Lecture Notes in Computer Science 11478

More information about this series at http://www.springer.com/series/7410

Yuval Ishai · Vincent Rijmen (Eds.)

Advances in Cryptology – EUROCRYPT 2019

38th Annual International Conference on the Theory and Applications of Cryptographic Techniques
Darmstadt, Germany, May 19–23, 2019
Proceedings, Part III

Editors
Yuval Ishai
Technion
Haifa, Israel

Vincent Rijmen
COSIC Group
KU Leuven
Heverlee, Belgium

ISSN 0302-9743 ISSN 1611-3349 (electronic)
Lecture Notes in Computer Science
ISBN 978-3-030-17658-7 ISBN 978-3-030-17659-4 (eBook)
https://doi.org/10.1007/978-3-030-17659-4

LNCS Sublibrary: SL4 – Security and Cryptology

This Springer imprint is published by the registered company Springer Nature Switzerland AG
The registered company address is: Gewerbestrasse 11, 6330 Cham, Switzerland

Preface

Eurocrypt 2019, the 38th Annual International Conference on the Theory and Applications of Cryptographic Techniques, was held in Darmstadt, Germany, during May 19–23, 2019. The conference was sponsored by the International Association for Cryptologic Research (IACR). Marc Fischlin (Technische Universität Darmstadt, Germany) was responsible for the local organization. He was supported by a local organizing team consisting of Andrea Püchner, Felix Günther, Christian Janson, and the Cryptoplexity Group. We are deeply indebted to them for their support and smooth collaboration.

The conference program followed the now established parallel track system where the works of the authors were presented in two concurrently running tracks. The invited talks and the talks presenting the best paper/best young researcher spanned over both tracks.

We received a total of 327 submissions. Each submission was anonymized for the reviewing process and was assigned to at least three of the 58 Program Committee members. Committee members were allowed to submit at most one paper, or two if both were co-authored. Submissions by committee members were held to a higher standard than normal submissions. The reviewing process included a rebuttal round for all submissions. After extensive deliberations the Program Committee accepted 76 papers. The revised versions of these papers are included in these three volume proceedings, organized topically within their respective track.

The committee decided to give the Best Paper Award to the paper "Quantum Lightning Never Strikes the Same State Twice" by Mark Zhandry. The runner-up was the paper "Compact Adaptively Secure ABE for NC^1 from k Lin" by Lucas Kowalczyk and Hoeteck Wee. The Best Young Researcher Award went to the paper "Efficient Verifiable Delay Functions" by Benjamin Wesolowski. All three papers received invitations for the *Journal of Cryptology.*

The program also included an IACR Distinguished Lecture by Cynthia Dwork, titled "Differential Privacy and the People's Data," and invited talks by Daniele Micciancio, titled "Fully Homomorphic Encryption from the Ground Up," and François-Xavier Standaert, titled "Toward an Open Approach to Secure Cryptographic Implementations."

We would like to thank all the authors who submitted papers. We know that the Program Committee's decisions can be very disappointing, especially rejections of very good papers that did not find a slot in the sparse number of accepted papers. We sincerely hope that these works eventually get the attention they deserve.

We are also indebted to the members of the Program Committee and all external reviewers for their voluntary work. The committee's work is quite a workload. It has been an honor to work with everyone. The committee's work was tremendously simplified by Shai Halevi's submission software and his support, including running the service on IACR servers.

Finally, we thank everyone else—speakers, session chairs, and rump-session chairs—for their contribution to the program of Eurocrypt 2019. We would also like to thank the many sponsors for their generous support, including the Cryptography Research Fund that supported student speakers.

May 2019

Yuval Ishai
Vincent Rijmen

Eurocrypt 2019

The 38th Annual International Conference on the Theory and Applications of Cryptographic Techniques

Sponsored by *the International Association for Cryptologic Research*

May 19–23, 2019
Darmstadt, Germany

General Chair

Marc Fischlin	Technische Universität Darmstadt, Germany

Program Co-chairs

Yuval Ishai	Technion, Israel
Vincent Rijmen	KU Leuven, Belgium and University of Bergen, Norway

Program Committee

Michel Abdalla	CNRS and ENS Paris, France
Adi Akavia	University of Haifa, Israel
Martin Albrecht	Royal Holloway, UK
Elena Andreeva	KU Leuven, Belgium
Paulo S. L. M. Barreto	University of Washington Tacoma, USA
Amos Beimel	Ben-Gurion University, Israel
Alex Biryukov	University of Luxembourg, Luxembourg
Nir Bitansky	Tel Aviv University, Israel
Andrej Bogdanov	Chinese University of Hong Kong, SAR China
Christina Boura	University of Versailles and Inria, France
Xavier Boyen	QUT, Australia
David Cash	University of Chicago, USA
Melissa Chase	MSR Redmond, USA
Kai-Min Chung	Academia Sinica, Taiwan
Dana Dachman-Soled	University of Maryland, USA
Ivan Damgård	Aarhus University, Denmark
Itai Dinur	Ben-Gurion University, Israel
Stefan Dziembowski	University of Warsaw, Poland
Serge Fehr	Centrum Wiskunde & Informatica (CWI) and Leiden University, The Netherlands
Juan A. Garay	Texas A&M University, USA
Sanjam Garg	UC Berkeley, USA

Christina Garman	Purdue University, USA
Siyao Guo	New York University Shanghai, China
Iftach Haitner	Tel Aviv University, Israel
Shai Halevi	IBM Research, USA
Brett Hemenway	University of Pennsylvania, USA
Justin Holmgren	Princeton University, USA
Stanislaw Jarecki	UC Irvine, USA
Dakshita Khurana	Microsoft Research New England, USA
Ilan Komargodski	Cornell Tech, USA
Gregor Leander	Ruhr-Universität Bochum, Germany
Huijia Lin	UCSB, USA
Atul Luykx	Visa Research, USA
Mohammad Mahmoody	University of Virginia, USA
Bart Mennink	Radboud University, The Netherlands
Tal Moran	IDC Herzliya, Israel
Svetla Nikova	KU Leuven, Belgium
Claudio Orlandi	Aarhus University, Denmark
Rafail Ostrovsky	UCLA, USA
Rafael Pass	Cornell University and Cornell Tech, USA
Krzysztof Pietrzak	IST Austria, Austria
Bart Preneel	KU Leuven, Belgium
Christian Rechberger	TU Graz, Austria
Leonid Reyzin	Boston University, USA
Guy N. Rothblum	Weizmann Institute, Israel
Amit Sahai	UCLA, USA
Christian Schaffner	QuSoft and University of Amsterdam, The Netherlands
Gil Segev	Hebrew University, Israel
abhi shelat	Northeastern University, USA
Martijn Stam	Simula UiB, Norway
Marc Stevens	CWI Amsterdam, The Netherlands
Stefano Tessaro	UCSB, USA
Mehdi Tibouchi	NTT, Japan
Frederik Vercauteren	KU Leuven, Belgium
Brent Waters	UT Austin, USA
Mor Weiss	Northeastern University, USA
David J. Wu	University of Virginia, USA
Vassilis Zikas	University of Edinburgh, UK

Additional Reviewers

Divesh Aggarwal
Shashank Agrawal
Gorjan Alagic
Abdelrahaman Aly
Andris Ambainis
Prabhanjan Ananth
Gilad Asharov
Tomer Ashur
Arash Atashpendar
Benedikt Auerbach
Christian Badertscher
Saikrishna Badrinarayanan
Shi Bai
Josep Balasch

Marshall Ball
James Bartusek
Balthazar Bauer
Carsten Baum
Christof Beierle
Fabrice Benhamouda
Iddo Bentov
Mario Berta
Ward Beullens
Ritam Bhaumik
Jean-François Biasse
Koen de Boer
Dan Boneh
Xavier Bonnetain
Charlotte Bonte
Carl Bootland
Jonathan Bootle
Joppe Bos
Adam Bouland
Florian Bourse
Benedikt Bünz
Wouter Castryck
Siu On Chan
Nishanth Chandran
Eshan Chattopadhyay
Yi-Hsiu Chen
Yilei Chen
Yu Long Chen
Jung-Hee Cheon
Mahdi Cheraghchi
Celine Chevalier
Nai-Hui Chia
Ilaria Chillotti
Chongwon Cho
Wutichai Chongchitmate
Michele Ciampi
Ran Cohen
Sandro Coretti
Ana Costache
Jan Czajkowski
Yuanxi Dai
Deepesh Data
Bernardo David
Alex Davidson
Thomas Debris-Alazard
Thomas De Cnudde
Thomas Decru
Luca De Feo
Akshay Degwekar
Cyprien Delpech de Saint Guilhem
Ioannis Demertzis
Ronald de Wolf
Giovanni Di Crescenzo
Christoph Dobraunig
Jack Doerner
Javad Doliskani
Leo Ducas
Yfke Dulek
Nico Döttling
Aner Ben Efraim
Maria Eichlseder
Naomi Ephraim
Daniel Escudero
Saba Eskandarian
Thomas Espitau
Pooya Farshim
Prastudy Fauzi
Rex Fernando
Houda Ferradi
Dario Fiore
Ben Fisch
Mathias Fitzi
Cody Freitag
Georg Fuchsbauer
Benjamin Fuller
Tommaso Gagliardoni
Steven Galbraith
Nicolas Gama
Chaya Ganesh
Sumegha Garg
Romain Gay
Peter Gazi
Craig Gentry
Marios Georgiou
Benedikt Gierlichs
Huijing Gong
Rishab Goyal
Lorenzo Grassi
Hannes Gross
Jens Groth
Paul Grubbs
Divya Gupta
Felix Günther
Helene Haagh
Björn Haase
Mohammad Hajiabadi
Carmit Hazay
Pavel Hubáček
Andreas Huelsing
Ilia Iliashenko
Muhammad Ishaq
Joseph Jaeger
Eli Jaffe
Aayush Jain
Abhishek Jain
Stacey Jeffery
Zhengfeng Ji
Yael Kalai
Daniel Kales
Chethan Kamath
Nathan Keller
Eike Kiltz
Miran Kim
Sam Kim
Taechan Kim
Karen Klein
Yash Kondi
Venkata Koppula
Mukul Kulkarni
Ashutosh Kumar
Ranjit Kumaresan
Rio LaVigne
Virginie Lallemand
Esteban Landerreche
Brandon Langenberg
Douglass Lee
Eysa Lee
François Le Gall
Chaoyun Li
Wei-Kai Lin
Qipeng Liu
Tianren Liu
Alex Lombardi
Julian Loss
Yun Lu
Vadim Lyubashevsky
Fermi Ma

Saeed Mahloujifar
Christian Majenz
Rusydi Makarim
Nikolaos Makriyannis
Nathan Manohar
Antonio Marcedone
Daniel Masny
Alexander May
Noam Mazor
Willi Meier
Rebekah Mercer
David Mestel
Peihan Miao
Brice Minaud
Matthias Minihold
Konstantinos Mitropoulos
Tarik Moataz
Hart Montgomery
Andrew Morgan
Pratyay Mukherjee
Luka Music
Michael Naehrig
Gregory Neven
Phong Nguyen
Jesper Buus Nielsen
Ryo Nishimaki
Daniel Noble
Adam O'Neill
Maciej Obremski
Sabine Oechsner
Michele Orrù
Emmanuela Orsini
Daniel Ospina
Giorgos Panagiotakos
Omer Paneth
Lorenz Panny
Anat Paskin-Cherniavsky
Alain Passelègue
Kenny Paterson
Chris Peikert
Geovandro Pereira
Léo Perrin
Edoardo Persichetti
Naty Peter
Rachel Player
Oxana Poburinnaya
Yuriy Polyakov
Antigoni Polychroniadou
Eamonn Postlethwaite
Willy Quach
Ahmadreza Rahimi
Sebastian Ramacher
Adrián Ranea
Peter Rasmussen
Shahram Rasoolzadeh
Ling Ren
Joao Ribeiro
Silas Richelson
Thomas Ricosset
Tom Ristenpart
Mike Rosulek
Dragos Rotaru
Yann Rotella
Lior Rotem
Yannis Rouselakis
Arnab Roy
Louis Salvail
Simona Samardziska
Or Sattath
Guillaume Scerri
John Schanck
Peter Scholl
André Schrottenloher
Sruthi Sekar
Srinath Setty
Brian Shaft
Ido Shahaf
Victor Shoup
Jad Silbak
Mark Simkin
Shashank Singh
Maciej Skórski
Caleb Smith
Fang Song
Pratik Soni
Katerina Sotiraki
Florian Speelman
Akshayaram Srinivasan
Uri Stemmer
Noah Stephens-Davidowitz
Alan Szepieniec
Gelo Noel Tabia
Aishwarya Thiruvengadam
Sergei Tikhomirov
Rotem Tsabary
Daniel Tschudy
Yiannis Tselekounis
Aleksei Udovenko
Dominique Unruh
Cédric Van Rompay
Prashant Vasudevan
Muthu Venkitasubramaniam
Daniele Venturi
Benoît Viguier
Fernando Virdia
Ivan Visconti
Giuseppe Vitto
Petros Wallden
Alexandre Wallet
Qingju Wang
Bogdan Warinschi
Gaven Watson
Hoeteck Wee
Friedrich Wiemer
Tim Wood
Keita Xagawa
Sophia Yakoubov
Takashi Yamakawa
Arkady Yerukhimovich
Eylon Yogev
Nengkun Yu
Yu Yu
Aaram Yun
Thomas Zacharias
Greg Zaverucha
Liu Zeyu
Mark Zhandry
Chen-Da Liu Zhang

Abstracts of Invited Talks

Differential Privacy and the People's Data

IACR Distinguished Lecture

Cynthia Dwork[1]

Harvard University
dwork@seas.harvard.edu

Abstract. Differential Privacy will be the confidentiality protection method of the 2020 US Decennial Census. We explore the technical and social challenges to be faced as the technology moves from the realm of information specialists to the large community of consumers of census data.

Differential Privacy is a definition of privacy tailored to the statistical analysis of large datasets. Roughly speaking, differential privacy ensures that anything learnable about an individual could be learned independent of whether the individual opts in or opts out of the data set under analysis. The term has come to denote a field of study, inspired by cryptography and guided by theoretical lower bounds and impossibility results, comprising algorithms, complexity results, sample complexity, definitional relaxations, and uses of differential privacy when privacy is not itself a concern.

From its inception, a motivating scenario for differential privacy has been the US Census: data of the people, analyzed for the benefit of the people, to allocate the people's resources (hundreds of billions of dollars), with a legal mandate for privacy. Over the past 4–5 years, differential privacy has been adopted in a number of industrial settings by Google, Microsoft, Uber, and, with the most fanfare, by Apple. In 2020 it will be the confidentiality protection method for the US Decennial Census.

Census data are used throughout government and in thousands of research studies every year. This mainstreaming of differential privacy, the transition from the realm of technically sophisticated information specialists and analysts into much broader use, presents enormous technical and social challenges. The Fundamental Theorem of Information Reconstruction tells us that overly accurate estimates of too many statistics completely destroys privacy. Differential privacy provides a measure of privacy loss that permits the tracking and control of cumulative privacy loss as data are analyzed and re-analyzed. But provably no method can permit the data to be explored without bound. How will the privacy loss "budget" be allocated? Who will enforce limits?

More pressing for the scientific community are questions of how the multitudes of census data consumers will interact with the data moving forward. The Decennial Census is simple, and the tabulations can be handled well with existing technology. In contrast, the annual American Community Survey, which covers only a few million households yearly, is rich in personal details on subjects from internet access in the home to employment to ethnicity, relationships among persons in the home, and fertility. We are not (yet?) able to

[1] Supported in part by NSF Grant 1763665 and the Sloan Foundation.

offer differentially private algorithms for every kind of analysis carried out on these data. Historically, confidentiality has been handled by a combination of data summaries, restricted use access to the raw data, and the release of public-use microdata, a form of noisy individual records. Summary statistics are the bread and butter of differential privacy, but giving even trusted and trustworthy researchers access to raw data is problematic, as their published findings are a vector for privacy loss: think of the researcher as an arbitrary non-differentially private algorithm that produces outputs in the form of published findings. The very *choice* of statistic to be published is inherently not privacy-preserving! At the same time, past microdata noising techniques can no longer be considered to provide adequate privacy, but generating synthetic public-use microdata while ensuring differential privacy is a computationally hard problem. Nonetheless, combinations of exciting new techniques give reason for optimism.

Towards an Open Approach to Secure Cryptographic Implementations

François-Xavier Standaert[1]

UCL Crypto Group, Université Catholique de Louvain, Belgium

Abstract. In this talk, I will discuss how recent advances in side-channel analysis and leakage-resilience could lead to both stronger security properties and improved confidence in cryptographic implementations. For this purpose, I will start by describing how side-channel attacks exploit physical leakages such as an implementation's power consumption or electromagnetic radiation. I will then discuss the definitional challenges that these attacks raise, and argue why heuristic hardware-level countermeasures are unlikely to solve the problem convincingly. Based on these premises, and focusing on the symmetric setting, securing cryptographic implementations can be viewed as a tradeoff between the design of modes of operation, underlying primitives and countermeasures.

Regarding modes of operation, I will describe a general design strategy for leakage-resilient authenticated encryption, propose models and assumptions on which security proofs can be based, and show how this design strategy encourages so-called leveled implementations, where only a part of the computation needs strong (hence expensive) protections against side-channel attacks.

Regarding underlying primitives and countermeasures, I will first emphasize the formal and practically-relevant guarantees that can be obtained thanks to masking (i.e., secret sharing at the circuit level), and how considering the implementation of such countermeasures as an algorithmic design goal (e.g., for block ciphers) can lead to improved performances. I will then describe how limiting the leakage of the less protected parts in a leveled implementations can be combined with excellent performances, for instance with respect to the energy cost.

I will conclude by putting forward the importance of sound evaluation practices in order to empirically validate (by lack of falsification) the assumptions needed both for leakage-resilient modes of operation and countermeasures like masking, and motivate the need of an open approach for this purpose. That is, by allowing adversaries and evaluators to know implementation details, we can expect to enable a better understanding of the fundamentals of physical security, therefore leading to improved security and efficiency in the long term.

[1] The author is a Senior Research Associate of the Belgian Fund for Scientific Research (FNRS-F.R.S.). This work has been funded in part by the ERC Project 724725.

Fully Homomorphic Encryption from the Ground Up

Daniele Micciancio

University of California, Mail Code 0404, La Jolla,
San Diego, CA, 92093, USA
daniele@cs.ucsd.edu
http://cseweb.ucsd.edu/~daniele/

Abstract. The development of fully homomorphic encryption (FHE), i.e., encryption schemes that allow to perform arbitrary computations on encrypted data, has been one of the main achievements of theoretical cryptography of the past 20 years, and probably the single application that brought most attention to lattice cryptography. While lattice cryptography, and fully homomorphic encryption in particular, are often regarded as a highly technical topic, essentially all constructions of FHE proposed so far are based on a small number of rather simple ideas. In this talk, I will try highlight the basic principles that make FHE possible, using lattices to build a simple private key encryption scheme that enjoys a small number of elementary, but very useful properties: a simple decryption algorithm (requiring, essentially, just the computation of a linear function), a basic form of circular security (i.e., the ability to securely encrypt its own key), and a very weak form of linear homomorphism (supporting only a bounded number of addition operations.)

All these properties are easily established using simple linear algebra and the hardness of the Learning With Errors (LWE) problem or standard worst-case complexity assumptions on lattices. Then, I will use this scheme (and its abstract properties) to build in a modular way a tower of increasingly more powerful encryption schemes supporting a wider range of operations: multiplication by arbitrary constants, multiplication between ciphertexts, and finally the evaluation of arithmetic circuits of arbitrary, but a-priory bounded depth. The final result is a *leveled*[1] FHE scheme based on standard lattice problems, i.e., a scheme supporting the evaluation of arbitrary circuits on encrypted data, as long as the depth of the circuit is provided at key generation time. Remarkably, lattices are used only in the construction (and security analysis) of the basic scheme: all the remaining steps in the construction do not make any direct use of lattices, and can be expressed in a simple, abstract way, and analyzed using solely the weakly homomorphic properties of the basic scheme.

Keywords: Lattice-based cryptography · Fully homomorphic encryption · Circular security · FHE bootstrapping

---

[1] The "leveled" restriction in the final FHE scheme can be lifted using "circular security" assumptions that have become relatively standard in the FHE literature, but that are still not well understood. Achieving (non-leveled) FHE from standard lattice assumptions is the main theoretical problem still open in the area.

Contents – Part III

Foundations I

On ELFs, Deterministic Encryption, and Correlated-Input Security 3
Mark Zhandry

New Techniques for Efficient Trapdoor Functions and Applications 33
Sanjam Garg, Romain Gay, and Mohammad Hajiabadi

Symbolic Encryption with Pseudorandom Keys 64
Daniele Micciancio

Efficient Secure Computation

Covert Security with Public Verifiability: Faster, Leaner, and Simpler 97
Cheng Hong, Jonathan Katz, Vladimir Kolesnikov, Wen-jie Lu, and Xiao Wang

Efficient Circuit-Based PSI with Linear Communication 122
Benny Pinkas, Thomas Schneider, Oleksandr Tkachenko, and Avishay Yanai

An Algebraic Approach to Maliciously Secure Private Set Intersection 154
Satrajit Ghosh and Tobias Nilges

Quantum II

On Finding Quantum Multi-collisions 189
Qipeng Liu and Mark Zhandry

On Quantum Advantage in Information Theoretic Single-Server PIR 219
Dorit Aharonov, Zvika Brakerski, Kai-Min Chung, Ayal Green, Ching-Yi Lai, and Or Sattath

Verifier-on-a-Leash: New Schemes for Verifiable Delegated Quantum Computation, with Quasilinear Resources 247
Andrea Coladangelo, Alex B. Grilo, Stacey Jeffery, and Thomas Vidick

Signatures I

Ring Signatures: Logarithmic-Size, No Setup—from Standard Assumptions 281
Michael Backes, Nico Döttling, Lucjan Hanzlik, Kamil Kluczniak, and Jonas Schneider

Group Signatures Without NIZK: From Lattices in the Standard Model 312
Shuichi Katsumata and Shota Yamada

A Modular Treatment of Blind Signatures from Identification Schemes 345
Eduard Hauck, Eike Kiltz, and Julian Loss

Best Paper Awards

Efficient Verifiable Delay Functions 379
Benjamin Wesolowski

Quantum Lightning Never Strikes the Same State Twice 408
Mark Zhandry

Information-Theoretic Cryptography

Secret-Sharing Schemes for General and Uniform Access Structures 441
Benny Applebaum, Amos Beimel, Oriol Farràs, Oded Nir, and Naty Peter

Towards Optimal Robust Secret Sharing with Security Against a Rushing Adversary 472
Serge Fehr and Chen Yuan

Simple Schemes in the Bounded Storage Model 500
Jiaxin Guan and Mark Zhandary

Cryptanalysis

From Collisions to Chosen-Prefix Collisions Application to Full SHA-1 527
Gaëtan Leurent and Thomas Peyrin

Preimage Attacks on Round-Reduced KECCAK-224/256 via an Allocating Approach 556
Ting Li and Yao Sun

BISON Instantiating the Whitened Swap-Or-Not Construction 585
Anne Canteaut, Virginie Lallemand, Gregor Leander, Patrick Neumann, and Friedrich Wiemer

Foundations II

Worst-Case Hardness for LPN and Cryptographic Hashing via Code Smoothing 619
Zvika Brakerski, Vadim Lyubashevsky, Vinod Vaikuntanathan, and Daniel Wichs

New Techniques for Obfuscating Conjunctions 636
James Bartusek, Tancrède Lepoint, Fermi Ma, and Mark Zhandry

Distributional Collision Resistance Beyond One-Way Functions 667
Nir Bitansky, Iftach Haitner, Ilan Komargodski, and Eylon Yogev

Signatures II

Multi-target Attacks on the Picnic Signature Scheme and Related Protocols 699
Itai Dinur and Niv Nadler

Durandal: A Rank Metric Based Signature Scheme 728
Nicolas Aragon, Olivier Blazy, Philippe Gaborit, Adrien Hauteville, and Gilles Zémor

SeaSign: Compact Isogeny Signatures from Class Group Actions 759
Luca De Feo and Steven D. Galbraith

Author Index 791

Foundations I

On ELFs, Deterministic Encryption, and Correlated-Input Security

Mark Zhandry(✉)

Princeton University, Princeton, USA
mzhandry@princeton.edu

Abstract. We construct deterministic public key encryption secure for any *constant* number of *arbitrarily correlated computationally unpredictable* messages. Prior works required either random oracles or non-standard knowledge assumptions. In contrast, our constructions are based on the exponential hardness of DDH, which is plausible in elliptic curve groups. Our central tool is a new *trapdoored extremely lossy function*, which modifies extremely lossy functions by adding a trapdoor.

1 Introduction

The Random Oracle Model [7] is a useful model whereby one models a hash function as a truly random function. Random oracles have many useful properties, such as collision resistance, pseudorandomness, correlation intractability, extractability, and more. Unfortunately, random oracles do not exist in the real world, and some random oracle properties are uninstantiable by concrete hash functions [11]. This has lead to a concerted effort in the community toward constructing hash functions with various strong security properties from standard, well-studied, and widely-accepted assumptions.

Correlated Input Security. In this work, we focus on one particular property satisfied by random oracles, namely correlated input security. Here, the adversary is given $y_i = f(x_i)$ for inputs $x_1, \ldots, x_k$ which may come from highly non-uniform and highly correlated distributions. At the simplest level, we ask that the adversary cannot guess any of the x_i, though stronger requirements such as the pseudorandomness of the y_i are possible. Correlated input security has applications to password hashing and searching on encrypted data [16] and is closely related to related-key security [4]. It is also a crucial security requirement for deterministic public key encryption [3], which is essentially a hash function with a trapdoor.

Correlated input secure functions follow trivially in the random oracle model, and standard-model constructions for specific classes of functions such as low-degree polynomials [16] or "block sources" [5,8,9,14,19] are known. However, there has been little progress toward attaining security for *arbitrary* correlations from standard assumptions, even in the case of just *two* correlated inputs.

Y. Ishai and V. Rijmen (Eds.): EUROCRYPT 2019, LNCS 11478, pp. 3–32, 2019.
https://doi.org/10.1007/978-3-030-17659-4_1

This Work. In this work, we construct hash functions and deterministic public key encryption (DPKE) with security for any *constant* number of *arbitrarily* correlated sources. In addition, we only require computational unpredictability for our sources, and our DPKE scheme even achieves CCA security. Our main new technical tool is a new construction of extremely lossy functions (ELFs) [22] that admit a trapdoor. Our construction is secure, assuming that DDH (or more generally k-Lin) is *exponentially* hard to solve. Such an assumption is plausible on elliptic curves.

1.1 Details

We now give an overview of our results and our approach. We start with correlated-input security for one-way functions, and gradually build up to our ultimate goal of deterministic public key encryption.

Correlated Input Secure OWFs. First, we observe that Zhandry's Extremely Lossy Functions (ELFs) [22] already give correlated-input secure one-way functions for any *constant* number of inputs. Recall that an ELF is a variant of a lossy trapdoor function (LTDF), which were introduced by Peikert and Waters [18]. LTDFs are functions with two modes, an injective mode that contains a secret trapdoor for inversion, and a lossy mode that is information-theoretically uninvertible. The security requirement is that these modes are computationally indistinguishable, if you do not know the trapdoor. LTDFs have many applications, including CCA-secure public key encryption [18], deterministic public key encryption [8], and more.

Similarly, and ELF also has two modes, injective and lossy similar to above. However the key difference is that in the lossy mode, the image size is so small that it is actually *polynomial.* Clearly, such a lossy mode can be distinguished from injective by an adversary whose running time is a slightly larger polynomial. So ELFs actually have a spectrum of lossy modes of differing polynomial image sizes, and the exact image size is chosen *based on the adversary* just large enough to fool it. The other main difference between ELFs and LTDFs is that, due to the particulars of Zhandry's construction, the injective mode for ELFs does not contain a trapdoor. Zhandry constructs ELFs based on the *exponential hardness* of the DDH assumption, or more generally exponential k-Lin.

Let f be an injective mode ELF. Consider a source S of d correlated inputs $x_1, \ldots, x_d$ as well as auxiliary information aux. Our goal is to show that, given aux and $f(x_1), \ldots, f(x_d)$, it is computationally infeasible to find any of the x_i. A necessary condition on S is that each x_i are computationally unpredictable given aux alone. Note that we *will* allow sources for which x_i *is* predictable given some of the other x_j. Note that such a source captures the setting where, say, $x_2 = x_1 + 1$, etc.

We now prove that f is one-way for any such computationally unpredictable source. To prove security, we first switch f to be a lossy mode with polynomial image size p. Since d is assumed to be constant, the number of possible value for the vector $f(x_1), \ldots, f(x_d)$ is p^d, also a polynomial. Therefore, this value can be guessed by the adversary with inverse polynomial probability. As such, if x_i can be guessed given aux and $f(x_1), \ldots, f(x_d)$ with non-negligible probability ϵ, it can also be guessed given just aux with probability at least ϵ/p^d, contradicting the unpredictability of S.

Correlated Input Secure PRGs. Next, we turn to the much harder task of constructing a PRG G for a constant number of correlated inputs. Here, the adversary is given $\mathsf{aux}, y_1, \ldots, y_d$ where either (1) $y_i = G(x_i)$ for all i or (2) y_i is chosen at random in the domain of G. The adversary tries to distinguish the two cases. In order for security to be possible at all, we need to place some minimal restrictions on the source S:

- As in the case of one-wayness, we must require that S is computationally unpredictable
- All the x_i must be distinct, with high probability. Otherwise, the adversary identify the $y_i = G(x_i)$ case by simply testing the equality of the y_i.

In this paper, in order to match notation from Zhandry [22], we will call a function G satisfying indistinguishability a *hardcore function* for computationally unpredictable sources on d inputs.

ELFs are not alone guaranteed to be such hardcore functions, as the outputs are not guaranteed to be random. Instead we build G by starting from Zhandry's hardcore function, which works in the case $d = 1$; that is, for single computationally unpredictable sources. Zhandry's construction is built from ELFs, but requires more machinery to prove pseudorandomness.

The core idea of Zhandry's hardcore function G is the following: first extract *many* Goldreich-Leving hardcore bits, far too many to be simultaneously hardcore. These cannot be output in the clear, as they would allow for trivial inversion. Instead, the bits are scrambled by feeding them through an ELF-based circuit. Zhandry shows (1) that the GL bits can be replaced with random without detection, and (2) if the GL bits are replaced with random, then the output of the circuit is random.

Unfortunately, for correlated sources, the GL bits for different inputs will be correlated: for example if the two inputs differ in a single bit, then if the parity function computing the GL bit is 0 in that position, the two GL bits will be identical. Therefore, the inputs to step (2) in Zhandry's proof may be highly correlated, and his circuit does not guarantee security against correlated inputs.

To mitigate this issue, we carefully modify Zhandry's function G. The idea is, rather than having fixed GL parities, we generate the GL parities as functions of the input itself. Different inputs will hopefully map to independent parities, leading to independent GL bits. We have to be careful, however, in order to avoid any circularities in the analysis, since we need the GL parities to be (pseudo)random and independent (in order to apply the GL theorem), but generating such random independent parities already seems to require extracting pseudorandom strings for arbitrarily correlated sources, leaving us back where we started.

Our construction works as follows: we have another ELF instance, which is applied to the input x, resulting in an value w. Then we apply a d-wise independent function R to w to generate the actual parities. Zhandry shows that this composition of an ELF and a d-wise independent function is collision-resistant for $d \geq 2$, meaning the d different x_i will indeed map to distinct parities, and in fact distinct w_i. Next, in the lossy mode for the ELF, there are only a polynomial number of w; since d is constant, there are also a polynomial number of possible d-tuples of $(w_1, \ldots, w_d)$. Therefore, we can actually *guess* the $(w_1, \ldots, w_d)$ vector that will result from applying the ELF to the x_i with inverse-polynomial probability. Next, since R is d-wise independent, we can take d independent sets of GL parities and program R to output these parities on the corresponding d values of w. This is not quite enough to show that the GL parities are themselves pseudorandom for correlated sources (since we only successfully program with inverse-polynomial probability), but with a careful proof we show that it is sufficient to prove the pseudorandomness of the overall construction.

Deterministic Public Key Encryption. Next, we turn to constructing deterministic public key encryption (DPKE). A DPKE protocol consists of 3 algorithms, $(\mathsf{DPKE.Gen}, \mathsf{DPKE.Enc}, \mathsf{DPKE.Dec})$. $\mathsf{DPKE.Gen}$ creates a secret/public key pair sk, pk. $\mathsf{DPKE.Enc}$ is a deterministic procedure that uses the public key pk to scramble a message m, arriving at a ciphertext c. $\mathsf{DPKE.Dec}$ is also deterministic, and maps the ciphertext c back to m.

We first consider security in the single-input setting; we note that it was previously open to construction DPKE for even a single arbitrary computationally unpredictable source. The canonical way to build DPKE [3] is to use an ordinary *randomized* public key encryption scheme with CPA security. The idea is to hash the message m using a hash function H, and use $H(m)$ as the randomness r: $\mathsf{DPKE.Enc}(\mathsf{pk}, m) = \mathsf{DPKE.Enc}(\mathsf{pk}, m; H(r))$ where $\mathsf{DPKE.Enc}$ is the randomized PKE encryption algorithm. In the random oracle model for H, Bellare, Boldyreva and O'Neill [3] show that this scheme obtains the strongest possible notion of security. One may hope that some ELF-based hash function H might be sufficient.

Unfortunately, Brzuska, Farshim and Mittelbach [10] give strong evidence that this scheme cannot be proven secure in the standard model, even under very strong assumptions. In particular, they devise a public key encryption scheme $\mathsf{PKE.Enc}$ such that, for any concrete hash function H, $\mathsf{DPKE.Enc}$ will be insecure. Their construction uses indistinguishability obfuscation [2] (iO), which is currently one of the more speculative tools used in cryptography. Nonetheless, in order to give a standard model construction of DPKE, one must either deviate from the scheme above, or else prove conclusively that iO does not exist.

On the other hand, lossy trapdoor functions have proven useful for building DPKE in the standard model (e.g. [8,9]). One limitation of these techniques, however, is that since the image size of a LTDF is always at least sub-exponential, constructions based on LTDFs tend to require high min-entropy/computational unpredictability requirements.

Our First Construction. We start by abstracting the constructions of Brakerski and Segev [9]. They construct DPKE for sub-exponentially unpredictable sources by essentially analyzing specific constructions of Lossy Trapdoor Functions (LTDFs), and showing that they satisfy the desired security experiment.

Our first construction abstracts their construction to work with arbitrary LTDFs. Our construction is the following, based on a semantically secure public key encryption scheme $\mathsf{PKE.Enc}$, a special kind of pseudorandom generator G, and a LTDF f generated in the injective mode:

$$\mathsf{DPKE.Enc}(\mathsf{pk}, m) = \mathsf{PKE.Enc}(\mathsf{pk}, f(m); G(m))$$

To prove security, we first switch to f being in the lossy mode. Now, notice that if m can be predicted with probability p, then it can still be predicted with probability p/r even given $f(m)$, by simply guessing the value of $f(m)$, which will be correct with probability $1/r$. In particular, if p is sub-exponentially small and r is sub-exponential, then p/r is also sub-exponential. Any LTDF can be set to have a sub-exponential-sized lossy mode by adjusting the security parameter accordingly.

Next, we observe that if G is hardcore for sub-exponentially unpredictable sources, then $G(m)$ will be pseudorandom given $f(m)$. Such a G can be built by extracting a sufficiently small polynomial-number of Goldreich-Levin [15] bits, and then expanding using a standard PRG.

At this point, we can replace $G(m)$ with a random bitstring, and then rely on the semantic security of $\mathsf{PKE.Enc}$ to show security, completing the security proof.

But what about *arbitrary* computationally unpredictable sources, which may not be sub-exponentially secure? Intuitively, all we need is that (1) r can be taken to be an arbitrarily small super-polynomial, and (2) that G is secure for arbitrary unpredictable sources, instead of just sub-exponential sources. We then recall that Zhandry's [22] construction of G already satisfies (2), and that ELF's themselves *almost* satisfy (1). Unfortunately, the resulting scheme is not an encryption scheme: Zhandry's ELFs do not have a trapdoor in the injective mode, meaning there is no way to decrypt.

Therefore, we propose the notion of a trapdoor ELFs, which combines the functionality of ELFs and LTDFs by allowing for both a polynomial image size *and* a trapdoor. Using a trapdoor ELF, the above construction becomes a secure DPKE scheme for any computationally unpredictable source. For now we will simply assume such trapdoor ELFs; we discuss constructing such functions below.

CCA Security. Next we turn to achieving CCA security for DPKE. CCA security has received comparatively less attention in the deterministic setting, though some standard-model constructions are known [8,17,19]. In particular, we are not aware of any constructions for computationally unpredictable sources, sub-exponentially hard or otherwise.

We observe that by combining techniques for building CCA-secure encryption from LTDFs [8,18] with our abstraction of Brakerski and Segev [9], we can achieve CCA security for sub-exponentially hard sources. The idea is to use all-but-one LTDFs, a generalization of LTDFs where the function f has many branches. In the injective mode, each branch is injective. In the lossy mode, a single branch is lossy, and the inverse function works for all other modes. The adversary cannot tell injective from lossy, even if it knowns the branch b^*. Peikert and Waters [18] show how to generically construct such ABO LTDFs from any LTDF.

First, we modify the definition to require that indistinguishability from injective and lossy holds even if the adversary can make inversion queries on all branches other than b^*. The generic construction from standard LTDFs satisfies this stronger notion.

Then, our CCA-secure construction can be seen as combining our construction above with the construction of [8]. We encrypt using the algorithm $\mathsf{DPKE.Enc}(\mathsf{pk}, m) = (b = G_0(m), \mathsf{PKE.Enc}(\mathsf{pk}, f(b, m); G_1(m)))$ where G_0, G_1 are strong pseudorandom generators, and $f(b, m)$ is the ABO LTDF evaluation on branch b. Here, we require $\mathsf{PKE.Enc}$ to be a CCA-secure PKE scheme.

Intuitively, G_0 determines the branch, and if it is injective, then each message has its own branch. Once the branch is fixed, the rest of the scheme basically becomes our basic scheme from above. The challenge ciphertext will be set to be the lossy branch, which can be proven to hide the message following the same proof as our basic scheme. We will need to simulate CCA queries, which can be handled by using the CCA-security of $\mathsf{PKE.Enc}$ and the security of f under inversion queries.

Using standard LTDFs, we thus get the first CCA-secure scheme for sub-exponentially hard computationally unpredictable sources.

Turning to the setting of arbitrary unpredictable sources, we need to replace the ABO LTDF with an ABO trapdoor ELF, which works. Unfortunately, as discussed below, the generic construction of ABO LTDF in [18] does not apply to trapdoor ELFs, so we need a different approach to construct an ABO trapdoor ELF. Our approach is outlined below when we discuss our ELF constructions.

Correlated Inputs. Next, we turn to constructing DPKE for correlated inputs. Here, we require essentially the same security notion as for hardcore functions; the only difference is in functionality, since there is a trapdoor for inversion.

In the case of CPA security, security trivially follows if we replace G with our hardcore function for correlated inputs. We therefore easily get the first DPKE scheme secure for a constant number of correlated sources. We also note that if the source is sub-exponentially unpredictable, our scheme can be based on standard LDTFs.

We can similarly extend this idea to get CCA security. Except here, we will need a trapdoor ELF with several lossy branches, one for each challenge ciphertext.

Constructing Trapdoor ELFs. Finally, we turn to actually constructing trapdoor ELFs. Our trapdoor ELFs will be based on Zhandry's ELFs, which are in turn based on constructions of LTDFs [13]. But unfortunately, Zhandry's ELFs lose the trapdoor from the LTDFs. Here, we show how to resurrect the trapdoor.

Zhandry's construction basically iterates Freeman et al.'s [13] LTDF at many security levels. Freeman et al.'s construction expands the inputs by a modest factor. Thus, Zhandry needs to compress the outputs of each iteration in order for the size to not grow exponentially. Unfortunately, this compression results in the trapdoor being lost, since it is un-invertible.

Instead, we opt to avoid compression by more carefully choosing the security parameters being iterated. Zhandry chooses powers of 2 from 2 up to the global security parameter. Instead, we choose double exponentials 2^{2^i}. We still cannot go all the way to the global security parameter, but we show that we can go high enough in order to capture any polynomial. Thus, we obtain ELFs that admit a trapdoor.

For our application to CCA-security, we need to introduce branches into our trapdoor ELFs. Unfortunately, the approach of Peikert and Waters [18] is insufficient for our purposes. In particular, they introduce branching by applying many different LTDFs in parallel to the same input, outputting all images. The overall image size is then roughly the product of the image sizes of each underlying LTDF. The branch specifies which subsets of LTDFs are applied; the LTDFs corresponding to the lossy branch are all set to be lossy. In this way, the lossy branch will indeed be lossy. On the other hand, any other branch will have at least one LTDF which is injective, meaning the overall function is injective. Unfortunately for us, this approach results in an exponential blowup in the size of the image space for the lossy branch, even if the original image size was polynomial. Hence, applying this transformation to an ELF would not result in an ABO ELF.

Instead, we opt for a direct construction though still based on Freeman et al.'s scheme. Recall Freeman et al.'s scheme: the function f^{-1} is specified by an $n \times n$ matrix $\mathbf{A}$ over $\mathbb{Z}_q$, and the function f is specified by $\mathbf{A}$, but encoded in the exponent of a cryptographic group over order q: $g^{\mathbf{A}}$. The function f takes an input $\mathbf{x} \in \{0,1\}^n$, and maps it to $g^{\mathbf{A}\cdot\mathbf{x}}$ by carrying out appropriate group operations on $g^{\mathbf{A}}$. The inverse function f^{-1} uses $\mathbf{A}^{-1}$ to recover $g^{\mathbf{x}}$ from $g^{\mathbf{A}\cdot\mathbf{x}}$, and then solves for $\mathbf{x}$, which is efficient since $\mathbf{x}$ is 0/1.

In the lossy mode, $\mathbf{A}$ is set to be a matrix of rank 1. By DDH, this is indistinguishable from full rank when just given $g^{\mathbf{A}}$. On the other hand, now the image size of f is only q. By setting $2^n \gg q$, this function will now be lossy.

We now give a direct construction of an ABO trapdoor ELF. Our idea is to make the matrices tall, say $2n$ rows and n columns. Note that *any* left inverse of $\mathbf{A}$ will work for inverting the function, and there are many.

Our actual construction is the following. For branches in $\{0,1\}^a$, f^{-1} will be specified by $2a+1$ matrices $\mathbf{B}, \mathbf{A}_{i,t}$ for $i \in [a], t \in \{0,1\}$. The description of f will simply be the corresponding encoded values of $\mathbf{B}, \mathbf{A}_{i,t}$. The branch $b \in \{0,1\}^a$ corresponds to the matrix $\mathbf{A}_b = \mathbf{B} + \sum_i \mathbf{A}_{i,b_i}$.

For a lossy mode with branch b, we set $\mathbf{A}_b$ to be rank 1. Then we choose $\mathbf{A}_{i,t}$ at random and set $\mathbf{B} = \mathbf{A}_b - \sum_i \mathbf{A}_{i,b_i}$.

We would now like to prove security. For a given branch b^*, suppose an adversary can distinguish the injective mode from the mode where b^* is lossy.

We now show how to use such an adversary to distinguish $g^{\mathbf{C}}$ for a full-rank $n \times n$ $\mathbf{C}$ from a random rank-1 $\mathbf{C}$.

First, we will set $\mathbf{A}_{b^*}$ to be the matrix $\mathbf{C}$, except with n more rows appended, all of which are zero. We can easily construct $g^{\mathbf{A}_{b^*}}$ from $g^{\mathbf{C}}$ without knowing $\mathbf{C}$. Then we choose random $\mathbf{A}_{i,t}$. Finally, we set $\mathbf{B} = \mathbf{A}_{b^*} - \sum_i \mathbf{A}_{i,b_i}$. We can easily construct $g^{\mathbf{B}}$ given $g^{\mathbf{C}}$, again without knowing $\mathbf{C}$.

Now notice that for each branch b, we know the bottom n rows of $\mathbf{A}_b$, and moreover for $b \neq b^*$ they are full rank. Therefore, we can invert on any branch other than b^*, allowing us to simulate the adversary's queries.

Unfortunately, the distribution simulated is not indistinguishable from the correct distribution. After all, $\mathbf{A}_{b^*}$ is all zeros on the bottom n rows, which is easily detectable by the adversary. In order to simulate the correct distribution, we actually left-multiply all the matrices $\mathbf{B}, \mathbf{A}_{i,t}$ by a random matrix $\mathbf{R} \in \mathbb{Z}_q^{2n\times 2n}$. This can easily be done in the exponent. Moreover, now in the case where $\mathbf{C}$ is random, the matrices $\mathbf{B}, \mathbf{A}_{i,t}$ are actually random. On the other hand if $\mathbf{C}$ is rank 1, we correctly simulate the case where b^* is lossy.

Our construction above can easily be extended to multiple lossy branches by iterating the construction several times, one for each branch that needs to be lossy. Then, we notice that we actually achieve a polynomial image size by setting q to be a polynomial, and then relying on the exponential hardness of DDH to prove indistinguishability. Thus, we achieve trapdoor ELFs with multiple lossy branches, as needed for our construction.

1.2 Discussion

Of course, one way to achieve a hash function with security for correlated inputs—or more generally any security property—is to simply make the "tautological" assumption that a given hash function such as SHA has the property. Assuming the hash function is well designed, such an assumption may seem plausible. In fact, for practical reasons this is may be the preferred approach.

However, in light of the impossibility of instantiating random oracles in general [11], it is a priori unclear which hash function properties are achievable in the standard model. It could be, for example, that certain correlations amongst inputs will always be trivially insecure, even for the best-designed hash functions. The *only* way to gain confidence that a particular hash function property is plausible at all is to give a construction provably satisfying the property under well-studied and widely accepted computational assumptions. Our correlated-input secure PRG G does exactly this.

On Exponential Hardness. Our constructions rely on the exponential hardness of DDH, which is plausible in elliptic curve groups based on the current state of knowledge. Elliptic curves have been studied for some time, and to date no attack has been found that violates the exponential hardness in general elliptic curves.

In fact, exponential hardness is exactly what makes elliptic curves desirable for practical cryptographic applications today. DDH over finite fields is solvable in *subexponential* time, meaning parameters must be set much larger to block attacks. This leads to much less efficient schemes. Some of the most efficient protocols in use today rely on elliptic curves, precisely because we can set parameters aggressively and still remain secure. Thus, the exponential hardness of DDH in elliptic curve groups is widely assumed for real-world schemes.

We also remark that, as explained by Zhandry [22], polynomial-time and even sub-exponential-time hardness are insufficient for one-way functions secure for arbitrary min-entropy sources, which in particular are implied by our correlated-input secure constructions. Therefore, some sort of extremely strong hardness is inherent in our applications.

Concretely, security for arbitrary min-entropy sources implies the following: for any super-logarithmic function $t(n)$, there is a problem in NP that (1) only requires $t(n)$ bits of non-determinism, but (2) is still not contained in P. Put another way, the problem can be brute-forced in very slightly super-polynomial time, but is not solvable by *any* algorithm in polynomial time, showing that brute-force is essentially optimal. This can be seen as a scaled-down version of the exponential time hypothesis. Thus, while exponential hardness may not be required for the applications, a scaled-down version of exponential hardness *is* required.

Common Random String. Our constructions are based on Zhandry's ELFs, which require a common random string (crs); this crs is just the description of the injective-mode function. Thus our hardcore functions require a crs, and moreover, we only obtain security if the crs is sampled independently of the inputs. A natural question is whether this is required. Indeed, the following simple argument shows that pseudorandomness for even a *single information-theoretically unpredictable* source is impossible without a crs. After all, for a fixed function G, let S sample a random input x conditioned on the first bit of $G(x)$ being 0. Then the first bit of $G(x)$ will always be zero, whereas the first bit of a random string will only be zero half the time. This argument also easily extends to the setting of a crs, but where the sampler *depends* on a crs. It also extends for security for DPKE schemes where the messages depend on the public key, as noted in [19].

Even if we restrict to inputs that are statistically close to uniform, but allow two inputs to be slightly correlated, a crs is still required for pseudorandomness. Indeed, for a function G that outputs n-bit strings, consider the following sampler: choose two inputs x_0, x_1 at random, conditioned on the first bit of $G(x_0) \oplus G(x_1)$ being 0.

In the case of one-wayness, the above does not quite apply (since $G(x)$ may still hide x), but we can show that one-wayness without a crs is impossible for any super-constant number of correlated inputs. Basically, for d inputs, the sampler S will choose a random $(d-1)\log\lambda$-bit string x_1, which has super-logarithmic min-entropy since d is super-constant. Then it will divide x into $d-1$ blocks of $\log\lambda$ bits $z_2, \dots, z_d$. It will then sample random $x_2, \dots, x_d$ such that the first $\log\lambda$ bits of $G(x_i)$ are equal to z_i (which requires $O(\lambda)$ evaluations of G). Finally, it outputs $x_1, \dots, x_d$. Given the outputs $y_1, \dots, y_d$, it is easy to reconstruct x_1.

Of course, we only achieve security for a constant number of correlated inputs *with* a crs, so this leaves open the interesting problem of constructing a one-way function for a constant number of correlated inputs *without* using a crs.

Barriers to Correlated-Input Security. Even with a crs, correlated-input security has been difficult to achieve. The following informal argument from Wichs [21] gives some indication why this is the case. Let P_1, P_2 be two functions. Consider correlated x_1, x_2 sampled as $x_1 = P_1(r), x_2 = P_2(r)$, *for the same choice of random* r. Now, a reduction showing correlated-input security would need to transform an attacker A for the correlated inputs into an algorithm B for some presumably hard problem. But it seems that B needs to some how feed into A a valid input $G(x_1), G(x_2)$, and then use A's attack in order to solve it's problem. But the only obvious way to generate a valid input for general P_1, P_2 is to choose a random r and set $x_1 = P_1(r), x_2 = P_2(r)$. But then B already knows what A would do, making A's attack useless.

The standard way (e.g. [1]) to get around this argument is to use G that are lossy, and this is the approach we use, exploiting the two modes of the ELF. Our results show that it is possible to attain security for a constant number of inputs.

What about larger numbers of correlated inputs? Wichs [21] shows that proofs relative to polynomial-time falsifiable assumptions that make *black-box* use of the adversary are impossible for any *super-logarithmic* number of correlated messages. Note that the impossibility does *not* apply to our results for three reasons:

- Our reduction requires knowing the adversary's success probability and running time, and is therefore very slightly non-black box. In the language of [12], our reduction is "non-uniform"
- We require exponential hardness, not polynomial-time hardness
- We only achieve a constant number of correlated messages.

Nonetheless, Wichs impossibility represents a barrier to significantly improving our results.

Deterministic Public Key Encryption. Deterministic public key encryption can be thought of as an injective hash function that also has a trapdoor. As a result, the definitions of security for DPKE are related to strong security notions for hash functions such as correlated-input security. We note that [6] construct correlated-input secure DPKE for an *arbitrary* number of correlated min-entropy sources. Their underlying building blocks are LTDFs and *universal computational extractors* (UCE's). Note that UCE's are a strong "uber" type assumption on hash functions that includes many different security properties, including correlated-input security. Therefore, the main difficulty in their work is showing how to take a hash function that already attains the security notion they want (and then some) and then building from it a function that also has a trapdoor.

Our correlated-input secure hash function is likely not a UCE. In particular, in light of Wich's impossibility results discussed above, we don't expect to be able to prove that our construction is correlated-input secure for a large number of inputs. More we do not expect to be able to prove all UCE security properties for our assumption. Therefore, we cannot simply plug our hash function construction into [6] to get a DPKE scheme.

2 Preliminaries

Definition 1. *Consider a distribution D on pairs (x, aux), indexed by the security parameter λ. We say that D is* computationally unpredictable *if, for any probabilistic polynomial time adversary A, there is a negligible function ϵ such that*

$$\Pr[A(\mathsf{aux}) = x : (x, \mathsf{aux}) \leftarrow D(\lambda)] < \epsilon(\lambda)$$

In other words, A cannot efficient guess x given aux.

Lemma 1. *Let D be a source of tuples (x, aux, z) such that (x, aux) is computationally unpredictable. Let $\mathcal{F}$ be a distribution over functions f with the following property. $f(\mathsf{aux}, x, z)$ is function such that, for any* aux, *$f(\mathsf{aux}, \cdot, z)$ has polynomial image size, and that it is possible to efficiently compute the polynomial-sized image. Then D' which samples $(x, \mathsf{aux}' = (\mathsf{aux}, f, f(\mathsf{aux}, x, z)))$ is computationally unpredictable.*

Proof. If there is an A adversary for D', we can simply make a random guess for the value of $f(\mathsf{aux}, x, z)$, which will be right with inverse polynomial probability. In this case, we correctly simulate the view of A, meaning A outputs x with non-negligible probability. Overall, we break the computational unpredictability of x with inverse polynomial probability. □

We will also consider a notion of computationally unpredictable sources on multiple correlated inputs:

Definition 2. *Consider a distribution D on tuples $(x_1 \dots, x_d, \mathsf{aux})$, indexed by the security parameter λ. We say that D is* computationally unpredictable *if the following hold:*

- *For any $i \neq j$, $\Pr[x_i = x_j]$ is negligible.*
- *For any probabilistic polynomial time adversary A, there is a negligible function ϵ such that*

$$\Pr[A(\mathsf{aux}) \in \{x_1, \dots, x_d\} : (x_1, \dots, x_d, \mathsf{aux}) \leftarrow D(\lambda)] < \epsilon(\lambda)$$

In other words, each distribution (x_i, aux) is computationally unpredictable.

Notice we do *not* require x_i to be unpredictable given $x_j, j \neq i$. As such, distributions such as $x, x+1, x+2, \mathsf{aux} = \emptyset$ are considered unpredictable.

We now consider hardcore functions:

Definition 3. *Let $\mathcal{G}$ be a sampling procedure for deterministic functions G on $n = n(\lambda)$ bits with $m = m(\lambda)$ bit outputs. We say that $\mathcal{G}$ is* hardcore for any computationally unpredictable source *if for any computationally unpredictable source D for $x \in \{0,1\}^n$, and any adversary A, there is a negligible function ϵ such that:*

$$|\Pr[A(G, G(x), \mathsf{aux}) = 1] - \Pr[A(G, R, \mathsf{aux})]| < \epsilon(\lambda)$$

where $G \leftarrow \mathcal{G}$, $(x, \mathsf{aux}) \leftarrow D(\lambda)$ and R is random in $\{0,1\}^m$. In other words, $G(x)$ is pseudorandom even given aux.

Definition 4. *Let $\mathcal{G}$ be a sampling procedure for deterministic functions G on $n = n(\lambda)$ bits with $m = m(\lambda)$ bit outputs. We say that $\mathcal{G}$ is* hardcore for any computationally unpredictable source *over d-inputs if for any computationally unpredictable source D for d inputs $x_1 \ldots, x_d \in \{0,1\}^n$, and any adversary A, there is a negligible function ϵ such that:*

$$|\Pr[A(G, G(x_1), \ldots, G(x_d), \mathsf{aux}) = 1] - \Pr[A(G, R_1, \ldots, R_d, \mathsf{aux})]| < \epsilon(\lambda)$$

where $G \leftarrow \mathcal{G}$, $(x_1, \ldots, x_d, \mathsf{aux}) \leftarrow D(\lambda)$ and $R_1 \ldots, R_d$ are random in $\{0,1\}^m$. In other words, $G(x)$ is pseudorandom even given aux *and the correlated inputs.*

2.1 Deterministic Public Key Encryption

A deterministic public key encryption scheme is a tuple of efficient algorithms (DPKE.Gen, DPKE.Enc, DPKE.Dec), where DPKE.Enc, DPKE.Dec are deterministic maps between messages and ciphertexts, and DPKE.Gen is randomized procedure for producing secret and public key pairs.

For security, we consider several possible notions. Security for arbitrary computational sources means that $(\mathsf{pk}, c^* = \mathsf{DPKE.Enc}(\mathsf{pk}, m), \mathsf{aux})$ is computationally indistinguishable from $(\mathsf{pk}, c^* = \mathsf{DPKE.Enc}(\mathsf{pk}, R), \mathsf{aux})$, where (m, aux) is sampled from an arbitrary computationally unpredictable source and R is uniformly random, and $(\mathsf{sk}, \mathsf{pk}) \leftarrow \mathsf{DPKE.Gen}(\lambda)$. CCA security means the same holds even if the adversary can later ask for decryption queries on ciphertexts other that c^*. Security for arbitrary correlated sources means that $(\mathsf{pk}, \mathsf{DPKE.Enc}(\mathsf{pk}, m_1), \ldots, \mathsf{DPKE.Enc}(\mathsf{pk}, m_1), \mathsf{aux})$ is indistinguishable from $(\mathsf{pk}, \mathsf{DPKE.Enc}(\mathsf{pk}, R_1), \ldots, \mathsf{DPKE.Enc}(\mathsf{pk}, R_d), \mathsf{aux})$ for arbitrary computationally unpredictable sources on d inputs.

2.2 ELFs

We recall the basic definition of Extremely Lossy Functions (ELFs) from Zhandry. We slightly change notation, but the definition is equivalent.

A Lossy Trapdoor Function, or LTDF [18], is a function family with two modes: an injective mode where the function is injective and there is a trapdoor for inversion, and a lossy mode where the image size of the function is much smaller than the domain. The security requirement is that no polynomial-time adversary can distinguish the two modes. An Extremely Lossy Function, or ELF [22], is a related notion without a trapdoor in the injective mode, but with a more powerful lossy mode. In particular, in the lossy mode the image size can be taken to be a polynomial r. One fixed polynomial r is insufficient (since then the lossy mode could easily be distinguished from injective), but instead, the polynomial r is tuned based on the adversary in question to be just large enough to fool the adversary.

Definition 5 (Zhandry [22]). *An ELF consists of two algorithms* $\mathsf{ELF.GenInj}$ *and* $\mathsf{ELF.GenLossy}$*, as well as a function* $N = N(M)$ *such that* $\log N$ *is polynomial in* $\log M$*.* $\mathsf{ELF.GenInj}$ *takes as input an integer* M*, and outputs the description of a function* $f : [M] \to [N]$ *such that:*

- *f is computable in time polynomial in the bit-length of their input, namely* $\log M$.
- *With overwhelming probability (in* $\log M$*),* f *is injective.*

$\mathsf{ELF.GenLossy}$ *on the other hand takes as input integers* M *and* $r \in [M]$*. It outputs the description of a function* $f : [M] \to [N]$ *such that:*

- *For all* $r \in [M]$, $|f([M])| \leq r$ *with overwhelming probability. That is, the function* f *has image size at most* r.
- *For any polynomial* p *and inverse polynomial function* δ *(in* $\log M$*), there is a polynomial* q *such that: for any adversary* $\mathcal{A}$ *running in time at most* p*, and any* $r \in [q(\log M), M]$*, we have that*

$$|\Pr[\mathcal{A}(f) = 1 : f \leftarrow \mathsf{ELF.GenInj}(M)] - \Pr[\mathcal{A}(f) = 1 : f \leftarrow \mathsf{ELF.GenLossy}(M, r)| < \delta$$

In other words, no polynomial-time adversary $\mathcal{A}$ *can distinguish an injective* f *from an* f *with polynomial image size.*

3 Correlated-Input Hardcore Functions

In this section, we build our correlated-input hardcore function. First, we recall Zhandry's [22] construction of hardcore functions for arbitrarily uncorrelated sources. The following description is taken essentially verbatim from Zhandry.

Construction 1. *Let* q *be the input length and* m *be the output length. Let* λ *be a security parameter. We will consider inputs* x *as* q*-dimensional vectors* $\mathbf{x} \in \mathbb{F}_2^q$*. Let* ELF *be an ELF. Let* $M = 2^{m+\lambda+1}$*, and let* n *be the bit-length of the ELF on input* $m+1$*. Set* $N = 2^n$*. Let* ℓ *be some polynomial in* m, λ *to be determined later. First, we will construct a function* H' *as follows.*

Choose random $f_1, \dots, f_\ell \leftarrow \mathsf{ELF.GenInj}(M)$ *where* $f_i : [M] \to [N]$*, and let* $h_1, \dots, h_{\ell-1} : [N] \to [M/2] = [2^{m+\lambda}]$ *and* $h_\ell : [N] \to [2^m]$ *be sampled from pairwise independent and uniform function families. Define* $\mathbf{f} = \{f_1, \dots, f_\ell\}$ *and* $\mathbf{h} = \{h_1, \dots, h_\ell\}$*. Define* $H'_i : \{0,1\}^i \to [M/2]$ *(and* $H'_\ell : \{0,1\}^\ell \to [2^m]$*) as follows:*

- $H'_0() = 1 \in [2^{m+\lambda}]$
- $H'_i(\mathbf{b}_{[1,i-1]}, b_i)$: *compute* $y_i = H'_{i-1}(\mathbf{b}_{[1,i-1]})$, $z_i \leftarrow f_i(y_i || b_i)$, *and output* $y_{i+1} \leftarrow h_i(z_i)$.

Then we set $H' = H'_\ell$. *Then to define* H, *choose a random matrix* $\mathbf{R} \in \mathbb{F}_2^{\ell \times q}$. *The description of* H *consists of* $\mathbf{f}, \mathbf{h}, \mathbf{R}$. *Then set* $H(x) = H'(\mathbf{R} \cdot \mathbf{x})$. *A diagram of* H *is given in Fig. 1.*

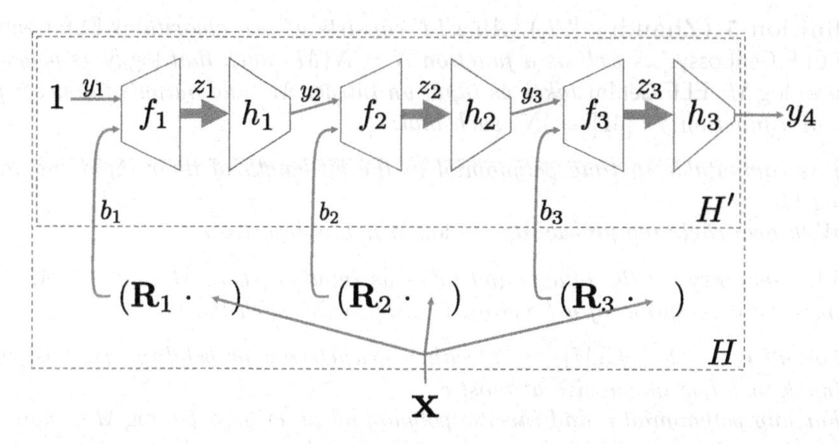

Fig. 1. An example taken from Zhandry [22] for $\ell = 3$. Notice that each iteration is identical, except for the final iteration, where h_ℓ has a smaller output.

Our Construction. We will modify Zhandry's construction as follows. Sample $\mathbf{f}, \mathbf{h}$ as in Construction 1. Then define the function $H_{\mathbf{R}}(\mathbf{x})$ to be the function H using Goldreich-Levin parities $\mathbf{R}$.

Our modification will be to generate $\mathbf{R}$ as a function of $\mathbf{x}$, and then apply $H_{\mathbf{R}}(\mathbf{x})$. In particular, we will set $\mathbf{R} = u(v(\mathbf{x}))$ where $v \leftarrow \mathsf{ELF.GenInj}(M)$ and u is a d-wise independent function. Actually, we need a stronger property of u: that each row of $\mathbf{R}$ is specified by an independent d-wise independent function u_i.

Theorem 2. *If* ELF *is a secure ELF, then* $H_{u(v(\mathbf{x}))}(\mathbf{x}) = H'(u(v(\mathbf{x})) \cdot \mathbf{x})$ *is a hardcore function for computationally unpredictable sources on* d *inputs, for any constant* d.

Proof. First, we recall some basic facts proved by Zhandry:

Claim. If $\ell \geq m + \lambda$, and if $\mathbf{b}$ is drawn uniformly at random, then $(H', H'(\mathbf{b}))$ is statistically close to (H', R) where R is uniformly random in $[2^m]$.

Therefore, given a source D which samples messages $m_1, \dots, m_d$ and auxiliary information aux, it is sufficient to prove the following are indistinguishable:

$(\mathbf{f}, \mathbf{h}, u, v, \mathsf{aux}, \{H'(u(v(\mathbf{x}_i)) \cdot \mathbf{x}_i)\}_i)$ and $(\mathbf{f}, \mathbf{h}, u, v, \mathsf{aux}, \{H'(\mathbf{b}_i)\}_i)$ for uniformly random $\mathbf{b}_i$.

Our proof will follow the same high-level idea as in Zhandry, but make adjustments along the way in order to prove security for correlated sources. Let $\mathcal{A}$ be an adversary with non-negligible advantage ϵ in distinguishing the two cases. We will assume it always checks that the images $v(\mathbf{x}_i)$ are all distinct and rejects if they are; by the property of the source D and the injectivity of v, this check will never trigger if sampled as above. Nonetheless, if the check triggers, we assume A outputs a random bit and aborts.

Let $\mathbf{R}_i = u(v(\mathbf{x}_i))$. Define $\mathbf{b}_i^{(j)}$ so that the first j bits of $\mathbf{b}_i^{(j)}$ are equal to the first j bits of $\mathbf{R}_i \cdot \mathbf{x}_i$, and the last $\ell - i$ bits are uniformly random and independent of $\mathbf{x}_1, \ldots, \mathbf{x}_d$.

We now define a sequence of hybrids. In **Hybrid** j, $\mathcal{A}$ is given the distribution $(\mathbf{f}, \mathbf{h}, u, v, \mathsf{aux}, \{H'(\mathbf{b}_i^{(j)})\}_i)$. Then $\mathcal{A}$ distinguishes **Hybrid 0** from **Hybrid** ℓ with probability ϵ. Now we choose an j at random from $[\ell]$. The adversary distinguishes **Hybrid** $j-1$ from **Hybrid** j with expected advantage at least ϵ/ℓ. Next, observe that since bits $j+1$ through t are random in either case, they can be simulated independently of the challenge. Moreover, $H'(\mathbf{b})$ can be computed given $H'_{j-1}(\mathbf{b}_{[j-1]})$, the bit b_j (be it random or equal to $\mathbf{R} \cdot \mathbf{x}$), and the random $b_{j+1}, \ldots, b_\ell$. Also, the d-wise independent functions $u_{j+1}, \ldots, u_\ell$ are never evaluated on the $\mathbf{x}_i$, so they can be simulated as well. Let $u_{[j]}(x)$ denote the output $(u_1(x), \ldots, u_j(x))$.

Thus, we can construct an adversary $\mathcal{A}'$ that distinguishes the following distributions:

$$(j, \mathbf{f}, \mathbf{h}, u_1, \ldots, u_j, v, \mathsf{aux}, \{H'_{j-1}(u_{[j-1]}(v(\mathbf{x}_i)) \cdot \mathbf{x}_i), u_j(v(\mathbf{x}_i)) \cdot \mathbf{x}_i\}_i) \text{ and}$$
$$(j, \mathbf{f}, \mathbf{h}, u_1, \ldots, u_j, v, \mathsf{aux}, \{H'_{j-1}(u_{[j-1]}(v(\mathbf{x}_i)) \cdot \mathbf{x}_i), b_i\}_i)$$

with advantage ϵ/ℓ, where j is chosen randomly in $[\ell]$, where b_i are random bits.

Next, notice that $\epsilon/5\ell$ is non-negligible, meaning there is an inverse polynomial δ such that $\epsilon/5\ell \geq \delta$ infinitely often. Then, there is a polynomial r such $\mathcal{A}'$ cannot distinguish f_i generated as $\mathsf{ELF.GenLossy}(M, r)$ from the honest f_i generated from $\mathsf{ELF.GenInj}(M)$, except with probability at most δ. Similarly we'll generate v by $\mathsf{ELF.GenLossy}(M, r)$.

This means, if we generate $f_i, v \leftarrow \mathsf{ELF.GenLossy}(M, r)$, we have that $\mathcal{A}'$ still distinguishes the distributions

$$(j, \mathbf{f}, \mathbf{h}, u_1, \ldots, u_j, v, \mathsf{aux}, \{H'_{j-1}(u_{[j-1]}(v(\mathbf{x}_i)) \cdot \mathbf{x}_i), u_j(v(\mathbf{x}_i)) \cdot \mathbf{x}_i\}_i) \text{ and}$$
$$(j, \mathbf{f}, \mathbf{h}, u_1, \ldots, u_j, v, \mathsf{aux}, \{H'_{j-1}(u_{[j-1]}(v(\mathbf{x}_i)) \cdot \mathbf{x}_i), b_i\}_i)$$

with advantage $\epsilon' = \epsilon/\ell - 4\delta$.

Next, we define new hybrids $J_0, \ldots, J_d$, where J_k is the distribution:

$$(j, \mathbf{f}, \mathbf{h}, u_1, \ldots, u_j, v, \mathsf{aux}, \{H'_{i-1}(u_{[j-1]}(v(\mathbf{x}_i)) \cdot \mathbf{x}_i), q_i\}_i)$$

where $q_i = u_j(v(\mathbf{x}_i)) \cdot \mathbf{x}_i$ for $i \leq k$ and q_i is uniformly random for $i > k$. Notice that J_0 and J_d are the two distributions distinguished with probability ϵ'.

Therefore, for a random $k \in [d]$, the expected distinguishing advantage between J_{i-1} and J_i is ϵ'/d. Thus, $\mathcal{A}'$ can be used to construct an adversary $\mathcal{A}''$ that distinguishes the two distributions:

$$\begin{pmatrix} j, k, \mathbf{f}, \mathbf{h}, \{u_i\}_i, v, \mathsf{aux}, \\ \{H'_{j-1}(u_{[j-1]}(v(\mathbf{x}_i)) \cdot \mathbf{x}_i)\}_{i \in [d]}, \{u_j(v(\mathbf{x}_i)) \cdot \mathbf{x}_i\}_{i<k}, u_j(v(\mathbf{x}_k)) \cdot \mathbf{x}_k \end{pmatrix} \text{ and}$$
$$\begin{pmatrix} j, k, \mathbf{f}, \mathbf{h}, \{u_i\}_i, v, \mathsf{aux}, \\ \{H'_{j-1}(u_{[j-1]}(v(\mathbf{x}_i)) \cdot \mathbf{x}_i)\}_{i \in [d]}, \{u_j(v(\mathbf{x}_i)) \cdot \mathbf{x}_i\}_{i<k}, b_k \end{pmatrix}$$

with advantage $\epsilon'/4$. Next, we devise an adversary $\mathcal{A}'''$ which distinguishes

$$\begin{pmatrix} j, k, \mathbf{f}, \mathbf{h}, \{u_i\}_i, v, \mathsf{aux}, \\ \{v(\mathbf{x}_i), H'_{i-1}(u_{[j-1]}(v(\mathbf{x}_i)) \cdot \mathbf{x}_i)\}_i, \{\mathbf{r}_i \cdot \mathbf{x}_i\}_{i<k}, \mathbf{r}_k \cdot \mathbf{x}_k \end{pmatrix} \text{ and}$$
$$\begin{pmatrix} j, k, \mathbf{f}, \mathbf{h}, \{u_i\}_i, v, \mathsf{aux}, \\ \{v(\mathbf{x}_i), H'_{i-1}(u_{[j-1]}(v(\mathbf{x}_i)) \cdot \mathbf{x}_i)\}_i, \{\mathbf{r}_i \cdot \mathbf{x}_i\}_{i<k}, b_k \end{pmatrix}$$

We recall that our adversary aborts if $v(\mathbf{x}_i)$ are not distinct. In the case where they are distinct, given one of the samples in the preceding equations, $\mathcal{A}'''$ samples u_j such that $u_j(v(\mathbf{x}_i)) = \mathbf{r}_i$ and that u_j is sampled uniformly according to the d-wise independent sampling procedure. Then $\mathcal{A}'''$ simulates the samples expected by $\mathcal{A}''$. The result is $\mathcal{A}'''$ distinguishes the two cases with probability ϵ'/d.

Now fix $\mathbf{f}, \mathbf{h}, u_1, \ldots, u_{j-1}, v$, which fixes H'_{i-1}. Let $y_i^{(j)} = H'_{j-1}(u_{[j-1]}(v(\mathbf{x}_i)) \cdot \mathbf{x}_i)$. Notice that since $\mathbf{f}, \mathbf{h}, u_1, \ldots, u_{j-1}, v$ are fixed and H'_{j-1} has image size at most r, there are at most r^d possible values for the vector $(y_1^{(j)}, \ldots, y_d^{(j)})$, and recall that r is a polynomial. If d is constant, then r^d is still polynomial. Moreover, there are at most r^d values for the vector $(v(\mathbf{x}_1), \ldots, v(\mathbf{x}_d))$.

Now, we use Lemma 1. Since $\mathbf{x}_k, \mathsf{aux}$ is computationally unpredictable and since there are only a polynomial number of images of v and H'_{j-1}, we have

$$(\mathbf{x}_k, (j, k, \mathbf{f}, \mathbf{h}, u_1, \ldots, u_{j-1}, v, \mathsf{aux}, \{v(\mathbf{x}_i), H'_{j-1}(u_{[j-1]}(v(\mathbf{x}_i)) \cdot \mathbf{x}_i)\}_i))$$

is computationally unpredictable as well. Even more, it must be that

$$\left(\mathbf{x}_k, \mathsf{aux}_k = \begin{pmatrix} j, k, \mathbf{f}, \mathbf{h}, \{u_i\}_i, v, \mathsf{aux}, \\ \{v(\mathbf{x}_i), H'_{j-1}(u_{[j-1]}(v(\mathbf{x}_i)) \cdot \mathbf{x}_i)\}_i, \{\mathbf{r}_i, \mathbf{r}_i \cdot \mathbf{x}_i\}_{i<k} \end{pmatrix}\right)$$

is computationally unpredictable, since there are only 2^d possible values to guess for $\mathbf{r}_i \cdot \mathbf{x}_i$.

Therefore, by Goldreich-Levin, we have that $(\mathsf{aux}_k, \mathbf{r}_k, \mathbf{r}_k \cdot \mathbf{x}_k)$ is computationally indistinguishable from $(\mathsf{aux}_k, \mathbf{r}_k, b_k)$ for random b_k. Putting this together in a simple hybrid argument, we have that the following are indistinguishable:

$$\begin{pmatrix} j, k, \mathbf{f}, \mathbf{h}, u_1, \ldots, u_{j-1}, v, \mathsf{aux}, \\ \{v(\mathbf{x}_i), H'_{i-1}(u_{[j-1]}(v(\mathbf{x}_i)) \cdot \mathbf{x}_i)\}_i, \{\mathbf{r}_i \cdot \mathbf{x}_i\}_{i<k}, \mathbf{r}_k \cdot \mathbf{x}_k \end{pmatrix} \text{ and}$$
$$\begin{pmatrix} j, k, \mathbf{f}, \mathbf{h}, u_1, \ldots, u_{j-1}, v, \mathsf{aux}, \\ \{v(\mathbf{x}_i), H'_{i-1}(u_{[j-1]}(v(\mathbf{x}_i)) \cdot \mathbf{x}_i)\}_i, \{\mathbf{r}_i \cdot \mathbf{x}_i\}_{i<k}, b_k \end{pmatrix}$$

But these are exactly the distributions distinguished by $\mathcal{A}'''$. Therefore, we must have ϵ'/d, and hence ϵ', is negligible. But since $\epsilon' = \epsilon/\ell - 4\delta$ and $\delta \leq \epsilon/5\ell$, we have that ϵ' is lower bounded by $\delta/5/\ell$ infinitely often, a contradiction. This completes the proof. □

4 Trapdoor ELFs

Here, we define and construct ELFs with a trapdoor, combining the features of LDTFs and ELFs.

Definition 6. *An Trapdoor ELF consists of two algorithms* $\mathsf{TELF.GenInj}$ *and* $\mathsf{TELF.GenLossy}$, *as well as a function* $N = N(M)$ *such that* $\log N$ *is polynomial in* $\log M$. $\mathsf{TELF.GenInj}$ *takes as input an integer* M, *and outputs the description of two functions* $f : [M] \to [N]$ *and* $f^{-1} : [N] \to [M] \cup \{\perp\}$ *such that:*

- *f, f^{-1} are computable in time polynomial in the bit-length of their input, namely* $\log M$.
- *With overwhelming probability (in* $\log M$*),* $f^{-1}(f(x)) = x$ *for all* $x \in [M]$. *In particular* f *is injective.*

$\mathsf{TELF.GenLossy}$ *on the other hand takes as input integers* M *and* $r \in [M]$. *It outputs the description of a function* $f : [M] \to [N]$ *such that:*

- *For all* $r \in [M]$, $|f([M])| \leq r$ *with overwhelming probability. That is, the function* f *has image size at most* r.
- *For any polynomial* p *and inverse polynomial function* δ *(in* $\log M$*), there is a polynomial* q *such that: for any adversary* $\mathcal{A}$ *running in time at most* p, *and any* $r \in [q(\log M), M]$, *we have that*

$$|\Pr[\mathcal{A}(f) = 1 : (f, f^{-1}) \leftarrow \mathsf{TELF.GenInj}(M)] - \Pr[\mathcal{A}(f) = 1 : f \leftarrow \mathsf{TELF.GenLossy}(M, r)| < \delta$$

In other words, no polynomial-time adversary $\mathcal{A}$ *can distinguish an injective* f *from an* f *with polynomial image size, in the case that* $\mathcal{A}$ *does not get the trapdoor for* f.

We also consider *all-but-some* Trapdoor ELFs, which contain many branches, some of which are lossy:

Definition 7. *An All-but-one Trapdoor ELF consists of algorithms* $\mathsf{TELF.GenInj}$ *and* $\mathsf{TELF.GenLossy}$, *as well as a function* $B = B(M), N = N(M)$ *such that* $\log B, \log N$ *are polynomial in* $\log M$. $\mathsf{TELF.GenInj}$ *takes as input an integer* M, *and outputs the description of two functions* $f : [B] \times [M] \to [N]$ *and* $f^{-1} : [B] \times [N] \to [M] \cup \{\perp\}$ *such that:*

- *f, f^{-1} are computable in time polynomial in the bit-length of their input, namely* $\log M$.

- *With overwhelming probability (in* $\log M$*), for all branches* $b \in [B]$*, we have that* $f^{-1}(b, f(b, x)) = x$ *for all* $x \in [M]$*. In particular* $f(b, \cdot)$ *is injective.*

$\mathsf{TELF.GenLossy}$ *on the other hand takes as input integers* M *and* $r \in [M]$*, and a branch* $b^* \in [B]$*. It outputs the description of functions* $f : [B] \times [M] \to [N]$ *and* $f^{-1} : [B] \times [N] \to [M] \cup \{\perp\}$ *such that:*

- *For all* $r \in [M]$*,* $|f(b, [M])| \leq r$ *with overwhelming probability. That is, the function* $f(b^*, \cdot)$ *has image size at most* r*.*
- *With overwhelming probability (in* $\log M$*), for all branches* $b \in [B] \setminus \{b^*\}$*,* $f^{-1}(b, f(b, x)) = x$ *for all* $x \in [M]$*.*
- *For any polynomial* p *and inverse polynomial function* δ *(in* $\log M$*), there is a polynomial* q *such that: for any adversary* $\mathcal{A}$ *running in time at most* p *and playing the following game, it's advantage is at most* δ*:*
  - *First,* $\mathcal{A}$ *chooses a branch* b^**, which is sends to the challenger.*
  - *The challenger then either runs* $(f, f^{-1}) \leftarrow \mathsf{TELF.GenInj}(M)$ *or runs* $(f, f^{-1} \leftarrow \mathsf{TELF.GenLossy}(M, b^*, r)$*, and sends* f *to* $\mathcal{A}$*.*
  - $\mathcal{A}$ *can make queries to* f^{-1} *on all branches other than* b^**.*
  - $\mathcal{A}$ *outputs a guess* b *for which* f *it was given*

  $\mathcal{A}$*'s advantage is defined to be the difference*

$$|\Pr[\mathcal{A}(f) = 1 : (f, f^{-1}) \leftarrow \mathsf{TELF.GenInj}(M)] \\ - \Pr[\mathcal{A}(f) = 1 : (f, f^{-1}) \leftarrow \mathsf{TELF.GenLossy}(M, b^*, r)|$$

In other words, no polynomial-time adversary $\mathcal{A}$ *can distinguish an injective* f *from an* f *where branch* b^* *has polynomial image size, in the case that* $\mathcal{A}$ *does not get the trapdoor for* f*.*

An all-but-some Trapdoor ELF generalizes the above to allow the lossy mode to contain multiple lossy branches. We omit the details of the definition.

4.1 Constructing Trapdoor ELFs

Here, we construct Trapdoor ELFs from exponentially-hard DDH, which is plausible on certain elliptic curve groups. Our construction will follow mostly Zhandry's [22] construction of ELFs, with some modifications to obtain a trapdoor.

Zhandry's scheme works as follows: first, he considers a *bounded adversary* ELF, which is secure against only adversaries of an a priori bounded running time. This scheme more or less follows from lossy trapdoor functions in the literature, just pushed into extreme parameter regimes. Then, he iterates the scheme many times, for many different bounds on the adversaries running time. ELF security follows by invoking security for the bounded adversary ELF that is just large enough to fool the given adversary.

We will adopt the same approach. In particular, we will construct a bounded adversary Trapdoor ELF following the LTDFs from the literature. We will trivially inherit the trapdoors from these schemes. Then, we will iterate the construction. Zhandry's construction, in order to remain efficient, must compress the image every after every iteration. This unfortunately means Zhandry's construction does not have a functioning trapdoor. We therefore devise a way to avoid compressing the input, allowing the trapdoor to remain intact.

Bounded Adversary Trapdoor ELFs. Here, we define a bounded adversary Trapdoor ELF, which is a Trapdoor ELF where security is guaranteed only against a prior bounded adversaries. The definition follows almost immediately from adapting Zhandry's bounded adversary ELF definition by adding a trapdoor.

Informally, in an ordinary Trapdoor ELF, r can be chosen based on the adversary to be just high enough to fool it. In contrast, in a bounded adversary Orf, r must be chosen independent of the adversary, and then security only applies to adversaries with running time sufficiently smaller than r. Moreover, the adversary gets to learn r.

Definition 8. *An* bounded adversary *Trapdoor ELF consists of two algorithms* $\mathsf{TELF.GenInj}'$ *and* $\mathsf{TELF.GenLossy}'$*, and a function* $N = N(M, r)$. $\mathsf{TELF.GenInj}'$ *takes as input an integer* M *and integer* $r \in [M]$ *and outputs the description of two functions* $f : [M] \to [N]$ *and* $f^{-1} : [N] \to [M] \cup \{\perp\}$ *such that:*

- f, f^{-1} *are computable in time polynomial in the bit-length of their input, namely* $\log M$.
- *With overwhelming probability (in* $\log M$*),* $f^{-1}(f(x)) = x$ *for all* $x \in [M]$*. In particular* f *is injective.*

$\mathsf{TELF.GenLossy}'$ *also takes as input integers* M *and* $r \in [M]$*. It outputs the description of a function* $f : [M] \to [N]$ *such that:*

- *For all* $r \in [M]$, $|f([M])| \leq r$ *with overwhelming probability. That is, the function* f *has image size at most* r.
- *For any polynomial* p *and inverse polynomial function* δ *(in* $\log M$*), there is a polynomial* q *such that: for any adversary* $\mathcal{A}$ *running in time at most* p*, and any* $r \in [q(\log M), M]$*, we have that*

$$|\Pr[\mathcal{A}(r, f) = 1 : (f, f^{-1}) \leftarrow \mathsf{TELF.GenInj}'(M)] - \Pr[\mathcal{A}(r, f) = 1 : f \leftarrow \mathsf{TELF.GenLossy}'(M, r)| < \delta$$

In other words, no polynomial-time adversary $\mathcal{A}$ *can distinguish an injective* f *from an* f *with polynomial image size, in the case that* $\mathcal{A}$ *does not get the trapdoor for* f*. Unlike an ordinary Trapdoor ELF, this holds* even if the adversary knows r.

Constructing Bounded Adversary Trapdoor ELFs. Our construction of bounded adversary Trapdoor ELFs, like Zhandry's ELFs, is based on the DDH-based lossy *trapdoor* functions of Peikert and Waters [18] and Freeman et al. [13]. In fact, since Zhandry did not need the trapdoor of prior constructions, the construction for ELFs was very slightly simplified. In contrast, our construction almost verbatim matches the construction Freeman et al., except that the group size is set to be much smaller, in particular polynomial. In order to maintain security in this regime, we must rely on the exponential hardness of the group.

Cryptographic Groups. The following definitions and notation are almost verbatim from Zhandry [22].

Definition 9. *A* cryptographic group *consists of an algorithm* Group.Gen *which takes in a security parameter* λ*, and produces a (description of a) cyclic group* $\mathbb{G}$ *of prime order* $p \in [2^\lambda, 2 \times 2^\lambda)$*, and a generator* g *for* $\mathbb{G}$ *such that:*

- *The group operation* $\times : \mathbb{G}^2 \to \mathbb{G}$ *is polynomial-time computable in* λ*.*
- *Exponentiation by elements in* $\mathbb{Z}_p$ *is polynomial-time computable in* λ*.*
- *The representation of a group element* h *has size polynomial in* λ*.*

For some notation: given a matrix $\mathbf{A} \in \mathbb{Z}_p^{m \times n}$, we write $g^{\mathbf{A}} \in \mathbb{G}^{m \times n}$ to be the $m \times n$ matrix of group elements $g^{A_{i,j}}$. Analogously define $g^{\mathbf{w}}$ for a vector $\mathbf{w} \in \mathbb{Z}_p^n$. Given a matrix $\hat{\mathbf{A}} \in \mathbb{G}^{m \times n}$ of group elements and a vector $\mathbf{v} \in \mathbb{Z}_p^n$, write $\hat{\mathbf{A}} \cdot \mathbf{v}$ to mean $\hat{\mathbf{w}} \in \mathbb{G}^m$ where $\hat{w}_i = \prod_{j=1}^n \hat{A}_{i,j}^{v_j}$. Using this notation, $(g^{\mathbf{A}}) \cdot \mathbf{v} = g^{\mathbf{A} \cdot \mathbf{v}}$. Therefore, the map $g^{\mathbf{A}}, \mathbf{v} \mapsto g^{\mathbf{A} \cdot \mathbf{v}}$ is efficiently computable.

Definition 10. *The* exponential decisional k-linear assumption *(k-eLin) on a cryptographic group specified by* Group.Gen *holds if there is a polynomial* $q(\cdot\cdot)$ *such that the following is true. For any time bound* t *and probability* ϵ*, let* $\lambda = \log q(t, 1/\epsilon)$*. Then for any adversary* $\mathcal{A}$ *running in time at most* t*, the following two distributions are indistinguishable, except with advantage at most* ϵ*:*

$$(\mathbb{G}, g, g^{a_1}, \ldots, g^{a_k}, g^c, g^{a_1 b_1}, \cdots g^{a_k b_k}) : \begin{matrix} (\mathbb{G}, g, p) \leftarrow \mathsf{Group.Gen}(\lambda) \\ a_i, b_i, c \leftarrow \mathbb{Z}_p \end{matrix}, \textit{ and}$$

$$(\mathbb{G}, g, g^{a_1}, \ldots, g^{a_k}, g^{\sum_{i=1}^k b_i}, g^{a_1 b_1}, \cdots g^{a_k b_k}) : \begin{matrix} (\mathbb{G}, g, p) \leftarrow \mathsf{Group.Gen}(\lambda) \\ a_i, b_i \leftarrow \mathbb{Z}_p \end{matrix}$$

$k = 1$ corresponds to the eDDH assumption above.

As a special case, $k = 1$ corresponds to the exponential DDH assumption. A plausible candidate for a cryptographic group supporting the eDDH assumption or k-linear assumption are groups based on elliptic curves. Despite over a decade or research, the best attacks on many elliptic curves are generic attacks which require exponential time. Therefore, the eDDH assumption on these groups appears to be a very reasonable assumption.

Construction. Our construction is as follows, and will be parameterized by k. $\mathsf{TELF.GenInj}'_k(M, r)$ does the following.

- Let λ be the largest integer such that $(2 \times 2^\lambda)^k < r$. Run $(\mathbb{G}, g, p) \leftarrow$ Group.Gen(λ).

- Let m be the smallest integer such that $2^m \geq M$. Let R be an efficiently invertible function from $[M]$ into $\{0,1\}^m$.
- Let $n \geq m$ (e.g. $m = 2n$) be chosen such that a random matrix sampled from $\mathbb{Z}_p^{n\times m}$ has rank m with overwhelming probability. Note that a random *square* matrix will be singular with probability $1/p$, and in our case, p is polynomial. Hence we require m somewhat larger than n.
- Choose a random matrix $n \times m$ matrix $\mathbf{A}$ of elements in $\mathbb{Z}_p^{n\times m}$. Set $\hat{\mathbf{A}} = g^{\mathbf{A}}$.
- Output functions f, f^{-1}. f is defined as $f(x) = \hat{\mathbf{A}} \cdot (R(x))$. The description of f will consist of $(\mathbb{G}, p, \hat{\mathbf{A}}, R, m, n)$.
  f^{-1} is defined as follows. Let $\mathbf{B} \in \mathbb{Z}_p^{m\times n}$ such that $\mathbf{B}\cdot\mathbf{A}$ is the identity. Given a vector $\mathbf{v} \in \mathbb{G}^n$, compute $\mathbf{w} = \mathbf{B} \cdot \mathbf{v}$. Then, try to compute the discrete log of each component by testing if the component is g^0 or g^1. If any of the discrete log computations fail, then output $\perp$. Otherwise, let $\mathbf{y}$ be the vector of exponents obtained. Invert R on y to obtain x. If inversion fails, output $\perp$. Otherwise, output x. The description of f^{-1} will consist of $(\mathbb{G}, p, \mathbf{B}, R, m, n)$.
  $\mathsf{TELF.GenLossy}'_k(M, r)$ is identical to $\mathsf{TELF.GenInj}'_k(M, r)$, except the matrix $\mathbf{A}$ is chosen to be random of rank k, rather than full rank. In this case, there is no $\mathbf{B}$ and hence no function f^{-1}.

Theorem 3. *If* Group.Gen *is a group where the k-eLin assumption holds for some constant k, then* $(\mathsf{TELF.GenInj}'_k, \mathsf{TELF.GenLossy}'_k)$ *is a bounded adversary Trapdoor ELF.*

Proof. For correctness, notice that $\mathbf{w}$ computed by f^{-1} is equal to $\mathbf{B}\cdot\mathbf{v} = \mathbf{B}\cdot\hat{\mathbf{A}}\cdot R(x) = g^{\mathbf{B}\cdot\mathbf{A}\cdot R(x)} = g^{R(x)}$. Therefore, when f^{-1} is given a valid output of f, it will recover $g^{R(x)}$, and the discrete log computations will yield $R(x)$ and the final inversion of R will yield x, as desired.

Security follows from an almost identical argument to the security of bounded adversary ELFs in Zhandry [22], and we only sketch the details here. All that needs to be shown is that $g^{\mathbf{A}}$ for a random matrix is indistinguishable from $g^{\mathbf{A}}$ for a random rank-k matrix. This follows by standard hybrid arguments (e.g. [20]) and the assumed k-linear assumption. □

Constructing Ordinary Trapdoor ELFs. We now turn to using bounded adversary Trapdoor ELFs to construct ordinary Trapdoor ELFs. Here, we depart slightly from Zhandry [22]. Zhandry's idea is to iterate many bounded adversary functions as r ranges over the powers of 2. The injective mode just sets all the bounded adversary functions to be injective. For the lossy mode, a single function is set to be lossy, namely the function that is big enough to fool the adversary in question.

One issue that immediately becomes apparent in the above approach is that the bounded adversary functions are expanding. As such, the overall domain will grow exponentially with the number of iterations, leading to an inefficient scheme. Zhandry gets around this by applying a pairwise independent function between each bounded adversary function to compress the output and keep it polynomial in size. Unfortunately, this compression destroys any trapdoor in the bounded adversary function.

Instead, our approach is to *not* compress the outputs, but be very careful about which r we choose for our bounded adversary Trapdoor ELFs. In particular, notice that our bounded adversary Trapdoor ELFs expand the input by a factor of $C_k \times \log r$, for some constant C that depends on k. Therefore, if our construction uses a sequence $r_1, \ldots, r_t$ of r's, the overall expansion is $C_k^t \prod \log r_i$. We need this expansion factor to be polynomial in size.

Notice that the powers of 2, namely $r_i = 2^i$, used by Zhandry do not work, as the overall expansion will be $C_k^t t!$. We need $r_t = 2^t$ to be larger than any polynomial in our security parameter $\log M$ (so that we can set r based on any adversary), meaning the overall expansion factor will be at least $C_k^{\log\log M}(\log\log M)!$. Notice that $(\log\log M)!$ is super-polynomial in $\log M$, leading to an inefficient scheme.

Instead, we choose $r_i = 2^{2^i}$, and let i go from 1 to $t = \sqrt{\log\log M}$. We see that the overall expansion factor is:

$$
\begin{aligned}
C_k^t \prod_{i=1}^t \log r_i &= C_k^{\sqrt{\log\log M}} \prod_{i=1}^t 2^i \\
&\leq C_k^{\log\log M} \prod_{i=1}^t 2^i = (\log M)^{\log C_k} \prod_{i=1}^t 2^i \\
&= (\log M)^{\log C_k} 2^{\sum_{i=1}^t i} \leq= (\log M)^{\log C_k} 2^{t^2} = (\log M)^{1+\log C_k}
\end{aligned}
$$

We also note that $r_t = 2^{2^{\sqrt{\log\log M}}}$ is larger than any polynomial in $\log M$. This means that for any polynomial p, we can always choose i so that r_i will be at most approximately p^2. This is exactly what we need to argue security.

In more detail, our construction does the following. Assume for the bounded adversary Trapdoor ELF that $N = N(M, r)$ satisfies $\log N \leq C(\log M)(\log r)$ for some universal constant C, as in our bounded adversary construction. Then $\mathsf{TELF.GenInj}(M)$ does the following:

- Let t be the smallest integer such that $2^{2^{t^2}} \geq M$.
- Let $M_1 = M$.
- For $i = 1, \ldots, t$, Run $(f_i, f_i^{-1}) \leftarrow \mathsf{TELF.GenInj}'(M_i, r_i)$ for $r_i = 2^{2^i}$. Let N_i be the output space of f_i, and set $M_{i+1} = N_i$.
- Let $f : [M] \to [N_t]$ be $f_t \circ f_{t-1} \circ \cdots \circ f_1$. Let f^{-1} attempt to compute $f_1^{-1} \circ \ldots f_t^{-1}$, and output $\perp$ if any of the inversions fail.
- Output (f, f^{-1}).

$\mathsf{TELF.GenLossy}(M, r)$ is the same as $\mathsf{TELF.GenInj}$, except that it lets i^* be the largest integer such that $r_{i^*} \leq r$ and $i^* \leq t$. It then computes $f_{i^*} \leftarrow \mathsf{TELF.GenLossy}'(M_{i^*}, r_{i^*})$ instead of using $\mathsf{TELF.GenInj}'$. It lets f be defined as above, and outputs f (but no f^*).

Theorem 4. *If* $(\mathsf{TELF.GenInj}', \mathsf{TELF.GenLossy}')$ *is a bounded-adversary Trapdoor ELF satisfying* $\log N \leq C(\log M)(\log r)$ *for some constant* C*, then we have that* $(\mathsf{TELF.GenInj}, \mathsf{TELF.GenLossy})$ *is an ordinary Trapdoor ELF.*

Proof. The image size in the lossy mode is guaranteed by how we chose i^*. Namely, the image size on input r is at most r_{i^*} which is at most r.

It remains to prove security. Let p be a polynomial and σ be an inverse polynomial in $\log M$. Let p' be p plus the running time of $\mathsf{TELF.GenInj}$. Let q be the polynomial guaranteed by $(\mathsf{TELF.GenInj}', \mathsf{TELF.GenLossy}')$ for p' and σ.

Notice that q will be a polynomial in the $\log M_i$, the domain for the functions $(\mathsf{TELF.GenInj}', \mathsf{TELF.GenLossy}')$, and not in $\log M$. Nonetheless, we can redefine q to be a polynomial in $\log M$ since $\log M_i$ is polynomial in $\log M$.

Consider any adversary A for $(\mathsf{TELF.GenInj}, \mathsf{TELF.GenLossy})$ running in time at most p. Let $r = r(M)$ be a computable function of M such that $r \in (q(\log M), M]$. Our goal is to show that A distinguishes f from $\mathsf{TELF.GenInj}(M)$ from $\mathsf{TELF.GenLossy}(M, r)$ with advantage less than δ.

Toward that goal, let i^* be the largest integer such that $r_{i^*} = 2^{2^{i^*}} \leq r$ and $i^* \leq t$. We construct an adversary A' for $(\mathsf{TELF.GenInj}', \mathsf{TELF.GenLossy}')$ with $r = r_{i^*}$. Let f_{i^*} be the f that A' receives, where f_{i^*} is either $\mathsf{TELF.GenInj}'(M, r_{i^*})$ or $\mathsf{TELF.GenLossy}'(M, r_{i^*})$. Then A' simulates the rest of f for itself, setting $(f_i, f_i^{-1}) \leftarrow \mathsf{TELF.GenInj}'(M_i, r_i)$ for $i \neq i^*$. A' then runs A on the simulated f. Notice that A' runs in time at most p'. Thus by the bounded-adversary security of $(\mathsf{TELF.GenInj}', \mathsf{TELF.GenLossy}')$, A' cannot distinguish injective or lossy mode, except with advantage σ. Moreover, if $f_{i^*} \leftarrow \mathsf{TELF.GenInj}'(M, r_{i^*})$, then this corresponds to $\mathsf{TELF.GenInj}$, and if $f_{i^*} \leftarrow \mathsf{TELF.GenLossy}'(M, r_{i^*})$, then this corresponds to $\mathsf{TELF.GenLossy}(M, r)$. Thus, A' and A have the same distinguishing advantage, and therefore A cannot distinguish the two cases except with probability less than σ. □

4.2 Constructing All-but-some Trapdoor ELFs

We now turn to constructing All-but-some Trapdoor ELFs. It is sufficient to construct a bounded adversary version of All-but-some Trapdoor ELFs, which can then be converted into full All-but-some Trapdoor ELFs using the conversion in the preceding section. Here, we describe how to do this. We focus on the all-but-one case, the all-but-some being a simple generalization.

Construction. Our construction is as follows, and will be parameterized by k. The branch set B will be interpreted as $\{0,1\}^a$ for some polynomial a. $\mathsf{TELF.GenInj}_k'(M, b, r)$ does the following.

- Let λ be the largest integer such that $(2 \times 2^\lambda)^k < r$. Run $(\mathbb{G}, g, p) \leftarrow \mathsf{Group.Gen}(\lambda)$.
- Let m be the smallest integer such that $2^m \geq M$. Let R be an efficiently invertible function from $[M]$ into $\{0,1\}^m$.

- Let $n \geq m$ be chosen such that a random matrix sampled from $\mathbb{Z}_p^{n\times m}$ has rank m with overwhelming probability. Note that a random *square* matrix will be singular with probability $1/p$, and in our case, p is polynomial. Therefore, we need to choose an n somewhat larger than m. It suffices to set $n = 2m$.
- Choose $2a+1$ random $2n \times m$ matrices $\mathbf{B}, \mathbf{A}_{i,t}$ in $\mathbb{Z}_q^{2n\times m}$, and let $\hat{\mathbf{A}}_{i,t} = g^{\mathbf{A}_{i,t}}, \hat{\mathbf{B}} = g^{\mathbf{B}}$.
  Define $\mathbf{A}_b = \mathbf{B} + \sum_i \mathbf{A}_{i,b_i}$.
- Output functions f, f^{-1}. f is defined as $f(b,x) = \hat{\mathbf{A}}_b \cdot (R(x))$. Note that $\hat{\mathbf{A}}_b$ can be computed from $\hat{\mathbf{A}}_{i,t}, \hat{\mathbf{B}}$. The description of f will consist of $(\mathbb{G}, p, \hat{\mathbf{B}}, \{\hat{\mathbf{A}}_{i,t}\}, R, m, n)$.
  $f^{-1}(b,v)$ is defined as follows. Let $\mathbf{A}_b^{-1} \in \mathbb{Z}_p^{m\times 2n}$ such that $\mathbf{A}_b^{-1}\cdot\mathbf{A}_b$ is the identity. Given a vector $\mathbf{v}\in\mathbb{G}^n$, compute $\mathbf{w} = \mathbf{A}_b^{-1}\cdot\mathbf{v}$. Then, try to compute the discrete log of each component by testing if the component is g^0 or g^1. If any of the discrete log computations fail, then output $\perp$. Otherwise, let $\mathbf{y}$ be the vector of exponents obtained. Invert R on y to obtain x. If inversion fails, output $\perp$. Otherwise, output x. The description of f^{-1} will consist of $(\mathbb{G}, p, \mathbf{B}, \{\mathbf{A}_{i,t}\}, R, m, n)$.
  $\mathsf{TELF.GenLossy}'_k(M,b,r)$ is identical to $\mathsf{TELF.GenInj}'_k(M,b,r)$, except the matrix $\mathbf{A}_b$ is chosen to be random of rank k, rather than full rank. Then $\mathbf{B}$ is set to $\mathbf{A}_b - \sum_i \mathbf{A}_{i,b_i}$.

Theorem 5. *If* Group.Gen *is a group where the k-eLin assumption holds for some constant k, then* $(\mathsf{TELF.GenInj}'_k, \mathsf{TELF.GenLossy}'_k)$ *is a bounded adversary all-but-one Trapdoor ELF.*

Proof. We just need to show, given a branch b^*, how to embed a challenge $g^{\mathbf{C}}$ into the description of f so that:

- If $\mathbf{C}$ is full rank, $\mathbf{B}, \mathbf{A}_{i,t}$ is distributed as in the injective mode, namely uniformly random.
- If $\mathbf{C}$ has rank k, then $\mathbf{B}, \mathbf{A}_{i,t}$ is distributed as in the lossy mode for branch b^*, namely $\mathbf{A}_{b^*}$ is random of rank k.
- We can simulate inversion queries on all other branches.

To do so, we exploit the fact that we have some extra rows to work with. We will assume the challenge $g^{\mathbf{C}}$ is $n\times m$. We will choose a uniformly random matrix $\mathbf{S}\in\mathbb{Z}_p^{2n\times 2n}$. We will set $\mathbf{A}'_{b^*}$ to be the block matrix with $\mathbf{C}$ on top, and $0^{2n\times m}$ on bottom. Then we will set $\mathbf{A}_{b^*} = \mathbf{S}\cdot\mathbf{A}'_{b^*}$.

We will choose $\mathbf{A}'_{i,t}$ as random $2n\times m$ matrices, and then set $\mathbf{A}_{i,t} = \mathbf{S}\cdot\mathbf{A}'_{i,t}$. Finally, we will set $\mathbf{B} = \mathbf{A}_{b^*} - \sum_i \mathbf{A}_{i,b_i^*} = \mathbf{S}\cdot\left(\mathbf{A}'_{b^*} - \sum_i \mathbf{A}'_{i,b_i^*}\right)$.

We can now compute $g^{\mathbf{A}_{i,t}}$ using our knowledge of $\mathbf{A}_{i,t}$, and $g^{\mathbf{B}}$ using our knowledge of $\mathbf{A}_{i,t}$, $\mathbf{S}$, and $g^{\mathbf{C}}$.

It is straightforward to show that if $\mathbf{C}$ is a uniformly random matrix, then so are all the matrices $\mathbf{B}, \mathbf{A}_{i,t}$. Moreover, if $\mathbf{C}$ is random of rank k, is is straightforward that the matrices are random, subject to $\mathbf{A}_{b^*}$ being rank k, as desired.

It remains to prove that we can answer inversion queries. Here, we simply use the fact that we know the bottom $n \times m$ matrices in the clear, meaning we can perform the inversion operation as in standard Trapdoor ELFs. As the last step, we just verify our inversion by evaluating the Trapdoor ELF on the derived pre-image, ensuring that it matches the provided image point. $\square$

We can easily use the above techniques to extend to ℓ lossy branches in several ways. One way is to simply evaluate ℓ different Trapdoor ELFs in sequence; to set the ℓ different branches, simply assign one branch to each of the Trapdoor ELFs.

5 DPKE for Computationally Unpredictable Sources

In this section, we show our basic DPKE construction, a deterministic public key encryption scheme (DPKE) for arbitrary computational sources.

The Construction. The message space for our scheme is $[M]$. We will use a hard-core function $\mathcal{G}$ with domain $[M]$, a PKE scheme $(\mathsf{PKE.Gen}, \mathsf{PKE.Enc}, \mathsf{PKE.Dec})$, and a trapdoor ELF $(\mathsf{TELF.GenInj}, \mathsf{TELF.GenLossy})$.

- $\mathsf{DPKE.Gen}$ runs $(\mathsf{sk}', \mathsf{pk}') \leftarrow \mathsf{PKE.Gen}(\lambda)$, $(f, f^{-1}) \leftarrow \mathsf{TELF.GenInj}(M)$, and $G \leftarrow \mathcal{G}$. It outputs $\mathsf{sk} = (\mathsf{sk}', f^{-1})$ and $\mathsf{pk} = (\mathsf{pk}', f, G)$.
- $\mathsf{DPKE.Enc}(\mathsf{pk}, m)$ runs $\mathsf{PKE.Enc}(\mathsf{pk}', f(m); G(m))$. That is, it encrypts $f(m)$ under the semantically secure encryption scheme, using random coins $G(m)$
- $\mathsf{DPKE.Dec}(\mathsf{sk}, c)$: run $y \leftarrow \mathsf{PKE.Dec}(\mathsf{sk}', c)$. If $y = \perp$ output $\perp$. Otherwise run $m \leftarrow f^{-1}(y)$ and output m.

Correctness of the scheme is immediate. For security, we have the following theorem:

Theorem 6. *For any constant d, if $\mathcal{G}$ is hardcore for arbitrary computationally unpredictable sources on d inputs,* $(\mathsf{PKE.Gen}, \mathsf{PKE.Enc}, \mathsf{PKE.Dec})$ *is semantically secure, and* $(\mathsf{TELF.GenInj}, \mathsf{TELF.GenLossy})$ *is a secure Trapdoor ELF, then* $(\mathsf{DPKE.Gen}, \mathsf{DPKE.Enc}, \mathsf{DPKE.Dec})$ *is a secure deterministic public key encryption scheme for arbitrary single computationally unpredictable sources on d inputs. If* $(\mathsf{PKE.Gen}, \mathsf{PKE.Enc}, \mathsf{PKE.Dec})$ *has pseudorandom ciphertexts, then so does* $(\mathsf{DPKE.Gen}, \mathsf{DPKE.Enc}, \mathsf{DPKE.Dec})$.

Proof. Consider an arbitrary computationally unpredictable source D, sampling messages $m_1, \dots, m_d$ and auxiliary information aux. We will prove the pseudorandom ciphertext case, the other case being analogous. We need to prove that $(\mathsf{pk}, \mathsf{DPKE.Enc}(\mathsf{pk}, m_1), \dots, \mathsf{DPKE.Enc}(\mathsf{pk}, m_d), \mathsf{aux})$ is computationally indistinguishable from $(\mathsf{pk}, C_1, \dots, C_d, \mathsf{aux})$, where $(\mathsf{sk}, \mathsf{pk}) \leftarrow \mathsf{DPKE.Gen}(\lambda)$, $(m_1, \dots, m_d, \mathsf{aux}) \leftarrow D$, and C_i are chosen uniformly random from the ciphertext space.

Suppose toward contradiction that we have an adversary A which distinguishes the two distributions with advantage ϵ. Let p be a polynomial such that $1/p \geq \epsilon$ infinitely often. We prove security through a sequence of hybrids:

- H_0. In this hybrid, the adversary is given $(\mathsf{pk}, c_1, \ldots, c_d, \mathsf{aux})$ where $\mathsf{pk} = (\mathsf{pk}', f, G)$, $(\mathsf{sk}', \mathsf{pk}') \leftarrow \mathsf{PKE.Gen}(\lambda)$, $(f, f^{-1}) \leftarrow \mathsf{TELF.GenInj}(M)$, $G \leftarrow \mathcal{G}$, and $c_i = \mathsf{DPKE.Enc}(\mathsf{pk}, m_i) = \mathsf{PKE.Enc}(\mathsf{pk}', f(m_i); G(m_i))$.
- H_1. In this hybrid, we change f to be lossy. That is we choose r so that A cannot distinguish $f \leftarrow \mathsf{TELF.GenLossy}(M, r)$ from f, except with probability $1/3p$. We then replace f with $f \leftarrow \mathsf{TELF.GenLossy}(M, r)$.
- H_2. In this hybrid, we change $c_i = \mathsf{PKE.Enc}(\mathsf{pk}', f(m_i); G(m_i))$ to $c_i = \mathsf{PKE.Enc}(\mathsf{pk}', f(m_i); R_i)$. That is, we replace $G(m_i)$ with R_i. We now claim that A distinguishes H_1 from H_2 with negligible probability.
  To prove this, notice that by Lemma 1 and the fact that d is constant, we have that $(m_1, \ldots, m_d,\ (\mathsf{aux}, f, f(m_1), \ldots, f(m_d)))$ is also computationally unpredictable. Then by the hardcore-ness of $\mathcal{G}$, we have that
  $$(G(m_1), \ldots, G(m_d),\ (\mathsf{aux}, G, f, f(m_1), \ldots, f(m_d)))$$
  is indistinguishable from
  $$(R_1, \ldots, R_d,\ (\mathsf{aux}, G, f, f(m_1), \ldots, f(m_d)))$$
  Finally by post-processing with $\mathsf{PKE.Enc}$, we have that
  $$(\{\mathsf{PKE.Enc}(\mathsf{pk}, f(m_i); G(m_i))\}, \mathsf{aux}, G, f, \{f(m_i)\}, \mathsf{pk})$$
  is indistinguishable from
  $$(\{\mathsf{PKE.Enc}(\mathsf{pk}, f(m_i); R_i)\}, \mathsf{aux}, G, f, \{f(m_i)\}, \mathsf{pk})$$
  The first case is H_1, and the second is H_2, proving their indistinguishability.
- H_3. Now we just change each c_i to be a uniformly random ciphertext C_i. The indistinguishability from H_2 follows from the pseudorandomness of $\mathsf{PKE.Enc}$.
- H_4. Finally, we change f back to the injective mode, generating $(f, f^{-1}) \leftarrow \mathsf{TELF.GenInj}(M)$ By analagous arguments, A distinguishes H_4 from H_3 with advantage $1/3p$. The result is that the adversary now sees $(\mathsf{pk}, C, \mathsf{aux})$.

Putting it all together, A distinguishes H_0 from H_4 with advantage at most $2/3p - \mathsf{negl} \leq 1/p \leq \epsilon$, a contradiction. □

6 Achieving CCA Security

In this section, we turn to building CCA-secure DPKE for computationally unpredictable sources.

We will loosely follow Peikert and Waters [18], who build CCA-secure public key encryption from lossy trapdoor functions (LTDFs). The main difficulty is that we want to switch to lossy mode in order to prove the security of the challenge ciphertext, but need to maintain the ability to decrypt all other ciphertexts. Their core idea is to devise a LTDF with many "branches", each ciphertext using a different branch. The challenge ciphertext is set to be encrypted using a lossy, and all others are injective.

We will use this idea, but the technical implementation will be somewhat different, and of course we will use a Trapdoor ELF with branches instead of an LTDF. The details are below.

6.1 Our Construction

Our building blocks will be a pseudorandom generator G, a CCA-secure public key encryption scheme $(\mathsf{PKE.Gen}, \mathsf{PKE.Enc}, \mathsf{PKE.Dec})$, and an all-but-one Trapdoor ELF $(\mathsf{TELF.GenInj}', \mathsf{TELF.GenLossy}')$.

- $\mathsf{DPKE.Gen}$ runs $(\mathsf{sk}', \mathsf{pk}') \leftarrow \mathsf{PKE.Gen}(\lambda)$, $(f, f^{-1}) \leftarrow \mathsf{TELF.GenInj}'(M)$, and $G_0, G_1 \leftarrow \mathcal{G}$. It outputs $\mathsf{sk} = (\mathsf{sk}', f^{-1})$ and $\mathsf{pk} = (\mathsf{pk}', f, G_0, G_1)$.
- $\mathsf{DPKE.Enc}(\mathsf{pk}, m)$ runs $b \leftarrow G_0(m)$ to select a branch. Then it applied our scheme from Sect. 5, using the branch b. Namely, it computes $d \leftarrow \mathsf{PKE.Enc}(\mathsf{pk}', f(b, m); G_1(m))$. The output is the ciphertext $c = (b, d)$.
- $\mathsf{DPKE.Dec}(\mathsf{sk}, c)$: run $y \leftarrow \mathsf{PKE.Dec}(\mathsf{sk}', c')$. If $y = \perp$ output $\perp$. Otherwise, it runs $m \leftarrow f^{-1}(b, y)$. Finally, it checks that the ciphertext is well-formed by re-encrypting m. Namely, it verifies that $b = G_0(m)$ and $d = \mathsf{PKE.Enc}(\mathsf{pk}, f(b, m); G_1(m))$. If the checks fail, it outputs $\perp$. Otherwise, it outputs m.

The completeness of the scheme is immediate. Next, we prove security.

Theorem 7. *For any constant d, if $\mathcal{G}$ is an injective hardcore function for any computationally unpredictable sources on d inputs,* $(\mathsf{PKE.Gen}, \mathsf{PKE.Enc}, \mathsf{PKE.Dec})$ *is CCA-secure,* $(\mathsf{TELF.GenInj}, \mathsf{TELF.GenLossy})$ *is a secure all-but-d Trapdoor ELF, then* $(\mathsf{DPKE.Gen}, \mathsf{DPKE.Enc}, \mathsf{DPKE.Dec})$ *is a CCA-secure deterministic public key encryption scheme for arbitrary computationally unpredictable sources on d inputs. If* $(\mathsf{PKE.Gen}, \mathsf{PKE.Enc}, \mathsf{PKE.Dec})$ *has pseudorandom ciphertexts, then so does* $(\mathsf{DPKE.Gen}, \mathsf{DPKE.Enc}, \mathsf{DPKE.Dec})$.

Proof. For simplicity, we prove the case $d = 1$, the more general case being a straightforward adaptation. Consider an arbitrary computationally unpredictable source D, sampling messages m and auxiliary information aux. We will prove the pseudorandom ciphertext case, the other case being analogous. We need to prove that $(\mathsf{pk}, \mathsf{DPKE.Enc}(\mathsf{pk}, m), \mathsf{aux})$ is computationally indistinguishable from $(\mathsf{pk}, C, \mathsf{aux})$, where $(\mathsf{sk}, \mathsf{pk}) \leftarrow \mathsf{DPKE.Gen}(\lambda)$, $(m, \mathsf{aux}) \leftarrow D$, and C is chosen uniformly random from the ciphertext space. This must hold even if an adversary can make decryption queries on any ciphertext except the challenge.

Suppose toward contradiction that we have an adversary A which distinguishes the two distributions with advantage ϵ. Let p be a polynomial such that $1/p \geq \epsilon$ infinitely often. We prove security through a sequence of hybrids:

- H_0. Here, we give the adversary $(\mathsf{pk}, c^*, \mathsf{aux})$ where $\mathsf{pk} = (\mathsf{pk}', f, G_0, G_1)$, $(\mathsf{sk}', \mathsf{pk}') \leftarrow \mathsf{PKE.Gen}(\lambda)$, and $(f, f^{-1}) \leftarrow \mathsf{TELF.GenInj}(M)$, and $G_0, G_1 \leftarrow \mathcal{G}$. Also, we set $c^* = \mathsf{DPKE.Enc}(\mathsf{pk}, m) = (b^*, d^*)$ where $b^* = G_0(m)$ and $d^* = \mathsf{PKE.Enc}(\mathsf{pk}', f(b^*, m); G_1(m))$.
- H_1. In this hybrid, we change f to be lossy on the branch b^*. That is, $(f, f^{-1}) \leftarrow \mathsf{TELF.GenInj}(M, b^*, r)$, where r is chosen so that A cannot distinguish this change except with advantage $1/3p$.

  We need to make sure that we can still answer CCA queries. For this, we just need that G_0 is injective, so that any other valid ciphertext will correspond to a different branch.

- H_2. In this hybrid, we replace $G_1(m)$ with random. We now claim that this change is indistinguishable to the adversary.
  Toward that end, first observe that since G_0 is hardcore, we have that $(G_0, G_0(m), \mathsf{aux})$ is indistinguishable from (G_0, S, aux) for a uniformly random S. This means that $(m, (\mathsf{aux}, G_0, G_0(m)))$ is computationally unpredictable. But then by Lemma 1, we also have that $(m, (\mathsf{aux}, G_0, b^*, f, f^{-1}, f(b^*, m)))$ is computationally unpredictable, where $(f, f^{-1}) \leftarrow \mathsf{TELF.GenLossy}(M, b^*, r)$ for $b^* = G_0(m)$. Finally, by the hardcore property of G_1, we have that the distribution $(G_1, G_1(m), \mathsf{aux}, G_0, b^*, f, f^{-1}, f(b^*, m))$ is indistinguishable from $(G_1, R, \mathsf{aux}, G_0, b^*, f, f^{-1}, f(b^*, m))$ for a random R.
  Now notice that an adversary given $(G_1, R, \mathsf{aux}, G_0, b^*, f, f^{-1}, f(b^*, m))$ for $R = G_1(m)$ (resp. random) can easily simulate the view of A in H_1 (resp. H_2) by using f^{-1} to answer decryption queries. Therefore, if A distinguishes the two hybrids, we can easily create a distinguisher for these two distribution, arriving at a contradiction.
- H_3. Now we just change c to be a uniformly random ciphertext C. The indistinguishability from H_2 follows from the CCA-secure pseudorandomness of PKE.Enc.
  Now notice that the d^* portion of the adversary's view is completely independent of m.
- H_4. Now we invoke the hardcore-ness of G_0 one more time to replace $G_0(m)$ with a random b^*.
- H_5. Finally, we change f back to the injective mode, generating $(f, f^{-1}) \leftarrow \mathsf{TELF.GenInj}(M)$ By analagous arguments, A distinguishes H_5 from H_4 with advantage $1/3p$. The result is that the adversary now sees $(\mathsf{pk}, C, \mathsf{aux})$.

Putting it all together, A distinguishes H_0 from H_5 with advantage at most $2/3p - \mathsf{negl} \leq 1/p \leq \epsilon$, a contradiction. □

References

1. Alwen, J., Dodis, Y., Wichs, D.: Survey: leakage resilience and the bounded retrieval model. In: Kurosawa, K. (ed.) ICITS 2009. LNCS, vol. 5973, pp. 1–18. Springer, Heidelberg (2010). https://doi.org/10.1007/978-3-642-14496-7_1
2. Barak, B., et al.: On the (im)possibility of obfuscating programs. In: Kilian, J. (ed.) CRYPTO 2001. LNCS, vol. 2139, pp. 1–18. Springer, Heidelberg (2001). https://doi.org/10.1007/3-540-44647-8_1
3. Bellare, M., Boldyreva, A., O'Neill, A.: Deterministic and efficiently searchable encryption. In: Menezes, A. (ed.) CRYPTO 2007. LNCS, vol. 4622, pp. 535–552. Springer, Heidelberg (2007). https://doi.org/10.1007/978-3-540-74143-5_30
4. Bellare, M., Cash, D., Miller, R.: Cryptography secure against related-key attacks and tampering. In: Lee, D.H., Wang, X. (eds.) ASIACRYPT 2011. LNCS, vol. 7073, pp. 486–503. Springer, Heidelberg (2011). https://doi.org/10.1007/978-3-642-25385-0_26
5. Bellare, M., Fischlin, M., O'Neill, A., Ristenpart, T.: Deterministic encryption: definitional equivalences and constructions without random oracles. In: Wagner, D. (ed.) CRYPTO 2008. LNCS, vol. 5157, pp. 360–378. Springer, Heidelberg (2008). https://doi.org/10.1007/978-3-540-85174-5_20

6. Bellare, M., Hoang, V.T.: Resisting randomness subversion: fast deterministic and hedged public-key encryption in the standard model. In: Oswald, E., Fischlin, M. (eds.) EUROCRYPT 2015, Part II. LNCS, vol. 9057, pp. 627–656. Springer, Heidelberg (2015). https://doi.org/10.1007/978-3-662-46803-6_21
7. Bellare, M., Rogaway, P.: Random oracles are practical: a paradigm for designing efficient protocols. In: Ashby, V. (ed.) ACM CCS 1993, pp. 62–73. ACM Press, November 1993
8. Boldyreva, A., Fehr, S., O'Neill, A.: On notions of security for deterministic encryption, and efficient constructions without random oracles. In: Wagner, D. (ed.) CRYPTO 2008. LNCS, vol. 5157, pp. 335–359. Springer, Heidelberg (2008). https://doi.org/10.1007/978-3-540-85174-5_19
9. Brakerski, Z., Segev, G.: Better security for deterministic public-key encryption: the auxiliary-input setting. In: Rogaway, P. (ed.) CRYPTO 2011. LNCS, vol. 6841, pp. 543–560. Springer, Heidelberg (2011). https://doi.org/10.1007/978-3-642-22792-9_31
10. Brzuska, C., Farshim, P., Mittelbach, A.: Indistinguishability obfuscation and UCEs: the case of computationally unpredictable sources. In: Garay, J.A., Gennaro, R. (eds.) CRYPTO 2014, Part I. LNCS, vol. 8616, pp. 188–205. Springer, Heidelberg (2014). https://doi.org/10.1007/978-3-662-44371-2_11
11. Canetti, R., Goldreich, O., Halevi, S.: The random oracle methodology, revisited (preliminary version). In: 30th ACM STOC, pp. 209–218. ACM Press, May 1998
12. Chung, K.-M., Lin, H., Mahmoody, M., Pass, R.: On the power of nonuniformity in proofs of security. In: Kleinberg, R.D. (ed.) ITCS 2013, pp. 389–400. ACM, January 2013
13. Freeman, D.M., Goldreich, O., Kiltz, E., Rosen, A., Segev, G.: More constructions of lossy and correlation-secure trapdoor functions. In: Nguyen, P.Q., Pointcheval, D. (eds.) PKC 2010. LNCS, vol. 6056, pp. 279–295. Springer, Heidelberg (2010). https://doi.org/10.1007/978-3-642-13013-7_17
14. Fuller, B., O'Neill, A., Reyzin, L.: A unified approach to deterministic encryption: new constructions and a connection to computational entropy. In: Cramer, R. (ed.) TCC 2012. LNCS, vol. 7194, pp. 582–599. Springer, Heidelberg (2012). https://doi.org/10.1007/978-3-642-28914-9_33
15. Goldreich, O., Levin, L.A.: A hard-core predicate for all one-way functions. In: 21st ACM STOC, pp. 25–32. ACM Press, May 1989
16. Goyal, V., O'Neill, A., Rao, V.: Correlated-input secure hash functions. In: Ishai, Y. (ed.) TCC 2011. LNCS, vol. 6597, pp. 182–200. Springer, Heidelberg (2011). https://doi.org/10.1007/978-3-642-19571-6_12
17. Matsuda, T., Hanaoka, G.: Chosen ciphertext security via UCE. In: Krawczyk, H. (ed.) PKC 2014. LNCS, vol. 8383, pp. 56–76. Springer, Heidelberg (2014). https://doi.org/10.1007/978-3-642-54631-0_4
18. Peikert, C., Waters, B.: Lossy trapdoor functions and their applications. In: Ladner, R.E., Dwork, C. (eds.) 40th ACM STOC, pp. 187–196. ACM Press, May 2008
19. Raghunathan, A., Segev, G., Vadhan, S.P.: Deterministic public-key encryption for adaptively chosen plaintext distributions. In: Johansson, T., Nguyen, P.Q. (eds.) EUROCRYPT 2013. LNCS, vol. 7881, pp. 93–110. Springer, Heidelberg (2013). https://doi.org/10.1007/978-3-642-38348-9_6
20. Villar, J.L.: Optimal reductions of some decisional problems to the rank problem. In: Wang, X., Sako, K. (eds.) ASIACRYPT 2012. LNCS, vol. 7658, pp. 80–97. Springer, Heidelberg (2012). https://doi.org/10.1007/978-3-642-34961-4_7

21. Wichs, D.: Barriers in cryptography with weak, correlated and leaky sources. In: Kleinberg, R.D. (ed.) ITCS 2013, pp. 111–126. ACM, January 2013
22. Zhandry, M.: The magic of ELFs. In: Robshaw, M., Katz, J. (eds.) CRYPTO 2016, Part I. LNCS, vol. 9814, pp. 479–508. Springer, Heidelberg (2016). https://doi.org/10.1007/978-3-662-53018-4_18

New Techniques for Efficient Trapdoor Functions and Applications

Sanjam Garg[1(✉)], Romain Gay[2,3], and Mohammad Hajiabadi[1,4]

[1] University of California, Berkeley, USA
{sanjamg,mdhajiabadi}@berkeley.edu
[2] Département informatique de l'ENS, École normale supérieure, CNRS, PSL University, 75005 Paris, France
romain.gay@ens.fr
[3] INRIA, Paris, France
[4] University of Virginia, Charlottesville, USA

Abstract. We develop techniques for constructing trapdoor functions (TDFs) with short image size and advanced security properties. Our approach builds on the recent framework of Garg and Hajiabadi [CRYPTO 2018]. As applications of our techniques, we obtain

- The first construction of deterministic-encryption schemes for block-source inputs (both for the CPA and CCA cases) based on the Computational Diffie-Hellman (CDH) assumption. Moreover, by applying our efficiency-enhancing techniques, we obtain CDH-based schemes with ciphertext size linear in plaintext size.
- The first construction of lossy TDFs based on the Decisional Diffie-Hellman (DDH) assumption with image size linear in input size, while retaining the lossiness rate of [Peikert-Waters STOC 2008].

Prior to our work, all constructions of deterministic encryption based even on the stronger DDH assumption incurred a quadratic gap between the ciphertext and plaintext sizes. Moreover, all DDH-based constructions of lossy TDFs had image size quadratic in the input size.

At a high level, we break the previous quadratic barriers by introducing a novel technique for encoding input bits via hardcore output bits with the use of erasure-resilient codes. All previous schemes used group elements for encoding input bits, resulting in quadratic expansions.

1 Introduction

Trapdoor functions (TDFs) are a fundamental primitive in cryptography and are typically used as a fundamental building block in the construction of advanced primitives such as CCA2-secure public-key encryption (PKE). Introduced in the 70s [DH76, RSA78], TDFs are a family of functions, where each individual function in the family is easy to compute, and also easy to invert if one posses an additional trapdoor key. The basic security requirement is that of one-wayness, requiring that a randomly chosen function from the family be one-way.

The usefulness of TDFs stems from the fact that the inversion algorithm recovers the entire input. This stands in sharp contrast to PKE, wherein the

Y. Ishai and V. Rijmen (Eds.): EUROCRYPT 2019, LNCS 11478, pp. 33–63, 2019.
https://doi.org/10.1007/978-3-030-17659-4_2

decryption algorithm may not recover the underlying randomness. This input recovery feature of TDFs is what makes them a useful tool, especially in applications where proofs of well-formedness are required.

On the other hand, building TDFs turns out to be much more difficult than building PKE, mostly due to the requirement of recovering the entire input, which in turn is the reason behind the lack of black-box transformations from PKE to TDFs [GMR01]. Specifically, in groups with discrete-log based hardness assumptions, this restricts the use of operations such as *exponentiation*, for which we do not have a generic trapdoor. Furthermore, in some applications we need TDFs to be *robust*, providing enhanced security properties rather than mere one-wayness (e.g., [BBO07,PW08,PW11,PVW08,BFOR08,RS09]).

Recently, Garg and Hajiabadi [GH18] introduced a new approach for building TDFs, obtaining the first construction of TDFs from the Computational Diffie-Hellman (CDH) assumption. Although their approach gives new feasibility results, their constructed TDFs are limited in certain ways: (a) Their TDFs are not robust enough—for example, it is not clear how to go beyond one-wayness, obtaining more advanced properties such as those required by deterministic encryption [BBO07,BFOR08,BFO08] or CCA2 security; and (b) The length of their TDF images grows (at least) quadratically with the length of the input.

We stress that Point (b) is not just an artifact of the construction of [GH18]. In fact, we do not know of any TDF constructions (even based on the stronger decisional Diffie-Hellman (DDH) assumption) with advanced properties, such as deterministic-encryption security, with images growing linearly in their inputs.[1] Since TDFs are typically used as building blocks in more advanced primitives, designing more efficient TDFs translates into the same features in target applications. For example, lossy TDFs [PW08,PW11] are an extremely versatile primitive with a long list of applications; e.g., [BFOR08,BHY09,BBN+09,MY10, BCPT13].

1.1 Our Results

We develop techniques for constructing efficient and robust TDFs. As concrete applications of our new techniques, we obtain the first construction of deterministic encryption for block sources (in the sense of [BFO08]) under the CDH assumption. We give both CPA and CCA2 versions of our constructions. We stress that prior to our work we knew how to build (even) CPA-secure deterministic encryption only from decisional assumptions, including DDH, QR, DCR and LWE [BFO08,Wee12]. Thus, in addition, we also obtain instantiations under the hardness of factoring assumption.

Furthermore, we show how to use our efficiency techniques to obtain:

[1] We note that building a TDF providing mere one-wayness with linear-size images is simple: if $\mathsf{TDF.F}(\mathsf{ik}, \cdot)$ maps n-bit inputs to n^c-bit outputs, define $\mathsf{TDF.F}'(\mathsf{ik}, \mathsf{x}||\mathsf{x}')$, where $|\mathsf{x}| = n$ and $|\mathsf{x}'| = n^c$, as $\mathsf{TDF.F}(\mathsf{ik}, \mathsf{x})||\mathsf{x}'$. Although this transformation results in TDFs with linear-image size, it destroy more advanced properties such as CCA2 security, deterministic-encryption security and the lossiness rate.

1. The first CDH-based deterministic encryption schemes with ciphertext size linear in plaintext size. Additionally, our CDH-based deterministic-encryption schemes beat all the previous DDH-based schemes in terms of ciphertext size. The sizes of other parameters (e.g., the secret key and public key) remain the same. See Table 1 for a comparison.
2. The first construction of lossy TDFs ([PW08,PW11]) from DDH with image size linear in input size. Our DDH-based lossy TDFs achieve the same lossiness rate as in [PW08,PW11]. All previous DDH-based lossy TDF constructions (achieving non-trivial lossiness rates) resulted in images quadratically large in their inputs.

Table 1. Bit complexity: p is the order of the group and n is the bit size of the TDF input. Here (n, k)-LTDF means lossy TDFs where in lossy mode the image-space size is at most 2^k. We call $1 - k/n$ the lossiness rate.

work	assumption	primitive	index key	trapdoor key	image
ours	CDH	CCA2 DE	$\Theta(n^2 \log p)$	$\Theta(n^2 \log p)$	$\log p + \Theta(n)$
[BFO08]	DDH	CCA2 DE	$\Theta(n^2 \log p)$	$\Theta(n^2 \log p)$	$\Theta(n \log p)$
ours	DDH	$(n, \log p)$-LTDF	$\Theta(n^2 \log p)$	$\Theta(n^2 \log p)$	$\log p + \Theta(n)$
[PW08, PW11, FGK+10]	DDH	$(n, \log p)$-LTDF	$\Theta(n^2 \log p)$	$\Theta(n^2 \log p)$	$\Theta(n \log p)$

1.2 Technical Overview

In this section we give an overview of our techniques for constructing robust and efficient TDFs. We will build TDFs with several abstract properties, and we will apply these techniques to the setting of deterministic encryption and lossy TDFs as concrete applications.

Our constructions rely on the same primitive of *recyclable one-way function with encryption (OWFE)* used by [GH18], so we first review this notion. An OWFE consists of a one-way function $\mathsf{f}(\mathsf{pp}, \cdot)\colon \{0,1\}^n \to \{0,1\}^\nu$, where pp is a public parameter, along with encapsulation/decapsulation algorithms (E, D). Specifically, E takes as input pp, an image $\mathsf{y} \in \{0,1\}^\nu$ of $\mathsf{f}(\mathsf{pp}, \cdot)$, a target index $i \in [n]$ and a target bit $b \in \{0,1\}$, and produces an *encapsulated ciphertext* ct and a corresponding *key bit* $\mathsf{e} \in \{0,1\}$. The algorithm D allows us to retrieve e from ct using any pre-image x of y, if $\mathsf{x}_i = b$. For security, letting $\mathsf{y} := \mathsf{f}(\mathsf{pp}, \mathsf{x})$, we require that if $(\mathsf{ct}, \mathsf{e}) \xleftarrow{\$} \mathsf{E}(\mathsf{pp}, \mathsf{y}, (i, 1 - \mathsf{x}_i))$, then even knowing both x and ct, one cannot distinguish e from a truly random bit. Finally, letting E_1 and E_2 refer to the first and second output pars of E, the recyclability requirement says that the output of E_1 does not depend on y, namely, we have: $\mathsf{ct} = \mathsf{E}_1(\mathsf{pp}, (i, b))$ and $\mathsf{e} = \mathsf{E}_2(\mathsf{pp}, \mathsf{y}, (i, b))$. (See Definition 4.) The work of [GH18] gives CDH instantiations of this notion.

Approach of [GH18]. A property implied by recyclable OWFE is the following: given $\mathsf{x} \in \{0,1\}^n$ and two fresh encapsulated ciphertexts $(\mathsf{ct}_0, \mathsf{ct}_1)$ made w.r.t. $\mathsf{y} := \mathsf{f}(\mathsf{pp}, \mathsf{x})$ and an arbitrary target index i and target bits 0 and 1 (respectively), one cannot distinguish the values of the corresponding two key bits $(\mathsf{e}_0, \mathsf{e}_1)$ from a pair in which we replace $\mathsf{e}_{1-\mathsf{x}_i}$ with a random bit. Exploiting this property, [GH18] set their index key to contain encapsulated ciphertexts $\mathsf{ct}_{i,b}$ made w.r.t. each value of $i \in [n]$ and $b \in \{0,1\}$—they put all the corresponding randomness values $[r_{i,b}]$ in the trapdoor key. The input to their TDF contains $\mathsf{x} \in \{0,1\}^n$ and the output u consists of $\mathsf{y} := \mathsf{f}(\mathsf{pp}, \mathsf{x})$ as well as a $2 \times n$ matrix $\mathbf{M}$ of bits $(\mathsf{e}_{i,b})_{i \in [n], b \in \{0,1\}}$, where for all i, they set $\mathsf{e}_{i,\mathsf{x}_i} := \mathsf{D}(\mathsf{pp}, \mathsf{x}, \mathsf{ct}_{i,\mathsf{x}_i})$ and set $\mathsf{e}_{i,1-\mathsf{x}_i}$ to be a random bit. Since TDFs are not allowed to make use of randomness, they draw $\mathsf{e}_{i,1-\mathsf{x}_i}$ for all i from an additional part of their input which they call the *blinding* part. For inverting $\mathsf{u} := (\mathsf{y}, \mathbf{M})$, the inverter may make use of its knowledge of all the randomness values underlying $\mathsf{ct}_{i,b}$'s to form the corresponding key bits w.r.t. y. Then the inverter may check each column of the resulting matrix, $\mathbf{M}'$, against the corresponding column of the matrix $\mathbf{M}$, and look for a matched coordinate. This would enable recovering half of the input bits (on average). The one-wayness of their scheme follows by the property alluded to above. Namely, for any $i \in [n]$, we may switch $\mathsf{e}_{1-\mathsf{x}_i}$ from uniformly random to $\mathsf{E}_2(\mathsf{pp}, \mathsf{y}, (i, 1-\mathsf{x}_i); r_{i,1-\mathsf{x}_i})$. Consequently, the image of the trapdoor function becomes: $(\mathsf{y}, (\mathsf{e}_{i,b})_{i,b})$, where $\mathsf{e}_{i,b} := \mathsf{E}_2(\mathsf{pp}, \mathsf{y}, (i, b); r_{i,b})$ for all $i \in [n]$ and $b \in \{0,1\}$. In other words, the entire view of a TDF adversary may be computed from y alone. At this point, the one-wayness of the TDF follows from the one-wayness of the underlying OWFE. Finally, [GH18] boosts correctness by repeating the above process in parallel. For future reference, we call the above initial TDF (which enables the recovery of half of the bits) *TDF gadget.*

Lack of perfect correctness in [GH18]. The TDF of [GH18] only achieves a weak form of correctness, under which the inversion algorithm may fail w.r.t. *any* index/trapdoor keys for a negligible fraction of the inputs. This severely restricts the applicability of CCA2-enhancing techniques, such as those of [RS09,KMO10], for obtaining CCA2 secure primitives. Even for the CPA case, the lack of perfect correctness hindered the construction of CPA-secure deterministic encryption schemes. Deterministic public-key encryption schemes [BBO07] are TDFs which hide information about plaintexts drawn from high min-entropy sources. There are various forms of this definition, e.g., [BBO07,BFO08,BFOR08,BS11,MPRS12]. Strong versions of this notion have so far been realized in the random oracle model [BBO07] and are subject to impossibility results [Wic13]. Boldyreva, Fehr and O'Neill [BFO08] formulated a relaxation of this notion (called *block-source security*), and showed how to realize this relaxed notion under standard assumptions such as DDH and LWE. Informally, block-source security requires that the (deterministic) encryptions of any two sources with high min entropy (more than a threshold k) remain computationally indistinguishable.[2] Ideally, we want $k << n$, where n is plaintext size.

The TDF of [GH18] does not achieve block-source security for the same reason that degraded their correctness property: The TDF input contains a blinding part, which in turn is copied in the clear in the output (but in hidden spots). To see how this breaks security, consider two sources, where the first one fixes the blinding part to all zeros, and the second one fixes it to all ones. Then it would be easy to distinguish between the outputs of the TDF w.r.t. these two sources, even though they may have high min entropy.

Enhancing to perfect correctness. We fix the imperfect correctness of [GH18] via a *mirroring* technique. Recall that the bits of the blinding input were previously used to form the values of $\mathsf{e}_{i,1-\mathsf{x}_i}$. Now, instead of having a blinding part in the input for making up the values of $\mathsf{e}_{i,1-\mathsf{x}_i}$, we set $\mathsf{e}_{i,1-\mathsf{x}_i} := \mathsf{a}_i - \mathsf{e}_{i,\mathsf{x}_i}$, where $\mathbf{a} := (\mathsf{a}_1, \ldots, \mathsf{a}_n) \in \{0,1\}^n$ is a random vector that comes from the index key. This way we get rid of inclusion of blinders as part of the input—the input now solely consists of a string $\mathsf{x} \in \{0,1\}^n$. We show that this method improves correctness: Our TDF is now perfectly correct for all but a negligible fraction of index/trapdoor keys; see Remark 1.

Lossy-like properties of our TDF toward obtaining deterministic encryption. So far, we showed how to fix the imperfect-correctness problem of [GH18], but this by itself does not guarantee deterministic-encryption security. Toward this goal, we show that the mirroring technique allows us to establish a *lossy-like* property for our TDFs, which in turn gives us block-source security.[3] Specifically, let y be an image point of $\mathsf{f}(\mathsf{pp}, \cdot)$ of the OWFE scheme, and let S be the set of all pre-images of y (which can be of exponential size under our CDH instantiation). We can now set the index key as ik_y, where (a) ik_y loses information w.r.t. all pre-images of y: for all $\mathsf{x}, \mathsf{x}' \in \mathsf{S}$ we have $\mathsf{TDF.F}(\mathsf{ik}_\mathsf{y}, \mathsf{x}) = \mathsf{TDF.F}(\mathsf{ik}_\mathsf{y}, \mathsf{x}')$ and (b) ik_y is computationally indistinguishable from an honestly generated ik. We exploit this property to prove block-source security for our TDFs.

Having achieved block-source CPA security, we may boost this scheme into a CCA2-secure deterministic-encryption scheme using the techniques of [RS09, KMO10].[4] Specifically, we show how to use our lossiness property to prove *k-repetition* security (introduced by [RS09]) for our TDF. Intuitively, k-repetition security requires one-wayness to hold even if the given input is evalu-

---

[2] This is the indistinguishability-based, single-message version of their notion, which as they show, is equivalent to the multiple-message version both for the indistinguishability- and simulation-based definitions.

[3] We note that this lossiness property is weaker than the one of [PW08, PW11], but it can be realized under CDH. We will later show efficient DDH-based instantiations of lossiness in the sense of [PW08, PW11].

[4] We mention that the transformation of [RS09] results in CCA-secure PKE schemes which use randomness, but this can be avoided by using the techniques of [KMO10] to get CCA2-secure TDFs.

ated under k-randomly chosen functions. The scheme of [GH18] fails to achieve k-repetition security, exactly because of the presence of blinders.

Finally, we mention that based on CDH we do not get lossiness in the sense of [PW08, PW11] as the amount of information we lose is negligible over the entire input space. Nevertheless, our weak lossiness property which can be realized under CDH suffices for our deterministic-encryption application, and may find other applications later.

Efficiency of our TDFs so far: quadratically large images. Under CDH instantiations of the above approach, for plaintexts of n bits, the bit-size of the ciphertext is $\Theta(n^2)$ in the CPA case, and $\Theta(n^2\omega(\log n))$ in the CCA case. In contrast, the DDH-based constructions of [BFO08] give ciphertext size $\Theta(n^2)$ both for the CPA and CCA cases.

Sources of inefficiency. Recall our TDF gadget has image size $\Theta(n)$. This TDF gadget may fail to recover any given bit of the input with probability 1/2. Thus, we ran many TDF gadgets in parallel, resulting in $\Theta(n^2)$ image size. We refer to this as *correctness repetition.* For the CCA2 case, since we relied on techniques of [RS09, KMO10] we needed to perform yet another repetition, which we call *CCA2 repetition.* This justifies the further blowup in CCA2 image size.

We develop techniques for avoiding both these repetitions, sketched below.

Erasure-resilient codes to the rescue: linear-image TDFs. We give techniques involving the use of erasure-resilient codes for making the size of our TDF images linear, while preserving other properties. Recall that under our TDF gadget, for a randomly chosen input $\mathsf{x} \in \{0,1\}^n$ and for any index $i \in [n]$, the inversion algorithm either recovers x_i correctly, or outputs $\perp$ for this bit position (with probability 1/2). Notice that the inversion process has a *local property,* in the sense that each bit individually may be recovered or not with probability 1/2.

Now instead of performing parallel repetition which results in a quadratic-size blowup, we boost correctness through the use of erasure-resilient codes. Suppose $(\mathsf{Encode}, \mathsf{Decode})$ is an erasure-resilient code, where $\mathsf{Encode}\colon \{0,1\}^n \to \{0,1\}^m$ (for $m = cn \in O(n)$), and where Decode only needs $c_1 n$ (noise-free) bits of a codeword $\mathsf{Encode}(\mathsf{x})$—for c_1 sufficiently smaller than c—in order to recover x. Such codes may be built from Reed-Solomon codes by adapting them to bit strings; see Definition 8.

The starting point of our technique is the following: On input $\mathsf{x} \in \{0,1\}^n$, apply the TDF gadget on the encoded input $\mathsf{z} := \mathsf{Encode}(\mathsf{x})$. To invert, we no longer need to recover *all* the m bits of z; recovering $c_1 n$ of them will do. Unfortunately, for codes over binary alphabets, the value of c_1/c is much greater than 1/2 and our TDF gadget may be incapable of recovering that many bits. We get around this issue by doing repetition but for a constant number of times: Instead of applying the TDF gadget to $\mathsf{z} := \mathsf{z}_1 \ldots \mathsf{z}_m$, apply it to the t-repetition copy of z where we repeat each bit of z t times. By choosing the underlying constants appropriately and using the mirroring idea, we can ensure perfect correctness for all but a negligible fraction of index/trapdoor keys. This way, images will grow linearly. The proof of CPA block-source security follows almost similarly as before.

Concretely, under CDH instantiations, CPA ciphertexts (for plaintexts of n bits) consist of one group element and a constant number of bits.[5] This substantially improves the ciphertext size of previous DDH-based schemes under which a ciphertext consists of n group elements.

Keeping image size linear in CCA-like applications. So far, we showed how to build linear-image TDFs with additional properties (e.g., block-source CPA security). Typically, TDFs with enhanced properties (such as k-repetition security [RS09] or lossy properties [PW08,PW11]) can be boosted into CCA2 primitives, but this requires "parallel repetition" of the base object, increasing the sizes. Our linear-image TDF turns out to be k-repetition secure, but we cannot afford to use previous techniques for getting CCA2 security, because we would like to keep image size linear. Here is where our other techniques come into play: We develop a half-simulation method for proving CCA2 security for our *same* TDF scheme without any further modifications. For this, we just need to choose the constant c in $m = cn$ big enough. Our CCA techniques are different from those of [PW08,PW11,RS09], which implicitly or explicitly relied on repetition.

As an application, we will get a CDH-based block-source CCA-secure deterministic encryption, beating the ciphertext size of DDH-based schemes. We now sketch our techniques.

Let $(\mathsf{Encode}, \mathsf{Decode})$ be a code obtained by repeating the output bit of a codeword (which in turn is obtained based on a linear-size-output error-correcting code) t times for a constant t. See Definition 8 and the two paragraphs afterward. A codeword z may be thought of as a string of m/t blocks, each consisting of entirely either t zeros or t ones.

Recall that a trapdoor key tk consists of all randomness values $r_{i,b}$'s used to form $\mathsf{ct}_{i,b}$'s (which are in turn fixed in the index key ik). On input $\mathsf{x} \in \{0,1\}^n$ we form $\mathsf{z} := \mathsf{Encode}(\mathsf{x}) \in \{0,1\}^m$ and return $\mathsf{u} := (\mathsf{y}, \mathbf{M} := \binom{\mathsf{e}_{1,0},\ldots,\mathsf{e}_{m,0}}{\mathsf{e}_{1,1},\ldots,\mathsf{e}_{m,1}})$, where $\mathsf{y} := \mathsf{f}(\mathsf{pp}, \mathsf{z})$, $\mathsf{e}_{i,\mathsf{z}_i} = \mathsf{D}(\mathsf{pp}, \mathsf{z}, \mathsf{ct}_{i,\mathsf{z}_i})$ and $\mathsf{e}_{i,1-\mathsf{z}_i} = \mathsf{a}_i - \mathsf{e}_{i,\mathsf{z}_i}$, where $\mathbf{a} := (\mathsf{a}_1, \ldots, \mathsf{a}_m) \in \{0,1\}^m$ is sampled in ik. The inversion algorithm will recover the ith bit of z iff $\mathsf{e}_{i,1-\mathsf{z}_i} = 1 - \mathsf{E}_2(\mathsf{pp}, \mathsf{y}, (i, 1-\mathsf{z}_i); r_{i,1-\mathsf{z}_i})$. Say the ith column of $\mathbf{M}$ is *hung* if $\mathsf{e}_{i,1-\mathsf{z}_i} = \mathsf{E}_2(\mathsf{pp}, \mathsf{y}, (i, 1-\mathsf{z}_i); r_{i,1-\mathsf{z}_i})$—if this happens, then the inverter cannot decide on the ith bit of z.

Let us argue CCA2 security w.r.t. two sources $\mathcal{S}_0$ and $\mathcal{S}_1$: The adversary should distinguish $(\mathsf{ik}, \mathsf{TDF.F}(\mathsf{ik}, \mathsf{x}_0))$ from $(\mathsf{ik}, \mathsf{TDF.F}(\mathsf{ik}, \mathsf{x}_1))$, where $\mathsf{x}_b \xleftarrow{\$} \mathcal{S}_b$. For deterministic encryption we may assume all CCA queries happen after seeing the index key and challenge ciphertext (Definition 2).

Our CCA2 simulation is based on a *half-trapdoor simulation* technique under which we forget one randomness value from each pair $(r_{i,0}, r_{i,1})$ in the trapdoor. Specifically, letting $\mathsf{x}^\star$ be the challenge plaintext, imagine a half-trapdoor key obtained based on $\mathsf{x}^\star$ from tk as $\mathsf{tk}_{\mathrm{rd},\mathsf{z}^\star} := (r_{1,\mathsf{z}_1^\star}, \ldots, r_{m,\mathsf{z}_m^\star})$, where $\mathsf{z}^\star = \mathsf{Encode}(\mathsf{x}^\star)$. We perform half-trapdoor inversion of a given point

[5] We have not yet optimized nor tried to get some upper bounds on the constants.

$\mathsf{u} := \left(\mathsf{y}, \binom{\mathsf{e}_{1,0},\ldots,\mathsf{e}_{m,0}}{\mathsf{e}_{1,1},\ldots,\mathsf{e}_{m,1}}\right)$ w.r.t. $\mathsf{tk}_{\mathrm{rd},\mathsf{z}^\star}$ as follows: Build (potentially) a noisy codeword z as follows: in order to recover the bits of the jth block, if at least for *one* index i in this block we have $\mathsf{e}_{i,\mathsf{z}_i^\star} = 1 - \mathsf{E}_2(\mathsf{pp}, \mathsf{y}, (i, \mathsf{z}_i^\star); r_{i,\mathsf{z}_i^\star})$, set all the bits of z in this block to $1 - \mathsf{z}_i^\star$; otherwise, set all those bits to the corresponding bit values of $\mathsf{z}^\star$ in those coordinates. Once z is formed, decode it to get a string and check if the re-encryption of that string gives back u. If so, return the string.

Letting $\mathsf{x}^\star$ be the challenge plaintext (and recalling that all CCA queries are post-challenge), we first show we may use $\mathsf{tk}_{\mathrm{rd},\mathsf{z}^\star}$ (instead of the full key tk) to reply to the CCA queries, without the adversary noticing any difference, using the following two facts. First, for a queried point u, if u is not a valid image (i.e., it does not have a pre-image), then both (full and half) inversions return $\perp$. This is because at the end of either inversion we re-encrypt the result to see whether we get the given image point back. So suppose for the queried $\mathsf{u} := \left(\mathsf{y}, \mathbf{M} := \binom{\mathsf{e}_{1,0},\ldots,\mathsf{e}_{m,0}}{\mathsf{e}_{1,1},\ldots,\mathsf{e}_{m,1}}\right)$ we have $\mathsf{u} := \mathsf{TDF.F}(\mathsf{ik}, \mathsf{x})$ for some $\mathsf{x} \in \{0,1\}^n \backslash \{\mathsf{x}^\star\}$. (If $\mathsf{x} = \mathsf{x}^\star$, then u will be the challenge ciphertext itself and hence not a permitted query.) Let $\mathsf{S} \subseteq [m/t]$ contain the indices of those blocks on which z and $\mathsf{z}^\star$ differ, where $\mathsf{z} := \mathsf{Encode}(\mathsf{x})$. Note the half-trapdoor inversion w.r.t. $\mathsf{tk}_{\mathrm{rd},\mathsf{z}^\star}$ will correctly recover *all* the bits of z that correspond to the blocks which are not in S.

For the blocks in S, we show that by choosing the constants appropriately, then for sufficiently-many indices $j \in \mathsf{S}$, the jth block of $\mathbf{M}$ is *not* hung; namely, for at least one index i in this block we have $\mathsf{e}_{i,1-\mathsf{z}_i} = 1 - \mathsf{E}_2(\mathsf{pp}, \mathsf{y}, (i, 1 - \mathsf{z}_i); r_{i,1-\mathsf{z}_i})$. For any index $j \in \mathsf{S}$ such that the above holds, the half-inversion process (w.r.t. $\mathsf{tk}_{\mathrm{rd},\mathsf{z}^\star}$) will recover the jth block of z (by definition). We will use these facts to argue we will have enough correctly generated bits in order to able to do error correction.

Once we solely use $\mathsf{tk}_{\mathrm{rd},\mathsf{z}^\star}$ to reply to decryption queries, letting $\mathsf{u}^\star := \left(\mathsf{y}^\star, \begin{pmatrix}\mathsf{e}^\star_{1,0},\ldots,\mathsf{e}^\star_{m,0}\\ \mathsf{e}^\star_{1,1},\ldots,\mathsf{e}^\star_{m,1}\end{pmatrix}\right)$ be the corresponding challenge ciphertext, we may replace each $\mathsf{e}^\star_{i,1-\mathsf{z}_i^\star}$ with $\mathsf{E}_2(\mathsf{pp}, \mathsf{y}^\star, (i, 1 - \mathsf{z}_i^\star); r_{i,1-\mathsf{z}_i^\star})$, and simultaneously set the ith bit of the vector $\mathbf{a}$ of the index key as $\mathsf{a}_i := \mathsf{E}_2(\mathsf{pp}, \mathsf{y}^\star, (i, \mathsf{z}_i^\star); r_{i,\mathsf{z}_i^\star}) + \mathsf{E}_2(\mathsf{pp}, \mathsf{y}^\star, (i, 1 - \mathsf{z}_i^\star); r_{i,1-\mathsf{z}_i^\star})$. This change goes unnoticed by the security of the OWFE. At this point the challenge ciphertext and index key only depend on $\mathsf{y}^\star$ and we only use $\mathsf{z}^\star$ to decide which randomness value from each pair of tk to forget. We will now switch back to using the full trapdoor, with analysis similar to before. At this point, the entire view of the adversary may be simulated using $\mathsf{y}^\star := \mathsf{f}(\mathsf{pp}, \mathsf{z}^\star)$, and thus we have block-source security in this hybrid similar to the CPA case.

Lossy TDFs. Recall that a TDF is lossy [PW08, PW11] if one may generate index keys in a lossy way which is (1) indistinguishable from honestly generated index keys and (2) which results in statistical loss of information if used during the evaluation algorithm. We show we can adapt the trapdoor functions of [PW08, PW11] using our erasure-resilient code based technique for encoding input bits via hardcore output bits. This allows us to obtain lossy TDFs based on DDH with image size linear in input size. All previous DDH-based constructions of lossy TDFs incur a quadratic blowup in image size [PW08, PW11, FGK+10, Wee12].

We defer the reader to Sect. 6 for details. We leave open the exciting problem of constructing lossy trapdoor functions from CDH.

Other related work. OWFE is a relaxation of the notion of (chameleon) hash encryption and its variants, which in turn imply strong primitives such as laconic oblivious transfer and identity-based encryption (IBE) in a non-black-box way [CDG+17, DG17b, DG17a, BLSV18, DGHM18].

Freeman et al. [FGK+10] give additional constructions and simplifications to the TDF construction of [PW08, PW11]. Further constructions of (lossy) TDFs from various assumptions are given in [Wee12, HO12, HO13]. As for efficient TDFs, Boyen and Waters show that in the bilinear setting one may drastically shorten the index-key size of the Peikert-Waters lossy-TDF construction from a quadratic number of group elements to linear [BW10].

Concurrent Work. In an exciting independent and concurrent work, Koppula and Waters [KW18] show that TDF techniques can be used to upgrade any attribute-based encryption or predicate encryption scheme to its CCA secure variant. Similarly to this work, Koppula and Waters build on the ideas from the CDH-based TDF construction of Garg and Hajiabadi [GH18]. In particular, Koppula and Waters [KW18] independently came up with a similar version of the mirroring technique along the way, which we also developed in this paper. However, the focus of our work is very different from that of Koppula and Waters. In particular, we develop efficient techniques for applications such as TDFs, deterministic encryption and lossy trapdoor functions.

Paper organization. We give standard definitions and lemmas in Sect. 2 and OWFE-related definitions in Sect. 3. We give our (inefficient) construction of TDFs with deterministic-encryption security in Sect. 4 and give our efficient construction in Sect. 5. Finally, we give our DDH-based lossy TDF construction with linear image size in Sect. 6.

2 Preliminaries

Notation. We use λ for the security parameter. We use $\stackrel{c}{\equiv}$ to denote computational indistinguishability between two distributions and use $\equiv$ to denote two distributions are identical. For any $\varepsilon > 0$, we write $\approx_\varepsilon$ to denote that two distributions are statistically close, within statistical distance ε, and use $\stackrel{s}{\equiv}$ for statistical indistinguishability. For a distribution $\mathcal{S}$ we use $x \xleftarrow{\$} \mathcal{S}$ to mean x is sampled according to $\mathcal{S}$ and use $y \in \mathcal{S}$ to mean $y \in \sup(\mathcal{S})$, where sup denotes the support of a distribution. For a set S we overload the notation to use $x \xleftarrow{\$} \mathsf{S}$ to indicate that x is chosen uniformly at random from S. If $\mathsf{A}(x_1, \ldots, x_n)$ is a randomized algorithm, then $\mathsf{A}(a_1, \ldots, a_n)$, for deterministic inputs $a_1, \ldots, a_n$, denotes the random variable obtained by sampling random coins r uniformly at random and returning $\mathsf{A}(a_1, \ldots, a_n; r)$.

The min-entropy of a distribution $\mathcal{S}$ is defined as $\mathsf{H}_\infty(\mathcal{S}) \triangleq -\log(\max_{\mathsf{x}} \Pr[\mathcal{S} = \mathsf{x}])$. We call a distribution $\mathcal{S}$ a (k,n)-source if $\mathsf{H}_\infty(\mathcal{S}) \geq k$ and $\sup(\mathcal{S}) \subseteq \{0,1\}^n$.

2.1 Standard Definitions

Definition 1 (Trapdoor functions (TDFs)). *Let $n = n(\lambda)$ be a polynomial. A family of trapdoor functions* TDF *with domain $\{0,1\}^n$ consists of three PPT algorithms* $\mathsf{TDF.KG}$, $\mathsf{TDF.F}$ *and* $\mathsf{TDF.F}^{-1}$ *with the following syntax and security properties.*

- $\mathsf{TDF.KG}(1^\lambda)$: *Takes the security parameter 1^λ and outputs a pair* $(\mathsf{ik}, \mathsf{tk})$ *of index/trapdoor keys.*
- $\mathsf{TDF.F}(\mathsf{ik}, \mathsf{x})$: *Takes an index key* ik *and a domain element* $\mathsf{x} \in \{0,1\}^n$ *and deterministically outputs an image element* u.
- $\mathsf{TDF.F}^{-1}(\mathsf{tk}, \mathsf{u})$: *Takes a trapdoor key* tk *and an image element* u *and outputs a value* $\mathsf{x} \in \{0,1\}^n \cup \{\perp\}$.

We require the following properties.

- ***Correctness:***

$$\Pr_{(\mathsf{ik},\mathsf{tk})}\left[\exists \mathsf{x} \in \{0,1\}^n \ s.t.\ \mathsf{TDF.F}^{-1}(\mathsf{tk}, \mathsf{TDF.F}(\mathsf{ik}, \mathsf{x})) \neq \mathsf{x}\right] = \mathsf{negl}(\lambda), \quad (1)$$

where the probability is taken over $(\mathsf{ik}, \mathsf{tk}) \xleftarrow{\$} \mathsf{TDF.KG}(1^\lambda)$.
- ***One-wayness:*** *For any PPT adversary $\mathcal{A}$, we have* $\Pr[\mathcal{A}(\mathsf{ik}, \mathsf{u}) = \mathsf{x}] = \mathsf{negl}(\lambda)$, *where* $(\mathsf{ik}, \mathsf{tk}) \xleftarrow{\$} \mathsf{TDF.KG}(1^\lambda)$, $\mathsf{x} \xleftarrow{\$} \{0,1\}^n$ *and* $\mathsf{u} := \mathsf{TDF.F}(\mathsf{ik}, \mathsf{x})$.

Remark 1. *The work of Garg and Hajiabadi [GH18] builds a TDF with a weaker correctness guarantee, under which for any choice of* $(\mathsf{ik}, \mathsf{tk})$, *we are allowed to have a negligible inversion error (over the choice of* $\mathsf{x} \xleftarrow{\$} \{0,1\}^n$*). Although the correctness condition of [GH18] implies that for a randomly chosen* $(\mathsf{ik}, \mathsf{tk})$ *and a randomly chosen* x, *the probability of an inversion error is negligible, it falls short in certain applications, such as CCA2 constructions, for which a stronger correctness condition, as that given in Definition 1, is needed.*

We will now define a single-message-based notion of indistinguishability for deterministic encryption of block sources, which as proved in [BFO08], is equivalent to both the simulation-based and indistinguishability-based multiple-message notions.

Definition 2 (Deterministic-encryption security [BFO08]**).** *Let* $\mathsf{TDF} = (\mathsf{TDF.KG}, \mathsf{TDF.F}, \mathsf{TDF.F}^{-1})$ *be as in Definition 1. We say that* TDF *is (k,n)-CPA-indistinguishable if for any two (k,n)-sources $\mathcal{S}_1$ and $\mathcal{S}_2$ we have* $(\mathsf{ik}, \mathsf{TDF.F}(\mathsf{ik}, \mathcal{S}_1)) \stackrel{c}{\equiv} (\mathsf{ik}, \mathsf{TDF.F}(\mathsf{ik}, \mathcal{S}_2))$, *where* $(\mathsf{ik}, *) \xleftarrow{\$} \mathsf{TDF.KG}(1^\lambda)$.

We say that TDF *is* (k,n)*-CCA2-indistinguishable if for any two* (k,n)*-sources* $\mathcal{S}_0$ *and* $\mathcal{S}_1$*, and any PPT adversary* $\mathcal{A}$ *the following probability is negligible:*

$$\Pr\left[b=b' : \begin{array}{l}(\mathsf{ik},\mathsf{tk}) \xleftarrow{\$} \mathsf{TDF.KG}(1^\lambda), b \xleftarrow{\$} \{0,1\}, \mathsf{x}^\star \xleftarrow{\$} \mathcal{S}_b, \mathsf{u}^\star := \mathsf{TDF.F}(\mathsf{ik},\mathsf{x}^\star) \\ b' \leftarrow \mathcal{A}^{\mathcal{O}_{\mathsf{Dec}}}(\mathsf{ik},\mathsf{u}^\star)\end{array}\right] - \frac{1}{2}$$

where on input u*, the decryption oracle* $\mathcal{O}_{\mathsf{Dec}}$ *returns* $\mathsf{TDF.F}^{-1}(\mathsf{tk},\mathsf{u})$ *if* $\mathsf{u} \neq \mathsf{u}^\star$*, and* $\perp$ *otherwise.*

We remark that considering only CCA2 queries (as opposed to both CCA1 and CCA2 queries) in the CCA2-indistinguishability definition for deterministic encryption is without loss of generality, since the plaintexts are not chosen by the adversary. See [BFO08] for further explanation.

Definition 3 (Computational Diffie-Hellman (CDH) assumption). *Let* G *be a group-generator scheme, which on input* 1^λ *outputs* $(\mathbb{G},p,g)$*, where* $\mathbb{G}$ *is the description of a group,* p *is the order of the group which is always a prime number and* g *is a generator of the group. We say that* G *is CDH-hard if for any PPT adversary* $\mathcal{A}$*:* $\Pr[\mathcal{A}(\mathbb{G},p,g,g^{a_1},g^{a_2}) = g^{a_1 a_2}] = \mathsf{negl}(\lambda)$*, where* $(\mathbb{G},p,g) \xleftarrow{\$} \mathsf{G}(1^\lambda)$ *and* $a_1, a_2 \xleftarrow{\$} \mathbb{Z}_p$*.*

2.2 Standard Lemmas

Lemma 1 (Chernoff inequality). *Let* $\mathcal{X}_1,\dots,\mathcal{X}_m$ *be independent Boolean variables each of expected value at least* p*. Then, for all* $\varepsilon > 0$*:*

$$\Pr\left[\frac{1}{m}\sum_{i=1}^{m}\mathcal{X}_i < p - \varepsilon\right] < e^{-2\varepsilon^2 m}.$$

Lemma 2 (Leftover hash lemma [ILL89]). *Let* $\mathcal{X}$ *be a random variable over* X *and* $h : \mathsf{S} \times \mathsf{X} \to \mathsf{Y}$ *be a 2-universal hash function, where* $|\mathsf{Y}| \leq 2^m$ *for some* $m > 0$*. If* $m \leq \mathsf{H}_\infty(\mathcal{X}) - 2\log\left(\frac{1}{\varepsilon}\right)$*, then* $(h(\mathcal{S},\mathcal{X}),\mathcal{S}) \approx_\varepsilon (\mathcal{U},\mathcal{S})$*, where* $\mathcal{S}$ *is uniform over* S *and* $\mathcal{U}$ *is uniform over* Y*.*

3 Smooth Recyclable OWFE

We recall the definition of recyclable one-way function with encryption from [GH18]. We adapt the definition to a setting in which the underlying input distribution is not necessarily uniform. We will also define a smoothness notion, which generalizes the one-wayness notion.

Definition 4 (Recyclable one-way function with encryption (OWFE)). *A recyclable* (k,n)*-OWFE scheme consists of the PPT algorithms* K*,* f*,* E_1*,* E_2 *and* D *with the following syntax.*

- $\mathsf{K}(1^\lambda)$: *Takes the security parameter* 1^λ *and outputs a public parameter* pp *(by tossing coins) for a function* $\mathsf{f}(\mathsf{pp}, \cdot)$ *from* n *bits to* ν *bits.*
- $\mathsf{f}(\mathsf{pp}, \mathsf{x})$: *Takes a public parameter* pp *and a preimage* $\mathsf{x} \in \{0,1\}^n$, *and deterministically outputs* $\mathsf{y} \in \{0,1\}^\nu$.
- $\mathsf{E}_1(\mathsf{pp}, (i, b); \rho)$: *Takes a public parameter* pp, *an index* $i \in [n]$, *a bit* $b \in \{0,1\}$ *and randomness* ρ, *and outputs a ciphertext* ct.[6]
- $\mathsf{E}_2(\mathsf{pp}, \mathsf{y}, (i, b); \rho)$: *Takes a public parameter* pp, *a value* y, *an index* $i \in [n]$, *a bit* $b \in \{0,1\}$ *and randomness* ρ, *and outputs a bit* e. *Notice that unlike* E_1, *which does not take* y *as input, the algorithm* E_2 *does take* y *as input.*
- $\mathsf{D}(\mathsf{pp}, \mathsf{ct}, \mathsf{x})$: *Takes a public parameter* pp, *a ciphertext* ct *and a preimage* $\mathsf{x} \in \{0,1\}^n$, *and deterministically outputs a bit* e.

We require the following properties.

- ***Correctness.*** *For any choice of* $\mathsf{pp} \in \mathsf{K}(1^\lambda)$, *any index* $i \in [n]$, *any preimage* $\mathsf{x} \in \{0,1\}^n$ *and any randomness value* ρ, *the following holds: letting* $\mathsf{y} := \mathsf{f}(\mathsf{pp}, \mathsf{x})$, $b := \mathsf{x}_i$ *and* $\mathsf{ct} := \mathsf{E}_1(\mathsf{pp}, (i, \mathsf{x}_i); \rho)$, *we have* $\mathsf{E}_2(\mathsf{pp}, \mathsf{y}, (i, \mathsf{x}_i); \rho) = \mathsf{D}(\mathsf{pp}, \mathsf{ct}, \mathsf{x})$.
- (k, n)***-One-wayness:*** *For any* (k, n) *source* $\mathcal{S}$ *and any PPT adversary* $\mathcal{A}$:

$$\Pr[\mathsf{f}(\mathsf{pp}, \mathcal{A}(\mathsf{pp}, \mathsf{y})) = \mathsf{y}] = \mathsf{negl}(\lambda),$$

  where $\mathsf{pp} \xleftarrow{\$} \mathsf{K}(1^\lambda)$, $\mathsf{x} \xleftarrow{\$} \mathcal{S}$ *and* $\mathsf{y} := \mathsf{f}(\mathsf{pp}, \mathsf{x})$.
- ***Security for encryption:*** *For any* $i \in [n]$ *and* $\mathsf{x} \in \{0,1\}^n$:

$$(\mathsf{x}, \mathsf{pp}, \mathsf{ct}, \mathsf{e}) \stackrel{c}{\equiv} (\mathsf{x}, \mathsf{pp}, \mathsf{ct}, \mathsf{e}')$$

  where $\mathsf{pp} \xleftarrow{\$} \mathsf{K}(1^\lambda)$, $\rho \xleftarrow{\$} \{0,1\}^*$, $\mathsf{ct} := \mathsf{E}_1(\mathsf{pp}, (i, 1 - \mathsf{x}_i); \rho)$, $\mathsf{e} := \mathsf{E}_2(\mathsf{pp}, \mathsf{f}(\mathsf{pp}, \mathsf{x}), (i, 1 - \mathsf{x}_i); \rho)$ *and* $\mathsf{e}' \xleftarrow{\$} \{0,1\}$.

Whenever we say an OWFE scheme (without specifying the parameters), we mean $k = n$.

Notation 2. *We define* $\mathsf{E}(\mathsf{pp}, \mathsf{y}, (i, b); \rho) \triangleq (\mathsf{E}_1(\mathsf{pp}, (i, b); \rho), \mathsf{E}_2(\mathsf{pp}, \mathsf{y}, (i, b); \rho))$.

We will now define the notion of smoothness which extends the one-wayness property to an indistinguishability-based property.

Definition 5 (Smoothness). *Let* $(\mathsf{K}, \mathsf{f}, \mathsf{E}, \mathsf{D})$ *be as in Definition 4. We say that* $(\mathsf{K}, \mathsf{f}, \mathsf{E}, \mathsf{D})$ *is* (k, n)*-smooth if for any two* (k, n)*-sources* $\mathcal{S}_1$ *and* $\mathcal{S}_2$ *we have* $(\mathsf{pp}, \mathsf{f}(\mathsf{pp}, \mathsf{x}_1)) \stackrel{c}{\equiv} (\mathsf{pp}, \mathsf{f}(\mathsf{pp}, \mathsf{x}_2))$, *where* $\mathsf{pp} \xleftarrow{\$} \mathsf{K}(1^\lambda)$, $\mathsf{x}_1 \xleftarrow{\$} \mathcal{S}_1$ *and* $\mathsf{x}_2 \xleftarrow{\$} \mathcal{S}_2$.

In the full version of this paper, we show that the recyclable OWFE from [GH18] based on CDH is (k, n)-smooth, for any $k \geq \log p + \omega(\log \lambda)$, where p is the order of the underlying CDH-hard group.

[6] ct is assumed to contain (i, b).

4 Strong TDFs from Smooth Recyclable OWFE

In this section we show that recyclable OWFE implies the existence of TDFs with almost-perfect correctness in the sense of Definition 1. This improves the correctness property of [GH18]; see Remark 1. Moreover, we show that if the base recyclable OWFE scheme is smooth (Definition 5), then the resulting TDF satisfies the notions of security for deterministic encryption (Definition 2). We will then use this statement along with the CDH-based OWFE from [GH18] to obtain the first deterministic-encryption scheme based on CDH. In particular, the existence of deterministic encryption (even) with CPA security from CDH has been open until now.

A central new tool developed in this work is a *mirroring* technique, which we will describe below. As notation, for a matrix $\mathbf{M} \in \mathbb{Z}_2^{k \times n}$, we define $\mathsf{RSum}(\mathbf{M}) \triangleq \mathbf{M}_1 + \cdots + \mathbf{M}_k \in \mathbb{Z}_2^n$, where $\mathbf{M}_i$ for $i \in [k]$ denotes the ith row of $\mathbf{M}$.

Definition 6. *(The mirror Function* Mir*) Let* $(\mathsf{K}, \mathsf{f}, \mathsf{E}_1, \mathsf{E}_2, \mathsf{D})$ *be a recyclable OWFE scheme. For a public parameter* pp*, a value* $\mathsf{x} \in \{0,1\}^n$*, a matrix* $\mathbf{CT} := \begin{pmatrix} \mathsf{ct}_{1,0}, \mathsf{ct}_{2,0}, \ldots, \mathsf{ct}_{n,0} \\ \mathsf{ct}_{1,1}, \mathsf{ct}_{2,1}, \ldots, \mathsf{ct}_{n,1} \end{pmatrix}$ *of ciphertexts outputted by* E_1*, and a vector* $\mathbf{a} \in \{0,1\}^n$*, the function* $\mathsf{Mir}(\mathsf{pp}, \mathsf{x}, \mathbf{CT}, \mathbf{a})$ *outputs a matrix* $\mathbf{M} \in \mathbb{Z}_2^{2 \times n}$*, where* $\mathbf{M} := \begin{pmatrix} \mathsf{b}_{1,0}, \mathsf{b}_{2,0}, \ldots, \mathsf{b}_{n,0} \\ \mathsf{b}_{1,1}, \mathsf{b}_{2,1}, \ldots, \mathsf{b}_{n,1} \end{pmatrix}$ *is formed deterministically and uniquely according to the following two rules:*

1. *for all* $i \in [n]$*:* $\mathsf{b}_{i,\mathsf{x}_i} = \mathsf{D}(\mathsf{pp}, \mathsf{ct}_{i,\mathsf{x}_i}, \mathsf{x})$*; and*
2. $\mathsf{RSum}(\mathbf{M}) = \mathbf{a}$.

Note that the above computation is deterministic and can be done efficiently.

Construction 3 (TDF construction). *We now present our TDF construction.*

Base primitive. *A recyclable OWFE scheme* $\mathcal{E} = (\mathsf{K}, \mathsf{f}, \mathsf{E}, \mathsf{D})$*. Let* Rand *be the randomness space of the algorithm* E*.*

Construction. *The construction is parameterized over two parameters* $n = n(\lambda)$ *and* $r = r(\lambda)$*, where* n *is the input length to the function* $\mathsf{f}(\mathsf{pp}, \cdot)$*, and* r *will be instantiated in the correctness proof. The input space of the TDF is* $\{0,1\}^n$*.*

- $\mathsf{TDF.KG}(1^\lambda)$*:*
  1. *Sample* $\mathsf{pp} \leftarrow \mathsf{K}(1^\lambda)$*.*
  2. *For each* $h \in [r]$*:*

$$\mathbf{P}_h := \begin{pmatrix} \rho_{1,0}^{(h)}, \rho_{2,0}^{(h)}, \ldots, \rho_{n,0}^{(h)} \\ \rho_{1,1}^{(h)}, \rho_{2,1}^{(h)}, \ldots, \rho_{n,1}^{(h)} \end{pmatrix} \xleftarrow{\$} \mathsf{Rand}^{2 \times n}, \tag{2}$$

$$\mathbf{CT}_h := \begin{pmatrix} \mathsf{E}_1(\mathsf{pp}, (1,0); \rho_{1,0}^{(h)}), \mathsf{E}_1(\mathsf{pp}, (2,0); \rho_{2,0}^{(h)}), \ldots, \mathsf{E}_1(\mathsf{pp}, (n,0); \rho_{n,0}^{(h)}) \\ \mathsf{E}_1(\mathsf{pp}, (1,1); \rho_{1,1}^{(h)}), \mathsf{E}_1(\mathsf{pp}, (2,1); \rho_{2,1}^{(h)}), \ldots, \mathsf{E}_1(\mathsf{pp}, (n,1); \rho_{n,1}^{(h)}) \end{pmatrix}. \tag{3}$$

3. *For* $h \in [r]$ *sample* $\mathbf{a}_h \xleftarrow{\$} \{0,1\}^n$.
4. *Form the index key* ik *and the trapdoor key* tk *as follows:*

$$\mathsf{ik} := (\mathsf{pp}, \mathbf{CT}_1, \ldots, \mathbf{CT}_r, \mathbf{a}_1, \ldots, \mathbf{a}_r), \tag{4}$$

$$\mathsf{tk} := (\mathsf{pp}, \mathbf{P}_1, \ldots, \mathbf{P}_r). \tag{5}$$

– $\mathsf{TDF.F}(\mathsf{ik}, \mathsf{x})$: *Parse* ik *as in Eq. 4. Set* $\mathsf{y} := \mathsf{f}(\mathsf{pp}, \mathsf{x})$. *Return*

$$\mathsf{u} := (\mathsf{y}, \mathsf{Mir}(\mathsf{pp}, \mathsf{x}, \mathbf{CT}_1, \mathbf{a}_1), \ldots, \mathsf{Mir}(\mathsf{pp}, \mathsf{x}, \mathbf{CT}_r, \mathbf{a}_r)). \tag{6}$$

– $\mathsf{TDF.F}^{-1}(\mathsf{tk}, \mathsf{u})$:
1. *Parse* $\mathsf{tk} := (\mathsf{pp}, \mathbf{P}_1, \ldots, \mathbf{P}_r)$ *and parse* $\mathbf{P}_h$ *for* $h \in [r]$ *as in Eq. 2.*
2. *Parse* $\mathsf{u} := (\mathsf{y}, \mathbf{M}_1, \ldots, \mathbf{M}_r)$, *where for all* $h \in [r]$, $\mathbf{M}_h \in \mathbb{Z}_2^{2\times n}$.
3. *Reconstruct* $\mathsf{x} := \mathsf{x}_1 \cdots \mathsf{x}_n \in \{0,1\}^n$ *bit-by-bit as follows. To recover the* i*th bit of* x*:*
   (a) *If for some* $h \in [r]$, $\mathbf{M}_h[i] = \begin{pmatrix} \mathsf{E}_2(\mathsf{pp}, \mathsf{y}, (i,0); \rho_{i,0}^{(h)}) \\ 1-\mathsf{E}_2(\mathsf{pp}, \mathsf{y}, (i,1); \rho_{i,1}^{(h)}) \end{pmatrix}$, *set* $\mathsf{x}_i = 0$. *Here* $\mathbf{M}_h[i]$ *denotes the* i*th column of* $\mathbf{M}_h$.
   (b) *Else, if for some* $h \in [r]$, $\mathbf{M}_h[i] = \begin{pmatrix} 1-\mathsf{E}_2(\mathsf{pp}, \mathsf{y}, (i,0); \rho_{i,0}^{(h)}) \\ \mathsf{E}_2(\mathsf{pp}, \mathsf{y}, (i,1); \rho_{i,1}^{(h)}) \end{pmatrix}$, *set* $\mathsf{x}_i = 1$.
   (c) *Otherwise, halt and return* $\perp$.
4. *Return* x.

Lemma 3 (TDF correctness). *We have*

$$\Pr_{(\mathsf{ik},\mathsf{tk})}[\exists \mathsf{x} \in \{0,1\}^n \text{ s.t. } \mathsf{TDF.F}^{-1}(\mathsf{tk}, (\mathsf{TDF.F}(\mathsf{ik}, \mathsf{x}))) \neq \mathsf{x}] \leq \frac{n2^n}{2^r}, \tag{7}$$

where the probability is taken over $(\mathsf{ik}, \mathsf{tk}) \xleftarrow{\$} \mathsf{TDF.KG}(1^\lambda)$. *For instance, setting:* $r = n + \omega(\log \lambda)$ *gives a negligible inversion error.*

Lemma 4 (TDF one-wayness and CPA-indistinguishability security). *Assuming* $\mathcal{E}$ *is an OWFE scheme (i.e., an* (n,n)*-OWFE scheme), the TDF* $(\mathsf{TDF.KG}, \mathsf{TDF.F}, \mathsf{TDF.F}^{-1})$ *given in Construction 3 is one-way. That is, for any PPT adversary* $\mathcal{A}$

$$\Pr[\mathcal{A}(\mathsf{ik}, \mathsf{TDF.F}(\mathsf{ik}, \mathsf{x})) = \mathsf{x}] = \mathsf{negl}(\lambda), \tag{8}$$

where $(\mathsf{ik}, \mathsf{tk}) \xleftarrow{\$} \mathsf{TDF.KG}(1^\lambda)$ *and* $\mathsf{x} \xleftarrow{\$} \{0,1\}^n$. *Moreover, if* $\mathcal{E}$ *is* (k,n)*-smooth (Definition 5), the constructed TDF is* (k,n)*-CPA-indistinguishable (Definition 2).*

We may now combine the CDH-based OWFE from [GH18], Lemmas 3 and 4 to get the first CPA-secure deterministic encryption scheme from CDH.

Corollary 1 (CDH implies deterministic encryption). *Let* G *be a CDH-hard group scheme. For any* $k \geq \log p + \omega(\log \lambda)$ *and any* $n \geq k$ *(where* p *is the order of the underlying group), there exists a* (k,n)*-CPA-indistinguishable deterministic encryption scheme with plaintext size* n *(in bits) and ciphertext size* $\Theta(n^2)$.

4.1 Proof of Correctness: Lemma 3

Proof. We will use notation given in Construction 3. Note that for a given $\mathsf{x} \in \{0,1\}^n$, the inversion succeeds unless there exists an index $i \in [n]$ for which the following bad event happens.

- $\mathsf{Bad}_{\mathsf{x},i}$: for all $h \in [r]$, $\mathbf{a}_h[i] = \mathsf{E}_2(\mathsf{pp}, \mathsf{y}, (i,0); \rho_{i,0}^{(h)}) + \mathsf{E}_2(\mathsf{pp}, \mathsf{y}, (i,1); \rho_{i,0}^{(h)}) \in \mathbb{Z}_2$, where $\mathbf{a}_h[i]$ denotes the i'th coordinate of $\mathbf{a}_h \in \{0,1\}^n$.

Since the bits $\mathbf{a}_h[i]$ for all $h \in [r]$ are chosen uniformly at random (independently of pp and $\rho_{i,b}$'s), we have: $\Pr[\mathsf{Bad}_{\mathsf{x},i}] = 2^{-r}$. Doing a union bound over all column $i \in [n]$ gives the probability $n \cdot 2^{-r}$ of an inversion error for a given x. We conclude using a union bound over all $\mathsf{x} \in \{0,1\}^n$. □

4.2 Proof of One-Wayness and CPA Security: Lemma 4

To prove Lemma 4 we first give a simulated way of sampling an index key together with an image point for a target input value.

Definition 7 (Simulated distribution Sim). *Let $\mathcal{E} = (\mathsf{K}, \mathsf{f}, \mathsf{E}, \mathsf{D})$ be the underlying recyclable OWFE scheme. Fix $\mathsf{x} \in \{0,1\}^n$ and let $\mathsf{y} := \mathsf{f}(\mathsf{pp}, \mathsf{x})$. We define a simulator $\mathsf{Sim}(\mathsf{pp}, n, \mathsf{y})$, which samples a simulated index key $\mathsf{ik}_{\mathrm{sim}}$ with a corresponding simulated TDF image $\mathsf{u}_{\mathrm{sim}}$ for x, as follows. For $h \in [r]$ sample $r_{i,b}^h \xleftarrow{\$} \{0,1\}^*$ for all $(i,b) \in [n] \times \{0,1\}$, and set*

$$(\mathbf{CT}_h, \mathbf{M}_h) \xleftarrow{\$} \begin{pmatrix} \mathsf{E}_1(\mathsf{pp}, (1,0); r_{1,0}^h), \ldots, \mathsf{E}_1(\mathsf{pp}, (n,0); r_{n,0}^h) \\ \mathsf{E}_1(\mathsf{pp}, (1,1); r_{1,1}^h), \ldots, \mathsf{E}_1(\mathsf{pp}, (n,1); r_{n,1}^h) \end{pmatrix}, \begin{pmatrix} \mathsf{E}_2(\mathsf{pp}, \mathsf{y}, (1,0); r_{1,0}^h), \ldots, \mathsf{E}_2(\mathsf{pp}, \mathsf{y}, (n,0); r_{n,0}^h) \\ \mathsf{E}_2(\mathsf{pp}, \mathsf{y}, (1,1); r_{1,1}^h), \ldots, \mathsf{E}_2(\mathsf{pp}, \mathsf{y}, (n,1); r_{n,1}^h) \end{pmatrix}. \quad (9)$$

Let

$$\mathsf{ik}_{\mathrm{sim}} := (\mathsf{pp}, \mathbf{CT}_1, \ldots, \mathbf{CT}_r, \mathsf{RSum}(\mathbf{M}_1), \ldots, \mathsf{RSum}(\mathbf{M}_r))$$
$$\mathsf{u}_{\mathrm{sim}} := (\mathsf{y}, \mathbf{M}_1, \ldots, \mathbf{M}_r).$$

Equipped with the above definition, we now give of the proof of Lemma 4.

Proof of Lemma 4. For any distribution $\mathcal{S}$ over $\{0,1\}^n$, we show that the sole security-for-encryption requirement of the recyclable OWFE implies that

$$(\mathsf{x}, \mathsf{ik}, \mathsf{TDF.F}(\mathsf{ik}, \mathsf{x})) \stackrel{c}{\equiv} (\mathsf{x}, \mathsf{Sim}(\mathsf{pp}, n, \mathsf{y})), \quad (10)$$

where $\mathsf{x} \xleftarrow{\$} \mathcal{S}$, $\mathsf{pp} \xleftarrow{\$} \mathsf{K}(1^\lambda)$, $(\mathsf{ik}, *) \xleftarrow{\$} \mathsf{TDF.KG}(1^\lambda)$ and $\mathsf{y} := \mathsf{f}(\mathsf{pp}, \mathsf{x})$.

We first show how to use Eq. 10 to derive the one-wayness and indistinguishability security of the resulting TDF from the corresponding one-wayness and smoothness of the underlying OWFE scheme, and will then prove Eq. 10.

For one-wayness, if there exists an inverter $\mathcal{A}$ that with non-negligible probability can compute x from $(\mathsf{ik}, \mathsf{TDF.F}(\mathsf{ik}, \mathsf{x})$—where $(\mathsf{ik}, *) \xleftarrow{\$} \mathsf{TDF.KG}(1^\lambda)$ and

$\mathsf{x} \xleftarrow{\$} \{0,1\}^n$—then Eq. 10 implies that with non-negligible probability the adversary $\mathcal{A}$ can compute x from $\mathsf{Sim}(\mathsf{pp}, n, \mathsf{y})$, where $\mathsf{y} := \mathsf{f}(\mathsf{pp}, \mathsf{x})$. However, this latter violates the one-wayness of f, because the computation of $\mathsf{Sim}(\mathsf{pp}, n, \mathsf{y})$ may be done efficiently with knowledge of pp, n and y.

For indistinguishability security (Definition 2) let $\mathcal{S}_0$ and $\mathcal{S}_1$ be two (k, n) sources and assume that the recyclable OWFE scheme is k-smooth (Definition 5).

Letting $(\mathsf{ik}, *) \xleftarrow{\$} \mathsf{TDF.KG}(1^\lambda)$, $\mathsf{x}_0 \xleftarrow{\$} \mathcal{S}_0$, $\mathsf{x}_1 \xleftarrow{\$} \mathcal{S}_1$, $\mathsf{y}_0 := \mathsf{f}(\mathsf{pp}, \mathsf{x}_0)$ and $\mathsf{y}_1 := \mathsf{f}(\mathsf{pp}, \mathsf{x}_1)$, by Eq. 10 we have

$$(\mathsf{ik}, \mathsf{TDF.F}(\mathsf{ik}, \mathsf{x}_0)) \stackrel{c}{\equiv} \mathsf{Sim}(\mathsf{pp}, n, \mathsf{y}_0) \stackrel{c}{\equiv} \mathsf{Sim}(\mathsf{pp}, n, \mathsf{y}_1) \stackrel{c}{\equiv} (\mathsf{ik}, \mathsf{TDF.F}(\mathsf{ik}, \mathsf{x}_1)),$$

where the second indistinguishability follows from the k-smoothness of the recyclable OWFE scheme, which states $(\mathsf{pp}, \mathsf{y}_0) \stackrel{c}{\equiv} (\mathsf{pp}, \mathsf{y}_1)$.

We are left to prove Eq. 10. Fix the distribution $\mathcal{S}$ for which we want to prove Eq. 10. To this end, we change the simulator Sim given in Definition 7 to define a new simulator Sim' which on input $\mathsf{Sim}'(\mathsf{pp}, \mathsf{x})$ samples a pair $(\mathsf{ik}'_{\mathrm{sim}}, \mathsf{u}'_{\mathrm{sim}})$ as follows. Let $\mathsf{y} := \mathsf{f}(\mathsf{pp}, \mathsf{x})$. For all $h \in [r]$, let $\mathbf{CT}_h$ be sampled as in $\mathsf{Sim}(\mathsf{pp}, n, \mathsf{y})$, but with the following modification to $\mathbf{M}_h$:

- Letting $\mathbf{M}_h := \begin{pmatrix} \mathsf{e}^{(h)}_{1,0}, \ldots, \mathsf{e}^{(h)}_{n,0} \\ \mathsf{e}^{(h)}_{1,1}, \ldots, \mathsf{e}^{(h)}_{n,1} \end{pmatrix}$ be formed as in $\mathsf{Sim}(\mathsf{pp}, \mathsf{y})$, for any $i \in [n]$ change $\mathsf{e}^{(h)}_{i,1-\mathsf{x}_i}$ to a random bit (fresh for each index).

Having defined how $\mathbf{CT}_h$ and $\mathbf{M}_h$ are sampled for $h \in [r]$ during $\mathsf{Sim}'(\mathsf{pp}, \mathsf{x})$, form $(\mathsf{ik}'_{\mathrm{sim}}, \mathsf{u}'_{\mathrm{sim}})$ exactly as how $(\mathsf{ik}_{\mathrm{sim}}, \mathsf{u}_{\mathrm{sim}})$ is formed during $\mathsf{Sim}(\mathsf{pp}, n, \mathsf{y})$.

The security-for-encryption requirement of the OWFE scheme implies that $(\mathsf{x}, \mathsf{ik}_{\mathrm{sim}}, \mathsf{u}_{\mathrm{sim}}) \stackrel{c}{\equiv} (\mathsf{x}, \mathsf{ik}'_{\mathrm{sim}}, \mathsf{u}'_{\mathrm{sim}})$, where $\mathsf{x} \xleftarrow{\$} \mathcal{S}$, $\mathsf{y} := \mathsf{f}(\mathsf{pp}, \mathsf{x})$, $(\mathsf{ik}_{\mathrm{sim}}, \mathsf{u}_{\mathrm{sim}}) \xleftarrow{\$} \mathsf{Sim}(\mathsf{pp}, n, \mathsf{y})$ and $(\mathsf{ik}'_{\mathrm{sim}}, \mathsf{u}'_{\mathrm{sim}}) \xleftarrow{\$} \mathsf{Sim}'(\mathsf{pp}, \mathsf{x})$. Moreover, it is easy to verify that $(\mathsf{x}, \mathsf{ik}'_{\mathrm{sim}}, \mathsf{u}'_{\mathrm{sim}})$ is identically distributed to $(\mathsf{x}, \mathsf{ik}, \mathsf{TDF.F}(\mathsf{ik}, \mathsf{x}))$, where $(\mathsf{ik}, \mathsf{tk}) \xleftarrow{\$} \mathsf{TDF.KG}(1^\lambda)$. The proof is now complete. □

The TDF given in Construction 3 is CPA secure (in a deterministic-encryption sense), but it is not hard to show that the construction is not CCA2 secure. However, we show in the full version of this paper that using techniques of [RS09,KMO10] one may use the TDF of Construction 3 to build another TDF which is CCA2 secure. This upgrading further increases the ciphertext size, resulting in ciphertext size $\Theta(n^3)$ (for the CDH-based instantiation), where n is the plaintext size.

5 Efficient Strong TDFs from Smooth OWFE

The TDF and deterministic encryption presented in Sect. 4 have the drawback that the output size grows at least quadratically with the input size. The reason behind this blowup is that we had to do "repetitions," resulting in $\Theta(n/2)$ output bits for every single bit of the input. In this section we show how to do away with excessive use of repetition, and to obtain TDFs (and deterministic encryption) whose image/ciphertext size grows linearly with input size. Our main idea involves the use of error-correcting codes, taking advantage of the local inversion property of our basic TDF. As a result, we will obtain the first CPA-secure deterministic encryption scheme with linear ciphertext size based on CDH. We stress that, even relying on DDH, previous DDH-based deterministic-encryption and TDF schemes resulted in quadratically large ciphertexts.

Definition 8 ($(m, n, d)_2$-Codes). *We recall the notion of $(m, n, d)_2$ error-correcting codes. Such a code is given by efficiently computable functions* $(\mathsf{Encode}, \mathsf{Decode})$, *where* $\mathsf{Encode} : \{0,1\}^n \to \{0,1\}^m$, *and where*

1. ***Distance.*** *For any two distinct* $\mathsf{x}_1, \mathsf{x}_2 \in \{0,1\}^n$ *we have* $\mathsf{H}_{\mathsf{dst}}(\mathsf{Encode}(\mathsf{x}_1), \mathsf{Encode}(\mathsf{x}_2)) \geq d$, *where* $\mathsf{H}_{\mathsf{dst}}$ *denotes the Hamming distance.*
2. ***Erasure correction.*** *For any* $\mathsf{x} \in \{0,1\}^n$, *letting* $\mathsf{z} := \mathsf{Encode}(\mathsf{x})$, *given any string* $\mathsf{z}' \in \{0,1,\perp\}^m$, *which has at most* $d-1$ $\perp$ *symbols, and whose all non-*$\perp$ *symbols agree with* z, *we have* $\mathsf{Decode}(\mathsf{z}') = \mathsf{x}$.
3. ***Error correction.*** *For any* $\mathsf{x} \in \{0,1\}^n$, *letting* $\mathsf{z} := \mathsf{Encode}(\mathsf{x})$, *given any* $\mathsf{z}' \in \{0,1\}^m$ *such that* $\mathsf{H}_{\mathsf{dst}}(\mathsf{z}, \mathsf{z}') < d/2$, *we have* $\mathsf{Decode}(\mathsf{z}') = \mathsf{x}$.

We are interested in binary codes with constant rate, constant relative distance, that is: $m = cn$, and $d = c_1 n$. Such codes can be obtained by concatenating codes with constant rate and constant relative distance over large fields—such as Reed-Solomon codes—with codes with constant rate and binary alphabet. See for instance binary Justesen codes [Jus72].

Definition 9 (rECC code). *We define a code that suites our purposes, which is the concatenation of an ECC code with a repetition code. Specifically, for a repetition constant* t, *a* t*-rECC code* $(\mathsf{Encode}, \mathsf{Decode})$ *consists of* $\mathsf{Encode}\colon \{0,1\}^n \to \{0,1\}^m$, *which is obtained by first applying a* $(cn, n, c_1 n)_2$ *code and then repeating each bit of the* cn *bit codeword* t *times. Thus,* $m = tcn$. *Note that this code is now a* $(tcn, n, tc_1 n)_2$*-code.*

Looking ahead, we remark that the use of these repetition codes makes decoding later easier. Specifically, with this repetition, an m bit codeword can be viewed as having cn *blocks* of t bits each. Furthermore, for decoding it is enough to recover one bit per block for at least $cn - c_1 n + 1$ blocks.

In our constructions, it is instructive to think of $c = 200$, $c_1 = 20$ and $t = 9$ for convenience in proofs.[7]

Block index versus bit index. Having codes given as above based on repetition, for a codeword $\mathsf{z} \in \{0,1\}^m$ we talk about a jth block of z for $j \in [m/t]$ to refer to the collections of the bits with indices $\{(j-1)t + 1, \ldots, jt\}$.

[7] The choices of the constants were made as above so to have slackness in proofs—they have not been optimized for efficiency.

Construction 4 (TDF construction). *We now describe our TDF construction.*

Base primitive. *A t-rECC code* (Encode, Decode)*, where* Encode: $\{0,1\}^n \to \{0,1\}^m$*, and a recyclable OWFE scheme* $\mathcal{E} = (\mathsf{K}, \mathsf{f}, \mathsf{E}, \mathsf{D})$*, where* f*'s input space is* $\{0,1\}^m$*. We will instantiate the value of constant t in the correctness proof. Let* Rand *be the randomness space of the encapsulation algorithm* E*.*

Construction.

- $\mathsf{TDF.KG}(1^\lambda)$:
  1. *Sample* $\mathsf{pp} \leftarrow \mathsf{K}(1^\lambda)$ *and*

$$\mathbf{P} := \begin{pmatrix} \rho_{1,0}, \rho_{2,0}, \ldots, \rho_{m,0} \\ \rho_{1,1}, \rho_{2,1}, \ldots, \rho_{m,1} \end{pmatrix} \xleftarrow{\$} \mathsf{Rand}^{2\times m}, \tag{11}$$

$$\mathbf{CT} := \begin{pmatrix} \mathsf{ct}_{1,0}, \mathsf{ct}_{2,0}, \ldots, \mathsf{ct}_{m,0} \\ \mathsf{ct}_{1,1}, \mathsf{ct}_{2,1}, \ldots, \mathsf{ct}_{m,1} \end{pmatrix}, \tag{12}$$

  where for all $i \in [m]$ *and* $b \in \{0,1\}$, $\mathsf{ct}_{i,b} := \mathsf{E}_1(\mathsf{pp}, (i,b); \rho_{i,b})$.
  2. *Sample* $\mathbf{a} \xleftarrow{\$} \{0,1\}^m$.
  3. *Form the index key* ik *and the trapdoor key* tk *as follows:*

$$\mathsf{ik} := (\mathsf{pp}, \mathbf{a}, \mathbf{CT}) \qquad \mathsf{tk} := (\mathsf{pp}, \mathbf{a}, \mathbf{P}). \tag{13}$$

- $\mathsf{TDF.F}(\mathsf{ik}, \mathsf{x})$: *Parse* $\mathsf{ik} := (\mathsf{pp}, \mathbf{a}, \mathbf{CT})$. *Let* $\mathsf{z} := \mathsf{Encode}(\mathsf{x})$ *and* $\mathsf{y} := \mathsf{f}(\mathsf{pp}, \mathsf{z})$. *Return*

$$\mathsf{u} := (\mathsf{y}, \mathsf{Mir}(\mathsf{pp}, \mathsf{z}, \mathbf{CT}, \mathbf{a})). \tag{14}$$

- $\mathsf{TDF.F}^{-1}(\mathsf{tk}, \mathsf{u})$:
  1. *Parse* $\mathsf{tk} := (\mathsf{pp}, \mathbf{a}, \mathbf{P})$ *and parse* $\mathbf{P}$ *as in Equation* (11)*. Parse* $\mathsf{u} := (\mathsf{y}, \mathbf{M})$*, where* $\mathbf{M} \in \mathbb{Z}_2^{2\times m}$*. If* $\mathsf{RSum}(\mathbf{M}) \neq \mathbf{a}$*, then return* $\perp$.
  2. *Construct* $\mathsf{z}' := \mathsf{z}'_1 \cdots \mathsf{z}'_m$ *bit-by-bit as follows. To recover the ith bit of* z':
     (a) *If* $\mathbf{M}[i] = \begin{pmatrix} \mathsf{E}_2(\mathsf{pp},\mathsf{y},(i,0);\rho_{i,0}) \\ 1-\mathsf{E}_2(\mathsf{pp},\mathsf{y},(i,1);\rho_{i,1}) \end{pmatrix}$*, set* $\mathsf{z}'_i = 0$*. Here* $\mathbf{M}[i]$ *denotes the ith column of* $\mathbf{M}$.
     (b) *Else if* $\mathbf{M}[i] = \begin{pmatrix} 1-\mathsf{E}_2(\mathsf{pp},\mathsf{y},(i,0);\rho_{i,0}) \\ \mathsf{E}_2(\mathsf{pp},\mathsf{y},(i,1);\rho_{i,1}) \end{pmatrix}$*, set* $\mathsf{z}'_i = 1$.
     (c) *Else, set* $\mathsf{z}'_i = \perp$.
  3. *Letting* $\mathsf{x} := \mathsf{Decode}(\mathsf{z}')$*, if* $\mathsf{TDF.F}(\mathsf{ik}, \mathsf{x}) = \mathsf{u}$*, then return* x*. Otherwise, return* $\perp$.

We will now give the correctness and security statements about our TDF, and will prove them in the subsequent subsections.

Lemma 5 (Correctness). *Using a t-rECC code* (Encode, Decode) *with parameters* $(tcn, n, tc_1n)_2$ *(Definition 9), we have*

$$\Pr_{(\mathsf{ik},\mathsf{tk})}[\exists \mathsf{x} \in \{0,1\}^n \ s.t.\ \mathsf{TDF.F}^{-1}(\mathsf{tk}, (\mathsf{TDF.F}(\mathsf{ik}, \mathsf{x}))) \neq \mathsf{x}] \leq 2^n \cdot \mathsf{e}^{-\frac{(2^t c_1 - c)^2 n}{2^{2t-1} c}}. \tag{15}$$

In particular, by choosing the repetition constant t based on c and c_1 *in such a way that* $2^t c_1 > c$ *and that* $\frac{(2^t c_1 - c)^2}{2^{2t-1} c} \geq 0.7$*, we will have a negligible error.*

Lemma 6 (TDF one-wayness and CPA-indistinguishability security). *Assuming $\mathcal{E}$ is an (n, m)-OWFE scheme, the TDF* $(\mathsf{TDF.KG}, \mathsf{TDF.F}, \mathsf{TDF.F}^{-1})$ *given in Construction 3 is one-way. That is, for any PPT adversary $\mathcal{A}$*

$$\Pr[\mathcal{A}(\mathsf{ik}, \mathsf{TDF.F}(\mathsf{ik}, \mathsf{x})) = \mathsf{x}] = \mathsf{negl}(\lambda), \tag{16}$$

where $(\mathsf{ik}, \mathsf{tk}) \xleftarrow{\$} \mathsf{TDF.KG}(1^\lambda)$ *and* $\mathsf{x} \xleftarrow{\$} \{0,1\}^n$. *Moreover, assuming that the underlying OWFE scheme is (k, m)-smooth (Definition 5), the constructed TDF is (k, n)-indistinguishable (Definition 2).*

Theorem 5 (CCA2-indistinguishability security). *Assuming that the underlying OWFE scheme is (k, m)-smooth and by appropriately choosing the parameters (in particular we will have $t, c, c_1 \in O(1)$), the constructed TDF is (k, n)-CCA2-indistinguishable.*

We may now combine the CDH-based OWFE from [GH18], 5 with Theorem 5 to get the following corollary.

Corollary 2 (CDH implies efficient deterministic encryption). *Let G be a CDH-hard group scheme. For any $k \geq \log p + \omega(\log \lambda)$ and any $n \geq k$ (where p is the order of the underlying group), there exists a (k, n)-CCA2-indistinguishable deterministic encryption scheme with plaintext size n (in bits) and ciphertext size* $\log p + O(n)$.

We prove Lemmas 5 and 6 in the full version of this paper.

5.1 Proof of CCA2 Security: Theorem 5

We give the proof of Theorem 5 via a series of lemmas. We first start with the following notation.

Notation 6. *For an OWFE scheme* $(\mathsf{K}, \mathsf{f}, \mathsf{E}_1, \mathsf{E}_2, \mathsf{D})$, *letting* $\mathbf{P} := \binom{\rho_{1,0}, \rho_{2,0}, \ldots, \rho_{m,0}}{\rho_{1,1}, \rho_{2,1}, \ldots, \rho_{m,1}}$ *we define*

$$\mathsf{E}(\mathsf{pp}, \mathsf{y}, \mathbf{P}) \triangleq \begin{pmatrix} \mathsf{E}_1(\mathsf{pp}, (1,0); \rho_{1,0}), \ldots, \mathsf{E}_1(\mathsf{pp}, (m,0); \rho_{m,0}) \\ \mathsf{E}_1(\mathsf{pp}, (1,1); \rho_{1,1}), \ldots, \mathsf{E}_1(\mathsf{pp}, (m,1); \rho_{m,1}) \end{pmatrix},$$
$$\begin{pmatrix} \mathsf{E}_2(\mathsf{pp}, \mathsf{y}, (1,0); \rho_{1,0}), \ldots, \mathsf{E}_2(\mathsf{pp}, \mathsf{y}, (m,0); \rho_{m,0}) \\ \mathsf{E}_2(\mathsf{pp}, \mathsf{y}, (1,1); \rho_{1,1}), \ldots, \mathsf{E}_2(\mathsf{pp}, \mathsf{y}, (m,1); \rho_{m,1}) \end{pmatrix}.$$

Half-trapdoor keys. In the proof of Theorem 5 we will make use of an alternative way of inversion which works with respect to knowledge of half of all the randomness values that were fixed in the trapdoor key. We refer to such trapdoor keys as *half* trapdoor keys (or simulated trapdoor keys). Recall that a real trapdoor key is of the form

$$(\mathsf{pp}, \mathbf{a}, (\rho_{1,0}, \rho_{1,1}), \ldots, (\rho_{m,0}, \rho_{m,1})). \tag{17}$$

A half-trapdoor key is a reduced version of a full trapdoor key in that we forget one randomness value from each pair, while remembering whether we chose to keep the first or the second coordinate of that pair. Formally, given a full trapdoor key as in Equation (17), a half trapdoor key is obtained based on a string $\mathsf{s} \in \{0,1\}^m$ as $\mathsf{tk_{rd}} := (\mathsf{pp}, \mathbf{a}, \mathsf{s}, (\rho_1, \ldots, \rho_m))$, where $\rho_i = \rho_{i,\mathsf{s}_i}$. (The subscript rd stands for "reduced.")

We will now define how to perform inversion w.r.t. half-trapdoor keys.

Definition 10 (Half-trapdoor inversion $\mathsf{TDF.F}_{\mathrm{rd}}^{-1}$). *For an image* $\mathsf{u} := (\mathsf{y}, \mathbf{M})$ *of our constructed TDF and a half-trapdoor key* $\mathsf{tk_{rd}} := (\mathsf{pp}, \mathbf{a}, \mathsf{s}, (\rho_1, \ldots, \rho_m))$ *we define* $\mathsf{TDF.F}_{\mathrm{rd}}^{-1}(\mathsf{tk_{rd}}, \mathsf{u})$ *as follows:*

1. *If* $\mathsf{RSum}(\mathbf{M}) \neq \mathbf{a} \in \mathbb{Z}_2^m$*, then return* $\perp$.
2. *Construct* $\mathsf{z}' \in \{0,1\}^m$ *bit by bit as follows. For all* $i \in [m]$*, we denote by* $\mathbf{M}[i] = \binom{\mathsf{e}_{i,0}}{\mathsf{e}_{i,1}}$ *the* i*'th column of* $\mathbf{M}$*. If* $\mathsf{e}_{i,\mathsf{s}_i} = 1 - \mathsf{E}_2(\mathsf{pp}, \mathsf{y}, (i, \mathsf{s}_i); \rho_i)$*, then set* $\mathsf{z}'_i = 1 - \mathsf{s}_i$*; otherwise set* $\mathsf{z}'_i = \perp$.
3. *For all* $j \in [cn]$*, if* $\exists i^* \in \{(j-1)t+1, \ldots, jt\}$ *such that* $\mathsf{z}'_{i^*} \neq \perp$ *then for all* $i \in \{(j-1)t+1, \ldots, jt\}$ *set* $\mathsf{z}''_i = \mathsf{z}'_{i^*}$*; else set* $\mathsf{z}''_i = \mathsf{s}_i$
4. *Letting* $\mathsf{x} := \mathsf{Decode}(\mathsf{z}'')$*, if* $\mathsf{TDF.F}(\mathsf{ik}, \mathsf{x}) = \mathsf{u}$*, return* x*. Otherwise, return* $\perp$.

As terminology, we say that $\mathsf{TDF.F}_{\mathrm{rd}}^{-1}(\mathsf{tk_{rd}}, \mathsf{u})$ *is able to open the ith column of* $\mathbf{M}$ *if* $\mathsf{z}'_i \neq \perp$ *(i.e., if* $\mathsf{e}_{i,\mathsf{s}_i} = 1 - \mathsf{E}_2(\mathsf{pp}, \mathsf{y}, (i, \mathsf{s}_i); \rho_i)$*).*

We first fix some notation and will then prove a useful property about half-inversion simulation, which in turn will be used in the CCA2 proof.

Notation 7 (Half trapdoor keys). *For a given* $\mathsf{tk} := (\mathsf{pp}, \mathbf{a}, (\rho_{1,0}, \rho_{1,1}), \ldots, (\rho_{m,0}, \rho_{m,1}))$ *and* $\mathsf{z} \in \{0,1\}^m$ *we define* $\mathsf{tk}/\mathsf{z} \triangleq (\mathsf{pp}, \mathbf{a}, \mathsf{z}, \rho_{1,\mathsf{z}_1}, \ldots, \rho_{m,\mathsf{z}_m})$.

We now give the following lemma about the effectiveness of the half-trapdoor inversion procedure.

Lemma 7 (Half-trapdoor inversion). *Fix* $\mathsf{x} \in \{0,1\}^n$ *and let* $\mathsf{z} := \mathsf{Encode}(\mathsf{x})$*. Using code* $(\mathsf{Encode}, \mathsf{Decode})$ *and setting* t *such that* $1 - 2^{-t} \geq \frac{1}{2} + \frac{c_1}{2c} - \frac{2}{c_1}$*, we have*

$$\Pr_{(\mathsf{ik},\mathsf{tk})}[\exists \mathsf{x}' \in \{0,1\}^n \backslash \{\mathsf{x}\} \text{ s.t. } \mathsf{TDF.F}^{-1}(\mathsf{tk}, \mathsf{u}') \neq \mathsf{TDF.F}_{\mathrm{rd}}^{-1}(\mathsf{tk_{rd}}, \mathsf{u}')] = 2^n \mathsf{e}^{-\frac{(c_1-4)^2 n}{2c_1}},$$

where $(\mathsf{ik}, \mathsf{tk}) \xleftarrow{\$} \mathsf{TDF.KG}(1^\lambda)$*,* $\mathsf{u}' := \mathsf{TDF.F}(\mathsf{ik}, \mathsf{x}')$ *and* $\mathsf{tk_{rd}} := \mathsf{tk}/\mathsf{z}$*. Thus, by appropriately choosing* c_1 *and* c *(and* t *based on these two values) the above probability will be negligible.*

Proof. Fix $\mathsf{x} \in \{0,1\}^n$ and let $\mathsf{z} := \mathsf{Encode}(\mathsf{x})$. For a sampled $(\mathsf{ik}, \mathsf{tk})$ we define the event Bad as

$$\mathrm{Bad} := \exists \mathsf{x}' \in (\{0,1\}^n \setminus \{\mathsf{x}\}) \text{ s.t. } \mathsf{TDF.F}^{-1}(\mathsf{tk}, \mathsf{u}') \neq \mathsf{TDF.F}_{\mathrm{rd}}^{-1}(\mathsf{tk_{rd}}, \mathsf{u}'),$$

where $\mathsf{u}' := \mathsf{TDF.F}(\mathsf{ik}, \mathsf{x}')$ and $\mathsf{tk_{rd}} := \mathsf{tk}/\mathsf{z}$.

First, note that if $\mathsf{TDF.F}^{-1}(\mathsf{tk}, \mathsf{TDF.F}(\mathsf{ik}, \mathsf{x}')) = \bot$, then $\mathsf{TDF.F}_{\mathrm{rd}}^{-1}(\mathsf{tk}_{\mathrm{rd}}, \mathsf{TDF.F}(\mathsf{ik}, \mathsf{x}')) = \bot$, and if $\mathsf{TDF.F}_{\mathrm{rd}}^{-1}(\mathsf{tk}_{\mathrm{rd}}, \mathsf{TDF.F}(\mathsf{ik}, \mathsf{x}')) \neq \bot$, then $\mathsf{TDF.F}^{-1}(\mathsf{tk}, \mathsf{TDF.F}(\mathsf{ik}, \mathsf{x}')) = \mathsf{x}' = \mathsf{TDF.F}_{\mathrm{rd}}^{-1}(\mathsf{tk}_{\mathrm{rd}}, \mathsf{TDF.F}(\mathsf{ik}, \mathsf{x}'))$. This follows from the descriptions of $\mathsf{TDF.F}_{\mathrm{rd}}^{-1}$ and $\mathsf{TDF.F}^{-1}$, and from the correctness property of $\mathsf{TDF.F}^{-1}$.

Thus, defining

$$\mathrm{Bad}' \colon \exists \mathsf{x}' \in (\{0,1\}^n \setminus \{\mathsf{x}\}) \text{ s.t. } \left(\mathsf{TDF.F}_{\mathrm{rd}}^{-1}(\mathsf{tk}_{\mathrm{rd}}, (\mathsf{TDF}(\mathsf{ik}, \mathsf{x}'))) = \bot\right),$$

we have $\Pr[\mathrm{Bad}] \leq \Pr[\mathrm{Bad}']$. In what follows, for any fixed $\mathsf{x}' \in \{0,1\}^n$ we will show

$$\Pr[\mathrm{Bad}_{\mathsf{x}'}] \leq \mathsf{e}^{-\frac{(c_1-4)^2 n}{2c_1}},$$

where we define $\mathrm{Bad}_{\mathsf{x}'} := \mathsf{TDF.F}_{\mathrm{rd}}^{-1}(\mathsf{tk}_{\mathrm{rd}}, (\mathsf{TDF}(\mathsf{ik}, \mathsf{x}'))) = \bot$. This will complete the proof.

For the fixed $\mathsf{x}' \in \{0,1\}^n$, let $\mathsf{u}' := \mathsf{TDF.F}(\mathsf{ik}, \mathsf{x}')$ and $\mathsf{z}' := \mathsf{Encode}(\mathsf{x}')$. Parse $\mathsf{u}' := (\mathsf{y}', \mathbf{M}')$.

In order to argue about the correctness of the output of $\mathsf{TDF.F}_{\mathrm{rd}}^{-1}(\mathsf{tk}_{\mathrm{rd}}, \mathsf{u}')$, let z^* denote the string that is constructed bit-by-bit during the execution of $\mathsf{TDF.F}_{\mathrm{rd}}^{-1}(\mathsf{tk}_{\mathrm{rd}}, \mathsf{u}')$. We will show that the fractional distance $\frac{\mathsf{H}_{\mathsf{dst}}(\mathsf{z}^*, \mathsf{z}')}{m} \leq \frac{c_1}{2c}$, and thus by the error-correction property of the underlying code (Item 3 of Definition 8) we will have $\mathsf{TDF.F}_{\mathrm{rd}}^{-1}(\mathsf{tk}_{\mathrm{rd}}, \mathsf{u}') = \mathsf{x}'$, as desired.

Let $\mathsf{S} \subseteq [cn]$ be the set of block indices on which z and z' are different. (Recall the notion of block index from the paragraphs after Definition 8.) Suppose $|\mathsf{S}| = v$ and let $\mathsf{S} := \{u_1, \ldots, u_v\}$. Note that $v \geq c_1 n$. We have

1. For any block index $j \in [cn] \setminus \mathsf{S}$, all those t bits of z' which come from its jth block will be equal to those of z^*. Namely, for all $i \in \{(j-1)t+1, \ldots, jt\}$ we have $\mathsf{z}'_i = \mathsf{z}^*_i$.
2. For any block index $j \in \mathsf{S}$, if the jth block of z' is different from that of z^*, then *all* the columns of the jth block of $\mathbf{M}'$ are hung; Namely, for all $i \in \{(j-1)t+1, \ldots, jt\}$, $\mathbf{M}'[i]$ is hung. This fact follows easily by inspection.

With the above intuition in mind, for $j \in [v]$ let W_j be a Boolean random variable where $\mathsf{W}_j = 0$ if the entire u_j'th block of $\mathbf{M}'$ is hung (i.e., all the corresponding t columns are hung), and $\mathsf{W}_j = 1$, otherwise. Note that for all j: $\Pr[\mathsf{W}_j = 1] = 1 - 2^{-t}$; this follows from the random choice of the vector $\mathbf{a}$ which is fixed in ik. Thus, by the bounds fixed in the lemma we have $\Pr[\mathsf{W}_j = 1] \geq \frac{1}{2} + \frac{c_1}{2c} - \frac{2}{c_1}$. Let $p := 1 - 2^{-t}$. We have

$$\begin{aligned}\Pr[\mathrm{Bad}_{\mathsf{x}'}] &\leq \Pr[\frac{1}{v}\sum_{j=1}^{v} \mathsf{W}_j < \frac{c_1}{2c}] \leq \Pr[\frac{1}{v}\sum_{j=1}^{v} \mathsf{W}_j < p - (\frac{1}{2} - \frac{2}{c_1})] \leq^* \mathsf{e}^{-2(\frac{1}{2}-\frac{2}{c_1})^2 v} \\ &\leq \mathsf{e}^{-2(\frac{1}{2}-\frac{2}{c_1})^2 c_1 n} = \mathsf{e}^{-\frac{2(c_1-4)^2 c_1 n}{4c_1^2}} = \mathsf{e}^{-\frac{(c_1-4)^2 n}{2c_1}},\end{aligned} \tag{18}$$

where the probability marked with $*$ follows from the Chernoff bounds. The proof is now complete. ∎

Our CCA2 hybrids will also make use of a simulated way of producing index/trapdoor keys. This procedure is described below.

Definition 11 (Simulated TDF key generation). *We define a simulated key-generation algorithm for the TDF given in Construction 4. Let* $(\mathsf{K}, \mathsf{f}, \mathsf{E}_1, \mathsf{E}_2, \mathsf{D})$ *be the underlying OWFE scheme. The simulated key generation algorithm* $\mathsf{TDF.KG_{sim}}(\mathsf{pp}, \mathsf{y})$ *takes* pp *and an image* y *of the function* f *as input, and outputs* $(\mathsf{ik}, \mathsf{tk})$ *formed as follows. Sample* $\mathbf{P} \xleftarrow{\$} \mathsf{Rand}^{2 \times m}$ *and set* $(\mathbf{CT}, \mathbf{M}) := \mathsf{E}(\mathsf{pp}, \mathsf{y}, \mathbf{P})$. *(See Notation 6.) Set* $\mathsf{ik} := (\mathsf{pp}, \mathsf{RSum}(\mathbf{M}), \mathbf{CT})$ *and* $\mathsf{tk} := (\mathsf{pp}, \mathsf{RSum}(\mathbf{M}), \mathbf{P})$.

We will now describe the hybrids for proving CCA2 security of the deterministic encryption scheme. We define the hybrids with respect to a distribution $\mathcal{D}$ and will then instantiate the distribution in the subsequent lemmas.

Hybrid $\mathsf{H}_0[\mathcal{D}]$*: real game.*

- **Index/trapdoor keys.** Sample $(\mathsf{ik}, \mathsf{tk}) \xleftarrow{\$} \mathsf{TDF.KG}(1^\lambda)$.
- **Challenge ciphertext.** Set $\mathsf{u} := \mathsf{TDF.F}(\mathsf{ik}, \mathsf{x})$, where $\mathsf{x} \leftarrow \mathcal{D}$.
- **CCA2 inversion queries.** Reply to each inversion query $\mathsf{u}' \neq \mathsf{u}$ with $\mathsf{TDF.F}^{-1}(\mathsf{tk}, \mathsf{u}')$.

Hybrid $\mathsf{H}_1[\mathcal{D}]$*: half-trapdoor inversion.* Same as H_0 except we reply to inversion queries using a half trapdoor and by using the algorithm $\mathsf{TDF.F}_{\mathrm{rd}}^{-1}$.

- **Index/trapdoor keys.** Sample $(\mathsf{ik}, \mathsf{tk}) \xleftarrow{\$} \mathsf{TDF.KG}(1^\lambda)$. Set the index key to be ik and form the trapdoor key as follows: sample $\mathsf{x} \leftarrow \mathcal{D}$, let $\mathsf{z} := \mathsf{Encode}(\mathsf{x})$ and set the trapdoor key to be $\mathsf{tk}_{\mathrm{rd}} := \mathsf{tk}/\mathsf{z}$ (Notation 7).
- **Challenge ciphertext.** Return $\mathsf{u} := \mathsf{TDF.F}(\mathsf{ik}, \mathsf{x})$, where recall that x was sampled in the previous step.
- **CCA2 inversion queries.** Reply to each inversion query $\mathsf{u}' \neq \mathsf{u}$ with $\mathsf{TDF.F}_{\mathrm{rd}}^{-1}(\mathsf{tk}_{\mathrm{rd}}, \mathsf{u}')$.

Hybrid $\mathsf{H}_2[\mathcal{D}]$*: half-trapdoor inversion with a simulated index key.* Same as $\mathsf{H}_1[\mathcal{D}]$ except that we sample the index key and the challenge ciphertext jointly in a simulated way.

- **Index/trapdoor keys:**
  1. Sample $\mathsf{x} \leftarrow \mathcal{D}$, and let $\mathsf{z} := \mathsf{Encode}(\mathsf{x})$. Set $\mathsf{y} := \mathsf{f}(\mathsf{pp}, \mathsf{z})$.
  2. Sample $(\mathsf{ik}, \mathsf{tk}) \xleftarrow{\$} \mathsf{TDF.KG_{sim}}(\mathsf{pp}, \mathsf{y})$.
  3. Set the index key to be ik and the trapdoor key to be $\mathsf{tk}_{\mathrm{rd}} := \mathsf{tk}/\mathsf{z}$.
- **Challenge ciphertext.** Return $\mathsf{u} := \mathsf{TDF.F}(\mathsf{ik}, \mathsf{x})$, where recall that x was sampled above.
- **CCA2 inversion queries.** Reply to each inversion query $\mathsf{u}' \neq \mathsf{u}$ with $\mathsf{TDF.F}_{\mathrm{rd}}^{-1}(\mathsf{tk}_{\mathrm{rd}}, \mathsf{u}')$.

Hybrid $\mathsf{H}_3[\mathcal{D}]$*: Full trapdoor inversion with a simulated index key.* Same as $\mathsf{H}_2[\mathcal{D}]$ except we use tk as the trapdoor key (instead of $\mathsf{tk}_{\mathrm{rd}}$) and will reply to each CCA2 inversion query $\mathsf{u}' \neq \mathsf{u}$ with $\mathsf{TDF.F}^{-1}(\mathsf{tk}, \mathsf{u}')$. That is:

- **Index/trapdoor keys:**
  1. Sample $\mathsf{x} \leftarrow \mathcal{D}$, and let $\mathsf{z} := \mathsf{Encode}(\mathsf{x})$. Set $\mathsf{y} := \mathsf{f}(\mathsf{pp}, \mathsf{z})$.
  2. Let the index/trapdoor key be $(\mathsf{ik}, \mathsf{tk}) \stackrel{\$}{\leftarrow} \mathsf{TDF.KG}_{\mathrm{sim}}(\mathsf{pp}, \mathsf{y})$.
- **Challenge ciphertext.** Return $\mathsf{u} := \mathsf{TDF.F}(\mathsf{ik}, \mathsf{x})$.
- **CCA2 inversion queries.** Reply to each inversion query $\mathsf{u}' \neq \mathsf{u}$ with $\mathsf{D}(\mathsf{tk}, \mathsf{u}')$.

The above concludes the description of the hybrids. We now define some notation and will then prove some lemmas.

Notation. For $i \in \{0, 1, 2, 3\}$ we use $\mathsf{out}_i[\mathcal{D}]$ to denote the output bit of an underlying adversary in hybrid $\mathsf{H}_i[\mathcal{D}]$. For $i, j \in \{0, 1, 2, 3\}$ and two distributions $\mathcal{S}_0$ and $\mathcal{S}_1$, we write $\mathsf{H}_i[\mathcal{S}_0] \stackrel{c}{\equiv} \mathsf{H}_j[\mathcal{S}_1]$ to mean that for all PPT adversaries $\mathcal{A}$ we have $|\Pr[\mathsf{out}_i[\mathcal{S}_0] = 1] - \Pr[\mathsf{out}_j[\mathcal{S}_1] = 1]| = \mathsf{negl}(\lambda)$.

The proof of Theorem 5 follows from the following lemmas.

Lemma 8 (Indistinguishability of Hybrids H_0 and H_1). *By appropriately choosing the parameters for c, c_1 and t, for any PPT adversary $\mathcal{A}$ we have* $|\Pr[\mathsf{out}_0[\mathcal{D}] = 1] - \Pr[\mathsf{out}_1[\mathcal{D}] = 1]| = \mathsf{negl}(\lambda)$.

Lemma 9 (Indistinguishability of Hybrids H_1 and H_2). *If the OWFE satisfies the security-for-encryption property, then for any distribution $\mathcal{D}$ and any PPT adversary $\mathcal{A}$, we have* $|\Pr[\mathsf{out}_1[\mathcal{D}] = 1] - \Pr[\mathsf{out}_2[\mathcal{D}] = 1]| = \mathsf{negl}(\lambda)$.

Lemma 10 (Indistinguishability of Hybrids H_2 and H_3). *If the OWFE satisfies the security-for-encryption property and by choosing the parameters appropriately, then for any distribution $\mathcal{D}$ and any PPT adversary $\mathcal{A}$, we have* $|\Pr[\mathsf{out}_2[\mathcal{D}] = 1] - \Pr[\mathsf{out}_3[\mathcal{D}] = 1]| = \mathsf{negl}(\lambda)$.

Lemma 11 (CCA2 Security in H_3). *If the OWFE is (k, m)-smooth, then for any two (k, n) sources $\mathcal{S}_0$ and $\mathcal{S}_1$ and any PPT adversary $\mathcal{A}$, we have* $|\Pr[\mathsf{out}_3[\mathcal{S}_0] = 1] - \Pr[\mathsf{out}_3[\mathcal{S}_1] = 1]| = \mathsf{negl}(\lambda)$.

Proof of Theorem 5. By applying the above lemmas, for any (k, n)-sources $\mathcal{S}_0$ and $\mathcal{S}_1$, we have:

$$\mathsf{H}_0[\mathcal{S}_0] \stackrel{c}{\equiv} \mathsf{H}_1[\mathcal{S}_0] \stackrel{c}{\equiv} \mathsf{H}_2[\mathcal{S}_0] \stackrel{c}{\equiv} \mathsf{H}_3[\mathcal{S}_0] \stackrel{c}{\equiv} \mathsf{H}_3[\mathcal{S}_1] \stackrel{c}{\equiv} \mathsf{H}_2[\mathcal{S}_1] \stackrel{c}{\equiv} \mathsf{H}_1[\mathcal{S}_1] \stackrel{c}{\equiv} \mathsf{H}_0[\mathcal{S}_1].$$

We prove these lemmas in the full version of the paper.

6 Lossy TDFs with Linear-Image Size

In this section, using our erasure-resilient code techniques, we show how to adapt a variant of the TDFs from [PW08,PW11] to obtain the first lossy trapdoor functions with images growing linearly in their inputs, based on the DDH assumption. This improves upon the lossy TDFs from [PW08,PW11], whose output size is quadratic in the input size. We first recall the definition of lossy TDF from [PW08,PW11].

Definition 12 (Lossy TDFs [PW08,PW11]**).** *An (n, k)-lossy TDF ((n, k)-LTDF) is given by four PPT algorithms* $\mathsf{TDF.KG}$, $\mathsf{TDF.KG_{ls}}$, $\mathsf{TDF.F}$, $\mathsf{TDF.F}^{-1}$, *where* $\mathsf{TDF.KG_{ls}}(1^\lambda)$ *only outputs a single key (as opposed to a pair of keys), and where the following properties hold:*

- ***Correctness in real mode.*** *The TDF* $(\mathsf{TDF.KG}, \mathsf{TDF.F}, \mathsf{TDF.F}^{-1})$ *satisfies correctness in the sense of Definition 1.*
- ***k-Lossiness.*** *For all but negligible probability over the choice of* $\mathsf{ik_{ls}} \xleftarrow{\$} \mathsf{TDF.KG_{ls}}(1^\lambda)$, *we have* $|\mathsf{TDF.F}(\mathsf{ik_{ls}}, \{0,1\}^n)| \leq 2^k$, *where we use* $\mathsf{TDF.F}(\mathsf{ik_{ls}}, \{0,1\}^n)$ *to denote the set of all images of* $\mathsf{TDF.F}(\mathsf{ik_{ls}}, \cdot)$.
- ***Indistinguishability of real and lossy modes.*** *We have* $\mathsf{ik} \stackrel{c}{\equiv} \mathsf{ik_{ls}}$, *where* $(\mathsf{ik}, *) \xleftarrow{\$} \mathsf{TDF.KG}(1^\lambda)$ *and* $\mathsf{ik_{ls}} \xleftarrow{\$} \mathsf{TDF.KG_{ls}}(1^\lambda)$.

Lossiness rate. In the definition above, we refer to the fraction $1 - k/n$ as the *lossiness rate*, describing the fraction of the bits lost. Ideally, we want this fraction to be as close to 1 as possible, e.g., $1 - o(1)$.

Our LTDF construction makes use of a balanced predicate, defined below.

6.1 Lossy TDF from DDH

Our LTDF construction makes use of the following notation.

Notation. Letting $\mathsf{x} \in \{0,1\}^n$ and $\mathbf{M} := \left(\begin{smallmatrix} g_{1,0}, g_{2,0} \dots, g_{n,0} \\ g_{1,1}, g_{2,1}, \dots, g_{n,1} \end{smallmatrix}\right)$ we define $\mathsf{x} \odot \mathbf{M} = \prod_{j \in [n]} g_{j,\mathsf{x}_j}$. For $i \in [n]$, $b \in \{0,1\}$ and $\mathbf{M}$ as above, we define the matrix $\mathbf{M}' := (\mathbf{M} \xrightarrow[(i,b)]{} g')$ to be the same as $\mathbf{M}$ except that instead of $g_{i,b}$ we put g' in $\mathbf{M}'$. If $\mathbf{M}$ is matrix of group elements, then $\mathbf{M}^r$ denotes entry-wise exponentiation to the power of r.

Overview of the construction and techniques. Let us first demonstrate the idea for retrieving the first bit of the input. Imagine two $2 \times n$ matrices $\mathbf{M}$ and $\mathbf{M}'$, where $\mathbf{M} := \left(\begin{smallmatrix} g_{1,0}, \dots, g_{n,0} \\ g_{1,1}, \dots, g_{n,1} \end{smallmatrix}\right)$ is chosen at random and where $\mathbf{M}' := (\mathbf{M}^r \xrightarrow[(1,b)]{} g_1)$, where $r \xleftarrow{\$} \mathbb{Z}_p$, $b \xleftarrow{\$} \{0,1\}$ and $g_1 \xleftarrow{\$} \mathbb{G}$. That is, $\mathbf{M}'$ is a perturbed rth power of $\mathbf{M}$, in that we replace one of the two elements of the first column of the exponentiated matrix with a random group element.

Think of $(\mathbf{M}, \mathbf{M}')$ as the index key. Suppose an evaluator $\mathsf{TDF.F}$ with input $\mathsf{x} \in \{0,1\}^n$ wants to use $(\mathbf{M}, \mathbf{M}')$ to communicate her first bit x_1 to an inverter who has knowledge of b, g_1 and r. A first attempt for $\mathsf{TDF.F}$ would be to output two group elements $(\tilde{g}, g_1') := (\mathsf{x} \odot \mathbf{M}, \mathsf{x} \odot \mathbf{M}')$. Given $(\tilde{g}, g_1')$, if $g_1' = \tilde{g}^r$, then $\mathsf{x}_1 = 1 - b$; otherwise, $\mathsf{x}_1 = b$—hence allowing $\mathsf{TDF.F}^{-1}$ to recover x_1.[8]

The above method is in fact what used (implicitly) in all previous approaches [PW08, FGK+10, PW11]. However, the cost paid is high: for communicating one bit of information we need to output (at least) one group element.

We will now illustrate our main idea. Let $\mathsf{BL} : \mathbb{G} \rightarrow \{0,1\}$ be a *balanced* predicate, meaning that $\mathsf{BL}(g^*)$ on a randomly generated g^* is a completely random bit. (We will show how to build this object unconditionally.) Returning to the above idea, instead of sending $(\tilde{g}, g_1')$ we will send $(\tilde{g}, \mathsf{BL}(g_1')) \in (\mathbb{G}, \{0,1\})$. Before arguing correctness and security, note that this method yields linear image size for the whole input, because the group element $\tilde{g}$ can be re-used across all indices.

To argue correctness, let us see how $\mathsf{TDF.F}^{-1}$—given b, r and g_1—may invert an encoding $(\tilde{g}, b')$ of the first bit of x. To this end, note the following two facts:

1. If $\mathsf{x}_1 = 1 - b$, then $b' = \mathsf{BL}(\tilde{g}^r)$.
2. If $\mathsf{x}_1 = b$, then $b' = \mathsf{BL}(\frac{\tilde{g}^r g_1}{g_{1,b}^r})$.

Thus, if $\mathsf{BL}(\tilde{g}^r) \neq \mathsf{BL}(\frac{\tilde{g}^r g_1'}{g_{1,b}^r})$, then we can determine the value of x_1. This is because in this case we either have

- $b' = \mathsf{BL}(\tilde{g}^r)$ and $b' \neq \mathsf{BL}(\frac{\tilde{g}^r g_1}{g_{1,b}^r})$: which implies $\mathsf{x}_1 = 1 - b$;
- $b' \neq \mathsf{BL}(\tilde{g}^r)$ and $b' = \mathsf{BL}(\frac{\tilde{g}^r g_1}{g_{1,b}^r})$: which implies $\mathsf{x}_1 = b$.

Summing up the above discussion, we are unable to determine the value of x_1 only when $\mathsf{BL}(\tilde{g}^r) = \mathsf{BL}(\frac{\tilde{g}^r g_1}{g_{1,b}^r})$. This happens with probability $1/2$ because the predicate BL is balanced. For any constant c, we may reduce this probability to $(1/2)^c$ via repetition for c times. Thus, by choosing the constant c appropriately, and doing the above procedure for every index, we will be able to retrieve a good fraction of all the bits of x, which will make the rest retrievable using error correction.

For security, we will show that this method admits a simple lossy way of generating public keys.

We now formally define the notion of balanced predicates, which will be used in our LTDF construction.

Definition 13 (Balanced predicates). *We say a randomized predicate* $\mathsf{P} : \mathsf{S} \times \{0,1\}^* \rightarrow \{0,1\}$ *is balanced over set* S *if* $\Pr[\mathsf{P}(x; r) = 0] = 1/2$*, where* $x \xleftarrow{\$} \mathsf{S}$ *and* $r \xleftarrow{\$} \{0,1\}^*$.

[8] For simplicity assume $g_1 \neq g_{1,b}^r$, hence we will not have a hung situation.

In the above definition, if $\mathsf{S} = \{0,1\}^n$, then we have a trivial predicate, one which returns, say, the first bit of its input. For our LTDF construction, we require the existence of a predicate for the underlying group $\mathbb{G}$. Assuming any 1-1 mapping $\mathbb{G} \to \{0,1\}^n$ (which may not be surjective), we may define the predicate P as the inner product function over $\mathbb{F}_2$: i.e., $\mathsf{P}(x,r) = \langle x, r\rangle$.

Construction 8 (Linear-image lossy TDF). *Let* G *be a group scheme and let* $(\mathsf{Encode}, \mathsf{Decode})$ *be an erasure code, where* $\mathsf{Encode}\colon \{0,1\}^n \to \{0,1\}^m$ *(Definition 8). Also, let* BL *be a balanced predicate for the underlying group (Definition 13).*

We define our LTDF construction $(\mathsf{TDF.KG}, \mathsf{TDF.KG_{ls}}, \mathsf{TDF.F}, \mathsf{TDF.F}^{-1})$ *as follows.*

- $\mathsf{TDF.KG}(1^\lambda)$*:*
  1. *Sample* $(\mathbb{G}, p, g) \xleftarrow{\$} \mathsf{G}(1^\lambda)$*, and*

$$\mathbf{M} := \begin{pmatrix} g_{1,0}, g_{2,0}, \ldots, g_{m,0} \\ g_{1,1}, g_{2,1}, \ldots, g_{m,1} \end{pmatrix} \xleftarrow{\$} \mathbb{G}^{2\times m}. \tag{19}$$

  2. *For all* $i \in [m]$*, sample* $g_i \xleftarrow{\$} \mathbb{G}$*,* $\rho_i \xleftarrow{\$} \mathbb{Z}_p$ *and* $b_i \xleftarrow{\$} \{0,1\}$*.*
  3. *Sample random coins* r *for the underlying function* BL*.*
  4. *Set the index and trapdoor keys as*

$$\mathsf{ik} := (\mathbf{M}, (\mathbf{M}^{\rho_1} \xrightarrow[(1,b_1)]{} g_1), \ldots, (\mathbf{M}^{\rho_m} \xrightarrow[(m,b_m)]{} g_m), r) \tag{20}$$

$$\mathsf{tk} := (\mathbf{M}, (\rho_1, b_1, g_1), \ldots, (\rho_m, b_m, g_m), r). \tag{21}$$

- $\mathsf{TDF.KG_{ls}}(1^\lambda)$*: Return* $\mathsf{ik_{ls}} := (\mathbf{M}, \mathbf{M}^{\rho_1}, \ldots, \mathbf{M}^{\rho_m}, r)$*, where* $\mathbf{M}$ *and* ρ_i *for* $i \in [m]$ *and* r *are sampled as above.*
- $\mathsf{TDF.F}(\mathsf{ik}, \mathsf{x} \in \{0,1\}^n)$*: Parse* $\mathsf{ik} := (\mathbf{M}, \mathbf{M}_1, \ldots, \mathbf{M}_m, r)$*. Set* $\mathsf{z} := \mathsf{Encode}(\mathsf{x})$ *and return*

$$\mathsf{u} := (\mathsf{z} \odot \mathbf{M}, \mathsf{BL}(\mathsf{z} \odot \mathbf{M}_1; r), \ldots, \mathsf{BL}(\mathsf{z} \odot \mathbf{M}_m; r)), \tag{22}$$

- $\mathsf{TDF.F}^{-1}(\mathsf{tk}, \mathsf{u})$*:*
  1. *Parse* $\mathsf{tk} := (\mathbf{M}, (\rho_1, b_1, g_1), \ldots, (\rho_m, b_m, g_m), r)$ *and* $\mathsf{u} := (g_c, b'_1, \ldots, b'_m)$*. Parse* $\mathbf{M}$ *as in Eq. 19.*
  2. *Construct* $\mathsf{z}' := \mathsf{z}'_1 \cdots \mathsf{z}'_m \in \{0,1,\perp\}^m$ *as follows. For* $i \in [m]$*:*
     (a) *Set* $g'_i := g_c^{\rho_i}$ *and* $g''_i := \frac{g_c^{\rho_i}}{g_{i,b_i}^{\rho_i}} \cdot g_i$*. Then*
        i. *If* $\mathsf{BL}(g'_i; r) = \mathsf{BL}(g''_i; r)$*, set* $\mathsf{z}'_i = \perp$*;*
        ii. *Else, if* $b'_i = \mathsf{BL}(g'_i; r)$*, set* $\mathsf{z}_i = 1 - b_i$*. Else (i.e.,* $b'_i = \mathsf{BL}(g''_i; r)$*), set* $\mathsf{z}_i = b_i$*.*
  3. *Return* $\mathsf{Decode}(\mathsf{z}')$*.*

The following theorem gives the lossiness property of the scheme.

Theorem 9 (Linear-image LTDF from DDH). *Using code* (Encode, Decode) *such that* $4^t c_1 > 3^t c$ *and if* $\frac{(4^t c_1 - 3^t c)^2}{2^{4t-1} c} \geq 0.7$*, the LTDF of Construction 8 is* $(n, \log p)$*-lossy with image size* $\log p + cn \in \Theta(n)$*. By setting* $n \in \omega(\log p)$ *we obtain* $1 - o(1)$ *lossiness rate.*

We prove all the required properties below.

Lemma 12 ($\log p$**-Lossiness).** *For any* $\mathsf{ik}_{\mathsf{ls}} \in \mathsf{TDF.KG}_{\mathsf{ls}}(1^\lambda)$ *we have*

$$|\mathsf{TDF.F}(\mathsf{ik}_{\mathsf{ls}}, \{0,1\}^n)| \leq p,$$

where recall that p *is the order of the underlying group.*

Proof. Parse $\mathsf{ik}_{\mathsf{ls}} := (\mathbf{M}, \mathbf{M}_1, \ldots, \mathbf{M}_m)$. It is easy to verify that for any two messages $\mathsf{x}, \mathsf{x}' \in \{0,1\}^n$ we have

$$\mathsf{TDF.F}(\mathsf{ik}_{\mathsf{ls}}, \mathsf{x}) \neq \mathsf{TDF.F}(\mathsf{ik}_{\mathsf{ls}}, \mathsf{x}') \Longleftrightarrow \mathsf{x} \odot \mathbf{M} \neq \mathsf{x}' \odot \mathbf{M}. \tag{23}$$

The statement of the lemma now follows, since $\{\mathsf{x} \odot \mathbf{M} \mid \mathsf{x} \in \{0,1\}^n\} \subseteq \mathbb{G}$, and thus we have $|\{\mathsf{x} \odot \mathbf{M} \mid \mathsf{x} \in \{0,1\}^n\}| \leq p$. □

Lemma 13 (Indistinguishability of real and lossy modes). *We have* $\mathsf{ik} \stackrel{c}{\equiv} \mathsf{ik}_{\mathsf{ls}}$*, where* $(\mathsf{ik}, *) \xleftarrow{\$} \mathsf{TDF.KG}(1^\lambda)$ *and* $\mathsf{ik}_{\mathsf{ls}} \xleftarrow{\$} \mathsf{TDF.KG}_{\mathsf{ls}}(1^\lambda)$.

Proof. Immediate by the DDH assumption using standard techniques. □

Lemma 14 (Correctness in real mode). *Using code* (Encode, Decode)*, we have*

$$\Pr_{(\mathsf{ik},\mathsf{tk})}[\exists \mathsf{x} \in \{0,1\}^n \ s.t.\ \mathsf{TDF.F}^{-1}(\mathsf{tk}, (\mathsf{TDF.F}(\mathsf{ik}, \mathsf{x}))) \neq \mathsf{x}] \leq 2^n \cdot \mathsf{e}^{-\frac{(2^t c_1 - c)^2}{2^{2t-1} c} n}. \tag{24}$$

In particular, by choosing the repetition constant t *such that* $2^t c_1 > c$ *and that* $\frac{(2^t c_1 - c)^2}{2^{2t-1} c} \geq 0.7$*, the probability in Equation* (24) *will be negligible.*

Proof. Fix $\mathsf{x} \in \{0,1\}^n$ and let $\mathsf{z} := \mathsf{Encode}(\mathsf{x})$. All probabilities below are taken over the random choice of $(\mathsf{ik}, \mathsf{tk})$. Parse

$$\mathsf{tk} := \left(\mathbf{M} := \begin{pmatrix} g_{1,0}, g_{2,0}, \ldots, g_{m,0} \\ g_{1,1}, g_{2,1}, \ldots, g_{m,1} \end{pmatrix}, (\rho_1, b_1, g_1), \ldots, (\rho_m, b_m, g_m), r \right).$$

For input $\mathsf{x} \in \{0,1\}^n$, let Fail_x be the event that $\mathsf{TDF.F}^{-1}(\mathsf{tk}, \mathsf{TDF.F}(\mathsf{ik}, \mathsf{x})) \neq \mathsf{x}$. Fix $\mathsf{x} \in \{0,1\}^n$ and let $\mathsf{z} := \mathsf{Encode}(\mathsf{x}) \in \{0,1\}^m$. Also, let $\mathsf{u} := \mathsf{TDF}(\mathsf{ik}, \mathsf{x}) := (g_c, b'_1, \ldots, b'_m)$.

Recall that z consists of cn blocks, where each block consists of t identical bits (Definition 9). For each block index $j \in [cn]$ we define an event Bad_j, which corresponds to the event that the inversion algorithm fails to recover the bit that corresponds to the jth block. That is, Bad_j occurs if all the t repetitions inside block j leads to failure during inversion. More formally:

– Bad_j: The event that for all $i \in \{(j-1)t+1, \ldots, jt\}$: $\mathsf{BL}(g'_i; r) = \mathsf{BL}(g''_i; r)$, where $g'_i := g_c^{\rho_i}$ and $g''_i := \frac{g_c^{\rho_i}}{g_{i,b_i}^{\rho_i}} \cdot g_i$.

Note that all Bad_j are i.i.d. events and we have $\Pr[\mathsf{Bad}_j] = (1/2)^t$. The reason for this is that all of $(g_1, \ldots, g_n)$ are sampled uniformly at random independently of all other values, and thus the two group elements g'_i and g''_i are uniform and independent.

Let $\mathsf{Good}_j = \overline{\mathsf{Bad}_j}$ and note that $\Pr[\mathsf{Good}_j] = 1 - (1/2)^t$. We now have

$$\Pr[\mathsf{Fail}_{\mathsf{x}}] \leq \Pr[\sum_{j=1}^{cn} \mathsf{Good}_j \leq cn - c_1 n] = \Pr[\frac{1}{cn}\sum_{j=1}^{cn} \mathsf{Good}_j \leq 1 - \frac{c_1}{c}]$$
$$= \Pr[\frac{1}{cn}\sum_{j=1}^{cn} \mathsf{Good}_j \leq 1 - \frac{1}{2^t} - (\frac{c_1}{c} - \frac{1}{2^t})] \leq^* \mathsf{e}^{-2cn(\frac{2^t c_1 - c}{2^t c})^2} \leq \mathsf{e}^{-\frac{(2^t c_1 - c)^2}{2^{2t-1} c} n}, \tag{25}$$

where the inequality marked with * follows from the Chernoff inequality (Theorem 1 with $p = 1 - 1/2^t$ and $\varepsilon = c_1/c - 1/2^t$. Note that since we must have $\epsilon > 0$, we should have $2^t c_1 > c$.)

We conclude using a union bound over all $\mathsf{x} \in \{0,1\}^n$. □

Acknowledgements. We thank Xiao Liang for suggesting a simplification to Construction 3. We are also grateful to the anonymous reviewers for their comments. Research supported in part from DARPA/ARL SAFEWARE Award W911NF15C0210, AFOSR Award FA9550-15-1-0274, AFOSR YIP Award, DARPA and SPAWAR under contract N66001-15-C-4065, a Hellman Award and research grants by the Okawa Foundation, Visa Inc., and Center for Long-Term Cybersecurity (CLTC, UC Berkeley), and a Google PhD fellowship. The views expressed are those of the authors and do not reflect the official policy or position of the funding agencies.

References

[BBN+09] Bellare, M., et al.: Hedged public-key encryption: how to protect against bad randomness. In: Matsui, M. (ed.) ASIACRYPT 2009. LNCS, vol. 5912, pp. 232–249. Springer, Heidelberg (2009). https://doi.org/10.1007/978-3-642-10366-7_14

[BBO07] Bellare, M., Boldyreva, A., O'Neill, A.: Deterministic and efficiently searchable encryption. In: Menezes, A. (ed.) CRYPTO 2007. LNCS, vol. 4622, pp. 535–552. Springer, Heidelberg (2007). https://doi.org/10.1007/978-3-540-74143-5_30

[BCPT13] Birrell, E., Chung, K.-M., Pass, R., Telang, S.: Randomness-dependent message security. In: Sahai, A. (ed.) TCC 2013. LNCS, vol. 7785, pp. 700–720. Springer, Heidelberg (2013). https://doi.org/10.1007/978-3-642-36594-2_39

[BFO08] Boldyreva, A., Fehr, S., O'Neill, A.: On notions of security for deterministic encryption, and efficient constructions without random oracles. In: Wagner, D. (ed.) CRYPTO 2008. LNCS, vol. 5157, pp. 335–359. Springer, Heidelberg (2008). https://doi.org/10.1007/978-3-540-85174-5_19

[BFOR08] Bellare, M., Fischlin, M., O'Neill, A., Ristenpart, T.: Deterministic encryption: definitional equivalences and constructions without random oracles. In: Wagner, D. (ed.) CRYPTO 2008. LNCS, vol. 5157, pp. 360–378. Springer, Heidelberg (2008). https://doi.org/10.1007/978-3-540-85174-5_20

[BHY09] Bellare, M., Hofheinz, D., Yilek, S.: Possibility and impossibility results for encryption and commitment secure under selective opening. In: Joux, A. (ed.) EUROCRYPT 2009. LNCS, vol. 5479, pp. 1–35. Springer, Heidelberg (2009). https://doi.org/10.1007/978-3-642-01001-9_1

[BLSV18] Brakerski, Z., Lombardi, A., Segev, G., Vaikuntanathan, V.: Anonymous IBE, leakage resilience and circular security from new assumptions. In: Nielsen, J.B., Rijmen, V. (eds.) EUROCRYPT 2018. LNCS, vol. 10820, pp. 535–564. Springer, Cham (2018). https://doi.org/10.1007/978-3-319-78381-9_20

[BS11] Brakerski, Z., Segev, G.: Better security for deterministic public-key encryption: the auxiliary-input setting. In: Rogaway, P. (ed.) CRYPTO 2011. LNCS, vol. 6841, pp. 543–560. Springer, Heidelberg (2011). https://doi.org/10.1007/978-3-642-22792-9_31

[BW10] Boyen, X., Waters, B.: Shrinking the keys of discrete-log-type lossy trapdoor functions. In: Zhou, J., Yung, M. (eds.) ACNS 2010. LNCS, vol. 6123, pp. 35–52. Springer, Heidelberg (2010). https://doi.org/10.1007/978-3-642-13708-2_3

[CDG+17] Cho, C., Döttling, N., Garg, S., Gupta, D., Miao, P., Polychroniadou, A.: Laconic oblivious transfer and its applications. In: Katz, J., Shacham, H. (eds.) CRYPTO 2017. LNCS, vol. 10402, pp. 33–65. Springer, Cham (2017). https://doi.org/10.1007/978-3-319-63715-0_2

[DG17a] Döttling, N., Garg, S.: From selective IBE to Full IBE and selective HIBE. In: Kalai, Y., Reyzin, L. (eds.) TCC 2017. LNCS, vol. 10677, pp. 372–408. Springer, Cham (2017). https://doi.org/10.1007/978-3-319-70500-2_13

[DG17b] Döttling, N., Garg, S.: Identity-based encryption from the Diffie-Hellman assumption. In: Katz, J., Shacham, H. (eds.) CRYPTO 2017. LNCS, vol. 10401, pp. 537–569. Springer, Cham (2017). https://doi.org/10.1007/978-3-319-63688-7_18

[DGHM18] Döttling, N., Garg, S., Hajiabadi, M., Masny, D.: New constructions of identity-based and key-dependent message secure encryption schemes. In: Abdalla, M., Dahab, R. (eds.) PKC 2018. LNCS, vol. 10769, pp. 3–31. Springer, Cham (2018). https://doi.org/10.1007/978-3-319-76578-5_1

[DH76] Diffie, W., Hellman, M.E.: New directions in cryptography. IEEE Trans. Inf. Theory **22**(6), 644–654 (1976)

[FGK+10] Freeman, D.M., Goldreich, O., Kiltz, E., Rosen, A., Segev, G.: More constructions of lossy and correlation-secure trapdoor functions. In: Nguyen, P.Q., Pointcheval, D. (eds.) PKC 2010. LNCS, vol. 6056, pp. 279–295. Springer, Heidelberg (2010). https://doi.org/10.1007/978-3-642-13013-7_17

[GH18] Garg, S., Hajiabadi, M.: Trapdoor functions from the computational Diffie-Hellman assumption. In: Shacham, H., Boldyreva, A. (eds.) CRYPTO 2018. LNCS, vol. 10992, pp. 362–391. Springer, Cham (2018). https://doi.org/10.1007/978-3-319-96881-0_13

[GMR01] Gertner, Y., Malkin, T., Reingold, O.: On the impossibility of basing trapdoor functions on trapdoor predicates. In: 42nd FOCS, Las Vegas, NV, USA, 14–17 October 2001, pp. 126–135. IEEE Computer Society Press (2001)

[HO12] Hemenway, B., Ostrovsky, R.: Extended-DDH and lossy trapdoor functions. In: Fischlin, M., Buchmann, J., Manulis, M. (eds.) PKC 2012. LNCS, vol. 7293, pp. 627–643. Springer, Heidelberg (2012). https://doi.org/10.1007/978-3-642-30057-8_37

[HO13] Hemenway, B., Ostrovsky, R.: Building lossy trapdoor functions from lossy encryption. In: Sako, K., Sarkar, P. (eds.) ASIACRYPT 2013. LNCS, vol. 8270, pp. 241–260. Springer, Heidelberg (2013). https://doi.org/10.1007/978-3-642-42045-0_13

[ILL89] Impagliazzo, R., Levin, L.A., Luby, M.: Pseudo-random generation from one-way functions (extended abstracts). In: 21st ACM STOC, Seattle, WA, USA, 15–17 May 1989, pp. 12–24. ACM Press (1989)

[Jus72] Justesen, J.: Class of constructive asymptotically good algebraic codes. IEEE Trans. Inf. Theory **18**(5), 652–656 (1972)

[KMO10] Kiltz, E., Mohassel, P., O'Neill, A.: Adaptive trapdoor functions and chosen-ciphertext security. In: Gilbert, H. (ed.) EUROCRYPT 2010. LNCS, vol. 6110, pp. 673–692. Springer, Heidelberg (2010). https://doi.org/10.1007/978-3-642-13190-5_34

[KW18] Koppula, V., Waters, B.: Realizing chosen ciphertext security generically in attribute-based encryption and predicate encryption. Cryptology ePrint Archive, Report 2018/847 (2018). https://eprint.iacr.org/2018/847

[MPRS12] Mironov, I., Pandey, O., Reingold, O., Segev, G.: Incremental deterministic public-key encryption. In: Pointcheval, D., Johansson, T. (eds.) EUROCRYPT 2012. LNCS, vol. 7237, pp. 628–644. Springer, Heidelberg (2012). https://doi.org/10.1007/978-3-642-29011-4_37

[MY10] Mol, P., Yilek, S.: Chosen-ciphertext security from slightly lossy trapdoor functions. In: Nguyen, P.Q., Pointcheval, D. (eds.) PKC 2010. LNCS, vol. 6056, pp. 296–311. Springer, Heidelberg (2010). https://doi.org/10.1007/978-3-642-13013-7_18

[PVW08] Peikert, C., Vaikuntanathan, V., Waters, B.: A framework for efficient and composable oblivious transfer. In: Wagner, D. (ed.) CRYPTO 2008. LNCS, vol. 5157, pp. 554–571. Springer, Heidelberg (2008). https://doi.org/10.1007/978-3-540-85174-5_31

[PW08] Peikert, C., Waters, B.: Lossy trapdoor functions and their applications. In: 40th ACM STOC, Victoria, British Columbia, Canada, 17–20 May 2008, pp. 187–196. ACM Press (2008)

[PW11] Peikert, C., Waters, B.: Lossy trapdoor functions and their applications. SIAM J. Comput. **40**(6), 1803–1844 (2011)

[RS09] Rosen, A., Segev, G.: Chosen-ciphertext security via correlated products. In: Reingold, O. (ed.) TCC 2009. LNCS, vol. 5444, pp. 419–436. Springer, Heidelberg (2009). https://doi.org/10.1007/978-3-642-00457-5_25

[RSA78] Rivest, R.L., Shamir, A., Adleman, L.M.: A method for obtaining digital signature and public-key cryptosystems. Commun. Assoc. Comput. Mach. **21**(2), 120–126 (1978)

[Wee12] Wee, H.: Dual projective hashing and its applications—lossy trapdoor functions and more. In: Pointcheval, D., Johansson, T. (eds.) EUROCRYPT 2012. LNCS, vol. 7237, pp. 246–262. Springer, Heidelberg (2012). https://doi.org/10.1007/978-3-642-29011-4_16

[Wic13] Wichs, D.: Barriers in cryptography with weak, correlated and leaky sources. In: ITCS 2013, Berkeley, CA, USA, 9–12 January 2013, pp. 111–126. ACM (2013)

Symbolic Encryption with Pseudorandom Keys

Daniele Micciancio(✉)

University of California, San Diego, Mail Code 0404, La Jolla, CA 92093, USA
daniele@cs.ucsd.edu
http://cseweb.ucsd.edu/~daniele/

Abstract. We give an efficient decision procedure that, on input two (acyclic) expressions making arbitrary use of common cryptographic primitives (namely, encryption and pseudorandom generators), determines (in polynomial time) if the two expressions produce computationally indistinguishable distributions for any cryptographic instantiation satisfying the standard security notions of pseudorandomness and indistinguishability under chosen plaintext attack. The procedure works by mapping each expression to a symbolic pattern that captures, in a fully abstract way, the information revealed by the expression to a computationally bounded observer. Our main result shows that if two expressions are mapped to different symbolic patterns, then there are secure pseudorandom generators and encryption schemes for which the two distributions can be distinguished with overwhelming advantage. At the same time if any two (acyclic) expressions are mapped to the same pattern, then the associated distributions are indistinguishable.

Keywords: Symbolic security · Greatest fixed points · Computational soundness · Completeness · Pseudorandom generators · Information leakage

1 Introduction

Formal methods for security analysis (e.g., [1,9,13,21,33,34]) typically adopt an all-or-nothing approach to modeling adversarial knowledge. For example, the adversary either knows a secret key or does not have any partial information about it. Similarly, either the message underlying a given ciphertext can be recovered, or it is completely hidden. In the computational setting, commonly used in modern cryptography for its strong security guarantees, the situation is much different: cryptographic primitives usually leak partial information about their inputs, and in many cases this cannot be avoided. Moreover, it is well known that computational cryptographic primitives, if not used properly, can easily

Research supported in part by NSF under grant CNS-1528068. Any opinions, findings, and conclusions or recommendations expressed in this material are those of the author and do not necessarily reflect the views of the National Science Foundation.

Y. Ishai and V. Rijmen (Eds.): EUROCRYPT 2019, LNCS 11478, pp. 64–93, 2019.
https://doi.org/10.1007/978-3-030-17659-4_3

lead to situations where individually harmless pieces of partial information can be combined to recover a secret in full. This is often the case when, for example, the same key or randomness is used within different cryptographic primitives.

Starting with the seminal work of Abadi and Rogaway [3], there has been considerable progress in combining the symbolic and computational approaches to security protocol design and analysis, with the goal of developing methods that are both easy to apply (e.g., through the use of automatic verification tools) and provide strong security guarantees, as offered by the computational security definitions. Still, most work in this area applies to scenarios where the use of cryptography is sufficiently restricted that the partial information leakage of computational cryptographic primitives is inconsequential. For example, [3] studies expressions that use a single encryption scheme as their only cryptographic primitive. In this setting, the partial information about a key k revealed by a ciphertext $\{\!|m|\!\}_k$ is of no use to an adversary (except, possibly, for identifying when two different ciphertexts are encrypted under the same, unknown, key), so one can treat k as if it were completely hidden. Other works [4,28] combine encryption with other cryptographic primitives (like pseudorandom generation and secret sharing,) but bypass the problem of partial information leakage simply by assuming that all protocols satisfy sufficiently strong syntactic restrictions to guarantee that different cryptographic primitives do not interfere with each other.

1.1 Our Results

In this paper we consider cryptographic expressions that make arbitrary (nested) use of encryption and pseudorandom generation, without imposing any syntactic restrictions on the messages transmitted by the protocols. In particular, following [3], we consider cryptographic expressions like $(\{\!|m|\!\}_k, \{\!|\{\!|k|\!\}_{k'}|\!\}_{k''})$, representing a pair of ciphertexts: the encryption of a message m under a session key k, and a double (nested) encryption of the session key k under two other keys k', k''. But, while in [3] key symbols represent independent randomly chosen keys, here we allow for derived keys obtained using a length doubling pseudorandom generator $k \mapsto \mathbb{G}_0(k); \mathbb{G}_1(k)$ that on input a single key k outputs a pair of (statistically correlated, but computationally indistinguishable) keys $\mathbb{G}_0(k)$ and $\mathbb{G}_1(k)$. The output of the pseudorandom generator can be used anywhere a key is allowed. In particular, pseudorandom keys $\mathbb{G}_0(k)$, $\mathbb{G}_1(k)$ can be used to encrypt messages, or as messages themselves (possibly encrypted under other random or pseudorandom keys), or as input to the pseudorandom generator. So, for example, one can iterate the application of the pseudorandom generator to produce an arbitrary long sequence of keys $\mathbb{G}_1(r), \mathbb{G}_1(\mathbb{G}_0(r)), \mathbb{G}_1(\mathbb{G}_0(\mathbb{G}_0(r))), \ldots$.

We remark that key expansion using pseudorandom generators occurs quite often in real world cryptography. In fact, the usefulness of pseudorandom generators is not limited to reducing the amount of randomness needed by cryptographic algorithms, and pseudorandom generators are often used as an essential tool in secure protocol design. For example, they are used in the design of *forward-secure* cryptographic functions to refresh a user private key [7,25], they are used in the best known (in fact, optimal [30]) multicast key distribution protocols [10] to compactly communicate (using a seed) a long sequence of pseudorandom keys, and they play an important role in Yao's classic garbled circuit construction for secure two-party computation to mask and selectively open part of a hidden circuit evaluation [24,35].

Pseudorandom generators (like any deterministic cryptographic primitive) inevitably leak partial information about their input key.[1] Similarly, a ciphertext $\{e\}_k$ may leak partial information about k if, for example, decryption succeeds (with high probability) only when the right key is used for decryption. As we consider the unrestricted use of encryption and pseudorandom generation, we need to model the possibility that given different pieces of partial information about a key, an adversary may be able to recover that key completely. Our main result shows how to do all this within a fairly simple symbolic model of computation, and still obtain strong computational soundness guarantees. Our treatment of partial information is extremely simple and in line with the spirit of formal methods and symbolic security analysis: we postulate that, given any two distinct pieces of partial information about a key, an adversary can recover the key in full. Perhaps not surprisingly, we demonstrate (Theorem 3) that the resulting symbolic semantics for cryptographic expressions is computationally sound, in the sense that if two (acyclic[2]) expressions are symbolically equivalent, then for any (length regular) semantically secure encryption scheme and (length doubling) pseudorandom generator the probability distributions naturally associated to the two expressions are computationally indistinguishable. More interestingly, we justify our symbolic model by proving a corresponding completeness theorem (Theorem 2), showing that if two cryptographic expressions are not symbolically equivalent (according to our definition), then there is an instantiation of the cryptographic primitives (satisfying the standard security notion of indistinguishability) such that the probability distributions corresponding to the two expressions can be efficiently distinguished with almost perfect advantage. In other words, if we want the symbolic semantics to be computationally sound with respect to any standard implementation of the cryptographic primitives, then our computationally sound symbolic semantics is essentially optimal. Moreover, our completeness theorem concretely shows what could go wrong when encrypting messages under related keys, even under a simple eavesdropping (passive) attack.

1.2 Techniques

A key technical contribution of our paper is a syntactic characterization of independent keys that exactly matches its computational counterpart, and a

[1] For example, $\mathbb{G}_0(k)$ gives partial information about k because it allows to distinguish k from any other key k' chosen independently at random: all that the distinguisher has to do is to compute $\mathbb{G}_0(k')$ and compare the result to $\mathbb{G}_0(k)$.

[2] For cyclic expressions, i.e., expressions containing encryption cycles, our soundness theorem still holds, but with respect to a slightly stronger "co-inductive" adversarial model based on greatest fixed point computations [26].

corresponding notion of computationally sound key renaming (Corollary 1). Our syntactic definition of independence is simple and intuitive: a set of keys $k_1, \ldots, k_n$ is symbolically independent if no key k_i can be obtained from another k_j via the syntactic application of the pseudorandom generator. We show that this simple definition perfectly captures the intuition behind the computational notion of pseudorandomness: we prove (Theorem 1) that our definition is both computationally sound and complete, in the sense that the keys $k_1, \ldots, k_n$ are symbolically independent *if and only if* the associated probability distribution is indistinguishable from a sequence of truly independent uniformly random keys. For example, although the probability distributions associated to pseudorandom keys $\mathbb{G}_0(k)$ and $\mathbb{G}_1(k)$ are not independent in a strict information theoretic sense, the dependency between these distributions cannot be efficiently recognized when k is not known because the joint distribution associated to the pair $(\mathbb{G}_0(k), \mathbb{G}_1(k))$ is indistinguishable from a pair of independent random values.

A key component of our completeness theorem is a technical construction of a secure pseudorandom generator $\mathbb{G}$ and encryption scheme $\{\!|\cdot|\!\}_k$ satisfying some very special properties (Lemma 4) that may be of independent interest. The properties are best described in terms of pseudorandom functions. Let f_k be the pseudorandom function obtained from the length-doubling pseudorandom generator $\mathbb{G}$ using the classic construction of [17]. We give an algorithm that on input any string w and two ciphertexts $c_0 = \{\!|m_0|\!\}_{k_0}$ and $c_1 = \{\!|m_1|\!\}_{k_1}$ (for arbitrarily chosen, and unknown messages m_0, m_1) determines if $k_1 = f_{k_0}(w)$, and, if so, completely recovers the value of the keys k_0 and k_1 with overwhelming probability. Building on this lemma, we define the symbolic semantics by means of an abstract adversary that is granted the ability to recover the keys k_0, k_1 whenever it observes two ciphertext encrypted under them. Our completeness theorem offers a precise technical justification for such strong symbolic adversary.

1.3 Active Attacks and Other Cryptographic Primitives

Our work focuses on security definitions with respect to passive attacks for two reasons. First, indistinguishability is essentially[3] the only notion of security applicable to primitives as simple as pseudorandom generators. Second, using passive security definitions only makes our main result (Theorem 2) stronger: our completeness theorem shows that if two expressions map to different symbolic patterns, then security can be completely subverted even under a simple eavesdropping attack. Still, we remark that our definitions and techniques could be useful also for the analysis of security under more realistic attacks in the presence of active adversaries, e.g., if combined together with other soundness results [5,6,11,19,32]. Also, our results immediately extend to other cryptographic primitives (e.g., non-interactive commitment schemes) which can be

[3] Active attacks against pseudorandom generators may be considered in the context of leakage resilient cryptography, fault injection analysis, and other side-channel attacks, which are certainly interesting, but also much more specialized models than those considered in this paper.

modeled as a weakening of public key encryption. Possible extension to other cryptographic primitives, e.g., using the notion of *deduction soundness* [8,12] is also an interesting possibility. However, such extensions are outside the scope of this paper, and they are left to future work.

1.4 Related Work

Cryptographic expressions with pseudorandom keys, as those considered in this paper, are used in the symbolic analysis of various cryptographic protocols, including multicast key distribution [28–30], cryptographically controlled access to XML documents [4], and (very recently) the symbolic analysis of Yao's garbled circuit construction for secure two party computation [24]. However, these works (with the exception of [24], which builds on the results from a preliminary version of our paper [27]) use ad-hoc methods to deal with pseudorandom keys by imposing syntactic restrictions on the way the keys are used. Even more general (so called "composed") encryption keys are considered in [23], but only under the random oracle heuristics. We remark that the use of such general composed keys is unjustified in the standard model of computation, and the significance of the results of [23] outside the random oracle model is unclear. In fact, our completeness results clearly show that modeling key expansion as new random keys is not sound with respect to computationally secure pseudorandom generators in the standard model.

The problem of defining a computationally sound and complete symbolic semantics for cryptographic expressions has already been studied in several papers before, e.g., [3,14,31]. However, to the best of our knowledge, our is the first paper to prove soundness and completeness results with respect to the standard notion of computationally secure encryption [18]. In the pioneering work [3], Abadi and Rogaway proved the first soundness theorem for basic cryptographic expressions. Although in their work they mention various notions of security, they focus on a (somehow unrealistic) variant of the standard security definition that requires the encryption scheme to completely hide both the key and the message being encrypted, including its length. This is the notion of security used in many other works, including [22]. The issue of completeness was first raised by Micciancio and Warinschi [31] who proved that the logic of Abadi and Rogaway is both sound and complete if one assumes the encryption scheme satisfies a stronger security property called confusion freeness (independently defined also in [2], and subsequently weakened in [14]). We remark that most symbolic models are trivially complete for trace properties. However, the same is not true for indistinguishability security properties.

The notion of completeness used in [2,14,31] is different from the one studied in this paper. The works [2,14,31] consider restricted classes of encryption schemes (satisfying stronger security properties) such that the computational equivalence relation induced on expressions is the same for all encryption schemes in the class. In other words, if two expressions can be proved not equivalent within the logic framework, then the probability distributions associated to the two expressions by evaluating them according to *any* encryption scheme (from the given class) are computationally distinguishable. It can be shown that no such notion of completeness can be achieved by the standard security definition of indistinguishability under chosen plaintext attack, as considered in this paper, i.e., different encryption schemes (all satisfying this standard notion of security) can define different equivalence relations. In this paper we use a different approach: instead of strengthening the computational security definitions to match the symbolic model of [3], we relax the symbolic model in order to match the standard computational security definition of [18]. Our relaxed symbolic model is still complete, in the sense that if two expressions evaluate to computationally equivalent distributions for any encryption scheme satisfying the standard security definition, then the equality between the two expressions can be proved within the logic. In other words, if two expressions are not equivalent in our symbolic model, then the associated probability distributions are not computationally equivalent for some (but not necessarily all) encryption scheme satisfying the standard computational security notion.

1.5 Organization

The rest of the paper is organized as follows. In Sect. 2 we review basic notions from symbolic and computational cryptography as used in this paper. In Sect. 3 we present our basic results on the computational soundness of pseudorandom keys, and introduce an appropriate notion of key renaming. In Sect. 4 we present our symbolic semantics for cryptographic expressions with pseudorandom keys. In Sect. 5, we present our main result: a completeness theorem which justifies the definitional choices made in Sect. 4. A corresponding soundness theorem is given in Sect. 6. Section 7 concludes the paper with some closing remarks.

2 Preliminaries

In this section we review standard notions and notation from symbolic and computational cryptography used in the rest of the paper. The reader is referred to [3,26] for more background on the symbolic model, and [15,16,20] (or any other modern cryptography textbook) for more information about the computational model, cryptographic primitives and their security definitions.

We write $\{0,1\}^*$ to denote the set of all binary strings, $\{0,1\}^n$ for the set of all strings of length n, $|x|$ for the bitlength of a string x, ϵ for the empty string, and ";" (or simple juxtaposition) for the string concatenation operation mapping $x \in \{0,1\}^n$ and $y \in \{0,1\}^m$ to $x;y \in \{0,1\}^{n+m}$. We also write $x \preceq y$ if x is a *suffix* of y, i.e., $y = zx$ for some $z \in \{0,1\}^*$. As usual, $x \prec y$ is $x \preceq y$ and $x \neq y$. The powerset of a set A is denoted $\wp(A)$.

As a general convention, we use bold uppercase names ($\mathbf{Exp}, \mathbf{Pat}$, etc.) for standard sets of symbolic expressions, bold lowercase names ($\mathbf{keys}, \mathbf{parts}$) for functions that return sets of symbolic expressions, and regular (non-bold) names ($\mathtt{shape}, \mathtt{norm}$) for functions returning a single symbolic expression. We also use uppercase letters (e.g., A, S) for set-valued variables, and lowercase letters (x, y)

for other variables. Calligraphic letters ($\mathcal{A}, \mathcal{G}, \mathcal{E}$, etc.) are reserved for probability distributions and algorithms in the computational setting.

2.1 Symbolic Cryptography

In the symbolic setting, messages are described by abstract terms. For any given sets of key and data terms **Keys**, **Data**, define **Exp** as the set of cryptographic expressions generated by the grammar

$$\mathbf{Exp} ::= \mathbf{Data} \mid \mathbf{Keys} \mid (\mathbf{Exp}, \mathbf{Exp}) \mid \{\!|\mathbf{Exp}|\!\}_{\mathbf{Keys}}, \tag{1}$$

where (e_1, e_2) denotes the ordered pair of subexpressions e_1 and e_2, and $\{\!|e|\!\}_k$ denotes the encryption of e under k. We write $\mathbf{Exp}[\mathbf{Keys}, \mathbf{Data}]$ (and, similarly, for patterns $\mathbf{Pat}[\mathbf{Keys}, \mathbf{Data}]$ later on) to emphasize that the definition of **Exp** depends on the underlying sets **Keys** and **Data**. As a notational convention, we assume that the pairing operation is right associative, and omit unnecessary parentheses. E.g., we write $\{\!|d_1, d_2, d_3|\!\}_k$ instead of $\{\!|(d_1, (d_2, d_3))|\!\}_k$. All ciphertexts in our symbolic expressions represent independent encryptions (each using fresh randomness in the computational setting), even when carrying the same message. This is so that an adversary cannot distinguish between, say, $(\{\!|0|\!\}_k, \{\!|0|\!\}_k)$ and $(\{\!|0|\!\}_k, \{\!|1|\!\}_k)$. Sometimes (e.g., when adding an equality predicate "$\mathbf{Exp} = \mathbf{Exp}$" to the language of expressions) it is desirable for equality of symbolic terms to correspond to equality of their computational interpretations. This can be easily achieved by decorating symbolic ciphertexts with a "randomness" tag, so that identical expressions $\{\!|m|\!\}_k^r = \{\!|m|\!\}_k^r$ correspond to identical ciphertexts, while independent encryptions (of possibly identical messages) are represented by different symbolic expressions $\{\!|m|\!\}_k^r \neq \{\!|m|\!\}_k^{r'}$. An alternative (and syntactically cleaner) method to represent identical ciphertexts is to extend the symbolic syntax with a *variable assignment* operation, like

$$\texttt{let } c := \{\!|\mathbf{Exp}|\!\}_{\mathbf{Keys}} \texttt{ in } \mathbf{Exp}',$$

where the bound variable c may appear (multiple times) in the second expression $\mathbf{Exp}'$. Here, each "let" expression implicitly encrypts using independent randomness, and identical ciphertexts are represented using bound variables. All our definitions and results are easily adapted to these extended expressions with explicit randomness tags or bound variables.

In [3,26], $\mathbf{Keys} = \{k_1, \ldots, k_n\}$ and $\mathbf{Data} = \{d_1, \ldots, d_n\}$ are two flat sets of atomic keys and data blocks. In this paper, we consider pseudorandom keys, defined according to the grammar

$$\mathbf{Keys} ::= \mathbf{Rand} \mid \mathbb{G}_0(\mathbf{Keys}) \mid \mathbb{G}_1(\mathbf{Keys}), \tag{2}$$

where $\mathbf{Rand} = \{r_1, r_2, \ldots\}$ is a set of atomic key symbols (modeling truly random and independent keys), and $\mathbb{G}_0$, $\mathbb{G}_1$ represent the left and right half of a length doubling pseudorandom generator $k \mapsto \mathbb{G}_0(k); \mathbb{G}_1(k)$. Notice that grammar (2) allows for the iterated application of the pseudorandom generator, so that from any key $r \in \mathbf{Rand}$, one can obtain keys of the form

$$\mathbb{G}_{b_1}(\mathbb{G}_{b_2}(\ldots(\mathbb{G}_{b_n}(r))\ldots))$$

for any $n \geq 0$, which we abbreviate as $\mathbb{G}_{b_1 b_2 \ldots b_n}(r)$. (As a special case, for $n = 0$, $\mathbb{G}_\epsilon(r) = r$.) For any set of keys $S \subseteq \mathbf{Keys}$, we write $\mathbb{G}^*(S)$ and $\mathbb{G}^+(S)$ to denote the sets

$$\mathbb{G}^*(S) = \{\mathbb{G}_w(k) \mid k \in S, w \in \{0,1\}^*\}$$
$$\mathbb{G}^+(S) = \{\mathbb{G}_w(k) \mid k \in S, w \in \{0,1\}^*, w \neq \epsilon\}$$

of keys which can be obtained from S through the repeated application of the pseudorandom generator functions $\mathbb{G}_0$ and $\mathbb{G}_1$, zero, one or more times. Using this notation, the set of keys generated by the grammar (2) can be written as $\mathbf{Keys} = \mathbb{G}^*(\mathbf{Rand})$. It is also convenient to define the set

$$\mathbb{G}^-(S) = \{k \mid \mathbb{G}^+(k) \cap S \neq \emptyset\} = \bigcup_{k' \in S} \{k \mid k' \in \mathbb{G}^+(k)\}.$$

Notice that, for any two keys k, k', we have $k \in \mathbb{G}^-(k')$ if and only if $k' \in \mathbb{G}^+(k)$, i.e., $\mathbb{G}^-$ corresponds to the inverse relation of $\mathbb{G}^+$.

The *shape* of an expression is obtained by replacing elements from **Data** and **Keys** with special symbols $\Box$ and $\circ$. Formally, shapes are defined as expressions over these dummy key/data symbols:

$$\mathbf{Shapes} = \mathbf{Exp}[\{\circ\}, \{\Box\}].$$

For notational simplicity, we omit the encryption keys $\circ$ in shapes and write $\{\!|s|\!\}$ instead of $\{\!|s|\!\}_\circ$. Shapes are used to model partial information (e.g., message size) that may be leaked by ciphertexts, even when the encrypting key is not known. (See Lemma 5 for a computational justification.)

The symbolic semantics of cryptographic expressions is defined by mapping them to *patterns*, which are expressions containing subterms of the form $\{\!|s|\!\}_k$, where $s \in \mathbf{Shapes}$ and $k \in \mathbf{Keys}$, representing undecryptable ciphertexts. Formally, the set of patterns $\mathbf{Pat}[\mathbf{Keys}, \mathbf{Data}]$ is defined as

$$\mathbf{Pat} ::= \mathbf{Data} \mid \mathbf{Keys} \mid (\mathbf{Pat}, \mathbf{Pat}) \mid \{\!|\mathbf{Pat}|\!\}_{\mathbf{Keys}} \mid \{\!|\mathbf{Shapes}|\!\}_{\mathbf{Keys}}. \quad (3)$$

Since expressions are also patterns, and patterns can be regarded as expressions over the extended sets $\mathbf{Keys} \cup \{\circ\}$, $\mathbf{Data} \cup \{\Box\}$, we use the letter e to denote expressions and patterns alike. We define a subterm relation $\sqsubseteq$ on $\mathbf{Pat}[\mathbf{Keys}, \mathbf{Data}]$ as the smallest reflexive transitive binary relation such that

$$e_1 \sqsubseteq (e_1, e_2), \qquad e_2 \sqsubseteq (e_1, e_2), \qquad \text{and} \qquad e \sqsubseteq \{\!|e|\!\}_k \quad (4)$$

for all $e, e_1, e_2 \in \mathbf{Pat}[\mathbf{Keys}, \mathbf{Data}]$ and $k \in \mathbf{Keys}$. The *parts* of a pattern $e \in \mathbf{Pat}$ are all of its subterms:

$$\mathbf{parts}(e) = \{e' \in \mathbf{Pat} \mid e' \sqsubseteq e\}. \quad (5)$$

The keys and shape of a pattern are defined by structural induction according to the obvious rules

$$\begin{aligned} \mathbf{keys}(d) &= \emptyset & \mathbf{shape}(d) &= \square \\ \mathbf{keys}(k) &= \{k\} & \mathbf{shape}(k) &= \circ \\ \mathbf{keys}(e_1, e_2) &= \mathbf{keys}(e_1) \cup \mathbf{keys}(e_2) & \mathbf{shape}(e_1, e_2) &= (\mathbf{shape}(e_1), \mathbf{shape}(e_2)) \\ \mathbf{keys}(\{\!|e|\!\}_k) &= \{k\} \cup \mathbf{keys}(e) & \mathbf{shape}(\{\!|e|\!\}_k) &= \{\!|\mathbf{shape}(e)|\!\} \end{aligned}$$

where $d \in \mathbf{Data}$, $k \in \mathbf{Keys}$, $e, e_1, e_2 \in \mathbf{Pat}[\mathbf{Keys}, \mathbf{Data}]$, and $\mathbf{shape}(s) = s$ for all shapes $s \in \mathbf{Shapes}$. Notice that, according to these definitions, $\mathbf{keys}(e)$ includes both the keys appearing in e as a message, and those appearing as an encryption key. On the other hand, $\mathbf{parts}(e)$ only includes the keys that are used as a message. As an abbreviation, we write

$$\mathbf{pkeys}(e) = \mathbf{parts}(e) \cap \mathbf{keys}(e)$$

for the set of keys that appear in e as a message. So, for example, if $e = (k, \{\!|0|\!\}_{k'}, \{\!|k''|\!\}_k)$ then $\mathbf{keys}(e) = \{k, k', k''\}$, but $\mathbf{pkeys}(e) = \{k, k''\}$. This is an important distinction to model the fact that an expression e only provides partial information about the keys in $\mathbf{keys}(e) \setminus \mathbf{parts}(e) = \{k'\}$.

2.2 Computational Model

We assume that all algorithms and constructions take as an implicit input a (positive integer) security parameter ℓ, which we may think as fixed at the outset. We use calligraphic letters, $\mathcal{A}, \mathcal{B}$, etc., to denote randomized algorithms or the probability distributions defined by their output. We write $x \leftarrow \mathcal{A}$ for the operation of drawing x from a probability distribution $\mathcal{A}$, or running a probabilistic algorithm $\mathcal{A}$ with fresh randomness and output x. The uniform probability distribution over a finite set S is denoted by $\mathcal{U}(S)$, and we write $x \leftarrow S$ as an abbreviation for $x \leftarrow \mathcal{U}(S)$. Technically, since algorithms are implicitly parameterized by the security parameter ℓ, each $\mathcal{A}$ represents a *distribution ensemble*, i.e., a sequence of probability distributions $\{\mathcal{A}(\ell)\}_{\ell \geq 0}$ indexed by ℓ. For brevity, we will informally refer to probability ensembles $\mathcal{A}$ simply as probability distributions, thinking of the security parameter ℓ as fixed. We use standard asymptotic notation $O(f)$, $\omega(f)$, etc., and write $f \approx g$ if the function $\epsilon(\ell) = f(\ell) - g(\ell) = \ell^{-\omega(1)}$ is negligible. Two probability distributions $\mathcal{A}_0$ and $\mathcal{A}_1$ are *computationally indistinguishable* (written $\mathcal{A}_0 \approx \mathcal{A}_1$) if for any efficiently computable predicate $\mathcal{D}$, $\Pr\{\mathcal{D}(x) : x \leftarrow \mathcal{A}_0\} \approx \Pr\{\mathcal{D}(x) : x \leftarrow \mathcal{A}_1\}$.

Cryptographic Primitives. In the computational setting, cryptographic expressions evaluate to probability distributions over binary strings, and two expressions are equivalent if the associated distributions are computationally indistinguishable. We consider cryptographic expressions that make use of two standard cryptographic primitives: pseudorandom generators, and (public or private key) encryption schemes.

A *pseudorandom generator* is an efficient algorithm $\mathcal{G}$ that on input a string $x \in \{0,1\}^\ell$ (the *seed*, of length equal to the security parameter ℓ) outputs a string $\mathcal{G}(x)$ of length bigger than ℓ, e.g., 2ℓ. We write $\mathcal{G}_0(x)$ and $\mathcal{G}_1(x)$ for the first and second half of the output of a (length doubling) pseudorandom generator, i.e., $\mathcal{G}(x) = \mathcal{G}_0(x); \mathcal{G}_1(x)$ with $|\mathcal{G}_0(x)| = |\mathcal{G}_1(x)| = |x| = \ell$. A pseudorandom generator $\mathcal{G}$ is computationally *secure* if the output distribution $\{\mathcal{G}(x) \colon x \leftarrow \{0,1\}^\ell\}$ is computationally indistinguishable from the uniform distribution $\mathcal{U}(\{0,1\}^{2\ell}) = \{y \colon y \leftarrow \{0,1\}^{2\ell}\}$.

A (private key) *encryption scheme* is a pair of efficient (randomized) algorithms $\mathcal{E}$ (for encryption) and $\mathcal{D}$ (for decryption) such that $\mathcal{D}(k, \mathcal{E}(k,m)) = m$ for any message m and key $k \in \{0,1\}^\ell$. The encryption scheme is *secure* if it satisfies the following definition of *indistinguishability under chosen plaintext attack*. More technically, for any probabilistic polynomial time adversary $\mathcal{A}$, the following must hold. Choose a bit $b \in \{0,1\}$ and a key $k \in \{0,1\}^\ell$ uniformly at random, and let $O_b(m)$ be an encryption oracle that on input a message m outputs $\mathcal{E}(k,m)$ if $b = 1$, or $\mathcal{E}(k, 0^{|m|})$ if $b = 0$, where $0^{|m|}$ is a sequence of 0s of the same length as m. The adversary $\mathcal{A}$ is given oracle access to $O_b(\cdot)$, and attempts to guess the bit b. The encryption scheme is secure if $\Pr\{\mathcal{A}^{O_b(\cdot)} = b\} \approx 1/2$. For notational convenience, the encryption $\mathcal{E}(k,m)$ of a message m under a key k is often written as $\mathcal{E}_k(m)$. Public key encryption is defined similarly. All our results hold for private and public key encryption algorithms, with hardly any difference in the proofs. So, for simplicity, we will focus the presentation on private key encryption, but we observe that adapting the results to public key encryption is straightforward.

In some of our proofs, it is convenient to use a seemingly stronger (but equivalent) security definition for encryption, where the adversary is given access to several encryption oracles, each encrypting under an independently chosen random key. More formally, the adversary $\mathcal{A}$ in the security definition is given access to a (stateful) oracle $O_b(i,m)$ that takes as input both a message m and a key index i. The first time $\mathcal{A}$ makes a query with a certain index i, the encryption oracle chooses a key $k_i \leftarrow \{0,1\}^\ell$ uniformly at random. The query $O_b(i,m)$ is answered using key k_i as in the previous definition: if $b = 1$ then $O_b(i,m) = \mathcal{E}(k_i, m)$, while if $b = 0$ then $O_b(i,m) = \mathcal{E}(k_i, 0^{|m|})$.

Computational evaluation. In order to map a cryptographic expression from **Exp** to a probability distribution, we need to pick a length doubling pseudorandom generator $\mathcal{G}$, a (private key) encryption scheme $\mathcal{E}$, a string representation γ_d for every data block $d \in$ **Data**, and a binary operation[4] π used to encode pairs of strings.

[4] We do not assume any specific property about π, other than invertibility and efficiency, i.e., $\pi(w_1, w_2)$ should be computable in polynomial (typically linear) time, and the substrings w_1 and w_2 can be uniquely recovered from $\pi(w_1, w_2)$, also in polynomial time. In particular, $\pi(w_1, w_2)$ is not just the string concatenation operation $w_1; w_2$ (which is not invertible), and the strings $\pi(w_1, w_2)$ and $\pi(w_2, w_1)$ may have different length. For example, $\pi(w_1, w_2)$ could be the string concatenation of a prefix-free encoding of w_1, followed by w_2.

Since encryption schemes do not hide the length of the message being encrypted, it is natural to require that all functions operating on messages are length-regular, i.e., the length of their output depends only on the length of their input. For example, $\mathcal{G}$ is length regular by definition, as it always maps strings of length ℓ to strings of length 2ℓ. Throughout the paper we assume that all keys have length ℓ equal to the security parameter, and the functions $d \mapsto \gamma_d$, π and $\mathcal{E}$ are length regular, i.e., $|\gamma_d|$ is the same for all $d \in \mathbf{Data}$, $|\pi(x_1, x_2)|$ depends only on $|x_1|$ and $|x_2|$, and $|\mathcal{E}(k, x)|$ depends only on ℓ and $|x|$.

Definition 1. *A computational interpretation is a tuple $(\mathcal{G}, \mathcal{E}, \gamma, \pi)$ consisting of a length-doubling pseudorandom generator $\mathcal{G}$, a length regular encryption scheme $\mathcal{E}$, and length regular functions γ_d and $\pi(x_1, x_2)$. If $\mathcal{G}$ is a secure pseudorandom generator, and $\mathcal{E}$ is a secure encryption scheme (satisfying indistinguishability under chosen plaintext attacks, as defined in the previous paragraphs), then we say that $(\mathcal{G}, \mathcal{E}, \gamma, \pi)$ is a* secure *computational interpretation.*

Computational interpretations are used to map symbolic expressions in **Exp** to probability distributions in the obvious way. We first define the evaluation $\sigma[\![e]\!]$ of an expression $e \in \mathbf{Exp}[\mathbf{Keys}, \mathbf{Data}]$ with respect to a fixed key assignment $\sigma\colon \mathbf{Keys} \to \{0,1\}^\ell$. The value $\sigma[\![e]\!]$ is defined by induction on the structure of the expression e by the rules $\sigma[\![d]\!] = \gamma_d$, $\sigma[\![k]\!] = \sigma(k)$, $\sigma[\![(e_1, e_2)]\!] = \pi(\sigma[\![e_1]\!], \sigma[\![e_2]\!])$, and $\sigma[\![\{e\}_k]\!] = \mathcal{E}(\sigma(k), \sigma[\![e]\!])$. All ciphertexts in a symbolic expressions are evaluated using fresh independent encryption randomness. The computational evaluation $[\![e]\!]$ of an expression e is defined as the probability distribution obtained by first choosing a random key assignment σ (as explained below) and then computing $\sigma[\![e]\!]$. When $\mathbf{Keys} = \mathbb{G}^*(\mathbf{Rand})$ is a set of pseudorandom keys, σ is selected by first choosing the values $\sigma(r) \in \{0,1\}^\ell$ (for $r \in \mathbf{Rand}$) independently and uniformly at random, and then extending σ to pseudorandom keys in $\mathbb{G}^+(\mathbf{Rand})$ using a length doubling pseudorandom generator $\mathcal{G}$ according to the rule

$$\mathcal{G}(\sigma(k)) = \sigma(\mathbb{G}_0(k)); \sigma(\mathbb{G}_1(k)).$$

It is easy to see that any two expressions $e, e' \in \mathbf{Exp}$ with the same shape $s = \mathtt{shape}(e) = \mathtt{shape}(e')$ always map to strings of exactly the same length, denoted $|[\![s]\!]| = |\sigma[\![e]\!]| = |\sigma'[\![e']\!]|$. The computational evaluation function $\sigma[\![e]\!]$ is extended to patterns by defining $\sigma[\![s]\!] = 0^{|[\![s]\!]|}$ for all shapes $s \in \mathbf{Shapes}$. Again, we have $|\sigma[\![e]\!]| = |[\![\mathtt{shape}(e)]\!]|$ for all patterns $e \in \mathbf{Pat}$, i.e., all patterns with the same shape evaluate to strings of the same length.

Notice that each expression e defines a probability ensemble $[\![e]\!]$, indexed by the security parameter ℓ defining the key length of $\mathcal{G}$ and $\mathcal{E}$. Two symbolic expressions (or patterns) e, e' are computationally equivalent (with respect to a given computational interpretation $(\mathcal{G}, \mathcal{E}, \gamma, \pi)$) if the corresponding probability ensembles $[\![e]\!]$ and $[\![e']\!]$ are computationally indistinguishable. An equivalence relation R on symbolic expressions is computationally *sound* if for any two equivalent expressions $(e, e') \in R$ and any secure computational interpretation, the distributions $[\![e]\!]$ and $[\![e']\!]$ are computationally indistinguishable. Conversely, we say that a relation R is *complete* if for any two unrelated expressions $(e, e') \notin R$, there is a secure computational interpretation such that $[\![e]\!]$ and $[\![e']\!]$ can be efficiently distinguished.

3 Symbolic Model for Pseudorandom Keys

In this section we develop a symbolic framework for the treatment of pseudorandom keys, and prove that it is computationally sound and complete. Specifically, we give a symbolic criterion for a set of keys which is satisfied *if and only if* the joint distribution associated to the set of keys is computationally indistinguishable from the uniform distribution. Before getting into the technical details we provide some intuition.

Symbolic keys are usually regarded as bound names, up to renaming. In the computational setting, this corresponds to the fact that changing the names of the keys does not alter the probability distribution associated to them. When pseudorandom keys are present, some care has to be exercised in defining an appropriate notion of key renaming. For example, swapping r and $\mathbb{G}_0(r)$ should not be considered a valid key renaming because the probability distributions associated to $(r, \mathbb{G}_0(r))$ and $(\mathbb{G}_0(r), r)$ can be easily distinguished.[5] A conservative approach would require a key renaming μ to act simply as a permutation over the set of atomic keys **Rand**. However, this is overly restrictive. For example, renaming $(\mathbb{G}_0(r), \mathbb{G}_1(r))$ to (r_0, r_1) should be allowed because $(\mathbb{G}_0(r), \mathbb{G}_1(r))$ represents a pseudorandom string, which is computationally indistinguishable from the truly random string given by (r_0, r_1). The goal of this section is to precisely characterize which key renamings can be allowed, and which cannot, to preserve computational indistinguishability.

The rest of the section is organized as follows. First, in Sect. 3.1, we introduce a symbolic notion of independence for pseudorandom keys. Informally, two (symbolic) keys are independent if neither of them can be derived from the other through the application of the pseudorandom generator. We give a computational justification for this notion by showing (see Theorem 1) that the standard (joint) probability distribution associated to a sequence of symbolic keys $k_1, \ldots, k_n \in \mathbf{Keys}$ in the computational model is pseudorandom precisely when the keys $k_1, \ldots, k_n$ are symbolically independent. Then, in Sect. 3.2, we use this definition of symbolic independence to define a computationally sound notion of key renaming. Intuitively, in order to be computationally sound and achieve other desirable properties, key renamings should map independent sets to independent sets. In Corollary 1 we prove that, under such restriction, applying a renaming to cryptographic expressions yields computationally *indistinguishable* distributions. This should be contrasted with the standard notion of key renaming used in the absence of pseudorandom keys, where equivalent expressions evaluate to *identical* probability distributions.

[5] All that the distinguisher has to do, on input a pair of keys (σ_0, σ_1), is to compute $\mathcal{G}_0(\sigma_1)$ and check if the result equals σ_0.

3.1 Independence

In this section we define a notion of independence for symbolic keys, and show that it is closely related to the computational notion of pseudorandomness.

Definition 2. *For any two keys* $k_1, k_2 \in \mathbf{Keys}$, *we say that* k_1 yields k_2 *(written* $k_1 \preceq k_2$*) if* $k_2 \in \mathbb{G}^*(k_1)$, *i.e.,* k_2 *can be obtained by repeated application of* $\mathbb{G}_0$ *and* $\mathbb{G}_1$ *to* k_1. *Two keys* k_1, k_2 *are* independent *(written* $k_1 \perp k_2$*) if neither* $k_1 \preceq k_2$ *nor* $k_2 \preceq k_1$. *We say that the keys* $k_1, \ldots, k_n$ *are independent if* $k_i \perp k_j$ *for all* $i \neq j$.

Notice that any two keys satisfy $\mathbb{G}_{w_0}(r_0) \preceq \mathbb{G}_{w_1}$ if and only if $r_0 = r_1$ and $w_0 \preceq w_1$. As an example, they keys $\mathbb{G}_0(r) \perp \mathbb{G}_{01}(r)$ are independent, but the keys $\mathbb{G}_0(r) \preceq \mathbb{G}_{10}(r)$ are not. As usual, we write $k_1 \prec k_2$ as an abbreviation for $(k_1 \preceq k_2) \wedge (k_1 \neq k_2)$. Notice that $(\mathbf{Keys}, \preceq)$ is a partial order, i.e., the relation $\preceq$ is reflexive, antisymmetric and transitive. Pictorially a set of keys $S \subseteq \mathbf{Keys}$ can be represented by the Hasse diagram[6] of the induced partial order $(S, \preceq)$. (See Fig. 1 for an example.) Notice that this diagram is always a forest, i.e., the union of disjoint trees with roots $\mathbf{roots}(S) = S \setminus \mathbb{G}^+(S)$. S is an independent set if and only if $S = \mathbf{roots}(S)$, i.e., each tree in the forest associated to S consists of a single node, namely its root.

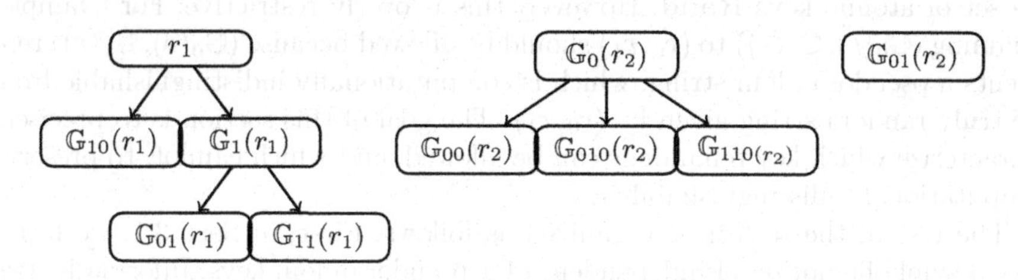

Fig. 1. Hasse diagram associated to the set of keys $S = \{r_1, \mathbb{G}_{10}(r_1), \mathbb{G}_1(r_1), \mathbb{G}_{01}(r_1), \mathbb{G}_{11}(r_1), \mathbb{G}_0(r_2), \mathbb{G}_{00}(r_2), \mathbb{G}_{010}(r_2), \mathbb{G}_{110}(r_2), \mathbb{G}_{01}(r_2)\}$. For any two keys, $k_1 \preceq k_2$ if there is a directed path from k_1 to k_2. The keys $\{\mathbb{G}_0(r_2), \mathbb{G}_{01}(r_2)\}$ form an independent set because neither $\mathbb{G}_0(r_2) \preceq \mathbb{G}_{01}(r_2)$, nor $\mathbb{G}_{01}(r_2) \preceq \mathbb{G}_0(r_2)$. The Hasse diagram of S is a forest consisting of 3 trees with roots $\mathbf{roots}(S) = \{r_1, \mathbb{G}_0(r_2), \mathbb{G}_{01}(r_2)\}$.

We consider the question of determining, symbolically, when (the computational evaluation of) a sequence of pseudorandom keys $k_1, \ldots, k_n$ is pseudorandom, i.e., it is computationally indistinguishable from n truly random independently chosen keys. The following lemma shows that our symbolic notion of independence corresponds exactly to the standard cryptographic notion of computational pseudorandomness. We remark that the correspondence proved in the lemma is *exact*, in the sense that the symbolic condition is both *necessary*

[6] The Hasse diagram of a partial order relation $\preceq$ is the graph associated to the transitive reduction of $\preceq$, i.e., the smallest relation R such that $\preceq$ is the symmetric transitive closure of R.

and *sufficient* for symbolic equivalence. This should be contrasted with typical computational soundness results [3], that only provide sufficient conditions for computational equivalence, and require additional work/assumptions to establish the completeness of the symbolic criterion [14,31].

Theorem 1. *Let* $k_1, \ldots, k_n \in \mathbf{Keys} = \mathbb{G}^*(\mathbf{Rand})$ *be a sequence of symbolic keys. Then, for any secure (length doubling) pseudorandom generator* $\mathcal{G}$*, the probability distributions* $[\![k_1, \ldots, k_n]\!]$ *and* $[\![r_1, \ldots, r_n]\!]$ *(where* $r_1, \ldots, r_n \in \mathbf{Rand}$ *are distinct atomic keys) are computationally indistinguishable if and only if the keys* $k_1, \ldots, k_n$ *are (symbolically) independent, i.e.,* $k_i \perp k_j$ *for all* $i \neq j$*.*

Proof. We first prove the "only if" direction of the equivalence, i.e., independence is a necessary condition for the indistinguishability of $[\![r_1, \ldots, r_n]\!]$ and $[\![k_1, \ldots, k_n]\!]$. Assume the keys in $(k_1, \ldots, k_n)$ are not independent, i.e., $k_i \preceq k_j$ for some $i \neq j$. By definition, $k_j = \mathbb{G}_w(k_i)$ for some $w \in \{0,1\}^*$. This allows to deterministically compute $[\![k_j]\!] = \mathcal{G}_w([\![k_i]\!])$ from $[\![k_i]\!]$ using the pseudorandom generator. The distinguisher between $[\![r_1, \ldots, r_n]\!]$ and $[\![k_1, \ldots, k_n]\!]$ works in the obvious way: given a sample $(\sigma_1, \ldots, \sigma_n)$, compute $\mathcal{G}_w(\sigma_i)$ and compare the result to σ_j. If the sample comes from $[\![k_1, \ldots, k_n]\!]$, then the test is satisfied with probability 1. If the sample comes from $[\![r_1, \ldots, r_n]\!]$, then the test is satisfied with exponentially small probability because $\sigma_i = [\![r_i]\!]$ is chosen at random independently from $\sigma_j = [\![r_j]\!]$. This concludes the proof for the "only if" direction.

Let us now move to the "if" direction, i.e., prove that independence is a sufficient condition for the indistinguishability of $[\![r_1, \ldots, r_n]\!]$ and $[\![k_1, \ldots, k_n]\!]$. Assume the keys in $(k_1, \ldots, k_n)$ are independent, and let m be the number of applications of $\mathbb{G}_0$ and $\mathbb{G}_1$ required to obtain $(k_1, \ldots, k_n)$ from the basic keys in **Rand**. We define $m+1$ tuples $K^i = (k_1^i, \ldots, k_n^i)$ of independent keys such that

- $K^0 = (k_1, \ldots, k_n)$
- $K^m = (r_1, \ldots, r_n)$, and
- for all i, the distributions $[\![K^i]\!]$ and $[\![K^{i+1}]\!]$ are computationally indistinguishable.

It follows by transitivity that $[\![K^0]\!] = [\![k_1, \ldots, k_n]\!]$ is computationally indistinguishable from $[\![K^m]\!] = [\![r_1, \ldots, r_n]\!]$. More precisely, any adversary that distinguishes $[\![k_1, \ldots, k_n]\!]$ from $[\![r_1, \ldots, r_n]\!]$ with advantage δ, can be efficiently transformed into an adversary that breaks the pseudorandom generator $\mathcal{G}$ with advantage at least δ/m. Each tuple K^{i+1} is defined from the previous one K^i as follows. If all the keys in $K^i = \{k_1^i, \ldots, k_n^i\}$ are random (i.e., $k_j^i \in \mathbf{Rand}$ for all $j = 1, \ldots, n$), then we are done and we can set $K^{i+1} = K^i$. Otherwise, let $k_j^i = \mathbb{G}_w(r) \in \mathbf{Keys} \setminus \mathbf{Rand}$ be a pseudorandom key in K^i, with $r \in \mathbf{Rand}$ and $w \neq \epsilon$. Since the keys in K^i are independent, we have $r \notin K^i$. Let $r', r'' \in \mathbf{Rand}$ be two new fresh key symbols, and define $K^{i+1} = \{k_1^{i+1}, \ldots, k_n^{i+1}\}$ as follows:

$$k_h^{i+1} = \begin{cases} \mathbb{G}_s(r') & \text{if } k_h^i = \mathbb{G}_s(\mathbb{G}_0(r)) \text{ for some } s \in \{0,1\}^* \\ \mathbb{G}_s(r'') & \text{if } k_h^i = \mathbb{G}_s(\mathbb{G}_1(r)) \text{ for some } s \in \{0,1\}^* \\ k_h^i & \text{otherwise} \end{cases}$$

It remains to prove that any distinguisher $\mathcal{D}$ between $[\![K^i]\!]$ and $[\![K^{i+1}]\!]$ can be used to break (with the same success probability) the pseudorandom generator $\mathcal{G}$. The distinguisher $\mathcal{D}'$ for the pseudorandom generator $\mathcal{G}$ is given as input a pair of strings (σ', σ'') chosen either uniformly (and independently) at random or running the pseudorandom generator $(\sigma', \sigma'') = \mathcal{G}(\sigma)$ on a randomly chosen seed σ. $\mathcal{D}'(\sigma', \sigma'')$ computes n strings $(\sigma_1, \ldots, \sigma_n)$ by evaluating $(k_1^{i+1}, k_2^{i+1}, \ldots, k_n^{i+1})$ according to an assignment that maps r' to σ', r'' to σ'', and all other base keys $r \in \mathbf{Rand}$ to independent uniformly chosen values. The output of $\mathcal{D}'(\sigma', \sigma'')$ is $\mathcal{D}(\sigma_1, \ldots, \sigma_n)$. Notice that if σ' and σ'' are chosen uniformly and independently at random, then $(\sigma_1, \ldots, \sigma_n)$ is distributed according to $[\![K^{i+1}]\!]$, while if $(\sigma', \sigma'') = \mathcal{G}(\sigma)$, then $(\sigma_1, \ldots, \sigma_n)$ is distributed according to $[\![K^i]\!]$. Therefore the success probability of $\mathcal{D}'$ in breaking $\mathcal{G}$ is exactly the same as the success probability of $\mathcal{D}$ in distinguishing $[\![K^i]\!]$ from $[\![K^{i+1}]\!]$. □

3.2 Renaming Pseudorandom Keys

We will show that key renamings are compatible with computational indistinguishability as long as they preserve the action of the pseudorandom generator, in the sense specified by the following definition.

Definition 3 (pseudo-renaming). *For any set of keys $S \subseteq$ **Keys**, a renaming $\mu\colon S \to$ **Keys** is compatible with the pseudorandom generator $\mathbb{G}$ if for all $k_1, k_2 \in S$ and $w \in \{0,1\}^*$,*

$$k_1 = \mathbb{G}_w(k_2) \qquad \text{if and only if} \qquad \mu(k_1) = \mathbb{G}_w(\mu(k_2)).$$

For brevity, we refer to renamings satisfying this property as pseudo-renamings.

Notice that the above definition does not require the domain of μ to be the set of all keys **Keys**, or even include all keys in **Rand**. So, for example, the function mapping $(\mathbb{G}_0(r_0), \mathbb{G}_1(r_0))$ to $(r_0, \mathbb{G}_{001}(r_1))$ is a valid pseudo-renaming, and it does not act as a permutation over **Rand**. The following lemmas show that Definition 3 is closely related to the notion of symbolic independence.

Lemma 1. *Let μ be a pseudo-renaming with domain $S \subseteq$ **Keys**. Then μ is a bijection from S to $\mu(S)$. Moreover, S is an independent set if and only if $\mu(S)$ is an independent set.*

Proof. Let $\mu\colon S \to$ **Keys** be a pseudo-renaming. Then μ is necessarily injective, because for all $k_1, k_2 \in S$ such that $\mu(k_1) = \mu(k_2)$, we have $\mu(k_1) = \mu(k_2) = \mathbb{G}_\epsilon(\mu(k_2))$. By definition of pseudo-renaming, this implies $k_1 = \mathbb{G}_\epsilon(k_2) = k_2$. This proves that μ is a bijection from S to $\mu(S)$.

Now assume S is *not* an independent set, i.e., $k_1 = \mathbb{G}_w(k_2)$ for some $k_1, k_2 \in S$ and $w \neq \epsilon$. By definition of pseudo-renaming, we also have $\mu(k_1) = \mathbb{G}_w(\mu(k_2))$. So, $\mu(S)$ is not an independent set either. Similarly, if $\mu(S)$ is *not* an independent set, then there exists keys $\mu(k_1), \mu(k_2) \in \mu(S)$ (with $k_1, k_2 \in S$) such that $\mu(k_1) = \mathbb{G}_w(\mu(k_2))$ for some $w \neq \epsilon$. Again, by definition of pseudo-renaming, $k_1 = \mathbb{G}_w(k_2)$, and S is not an independent set. □

In fact, pseudo-renamings can be equivalently defined as the natural extension of bijections between two independent sets of keys.

Lemma 2. *Any pseudo-renaming μ with domain S can be uniquely extended to a pseudo-renaming $\bar{\mu}$ with domain $\mathbb{G}^*(S)$. In particular, any pseudo-renaming can be (uniquely) specified as the extension $\bar{\mu}$ of a bijection $\mu\colon A \to B$ between two independent sets $A = \mathbf{roots}(S)$ and $B = \mu(A)$.*

Proof. Let $\mu\colon S \to \mathbf{Keys}$ be a pseudo-renaming. For any $w \in \{0,1\}^*$ and $k \in S$, define $\bar{\mu}(\mathbb{G}_w(k)) = \mathbb{G}_w(\mu(k))$. This definition is well given because μ is a pseudo-renaming, and therefore for any two representations of the same key $\mathbb{G}_w(k) = \mathbb{G}_{w'}(k') \in \mathbb{G}^*(S)$ with $k, k' \in S$, we have $\mathbb{G}_w(\mu(k)) = \mu(\mathbb{G}_w(k)) = \mu(Gen_{w'}(k')) = \mathbb{G}_{w'}(\mu(k'))$. Moreover, it is easy to check that $\bar{\mu}$ is a pseudo-renaming, and any pseudo-renaming that extends μ must agree with $\bar{\mu}$. We now show that pseudo-renamings can be uniquely specified as bijections between two independent sets of keys. Specifically, for any pseudo-renaming μ with domain S, consider the restriction μ_0 of μ to $A = \mathbf{roots}(S)$. By Lemma 1, μ_0 is a bijection between independent sets A and $B = \mu_0(A)$. Consider the extensions of μ and μ_0 to $\mathbb{G}^*(S) = \mathbb{G}^*(\mathbf{roots}(S)) = \mathbb{G}^*(A)$. Since μ and μ_0 agree on $A = \mathbf{roots}(S)$, both $\bar{\mu}$ and $\bar{\mu}_0$ are extensions of μ_0. By uniqueness of this extension, we get $\bar{\mu}_0 = \bar{\mu}$. Restricting both functions to S, we get that the original pseudo-renaming μ can be expressed as the restriction of $\bar{\mu}_0$ to S. In other words, μ can be expressed as the extension to S of a bijection μ_0 between two independent sets of keys $A = \mathbf{roots}(S)$ and $B = \mu(A)$. □

We remark that a pseudo-renaming $\mu\colon S \to \mathbf{Keys}$ cannot, in general, be extended to one over the set $\mathbf{Keys} = \mathbb{G}^*(\mathbf{Rand})$ of all keys. For example, $\mu\colon \mathbb{G}_0(r_0) \mapsto r_1$ is a valid pseudo-renaming, but it cannot be extended to include r_0 in its domain.

The next lemma gives one more useful property of pseudo-renamings: they preserve the root keys.

Lemma 3. *For any pseudo-renaming $\mu\colon A \to \mathbf{Keys}$, we have $\mu(\mathbf{roots}(A)) = \mathbf{roots}(\mu(A))$.*

Proof. By Lemma 1, μ is injective. Therefore, $\mu(\mathbf{roots}(A))$ equals $\mu(A \setminus \mathbb{G}^+(A)) = \mu(A) \setminus \mu(\mathbb{G}^+(A))$. From the defining property of pseudo-renamings we also easily get that $\mu(\mathbb{G}^+(A)) = \mathbb{G}^+(\mu(A))$. Therefore, $\mu(\mathbf{roots}(A)) = \mu(A) \setminus \mathbb{G}^+(\mu(A)) = \mathbf{roots}(\mu(A))$. □

Using Lemma 2, throughout the paper we specify pseudo-renamings as bijections between two independent sets of keys. Of course, in order to apply $\mu\colon S \to \mu(S)$ to an expression e, the key set $\mathbf{keys}(e)$ must be contained in $\mathbb{G}^*(S)$. Whenever we apply a pseudo-renaming $\mu\colon S \to \mathbf{Keys}$ to an expression or pattern e, we implicitly assume that $\mathbf{keys}(e) \subset \mathbb{G}^*(S)$. (Typically, $S = \mathbf{roots}(\mathbf{keys}(e))$, so that $\mathbf{keys}(e) \subset \mathbb{G}^*(\mathbf{roots}(\mathbf{keys}(e))) = \mathbb{G}^*(S)$ is always satisfied.) Formally, the result of applying a pseudo-renaming μ to an expression or pattern $e \in \mathbf{Pat}(\mathbf{Keys}, \mathbf{Data})$ is defined as

$$\mu(d) = d \qquad \mu(\{\!\!\{e\}\!\!\}_k) = \{\!\!\{\mu(e)\}\!\!\}_{\bar{\mu}(k)}$$
$$\mu(k) = \bar{\mu}(k) \qquad \mu(s) = s$$
$$\mu(e_1, e_2) = (\mu(e_1), \mu(e_2))$$

for all $d \in \mathbf{Data}$, $k \in \mathbf{Keys}$, $e, e_1, e_2 \in \mathbf{Pat}(\mathbf{Keys}, \mathbf{Data})$ and $s \in \mathbf{Shapes}$. We can now define an appropriate notion of symbolic equivalence up to renaming.

Definition 4. *Two expressions or patterns $e_1, e_2 \in \mathbf{Pat}(\mathbf{Keys}, \mathbf{Data})$ are* equivalent up to pseudo-renaming *(written $e_1 \cong e_2$), if there is a pseudo-renaming μ such that $\bar{\mu}(e_1) = e_2$. Equivalently, by Lemma 2, $e_1 \cong e_2$ if there is a bijection $\mu\colon \mathbf{roots}(\mathbf{keys}(e_1)) \to \mathbf{roots}(\mathbf{keys}(e_2))$ such that $\bar{\mu}(e_1) = e_2$.*

It easily follows from the definitions and Theorem 1 that $\cong$ is an equivalence relation, and expressions that are equivalent up to pseudo-renaming are computationally equivalent.

Corollary 1. *The equivalence relation $\cong$) is computationally sound, i.e., for any two patterns $e_1, e_2 \in \mathbf{Pat}(\mathbf{Keys}, \mathbf{Data})$ such that $e_1 \cong e_2$, the distributions $[\![e_1]\!]$ and $[\![e_2]\!]$ are computationally indistinguishable.*

Proof. Assume $e_1 \cong e_2$, i.e., there exists a bijection $\mu : \mathbf{roots}(\mathbf{keys}(e_1)) \to \mathbf{roots}(\mathbf{keys}(e_2))$ such that $\bar{\mu}(e_1) = e_2$. Let n be the size of $A_1 = \mathbf{roots}(\mathbf{keys}(e_1))$ and $A_2 = \mathbf{roots}(\mathbf{keys}(e_2)) = \mu(A_1)$. We show that any distinguisher $\mathcal{D}$ between $[\![e_1]\!]$ and $[\![e_2]\!] = [\![\bar{\mu}(e_1)]\!]$ can be efficiently transformed into a distinguisher $\mathcal{A}$ between $[\![A_1]\!]$ and $[\![A_2]\!]$ with the same advantage as $\mathcal{D}$. Since A_1 and A_2 are independent sets of size n, by Theorem 1 the probability distributions $[\![A_1]\!]$ and $[\![A_2]\!]$ are indistinguishable from $[\![r_1, \ldots, r_n]\!]$. So, $[\![A_1]\!]$ and $[\![A_2]\!]$ must be indistinguishable from each other, and $\mathcal{A}$'s advantage must be negligible. We now show how to build $\mathcal{A}$ from $\mathcal{D}$. The distinguisher $\mathcal{A}$ takes as input a sample σ coming from either $[\![A_1]\!]$ or $[\![A_2]\!]$. $\mathcal{A}$ evaluates e_1 according to the key assignment $A_1 \mapsto \sigma$, and outputs $\mathcal{D}(\sigma[\![e_1]\!])$. By construction, $\sigma[\![e_1]\!]$ is distributed according to $[\![e_1]\!]$ when $\sigma = [\![A_1]\!]$, while it is distributed according to $[\![e_2]\!] = [\![\bar{\mu}(e_1)]\!]$ when $\sigma = [\![A_2]\!] = [\![\mu(A_1)]\!]$. It follows that $\mathcal{A}$ has exactly the same advantage as $\mathcal{D}$. □

Based on the previous corollary, it is convenient to define a notion of "normal pattern", where the keys have been renamed in some standard way.

Definition 5. *The normalization of $e \in \mathbf{Pat}$ is the pattern $\mathtt{norm}(e) = \mu(e)$ obtained by applying the pseudo-renaming $\mu(k_i) = r_i$, where $K = \{k_1, \ldots, k_n\} = \mathbf{roots}(\mathbf{keys}(e))$ and $r_1, \ldots, r_n \in \mathbf{Rand}$.*

It immediately follows from the definition that $\mathtt{norm}(e) \cong e$, and that any two patterns e_0, e_1 are equivalent up to renaming ($e_0 \cong e_1$) if and only if their normalizations $\mathtt{norm}(e_0) = \mathtt{norm}(e_1)$ are identical.

4 Symbolic Semantics

Following [3,26], the symbolic semantics of an expression $e \in \mathbf{Exp}$ is defined by specifying the set of keys $S \subseteq \mathbf{keys}(e)$ recoverable from e by an adversary, and a corresponding pattern $\mathtt{proj}(e, S)$, which, informally, represents the adversary's view of e when given the ability to decrypt only under the keys in S. Informally, $\mathtt{proj}(e, S)$ can be thought as the projection of e onto the subset of expressions that use only keys in S for encryption. More specifically, $\mathtt{proj}(e, S)$ is obtained from e by replacing all undecryptable subexpression $\{\!|e'|\!\}_k \sqsubseteq e$ (where $k \notin S$) with a pattern $\{\!|\mathtt{shape}(e')|\!\}_k$ that reveals only the shape of the encrypted message. The formal definition of $\mathtt{proj}$ is given in Fig. 2.

We remark that the definition of $\mathtt{proj}$ is identical to previous work [3,26], as it treats pseudo-random keys $\mathbf{Keys} = \mathbb{G}^*(\mathbf{Rand})$ just as regular keys, disregarding their internal structure. (Relations between pseudorandom keys will be taken into account when defining the set of keys S known to the adversary.) In particular, as shown in [3,26], this function satisfies the following properties[7]

$$\mathtt{proj}(e, \mathbf{Keys}) = e \tag{6}$$

$$\mathtt{proj}(\mathtt{proj}(e, S), T) = \mathtt{proj}(e, S \cap T). \tag{7}$$

In order to define S, we need to specify the set of keys $\mathbf{rec}(e) \subseteq \mathbf{keys}(e)$ that an adversary may (potentially) extract from all the parts of an expression (or pattern) e. In the standard setting, where keys are atomic symbols, and encryption is the only cryptographic primitive, $\mathbf{rec}(e)$ can be simply defined as the set of keys appearing in e as a message. This is because the partial information about a key k revealed by a ciphertext $\{\!|m|\!\}_k$ is of no use to an adversary, except possibly for telling when two ciphertexts are encrypted under the same key. When dealing with expressions that make use of possibly related pseudorandom keys and multiple cryptographic primitives, one needs to take into account the possibility that an adversary may combine different pieces of partial information about the keys in mounting an attack. To this end, we define $\mathbf{rec}(e)$ to include all keys k such that either

1. e contains k as a message (directly revealing the value of k), or
2. e contains both a message encrypted under k (providing partial information about k) and some other related key k' (providing an additional piece of information about k).

In other words, our definition postulates that the symbolic adversary can fully recover a key k whenever it is given two distinct pieces of partial information about it. In addition, $\mathbf{rec}(e)$ contains all other keys that can be derived using the pseudorandom generator $\mathbb{G}$.

[7] Notice that by (7), the functions $\mathtt{proj}(\cdot, S)$ and $\mathtt{proj}(\cdot, T)$ commute, i.e., $\mathtt{proj}(\mathtt{proj}(e, S), T) = \mathtt{proj}(\mathtt{proj}(e, T), S)$ for any expression e. Indeed, for example, if $S = \{k_1\}$, $T = \{k_2\}$ and $e = \{\!|\{\!|m|\!\}_{k_1}|\!\}_{k_2}$, then $\mathtt{proj}(e, \{k_1\}) = \{\!|\{\!|\square|\!\}|\!\}_{k_2}$, $\mathtt{proj}(e, \{k_2\}) = \{\!|\{\!|\square|\!\}_{k_1}|\!\}_{k_2}$, and $\mathtt{proj}(\mathtt{proj}(e, \{k_1\}), \{k_2\}) = \mathtt{proj}(\mathtt{proj}(e, \{k_2\}), \{k_1\}) = \mathtt{proj}(e, \emptyset) = \{\!|\{\!|\square|\!\}|\!\}_{k_2}$.

Definition 6. *For any pattern e, let* $\mathbf{rec}(e) = \mathbf{keys}(e) \cap \mathbb{G}^*(K)$ *where*

$$K = \mathbf{keys}(e) \cap (\mathbf{parts}(e) \cup \mathbb{G}^-(\mathbf{keys}(e)))$$
$$= \mathbf{pkeys}(e) \cup (\mathbf{keys}(e) \cap \mathbb{G}^-(\mathbf{keys}(e))).$$

The expression $\mathbf{keys}(e) \cap \mathbb{G}^*(K)$ simply extends the set of known keys K using the pseudorandom generator. The interesting part of Definition 6 is the set K, which captures the key recovery capabilities of the adversary: $\mathbf{pkeys}(e)$ are all the keys that appear in e as a message, and $\mathbf{keys}(e) \cap \mathbb{G}^-(\mathbf{keys}(e))$ are the keys for which the adversary can obtain some additional partial information.[8] Our definition may seem overly conservative, as it postulates, for example, that a key k can be completely recovered simply given two ciphertexts $\{\!|\Box|\!\}_k$ and $\{\!|\Box|\!\}_{k'}$ where $k' = \mathbb{G}_{101}(k)$ is derived form k using the (one-way) functions $\mathbb{G}_0, \mathbb{G}_1$. In Sect. 5 we justify our definition by showing that there are encryption schemes and pseudorandom generators for which this is indeed possible, and proving a completeness theorem for the symbolic sematics associated to Definition 6. Specifically, if our definition enables a symbolic attacker to distinguish between two expressions e and e', then there is also an efficient computational adversary that distinguishes between the corresponding probability distributions for some valid computational interpretation of the cryptographic primitives.

The functions $\mathtt{proj}$ and $\mathbf{rec}$ are used to associate to each expression e a corresponding *key recovery map* $\mathbb{F}_e$, which, on input a set of keys S, outputs the set of keys $\mathbb{F}_e(S)$ potentially recoverable from e when using the keys in S for decryption.

$$\mathbb{F}_e \colon \mathbf{keys}(e) \to \mathbf{keys}(e) \qquad \text{where} \qquad \mathbb{F}_e(S) = \mathbf{rec}(\mathtt{proj}(e, S)). \tag{8}$$

A symbolic adversary that intercepts the expression e, and whose initial knowledge is the empty set of keys $S_0 = \emptyset$, can obtain more and more keys $S_1 = \mathbb{F}_e(S_0)$, $S_2 = \mathbb{F}_e(S_1), \ldots, S_{i+1} = \mathbb{F}_e(S_i) = \mathbb{F}_e^{i+1}(\emptyset)$, and ultimately recover all the keys in the set[9]

$$\text{fix}(\mathbb{F}_e) = \bigcup_{n \geq 0} S_n = \bigcup_{n \geq 0} \mathbb{F}_e^n(\emptyset). \tag{9}$$

In summary, the symbolic semantics of an expression e can be defined as follows.

Definition 7. *The (least fixed point) symbolic semantics of a cryptographic expression e is the pattern*

$$\mathtt{pattern}(e) = \mathtt{norm}(\mathtt{proj}(e, \mathit{fix}(\mathbb{F}_e)))$$

where $\mathit{fix}(\mathbb{F}_e) = \bigcup_{n \geq 0} \mathbb{F}_e^n(\emptyset)$.

[8] By symmetry, and the final application of $\mathbb{G}$ in the definition of $\mathbf{rec}$, keys recoverable from partial information of type $\mathbf{keys}(e) \cap \mathbb{G}^+(\mathbf{keys}(e))$ are also captured implicitly by the this definition, simply by swapping the role of the two keys.

[9] As we will see, the key recovery map $\mathbb{F}_e$ is monotone, i.e., if $S \subseteq S'$, then $\mathbb{F}_e(S) \subseteq \mathbb{F}_e(S')$, for any two sets of keys S, S'. Therefore, $\mathbb{F}_e$ defines a monotonically increasing sequence of known sets of keys $S_0 \subset S_1 \subset S_2 \subset \ldots S_n = S_{n+1}$ and the set of keys recoverable by the adversary $S_n = \text{fix}(\mathbb{F}_e)$ is precisely the least fixed point of $\mathbb{F}_e$, i.e., the smallest set S such that $\mathbb{F}_e(S) = S$.

$$\texttt{proj}(d, S) = d \qquad \texttt{proj}((e_1, e_2), S) = (\texttt{proj}(e_1, S), \texttt{proj}(e_2, S))$$
$$\texttt{proj}(k, S) = k \qquad \texttt{proj}(\{\!\!\{e\}\!\!\}_k, S) = \begin{cases} \{\!\!\{\texttt{shape}(e)\}\!\!\}_k & \text{if } k \notin S \\ \{\!\!\{\texttt{proj}(e, S)\}\!\!\}_k & \text{if } k \in S \end{cases}$$

Fig. 2. The pattern function $\texttt{proj}: \mathbf{Pat}[\mathbf{Keys}, \mathbf{Data}] \times \wp(\mathbf{Keys}) \rightarrow \mathbf{Pat}[\mathbf{Keys}, \mathbf{Data}]$ where $k \in \mathbf{Keys}$, $d \in \mathbf{Data}$, and $(e_1, e_2), \{\!\!\{e\}\!\!\}_k \in \mathbf{Pat}[\mathbf{Keys}, \mathbf{Data}]$. Intuitively, $\texttt{proj}(e, S)$ is the observable pattern of e, when using the keys in S for decryption.

In the above definition, $S = \text{fix}(\mathbb{F}_e)$ is the set of all keys recoverable by an adversary that intercepts e, $\texttt{proj}(e, S)$ is (the symbolic representation of) what part of e can be decrypted by the adversary, and the final application of $\texttt{norm}$ takes care of key renamings.

We conclude this section by observing that the function **rec** satisfies the fundamental property

$$\mathbf{rec}(\texttt{proj}(e, S)) \subseteq \mathbf{rec}(e) \tag{10}$$

which, informally, says that projecting an expression (or pattern) e does not increase the amount of information recoverable from it. In fact, for any pattern e, the set $\mathbf{rec}(e)$ depends only on the sets $\mathbf{keys}(e)$ and $\mathbf{pkeys}(e)$. Moreover, this dependence is monotone. Since we have $\mathbf{keys}(\texttt{proj}(e, S)) \subseteq \mathbf{keys}(e)$ and $\mathbf{pkeys}(\texttt{proj}(e, S)) \subseteq \mathbf{pkeys}(e)$, by monotonicity we get $\mathbf{rec}(\texttt{proj}(e, S)) \subseteq \mathbf{rec}(e)$.

As an application, [26, Theorem 1] shows that for any functions $\texttt{proj}, \mathbf{rec}$ satisfying properties (6), (7) and (10), the function $\mathbb{F}_e(S) = \mathbf{rec}(\texttt{proj}(e, S))$ is monotone, i.e., if $S \subseteq T$, then $\mathbb{F}_e(S) \subseteq \mathbb{F}_e(T)$.

5 Completeness

In this section we prove that the symbolic semantics defined in Sect. 4 is complete, i.e., if two cryptographic expressions map to different symbolic patterns (as specified in Definition 7), then the corresponding probability distributions can be efficiently distinguished. More specifically, we show that for any two such symbolic expressions e_0, e_1, there is a secure computational interpretation $[\![\cdot]\!]$ (satisfying the standard computational notions of security for pseudorandom generators and encryption schemes) and an efficiently computable predicate $\mathcal{D}$ such that $\Pr\{\mathcal{D}([\![e_0]\!])\} \approx 0$ and $\Pr\{\mathcal{D}([\![e_1]\!])\} \approx 1$.

The core of our completeness theorem is the following lemma, which shows that computationally secure encryption schemes and pseudorandom generators can leak enough partial information about their keys, so to make the keys completely recoverable whenever two keys satisfying a nontrivial relation are used to encrypt. The key recovery algorithm $\mathcal{A}$ described in Lemma 4 provides a tight computational justification for the symbolic key recovery function **rec** described in Definition 6.

Lemma 4. *If pseudorandom generators and encryption schemes exist at all, then there is a secure computational interpretation* $(\mathcal{G}, \mathcal{E}, \gamma, \pi)$ *and a deterministic polynomial time key recovery algorithm* $\mathcal{A}$ *such that the following holds. For any (symbolic) keys* $k_0, k_1 \in \mathbf{Keys}$, *messages* m_0, m_1, *and binary string* $w \neq \epsilon$,

- *if* $k_1 = \mathbb{G}_w(k_0)$, *then* $\mathcal{A}(\mathcal{E}_{\sigma(k_0)}(m_0), \mathcal{E}_{\sigma(k_1)}(m_1), w) = \sigma(k_0)$ *for any key assignment* σ; *and*
- *if* $k_1 \neq \mathbb{G}_w(k_0)$, *then* $\mathcal{A}(\mathcal{E}_{\sigma(k_0)}(m_0), \mathcal{E}_{\sigma(k_1)}(m_1), w) = \bot$ *outputs a special symbol* $\bot$ *denoting failure, except with negligible probability over the random choice of the key assignment* σ.

Proof. We show how to modify any (length doubling) pseudorandom generator $\mathcal{G}'$ and encryption scheme $\mathcal{E}'$ to satisfy the properties in the lemma. Before describing the actual construction, we provide some intuition. The idea is to use an encryption scheme that splits the key $k = (k[0]; k[1])$, uses half of the key (say, $k[1]$) and leaks the first half $k[0]$ as part of the ciphertext. Notice that this is already enough to tell if two ciphertexts are encrypted under the same key, as exposed by patterns like $(\{\!|\Box|\!\}_k, \{\!|\Box|\!\}_{k'})$. But, still, this does not leak any information about the messages, which are well protected by the undisclosed portion of the keys. In order to prove the lemma, we need an appropriate pseudorandom generator which, when combined with the encryption scheme, leads to a key recovery attack. Similarly to the encryption scheme, the pseudorandom generator uses only $k[0]$ (which is expanded by a factor 4, to obtain a string twice as long as the original k), and uses the result to "mask" the second part $k[1]$. Specifically, each half of the output $\mathcal{G}_b(k)$ equals $(\mathcal{G}'_{0b}(k[0]), \mathcal{G}'_{1b}(k[0]) \oplus k[1])$. Now, given an encryption under k (which leaks $k[0]$), and a one-way function $\mathcal{G}_b(k)$ (for any bit b) of the key, one can recover $k[1]$ as follows: expand $k[0]$ to $(\mathcal{G}'_{0b}(k[0]), \mathcal{G}'_{1b}(k[0]))$ and use the result to unmask $\mathcal{G}_b(k)$, to reveal $(0, k[1])$. The same argument is easily adapted to work for any one-way function $\mathcal{G}_w(k)$ corresponding to an arbitrary sequence of applications w of the pseudorandom generator. The problem with this intuitive construction is that it requires to see the full output of $\mathcal{G}_b(k)$. If, instead, we are given only two ciphertexts (encrypted under k and $\mathcal{G}_b(k)$) one gets to learn only the first half $\mathcal{G}_b(k)$, which is not enough to recover $k[1]$. An easy fix to this specific problem is to let $\mathcal{G}_b(k)$ to mask $(k[1], k[1])$ instead of $(0, k[1])$. But this would not allow the attack to carry over to longer applications $\mathcal{G}_w(k)$ of the pseudorandom generator. So, the actual construction required to prove the lemma is a bit more complex, and splits the key into three parts.

The new $\mathcal{E}$ and $\mathcal{G}$ use keys that are three times as long as those of $\mathcal{E}'$ and $\mathcal{G}'$. Specifically, each new key $\sigma(k)$ consists of three equal length blocks which we denote as $\sigma(k)[0], \sigma(k)[1]$ and $\sigma(k)[2]$, where each block can be used as a seed or encryption key for the original $\mathcal{G}'$ and $\mathcal{E}'$. Alternatively, we may think of k as consisting of three atomic symbolic keys $k = (k[0], k[1], k[2])$, each corresponding to ℓ bits of $\sigma(k)$. For notational simplicity, in the rest of the proof, we fix a random key assignment σ, and, with slight abuse of notation, we identify the symbolic keys $k[i]$ with the corresponding ℓ-bit strings $\sigma(k)[i]$. So, for example, we will write k and $k[i]$ instead of $\sigma(k)$ and $\sigma(k)[i]$. Whether each $k[i]$ should be interpreted as a symbolic expression or as a bitstring will always be clear from the context.

The new encryption scheme $\mathcal{E}(k,m) = k[0]; k[1]; \mathcal{E}'(k[2],m)$ simply leaks the first two blocks of the key, and uses the third block to perform the actual encryption. It is easy to see that if $\mathcal{E}'$ is secure against chosen plaintext attacks, then $\mathcal{E}'$ is also secure. Moreover, $\mathcal{E}$ can be made length regular simply by padding the output of $\mathcal{E}'$ to its maximum length.

For the pseudo-random generator, assume without loss of generality that $\mathcal{G}'$ is length doubling, mapping strings of length ℓ to strings of length 2ℓ. We need to define a new $\mathcal{G}$ mapping strings of length 3ℓ to strings of length 6ℓ. On input $k = k[0]; k[1]; k[2]$, the new $\mathcal{G}$ stretches $k[0]$ to a string of length 6ℓ corresponding to the symbolic expression

$$(\mathbb{G}_{00}(k[0]), \mathbb{G}_{010}(k[0]), \mathbb{G}_{110}(k[0]), \mathbb{G}_{01}(k[0]), \mathbb{G}_{011}(k[0]), \mathbb{G}_{111}(k[0])) \qquad (11)$$

and outputs the exclusive-or of this string with $(0; k[2]; k[2]; 0; k[2]; k[2])$. The expression (11) is evaluated using $\mathcal{G}'$. Since $\mathcal{G}'$ is a secure length doubling pseudorandom generator, and the keys in (11) are symbolically independent, by Theorem 1 expression (11) is mapped to a pseudorandom string of length 6ℓ. Finally, since taking the exclusive-or with any fixed string $(0; k[2]; k[2]; 0; k[2]; k[2])$ maps the uniform distribution to itself, the output of $\mathcal{G}$ is also computationally indistinguishable from a uniformly random string of length 6ℓ. This proves that $\mathcal{G}$ is a secure length doubling pseudorandom generator as required. It will be convenient to refer to the first and second halves of this pseudorandom generator $\mathcal{G}(k) = \mathcal{G}_0(k); \mathcal{G}_1(k)$. Using the definition of $\mathcal{G}$, we see that for any bit $b \in \{0,1\}$, the corresponding half of the output consists of the following three blocks:

$$\mathcal{G}_b(k)[0] = [\![\mathbb{G}_{0b}(k[0])]\!] \qquad (12)$$

$$\mathcal{G}_b(k)[1] = [\![\mathbb{G}_{01b}(k[0])]\!] \oplus k[2] \qquad (13)$$

$$\mathcal{G}_b(k)[2] = [\![\mathbb{G}_{11b}(k[0])]\!] \oplus k[2]. \qquad (14)$$

Next, we describe the key recovery algorithm $\mathcal{A}$. This algorithm takes as input two ciphertexts $\mathcal{E}_{k_0}(m_0)$, $\mathcal{E}_{k_1}(m_1)$ and a binary string w. The two ciphertexts are only used for the purpose of recovering the partial information about the keys $k_0[0], k_0[1], k_1[0], k_1[1]$ leaked by $\mathcal{E}$. So, we assume $\mathcal{A}$ is given $k_0[0], k_0[1]$ and $k_1[0], k_1[1]$ to start with. Let $w = w_n \ldots w_1$ be any bitstring of length n, and define the sequence of keys $k^i = (k^i[0], k^i[1], k^i[2])$ by induction as

$$k^0 = k_0, \qquad k^{i+1} = \mathcal{G}_{w_{i+1}}(k^i)$$

for $i = 0, \ldots, n-1$. Notice that, if k_0 and k_1 are symbolically related by $k_1 = \mathbb{G}_w(k_0)$, then the last key in this sequence equals $k^n = k_1$ as a string in $\{0,1\}^{3\ell}$.

Using (12), the first block of these keys can be expressed symbolically as

$$k^i[0] = [\![\mathbb{G}_{u_i}(k_0[0])]\!] \qquad \text{where} \qquad u_i = 0w_i0w_{i-1}\ldots 0w_1.$$

So, Algorithm $\mathcal{A}(k_0[0], k_0[1], k_1[0], k_1[1], w)$ begins by computing the value of all $k^i[0] = [\![\mathbb{G}_{u_i}(k_0[0])]\!]$ (for $i = 0, \ldots, n$) starting from the input value $k_0[0]$ and applying the pseudorandom generator $\mathcal{G}'$ as directed by u_i. At this point, $\mathcal{A}$ may compare $k^n[0]$ with its input $k_1[0]$, and expect these two values to be equal. If the values differ, $\mathcal{A}$ immediately terminates with output $\perp$. We will prove later on that if $k_1 \neq \mathbb{G}_w(k_0)$, then $k^n[0] \neq k_1[0]$ with high probability, and $\mathcal{A}$ correctly outputs $\perp$. But for now, let us assume that $k_1 = \mathbb{G}_w(k_0)$, so that $k_1 = [\![\mathbb{G}_w(k_0)]\!] = k^n$ and the condition $k^n[0] = k_1[0]$ is satisfied. In this case, $\mathcal{A}$ needs to recover and output the key k_0. Since algorithm $\mathcal{A}$ is already given $k_0[0]$ and $k_0[1]$ as part of its input, all we need to do is to recover the last block $k_0[2]$ of the key. To this end, $\mathcal{A}$ first uses (13) to compute $k^{n-1}[2]$ as

$$\begin{aligned} k_1[1] \oplus \mathcal{G}_0(\mathcal{G}_1(\mathcal{G}_{w_n}(k^{n-1}[0]))) &= k^n[1] \oplus [\![\mathbb{G}_{01w_n}(k^{n-1}[0])]\!] \\ &= k^n[1] \oplus (\mathcal{G}_{w_n}(k^{n-1})[1] \oplus k^{n-1}[2]) \\ &= k^n[1] \oplus (k^n[1] \oplus k^{n-1}[2]) = k^{n-1}[2]. \end{aligned}$$

Similarly, starting from $k^{n-1}[2]$, $\mathcal{A}$ uses (14) to compute $k^i[2]$ for $i = n-2, n-3, \ldots, 0$ as

$$\begin{aligned} k^{i+1}[2] \oplus \mathcal{G}_1(\mathcal{G}_1(\mathcal{G}_{w_{i+1}}(k^i[0]))) &= k^{i+1}[2] \oplus [\![\mathbb{G}_{11w_{i+1}}(k^i[0])]\!] \\ &= k^{i+1}[2] \oplus (\mathcal{G}_{w_{i+1}}(k^i)[2] \oplus k^i[2]) \\ &= k^{i+1}[2] \oplus (k^{i+1}[2] \oplus k^i[2]) = k^i[2]. \end{aligned}$$

At this point, $\mathcal{A}$ can output $(k_0[0], k_0[1], k^0[2]) = (k_0[0], k_0[1], k_0[2]) = k_2$. This completes the analysis for the case $k_1 = \mathbb{G}_w(k_0)$.

We need to show that if $k_1 \neq \mathbb{G}_w(k_0)$, then the probability that $k^n[0] = k_1[0]$ is negligible, so that $\mathcal{A}$ correctly outputs $\perp$. Since we are interested only in the first blocks $k^n[0], k_1[0]$ of the keys, we introduce some notation. For any bitstring $v = v_1 \ldots v_m$, let $0|v = 0v_1 0v_2 \ldots 0v_m$ be the result of shuffling v with a string of zeros of equal length. If we express $k^n[0] = \mathbb{G}_{0|w}(k_0[0])$ in terms of $k_0[0]$, the goal becomes to prove that $\mathbb{G}_{0|w}(k_0[0])$ and $k_1[0]$ evaluate to different strings with overwhelming probability. The proof proceeds by cases, depending on whether $k_0 \perp k_1$, $k_0 \prec k_1$, or $k_1 \preceq k_0$, and makes use of the symbolic characterization of computational independence from Sect. 3.

Case 1. If $k_0 \perp k_1$, then $k_0 = \mathbb{G}_{v_0}(r_0)$ and $k_1 = \mathbb{G}_{v_1}(r_1)$ for some r_0, r_1, v_0, v_1 such that either $r_0 \neq r_1$, or v_0, v_1 are not one a suffix of the other. It follows that $k_0[0] = \mathbb{G}_{0|v_0}(r_0)$ and $k_1[0] = \mathbb{G}_{0|v_1}(r_1)$ are also symbolically independent because either $r_0 \neq r_1$, or $(0|v_0), (0|v_1)$ are not one a suffix of the other. In this case, also $k^n[0] = \mathbb{G}_{0|w}(k_0[0])$ and $k_1[0]$ are symbolically independent. It follows, from Theorem 1, that the distribution $[\![\mathbb{G}_{0|w}(k_0[0]), k_1[0]]\!]$ is computationally indistinguishable from the evaluation $[\![r_0, r_1]\!]$ of two independent uniformly random keys. In particular, since r_0 and r_1 evaluate to the same bitstring with exponentially small probability $2^{-\ell}$, the probability that $k^n[0] = \mathbb{G}_{0|w}(k_0[0])$ and $k_1[0]$ evaluate to the same string is also negligible.

Case 2. If $k_1 \prec k_0$, then $k_0 = \mathbb{G}_v(k_1)$ for some string $v \neq \epsilon$, and $k_0[0] = \mathbb{G}_{0|v}(k_1[0])$. Then, the pair of keys $(k^n[0], k_1[0])$ where

$$k^n[0] = \mathbb{G}_{0|w}(k_0[0]) = \mathbb{G}_{0|w}(\mathbb{G}_{0|v}(k_1[0])) = \mathbb{G}_{0|wv}(k_1[0])$$

is symbolically equivalent to $(\mathbb{G}_u(r), r)$ for some $u = (0|wv) \neq \epsilon$. So, by Theorem 1, we can equivalently bound the probability δ (over the random choice of σ) that $[\![\mathbb{G}_u(r)]\!]_\sigma$ evaluates to $[\![r]\!]_\sigma$. The trivial (identity) algorithm $\mathcal{I}(y) = y$ inverts the function defined by $\mathbb{G}_u$ with probability at least δ. Since $u \neq \epsilon$, $\mathbb{G}_u$ defines a one-way function, and δ must be negligible.

Case 3. Finally, if $k_0 \preceq k_1$, then $k_1 = \mathbb{G}_v(k_0)$ for some string $v \neq w$, and $k_1[0] = \mathbb{G}_{0|v}(k_0[0])$. This time, we are given a pair of keys

$$(k^n[0], k_1[0]) = (\mathbb{G}_{0|w}(k_0[0]), \mathbb{G}_{0|v}(k_0[0]))$$

which are symbolically equivalent to $(\mathbb{G}_{0|w}(r), \mathbb{G}_{0|v}(r))$. As before, by Theorem 1, it is enough to evaluate the probability δ that $\mathbb{G}_{0|w}(r)$ and $\mathbb{G}_{0|v}(r)$ evaluate to the same bitstring. If v is a (strict) suffix of w or w is a (strict) suffix of v, then δ must be negligible by the same argument used in Case 2. Finally, if v and w are not one a suffix of the other, then $\mathbb{G}_{0|w}(r)$ and $\mathbb{G}_{0|v}(r)$ are symbolically independent, and δ must be negligible by the same argument used in Case 1.

We have shown that in all three cases, the probability δ that $\mathbb{G}_{u_n}(k_0[0])$ and $k_1[0]$ evaluate to the same bitstring is negligible. So, the test performed by $\mathcal{A}$ fails (expect with negligible probability) and $\mathcal{A}$ outputs $\perp$ as required by the lemma. □

We use Lemma 4 to distinguish between expressions that have the same shape. Expressions with different shapes can be distinguished more easily simply by looking at their bitsize. Recall that for any (length regular) instantiation of the cryptographic primitives, the length of all strings in the computational interpretation of a pattern $[\![e]\!]$ (denoted $|[\![e]\!]|$) depends only on $\mathtt{shape}(e)$. In other words, for any two patterns e_0, e_1, if $\mathtt{shape}(e_0) = \mathtt{shape}(e_1)$, then $|[\![e_0]\!]| = |[\![e_1]\!]|$. The next lemma provides a converse of this property, showing that whenever two patterns have different shape, they may evaluate to strings of different length. So, secure computational interpretations are not guaranteed to protect any piece of partial information about the shape of symbolic expressions.

Lemma 5. *If pseudorandom generators and encryption schemes exist at all, then for any two expressions e_0 and e_1 with $\mathtt{shape}(e_0) \neq \mathtt{shape}(e_1)$, there exists a secure computational interpretation $(\mathcal{G}, \mathcal{E}, \gamma, \pi)$ such that $|[\![e_0]\!]| \neq |[\![e_1]\!]|$.*

Proof. We show how to modify any secure computational interpretation simply by padding the output length, so that the lemma is satisfied. More specifically, we provide a computational interpretation such that the length of $[\![e_0]\!]$ is different from the length of any expression with different shape. Let $S = \{\mathtt{shape}(e) \mid e \in \mathbf{parts}(e_0)\}$ be the set of all shapes of subexpressions of e_0, and let $n = |S| + 1$. Associate to each shape $s \in S$ a unique number $\varphi(s) \in \{1, \ldots, n-1\}$, and define $\varphi(s) = 0$ for all shapes $s \notin S$. Data blocks and keys are padded to bit-strings of length congruent to $\varphi(\square)$ and $\varphi(\circ)$ modulo n, respectively. The encryption function first applies an arbitrary encryption scheme, and then pads the ciphertext $\mathcal{E}(m)$ so that its length modulo n equals $\varphi(\{\!|s|\!\})$, for some shape s such that $|m| = \varphi(s)$. The pairing function π is defined similarly: if the two strings being combined in a pair have length $|m_0| = \varphi(s_0) \pmod n$ and $|m_1| = \varphi(s_1) \pmod n$, then the string encoding the pair (m_0, m_1) is padded so that its length equals $\varphi(s_0, s_1)$ modulo n. It is easy to check that all patterns e are evaluated to strings of length $|[\![e]\!]| = \varphi(\mathtt{shape}(e)) \pmod n$. Since $\mathtt{shape}(e_0) \in S$ and $\mathtt{shape}(e_1) \notin S$, we get $|[\![e_0]\!]| \neq 0 \pmod n$ and $|[\![e_1]\!]| = 0 \pmod n$. In particular, $|[\![e_0]\!]| \neq |[\![e_1]\!]|$. □

We are now ready to prove our completeness theorem, and establish the optimality of our symbolic semantics.

Theorem 2. *For any two expressions e_0 and e_1, if $\mathtt{pattern}(e_0) \neq \mathtt{pattern}(e_1)$, then there exists a secure computational interpretation $(\mathcal{G}, \mathcal{E}, \gamma, \pi)$ and a polynomial time computable predicate $\mathcal{D}$ such that $\Pr\{\mathcal{D}([\![e_0]\!])\} \approx 0$ and $\Pr\{\mathcal{D}([\![e_1]\!])\} \approx 1$, i.e., the distributions $[\![e_0]\!]$ and $[\![e_1]\!]$ can be distinguished with negligible probability of error.*

Proof. We consider two cases, depending on the shapes of the expressions. If $\mathtt{shape}(e_0) \neq \mathtt{shape}(e_1)$, then let $[\![\cdot]\!]$ be the computational interpretation defined in Lemma 5. Given a sample α from one of the two distributions, the distinguisher $\mathcal{D}$ simply checks if $|\alpha| = |[\![\mathtt{shape}(e_1)]\!]|$. If they are equal, it accepts. Otherwise it rejects. It immediately follows from Lemma 5 that this distinguisher is always correct, accepting all samples α from $[\![e_1]\!]$, and rejecting all samples α from $[\![e_0]\!]$.

The more interesting case is when $\mathtt{shape}(e_0) = \mathtt{shape}(e_1)$. This time the difference between the two expressions is not in their shape, but in the value of the keys and data. This time we use the computational interpretation $[\![\cdot]\!]$ defined in Lemma 4, and show how to distinguish between samples from $[\![e_0]\!]$ and samples from $[\![e_1]\!]$, provided $\mathtt{pattern}(e_0) \neq \mathtt{pattern}(e_1)$.

Let $S_b^i = \mathbb{F}_{e_b}^i(\emptyset)$ be the sequence of sets of keys defined by e_b. We know that $\emptyset = S_b^0 \subseteq S_b^1 \subseteq S_b^2 \subseteq \cdots \subseteq S_b^n = \mathrm{fix}(\mathbb{F}_{e_b})$ for some integer n. Let $e_b^i = \mathbf{Pat}(e_b, S_b^i)$ be the sequence of patterns defined by the sets S_b^i. Since $\mathtt{pattern}(e_0) \neq \mathtt{pattern}(e_1)$, we have $e_0^n \not\cong e_1^n$. Let i the smallest index such that $e_0^i \not\cong e_1^i$. We will give a procedure that iteratively recovers all the keys in the sets $S_b^0, S_b^1, \ldots, S_b^{i-1}$, and then distinguishes between samples coming from the two distributions associated to e_0 and e_1.

The simplest case is when $i = 0$, i.e., $e_0^0 \not\cong e_1^0$. In this case $S_0^0 = \emptyset = S_1^0$, and we do not need to recover any keys. Since e_0 and e_1 have the same shape, $\mathcal{D}$ can unambiguously parse α as a concatenation of data blocks d, keys k and ciphertexts of type $\{\!|s|\!\}_k$, without knowing if α comes from $[\![e_0]\!]$ or $[\![e_1]\!]$. If the two patterns e_0^0, e_1^0 differ in one of the data blocks, then $\mathcal{D}$ can immediately tell if α comes from e_0^0 or e_1^0 by looking at the value of that piece of data. So, assume all data blocks are identical, and e_0^0 and e_1^0 differ only in the values of the keys. Consider the set P of all key positions in e_0^0 (or, equivalently, in e_1^0), and for every position $p \in P$, let k_b^p be the key in e_b^0 at position p. (Positions include both plain keys k_b^p and ciphertexts $\{\!|s_p|\!\}_{k_b^p}$.) For any two positions p, p', define the relation $r_b(p, p')$ between the keys k_b^p and $k_b^{p'}$ to be

$$r_b(p, p') = \begin{cases} +w \text{ if } k_b^{p'} = \mathbb{G}_w(k_b^p) \text{ for some } w \in \{0,1\}^+ \\ -w \text{ if } k_b^p = \mathbb{G}_w(k_b^{p'}) \text{ for some } w \in \{0,1\}^+ \\ 0 \quad \text{if } k_b^p = k_b^{p'} \\ \bot \quad \text{otherwise.} \end{cases}$$

Notice that if $r_0(p, p') = r_1(p, p')$ for all positions $p, p' \in P$, then the map $\mu(k_0^p) = k_1^p$ is a valid pseudo-renaming. Since $\mu(e_0^0) = e_1^0$, this would show that $e_0^0 \cong e_1^0$, a contradiction. So, there must be two positions p, p' such that $r_0(p, p') \neq r_1(p, p')$, i.e., the keys at positions p and p' in the two expressions e_0^0 and e_1^0 satisfy different relations. At this point we distinguish two cases:

- If two keys are identical ($r_b(p, p') = 0$) and the other two keys are unrelated ($r_{1-b}(p, p') = \bot$), then we can determine the value of b simply by checking if the corresponding keys recovered from the sample α are identical or not. Notice that even if the subexpression at position p (or p') is a ciphertext, the encryption scheme defined in Lemma 4 still allows to recover the first 2ℓ bits of the keys, and this is enough to tell if two keys are identical or independent with overwhelming probability.
- Otherwise, it must be the case that one of the two relations is $r_b(p, p') = \pm w$ for some string w. By possibly swapping p and p', and e_0 and e_1, we may assume that $r_0(p, p') = +w$ while $r_1(p, p') \neq +w$. In other words, $k_0^{p'} = \mathbb{G}_w(k_0^p)$, while $k_1^{p'} \neq \mathbb{G}_w(k_0^p)$. We may also assume that the subexpressions at position p and p' are ciphertexts. (If the subexpression at one of these positions is a key, we can simply use it to encrypt a fixed message m, and obtain a corresponding ciphertext.) Let α_0, α_0' be the ciphertexts extracted from α corresponding to positions p and p'. We invoke the algorithm $\mathcal{A}(\alpha_0, \alpha_0', w)$ from Lemma 4 and check if it outputs a key or the special failure symbol $\bot$. The distinguisher accepts if and only if $\mathcal{A}(\alpha_0, \alpha_0', w) = \bot$. By Lemma 4, if α was sampled from $[\![e_0]\!]$, then $\mathcal{A}(\alpha_0, \alpha_0', w)$ will recover the corresponding key with probability 1, and $\mathcal{D}$ rejects the sample α. On the other hand, if α was sampled from $[\![e_1]\!]$, then $\mathcal{A}(\alpha_0, \alpha_0', w) = \bot$ with overwhelming probability, and $\mathcal{D}$ accepts the sample α.

This completes the description of the decision procedure $\mathcal{D}$ when $i = 0$. When $i > 1$, we first use Lemma 4 to recover the keys in S_b^1. Then we use these keys to decrypt the corresponding subexpressions in α, and use Lemma 4 again to recover all the keys in S_b^2. We proceed in a similar fashion all the way up to S_b^{i-1}. Notice that since all the corresponding patterns $e_0^j \cong e_1^j$ (for $j \leq i$) are equivalent up to renaming, all the keys at similar positions p, p' satisfy the same relations $r_0(p, p') = r_1(p, p')$, and we can apply Lemma 4 identically, whether the sample α comes from $[\![e_0]\!]$ or $[\![e_1]\!]$. This allows to recover the keys in S_b^i, at which point we can parse (and decrypt) α to recover all the data blocks, keys and ciphertexts appearing in e_b^i. Finally, using the fact that $e_0^i \not\cong e_1^i$, we proceed as in the case $i = 0$ to determine the value of b. □

6 Computational Soundness

Computational soundness results for symbolic cryptography usually forbid encryption cycles, e.g., collections of ciphertexts where k_i is encrypted under k_{i+1} for $i = 1, \ldots, n-1$, and k_n is encrypted under k_1. Here we follow an alternative approach, put forward in [26], which defines the adversarial knowledge as the *greatest fixed point* of $\mathbb{F}_e$, i.e., the *largest* set S such that $\mathbb{F}_e(S) = S$. Interestingly, [26] shows that under this "co-inductive" definition of the set of known keys, soundness can be proved in the presence of encryption cycles, offering a tight connection between symbolic and computational semantics. At the same time, [26, Theorem 2] also shows that if e has no encryption cycles, then $\mathbb{F}_e$ has a unique fixed point, and therefore $\mathrm{fix}(\mathbb{F}_e) = \mathrm{FIX}(\mathbb{F}_e)$. So, computational soundness under the standard "least fixed point" semantics for acyclic expressions follows as a corollary. We remark that $\mathbb{F}_e$ may have a unique fixed point even if e contains encryption cycles. So, based on [26, Theorem 2], we generalize the definition of acyclic expressions to include all expressions e such that $\mathbb{F}_e$ has a unique fixed point $\mathrm{fix}(\mathbb{F}_e) = \mathrm{FIX}_e(\mathbb{F}_e)$.

In this section we extend the results of [26] to expressions with pseudorandom keys. But, before doing that, we explain the intuition behind the co-inductive (greatest fixed point) semantics. Informally, using the greatest fixed point corresponds to working by induction on the set of keys that are *hidden* from the adversary, starting from the empty set (i.e., assuming that no key is hidden a-priori), and showing that more and more keys are provably secure. Formulating this process in terms of the complementary set of *potentially known* keys, one starts from the set of all keys $K = \mathbf{keys}(e)$, and repeatedly applies $\mathbb{F}_e$ to it. By monotonicity of $\mathbb{F}_e$ the result is a sequence of smaller and smaller sets

$$K \supset \mathbb{F}_e(K) \supset \mathbb{F}_e^2(K) \supset \mathbb{F}_e^3(K) \supset \cdots$$

of potentially known keys, which converges to the greatest fixed point

$$\mathrm{FIX}(\mathbb{F}_e) = \bigcap_n \mathbb{F}_e^n(\mathbf{keys}(e)).$$

We emphasize $\mathrm{FIX}(\mathbb{F}_e)$ should be interpreted as the set of keys that are only *potentially* recoverable by an adversary. Depending on the details of the encryption scheme (e.g., if it provides some form of key dependent message security), an adversary may or may not be able to recover all the keys in $\mathrm{FIX}(\mathbb{F}_e)$. On the other hand, all keys in the complementary set $\mathbf{Keys}(e) \setminus \mathrm{FIX}(\mathbb{F}_e)$ are *provably secret*, for any encryption scheme providing the minimal security level of indistinguishability under chosen message attack. Using the greatest fixed point, one can define an alternative symbolic semantics for cryptographic expressions,

$$\mathtt{PATTERN}(e) = \mathtt{norm}(\mathtt{proj}(e, \mathrm{FIX}(\mathbb{F}_e))). \tag{15}$$

In general, $\mathrm{fix}(\mathbb{F}_e)$ can be a strict subset of $\mathrm{FIX}(\mathbb{F}_e)$, so (15) may be different from the patterns defined in Sect. 4. However, if e is acyclic, then $\mathrm{FIX}(\mathbb{F}_e) = \mathrm{fix}_e(\mathbb{F}_e)$, and therefore $\mathtt{PATTERN}(e) = \mathtt{pattern}(e)$.

Theorem 3. *For any secure computational interpretation* $(\mathcal{G}, \mathcal{E}, \gamma, \pi)$ *and any expression* e*, the distributions* $[\![e]\!]$ *and* $[\![\mathtt{PATTERN}(e)]\!]$ *are computationally indistinguishable. In particular, if* $\mathtt{PATTERN}(e_0) = \mathtt{PATTERN}(e_1)$*, then* $[\![e_0]\!] \approx [\![e_1]\!]$*, i.e., the equivalence relation induced by* PATTERN *is computationally sound.*

Corollary 2. *If* e_0, e_1 *are acyclic expressions, and* $\mathtt{pattern}(e_0) = \mathtt{pattern}(e_1)$*, then* $[\![e_0]\!]$ *and* $[\![e_1]\!]$ *are computationally indistinguishable.*

The proof of the soundness theorem is pretty standard, and similar to previous work, and can be found in the full version of the paper [27].

7 Conclusion

We presented a generalization of the computational soundness result of Abadi and Rogaway [3] (or, more precisely, its co-inductive variant put forward in [26]) to expressions that mix encryption with a pseudo-random generator. Differently from previous work in the area of multicast key distribution protocols [28–30], we considered unrestricted use of both cryptographic primitives, which raises new issues related partial information leakage that had so far been dealt with using ad-hoc methods. We showed that partial information can be adequately taken into account in a simple symbolic adversarial model where the attacker can fully recover a key from any two pieces of partial information. While, at first, this attack model may seem unrealistically strong, we proved, as our main result, a completeness theorem showing that the model is essentially optimal.

A slight extension of our results (to include the random permutation of ciphertexts) has recently been used in [24], which provides a computationally sound symbolic analysis of Yao's garbled circuit construction for secure two party computation. The work of [24] illustrates the usefulness of the methods developed in this paper to the analysis of moderately complex protocols, and also provides an implementation showing that our symbolic semantics can be evaluated extremely fast even on fairly large expressions, e.g., those describing garbled circuits with thousands of gates. Our results can be usefully generalized even further, to include richer collections of cryptographic primitives, e.g., different types of (private and public key) encryption, secret sharing schemes (as used in [4]), and more. Extensions to settings involving active attacks are also possible [19,32], but probably more challenging.

Acknowledgments. The author thanks the anonymous Eurocrypt 2019 referees for their useful comments.

References

1. Abadi, M., Gordon, A.: A calculus for cryptogaphic protocols: the spi calculus. Inf. Comput. **148**(1), 1–70 (1999). https://doi.org/10.1006/inco.1998.2740
2. Abadi, M., Jürjens, J.: Formal eavesdropping and its computational interpretation. In: Kobayashi, N., Pierce, B.C. (eds.) TACS 2001. LNCS, vol. 2215, pp. 82–94. Springer, Heidelberg (2001). https://doi.org/10.1007/3-540-45500-0_4

3. Abadi, M., Rogaway, P.: Reconciling two views of cryptography (the computational soundness of formal encryption). J. Cryptol. **15**(2), 103–127 (2002). https://doi.org/10.1007/s00145-007-0203-0
4. Abadi, M., Warinschi, B.: Security analysis of cryptographycally controlledaccess to XML documents. J. ACM **55**(2), 1–29 (2008). https://doi.org/10.1145/1346330.1346331. Prelim. version in PODS'05
5. Backes, M., Pfitzmann, B.: Symmetric encryption in a simulatable Dolev-Yao style cryptographic library. In: CSFW 2004, pp. 204–218. IEEE (2004). https://doi.org/10.1109/CSFW.2004.20
6. Backes, M., Pfitzmann, B., Waidner, M.: Symmetric authentication in a simulatable Dolev-Yao-style cryptographic library. Int. J. Inf. Secur. **4**(3), 135–154 (2005). https://doi.org/10.1007/s10207-004-0056-6
7. Bellare, M., Miner, S.K.: A forward-secure digital signature scheme. In: Wiener, M. (ed.) CRYPTO 1999. LNCS, vol. 1666, pp. 431–448. Springer, Heidelberg (1999). https://doi.org/10.1007/3-540-48405-1_28
8. Böhl, F., Cortier, V., Warinschi, B.: Deduction soundness: prove one, get five for free. In: CCS 2013, pp. 1261–1272. ACM (2013). https://doi.org/10.1145/2508859.2516711
9. Burrows, M., Abadi, M., Needham, R.M.: A logic of authentication. ACM Trans. Comput. Syst. **8**(1), 18–36 (1990). https://doi.org/10.1145/77648.77649
10. Canetti, R., Garay, J., Itkis, G., Micciancio, D., Naor, M., Pinkas, B.: Multicast security: a taxonomy and some efficient constructions. In: INFOCOM 1999, pp. 708–716 (1999). https://doi.org/10.1109/INFCOM.1999.751457
11. Comon-Lundh, H., Cortier, V.: Computational soundness of observational equivalence. In: CCS 2008, pp. 109–118. ACM (2008). https://doi.org/10.1145/1455770.1455786
12. Cortier, V., Warinschi, B.: A composable computational soundness notion. In: CCS 2011, pp. 63–74. ACM (2011). https://doi.org/10.1145/2046707.2046717
13. Dolev, D., Yao, A.: On the security of public key protocols. IEEE Trans. Inf. Theory **29**(2), 198–208 (1983). https://doi.org/10.1109/TIT.1983.1056650
14. Horvitz, O., Gligor, V.: Weak key authenticity and the computational completeness of formal encryption. In: Boneh, D. (ed.) CRYPTO 2003. LNCS, vol. 2729, pp. 530–547. Springer, Heidelberg (2003). https://doi.org/10.1007/978-3-540-45146-4_31
15. Goldreich, O.: Foundations of Cryptography, Volume I - Basic Tools. Cambridge Unievrsity Press, Cambridge (2001)
16. Goldreich, O.: Foundation of Cryptography, Volume II - Basic Applications. Cambridge Unievrsity Press, Cambridge (2004)
17. Goldreich, O., Goldwasser, S., Micali, S.: How to construct random functions. J. ACM **33**(4), 792–807 (1986). https://doi.org/10.1145/6490.6503
18. Goldwasser, S., Micali, S.: Probabilistic encryption. J. Comput. Syst. Si. **28**(2), 270–299 (1984). https://doi.org/10.1016/0022-0000(84)90070-9
19. Hajiabadi, M., Kapron, B.M.: Computational soundness of coinductive symbolic security under active attacks. In: Sahai, A. (ed.) TCC 2013. LNCS, vol. 7785, pp. 539–558. Springer, Heidelberg (2013). https://doi.org/10.1007/978-3-642-36594-2_30
20. Katz, J., Lindell, Y.: Introduction to Modern Cryptography (Cryptography and Network Security Series). Chapman & Hall/CRC, Boca Raton (2007)
21. Kemmerer, R.A., Meadows, C.A., Millen, J.K.: Three system for cryptographic protocol analysis. J. Cryptology **7**(2), 79–130 (1994). https://doi.org/10.1007/BF00197942

22. Laud, P.: Encryption cycles and two views of cryptography. In: NORDSEC 2002, pp. 85–100. No. 2002:31 in Karlstad University Studies (2002)
23. Laud, P., Corin, R.: Sound computational interpretation of formal encryption with composed keys. In: Lim, J.-I., Lee, D.-H. (eds.) ICISC 2003. LNCS, vol. 2971, pp. 55–66. Springer, Heidelberg (2004). https://doi.org/10.1007/978-3-540-24691-6_5
24. Li, B., Micciancio, D.: Symbolic security of garbled circuits. In: Computer Security Foundations Symposium - CSF 2018, pp. 147–161. IEEE (2018). https://doi.org/10.1109/CSF.2018.00018
25. Malkin, T., Micciancio, D., Miner, S.: Efficient generic forward-secure signatures with an unbounded number of time periods. In: Knudsen, L.R. (ed.) EUROCRYPT 2002. LNCS, vol. 2332, pp. 400–417. Springer, Heidelberg (2002). https://doi.org/10.1007/3-540-46035-7_27
26. Micciancio, D.: Computational soundness, co-induction, and encryption cycles. In: Gilbert, H. (ed.) EUROCRYPT 2010. LNCS, vol. 6110, pp. 362–380. Springer, Heidelberg (2010). https://doi.org/10.1007/978-3-642-13190-5_19
27. Micciancio, D.: Symbolic encryption with pseudorandom keys. Report 2009/249, IACR ePrint archive (2018). https://eprint.iacr.org/2009/249. version 201800223:160820
28. Micciancio, D., Panjwani, S.: Adaptive security of symbolic encryption. In: Kilian, J. (ed.) TCC 2005. LNCS, vol. 3378, pp. 169–187. Springer, Heidelberg (2005). https://doi.org/10.1007/978-3-540-30576-7_10
29. Micciancio, D., Panjwani, S.: Corrupting one vs. corrupting many: the case of broadcast and multicast encryption. In: Bugliesi, M., Preneel, B., Sassone, V., Wegener, I. (eds.) ICALP 2006. LNCS, vol. 4052, pp. 70–82. Springer, Heidelberg (2006). https://doi.org/10.1007/11787006_7
30. Micciancio, D., Panjwani, S.: Optimal communication complexity of generic multicast key distribution. IEEE/ACM Trans. Network. **16**(4), 803–813 (2008). https://doi.org/10.1109/TNET.2007.905593
31. Micciancio, D., Warinschi, B.: Completeness theorems for the Abadi-Rogaway logic of encrypted expressions. J. Comput. Secur. **12**(1), 99–129 (2004). https://doi.org/10.3233/JCS-2004-12105
32. Micciancio, D., Warinschi, B.: Soundness of formal encryption in the presence of active adversaries. In: Naor, M. (ed.) TCC 2004. LNCS, vol. 2951, pp. 133–151. Springer, Heidelberg (2004). https://doi.org/10.1007/978-3-540-24638-1_8
33. Millen, J.K., Clark, S.C., Freedman, S.B.: The Interrogator: protocol security analysis. IEEE Trans. Softw. Eng. **SE–13**(2), 274–288 (1987). https://doi.org/10.1109/TSE.1987.233151
34. Paulson, L.C.: The inductive approach to verifying cryptographic protocols. J. Comput. Secur. **6**(1–2), 85–128 (1998). https://doi.org/10.3233/JCS-1998-61-205
35. Yao, A.C.: Protocols for secure computations (extended abstract). In: FOCS 1982, pp. 160–164 (1982). https://doi.org/10.1109/SFCS.1982.38

Efficient Secure Computation

Covert Security with Public Verifiability: Faster, Leaner, and Simpler

Cheng Hong[1], Jonathan Katz[2], Vladimir Kolesnikov[3], Wen-jie Lu[4], and Xiao Wang[5,6](✉)

[1] Alibaba, Hangzhou, China
vince.hc@alibaba-inc.com
[2] University of Maryland, College Park, USA
jkatz@cs.umd.edu
[3] Georgia Tech, Atlanta, USA
kolesnikov@gatech.edu
[4] University of Tsukuba, Tsukuba, Japan
riku@mdl.cs.tsukuba.ac.jp
[5] MIT, Cambridge, USA
wangxiao@northwestern.edu
[6] BU, Boston, USA

Abstract. The notion of *covert security* for secure two-party computation serves as a compromise between the traditional semi-honest and malicious security definitions. Roughly, covert security ensures that cheating behavior is detected by the honest party with reasonable probability (say, 1/2). It provides more realistic guarantees than semi-honest security with significantly less overhead than is required by malicious security.

The rationale for covert security is that it dissuades cheating by parties that care about their reputation and do not want to risk being caught. But a much stronger disincentive is obtained if the honest party can generate a publicly verifiable *certificate* when cheating is detected. While the corresponding notion of publicly verifiable covert (PVC) security has been explored, existing PVC protocols are complex and less efficient than the best covert protocols, and have impractically large certificates.

We propose a novel PVC protocol that significantly improves on prior work. Our protocol uses only "off-the-shelf" primitives (in particular, it avoids signed oblivious transfer) and, for deterrence factor 1/2, has only 20–40% overhead compared to state-of-the-art *semi-honest* protocols. Our protocol also has, for the first time, *constant-size* certificates of cheating (e.g., 354 bytes long at the 128-bit security level).

As our protocol offers strong security guarantees with low overhead,

J. Katz—Work supported in part by a grant from Alibaba.
V. Kolesnikov—Work supported in part by Sandia National Laboratories, a multi-mission laboratory managed and operated by National Technology and Engineering Solutions of Sandia, LLC., a wholly owned subsidiary of Honeywell International, Inc., for the U.S. Department of Energys National Nuclear Security Administration under contract DE-NA-0003525.

Y. Ishai and V. Rijmen (Eds.): EUROCRYPT 2019, LNCS 11478, pp. 97–121, 2019.
https://doi.org/10.1007/978-3-030-17659-4_4

we suggest that it is the best choice for many practical applications of secure two-party computation.

1 Introduction

Secure two-party computation allows two mutually distrusting parties $\mathsf{P_A}$ and $\mathsf{P_B}$ to evaluate a function of their inputs without requiring either party to reveal their input to the other. Traditionally, two security notions have been considered [7]. Protocols with *semi-honest security* can be quite efficient, but only protect against passive attackers who do not deviate from the prescribed protocol. *Malicious security*, in contrast, categorically prevents an attacker from gaining any advantage by deviating from the protocol; unfortunately, despite many advances over the past few years, protocols achieving malicious security are still noticeably less efficient than protocols with semi-honest security.

The notion of *covert security* [3] was proposed as a compromise between semi-honest and malicious security. Roughly, covert security ensures that while a cheating attacker may be successful with some small probability, the attempted cheating will fail *and be detected by the other party* with the remaining probability. The rationale for covert security is that it dissuades cheating by parties that care about their reputation and do not want to risk being caught. Covert security thus provides stronger guarantees than semi-honest security; it can also be achieved with better efficiency than malicious security [3,6,9,17].

Nevertheless, the guarantee of covert security is not fully satisfactory. Covert security only ensures that when cheating is unsuccessful, the honest party detects the fact that cheating took place—but it provides no mechanism for the honest party to *prove* this fact to anyone else (e.g., a judge or the public) and, indeed, existing covert protocols do not provide any such mechanism. Thus, a cheating attacker only risks harming its reputation with one other party; even if the honest party publicly announces that it caught the other party cheating, the cheating party can simply counter that it is being falsely accused.

Motivated by this limitation of covert security, Asharov and Orlandi [2] proposed the stronger notion of *publicly verifiable* covert (PVC) security. As in the covert model, any attempted cheating is detected with some probability; now, however, when cheating is detected the honest party can generate a *publicly verifiable certificate* of that fact. This small change would have a significant impact in practice, as a cheating attacker now risks having its reputation publicly and permanently damaged if it is caught. Alternatively (or additionally), the cheating party can be brought to court and fined for its misbehavior; the parties may even sign a contract in advance that describes the penalties to be paid if either party is caught. Going further, the parties could execute a "smart contract" in advance of the protocol execution that would automatically pay out if a valid certificate of cheating is posted on a blockchain. All these consequences are infeasible in the original covert model and, overall, the PVC model seems to come closer to the original goal of covert security.

Asharov and Orlandi [2] mainly focus on feasibility; although their protocol is implementable, it is not competitive with state-of-the-art semi-honest protocols since, in particular, it requires a stronger variant of oblivious transfer (OT) called *signed OT* and thus is not directly compatible with OT extension. Subsequent work by Kolesnikov and Malozemoff [13] shows various efficiency improvements to the Asharov-Orlandi protocol, with the primary gain resulting from a new, dedicated protocol for signed-OT extension. (Importantly, signed-OT extension does not follow generically from standard OT extension, and so cannot take advantage of the most-efficient recent constructions of the latter.)

Unfortunately, existing PVC protocols [2,13] seem not to have attracted much attention; for example, to the best of our knowledge, they have never been implemented. We suggest this is due to a number of considerations:

- **High overhead.** State-of-the-art PVC protocols still incur a significant overhead compared to known semi-honest protocols, and even existing covert protocols. (See Sect. 6.)
- **Large certificates.** Existing PVC protocols have certificates of size at least $\kappa \cdot |\mathcal{C}|$ bits, where κ is the (computational) security parameter and $|\mathcal{C}|$ is the circuit size.[1] Certificates this large are prohibitively expensive to propagate and are incompatible with some of the applications mentioned above (e.g., posting a certificate on a blockchain).
- **Complexity.** Existing PVC protocols rely on signed OT, a non-standard primitive that is less efficient than standard OT, is not available in existing secure-computation libraries, and is somewhat complicated to realize (especially for signed-OT extension).

1.1 Our Contributions

In this work we put forward a new PVC protocol in the random oracle model that addresses the issues mentioned above. Specifically:

- **Low overhead.** We improve on the efficiency of prior work by roughly a factor of 3× for deterrence factor 1/2, and even more for larger deterrence. (An exact comparison depends on a number of factors; we refer to Sect. 6 for a detailed discussion.) Strikingly, our PVC protocol (with deterrence factor 1/2) incurs *only 20–40% overhead compared to state-of-the-art* **semi-honest** *protocols based on garbled circuits.*
- **Small certificates.** We achieve, for the first time, *constant-size* certificates (i.e., independent of the circuit size or the lengths of the parties' inputs). Concretely, our certificates are small: at the 128-bit security level, they are only 354 bytes long.
- **Simplicity.** Our protocol avoids entirely the need for signed OT, and relies only on standard building blocks such as (standard) OT and circuit garbling.

[1] We observe that the certificate size in [13] can be improved to $O(\kappa \cdot n)$ bits (where n is the parties' input lengths) by carefully applying ideas from the literature. In many cases, this is still unacceptably large.

We also dispense with the XOR-tree technique for preventing selective-failure attacks; this allows us to avoid increasing the number of effective OT inputs. This reduction in complexity allowed us to produce a simple and efficient (and, to our knowledge, the first) implementation of a PVC protocol.

Overview of the paper. In Sect. 2 we provide an overview of prior PVC protocols and explain the intuition behind the construction of our protocol. After some background in Sect. 3, we present the description of our protocol in Sect. 4 and prove security in Sect. 5. Section 6 gives an experimental evaluation of our protocol and a comparison to prior work.

2 Technical Overview

We begin by providing an overview of the approach taken in prior work designing PVC protocols. Then we discuss the intuition behind our improved protocol.

2.1 Overview of Prior Work

At a high level, both previous works constructing PVC protocols [2,13] rely on the standard cut-and-choose paradigm [18] using a small number of garbled circuits, with some additional complications to achieve public verifiability. Both works rely crucially on a primitive called *signed OT*; this is a functionality similar to OT but where the receiver additionally learns the sender's signatures on all the values it obtains. Roughly, prior protocols proceed as follows:

1. Let λ be a parameter that determines the deterrence factor (i.e., the probability of detecting misbehavior). $\mathsf{P_A}$ picks random seeds $\{\mathsf{seed}_j\}_{j=1}^{\lambda}$ and $\mathsf{P_B}$ chooses a random index $\hat{\jmath} \in \{1, \ldots, \lambda\}$ that will serve as the "evaluation index" while the $j \neq \hat{\jmath}$ will be "check indices." The parties run signed OT using these inputs, which allows $\mathsf{P_B}$ to learn $\{\mathsf{seed}_j\}_{j \neq \hat{\jmath}}$ along with signatures of $\mathsf{P_A}$ on all those values.
2. $\mathsf{P_A}$ generates λ garbled circuits, and then sends signed commitments to those garbled circuits (along with the input-wire labels corresponding to $\mathsf{P_A}$'s input wires). Importantly, seed_j is used to derive the (pseudo)randomness for the jth garbling as well as the jth commitment.
   The parties also use signed OT so that $\mathsf{P_B}$ can obtain the input-wire labels for its inputs across all the circuits.
3. For all $j \neq \hat{\jmath}$, party $\mathsf{P_B}$ checks that the commitment to the jth garbled circuit is computed correctly based on seed_j and that the input-wire labels it received are correct; if this is not the case, then $\mathsf{P_B}$ can generate a certificate of cheating that consists of the inconsistent values plus their signatures.
4. Assuming no cheating was detected, $\mathsf{P_B}$ reveals $\hat{\jmath}$ to $\mathsf{P_A}$, who then sends the $\hat{\jmath}$th garbled circuit and the input-wire labels corresponding to its own inputs for that circuit. $\mathsf{P_B}$ can then evaluate the garbled circuit as usual.

Informally, we refer to the jth garbled circuit and commitment as the *jth instance of the protocol.* If $\mathsf{P_A}$ cheats in the jth instance of the protocol, then it is caught with probability at least $1 - \frac{1}{\lambda}$ (i.e., if j is a check index). Moreover, if $\mathsf{P_A}$ is caught, then $\mathsf{P_B}$ has a signed seed (which defines what $\mathsf{P_A}$ was supposed to do in the jth instance) and also a signed commitment to an incorrect garbled circuit or incorrect input-wire labels. These values allow $\mathsf{P_B}$ to generate a publicly verifiable certificate that $\mathsf{P_A}$ cheated.

As described, the protocol still allows $\mathsf{P_A}$ to carry out a selective-failure attack when transferring garbled labels for $\mathsf{P_B}$'s input wires. Specifically, it may happen that a malicious $\mathsf{P_A}$ corrupts a single input-wire label (used as input to the OT protocol) for the $\hat{j}$th garbled circuit—say, the label corresponding to a '1' input on some wire. If $\mathsf{P_B}$ aborts, then $\mathsf{P_A}$ learns that $\mathsf{P_B}$'s input on that wire was equal to 1. Such selective-failure attacks can be prevented using the XOR-tree approach [18].[2] This approach introduces significant overhead because it increases the number of effective inputs, which in turn requires additional signed OTs. The analysis in prior work [2,3,13] shows that to achieve deterrence factor (i.e., probability of being caught cheating) 1/2, a replication factor of $\lambda = 3$ with $\nu = 3$ is needed. More generally, the deterrence factor as a function of λ and the XOR-tree expansion factor ν is $(1 - \frac{1}{\lambda}) \cdot (1 - 2^{-\nu+1})$.

Practical performance. Several aspects of the above protocol are relatively inefficient. First, the dependence of the deterrence factor on the replication factor λ is not optimal due to the XOR tree, e.g., to achieve deterrence factor 1/2 at least $\lambda = 3$ garbled circuits are needed (unless ν is impractically large); the issue becomes even more significant when a larger deterrence factor is desired. In addition, the XOR-tree approach used in prior work increases the effective input length by at least a factor of 3, which necessitates $3\times$ more signed OTs; recall these are relatively expensive since signed-OT extension is. Finally, prior protocols have large certificates. This seems inherent in the more efficient protocol of [13] due to the way they do signed-OT extension. (Avoiding signed-OT extension would result in a much less efficient protocol overall.)

2.2 Our Solution

The reliance of prior protocols on signed OT and their approach to preventing selective-failure attacks affect both their efficiency as well as the size of their certificates. We address both these issues in the protocol we design.

As in prior work, we use the cut-and-choose approach and have $\mathsf{P_B}$ evaluate one garbled circuit while checking the rest, and we realize this by having $\mathsf{P_A}$ choose seeds for each of λ executions and then allowing $\mathsf{P_B}$ to obliviously learn all-but-one of those seeds. One key difference in our protocol is that we utilize the seeds chosen by $\mathsf{P_A}$ not only to "derandomize" the garbled-circuit generation and commitments, but *also to derandomize the entire remainder of $\mathsf{P_A}$'s execution,*

[2] For reasonable values of the parameters, the XOR-tree approach will be more efficient than a coding-theoretic approach [18].

and in particular its execution of the OT protocol used to transfer $\mathsf{P_B}$'s input-wire labels to $\mathsf{P_B}$. This means that after $\mathsf{P_B}$ obliviously learns all-but-one of the seeds of $\mathsf{P_A}$, the rest of $\mathsf{P_A}$'s execution is entirely deterministic; thus, $\mathsf{P_B}$ can verify correct execution of $\mathsf{P_A}$ during the entire rest of the protocol for all-but-one of the seeds. Not only does this eliminate the need for signed OT for the input-wire labels, but it also defends against the selective-failure attack described earlier *without* the need to increase the effective input length at all.

As described, the above allows $\mathsf{P_B}$ to *detect* cheating by $\mathsf{P_A}$ but does not yet achieve public verifiability. For this, we additionally require $\mathsf{P_A}$ to sign its protocol messages; if $\mathsf{P_A}$ cheats, $\mathsf{P_B}$ can generate a certificate of cheating from the seed and the corresponding signed inconsistent transcript.

Thus far we have focused on the case where $\mathsf{P_A}$ is malicious. We must also consider the case of a malicious $\mathsf{P_B}$ attempting to frame an honest $\mathsf{P_A}$. We address this by also having $\mathsf{P_B}$ commit in advance to *its* randomness[3] for each of the λ protocol instances. The resulting commitments will be included in $\mathsf{P_A}$'s signature, and will ensure that a certificate will be rejected if it corresponds to an instance in which $\mathsf{P_B}$ deviated from the protocol.

Having $\mathsf{P_B}$ commit to its randomness also allows us to avoid the need for signed OT in the first step, when $\mathsf{P_B}$ learns all-but-one of $\mathsf{P_A}$'s seeds. This is because those seeds can be reconstructed from $\mathsf{P_B}$'s view of the protocol, i.e., from the transcript of the (standard) OT protocol used to transfer those seeds plus $\mathsf{P_B}$'s randomness. Having $\mathsf{P_A}$ sign the transcripts of those OT executions serves as publicly verifiable evidence of the seeds used by $\mathsf{P_A}$.

We refer to Sect. 4 for further intuition behind our protocol, as well as its formal specification.

3 Covert Security with Public Verifiability

Before defining the notion of PVC security, we review the (plain) covert model [3] it extends. We focus on the strongest formulation of covert security, namely the *strong explicit cheat* formulation. This notion is formalized via an ideal functionality that explicitly allows an adversary to specify an attempt at cheating; in that case, the ideal functionality allows the attacker to successfully cheat with probability $1-\epsilon$, but the attacker is caught by the other party with probability ϵ; see Fig. 1. (As in [2], we also allow an attacker to "blatantly cheat," which guarantees that it will be caught.) For simplicity, we adapt the functionality such that only $\mathsf{P_A}$ has this option (since this is what is achieved by our protocol). For conciseness, we refer to a protocol realizing this functionality (against malicious adversaries) as having *covert security with deterrence* ϵ.

The PVC model extends the above to consider a setting wherein, before execution of the protocol, $\mathsf{P_A}$ has generated keys (pk, sk) for a digital-signature scheme, with the public key pk known to $\mathsf{P_B}$. We *do not* require that (pk, sk) is

[3] As an optimization, we have $\mathsf{P_B}$ commit to seeds, just like $\mathsf{P_A}$, and then use those seeds to generate the (pseudo)randomness to use in each instance. (This optimization is critical for realizing constant-size certificates.).

Functionality $\mathcal{F}$

$\mathsf{P_A}$ sends $x \in \{0,1\}^{n_1} \cup \{\perp, \mathsf{blatantCheat}, \mathsf{cheat}\}$ and $\mathsf{P_B}$ sends $y \in \{0,1\}^{n_2}$.

1. If $x \in \{0,1\}^{n_1}$ then compute $f(x,y)$ and send it to $\mathsf{P_B}$.
2. If $x = \perp$ then send $\perp$ to both parties.
3. If $x = \mathsf{blatantCheat}$, then send $\mathsf{corrupted}$ to both parties.
4. If $x = \mathsf{cheat}$ then:
   - With probability ϵ, send $\mathsf{corrupted}$ to both parties.
   - With probability $1-\epsilon$, send $(\mathsf{undetected}, y)$ to $\mathsf{P_A}$. Then wait to receive $z \in \{0,1\}^{n_3}$ from $\mathsf{P_A}$, and send z to $\mathsf{P_B}$.

Fig. 1. Functionality $\mathcal{F}$ for covert security with deterrence ϵ for two-party computation of a function f.

honestly generated, or that $\mathsf{P_A}$ gives any proof of knowledge of the secret key sk corresponding to the public key pk. In addition, the protocol is augmented with two additional algorithms, Blame and Judge. The Blame algorithm is run by $\mathsf{P_B}$ when it outputs $\mathsf{corrupted}$. This algorithm takes as input $\mathsf{P_B}$'s view of the protocol execution thus far, and outputs a certificate cert which is then sent to $\mathsf{P_A}$. The Judge algorithm takes as input $\mathsf{P_A}$'s public key pk, (a description of) the circuit $\mathcal{C}$ being evaluated, and a certificate cert, and outputs 0 or 1.

A protocol Π along with algorithms $\mathsf{Blame}, \mathsf{Judge}$ is said to be *publicly verifiable covert with deterrence* ϵ for computing a circuit $\mathcal{C}$ if the following hold:

Covert security: The protocol Π has covert security with deterrence ϵ. (Since the protocol includes the step of possibly sending cert to $\mathsf{P_A}$ if $\mathsf{P_B}$ outputs $\mathsf{corrupted}$, this ensures that cert itself does not violate privacy of $\mathsf{P_B}$.)

Public verifiability: If the honest $\mathsf{P_B}$ outputs cert in an execution of the protocol, then we know $\mathsf{Judge}(pk, \mathcal{C}, \mathsf{cert}) = 1$, except with negligible probability.

Defamation freeness: If $\mathsf{P_A}$ is honest, then the probability that a malicious $\mathsf{P_B}$ generates a certificate cert for which $\mathsf{Judge}(pk, \mathcal{C}, \mathsf{cert}) = 1$ is negligible.[4]

As in prior work on the PVC model, we assume the Judge algorithm learns the circuit $\mathcal{C}$ through some "out-of-band" mechanism; in particular, we do not include $\mathcal{C}$ as part of the certificate. In some applications (such as the smart-contract example), it may indeed be the case that the party running the Judge algorithm is aware of the circuit being computed in advance. When this is not the case, a description of $\mathcal{C}$ must be included as part of the certificate. However, we stress that the description of a circuit may be much shorter than the full circuit; for example, specifying a circuit for computing the Hamming distance between two 10^6-bit vectors requires only a few lines of high-level code in modern secure-computation platforms even though the circuit itself may have millions of gates.

[4] Note that defamation freeness implies that the protocol is also non-halting detection accurate [3].

Alternately, there may be a small set of commonly used "reference circuits" that can be identified by ID number rather than by their complete wiring diagram.

4 Our PVC Protocol

4.1 Preliminaries

We let $[n] = \{1, \ldots, n\}$. We use κ for the (computational) security parameter, but for compactness in the protocol description we let κ be an implicit input to our algorithms. For a boolean string y, we let $y[i]$ denote the ith bit of y.

We let Com denote a commitment scheme. We assume for simplicity that it is non-interactive, but this restriction can easily be removed. The decommitment decom is simply the random coins used during commitment. H is a hash function with 2κ-bit output length.

We say a party "uses randomness derived from seed" to mean that the party uses a pseudorandom function (with seed as the key) in CTR mode to obtain sufficiently many pseudorandom values that it then uses as its random coins. If $m_1, m_2, \ldots$ is a transcript of an execution of a two-party protocol (where the parties alternate sending the messages), the *transcript hash* of the execution is defined to be $\mathcal{H} = (H(m_1), H(m_2), \ldots)$.

We let Π_{OT} be an OT protocol realizing a parallel version of the OT functionality, as in Fig. 2.

Functionality $\mathcal{F}_{\mathsf{OT}}$

Private inputs: $\mathsf{P_A}$ has input $\{(B_{i,0}, B_{i,1})\}_{i=1}^{n_2}$ and $\mathsf{P_B}$ has input $y \in \{0,1\}^{n_2}$.

1. Upon receiving $\{(B_{i,0}, B_{i,1})\}_{i=1}^{n_2}$ from $\mathsf{P_A}$ and y from $\mathsf{P_B}$, send $\{B_{i,y[i]}\}_{i=1}^{n_2}$ to $\mathsf{P_B}$.

Fig. 2. Functionality $\mathcal{F}_{\mathsf{OT}}$ for parallel oblivious transfer.

Garbling. Our protocol relies on a (circuit) *garbling scheme*. For our purposes, a garbling scheme is defined by algorithms (Gb, Eval) having the following syntax:

- Gb takes as input the security parameter 1^κ and a circuit C with $n = n_1 + n_2$ input wires and n_3 output wires. It outputs *input-wire labels* $\{X_{i,0}, X_{i,1}\}_{i=1}^{n}$, a *garbled circuit* GC, and *output-wire labels* $\{Z_{i,0}, Z_{i,1}\}_{i=1}^{n_3}$.
- Eval is a deterministic algorithm that takes as input a set of input-wire labels $\{X_i\}_{i=1}^{n}$ and a garbled circuit GC. It outputs a set of output-wire labels $\{Z_i\}_{i=1}^{n_3}$.

Correctness is defined as follows: For any circuit $\mathcal{C}$ as above and any input $w \in \{0,1\}^n$, consider the experiment in which we first run $(\{X_{i,0}, X_{i,1}\}_{i=1}^n, \mathsf{GC}, \{Z_{i,0}, Z_{i,1}\}_{i=1}^{n_3}) \leftarrow \mathsf{Gb}(1^\kappa, \mathcal{C})$ followed by $\{Z_i\} := \mathsf{Eval}(\{X_{i,w[i]}\}, \mathsf{GC})$. Then, except with negligible probability, it holds that $Z_i = Z_{i,y[i]}$ and $Z_i \neq Z_{i,1-y[i]}$ for all i, where $y = \mathcal{C}(w)$.

A garbling scheme can be used by (honest) parties $\mathsf{P_A}$ and $\mathsf{P_B}$ to compute $\mathcal{C}$ in the following way: first, $\mathsf{P_A}$ computes $(\{X_{i,0}, X_{i,1}\}_{i=1}^n, \mathsf{GC}, \{Z_{i,0}, Z_{i,1}\}_{i=1}^{n_3}) \leftarrow \mathsf{Gb}(1^\kappa, \mathcal{C})$ and sends $\mathsf{GC}, \{Z_{i,0}, Z_{i,1}\}_{i=1}^{n_3}$ to $\mathsf{P_B}$. Next, $\mathsf{P_B}$ learns the input-wire labels $\{X_{i,w[i]}\}$ corresponding to some input w. (In a secure-computation protocol, $\mathsf{P_A}$ would send $\mathsf{P_B}$ the input-wire labels corresponding to its own portion of the input, while the parties would use OT to enable $\mathsf{P_B}$ to learn the input-wire labels corresponding to $\mathsf{P_B}$'s portion of the input.) Then $\mathsf{P_B}$ computes $\{Z_i\} := \mathsf{Eval}(\{X_{i,w[i]}\}, \mathsf{GC})$. Finally, $\mathsf{P_B}$ sets $y[i]$, for all i, to be the (unique) bit for which $Z_i = Z_{i,y[i]}$; the output is y.

We assume the garbling scheme satisfies the standard security definition [10, 15]. That is, we assume there is a simulator $\mathcal{S}_{\mathsf{Gb}}$ such that for all $\mathcal{C}, w$, the distribution $\left\{\mathcal{S}_{\mathsf{Gb}}(1^\kappa, \mathcal{C}, \mathcal{C}(w))\right\}$ is computationally indistinguishable from

$$\left\{(\{X_{i,0}, X_{i,1}\}_{i=1}^n, \mathsf{GC}, \{Z_{i,0}, Z_{i,1}\}_{i=1}^{n_3}) \leftarrow \mathsf{Gb}(1^\kappa, \mathcal{C}) \;:\; (\{X_{i,w[i]}\}, \mathsf{GC}, \{Z_{i,0}, Z_{i,1}\}_{i=1}^{n_3})\right\}.$$

As this is the "minimal" security notion for garbling, it is satisfied by garbling schemes including all state-of-the-art optimizations [4, 14, 20].

4.2 Our Scheme

We give a high-level description of our protocol below; a formal definition of the protocol is provided in Fig. 3. The Blame algorithm is included as part of the protocol description (cf. Step 6) for simplicity. The Judge algorithm is specified in Fig. 5.

We use a signature scheme $(\mathsf{Gen}, \mathsf{Sign}, \mathsf{Vrfy})$. Before executing the protocol, $\mathsf{P_A}$ runs Gen to obtain public key pk and private key sk; we assume that $\mathsf{P_B}$ knows pk before running the protocol. As noted earlier, if $\mathsf{P_A}$ is malicious then it may choose pk arbitrarily.

The main idea of the protocol is to run λ parallel instances of a "basic" garbled-circuit protocol that is secure against a semi-honest $\mathsf{P_A}$ and a malicious $\mathsf{P_B}$. Of these instances, $\lambda - 1$ will be checked by $\mathsf{P_B}$, while a random one (the $\hat{\jmath}$th) will be evaluated by $\mathsf{P_B}$ to learn its output. To give $\mathsf{P_B}$ the ability to verify honest behavior in the check instances, we make all the executions deterministic by having $\mathsf{P_A}$ use (pseudo)randomness derived from corresponding seeds $\{\mathsf{seed}_j^A\}_{j \in [\lambda]}$. That is, $\mathsf{P_A}$ will uniformly sample each seed seed_j^A and use it to generate (pseudo)randomness for its jth instance. Then $\mathsf{P_A}$ and $\mathsf{P_B}$ run an OT protocol Π_{OT} (with malicious security) that allows $\mathsf{P_B}$ to learn $\lambda - 1$ of those seeds. Since $\mathsf{P_A}$'s behavior in those $\lambda - 1$ instances is completely determined by $\mathsf{P_B}$'s messages and those seeds, it is possible for $\mathsf{P_B}$ to check $\mathsf{P_A}$'s behavior in those instances.

Protocol Π_{pvc}

Private inputs: $\mathsf{P_A}$ has input $x \in \{0,1\}^{n_1}$ and keys (pk, sk) for a signature scheme. $\mathsf{P_B}$ has input $y \in \{0,1\}^{n_2}$ and knows pk.
Public inputs: Both parties also agree on a circuit $\mathcal{C}$ and parameters κ, λ.
Protocol:

1. $\mathsf{P_B}$ chooses uniform κ-bit strings $\{\mathsf{seed}_j^B\}_{j\in[\lambda]}$, sets $h_j \leftarrow \mathsf{Com}(\mathsf{seed}_j^B)$ for all j, and sends $\{h_j\}_{j\in[\lambda]}$ to $\mathsf{P_A}$.
2. $\mathsf{P_A}$ chooses uniform κ-bit strings $\{\mathsf{seed}_j^A, \mathsf{witness}_j\}_{j\in[\lambda]}$, while $\mathsf{P_B}$ chooses uniform $\hat{\jmath} \in [\lambda]$ and sets $b_{\hat{\jmath}} := 1$ and $b_j := 0$ for $j \neq \hat{\jmath}$.
   $\mathsf{P_A}$ and $\mathsf{P_B}$ run λ executions of Π_{OT}, where in the jth execution $\mathsf{P_A}$ uses $(\mathsf{seed}_j^A, \mathsf{witness}_j)$ as input, and $\mathsf{P_B}$ uses b_j as input and randomness derived from seed_j^B. Upon completion, $\mathsf{P_B}$ obtains $\{\mathsf{seed}_j^A\}_{j\neq\hat{\jmath}}$ and $\mathsf{witness}_{\hat{\jmath}}$. Let trans_j be the transcript of the jth execution of Π_{OT}.
3. For each $j \in [\lambda]$, $\mathsf{P_A}$ garbles $\mathcal{C}$ using randomness derived from seed_j^A. Denote the jth garbled circuit by GC_j, the input-wire labels of $\mathsf{P_A}$ by $\{A_{j,i,b}\}_{i\in[n_1],b\in\{0,1\}}$, the input-wire labels of $\mathsf{P_B}$ by $\{B_{j,i,b}\}_{i\in[n_2],b\in\{0,1\}}$, and the output-wire labels by $\{Z_{j,i,b}\}_{i\in[n_3],b\in\{0,1\}}$.
   $\mathsf{P_A}$ and $\mathsf{P_B}$ then run λ executions of Π_{OT}, where in the jth execution $\mathsf{P_A}$ uses $\{(B_{j,i,0}, B_{j,i,1})\}_{i=1}^{n_2}$ as input, and $\mathsf{P_B}$ uses y as input if $j = \hat{\jmath}$ and 0^{n_2} otherwise. The parties use seed_j^A and seed_j^B, respectively, to derive all their randomness in the jth execution. In this way, $\mathsf{P_B}$ obtains $\{B_{\hat{\jmath},i,y[i]}\}_{i\in[n_2]}$. We let $\mathcal{H}_j$ denote the transcript hash for the jth execution of Π_{OT}.
4. $\mathsf{P_A}$ computes commitments $h_{j,i,b}^A \leftarrow \mathsf{Com}(A_{j,i,b})$ for all j, i, b, and then computes the commitments $\mathsf{c}_j \leftarrow \mathsf{Com}\ \mathsf{GC}_j, \{h_{j,i,b}^A\}_{i\in[n_1],b\in\{0,1\}}, \{Z_{j,i,b}\}_{i\in[n_3],b\in\{0,1\}}$ for all j, where each pair $(h_{j,i,0}^A, h_{j,i,1}^A)$ is randomly permuted. All randomness in the jth instance is derived from seed_j^A. Finally, $\mathsf{P_A}$ sends $\{\mathsf{c}_j\}_{j\in[\lambda]}$ to $\mathsf{P_B}$.
5. For each $j \in [\lambda]$, $\mathsf{P_A}$ computes $\sigma_j \leftarrow \mathsf{Sign}_{sk}(\mathcal{C}, j, h_j, \mathsf{trans}_j, \mathcal{H}_j, \mathsf{c}_j)$ and sends σ_j to $\mathsf{P_B}$. Then $\mathsf{P_B}$ checks that σ_j is a valid signature for all j, and aborts with output $\perp$ if not.
6. For each $j \neq \hat{\jmath}$, $\mathsf{P_B}$ uses seed_j^A and the messages it sent to simulate $\mathsf{P_A}$'s computation in steps 3 and 4, and in particular computes $\hat{\mathcal{H}}_j, \hat{\mathsf{c}}_j$. It then checks that $(\hat{\mathcal{H}}_j, \hat{\mathsf{c}}_j) = (\mathcal{H}_j, \mathsf{c}_j)$. If the check fails for some $j \neq \hat{\jmath}$, then $\mathsf{P_B}$ chooses a uniform such j, outputs corrupted, sends $\mathsf{cert} := (j, \mathsf{trans}_j, \mathcal{H}_j, \mathsf{c}_j, \sigma_j, \mathsf{seed}_j^B, \mathsf{decom}_j^B)$ to $\mathsf{P_A}$, and halts.

Fig. 3. Full description of our PVC protocol (part I).

The above idea allows $\mathsf{P_B}$ to catch a cheating $\mathsf{P_A}$, but not to generate a publicly verifiable certificate that $\mathsf{P_A}$ has cheated. To add this feature, we have $\mathsf{P_A}$ sign the transcripts of each instance, including the transcript of the execution of the OT protocol by which $\mathsf{P_B}$ learned the corresponding seed. If $\mathsf{P_A}$ cheats in, say, the jth instance ($j \neq \hat{\jmath}$) and is caught, then $\mathsf{P_B}$ can output a certificate that includes $\mathsf{P_B}$'s view (including its randomness) in the execution of the jth OT protocol (from which seed_j^A can be recomputed) and the transcript of the

Protocol Π_{pvc}

7. $\mathsf{P_B}$ sends $(\hat{\jmath}, \{\mathsf{seed}_j^A\}_{j \neq \hat{\jmath}}, \mathsf{witness}_{\hat{\jmath}})$ to $\mathsf{P_A}$, who checks that $\{\mathsf{seed}_j^A\}_{j \neq \hat{\jmath}}, \mathsf{witness}_{\hat{\jmath}}$ are all correct and aborts if not.
8. $\mathsf{P_A}$ sends $\mathsf{GC}_{\hat{\jmath}}$, $\{A_{\hat{\jmath},i,x[i]}\}_{i \in [n_1]}$, $\{h^A_{\hat{\jmath},i,b}\}_{i \in [n_1], b \in \{0,1\}}$ (in the same permuted order as before), and $\{Z_{\hat{\jmath},i,b}\}_{i \in [n_3], b \in \{0,1\}}$ to $\mathsf{P_B}$, along with decommitments $\mathsf{decom}_{\hat{\jmath}}$ and $\{\mathsf{decom}^A_{\hat{\jmath},i,x[i]}\}$. If $\mathsf{Com}(\mathsf{GC}_{\hat{\jmath}}, \{h^A_{\hat{\jmath},i,b}\}, \{Z_{\hat{\jmath},i,b}\}; \mathsf{decom}_{\hat{\jmath}}) \neq \mathsf{c}_{\hat{\jmath}}$ or $\mathsf{Com}(A_{\hat{\jmath},i,x[i]}; \mathsf{decom}^A_{\hat{\jmath},i,x[i]}) \notin \{h^A_{\hat{\jmath},i,b}\}_{b \in \{0,1\}}$ for some i, then $\mathsf{P_B}$ aborts with output $\perp$.
   Otherwise, $\mathsf{P_B}$ evaluates $\mathsf{GC}_{\hat{\jmath}}$ using $\{A_{\hat{\jmath},i,x[i]}\}_{i \in [n_1]}$ and $\{B_{\hat{\jmath},i,y[i]}\}_{i \in [n_2]}$ to obtain output-wire labels $\{Z_i\}_{i \in [n_3]}$. For each $i \in [n_3]$, if $Z_i = Z_{\hat{\jmath},i,0}$, set $z[i] := 0$; if $Z_i = Z_{\hat{\jmath},i,1}$, set $z[i] := 1$. (If $Z_i \notin \{Z_{\hat{\jmath},i,0}, Z_{\hat{\jmath},i,1}\}$ for some i, then abort with output $\perp$.) Output z.

Fig. 4. Full description of our PVC protocol (part II).

$\jmath$th instance, along with $\mathsf{P_A}$'s signature on the transcripts. Note that, given the randomness of both $\mathsf{P_A}$ and $\mathsf{P_B}$, the entire transcript of the instance can be recomputed and anyone can then check whether it is consistent with seed_j^A. We remark that nothing about $\mathsf{P_B}$'s inputs is revealed by a certificate since $\mathsf{P_B}$ uses a dummy input in all the check instances.

There still remains the potential issue of *defamation*. Indeed, an honest $\mathsf{P_A}$'s messages might be deemed inconsistent if $\mathsf{P_B}$ includes in the certificate fake messages different from those sent by $\mathsf{P_B}$ in the real execution. We prevent this by having $\mathsf{P_B}$ commit to its randomness for each instance at the beginning of the protocol, and having $\mathsf{P_A}$ sign those commitments. Consistency of $\mathsf{P_B}$'s randomness and the given transcript can then be checked as part of verification of the certificate.

As described, the above would result in a certificate that is linear in the length of $\mathsf{P_B}$'s inputs, since there are that many OT executions (in each instance) for which $\mathsf{P_B}$ must generate randomness. We compress this to a constant-size certificate by having $\mathsf{P_B}$ also generate its (pseudo)randomness from a short seed.

The above description conveys the main ideas of the protocol, though various other modifications are needed for the proof of security. We refer the reader to Figs. 3 and 5 for the details.

4.3 Optimizations

Our main protocol is already quite efficient, but we briefly discuss some additional optimizations that can be applied.

Commitments in the random-oracle model. When standard garbling schemes are used, all the values committed during the course of the protocol have high entropy; thus, commitment to a string r can be done by simply computing

Algorithm Judge

Inputs: A public key pk, a circuit $\mathcal{C}$, and a certificate cert.

1. Parse cert as $(j, \mathsf{trans}_j, \mathcal{H}_j, \mathsf{c}_j, \sigma_j, \mathsf{seed}_j^B, \mathsf{decom}_j^B)$. Compute $h_j := \mathsf{Com}(\mathsf{seed}_j^B; \mathsf{decom}_j^B)$.
2. If $\mathsf{Vrfy}_{pk}((\mathcal{C}, j, h_j, \mathsf{trans}_j, \mathcal{H}_j, \mathsf{c}_j), \sigma_j) = 0$, output 0.
3. Simulate an execution of Π_{OT} by $\mathsf{P_B}$, where $\mathsf{P_B}$'s input is 0, its randomness is derived from seed_j^B, and $\mathsf{P_A}$'s messages are those included in trans_j. Check that all of $\mathsf{P_B}$'s messages generated in this simulation are consistent with trans_j; terminate with output 0 if not. Otherwise, let seed_j^A denote the output of $\mathsf{P_B}$ from the simulated execution of Π_{OT}.
4. Use seed_j^A and seed_j^B to simulate an honest execution of steps 3 and 4 of the protocol, and in particular compute $\hat{\mathcal{H}}_j, \hat{\mathsf{c}}_j$.
5. Do:
   (a) If $(\hat{\mathcal{H}}_j, \hat{\mathsf{c}}_j) = (\mathcal{H}_j, \mathsf{c}_j)$ then output 0.
   (b) If $\hat{\mathsf{c}}_j \neq \mathsf{c}_j$ then output 1.
   (c) Find the first message for which $\hat{\mathcal{H}}_j \neq \mathcal{H}_j$. If this corresponds to a message sent by $\mathsf{P_A}$, output 1; otherwise, output 0.

Fig. 5. The Judge algorithm.

$H(r)$ (if H is modeled as a random oracle) and decommitment requires only sending r.

Using correlated oblivious transfer. One optimization introduced by Asharov et al. [1] is using correlated OT for transferring $\mathsf{P_B}$'s input-wire labels when garbling is done using the free-XOR approach [14]. This optimization is compatible with our protocol in a straightforward manner.

Avoiding committing to the input-wire labels. In our protocol, we have $\mathsf{P_A}$ commit to its input-wire labels (along with the rest of the garbled circuit). This is done to prevent $\mathsf{P_A}$ from sending incorrect input-wire labels in the final step. We observe that this is unnecessary if the garbling scheme has the additional property that it is infeasible to generate a garbled circuit along with incorrect input-wire labels that result in a valid output when evaluated. (We omit a formal definition.) Many standard garbling schemes have this property.

5 Proof of Security

The remainder of this section is devoted to a proof of the following result:

Theorem 1. *Assume* Com *is computationally hiding/binding, H is collision-resistant, the garbling scheme is secure, Π_{OT} realizes $\mathcal{F}_{\mathsf{OT}}$, and the signature scheme is existentially unforgeable under a chosen-message attack. Then protocol Π_{pvc} along with* Blame *as in step 6 and* Judge *as in Fig. 5 is publicly verifiable covert with deterrence $\epsilon = 1 - \frac{1}{\lambda}$.*

Proof. We separately prove covert security with ϵ-deterrence (handling the cases where either $\mathsf{P_A}$ or $\mathsf{P_B}$ is corrupted), public verifiability, and defamation freeness.

Covert Security—Malicious $\mathsf{P_A}$

Let $\mathcal{A}$ be an adversary corrupting $\mathsf{P_A}$. We construct the following simulator $\mathcal{S}$ that holds pk and runs $\mathcal{A}$ as a subroutine, while playing the role of $\mathsf{P_A}$ in the ideal world interacting with $\mathcal{F}$:

1. Choose uniform κ-bit strings $\{\mathsf{seed}_j^B\}_{j\in[\lambda]}$, set $h_j \leftarrow \mathsf{Com}(\mathsf{seed}_j^B)$ for all j, and send $\{h_j\}_{j\in[\lambda]}$ to $\mathcal{A}$.
2. For all $j \in [\lambda]$, run Π_{OT} with $\mathcal{A}$, using input 0 and randomness derived from seed_j^B. In this way, $\mathcal{S}$ obtains $\{\mathsf{seed}_j^A\}_{j\in[\lambda]}$. Let trans_j denote the transcript of the jth execution.
3. For $j \in [\lambda]$, run an execution of Π_{OT} with $\mathcal{A}$, using input 0^{n_2} and randomness derived from seed_j^B. Let $\mathcal{H}_j$ denote the transcript hash of the jth execution.
4. Receive $\{\mathsf{c}_j\}_{j\in[\lambda]}$ from $\mathcal{A}$.
5. Receive $\{\sigma_j\}$ from $\mathcal{A}$. If any of the signatures are invalid, send $\perp$ to $\mathcal{F}$ and halt.
6. For all $j \in [\lambda]$, use seed_j^A and the messages sent previously to simulate the computation of an honest $\mathsf{P_A}$ in steps 3 and 4, and in particular compute $\hat{\mathcal{H}}_j, \hat{\mathsf{c}}_j$. Let J be the set of indices for which $(\hat{\mathcal{H}}_j, \hat{\mathsf{c}}_j) \neq (\mathcal{H}_j, \mathsf{c}_j)$.
   There are now three cases:
   - If $|J| \geq 2$ then send $\mathsf{blatantCheat}$ to $\mathcal{F}$, send $\mathsf{cert} := (j, \mathsf{trans}_j, \mathcal{H}_j, \mathsf{c}_j, \sigma_j, \mathsf{seed}_j^B, \mathsf{decom}_j^B)$ to $\mathcal{A}$ (for uniform $j \in J$), and halt.
   - If $|J| = 1$ then send cheat to $\mathcal{F}$. If $\mathcal{F}$ returns $\mathsf{corrupted}$ then set $\mathsf{caught} := \mathsf{true}$; if $\mathcal{F}$ returns $(\mathsf{undetected}, y)$, set $\mathsf{caught} := \mathsf{false}$. In either case, continue below.
   - If $|J| = 0$ then set $\mathsf{caught} := \perp$ and continue below.

0′. Rewind $\mathcal{A}$ and run steps 1′–6′ below until[5] $|J'| = |J|$ and $\mathsf{caught}' = \mathsf{caught}$.

1′. Choose uniform $\hat{\jmath} \in [\lambda]$. For $j \neq \hat{\jmath}$, choose uniform κ-bit strings $\{\mathsf{seed}_j^B\}$ and set $h_j \leftarrow \mathsf{Com}(\mathsf{seed}_j^B)$. Set $h_{\hat{\jmath}} \leftarrow \mathsf{Com}(0^\kappa)$. Send $\{h_j\}_{j\in[\lambda]}$ to $\mathcal{A}$.

2′. For all $j \neq \hat{\jmath}$, run Π_{OT} with $\mathcal{A}$, using input 0 and randomness derived from seed_j^B. In this way, $\mathcal{S}$ obtains $\{\mathsf{seed}_j^A\}_{j\neq\hat{\jmath}}$. For the $\hat{\jmath}$th execution, use the simulator $\mathcal{S}_{\mathsf{OT}}$ for protocol Π_{OT}, thus extracting both $\mathsf{seed}_{\hat{\jmath}}^A$ and $\mathsf{witness}_{\hat{\jmath}}$. Let trans_j denote the transcript of the jth execution.

3′. For all $j \neq \hat{\jmath}$, run Π_{OT} with $\mathcal{A}$, using input 0^{n_2} and randomness derived from seed_j^B. For $j = \hat{\jmath}$, use the simulator $\mathcal{S}_{\mathsf{OT}}$ for protocol Π_{OT}, thus extracting $\{B_{\hat{\jmath},i,b}\}_{i\in[n_2],b\in\{0,1\}}$. Let $\mathcal{H}_j$ denote the transcript hash of the jth execution.

4′. Receive $\{\mathsf{c}_j\}_{j\in[\lambda]}$ from $\mathcal{A}$.

5′. Receive $\{\sigma_j\}$ from $\mathcal{A}$. If any of the signatures are invalid, then return to step 1′.

[5] We use standard techniques [8,16] to ensure that $\mathcal{S}$ runs in expected polynomial time; details are omitted for the sake of the exposition.

6′. For all $j \in [\lambda]$, use seed_j^A and the messages sent previously to simulate the computation of an honest $\mathsf{P_A}$ in steps 3′ and 4′, and in particular compute $\hat{\mathcal{H}}_j, \hat{\mathsf{c}}_j$. Let J' be the set of indices for which $(\hat{\mathcal{H}}_j, \hat{\mathsf{c}}_j) \neq (\mathcal{H}_j, \mathsf{c}_j)$. If $|J'| = 1$ and $\hat{\jmath} \notin J'$ then set $\mathsf{caught}' := \mathsf{true}$. If $|J'| = 1$ and $\hat{\jmath} \in J'$ then set $\mathsf{caught}' := \mathsf{false}$. If $|J'| = 0$ then set $\mathsf{caught}' := \perp$.

7. If $|J'| = 1$ and $\mathsf{caught}' = \mathsf{true}$, then send $\mathsf{cert} := (j, \mathsf{trans}_j, \mathcal{H}_j, \mathsf{c}_j, \sigma_j, \mathsf{seed}_j^B)$ to $\mathcal{A}$ (where j is the unique index in J') and halt.
   Otherwise, send $(\hat{\jmath}, \{\mathsf{seed}_j^A\}_{j \neq \hat{\jmath}}, \mathsf{witness}_{\hat{\jmath}})$ to $\mathcal{A}$.
8. Receive GC, $\{A_i\}_{i \in [n_1]}$, $\{h_{i,b}^A\}_{i \in [n_1], b \in \{0,1\}}$, $\{Z_{i,b}\}_{i \in [n_3], b \in \{0,1\}}$, and the corresponding decommitments from $\mathcal{A}$. If any of the decommitments are incorrect, send $\perp$ to $\mathcal{F}$ and halt.
   Otherwise, there are two possibilities:
   - If $|J'| = 1$ and $\mathsf{caught}' = \mathsf{false}$, then use $\{B_{\hat{\jmath},i,b}\}_{i \in [n_2], b \in \{0,1\}}$ and the value y received from $\mathcal{F}$ to compute an output z exactly as an honest $\mathsf{P_B}$ would. Send z to $\mathcal{F}$ and halt.
   - If $|J'| = 0$, then compute an effective input $x \in \{0,1\}^{n_1}$ using $\mathsf{seed}_{\hat{\jmath}}^A$ and the input-wire labels $\{A_i\}_{i \in [n_1]}$. Send x to $\mathcal{F}$ and halt.

We now show that the joint distribution of the view of $\mathcal{A}$ and the output of $\mathsf{P_B}$ in the ideal world is computationally indistinguishable from the joint distribution of the view of $\mathcal{A}$ and the output of $\mathsf{P_B}$ in a real protocol execution. We prove this by considering a sequence of experiments, where the output of each is defined to be the view of $\mathcal{A}$ and the output of $\mathsf{P_B}$, and showing that the output of each is computationally indistinguishable from the output of the next one.

Expt$_0$. This is the ideal-world execution between $\mathcal{S}$ (as described above) and the honest $\mathsf{P_B}$ holding some input y, both interacting with functionality $\mathcal{F}$.

By inlining the actions of $\mathcal{S}, \mathcal{F}$, and $\mathsf{P_B}$, we may rewrite the experiment as follows:

1. Choose uniform κ-bit strings $\{\mathsf{seed}_j^B\}_{j \in [\lambda]}$, set $h_j \leftarrow \mathsf{Com}(\mathsf{seed}_j^B)$ for all j, and send $\{h_j\}_{j \in [\lambda]}$ to $\mathcal{A}$.
2. For all $j \in [\lambda]$, run Π_{OT} with $\mathcal{A}$, using input 0 and randomness derived from seed_j^B. Obtain $\{\mathsf{seed}_j^A\}_{j \in [\lambda]}$ as the outputs. Let trans_j denote the transcript of the jth execution.
3. For $j \in [\lambda]$, run an execution of Π_{OT} with $\mathcal{A}$, using input 0^{n_2} and randomness derived from seed_j^B. Let $\mathcal{H}_j$ denote the transcript hash of the jth execution.
4. Receive $\{\mathsf{c}_j\}_{j \in [\lambda]}$ from $\mathcal{A}$.
5. Receive $\{\sigma_j\}$ from $\mathcal{A}$. If any of the signatures are invalid, then $\mathsf{P_B}$ outputs $\perp$ and the experiment halts.
6. For all $j \in [\lambda]$, use seed_j^A and the messages sent previously to $\mathcal{A}$ to simulate the computation of an honest $\mathsf{P_A}$ in steps 3 and 4, and in particular compute $\hat{\mathcal{H}}_j, \hat{\mathsf{c}}_j$. Let J be the set of indices for which $(\hat{\mathcal{H}}_j, \hat{\mathsf{c}}_j) \neq (\mathcal{H}_j, \mathsf{c}_j)$.
   There are now three cases:
   - If $|J| \geq 2$, send $\mathsf{cert} := (j, \mathsf{trans}_j, \mathcal{H}_j, \mathsf{c}_j, \sigma_j, \mathsf{seed}_j^B)$ to $\mathcal{A}$ (for uniform $j \in J$). Then $\mathsf{P_B}$ outputs $\mathsf{corrupted}$ and the experiment halts.

- If $|J| = 1$ then with probability ϵ set $\mathsf{caught} := \mathsf{true}$ and with the remaining probability set $\mathsf{caught} := \mathsf{false}$. If $\mathsf{caught} = \mathsf{true}$ then $\mathsf{P_B}$ outputs $\mathsf{corrupted}$ (but the experiment continues below in either case).
- If $|J| = 0$ then set $\mathsf{caught} := \perp$ and continue below.

0′. Rewind $\mathcal{A}$ and run steps 1′–6′ below until $|J'| = |J|$ and $\mathsf{caught}' = \mathsf{caught}$ (using standard techniques [8,16] to ensure the experiment runs in expected polynomial time).

1′. Choose uniform $\hat{\jmath} \in [\lambda]$. For $j \neq \hat{\jmath}$, choose uniform κ-bit strings $\{\mathsf{seed}_j^B\}$ and set $h_j \leftarrow \mathsf{Com}(\mathsf{seed}_j^B)$. Set $h_{\hat{\jmath}} \leftarrow \mathsf{Com}(0^\kappa)$. Send $\{h_j\}_{j \in [\lambda]}$ to $\mathcal{A}$.

2′. For all $j \neq \hat{\jmath}$, run Π_{OT} with $\mathcal{A}$, using input 0 and randomness derived from seed_j^B. Obtain $\{\mathsf{seed}_j^A\}_{j \neq \hat{\jmath}}$ as the outputs of these executions. For the $\hat{\jmath}$th execution, use the simulator $\mathcal{S}_{\mathsf{OT}}$ for protocol Π_{OT}, thus extracting both $\mathsf{seed}_{\hat{\jmath}}^A$ and $\mathsf{witness}_{\hat{\jmath}}$. Let trans_j denote the transcript of the jth execution.

3′. For $j \neq \hat{\jmath}$, run an execution of Π_{OT} with $\mathcal{A}$ using input 0^{n_2} and randomness derived from seed_j^B. For $j = \hat{\jmath}$, use the simulator $\mathcal{S}_{\mathsf{OT}}$ for protocol Π_{OT}, thus extracting $\{B_{\hat{\jmath},i,b}\}_{i \in [n_2], b \in \{0,1\}}$. Let $\mathcal{H}_j$ denote the transcript hash of the jth execution.

4′. Receive $\{\mathsf{c}_j\}_{j \in [\lambda]}$ from $\mathcal{A}$.

5′. Receive $\{\sigma_j\}$ from $\mathcal{A}$. If any of the signatures are invalid, then return to step 1′.

6′. For all $j \in [\lambda]$, use seed_j^A and the messages sent previously to simulate the computation of an honest $\mathsf{P_A}$ in steps 3′ and 4′, and in particular compute $\hat{\mathcal{H}}_j, \hat{\mathsf{c}}_j$. Let J' be the set of indices for which $(\hat{\mathcal{H}}_j, \hat{\mathsf{c}}_j) \neq (\mathcal{H}_j, \mathsf{c}_j)$. If $|J'| = 1$ and $\hat{\jmath} \notin J'$ then set $\mathsf{caught}' := \mathsf{true}$. If $|J'| = 1$ and $\hat{\jmath} \in J'$ then set $\mathsf{caught}' := \mathsf{false}$. If $|J'| = 0$ then set $\mathsf{caught}' := \perp$.

7. If $|J'| = 1$ and $\mathsf{caught}' = \mathsf{true}$, then send $\mathsf{cert} := (j, \mathsf{trans}_j, \mathcal{H}_j, \mathsf{c}_j, \sigma_j, \mathsf{seed}_j^B)$ to $\mathcal{A}$ (where j is the unique index in J') and halt.
   Otherwise, send $(\hat{\jmath}, \{\mathsf{seed}_j^A\}_{j \neq \hat{\jmath}}, \mathsf{witness}_{\hat{\jmath}})$ to $\mathcal{A}$.
8. Receive GC, $\{A_i\}_{i \in [n_1]}$, $\{h_{i,b}^A\}_{i \in [n_1], b \in \{0,1\}}$, $\{Z_{i,b}\}_{i \in [n_3], b \in \{0,1\}}$, and the corresponding decommitments from $\mathcal{A}$. If any of the decommitments are incorrect, then $\mathsf{P_B}$ outputs $\perp$ and the experiment halts.
   Otherwise, there are two possibilities:
   - If $|J'| = 1$ and $\mathsf{caught}' = \mathsf{false}$ then use $\{B_{\hat{\jmath},i,b}\}_{i \in [n_2], b \in \{0,1\}}$ and y to compute z exactly as in the protocol. $\mathsf{P_B}$ outputs z and the experiment halts.
   - If $|J'| = |J| = 0$, compute an effective input $x \in \{0,1\}^{n_1}$ using $\mathsf{seed}_{\hat{\jmath}}^A$ and the input-wire labels $\{A_i\}_{i \in [n_1]}$. Then $\mathsf{P_B}$ outputs $f(x, y)$ and the experiment halts.

Expt$_1$. Here we modify the previous experiment in the following way: Choose a uniform $\hat{\jmath} \in [\lambda]$ at the outset of the experiment. Then in step 6:

- If $|J| \geq 2$ then send $\mathsf{cert} := (j, \mathsf{trans}_j, \mathcal{H}_j, \mathsf{c}_j, \sigma_j, \mathsf{seed}_j^B)$ to $\mathcal{A}$ for uniform $j \in J \setminus \{\hat{\jmath}\}$. Then $\mathsf{P_B}$ outputs $\mathsf{corrupted}$ and the experiment halts.
- if $|J| = 1$ set $\mathsf{caught} := \mathsf{true}$ if $\hat{\jmath} \notin J$ and set $\mathsf{caught} := \mathsf{false}$ if $\hat{\jmath} \in J$.

Since $\hat{\jmath} \notin J$ with probability ϵ when $|J| = 1$, the outputs of $\mathbf{Expt}_1$ and $\mathbf{Expt}_0$ are identically distributed.

$\mathbf{Expt}_2$. The previous experiment is modified as follows: In step 1, do not choose $\mathsf{seed}_{\hat{\jmath}}^B$. Instead, in step 1 set $h_{\hat{\jmath}} \leftarrow \mathsf{Com}(0^\kappa)$, and in steps 2 and 4 use true randomness in the $\hat{\jmath}$th execution of Π_{OT}.

It is immediate that the distribution of the output of $\mathbf{Expt}_2$ is computationally indistinguishable from the distribution of the output of $\mathbf{Expt}_1$.

$\mathbf{Expt}_3$. We change the previous experiment in the following way: In steps 2 and 4, use $\mathcal{S}_{\mathsf{OT}}$ to run the $\hat{\jmath}$th instances of Π_{OT}. In doing so, extract all of $\mathcal{A}$'s inputs in those executions.

It follows from security of Π_{OT} that the distribution of the output of $\mathbf{Expt}_3$ is computationally indistinguishable from the distribution of the output of $\mathbf{Expt}_2$.

$\mathbf{Expt}_{3a}$. Because steps 1′–4′ in $\mathbf{Expt}_3$ are identical to steps 1–4, we can "collapse" the rewinding and thus obtain the following experiment $\mathbf{Expt}_{3a}$ that is statistically indistinguishable from $\mathbf{Expt}_2$ (with the only difference occurring in case of an aborted rewinding in the latter):

1. Choose uniform $\hat{\jmath} \in [\lambda]$. For $j \neq \hat{\jmath}$, choose uniform κ-bit strings $\{\mathsf{seed}_j^B\}$ and set $h_j \leftarrow \mathsf{Com}(\mathsf{seed}_j^B)$. Set $h_{\hat{\jmath}} \leftarrow \mathsf{Com}(0^\kappa)$. Send $\{h_j\}_{j\in[\lambda]}$ to $\mathcal{A}$.
2. For all $j \neq \hat{\jmath}$, run Π_{OT} with $\mathcal{A}$, using input 0 and randomness derived from seed_j^B. Obtain $\{\mathsf{seed}_j^A\}_{j\neq\hat{\jmath}}$ as the outputs of these executions. For the $\hat{\jmath}$th execution, use the simulator $\mathcal{S}_{\mathsf{OT}}$ for protocol Π_{OT}, thus extracting both $\mathsf{seed}_{\hat{\jmath}}^A$ and $\mathsf{witness}_{\hat{\jmath}}$. Let trans_j denote the transcript of the jth execution.
3. For all $j \neq \hat{\jmath}$, run Π_{OT} with $\mathcal{A}$ using input 0^{n_2} and randomness derived from seed_j^B. For $j = \hat{\jmath}$, use the simulator $\mathcal{S}_{\mathsf{OT}}$ for protocol Π_{OT}, thus extracting $\{B_{\hat{\jmath},i,b}\}_{i\in[n_2],b\in\{0,1\}}$. Let $\mathcal{H}_j$ denote the transcript hash of the jth execution.
4. Receive $\{\mathsf{c}_j\}_{j\in[\lambda]}$ from $\mathcal{A}$.
5. Receive $\{\sigma_j\}$ from $\mathcal{A}$. If any of the signatures are invalid, then $\mathsf{P_B}$ outputs $\perp$ and the experiment halts.
6. For all $j \in [\lambda]$, use seed_j^A and the messages sent previously to $\mathcal{A}$ to simulate the computation of an honest $\mathsf{P_A}$ in steps 3 and 4, and in particular compute $\hat{\mathcal{H}}_j, \hat{\mathsf{c}}_j$. Let J be the set of indices for which $(\hat{\mathcal{H}}_j, \hat{\mathsf{c}}_j) \neq (\mathcal{H}_j, \mathsf{c}_j)$.
   There are now two cases:
   - If $|J| \geq 2$, or if $|J| = 1$ and $\hat{\jmath} \notin J$, then choose uniform $j \in J \setminus \{\hat{\jmath}\}$ and send $\mathsf{cert} := (j, \mathsf{trans}_j, \mathcal{H}_j, \mathsf{c}_j, \sigma_j, \mathsf{seed}_j^B)$ to $\mathcal{A}$. Then $\mathsf{P_B}$ outputs corrupted and the experiment halts.
   - If $|J| = 1$ and $\hat{\jmath} \in J$, or if $|J| = 0$, then continue below.
7. Send $(\hat{\jmath}, \{\mathsf{seed}_j^A\}_{j\neq\hat{\jmath}}, \mathsf{witness}_{\hat{\jmath}})$ to $\mathcal{A}$.
8. Receive GC, $\{A_i\}_{i\in[n_1]}$, $\{h_{i,b}^A\}_{i\in[n_1],b\in\{0,1\}}$, $\{Z_{i,b}\}_{i\in[n_3],b\in\{0,1\}}$, and the corresponding decommitments from $\mathcal{A}$. If any of the decommitments are incorrect, then $\mathsf{P_B}$ outputs $\perp$ and the experiment halts.
   Otherwise, there are two possibilities:
   - If $|J| = 1$ then $\mathsf{P_B}$ uses $\{B_{\hat{\jmath},i,b}\}_{i\in[n_2],b\in\{0,1\}}$ and y to compute z exactly as in the protocol. $\mathsf{P_B}$ outputs z and the experiment halts.

- If $|J| = 0$, then compute an effective input $x \in \{0,1\}^{n_1}$ using $\mathsf{seed}_{\hat{\jmath}}^A$ and the input-wire labels $\{A_i\}_{i\in[n_1]}$. Then $\mathsf{P_B}$ outputs $f(x, y)$ and the experiment halts.

Expt$_4$. We modify the previous experiment as follows: In step 8, if $|J| = 0$ (and $\mathsf{P_B}$ has not already output $\perp$ in that step), use y to compute z exactly as in the protocol. Then $\mathsf{P_B}$ outputs z and the experiment halts.

Since $|J| = 0$, we know that $\mathsf{c}_{\hat{\jmath}}$ is a commitment to a correctly computed garbled circuit along with commitments to (correctly permuted) input-wire labels $\{A_{\hat{\jmath},i,b}\}$ and output-wire labels. Thus—unless $\mathcal{A}$ has managed to violate the commitment property of Com—if $\mathsf{P_B}$ does not output $\perp$ in this step it must be the case that the values GC, $\{A_i\}_{i\in[n_1]}$, $\{h^A_{i,b}\}_{i\in[n_1],b\in\{0,1\}}$, and $\{Z_{i,b}\}_{i\in[n_3],b\in\{0,1\}}$ sent by $\mathcal{A}$ in step 8 are correct. Moreover, since $|J| = 0$ the execution of Π_{OT} in step 4 was run honestly by $\mathcal{A}$ using correct input-wire labels $\{B_{\hat{\jmath},i,b}\}$. Thus, evaluating GC using $\{A_i\}_{i\in[n_1]}$ and $\{B_{\hat{\jmath},i,y[i]}\}$ yields a result that is equal to $f(x, y)$ as computed in **Expt**$_3$.

Since Com is computationally binding, this means that the distribution of the output of **Expt**$_4$ is computationally indistinguishable from the distribution of the output of **Expt**$_{3a}$.

Expt$_5$. Here we change the previous experiment in the following way: The computation in step 6 is done only for $j \in [\lambda] \setminus \{\hat{\jmath}\}$; let $\hat{J} \subseteq [\lambda] \setminus \{\hat{\jmath}\}$ be the set of indices for which $(\hat{\mathcal{H}}_j, \hat{\mathsf{c}}_j) \neq (\mathcal{H}_j, \mathsf{c}_j)$. Then:

- If $\hat{J} \neq \emptyset$ choose uniform $j \in \hat{J}$ and send $\mathsf{cert} := (j, \mathsf{trans}_j, \mathcal{H}_j, \mathsf{c}_j, \sigma_j, \mathsf{seed}_j^B)$ to $\mathcal{A}$. Then $\mathsf{P_B}$ outputs $\mathsf{corrupted}$ and the experiment halts.
- If $\hat{J} = \emptyset$ then run steps 7 and 8 as in **Expt**$_4$.

Letting J be defined as in **Expt**$_4$, note that

$$|J| \geq 2 \text{ or } |J| = 1;\ \hat{\jmath} \notin J \iff \hat{J} \neq \emptyset$$

and

$$|J| = 1, \hat{\jmath} \in J \text{ or } |J| = 0 \iff \hat{J} = \emptyset.$$

Thus, the outputs of **Expt**$_4$ and **Expt**$_5$ are identically distributed.

Expt$_6$. We now modify the previous experiment by running the $\hat{\jmath}$th instances of Π_{OT} honestly in steps 2 and 4, using input 1 in step 2 and input y in step 4.

It follows from security of Π_{OT} that the distribution of the output of **Expt**$_6$ is computationally indistinguishable from the distribution of the output of **Expt**$_5$.

Expt$_7$. Finally, we modify the previous experiment so the $\hat{\jmath}$th instance of Π_{OT} in steps 2 and 4 uses pseudorandomness derived from a uniform seed $\mathsf{seed}_{\hat{\jmath}}^B$, and we compute $h_{\hat{\jmath}} \leftarrow \mathsf{Com}(\mathsf{seed}_{\hat{\jmath}}^B)$.

It is immediate that the distribution of the output of **Expt**$_7$ is computationally indistinguishable from the distribution of the output of **Expt**$_6$.

Since $\mathbf{Expt}_7$ corresponds to a real-world execution of the protocol between $\mathcal{A}$ and $\mathsf{P_B}$ holding input y, this completes the proof.

Covert Security—Malicious $\mathsf{P_B}$
Let $\mathcal{A}$ be an adversary corrupting $\mathsf{P_B}$. We construct the following simulator $\mathcal{S}$ that runs $\mathcal{A}$ as a subroutine while playing the role of $\mathsf{P_B}$ in the ideal world interacting with $\mathcal{F}$:

0. Run Gen to generate keys (pk, sk), and send pk to $\mathcal{A}$.
1. Receive $\{h_j\}_{j\in[\lambda]}$ from $\mathcal{A}$.
2. Use the simulator $\mathcal{S}_{\mathsf{OT}}$ for protocol Π_{OT} to interact with $\mathcal{A}$. In this way, $\mathcal{S}$ extracts $\mathcal{A}$'s inputs $\{b_j\}_{j\in[\lambda]}$; let $J := \{j : b_j = 1\}$. As part of the simulation, return uniform κ-bit strings $\{\mathsf{seed}_j^A\}_{j\notin J}$ and $\{\mathsf{witness}_j\}_{j\in J}$ as output to $\mathcal{A}$.
3. For each $j \notin J$, run this step exactly as an honest $\mathsf{P_A}$ would. For each $j \in J$ do:
   - If $|J| = 1$ then let $\hat{\jmath}$ be the unique index in J. Use $\mathcal{S}_{\mathsf{OT}}$ to interact with $\mathcal{A}$ in the $\hat{\jmath}$th execution of Π_{OT}. In this way, $\mathcal{S}$ extracts $\mathcal{A}$'s input y for that execution. Send y to $\mathcal{F}$, and receive in return a value z. Compute
   $$(\{A_{\hat{\jmath},i}\}, \{B_{\hat{\jmath},i}\}, \mathsf{GC}_{\hat{\jmath}}, \{Z_{\hat{\jmath},i,b}\}) \leftarrow \mathcal{S}_{\mathsf{Gb}}(1^\kappa, \mathcal{C}, z),$$
   where we let $\{A_{\hat{\jmath},i}\}$ correspond to input wires of $\mathsf{P_A}$ and $\{B_{\hat{\jmath},i}\}$ correspond to input wires of $\mathsf{P_B}$. Return $\{B_{\hat{\jmath},i}\}$ as output to $\mathcal{A}$ from this execution of Π_{OT}.
   - If $|J| > 1$ then act as an honest $\mathsf{P_A}$ would but using true randomness.
4. For each $j \notin J$, compute c_j exactly as an honest $\mathsf{P_A}$ would. For each $j \in J$ do:
   - If $|J| = 1$ then compute $h^A_{\hat{\jmath},i,0} \leftarrow \mathsf{Com}(A_{\hat{\jmath},i})$ and let $h^A_{\hat{\jmath},i,1}$ be a commitment to the 0-string. Compute $\mathsf{c}_{\hat{\jmath}} \leftarrow \mathsf{Com}(\mathsf{GC}_{\hat{\jmath}}, \{h^A_{\hat{\jmath},i,b}\}, \{Z_{\hat{\jmath},i,b}\})$, where each pair $(h^A_{\hat{\jmath},i,0}, h^A_{\hat{\jmath},i,1}\}$ is in random permuted order.
   - If $|J| > 1$ then compute c_j exactly as an honest $\mathsf{P_A}$ would but using true randomness.

   Send $\{\mathsf{c}_j\}_{j\in[\lambda]}$ to $\mathcal{A}$.

5–6. Compute signatures $\{\sigma_j\}$ as an honest $\mathsf{P_A}$ would, and send them to $\mathcal{A}$.

7. If $|J| \neq 1$ then abort. Otherwise, receive $(\hat{\jmath}, \{\mathsf{seed}_j\}_{j\neq\hat{\jmath}}, \mathsf{witness}_{\hat{\jmath}})$ from $\mathcal{A}$ and verify these as an honest $\mathsf{P_A}$ would. (If verification fails, then abort.)
8. Send $\mathsf{GC}_{\hat{\jmath}}$, $\{A_{\hat{\jmath},i}\}$, $\{h^A_{\hat{\jmath},i,b}\}$ (in the same permuted order as before), and $\{Z_{\hat{\jmath},i,b}\}$ to $\mathcal{A}$, along with the corresponding decommitments. Then halt.

We show that the distribution of the view of $\mathcal{A}$ in the ideal world is computationally indistinguishable from its view in a real protocol execution. (Note that $\mathsf{P_A}$ has no output.) Let $\mathbf{Expt}_0$ be the ideal-world execution between $\mathcal{S}$ (as described above) and the honest $\mathsf{P_A}$ holding some input x, both interacting with functionality $\mathcal{F}$.

$\mathbf{Expt}_1$. Here we modify the previous experiment when $|J| = 1$ as follows. In step 3, compute

$$(\{A_{\hat{\jmath},i,b}\}, \{B_{\hat{\jmath},i,b}\}, \mathsf{GC}_{\hat{\jmath}}, \{Z_{\hat{\jmath},i,b}\}) \leftarrow \mathsf{Gb}(1^\kappa, \mathcal{C}),$$

and return the values $\{B_{\hat{\jmath},i,y[i]}\}$ as output to $\mathcal{A}$ from the simulated execution of Π_{OT} in that step. In steps 4 and 8, the values $A_{\hat{\jmath},i,x[i]}$ are used in place of $A_{\hat{\jmath},i}$.

It follows from security of the garbling scheme that the view of $\mathcal{A}$ in $\mathbf{Expt}_1$ is computationally indistinguishable from its view in $\mathbf{Expt}_0$.

$\mathbf{Expt}_2$. Now we change the previous experiment when $|J| = 1$ as follows: In step 3, compute $h^A_{\hat{\jmath},i,b} \leftarrow \mathsf{Com}(A_{\hat{\jmath},i,b})$ for all i, b. It follows from the hiding property of the commitment scheme that the view of $\mathcal{A}$ in $\mathbf{Expt}_2$ is computationally indistinguishable from its view in $\mathbf{Expt}_1$.

$\mathbf{Expt}_3$. This time, the previous experiment is modified by executing protocol Π_{OT} with $\mathcal{A}$ when $|J| = 1$ in step 3. Security of Π_{OT} implies that the view of $\mathcal{A}$ in $\mathbf{Expt}_3$ is computationally indistinguishable from its view in $\mathbf{Expt}_2$.

$\mathbf{Expt}_4$. The previous experiment is now modified in the following way. In step 2, also choose uniform $\{\mathsf{seed}^A_j\}_{j \in J}$ and $\{\mathsf{witness}^A_j\}_{j \notin J}$, and use pseudorandomness derived from $\{\mathsf{seed}^A_j\}_{j \in J}$ in steps 3 and 4 in place of true randomness. Also, in step 7 continue to run the protocol as an honest $\mathsf{P_A}$ would even in the case that $|J| \neq 1$.

It is not hard to show that when $|J| \neq 1$ then $\mathsf{P_A}$ aborts in $\mathbf{Expt}_4$ with all but negligible probability. Computational indistinguishability of $\mathcal{A}$'s view in $\mathbf{Expt}_4$ and $\mathbf{Expt}_3$ follows.

$\mathbf{Expt}_5$. Finally, we change the last experiment by executing protocol Π_{OT} in step 2. It follows from the security of Π_{OT} that the view of $\mathcal{A}$ in $\mathbf{Expt}_5$ is computationally indistinguishable from its view in $\mathbf{Expt}_4$.

Since $\mathbf{Expt}_5$ corresponds to a real-world execution of the protocol, this completes the proof.

Public Verifiability and Defamation Freeness

It is easy to check (by inspecting the protocol) that whenever an honest $\mathsf{P_B}$ outputs $\mathsf{corrupted}$ then it also outputs a valid certificate. Thus our protocol satisfies public verifiability. It is similarly easy to verify defamation freeness under the assumptions of the theorem.

6 Implementation and Evaluation

We implemented our PVC protocol using the optimizations from Sect. 4.3 and state-of-the-art techniques for garbling [4,20], oblivious transfer [5], and OT extension [12]. Our implementation uses SHA-256 for the hash function (modeled as a random oracle) and the standard ECDSA implementation provided by openssl as the signature scheme. We target $\kappa = 128$ in our implementation.

We evaluate our protocol in both LAN and WAN settings. In the LAN setting, the network bandwidth is 1 Gbps and the latency is less than 1 ms; in the WAN setting, the bandwidth is 200 Mbps and the latency is 75 ms. In either setting, the machines running the protocol have 32 cores, each running at 3.0 GHz. Due to pipelining, we never observe any issues with memory usage. All reported timing results are computed as the average of 10 executions.

6.1 Certificate Size

The size of the certificate in our protocol is independent of the circuit size or the lengths of the parties' inputs. The following figure gives a graphical decomposition of the certificate. (Note that since we instantiate Com by a random oracle as discussed in Sect. 4.3, we do not need to include an extra decommitment in the certificate.) In total, a certificate requires 354 bytes.

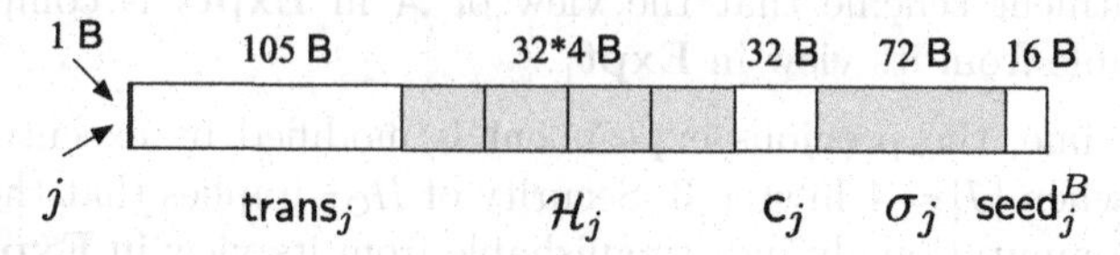

$\mathcal{H}_j$ contains 4 hash values, corresponding to a 4-round OT protocol obtained by piggybacking a 2-round OT-extension protocol with a 3-round base-OT protocol. The signature size varies from 70–72 bytes; we allocate 72 bytes for the signature so the total length of a certificate is fixed.

6.2 Comparison to Prior PVC Protocols

Because it enables signed-OT extension, the PVC protocol by Kolesnikov and Malozemoff [13] (the *KM15 protocol*) would be strictly more efficient than the original PVC protocol by Asharov and Orlandi [2]. We therefore focus our attention on the KM15 protocol. We compare our protocol to theirs in three respects.

Parameters. We briefly discuss the overhead needed to achieve deterrence factors larger than $\frac{1}{2}$ for each protocol. Recall that in the KM15 protocol the overall deterrence factor ϵ depends on both the garbled-circuit replication factor λ and the XOR-tree expansion factor ν as $\epsilon = (1 - \frac{1}{\lambda}) \cdot (1 - 2^{-\nu+1})$. For deterrence $\epsilon \approx 1 - \frac{1}{2^k}$, setting $\lambda = 2^{k+1}, \nu = k + 2$ gives the best efficiency. In contrast, our protocol achieves this deterrence with $\lambda = 2^k, \nu = 1$, which means garbling half as many circuits and avoiding the XOR-tree approach altogether. For example, to achieve deterrence $\epsilon = 7/8$, our protocol garbles 8 circuits, whereas prior work would need to garble 16 circuits. Additionally, prior work would need to execute 5× as many OTs. (Plus, in prior work each OT is actually a *signed* OT, which is more expensive than standard OT; see next.)

Signed OT vs. standard OT. Signed OT induces higher costs than standard OT in terms of both communication and computation. As an illustration, fix the deterrence factor to 1/2. In that case our protocol runs OT extension twice, where each is used for n_2 OTs on κ-bit strings. Compared to this, the KM15 protocol needs to run $3n_2$ OTs on 2κ-bit strings. The total communication complexity of the OT step (for the input-wire labels) is $4\kappa n_2$ bits in our protocol, while in the KM15 protocol it is $3 * 2 * 3\kappa n_2 + 3 * 2.6\kappa n_2 = 25.8\kappa n_2$ bits, more than 6× higher.

Moreover, signed OT also has a very high computational overhead:

– Signed-OT extension needs to use a wider matrix (by a factor of roughly 2.6×) compared to standard OT extension. Besides the direct penalty this

Table 1. Circuits used in our evaluation. The parties' input lengths are n_1 and n_2, and the output length is n_3. The number of AND gates in the circuit is denoted by $|\mathcal{C}|$.

Circuit	n_1	n_2	n_3	$\vert\mathcal{C}\vert$
AES-128	128	128	128	6800
SHA-128	256	256	160	37300
SHA-256	256	256	256	90825
Sorting	131072	131072	131072	10223K
Integer mult.	2048	2048	2048	4192K
Hamming dist.	1048K	1048K	22	2097K

incurs, a wider matrix means that the correlation-robust hash H cannot be based on fixed-key AES but must instead be based on a hash function like SHA-256. This impacts performance significantly.
- As part of signed-OT extension, $\mathsf{P_B}$ needs to reveal κ random columns in the matrix. Even with AVX operations, this incurs significant computational overhead.

Signed-OT extension [13] is complex, and we did not implement it in its entirety. However, we modified an existing (standard) OT-extension protocol to match the matrix width required by signed-OT extension; this can be used to give a conservative lower bound on the performance of signed-OT extension. Our results indicate that signed-OT extension requires roughly 5× more computation than state-of-the-art OT extension.

Certificate size. In the KM15 protocol, the certificate size is at least $2\kappa \cdot n_2$ bits. Even for AES (with only 128-bit input length), this gives a certificate roughly 10× larger than ours.

6.3 Comparing to Semi-Honest and Malicious Protocols

We believe our PVC protocol provides an excellent performance/security tradeoff that makes it the best choice for many applications of secure computation.

Performance. Our protocol is not much less efficient that the best known semi-honest protocols, and is significantly faster than the best known malicious protocols.

Security. The PVC model provides much more meaningful guarantees than the notion of semi-honest security, and may be appropriate for many (even if not all) applications of secure computation where full malicious security is overkill.

To support the first point, we compare the performance of our PVC protocol against state-of-the-art two-party computation protocols. The semi-honest protocol we compare against is a garbled-circuit protocol including all existing optimizations; for the malicious protocol we use the recent implementation of Wang et al. [19]. Our comparison uses the circuits listed in Table 1.

Table 2. Comparing the running times of our protocol and a semi-honest protocol in the LAN and WAN settings.

Circuit	LAN setting			WAN setting		
	Our PVC	Semi-honest	Slowdown	Our PVC	Semi-honest	Slowdown
AES-128	25 ms	15 ms	1.60×	960 ms	821 ms	1.17×
SHA-128	34 ms	25 ms	1.36×	1146 ms	977 ms	1.17×
SHA-256	48 ms	38 ms	1.27×	1252 ms	1080 ms	1.16×
Sort.	3468 ms	2715 ms	1.28×	13130 ms	12270 ms	1.07×
Mult.	1285 ms	1110 ms	1.16×	5707 ms	5462 ms	1.04×
Hamming	2585 ms	1550 ms	1.67×	11850 ms	6317 ms	1.69×

Running time. In Table 2 we compare the running time of our protocol to that of a semi-honest protocol. From the table, we see that over a LAN our protocol adds at most 36% overhead except in two cases: AES and Hamming-distance computation. For AES, the reason is that the circuit is small and so the overall time is dominated by the base OTs. For Hamming distance, the total input size is equal to the number of AND gates in the circuit; therefore, the cost of processing the inputs becomes more significant.

In the WAN setting, our PVC protocol incurs only 17% overhead except for the Hamming-distance example (for a similar reason as above).

The comparison between our PVC protocol and the malicious protocol is shown in Table 3. As expected, our PVC protocol achieves much better performance, by a factor of 4–18×.

Table 3. Comparing the running times of our protocol and a malicious protocol in the LAN and WAN settings.

Circuit	LAN setting			WAN setting		
	Our PVC	Malicious [19]	Speedup	Our PVC	Malicious [19]	Speedup
AES-128	25 ms	157 ms	6.41×	960 ms	11170 ms	11.6×
SHA-128	34 ms	319 ms	9.47×	1146 ms	13860 ms	12.1×
SHA-256	48 ms	612 ms	12.6×	1252 ms	17300 ms	13.8×
Sort.	3468 ms	45130 ms	13.0×	13130 ms	197900 ms	15.1×
Mult.	1285 ms	17860 ms	13.9×	5707 ms	99930 ms	17.5×
Hamming	2586ms	11380 ms	4.40×	11850 ms	76280 ms	6.44×

Table 4. Communication complexity in MB of our protocol with $\lambda = 2$ and other protocols.

	AES-128	SHA-128	SHA-256	Sort.	Mult.	Hamming
Semi-honest	0.2218	1.165	2.800	313.1	128.0	96.01
Malicious [11]	3.545	17.69	42.95	2953	1228	662.7
Our PVC	0.2427	1.205	2.844	325.1	128.2	144.2

Communication complexity. We also compare the communication complexity of our protocol to other protocols in a similar way; see Table 4. In this comparison we use the same semi-honest protocol as above, but use the more communication-efficient protocol by Katz et al. [11] as the malicious protocol. We see that, with the exception of the Hamming-distance example, the communication of our protocol is very close to the semi-honest case.

6.4 Higher Deterrence Factors

Another important aspect of our protocol is how the performance is affected by the deterrence factor. Recall that the deterrence factor ϵ is the probability that a cheating party is caught, and in our protocol $\epsilon = 1 - \frac{1}{\lambda}$ where λ is the garbled-circuit replication factor. The performance of our protocol as a function of ϵ is shown in Table 5. We see that when doubling the value of λ, the running time of the protocol increases by only ≈20% unless the circuit is very small (in which case the cost of the base OTs dominates the total running time). The running time when $\epsilon = 3/4$ (i.e., $\lambda = 4$) is still less than twice the running time of a semi-honest protocol.

Table 5. Running time in milliseconds of our protocol for different λ. $\epsilon = 1 - \frac{1}{\lambda}$.

		AES-128	SHA-128	SHA-256	Sort.	Mult.	Hamming
	$\lambda = 2$	25	34	49	3468	1285	2586
LAN	$\lambda = 4$	36	46	59	3554	1308	3156
	$\lambda = 8$	47	57	71	3954	1396	4856
	$\lambda = 16$	101	127	152	6238	2355	7143
	$\lambda = 32$	175	228	229	7649	2995	12984
	$\lambda = 2$	960	1146	1252	13130	5707	11850
WAN	$\lambda = 4$	1112	1375	1700	14400	5952	12899
	$\lambda = 8$	1424	1912	2436	16130	6167	19840
	$\lambda = 16$	1920	2094	2191	19087	7801	36270
	$\lambda = 32$	3228	3434	3535	25197	9229	64468

6.5 Scalability

Our protocol scales linearly in all parameters, and so can easily handle large circuits. To demonstrate this, we benchmarked our protocol with different input lengths, output lengths, and circuit sizes. Initially, the input and output lengths are all 128 bits, and the circuit size is 1024 AND gates. We then gradually increase one of the input/output lengths or circuit size (while holding everything else constant) and record the running time. Since the dependence is linear in all cases, we report only the marginal cost (i.e., the slope), summarized in Table 6.

Table 6. Scalability of our protocol.

	n_1 (μs/bit)	n_2 (μs/bit)	n_3 (μs/bit)	$\lvert\mathcal{C}\rvert$ (μs/gate)
LAN	0.20	0.88	0.23	0.29
WAN	0.61	3.13	0.62	1.10

References

1. Asharov, G., Lindell, Y., Schneider, T., Zohner, M.: More efficient oblivious transfer and extensions for faster secure computation. In: 20th ACM Conference on Computer and Communications Security (CCS), pp. 535–548. ACM Press (2013)
2. Asharov, G., Orlandi, C.: Calling out cheaters: covert security with public verifiability. In: Wang, X., Sako, K. (eds.) ASIACRYPT 2012. LNCS, vol. 7658, pp. 681–698. Springer, Heidelberg (2012). https://doi.org/10.1007/978-3-642-34961-4_41
3. Aumann, Y., Lindell, Y.: Security against covert adversaries: efficient protocols for realistic adversaries. J. Cryptol. **23**(2), 281–343 (2010)
4. Bellare, M., Hoang, V.T., Keelveedhi, S., Rogaway, P.: Efficient garbling from a fixed-key blockcipher. In: 2013 IEEE Symposium on Security & Privacy, pp. 478–492. IEEE (2013)
5. Chou, T., Orlandi, C.: The simplest protocol for oblivious transfer. In: Lauter, K., Rodríguez-Henríquez, F. (eds.) LATINCRYPT 2015. LNCS, vol. 9230, pp. 40–58. Springer, Cham (2015). https://doi.org/10.1007/978-3-319-22174-8_3
6. Damgård, I., Geisler, M., Nielsen, J.B.: From passive to covert security at low cost. In: Micciancio, D. (ed.) TCC 2010. LNCS, vol. 5978, pp. 128–145. Springer, Heidelberg (2010). https://doi.org/10.1007/978-3-642-11799-2_9
7. Goldreich, O.: Foundations of Cryptography, Volume 2: Basic Applications. Cambridge University Press, Cambridge (2004)
8. Goldreich, O., Kahan, A.: How to construct constant-round zero-knowledge proof systems for NP. J. Cryptol. **9**(3), 167–190 (1996)
9. Goyal, V., Mohassel, P., Smith, A.: Efficient two party and multi party computation against covert adversaries. In: Smart, N. (ed.) EUROCRYPT 2008. LNCS, vol. 4965, pp. 289–306. Springer, Heidelberg (2008). https://doi.org/10.1007/978-3-540-78967-3_17
10. Katz, J., Ostrovsky, R.: Round-optimal secure two-party computation. In: Franklin, M. (ed.) CRYPTO 2004. LNCS, vol. 3152, pp. 335–354. Springer, Heidelberg (2004). https://doi.org/10.1007/978-3-540-28628-8_21
11. Katz, J., Ranellucci, S., Rosulek, M., Wang, X.: Optimizing authenticated garbling for faster secure two-party computation. In: Shacham, H., Boldyreva, A. (eds.) CRYPTO 2018. LNCS, vol. 10993, pp. 365–391. Springer, Cham (2018). https://doi.org/10.1007/978-3-319-96878-0_13
12. Keller, M., Orsini, E., Scholl, P.: Actively secure OT extension with optimal overhead. In: Gennaro, R., Robshaw, M. (eds.) CRYPTO 2015. LNCS, vol. 9215, pp. 724–741. Springer, Heidelberg (2015). https://doi.org/10.1007/978-3-662-47989-6_35
13. Kolesnikov, V., Malozemoff, A.J.: Public verifiability in the covert model (almost) for free. In: Iwata, T., Cheon, J.H. (eds.) ASIACRYPT 2015. LNCS, vol. 9453, pp. 210–235. Springer, Heidelberg (2015). https://doi.org/10.1007/978-3-662-48800-3_9

14. Kolesnikov, V., Schneider, T.: Improved garbled circuit: free XOR gates and applications. In: Aceto, L., Damgård, I., Goldberg, L.A., Halldórsson, M.M., Ingólfsdóttir, A., Walukiewicz, I. (eds.) ICALP 2008. LNCS, vol. 5126, pp. 486–498. Springer, Heidelberg (2008). https://doi.org/10.1007/978-3-540-70583-3_40
15. Lindell, Y., Pinkas, B.: A proof of security of Yao's protocol for two-party computation. J. Cryptol. **22**(2), 161–188 (2009)
16. Lindell, Y.: A note on constant-round zero-knowledge proofs of knowledge. J. Cryptol. **26**(4), 638–654 (2013)
17. Lindell, Y.: Fast cut-and-choose-based protocols for malicious and covert adversaries. J. Cryptol. **29**(2), 456–490 (2016)
18. Lindell, Y., Pinkas, B.: An efficient protocol for secure two-party computation in the presence of malicious adversaries. In: Naor, M. (ed.) EUROCRYPT 2007. LNCS, vol. 4515, pp. 52–78. Springer, Heidelberg (2007). https://doi.org/10.1007/978-3-540-72540-4_4
19. Wang, X., Ranellucci, S., Katz, J.: Authenticated garbling and efficient maliciously secure two-party computation. In: 24th ACM Conference on Computer and Communications Security (CCS), pp. 21–37. ACM Press (2017)
20. Zahur, S., Rosulek, M., Evans, D.: Two halves make a whole - reducing data transfer in garbled circuits using half gates. In: Oswald, E., Fischlin, M. (eds.) EUROCRYPT 2015. LNCS, vol. 9057, pp. 220–250. Springer, Heidelberg (2015). https://doi.org/10.1007/978-3-662-46803-6_8

Efficient Circuit-Based PSI with Linear Communication

Benny Pinkas[1(✉)], Thomas Schneider[2], Oleksandr Tkachenko[2], and Avishay Yanai[1]

[1] Bar-Ilan University, Ramat Gan, Israel
benny@pinkas.net, ay.yanay@gmail.com
[2] TU Darmstadt, Darmstadt, Germany
{schneider,tkachenko}@encrypto.cs.tu-darmstadt.de

Abstract. We present a new protocol for computing a circuit which implements the private set intersection functionality (PSI). Using circuits for this task is advantageous over the usage of specific protocols for PSI, since many applications of PSI do not need to compute the intersection itself but rather functions based on the items in the intersection.

Our protocol is the *first circuit-based PSI protocol to achieve linear communication complexity*. It is also concretely more efficient than all previous circuit-based PSI protocols. For example, for sets of size 2^{20} it improves the communication of the recent work of Pinkas et al. (EUROCRYPT'18) by more than 10 times, and improves the run time by a factor of 2.8x in the LAN setting, and by a factor of 5.8x in the WAN setting.

Our protocol is based on the usage of a protocol for computing oblivious programmable pseudo-random functions (OPPRF), and more specifically on our technique to amortize the cost of batching together multiple invocations of OPPRF.

Keywords: Private Set Intersection · Secure computation

1 Introduction

The functionality of Private Set Intersection (PSI) enables two parties, P_1 and P_2, with respective input sets X and Y to compute the intersection $X \cap Y$ without revealing any information about the items which are not in the intersection. There exist multiple constructions of secure protocols for computing PSI, which can be split into two categories: (i) constructions that output the intersection itself and (ii) constructions that output the result of a function f computed on the intersection. In this work, we concentrate on the second type of constructions (see Sect. 1.2 for motivation). These constructions keep the intersection $X \cap Y$ secret from both parties and allow the function f to be securely computed on top of it, namely, yielding only $f(X \cap Y)$. Formally, denote by $\mathcal{F}_{\mathsf{PSI},f}$ the functionality $(X, Y) \mapsto (f(X \cap Y), f(X \cap Y))$.

Y. Ishai and V. Rijmen (Eds.): EUROCRYPT 2019, LNCS 11478, pp. 122–153, 2019.
https://doi.org/10.1007/978-3-030-17659-4_5

A functionality for computing $f(X \cap Y)$ can be naively implemented using generic MPC protocols by expressing the functionality as a circuit. However, naive protocols for computing $f(X \cap Y)$ have high communication complexity, which is of paramount importance for real-world applications. The difficulty in designing a circuit for computing the intersection is in deciding which pairs of items of the two parties need to be compared. We refer here to the number of comparisons computed by the circuit as the major indicator of the overhead, since it directly affects the amount of communication in the protocol (which is proportional to the number of comparisons, times the length of the representation of the items, times the security parameter). Since the latter factors (input length and security parameter) are typically given, and since the circuit computation mostly involves symmetric key operations, the goal is to minimize the communication overhead as a function of the input size. We typically state this goal as minimizing the number of comparisons computed in the circuit. The protocol presented in this paper is the first to achieve linear communication overhead, which is optimal.

Suppose that each party has an input set of n items. A naive circuit for this task compares all pairs and computes $O(n^2)$ comparisons. More efficient circuits are possible, assuming that the parties first order their respective inputs in specific ways. For example, if each party has sorted its input set then the intersection can be computed using a circuit which first computes, using a merge-sort network, a sorted list of the union of the two sets, and then compares adjacent items [HEK12]. This circuit computes only $O(n \log n)$ comparisons. The protocol of [PSSZ15] (denoted "Circuit-Phasing") has P_1 map its items to a table using Cuckoo hashing, and P_2 maps its items using simple hashing. The intersection is computed on top of these tables by a circuit with $O(n \log n/ \log\log n)$ comparisons. This protocol is the starting point of our work.

A recent circuit-based PSI construction [PSWW18] is based on a new hashing algorithm, denoted "two-dimensional Cuckoo hashing", which uses a table of size $O(n)$ and a stash of size $\omega(1)$. Each party inserts its inputs to a separate table, and the hashing scheme assures that each value in the intersection is mapped by both parties to exactly one mutual bin. Hence, a circuit which compares the items that the two parties mapped to each bin, and also compares all stash items to all items of the other party, computes the intersection in only $\omega(n)$ comparisons (namely, the overhead is slightly more than linear, although it can be made arbitrarily close to being linear).

Our work is based on the usage of an oblivious programmable pseudo-random function (OPPRF), which is a new primitive that was introduced in [KMP+17]. An OPRF—oblivious pseudo-random function (note, this is different than an OPPRF)—is a two-party protocol where one party has a key to a PRF F and the other party can privately query F at specific locations. An OPPRF is an extension of the protocol which lets the key owner "program" F so that it has specific outputs for some specific input values (and is pseudo-random on all other values). The other party which evaluates the OPPRF does not learn whether it learns a "programmed" output of F or just a pseudo-random value.

1.1 Overview of Our Protocol

The starting point for our protocols is the Circuit-Phasing PSI protocol of [PSSZ15], in which $O(n)$ bins are considered and the circuit computes $O(n \log n/ \log\log n)$ comparisons. Party P_1 uses Cuckoo hashing to map at most one item to each bin, whereas party P_2 maps its items to the bins using simple hashing (two times, once with each of the two functions used in the Cuckoo hashing of the first party). Thus, P_2 maps up to $S = O(\log n/ \log\log n)$ items to each bin. Since the parties have to hide the number of items that are mapped to each bin, they pad the bins with "dummy" items to the maximum bin size. That is, P_1 pads all bins so they all contain exactly one item and P_2 pads all bins so they all contain S items.

Both parties use the same hash functions, and therefore for each input element x that is owned by both parties there is exactly one bin to which x is mapped by both parties. Thus, it is only needed to check whether the item that P_1 places in a bin is among the items that are placed in this bin by P_2. This is essentially a private set membership (PSM) problem: As input, P_1 has a single item x and P_2 has a set Σ with $|\Sigma|$ items, where $S = |\Sigma|$. As for the output, if $x \in \Sigma$ then both parties learn the same random output, otherwise they learn independent random outputs. These outputs can then be fed to a circuit, which computes the intersection. The Circuit-Phasing protocol [PSSZ15] essentially computes the PSM functionality using a sub-circuit of the overall circuit that it computes. Namely, let $S = O(\log n/ \log\log n)$ be an upper bound on the number of items mapped by P_2 to a single bin. For each bin the sub-circuit receives one input from P_1 and S inputs from P_2, computes S comparisons, and feeds the result to the main part of the circuit which computes the intersection itself (and possibly some function on top of the intersection). Therefore the communication overhead is $O(nS) = O(n \log n/ \log\log n)$. A very recent work in [CO18] uses the same hashing method and computes the PSM using a specific protocol whose output is fed to the circuit. The circuit there computes only $\omega(n)$ comparisons but the PSM protocol itself incurs a communication overhead of $O(\log n/ \log\log n)$ and is run $O(n)$ times. Therefore, the communication overhead of [CO18] is also $O(n \log n/ \log\log n)$.

We diverge from the protocol of [PSSZ15] in the method for comparing the items mapped to each bin. In our protocol, the parties run an oblivious programmable PRF (OPPRF) protocol for each bin i, such that party P_2 chooses the PRF key and the programmed values, and the first party learns the output. The function is "programmed" to give the same output β_i for each of the $O(\log n/ \log\log n)$ items that P_2 mapped to this bin. Therefore, if there is any match in this bin then P_1 learns the same value β_i. Then, the parties evaluate a circuit, where for each bin i party P_1 inputs its output in the corresponding OPPRF protocol, and P_2 inputs β_i. This circuit therefore needs to compute only a *single* comparison per bin.

The communication overhead of an OPPRF is linear in the number of programmed values. Thus, a stand alone invocation of an OPPRF for every bin incurs an overall overhead of $O(n \log n/ \log\log n)$. We achieve linear overhead for comparing the items in all bins, by observing that although each bin is of maximal size $O(\log n / \log\log n)$ (and therefore naively requires to program this number of values in the OPPRF), the total number of items that need to be programmed in all bins is $O(n)$. We can amortize communication so that the total communication of computing all $O(n)$ OPPRFs is the same as the total number of items, which is $O(n)$.

In addition to comparing the items that are mapped to the hash tables, the protocol must also compare items that are mapped to the stash of the Cuckoo hashing scheme. Fixing a stash size $s = O(1)$, the probability that the stash does not overflow is $O(n^{-(s+1)})$ [KMW09]. It was shown in [GM11] that a stash of size $O(\log n)$ ensures a negligible failure probability (namely, a probability that is asymptotically smaller than any polynomial function). Each item that P_1 places in the stash must be compared to all items of P_2, and therefore a straightforward implementation of this step requires the circuit to compute $\omega(n)$ comparisons. However, we show an advanced variant of our protocol that computes all comparisons (including elements in the stash) with only $O(n)$ comparisons.

In addition to designing a generic $O(n)$ circuit-based PSI protocol, we also investigate an important and commonly used variant of the problem where each item is associated with some value ("payload"), and it is required to compute a function of the payloads of the items in the intersection. (For example, compute the sum of financial transactions associated with these items). The challenge is that each of the S items that the second party maps to a bin has a different payload and therefore it is hard to represent them using a single value. (The work in [PSSZ15, CO18], for example, did not consider payloads). We describe a variant of our PSI protocol which injects the correct payloads to the circuit while keeping the $O(n)$ overhead.

Overall, the work in this paper improves the state of the art in two dimensions:

- With regards to **asymptotic** performance, we show a protocol for circuit-based PSI which has only $O(n)$ communication. This cost is asymptotically smaller than that of all known circuit-based constructions of PSI, and matches the obvious lower bound on the number of comparisons that must be computed.
- With regards to **concrete** overhead, our most efficient protocols improve communication by a factor of 2.6x to 12.8x, and run faster by factor 2.8x to 5.8x compared to the previous best circuit-based PSI protocol of [PSWW18]. We demonstrate this both analytically and experimentally.

1.2 Motivation for Circuit-Based PSI

Most research on computing PSI focused on computing the intersection itself (see Sect. 1.4). On the other hand, many **applications** of PSI are based on computing arbitrary functions of the intersection. For example, Google reported a PSI-based application for measuring the revenues from *online* ad viewers who later perform a related *offline* transaction (namely, ad conversion rates) [Yun15,Kre17]. This computation compares the set of people who were shown an ad with the set of people who have completed a transaction. These sets are held by the advertiser, and by merchants, respectively. A typical use case is where the merchant inputs pairs of the customer-identity and the value of the transactions made by this customer, and the computation calculates the total revenue from customers who have seen an ad, namely customers in the intersection of the sets known to the advertiser and the merchant. Google reported implementing this computation using a Diffie-Hellman-based PSI cardinality protocol (for computing the cardinality of the intersection) and Paillier encryption (for computing the total revenues) [IKN+17,Kre18]. In fact, it was recently reported that Google is using such a "double-blind encryption" protocol in a beta version of their ads tool.[1] However, their protocol reveals the size of the intersection, and has substantially higher runtimes than our protocol as it uses public key operations, rather than efficient symmetric cryptographic operations (cf. Sect. 7.4).

Another motivation for running circuit-based PSI is **adaptability.** A protocol that is specific for computing the intersection, or a specific function such as the cardinality of the intersection, cannot be easily changed to compute another function of the intersection (say, the cardinality plus some noise to preserve differential privacy). Any change to a specialized protocol will require considerable cryptographic know-how, and might not even be possible. On the other hand, the task of writing a new circuit component which computes a different function of the intersection is rather trivial.

Circuit-based protocols also benefit from the **existing code base** for generic secure computation. Users only need to design the circuit to be computed, and can use available libraries of optimized code for secure computation, such as [HEKM11,EFLL12,DSZ15,LWN+15].

1.3 Computing Symmetric Functions

We focus in this work on constructing a circuit which computes the intersection. On top of that circuit it is possible to compose a circuit for computing any function that is based on the intersection. In order to preserve privacy, that function must be a symmetric function of the items in the intersection. Namely, the output of the function must not depend on the *order* of its inputs.

If the function that needs to be computed is non-symmetric, then the circuit for computing the intersection must shuffle its output, in order to place each item of the intersection in a location which is independent of the other values. The result is used as the input to the function. The size of this "shuffle" step is $O(n \log n)$, as is described in [HEK12], and it dominates the $O(n)$ size of the intersection circuit. We therefore focus on the symmetric case.[2]

---

[1] https://www.bloomberg.com/news/articles/2018-08-30/google-and-mastercard-cut-a-secret-ad-deal-to-track-retail-sales.

[2] Note that outputting the intersection is a *non-symmetric* function. Therefore in that case the order of the elements must be shuffled. However, it is unclear why a circuit-based protocol should be used for computing the intersection, since there are specialized protocols for this which are much more efficient, e.g. [KKRT16,PSZ18].

Most interesting functions of the intersection (except for the intersection itself) are symmetric. Examples of symmetric functions include:

- The size of the intersection, i.e., PSI cardinality (PSI-CA).
- A threshold function that is based on the size of the intersection. For example identifying whether the size of the intersection is greater than some threshold (PSI-CAT). An extension of PSI-CAT, where the intersection is revealed only if the size of the intersection is greater than a threshold, can be used for privacy-preserving ridesharing [HOS17]. Other public-key based protocols for this functionality appear in [ZC17, ZC18].
- A differentially private [Dwo06] value of the size of the intersection, which is computed by adding some noise to the exact count.
- The sum of values associated with the items in the intersection. This is used for measuring ad-generated revenues (cf. Sect. 1.2).

The circuits for computing all these functions are of size $O(n)$. Therefore, with our new construction the total size of the circuits for applying these functions to the intersection is $O(n)$.

1.4 Related Work

We classify previous works into dedicated protocols for *PSI*, generic protocols for *circuit-based PSI*, and dedicated protocols for *PSI cardinality*.

PSI. The first PSI protocols were based on public-key cryptography, e.g., on the Diffie Hellman function (e.g. [Mea86], with an earlier mention in [Sha80]), oblivious polynomial evaluation [FNP04], or blind RSA [DT10]. More recent protocols are based on oblivious transfer (OT) which can be efficiently instantiated using symmetric key cryptography [IKNP03, ALSZ13]: these protocols use either Bloom filters [FNP04] or hashing to bins [PSZ14, PSSZ15, KKRT16, PSZ18]. All these PSI protocols have super-linear complexity and many of them were compared experimentally in [PSZ18]. PSI protocols have also been evaluated on mobile devices, e.g., in [HCE11, ADN+13, CADT14, KLS+17]. PSI protocols with input sets of different sizes were studied in [KLS+17, PSZ18, RA18].

Circuit-based PSI. These protocols use secure evaluation of circuits for PSI. A trivial circuit for PSI computes $O(n^2)$ comparisons which result in $O(\sigma n^2)$ gates, where σ is the bit-length of the elements.

The sort-compare-shuffle (SCS) PSI circuit of [HEK12] computes $O(n \log n)$ comparisons and is of size $O(\sigma n \log n)$ gates (even without the final shuffle layer).

The Circuit-Phasing PSI circuit of [PSSZ15] uses Cuckoo hashing to $O(n)$ bins by one party and simple hashing by the other party which maps at most $O(\log n / \log \log n)$ elements per bin. Therefore, the Circuit-Phasing circuit has a size of $O(\sigma n \log n / \log \log n)$ gates.

The recent circuit-based PSI protocol of [CO18] applies a protocol based on OT extension to compute private set membership in each bin. The outputs of the invocations of this protocol are input to a comparison circuit. The circuit itself computes a linear number of comparisons, but the total communication complexity of the private set membership protocols is of the same order as that of the Circuit-Phasing circuit [PSSZ15] with $O(\sigma n \log n / \log \log n)$ gates.

Another recent circuit-based PSI protocol of [FNO18, Sect. 8] has communication complexity $O(\sigma n \log \log n)$. It uses hashing to $O(n)$ bins where each bin has multiple buckets and then runs the SCS circuit of [HEK12] to compute the intersection of the elements in the respective bins.

The two-dimensional Cuckoo hashing circuit of [PSWW18] uses a new variant of Cuckoo hashing in two dimensions and has an almost linear complexity of $\omega(\sigma n)$ gates.

In this work, we present the first circuit-based PSI protocol with a true linear complexity of $O(\sigma n)$ gates.

PSI Cardinality. Several protocols for securely computing the cardinality of the intersection, i.e., $|X \cap Y|$, were proposed in the literature. These protocols have linear complexity and are based on public-key cryptography, namely Diffie-Hellman [DGT12], the Goldwasser-Micali cryptosystem [DD15], or additively homomorphic encryption [DC17]. However, these protocols reveal the cardinality of the intersection to one of the parties. In contrast, circuit-based PSI protocols can easily be adapted to efficiently compute the cardinality and even functions of it using mostly symmetric cryptography.

1.5 Our Contributions

In summary, in this paper we present the following contributions:

- The first circuit-based PSI protocol with linear asymptotic communication overhead. We remark that achieving a linear overhead is technically hard since hashing to a table of linear size requires a stash of super-linear size in order to guarantee a negligible failure probability. It is hard to achieve linear overhead with objects of super-linear size.
- A circuit-based PSI protocol with small constants and an improved concrete overhead over the state of the art. As a special case, we consider a very common variant of PSI, namely threshold PSI, in which the intersection is revealed only if it is bigger/smaller than some threshold. Surprisingly, our protocol is 1–2 orders of magnitude more efficient than the state-of-the-art [ZC18] and has the same asymptotic communication complexity of $O(n)$, despite the fact that the protocol in [ZC18] is a special purpose protocol for threshold-PSI.
- Our protocol supports associating data ("payload") with each input (from both parties), and compute a function that depends on the data associated with the items in the intersection. This property was not supported by the Phasing circuit-based protocol in [PSSZ15]. It is important for applications that compute some function of data associated with the items in the intersection, e.g., aggregate revenues from common users (cf. Sect. 1.2).

- On a technical level, we present a new paradigm for handling $\omega(1)$ stash sizes and obtaining an overall overhead that is linear. This is achieved by running an extremely simple dual-execution variant of the protocol.
- Finally, with regards to concrete efficiency, we introduce a circuit-based PSI protocol with linear complexity. This is achieved by using Cuckoo hashing with $K = 3$ instead of $K = 2$ hash functions, and no stash. This protocol substantially reduces communication (by a factor of 2.6x to 12.8x) and runtime (by a factor of 2.8x to 5.8x) compared to the best previous circuit-based PSI protocol of [PSWW18].

2 Preliminaries

2.1 Setting

There are two parties, which we denote as P_1 (the "receiver") and P_2 (the "sender"). They have input sets, X and Y, respectively, each of which contains n items of bitlength λ. We assume that both parties agree on a function f and wish to securely compute $f(X \cap Y)$. They also agree on a circuit C that receives the items in the intersection as input and computes f. That is, C has $O(n\lambda)$ input wires if we consider a computation on the elements themselves or $O(n(\lambda + \rho))$ if we consider a computation on the elements and their associated payloads where the associated payload of each item has bitlength ρ. We denote the computational and statistical security parameters by κ and σ, respectively. Denote the set $1, \ldots, c$ by $[c]$. We use the notation $X(i)$ to denote the i-th element in the set X.

2.2 Security Model

This work, similar to most protocols for private set intersection, operates in the semi-honest model, where adversaries may try to learn as much information as possible from a given protocol execution but are not able to deviate from the protocol steps. This is in contrast to malicious adversaries which are able to deviate arbitrarily from the protocol. PSI protocols for the malicious setting exist, but they are less efficient than protocols for the semi-honest setting, e.g., [FNP04, DSMRY09, HN10, DKT10, FHNP16, RR17a, RR17b]. The only circuit-based PSI protocol that can be easily secured against malicious adversaries is the Sort-Compare-Shuffle protocol of [HEK12]: here a circuit of size $O(n)$ can be used to check that the inputs are sorted, resulting in an overall complexity of $O(n \log n)$. For the recent circuit-based PSI protocols that rely on Cuckoo hashing, ensuring that the hashing was done correctly remains the challenge.The semi-honest adversary model is appropriate for scenarios where execution of the intended software is guaranteed via software attestation or business restrictions, and yet an untrusted third party is able to obtain the transcript of the protocol after its execution, by stealing it or by legally enforcing its disclosure.

FUNCTIONALITY 1 (Two-Party Computation)

Parameters. The Boolean circuit C to be computed, with I_1, I_2 inputs and O_1, O_2 outputs associated with P_1 and P_2 resp.
Inputs. P_1 inputs bits $x_1, \ldots, x_{I_1}$ and P_2 inputs bits $y_1, \ldots, y_{I_2}$.
Outputs. The functionality computes the circuit C on the parties' inputs and returns the outputs to the parties.

2.3 Secure Two-Party Computation

There are two main approaches for generic secure two-party computation of Boolean circuits with security against semi-honest adversaries: (1) Yao's garbled circuit protocol [Yao86] has a constant round complexity and with today's most efficient optimizations provides free XOR gates [KS08], whereas securely evaluating an AND gate requires sending two ciphertexts [ZRE15]. (2) The GMW protocol [GMW87] also provides free XOR gates and also sends two ciphertexts per AND gate using OT extension [ALSZ13].

The main advantage of the GMW protocol is that *all* symmetric cryptographic operations can be pre-computed in a constant number of rounds in a setup phase, whereas the online phase is very efficient, but requires interaction for each layer of AND gates. In more detail, the setup phase is independent of the actual inputs and precomputes multiplication triples for each AND gate using OT extension in a constant number of rounds. The online phase begins when the inputs are provided and involves a communication round for each layer of AND gates. See [SZ13] for a detailed description and comparison between Yao and GMW.

In our protocol we make use of Functionality 1.

2.4 Cuckoo Hashing

Cuckoo hashing [PR01] uses two hash functions h_0, h_1 to map n elements to two tables T_0, T_1 which each contain $(1 + \varepsilon)n$ bins. (It is also possible to use a single table T with $2(1 + \varepsilon)n$ bins. The two versions are essentially equivalent). Each bin accommodates at most a single element. The scheme avoids collisions by relocating elements when a collision is found using the following procedure: Let $b \in \{0, 1\}$. An element x is inserted into a bin $h_b(x)$ in table T_b. If a prior item y exists in that bin, it is evicted to bin $h_{1-b}(y)$ in T_{1-b}. The pointer b is then assigned the value $1 - b$. The procedure is repeated until no more evictions are necessary, or until a threshold number of relocations has been reached. In the latter case, the last element is put in a special stash. It was shown in [KMW09] that for a stash of constant size s the probability that the stash overflows is at most $O(n^{-(s+1)})$. It was also shown in [GM11] that this failure probablity is negligilble when the stash is of size $O(\log n)$. An observation in [KM18] shows that this is also the case when $s = O(\omega(1) \cdot \frac{\log n}{\log \log n})$. After insertion, each item can be found in one of two locations or in the stash.

2.5 PSI Based on Hashing

Existing constructions for circuit-based PSI require the parties to reorder their inputs before inputting them to the circuit. In the sorting network-based circuit of [HEK12], the parties sort their inputs. In the hashing-based construction of [PSSZ15], the parties map their items to bins using a hashing scheme.

It was observed as early as [FNP04] that if the two parties agree on the same hash function and use it to assign their respective input to bins, then the items that one party maps to a specific bin only need to be compared to the items that the other party maps to the same bin. However, the parties must be careful not to reveal to each other the number of items that they mapped to each bin, since this leaks information about their input sets. Therefore, the parties agree beforehand on an upper bound m for the maximum number of items that can be mapped to a bin (such upper bounds are well known for common hashing algorithms, and can also be substantiated using simulation), and pad each bin with random dummy values until it has exactly m items in it. If both parties use the same hash algorithm, then this approach considerably reduces the overhead of the computation from $O(n^2)$ to $O(\beta \cdot m^2)$ where β is the number of bins.

When using a random hash function h to map n items to n bins such that x is mapped to bin $h(x)$, the most occupied bin has at most $m = \frac{\ln n}{\ln \ln n}(1 + o(1))$ items with high probability [Gon81]. For instance, for $n = 2^{20}$ and a desired error probability of 2^{-40}, a careful analysis shows that $m = 20$. Cuckoo hashing is much more promising, since it maps n items to $2n(1+\varepsilon)$ bins, where each bin stores at most $m = 1$ items.

It is tempting to let both parties, P_1 and P_2, map their items to bins using Cuckoo hashing, and then only compare the item that P_1 maps to a bin with the item that P_2 maps to the same bin. The problem is that P_1 might map x to $h_0(x)$ whereas P_2 might map it to $h_1(x)$. Unfortunately, they cannot use a protocol where P_1's value in bin $h_0(x)$ is compared to the two bins $h_0(x), h_1(x)$ in P_2's input, since this reveals that P_1 has an item which is mapped to these two locations. The solution used in [FHNP16, PSZ14, PSSZ15] is to let P_1 map its items to bins using Cuckoo hashing, and P_2 map its items using simple hashing. Namely, each item of P_2 is mapped to both bins $h_0(x), h_1(x)$. Therefore, P_2 needs to pad its bins to have exactly $m = O(\log n / \log \log n)$ items in each bin, and the total number of comparisons is $O(n \log n / \log \log n)$.

3 OPPRF – Oblivious Programmable PRF

Our protocol builds on a (batched) oblivious programmable pseudorandom function (OPPRF). In this section we gradually present the properties required by that kind of a primitive, by first describing simpler primitives, namely, Programmable PRF (and its batched version) and Oblivious PRF.

3.1 Oblivious PRF

An oblivious PRF (OPRF) [FIPR05] is a two-party protocol implementing a functionality between a sender and a receiver. Let F be a pseudo-random function (PRF) such that $F : \{0,1\}^\kappa \times \{0,1\}^\ell \rightarrow \{0,1\}^\ell$. The sender inputs a key k to F and the receiver inputs $q_1 \ldots, q_c$. The functionality outputs $F(k, q_1), \ldots, F(k, q_c)$ to the receiver and nothing to the sender. In another variant of oblivious PRF the sender is given a fresh random key k as an output from the functionality rather than choosing it on its own. In our protocol we will make use of a "one-time" OPRF functionality in which the receiver can query *a single query*, namely, the sender inputs nothing and the receiver inputs a query q; the functionality outputs to the sender a key k and to the receiver the result $F_k(q)$. Let us denote that functionality by $\mathcal{F}_{\mathsf{OPRF}}$.

3.2 (One-Time) Programmable PRF (PPRF)

A programmable PRF (PPRF) is similar to a PRF, with the additional property that on a certain "programmed" set of inputs the function outputs "programmed" values. Namely, for an arbitrary set X and a "target" multi-set T, where $|X| = |T|$ and each $t \in T$ is uniformly distributed[3], it is guaranteed that on input $X(i)$ the function outputs $T(i)$. Let $\mathcal{T}$ be a distribution of such multi-sets, which may be public to both parties.

The restriction of the PPRF to be only one-time comes from the fact that we allow the elements in T to be correlated. If the elements are indeed correlated then by querying it two times (on the correlated positions) it would be easy to distinguish it from a random function.

We capture the above notion by the following formal definition:

Definition 1. *An ℓ-bits PPRF is a pair of algorithms $\hat{F} = (\mathsf{Hint}, F)$ as follows:*

- $\mathsf{Hint}(k, X, T) \rightarrow \mathsf{hint}_{k,X,T}$: *Given a uniformly random key $k \in \{0,1\}^\kappa$, the set X where $|X(i)| = \ell$ for all $i \in [|X|]$ and a target multi-set T with $|T| = |X|$ and all elements in T are uniformly distributed (but may be correlated), output the hint $\mathsf{hint}_{k,X,T} \in \{0,1\}^{\kappa \cdot |X|}$.*
- $F(k, \mathsf{hint}, x) \rightarrow y^\star$. *Given a key $k \in \{0,1\}^\kappa$, a hint $\mathsf{hint} \in \{0,1\}^{\kappa \cdot |X|}$ and an input $x \in \{0,1\}^\ell$, output $y^\star \in \{0,1\}^\ell$.*

We consider two properties of a PPRF, correctness and security:

- **Correctness.** *For every k, T and X, and for every $i \in [|X|]$ we have:*

$$F(k, \mathsf{hint}, X(i)) = T(i).$$

- **Security.** *We say that an interactive machine M is a* PPRF oracle over $\hat{F}$ *if, when interacting with a "caller" $\mathcal{A}$, it works as follows:*
  1. *M is given a set X from $\mathcal{A}$.*
  2. *M samples a uniformly random $k \in \{0,1\}^\kappa$ and T from $\mathcal{T}$, invokes $\mathsf{hint} \leftarrow \mathsf{Hint}(k, X, T)$ and hands hint to $\mathcal{A}$.*

---

[3] We require that each element in T is uniformly random but the elements may be correlated.

CONSTRUCTION 2 (PPRF)

Let $F' : \{0,1\}^{\kappa} \times \{0,1\}^{\ell} \rightarrow \{0,1\}^{\ell}$ be a PRF.
- $\mathsf{Hint}(k, X, T)$. Interpolate a polynomial p over the points $\{(X(i)\ ,\ F'_k(X(i)) \oplus T(i))\}_{i \in [|X|]}$. Return p as the hint.
- $F(k, \mathsf{hint}, x)$. Interpret hint as a polynomial, denoted p. Return $F'_k(x) \oplus p(x)$.

3. M is given an input $x \in \{0,1\}^{\ell}$ from $\mathcal{A}$ and responds with $F(k, \mathsf{hint}, x)$.
4. M halts (any subsequent queries will be ignored).
The scheme $\hat{F}$ is said to be secure *if, for every X input by $\mathcal{A}$ (i.e. the caller), the interaction of $\mathcal{A}$ with M is computationally indistinguishable from the interaction with the* PPRF oracle $\mathcal{S}$, *where $\mathcal{S}$ outputs a uniformly random "hint" $\{0,1\}^{\kappa \cdot |X|}$ and a "PRF result" from $\{0,1\}^{\ell}$.*

The definition is reminiscent of a semantically secure encryption scheme. Informally, semantic security means that whatever is efficiently computable about the cleartext given the ciphertext, is also efficiently computable without the ciphertext. Also here, whatever can be efficiently computable given X is also efficiently computable given only $|X|$. That implicitely means that the interaction with a PPRF oracle M over (KeyGen, F) does not leak the elements in X.

Our security definition diverges from that of [KMP+17] in two aspects:

1. In [KMP+17], $\mathcal{A}$ has many queries to M in Step 3 of the interaction, whereas our definition allows only a single query. In the (n,t)-security definition in [KMP+17] this corresponds to setting $t = 1$. Our definition is weaker in this sense, but this is sufficient for our protocol as we invoke multiple instances of the one-time PPRF.
2. The definition in [KMP+17] compensates for the fact that $\mathcal{A}$ has many queries, by requiring that the function F outputs an *independent* target value for every $x \in X$. Our definition is stronger as it allows having correlated target elements in T. In the most extreme form of correlation all values in T are equal, which makes the task of the adversary "easier". We require the security property to hold even in this case.

We present in Construction 2 a polynomial-based PPRF scheme that is based on the construction in [KMP+17].

Theorem 3. *Construction 2 is a PPRF.*

Proof. It is easy to see that this construction is correct. For every k, X and T, let $p = \mathsf{Hint}(k, X, T)$, then for all $i \in |X|$ it holds that

$$\begin{aligned} F(k, p, X(i)) &= F'(k, X(i)) \oplus p(X(i)) \\ &= F'(k, X(i)) \oplus F'(k, X(i)) \oplus T(i) \\ &= T(i) \end{aligned}$$

as required.We now reduce the security of the scheme to the security of a PRF (i.e., to the standard PRF definition, with many oracle accesses). Let M be a PPRF oracle over $\hat{F}$ of Construction 2. Assume there exists a distinguisher $\mathcal{D}$ and a caller $\mathcal{A}$ such that $\mathcal{D}$ distinguishes between the output of M after interacting with $\mathcal{A}$, when $\mathcal{A}$ chooses X and x as its inputs, and the output of $\mathcal{S}(1^\kappa, |X|)$ (where $\mathcal{S}$ is the simulator described in Definition 1) with probability μ.

We present a distinguisher $\mathcal{D}'$ that has an oracle access to either a truly random function $R(\cdot)$ or a PRF $\tilde{F}(k, \cdot)$. The distinguisher $\mathcal{D}'$ runs as follows:

Given an oracle $\mathcal{O}$ to either $R(\cdot)$ or $\tilde{F}(k, \cdot)$, $\mathcal{D}'$ samples T from $\mathcal{T}$, then, for every $i \in [|X|]$ it queries the oracle on $X(i)$ and obtains $\mathcal{O}(X(i))$. It interpolates the polynomial p using the points $\{(X(i), \mathcal{O}(X(i)) \oplus T(i))\}_{i \in |X|}$ and provides p's coefficients to $\mathcal{D}$. For the query x, $\mathcal{D}'$ hands $\mathcal{D}$ the value $\mathcal{O}(x) \oplus p(x)$ and outputs whatever $\mathcal{D}$ outputs.

Observe that if $\mathcal{O}$ is truly random, then the values $\{R(X(i)) \oplus T(i))\}_{i \in [|X|]}$ are uniformly random and thus the polynomial p is uniformly random and independent of T. If $x \notin X$ then the value $R(x) \oplus p(x)$ is obviously random since $R(x)$ is independent of p. In addition, if $x = X(i)$ for some i, then the value $R(x) \oplus p(x)$ equals $T(i)$ for some $i \in [|X|]$, which is uniformly random since T is sampled from $\mathcal{T}$ and every $t \in T$ is distributed uniformly. Therefore, the pair $(p, R(x) \oplus p(x))$ is distributed identically to the output of $\mathcal{S}$. On the other hand, if $\mathcal{O}$ is a pseudorandom function, then the values $\{F_k(X(i)) \oplus T(i))\}_{i \in [|X|]}$ from which the polynomial p is interpolated, along with the second output $F_k(x) \oplus p(x)$, are distributed identically to the output of M upon an interaction with $\mathcal{A}$. This leads to the same distinguishing success probability μ, for both $\mathcal{D}$ and $\mathcal{D}'$, which must be negligible. □

3.3 Batch PPRF

Note that the size of the hint generated by algorithm KeyGen is $\kappa \cdot |X|$ (i.e., the polynomial is represented by $|X|$ coefficients, each of size κ bits). In our setting we use an independent PPRF per bin, where each bin contains at most $O(\log n / \log \log n)$ values. Therefore the hint for one bin is of size $O(\kappa \cdot \log n / \log \log n)$, and the size of all hints is $O(\kappa \cdot n \cdot \log n / \log \log n)$. However, we know that the total number of values in all P_2's bins is $2n$, since each value is stored in (at most) two locations of the table[4]. We next show that it is possible to combine the hints of *all* bins to a single hint of length $2n$, thus reducing the total communication for all hints to $O(n)$.

We first present a formal definition of the notion of batch PPRF.

Definition 2. *An ℓ-bits, β-bins PPRF (or (ℓ, β)-PPRF) is a pair of algorithms $\hat{F} = (\mathsf{KeyGen}, F)$ as follows:*

---

[4] In the actual implementation we use a more general variant of Cuckoo hashing with a parameter $K \in \{2, 3\}$ where each item is stored in K locations in the table. The size of the hint will be $K \cdot n$.

CONSTRUCTION 4 (Batched PPRF)

Let $F' : \{0,1\}^\kappa \times \{0,1\}^\ell \rightarrow \{0,1\}^\ell$ be a PRF.

- $\mathsf{Hint}(k, X, T)$.
  Given the keys $k = k_1, \ldots, k_\beta$, the sets $X = X_1, \ldots, X_\beta$ and the target multi-sets $T = T_1, \ldots, T_\beta$, interpolate the polynomial p using the points $\{(X_j(i), F'(k_j, X_j(i)) \oplus T_j(i))\}_{j \in \beta; i \in [|X_j|]}$. Return p as the hint.
- $F(k, \mathsf{hint}, x)$.
  Interpret hint as a polynomial, denoted p. Return $F'(k, x) \oplus p(x)$. (Same as in Construction 2.)

- *$\mathsf{Hint}(k, X, T) \rightarrow \mathsf{hint}_{k,X,T}$. Given a set of uniformly random and independent keys $k = k_1, \ldots, k_\beta \in \{0,1\}^\kappa$, the sets $X = X_1, \ldots, X_\beta$ where $|X_j(i)| = \ell$ for all $j \in [\beta]$ and $i \in [|X|]$ and a target multi-sets $T = T_1, \ldots, T_\beta$ where for every $j \in [\beta]$ it holds that $|T_j| = |X_j|$ and all elements in T_j are uniformly distributed (but, again, may be correlated), output the hint $\mathsf{hint}_{k,X,T} \in \{0,1\}^{\kappa \cdot N}$ where $N = \sum_{j=1}^{\beta} |X_j|$.*
- *$F(k, \mathsf{hint}, x) \rightarrow y^\star$. Given a key $k \in \{0,1\}^\kappa$, a hint $\mathsf{hint} \in \{0,1\}^{\kappa \cdot N}$ and an input $x \in \{0,1\}^\ell$, output $y^\star \in \{0,1\}^\ell$.*

As before, we want a batched PPRF to have the following properties:

- **Correctness.** *For every $k = k_1, \ldots, k_\beta$, $T = T_1, \ldots, T_\beta$ and $X = X_1, \ldots, X_\beta$ as above, we have*
$$F(k_j, \mathsf{hint}, X_j(i)) = T_j(i)$$
  for every $j \in [\beta]$ and $i \in [|X_j|]$.
- **Security.** *We say that an interactive machine M is a* batched PPRF oracle over $\hat{F}$ *if, when interacting with a "caller" $\mathcal{A}$, it works as follows:*
  1. *M is given $X = X_1, \ldots, X_\beta$ from $\mathcal{A}$.*
  2. *M samples uniformly random keys $k = k_1, \ldots, k_\beta$ and target multi-sets $T = T_1, \ldots, T_\beta$ from $\mathcal{T}$, and invokes $\mathsf{hint} \leftarrow \mathsf{Hint}(k, X, T)$ hands hint to $\mathcal{A}$.*
  3. *M is given β queries $x_1, \ldots, x_\beta$ from $\mathcal{A}$ and responds with $y_1^\star, \ldots, y_\beta^\star$ where $y_j^\star = F(k_j, \mathsf{hint}, x_j)$.*
  4. *M halts.*

  The scheme $\hat{F}$ is said to be secure if for every disjoint sets $X_1, \ldots, X_\beta$ (where $N = \sum_{j \in [\beta]} |X_j|$) input by a PPT machine $\mathcal{A}$, the output of M is computationally indistinguishable from the output of $\mathcal{S}(1^\kappa, N)$, such that $\mathcal{S}$ outputs a uniformly random $\mathsf{hint} \in \{0,1\}^{\kappa \cdot N}$ and a set of β uniformly random values from $\{0,1\}^\ell$.

Construction 4 is a batched version of Construction 2.

Theorem 5. *Construction 4 is a secure (ℓ, β)-PPRF.*

Proof. For correctness, note that for every $j \in [\beta]$ and $i \in [|X_j|]$ it holds that

$$\begin{aligned} F(k_j, p, X_j(i)) &= F'(k_j, X_j(i)) \oplus p(X_j(i)) \\ &= F'(k_i, X_j(i)) \oplus F'(k_i, X_j(i)) \oplus T_j(i) \\ &= T_j(i). \end{aligned}$$

The security of the scheme is reduced to the security of a batch PRF $\tilde{F}$. Informally, a batch PRF works as follows: Sample uniform keys $k_1, \ldots, k_\beta \in \{0,1\}^\kappa$ and for a query (j, x) respond with $\tilde{F}(k_j, x)$. One can easily show that a batch PRF is indistinguishable from a set of β truly random functions $R_1, \ldots, R_\beta$ where on query (j, x) the output is $R_j(x)$.

Let M be a batched PPRF oracle over $\hat{F}$ of Construction 4. Assume there exists a distinguisher $\mathcal{D}$ and a caller $\mathcal{A}$ such that $\mathcal{D}$ distinguishes between the output of M after interacting with $\mathcal{A}$, when $\mathcal{A}$ chooses $X_1, \ldots, X_\beta$ and $x_1, \ldots, x_\beta$ as its inputs, and the outputs of $\mathcal{S}(1^\kappa, N)$, where $\mathcal{S}$ is the simulator described in Definition 2.

We present a distinguisher $\mathcal{D}'$ that has an oracle access $\mathcal{O}$, to either a batch PRF $\tilde{F}(k_j, \cdot)$ or a set of truly random functions $R_j(\cdot)$ (where $j \in [\beta]$). The distinguisher $\mathcal{D}'$ works as follows: sample $T_1, \ldots, T_\beta$ from $\mathcal{T}$, interpolate a polynomial p with the points $\{(X_j(i), \mathcal{O}(j, X_j(i)) \oplus T_j(i))\}_{j \in [\beta]; i \in [|X_j|]}$ and hand p's coefficients to $\mathcal{D}$ as the hint. Then, for query x_j of $\mathcal{D}$, respond with $y_j^\star = \mathcal{O}(x_j) \oplus p(x_j)$. Finally, $\mathcal{D}'$ outputs whatever $\mathcal{D}$ outputs.

First note that if $\mathcal{O}$ is a set of truly random functions then the polynomial p is uniformly random and independent of $y_1^\star, \ldots, y_\beta^\star$ because all interpolation points are uniformly random. Now, if $x_j \notin X_j$ then the result is obviously uniformly random. Otherwise, if $x_j = X_j(i)$ for some i then note that the result is $T_j(i)$ which is uniformly random as well, since the other elements in T_j are unknown. Thus, this is distributed identically to the output of $\mathcal{S}(1^\kappa, N)$. On the other hand, if $\mathcal{O}$ is a batch PRF then the interpolation points $\{(X_j(i), \mathcal{O}(j, X_j(i)) \oplus T_j(i))\}_{j \in [\beta]; i \in [|X_j|]}$ along with $y_1^\star, \ldots, y_\beta^\star$ are distributed identically to the output of M upon an interaction with $\mathcal{A}$. This leads to the same distinguishing success probability for both $\mathcal{D}$ and $\mathcal{D}'$, which must be negligible. □

3.4 Batch *Oblivious* Programmable Pseudorandom Functions

In this section we define a two-party functionality for batched oblivious programmable pseudorandom function (Functionality 6), which is the main building block in our PSI protocols. The functionality is parametrised by a (ℓ, β)-PPRF $\hat{F} = (\mathsf{Hint}, F)$ and interacts with a sender, who programs $\hat{F}$ with β sets, and a receiver who queries $\hat{F}$ with β queries. The functionality guarantees that the sender does not learn what are the receiver's queries and the receiver does not learn what are the programmed points.

Given a protocol that realizes $\mathcal{F}_{\mathsf{OPRF}}$ and a secure (ℓ, β)-PPRF, the realization of Functionality 6 is simple and described in Protocol 7.

Theorem 8. *Given an (ℓ, β)-PPRF, Protocol 7 securely realizes Functionality 6 in the $\mathcal{F}_{\mathsf{OPRF}}$-hybrid model.*

FUNCTIONALITY 6 (Batch Oblivious PPRF)

Parameters. A (ℓ, β)-PPRF $\hat{F} = (\mathsf{Hint}, F)$.
Sender's inputs. These are the following values:
- Disjoint sets $X = X_1, \ldots, X_\beta$ where $|X_j(i)| \in \{0,1\}^\ell$ for every $j \in [\beta]$ and $i \in [|X_j|]$. Let the total number of elements in all sets be $N = \sum_j |X_j|$.
- The sets $T = T_1, \ldots, T_\beta$ sampled independently from $\mathcal{T}$.

Receiver's inputs. The queries $x_1, \ldots, x_\beta \in \{0,1\}^\ell$.

The functionality works as follows:
1. Sample uniformly random and independent keys $k = k_1, \ldots, k_\beta$.
2. Invoke $\mathsf{Hint}(k, X, T) \to \mathsf{hint}$.
3. Output hint to P_1 (P_2 can compute it on its own from k, X, T).
4. For every $j \in [\beta]$ output $F(k_j, \mathsf{hint}, x_j)$ to the receiver.

PROTOCOL 7 (Batch Oblivious PPRF)

The protocol is defined in the $\mathcal{F}_{\mathsf{OPRF}}$-hybrid model and receives an (ℓ, β)-PPRF $\hat{F} = (\mathsf{Hint}, F)$ as a parameter. The underlying PRF in both $\mathcal{F}_{\mathsf{OPRF}}$ and $\hat{F}$ is the same and denoted F'. The protocol proceeds as follows:
1. The parties invoke β instances of $\mathcal{F}_{\mathsf{OPRF}}$. In the $j \in [\beta]$ instance, P_2 inputs nothing and receives the key k_j, and P_1 inputs x_j and receives $F'(k_j, x_j)$.
2. Party P_2 invokes $p \leftarrow \mathsf{Hint}(k, X, T)$ and sends p to P_1.
3. For every $j \in [\beta]$, party P_1 outputs $F'(k_j, x_j) \oplus p(x_j)$.

Proof. Note that party P_2 receives nothing in the functionality but receives $k_1, \ldots, k_\beta$ in the real execution as output from $\mathcal{F}_{\mathsf{OPRF}}$. Therefore, P_2's view can be easily simulated with the simulator of $\mathcal{F}_{\mathsf{OPRF}}$.

As for the view of P_1, from the security of the PPRF it follows that it is indistinguishable from the output of $\mathcal{S}(1^\kappa, N)$ where $\mathcal{S}$ is the simulator from Definition 2. □

4 A Super-Linear Communication Protocol

4.1 The Basic Construction

Let $C_{a,b}$ be a Boolean circuit that has $2 \cdot a \cdot (b + \lambda)$ input wires, divided to a sections of $2b + \lambda$ inputs wires each. For each section, the first (resp. second) β input wires are associated with P_1 (resp. P_2). The last λ input wires are associated with P_1 as well. Denote the first (resp. second) β bits input to the i-th section by $u_{i,1}$ (resp. $v_{i,2}$) and the last λ bits by z_i. The circuit first compares $u_{i,1}$ to $v_{i,2}$ for every $i \in [\alpha]$ and produces $w_i = 1$ if $u_{i,1} = v_{i,2}$ and 0 otherwise.

PROTOCOL 9 (Private Set Intersection)

Inputs. P_1 has $X = \{x_1, \ldots, x_n\}$ and P_2 has $Y = \{y_1, \ldots, y_n\}$.
Protocol. The protocol proceeds in 3 steps as follows:

1. **Hashing.** The parties agree on hash functions $H_1, H_2 : \{0,1\}^\ell \rightarrow [\beta]$, which are used as follows:
   - P_1 uses H_1, H_2 in a Cuckoo hashing construction that maps $x_1, \ldots, x_n$ to a table Table_1 of $\beta = 2(1+\varepsilon)n$ entries, where input x_i is mapped to either entry $\mathsf{Table}_1[H_1(x_i)]$ or $\mathsf{Table}_1[H_2(x_i)]$ or the stash Stash (which is of size s)[a]. Since $\beta > n$, P_1 fills the empty entries in Table_1 with a uniformly random value.
   - P_2 maps $y_1, \ldots, y_n$ to Table_2 of β entries using both H_1 and H_2. That is, y_i is placed in both $\mathsf{Table}_2[H_1(y_i)]$ and $\mathsf{Table}_2[H_2(y_i)]$. (Obviously, some bins will have multiple items mapped to them. This is not an issue, and there is even no need to use a probabilistic upper bound on the occupancy of the bin.)
2. **Computing batch OPPRF.** P_2 samples uniformly random and independent target values $t_1, \ldots, t_\beta \in \{0,1\}^\kappa$. The parties invoke an (λ, β)-OPPRF (Functionality 6; recall that λ is the bit-length of the items). P_2 inputs $Y_1, \ldots, Y_\beta$ and $T_1, \ldots, T_\beta$ where $Y_j = \mathsf{Table}_2[j] = \{y||j \mid y \in Y \wedge j \in \{H_1(y), H_2(y)\}\}$ and T_j has $|Y_j|$ elements, all equal to t_j. If, $j = H_1(y) = H_2(y)$ for some $y \in Y$ then P_2 adds a uniformly random element to $\mathsf{Table}_2[j]$. P_1 inputs $\mathsf{Table}_1[1], \ldots, \mathsf{Table}_1[\beta]$ and receives $y_1^\star, \ldots, y_\beta^\star$. According to the definition of the OPPRF, if $\mathsf{Table}_1[j] \in \mathsf{Table}_2[j]$ then $y_j^\star = t_j$.
3. **Computing the circuit.** The parties use a two-party computation (Functionality 1) with the circuit $C_{\beta+s\cdot n,\gamma}$.[b] For section $j \in [\beta]$ of the circuit, party P_1 inputs the first γ bits of $y_j^\star$ and $\mathsf{Table}_1[j]$, and P_2 inputs the first γ bits t_j; for the $\beta + j$-th section P_1 inputs $\mathsf{Stash}[\lceil j/n \rceil + 1]$ and P_2 inputs $\mathsf{Table}[(j \bmod n) + 1]$.

[a] We discuss the value of s in Sect. 4.2 and the value of ε in Sect. 7.1.
[b] We discuss the value of γ in Sect. 4.2.

Then, the circuit computes and outputs $f(Z)$ where $Z = \{z_i \mid w_i = 1\}_{i \in [a]}$ and f is the function required to be computed in the $\mathcal{F}_{\mathsf{PSI},f}$ functionality.

Correctness. If $z \in X \cap Y$ then z is mapped to both $\mathsf{Table}_2[H_1(z)]$ and $\mathsf{Table}_2[H_2(z)]$ by P_2. There are two cases: (1) z is mapped to $\mathsf{Table}_1[H_b(z)]$ by P_1 for $b \in \{1,2\}$. (2) z is mapped to Stash by P_1. In the first case the match is found in section $H_b(z)$ of the circuit; in the second case the match is certainly found since every item in the Stash is compared to every item in Y.

Two items $x \in X$ and $y \in Y$ where $x \neq y$ will not be matched, since by the properties of the PPRF P_1 receives a pseudorandom output. Since the parties only input the first γ bits of the PPRF results, those values will be matched with probability $2^{-\gamma}$. See Sect. 4.2 for a discussion on limiting the failure probability.

Security. The security of the protocol follows immediately from the security of the OPPRF and the two-party computation functionalities.

4.2 Limiting the Failure Probability

Protocol 9 might fail due to two reasons:

- **Stash size.** For an actual implementation, one needs to fix s and ε so that the stash failure probability will be smaller than $2^{-\sigma}$. If the stash is overflowed (i.e., more than s items are mapped to it) then the protocol fails.[5] As discussed in Sect. 2, setting $s = O(\log n / \log\log n)$ makes the failure probability negligible.
- **Input encoding.** The circuit compares the first γ bits of $y_j^\star$ of P_1 to the first γ bits of t_j of P_2. Thus, the false positive error probability in each comparison equals $2^{-\gamma}$ (due to $F(x)$, for $x \notin Y$, being equal to the programmed output), and therefore the overall probability of a false positive is at most $\beta \cdot 2^{-\gamma} = 2(1+\varepsilon)n \cdot 2^{-\gamma}$.

4.3 Reducing Computation

A major computation task of the protocol is interpolating the polynomial which encodes the hint. If we use Cuckoo hashing with $K = O(1)$ hash functions then the polynomial encodes $O(n)$ items and is of degree $O(n)$. This section describes how to reduce the *asymptotic* overhead of *computing* the polynomial and therefore we will use asymptotic notation. The concrete overhead is discussed in Sect. 7.2.

The overhead of interpolating a polynomial of degree $O(n)$ over arbitrary points is $O(n^2)$ operations using Lagrange interpolation, or $O(n \log^2 n)$ operations using FFT. The overhead can be reduced by dividing the polynomial to several lower-degree polynomials. In particular, let us divide the $\beta = O(n)$ bins to B "mega-bins", each encompassing β/B bins. Suppose that we have an upper bound such that the number of items in a mega-bin is at most m, except with negligible probability. Then the protocol can invoke a batch OPPRF for each mega-bin, using a different hint polynomial. Each such polynomial is of degree m. Therefore the computation overhead is $O(B \cdot m \log^2 m)$. Ideally, the upper bound on the number of items in a mega-bin, m, is of the same order as the expected number of items in a mega-bin, $O(n/B)$. In this case the computation overhead is $O(n/B \cdot B \cdot \log^2(n/B)) = O(n \log^2(n/B))$ and will be minimized when the number of mega-bins B is maximal.

It is known that when mapping $O(n)$ items to $B = n/\log n$ (mega-)bins, then with high probability the most occupied bin has less than $m = O(n/B) = O(\log n)$ items. When interested in concrete efficiency we can use the analysis

[5] In that case either not all items are stored in the stash – resulting in the protocol ignoring part of the input and potentially computing the wrong output, or P_1 needs to inform P_2 that it uses a stash larger than s – resulting in a privacy breach.

in [PSZ18] to find the exact number of mega-bins to make the failure probability sufficiently small (see Sect. 7.2). When interested in asymptotic analysis, it is easy to deduce from the analysis in [PSZ18] that with $B = n/\log n$ mega-bins, the probability of having more than $\omega(\log n)$ items in a mega-bin is negligible. Therefore when using this number of mega-bins, the computation overhead is only $\omega((n/\log n) \cdot \log^2(n)) = \omega(n \log n)$ using Lagrange interpolation. Using FFT interpolation, the asymptotic overhead is reduced to $\omega((n/\log n) \log n (\log\log n)^2) = \omega(n \cdot (\log\log n)^2)$. But since we map relatively few items to each mega-bin the gain in practice of using FFT is marginal.

5 A Linear Communication Protocol

We describe here a protocol in which the circuit computes only $O(n)$ comparisons. This protocol outperforms the protocols in Sect. 4.1 or in [PSWW18, CO18] which have a circuit that computes $\omega(n)$ comparisons. A careful analysis reveals that those protocols require $O(n)$ comparisons to process all items that were mapped to the Cuckoo hash table, and an additional $s \cdot n$ comparisons to process the $s = \omega(1)$ items that were mapped to the stash. We note that the concurrent and independent work of [FNO18] proposes to use a PSI protocol for unbalanced set sizes, such as in the work of [KLS+17], to reduce the complexity of handling the stash from $\omega(n)$ to $O(n)$ in PSI protocols. However, their idea can only be applied when the output is the intersection itself. When the output is a function of the intersection then their protocol has communication complexity $O(n \log\log n)$, cf. Sect. 1.4). In contrast, we achieve $O(n)$ communication even when the output is a function of the intersection.

We present two different techniques to achieve a linear communication protocol with failure probability that is negligible in the statistical security parameter σ. The first technique (see Sect. 5.1) is implied by a mathematical analysis of the failure probability (as argued in Sect. 1.4). The second technique (see Sect. 5.2) is implied by the empirical analysis presented in [PSZ18].

5.1 Linear Communication via Dual Execution

We overcome the difficulty of handling the stash by running a modified version of the protocol in three phases. The first phase is similar to the basic protocol, but ignores the items that P_1 maps to the stash. Therefore this phase inputs to the circuit the $O(n)$ results of comparing P_1's input items (except those mapped to the stash) with all of P_2's items. The second phase reverses the roles of the parties, and in addition now P_1 inputs only the items that it previously mapped to the stash. In this phase P_2 uses Cuckoo hashing and might map some items to the stash. The last phase only compares the items that P_1 mapped to the stash in the first phase, to the items that P_2 mapped to the stash in the second phase, and therefore only needs to handle very few items. Below, we describe our protocol in more detail: In Protocol 10, we describe our protocol in more detail.

PROTOCOL 10 (PSI with Linear Communication)

Inputs. P_1 has $X = \{x_1, \dots, x_n\}$ and P_2 has $Y = \{y_1, \dots, y_n\}$.
Protocol. The protocol proceeds in 3 phases as follows:

1. Run steps 1-2 of Protocol 9. Denote the items mapped to P_1's table by X_T (i.e., excluding the items mapped to the stash). In the end of this phase, for every $j \in [\beta]$, P_1 holds the OPPRF result $y_j^\star$ and P_2 holds the target value t_j.
2. Reverse the roles of P_1 and P_2 and run steps 1-2 of Protocol 9 again, where P_1 inputs $X_S = X \setminus X_T$ (i.e., only the items that were previously mapped to the stash) and P_2 inputs Y. Since the roles are reversed then P_1 maps X_S using simple hashing and P_2 maps Y using Cuckoo hashing. Denote the items mapped to the table and stash of P_2 by Y_T and Y_S, respectively. In the end of this phase, for every $j \in [\beta]$, P_1 has the target value $\tilde{t}_j$ and P_2 has the OPPRF result $\tilde{y}_j^\star$.
3. The parties use secure two-party computation (Functionality 1) with the circuit $C_{2\beta+s^2,\gamma}$ (where s is the stash size). For section $j \in [\beta]$ of the circuit, P_1 and P_2 input the first γ bits of $y_j^\star$ and t_j resp. For section $j \in \{\beta+1, \dots, 2\beta\}$ of the circuit P_1 and P_2 input the first γ bits of $\tilde{t}_j$ and $\tilde{y}_j^\star$, respectively. Finally, for the rest s^2 sections of the circuit, the parties input every combination of $X_S \times Y_S$ (padded with uniformly random items so that $|X_S| = |Y_S| = s$).

Correctness & Efficiency. The protocol compares every pair in $X \times Y$ and therefore every item in the intersection is input to the circuit exactly once: Sections $1, \dots, \beta$ of the circuit cover all pairs in $X_T \times Y$, sections $\beta + 1, \dots, 2\beta$ cover all pairs in $X_S \times Y_T$ and sections $2\beta + 1, \dots, 2\beta + s^2$ covers all pairs in $X_S \times Y_S$. This implies that the result of the three-phase construction is exactly the intersection $X \cap Y$. The communication complexity in the first two steps of the protocol is $O(n \cdot \kappa)$ as they involve the execution of a OPPRF with at most $O(n)$ items to the parties. The communication complexity of the third step is $O(n \cdot \gamma)$ since it involves $2n + s^2$ comparisons of γ-bit elements. Since the stash size is $s = O(\log n)$, overall there are $O(n)$ comparisons.

Security. As in the basic protocol (see Sect. 4.1), the security of this protocol is implied by the security of the OPPRF and secure two-party computation.

5.2 Linear Communication via Stash-Less Cuckoo Hashing

The largest communication cost factor in our protocols is the secure evaluation of the circuit. The asymptotically efficient Protocol 10 requires computing at least two copies of the basic circuit (for Phases 1 and 2), and it is therefore preferable to implement a protocol which has better *concrete* efficiency. We design a protocol that requires no stash (while achieving a small failure probability of less than 2^{-40}), and hence uses no dual execution.

In order to be able to not use the stash, hashing is done with $K > 2$ hash functions. We take into account the results of [PSZ18], which ran an empirical evaluation for the failure probability of Cuckoo hashing (failure is defined as the event where an item cannot be stored in the table and must be stored in the stash). They run experiments for a failure probability of 2^{-30} with $K = 3, 4$ and 5 hash functions, and extrapolated the results to yield the minimum number of bins for achieving a failure probability of less than 2^{-40}. The results showed that $\beta = 1.27n, 1.09n$, and $1.05n$ bins are required for $K = 3, 4$, and 5, respectively.

The main obstacle in using more than two hash functions in previous works on PSI was that the communication was still linear in $O(\max_b \cdot \beta)$, where $\max_b$ is the maximal number of elements in a bin of the simple hash table. The value of $\max_b$ increases with K since each item is stored K times in the simple hash table. In our protocol the communication for the circuit is independent of $\max_b$, as it only depends on the number of bins β. The communication for sending the polynomials, whose size is $O(K \cdot n \cdot \kappa)$, is just a small fraction of the overall communication and was in our experiments always smaller than 3%. In this paper, we therefore use $K = 3$ hash functions for our stash-less protocol.

6 PSI with Associated Payload

In many cases, each input item of the parties has some "payload" data associated with it. For example, an input item might include an id which is a credit card number, and a payload which is a transaction that was paid using this credit card. The parties might wish to compute some function of the *payloads* of the items in the intersection (for example, the sum of the financial transactions associated with these items). However, a straightforward application of our techniques does not seem to support this type of computation: Recall that P_2 might map multiple items to each bin. The OPPRF associates a single output β to all these items, and this value is compared in the circuit with the output α of P_1. But if P_2 inserts a single item to the circuit, it seems that this item cannot encode the payloads of all items mapped to this bin.

The 2D Cuckoo hashing circuit-based PSI protocol of [PSWW18] handles payloads well, since each comparison involves only a single item from each party. While our basic protocol cannot handle payloads, we show here how it can be adapted to efficiently encode payloads in the input to the circuit.

Let Table_1 and Stash be P_1's table and stash after mapping its items using Cuckoo hashing and let Table_2 be P_2's table after mapping its items using simple hashing. In addition, denote by $U(x)$ and $V(y)$ the payloads associated with $x \in X$ and $y \in Y$ respectively and assume that all payloads have the same length δ. The parties invoke two instances of batch OPPRF as follows:

1. A batch OPPRF where P_1 inputs $\mathsf{Table}_1[1], \ldots, \mathsf{Table}_1[\beta]$ and P_2 inputs $\mathsf{Table}_2[1], \ldots, \mathsf{Table}_2[\beta]$ and $T_1, \ldots, T_\beta$ where T_j has $|\mathsf{Table}_2[j]|$ elements, all equal to a uniformly random and independent value $t_j \in \{0,1\}^\lambda$. This is the same invocation of a batch OPPRF as in Protocol 9. At the end, P_1 has the OPPRF results $y_1^\star, \ldots, y_\beta^\star$ and P_2 has the target values $t_1, \ldots, t_j$.

2. In the second batch OPPRF, P_2 chooses the target values such that the elements in the set T_j are not equal. Specifically, P_1 inputs $\mathsf{Table}_1[1], \ldots, \mathsf{Table}_1[\beta]$ and P_2 samples $\tilde{t}_1, \ldots, \tilde{t}_\beta$ uniformly, and inputs $\mathsf{Table}_2[1], \ldots, \mathsf{Table}_2[\beta]$ and $T_1, \ldots, T_\beta$ where $T_j(i) = \tilde{t}_j \oplus V(\mathsf{Table}_2[j](i))$. Denote the OPPRF results that P_1 obtains by $\tilde{y}_1^\star, \ldots, \tilde{y}_\beta^\star$.

Then, the circuit operates in the following way: For the j-th section, P_1 inputs $\mathsf{Table}_1[j], y_j^\star, \tilde{y}_j^\star$ and $U(\mathsf{Table}_1[j])$, and P_2 inputs t_j and $\tilde{t}_j$. The circuit compares $y_j^\star$ to t_j. If they are equal then it forwards to the sub-circuit that computes f the item $\mathsf{Table}_1[j]$ itself, P_1's payload $U(\mathsf{Table}_1[j])$ and P_2's payload $\tilde{y}_j^\star \oplus \tilde{t}_j$. This holds since if $\mathsf{Table}_1[j]$ is the i-th item in P_2's table, namely, $\mathsf{Table}_2[j](i)$, then the value $\tilde{y}_j^\star$ received by P_1 is $\tilde{y}_j^\star = \tilde{t}_j \oplus V(\mathsf{Table}_2[j](i))$. Thus, $\tilde{y}_j^\star \oplus \tilde{t}_j = V(\mathsf{Table}_2[j](i))$ as required.

Efficiency. The resulting protocol has the same asymptotic complexity as our initial protocols without payloads. The number of comparisons in the circuit is the same as in the basic circuit.

Table 1. The results of [PSZ18] for the required stash sizes s for $K = 2$ hash functions and $\beta = 2.4n$ bins, and the minimum OPPRF output bitlength γ to achieve failure probability $< 2^{-40}$ when mapping n elements into β bins with Cuckoo hashing. For $K > 2$ hash functions we choose a large enough number of bins β to achieve stash failure probability $< 2^{-40}$.

# Elements n		2^8	2^{12}	2^{16}	2^{20}	2^{24}
Stash size s for $K = 2$		12	6	4	3	2
OPPRF output length γ	$K = 2$, $\beta = 2.4n$	50	54	58	62	66
	$K = 3$, $\beta = 1.27n$, $s = 0$	49	53	57	61	65
	$K = 4$, $\beta = 1.09n$, $s = 0$	49	53	57	61	65
	$K = 5$, $\beta = 1.05n$, $s = 0$	49	53	57	61	65

7 Concrete Costs

In this section we evaluate the concrete costs of our protocol for concrete values of the security parameters. We set the computational security parameter to $\kappa = 128$, and the statistical security parameter to $\sigma = 40$.

7.1 Parameter Choices for Sufficiently Small Failure Probability

For $K = 2$ hash functions, following previous works on PSI (e.g., [PSSZ15, PSWW18]), we set the table size parameter for Cuckoo hashing to $\epsilon = 0.2$, and use a Cuckoo table with $\beta = 2n(1 + \epsilon) = 2.4n$ bins. The resulting stash sizes for mapping n elements into $\beta = 2.4n$ bins, as determined by the experiments in

[PSZ18], are summarized in Table 1. Note that we use here concrete values for the stash size, and are aiming for a failure probability smaller than 2^{-40}. This can either be achieved using the basic protocol of Sect. 4.1 with the right choice of the stash size, or by running the three rounds $O(n)$ complexity protocol of Sect. 5.

Another option is described in Sect. 5.2, where we use more than two hash functions (specifically, use $K = 3, 4$, or 5 functions), with the hash table being of size $\beta = 1.27n, 1.09n$, or $1.05n$, respectively. These parameters achieve a failure probability smaller than 2^{-40} according to the experimental analysis in [PSZ18].

As described in Sect. 4.2, even if there are no stash failures, the scheme can fail due to collisions in the output of the PRF, with probability $\beta \cdot 2^{-\gamma}$, where γ is the output bitlength of the OPPRF. To make this failure probability smaller than the statistical security parameter (which we set to 40), the output bitlength of the OPPRF must be $\gamma = 40 + \log_2 \beta$ bits.

7.2 Computing Polynomial Interpolation

We implemented interpolation of polynomials of degree d using an $O(d^2)$ algorithm based on Lagrange interpolation in a prime field where the prime is the Mersenne prime $2^{61} - 1$. The runtime for interpolating a polynomial of degree $d = 1024$ was 7 ms, measured on an Intel Core i7-4770K CPU with a single thread. The runtimefor different values of d behaved (very accurately) as a quadratic function of d. The actual algorithmsare those implemented in NTL v10.0 with field arithmetics replaced with our customized arithmetic operationsover the Mersenne prime $2^{61} - 1$. Most importantly, this field enables an order of magnitude faster multiplication of field elements: multiplying $x \cdot y$ with $|x|, |y| \leq 61$ is implemented by multiplying x and y over $\mathbb{Z}$ to obtain $z = xy$ with $|z| \leq 122$. Then the result is the sum of the element represented by the lower 61 bits of z with the element represented by the higher 61 bits of z (and therefore no expensive modular reduction is required). The Mersenne prime $2^{61} - 1$ allows the use of at least 40-bit statistical security for up to $n = 2^{20}$ elements for all our algorithms using permutation-based hashing (cf. [PSSZ15]). To use larger sets, we see two possible solutions: (i) using a larger Mersenne prime or (ii) reducing the statistical security parameter σ (e.g., using $\sigma = 38$ for achieving less than $2^{-\sigma}$ failure probability for $n = 2^{22}$ elements, $K = 3$ hash functions, and $\beta = 1.27n$ bins). The required minimum bit-length of the elements using permutation-based hashing with failure probability $2^{-\sigma}$ is computed as $\ell = \sigma + 2\log_2 n - \log_2 \beta$. The OPPRF output is also ≤ 61 bits in most cases as shown in Table 1.

For reducing the computation complexity of our protocol, we use the approach described in Sect. 4.3, where instead of interpolating a polynomial of degree $K \cdot n$, where K is the number of hash functions and n is the number of elements for PSI, we interpolate multiple smaller polynomials of degree at most $d = 1024$. We therefore have to determine the minimum number of mega-bins B such that when mapping $N = K \cdot n$ elements to B bins, the probability of having a bin with more than $\max_b = 1024$ elements is smaller than 2^{-40}. As in the analysis for simple hashing in [PSZ18], we use the formula from [MR95]:

$$P(\text{“}\exists\text{ bin with} \geq \max_b \text{ elements”}) \leq \sum_{i=1}^{B} P(\text{“bin } i \text{ has} \geq \max_b \text{ elements”})$$
$$= B \cdot \sum_{i=\max_b}^{N} \binom{N}{i} \cdot \left(\frac{1}{B}\right)^i \cdot \left(1 - \frac{1}{B}\right)^{N-i}.$$

We depict the corresponding numbers in Table 2. With these numbers and our experiments for polynomial interpolation described above, the estimated runtimes for the polynomial interpolation are $B \cdot 7$ ms. The hints (polynomials) that need to be sent have size $B \cdot \max_b \cdot \gamma$ bits which is only slightly larger than the ideal communication of $K \cdot n \cdot \gamma$ bits when using one large polynomial as shown in Table 2.

Note that in contrast to many PSI solutions whose main run-time bottleneck is already network bandwidth (which cannot be easily improved in many settings such as over the Internet), the run-time of our protocols can be improved by using multiple threads instead of one thread. Since the interpolation of polynomials for different mega-bins is independent of each other, the computation scales linearly in the number of physical cores and thus can be efficiently parallelized.

Table 2. Parameters for mapping $N = K \cdot n$ elements to B mega-bins s.t. each mega-bin has at most $\max_b \leq 1024$ elements with probability smaller than 2^{-40}. The lower half of the table contains the expected costs for the polynomial interpolations.

# hash functions	$K = 2$			$K = 3$		
Set size	$n = 2^{12}$	$n = 2^{16}$	$n = 2^{20}$	$n = 2^{12}$	$n = 2^{16}$	$n = 2^{20}$
# mega-bins B	11	165	2663	16	248	4002
Maximum number of elements $\max_b$	944	1021	1024	975	1021	1024
Polynomial interpolation [in milliseconds]	126	1815	29293	183	2809	45335
Size of hints [in bits]	560736	9770970	169068544	826800	14432856	249980928
Ideal size of hints for one polynomial [in bits]	436330	7505580	128477895	651264	11206656	191889408

7.3 Communication and Depth Comparison

We first compute the communication complexity of our basic construction from Sect. 4.1. The communication is composed of (a) the OPRF evaluations for each of the B bins, (b) the hints consisting of the polynomials, (c) the circuit for comparing the outputs of the OPPRFs in each bin, and (d) the circuit for comparing the s elements on the stash with the n elements of P_2.

With regards to (a), the OPRF protocol of [KKRT16], which was also used in [KMP+17], has an amortized communication of at most 450 bits per OPRF evaluation for set sizes up to $n = 2^{24}$ elements (cf. [KKRT16, Table 1]). This amounts to $B \cdot 450$ bits of communication.

With regards to (b), for the size of the hints in the OPPRF construction we use the values given in Table 2. These numbers represent the communication when using mega-bins, and are slightly larger than the ideal communication of $K \cdot n$ coefficients of size γ bits each, that would have been achieved by using a single polynomial for all values. However, it is preferable to use mega-bins since their usage substantially improves the computation complexity as described in Sect. 4.3, while the total communication for the hints is at most 3% of the total communication. (This also shows that any improvements of the size of the hints will have only a negligible effect on the total communication).

With regards to (c), the circuit compares B elements of bitlength γ, and hence requires $B \cdot (\gamma - 1)$ AND gates. With 256 bits per AND gate [ALSZ13, ZRE15] this yields $B \cdot (\gamma - 1) \cdot 256$ bits of communication.

With regards to (d), the final circuit consists of $s \cdot n$ comparisons of bitlength σ. This requires $sn(\sigma - 1) \cdot 256$ bits of communication.

We now analyze the communication complexity of our $O(n)$ protocol described in Sect. 5. The main difference compared to the basic protocol analyzed above is that a different method is used for comparing the elements of the stash, i.e., replacing step (d) above. The new method replaces this step by letting P_2 use Cuckoo hashing of its n elements into B bins and then evaluating OPRF for each of these bins. This requires $B \cdot 450$ bits of communication plus B comparisons of γ bit values. Overall, this amounts to $B \cdot (450 + (\gamma - 1) \cdot 256)$ bits of communication. For simplicity, we omit the communication for comparing the elements for phase 3 which compares the elements on the two stashes, as it is negligible.

Comparison to Previous Work. In Table 3, we compare the resulting communication of our protocols to those of previous circuit-based PSI protocols of [HEK12, PSSZ15, PSWW18, FNO18]. As can be seen from this table, our protocols improve communication by an integer factor, where the main advantage of our protocols is that their communication complexity is *independent* of the bitlength of the input elements. Namely, for arbitrary input bitlengths, our no-stash protocol improves the communication over the previous best protocol of [PSWW18] *by a factor of 12.8x for* $n = 2^{12}$ *to a factor of 10.1x for* $n = 2^{20}$. For fixed bitlength of $\sigma = 32$ bits, our no-stash protocol improves communication over [PSWW18] *by a factor of 5.7x for* $n = 2^{12}$ *to a factor of 2.6x for* $n = 2^{20}$.

Circuit Depth. For some secure circuit evaluation protocols like GMW [GMW87] the round complexity depends on the depth of the circuit. In Table 4, we depict the circuit depths for concrete parameters of our protocols and previous work, and show that our circuits have about the same low depth as the best previous works [PSSZ15, PSWW18]. In more detail, the Sort-Compare-Shuffle (SCS) circuit of [HEK12] has depth $\log_2 \sigma \cdot \log_2 n$ when using depth-optimized comparison circuits. The protocols of [PSSZ15, PSWW18] have depth

Table 3. Communication in MB for circuit-based PSI on n elements of fixed bitlength $\sigma = 32$ (left) and arbitrary bitlength hashed to $\sigma = 40 + 2\log_2(n) - 1$ bits (right). The numbers for previous protocols are based on the circuit sizes given in [PSWW18, Table 3] with 256 bit communication per AND gate. The best values are marked in bold.

Protocol $n =$	$\sigma = 32$			Arbitrary σ		
	2^{12}	2^{16}	2^{20}	2^{12}	2^{16}	2^{20}
SCS [HEK12]	104	2 174	42 976	205	4 826	106 144
Circuit-Phasing [PSSZ15]	130	1 683	21 004	320	5 552	97 708
Hashing + SCS [FNO18]	-	1 537	21 207	-	3 998	72 140
2D CH [PSWW18]	51	612	6 582	115	1 751	25 532
Ours Basic Sect. 4.1	41	550	8 123	65	870	12 731
Ours Advanced Sect. 5	35	604	10 277	35	604	10 277
Ours No-Stash Sect. 5.2	**9**	**149**	**2 540**	**9**	**149**	**2 540**
Breakdown:						
OPRF	0.3 (3%)	5 (3%)	72 (3%)	0.3 (3%)	5 (3%)	72 (3%)
Sending polynomials	0.1 (1%)	2 (1%)	30 (1%)	0.1 (1%)	2 (1%)	30 (1%)
Circuit	9 (96%)	142 (96%)	2438 (96%)	9 (96%)	142 (96%)	2438 (96%)
Improvement factor	5.7x	4.1x	2.6x	12.8x	11.8x	10.1x

$\log_2 \sigma$. A depth-optimized SCS circuit for the construction in [FNO18] has depth $\log_2(\sigma - \log_2(n/b)) \cdot \log_2((1+\delta)b)$, where concrete parameters for n, δ, b are given in [FNO18, Table 1]. Our protocols consist of circuits for comparing the elements on the stash of bitlength σ and the outputs of the OPPRFs of length γ and therefore have depth $\max(\log \sigma, \log \gamma) = \max(\log \sigma, \log_2(40 + 2\log_2(n) - 1))$.

Table 4. Circuit depth for circuit-based PSI on n elements of fixed bitlength $\sigma = 32$ (left) and arbitrary bitlength hashed to $\sigma = 40 + 2\log_2(n) - 1$ bits (right).

Protocol $n =$	$\sigma = 32$			Arbitrary σ		
	2^{12}	2^{16}	2^{20}	2^{12}	2^{16}	2^{20}
SCS [HEK12]	60	80	100	72	98	126
Circuit-Phasing [PSSZ15]	5	5	5	6	7	7
Hashing + SCS [FNO18]	-	42	36	-	54	51
2D CH [PSWW18]	5	5	5	6	7	7
Our Protocols	6	6	6	6	7	7

7.4 Runtime Comparison

In this section we compare the runtimes of different PSI protocols. In Sect. 7.2 we conducted experiments for polynomial interpolation, the main new part of our protocol, and we show below that this step takes only a small fraction of the total runtime. We also implemented our most efficient protocol (see Sect. 5.2).[6] In addition, we estimate the runtime of our less efficient basic protocol (see

[6] Our implementation is available at https://github.com/encryptogroup/OPPRF-PSI.

Sect. 4.1) and the protocol with linear communication overhead (see Sect. 5) based on the experiments of the interpolation procedure and rigorous estimations from previous works.

Previous Work. As we have seen in the analysis of the communication overhead in Sect. 7.3, our protocols provide better improvements to performance in the case of arbitrary bitlengths. The previous work of [PSWW18] gave runtimes only for fixed bitlength of 32 bits in [PSWW18, Table 4]. Therefore, we extrapolate the runtimes of the previous protocols from fixed bitlength to arbitrary bitlength based on the circuit sizes given in [PSWW18, Table 3]. The estimated runtimes are given in Table 5. The LAN setting is a 1 Gbit/s network with round-trip time of 1 ms and the WAN setting is a 100 Mbit/s network with round-trip time of 100 ms. Runtimes were not presented in [FNO18], but since their circuit sizes and depths are substantially larger than those of [PSWW18] (cf. Tables 3 and 4), their runtimes will also be substantially higher than those of [PSWW18].

Our Implementation. We implemented and benchmarked our most efficient no-stash OPPRF-based PSI protocol (see Sect. 5.2) on two commodity PCs with an Intel Core i7-4770K CPU. We instantiated our protocol with security parameter $\kappa = 128$ bits, $K = 3$ hash functions, $B = 1.27n$ bins, and no stash (see Sect. 5.2). Our OPPRF implementation is based on the OPRF protocol of [PSZ18].[7] For the secure circuit evaluation, we used the ABY framework [DSZ15]. The run-times are averaged over 50 executions. The results are described in Table 5.

Comparison with PSI Protocols. As a baseline, we compare our performance with specific protocols for computing the intersection itself. (However, as is detailed in Sect. 1.2, our protocol is circuit-based and therefore has multiple advantages compared to specific PSI protocols). Our best protocol is slower by a factor of 41x than today's fastest PSI protocol of [KKRT16] for $n = 2^{20}$ elements in the WAN setting (cf. Table 5).

Comparison with Public Key-Based PSI Variant Protocols. Our circuit-based protocol is substantially faster than previous public key-based protocols for computing variants of PSI, although they have similar asymptotic linear complexity. As an example, consider comparing whether the size of the intersection is greater than a threshold (PSI-CAT). In our protocol, we can compute the PSI-CAT functionality by extending the PSI circuit of Table 5 with a Hamming distance circuit (which, using the size-optimal construction of [BP06], adds less than n AND gates). The final comparison with the threshold adds another $\log_2 n$

[7] This OPRF protocol has communication that is higher by 10% to 15% than the communication of the OPRF protocol of [KKRT16]. But since OPRF requires less than 3% of the total communication, this additional cost is negligible in our protocol.

Table 5. Total run-times in ms for PSI variant protocols on n elements of arbitrary bitlength using GMW [GMW87] for secure circuit evaluation and one thread. Numbers for all but our protocols are based on [PSWW18]. The best values for generic circuit-based PSI protocols are marked in bold.

Network	LAN			WAN		
Protocol $n =$	2^{12}	2^{16}	2^{20}	2^{12}	2^{16}	2^{20}
Special-purpose PSI protocols (as baseline)						
DH/ECC PSI [Sha80, Mea86, DGT12]	3296	49010	7904054	4082	51866	8008771
BaRK-OPRF [KKRT16]	113	295	3882	540	1247	14604
Generic circuit-based PSI protocols						
Circuit-Phasing [PSSZ15]	7825	67292	1126848	37380	327976	4850571
2D CH [PSWW18]	5031	25960	336134	22796	129436	1512505
Ours Basic Sect. 4.1 (estimated)	2908	13767	182204	12934	63861	752695
Ours Advanced Sect. 5 (estimated)	1674	9763	148436	7372	43675	597885
Ours No-Stash Sect. 5.2, Total	**1 199**	**8 486**	**120 731**	**5 910**	**22 134**	**261 481**
Breakdown:						
OPRF	724 (60%)	1097 (13%)	5844 (5%)	2867 (49%)	4164 (19%)	26121 (10%)
Polynomial interpolation	183 (15%)	2809 (33%)	45335 (38%)	183 (3%)	2809 (13%)	45335 (17%)
Polynomial transmission	16 (1%)	145 (2%)	667 (0%)	816 (13%)	1079 (5%)	4012 (2%)
Polynomial evaluation	58 (5%)	1344 (16%)	21768 (18%)	58 (1%)	1344 (6%)	21768 (8%)
Circuit	218 (18%)	3091 (36%)	47117 (39%)	1986 (34%)	12738 (57%)	164245 (63%)
Improvement over [PSWW18]	4.2x	3.1x	2.8x	3.9x	5.8x	5.8x

AND gates [BPP00] which are negligible as well. For the PSI-CAT functionality, [ZC17] report runtimes of 779 s for $n = 2^{11}$ elements, [HOS17] report runtimes of 728 s for $n = 2^{11}$ elements, and [ZC18] report runtimes of at least 138 s for $n = 100$ elements, whereas our protocol requires 0.52 s for $n = 2^{11}$ elements and 0.34 s for $n = 100$ elements. Hence, we improve over [ZC17] by a factor of 1 498x, over [HOS17] by a factor of 1 400x, and over [ZC18] by a factor of 405x. As an example for computing PSI-CAT with larger set sizes, our protocol requires 124 s for $n = 2^{20}$ elements.

The protocol described by Google for computing ad revenues [Yun15, Kre17] (see Sect. 1.2) is based on the DH-based PSI protocol which is already 65x slower than our protocol for $n = 2^{20}$ elements over a LAN (cf. Table 5) and leaks the intersection cardinality as an intermediate result. Here, too, our circuit would be only slightly larger than the PSI circuit of Table 5.

Comparison with Circuit-Based PSI Protocols. As can be seen from Table 5, our no-stash protocol from Sect. 5.2 is substantially more efficient than our basic protocol and our linear asymptotic overhead protocol from Sect. 4.1 and Sect. 5, respectively. It improves over the best previous circuit-based PSI protocol from [PSWW18] by factors of 4.2x to 2.8x in the LAN setting, and by factors of 5.8x to 3.9x in the WAN setting. From the micro-benchmarks in Table 5, we also observe that the runtimes for the polynomial interpolation are a significant fraction of the total runtimes of our protocols (3% to 33% for the interpolation and 1% to 18% for the evaluation). Since polynomials are independent of each other, the interpolation and evaluation can be trivially parallelized for running with multiple threads, which would give this part of the computation a speed-up that is linear in the number of physical cores of the processor.

Acknowledgements. We thank Ben Riva and Udi Wieder for valuable discussions about this work. This work has been co-funded by the DFG within project E4 of the CRC CROSSING and by the BMBF and the HMWK within CRISP, by the BIU Center for Research in Applied Cryptography and Cyber Security in conjunction with the Israel National Cyber Bureau in the Prime Ministers Office, and by a grant from the Israel Science Foundation.

References

[ADN+13] Asokan, N., et al.: CrowdShare: secure mobile resource sharing. In: Jacobson, M., Locasto, M., Mohassel, P., Safavi-Naini, R. (eds.) ACNS 2013. LNCS, vol. 7954, pp. 432–440. Springer, Heidelberg (2013). https://doi.org/10.1007/978-3-642-38980-1_27

[ALSZ13] Asharov, G., Lindell, Y., Schneider, T., Zohner, M.: More efficient oblivious transfer and extensions for faster secure computation. In: CCS (2013)

[BP06] Boyar, J., Peralta, R.: Concrete multiplicative complexity of symmetric functions. In: Královič, R., Urzyczyn, P. (eds.) MFCS 2006. LNCS, vol. 4162, pp. 179–189. Springer, Heidelberg (2006). https://doi.org/10.1007/11821069_16

[BPP00] Boyar, J., Peralta, R., Pochuev, D.: On the multiplicative complexity of Boolean functions over the basis $(\wedge, \oplus, 1)$. TCS **235**(1), 43–57 (2000)

[CADT14] H. Carter, C. Amrutkar, I. Dacosta, and P. Traynor. For your phone only: custom protocols for efficient secure functionevaluation on mobile devices. Secur. Commun. Netw. **7**(7) (2014)

[CO18] Ciampi, M., Orlandi, C.: Combining private set-intersection with secure two-party computation. In: Catalano, D., De Prisco, R. (eds.) SCN 2018. LNCS, vol. 11035, pp. 464–482. Springer, Cham (2018). https://doi.org/10.1007/978-3-319-98113-0_25

[DC17] Davidson, A., Cid, C.: An efficient toolkit for computing private set operations. In: Pieprzyk, J., Suriadi, S. (eds.) ACISP 2017. LNCS, vol. 10343, pp. 261–278. Springer, Cham (2017). https://doi.org/10.1007/978-3-319-59870-3_15

[DD15] Debnath, S.K., Dutta, R.: Secure and efficient private set intersection cardinality using bloom filter. In: Lopez, J., Mitchell, C.J. (eds.) ISC 2015. LNCS, vol. 9290, pp. 209–226. Springer, Cham (2015). https://doi.org/10.1007/978-3-319-23318-5_12

[DGT12] De Cristofaro, E., Gasti, P., Tsudik, G.: Fast and private computation of cardinality of set intersection and union. In: Pieprzyk, J., Sadeghi, A.-R., Manulis, M. (eds.) CANS 2012. LNCS, vol. 7712, pp. 218–231. Springer, Heidelberg (2012). https://doi.org/10.1007/978-3-642-35404-5_17

[DKT10] De Cristofaro, E., Kim, J., Tsudik, G.: Linear-complexity private set intersection protocols secure in malicious model. In: Abe, M. (ed.) ASIACRYPT 2010. LNCS, vol. 6477, pp. 213–231. Springer, Heidelberg (2010). https://doi.org/10.1007/978-3-642-17373-8_13

[DSMRY09] Dachman-Soled, D., Malkin, T., Raykova, M., Yung, M.: Efficient robust private set intersection. In: Abdalla, M., Pointcheval, D., Fouque, P.-A., Vergnaud, D. (eds.) ACNS 2009. LNCS, vol. 5536, pp. 125–142. Springer, Heidelberg (2009). https://doi.org/10.1007/978-3-642-01957-9_8

[DSZ15] Demmler, D., Schneider, T., Zohner, M.: ABY - a framework for efficient mixed-protocol secure two-party computation. In: NDSS (2015)

[DT10] De Cristofaro, E., Tsudik, G.: Practical private set intersection protocols with linear complexity. In: Sion, R. (ed.) FC 2010. LNCS, vol. 6052, pp. 143–159. Springer, Heidelberg (2010). https://doi.org/10.1007/978-3-642-14577-3_13

[Dwo06] Dwork, C.: Differential privacy. In: ICALP (2006)

[EFLL12] Ejgenberg, Y., Farbstein, M., Levy, M., Lindell, Y.: SCAPI: the secure computation application programming interface. Cryptology ePrint Archive, Report 2012/629 (2012)

[FHNP16] Freedman, M.J., Hazay, C., Nissim, K., Pinkas, B.: Efficient set intersection with simulation-based security. J. Cryptol. **29**(1), 115–155 (2016)

[FIPR05] Freedman, M.J., Ishai, Y., Pinkas, B., Reingold, O.: Keyword search and oblivious pseudorandom functions. In: Kilian, J. (ed.) TCC 2005. LNCS, vol. 3378, pp. 303–324. Springer, Heidelberg (2005). https://doi.org/10.1007/978-3-540-30576-7_17

[FNO18] Falk, B.H., Noble, D., Ostrovsky, R.: Private set intersection with linear communication from general assumptions. Cryptology ePrint Archive, Report 2018/238 (2018)

[FNP04] Freedman, M.J., Nissim, K., Pinkas, B.: Efficient private matching and set intersection. In: Cachin, C., Camenisch, J.L. (eds.) EUROCRYPT 2004. LNCS, vol. 3027, pp. 1–19. Springer, Heidelberg (2004). https://doi.org/10.1007/978-3-540-24676-3_1

[GM11] Goodrich, M.T., Mitzenmacher, M.: Privacy-preserving access of outsourced data via oblivious RAM simulation. In: Aceto, L., Henzinger, M., Sgall, J. (eds.) ICALP 2011. LNCS, vol. 6756, pp. 576–587. Springer, Heidelberg (2011). https://doi.org/10.1007/978-3-642-22012-8_46

[GMW87] Goldreich, O., Micali, S., Wigderson, A.: How to play any mental game or a completeness theorem for protocols with honest majority. In: STOC (1987)

[Gon81] Gonnet, G.H.: Expected length of the longest probe sequence in hash code searching. J. ACM **28**(2), 289–304 (1981)

[HCE11] Huang, Y., Chapman, P., Evans, D.: Privacy-preserving applications on smartphones. In: Hot Topics in Security (HotSec) (2011)

[HEK12] Huang, Y., Evans, D., Katz, J.: Private set intersection: are garbled circuits better than custom protocols? In: NDSS (2012)

[HEKM11] Huang, Y., Evans, D., Katz, J., Malka, L.: Faster secure two-party computation using garbled circuits. In: USENIX Security (2011)

[HN10] Hazay, C., Nissim, K.: Efficient set operations in the presence of malicious adversaries. In: Nguyen, P.Q., Pointcheval, D. (eds.) PKC 2010. LNCS, vol. 6056, pp. 312–331. Springer, Heidelberg (2010). https://doi.org/10.1007/978-3-642-13013-7_19

[HOS17] Hallgren, P., Orlandi, C., Sabelfeld, A.: PrivatePool: privacy-preserving ridesharing. In: Computer Security Foundations Symposium (CSF) (2017)

[IKN+17] Ion, M., et al.: Private intersection-sum protocol with applications to attributing aggregate ad conversions. Cryptology ePrint Archive, Report 2017/738 (2017)

[IKNP03] Ishai, Y., Kilian, J., Nissim, K., Petrank, E.: Extending oblivious transfers efficiently. In: Boneh, D. (ed.) CRYPTO 2003. LNCS, vol. 2729, pp. 145–161. Springer, Heidelberg (2003). https://doi.org/10.1007/978-3-540-45146-4_9
[KKRT16] Kolesnikov, V., Kumaresan, R., Rosulek, M., Trieu, N.: Efficient batched oblivious PRF with applications to private set intersection. In: CCS (2016)
[KLS+17] Kiss, Á., Liu, J., Schneider, T., Asokan, N., Pinkas, B.: Private set intersection for unequal set sizes with mobile applications. PoPETs **2017**(4), 177–197 (2017)
[KM18] Kushilevitz, E., Mour, T.: Sub-logarithmic distributed oblivious RAM with small block size. CoRR, abs/1802.05145 (2018)
[KMP+17] Kolesnikov, V., Matania, N., Pinkas, B., Rosulek, M., Trieu, N.: Practical multi-party private set intersection from symmetric-key techniques. In: CCS (2017)
[KMW09] Kirsch, A., Mitzenmacher, M., Wieder, U.: More robust hashing: cuckoo hashing with a stash. SIAM J. Comput. **39**(4), 1543–1561 (2009)
[Kre17] Kreuter, B.: Secure multiparty computation at Google. In: RWC (2017)
[Kre18] Kreuter, B.: Techniques for Scalable Secure Computation Systems. Ph.D. thesis, Northeastern University (2018)
[KS08] Kolesnikov, V., Schneider, T.: Improved garbled circuit: free XOR gates and applications. In: Aceto, L., Damgård, I., Goldberg, L.A., Halldórsson, M.M., Ingólfsdóttir, A., Walukiewicz, I. (eds.) ICALP 2008. LNCS, vol. 5126, pp. 486–498. Springer, Heidelberg (2008). https://doi.org/10.1007/978-3-540-70583-3_40
[LWN+15] Liu, C., Wang, X. S., Nayak, K., Huang, Y., Shi, E.: ObliVM: a programming framework for secure computation. In: S&P (2015)
[Mea86] Meadows, C.: A more efficient cryptographic matchmaking protocol for use in the absence of a continuously available third party. In: S&P (1986)
[MR95] Motwani, R., Raghavan, P.: Randomized Algorithms (1995)
[PR01] Pagh, R., Rodler, F.F.: Cuckoo hashing. In: auf der Heide, F.M. (ed.) ESA 2001. LNCS, vol. 2161, pp. 121–133. Springer, Heidelberg (2001). https://doi.org/10.1007/3-540-44676-1_10
[PSSZ15] Pinkas, B., Schneider, T., Segev, G., Zohner, M.: Phasing: private set intersection using permutation-based hashing. In: USENIX Security (2015)
[PSWW18] Pinkas, B., Schneider, T., Weinert, C., Wieder, U.: Efficient circuit-based PSI via cuckoo hashing. In: Nielsen, J.B., Rijmen, V. (eds.) EUROCRYPT 2018. LNCS, vol. 10822, pp. 125–157. Springer, Cham (2018). https://doi.org/10.1007/978-3-319-78372-7_5
[PSZ14] Pinkas, B., Schneider, T., Zohner, M.: Faster private set intersection based on OT extension. In: USENIX Security (2014)
[PSZ18] Pinkas, B., Schneider, T., Zohner, M.: Scalable private set intersection based on OT extension. TOPS **21**(2), 7 (2018)
[RA18] Resende, A.C.D., Aranha, D.F.: Faster unbalanced private set intersection. In: FC (2018)
[RR17a] Rindal, P., Rosulek, M.: Improved private set intersection against malicious adversaries. In: Coron, J.-S., Nielsen, J.B. (eds.) EUROCRYPT 2017. LNCS, vol. 10210, pp. 235–259. Springer, Cham (2017). https://doi.org/10.1007/978-3-319-56620-7_9
[RR17b] Rindal, P., Rosulek, M.: Malicious-secure private set intersection via dual execution. In: CCS (2017)

[Sha80] Shamir, A.: On the power of commutativity in cryptography. In: de Bakker, J., van Leeuwen, J. (eds.) ICALP 1980. LNCS, vol. 85, pp. 582–595. Springer, Heidelberg (1980). https://doi.org/10.1007/3-540-10003-2_100

[SZ13] Schneider, T., Zohner, M.: GMW vs. Yao? Efficient secure two-party computation with low depth circuits. In: Sadeghi, A.-R. (ed.) FC 2013. LNCS, vol. 7859, pp. 275–292. Springer, Heidelberg (2013). https://doi.org/10.1007/978-3-642-39884-1_23

[Yao86] Yao, A.C.: How to generate and exchange secrets. In: FOCS (1986)

[Yun15] Yung, M.: From mental poker to core business: why and how to deploy secure computation protocols? In: CCS (2015)

[ZC17] Zhao, Y., Chow, S.S.M.: Are you the one to share? Secret transfer with access structure. PoPETs **2017**(1), 149–169 (2017)

[ZC18] Zhao, Y., Chow, S.S.M.: Can you find the one for me? Privacy-preserving matchmaking via threshold PSI. In: WPES (2018)

[ZRE15] Zahur, S., Rosulek, M., Evans, D.: Two halves make a whole: reducing data transfer in garbled circuits using half gates. In: Oswald, E., Fischlin, M. (eds.) EUROCRYPT 2015. LNCS, vol. 9057, pp. 220–250. Springer, Heidelberg (2015). https://doi.org/10.1007/978-3-662-46803-6_8

An Algebraic Approach to Maliciously Secure Private Set Intersection

Satrajit Ghosh[1(✉)] and Tobias Nilges[2]

[1] Department of Computer Science, Aarhus University, Aarhus, Denmark
satrajit@cs.au.dk

[2] ITK Engineering GmbH, Rülzheim, Germany

Abstract. Private set intersection (PSI) is an important area of research and has been the focus of many works over the past decades. It describes the problem of finding an intersection between the input sets of at least two parties without revealing anything about the input sets apart from their intersection.

In this paper, we present a new approach to compute the intersection between sets based on a primitive called Oblivious Linear Function Evaluation (OLE). On an abstract level, we use this primitive to efficiently add two polynomials in a randomized way while preserving the roots of the added polynomials. Setting the roots of the input polynomials to be the elements of the input sets, this directly yields an intersection protocol with optimal asymptotic communication complexity $O(m\kappa)$. We highlight that the protocol is information-theoretically secure against a malicious adversary assuming OLE.

We also present a natural generalization of the 2-party protocol for the fully malicious multi-party case. Our protocol does away with expensive (homomorphic) threshold encryption and zero-knowledge proofs. Instead, we use simple combinatorial techniques to ensure the security. As a result we get a UC-secure protocol with asymptotically optimal communication complexity $O((n^2+nm)\kappa)$, where n is the number of parties, m is the set size and κ is the security parameter. Apart from yielding an asymptotic improvement over previous works, our protocols are also conceptually simple and require only simple field arithmetic. Along the way we develop techniques that might be of independent interest.

1 Introduction

Private set intersection (PSI) has been the focus of research for decades and describes the following basic problem. Two parties, Alice and Bob, each have a

S. Ghosh—Supported by the European Union's Horizon 2020 research and innovation programme under grant agreement #669255 (MPCPRO), the European Union's Horizon 2020 research and innovation programme under grant agreement #731583 (SODA) and the Independent Research Fund Denmark project BETHE.

T. Nilges—Part of the research leading to these results was done while the author was at Aarhus University. Supported by the European Union's Horizon 2020 research and innovation programme under grant agreement #669255 (MPCPRO).

Y. Ishai and V. Rijmen (Eds.): EUROCRYPT 2019, LNCS 11478, pp. 154–185, 2019.
https://doi.org/10.1007/978-3-030-17659-4_6

set S_A and S_B, respectively, and want to find the intersection $S_\cap = S_\mathsf{A} \cap S_\mathsf{B}$ of their sets. This problem is non-trivial if both parties must not learn anything but the intersection. There are numerous applications for PSI from auctions [29] over advertising [32] to proximity testing [30].

Over the years several techniques for two-party PSI have been proposed, which can be roughly placed in four categories: constructions built from specific number-theoretic assumptions [8,9,21,23,28,38], using garbled circuits [20,32], based on oblivious transfer (OT) [10,26,31,33–36] and based on oblivious polynomial evaluation (OPE) [7,12,13,17,18]. There also exists efficient PSI protocols in server-aided model [24].

Some of these techniques translate to the multi-party setting. The first (passively secure) multi-party PSI (MPSI) protocol was proposed by Freedman et al. [13] based on OPE and later improved in a series of works [5,25,37] to achieve full malicious security. Recently, Hazay and Venkitasubramaniam [19] proposed new protocols secure against semi-honest and fully malicious adversaries. They improve upon the communication efficiency of previous works by designating a central party that runs a version of the protocol of [13] with all other parties and aggregates the results.

Given the state of the art, it remains an open problem to construct a protocol with asymptotically optimal communication complexity in the fully malicious multi-party setting. The main reason for this is the use of zero-knowledge proofs and expensive checks in previous works, which incur an asymptotic overhead over passively secure solutions.

In a concurrent and independent work, Kolesnikov et al. [27] presented a new paradigm for solving the problem of MPSI from oblivious programmable pseudorandom functions (OPPRF). Their approach yields very efficient protocols for multi-party PSI, but the construction achieves only passive security against $n-1$ corruptions. However, their approach to aggregate the intermediate results uses ideas similar to our masking scheme in the multi-party case.

1.1 Our Contribution

We propose a new approach to (multi-party) private set intersection based on oblivious linear function evaluation (OLE). OLE allows two mutually distrusting parties to evaluate a linear function $ax + b$, where the sender knows a and b, and the receiver knows x. Nothing apart from the result $ax + b$ is learned by the receiver, and the sender learns nothing about x. OLE can be instantiated in the OT-hybrid model from a wide range of assumptions with varying communication efficiency, like LPN [1], Quadratic/Composite Residuosity [22] and Noisy Encodings [14,22], or even unconditionally [22].

Our techniques differ significantly from previous works and follow a new paradigm which leads to conceptually simple and very efficient protocols. Our approach is particularly efficient if all input sets are of similar size. To showcase the benefits of our overall approach, we also describe how our MPSI protocol can be modified into a threshold MPSI protocol.

Concretely, we achieve the following:

- Two-party PSI with communication complexity $O(m\kappa)$ and computational cost of $O(m \log m)$ multiplications. The protocol is information theoretically secure against a malicious adversary in the OLE-hybrid model.
- UC-secure Multi-party PSI in fully malicious setting with communication complexity $O((n^2 + nm)\kappa)$ and computational complexity dominated by $O(nm \log m)$ multiplications for the central party and $O(m \log m)$ multiplications for other parties.
- A simple extension of the multi-party PSI protocol to threshold PSI, with the same complexity. To the best of our knowledge, this is the first actively secure threshold multi-party PSI protocol.[1]

In comparison to previous works which rely heavily on exponentiations in fields or groups, our protocols require only field addition and multiplication (and OWF in the case of MPSI).

If we compare our result with the asymptotically optimal 2-party PSI protocols by [8,23], which have linear communication and computation, our first observation is that although they only have a linear number of modular exponentiations, the number of field operations is not linear but rather in the order of $O(m\kappa)$, and further they need a ZK proof in the ROM for each exponentiation, which is also expensive. Additionally, their result is achieved with specific number-theoretic assumptions, so the parameter sizes are probably not favourable compared to our protocol, and the construction is not black-box. We provide a detailed comparison of the concrete efficiency of our result with the recent protocol by Rindal and Rosulek [36], which has very good concrete efficiency.

In the setting of MPSI, our techniques result in asymptotic efficiency improvements over previous works in both communication and computational complexity (cf. Table 1).

We want to emphasize that our efficiency claims hold *including* the communication and computation cost for the OLE, if the recent instantiation by Ghosh et al. [14] is used, which is based on noisy Reed-Solomon codes. This OLE protocol has a constant communication overhead per OLE if instantiated with an efficient OT-extension protocol like [31] and therefore does not influence the asymptotic efficiency of our result.

Our results may seem surprising in light of the information-theoretic lower bound of $O(n^2 m\kappa)$ in the communication complexity for multi-party PSI in the fully malicious UC setting. We circumvent this lower bound by considering a slightly modified ideal functionality, resulting in a UC-secure solution for multi-party PSI with asymptotically optimal communication overhead. By asymptotically optimal, we mean that our construction matches the optimal bounds in the plain model for $m > n$, even for passive security, where n is the number of parties, m is the size of the sets and κ is the security parameter. All of our protocols work over fields $\mathbb{F}$ that are exponential in the size of the security parameter κ.

[1] Please see the full version of the paper [15].

Table 1. Comparison of multi-party PSI protocols, where n is the number of parties, m the size of the input set and κ a security parameter. Here, THE denotes a threshold homomorphic encryption scheme, CRS a common reference string and OPPRF an oblivious programmable PRF. The computational cost is measured in terms of number of multiplications. Some of the protocols perform better if the sizes of the input sets differ significantly, or particular domains for inputs are used. The overhead described here assumes sets of similar size, with κ bit elements.

Protocol	Tools	Communication	Computation	Corruptions	Security
[27]	OPPRF	$O(nm\kappa)$	$O(n\kappa^2)$	$n-1$	semi-honest
[19]	THE	$O(nm\kappa)$	$O(nm\log m\kappa)$	$n-1$	semi-honest
[25]	THE, ZK	$O(n^2m^2\kappa)$	$O(n^2m + nm^2\kappa)$	$n-1$	malicious
[5]	THE, ZK	$O(n^2m\kappa)$	$O(n^2m + nm\kappa)$	$t < n/2$	malicious
[19]	CRS, THE	$O((n^2 + nm\log m)\kappa)$	$O(m^2\kappa)$	$n-1$	malicious
Ours+ [14]	OLE	$O((n^2 + nm)\kappa)$	$O(nm\log m)$	$n-1$	malicious[a]

[a]Our protocol is UC-secure in the fully malicious setting.

We believe that our approach provides an interesting alternative to existing solutions and that the techniques which we developed can find application in other settings as well.

1.2 Technical Overview

Abstractly, we let both parties encode their input set as a polynomial, such that the roots of the polynomials correspond to the inputs. This is a standard technique, but usually the parties then use OPE to obliviously evaluate the polynomials or some form of homomorphic encryption. Instead, we devise an OLE-based construction to add the two polynomials in an oblivious way, which results in an intersection polynomial. Kissner and Song [25] also create an intersection polynomial similar to ours, but encrypted under a layer of homomorphic encryption, whereas our technique results in a *plain* intersection polynomial. Since the intersection polynomial already hides everything but the intersection, one could argue that the layer of encryption in [25] incurs *additional overhead* in terms of expensive computations and complex checks.

In our case, both parties simply evaluate the intersection polynomial on their input sets and check if it evaluates to 0. This construction is information-theoretically secure in the OLE-hybrid model and requires only simple field operations. Conceptually, we compute the complete intersection in one step. In comparison to the naive OPE-based approach[2], our solution directly yields an asymptotic communication improvement in the input size. Another advantage is that our approach generalizes to the multi-party setting.

We start with a detailed overview of our constructions and technical challenges.

[2] Here we mean an OPE is used for each element of the receiver's input set. This can be circumvented by clever hashing strategies, e.g. [13,19].

Oblivious polynomial addition from OLE. Intuitively, OLE is the generalization of OT to larger fields, i.e. it allows a sender and a receiver to compute a linear function $c(x) = ax + b$, where the sender holds a, b and the receiver inputs x and obtains c. OLE guarantees that the receiver learns nothing about a, b except for the result c, while the sender learns nothing about x.

Based on this primitive, we define and realize a functionality OPA that allows to add two polynomials in such a way that the receiver cannot learn the sender's input polynomial, while the sender learns nothing about the receiver's polynomial or the output. We first describe a passively secure protocol. Concretely, assume that the sender has an input polynomial $\mathbf{a}$ of degree $2d$, and the receiver has a polynomial $\mathbf{b}$ of degree d. The sender additionally draws a uniformly random polynomial $\mathbf{r}$ of degree d. Both parties point-wise add and multiply their polynomials, i.e. they evaluate their polynomials over a set of $2d + 1$ distinct points $\alpha_1, \ldots, \alpha_{2d+1}$, resulting in $a_i = \mathbf{a}(\alpha_i), b_i = \mathbf{b}(\alpha_i)$ and $r_i = \mathbf{r}(\alpha_i)$ for $i \in [2d + 1]$. Then, for each of $2d + 1$ OLEs, the sender inputs r_i, a_i, while the receiver inputs b_i and thereby obtains $c_i = r_i b_i + a_i$. The receiver interpolates the polynomial $\mathbf{c}$ from the $2d+1$ (α_i, c_i) and outputs it. Since we assume semi-honest behaviour, the functionality is realized by this protocol.

The biggest hurdle in achieving active security for the above protocol lies in ensuring a non-zero $\mathbf{b}$ and $\mathbf{r}$ as input. Otherwise, e.g. if $\mathbf{b} = 0$, the receiver could learn $\mathbf{a}$. One might think that it is sufficient to perform a coin-toss and verify that the output satisfies the supposed relation, i.e. pick a random x, compute $\mathbf{a}(x), \mathbf{b}(x), \mathbf{r}(x)$ and $\mathbf{c}(x)$ and everyone checks if $\mathbf{b}(x)\mathbf{r}(x) + \mathbf{a}(x) = \mathbf{c}(x)$ and if $\mathbf{b}(x), \mathbf{r}(x)$ are non-zero[3]. For $\mathbf{r}(x) \neq 0$, the check is actually sufficient, because $\mathbf{r}$ must have degree at most d, otherwise the reconstruction fails, and only d points of $\mathbf{r}$ can be zero ($\mathbf{r} = 0$ would require $2d+1$ zero inputs). For $\mathbf{b} \neq 0$, however, just checking for $\mathbf{b}(x) \neq 0$ is not sufficient, because at this point, even if the input $\mathbf{b} \neq 0$, the receiver can input d zeroes in the OLE, which in combination with the check is sufficient to learn $\mathbf{a}$ completely. We resolve this issue by constructing an enhanced OLE functionality which ensures that the receiver input is non-zero. We believe that this primitive is of independent interest and describe it in more detail later in this section.

Two-party PSI from OLE. Let us first describe a straightforward two-party PSI protocol with one-sided output from the above primitive. Let S_A and S_B denote the inputs for Alice and Bob, respectively, where $|S_\mathsf{P}| = m$. Assuming that Bob is supposed to get the intersection, they pick random $\mathbf{p}_\mathsf{A}$ and $\mathbf{p}_\mathsf{B}$ with the restriction that $\mathbf{p}_\mathsf{P}(\gamma) = 0$ for $\gamma \in S_\mathsf{P}$. As they will use OPA, $\deg \mathbf{p}_\mathsf{A} = 2m$, while $\deg \mathbf{p}_\mathsf{B} = m$. Further, Alice picks a uniformly random polynomial $\mathbf{r}_\mathsf{A}$ of degree m and inputs $\mathbf{p}_\mathsf{A}, \mathbf{r}_\mathsf{A}$ into OPA. Bob inputs $\mathbf{p}_\mathsf{B}$, obtains $\mathbf{p}_\cap = \mathbf{p}_\mathsf{A} + \mathbf{p}_\mathsf{B}\mathbf{r}_\mathsf{A}$ and outputs all $\gamma_j \in S_\mathsf{B}$ for which $\mathbf{p}_\cap(\gamma_j) = 0$. Obviously, $\mathbf{r}_\mathsf{A}$ does not remove any of the roots of $\mathbf{p}_\mathsf{B}$, and therefore all points γ where $\mathbf{p}_\mathsf{B}(\gamma) = 0 = \mathbf{p}_\mathsf{A}(\gamma)$ remain in $\mathbf{p}_\cap$.

[3] Since this check leaks some information about the inputs, we have to perform the check in a secure manner.

However, as a stepping stone for multi-party PSI, we are more interested in protocols that provide output to both parties. If we were to use the above protocol and simply announce $\mathbf{p}_{\cap}$ to Alice, then Alice could learn Bob's input. Therefore we have to take a slightly different approach. Let $\mathbf{u}_{\mathsf{A}}$ be an additional random polynomial chosen by Alice. Instead of using her own input in the OPA, Alice uses $\mathbf{r}_{\mathsf{A}}, \mathbf{u}_{\mathsf{A}}$, which gives $\mathbf{s}_{\mathsf{B}} = \mathbf{u}_{\mathsf{A}} + \mathbf{p}_{\mathsf{B}}\mathbf{r}_{\mathsf{A}}$ to Bob. Then they run another OPA in the other direction, i.e. Bob inputs $\mathbf{r}_{\mathsf{B}}, \mathbf{u}_{\mathsf{B}}$ and Alice $\mathbf{p}_{\mathsf{A}}$. Now, both Alice and Bob have a randomized "share" of the intersection, namely $\mathbf{s}_{\mathsf{A}}$ and $\mathbf{s}_{\mathsf{B}}$, respectively. Adding $\mathbf{s}_{\mathsf{A}}$ and $\mathbf{s}_{\mathsf{B}}$ yields a masked but correct intersection. We still run into the problem that sending either $\mathbf{s}_{\mathsf{B}}$ to Alice or $\mathbf{s}_{\mathsf{A}}$ to Bob allows the respective party to learn the other party's input. We also have to use additional randomization polynomials $\mathbf{r}'_{\mathsf{A}}, \mathbf{r}'_{\mathsf{B}}$ to ensure privacy of the final result.

Our solution is to simply use the masks $\mathbf{u}$ to enforce the addition of the two shares. Let us fix Alice as the party that combines the result. Bob computes $\mathbf{s}'_{\mathsf{B}} = \mathbf{s}_{\mathsf{B}} - \mathbf{u}_{\mathsf{B}} + \mathbf{p}_{\mathsf{B}}\mathbf{r}'_{\mathsf{B}}$ and sends it to Alice. Alice computes $\mathbf{p}_{\cap} = \mathbf{s}'_{\mathsf{B}} + \mathbf{s}_{\mathsf{A}} - \mathbf{u}_{\mathsf{A}} + \mathbf{p}_{\mathsf{A}}\mathbf{r}'_{\mathsf{A}}$. This way, the only chance to get rid of the blinding polynomial $\mathbf{u}_{\mathsf{B}}$ is to add both values. But since each input is additionally randomized via the $\mathbf{r}$ polynomials, Alice cannot subtract her own input from the sum. Since the same also holds for Bob, Alice simply sends the result to Bob.

The last step is to check if the values that are sent and the intersection polynomial are consistent. We do this via a simple coin-toss for a random x, and the parties evaluate their inputs on x and can abort if the relation $\mathbf{p}_{\cap} = \mathbf{p}_{\mathsf{B}}(\mathbf{r}_{\mathsf{A}} + \mathbf{r}'_{\mathsf{B}}) + \mathbf{p}_{\mathsf{A}}(\mathbf{r}'_{\mathsf{A}} + \mathbf{r}_{\mathsf{B}})$ does not hold, i.e. $\mathbf{p}_{\cap}$ is computed incorrectly. This type of check enforces semi-honest behaviour, and was used previously e.g. in [2].

A note on the MPSI functionality. We show that by slightly modifying the ideal functionality for multi-party PSI we get better communication efficiency, without compromising the security at all. A formal definition is given in Sect. 6.1. Typically, it is necessary for the simulator to extract *all* inputs from the malicious parties, input them into the ideal functionality, and then continue the simulation with the obtained ideal intersection. In a fully malicious setting, however, this requires every party to communicate in $O(m\kappa)$ with every other party—otherwise the input is information-theoretically undetermined and cannot be extracted—which results in $O(n^2 m\kappa)$ communication complexity.

The crucial observation here is that in the setting of multi-party PSI, an intermediate intersection between a single malicious party and *all* honest parties is sufficient for simulation. This is due to the fact that inputs by additional malicious parties can only reduce the size of the intersection, and as long as we observe the additional inputs at some point, we can correctly reduce the intersection in the ideal setting before outputting it. On a technical level, we no longer need to extract all malicious inputs right away to provide a correct simulation of the intersection. Therefore, it is not necessary for every party to communicate in $O(m\kappa)$ with every other party. Intuitively, the intermediate intersection corresponds to the case where all malicious parties have the same input. We therefore argue that the security of this modified setting is identical to standard MPSI up to input substitution of the adversary.[4]

Multi-party PSI. The multi-party protocol is a direct generalization of the two-party protocol, with some small adjustments. We consider a network with a star topology, similar to the recent result of [19]. One party is set to be the central party, and all other parties (mainly) interact with this central party to compute the result. The main idea here is to delegate most of the work to the central party, which in turn allows to reduce the communication complexity. Since no party is supposed to get any intermediate intersections, we let each party create an additive sharing of their intersection with the central party.

First, consider the following (incorrect) toy example. Let each party P_i execute the two-party PSI as described above with P_0, up to the point where both parties have shares $\mathbf{s}^i_{P_0}, \mathbf{s}'_{P_i}$. All parties P_i send their shares $\mathbf{s}'_{P_i}$ to P_0, who adds all polynomials and broadcasts the output. By design of the protocols and the inputs, this yields the intersection of all parties. Further, the communication complexity is in $O(nm\kappa)$, which is optimal. However, this protocol also allows P_0 to learn all intermediate intersections with the other parties, which is not allowed. Previously, all maliciously secure multi-party PSI protocols used threshold encryption to solve this problem, and indeed it might be possible to use a similar approach to ensure active security for the above protocol. For example, a homomorphic threshold encryption would allow to add all these shares homomorphically, without leaking the intermediate intersections. But threshold encryption incurs a significant computational overhead (and increases the complexity of the protocol and its analysis) which we are unwilling to pay.

Instead, we propose a new solution which is conceptually very simple. We add another layer of masking on the shares $\mathbf{s}_{P_i}$, which forces P_0 to add *all* intermediate shares—at least those of the honest parties. For this we have to ensure that the communication complexity does not increase, so all parties exchange seeds (instead of sending random polynomials directly), which are used in a PRG to mask the intermediate intersections. This technique is somewhat reminiscent of the pseudorandom secret-sharing technique by Cramer et al. [6]. We emphasize that we do not need any public key operations.

Concretely, all parties exchange a random seed and use it to compute a random polynomial in such a way that every pair of parties P_i, P_j holds two polynomials $\mathbf{v}_{ij}, \mathbf{v}_{ji}$ with $\mathbf{v}_{ij} + \mathbf{v}_{ji} = 0$. Then, instead of sending $\mathbf{s}'_{P_i}$, each party P_i computes $\mathbf{s}''_{P_i} = \mathbf{s}'_{P_i} + \sum \mathbf{v}_{ij}$ and sends this value. If P_0 obtains this value, it has to add the values $\mathbf{s}''_{P_i}$ of all parties to remove the masks, otherwise $\mathbf{s}''_{P_i}$ will be uniformly random.

Finally, to ensure that the central party actually computed the aggregation, we add a check similar to two-party PSI, where the relation, i.e. computing the sum, is verified by evaluating the inputs on a random value x which is obtained by a multi-party coin-toss.

[4] Our multi-party PSI functionality uses similar idea as augmented semi-honest multi-party PSI as in previous works [27].

Threshold (M)PSI. We defer the threshold extensions to the full version of this paper [15] and only give a very brief technical overview.

First of all, we clarify the term *threshold PSI*. We consider the setting where all parties have m elements as their input, and the output is only revealed if the intersection of the inputs among all parties is bigger than a certain threshold ℓ. In [16] Hallgren et al. defined this notion for two party setting, and finds application whenever two entities are supposed to be matched once a certain threshold is reached, e.g. for dating websites or ride sharing.

We naturally extend the idea of threshold PSI from [16] to the multi-party setting and propose the first actively secure threshold multi-party PSI protocol. On a high level, our solution uses a similar idea to [16], but we use completely different techniques and achieve stronger security and better efficiency. The main idea is to use a robust secret sharing scheme, and the execution of the protocol basically transfers a subset of these shares to the other parties, one share for each element in the intersection. If the intersection is large enough, the parties can reconstruct the shared value.

Specifically, the trick is to modify the input polynomials of each party P_i for the MPSI protocol and add an additional check. Instead of simply setting $\mathbf{p}_i$ such that $\mathbf{p}_i(\gamma_j) = 0$ for all $\gamma_j \in S_i$, we set $\tilde{\mathbf{p}}_i(\gamma_j) = 1$. Further, for each of the random polynomials $\tilde{\mathbf{r}}_i, \tilde{\mathbf{r}}'_i$ we set $\tilde{\mathbf{r}}_i(\gamma_j) = \rho_j$ and $\tilde{\mathbf{r}}'_i(\gamma_j) = \rho'_j$, where $\rho_1, \ldots, \rho_n$, $\rho'_1, \ldots, \rho'_n$ are the shares of two robust (ℓ, n)-secret sharings of random values s_i^0 and s_i^1, respectively. Now, by computing the modified intersection polynomial $\tilde{\mathbf{p}}_\cap$ as before, each party obtains exactly $m_\cap = |S_\cap|$ shares, one for each $\gamma_j \in S_i$.

Now if $m_\cap \geq \ell$ then each party can reconstruct $r_\cap = \sum_{i=1}^{n} (s_i^0 + s_i^1)$. Otherwise the intersection remains hidden completely. We omitted some of the details due to the space constraints and refer to the full version [15].

A New Flavour of OLE. One of the main technical challenges in constructing our protocols is to ensure a non-zero input into the OLE functionality by the receiver. Recall that an OLE computes a linear function $ax + b$. We define an enhanced OLE functionality (cf. Sect. 3) which ensures that $x \neq 0$, otherwise the output is uniformly random. Our protocol which realises this functionality makes two black-box calls to a normal OLE and is otherwise purely algebraic.

Before we describe the solution, let us start with a simple observation. If the receiver inputs $x = 0$, an OLE returns the value b. Therefore, it is critical that the receiver cannot force the protocol to output b. One way to achieve this is by forcing the receiver to multiply b with some correlated value via an OLE, let's call it $\hat{x}$. Concretely, we can use an OLE where the receiver inputs $\hat{x}$ and a random s, while the sender inputs b and obtains $\hat{x}b + s$. Now if the sender uses $a + b\hat{x} + s, 0$ as input for another OLE, where the receiver inputs x, the receiver obtains $ax + b\hat{x}x + sx$. Which means that if $\hat{x} = x^{-1}$ then the receiver can extract the correct output. This looks like a step in the right direction, since for $x = 0$ or $\hat{x} = 0$, the output would not be b. On the other hand, the receiver can now force the OLE to output a by choosing $\hat{x} = 0$ and $x = 1$, so maybe we only shifted the problem.

The final trick lies in masking the output such that it is uniform for inconsistent inputs $x, \hat{x}$. We do this by splitting b into two shares that only add to b if $x \cdot \hat{x} = 1$. The complete protocol looks like this: the receiver plays the sender for one OLE with input x^{-1}, s, and the sender inputs a random u to obtain $t = x^{-1}u + s$. Then the sender plays the sender for the second OLE and inputs $t + a, b - u$, while the receiver inputs x and obtains $c' = (t+a)x + b - u = ux^{-1}x + sx + ax + b - u = ax + b + sx$, from which the receiver can subtract sx to get the result. A cheating receiver with inconsistent input $x^*, \hat{x}^*$ will get $ax + b + u(x^*\hat{x}^* - 1)$ as an output, which is uniform over the choice of u.

2 Preliminaries

We assume $|\mathbb{F}| \in \theta(2^\kappa)$, where κ is a statistical security parameter. Typically, $x \in \mathbb{F}$ denotes a field element, while $\mathbf{p} \in \mathbb{F}[X]$ denotes a polynomial. Let $\mathcal{M}_0(\mathbf{p})$ denote the zero-set for $\mathbf{p} \in \mathbb{F}[X]$, i.e. $\forall x \in \mathcal{M}_0(\mathbf{p}), \mathbf{p}(x) = 0$.

In the proofs, $\hat{x}$ denotes an element either extracted or simulated by the simulator, while x^* denotes an element sent by the adversary.

We slightly abuse notation and denote by $\mathbf{v} = \mathsf{PRG}_d(s)$ the deterministic pseudorandom polynomial of degree d derived from evaluating PRG on seed s.

2.1 Security Model

We prove our protocol in the Universal Composability (UC) framework [4]. In the framework, security of a protocol is shown by comparing a real protocol π in the real world with an ideal functionality $\mathcal{F}$ in the ideal world. $\mathcal{F}$ is supposed to accurately describe the security requirements of the protocol and is secure per definition. An environment $\mathcal{Z}$ is plugged either to the real protocol or the ideal protocol and has to distinguish the two cases. For this, the environment can corrupt parties. To ensure security, there has to exist a simulator in the ideal world that produces a protocol transcript indistinguishable from the real protocol, even if the environment corrupts a party. We say π UC-realises $\mathcal{F}$ if for all adversaries $\mathcal{A}$ in the real world there exists a simulator $\mathcal{S}$ in the ideal world such that all environments $\mathcal{Z}$ cannot distinguish the transcripts of the parties' outputs.

Oblivious Linear Function Evaluation. Oblivious Linear Function Evaluation (OLE) is the generalized version of OT over larger fields. The sender has as input two values $a, b \in \mathbb{F}$ that determine a linear function $f(x) = a \cdot x + b$ over $\mathbb{F}$, and the receiver gets to obliviously evaluate the linear function on input $x \in \mathbb{F}$. The receiver will learn only $f(x)$, and the sender learns nothing at all. The ideal functionality is shown in Fig. 1.

2.2 Technical Lemmas

We state several lemmas which are used to show the correctness of our PSI protocols later on.

Lemma 2.1. *Let* $\mathbf{p}, \mathbf{q} \in \mathbb{F}[X]$ *be non-trivial polynomials. Then,*

$$\mathcal{M}_0(\mathbf{p}) \cap \mathcal{M}_0(\mathbf{p}+\mathbf{q}) = \mathcal{M}_0(\mathbf{p}) \cap \mathcal{M}_0(\mathbf{q}) = \mathcal{M}_0(\mathbf{q}) \cap \mathcal{M}_0(\mathbf{p}+\mathbf{q}).$$

This lemma shows that the sum of two polynomials contains the intersection with respect to the zero-sets of both polynomials.

Functionality $\mathcal{F}_{\text{OLE}}$

1. Upon receiving a message $(\texttt{inputS}, (a, b))$ from the sender with $a, b \in \mathbb{F}$, verify that there is no stored tuple, else ignore that message. Store a and b and send a message $(\texttt{input})$ to $\mathcal{A}$.
2. Upon receiving a message $(\texttt{inputR}, x)$ from the receiver with $x \in \mathbb{F}$, verify that there is no stored tuple, else ignore that message. Store x and send a message $(\texttt{input})$ to $\mathcal{A}$.
3. Upon receiving a message $(\texttt{deliver}, \mathsf{S})$ from $\mathcal{A}$, check if both (a, b) and x are stored, else ignore that message. Send $(\texttt{delivered})$ to the sender.
4. Upon receiving a message $(\texttt{deliver}, \mathsf{R})$ from $\mathcal{A}$, check if both (a, b) and x are stored, else ignore that message. Set $y = a \cdot x + b$ and send $(\texttt{output}, y)$ to the receiver.

Fig. 1. Ideal functionality for oblivious linear function evaluation.

Proof. Let $\mathcal{M}_\cap = \mathcal{M}_0(\mathbf{p}) \cap \mathcal{M}_0(\mathbf{q})$.

"$\supseteq$": $\forall x \in \mathcal{M}_\cap$: $\mathbf{p}(x) = \mathbf{q}(x) = 0$. But $\mathbf{p}(x) + \mathbf{q}(x) = 0$, so $x \in \mathcal{M}_0(\mathbf{p}+\mathbf{q})$.

"$\subseteq$": It remains to show that there is no x such that $x \in \mathcal{M}_0(\mathbf{p}) \cap \mathcal{M}_0(\mathbf{p}+\mathbf{q})$ but $x \notin \mathcal{M}_\cap$, i.e. $\mathcal{M}_0(\mathbf{p}) \cap (\mathcal{M}_0(\mathbf{p}+\mathbf{q}) \setminus \mathcal{M}_\cap) = \emptyset$. Similarly, $\mathcal{M}_0(\mathbf{q}) \cap (\mathcal{M}_0(\mathbf{p}+\mathbf{q}) \setminus \mathcal{M}_\cap) = \emptyset$.

Assume for the sake of contradiction that $\mathcal{M}_0(\mathbf{p}) \cap (\mathcal{M}_0(\mathbf{p}+\mathbf{q}) \setminus \mathcal{M}_\cap) \neq \emptyset$. Let $x \in \mathcal{M}_0(\mathbf{p}) \cap (\mathcal{M}_0(\mathbf{p}+\mathbf{q}) \setminus \mathcal{M}_\cap)$. Then, $\mathbf{p}(x) = 0$, but $\mathbf{q}(x) \neq 0$, otherwise $x \in \mathcal{M}_\cap$. But this means that $\mathbf{p}(x) + \mathbf{q}(x) \neq 0$, i.e. $x \notin \mathcal{M}_0(\mathbf{p}+\mathbf{q})$. This contradicts our assumption, and we get that $\mathcal{M}_0(\mathbf{p}) \cap (\mathcal{M}_0(\mathbf{p}+\mathbf{q}) \setminus \mathcal{M}_\cap) = \emptyset$.

Symmetrically, we get that $\mathcal{M}_0(\mathbf{q}) \cap (\mathcal{M}_0(\mathbf{p}+\mathbf{q}) \setminus \mathcal{M}_\cap) = \emptyset$. The claim follows. □

Lemma 2.2. *Let* $d \in \mathsf{poly}(\log|\mathbb{F}|)$. *Let* $\mathbf{p} \in \mathbb{F}[X]$, $\deg(\mathbf{p}) = d$ *be a non-trivial random polynomial with* $\Pr[x \in \mathcal{M}_0(\mathbf{p})] \leq \mathsf{negl}(|\mathbb{F}|)$ *for all* x. *Then, for all* $\mathbf{q}_1, \ldots, \mathbf{q}_l \in \mathbb{F}[X]$ *with* $\deg(\mathbf{q}_i) \leq d$,

$$\Pr[(\mathcal{M}_0(\mathbf{p}) \cap \mathcal{M}_0(\sum_{i=1}^{l} \mathbf{q}_i + \mathbf{p})) \neq (\mathcal{M}_0(\mathbf{p}) \cap \bigcap_{i=1}^{l} \mathcal{M}_0(\mathbf{q}_i))] \leq \mathsf{negl}(|\mathbb{F}|).$$

This lemma is basically an extension of Lemma 2.1 and shows that the sum of several polynomials does not create new elements in the intersection unless the supposedly unknown zero-set of $\mathbf{p}$ can be guessed with non-negligible probability.

Proof. "$\subseteq$": We first observe that $\bigcap_{i=1}^{l} \mathcal{M}_0(\mathbf{q}_i) \subseteq \mathcal{M}_0(\sum_{i=1}^{l} \mathbf{q}_i)$: it holds that for all $x \in \bigcap_{i=1}^{l} \mathcal{M}_0(\mathbf{q}_i)$, $\mathbf{q}_i(x) = 0$ for $i \in [l]$. It follows that $\sum_{i=1}^{l} \mathbf{q}_i(x) = 0$, i.e. $x \in \mathcal{M}_0(\sum_{i=1}^{l} \mathbf{q}_i)$.

"$\supseteq$": Assume for the sake of contradiction that

$$(\mathcal{M}_0(\mathbf{p}) \cap \mathcal{M}_0(\sum_{i=1}^{l} \mathbf{q}_i) + \mathbf{p}) \neq (\mathcal{M}_0(\mathbf{p}) \cap \bigcap_{i=1}^{l} \mathcal{M}_0(\mathbf{q}_i))$$

with non-negligible probability ϵ. Let $\mathcal{M} = \mathcal{M}_0(\sum_{i=1}^{l} \mathbf{q}_i + \mathbf{p}) \setminus \bigcap_{i=1}^{l} \mathcal{M}_0(\mathbf{q}_i)$.

Then with probability at least ϵ, the set $\mathcal{M}$ is not empty. Further, we can bound $|\mathcal{M}| \leq d$. Pick a random $x \in \mathcal{M}$. It now holds that $\Pr[x \in \mathcal{M}_0(\mathbf{p})] \geq \epsilon/d$, which directly contradicts our assumption that for an unknown $\mathbf{p}$ the probability of guessing $x \in \mathcal{M}_0(\mathbf{p})$ is negligible over choice of $\mathbf{p}$. The claim follows. $\square$

Lemma 2.3. *Let* $d, d' \in \mathsf{poly}(\log |\mathbb{F}|)$. *Let* $\mathbf{r} \in \mathbb{F}[X]$, $\deg(\mathbf{r}) = d$ *be a uniformly random polynomial. For all non-trivial* $\mathbf{p} \in \mathbb{F}[X]$, $\deg(\mathbf{p}) = d'$,

$$\Pr_{\mathbf{r} \in \mathbb{F}[X]}[(\mathcal{M}_0(\mathbf{r}) \cap \mathcal{M}_0(\mathbf{p})) \neq \emptyset] \leq \mathsf{negl}(|\mathbb{F}|).$$

This lemma establishes that the intersection of a random polynomial with another polynomial is empty except with negligible probability.

Proof. This follows from the fundamental theorem of algebra, which states that a polynomial of degree d evaluates to 0 in a random point only with probability $d/|\mathbb{F}|$.

Since $\mathbf{r}$ (and therefore all $x \in \mathcal{M}_0(\mathbf{r})$) is uniformly random and $|\mathcal{M}_0(\mathbf{r})| = d$, while $|\mathcal{M}_0(\mathbf{p})| = d'$, we get that

$$\Pr[(\mathcal{M}_0(\mathbf{r}) \cap \mathcal{M}_0(\mathbf{p})) \neq \emptyset] \leq dd'/|\mathbb{F}|.$$

$\square$

Lemma 2.4. *Let* $d \in \mathsf{poly}(\log |\mathbb{F}|)$. *Let* $\mathbf{p} \in \mathbb{F}[X]$, $\deg(\mathbf{p}) = d$ *be a fixed but unknown non-trivial polynomial. Further let* $\mathbf{r} \in \mathbb{F}[X]$, $\deg(\mathbf{r}) = d$ *be a uniformly random polynomial. For all non-trivial* $\mathbf{q}, \mathbf{s} \in \mathbb{F}[X]$ *with* $\deg(\mathbf{q}) \leq d$ *and* $\deg(\mathbf{s}) \leq d$,

$$\Pr_{\mathbf{r} \in \mathbb{F}[X]}[(\mathcal{M}_0(\mathbf{p}) \cap \mathcal{M}_0(\mathbf{ps} + \mathbf{rq})) \neq (\mathcal{M}_0(\mathbf{p}) \cap \mathcal{M}_0(\mathbf{q}))] \leq \mathsf{negl}(|\mathbb{F}|).$$

This lemma shows that the multiplication of (possibly maliciously chosen) polynomials does not affect the intersection except with negligible probability, if one random polynomial is used.

Proof.

$$\begin{aligned}
\mathcal{M}_0(\mathbf{p}) \cap \mathcal{M}_0(\mathbf{ps} + \mathbf{rq}) &\overset{\text{Lemma 2.1}}{=} \mathcal{M}_0(\mathbf{p}) \cap (\mathcal{M}_0(\mathbf{ps}) \cap \mathcal{M}_0(\mathbf{qr})) \\
&= \mathcal{M}_0(\mathbf{p}) \cap ((\mathcal{M}_0(\mathbf{p}) \cup \mathcal{M}_0(\mathbf{s})) \cap (\mathcal{M}_0(\mathbf{q}) \cup \mathcal{M}_0(\mathbf{r}))) \\
&= \mathcal{M}_0(\mathbf{p}) \cap ((\mathcal{M}_0(\mathbf{p}) \cap \mathcal{M}_0(\mathbf{q})) \cup \underbrace{(\mathcal{M}_0(\mathbf{p}) \cap \mathcal{M}_0(\mathbf{r}))}_{T_1} \\
&\quad \cup \underbrace{(\mathcal{M}_0(\mathbf{s}) \cap \mathcal{M}_0(\mathbf{q}))}_{\subseteq \mathcal{M}_0(\mathbf{q})} \cup \underbrace{(\mathcal{M}_0(\mathbf{s}) \cap \mathcal{M}_0(\mathbf{r}))}_{T_2})
\end{aligned}$$

From Lemma 2.3 it follows that $\Pr[\mathcal{T}_1 \neq \emptyset] \leq d^2/|\mathbb{F}|$, and also $\Pr[\mathcal{T}_2 \neq \emptyset] \leq d^2/|\mathbb{F}|$. Since

$$\mathcal{M}_0(\mathbf{p}) \cap ((\mathcal{M}_0(\mathbf{p}) \cap \mathcal{M}_0(\mathbf{q})) \cup \mathcal{M}_0(\mathbf{q})) = \mathcal{M}_0(\mathbf{p}) \cap \mathcal{M}_0(\mathbf{q}),$$

we get

$$\Pr_{\mathbf{r} \in \mathbb{F}[X]}[(\mathcal{M}_0(\mathbf{p}) \cap \mathcal{M}_0(\mathbf{ps} + \mathbf{rq})) \neq (\mathcal{M}_0(\mathbf{p}) \cap \mathcal{M}_0(\mathbf{q}))] \leq 2d^2/|\mathbb{F}|.$$

□

3 Enhanced Oblivious Linear Function Evaluation $\mathcal{F}_{\mathrm{OLE}^+}$

In this section we present an enhanced version of the OLE functionality. The standard OLE functionality allows the sender to input a, b, while the receiver inputs x and obtains $ax + b$. For our applications, we do not want the receiver to be able to learn b, i.e. it has to hold that $x \neq 0$. Our approach is therefore to modify the OLE functionality in such a way that it outputs a random field element upon receiving an input $x = 0$ (cf. Fig. 2). A different approach might be to output a special abort symbol or 0, but crucially the output must *not* satisfy the relation $ax + b$. This is a particularly useful feature, as we will show in the next section.

$\mathcal{F}_{\mathrm{OLE}^+}$

1. Upon receiving a message $(\texttt{inputS}, (a, b))$ from the sender with $a, b \in \mathbb{F}$, verify that there is no stored tuple, else ignore that message. Otherwise, store (a, b) and send $(\texttt{input})$ to $\mathcal{A}$.
2. Upon receiving a message $(\texttt{inputR}, x)$ from the receiver with $x \in \mathbb{F}$, verify that there is no stored value, else ignore that message. Otherwise, store x and send $(\texttt{input})$ to $\mathcal{A}$.
3. Upon receiving a message $(\texttt{deliver})$ from $\mathcal{A}$, check if both (a, b) and x are stored, else ignore that message. If $x \neq 0$, set $c = ax + b$, otherwise pick a uniformly random $c \in \mathbb{F}$ and send $(\texttt{output}, c)$ to the receiver. Ignore all further messages.

Fig. 2. Ideal functionality for the enhanced oblivious linear function evaluation.

While it might be possible to modify existing OLE protocols in such a way that a non-zero input is guaranteed, we instead opt to build a protocol black-box from the standard OLE functionality $\mathcal{F}_{\mathrm{OLE}}$.

We refer to the introduction for an abstract overview and a description of the ideas of our construction. The formal description of the protocol is given in Fig. 3.

Lemma 3.1. *Π_{OLE^+} unconditionally UC-realizes $\mathcal{F}_{\mathrm{OLE}^+}$ in the $\mathcal{F}_{\mathrm{OLE}}$-hybrid model.*

Proof. The simulator against a corrupted sender simulates both instances of $\mathcal{F}_{\text{OLE}}$. Let α_1 be the sender's input in the first OLE, and (α_2, α_3) be the inputs into the second OLE. The simulator sets $\hat{b} = \alpha_1 + \alpha_3$ and $\hat{a} = \alpha_2 - \hat{t}$, where $\hat{t}$ is chosen as the uniformly random output to $\mathcal{A}_{\mathsf{S}}$ of the first OLE. The simulator simply inputs $(\texttt{inputS}, (\hat{a}, \hat{b}))$ into $\mathcal{F}_{\text{OLE}^+}$. Let us briefly argue that this simulation is indistinguishable from a real protocol run. The value $\hat{t}$ is indistinguishable from a valid t, since the receiver basically uses a one-time-pad s to mask the multiplication. Therefore, the sender can only change his inputs into the OLEs. Since his inputs uniquely determine both $\hat{a}$ and $\hat{b}$, the extraction by the simulator is correct and the simulation is indistinguishable from a real protocol run.

Π_{OLE^+}

1. **Receiver (Input $x \in \mathbb{F}$): Pick $s \in \mathbb{F}$ and send $(\texttt{inputS}, (x^{-1}, s))$ to the first $\mathcal{F}_{\text{OLE}}$.**
2. **Sender (Input $a, b \in \mathbb{F}$): Pick $u \in \mathbb{F}$ uniformly at random and send $(\texttt{inputR}, u)$ to the first $\mathcal{F}_{\text{OLE}}$ to learn $t = ux^{-1} + s$. Send $(\texttt{inputS}, (t + a, b - u))$ to the second $\mathcal{F}_{\text{OLE}}$.**
3. **Receiver: Send $(\texttt{inputR}, x)$ to the second $\mathcal{F}_{\text{OLE}}$ and obtain $c = ax + b + sx$. Output $c - sx$.**

Fig. 3. Protocol that realizes $\mathcal{F}_{\text{OLE}^+}$ in the $\mathcal{F}_{\text{OLE}}$-hybrid model.

Against a corrupted receiver, the simulator simulates the two instance of $\mathcal{F}_{\text{OLE}}$ and obtains the receiver's inputs (ξ_1, ξ_3) and ξ_2. If $\xi_1 \cdot \xi_2 = 1$, the simulator sets $\hat{x} = \xi_2$, sends $(\texttt{inputR}, \hat{x})$ to $\mathcal{F}_{\text{OLE}^+}$ and receives $(\texttt{output}, c)$. It forwards $c' = c + \xi_2\xi_3$ to $\mathcal{A}_{\mathsf{R}}$. If $\xi_1 \cdot \xi_2 \neq 1$, the simulator sends $(\texttt{inputR}, 0)$ to $\mathcal{F}_{\text{OLE}^+}$ and forwards the output c to the receiver. It remains to argue that this simulation is indistinguishable from the real protocol. From $\mathcal{A}$'s view, the output c is determined as

$$c = u\xi_1\xi_2 + a\xi_2 + b - u + \xi_2\xi_3 = a\xi_2 + b + u(\xi_1\xi_2 - 1) + \xi_2\xi_3.$$

We can ignore the last term, since it is known to $\mathcal{A}$. If $\xi_1\xi_2 \neq 1$, then $u(\xi_1\xi_2 - 1)$ does not vanish and the result will be uniform over the choice of u. Thus, by using ξ_2 as the correct input otherwise, we extract the correct value and the simulation is indistinguishable from the real protocol. □

4 Randomized Polynomial Addition from OLE

Concretely, we have two parties, the sender with a polynomial of degree $2d$ as input and the receiver with a polynomial of degree d as input. The goal is that the receiver obtains the sum of these two polynomials such that it cannot learn the sender's polynomial fully. We want to achieve this privacy property by using a randomization polynomial that prevents the receiving party from simply subtracting its input from the result. This functionality is defined in Fig. 4.

Notice that we have some additional requirements regarding the inputs of the parties. First, the degree of the inputs has to be checked, but the functionality also makes sure that the receiver does not input a 0 polynomial, because otherwise he might learn the input of the sender. Also note that the functionality leaks some information about the sender's polynomial. Looking ahead in the PSI protocol, where the input of the sender is always a uniformly random $2d$ degree polynomial, this leakage of the ideal functionality will not leak any non-trivial information in the PSI protocol.

$\mathcal{F}_{\mathrm{OPA}}$

Implicitly parameterized by d signifying the maximal input degree.

1. **Upon receiving a message $(\mathtt{inputS}, (\mathbf{a}, \mathbf{r}))$ from the sender where $\mathbf{a}, \mathbf{r} \in \mathbb{F}[X]$, check whether**
   - $\mathbf{r} \neq 0$
   - **$\deg(\mathbf{r}) \leq d$ and $\deg(\mathbf{a}) = 2d$ OR $\deg(\mathbf{r}) = d$ and $\deg(\mathbf{a}) \leq 2d$**

   and ignore that message if not. Store $(\mathbf{a}, \mathbf{r})$ and send $(\mathtt{input})$ to $\mathcal{A}$.
2. **Upon receiving a message $(\mathtt{inputR}, \mathbf{b})$ from the receiver where $\mathbf{b} \in \mathbb{F}[X]$, check whether $\deg(\mathbf{b}) \leq d$ and $\mathbf{b} \neq 0$. If not, ignore that message. Otherwise, retrieve $\mathbf{a}, \mathbf{r}$, compute $\mathbf{s} = \mathbf{r} \cdot \mathbf{b} + \mathbf{a}$ and send $(\mathtt{res}, \mathbf{s})$ to the receiver. Ignore all further messages.**

Fig. 4. Ideal functionality that allows to obliviously compute an addition of polynomials.

It is instructive to first consider a passively secure protocol. In the semi-honest case, both sender and receiver evaluate their input polynomials on a set of distinct points $\mathcal{P} = \{\alpha_1, \ldots, \alpha_{2d+1}\}$, where d is the degree of the input polynomials. The sender additionally picks a random polynomial $\mathbf{r} \in \mathbb{F}[X]$ of degree d and also evaluates it on $\mathcal{P}$.

Instead of using OLE in the "traditional" sense, i.e. instead of computing $\mathbf{ab} + \mathbf{r}$ where $\mathbf{r}$ blinds the multiplication of the polynomials, we basically compute $\mathbf{rb} + \mathbf{a}$. This means that the sender randomizes the polynomial of the receiver, and then adds his own polynomial. This prevents the receiver from simply subtracting his input polynomial and learning $\mathbf{a}$. In a little more detail, sender and receiver use $2d + 1$ OLEs to add the polynomials as follows: for each $i \in [2d + 1]$, the sender inputs (r_i, a_i) in OLE i, while the receiver inputs b_i and obtains $s_i = r_i b_i + a_i$. He then interpolates the resulting polynomial $\mathbf{s}$ of degree $2d$ using the $2d + 1$ values s_i.

In going from passive to active security, we have to ensure that the inputs of the parties are correct. Here, the main difficulty obviously lies in checking for $\mathbf{b} = 0$. In fact, since $\mathcal{F}_{\mathrm{OPA}}$ does not even leak a single point a_i we have to make sure that all $b_i \neq 0$. However, this can easily be achieved by using $\mathcal{F}_{\mathrm{OLE}^+}$

instead of $\mathcal{F}_{\text{OLE}}$. We also have to verify that the inputs are well-formed via a simple polynomial check. For a more detailed overview we refer the reader to the introduction.

The complete actively secure protocol is shown in Fig. 5. Here, we use two instances of $\mathcal{F}_{\text{OLE}}$ that implement a commitment and a check. We named the first OLE that is used for a commitment to a blinding value u $\mathcal{F}_{\text{OLE}}^{\text{com}}$. The check is performed by comparing the blinded reconstructed polynomial $\mathbf{s}$ evaluated in x_S with the inputs in this location using the second OLE denoted by $\mathcal{F}_{\text{OLE}}^{\text{check}}$.[5]

Π_{OPA}

Let $\mathcal{P} = \{\alpha_1, \ldots, \alpha_{2d+1}\}, \alpha_i \in \mathbb{F}$ be a set of distinct points and let $\mathcal{F}_{\text{OLE}}^{\text{check}}, \mathcal{F}_{\text{OLE}}^{\text{com}}$ be instances of $\mathcal{F}_{\text{OLE}}$.

1. **Sender (Input $\mathbf{a}, \mathbf{r} \in \mathbb{F}[X]$, $\deg(\mathbf{a}) \leq 2d$, $\deg(\mathbf{r}) = d$): Evaluate $\mathbf{a}, \mathbf{r}$ on $\mathcal{P}$ to obtain $(a_i, r_i), i \in [2d+1]$. Input (r_i, a_i) into $\mathcal{F}_{\text{OLE}^+}^{(\text{i})}$.**
2. **Receiver (Input $\mathbf{b} \in \mathbb{F}[X]$, $\deg(\mathbf{b}) \leq d$): Evaluate $\mathbf{b}$ on $\mathcal{P}$ and obtain $b_i, i \in [2d+1]$. Input b_i into $\mathcal{F}_{\text{OLE}^+}^{(\text{i})}$ and receive $s_i = a_i + b_i \cdot r_i$. Reconstruct $\mathbf{s}$ from the s_i and check if $\deg(\mathbf{s}) \leq 2d$, otherwise abort.**
3. **Consistency check: Sender picks a random $x_s \in \mathbb{F}$ and send it to the receiver.**
   - **Receiver: Pick random $u, v \in \mathbb{F}$ and input them into $\mathcal{F}_{\text{OLE}}^{\text{com}}$. Further input $(\mathbf{b}(x_s), -\mathbf{s}(x_s) + u)$ into $\mathcal{F}_{\text{OLE}}^{\text{check}}$.**
   - **Sender: Pick a random $t \in \mathbb{F}$ and input it into $\mathcal{F}_{\text{OLE}}^{\text{com}}$ and obtain $c = ut + v$. Input $\mathbf{r}(\mathbf{x_s})$ into $\mathcal{F}_{\text{OLE}}^{\text{check}}$ and obtain $\mathbf{r}(x_s)\mathbf{b}(\mathbf{x_s}) - \mathbf{s}(x_s) + u + \mathbf{a}(x_s) = u'$. Send u' to the receiver.**
   - **Receiver: If $u' \neq u$ abort, otherwise send v to the sender.**
   - **Sender: If $u't + v \neq c$ abort.**
4. **Receiver picks $x_r \in \mathbb{F}$ and runs similar consistency check with the sender.**

Fig. 5. Protocol that realizes $\mathcal{F}_{\text{OPA}}$ in the $(\mathcal{F}_{\text{OLE}^+}, \mathcal{F}_{\text{OLE}})$-hybrid model.

Lemma 4.1. *Π_{OPA} unconditionally UC-realizes $\mathcal{F}_{\text{OPA}}$ in the $\mathcal{F}_{\text{OLE}^+}$-hybrid model.*

Proof (Sketch). **Corrupted Sender.** The simulator $\mathcal{S}_S$ against a corrupted sender proceeds as follows. It simulates $\mathcal{F}_{\text{OLE}^+}^{(\text{i})}$ and thereby obtains (r_i^*, a_i^*) for all $i \in [2d+1]$. From these values, the simulator reconstructs $\hat{\mathbf{r}}$ and $\hat{\mathbf{a}}$. It aborts in Step 3 if $\deg(\hat{\mathbf{r}}) > d$ or $\deg(\hat{\mathbf{a}}) > 2d$. It also aborts if $\hat{\mathbf{a}}$ or $\hat{\mathbf{r}}$ are zero, and otherwise sends $(\texttt{inputS}, (\hat{\mathbf{a}}, \hat{\mathbf{r}}))$ to $\mathcal{F}_{\text{OPA}}$.

The extraction of the corrupted sender's inputs is correct if his inputs $\mathbf{r}^*$ corresponds to a polynomial of degree d and $\mathbf{a}^*$ corresponds to a polynomial of degree $2d$. Thus, the only possibility for an environment to distinguish between

[5] The commitment we implicitly use has been used previously in [11], as has the check sub-protocol.

the simulation and the real protocol is by succeeding in answering the check while using a malformed input, i.e. a polynomial of incorrect degree or 0-polynomials. If the polynomials have degree greater than d and $2d$, respectively, the resulting polynomial $\mathbf{s}$ has degree $2d+1$ instead of $2d$, i.e. the receiver cannot reconstruct the result from $2d+1$ points. Since the sender learns nothing about the receiver's inputs, the thus incorrectly reconstructed polynomial will be uniformly random from his point of view and the probability that his response to the challenge is correct is $1/|\mathbb{F}|$. Also, both $\hat{\mathbf{a}}$ and $\hat{\mathbf{r}}$ have to be non-zero, because in each case the polynomials are evaluated in $2d+1$ points, and it requires $2d+1$ zeros as a_i, r_i to get a 0 polynomial. But since both $\mathbf{a}, \mathbf{r}$ have degree at most $2d$, there are at most $2d$ roots of these polynomials. Therefore, in order to pass the check, $\mathbf{a}(x)$ and $\mathbf{b}(x)$ would need to be 0, which is also checked for.

Corrupted Receiver. The simulator $\mathcal{S}_\mathsf{R}$ against a corrupted receiver simulates $\mathcal{F}^{(\mathrm{i})}_{\mathrm{OLE}^+}$ and obtains $\mathbf{b}_i^*$ for all $i \in [2d+1]$. It reconstructs $\hat{\mathbf{b}}$ and aborts the check in Step 3 if $\deg(\hat{\mathbf{b}}) > d$. The simulator sends $(\mathtt{inputR}, \hat{\mathbf{b}})$ to $\mathcal{F}_{\mathrm{OPA}}$ and receives $(\mathtt{res}, \hat{\mathbf{s}})$. It evaluates $\hat{\mathbf{s}}$ on $\mathcal{P}$ and returns s_i for the corresponding OLEs. $\mathcal{S}_\mathsf{R}$ simulates the rest according to the protocol.

Clearly, if the corrupted receiver $\mathcal{A}_\mathsf{R}$ inputs a degree d polynomial, the simulator will extract the correct polynomial. In order to distinguish the simulation from the real protocol, the adversary can either input 0 in an OLE or has to input a polynomial of higher degree, while still passing the check. In the first case, assume w.l.o.g. that $\mathcal{A}_\mathsf{R}$ cheats in $\mathcal{F}^{(\mathrm{j})}_{\mathrm{OLE}^+}$ for some j. This means $\mathcal{A}_\mathsf{R}$ receives a value $\hat{s}_i$, which is uniformly random. This means that only with probability $1/|\mathbb{F}|$ will $\hat{s}_i$ satisfy the relation $\mathbf{rb}+\mathbf{a}$ and the check will fail, i.e. he can lie about u, but the commitment to u cannot be opened without knowing t. In the second case, the resulting polynomial would be of degree $2d+1$, while the receiver only gets $2d+1$ points of the polynomial. Therefore the real polynomial is underdetermined and $\mathcal{A}$ can only guess the correct value $\hat{\mathbf{s}}(x)$, i.e. the check will fail with overwhelming probability. □

5 Maliciously Secure Two-Party PSI

In this section we provide a maliciously secure two-party PSI protocol with output for *both* parties, i.e. we realize $\mathcal{F}_{\mathrm{PSI}}$ as described in Fig. 6.

$\mathcal{F}_{\mathrm{PSI}}$

1. Upon receiving a message $(\mathtt{input}, P, S_P)$ from party $P \in \{\mathsf{A}, \mathsf{B}\}$, store the set S_P. Once all inputs are given, set $S_\cap = S_\mathsf{A} \cap S_\mathsf{B}$ and send $(\mathtt{output}, S_\cap)$ to $\mathcal{A}$.
2. Upon receiving a message $(\mathtt{deliver})$ from $\mathcal{A}$, send $(\mathtt{output}, S_\cap)$ to the honest party.

Fig. 6. Ideal functionality $\mathcal{F}_{\mathrm{PSI}}$ for two-party PSI.

$\Pi_{2\text{PSI}}$

Let $m = \max_i |S_i| + 1$.

Computation of Intersection

1. Alice (Input S_A): Pick a random polynomial $\mathbf{p}_\mathsf{A}$ of degree m such that $\mathbf{p}_\mathsf{A}(\gamma_j) = 0$ for all $\gamma_j \in S_\mathsf{A}$. Generate two random polynomials $\mathbf{r}_\mathsf{A}, \mathbf{r}'_\mathsf{A}$ of degree m and a random polynomial $\mathbf{u}_\mathsf{A}$ of degree $2m$.
   - Input $\mathbf{r}_\mathsf{A}, \mathbf{u}_\mathsf{A}$ into $\mathcal{F}^{(1)}_{\text{OPA}}$.
   - Input $\mathbf{p}_\mathsf{A}$ into $\mathcal{F}^{(2)}_{\text{OPA}}$ and obtain $\mathbf{s}_\mathsf{A} = \mathbf{p}_\mathsf{A}\mathbf{r}_\mathsf{B} + \mathbf{u}_\mathsf{B}$.
   - Set $\mathbf{s}'_\mathsf{A} = \mathbf{s}_\mathsf{A} - \mathbf{u}_\mathsf{A} + \mathbf{p}_\mathsf{A}\mathbf{r}'_\mathsf{A}$ and send it to Bob.
2. P_B (Input S_B): Pick a random polynomial $\mathbf{p}_\mathsf{B}$ of degree m such that $\mathbf{p}_\mathsf{B}(\gamma_j) = 0$ for all $\gamma_j \in S_\mathsf{B}$. Generate two random polynomials $\mathbf{r}_\mathsf{B}, \mathbf{r}'_\mathsf{B}$ of degree m and a random polynomial $\mathbf{u}_\mathsf{B}$ of degree $2m$.
   - Input $\mathbf{r}_\mathsf{B}, \mathbf{u}_\mathsf{B}$ into $\mathcal{F}^{(2)}_{\text{OPA}}$.
   - Input $\mathbf{p}_\mathsf{B}$ into $\mathcal{F}^{(1)}_{\text{OPA}}$ and obtain $\mathbf{s}_\mathsf{B} = \mathbf{p}_\mathsf{B}\mathbf{r}_\mathsf{A} + \mathbf{u}_\mathsf{A}$.
   - Upon receiving $\mathbf{s}'_\mathsf{A}$, compute $\mathbf{p}_\cap = \mathbf{s}'_\mathsf{A} + \mathbf{s}_\mathsf{B} + \mathbf{p}_\mathsf{B}\mathbf{r}'_\mathsf{B} - \mathbf{u}_\mathsf{B}$ and send it to Alice.

Output Verification

3. Alice: Pick a random $x_\mathsf{A} \in \mathbb{F}$ and send it to Bob.
4. Bob: Set $\alpha_\mathsf{B} = \mathbf{p}_\mathsf{B}(x_\mathsf{A})$, $\beta_\mathsf{B} = \mathbf{r}_\mathsf{B}(x_\mathsf{A})$ and $\delta_\mathsf{B} = \mathbf{r}'_\mathsf{B}(x_\mathsf{A})$. Pick a random $x_\mathsf{B} \in \mathbb{F}$ and send $(x_\mathsf{B}, \alpha_\mathsf{B}, \beta_\mathsf{B}, \delta_\mathsf{B})$ to Alice.
5. Alice: Check if $\mathbf{p}_\mathsf{A}(x_\mathsf{A})(\beta_\mathsf{B} + \mathbf{r}'_\mathsf{A}(x_\mathsf{A})) + \alpha_\mathsf{B}(\mathbf{r}_\mathsf{A}(x_\mathsf{A}) + \delta_\mathsf{B}) = \mathbf{p}_\cap(x_\mathsf{A})$, otherwise abort. For each $\gamma_j \in S_\mathsf{A}$: If $\mathbf{p}_\cap(\gamma_j) = 0$, add γ_j to $S_\cap$. Send $\alpha_\mathsf{A} = \mathbf{p}_\mathsf{A}(x_\mathsf{B})$, $\beta_\mathsf{A} = \mathbf{r}_\mathsf{A}(x_\mathsf{B})$ and $\delta_\mathsf{A} = \mathbf{r}'_\mathsf{A}(x_\mathsf{B})$ to Bob. Output $S_\cap$.
6. Bob: Check if $\alpha_\mathsf{A}(\mathbf{r}_\mathsf{B}(x_\mathsf{B}) + \delta_\mathsf{A}) + \mathbf{p}_\mathsf{B}(x_\mathsf{B})(\beta_\mathsf{A} + \mathbf{r}'_\mathsf{B}(x_\mathsf{B})) = \mathbf{p}_\cap(x_\mathsf{B})$, otherwise abort. For each $\gamma_j \in S_\mathsf{B}$: If $\mathbf{p}_\cap(\gamma_j) = 0$, add γ_j to $S_\cap$. Output $S_\cap$.

Fig. 7. Protocol $\Pi_{2\text{PSI}}$ UC-realises $\mathcal{F}_{\text{PSI}}$ in the $\mathcal{F}_{\text{OPA}}$-hybrid model.

We briefly sketch the protocol in the following; a more detailed overview can be found in the introduction. First, Alice and Bob simply transform their input sets into polynomials. Then, both compute a randomized share of the intersection via our previously defined OPA in such a way that Alice can send her share to Bob without him being able to learn her input. This can be achieved by adding a simple mask to the intermediate share. Bob adds both shares and sends the output to Alice. The protocol only requires two OPA and a simple check which ensures semi-honest behaviour, and no computational primitives. A formal description is given in Fig. 7.

Theorem 5.1. *The protocol $\Pi_{2\text{PSI}}$ UC-realises $\mathcal{F}_{\text{PSI}}$ in the $\mathcal{F}_{\text{OPA}}$-hybrid model with communication complexity $O(m\kappa)$.*

Proof. Let us argue that $\mathbf{p}_\cap = \mathbf{p}_\mathsf{A}(\mathbf{r}'_\mathsf{A} + \mathbf{r}_\mathsf{B}) + \mathbf{p}_\mathsf{B}(\mathbf{r}_\mathsf{A} + \mathbf{r}'_\mathsf{B})$ actually hides the inputs. The main observation here is that $\mathbf{r}'_\mathsf{P} + \mathbf{r}_{\bar{\mathsf{P}}}$ is uniformly random as long as one party is honest. Since $\mathbf{p}_\mathsf{A} + \mathbf{p}_\mathsf{B}$ validly encodes the intersection

Simulator $\mathcal{S}_{\mathsf{A}}$

1. Extract the inputs $\hat{\mathbf{p}}_{\mathsf{A}}, \hat{\mathbf{r}}_{\mathsf{A}}, \hat{\mathbf{u}}_{\mathsf{A}}$ by simulating $\mathcal{F}_{\mathrm{OPA}}$.
2. Find the roots $\hat{\gamma}_1, \ldots, \hat{\gamma}_m$ of $\hat{\mathbf{p}}_{\mathsf{A}}$ and thereby the set $\hat{S}_{\mathsf{A}} = \{\hat{\gamma}_1, \ldots, \hat{\gamma}_m\}$.
3. Send $(\mathtt{input}, \mathsf{A}, \hat{S}_{\mathsf{A}})$ to $\mathcal{F}_{\mathrm{PSI}}$.
4. Upon receiving $(\mathtt{output}, \hat{S}_{\cap})$ from $\mathcal{F}_{\mathrm{PSI}}$, pick a random degree m polynomial $\hat{\mathbf{p}}_{\mathsf{B}}$ such that $\hat{\mathbf{p}}_{\mathsf{B}}(\gamma) = 0$ for all $\gamma \in \hat{S}_{\cap}$.
5. Use $\hat{\mathbf{p}}_{\mathsf{B}}$ as input for simulating the $\mathcal{F}_{\mathrm{OPA}}$ together with random polynomials $\hat{\mathbf{r}}_{\mathsf{B}}$ and $\hat{\mathbf{u}}_{\mathsf{B}}$, i.e. keep $\hat{\mathbf{s}}_{\mathsf{B}} = \hat{\mathbf{p}}_{\mathsf{B}} \cdot \hat{\mathbf{r}}_{\mathsf{A}} + \hat{\mathbf{u}}_{\mathsf{A}}$ and send $\hat{\mathbf{s}}_{\mathsf{A}} = \hat{\mathbf{p}}_{\mathsf{A}} \cdot \hat{\mathbf{r}}_{\mathsf{B}} + \hat{\mathbf{u}}_{\mathsf{B}}$ to $\mathcal{A}$.
6. Simulate the rest according to Π_{2PSI}, but abort in Step 6, if after setting

$$\hat{\mathbf{r}}'_{\mathsf{A}} = (\mathbf{s}'^{*}_{\mathsf{A}} + \hat{\mathbf{u}}_{\mathsf{A}} - \hat{\mathbf{u}}_{\mathsf{B}} - \hat{\mathbf{p}}_{\mathsf{A}}\hat{\mathbf{r}}_{\mathsf{B}})/\hat{\mathbf{p}}_{\mathsf{A}},$$

$\alpha^*_{\mathsf{A}} \neq \hat{\mathbf{p}}_{\mathsf{A}}(x)$, $\beta^*_{\mathsf{A}} \neq \hat{\mathbf{r}}_{\mathsf{A}}(x)$ or $\delta^*_{\mathsf{A}} \neq \hat{\mathbf{r}}'_{\mathsf{A}}(x)$, even though the check in Step 6 would pass.

Fig. 8. Simulator $\mathcal{S}_{\mathsf{A}}$ against a corrupted Alice.

(see Lemma 2.1), $\mathbf{p}_{\cap}$ is uniformly random over the choice of the randomization polynomials $\mathbf{r}_{\mathsf{A}}, \mathbf{r}'_{\mathsf{A}}, \mathbf{r}_{\mathsf{B}}$ and $\mathbf{r}'_{\mathsf{B}}$, except for the roots denoting the intersection.

Corrupted Alice. We show the indistinguishability of the simulation of $\mathcal{S}_{\mathsf{A}}$ (cf. Fig. 8). The simulator extracts Alice's inputs and then checks for any deviating behaviour. If such behaviour is detected, it aborts, even if the protocol would succeed. Proving indistinguishability of the simulation shows that the check in the protocol basically enforces semi-honest behaviour by Alice, up to input substitution.

Consider the following series of hybrid games.

Hybrid 0: $\mathsf{Real}^{\mathcal{A}_{\mathsf{A}}}_{\Pi_{\mathrm{2PSI}}}$.
Hybrid 1: Identical to Hybrid 0, except that $\mathcal{S}_1$ simulates $\mathcal{F}_{\mathrm{OPA}}$, learns all inputs and aborts if $\alpha^*_{\mathsf{A}} \neq \hat{\mathbf{p}}_{\mathsf{A}}(x)$ or $\beta^*_{\mathsf{A}} \neq \hat{\mathbf{r}}_{\mathsf{A}}(x)$, but the check is passed. Let $\alpha^*_{\mathsf{A}} = \alpha_{\mathsf{A}} + e$ be $\mathcal{A}_{\mathsf{A}}$'s check value. Then the check in Step 6 will fail with overwhelming probability. Let σ denote the outcome of the check. If $\mathcal{A}_{\mathsf{A}}$ behaves honestly, then

$$\sigma = \alpha^*_{\mathsf{A}}(\mathbf{r}_{\mathsf{B}}(x) + \delta^*_{\mathsf{A}}) + \mathbf{p}_{\mathsf{B}}(x)(\beta^*_{\mathsf{A}} + \mathbf{r}'_{\mathsf{B}}(x)) - \mathbf{p}_{\cap}(x) = 0.$$

Using $\alpha^*_{\mathsf{A}} = \alpha_{\mathsf{A}} + e$, however, we get

$$\sigma' = (\alpha_{\mathsf{A}}+e)(\mathbf{r}_{\mathsf{B}}(x)+\delta^*_{\mathsf{A}})+\mathbf{p}_{\mathsf{B}}(x)(\beta^*_{\mathsf{A}}+\mathbf{r}'_{\mathsf{B}}(x))-\mathbf{p}_{\cap}(x) = e\cdot(\mathbf{r}_{\mathsf{B}}(x)+\delta^*_{\mathsf{A}}) \neq \mathsf{const}.$$

This means that the outcome of the check is uniformly random from $\mathcal{A}_{\mathsf{A}}$'s view over the choice of $\mathbf{r}_{\mathsf{B}}$ (or $\mathbf{p}_{\mathsf{B}}$ for $\beta^*_{\mathsf{A}} \neq \mathbf{r}_{\mathsf{A}}(x)$). Therefore, the check will fail except with probability $2/|\mathbb{F}|$ and Hybrids 0 and 1 are statistically close.

Hybrid 2: Identical to Hybrid 1, except that $\mathcal{S}_2$ aborts according to Step 6 in Fig. 8.
An environment distinguishing Hybrids 1 and 2 must manage to send $\mathbf{s}'^*_{\mathsf{A}}$ such that

$$\mathbf{s}'^*_{\mathsf{A}} + \hat{\mathbf{u}}_{\mathsf{A}} - \hat{\mathbf{u}}_{\mathsf{B}} \neq \hat{\mathbf{p}}_{\mathsf{A}} \cdot (\hat{\mathbf{r}}_{\mathsf{B}} + \hat{\mathbf{r}}'_{\mathsf{A}})$$

while passing the check in Step 6 with non-negligible probability.
Let $\mathbf{f} = \mathbf{s}'^*_{\mathsf{A}} + \hat{\mathbf{u}}_{\mathsf{A}} - \hat{\mathbf{u}}_{\mathsf{B}} - \hat{\mathbf{p}}_{\mathsf{A}} \cdot (\hat{\mathbf{r}}_{\mathsf{B}} + \hat{\mathbf{r}}'_{\mathsf{A}})$. We already know that $\mathbf{f}(x) = 0$, otherwise we have $\alpha^*_{\mathsf{A}} = \alpha_{\mathsf{A}} + \mathbf{f}(x) \neq \alpha_{\mathsf{A}}$ (or an invalid β^*_{A}), and the check fails. But since x is uniformly random, the case that $\mathbf{f}(x) = 0$ happens only with probability $m/|\mathbb{F}|$, which is negligible. Therefore, Hybrid 1 and Hybrid 2 are statistically close.
Hybrid 3: Identical to Hybrid 2, except that $\mathcal{S}_3$ generates the inputs $\hat{\mathbf{s}}_{\mathsf{A}}, \hat{\mathbf{s}}_{\mathsf{B}}$ according to Step 5 in Fig. 8 and adjusts the output. This corresponds to $\mathsf{Ideal}^{\mathcal{S}_{\mathsf{A}}}_{\mathcal{F}_{\mathrm{PSI}}}$.
The previous hybrids established that the inputs $\hat{\mathbf{p}}_{\mathsf{A}}, \hat{\mathbf{r}}_{\mathsf{A}}$ are extracted correctly. Therefore, by definition, $\hat{S}_{\mathsf{A}} = \mathcal{M}_0(\hat{\mathbf{p}}_{\mathsf{A}})$. It remains to argue that the simulated outputs are indistinguishable. First, note that the received intersection $\hat{S}_{\cap} = \mathcal{M}_0(\hat{\mathbf{p}}_{\mathsf{B}})$ defines $\hat{\mathbf{p}}_{\mathsf{B}}$. From Lemma 2.4 it follows that $\mathcal{M}_0(\mathbf{p}_{\cap}) = \mathcal{M}_0(\hat{\mathbf{p}}_{\mathsf{A}}) \cap \mathcal{M}_0(\hat{\mathbf{p}}_{\mathsf{B}}) = \hat{S}_{\cap}$ w.r.t. $\mathcal{M}_0(\hat{\mathbf{p}}_{\mathsf{B}})$, even for a maliciously chosen $\hat{\mathbf{r}}_{\mathsf{A}}$, i.e. the $\mathcal{A}_{\mathsf{A}}$ cannot increase the intersection even by a single element except with negligible probability.
Further, note that $\hat{\mathbf{s}}_{\mathsf{A}} = \hat{\mathbf{p}}_{\mathsf{A}} \cdot \hat{\mathbf{r}}_{\mathsf{B}} + \hat{\mathbf{u}}_{\mathsf{B}}$ is uniformly distributed over the choice of $\hat{\mathbf{u}}_{\mathsf{B}}$, and $\hat{\mathbf{p}}_{\cap}$ is uniform over the choice of $\hat{\mathbf{r}}_{\mathsf{B}}, \hat{\mathbf{r}}'_{\mathsf{B}}$.
Finally, since $\hat{\mathbf{r}}_{\mathsf{B}}, \hat{\mathbf{r}}'_{\mathsf{B}}$ are uniformly random and the degree of $\hat{\mathbf{p}}_{\mathsf{B}}$ is m, i.e. $\max_i |\mathcal{S}_i| + 1$, the values $\hat{\alpha}_{\mathsf{B}}$, $\hat{\beta}_{\mathsf{B}}$ and $\hat{\delta}_{\mathsf{B}}$ are uniformly distributed as well. In conclusion, the Hybrids 2 and 3 are statistically close.

As a result we get that for all environments $\mathcal{Z}$,

$$\mathsf{Real}^{\mathcal{A}_{\mathsf{A}}}_{\Pi_{2\mathrm{PSI}}}(\mathcal{Z}) \approx_s \mathsf{Ideal}^{\mathcal{S}_{\mathsf{A}}}_{\mathcal{F}_{\mathrm{PSI}}}(\mathcal{Z}).$$

Corrupted Bob. The simulator against a corrupted Bob is essentially the same as the one against a corrupted Alice, except for a different way to check his inputs.For the full proof we refer the reader to the full version [15] of the paper.

Efficiency. The protocol makes two calls to OPA, which in turn is based on OLE. Overall, $2m$ calls to OLE are necessary in OPA. Given the recent constant overhead OLE of Ghosh et al. [14], the communication complexity of $\Pi_{2\mathrm{PSI}}$ lies in $O(m\kappa)$.

The computational cost of the protocol is dominated by multi-point evaluation of polynomials of degree m, which requires $O(m \log m)$ multiplications using fast modular transform [3]. Note that this cost includes computational cost of the OLE instantiation from [14]. This concludes the proof. □

6 Maliciously Secure Multi-party PSI

6.1 Ideal Functionality

The ideal functionality for multi-party private set intersection $\mathcal{F}^*_{\mathrm{MPSI}}$ simply takes the inputs from all parties and computes the intersection of these inputs. Our functionality $\mathcal{F}^*_{\mathrm{MPSI}}$ in Fig. 9 additionally allows an adversary to learn the intersection and then possibly update the result to be only a subset of the original result.

$\mathcal{F}^*_{\mathrm{MPSI}}$

Let A denote the set of malicious parties, and H the set of honest parties.

1. Upon receiving a message $(\texttt{input}, P_i, S_i)$ from party P_i, store the set S_i. Once all inputs $i \in [n]$ are input, set $S_\cap = \bigcap_{i=1}^{n} S_i$ and send $(\texttt{output}, S_\cap)$ to $\mathcal{A}$.
2. Upon receiving a message $(\texttt{deliver}, S'_\cap)$ from $\mathcal{A}$, check if $S'_\cap \subseteq S_\cap$. If not, set $S'_\cap = \perp$. Send $(\texttt{output}, S'_\cap)$ to H.

Fig. 9. Ideal functionality $\mathcal{F}^*_{\mathrm{MPSI}}$ for multi-party PSI.

Let us briefly elaborate on why we chose to use this modified functionality. In the UC setting, in order to extract the inputs of all malicious parties, any honest party has to communicate with all malicious parties. In particular, since the simulator has to extract the complete input, this requires at least $O(nm)$ communication per party for the classical MPSI functionality. In turn, for the complete protocol, this means that the communication complexity lies in $O(n^2m)$.

Instead, we want to take an approach similar to the recent work of Hazay et al. [19], i.e. we have one central party, and some of the work is delegated to this party. This removes the need for the other parties to extensively communicate with each other and potentially allows communication complexity $O(mn)$, which is asymptotically optimal in any setting. However, if we assume that the central party and at least one additional party are corrupted, the honest party does not (extensively) interact with this additional party and does not learn its inputs; it can only learn the input of the central party. If the input set of the other malicious party is the same as the one of the central party, the output remains the same. If this input is different, however, the actual intersection might be smaller. One might argue that this case simply corresponds to input substitution by the malicious party, but for any type of UC simulation this poses a problem, since the output of the honest party in the protocol might be different from the intersection in the ideal world. Thus, $\mathcal{F}^*_{\mathrm{MPSI}}$ allows a malicious party to modify the output. Crucially, the updated intersection can only be smaller and may not changed arbitrarily by the adversary. We believe that this weaker multiparty PSI functionality is sufficient for most scenarios.

6.2 Multi-party PSI from OLE

Our multi-party PSI protocol uses the same techniques that we previously employed to achieve two-party PSI. This is similar in spirit to the approach taken in [19], who employ techniques from the two-party PSI of [13] and apply them in the multi-party setting. We also adopt the idea of a designated central party that performs a two-party PSI with the remaining parties, because this allows to delegate most of the computation to this party and saves communication. Apart from that, our techniques differ completely from [19]. Abstractly, they run the two-party PSI with each party and then use threshold encryption and zero-knowledge proofs to ensure the correctness of the computation. These tools inflict a significant communication and computation penalty.

In our protocol (cf. Fig. 10) we run our two-party PSI between the central party and every other party, but we ensure privacy of the aggregation not via threshold encryption and zero-knowledge proofs, but instead by a simple masking of the intermediate values and a polynomial check. This masking is created in a setup phase, where every pair of parties exchanges a random seed that is used to create two random blinding polynomials which cancel out when added.

Once the central party receives all shares of the computation, it simply add these shares, thereby removing the random masks. The central party broadcasts the result to all parties. Then, all parties engage in a multi-party coin-toss and obtain a random value x. Since all operations in the protocol are linear operations on polynomials, the parties evaluate their input polynomials on x and broadcast the result. This allows every party to locally verify the relation and as a consequence also the result. Here we have to ensure that a rushing adversary cannot cheat by waiting for all answers before providing its own answer. We solve this issue by simply committing to the values first, and the unveiling them in the next step. This leads to malleability problems, i.e. we have to use non-malleable commitments[6].

Theorem 6.1. *The protocol Π_{MPSI} computationally UC-realises $\mathcal{F}^*_{\text{MPSI}}$ in the $\mathcal{F}_{\text{OPA}}$-hybrid model with communication complexity in $O((n^2 + nm)\kappa)$.*

Proof. We have to distinguish between the case where the central party is malicious and the case where it is honest. We show UC-security of Π_{MPSI} by defining a simulator $\mathcal{S}$ for each case which produces an indistinguishable simulation of the protocol to any environment $\mathcal{Z}$ trying to distinguish the ideal world from the real world. The approach of the simulation is straightforward: the simulator extracts the input polynomials into $\mathcal{F}_{\text{OPA}}$ and thus obtains an intersection of the adversary's inputs.

In the case of an honest central party, all parties communicate with this party, i.e. the simulator can extract all inputs of all malicious parties. In the case where P_0 is malicious, however, the simulator can at most learn the central party's input at the beginning. He inputs this result into the ideal functionality

[6] In order to achieve our claimed efficiency we actually use UC commitments, but non-malleable commitments are sufficient for the security of the protocol.

Π_{MPSI}

Let $m = \max_i \{|S_i|\} + 1$ and NMCOM be a bounded-concurrent non-malleable commitment scheme against synchronized adversaries. $\mathcal{F}_{\text{OPA}}^{(\text{i,j})}$ denotes the jth instance for parties P_0 and P_i.

Setup

1. All parties P_i and P_j for $i, j \in \{1, \ldots, n-1\}$ exchange a random polynomial as follows. For all $j \neq i$, if $\mathbf{v}_{ij} = \perp$, P_i picks seed_{ij} uniformly at random and sets $\mathbf{v}_{ij} = \mathsf{PRG}_{2m}(\mathsf{seed}_{ij})$. It sends seed_{ij} to P_j, who sets $\mathbf{v}_{ji} = -\mathsf{PRG}_{2m}(\mathsf{seed}_{ij})$.

Share Computation

2. P_0 (Input S_0): Compute a polynomial $\mathbf{p}_0$ of degree m s.t. $\mathbf{p}_0(\gamma_j) = 0$ for all $\gamma_j \in S_0$. Generate n random polynomials $\mathbf{r}_0^i \in \mathbb{F}[X], i \in \{1, \ldots, n-1\}$ and $\mathbf{r}_0' \in \mathbb{F}[X]$ of degree m each and $n-1$ random polynomials $\mathbf{u}_0^i \in \mathbb{F}[X], i \in \{1, \ldots, n-1\}$ of degree $2m$. For $i \in [n-1]$
   - Input $\mathbf{r}_0^i, \mathbf{u}_0^i$ into $\mathcal{F}_{\text{OPA}}^{(\text{i,1})}$ for each $i \in \{1, \ldots, n-1\}$.
   - Input $\mathbf{p}_0$ into $\mathcal{F}_{\text{OPA}}^{(\text{i,2})}$ and obtain $\mathbf{s}_0^i = \mathbf{p}_0 \cdot \mathbf{r}_i + \mathbf{u}_i$.
3. P_i (Input S_i): Compute a polynomial $\mathbf{p}_i$ of degree m s.t. $\mathbf{p}_i(\gamma_j) = 0$ for all $\gamma_j \in S_i$. Additionally, pick $\mathbf{r}_i, \mathbf{r}_i' \in \mathbb{F}[X]$ uniformly of degree m and $\mathbf{u}_i \in \mathbb{F}[X]$ uniformly of degree $2m$.
   - Input $\mathbf{p}_i$ into $\mathcal{F}_{\text{OPA}}^{(\text{i,1})}$ and obtain $\mathbf{s}_i = \mathbf{p}_i \cdot \mathbf{r}_0^i + \mathbf{u}_0^i$.
   - Input $\mathbf{r}_i, \mathbf{u}_i$ into $\mathcal{F}_{\text{OPA}}^{(\text{i,2})}$.
   - Set $\mathbf{s}_i' = \mathbf{s}_i - \mathbf{u}_i + \mathbf{p}_i \mathbf{r}_i' + \sum_{i \neq j} \mathbf{v}_{ij}$ and send it to P_0.

Output Aggregation and Verification

4. P_0: Compute $\mathbf{p}_\cap = \sum_{i=1}^{n-1} (\mathbf{s}_i' + \mathbf{s}_0^i - \mathbf{u}_0^i + \mathbf{p}_0 \mathbf{r}_0')$ and broadcast $\mathbf{p}_\cap$.
5. *All parties*:
   - Run a multiparty coin-toss protocol Π_{CT} to obtain a random $x \in \mathbb{F}$.
   - Evaluate $\alpha_i = \mathbf{p}_i(x)$, $\beta_i = \mathbf{r}_i(x)$, $\delta_i = \mathbf{r}_i'(x)$ and compute $(\mathsf{com}_i, \mathsf{unv}_i) = \mathsf{NMCOM.Commit}(\alpha_i, \beta_i, \delta_i)$. Broadcast com_i.
   - Once all commitments are received, broadcast unv_i and $(\alpha_i, \beta_i, \delta_i)$. Abort if $\sum \alpha_0 \cdot (\beta_i + \delta_0) + \alpha_i \cdot (\beta_0 + \delta_i) \neq \mathbf{p}_\cap(x)$ or $\mathsf{NMCOM.Open}(\mathsf{com}_i, \mathsf{unv}_i, (\alpha_i, \beta_i, \delta_i)) \neq 1$.
   - For each $\gamma_j \in S_i$: if $\mathbf{p}_\cap(\gamma_j) = 0$, add γ_j to $S_\cap$. Output $S_\cap$.

Fig. 10. Protocol Π_{MPSI} UC-realises $\mathcal{F}_{\text{MPSI}}^*$ in the $\mathcal{F}_{\text{OPA}}$-hybrid model.

and uses the intermediate result for the simulation. The malicious central party can later "simulate" the other malicious parties and thereby possibly change the intersection for the honest parties. We show that $\mathcal{A}$ can only reduce the intersection unless it already knows $x \in S_j$ for at least one $j \in \mathsf{H}$, i.e. we assume that $\mathcal{A}$ cannot predict a single element of the set of an honest party except with negligible probability. This reduced intersection can be passed by the simulator to the ideal functionality.

Simulator $\mathcal{S}_{P_0}$

Let $\mathsf{A} = \{i | P_i \text{ is malicious}\}$ denote the index set of corrupted parties, where $|\mathsf{A}| = t \leq n - 1$. Further let H denote the index set of honest parties.

1. Simulate the setup and obtain all $\mathbf{v}^*_{ij}$ for $i \in \mathsf{A}$ and $j \in \mathsf{H}$. Pick uniformly random $\hat{\mathbf{v}}_{jl} \in \mathbb{F}[X]$ of degree $2m$ for $j, l \in \mathsf{H}$ and set $\hat{\mathbf{v}}_{lj} = -\hat{\mathbf{v}}_{jl}$.
2. Extract the inputs $(\hat{\mathbf{p}}^j_0, \hat{\mathbf{r}}^j_0, \hat{\mathbf{u}}^j_0)$ of P_0 for all $j \in \mathsf{H}$ by simulating $\mathcal{F}_{\text{OPA}}$.
3. Abort in Step 5 of Π_{MPSI} if the $\hat{\mathbf{p}}^j_0$ are not all identical. Set $\hat{\mathbf{p}}_{\mathsf{A}} = \hat{\mathbf{p}}^j_0$ for a random $j \in \mathsf{H}$, and find the roots $\hat{\gamma}_1, \dots, \hat{\gamma}_{2m}$ of $\hat{\mathbf{p}}_{\mathsf{A}}$ and thereby the set $\hat{S}_{\mathsf{A}} = \{\hat{\gamma}_1, \dots, \hat{\gamma}_{2m}\}$.
4. Send $(\texttt{input}, P_i, \hat{S}_{\mathsf{A}})$ to $\mathcal{F}^*_{\text{MPSI}}$ for all parties $i \in \mathsf{A}$.
5. Upon receiving $(\texttt{output}, \hat{S}_{\cap})$ from $\mathcal{F}^*_{\text{MPSI}}$, pick $n-t$ random degree m polynomials $\hat{\mathbf{p}}_j$ such that $\hat{\mathbf{p}}_j(\gamma) = 0$ for all $\gamma \in \hat{S}_{\cap}, j \in \mathsf{H}$.
6. Use the $\hat{\mathbf{p}}_j$ as input for each instance of $\mathcal{F}_{\text{OPA}}$ together with random polynomials $\hat{\mathbf{r}}_j$ and $\hat{\mathbf{u}}_j$ for $j \in \mathsf{H}$, i.e. send $\hat{\mathbf{s}}^j_0 = \hat{\mathbf{p}}^j_0 \cdot \hat{\mathbf{r}}_j + \hat{\mathbf{u}}_j$ to $\mathcal{A}$ and keep $\hat{\mathbf{s}}_j = \hat{\mathbf{p}}_j \cdot \hat{\mathbf{r}}^j_0 + \hat{\mathbf{u}}^j_0$.
7. Simulate the rest according to Π_{MPSI}, but abort in Step 5, if the extracted $\hat{\mathbf{p}}_{\mathsf{A}}(x) \neq \alpha_0$ or $\hat{\mathbf{r}}^j_0(x) \neq \beta^j_0$ for any $j \in \mathsf{H}$, even if the check passes otherwise.
8. Upon receiving $\mathbf{p}^*_{\cap}$ from $\mathcal{A}$, if the check in Step 5 of Π_{MPSI} passes, test for all $s \in \hat{S}_{\cap}$ if $\mathbf{p}^*_{\cap}(s) = 0$. If yes, set $\hat{S}'_{\cap} = \hat{S}'_{\cap} \cup s$. Send $(\texttt{deliver}, \hat{S}'_{\cap})$ to $\mathcal{F}^*_{\text{MPSI}}$.

Fig. 11. Simulator $\mathcal{S}_{P_0}$ for $P_0 \in \mathsf{A}$.

P_0 is malicious: Consider the simulator in Fig. 11.

We show the indistinguishability of the simulation and the real protocol through the following hybrid games. In the following, let $\mathcal{A}$ denote the dummy adversary controlled by $\mathcal{Z}$.

Hybrid 0: $\mathsf{Real}^{\mathcal{A}}_{\Pi_{\text{MPSI}}}$.
Hybrid 1: Identical to Hybrid 0, except that $\mathcal{S}_1$ simulates $\mathcal{F}_{\text{OPA}}$ and learns all inputs.
Hybrid 2: Identical to Hybrid 1, except that $\mathcal{S}_2$ aborts according to Step 7 in Fig. 11.
Hybrid 3: Identical to Hybrid 2, except that $\mathcal{S}_3$ aborts if the extracted $\hat{\mathbf{p}}_0$ are not identical, but the check is passed.
Hybrid 4: Identical to Hybrid 3, except that $\mathcal{S}_4$ replaces the $\mathbf{v}_{jl}$ between honest parties j, l by uniformly random polynomials.
Hybrid 5: Identical to Hybrid 4, except that $\mathcal{S}_5$ generates the inputs $\hat{\mathbf{s}}^j_0, \hat{\mathbf{s}}_j$ according to Step 6 in Fig. 11 and adjusts the output. This corresponds to $\mathsf{Ideal}^{\mathcal{S}_{P_0}}_{\mathcal{F}^*_{\text{MPSI}}}$.

Hybrids 0 and 1 are trivially indistinguishable. We show that Hybrid 1 and Hybrid 2 are computationally indistinguishable in Lemma 6.1.1. This step ensures that the correct $\hat{p}_0$ was extracted, and that all the intermediate values of the honest parties are added up. Hybrids 2 and 3 are indistinguishable due to the security of the coin-toss. This is formalized in Lemma 6.1.2. As an intermediate step to complete the full simulation, we replace all pseudorandom polynomials $\mathbf{v}_{jl}$ between honest parties j, l by uniformly random ones. Computational indistinguishability of Hybrid 3 and Hybrid 4 follows from a straightforward reduction to the pseudorandomness of PRG. We establish the statistical indistinguishability of Hybrids 4 and 5 in Lemma 6.1.3. As a result we get that for all PPT environments $\mathcal{Z}$,

$$\mathsf{Real}^{\mathcal{A}}_{\Pi_{\mathrm{MPSI}}}(\mathcal{Z}) \approx_c \mathsf{Ideal}^{\mathcal{S}_{P_0}}_{\mathcal{F}^*_{\mathrm{MPSI}}}(\mathcal{Z}).$$

Lemma 6.1.1. *Assume that* NMCOM *is a bounded-concurrent non-malleable commitment scheme against synchronizing adversaries. Then Hybrid 1 and Hybrid 2 are computationally indistinguishable.*

Proof. The only difference between Hybrid 1 and Hybrid 2 lies in the fact that $\mathcal{S}_2$ aborts if the extracted $\hat{\mathbf{p}}_{\mathsf{A}}$ evaluated on x does not match the check value α_0, but the check is still passed. Therefore, in order for $\mathcal{Z}$ to distinguish both hybrids, it has to be able to produce a value $\alpha_0^* \neq \hat{\mathbf{p}}_{\mathsf{A}}(x)$ and pass the check with non-negligible probability ϵ. W.l.o.g. it is sufficient that α_0^* is incorrect for only one $\hat{\mathbf{p}}_0$. We show that such a $\mathcal{Z}$ breaks the non-malleability property of NMCOM.

Let σ denote the outcome of the check. If $\mathcal{A}$ is honest, i.e. $\alpha_0 = \hat{\mathbf{p}}_0(x)$ and $\beta_0^i = \hat{\mathbf{r}}_0^i(x)$, then

$$\sigma = \sum_{i=0}^{n} (\alpha_0(\beta_i + \delta_0) + \alpha_i(\beta_0^{ii} + \delta_i)) - \mathbf{p}_\cap(x) = 0, \tag{1}$$

where

$$\mathbf{p}_\cap = \sum_{i \in \mathsf{A}} (\mathbf{s}_i + \mathbf{s}_0^i) + \sum_{j \in \mathsf{H}} (\mathbf{s}_j + \mathbf{s}_0^j).$$

We first observe that $\sum_{j \in \mathsf{H}} (\mathbf{s}_j + \mathbf{s}_0^j) = \sum_{j \in \mathsf{H}} \hat{\mathbf{p}}_j(\hat{\mathbf{r}}_0^j + \hat{\mathbf{r}'_j}) + \hat{\mathbf{p}}_0(\hat{\mathbf{r}'_0} + \hat{\mathbf{r}}_j)$ is uniform over the choice of the $\hat{\mathbf{r}}_j, \hat{\mathbf{r}}'_j$. Therefore, if $\mathcal{A}$ uses $\mathbf{p}_\cap^*$ without adding $\sum_{j \in \mathsf{H}} (\mathbf{s}_j + \mathbf{s}_0^j)$, the check will fail with overwhelming probability.

Since $\mathcal{A}$ controls the inputs of the malicious parties $i \in \mathsf{A}$, in order to pass the check it is sufficient for $\mathcal{A}$ to satisfy the following simplification of Eq. (1).

$$\sigma' = \sum_{j \in \mathsf{H}} (\alpha_0(\beta_j + \delta_0) + \alpha_j(\beta_0^j + \delta_j)) - \sum_{j \in \mathsf{H}} (\mathbf{s}_j(x) + \mathbf{s}_0^j(x)) = \mathsf{const}$$

Here const is a fixed constant known to $\mathcal{A}$ (0 if $\mathcal{A}$ is honest) determined by setting the inputs α_i, β_i for $i \in \mathsf{A}$ accordingly. But if $\alpha_0^* \neq \hat{\mathbf{p}}_0(x)$, i.e. $\alpha_0^* = \alpha_0 + e$, then we get that

$$\begin{aligned}\sigma' &= \sum_{j \in \mathsf{H}} ((\alpha_0 + e)(\beta_j + \delta_0) + \alpha_j(\beta_0^j + \delta_j)) - \sum_{j \in \mathsf{H}} (\mathbf{s}_j(x) + \mathbf{s}_0^j(x)) \\ &= \sum_{j \in \mathsf{H}} (\alpha_0(\beta_j + \delta_0) + \alpha_j(\beta_0^j + \delta_j)) - \sum_{j \in \mathsf{H}} (\mathbf{s}_j(x) + \mathbf{s}_0^j(x)) + e \sum_{j \in \mathsf{H}} (\beta_j + \delta_0) \\ &= e \sum_{j \in \mathsf{H}} (\beta_j + \delta_0) \neq \mathsf{const}\end{aligned}$$

Similarly for $\beta_0^j \neq \hat{\mathbf{r}}_0^j(x)$ for any $j \in \mathsf{H}$. Thus, except for the case of $\alpha_0^* = \alpha_0 + e/\sum_{j \in \mathsf{H}} \beta_j$, the check will fail for $\alpha_0^* \neq \hat{\mathbf{p}}_0(x)$. But since we assumed that $\mathcal{A}$ passes the check with non-negligible probability, and NMCOM is statistically binding, $\mathcal{A}$ has to produce a valid commitment to $\tilde{\alpha}_0 = \alpha_0 + e/\sum_{j \in \mathsf{H}} (\beta_j + \delta_0)$ with the same probability.

Note, that $\mathcal{A}$ interacts in both the left and right session of NMCOM with the same party (actually all parties simultaneously, since everything is broadcast). But this means that $\mathcal{A}$ cannot let the left session finish before starting the right session, i.e. $\mathcal{A}$ is a synchronizing adversary against NMCOM. Concretely, in the left session, $\mathcal{S}_2$ commits to $(\hat{\mathbf{p}}_j(x), \hat{\mathbf{r}}_j(x), \hat{\mathbf{r}}_j'(x)) = (\alpha_j, \beta_j, \delta_j)$ for $j \in \mathsf{H}$, while $\mathcal{A}$ commits in the right session to $(\alpha_0, \{\beta_0^i\}_{i \in [n]}, \delta_0)$ and $(\alpha_i, \beta_i, \delta_i)$ for $i \in \mathsf{A}$ to $\mathcal{S}_2$. Further, the number of sessions that $\mathcal{A}$ can start is bounded in advance at $n-1$, i.e. it is sufficient to consider bounded-concurrency.

Consider the two views

$$\mathsf{Real} = \{\hat{\mathbf{s}}_j, \{\mathsf{com}_j\}\}_{j \in \mathsf{H}}, \qquad \mathsf{Rand} = \{\hat{\mathbf{s}}_j, \{\widehat{\mathsf{com}}_j\}\}_{j \in \mathsf{H}},$$

where $\mathsf{com}_j \leftarrow \mathsf{NMCOM.Commit}(\alpha_j, \beta_j)$ and $\widehat{\mathsf{com}}_j \leftarrow \mathsf{NMCOM.Commit}(0)$. Real corresponds to a real protocol view of $\mathcal{A}$ before committing itself[7].

Obviously, $\mathsf{Real} \approx_c \mathsf{Rand}$ if NMCOM is non-malleable. However, we will argue that $\mathcal{A}$ cannot output a valid commitment on $\tilde{\alpha}_0$ except with negligible probability, i.e.

$$\Pr[(\mathsf{com}_0^*, \mathsf{unv}_0^*, (\tilde{\alpha}_0, \{\tilde{\beta_0^i}\}_{i \in [n]}, \tilde{\delta}_0) \leftarrow \mathcal{A}(\mathsf{Rand}) \wedge \mathsf{valid}] \leq \mathsf{negl}(\kappa),$$

where valid is the event that $\mathsf{NMCOM.Open}(\mathsf{com}_0^*, \mathsf{unv}_0^*, (\tilde{\alpha}_0, \{\tilde{\beta_0^i}\}_{i \in [n]}, \delta_0) = 1$. We first observe that $\hat{\mathbf{p}}_j$ and $\hat{\mathbf{r}}_j$ for $j \in \mathsf{H}$ cannot be obtained by $\mathcal{A}$ via $\hat{\mathbf{s}}_j = \hat{\mathbf{p}}_j \cdot \hat{\mathbf{r}}_0^j - \hat{\mathbf{u}}_j$. The polynomial $\hat{\mathbf{s}}_j$ itself is uniformly random over the choice of $\hat{\mathbf{u}}_j$, and the only equation that $\mathcal{A}$ has is $\hat{\mathbf{p}}_\cap = \sum_{i \in \mathsf{A}} (\mathbf{s}_i + \mathbf{s}_0^i) + \sum_{j \in \mathsf{H}} (\mathbf{s}_j + \mathbf{s}_0^j) = \sum_{i \in \mathsf{A}} (\hat{\mathbf{p}}_0 \cdot (\hat{\mathbf{r}}_i + \hat{\mathbf{r}}_0') + \hat{\mathbf{p}}_i \cdot (\hat{\mathbf{r}}_0^i + \hat{\mathbf{r}}_i')) + \sum_{j \in \mathsf{H}} (\hat{\mathbf{p}}_0 \cdot (\hat{\mathbf{r}}_j + \hat{\mathbf{r}}_0') + \hat{\mathbf{p}}_j \cdot (\hat{\mathbf{r}}_0^j + \hat{\mathbf{r}}_j'))$. Note, that the honest $\hat{\mathbf{r}}_j, \hat{\mathbf{r}}_j'$ have degree d and therefore hide $\hat{\mathbf{p}}_j$. Further, the commitments com_j contain the value 0 and are therefore independent of $\hat{\mathbf{p}}_j$ and $\hat{\mathbf{r}}_j$. Thus, the probability that $\mathcal{A}$ obtains a commitment on $\tilde{\alpha}_0$ is negligible.

[7] For ease of notation, here we assume that the commitments are completely sent before $\mathcal{A}$ commits himself. The very same argument also holds if $\mathcal{A}$ only received synchronized messages of com_j and has to start committing concurrently.

But since $\mathsf{Real} \approx_c \mathsf{Rand}$, we also get that

$$\Pr[(\mathsf{com}_0^*, \mathsf{unv}_0^*, (\tilde{\alpha}_0, \{\tilde{\beta}_0^i\}_{i\in[n]}, \tilde{\delta}_0) \leftarrow \mathcal{A}(\mathsf{Real}) \wedge \mathsf{valid}] \leq \mathsf{negl}(\kappa),$$

which contradicts our assumption that $\mathcal{A}$ produces the commitment with non-negligible probability ϵ.

In conclusion, Hybrid 1 and Hybrid 2 are computationally indistinguishable. □

Lemma 6.1.2. *Assume that Π_{CT} provides a uniformly random x with computational security. Then Hybrid 2 and Hybrid 3 are computationally indistinguishable.*

Proof. Assume that there exists an environment $\mathcal{Z}$ that distinguishes Hybrids 2 and 3 with non-negligible probability ϵ. In order to distinguish Hybrid 2 and Hybrid 3 $\mathcal{Z}$ has to provide two distinct polynomials for a malicious P_0 and still pass the check in the protocol. Then we can construct from $\mathcal{Z}$ an adversary $\mathcal{B}$ that predicts the outcome of Π_{CT} with non-negligible probability.

Let $\mathcal{A}$ input w.l.o.g. two polynomials $\hat{\mathbf{p}}_0^1 \neq \hat{\mathbf{p}}_0^2$. The check with the random challenge x allows $\mathcal{A}$ to send only one value α_0^*, but from Lemma 6.1.1 we know that it has to hold that $\alpha_0^* = \hat{\mathbf{p}}_0^1(x) = \hat{\mathbf{p}}_0^2(x)$, or the check will fail. First note that two polynomials of degree m agree in a random point x over $\mathbb{F}$ only with probability $m/|\mathbb{F}|$, which is negligible in our case.

Our adversary $\mathcal{B}$ proceeds as follows. It simulates the protocol for $\mathcal{Z}$ according to $\mathcal{S}_1$ up to the point where $\mathcal{S}_1$ learns the polynomials $\hat{\mathbf{p}}_0^1 \neq \hat{\mathbf{p}}_0^2$. $\mathcal{B}$ sets $\mathbf{f} = \hat{\mathbf{p}}_0^1 - \hat{\mathbf{p}}_0^2$ and computes the roots $\gamma_1, \ldots, \gamma_m$ of $\mathbf{f}$. One of these roots has to be the random point x, otherwise $\hat{\mathbf{p}}_0^1(x) - \hat{\mathbf{p}}_0^2(x) \neq 0$ and the check in Π_{MPSI} fails (since there is only one α_0^*). $\mathcal{B}$ picks a random index $l \in [m]$ and predicts the output of the coin-flip as γ_l. Thus, $\mathcal{B}$ predicts the outcome of the coin-toss correctly with probability ϵ/m, which is non-negligible. This contradicts the security of Π_{CT}.

This establishes the indistinguishability of Hybrid 2 and Hybrid 3. □

Lemma 6.1.3. *Hybrid 4 and Hybrid 5 are statistically close.*

Proof. A malicious environment $\mathcal{Z}$ can distinguish Hybrid 4 and Hybrid 5 if (a) the extracted inputs are incorrect or if (b) the simulated messages can be distinguished from real ones.

Concerning (a), if the inputs were not correctly extracted, $\mathcal{Z}$ would receive different outputs in the two hybrids. We already established that the extracted polynomial $\hat{\mathbf{p}}_0$ is correct. Similarly, the extracted $\hat{\mathbf{r}}_0^j$ are also correct. By implication this also ensures that the intermediate intersection is computed correctly.

We argue that the correction of the intersection is also correct, i.e. the set $\hat{S}'_\cap$ is computed correctly and in particular it holds that $(\mathcal{M}_0(\mathbf{p}_\cap^*) \cap \mathcal{M}_0(\hat{\mathbf{p}}_j)) \subseteq \hat{S}_\cap$. First of all, we have to show that the intermediate intersection polynomial $\hat{\mathbf{p}}_{\mathsf{int}}$ actually provides the intersection for all parties. For all P_j it holds with overwhelming probability:

$$
\begin{aligned}
\mathcal{M}_0(\hat{\mathbf{p}}_j) \cap \mathcal{M}_0(\hat{\mathbf{p}}_{\mathsf{int}}) &= \mathcal{M}_0(\hat{\mathbf{p}}_j) \cap \mathcal{M}_0(\sum_{j \in \mathsf{H}} (\hat{\mathbf{p}}_0 \cdot (\hat{\mathbf{r}}_j + \hat{\mathbf{r}}'_0) + \hat{\mathbf{p}}_j \cdot (\hat{\mathbf{r}}^j_0 + \hat{\mathbf{r}}'_j))) \\
&\overset{Lemma\ 2.2}{=} \mathcal{M}_0(\hat{\mathbf{p}}_j) \cap (\bigcap_{j \in \mathsf{H}} \mathcal{M}_0((\hat{\mathbf{p}}_0 \cdot (\hat{\mathbf{r}}_j + \hat{\mathbf{r}}'_0) + \hat{\mathbf{p}}_j \cdot (\hat{\mathbf{r}}^j_0 + \hat{\mathbf{r}}'_j))) \\
&\overset{Lemma\ 2.4}{=} \mathcal{M}_0(\hat{\mathbf{p}}_j) \cap (\bigcap_{j \in \mathsf{H}} \mathcal{M}_0(\hat{\mathbf{p}}_0) \cap \mathcal{M}_0(\hat{\mathbf{p}}_j)) \\
&= \hat{S}_\cap
\end{aligned}
$$

Once the intermediate intersection is computed, the adversary can only add an update polynomial $\hat{\mathbf{p}}_{\mathsf{upt}}$ to get the final intersection polynomial $\mathbf{p}^*_\cap$. It remains to show that this final intersection does not include any points that were not already in the intermediate intersection for any of the parties' polynomials $\hat{\mathbf{p}}_j$.

For this, we consider the intersection of every honest party's (unknown) input $\mathbf{p}_j$ with the intersection. It has to hold that $\hat{S}'_\cap \subseteq \hat{S}_\cap$ for all P_j except with negligible probability. Here we require that $\Pr[x \in \mathcal{M}_0(\mathbf{p}_j)] \leq \mathsf{negl}(|\mathbb{F}|)$ for all x, i.e. the adversary can only guess an element of P_j's input set.

$$
\begin{aligned}
\mathcal{M}_0(\hat{\mathbf{p}}_j) \cap \mathcal{M}_0(\mathbf{p}^*_\cap) &= \mathcal{M}_0(\hat{\mathbf{p}}_j) \cap (\mathcal{M}_0(\hat{\mathbf{p}}_{\mathsf{int}} + \hat{\mathbf{p}}_{\mathsf{upt}})) \\
&\overset{Lemma\ 2.2}{=} \mathcal{M}_0(\hat{\mathbf{p}}_j) \cap (\mathcal{M}_0(\hat{\mathbf{p}}_{\mathsf{int}}) \cap \mathcal{M}_0(\hat{\mathbf{p}}_{\mathsf{upt}})) \\
&= \mathcal{M}_0(\hat{\mathbf{p}}_j) \cap (\hat{S}_\cap \cap \mathcal{M}_0(\hat{\mathbf{p}}_{\mathsf{upt}})) \\
&\subseteq \mathcal{M}_0(\hat{\mathbf{p}}_j) \cap \hat{S}_\cap = \hat{S}_\cap
\end{aligned}
$$

Therefore, $\hat{S}'_\cap \subseteq \hat{S}_\cap$, and the output in both hybrids is identical.

Regarding (b), we make the following observations. Since $\mathcal{S}_4$ sends $\hat{\mathbf{s}}'_j = \hat{\mathbf{s}}_j - \mathbf{u}_j + \sum_{i \neq j} \mathbf{v}_{ij}$, the value $\hat{\mathbf{s}}'_j$ is uniformly random over the choice of $\mathbf{u}_j$ (and over $\sum \mathbf{v}_{ij}$, if $t \leq n-2$). Therefore, the simulation of $\hat{\mathbf{s}}'_j$ is identically distributed to Hybrid 4.

Similarly, we have:

$$
\sum_{j \in \mathsf{H}} (\hat{\mathbf{s}}'_j + \hat{\mathbf{s}}^j_0) = \sum_{j \in \mathsf{H}} (\hat{\mathbf{p}}_0 \cdot (\hat{\mathbf{r}}_j + \hat{\mathbf{r}}'_0) + \hat{\mathbf{p}}_j \cdot (\hat{\mathbf{r}}^j_0 + \hat{\mathbf{r}}'_j)) \quad [+ \sum_{i \in \mathsf{A}, j \in \mathsf{H}} \mathbf{v}_{ij}]
$$

We can ignore the $\mathbf{v}_{ij}$ values, since these are known to $\mathcal{A}$. The sum is uniform over the choice of the $\hat{\mathbf{r}}_j, \hat{\mathbf{r}}'_j$ apart from the points $\gamma \in \hat{S}_\cap$ (since $\mathcal{F}_{\mathrm{OPA}}$ guarantees that $\hat{p}_0 \neq 0$) and therefore identically distributed to Hybrid 5, since the extraction in correct.

□

P_0 **is honest:** The proof itself is very similar to the proof of a corrupted P_0. It is actually easier to simulate in the sense that $\mathcal{S}_{\bar{P}_0}$ observes the inputs of all malicious parties. In this sense, Π_{MPSI} actually realises $\mathcal{F}_{\mathrm{MPSI}}$ if P_0 is honest, since no adjustment of the output is necessary. We refer to the full version [15] of the paper for the proof.

Efficiency. The setup, i.e. the distribution of seeds, has communication complexity $O(n^2\kappa)$. The oblivious addition of the polynomials has communication overhead of $O(nm\kappa)$. The check phase first requires a multi-party coin-toss.

In the full version of this paper [15], we sketch a coin-tossing protocol in combination with an OLE-based commitment scheme (replacing the non-malleable commitment for better efficiency) that results in an asymptotic communication overhead of $O(n^2\kappa)$ for the check and the coin-toss phase. Combining this with the above observations, Π_{MPSI} has communication complexity $O((n^2 + nm)\kappa)$ in the $\mathcal{F}_{\text{OLE}}$-hybrid model.

For concrete instantiations of $\mathcal{F}_{\text{OLE}}$, the OLE protocol of Ghosh et al. [14] has a constant communication overhead per OLE. In summary, the complete protocol has communication complexity $O((n^2+nm)\kappa)$, which is asymptotically optimal for $m \geq n$.

Similar to the two-party case, the computational cost is dominated by the cost of polynomial interpolation. In particular, the central party has to run the two-party protocol n times, which leads to a computational overhead of $O(nm \log m)$ multiplications. The other parties basically have the same computational overhead as in the two-party case. □

7 Performance Analysis

In this section, we give an estimation of the communication efficiency with concrete parameters and provide a comparison with existing results. For this, we simply count the number of field elements that have to be sent for the protocols. We first look at the communication overhead of the OLE primitive. Instantiated with the result by Ghosh et al. [14], each OLE has an overhead of 64 field elements including OT extension (32 without), which translates to 256 field elements per item per OPA. The factor 4 stems from the fact that OPA needs $2d$ OLE to compute a degree d output, and OLE+ requires two OLE per instance.

Table 2. Comparison of two-party PSI protocols from [36] for input-size $m = \{2^{16}, 2^{20}\}$, where κ denotes statistical security parameter, σ denotes size of each item in bits, SM denotes standard model, ROM denotes random oracle model.

Protocol	Communication cost	
	$m = 2^{16}$	$m = 2^{20}$
[36] (EC-ROM)	79 MB ($\kappa = 40$)	1.32 GB ($\kappa = 40$)
[36] (DE-ROM)	61 MB ($\kappa = 40$)	1.07 GB ($\kappa = 40$)
[36] (SM, $\sigma = 40$)	451 MB ($\kappa = 40$)	>7.7 GB ($\kappa = 40$)
[36] (SM, $\sigma = 64$)	1.29 GB ($\kappa = 40$)	22.18 GB ($\kappa = 40$)
Ours ($\sigma = 40$)	80 MB ($\kappa = 40$)	1.25 GB ($\kappa = 40$)
Ours ($\sigma = 64$)	128 MB ($\kappa = 64$)	2 GB ($\kappa = 64$)

2-party PSI. To get a feeling for the concrete communication efficiency of our two-party protocol, we compare it with the recent maliciously UC-secure protocols from [36]. These protocols give only one-sided output, whereas our protocol gives two-sided output. However, OPA is sufficient for one-sided PSI, consequently a one-sided PSI would cost 256 field elements per item in our case.

Table 2 clearly shows that the communication overhead of our protocol is significantly less than the standard model (SM) protocol from [36]. Note that our instantiation is also secure in SM, given $O(\kappa)$ base OTs. Like [36] we use the OT-extension protocol from [31] for the instantiation. Even if we compare our result to the ROM approach of [36], we achieve fairly competitive communication efficiency.

One should consider that in the ROM there exist other PSI protocols with linear communication complexity [8,23]. The concrete bandwidth of those protocols are much less than our specific instantiation, for example for sets of 2^{20} elements the total communication cost of [8] is about 213 MB[8]. Further [23] has lower bandwidth than [8]. However, in both the cases communication efficiency comes at the cost of huge computational expenses due to lots of public key operations. We believe that the simple field arithmetic of our protocols (including the cost of the OLE of [14]) does not incur such a drawback in practice.

Table 3. Comparison of communication overhead per party of MPSI protocol with [27] for 2^{20} elements with 40 bit statistical security, without the cost for OT extension.

Protocol	Parties	Corr.	Comm. 2^{20} elements
[27] (passive)	n	$n-1$	$(n-1) \cdot 467$ MB
Ours (active)	n	$n-1$	≈ 2.5 GB
Ours (passive)	n	$n-1$	≈ 1.25 GB

Multi-party PSI. To the best of our knowledge, there are currently no maliciously secure MPSI implementations with which we could compare our result. A direct comparison with the *passively* secure MPSI from [27], however, directly shows the difference in asymptotic behaviour to our result. Their communication costs per party increase with the number of parties, whereas it remains constant in our case (except for the central party). If we average over all parties, the central party's overhead can be distributed over all parties, which at most doubles the average communication cost per party (cf. Table 3). We can upper bound the communication cost per party by 2.5 GB for 2^{20} elements (excluding the cost for OT extension in order to get comparable results to [27]). From the table we can deduce that with only 6 parties, our actively secure protocol is more efficient than their passive one. Replacing the actively secure OPA in our MPSI protocol with the passively secure one yields a passively secure MPSI protocol. We gain

[8] For reference see Figure 8 of [36].

another factor of 2 in communication efficiency and our construction is more efficient than [27] starting from 4 parties.

References

1. Applebaum, B., Damgård, I., Ishai, Y., Nielsen, M., Zichron, L.: Secure arithmetic computation with constant computational overhead. In: Katz, J., Shacham, H. (eds.) CRYPTO 2017. LNCS, vol. 10401, pp. 223–254. Springer, Cham (2017). https://doi.org/10.1007/978-3-319-63688-7_8
2. Ben-Sasson, E., Fehr, S., Ostrovsky, R.: Near-linear unconditionally-secure multiparty computation with a dishonest minority. In: Safavi-Naini, R., Canetti, R. (eds.) CRYPTO 2012. LNCS, vol. 7417, pp. 663–680. Springer, Heidelberg (2012). https://doi.org/10.1007/978-3-642-32009-5_39
3. Borodin, A., Moenck, R.: Fast modular transforms. J. Comput. Syst. Sci. **8**(3), 366–386 (1974). http://dx.doi.org/10.1016/S0022-0000(74)80029-2
4. Canetti, R.: Universally composable security: a new paradigm for cryptographic protocols. In: 42nd FOCS, pp. 136–145. IEEE Computer Society Press, October 2001
5. Cheon, J.H., Jarecki, S., Seo, J.H.: Multi-party privacy-preserving set intersection with quasi-linear complexity. IEICE Trans. **95–A**(8), 1366–1378 (2012)
6. Cramer, R., Damgård, I., Ishai, Y.: Share conversion, pseudorandom secret-sharing and applications to secure computation. In: Kilian, J. (ed.) TCC 2005. LNCS, vol. 3378, pp. 342–362. Springer, Heidelberg (2005). https://doi.org/10.1007/978-3-540-30576-7_19
7. Dachman-Soled, D., Malkin, T., Raykova, M., Yung, M.: Efficient robust private set intersection. In: Abdalla, M., Pointcheval, D., Fouque, P.-A., Vergnaud, D. (eds.) ACNS 2009. LNCS, vol. 5536, pp. 125–142. Springer, Heidelberg (2009). https://doi.org/10.1007/978-3-642-01957-9_8
8. De Cristofaro, E., Kim, J., Tsudik, G.: Linear-complexity private set intersection protocols secure in malicious model. In: Abe, M. (ed.) ASIACRYPT 2010. LNCS, vol. 6477, pp. 213–231. Springer, Heidelberg (2010). https://doi.org/10.1007/978-3-642-17373-8_13
9. De Cristofaro, E., Tsudik, G.: Practical private set intersection protocols with linear complexity. In: Sion, R. (ed.) FC 2010. LNCS, vol. 6052, pp. 143–159. Springer, Heidelberg (2010). https://doi.org/10.1007/978-3-642-14577-3_13
10. Dong, C., Chen, L., Wen, Z.: When private set intersection meets big data: an efficient and scalable protocol. In: Sadeghi, A.R., Gligor, V.D., Yung, M. (eds.) ACM CCS 2013, pp. 789–800. ACM Press, November 2013
11. Döttling, N., Ghosh, S., Nielsen, J.B., Nilges, T., Trifiletti, R.: TinyOLE: efficient actively secure two-party computation from oblivious linear function evaluation. In: Thuraisingham, B.M., Evans, D., Malkin, T., Xu, D. (eds.) ACM CCS 2017, pp. 2263–2276. ACM Press, October/November 2017
12. Freedman, M.J., Hazay, C., Nissim, K., Pinkas, B.: Efficient set intersection with simulation-based security. J. Cryptol. **29**(1), 115–155 (2016)
13. Freedman, M.J., Nissim, K., Pinkas, B.: Efficient private matching and set intersection. In: Cachin, C., Camenisch, J.L. (eds.) EUROCRYPT 2004. LNCS, vol. 3027, pp. 1–19. Springer, Heidelberg (2004). https://doi.org/10.1007/978-3-540-24676-3_1

14. Ghosh, S., Nielsen, J.B., Nilges, T.: Maliciously secure oblivious linear function evaluation with constant overhead. In: Takagi, T., Peyrin, T. (eds.) ASIACRYPT 2017. LNCS, vol. 10624, pp. 629–659. Springer, Cham (2017). https://doi.org/10.1007/978-3-319-70694-8_22
15. Ghosh, S., Nilges, T.: An algebraic approach to maliciously secure private set intersection. Cryptology ePrint Archive, Report 2017/1064 (2017). https://eprint.iacr.org/2017/1064
16. Hallgren, P.A., Orlandi, C., Sabelfeld, A.: Privatepool: privacy-preserving ridesharing. In: 30th IEEE Computer Security Foundations Symposium, CSF 2017, 21–25 August 2017, Santa Barbara, CA, USA, pp. 276–291 (2017)
17. Hazay, C.: Oblivious polynomial evaluation and secure set-intersection from algebraic PRFs. In: Dodis, Y., Nielsen, J.B. (eds.) TCC 2015. LNCS, vol. 9015, pp. 90–120. Springer, Heidelberg (2015). https://doi.org/10.1007/978-3-662-46497-7_4
18. Hazay, C., Nissim, K.: Efficient set operations in the presence of malicious adversaries. J. Cryptol. **25**(3), 383–433 (2012)
19. Hazay, C., Venkitasubramaniam, M.: Scalable multi-party private set-intersection. In: Fehr, S. (ed.) PKC 2017. LNCS, vol. 10174, pp. 175–203. Springer, Heidelberg (2017). https://doi.org/10.1007/978-3-662-54365-8_8
20. Huang, Y., Evans, D., Katz, J.: Private set intersection: are garbled circuits better than custom protocols? In: NDSS 2012. The Internet Society, February 2012
21. Huberman, B.A., Franklin, M., Hogg, T.: Enhancing privacy and trust in electronic communities. In: Proceedings of the 1st ACM Conference on Electronic Commerce, EC 1999, pp. 78–86 (1999)
22. Ishai, Y., Prabhakaran, M., Sahai, A.: Secure arithmetic computation with no honest majority. In: Reingold, O. (ed.) TCC 2009. LNCS, vol. 5444, pp. 294–314. Springer, Heidelberg (2009). https://doi.org/10.1007/978-3-642-00457-5_18
23. Jarecki, S., Liu, X.: Fast secure computation of set intersection. In: Garay, J.A., De Prisco, R. (eds.) SCN 2010. LNCS, vol. 6280, pp. 418–435. Springer, Heidelberg (2010). https://doi.org/10.1007/978-3-642-15317-4_26
24. Kamara, S., Mohassel, P., Raykova, M., Sadeghian, S.: Scaling private set intersection to billion-element sets. In: Christin, N., Safavi-Naini, R. (eds.) FC 2014. LNCS, vol. 8437, pp. 195–215. Springer, Heidelberg (2014). https://doi.org/10.1007/978-3-662-45472-5_13
25. Kissner, L., Song, D.: Privacy-preserving set operations. In: Shoup, V. (ed.) CRYPTO 2005. LNCS, vol. 3621, pp. 241–257. Springer, Heidelberg (2005). https://doi.org/10.1007/11535218_15
26. Kolesnikov, V., Kumaresan, R., Rosulek, M., Trieu, N.: Efficient batched oblivious PRF with applications to private set intersection. In: Weippl, E.R., Katzenbeisser, S., Kruegel, C., Myers, A.C., Halevi, S. (eds.) ACM CCS 2016, pp. 818–829. ACM Press, October 2016
27. Kolesnikov, V., Matania, N., Pinkas, B., Rosulek, M., Trieu, N.: Practical multi-party private set intersection from symmetric-key techniques. In: Thuraisingham, B.M., Evans, D., Malkin, T., Xu, D. (eds.) ACM CCS 2017, pp. 1257–1272. ACM Press, October/November 2017
28. Meadows, C., Mutchler, D.: Matching secrets in the absence of a continuously available trusted authority. IEEE Trans. Softw. Eng. **SE-13**(2), 289–292 (1987)
29. Naor, M., Pinkas, B., Sumner, R.: Privacy preserving auctions and mechanism design. In: EC, pp. 129–139 (1999)
30. Narayanan, A., Thiagarajan, N., Lakhani, M., Hamburg, M., Boneh, D.: Location privacy via private proximity testing. In: NDSS 2011. The Internet Society, February 2011

31. Orrù, M., Orsini, E., Scholl, P.: Actively secure 1-out-of-N OT extension with application to private set intersection. In: Handschuh, H. (ed.) CT-RSA 2017. LNCS, vol. 10159, pp. 381–396. Springer, Cham (2017). https://doi.org/10.1007/978-3-319-52153-4_22
32. Pinkas, B., Schneider, T., Segev, G., Zohner, M.: Phasing: private set intersection using permutation-based hashing. In: 24th USENIX Security Symposium, USENIX Security 2015, 12–14 August 2015, Washington, D.C., USA, pp. 515–530 (2015)
33. Pinkas, B., Schneider, T., Zohner, M.: Faster private set intersection based on OT extension. In: Proceedings of the 23rd USENIX Security Symposium, 20–22 August 2014, San Diego, CA, USA, pp. 797–812 (2014)
34. Pinkas, B., Schneider, T., Zohner, M.: Scalable private set intersection based on OT extension. ACM Trans. Priv. Secur. **21**(2), 7:1–7:35 (2018)
35. Rindal, P., Rosulek, M.: Improved private set intersection against malicious adversaries. In: Coron, J.-S., Nielsen, J.B. (eds.) EUROCRYPT 2017. LNCS, vol. 10210, pp. 235–259. Springer, Cham (2017). https://doi.org/10.1007/978-3-319-56620-7_9
36. Rindal, P., Rosulek, M.: Malicious-secure private set intersection via dual execution. In: Thuraisingham, B.M., Evans, D., Malkin, T., Xu, D. (eds.) ACM CCS 2017, pp. 1229–1242. ACM Press, October/November 2017
37. Sang, Y., Shen, H.: Privacy preserving set intersection based on bilinear groups. In: Proceedings of the Thirty-First Australasian Conference on Computer Science, ACSC 2008, vol. 74. pp. 47–54 (2008)
38. Shamir, A.: On the power of commutativity in cryptography. In: de Bakker, J., van Leeuwen, J. (eds.) ICALP 1980. LNCS, vol. 85, pp. 582–595. Springer, Heidelberg (1980). https://doi.org/10.1007/3-540-10003-2_100

Quantum II

On Finding Quantum Multi-collisions

Qipeng Liu(✉) and Mark Zhandry

Princeton University, Princeton, NJ 08544, USA
qipengl@princeton.edu

Abstract. A k-collision for a compressing hash function H is a set of k distinct inputs that all map to the same output. In this work, we show that for any constant k, $\Theta\left(N^{\frac{1}{2}(1-\frac{1}{2^k-1})}\right)$ quantum queries are both necessary and sufficient to achieve a k-collision with constant probability. This improves on both the best prior upper bound (Hosoyamada et al., ASIACRYPT 2017) and provides the first non-trivial lower bound, completely resolving the problem.

1 Introduction

Collision resistance is one of the central concepts in cryptography. A *collision* for a hash function $H : \{0,1\}^m \to \{0,1\}^n$ is a pair of distinct inputs $x_1 \neq x_2$ that map to the same output: $H(x_1) = H(x_2)$.

Multi-collisions. Though receiving comparatively less attention in the literature, multi-collision resistance is nonetheless an important problem. A k-collision for H is a set of k distinct inputs $\{x_1, \ldots, x_k\}$ such that $x_i \neq x_j$ for $i \neq j$ where $H(x_i) = H(x_j)$ for all i, j.

Multi-collisions frequently surface in the analysis of hash functions and other primitives. Examples include MicroMint [RS97], RMAC [JJV02], chopMD [CN08], Leamnta-LW [HIK+11], PHOTON and Parazoa [NO14], Keyed-Sponge [JLM14], all of which assume the multi-collision resistance of a certain function. Multi-collisions algorithms have also been used in attacks, such as MDC-2 [KMRT09], HMAC [NSWY13], Even-Mansour [DDKS14], and LED [NWW14]. Multi-collision resistance for polynomial k has also recently emerged as a theoretical way to avoid keyed hash functions [BKP18,BDRV18], or as a useful cryptographic primitives, for example, to build statistically hiding commitment schemes with succinct interaction [KNY18].

Quantum. Quantum computing stands to fundamentally change the field of cryptography. Importantly for our work, Grover's algorithm [Gro96] can speed up brute force searching by a quadratic factor, greatly increasing the speed of pre-image attacks on hash functions. In turn, Grover's algorithm can be used to find ordinary collisions ($k = 2$) in time $O(2^{n/3})$, speeding up the classical "birthday"

Y. Ishai and V. Rijmen (Eds.): EUROCRYPT 2019, LNCS 11478, pp. 189–218, 2019.
https://doi.org/10.1007/978-3-030-17659-4_7

attack which requires $O(2^{n/2})$ time. It is also known that, in some sense (discussed below), these speedups are optimal [AS04, Zha15a]. These attacks require updated symmetric primitives with longer keys in order to make such attacks intractable.

1.1 This Work: Quantum Query Complexity of Multi-collision Resistance

In this work, we consider *quantum* multi-collision resistance. Unfortunately, little is known of the difficulty of finding *multi*-collisions for $k \geq 3$ in the quantum setting. The only prior work on this topic is that of Hosoyamada et al. [HSX17], who give a $O(2^{4n/9})$ algorithm for 3-collisions, as well as algorithms for general constant k. On the lower bounds side, the $\Omega(2^{n/3})$ from the $k = 2$ case applies as well for higher k, and this is all that is known.

We completely resolve this question, giving tight upper and lower bounds for any constant k. In particular, we consider the quantum query complexity of multi-collisions. We will model the hash function H as a *random oracle.* This means, rather than getting concrete code for a hash function H, the adversary is given black box access to a function H chosen uniformly at random from the set of all functions from $\{0,1\}^m$ into $\{0,1\}^n$. Since we are in the quantum setting, black box access means the adversary can make quantum queries to H. Each query will cost the adversary 1 time step. The adversary's goal is to solve some problem—in our case find a k-collision—with the minimal cost. Our results are summarized in Table 1. Both our upper bounds and lower bounds improve upon the prior work for $k \geq 3$; for example, for $k = 3$, we show that the quantum query complexity is $\Theta(2^{3n/7})$.

Table 1. Quantum query complexity results for k-collisions. k is taken to be a constant, and all Big O and Ω notations hide constants that depend on k. In parenthesis are the main restrictions for the lower bounds provided. We note that in the case of 2-to-1 functions, $m \leq n+1$, so implicitly these bounds only apply in this regime. In these cases, m characterizes the query complexity. On the other hand, for random or arbitrary functions, n is the more appropriate way to measure query complexity. We also note that for arbitrary functions, when $m \leq n + \log(k-1)$, it is possible that H contains no k-collisions, so the problem becomes impossible. Hence, $m \geq n + \log k$ is essentially tight. For random functions, there will be no collisions w.h.p unless $m \gtrsim (1 - \frac{1}{k})n$, so algorithms on random functions must always operate in this regime.

	Upper Bound (Algorithm)	Lower Bound
[BHT98]	$O(2^{m/3})$ for $k = 2$ (2-to-1)	
[AS04]		$\Omega(2^{m/3})$ for $k = 2$ (2-to-1)
[Zha15a]	$O(2^{n/3})$ for $k = 2$ (Random, $m \geq n/2$)	$\Omega(2^{n/3})$ for $k = 2$ (Random)
[HSX17]	$O\left(2^{\frac{1}{2}(1-\frac{1}{3^{k-1}})n}\right)$ $(m \geq n + \log k)$	
This Work	$\mathbf{O}\left(2^{\frac{1}{2}(1-\frac{1}{2^k-1})n}\right)$ $\mathbf{(m \geq n + \log k)}$	$\mathbf{\Omega}\left(2^{\frac{1}{2}(1-\frac{1}{2^k-1})n}\right)$ **(Random)**

1.2 Motivation

Typically, the parameters of a hash function are set to make finding collisions intractable. One particularly important parameter is the output length of the hash function, since the output length in turn affects storage requirements and the efficiency of other parts of a cryptographic protocol.

Certain attacks, called *generic* attacks, apply regardless of the implementation details of the hash function H, and simply work by evaluating H on several inputs. For example, the birthday attack shows that it is possible to find a collision in time approximately $2^{n/2}$ by a classical computer. Generalizations show that k-collisions can be found in time $\Theta(2^{(1-1/k)n})$[1].

These are also known to be optimal among *classical* generic attacks. This is demonstrated by modeling H as an oracle, and counting the number of queries needed to find (k-)collisions in an arbitrary hash function H. In cryptographic settings, it is common to model H as a *random* function, giving stronger *average case* lower bounds.

Understanding the effect of generic attacks is critical. First, they cannot be avoided, since they apply no matter how H is designed. Second, other parameters of the function, such as the number of iterations of an internal round function, can often be tuned so that the best known attacks are in fact generic. Therefore, for many hash functions, the complexity of generic attacks accurately represents the actual cost of breaking them.

Therefore, for "good" hash functions where generic attacks are optimal, in order to achieve security against classical adversaries n must be chosen so that $t = 2^{n/2}$ time steps are intractable. This often means setting $t = 2^{128}$, so $n = 256$. In contrast, generic classical attacks can find k-collisions in time $\Theta(2^{(1-1/k)n})$. For example, this means that n must be set to 192 to avoid 3-collisions, or 171 to avoid 4-collisions.

Once quantum computers enter the picture, we need to consider quantum queries to H in order to model actual attacks that evaluate H in superposition. This changes the query complexity, and makes proving bounds much more difficult. Just as understanding query complexity in the classical setting was crucial to guide parameter choices, it will be critical in the quantum world as well.

We also believe that quantum query complexity is an important study in its own right, as it helps illuminate the effects quantum computing will have on various areas of computer science. It is especially important to cryptography, as many of the questions have direct implications to the post-quantum security of cryptosystems. Even more, the techniques involved are often closely related to proof techniques in post-quantum cryptography. For example, bounds for the quantum query complexity of finding collisions in random functions [Zha15a], as well as more general functions [EU17,BES17], were developed from techniques for proving security in the quantum random oracle model [BDF+11,Zha12,TU16]. Similarly, the lower bounds in this work build on techniques for proving quantum indifferentiability [Zha18]. On the other hand,

[1] Here, the Big Theta notation hides a constant that depends on k.

proving the security of MACs against superposition queries [BZ13] resulted in new lower bounds for the quantum oracle interrogation problem [van98] and generalizations [Zha15b].

Lastly, multi-collision finding can be seen as a variant of k-distinctness, which is essentially the problem of finding a k-collision in a function $H : \{0,1\}^n \to \{0,1\}^n$, where the k-collision may be unique and all other points are distinct. The quantum query complexity of k-distinctness is currently one of the main open problems in quantum query complexity. An upper bound of $(2^n)^{\frac{3}{4}-\frac{1}{4(2^k-1)}}$ was shown by Belovs [Bel12]. The best known lower bound is $\Omega((2^n)^{\frac{3}{4}-\frac{1}{2k}})$ [BKT18]. Interestingly, the dependence of the exponent on k is exponential for the upper bound, but polynomial for the lower bound, suggesting a fundamental gap our understanding of the problem.

Note that our results do not immediately apply in this setting, as our algorithm operates only in a regime where there are many ($\leq k$-)collisions, whereas k-distinctness applies even if the k-collision is unique and all other points are distinct (in particular, no $(k-1)$-collisions). On the other hand, our lower bound is always lower than $2^{n/2}$, which is trivial for this problem. Nonetheless, both problems are searching for the same thing—namely a k-collisions—just in different settings. We hope that future work may be able to extend our techniques to solve the problem of k-distinctness.

1.3 The Reciprocal Plus 1 Rule

For many search problems over random functions, such as pre-image search, collision finding, k-sum, quantum oracle interrogation, and more, a very simple folklore rule of thumb translates the classical query complexity into quantum query complexity.

In particular, let $N = 2^n$, all of these problems have a classical query complexity $\Theta(N^{1/\alpha})$ for some rational number α. Curiously, the quantum query complexity of all these problems is always $\Theta(N^{\frac{1}{\alpha+1}})$.

In slightly more detail, for all of these problems the best classical q-query algorithm solves the problem with probability $\Theta(q^c/N^d)$ for some constants c, d. Then the classical query complexity is $\Theta(N^{d/c})$. For this class of problems, the success probability of the best q query quantum algorithm is obtained simply by increasing the power of q by d. This results in a quantum query complexity of $\Theta(N^{d/(c+d)})$. Examples:

- Grover's pre-image search [Gro96] improves success probability from q/N to q^2/N, which is known to be optimal [BBBV97]. The result is a query complexity improvement from $N = N^{1/1}$ to $N^{1/2}$.
  Similarly, finding, say, 2 pre-images has classical success probability q^2/N^2; it is straightforward to adapt known techniques to prove that the best quantum success probability is q^4/N^2. Again, the query complexity goes from N to $N^{1/2}$. Analogous statements hold for any *constant* number of pre-images.
- The BHT collision finding algorithm [BHT98] finds a collision with probability q^3/N, improving on the classical birthday attack q^2/N. Both of these are

known to be optimal [AS04,Zha15a]. Thus quantum algorithms improve the query complexity from $N^{1/2}$ to $N^{1/3}$.
Similarly, finding, say, 2 distinct collisions has classical success probability q^4/N^2, whereas we show that the quantum success probability is q^6/N^2. More generally, any *constant* number of distinct collisions conforms to the Reciprocal Plus 1 Rule.
- k-sum asks to find a set of k inputs such that the sum of the outputs is 0. This is a different generalization of collision finding than what we study in this work. Classically, the best algorithm succeeds with probability q^k/N. Quantumly, the best algorithm succeeds with probability q^{k+1}/N [BS13,Zha18]. Hence the query complexity goes from $N^{1/k}$ to $N^{1/(k+1)}$.
Again, solving for any *constant* number of distinct k-sum solutions also conforms to the Reciprocal Plus 1 Rule.
- In the oracle interrogation problem, the goal is to compute $q+1$ input/output pairs, using only q queries. Classically, the best success probability is clearly $1/N$. Meanwhile, Boneh and Zhandry [BZ13] give a quantum algorithm with success probability roughly q/N, which is optimal.

Some readers may have noticed that Reciprocal Plus 1 (RP1) rule does not immediately appear to apply the Element Distinctness. The Element Distinctness problem asks to find a collision in $H : [M] \to [N]$ where the collision is unique. Classically, the best algorithm succeeds with probability $\Theta(q^2/M^2)$. On the other hand, quantum algorithms can succeed with probability $\Theta(q^3/M^2)$, which is optimal [Amb04,Zha15a]. This does not seem to follow the prediction of the RP1 rule, which would have predicted q^4/M^2. However, we note that unlike the settings above which make sense when $N \ll M$, and where the complexity is characterized by N, the Element Distinctness problem requires $M \leq N$ and the complexity is really characterized by the domain size M. Interestingly, we note that for a random expanding function, when $N \approx M^2$, there will with constant probability be exactly one collision in H. Thus, in this regime the collision problem matches the Element Distinctness problem, and the RP1 rule gives the right query complexity!

Similarly, the quantum complexity for k-sum is usually written as $M^{k/(k+1)}$, not $N^{1/(k+1)}$. But again, this is because most of the literature considers H for which there is a unique k-sum and H is non-compressing, in which case the complexity is better measured in terms of M. Notice that a random function will contain a unique k collision when $N \approx M^k$, in which case the bound we state (which follows the RP1 rule) exactly matches the statement usually given.

On the other hand, the RP1 rule does not give the right answer for k-distinctness for $k \geq 3$, since the RP1 rule would predict the exponent to approach $1/2$ for large k, whereas prior work shows that it approaches $3/4$ for large k. That RP1 does not apply perhaps makes sense, since there is no setting of N, M where a random function will become an instance of k-distinctness: for any setting of parameters where a random function has a k-collision, it will also most likely have many $(k-1)$-collisions.

The takeaway is that the RP1 Rule seems to apply for natural search problems that make sense on random functions when $N \ll M$. Even for problems that do not immediately fit this setting such as Element Distinctness, the rule often still gives the right query complexity by choosing M, N so that a random function is likely to give an instance of the desired problem.

Enter k-collisions. In the case of k-collisions, the classical best success probability is $q^k/N^{(k-1)}$, giving a query complexity of $N^{(k-1)/k} = N^{1-1/k}$. Since the k-collision problem is a generalization of collision finding, is similar in spirit to the problems above, and applies to compressing random functions, one may expect that the Reciprocal Plus 1 Rule applies. If true, this would give a quantum success probability of q^{2k-1}/N^{k-1}, and a query complexity of $N^{(k-1)/(2k-1)} = N^{\frac{1}{2}(1-\frac{1}{2k-1})}$.

Even more, for small enough q, it is straightforward to find a k-collision with probability $O(q^{2k-1}/N^{k-1})$ as desired. In particular, divide the q queries into $k-1$ blocks. Using the first $q/(k-1)$ queries, find a 2-collision with probability $(q/(k-1))^3/N = O(q^3/N)$. Let y be the image of the collision. Then, for each of the remaining $(k-2)$ blocks of queries, find a pre-image of y with probability $(q/(k-1))^2/N = O(q^2/N)$ using Grover search. The result is k colliding inputs with probability $O(q^{3+2(k-2)}/N^{k-1}) = O(q^{2k-1}/N^{k-1})$. It is also possible to prove that this is a lower bound on the success probability (see lower bound discussion below). Now, this algorithm works as long $q \leq N^{1/3}$, since beyond this range the 2-collision success probability is bounded by $1 < q^3/N$. Nonetheless, it is asymptotically tight in the regime for which it applies. This seems to suggest that the limitation to small q might be an artifact of the algorithm, and that a more clever algorithm could operate beyond the $N^{1/3}$ barrier. In particular, this strongly suggests k-collisions conforms to the Reciprocal Plus 1 Rule.

Note that the RP1 prediction gives an exponent that depends polynomially on k, asymptotically approaching 1/2. In contrast, the prior work of [HSX17] approaches 1/2 *exponentially fast* in k. Thus, prior to our work we see an exponential vs polynomial gap for k-collisions, similar to the case of k-distinctness.

Perhaps surprisingly given the above discussion[2], our work demonstrates that the right answer is in fact exponential, refuting the RP1 rule for k-collisions.

As mentioned above, our results do not immediately give any indication for the query complexity of k-distinctness. However, our results may hint that k-distinctness also exhibits an exponential dependence on k. We hope that future work, perhaps building on our techniques, will be able to resolve this question.

1.4 Technical Details

The Algorithm. At their heart, the algorithms for pre-image search, collision finding, k-sum, and the recent algorithm for k-collision, all rely on Grover's algorithm. Let $f : \{0,1\}^n \to \{0,1\}$ be a function with a fraction δ of accepting inputs. Grover's algorithm finds the input with probability $O(\delta q^2)$ using q

[2] At least, the authors found it surprising!.

quantum queries to f. Grover's algorithm finds a pre-image of a point y in H by setting $f(x)$ to be 1 if and only if $H(x) = y$.

The BHT algorithm [BHT98] uses Grover's to find a collision in H. First, it queries H on $q/2 = O(q)$ random points, assembling a database D. As long as $q \ll N^{1/2}$, all the images in D will be distinct. Now, it lets $f(x)$ be the function that equals 1 if and only if $H(x)$ is found amongst the images in D, and x is not among the pre-images. By finding an accepting input to f, one immediately finds a collision. Notice that the fraction of accepting inputs is approximately q/N.

By running Grover's for $q/2 = O(q)$ steps, one obtains a such a pre-image, and hence a collision, with probability $O((q/N)q^2) = O(q^3/N)$.

Hosoyamada et al. show how this idea can be recursively applied to find multi-collisions. For $k = 3$, the first step is to find a database D_2 consisting of r distinct 2-collisions. By recursively applying the BHT algorithm, each 2-collision takes time $N^{1/3}$. Then, to find a 3 collision, set up f as before: $f(x) = 1$ if and only if $H(x)$ is amongst the images in D and x is not among the pre-images. The fraction of accepting inputs is approximately r/N, so Grover's algorithm will find a 3-collision in time $(N/r)^{1/2}$. Setting r to be $N^{1/9}$ optimizes the total query count as $N^{4/9}$. For $k = 4$, recursively build a table D_3 of 3-collisions, and set up f to find a collision with the database.

The result is an algorithm for k-collisions for any constant k, using $O(N^{\frac{1}{2}(1-\frac{1}{3^k-1})})$ queries.

Our algorithm improves on Hosoyamada et al.'s, yielding a query complexity of $O(N^{\frac{1}{2}(1-\frac{1}{2^k-1})})$. Note that for Hosoyamada et al.'s algorithm, when constructing D_{k-1}, many different D_{k-2} databases are being constructed, one for each entry in D_{k-1}. Our key observation is that a single database can be re-used for the different entries of D_{k-1}. This allows us to save on some of the queries being made. These extra queries can then be used in other parts of the algorithm to speed up the computation. By balancing the effort correctly, we obtain our algorithm. Put another way, the cost of finding many (k-)collisions can be amortized over many instances, and then recursively used for finding collisions with higher k. Since the recursive steps involve solving many instances, this leads to an improved computational cost.

In more detail, we iteratively construct databases $D_1, D_2, \ldots, D_k$. Each D_i will have r_i i-collisions. We set $r_k = 1$, indicating that we only need a single k-collision. To construct database D_1, simply query on r_1 arbitrary points. To construct database $D_i, i \geq 2$, define the function f_i that accepts inputs that collide with D_{i-1} but are not contained in D_{i-1}. The fraction of points accepted by f_i is approximately r_{i-1}/N. Therefore, Grover's algorithm returns an accepting input in time $(N/r_{i-1})^{1/2}$. We simply run Grover's algorithm r_i times *using the same database* D_{i-1} to construct D_i in time $r_i(N/r_{i-1})^{1/2}$.

Now we just optimize $r_1, \ldots, r_{k-1}$ by setting the number of queries to construct each database to be identical. Notice that $r_1 = O(q)$, so solving for r_i gives us

$$r_k = O\left(\frac{q^{2-\frac{1}{2^k-1}}}{N^{1-\frac{1}{2^k-1}}}\right)$$

Setting $r_k = 1$ and solving for q gives the desired result. In particular, in the case $k = 3$, our algorithm finds a collision in time $O(N^{3/7})$.

The Lower Bound. Notice that our algorithm fails to match the result one would get by applying the "Reciprocal Plus 1 Rule". Given the discussion above, one may expect that our iterative algorithm could potentially be improved on even more. To the contrary we prove that, in fact, our algorithm is asymptotically optimal for any constant k.

Toward that end, we employ a recent technique developed by Zhandry [Zha18] for analyzing quantum queries to *random* functions. We use this technique to show that our algorithm is tight for random functions, giving an average-case lower bound.

Zhandry's "Compressed Oracles." Zhandry demonstrates that the information an adversary knows about a random oracle H can be summarized by a database D^* of input/output pairs, which is updated according to some special rules. In Zhandry's terminology, D^* is the "compressed standard/phase oracle".

This D^* is not a classical database, but technically a superposition of all databases, meaning certain amplitudes are assigned to each possible database. D^* can be measured, obtaining an actual classical database D with probability equal to its amplitude squared. In the following discussion, we will sometimes pretend that D^* is actually a classical database. While inaccurate, this will give the intuition for the lower bound techniques we employ. In the Sect. 4 we take care to correctly analyze D^* as a superposition of databases.

Zhandry shows roughly the following:

- Consider any "pre-image problem", whose goal is to find a set of pre-images such that the images satisfy some property. For example, k-collision is the problem of finding k pre-images such that the corresponding images are all the same.
  Then after q queries, consider measuring D^*. The adversary can only solve the pre-image problem after q queries if the measured D^* has a solution to the pre-image problem.
  Thus, we can always upper bound the adversary's success probability by upper bounding the probability D^* contains a solution.
- D^* starts off empty, and each query can only add one point to the database.
- For any image point y, consider the amplitude on databases containing y as a function of q (remember that amplitude is the square root of the probability). Zhandry shows that this amplitude can only increase by $O(\sqrt{1/N})$ from one query to the next. More generally, for a set S of r different images, the amplitude on databases containing any point in S can only increase by $O(\sqrt{|S|/N})$.

The two results above immediately imply the optimality of Grover's search. In particular, the amplitude on databases containing y is at most $O(q\sqrt{1/N})$ after q queries, so the probability of obtaining a solution is the square of this amplitude, or $O(q^2/N)$. This also readily gives a lower bound for the collision problem. Namely, in order to introduce a collision to D^*, the adversary must add a point that collides with one of the existing points in D^*. Since there are at most q such points, the amplitude on such D^* can only increase by $O(\sqrt{q/N})$. This means the overall amplitude after q queries is at most $O(q^{3/2}/N^{1/2})$. Squaring to get a probability gives the correct lower bound.

A First Attempt. Our core idea is to attempt a lower bound for k-collision by applying these ideas recursively. The idea is that, in order to add, say, a 3-collision to D^*, there must be an existing 2-collision in the database. We can then use the 2-collision lower bound to bound the increase in amplitude that results from each query.

More precisely, for very small q, we can bound the amplitude on databases containing ℓ distinct 2-collisions as $O(\ (q^{3/2}/N^{1/2})^\ell)$. If $q \ll N^{1/3}$, ℓ must be a constant else this term is negligible. So we can assume for $q < N^{1/3}$ that ℓ is a constant.

Then, we note that in order to introduce a 3-collision, the adversary's new point must collide with one of the existing 2-collisions. Since there are at most ℓ, we know that the amplitude increases by at most $O(\sqrt{\ell}/N^{1/2}) = O(1/N^{1/2})$ since ℓ is a constant. This shows that the amplitude on databases with 3-collisions is at most $q/N^{1/2}$.

We can bound the amplitude increase even smaller by using not only the fact that the database contains at most ℓ 2-collisions, but the fact that the amplitude on databases containing even a single 2-collision is much less than 1. In particular, it is $O(q^{3/2}/N^{1/2})$ as demonstrated above. Intuitively, it turns out we can actually just multiply the $1/N^{1/2}$ amplitude increase in the case where the database contains a 2-collision by the $q^{3/2}/N^{1/2}$ amplitude on databases containing any 2-collision to get an overall amplitude increase of $q^{3/2}/N$.

Overall then, we upper bound the amplitude after $q < N^{1/3}$ queries by $O(q^{5/2}/N)$, given an upper bound of $O(q^5/N^2)$ on the probability of finding a 3-collision. This lower bound can be extended recursively to any constant k-collisions, resulting in a bound that exactly matches the Reciprocal Plus 1 Rule, as well as the algorithm for small q! This again seems to suggest that our algorithm is not optimal.

Our Full Proof. There are two problems with the argument above that, when resolved, actually do show our algorithm is optimal. First, when $q \geq N^{1/3}$, the $O(q^{3/2}/N^{1/2})$ part of the amplitude bound becomes vacuous, as amplitudes can never be more than 1. Second, the argument fails to consider algorithms that find many 2-collisions, which is possible when $q > N^{1/3}$. Finding many 2-collisions of course takes more queries, but then it makes extending to 3-collisions easier, as there are more collisions in the database to match in each iteration.

In our full proof, we examine the amplitude on the databases containing a 3-collision as well as r 2-collisions, after q queries. We call this amplitude $g_{q,r}$. We show a careful recursive formula for bounding g using Zhandry's techniques, which we then solve.

More generally, for any constant k, we let $g_{q,r,s}^{(k)}$ be the amplitude on databases containing exactly r distinct $(k-1)$-collisions and at least s distinct k-collisions after q queries. We develop a multiply-recursive formula for the $g^{(k)}$ in terms of the $g^{(k)}$ and $g^{(k-1)}$. We then recursively plug in our solution to $g^{(k-1)}$ so that the recursion is just in terms of $g^{(k)}$, which we then solve using delicate arguments.

Interestingly, this recursive structure for our lower bound actually closely matches our algorithm. Namely, our proof lower bounds the difficulty of adding an i-collision to a database D^* containing many $i-1$ collisions, exactly the problem our algorithm needs to solve. Our techniques essentially show that every step of our algorithm is tight, resulting in a lower bound of $\boldsymbol{\Omega}\left(\mathbf{N}^{\frac{1}{2}(1-\frac{1}{2^k-1})}\right)$, exactly matching our algorithm. Thus, we solve the quantum query complexity of k-collisions.

1.5 Other Related Work

Most of the related work has been mentioned earlier. Recently, in [HSTX18], Hosoyamada, Sasaki, Tani and Xagawa gave the same improvement. And they also showed that, their algorithm can also find a multi-collision for a more general setting where $|X| \geq \frac{l}{c_N} \cdot |Y|$ for any positive value $c_N \geq 1$ which is in $o(N^{\frac{1}{2^l-1}})$ and find a multiclaw for random functions with the same query complexity. They also noted that our improved collision finding algorithm for the case $|X| \geq l \cdot |Y|$ was reported in the Rump Session of AsiaCrypt 2017. They did not give an accompanying lower bound.

2 Preliminaries

Here, we recall some basic facts about quantum computation, and review the relevant literature on quantum search problems.

2.1 Quantum Computation

A quantum system Q is defined over a finite set B of classical states. In this work we will consider $B = \{0,1\}^n$. A **pure state** over Q is a unit vector in $\mathbb{C}^{|B|}$, which assigns a complex number to each element in B. In other words, let $|\phi\rangle$ be a pure state in Q, we can write $|\phi\rangle$ as:

$$|\phi\rangle = \sum_{x \in B} \alpha_x |x\rangle$$

where $\sum_{x\in B} |\alpha_x|^2 = 1$ and $\{|x\rangle\}_{x\in B}$ is called the "**computational basis**" of $\mathbb{C}^{|B|}$. The computational basis forms an orthonormal basis of $\mathbb{C}^{|B|}$.

Given two quantum systems Q_1 over B_1 and Q_2 over B_2, we can define a **product** quantum system $Q_1 \otimes Q_2$ over the set $B_1 \times B_2$. Given $|\phi_1\rangle \in Q_1$ and $|\phi_2\rangle \in Q_2$, we can define the product state $|\phi_1\rangle \otimes |\phi_2\rangle \in Q_1 \otimes Q_2$.

We say $|\phi\rangle \in Q_1 \otimes Q_2$ is **entangled** if there does not exist $|\phi_1\rangle \in Q_1$ and $|\phi_2\rangle \in Q_2$ such that $|\phi\rangle = |\phi_1\rangle \otimes |\phi_2\rangle$. For example, consider $B_1 = B_2 = \{0, 1\}$ and $Q_1 = Q_2 = \mathbb{C}^2$, $|\phi\rangle = \frac{|00\rangle + |11\rangle}{\sqrt{2}}$ is entangled. Otherwise, we say $|\phi\rangle$ is un-entangled.

A pure state $|\phi\rangle \in Q$ can be manipulated by a unitary transformation U. The resulting state $|\phi'\rangle = U|\phi\rangle$.

We can extract information from a state $|\phi\rangle$ by performing a **measurement**. A measurement specifies an orthonormal basis, typically the computational basis, and the probability of getting result x is $|\langle x|\phi\rangle|^2$. After the measurement, $|\phi\rangle$ "collapses" to the state $|x\rangle$ if the result is x.

For example, given the pure state $|\phi\rangle = \frac{3}{5}|0\rangle + \frac{4}{5}|1\rangle$ measured under $\{|0\rangle, |1\rangle\}$, with probability 9/25 the result is 0 and $|\phi\rangle$ collapses to $|0\rangle$; with probability 16/25 the result is 1 and $|\phi\rangle$ collapses to $|1\rangle$.

We finally assume a quantum computer can implement any unitary transformation (by using these basic gates, Hadamard, phase, CNOT and $\frac{\pi}{8}$ gates), especially the following two unitary transformations:

- **Classical Computation:** Given a function $f : X \to Y$, one can implement a unitary U_f over $\mathbb{C}^{|X|\cdot|Y|} \to \mathbb{C}^{|X|\cdot|Y|}$ such that for any $|\phi\rangle = \sum_{x\in X, y\in Y} \alpha_{x,y}|x, y\rangle$,

$$U_f|\phi\rangle = \sum_{x\in X, y\in Y} \alpha_{x,y}|x, y \oplus f(x)\rangle$$

  Here, $\oplus$ is a commutative group operation defined over Y.
- **Quantum Fourier Transform:** Let $N = 2^n$. Given a quantum state $|\phi\rangle = \sum_{i=0}^{2^n-1} x_i|i\rangle$, by applying only $O(n^2)$ basic gates, one can compute $|\psi\rangle = \sum_{i=0}^{2^n-1} y_i|i\rangle$ where the sequence $\{y_i\}_{i=0}^{2^n-1}$ is the sequence achieved by applying the classical Fourier transform QFT_N to the sequence $\{x_i\}_{i=0}^{2^n-1}$:

$$y_k = \frac{1}{\sqrt{N}} \sum_{i=0}^{2^n-1} x_i \omega_n^{ik}$$

  where $\omega_n = e^{2\pi i/N}$, i is the imaginary unit.
  One interesting property of QFT is that by preparing $|0^n\rangle$ and applying QFT_2 to each qubit, $(\mathsf{QFT}_2|0\rangle)^{\otimes n} = \frac{1}{\sqrt{2^n}} \sum_{x\in\{0,1\}^n} |x\rangle$ which is a uniform superposition over all possible $x \in \{0, 1\}^n$.

For convenience, we sometimes ignore the normalization of a pure state which can be calculated from the context.

2.2 Grover's Algorithm and BHT Algorithm

Definition 1 (Database Search Problem). *Suppose there is a function/database encoded as $F : X \to \{0,1\}$ and $F^{-1}(1)$ is non-empty. The problem is to find $x^* \in X$ such that $F(x^*) = 1$.*

We will consider adversaries with quantum access to F, meaning they submit queries as $\sum_{x\in X, y\in\{0,1\}} \alpha_{x,y}|x, y\rangle$ and receive in return $\sum_{x\in X, y\in\{0,1\}} \alpha_{x,y}|x, y \oplus F(x)\rangle$. Grover's algorithm [Gro96] finds a pre-image using an optimal number of queries:

Theorem 1 ([Gro96,BBHT98]). *Let F be a function $F : X \to \{0,1\}$. Let $t = |F^{-1}(1)| > 0$ be the number of pre-images of 1. There is a quantum algorithm that finds $x^* \in X$ such that $F(x^*) = 1$ with an expected number of quantum queries to F at most $O\left(\sqrt{\frac{|X|}{t}}\right)$ even without knowing t in advance.*

We will normally think of the number of queries as being fixed, and consider the probability of success given the number of queries. The algorithm from Theorem 1, when runs for q queries, can be shown to have a success probability $\min(1, O(q^2/(|X|/t)))$. For the rest of the paper, "Grover's algorithm" will refer to this algorithm.

Now let us look at another important problem: 2-collision finding problem on 2-to-1 functions.

Definition 2 (Collision Finding on 2-to-1 Functions). *Assume $|X| = 2|Y| = 2N$. Consider a function $F : X \to Y$ such that for every $y \in Y$, $|F^{-1}(y)| = 2$. In other words, every image has exactly two pre-images. The problem is to find $x \neq x'$ such that $F(x) = F(x')$.*

Brassard, Høyer and Tapp proposed a quantum algorithm [BHT98] that solved the problem using only $O(N^{1/3})$ quantum queries. The idea is the following:

- Prepare a list of input and output pairs, $L = \{(x_i, y_i = F(x_i)\}_{i=1}^t$ where x_i is drawn uniformly at random and $t = N^{1/3}$;
- If there is a 2-collision in L, output that pair. Otherwise,
- Run Grover's algorithm on the following function F': $F'(x) = 1$ if and only if there exists $i \in \{1, 2, \cdots, t\}$, $F(x) = y_i = F(x_i)$ and $x \neq x_i$. Output the solution x, as well as whatever x_i it collides with.

This algorithm takes $O(t + \sqrt{N/t})$ quantum queries and when $t = \Theta(N^{1/3})$, the algorithm finds a 2-collision with $O(N^{1/3})$ quantum queries.

2.3 Multi-collision Finding and [HSX17]

Hosoyamada, Sasaki and Xagawa proposed an algorithm for k-collision finding on any function $F : X \to Y$ where $|X| \geq k|Y|$ (k is a constant). They generalized the idea of [BHT98] and gave the proof for even arbitrary functions. We now briefly talk about their idea. For simplicity in this discussion, we assume F is a k-to-1 function.

The algorithm prepares t pairs of 2-collisions $(x_1, x_1'), \cdots, (x_t, x_t')$ by running the BHT algorithm t times. If two pairs of 2-collisions collide, there is at least a 3-collision (possibly a 4-collision). Otherwise, it uses Grover's algorithm to find a $x'' \neq x_i$, $x'' \neq x_i'$ and $f(x'') = f(x_i) = f(x_i')$. The number of queries is $O(tN^{1/3} + \sqrt{N/t})$. When $t = \Theta(N^{1/9})$, the query complexity is $O(N^{4/9})$.

By induction, finding a $(k-1)$-collision requires $O(N^{(3^{k-1}-1)/(2\cdot 3^{k-1})})$ quantum queries. By preparing t $(k-1)$-collisions and applying Grover's algorithm to it, it takes $O(tN^{(3^{k-1}-1)/(2\cdot 3^{k-1})} + \sqrt{\frac{N}{t}})$ quantum queries to get one k-collision. It turns out that $t = \Theta(N^{1/3^k})$ and the complexity of finding k-collision is $O(N^{(3^k-1)/(2\cdot 3^k)})$.

2.4 Compressed Fourier Oracles and Compressed Phase Oracles

In [Zha18], Zhandry showed a new technique for analyzing cryptosystems in the random oracle model. He also showed that his technique can be used to re-prove several known quantum query lower bounds. In this work, we will extend his technique in order to prove a new optimal lower bound for multi-collisions.

The basic idea of Zhandry's technique is the following: assume $\mathcal{A}$ is making a query to a random oracle H and the query is $\sum_{x,u,z} a_{x,u,z}|x, u, z\rangle$ where x is the query register, u is the response register and z is its private register. Instead of only considering the adversary's state $\sum_{x,u,z} a_{x,u,z}|x, u + H(x), z\rangle$ for a random oracle H, we can actually treat the whole system as

$$\sum_{x,u,z} \sum_{H} a_{x,u,z}|x, u + H(x), z\rangle \otimes |H\rangle$$

where $|H\rangle$ is the truth table of H. By looking at random oracles that way, Zhandry showed that these five random oracle models/simulators are equivalent:

1. **Standard Oracles:**

$$\mathsf{StO} \sum_{x,u,z} a_{x,u,z}|x, u, z\rangle \otimes \sum_{H} |H\rangle \Rightarrow \sum_{x,u,z} \sum_{H} a_{x,u,z}|x, u + H(x), z\rangle \otimes |H\rangle$$

2. **Phase Oracles:**

$$\mathsf{PhO} \sum_{x,u,z} a_{x,u,z}|x, u, z\rangle \otimes \sum_{H} |H\rangle \Rightarrow \sum_{x,u,z} a_{x,u,z}|x, u, z\rangle \otimes \sum_{H} \omega_n^{H(x)\cdot u}|H\rangle$$

where $\omega_n = e^{2\pi i/N}$ and $\mathsf{PhO} = (I \otimes \mathsf{QFT}^\dagger \otimes I) \cdot \mathsf{StO} \cdot (I \otimes \mathsf{QFT} \otimes I)$. In other words, it first applies the QFT to the u register, applies the standard query, and then applies $\mathsf{QFT}^\dagger$ one more time.

3. **Fourier Oracles:** We can view $\sum_H |H\rangle$ as $\mathsf{QFT}|0^N\rangle$. In other words, if we perform Fourier transform on a function that always outputs 0, we will get a uniform superposition over all the possible functions $\sum_H |H\rangle$.
Moreover, $\sum_H \omega^{H(x)\cdot u}|H\rangle$ is equivalent to $\mathsf{QFT}|0^N \oplus (x,u)\rangle$. Here $\oplus$ means updating (xor) the x-th entry in the database with u.
So in this model, we start with $\sum_{x,u,z} a^0_{x,u,z}|x,u,z\rangle \otimes \mathsf{QFT}|D_0\rangle$ where D_0 is an all-zero function. By making the i-th query, we have
$$\mathsf{PhO} \sum_{x,u,z,D} a^{i-1}_{x,u,z,D}|x,u,z\rangle \otimes \mathsf{QFT}|D\rangle \Rightarrow \sum_{x,u,z,D} a^{i-1}_{x,u,z,D}|x,u,z\rangle \otimes \mathsf{QFT}|D \oplus (x,u)\rangle$$
The Fourier oracle incorporates QFT and operates directly on the D registers:
$$\mathsf{FourierO} \sum_{x,u,z,D} a^{i-1}_{x,u,z,D}|x,u,z\rangle \otimes |D\rangle \Rightarrow \sum_{x,u,z,D} a^{i-1}_{x,u,z,D}|x,u,z\rangle \otimes |D \oplus (x,u)\rangle$$
4. **Compressed Fourier Oracles:** The idea is basically the same as Fourier oracles. But when the algorithm only makes q queries, the database D with non-zero weight contains at most q non-zero entries.
So to describe D, we only need at most q different (x_i, u_i) pairs $(u_i \neq 0)$ which says the database outputs u_i on x_i and 0 everywhere else. And $D \oplus (x,u)$ is doing the following: (1) if x is not in the list D and $u \neq 0$, put (x,u) in D; (2) if (x,u') is in the list D and $u' \neq u$, update u' to $u' \oplus u$ in D; (3) if (x,u') is in the list and $u' = u$, remove (x,u') from D.
In the model, we start with $\sum_{x,u,z} a^0_{x,u,z}|x,u,z\rangle \otimes |D_0\rangle$ where D_0 is an empty list. After making the i-th query, we have
$$\mathsf{CFourierO} \sum_{x,u,z,D} a^{i-1}_{x,u,z,D}|x,u,z\rangle \otimes |D\rangle \Rightarrow \sum_{x,u,z,D} a^{i-1}_{x,u,z,D}|x,u,z\rangle \otimes |D \oplus (x,u)\rangle$$
5. **Compressed Standard/Phase Oracles:** These two models are essentially equivalent up to an application of QFT applied to the query response register. From now on we only consider compressed phase oracles.
By applying QFT on the u entries of the database registers of a compressed Fourier oracle, we get a compressed phase oracle.
In this model, D contains all the pair (x_i, u_i) which means the oracle outputs u_i on x_i and uniformly at random on other inputs. When making a query on $|x,u,z,D\rangle$,
   - if (x,u') is in the database D for some u', a phase $\omega_n^{uu'}$ will be added to the state; it corresponds to update w to $w+u$ in the compressed Fourier oracle model where $w = D(x)$ in the compressed Fourier database.
   - otherwise a superposition is appended to the state $|x\rangle \otimes \sum_{u'} \omega_n^{uu'}|u'\rangle$; it corresponds to put a new pair (x,u) in the list of the compressed Fourier oracle model;
   - also make sure that the list will never have an $(x,0)$ pair in the compressed Fourier oracle model (in other words, it is $|x\rangle \otimes \sum_y |y\rangle$ in the compressed phase oracle model); if there is one, delete that pair;
   - All the 'append' and 'delete' operations above mean applying QFT.

3 Algorithm for Multi-collision Finding

In this section, we give an improved algorithm for k-collision finding. We use the same idea from [HSX17] but carefully reorganize the algorithm to reduce the number of queries.

As a warm-up, let us consider the case $k = 3$ and the case where $F : X \to Y$ is a 3-to-1 function, $|X| = 3|Y| = 3N$. They gives an algorithm with $O(N^{4/9})$ quantum queries. Here is our algorithm with only $O(N^{3/7})$ quantum queries:

- Prepare a list $L = \{(x_i, y_i = F(x_i))\}_{i=1}^{t_1}$ where x_i are distinct and $t_1 = N^{3/7}$. This requires $O(N^{3/7})$ classical queries on random points.
- Define the following function F' on X:

$$F'(x) = \begin{cases} 1, & x \notin \{x_1, x_2, \cdots, x_{t_1}\} \text{ and } F(x) = y_j \text{ for some } j \\ 0, & \text{otherwise} \end{cases}$$

  Run Grover's algorithm on function F'. Wlog (by reordering L), we find x_1' such that $x_1' \neq x_1$ and $F(x_1') = F(x_1)$ using $O(\sqrt{N/N^{3/7}}) = O(N^{2/7})$ quantum queries.
- Repeat the last step $t_2 = N^{1/7}$ times, we will have $N^{1/7}$ 2-collisions $L' = \{(x_i, x_i', y_i)\}_{i=1}^{t_2}$. This takes $O(N^{1/7} \cdot \sqrt{N/N^{3/7}}) = O(N^{3/7})$ quantum queries.
- If two elements in L' collide, simply output a 3-collision. Otherwise, run Grover's on function G:

$$G(x) = \begin{cases} 1, & x \notin \{x_1, x_2, \cdots, x_{t_2}, x_1', \cdots, x_{t_2}'\} \text{ and } F(x) = y_j \text{ for some } j \\ 0, & \text{otherwise} \end{cases}$$

  A 3-collision will be found when Grover's algorithm finds a pre-image of 1 on G. It takes $O(\sqrt{N/N^{1/7}}) = O(N^{3/7})$ quantum queries.

Overall, the algorithm finds a 3-collision using $O(N^{3/7})$ quantum queries.

The similar algorithm and analysis works for any constant k and any k-to-1 function which only requires $O(N^{(2^{k-1}-1)/(2^k-1)})$ quantum queries. Let $t_1 = N^{(2^{k-1}-1)/(2^k-1)}, t_2 = N^{(2^{k-2}-1)/(2^k-1)}, \cdots, t_i = N^{(2^{k-i}-1)/(2^k-1)}, \cdots, t_{k-1} = N^{1/(2^k-1)}$. The algorithm works as follows:

- Assume $F : X \to Y$ is a k-to-1 function and $|X| = k|Y| = kN$.
- Prepare a list L_1 of input-output pairs of size t_1. With overwhelming probability $(1 - N^{-1/2^k})$, L_1 does not contain a collision. By letting $t_0 = N$, this step makes $t_1\sqrt{N/t_0}$ quantum queries.
- Define a function $F_2(x)$ that returns 1 if the input x is not in L_1 but the image $F(x)$ collides with one of the images in L_1, otherwise it returns 0. Run Grover's on F_2 t_2 times. Every time Grover's algorithm outputs x', it gives a 2-collision. With probability $1 - O(N^{-1/2^k})$ (explained below), all these t_2 collisions do not collide. So we have a list L_2 of t_2 different 2-collisions. This step makes $t_2\sqrt{N/t_1}$ quantum queries.

- For $2 \leq i \leq k-1$, define a function $F_i(x)$ that returns 1 if the input x is not in L_{i-1} but the image $F(x)$ collides with one of the images of $(i-1)$-collisions in L_{i-1}, otherwise it returns 0. Run Grover's algorithm on F_i t_i times. Every time Grover's algorithm outputs x', it gives an i-collision. With probability $1 - O(t_i^2/t_{i-1}) = 1 - O(N^{-1/2^k})$, all these t_i collisions do not collide. So we have a list L_i of t_i different i-collisions. This step makes $t_i\sqrt{N/t_{i-1}}$ quantum queries.
- Finally given t_{k-1} $(k-1)$-collisions, using Grover's to find a single x' that makes a k-collision with one of the $(k-1)$-collision in L_{k-1}. This step makes $t_k\sqrt{N/t_{k-1}}$ quantum queries by letting $t_k = 1 = N^{(2^{k-k}-1)/(2^k-1)}$.

The number of quantum queries made by the algorithm is simply:

$$\begin{aligned}\sum_{i=0}^{k-1} t_{i+1}\sqrt{N/t_i} &= \sum_{i=0}^{k-1}\sqrt{N\frac{t_{i+1}^2}{t_i}} \\ &= \sum_{i=0}^{k-1}\sqrt{N \cdot N^{\frac{2\cdot\left(2^{k-(i+1)}-1\right)-(2^{k-i}-1)}{2^k-1}}} \\ &= k \cdot N^{(2^{k-1}-1)/(2^k-1)}\end{aligned}$$

So we have the following theorem:

Theorem 2. *For any constant k, any k-to-1 function $F : X \to Y$ ($|X| = k|Y| = kN$), the algorithm above finds a k-collision using $O(N^{(2^{k-1}-1)/(2^k-1)})$ quantum queries.*

We now show the above conclusion holds for an arbitrary function $F : X \to Y$ as long as $|X| \geq k|Y| = kN$. To prove this, we use the following lemma:

Lemma 1. *Let $F : X \to Y$ be a function and $|X| = k|Y| = kN$. Let $\mu_F = \Pr_x\left[|F^{-1}(F(x))| \geq k\right]$ be the probability that if we choose x uniformly at random and $y = F(x)$, the number of pre-images of y is at least k. We have $\mu_F \geq \frac{1}{k}$.*

Proof. We say an input or a collision is good if its image has at least k pre-images.

To make the probability as small as possible, we want that if y has less than k pre-images, y should have exactly $k-1$ pre-images. So the probability is at least

$$\mu_F = \frac{|\{x \,|\, x \text{ is good}\}|}{|X|} \geq \frac{kN - (k-1)N}{kN} = \frac{1}{k}$$

□

Theorem 3. *Let $F : X \to Y$ be a function and $|X| \geq k|Y| = kN$. The above algorithm finds a k-collision using $O(N^{(2^{k-1}-1)/(2^k-1)})$ quantum queries with overwhelming probability.*

Proof. We prove the case $|X| = k|Y|$. The case $|X| > k|Y|$ follows readily by choosing an arbitrary subset $X' \subseteq X$ such that $|X'| = k|Y|$ and restrict the algorithm to the domain X'.

As what we did in the previous algorithm, in the list L_1, with overwhelming probability, there are $0.999\mu_F \cdot t_1$ good inputs by Chernoff bound because every input is good with probability μ_F. Then every 2-collision in L_2 has probability $0.999\mu_F$ to be good. So by Chernoff bound, L_2 contains at least $0.999^2\mu_F t_2$ good 2-collisions with overwhelming probability. By induction, in the final list L_{k-1}, with overwhelming probability, there are $0.999^{k-1}\mu_F \cdot t_{k-1}$ good $(k-1)$-collisions. Finally, the algorithm outputs a k-collision with probability 1, by making at most $O(\sqrt{N/(0.99^{k-1}\mu_F t_{k-1})})$ quantum queries.

As long as k is a constant, the coefficients before t_i are all constants. The number of quantum queries is scaled by a constant and is still $O(N^{(2^{k-1}-1)/(2^k-1)})$ and the algorithm succeeds with overwhelming probability. □

4 Lower Bound for Multi-collision Finding

4.1 Idea in [Zha18]

We will first show how Zhandry re-proved the lower bound of 2-collision finding using compressed oracle technique. The idea is that when we are working under compressed phase/standard oracle model, a query made by the adversary (x, u) can be recorded in the compressed oracle database.

Suppose before making the next quantum query, the current joint state is the following

$$|\phi\rangle = \sum_{x,u,z,D} a_{x,u,z,D}|x, u, z\rangle \otimes |D\rangle$$

where x is the query register, u is the response register, z is the private storage of the adversary and D is the database in the compressed phase oracle model. Consider measuring D after running the algorithm. Because the algorithm only has information about the points in the database D, the only way to have a non-trivial probability of finding a collision is for the D that results from measurement to have a collision. More formally, here is a lemma from [Zha18].

Lemma 2 (Lemma 5 from [Zha18]**).** *Consider a quantum algorithm A making queries to a random oracle H and outputting tuples $(x_1, \cdots, x_k, y_1, \cdots, y_k, z)$. Let R be a collection of such tuples. Suppose with probability p, A outputs a tuple such that (1) the tuple is in R and (2) $H(x_i) = y_i$ for all i. Now consider running A with compressed standard/phase oracle, and suppose the database D is measured after A produces its output. Let p' be the probability that (1) the tuple is in R, and (2) $D(x_i) = y_i$ for all i (and in particular $D(x_i) \neq \bot$). Then*

$$\sqrt{p} \leq \sqrt{p'} + \sqrt{k/2^n}$$

As long as k is small, the difference is negligible. So we can focus on bounding the probability p'.

Let $\tilde{P}_1$ be a projection spanned by all the states with z, D containing at least one collision in the compressed phase oracles. In other words, z contains $x \neq x'$ such that $D(x) \neq \bot$, $D(x') \neq \bot$ and $D(x) = D(x')$.

$$\tilde{P}_1 = \sum_{\substack{x,u,z \\ z,D:\geq 1 \text{ collision}}} |x,u,z,D\rangle\langle x,u,z,D|$$

We care about the amplitude (square root of the probability) $\left|\tilde{P}_1|\phi\rangle\right|$. As in the above lemma, $\left|\tilde{P}_1|\phi\rangle\right| = \sqrt{p'}$ and $k = 2$. Moreover, we can bound the amplitude of the following measurement.

$$P_1 = \sum_{\substack{x,u,z \\ D:\geq 1 \text{ collision}}} |x,u,z,D\rangle\langle x,u,z,D|$$

Here "$D :\geq 1$ collision" meaning D as a compressed phase oracle, it has a pair of $x \neq x'$ such that $D(x) = D(x')$. It is easy to see $|P_1|\phi\rangle| \geq |\tilde{P}_1|\phi\rangle|$. So we will focus on bounding $|P_1|\phi\rangle|$ in the rest of the paper.

For every $|x,u,z,D\rangle$, after making one quantum query, the size of D will increase by at most 1. Let $|\phi_i\rangle$ be the state before making the $(i+1)$-th quantum query and $|\phi_i'\rangle$ be the state after it. Let O be the unitary over the joint system corresponding to an oracle query, in other words, $|\phi_i'\rangle = O|\phi_i\rangle$. By making q queries, the computation looks like the following:

- At the beginning, it has $|\phi_0\rangle$;
- For $1 \leq i \leq q$, it makes a quantum query; the state $|\phi_{i-1}\rangle$ becomes $|\phi_{i-1}'\rangle$; and it applies a unitary on its registers $U^i \otimes \mathsf{id}$ to get $|\phi_i\rangle$ where U^i is some unitary defined over the registers x, u, z.
- Finally measure it using P_1, the probability of finding a collision (in the compressed phase oracle) is at most $|P_1|\phi_q\rangle|^2$

We have the following two lemmas:

Lemma 3. *For any unitary* U^i,

$$|P_1|\phi_{i-1}'\rangle| = |P_1 \cdot (U^i \otimes \mathsf{id}) \cdot |\phi_{i-1}'\rangle| = |P_1|\phi_i\rangle|$$

Proof. Intuitively, P_1 is a measurement on the oracle's register and U^i is a unitary on the adversary's registers, applying the unitary does not affect the measurement P_1.

Because U^i is a unitary defined over the registers x, u, z and P_1 is a projective measurement defined over the database register D, we have

$$\begin{aligned} |P_1 \cdot (U^i \otimes \mathsf{id}) \cdot |\phi_{i-1}'\rangle| &= \left| P_1 \cdot (U^i \otimes \mathsf{id}) \cdot \sum_{x,u,z,D} \alpha_{x,u,z,D} |x,u,z,D\rangle \right| \\ &= \left| P_1 \cdot (U^i \otimes \mathsf{id}) \cdot \sum_D |\psi_D\rangle \otimes |D\rangle \right| \end{aligned}$$

$$= \sqrt{\sum_{\text{D}\geq 1 \text{ collision}} |U^i|\psi_D\rangle|^2} = \sqrt{\sum_{\text{D}\geq 1 \text{ collision}} ||\psi_D\rangle|^2}$$

which is the same as $|P_1|\phi'_{i-1}\rangle|$. □

Lemma 4. $|P_1|\phi'_i\rangle| \leq |P_1|\phi_i\rangle| + \frac{\sqrt{i}}{\sqrt{N}}$.

Proof. We have

$$\begin{aligned} |P_1|\phi'_i\rangle| &= |P_1 O|\phi_i\rangle| \\ &= |P_1 O\left(P_1|\phi_i\rangle + (I - P_1)|\phi_i\rangle\right)| \\ &\leq |P_1 O P_1|\phi_i\rangle| + |P_1 O(I - P_1)|\phi_i\rangle| \\ &\leq |P_1|\phi_i\rangle| + |P_1 O(I - P_1)|\phi_i\rangle| \end{aligned}$$

$|P_1 O P_1|\phi_i\rangle| \leq |P_1|\phi_i\rangle|$ is because $P_1|\phi_i\rangle$ contains only D with collisions. By making one more query, the total magnitude will not increase.

So we only need to bound the second term $|P_1 O(I - P_1)|\phi_i\rangle|$. $(I - P_1)|\phi\rangle$ contains only states $|x, u, z, D\rangle$ that D has no collision. If after applying O to a state $|x, u, z, D\rangle$, the size of D does not increase (stays the same or becomes smaller), the new database still does not contain any collision. Otherwise, it becomes $\sum_{u'} \omega_n^{uu'}|x, u, z, D \oplus (x, u')\rangle$. And only $|D| \leq i$ out of N possible $D \oplus (x, u')$ contain a collision.

$$\begin{aligned} |P_1 O(I - P_1)|\phi_i\rangle| &= \left| P_1 O \sum_{\substack{x,u,z,D \\ D:\text{ no collision}}} a_{x,u,z,D}|x, u, z, D\rangle \right| \\ &= \left| P_1 \sum_{\substack{x,u,z,D \\ D:\text{ no collision}}} \frac{1}{\sqrt{N}} \sum_{u'} \omega_n^{uu'} a_{x,u,z,D}|x, u, z, D \oplus (x, u')\rangle \right| \\ &\leq \left(\sum_{\substack{x,u,z,D \\ D:\text{ no collision}}} \frac{i}{N} \cdot a^2_{x,u,z,D} \right)^{1/2} \leq \frac{\sqrt{i}}{\sqrt{N}} \end{aligned}$$

□

By combining Lemmas 3 and 4, we have that $|P_1|\phi_i\rangle| \leq \sum_{j=1}^{i-1} \frac{\sqrt{j}}{\sqrt{N}} = O(i^{3/2}/N^{1/2})$. So we re-prove the following theorem:

Theorem 4. *For any quantum algorithm, given a random function $f : X \to Y$ where $|Y| = N$, it needs to make $\Omega(N^{1/3})$ quantum queries to find a 2-collision with constant probability.*

4.2 Intuition for Generalizations

Here is the intuition for $k = 3$: as we have seen in the proof for $k = 2$, after $T_1 = O(N^{1/3})$ quantum queries, the database has high probability to contain a 2-collision. Following the same formula, after making T_2 queries, the amplitude that it contains two 2-collisions is about

$$\sum_{T_1+1}^{T_2} \frac{\sqrt{i}}{\sqrt{N}} = O\left(\frac{T_2^{3/2} - T_1^{3/2}}{\sqrt{N}}\right) \Rightarrow T_2 = O(2^{2/3} N^{1/3})$$

And similarly after $T_i = O(i^{2/3} N^{1/3})$, the database will contain i 2-collisions. Now we just assume between the $(T_{i-1} + 1)$-th query and T_i-th query, the database contains exactly $(i - 1)$ 2-collisions.

Every time a quantum query is made to a database with i 2-collisions, with probability at most i/N, the new database will contain a 3-collision. Similar to the Lemma 4, when we make queries until the database contains m 2-collisions, the amplitude that it contains a 3-collision in the database is at most

$$\sum_{i=1}^{m} \frac{\sqrt{i}}{\sqrt{N}} (T_i - T_{i-1}) \approx \int_1^m \frac{x^{1/6}}{N^{1/6}} \mathrm{d}x \approx x^{7/6}/N^{1/6}$$

which gives us that the number of 2-collisions is $m = N^{1/7}$. And the total number of quantum queries is $T_m = m^{2/3} \cdot N^{1/3} = N^{3/7}$ which is what we expected.

In the following sections, we will show how to bound the probability/amplitude of finding a $k = 2, 3, 4$-collision and any constant k-collision with constant probability. All the proof ideas are explained step by step through the proof for $k = 2, 3, 4$. The proof for any constant k is identical to the proof for $k = 4$ but every parameter is replaced with functions of k.

4.3 Lower Bound for 2-Collisions

Let $P_{2,j}$ be a projection spanned by all the states with D containing at least j distinct 2-collisions in the compressed phase oracle model.

$$P_{2,j} = \sum_{\substack{x,u,z \\ D:\geq j \text{ 2-collisions}}} |x, u, z, D\rangle\langle x, u, z, D|$$

Let the current joint state be $|\phi\rangle$ (after making i quantum queries but before the $(i + 1)$-th query), and $|\phi'\rangle$ be the state after making the $(i + 1)$-th quantum query.

$$|\phi\rangle = \sum_{x,u,z,D} a_{x,u,z,D} |x, u\rangle \otimes |z, D\rangle$$

We have the relation following from Lemma 4:

$$|P_{2,1}|\phi'\rangle| \leq |P_{2,1}|\phi\rangle| + \frac{\sqrt{i}}{\sqrt{N}}$$

$$|P_{2,j}|\phi'\rangle| \leq |P_{2,j}|\phi\rangle| + \frac{\sqrt{i}}{\sqrt{N}} |P_{2,j-1}|\phi'\rangle| \text{ for all } j > 0$$

Let $|\phi_0\rangle, |\phi_1\rangle, \cdots, |\phi_i\rangle$ be the state after making $0, 1, \cdots, i$ quantum queries respectively. Let $f_{i,j} = |P_{2,j}|\phi_i\rangle|$. We rewrite the relations using $f_{i,j}$:

$$f_{i,1} \leq f_{i-1,1} + \frac{\sqrt{i-1}}{\sqrt{N}} = \sum_{0\leq l<i} \frac{\sqrt{l}}{\sqrt{N}} < \frac{i^{3/2}}{\sqrt{N}}$$

$$\begin{aligned} f_{i,j} &\leq f_{i-1,j} + \frac{\sqrt{i-1}}{N} f_{i-1,j-1} \\ &= \sum_{0\leq l_1<i} \frac{\sqrt{l_1}}{\sqrt{N}} f_{l_1,j-1} \\ &= \sum_{0\leq l_j<l_{j-1}<\cdots<l_2<l_1<i} \prod_{k=1}^{j} \left(\frac{\sqrt{l_k}}{\sqrt{N}}\right) \\ &< \frac{1}{j!} \sum_{0\leq l_j, l_{j-1},\cdots,l_2,l_1<i} \prod_{k=1}^{j} \left(\frac{\sqrt{l_k}}{\sqrt{N}}\right) \\ &= \frac{1}{j!} (f_{i,1})^j < \left(\frac{e\cdot i^{3/2}}{j\sqrt{N}}\right)^j \end{aligned}$$

We observe that when $i = o(j^{2/3}N^{1/3})$, $f_{i,j} = o(1)$.

Corollary 1. *For any quantum algorithm, given a random function $f: X \to Y$ where $|Y| = N$, by making i queries, the probability of finding constant j 2-collisions is at most $O\left((\frac{i^3}{N})^j\right)$.*

Theorem 5. *For any quantum algorithm, given a random function $f: X \to Y$ where $|Y| = N$, it needs to make $\Omega(j^{2/3}N^{1/3})$ quantum queries to find j 2-collisions with constant probability.*

4.4 Lower Bound for 3-Collisions

Let $P_{3,k}$ be a projection spanned by all the states with D containing at least k distinct 3-collisions in the compressed phase model. And let $P_{3,j,k}$ be a projection spanned by all the states with D containing **exactly** j distinct 2-collisions and at least k 3-collisions.

Let the current joint state be $|\phi\rangle$ (after making i quantum queries but before the $(i+1)$-th query), and $|\phi'\rangle$ be the state after making the $(i+1)$-th quantum query. We have the following relation similar to Lemma 4:

$$\begin{aligned} &|P_{3,k}|\phi'\rangle| \leq |P_{3,k}|\phi\rangle| \\ &\quad + \left| P_{3,k} \sum_{l\geq 0} \sum_{\substack{x,u,z \\ D:\text{ exactly } l \text{ 2-collisions} \\ \text{exactly } k-1 \text{ 3-collision}}} \frac{1}{\sqrt{N}} \sum_{u'} \omega_n^{uu'} \cdot \alpha_{x,u,z,D} |x,u,z,D \oplus (x,u')\rangle \right| \end{aligned}$$

where the first term means D already contains at least k 3-collisions before the query; and the second term is the case where a new 3-collision is added into the database. Similar to Lemma 4, only l out of N u' will make $D \oplus (x, u')$ contain k 3-collisions. So we have,

$$|P_{3,k}|\phi'\rangle| \leq |P_{3,k}|\phi\rangle| + \sqrt{\sum_{l\geq 0} \frac{l}{N} \sum_{\substack{x,u,z \\ D:\text{ exactly } l \text{ 2-collisions} \\ \text{exactly } k-1 \text{ 3-collision}}} |\alpha|^2_{x,u,z,D}}$$

$$\leq |P_{3,k}|\phi\rangle| + \sqrt{\sum_{l\geq 0} \frac{l}{N} |P_{3,l,k-1}|\phi\rangle|^2}$$

Let $g_{i,k}$ be the amplitude $|P_{3,k}|\phi_i\rangle|$ and $g_{i,j,k} = |P_{3,j,k}|\phi_i\rangle|$. It is easy to see $g_{i,0} \leq 1$ for any $i \geq 0$ since it is an amplitude. We have the following:

$$g_{i,k} \leq g_{i-1,k} + \sqrt{\sum_{l\geq 0} \frac{l}{N} \cdot g^2_{i-1,l,k-1}}$$

Let $f_{i,j} = |P_{2,j}|\phi_i\rangle|$. Define $h_3(i) = \max\{2e \cdot \frac{i^{3/2}}{\sqrt{N}}, 10N^{1/8}\}$. We have the following lemma:

Lemma 5.

$$g_{i,k} \leq g_{i-1,k} + \sqrt{\frac{h_3(i-1)}{N}} g_{i-1,k-1} + f_{i-1,h_3(i-1)}$$

Proof.

$$g_{i,k} \leq g_{i-1,k} + \sqrt{\sum_{l\geq 0} \frac{l}{N} \cdot g^2_{i-1,l,k-1}}$$

$$\leq g_{i-1,k} + \sqrt{\sum_{0\leq l\leq h_3(i-1)} \frac{l}{N} \cdot g^2_{i-1,l,k-1}} + \sqrt{\sum_{l>h_3(i-1)} 1 \cdot g^2_{i-1,l,k-1}}$$

$$\leq g_{i-1,k} + \sqrt{\frac{h_3(i-1)}{N}} \cdot \sqrt{\sum_{l\geq 0} g^2_{i-1,l,k-1}} + \sqrt{\sum_{l>h_3(i-1)} g^2_{i-1,l,k-1}}$$

$$\leq g_{i-1,k} + \sqrt{\frac{h_3(i-1)}{N}} \cdot g_{i-1,k-1} + f_{i-1,h_3(i-1)}$$

Here, in the last line, we used the fact that $\sum_{l\geq 0} g^2_{i-1,l,k-1}$ represents the total probability of the database having $k-1$ distinct 3-collisions, and so is equal to $g^2_{i-1,k-1}$. Similarly, we used that $\sum_{l>h_3(i-1)} g^2_{i-1,l,k-1}$ represents the total probability of having at least $k-1$ distinct 3-collisions *and* at least $h_3(i-1)$ distinct 2-collisions. This probability is bounded above by the probability of just having at least $h_3(i-1)$ distinct 2-collisions, which is $f^2_{i-1,h_3(i-1)}$. □

Lemma 6. *Define* $A_i = \sum_{l=0}^{i-1} \sqrt{\frac{h_3(l)}{N}}$. *Then* $g_{i,k}$ *can be bounded as the following:*

$$g_{i,k} \leq \frac{A_i^k}{k!} + 2^{-N^{1/8}} \qquad \text{for all } i \leq N^{1/2}, 1 \leq k \leq N^{1/8}$$

Proof. If we expand Lemma 5, we have

$$\begin{aligned}
g_{i,k} &\leq g_{i-1,k} + \sqrt{\frac{h_3(i-1)}{N}} g_{i-1,k-1} + f_{i-1,h_3(i-1)} \\
&\leq g_{i-2,k} + \sum_{l=i-2}^{i-1} \left(\sqrt{\frac{h_3(l)}{N}} g_{l,k-1} + f_{l,h_3(l)} \right) \\
&\vdots \\
&\leq g_{0,k} + \sum_{l=0}^{i-1} \left(\sqrt{\frac{h_3(l)}{N}} g_{l,k-1} + f_{l,h_3(l)} \right)
\end{aligned}$$

where if $k \geq 1$, $g_{0,k} = 0$. Next,

$$\begin{aligned}
g_{i,k} &\leq \sum_{0 \leq l < i} \sqrt{\frac{h_3(l)}{N}} g_{l,k-1} + \sum_{0 \leq l < i} f_{l,h_3(l)} \\
&\leq \sum_{0 \leq l < i} \sqrt{\frac{h_3(l)}{N}} g_{l,k-1} + N^{1/2} \left(\frac{1}{2} \right)^{10N^{1/8}} \\
&\leq \sum_{0 \leq l < i} \sqrt{\frac{h_3(l)}{N}} g_{l,k-1} + 2^{-9.5N^{1/8}}
\end{aligned}$$

By recursively expanding the inequality, let $C = 2^{-9.5N^{1/8}}$, we will get

$$\begin{aligned}
g_{i,k} &\leq \sum_{0 \leq l_1 < i} \sqrt{\frac{h_3(l_1)}{N}} g_{l_1,k-1} + C \\
&\leq \sum_{0 \leq l_1 < i} \sqrt{\frac{h_3(l_1)}{N}} \left(\sum_{0 \leq l_2 < l_1} \sqrt{\frac{h_3(l_2)}{N}} g_{l_2,k-2} + C \right) + C \\
&\leq \sum_{0 \leq l_1 < i} \sqrt{\frac{h_3(l_1)}{N}} \left(\sum_{0 \leq l_2 < l_1} \sqrt{\frac{h_3(l_2)}{N}} \left(\sum_{0 \leq l_3 < l_2} \sqrt{\frac{h_3(l_3)}{N}} \cdots \right) + C \right) + C \\
&= \sum_{0 \leq l_k < \cdots < l_1 < i} \prod_{j=1}^{k} \left(\sqrt{\frac{h_3(l_j)}{N}} \right) + C \sum_{t=0}^{k-1} \sum_{0 \leq l_t < \cdots < l_1 < i} \prod_{j=1}^{t} \left(\sqrt{\frac{h_3(l_j)}{N}} \right) \\
&< \frac{A_i^k}{k!} + \sum_{t=0}^{k-1} \frac{A_i^t}{t!} \cdot C \\
&< \frac{A_i^k}{k!} + e^{A_i} \cdot 2^{-9.5N^{1/8}}
\end{aligned}$$

We then bound A_i for all $i \leq N^{1/2}$ (we can always assume $i = o(N^{1/2})$, because finding any constant-collision using $O(N^{1/2})$ quantum queries is easy by a quantum computer, just repeatedly applying Grover's algorithm):

$$A_i \leq \sum_{l=1}^{i} \frac{\sqrt{2e} \cdot l^{3/2}}{N^{3/4}} + \sum_{l:h_3(l)=10N^{1/8}} \frac{\sqrt{10N^{1/8}}}{N^{1/2}}$$
$$\leq \sqrt{2e} \cdot \frac{i^{7/4}}{N^{3/4}} + O\left(N^{-1/48}\right)$$

Which implies $A_i < 2e \cdot N^{1/8}$ (by letting $i = \sqrt{N}$). So we complete the proof:

$$g_{i,k} \leq \frac{A_i^k}{k!} + e^{A_i} \cdot 2^{-9.5N^{1/8}}$$
$$\leq \frac{A_i^k}{k!} + e^{2e \cdot N^{1/8}} \cdot 2^{-9.5N^{1/8}}$$
$$< \frac{A_i^k}{k!} + 2^{-N^{1/8}}$$

□

Theorem 6. *For any quantum algorithm, given a random function $f : X \to Y$ where $|Y| = N$, it needs to make $\Omega(j^{4/7}N^{3/7})$ quantum queries to find j 3-collisions for any $j \leq N^{1/8}$ with constant probability.*

Proof. We have two cases:

- When j is a constant: If $i^* = o(N^{3/7})$, we have $g_{i^*,j} \leq o(1) + O(N^{-1/48})$.
- When j is not a constant: For any j, let i^* be the largest integer such that $A_{i^*} < \frac{1}{2e} \cdot j$. In this case, $i^* = O\left(j^{4/7}N^{3/7}\right)$. So the probability of having at least j 3-collisions is bounded by $g_{i^*,j}^2$ where $g_{i^*,j} \leq (eA_{i^*}/j)^j + 2^{-N^{1/8}} \leq 2^{-j+1} + 2^{-N^{1/8}} = o(1)$.

□

4.5 Lower Bound for 4-Collisions

Here we show the proof for lower bound of finding 4-collisions. The proof for arbitrary constant has the same structure but different parameters which is shown in the next section. We prove the case of 4-collisions here to give the idea before generalizing.

Let $f_{i,j}$ be the amplitude of the database containing at least j 3-collisions after making i quantum queries, $g_{i,j,k}$ be the amplitude of the database containing exactly j 3-collisions and at least k 4-collisions after i quantum queries, $g_{i,k}$ be the amplitude of containing at least k 4-collisions after i quantum queries.

As we have seen in the last proof, we have

$$g_{i,k} \leq g_{i-1,k} + \sqrt{\sum_{l \geq 0} \frac{l}{N} g_{i-1,l,k-1}^2}$$

Define $h_4(i) = \max\{(2e)^{3/2} \cdot \frac{i^{7/4}}{N^{3/4}}, 10N^{1/16}\}$. Again, we can bound $g_{i,k}$ by dividing the summation into two parts:

$$\begin{aligned} g_{i,k} &\le g_{i-1,k} + \sqrt{\sum_{l \le h_4(i-1)} \frac{l}{N} g^2_{i-1,l,k-1}} + \sqrt{\sum_{l > h_4(i-1)} 1 \cdot g^2_{i-1,l,k-1}} \\ &\le g_{i-1,k} + \sqrt{\frac{h_4(i-1)}{N}} g_{i-1,k-1} + f_{i-1,h_4(i-1)} \\ &\vdots \\ &\le \sum_{0 \le l < i} \sqrt{\frac{h_4(l)}{N}} g_{l,k-1} + \sum_{0 \le l < i} f_{l,h_4(l)} \end{aligned}$$

The second term can be bounded as the following (and we can safely assume $i < N^{1/2}$)

$$\begin{aligned} \sum_{0 \le l < i} f_{l,h_4(l)} &\le \sum_{0 \le l < i} \left(\frac{A_l^{h_4(l)}}{h_4(l)!} + 2^{-N^{1/8}} \right) \\ &\le \sum_{0 \le l < i} \left(\frac{eA_l}{h_4(l)} \right)^{h_4(l)} + N^{1/2} \cdot 2^{-N^{1/8}} \\ &\le \sum_{0 \le l < i} \left(\frac{1}{2} + o(1) \right)^{10N^{1/16}} + N^{1/2} \cdot 2^{-N^{1/8}} \\ &\le 2^{-9.5N^{1/16}} \end{aligned}$$

Let $B_i = \sum_{0 \le l < i} \sqrt{\frac{h_4(l)}{N}}$. And similarly, for all $i \le N^{1/2}$,

$$B_i \le (2e)^{3/4} \frac{i^{15/8}}{N^{7/8}} + O(N^{-\frac{1}{16} \cdot \frac{1}{14}})$$

The proof follows from the last proof for $k = 3$. A generalized version (for any constant) can be found in the next section. And B_i is bounded by $B_{\sqrt{N}}$ which is at most $2e \cdot N^{1/16}$.

Finally we have the following closed form:

$$g_{i,k} \le \frac{B_i^k}{k!} + \sum_{l=0}^{k-1} \frac{B_i^l}{l!} \cdot 2^{-9.5N^{1/16}} < \frac{B_i^k}{k!} + e^{B_{\sqrt{N}}} \cdot 2^{-9.5N^{1/16}} \le \frac{B_i^k}{k!} + 2^{-N^{1/16}}$$

So we can conclude the following theorem:

Theorem 7. *For any quantum algorithm, given a random function $f : X \to Y$ where $N = |Y|$, it needs to make $\Omega(j^{8/15} N^{7/15})$ quantum queries to find j 4-collisions for any $j \le N^{1/16}$ with constant probability.*

4.6 Lower Bound for Finding a Constant-Collision

In this section, we are going to show that the theorem can be generalized to any constant-collision. Let $f_{i,j}$ be the amplitude of the database containing at least j distinct s-collisions after i quantum queries, $g_{i,j,k}$ be the amplitude of the database containing exactly j distinct s-collisions and at least k distinct $(s+1)$-collisions after i quantum queries. Also let $g_{i,k}$ be the amplitude of the database with at least k distinct $(s+1)$-collisions after i quantum queries.

We assume $f_{i,j}$ is only defined for $i \leq \sqrt{N}, 1 \leq j \leq N^{1/2^s}$ and $g_{i,k}$ is only defined for $i \leq \sqrt{N}, 1 \leq k \leq N^{1/2^{s+1}}$. It holds for the base cases $s = 4$.

Define $h_s(i)$ (for any $s \geq 3$) as the following:

$$h_s(i) = \max\left\{(2e)^{\frac{2^{s-2}-1}{2^{s-3}}} \frac{i^{(2^{s-1}-1)/2^{s-2}}}{N^{(2^{s-2}-1)/2^{s-2}}}, 10 \cdot N^{1/2^s}\right\}$$

It holds for $s = 3, 4$ where $h_3(i) = \max\{(2e) \cdot i^{3/2}/N^{1/2}, 10N^{1/8}\}$ and $h_4(i) = \max\{(2e)^{3/2} \cdot i^{7/4}/N^{3/4}, 10N^{1/16}\}$.

Define $A_{i,s} = \sum_{l=0}^{i-1} \sqrt{\frac{h_s(l)}{N}}$. It is easy to see A_i and B_i in the last proof are $A_{i,3}$ and $A_{i,4}$. And we have $A_{i,s} \leq (2e)^{\frac{2^{s-2}-1}{2^{s-2}}} \frac{i^{(2^s-1)/2^{s-1}}}{N^{(2^{s-1}-1)/2^{s-1}}} + O(N^{-1/(2^s(2^s-2))})$.

Lemma 7. *$A_{i,s} \leq (2e)^{\frac{2^{s-2}-1}{2^{s-2}}} \frac{i^{(2^s-1)/2^{s-1}}}{N^{(2^{s-1}-1)/2^{s-1}}} + O(N^{-1/(2^s(2^s-2))})$ holds for all constant $s \geq 3$.*

The lemma is consistent with the cases where $s = 3, 4$.

Proof.

$$\begin{aligned}
A_{i,s} &= \sum_{l=0}^{i-1} \sqrt{\frac{h_s(l)}{N}} \\
&= \sum_{l:h_s(l)=10N^{1/2^s}} \sqrt{\frac{10N^{1/2^s}}{N}} + \sum_{l:h_s(l)>10N^{1/2^s}} \sqrt{\frac{h_s(l)}{N}} \\
&= \sum_{l:h_s(l)=10N^{1/2^s}} \sqrt{\frac{10N^{1/2^s}}{N}} + \sum_{l=0}^{i-1} (2e)^{\frac{2^{s-2}-1}{2^{s-2}}} \frac{l^{(2^{s-1}-1)/2^{s-1}}}{N^{(2^{s-2}-1)/2^{s-1}}} \cdot N^{-1/2}
\end{aligned}$$

where the second summation is at most $(2e)^{\frac{2^{s-2}-1}{2^{s-2}}} \frac{i^{(2^s-1)/2^{s-1}}}{N^{(2^{s-1}-1)/2^{s-1}}}$ and the first summation is at most

$$\begin{aligned}
\sum_{l:h_s(l)=10N^{1/2^s}} \sqrt{\frac{10N^{1/2^s}}{N}} &= \sqrt{\frac{10N^{1/2^s}}{N}} \cdot O\left(N^{\left(\frac{1}{2^s}+1-\frac{1}{2^{s-2}}\right)\cdot\frac{2^{s-2}}{2^{s-1}-1}}\right) \\
&\leq O\left(N^{-\frac{1}{2}+\frac{1}{2^{s+1}}} \cdot N^{\frac{2^s-3}{4(2^{s-1}-1)}}\right) \\
&\leq O\left(N^{-\frac{1}{2^s(2^s-2)}}\right)
\end{aligned}$$

which completes the proof. □

Finally, we assume $f_{i,j} \leq \frac{A_{i,s}^j}{j!} + O(2^{-N^{1/2^s}})$ which holds for both $s = 3, 4$. We are going to show it holds for $(s+1)$, in other words, $g_{i,k} \leq \frac{A_{i,s+1}^k}{k!} + O(2^{-N^{1/2^{s+1}}})$. And by induction, it holds for all constant s.

As we have seen in the last proof, we have the following inequality:

$$\begin{aligned} g_{i,k} &\leq g_{i-1,k} + \sqrt{\sum_{l\geq 0} \frac{l}{N} g_{i-1,l,k-1}^2} \\ &\leq g_{i-1,k} + \sqrt{\frac{h_{s+1}(i-1)}{N}} \cdot g_{i-1,k-1} + f_{i-1,h_{s+1}(i-1)} \end{aligned}$$

where as $i \leq N^{1/2}$, for sufficient large N, the last term $f_{i-1,h_{s+1}(i-1)}$ can be bounded as:

$$\begin{aligned} & f_{i-1,h_{s+1}(i-1)} \\ & \leq \frac{A_{i-1,s}^{h_{s+1}(i-1)}}{h_{s+1}(i-1)!} + O(2^{-N^{1/2^s}}) \\ & \leq \left(e \cdot \frac{(2e)^{\frac{2^{s-2}-1}{2^{s-2}}} \frac{(i-1)^{(2^s-1)/2^{s-1}}}{N^{(2^{s-1}-1)/2^{s-1}}} + O(N^{-1/(2^s(2^s-2))})}{\max\left\{ (2e)^{\frac{2^{s-1}-1}{2^{s-2}}} \frac{(i-1)^{(2^s-1)/2^{s-1}}}{N^{(2^{s-1}-1)/2^{s-1}}}, 10 \cdot N^{1/2^{s+1}} \right\}} \right)^{10N^{1/2^{s+1}}} + O(2^{-N^{1/2^s}}) \\ & \leq \left(\frac{1}{2} + o(1) \right)^{10N^{1/2^{s+1}}} + O(2^{-N^{1/2^s}}) \\ & < 2^{-9.8N^{1/2^{s+1}}} \end{aligned}$$

By expanding the inequality, we get

$$\begin{aligned} g_{i,k} &\leq \sum_{l=0}^{i-1} \sqrt{\frac{h_{s+1}(l)}{N}} g_{l,k-1} + N^{1/2} \cdot 2^{-9.8N^{1/2^{s+1}}} \\ &\leq \sum_{l=0}^{i-1} \sqrt{\frac{h_{s+1}(l)}{N}} g_{l,k-1} + 2^{-9.5N^{1/2^{s+1}}} \\ &\leq \frac{A_{i,s+1}^k}{k!} + \sum_{l=0}^{k-1} \frac{A_{i,s+1}^l}{l!} \cdot 2^{-9.5N^{1/2^{s+1}}} \\ &\leq \frac{A_{i,s+1}^k}{k!} + e^{A_{i,s+1}} \cdot 2^{-9.5N^{1/2^{s+1}}} \end{aligned}$$

Because $i \leq \sqrt{N}$, $A_{i,s+1} < 2eN^{1/2^{s+1}}$. Finally, we have

$$g_{i,k} \leq \frac{A_{i,s+1}^k}{k!} + 2^{-N^{1/2^{s+1}}}$$

which completes the induction. So we have the following theorem:

Corollary 2. *For any constant $s \geq 2$, let $f_{i,j}$ be the amplitude of the database containing at least j s-collisions after i quantum queries. For all $1 \leq j \leq N^{1/2^s}$, we have*

$$f_{i,j} \leq \frac{A_{i,s}^j}{j!} + O(2^{-N^{1/2^s}})$$

where

$$A_{i,s} \leq (2e)^{\frac{2^{s-2}-1}{2^{s-2}}} \frac{i^{(2^s-1)/2^{s-1}}}{N^{(2^{s-1}-1)/2^{s-1}}} + O(N^{-1/(2^s(2^s-2))})$$

Theorem 8. *For any quantum algorithm, given a random function $f : X \rightarrow Y$ where $N = |Y|$, it needs to make $\Omega(j^{2^{s-1}/(2^s-1)} N^{(2^{s-1}-1)/(2^s-1)})$ quantum queries to find j s-collisions for any $j \leq N^{1/2^s}$.*

Moreover, for any quantum algorithm, given a random function $f : X \rightarrow Y$ where $N = |Y|$, it needs to make $\Omega(N^{(2^{s-1}-1)/(2^s-1)})$ quantum queries to find one s-collision.

Acknowledgements. This work is supported in part by NSF. Opinions, findings and conclusions or recommendations expressed in this material are those of the author(s) and do not necessarily reflect the views of NSF.

References

[Amb04] Ambainis, A.: Quantum walk algorithm for element distinctness. In: 45th FOCS, pp. 22–31. IEEE Computer Society Press, October 2004

[AS04] Aaronson, S., Shi, Y.: Quantum lower bounds for the collision and the element distinctness problems. J. ACM **51**(4), 595–605 (2004)

[BBBV97] Bennett, C.H., Bernstein, E., Brassard, G., Vazirani, U.: Strengths and weaknesses of quantum computing. SIAM J. Comput. **26**(5), 1510–1523 (1997)

[BBHT98] Boyer, M., Brassard, G., Høyer, P., Tapp, A.: Tight bounds on quantum searching. Fortschritte der Physik: Progress Phys. **46**(4–5), 493–505 (1998)

[BDF+11] Boneh, D., Dagdelen, Ö., Fischlin, M., Lehmann, A., Schaffner, C., Zhandry, M.: Random oracles in a quantum world. In: Lee, D.H., Wang, X. (eds.) ASIACRYPT 2011. LNCS, vol. 7073, pp. 41–69. Springer, Heidelberg (2011). https://doi.org/10.1007/978-3-642-25385-0_3

[BDRV18] Berman, I., Degwekar, A., Rothblum, R.D., Vasudevan, P.N.: Multi-collision resistant hash functions and their applications. In: Nielsen, J.B., Rijmen, V. (eds.) EUROCRYPT 2018. LNCS, vol. 10821, pp. 133–161. Springer, Cham (2018). https://doi.org/10.1007/978-3-319-78375-8_5

[Bel12] Belovs, A.: Learning-graph-based quantum algorithm for k-distinctness. In: 53rd FOCS, pp. 207–216. IEEE Computer Society Press, October 2012

[BES17] Balogh, M., Eaton, E., Song, F.: Quantum collision-finding in non-uniform random functions. Cryptology ePrint Archive, Report 2017/688 (2017). http://eprint.iacr.org/2017/688

[BHT98] Brassard, G., Høyer, P., Tapp, A.: Quantum cryptanalysis of hash and claw-free functions. In: Lucchesi, C.L., Moura, A.V. (eds.) LATIN 1998. LNCS, vol. 1380, pp. 163–169. Springer, Heidelberg (1998). https://doi.org/10.1007/BFb0054319

[BKP18] Bitansky, N., Kalai, Y.T., Paneth, O.: Multi-collision resistance: a paradigm for keyless hash functions. In: Diakonikolas, I., Kempe, D., Henzinger, M. (eds.) 50th ACM STOC, pp. 671–684. ACM Press, June 2018

[BKT18] Bun, M., Kothari, R., Thaler, J.: The polynomial method strikes back: tight quantum query bounds via dual polynomials. In: Diakonikolas, I., Kempe, D., Henzinger, M. (eds.) 50th ACM STOC, pp. 297–310. ACM Press, June 2018

[BS13] Belovs, A., Spalek, R.: Adversary lower bound for the k-sum problem. In: Kleinberg, R.D. (ed.) ITCS 2013, pp. 323–328. ACM, January 2013

[BZ13] Boneh, D., Zhandry, M.: Quantum-secure message authentication codes. In: Johansson, T., Nguyen, P.Q. (eds.) EUROCRYPT 2013. LNCS, vol. 7881, pp. 592–608. Springer, Heidelberg (2013). https://doi.org/10.1007/978-3-642-38348-9_35

[CN08] Chang, D., Nandi, M.: Improved indifferentiability security analysis of chopMD hash function. In: Nyberg, K. (ed.) FSE 2008. LNCS, vol. 5086, pp. 429–443. Springer, Heidelberg (2008). https://doi.org/10.1007/978-3-540-71039-4_27

[DDKS14] Dinur, I., Dunkelman, O., Keller, N., Shamir, A.: Cryptanalysis of iterated Even-Mansour schemes with two keys. In: Sarkar, P., Iwata, T. (eds.) ASIACRYPT 2014. LNCS, vol. 8873, pp. 439–457. Springer, Heidelberg (2014). https://doi.org/10.1007/978-3-662-45611-8_23

[EU17] Ebrahimi, E., Unruh, D.: Quantum collision-resistance of non-uniformly distributed functions: upper and lower bounds. Cryptology ePrint Archive, Report 2017/575 (2017). http://eprint.iacr.org/2017/575

[Gro96] Grover, L.K.: A fast quantum mechanical algorithm for database search. In: 28th ACM STOC, pp. 212–219. ACM Press, May 1996

[HIK+11] Hirose, S., Ideguchi, K., Kuwakado, H., Owada, T., Preneel, B., Yoshida, H.: A lightweight 256-bit hash function for hardware and low-end devices: Lesamnta-LW. In: Rhee, K.-H., Nyang, D.H. (eds.) ICISC 2010. LNCS, vol. 6829, pp. 151–168. Springer, Heidelberg (2011). https://doi.org/10.1007/978-3-642-24209-0_10

[HSTX18] Hosoyamada, A., Sasaki, Y., Tani, S., Xagawa, K.: Improved quantum multicollision-finding algorithm. Cryptology ePrint Archive, Report 2018/1122 (2018). https://eprint.iacr.org/2018/1122

[HSX17] Hosoyamada, A., Sasaki, Y., Xagawa, K.: Quantum multicollision-finding algorithm. In: Takagi, T., Peyrin, T. (eds.) ASIACRYPT 2017. LNCS, vol. 10625, pp. 179–210. Springer, Cham (2017). https://doi.org/10.1007/978-3-319-70697-9_7

[JJV02] Jaulmes, É., Joux, A., Valette, F.: On the security of randomized CBC-MAC beyond the birthday paradox limit a new construction. In: Daemen, J., Rijmen, V. (eds.) FSE 2002. LNCS, vol. 2365, pp. 237–251. Springer, Heidelberg (2002). https://doi.org/10.1007/3-540-45661-9_19

[JLM14] Jovanovic, P., Luykx, A., Mennink, B.: Beyond $2^{c/2}$ security in sponge-based authenticated encryption modes. In: Sarkar, P., Iwata, T. (eds.) ASIACRYPT 2014. LNCS, vol. 8873, pp. 85–104. Springer, Heidelberg (2014). https://doi.org/10.1007/978-3-662-45611-8_5

[KMRT09] Knudsen, L.R., Mendel, F., Rechberger, C., Thomsen, S.S.: Cryptanalysis of MDC-2. In: Joux, A. (ed.) EUROCRYPT 2009. LNCS, vol. 5479, pp. 106–120. Springer, Heidelberg (2009). https://doi.org/10.1007/978-3-642-01001-9_6

[KNY18] Komargodski, I., Naor, M., Yogev, E.: Collision resistant hashing for paranoids: dealing with multiple collisions. In: Nielsen, J.B., Rijmen, V. (eds.) EUROCRYPT 2018. LNCS, vol. 10821, pp. 162–194. Springer, Cham (2018). https://doi.org/10.1007/978-3-319-78375-8_6

[NO14] Naito, Y., Ohta, K.: Improved indifferentiable security analysis of PHOTON. In: Abdalla, M., De Prisco, R. (eds.) SCN 2014. LNCS, vol. 8642, pp. 340–357. Springer, Cham (2014). https://doi.org/10.1007/978-3-319-10879-7_20

[NSWY13] Naito, Y., Sasaki, Y., Wang, L., Yasuda, K.: Generic state-recovery and forgery attacks on ChopMD-MAC and on NMAC/HMAC. In: Sakiyama, K., Terada, M. (eds.) IWSEC 2013. LNCS, vol. 8231, pp. 83–98. Springer, Heidelberg (2013). https://doi.org/10.1007/978-3-642-41383-4_6

[NWW14] Nikolić, I., Wang, L., Wu, S.: Cryptanalysis of round-reduced LED. In: Moriai, S. (ed.) FSE 2013. LNCS, vol. 8424, pp. 112–129. Springer, Heidelberg (2014). https://doi.org/10.1007/978-3-662-43933-3_7

[RS97] Rivest, R.L., Shamir, A.: PayWord and MicroMint: two simple micropayment schemes. In: Lomas, M. (ed.) Security Protocols 1996. LNCS, vol. 1189, pp. 69–87. Springer, Heidelberg (1997). https://doi.org/10.1007/3-540-62494-5_6

[TU16] Targhi, E.E., Unruh, D.: Post-quantum security of the Fujisaki-Okamoto and OAEP transforms. In: Hirt, M., Smith, A. (eds.) TCC 2016. LNCS, vol. 9986, pp. 192–216. Springer, Heidelberg (2016). https://doi.org/10.1007/978-3-662-53644-5_8

[van98] van Dam, W.: Quantum oracle interrogation: getting all information for almost half the price. In: 39th FOCS, pp. 362–367. IEEE Computer Society Press, November 1998

[Zha12] Zhandry, M.: Secure identity-based encryption in the quantum random oracle model. In: Safavi-Naini, R., Canetti, R. (eds.) CRYPTO 2012. LNCS, vol. 7417, pp. 758–775. Springer, Heidelberg (2012). https://doi.org/10.1007/978-3-642-32009-5_44

[Zha15a] Zhandry, M.: A note on the quantum collision and set equality problems. Quantum Inf. Comput. **15**(7&8) (2015)

[Zha15b] Zhandry, M.: Quantum oracle classification-the case of group structure. arXiv preprint arXiv:1510.08352 (2015)

[Zha18] Zhandry, M.: How to record quantum queries, and applications to quantum indifferentiability. Cryptology ePrint Archive, Report 2018/276 (2018). https://eprint.iacr.org/2018/276

On Quantum Advantage in Information Theoretic Single-Server PIR

Dorit Aharonov[1](✉), Zvika Brakerski[2], Kai-Min Chung[3], Ayal Green[1], Ching-Yi Lai[3], and Or Sattath[4]

[1] Hebrew University, Jerusalem, Israel
doria@cs.huji.ac.il
[2] Weizmann Institute of Science, Rehovot, Israel
[3] Academia Sinica, Taipei, Taiwan
[4] Ben-Gurion University, Beersheba, Israel

Abstract. In (single-server) Private Information Retrieval (PIR), a server holds a large database DB of size n, and a client holds an index $i \in [n]$ and wishes to retrieve $\mathtt{DB}[i]$ without revealing i to the server. It is well known that information theoretic privacy even against an "honest but curious" server requires $\Omega(n)$ communication complexity. This is true even if quantum communication is allowed and is due to the ability of such an adversarial server to execute the protocol on a superposition of databases instead of on a specific database ("input purification attack").

Nevertheless, there have been some proposals of protocols that achieve sub-linear communication and appear to provide some notion of privacy. Most notably, a protocol due to Le Gall (ToC 2012) with communication complexity $O(\sqrt{n})$, and a protocol by Kerenidis et al. (QIC 2016) with communication complexity $O(\log(n))$, and $O(n)$ shared entanglement.

We show that, in a sense, input purification is the only potent adversarial strategy, and protocols such as the two protocols above are secure in a restricted variant of the quantum honest but curious (a.k.a specious) model. More explicitly, we propose a restricted privacy notion called *anchored privacy*, where the adversary is forced to execute on a classical database (i.e. the execution is anchored to a classical database). We show that for measurement-free protocols, anchored security against honest adversarial servers implies anchored privacy even against specious adversaries.

Finally, we prove that even with (unlimited) pre-shared entanglement it is impossible to achieve security in the standard specious model with sub-linear communication, thus further substantiating the necessity of our relaxation. This lower bound may be of independent interest (in particular recalling that PIR is a special case of Fully Homomorphic Encryption).

1 Introduction

Private Information Retrieval (PIR), introduced by Chor et al. [CGKS95], is perhaps the most basic form of joint computation with privacy guarantee. PIR

Y. Ishai and V. Rijmen (Eds.): EUROCRYPT 2019, LNCS 11478, pp. 219–246, 2019.
https://doi.org/10.1007/978-3-030-17659-4_8

is concerned with privately retrieving an entry from a database, without revealing which entry has been accessed. Formally, a PIR protocol is a communication protocol between two parties, a server holding a large database DB containing n binary entries[1], and a client who wishes to retrieve the ith element of the database but without revealing the index i. Privacy can be defined using standard cryptographic notions such as indistinguishability or simulation (see [Gol04]). The simplicity of this primitive is since there is no privacy requirement for the database (i.e. we allow sending more information than necessary) and that the server is not required to produce any output in the end of the interaction, so functionality and privacy are *one sided.*

Clearly PIR is achievable by sending all of DB to the client. This will have communication complexity n and will be perfectly private under any plausible definition since the client sends no information. The absolute optimal result one could hope for is a protocol with logarithmic communication, matching the most communication efficient protocol without privacy constraints, in which the client sends the index i to the server and receives DB[i] in response.

Alas, [CGKS95] proved that linear (in n) communication complexity is *necessary* for PIR, and that this is the case even in the presence of arbitrary setup information.[2] Despite its pessimistic outlook, this lower-bound served (already in [CGKS95] itself) as starting point to two extremely prolific and influential lines of research, showing that the communication complexity can be vastly improved if we place some restrictions on the server. The first considered *multiple non-interacting* servers (see, e.g., [Efr12, DG15] and references therein), instead of just a single server, and the second considered *computationally bounded* servers and relying on *cryptographic assumptions* (see, e.g., [CMS99, Gen09, BV11]).

While our discussion so far referred to protocols executed by classical parties over classical communication channels, the focus of this work is on the quantum setting, where there is a quantum communication channel between the client and server, and where the parties themselves are capable of performing quantum operations. Importantly, we still only require functionality for a classical database and a classical index.

One could hope that introducing quantum channels could allow an information theoretic solution to a problem that classically can only be solved using cryptographic assumptions, as has been the case for quantum key distribution [BB84], quantum money [Wie83], quantum digital signatures [GC01], quantum coin-flipping [Moc07, CK09, ACG+16] and more [BS16]. Indeed, the notion of Quantum PIR (or QPIR) is quite a natural extension of its classical counterpart and has also been extensively studied in the literature. Nayak's famous result on the impossibility of random access codes [Nay99] implies a linear lower bound

[1] Throughout this work we will focus on the setting of binary database. We do note that there is vast literature concerned with optimizations for the case of larger alphabet.

[2] Setup refers to any information that is provided to the parties prior to the execution of the protocol by a trusted entity, but crucially one that does not depend on the parties' inputs. Shared randomness or shared entanglement are common examples.

for non-interactive protocols (ones that consists of only a single message from the server to the client), and implicitly, via extension of the same methods, also for multi-round protocols. Formal variants of this lower bound were proven also by Jain, Radhakrishnan and Sen [JRS09] (in terms of quantum mutual information) and by Baumeler and Broadbent [BB15]. Indeed, one could trace back all of these results to the notion of *adversary purification* which was used to show the impossibility of various cryptographic tasks in the information-theoretic quantum model starting as early as [Lo97,LC97,May97]. In the context of QPIR, it can be shown that executing a QPIR protocol with sub-linear communication on a superposition of databases instead of on a single database, will leave the server at the end of the execution with a state that reveals some information about the index i. This is made explicit in [JRS09, Section 3.1] and is also implicit in the proof of [BB15].

Most relevant to our work is the aforementioned [BB15], which provides an analysis from a cryptographic perspective and considers a well defined adversarial model known as privacy against specious adversaries, or *the specious model* for short. This adversarial model was introduced by Dupuis, Nielsen and Salvail [DNS10] as a quantum counterpart to the classical notion of *honest but curious* (a.k.a semi-honest) adversaries.[3] A specious adversary can be thought of as one that contains, as a part of its local state, a sub-state which is indistinguishable from that of the respective honest party, even when inspected jointly with the other party's local state.[4]

Let us provide a high level description of the specious model. We provide a general outline for two-party protocols, and not one that is specific to QPIR. Consider a protocol executed between parties A, B on input registers X, Y respectively. Let A, B also denote the local state of the parties at a given point in time. Then the state of an honest execution of the protocol on inputs XY can be described by the joint density matrix of the registers $XABY$. A specious adversarial strategy for party A can be thought of as one where at any point in time, the local state of the adversary is of the form $A'XA$ (i.e. the adversary is allowed to maintain additional information, possibly in superposition with other parts of the system), such that the reduced density matrix of $XABY$ is still indistinguishable from the one obtained in an honest execution. This provides a potential advantage to a specious adversary (compared to an honest A) since it is quite possible that together with A', the joint state is no longer honest. Thus the local view of the adversary, i.e. the registers $A'XA$, might in fact reveal information about B's input Y that was supposed to have been kept private.

In the QPIR setting, say taking A to be the server and B to be the client, the register X holds the database $\mathtt{DB}$, and Y holds the index i. Indeed, [BB15] shows

[3] As [DNS10] point out, their model is stronger, i.e. excludes a larger class of attacks, compared to the honest but curious model, even when restricted to a completely classical setting.

[4] More accurately, indistinguishability is required to hold even in the presence of an environment which can be arbitrary correlated (or entangled) with the parties' inputs. In the quantum setting this usually corresponds to the environment.

that it is sufficient that A' contains a purification of XA, where X is a uniform distribution over all databases. We call this the *purification attack*. Thus, while the adversary pretends to execute the protocol on a randomly sampled database, it is in fact executed on a superposition of all possible databases at the same time (indeed this is the case since A' contains a purification of X). As explained above, this methodology is not new, but [BB15] analyze and show that no meaningful notion of QPIR can be achieved against this class of adversaries.

While the negative results could leave us pessimistic as to the abilities of quantum techniques to improve the state of the art on single-server PIR, there is some optimism suggested by two works. Le Gall [LG12] proposed a protocol with sub-linear communication (specifically $O(\sqrt{n})$). Kerenidis et al. [KLGR16] proposed two protocols – an explicit one, with $O(\log n)$ communication, which requires linear pre-shared entanglement; and a second protocol, with poly-logarithmic communication (and does not require pre-shared entanglement). In terms of privacy, it is shown that in a perfectly honest execution of the protocol, client's privacy is preserved. It might not be immediately clear how to translate this proof of privacy to the existing security models and reconcile it with the negative results. It is explained in [LG12] that the protocol is not actually secure if the server deviates from the protocol. However, as [BB15] observed, even a specious attacker that purifies the adversary can violate the security of the protocol, and the privacy proof strongly hinges on the honest execution using a classical database.

Challenges. The state of affairs prior to this work, was that (non-trivial) QPIR was proven impossible even against fairly weak adversaries (namely, specious). Nevertheless, it appears that [LG12, KLGR16] achieve some non-trivial privacy guarantee using sub-linear communication. This privacy guarantee appears not to be captured by the existing security model. Lastly, we notice that all existing negative results are proven in a standalone model and did not consider protocols where the parties are allowed to share (honestly generated) setup information, such as the one by Kerenidis et al. [KLGR16]. In the quantum setting, a natural question is whether shared entanglement can help in achieving a stronger result.[5] The goal of this work is to address these challenges.

1.1 Our Results

Anchored Privacy. We start by formalizing a refinement of the standard notion of quantum privacy - one where the adversary is not allowed to purify its input register. We show anchored privacy against specious adversaries follows from anchored privacy against an honest party, if the protocol itself does not require

[5] We note that to the best of our understanding, even prior "entropic" results such as [JRS09] seem to fall short of capturing the potential additional power of shared entanglement. This is essentially due to the property that if AB are entangled, then it is possible that the reduced state of B will have (much) higher von Neumann entropy than the joint AB (whose entropy might even be 0).

parties to perform measurements (i.e. is measurement-free). Formally, using our notation from above, privacy in our model is only required to hold if the reduced density matrix of the register X is a standard basis element, i.e. a fixed classical value. We call our model *anchored* privacy as we can view our adversary as anchored to a specific value for its input X.

We observe that Le Gall's $O(\sqrt{n})$ protocol [LG12] and the two protocols mentioned above by Kerenidis et al. [KLGR16] are in fact private against honest servers. We prove that explicitly for the pre-shared entanglement protocol by Kerinidis et al. in Appendix B. Using our reduction we can deduce that these protocol are also anchored private against specious adversaries, namely that so long as the adversary does not attempt to execute the protocol on a superposition of databases (and is still specious in the manner explained above), privacy is guaranteed. In a sense, we formalize the folklore reliance on input purification to attack cryptographic schemes (and QPIR in particular), and show that in a model where input purification is impossible or prevented via some external restriction, it is possible to achieve security against specious adversaries.

We believe this model is interesting for three main reasons:

1. Conceptually, this model helps clarify the exact reason for the impossibility of QPIR - it is precisely because of the purification attack. Indeed, there is a formal sense in which some anchoring is necessary since we know that for any proposed protocol, allowing to execute on a superposition of inputs allows to violate security – see the preceding discussion in Sect. 1.
2. We view the anchored specious model as a stepping stone towards more robust notions. One intriguing future direction (mentioned briefly in our list of open problems) is to try to develop a malicious analog that still implements the ideology of "forbidden input purification", e.g. by forcing the adversary to "classically open" the database before or after the execution in a manner that is consistent with the client's output. Another interesting direction is to try to enforce anchoring using a two-server setting, thus achieving logarithmic two-server QPIR (which is currently still beyond reach).
3. We believe that our new model may be plausible in certain situations where one could certify that the server cannot employ a superposition on databases. We note that this model can be externally enforced, e.g. by conducting an inspection of the server's local computation device (with a very low probability) and making sure that it complies, and otherwise apply a heavy penalty. One could imagine such an inspection verifying that a copy of the database is stored on a macroscopic device that cannot be placed in superposition using available technology. Another example of a setting where the anchored model could be applicable is when the database contains information with some semantic meaning, so that the client can easily notice when a nonsense value has been used (this is somewhat similar to the setting considered in [GLM08]). We recall that semi-honest protocols are often used as building blocks, with additional external mechanisms that are employed to validate the assumptions of the model, and hope that our model can also be used in this way. Lastly, from a purely scientific perspective, we believe that formalizing and

pinpointing a non-trivial model where non-trivial QPIR is possible will allow to better understand this primitive and the relation between quantum privacy and its classical counterpart.

Improved Lower Bound. It would be instrumental to understand why the known QPIR lower bounds do not apply to our logarithmic protocol described above. Specifically, the protocol makes use of setup (pre-shared entanglement), and one could wonder whether this is the source of improvement, and perhaps with pre-shared entanglement it is possible to prove security even in the standard specious model. We show that this is not the case by providing a lower bound in the specious model even for the *one-sided communication* from the server to the client. Namely, we show that linear communication from the server to the client is necessary even if we allow arbitrary communication from the client to the server. In particular, this rules out the ability to use the setup to circumvent the lower bound, since the client (which is assumed to be honest) can generate the setup locally, and send the server's share across the channel at the beginning of the protocol. This completes the picture in terms of the impossibility of QPIR in the specious model and further justifies our relaxation of the model in order to achieve meaningful results.

Noting that PIR can be thought of as a special case of Fully Homomorphic Encryption (FHE), our lower bound implies that even a Quantum Fully Homomorphic Encryption (QFHE) with (even approximate) information theoretic security cannot have non-trivial communication complexity, even if the QFHE protocol is allowed to make use of shared entanglement between the server and the client. We thus generalize (to allow shared prior entanglement) the impossibility results for (even imperfect) QPIR of [BB15] (as well as those of [YPF14] which explicitly referred to QFHE).

1.2 Overview of Our Techniques

Anchored-Specious Security. Recall the notation introduced above for two party protocol (A, B) on inputs (X, Y), and recall that a specious adversary can be thought of as one where the local state of the adversary is of the form $A'XA$. Now let us consider the case of measurement-free protocols and also assume that the client's input Y is a pure state (this can be justified since otherwise we can apply our argument on the joint state of Y and its purifying environment instead of Y itself). In such an execution, it holds that at any stage $XABY$ is a pure superposition (i.e. its density matrix is of rank 1). Now let us consider the joint state together with the specious adversary's additional register, i.e. $A'XABY$. Since $(XABY)$ is pure, A' cannot be entangled with it, and therefore A' is in tensor product with the remainder of the state, namely $(XABY)$. It follows that the status of the register A' can be simulated at any point in time without any knowledge of the other components of the protocol. There is a delicate point here, since A' may indeed be in tensor product, but we must also argue that it is independent of Y. Intuitively, to see why such dependence on Y cannot occur consider, e.g., $Y = |y_1\rangle + |y_2\rangle$. Then YA' is in the sate $|y_1\rangle \otimes \rho_{A'} + |y_2\rangle \otimes \rho_{A'}$ (importantly the same $\rho_{A'}$ appears twice). However, this state is exactly the purification of executing the protocol either with $Y = |y_1\rangle$ or with $Y = |y_2\rangle$. We conclude that $\rho_{A'}$ must be the same in both settings, and by extension it can be shown to be the same for all Y.

After taking care of A', we need to consider the other part of the adversary's state, namely the register (XA). This register is, by definition, identical (or indistinguishable) from the state of an honest party during the execution. Recall that we assume our protocol is anchored private against honest servers. So the local honest state (XA) is guaranteed not to leak information about B's input. Add to that the conclusion about A' being in tensor product and independent of B's state, and we get that the entire local state of the specious adversary does not reveal any disallowed information.

As a conclusion, since we can show, e.g. in Le Gall's protocol or in our logarithmic protocol, that an honest execution with a classical X does not leak information about Y, this will also be the case in the anchored-specious setting.

Obviously many details are omitted from this high level overview. For example, a specious adversary is not required to make $(XABY)$ identical to an honest execution but rather only statistically close (in trace distance), which requires a more delicate analysis. Furthermore, the formal construction of a simulator for the adversary as required by the specious definition requires some care to detail. For the formal definitions and analysis see Sect. 3 below.

Our Lower Bound. We first note that previous lower bound proofs in [Nay99, BB15] bounded the *total* communication complexity by a reduction to quantum random access codes. It is not a-priori clear how to generalize this proof method to the presence of shared entanglement. To do so, we provide a new lower bound argument that establishes a linear lower bound on the *server's* communication complexity. Specifically, we show that the server needs to transmit at least roughly $n/2$ qubits to the client, no matter how many qubits is transmitted from the client to the server (assuming that the protocol has sufficiently small correctness and privacy error). As we mentioned above, such a lower bound trivially extends to hold with prior shared entanglement, since one can think of that the shared entanglement is established by the client sending messages to the server.

Our new lower bound argument is based on an interactive leakage chain rule in [LC18] and might even be considered conceptually simpler than previous methods. At a high level, we consider a server holding a uniformly random database $\mathbf{a} \in \{0,1\}^n$ and running a QPIR protocol with a client. Initially, from the client's point of view, the database $\mathbf{a}$ has n-bits of min-entropy, and the protocol execution can be viewed as an "interactive leakage" that leaks information about $\mathbf{a}$ to the client. Let m_A and m_B denote the server and the client's communication complexity in the protocol. The interactive leakage chain rule in [LC18] states that the min-entropy of $\mathbf{a}$ can only be decreased by at most $\min\{2m_A, m_A + m_B\}$. More precisely, let ρ_{AB} denote the states at the end of the protocol execution where the A register stores the (classical) random database $\mathbf{a}$ and B denotes the client's local register. The interactive leakage chain rule states that $H_{\min}(A|B)_\rho \geq n - \min\{2m_A, m_A + m_B\}$. By the operational meaning of quantum min-entropy, given the client's state ρ_B, one cannot predict the database correctly with probability higher than $2^{-(n-\min\{2m_A,m_A+m_B\})}$. On the other hand, suppose the protocol is secure against specious servers with sufficiently small correctness and privacy error. We can combine the by-now standard lower bound argument by Lo [Lo97] and gentle measurement [Win99, Aar04, ON07], we show that one can reconstruct the database $\mathbf{a}$ from the client's state ρ_B with a constant probability. Combining both claims allows us to establish lower bounds on both the server's and the total communication complexity in a unified way.

1.3 Remaining Open Problems

We proposed a new model and a new protocol which, we believe, resurfaces the question of what can be achieved in the context of QPIR. We believe that a number of intriguing questions still remain for future work.

1. As discussed above, our model is a relaxation of the specious model, which is by itself a semi-honest model. Such models are fairly restrictive in the sense that they make structural assumptions on the adversary (i.e. that it follows the protocol, or contains a part that follows the protocol). Obviously, if we hope for non-trivial results, any model that we formalize must preclude purification of input. It is thus an intriguing question whether it is possible to formulate *malicious* adversarial models that are still purification-free, and what can be said about the plausibility of QPIR in such models. The current definition of anchored privacy will need to be amended, since a malicious server is allowed to just ignore its prescribed input, so a different method of anchoring needs to be devised.
2. Another natural question is whether setup is necessary to achieve logarithmic QPIR in the anchored specious model. We know from Kerenidis et al.'s result that polylogarithmic communication is achievable even without setup. Is there a reason that one can only improve it when assuming a setup? Another surprising aspect is that the shared entanglement created during the setup is not consumed during the protocol, and can be used for other needs after the execution of the protocol (e.g., running another execution of PIR, or teleportation). A similar phenomenon occurs in quantum information: catalyst quantum states are useful for mapping one bi-partite state to another using LOCC, without consuming the catalyst state [JP99, Kli07]. The related notion of quantum embezzlement [vDH03] has a similar property, but in this case, the original shared state changes slightly. The authors are not aware of any other cryptographic protocol with this non-consumption property.
3. Most state of the art classical PIR protocols (both in the multi-server setting and in the computational cryptographic setting) only require one round of communication. That is, one message (query) from the client to the server (or servers) and one response message. All the existing sublinear QPIR protocols

have multiple rounds. Understanding the round complexity of QPIR in light of the classical state of the art is also an intriguing direction.

4. A main contribution of this work is to formalize the notion of anchored security and show it can be used to provide a non-trivial cryptographic primitive. It would be interesting to study the relevance of this notion (or adequately adapted versions) in the context of a variety of other cryptographic tasks. In particular, the question of whether it is possible to construct information theoretically secure fully homomorphic encryption (FHE) given quantum channels has received attention in recent years (see, e.g., [YPF14]). In homomorphic encryption, the server has a function f and the client has an input x, and the goal of the protocol is for the client to learn $f(x)$ without revealing any information about x. PIR and FHE functionalities are intimately related (think about a function $f_{\mathtt{DB}}(i) = \mathtt{DB}[i]$ for FHE, and about executing PIR with database equal to the truth table of some function), and it is thus intriguing whether the anchored model is applicable in the context of FHE as well.

1.4 Paper Organization

General preliminaries are provided in Sect. 2. We present our new model, and the proof that for pure protocols honest security implies anchored specious security in Sect. 3. Our new lower bound is stated and proven in Sect. 4. In Appendix B, we show that the protocol by Kerenidis et al. is anchored private against specious adversaries.

This work is also available on the arXiv eprint [ABC+19].

2 Preliminaries

Standard preliminaries regarding Hilbert spaces and quantum states can be found in Appendix A. We provide below background and definitions concerning two-party quantum protocols, specious adversaries and quantum private information retrieval.

2.1 Two-Party Quantum Protocols

As in [BB15], we base our definitions on the works of [GW07] and [DNS10]. However, we make slight adaptations to allow for prior entanglement between the parties.

Definition 2.1 (Two-party quantum protocol). *An s-round, two-party quantum protocol, denoted $\Pi = \{\mathscr{A}, \mathscr{B}, \rho_{joint}, s\}$ consists of:*

1. *input spaces $\mathcal{A}_0$ and $\mathcal{B}_0$ for parties $\mathscr{A}$ and $\mathscr{B}$ respectively,*
2. *initial spaces $\mathcal{A}_p$ and $\mathcal{B}_p$ (p for pre-shared state) for parties $\mathscr{A}$ and $\mathscr{B}$ respectively,*
3. *a joint initial state $\rho_{joint} \in \mathcal{A}_p \otimes \mathcal{B}_p$, split between the two parties,*

4. *memory spaces* $\mathcal{A}_1, \ldots, \mathcal{A}_s$ *for* $\mathscr{A}$ *and* $\mathcal{B}_1, \ldots, \mathcal{B}_s$ *for* $\mathscr{B}$, *and communication spaces* $\mathcal{X}_1, \ldots, \mathcal{X}_s$, $\mathcal{Y}_1, \ldots, \mathcal{Y}_{s-1}$,
5. *an s-tuple of quantum operations* $(\mathscr{A}_1, \ldots, \mathscr{A}_s)$ *for* $\mathscr{A}$, *where* $\mathscr{A}_1 : L(\mathcal{A}_0 \otimes \mathcal{A}_p) \mapsto L(\mathcal{A}_1 \otimes \mathcal{X}_1)$, *and* $\mathscr{A}_t : L(\mathcal{A}_{t-1} \otimes \mathcal{Y}_{t-1}) \mapsto L(\mathcal{A}_t \otimes \mathcal{X}_t)$ $(2 \le t \le s)$,
6. *an s-tuple of quantum operations* $(\mathscr{B}_1, \ldots, \mathscr{B}_s)$ *for* $\mathscr{B}$, *where* $\mathscr{B}_1 : L(\mathcal{B}_0 \otimes \mathcal{B}_p \otimes \mathcal{X}_1) \mapsto L(\mathcal{B}_1 \otimes \mathcal{Y}_1)$, $\mathscr{B}_t : L(\mathcal{B}_{t-1} \otimes \mathcal{X}_t) \mapsto L(\mathcal{B}_t \otimes \mathcal{Y}_t)$ $(2 \le t \le s-1)$, *and* $\mathscr{B}_s : L(\mathcal{B}_{s-1} \otimes \mathcal{X}_s) \mapsto L(\mathcal{B}_s)$.

Note that in order to execute a protocol as defined above, one has to specify the input, namely a quantum state $\rho_{in} \in S(\mathcal{A}_0 \otimes \mathcal{B}_0)$ from which the execution starts.

Definition 2.2 (Protocol Execution). *If* $\Pi = \{\mathscr{A}, \mathscr{B}, \rho_{joint}, s\}$ *is an s-round two-party protocol, then the state after the t-th step* $(1 \le t \le 2s)$, *and upon input state* $\rho_{in} \in S(\mathcal{A}_0 \otimes \mathcal{B}_0 \otimes \mathcal{R})$, *for any* $\mathcal{R}$, *is defined as*

$$\rho_t(\rho_{in}) := (\mathscr{A}_{(t+1)/2} \otimes I_{\mathcal{B}_{(t-1)/2}}) \ldots (\mathscr{B}_1 \otimes I_{\mathcal{A}_1})(\mathscr{A}_1 \otimes I_{\mathcal{B}_0, \mathcal{B}_p})(\rho_{in} \otimes \rho_{joint}),$$

for t odd, and

$$\rho_t(\rho_{in}) := (\mathscr{B}_{t/2} \otimes I_{\mathcal{A}_{t/2}}) \ldots (\mathscr{B}_1 \otimes I_{\mathcal{A}_1})(\mathscr{A}_1 \otimes I_{\mathcal{B}_0, \mathcal{B}_p})(\rho_{in} \otimes \rho_{joint}),$$

for t even. We define the final state of protocol $\Pi = \{\mathscr{A}, \mathscr{B}, \rho_{joint}, s\}$ *upon input state* $\rho_{in} \in S(\mathcal{A}_0 \otimes \mathcal{B}_0 \otimes \mathcal{R})$ *as:* $[\mathscr{A} \circledast \mathscr{B}](\rho_{in}) := \rho_{2s}(\rho_{in})$.

The communication complexity of a protocol is the number of qubits that are exchanged between the parties. Slightly more generally, we can consider the logarithm of the dimension of the message registers $\mathcal{X}_t$, $\mathcal{Y}_t$. The formal definition thus follows.

Definition 2.3 (Communication Complexity). *The* communication complexity *of a protocol as in Definition 2.1 is*

$$\sum_{t=1}^{s} \log \dim(\mathcal{X}_t) + \sum_{t=1}^{s-1} \log \dim(\mathcal{Y}_t).$$

We sometimes also refer to one-sided communication complexity, i.e. the total communication originating from one party to the other. The communication complexity of $\mathscr{A}$ *is defined to be the communication originating from* $\mathscr{A}$, *or formally* $\sum_{t=1}^{s} \log \dim(\mathcal{X}_t)$. *Symmetrically the communication complexity of* $\mathscr{B}$ *is* $\sum_{t=1}^{s-1} \log \dim(\mathcal{Y}_t)$.

2.2 Specious Adversary

Given a two-party quantum protocol $\Pi = \{\mathscr{A}, \mathscr{B}, \rho_{joint}, s\}$, an adversary $\tilde{\mathscr{A}}$ for $\mathscr{A}$ is an s-tuple of quantum operations $(\tilde{\mathscr{A}}_1, \ldots, \tilde{\mathscr{A}}_s)$, where $\tilde{\mathscr{A}}_1 : L(\tilde{\mathcal{A}}_0) \mapsto L(\mathcal{A}_1 \otimes \mathcal{X}_1)$ and $\tilde{\mathscr{A}}_t : L(\tilde{\mathcal{A}}_{t-1} \otimes \mathcal{Y}_{t-1}) \mapsto L(\tilde{\mathcal{A}}_t \otimes \mathcal{X}_t)$, $2 \le t \le s$, with $\tilde{\mathcal{A}}_1, \ldots, \tilde{\mathcal{A}}_s$ being $\tilde{\mathscr{A}}$'s memory spaces. The global state after the tth step of a protocol run with $\tilde{\mathscr{A}}$ is $\tilde{\rho}_t(\tilde{\mathscr{A}}, \rho_{in})$. An adversary $\tilde{\mathscr{B}}$ for $\mathscr{B}$ is similarly defined.

Definition 2.4 (Specious adversaries). *Let $\Pi = \{\mathscr{A}, \mathscr{B}, \rho_{joint}, s\}$ be an s-round two-party protocol. An adversary $\tilde{\mathscr{A}}$ for $\mathscr{A}$ is said to be γ-specious, if there exists a sequence of quantum operations (called recovery operators) $\mathscr{F}_1, \ldots, \mathscr{F}_{2s}$, such that for $1 \leq t \leq 2s$ and for all $\rho_{in} \in S(\mathcal{A}_0 \otimes \mathcal{B}_0 \otimes \mathcal{R})$:*

1. *For all t even, $\mathscr{F}_t : L(\tilde{\mathcal{A}}_{t/2}) \mapsto L(\mathcal{A}_{t/2})$.*
2. *For all t odd, $\mathscr{F}_t : L(\tilde{\mathcal{A}}_{(t+1)/2} \otimes \mathcal{X}_{(t+1)/2}) \mapsto L(\mathcal{A}_{(t+1)/2} \otimes \mathcal{X}_{(t+1)/2})$.*
3. *For every input state $\rho_{in} \in S(\mathcal{A}_0 \otimes \mathcal{B}_0 \otimes \mathcal{R})$, for any $\mathcal{R}$,*

$$\Delta\left((\mathscr{F}_t \otimes I_{\mathcal{B}_t,\mathcal{R}})\left(\tilde{\rho}_t(\tilde{\mathscr{A}}, \rho_{in})\right), \rho_t\left(\rho_{in}\right)\right) \leq \gamma. \tag{1}$$

A γ-specious adversary $\tilde{\mathscr{B}}$ for $\mathscr{B}$ is similarly defined.

2.3 Quantum Private Information Retrieval

We define QPIR similarly to [BB15].

Definition 2.5 (Quantum Private Information Retrieval). *An s-round, n-bit Quantum Private Information Retrieval protocol (QPIR) is a two-party protocol $\Pi_{QPIR} = \{\mathscr{A}, \mathscr{B}, \rho_{joint}, s\}$, where $\mathscr{A}$ is the server, $\mathscr{B}$ is the client, and ρ_{joint} is an initial state shared between them prior to the protocol. We call Π_{QPIR} $(1-\delta)$-correct if, for all inputs $\rho_{in} = |x\rangle\langle x|_{\mathcal{A}_0} \otimes |i\rangle\langle i|_{\mathcal{B}_0}$, with $x = x_1, \ldots, x_n \in \{0,1\}^n$ and $i \in \{1, \ldots, n\}$, there exists a measurement $\mathcal{M}$ acting on $\mathcal{B}_s$ with outcome 0 or 1, such that:*

$$\Pr\left\{\mathcal{M}\left(tr_{\mathcal{A}_s}\left[\mathscr{A} \circledast \mathscr{B}\right](\rho_{in})\right) = x_i\right\} \geq 1 - \delta.$$

If $\delta = 0$ we say that the protocol is perfectly correct.

We call Π_{QPIR} ϵ-private against a (possibly adversarial) server $\tilde{\mathscr{A}}$, if there exists a sequence of quantum operations (simulators) $\mathscr{S}_1, \ldots, \mathscr{S}_{s-1}$, where $\mathscr{S}_t : L(\mathcal{A}_0 \otimes \mathcal{A}_p) \mapsto L(\tilde{\mathcal{A}}_t \otimes \mathcal{Y}_t)$, such that for all $1 \leq t \leq s-1$ and for all $\rho_{in} \in S(\mathcal{A}_0 \otimes \mathcal{B}_0 \otimes \mathcal{R})$,

$$\Delta\left(tr_{\mathcal{B}_0}\left(\mathscr{S}_t \otimes \mathbb{I}_{\mathcal{B}_0,\mathcal{R}}(\rho_{in})\right), tr_{\mathcal{B}_t}(\tilde{\rho}_{2t}(\tilde{\mathscr{A}}, \rho_{in}))\right) \leq \epsilon. \tag{2}$$

If $\epsilon = 0$ we say that the protocol is perfectly private.

We say that a QPIR protocol is ϵ-private against a class of servers if it is ϵ-private against any server from this class.

We note that in the above definition privacy is required to hold also for adversarial input states for the client and server, which also includes inputs in superposition, and even for the case where the client and server (and possibly a third party) are entangled. Nayak [Nay99, ANTSV02] showed that a perfectly private QPIR protocol, even only against 0-specious servers, must have communication complexity at least $(1 - H(1-\delta))n$, where $H(p)$ is the binary entropy function. Baumeler and Broadbent [BB15] extended this lower bound to the case of $\epsilon > 0$ and presented a communication lower bound of

$$\left(1 - H\left(1 - \delta - 2\sqrt{\epsilon(2-\epsilon)}\right)\right) n. \tag{3}$$

3 Anchored Privacy Against Specious Adversaries

We now present our new restricted notion of privacy, that we call *anchored privacy.* A protocol is anchored private if it satisfies the standard definition of privacy with respect to classical inputs on the adversary's side. There is no privacy requirement for superposition input states on the adversary's side (and therefore this notion of privacy is weaker, and hence, easier to achieve). A formal definition follows.

Definition 3.1 (Anchored Privacy). *A QPIR protocol is anchored ϵ-private if Eq.* (2) *holds for all $\rho_{in} \in \mathcal{A}_0 \otimes \mathcal{B}_0 \otimes \mathcal{R}$ (for any $\mathcal{R}$), for which $\rho_{in}|_{\mathcal{A}_0} = |x\rangle\langle x|$ for some $x \in \{0,1\}^n$.*

We note that prior intuitive notions of security such as that implied by the analysis of Le Gall [LG12] in fact correspond to anchored privacy against honest servers. Our main theorem below shows that this type of privacy extends to the specious setting as well.

Theorem 3.2. *Let Π be a measurement-free QPIR protocol which is anchored ϵ-private against honest servers, then Π is anchored $(\epsilon + 3\sqrt{2\gamma})$-private against γ-specious servers.*

Critically, the theorem only holds for measurement-free QPIR protocols. To see this, consider the following protocol, which will be anchored-private against honest servers but not anchored-private against specious ones. Let Π be a QPIR protocol which is anchored-private against honest servers (e.g., Le-Gall's protocol [LG12]). Now consider the following protocol Π' which first generates a superposition over all possible databases, then measures this superposition to obtain a classical value for the database. It then runs Π on this measured database (with the client using its real input index). Finally, both parties toss out the output of this first execution, and run Π again, now using the actual input database.

Let us first see that Π' is anchored-private against honest servers. This follows since Π is secure against honest adversaries when executed over input states in which the server's input is classical, and hence so is Π' which just consists of two sequential executions of Π over classical databases. However, a purification of an honest server allows to execute a purification attack on the first execution of Π in a way that allows to recover the client's input, even though the database used as input for Π' is completely classical.

Warm-Up. We first give a proof under some simplifying assumptions: (i) $\gamma = \epsilon = 0$. (ii) the input is pure (iii) the purification space is trivial: $\mathcal{R} = \mathbb{C}$ and (iv) the specious server's quantum operations $\tilde{\mathscr{A}}_t$ are unitary. The main point that makes the analysis easier in this case is assumption (i).

Fix a step of the protocol t.

1. We claim that for every unitary γ-specious adversary, which is perfect (i.e. $\gamma = 0$) the entire state, (written in some *fixed* but maybe non standard basis), is of the form $|\eta\rangle_{S'} \otimes |\psi_t\rangle_{S,C}$ where $|\psi_t\rangle$ is the state that an honest server and

client would have when running on the same input. Here, and later, we use the notation $\mathcal{S}$ for all of the honest server registers at step t, $\mathcal{C}$ for all of the client's registers at step t and $\mathcal{S}'$ for the specious server's ancillary register at step t. Crucially, $|\eta\rangle$ is independent of the (server and client) input.

We now prove the above claim. By the specious property, we know that there exists a quantum operation $\mathscr{F}_t$ which maps the global state at the tth stage to the state $|\psi_t\rangle$. We know that the state in step t in the honest run is necessarily pure since Π is measurement free. W.l.o.g. we can assume that the operation $\mathscr{F}_t$ is a unitary U_t, followed by tracing out everything other then the $\mathcal{S}$ and $\mathcal{C}$ registers.

Let's assume towards contradiction that the state in the basis $U_t^\dagger$ is of the form $|\eta(input)\rangle \otimes |\psi_t\rangle$, where $|\eta(input)\rangle$ depends on the input (where here we mean both the client and the server's input). There must be two different input states such that running them would give $|\eta(1)\rangle \otimes |\psi_t(1)\rangle$ and $|\eta(2)\rangle \otimes |\psi_t(2)\rangle$ for which $|\eta(1)\rangle \neq |\eta(2)\rangle$. Since the honest runs are entirely unitary (by the measurement-free property) and have different inputs, necessarily, $|\psi_t(1)\rangle \neq |\psi_t(2)\rangle$. By running the specious adversary on a superposition of these two inputs, we get that after applying $\mathscr{F}_t$, the state becomes a mixture of the two states, $|\psi_t(1)\rangle$ and $|\psi_t(2)\rangle$. This contradicts the perfect specious property (see Eq. (1)) – which requires the state to be the pure (since all the operations of the client and honest servers are unitary, and their input in this case is pure).

2. By the perfect anchored-privacy against the honest server, the state $\rho_t = \mathrm{tr}_{\mathcal{C}}(|\psi_t\rangle\langle\psi_t|_{\mathcal{S},\mathcal{C}})$ is independent of the client's input, and therefore, could only depend on x – the server's input. To emphasize that independence on the client's input (and possible dependence on the server's input), we denote the state ρ_t by $\rho_t(x)$.

Our goal is to show the anchored-privacy property for the specious server. Indeed, the two points above show that the specious server's state (in the fixed basis we choose to work in) is $|\eta\rangle\langle\eta| \otimes \rho_t(x)$, which is independent of the client's input. Therefore the simulator can generate that state exactly by using the server's classical input x, as required (see Eq. (2)).

Outline of the General Proof. For each round t we construct a simulator for the server in the following way: we first construct a simulator $\tilde{\mathscr{I}}_t^{x_0,0}$ for input $|x_0\rangle \otimes |0\rangle$ where $|x_0\rangle$ is an input for the server and $|0\rangle$ is an input for the client. We construct this simulator using the simulator for the honest server along with the 'specious operator', and an ancillary state $|\sigma_{x_0,0}\rangle$. We then show that $|\sigma_{x_0,0}\rangle$ is also an appropriate ancillary state for any input $|x\rangle \otimes |\eta\rangle$. Using this, we show that $\tilde{\mathscr{I}}_t^{x,0}$ is indeed a simulator for any input $|x\rangle \otimes |\eta\rangle$, with slightly worse parameters.

We are now ready to give the proof in full generality:

Proof (Theorem 3.2 *(Proof)).* Let Π be a purified QPIR protocol which is anchored ϵ-private against honest servers, and let $\tilde{\mathscr{A}}$ be a γ-specious server for

Π. W.l.o.g we can assume that $\tilde{\mathscr{A}}$ is purified, namely, a unitary[6]. From now on, we will fix t. We can denote

$$|\psi_t^{\rho_{in}}\rangle\langle\psi_t^{\rho_{in}}| = \rho_t(\rho_{in}) \tag{4}$$

where $|\psi_t^{\rho_{in}}\rangle \in \mathcal{S} \otimes \mathcal{C} \otimes \mathcal{R}$ for some $\mathcal{R}$, and we use $\mathcal{S}$ to represent the server's registers $\mathcal{S} = \mathcal{A}_t \otimes \mathcal{Y}_t \otimes \mathcal{A}_p$ (for t odd. otherwise $\mathcal{S} = \mathcal{A}_t \otimes \mathcal{A}_p$), and $\mathcal{C}$ to represent the client's registers $\mathcal{C} = \mathcal{B}_t \otimes \mathcal{X}_t \otimes \mathcal{B}_p$ (for t even. otherwise $\mathcal{C} = \mathcal{B}_t \otimes \mathcal{B}_p$). Furthermore, w.l.o.g we assume the various recovery operators for $\tilde{\mathscr{A}}$ are purified. That is, there exist unitary operators $\hat{\mathscr{F}}_t$ such that $\mathscr{F}_t(\cdot) = \mathrm{tr}_{\mathcal{S}'}\left(\hat{\mathscr{F}}_t(\cdot)\right)$ for some purification space $\mathcal{S}'$ which is at the hands of the server (from now on, for the sake of this proof, where we say "recovery operators" we regard these unitary $\hat{\mathscr{F}}_t$ operators). Therefore we can denote

$$|\tilde{\psi}_t^{\rho_{in}}\rangle\langle\tilde{\psi}_t^{\rho_{in}}| = \tilde{\rho}_t\left(\tilde{\mathscr{A}}, \rho_{in}\right) \tag{5}$$

where w.l.o.g $|\tilde{\psi}_t^{\rho_{in}}\rangle \in S' \otimes S \otimes C \otimes R$. We note that all of the unitary operators - $\mathscr{A}_t, \mathscr{B}_t$ which are used in the original protocol (by either the server or the client), $\tilde{\mathscr{A}}_t$ which are used by the specious server $\tilde{\mathscr{A}}$, and the recovery $\hat{\mathscr{F}}_t$ operators are independent of both the client's and the server's inputs.

For each round t, we will start by constructing a simulator for $\tilde{\mathscr{A}}$ acting on $\rho_{in} = |x_0\rangle\langle x_0|_{\mathcal{A}_0} \otimes |0\rangle\langle 0|_{\mathcal{B}_0}$, where $x_0 \in \{0,1\}^n$ (in this specific input, $\mathcal{R}$ is trivial and is thus omitted). By γ-speciousness of $\tilde{\mathscr{A}}$, along with our purification assumptions, there exists a unitary recovery operator $\hat{\mathscr{F}}_{2t} : L(\tilde{A}_t) \mapsto L(\mathcal{S}' \otimes \mathcal{A}_t)$ such that

$$\Delta\left(\mathrm{tr}_{\mathcal{S}'}\left(\left(\hat{\mathscr{F}}_{2t} \otimes \mathbb{I}_{\mathcal{C}}\right)|\tilde{\psi}_{2t}^{|x_0\rangle\otimes|0\rangle}\rangle\right), |\psi_{2t}^{|x_0\rangle\otimes|0\rangle}\rangle\right) \leq \gamma \tag{6}$$

By Lemma A.1, this means that there exists a state $|\sigma_{x_0,0}\rangle \in \mathcal{S}'$ such that:

$$\Delta\left(\left(\hat{\mathscr{F}}_{2t} \otimes \mathbb{I}\right)|\tilde{\psi}_{2t}^{|x_0\rangle\otimes|0\rangle}\rangle, |\sigma_{x_0,0}\rangle \otimes |\psi_{2t}^{|x_0\rangle\otimes|0\rangle}\rangle\right) \leq \sqrt{\gamma} \tag{7}$$

We can now operate on Eq. (7) with $\hat{\mathscr{F}}_{2t}^{\dagger} \otimes \mathbb{I}$ to get:

$$\Delta\left(|\tilde{\psi}_{2t}^{|x_0\rangle\otimes|0\rangle}\rangle, \left(\hat{\mathscr{F}}_{2t}^{\dagger} \otimes \mathbb{I}\right)\left(|\sigma_{x_0,0}\rangle \otimes |\psi_{2t}^{|x_0\rangle\otimes|0\rangle}\rangle\right)\right) \leq \sqrt{\gamma} \tag{8}$$

The above connects the states derived from the execution with the specious server to that with the honest server. By anchored ϵ-privacy of Π against honest servers, there exists a simulator $\mathscr{I}_t : L(\mathcal{A}_0 \otimes \mathcal{A}_p) \mapsto L(\mathcal{A}_t \otimes \mathcal{X}_t)$ such that for all $x \in \{0,1\}^n$ and $|\alpha\rangle \in \mathcal{B}_0 \otimes \mathcal{R}$, for any $\mathcal{R}$,

$$\Delta\left(\mathrm{tr}_{\mathcal{B}_0,\mathcal{B}_p}\left(\left(\mathscr{I}_t \otimes \mathbb{I}_{\mathcal{B}_0,\mathcal{B}_p}\right) \circ \left(|x\rangle\langle x|_{\mathcal{A}_0} \otimes |\alpha\rangle\langle\alpha|_{\mathcal{R},\mathcal{B}_0} \otimes \rho_{joint}\right)\right), \mathrm{tr}_{\mathcal{B}_t}\left(|\psi_{2t}^{|x\rangle\otimes|\alpha\rangle}\rangle\right)\right) \leq \epsilon \tag{9}$$

[6] This is because we can include the purification register at any point, as the server could have included himself rather than throwing it away.

(In fact, the above holds for any mixture over such α's, by convexity). We can now define the simulator for ρ_{in} corresponding to input state $|x_0\rangle \otimes |0\rangle$ to be the following unitary embedding from $\mathcal{A}_0 \otimes \mathcal{A}_p$ to $\mathcal{S}' \otimes \mathcal{A}_0 \otimes \mathcal{A}_p$:

$$\tilde{\mathscr{I}}_t^{x_0,0}(\cdot) = \hat{\mathscr{F}}_{2t}^\dagger \circ \left(|\sigma_{x_0,0}\rangle\langle\sigma_{x_0,0}| \otimes \mathscr{I}_t(\cdot)\right) \tag{10}$$

To show that it indeed satisfies the requirements from a simulator, we combine Eqs. (8), (10), and (9) for $x = x_0$, $|\alpha\rangle = |0\rangle$, to get that

$$\Delta\left(tr_{\mathcal{B}_0,\mathcal{B}_p}\left(\left(\tilde{\mathscr{I}}_t^{x_0,0} \otimes \mathbb{I}_{\mathcal{B}_0,\mathcal{B}_p}\right) \circ \left(|x_0\rangle\langle x_0|_{\mathcal{A}_0} \otimes |0\rangle\langle 0|_{\mathcal{B}_0} \otimes \rho_{joint}\right)\right), \mathrm{tr}_{\mathcal{B}_t}\left(|\tilde{\psi}_{2t}^{|x_0\rangle\otimes|0\rangle}\rangle\right)\right) \leq \epsilon + \sqrt{\gamma} \tag{11}$$

We now define the simulator for any input to be this exact simulator:

$$\tilde{\mathscr{I}}_t(\cdot) = \tilde{\mathscr{I}}_t^{x_0,0}(\cdot); \tag{12}$$

In the remainder of the proof we show that $\tilde{\mathscr{I}}_t(\cdot)$ satisfies an inequality similar to Eq. (11) with respect to all classical server inputs $x \in \{0,1\}^n$ (not necessarily x_0) and any input state $|\alpha\rangle \in \mathcal{B}_0 \otimes \mathcal{R}$ for any $\mathcal{R}$, as well as for a mixture of such α's; this would imply anchored privacy for the specious server. To this end we show that also for this input, a similar inequality to Eq. (11) holds (with a slightly worse bound). Define

$$|x\alpha_+\rangle = \frac{1}{\sqrt{2}}|0\rangle_{\mathcal{R}'}|x_0\rangle_{\mathcal{A}_0}|0\rangle_{\mathcal{B}_0,\mathcal{R}} + \frac{1}{\sqrt{2}}|1\rangle_{\mathcal{R}'}|x\rangle_{\mathcal{A}_0}|\alpha\rangle_{\mathcal{B}_0,\mathcal{R}},$$

where we have added an additional (control) qubit in the space $\mathcal{R}'$. The specious adversary condition applies to this input state as well, and thus using the same derivation as for Eq. (8)) we get:

$$\Delta\left(|\tilde{\psi}_{2t}^{|x\alpha_+\rangle}\rangle, \left(\hat{\mathscr{F}}_{2t}^\dagger \otimes \mathbb{I}\right)\left(|\sigma_{x\alpha_+}\rangle \otimes |\psi_{2t}^{|x\alpha_+\rangle}\rangle\right)\right) \leq \sqrt{\gamma} \tag{13}$$

Using the fact that neither the server nor the client act on the $\mathcal{R}'$ register, we get:

$$|\psi_{2t}^{|x\alpha_+\rangle}\rangle = \frac{1}{\sqrt{2}}|0\rangle_{\mathcal{R}'} \otimes |\psi_{2t}^{|x_0\rangle\otimes|0\rangle}\rangle_{\mathcal{S},\mathcal{C},\mathcal{R}} + \frac{1}{\sqrt{2}}|1\rangle_{\mathcal{R}'} \otimes |\psi_{2t}^{|x\rangle\otimes|\alpha\rangle}\rangle_{\mathcal{S},\mathcal{C},\mathcal{R}} \tag{14}$$

Similarly, since the same is true for the adversarial run, we get:

$$|\tilde{\psi}_{2t}^{|x\alpha_+\rangle}\rangle = \frac{1}{\sqrt{2}}|0\rangle_{\mathcal{R}'} \otimes |\tilde{\psi}_{2t}^{|x_0\rangle\otimes|0\rangle}\rangle_{\mathcal{S},\mathcal{C},\mathcal{R}} + \frac{1}{\sqrt{2}}|1\rangle_{\mathcal{R}'} \otimes |\tilde{\psi}_{2t}^{|x\rangle\otimes|\alpha\rangle}\rangle_{\mathcal{S}',\mathcal{S},\mathcal{C},\mathcal{R}} \tag{15}$$

We plug Eqs. (14) and (15) into Eq. (13), and project the register $\mathcal{R}'$ in the resulting state onto $|1\rangle_{\mathcal{R}'}$ to get:

$$\Delta\left(\frac{1}{\sqrt{2}}|1\rangle_{\mathcal{R}'}\otimes|\tilde{\psi}_{2t}^{|x\rangle\otimes|\alpha\rangle}\rangle_{S,C},\left(\hat{\mathscr{F}}_{2t}^{\dagger}\otimes\mathbb{I}_{\mathcal{R},C}\right)\left(\frac{1}{\sqrt{2}}|1\rangle_{\mathcal{R}'}\otimes|\sigma_{x,\alpha_+}\rangle_{S'}\otimes|\psi_{2t}^{|x\rangle\otimes|\alpha\rangle}\rangle_{S,C}\right)\right)\leq\sqrt{\gamma} \tag{16}$$

Now we apply the fact that $\hat{\mathscr{F}}_{2t}^{\dagger}$ doesn't act on the client's input; the fact that a unitary operator doesn't change the distance between states; and the fact that tracing out doesn't increase that distance [AKN98], and Eq. (16) becomes:

$$\Delta\left(|\tilde{\psi}_{2t}^{|x\rangle\otimes|\alpha\rangle}\rangle,\left(\hat{\mathscr{F}}_{2t}^{\dagger}\otimes\mathbb{I}\right)\left(|\sigma_{x\alpha_+}\rangle\otimes|\psi_{2t}^{|x\rangle\otimes|\alpha\rangle}\rangle\right)\right)\leq\sqrt{2\gamma} \tag{17}$$

Similarly, by projecting onto $|0\rangle_{\mathcal{R}'}$ instead of $|1\rangle_{\mathcal{R}'}$ in the derivation of 16, we get

$$\Delta\left(|\tilde{\psi}_{2t}^{|x_0\rangle\otimes|0\rangle}\rangle,\left(\hat{\mathscr{F}}_{2t}^{\dagger}\otimes\mathbb{I}\right)\left(|\sigma_{x\alpha_+}\rangle\otimes|\psi_{2t}^{|x_0\rangle\otimes|0\rangle}\rangle\right)\right)\leq\sqrt{2\gamma} \tag{18}$$

We now want to apply the triangle inequality to (18), using Eq. (8). Applying yet again the same sequence of simple argument, namely the fact that unitary transformations preserve the trace distance and tracing out can only decrease it, we get

$$\Delta\left(|\sigma_{x_0,0}\rangle,|\sigma_{x\alpha_+}\rangle\right)\leq 2\sqrt{2\gamma} \tag{19}$$

And we can use Eq. (19) together with Eq. (17) to get:

$$\Delta\left(|\tilde{\psi}_{2t}^{|x\rangle\otimes|\alpha\rangle}\rangle,\left(\hat{\mathscr{F}}_{2t}^{\dagger}\otimes\mathbb{I}\right)\left(|\sigma_{x_0,0}\rangle\otimes|\psi_{2t}^{|x\rangle\otimes|\alpha\rangle}\rangle\right)\right)\leq 3\sqrt{2\gamma} \tag{20}$$

And finally combine Eqs. (20), (9) and (12) (in a similar way to how we derived Eq. (11)) to get:

$$\Delta\left(\mathrm{tr}_{\mathcal{B}_0}\left(\left(\tilde{\mathscr{I}}_t\otimes\mathbb{I}\right)(|x\rangle\langle x|\otimes|\alpha\rangle\langle\alpha|\otimes\rho_{joint})\right),\mathrm{tr}_{\mathcal{B}_t}\left(|\tilde{\psi}_t^{|x\rangle\otimes|\alpha\rangle}\rangle\right)\right)\leq\epsilon+3\sqrt{2\gamma}. \tag{21}$$

This finishes our proof. □

4 Linear Lower Bound in the Specious Model, Even with Prior Entanglement

In this section we show that in the standard specious model, even allowing arbitrarily long prior entanglement, it is still impossible to achieve QPIR with sublinear communication. We do so by presenting a new lower bound argument based on an interactive leakage chain rule in [LC18], which allows us to establish linear lower bounds on both the server's communication complexity and the total communication complexity in a unified way. Then we observe that the lower bound on the server's communication complexity extends trivially to the case with arbitrary prior entanglement. In the following, we state some useful preliminaries in Sect. 4.1 and present our lower bound in Sect. 4.2.

4.1 Quantum Information Theory Background

We first recall the notion of quantum min-entropy. Consider a bipartite quantum state ρ_{AB}. The quantum min-entropy of A conditioned on B is defined as

$$H_{\min}(A|B)_\rho = -\inf_{\sigma_B} \left\{\inf\left\{\lambda \in \mathbb{R} : \rho_{AB} \leq 2^\lambda I_A \otimes \sigma_B\right\}\right\}.$$

When ρ_{AB} is a cq-state (i.e., the A register is a classical state), the quantum min-entropy has a nice operational meaning in terms of guessing probability [KRS09]. Specifically, if $H_{\min}(A|B)_\rho = k$, then the optimal probability of predicting the value of A given ρ_B is exactly 2^{-k}.

In the following, we state the interactive leakage chain rule in [LC18]. Let $\rho = \rho_{AB}$ be a cq-state, that is, the system A is classical while B is quantum. The interactive leakage chain rule bounds how much the min-entropy $H_{\min}(A|B)_\rho$ can be decreased by an "interactive leakage" produced by applying a two-party protocol $\Pi = \{\mathscr{A}, \mathscr{B}, \rho_{joint}, s\}$ to ρ, where A is treated as a classical input to $\mathscr{A}$ and B is given to $\mathscr{B}$ as part of its initial state in ρ_{joint}.

Definition 4.1. *Let $\rho = \rho_{AB}$ be a cq-state. Let $\Pi = \{\mathscr{A}, \mathscr{B}, \rho_{joint}, s\}$ be a two-party protocol where ρ_{joint} contains ρ_B in the $\mathscr{B}_p$ system, and ρ_{in} be an input state where the classical state ρ_A is copied to $\mathcal{A}_0$ as the input for $\mathscr{A}$. (That is, $\mathcal{A}_0$ has an initial state $|0\rangle_{\mathcal{A}_0}$ and we do controlled NOT gates from ρ_A to $|0\rangle_{\mathcal{A}_0}$.) Consider the protocol execution $[\mathscr{A} \circledast \mathscr{B}](\rho_{in})$ and let σ_{AB_s} be the final state where A denotes the original classical state and B_s denotes the final state of $\mathscr{B}$. We say σ_{B_s} is an* interactive leakage *of A produced by Π.*

Theorem 4.2. *Let $\rho = \rho_{AB}$ be cq-state. Let σ_{AB_s} be the final state of a two-party protocol $\Pi = \{\mathscr{A}, \mathscr{B}, \rho_{joint}, s\}$ with certain input state ρ_{in}. Let m_A and m_B be the communication complexity of $\mathscr{A}$ and $\mathscr{B}$, respectively. We have*

$$H_{\min}(A|B_s)_\sigma \geq H_{\min}(A|B)_\rho - \min\{m_A + m_B, 2m_A\}, \tag{22}$$

We will also use the following lemma about gentle measurement, which is first proved by Winter [Win99] and improved by Ogawa and Nagaoka [ON07], and is also referred to as the almost-as-good-as-new Lemma by Aaronson [Aar04]. It says that the post-measurement state of an almost-sure measurement will remain close to its original. The following version is taken from Wilde's book [Wil13].

Lemma 4.3. *Suppose $0 \leq \Lambda \leq I$ is a measurement operator such that for a mixed state ρ,*

$$tr(\Lambda\rho) \geq 1 - \epsilon.$$

Then the post-measurement state $\tilde{\rho}$ is $\sqrt{\epsilon}$-close to the original state ρ:

$$||\tilde{\rho} - \rho||_{tr} \leq \sqrt{\epsilon}.$$

We will also need the following lemma, which can be proved by a standard argument using Uhlmann theorem and the Fuchs and van de Graaf inequality [FvdG99] (for a proof, see, e.g., [BB15]).

Lemma 4.4. *Suppose ρ_A, $\sigma_A \in \mathcal{A}$ are two quantum states with purifications $|\phi\rangle_{AB}$, $|\psi\rangle_{AB} \in \mathcal{A} \otimes \mathcal{B}$, respectively, and $||\rho_A - \sigma_A||_{tr} \leq \epsilon$. Then there exists a unitary $U_B \in L(\mathcal{B})$ such that*

$$|||\phi\rangle_{AB} - I_A \otimes U_B|\psi\rangle_{AB}||_{tr} \leq \sqrt{\epsilon(2-\epsilon)}.$$

4.2 Our Lower Bound

Theorem 4.5. *Let $\Pi = \{\mathscr{A}, \mathscr{B}, \rho_{joint} = |0\rangle\langle 0|, s\}$ be a QPIR protocol for the server's database of size n. Suppose Π is $(1-\delta)$-correct and ϵ-private against γ-specious servers with $\delta \leq n^{-4}/100, \epsilon \leq n^{-8}/100$. Then the server's communication complexity is at least $(n-1)/2$ and the total communication complexity is at least $n-1$.*

In the above theorem, we consider protocols with no prior setup, i.e., $\rho_{joint} = |0\rangle\langle 0|$. We observe that the lower bound for the server's communication complexity extends for general ρ_{joint}, since one can think of ρ_{joint} as prepared by the client, who sends the server's initial state to the server at the beginning of the protocol. This simple reduction does not increase the server's communication complexity and extends the lower bound on the server's communication complexity for arbitrary ρ_{joint}.

Corollary 4.6. *Let $\Pi = \{\mathscr{A}, \mathscr{B}, \rho_{joint}, s\}$ be a QPIR protocol for the server's database of size n with arbitrary ρ_{joint}. Suppose Π is $(1-\delta)$-correct and ϵ-private against γ-specious servers with $\delta \leq n^{-4}/100, \epsilon \leq n^{-8}/100$. Then the server's communication complexity is at least $(n-1)/2$.*

We now prove Theorem 4.5.

Proof. To establish communication complexity lower bound for Π, we consider a purified version $\bar{\Pi} = \{\bar{\mathscr{A}}, \bar{\mathscr{B}}, \rho_{joint}, s\}$ of Π, where both parties' operations are purified. Specifically, $\bar{\mathscr{A}}$ is modified from $\mathscr{A}$, where the sequence of quantum operations $\bar{\mathscr{A}}_1, \ldots, \bar{\mathscr{A}}_s$ are unitaries

$$\bar{\mathscr{A}}_1 : L(\mathcal{A}_0 \otimes \bar{\mathcal{A}}_0) \to L(\mathcal{A}_1 \otimes \bar{\mathcal{A}}_1 \otimes \mathcal{X}_1),$$
$$\bar{\mathscr{A}}_t : L(\mathcal{A}_{t-1} \otimes \bar{\mathcal{A}}_{t-1} \otimes \mathcal{Y}_{t-1}) \to L(\mathcal{A}_t \otimes \bar{\mathcal{A}}_t \otimes \mathcal{X}_t), t = 2, \ldots, s;$$

$\bar{\mathcal{A}}_0$ is of sufficiently large dimension and initialized to $|0\rangle$; $\bar{\mathcal{A}}_t$ are called purifying spaces and

$$\mathrm{tr}_{\bar{\mathcal{A}}_t}(\bar{\rho}_t(\rho_{in})) = \rho_t(\rho_{in})$$

for all $\rho \in \mathcal{A}_0 \otimes \mathcal{B}_0$. The purified $\bar{\mathscr{B}}$ for $\mathscr{B}$ is similarly defined.

By inspection, it is easy to verify that $\bar{\Pi}$ preserves the properties of Π, i.e., $\bar{\Pi}$ is also $(1-\delta)$-correct, ϵ-private against γ-specious servers, and has the same communication complexity as Π. Thus, communication complexity lower bound for $\bar{\Pi}$ implies that for Π. Also, note that $\bar{\mathscr{A}}$ is a 0-specious adversary for Π.

Now, let us consider an experiment that first samples a uniformly random database $\mathbf{a} \in \{0,1\}^n$, and use $\mathbf{a}$ as the server's database to run the protocol $\bar{\Pi}$

with an arbitrary fixed input of the client. Note that execution of the protocol can be viewed as producing an interactive leakage of $\mathbf{a}$. Let ρ_{AB} denote the final state where system A denotes the input $\mathbf{a}$ and system B has the client's final local state. By Theorem 4.2, we have

$$H(A|B)_\rho \geq H(A)_\rho - \min\{2m_A, m_A + m_B\},$$

where m_A, m_B denote the server and the client's communication complexities, respectively. The operational meaning of min-entropy says that given the client's state ρ_B, one cannot guess the random database $\mathbf{a}$ correctly with probability higher than $2^{-(H(A)_\rho - \min\{2m_A, m_A+m_B\})}$. To derive a lower bound on the communication complexity, we show a strategy to predict the database $\mathbf{a}$ with probability at least $1 - n^2\sqrt{\delta + 2\sqrt{\epsilon(1-\epsilon)}} > 1/2$, which gives the desired lower bound.

Let $\sigma_B^i = \mathrm{tr}_A[\bar{\mathscr{A}} \circledast \bar{\mathscr{B}}](|\mathbf{a}\rangle\langle\mathbf{a}|_{A_0} \otimes |i\rangle\langle i|_{B_0})$ and $\sigma_A^i = \mathrm{tr}_B[\bar{\mathscr{A}} \circledast \bar{\mathscr{B}}](|\mathbf{a}\rangle\langle\mathbf{a}|_{A_0} \otimes |i\rangle\langle i|_{B_0})$.

By the definition of privacy, there exists a quantum operation $\mathscr{F}$ such that

$$\Delta\left(\mathrm{tr}_{B_0}\mathscr{F}_0 \otimes I_{\bar{B}_0}\left(\rho_{in}^1\right), \sigma_A^1\right) \leq \epsilon. \tag{23}$$

Since $\mathrm{tr}_{B_0}\mathscr{F}_0 \otimes I_{\bar{B}_0}\left(\rho_{in}^1\right) = \mathrm{tr}_{B_0}\mathscr{F}_0 \otimes I_{\bar{B}_0}\left(\rho_{in}^i\right)$ for all i,

$$\Delta\left(\mathrm{tr}_{B_0}\mathscr{F}_0 \otimes I_{\bar{B}_0}\left(\rho_{in}^1\right) - \sigma_A^i\right) \leq \epsilon \tag{24}$$

We have, by triangle inequality,

$$\Delta\left(\sigma_A^1 - \sigma_A^i\right) \leq 2\epsilon.$$

for all i.

By Lemma 4.4, we have

$$\Delta\left(I_A \otimes U_B^{1\to i}|\psi^1\rangle_{\bar{A}\bar{B}}, |\psi^i\rangle_{\bar{A}\bar{B}}\right) \leq 2\sqrt{\epsilon(1-\epsilon)} \triangleq \epsilon', \tag{25}$$

where $|\phi\rangle_{\bar{A}\bar{B}}$ and $|\psi^i\rangle_{\bar{A}\bar{B}}$ are purifications of σ_A^1 and σ_A^i, respectively.

By the definition of correctness error, there exists measurement $\mathcal{M}_i$ such that

$$\Pr\left\{\mathcal{M}_i\left(\sigma_B^i\right) = a_i\right\} \geq 1 - \delta.$$

Let

$$\mathcal{M}_i' = \left(U_B^{1\to i}\right)^\dagger \mathcal{M}_i U_B^{1\to i}$$

for $i = 2, \ldots, n$. Thus we have by Eq. (25)

$$\Pr\left\{\mathcal{M}_i'\left(\sigma_B^1\right) = a_i\right\} \geq 1 - \delta - \epsilon'. \tag{26}$$

By Lemma 4.3, the client can recover $\tilde{\sigma}_B^{(i)}$ such that

$$\Delta\left(\tilde{\sigma}_B^{(i)}, \sigma_B^1\right) \leq \sqrt{\delta + \epsilon'}. \tag{27}$$

Now we construct a protocol for the client to learn all the bits $\mathbf{a} = a_1, \ldots, a_n$. First the client chooses input $|1\rangle\langle 1|$. Then he plays the protocol $\bar{\Pi}$ with Alice and obtains σ_B^1. Measuring σ_B^1 by $\mathcal{M}_1$, the client gets a_1 with probability at least $1 - \delta$. By Lemma 4.3, the client can recover $\tilde{\sigma}_B^1$ such that

$$\Delta\left(\tilde{\sigma}_B^1, \sigma_B^1\right) \leq \sqrt{\delta}.$$

Then the client measures $\mathcal{M}_2'$ on $\tilde{\sigma}_B^1$ and then recovers $\tilde{\sigma}_B^2$. Continue this process and $\tilde{\sigma}_B^k$ will be the state recovered from applying $\mathcal{M}_k'$ to $\tilde{\sigma}_B^{k-1}$. We claim that

$$\Delta\left(\tilde{\sigma}_B^k, \sigma_B^1\right) \leq k\sqrt{\delta + \epsilon'}. \tag{28}$$

Suppose this is true for $i = 2, \cdots, k$. If we measure $\mathcal{M}_{k+1}'$ on $\tilde{\sigma}_B^{k+1}$ and on σ_B^1, respectively, and recover $\tilde{\sigma}_B^{k+1}$ and $\tilde{\sigma}_B^{(k+1)}$, respectively, we have

$$\Delta\left(\tilde{\sigma}_B^{k+1}, \tilde{\sigma}_B^{(k+1)}\right) \leq \Delta\left(\tilde{\sigma}_B^k, \sigma_B^1\right) \leq k\sqrt{\delta + \epsilon'} \tag{29}$$

where the first inequality is because quantum operations do not increase trace distance. Now use the triangle inequality with Eqs. (27) and (29), and the claim follows by induction.

By Eqs. (26) and (28), the probability of recovering a_i by measuring $\mathcal{M}_i'$ on $\tilde{\sigma}_B^{i-1}$ is at least $1 - i\sqrt{\delta + \epsilon'}$. Therefore, the client learns $\mathbf{a}$ with probability at least

$$\prod_{i=1}^{n}\left(1 - i\sqrt{\delta + \epsilon'}\right) \geq 1 - n^2\sqrt{\delta + \epsilon'},$$

which is what we need to complete the proof. □

Acknowledgments. We thank the anonymous referees for presenting us with the work of Kerenidis et al. [KLGR16], and other valuable comments. ZB is supported by the Israel Science Foundation (Grant 468/14), Binational Science Foundation (Grants 2016726, 2014276), and by the European Union Horizon 2020 Research and Innovation Program via ERC Project REACT (Grant 756482) and via Project PROMETHEUS (Grant 780701). OS is supported by ERC Grant 280157, by the Israel Science Foundation (Grant 682/18), and by the Cyber Security Research Center at Ben-Gurion University. CYL is financially supported from the Young Scholar Fellowship Program by Ministry of Science and Technology (MOST) in Taiwan, under Grant MOST107-2636-E-009-005. KMC is partially supported by 2016 Academia Sinica Career Development Award under Grant No. 23-17 and the Ministry of Science and Technology, Taiwan under Grant No. MOST 103-2221- E-001-022-MY3. DA and AG were supported by ERC Grant 280157 for part of the work on this project, and are supported by the Israel Science Foundation (Grant 1721/17).

A Hilbert Spaces and Quantum States

The Hilbert space of a quantum system A is denoted by the corresponding calligraphic letter $\mathcal{A}$ and its dimension is denoted by $\dim(\mathcal{A})$. Let $L(\mathcal{A})$ be the space of linear operators on $\mathcal{A}$. A quantum state of system A is described by a *density operator* $\rho_A \in L(\mathcal{A})$ that is positive semidefinite and with unit trace ($\mathrm{tr}(\rho_A) = 1$). Let $S(\mathcal{A}) = \{\rho_A \in L(\mathcal{A}) : \rho_A \geq 0, \mathrm{tr}(\rho_A) = 1\}$ be the set of density operators on $\mathcal{A}$. When $\rho_A \in S(\mathcal{A})$ is of rank one, it is called a *pure* quantum state and we can write $\rho = |\psi\rangle\langle\psi|_A$ for some unit vector $|\psi\rangle_A \in \mathcal{A}$, where $\langle\psi| = |\psi\rangle^\dagger$ is the conjugate transpose of $|\psi\rangle$. If ρ_A is not pure, it is called a *mixed* state and can be expressed as a convex combination of pure quantum states.

The Hilbert space of a joint quantum system AB is the tensor product of the corresponding Hilbert spaces $\mathcal{A} \otimes \mathcal{B}$. For $\rho_{AB} \in S(\mathcal{A} \otimes \mathcal{B})$, its reduced density operator in system A is $\rho_A = \mathrm{tr}_B(\rho_{AB})$, where

$$\mathrm{tr}_B(\rho_{AB}) = \sum_i I_A \otimes \langle i|_B \left(\rho_{AB}\right) I_A \otimes |i\rangle_B$$

for an orthonormal basis $\{|i\rangle_B\}$ for $\mathcal{B}$. We sometimes use the equivalent notation,

$$\rho_{AB}|_A := tr_B(\rho_{AB}).$$

Suppose $\rho_A \in S(\mathcal{A})$ of finite dimension $\dim(\mathcal{A})$. Then there exists $\mathcal{B}$ of dimension $\dim(\mathcal{B}) \geq \dim(\mathcal{A})$ and $|\psi\rangle_{AB} \in \mathcal{A} \otimes \mathcal{B}$ such that

$$\mathrm{tr}_B|\psi\rangle\langle\psi|_{AB} = \rho_A.$$

The state $|\psi\rangle_{AB}$ is called a purification of ρ_A.

The trace distance between two quantum states ρ and σ is

$$\Delta(\rho, \sigma) = ||\rho - \sigma||_{\mathrm{tr}},$$

where $||X||_{\mathrm{tr}} = \frac{1}{2}\mathrm{tr}\sqrt{X^\dagger X}$ is the trace norm of X. Hence the trace distance between two pure states $|\alpha\rangle, |\beta\rangle$ is

$$\Delta(|\alpha\rangle\langle\alpha|, |\beta\rangle\langle\beta|) = \sqrt{1 - |\langle\alpha|\beta\rangle|^2}. \tag{30}$$

Lemma A.1. *Consider a quantum state ρ_{XY} over two registers X, Y, and denote $\rho_X = tr_Y(\rho_{XY})$. Then if there exists $\epsilon, |\varphi\rangle$ s.t. $\Delta(\rho_X, |\varphi\rangle\langle\varphi|) \leq \epsilon$, then there exists $\tilde{\rho}_Y$ s.t. $\Delta(\rho_{XY}, |\varphi\rangle\langle\varphi| \otimes \tilde{\rho}_Y) \leq \sqrt{\epsilon}$. Furthermore, if ρ_{XY} is pure then so is $\tilde{\rho}_Y$.*

Proof. It is sufficient w.l.o.g to prove for a pure ρ_{XY}, since it is always possible to purify ρ_{XY} by adding an additional register Z, and consider the pure state ρ_{XYZ}. The transitivity of the partial trace operation implies that if the theorem is true for $X, (YZ)$, then it is also true for X, Y. Also assume w.l.o.g that $|\varphi\rangle = |0\rangle$ (this is just a matter of choosing a basis elements).

Thus we will provide a proof in the case where the joint state of X, Y can be written as a superposition $|\alpha\rangle = \sum_{x,y} w_{x,y}|x\rangle|y\rangle$. Define $P_0 = \Pr[X = 0] = \sum_y |w_{0,y}|^2$, and note that it must be the case that $P_0 \geq 1 - \epsilon$. To see this, note that P_0 is the probability of measuring $X = 0$ in the experiment where we first trace out Y and then measuring X. Since $\Delta(\rho_X, |0\rangle\langle 0|) \leq \epsilon$, the probability of measuring $X = 0$ after tracing out Y is ϵ close to the probability of measuring $X = 0$ in $|0\rangle\langle 0|$, which is 1 (see, e.g., [AKN98]). The claim $P_0 \geq 1 - \epsilon$ follows.

Now define $|\beta\rangle = \frac{1}{\sqrt{P_0}} \sum_y w_{0,y}|y\rangle$, and let $\tilde{\rho}_Y = |\beta\rangle\langle\beta|$. Then

$$\Delta(\rho_{XY}, |0\rangle\langle 0| \otimes \tilde{\rho}_Y) = \Delta(|\alpha\rangle\langle\alpha|, |0\rangle\langle 0| \otimes |\beta\rangle\langle\beta|) = \sqrt{1 - |\langle\alpha|(0,\beta)\rangle|^2}. \quad (31)$$

We have

$$\langle\alpha|(0,\beta)\rangle = \frac{1}{\sqrt{P_0}} \sum_y |w_{0,y}|^2 = \sqrt{P_0}, \quad (32)$$

which implies that indeed $\Delta(\rho_{XY}, |0\rangle\langle 0| \otimes \tilde{\rho}_Y) = \sqrt{1 - P_0} \leq \sqrt{\epsilon}$. □

B Security Analysis of Kereneidis et al.'s Protocol

For completeness, we restate[7] the QPIR protocol with pre-shared entanglement by Kerenidis et al. [KLGR16, Section 6]. Given a database $\texttt{DB} \in \{0,1\}^n$ for some $n = 2^\ell$ as input to the server, and index $i \in [n]$ as input to the client (If the client's input is a superposition, the algorithm is run in superposition), we denote the protocol Π_n as follows.

The protocol Π_n is recursive and calls $\Pi_{n/2}$ as a subroutine. For the execution of Π_n, the parties are required to pre-share a pair of entangled state registers $\frac{1}{2^{n/4}} \sum_{\mathbf{r} \in \{0,1\}^{n/2}} |\mathbf{r}\rangle_R |\mathbf{r}\rangle_{R'}$, where R is held by the server and R' is held by the client. They also share an entangled state needed for the recursive application of the protocol $\Pi_{n/2}$ (and the recursive calls it entails). Unfolding the recursion, this means that for all $n' = 2^{\ell'}$ with $\ell' \in [\ell - 1]$, there is an entangled register of length n' shared between the client and the server.

The protocol execution is described in shorthand Fig. 1. In what follows we provide a detailed description and analyze the steps of the protocol to establish correctness and assert properties that will allow us to analyze privacy.

1. If $n = 1$ then the database contains a single value. In this case there is no need for shared entanglement, and the server sends a register F containing $|\texttt{DB}\rangle$ (the final response) to the client, and the protocol terminates. This is trivially secure and efficient. Otherwise proceed to the next steps.
2. The server denotes $\texttt{DB}_0, \texttt{DB}_1 \in \{0,1\}^{n/2}$ s.t. $\texttt{DB} = [\texttt{DB}_0 \| \texttt{DB}_1]$, i.e. the low-order and high-order bits of the database respectively. The server starts with two single-bit registers Q_0, Q_1 initialized to 0. The server CNOTs Q_b with the inner product of R and $\texttt{DB}_b$ so that it contains $|\mathbf{r} \cdot \texttt{DB}_b\rangle_{Q_b}$, and sends Q_0, Q_1 to the client.

[7] We make one minor adaptation – see Remark B.1.

At this point, the joint state between the client and (an honest) server is

$$\sum_{\mathbf{r}\in\{0,1\}^{n/2}} |\mathbf{r}\rangle_R|\mathbf{r}\rangle_{R'}|\mathbf{r}\cdot\mathtt{DB}_0\rangle_{Q_0}|\mathbf{r}\cdot\mathtt{DB}_1\rangle_{Q_1}.$$

In particular the reduced density matrix of the server's state is independent of the index i.

3. Let $b^* = \lfloor\frac{i-1}{n}\rceil$ denote the most significant bit of i. The client evaluates a Z gate on Q_{b^*}. It sends Q_0, Q_1 back to the server.
   At this point, the joint state between the client and (an honest) server is

$$\sum_{\mathbf{r}\in\{0,1\}^{n/2}} (-1)^{\mathbf{r}\cdot\mathtt{DB}_{b^*}}|\mathbf{r}\rangle_R|\mathbf{r}\rangle_{R'}|\mathbf{r}\cdot\mathtt{DB}_0\rangle_{Q_0}|\mathbf{r}\cdot\mathtt{DB}_1\rangle_{Q_1}.$$

   Importantly, the reduced density matrix of the server, which contains the registers R, Q_0, Q_1, is the diagonal matrix that corresponds to the classical distribution of sampling a random $\mathbf{r}$ in register R, and placing $\mathbf{r}\cdot\mathtt{DB}_0, \mathbf{r}\cdot\mathtt{DB}_1$ in Q_0, Q_1. This density matrix is independent of b^* and therefore of i.
4. The server again CNOTs Q_b with the inner product of R and $\mathtt{DB}_b$.
   At this point, the joint state between the client and (an honest) server is

$$\sum_{\mathbf{r}\in\{0,1\}^{n/2}} (-1)^{\mathbf{r}\cdot\mathtt{DB}_{b^*}}|\mathbf{r}\rangle_R|\mathbf{r}\rangle_{R'}|0\rangle_{Q_0}|0\rangle_{Q_1}.$$

   From this point on we disregard Q_0, Q_1 since they remain zero throughout. Since this step only involves a local unitary by the server, we are guaranteed that its reduced density matrix is still independent of i.
5. The server performs QFT on R and the client performs QFT on R'. The resulting state is

$$\frac{1}{2^{3n/4}}\sum_{\mathbf{r},\mathbf{y},\mathbf{w}\in\{0,1\}^{n/2}} (-1)^{\mathbf{r}\cdot(\mathtt{DB}_{b^*}\oplus\mathbf{y}\oplus\mathbf{w})}|\mathbf{y}\rangle_R|\mathbf{w}\rangle_{R'} = \frac{1}{2^{n/4}}\sum_{\mathbf{y}\in\{0,1\}^{n/2}} |\mathbf{y}\rangle_R|\underbrace{\mathbf{y}\oplus\mathtt{DB}_{b^*}}_{\mathbf{w}}\rangle_{R'}.$$

   Since we only performed local operations on the server and client side (without communication), the server's density matrix remains perfectly independent of b^* and thus of i.
6. Note that at this point, the joint state of the client and server is a "shifted" entangled state where the shift corresponds to the half-database $\mathtt{DB}_{b^*}$ that contains the element that the client wishes to retrieve. More explicitly, $\mathtt{DB}[i] = \mathtt{DB}_{b^*}[i^*]$ for $i^* = i \pmod{n/2}$ contains the $(\ell-1)$ least significant bits of i. Therefore, for all $\mathbf{y}, \mathbf{w}$ in the support of the joint state, it holds that $\mathtt{DB}[i] = \mathbf{w}[i^*]\oplus\mathbf{y}[i^*]$.
   The client will now ignore (temporarily) the register R' and execute $\Pi_{n/2}$ recursively on index i^*. The (honest) server will carry out the protocol with the value $\mathbf{y}$ from the register R serving as the server's database. Note that since the register R' is not touched, for the purposes of executing the protocol

the value $\mathbf{w}$ in R' is equivalent to have been measured, and the value $\mathbf{y}$ in R is equivalent to the deterministic register $\mathbf{w} \oplus \mathtt{DB}_{b^*}$.

We are recursively guaranteed that in the end of the execution of $\Pi_{n/2}$, the client receives a register F containing the value $\mathbf{y}[i^*] = \mathbf{w}[i^*] \oplus \mathtt{DB}_{b^*}[i^*] = \mathbf{w}[i^*] \oplus \mathtt{DB}[i]$. Since the client still maintains the original register R' containing $\mathbf{w}$, it can CNOT the value $\mathbf{w}[i^*]$ from F and obtain $|\mathtt{DB}[i]\rangle_A$. Namely, in the end of the execution, the register F indeed contains the desired value $\mathtt{DB}[i]$.

7. Lastly, if the client and server desire to "clean up" and restore the shared entanglement so that it can be reused in consequent executions, the client can copy the contents of the register F to a fresh register (which is possible since this register contains a classical value). Since the client and server are pure (i.e. do not measure) throughout the protocol, they can rewind the execution of the protocol to restore their initial joint entanglement.

If the final cleanup step is not executed then the total number of rounds of Π_n is $2\ell + 1$, and the total communication complexity is $4l + 1$ (recall that $\ell = \log(n)$). If the cleanup step is executed, the round complexity and communication complexity are doubled due to rewinding the execution.

Remark B.1. Note that in the original protocol by Kerenidis et al. step 7 does not appear, and it is not mentioned that the shared entanglement can be cleaned and reused.

Lemma B.2. *The protocol Π_n is a PIR protocol with perfect correctness and perfect anchored privacy against honest servers. It furthermore has communication complexity $O(\log n)$, and uses $O(n)$ bits of (reusable) shared entanglement.*

Proof. The analysis in the body of the protocol establishes that the local view of the adversary is independent of the input i, when i is treated as a fixed (classical) parameter. Next, we show that the server's local state is independent of the client's input, even when the input is an arbitrary quantum state.

We observe two facts: (i) since we are interested in the server's local view, the input register is traced out, (ii) the client interacts with its input qubits as control bits for Controlled operations only. By property (i), we can assume for the sake of the analysis that the qubits are measured just before tracing them out. Using property (ii), the entire protocol commutes with a measurement in the standard basis of the input register. Therefore, we can assume the server's local view would not be changed by adding a measurement in the standard basis of the input register at the very beginning of the protocol. By the argument in the previous paragraph, we already know that for a classical input, the server's local view is independent of the input. The measurement in the standard basis collapses the state a classical, and we conclude that the server's local view is independent of the input, for any input state.

In order to comply with the simulation based privacy definition (see Definition 2.5), we can define the simulators to be simulations of the protocol run with input $i = |0\rangle\langle 0|$. Since the server's state is independent of the input, we complete the proof that the protocol has perfect anchored privacy against honest servers.

Recursive QPIR with Logarithmic Communication

Server input: Database $\mathtt{DB} \in \{0,1\}^n$.
Client input: Index $i \in [n]$.
Desired output: Value $\mathtt{DB}[i]$ stored in register F on the client side.
Setup: Register R for server and R' for client in joint state $\frac{1}{2^{n/4}} \sum_{\mathbf{r} \in \{0,1\}^{n/2}} |\mathbf{r}\rangle_R |\mathbf{r}\rangle_{R'}$.
(This setup is for external recursion loop, internal loops require their own R, R' defined recursively.)

1. If $n = 1$, copy the (single-bit) database into a register and send to client, then terminate (or go to clean up step 7 below).
2. The server denotes $\mathtt{DB}_0, \mathtt{DB}_1 \in \{0,1\}^{n/2}$ s.t. $\mathtt{DB} = [\mathtt{DB}_0 \| \mathtt{DB}_1]$, i.e. the low-order and high-order bits of the database respectively. The server starts with two single-bit registers Q_0, Q_1 initialized to 0. The server CNOTs Q_b with the inner product of R and $\mathtt{DB}_b$ so that it contains $|\mathbf{r} \cdot \mathtt{DB}_b\rangle_{Q_b}$. It sends Q_0, Q_1 to the client.
3. Let $b^* = \lfloor \frac{i-1}{n} \rceil$ denote the most significant bit of i. The client evaluates a Z gate on Q_{b^*}. It sends Q_0, Q_1 back to the server.
4. The server again CNOTs Q_b with the inner product of R and $\mathtt{DB}_b$.
5. The server performs QFT on R and the client performs QFT on R'.
6. Call $\Pi_{n/2}$ recursively (with fresh R, R' obtained from the setup). The server input is the contents of the register R (of length $n/2$). The client input is $i^* = i \pmod{n/2} \in [n/2]$. The client receives a response register F as the output of the recursive call. It then CNOTs $R'[i^*]$ into F. Finally, F contains the output of the recursive execution.
7. If it is desired to restore the shared entanglement, copy the (classical) output into a fresh register and rewind the execution of the protocol.

Fig. 1. The QPIR protocol of Kerenidis et al.

The communication complexity and the amount of reusable shared entanglement needed in this protocol follow directly from the protocol. □

We can therefore apply Theorem 3.2 and conclude that Π is secure against anchored-specious adversaries.

Corollary B.3. *There exists a PIR protocol Π with logarithmic communication complexity assuming linear shared entanglement, which is perfectly correct and anchored $O(\sqrt{\gamma})$-private against γ-specious adversaries.*

References

[Aar04] Aaronson, S.: Limitations of quantum advice and one-way communication. In: Proceedings of the 19th IEEE Annual Conference on Computational Complexity, pp. 320–332, June 2004. https://doi.org/10.1109/CCC.2004.1313854

[ABC+19] Aharonov, D., Brakerski, Z., Chung, K.-M., Green, A., Lai, C.-Y., Sattath, O.: On quantum advantage in information theoretic single-server PIR (2019). arXiv:1902.09768

[ACG+16] Aharonov, D., Chailloux, A., Ganz, M., Kerenidis, I., Magnin, L.: A simpler proof of the existence of quantum weak coin flipping with arbitrarily small bias. SIAM J. Comput. **45**(3), 633–679 (2016). https://doi.org/10.1137/14096387X

[AKN98] Aharonov, D., Kitaev, A.Y., Nisan, N.: Quantum circuits with mixed states. In: Proceedings of the Thirtieth Annual ACM Symposium on the Theory of Computing, Dallas, Texas, USA, 23–26 May 1998, pp. 20–30 (1998). https://doi.org/10.1145/276698.276708

[ANTSV02] Ambainis, A., Nayak, A., Ta-Shma, A., Vazirani, U.: Dense quantum coding and quantum finite automata. JACM **49**(4), 496–511 (2002). https://doi.org/10.1145/581771.581773

[BB84] Bennett, C.H., Brassard, G.: Quantum cryptography: public key distribution and coin tossing. In: Proceedings of IEEE International Conference on Computers, Systems, and Signal Processing, p. 175 (1984)

[BB15] Baumeler, Ä., Broadbent, A.: Quantum private information retrieval has linear communication complexity. J. Cryptol. **28**(1), 161–175 (2015). https://doi.org/10.1007/s00145-014-9180-2

[BS16] Broadbent, A., Schaffner, C.: Quantum cryptography beyond quantum key distribution. Des. Codes Crypt. **78**(1), 351–382 (2016). https://doi.org/10.1007/s10623-015-0157-4

[BV11] Brakerski, Z., Vaikuntanathan, V.: Efficient fully homomorphic encryption from (standard) LWE. In: Ostrovsky, R. (ed.) FOCS, pp. 97–106. IEEE (2011). https://eprint.iacr.org/2011/344.pdf

[CGKS95] Chor, B., Goldreich, O., Kushilevitz, E., Sudan, M.: Private information retrieval. In: 36th Annual Symposium on Foundations of Computer Science, Milwaukee, Wisconsin, USA, 23–25 October 1995, pp. 41–50. IEEE Computer Society (1995). https://doi.org/10.1109/SFCS.1995.492461

[CK09] Chailloux, A., Kerenidis, I.: Optimal quantum strong coin flipping. In: 50th Annual IEEE Symposium on Foundations of Computer Science, FOCS 2009, Atlanta, Georgia, USA, 25–27 October 2009, pp. 527–533. IEEE Computer Society (2009). https://doi.org/10.1109/FOCS.2009.71

[CMS99] Cachin, C., Micali, S., Stadler, M.: Computationally private information retrieval with polylogarithmic communication. In: Stern, J. (ed.) EUROCRYPT 1999. LNCS, vol. 1592, pp. 402–414. Springer, Heidelberg (1999). https://doi.org/10.1007/3-540-48910-X_28

[DG15] Dvir, Z., Gopi, S.: 2-Server PIR with sub-polynomial communication. In: Servedio, R.A., Rubinfeld, R. (eds.) Proceedings of the Forty-Seventh Annual ACM on Symposium on Theory of Computing, STOC 2015, Portland, OR, USA, 14–17 June 2015, pp. 577–584. ACM (2015). https://doi.org/10.1145/2746539.2746546

[DNS10] Dupuis, F., Nielsen, J.B., Salvail, L.: Secure two-party quantum evaluation of unitaries against specious adversaries. In: Rabin, T. (ed.) CRYPTO 2010. LNCS, vol. 6223, pp. 685–706. Springer, Heidelberg (2010). https://doi.org/10.1007/978-3-642-14623-7_37

[Efr12] Efremenko, K.: 3-query locally decodable codes of subexponential length. SIAM J. Comput. **41**(6), 1694–1703 (2012). https://doi.org/10.1137/090772721

[FvdG99] Fuchs, C.A., van de Graaf, J.: Cryptographic distinguishability measures for quantum-mechanical states. IEEE Trans. Inf. Theory **45**(4), 1216–1227 (1999). https://doi.org/10.1109/18.761271

[GC01] Gottesman, D., Chuang, I.: Quantum digital signatures (2001). arXiv:quant-ph/0105032

[Gen09] Gentry, C.: A fully homomorphic encryption scheme. Ph.D. thesis. Stanford University (2009)

[GLM08] Giovannetti, V., Lloyd, S., Maccone, L.: Quantum private queries. Phys. Rev. Lett. **100**, 230502 (2008). https://doi.org/10.1103/PhysRevLett.100.230502

[Gol04] Goldreich, O.: The Foundations of Cryptography - Volume 2, Basic Applications. Cambridge University Press, Cambridge (2004)

[GW07] Gutoski, G., Watrous, J.: Toward a general theory of quantum games. In: Proceedings of the Thirty-Ninth Annual ACM Symposium on Theory of Computing, pp. 565–574. ACM (2007). https://doi.org/10.1145/1250790.1250873

[JP99] Jonathan, D., Plenio, M.B.: Entanglement-assisted local manipulation of pure quantum states. Phys. Rev. Lett. **83**, 3566–3569 (1999). https://doi.org/10.1103/PhysRevLett.83.3566

[JRS09] Jain, R., Radhakrishnan, J., Sen, P.: A property of quantum relative entropy with an application to privacy in quantum communication. J. ACM **56**(6), 33:1–33:32 (2009). https://doi.org/10.1145/1568318.1568323

[KLGR16] Kerenidis, I., Laurière, M., Gall, F.L., Rennela, M.: Information cost of quantum communication protocols. Quantum Inf. Comput. **16**(3&4), 181–196 (2016). http://www.rintonpress.com/xxqic16/qic-16-34/0181-0196.pdf

[Kli07] Klimesh, M.: Inequalities that collectively completely characterize the catalytic majorization relation (2007). arXiv:0709.3680

[KRS09] Konig, R., Renner, R., Schaffner, C.: The operational meaning of min- and max-entropy. IEEE Trans. Inf. Theory **55**(9), 4337–4347 (2009). https://doi.org/10.1109/TIT.2009.2025545

[LC97] Lo, H.-K., Chau, H.F.: Is quantum bit commitment really possible? Phys. Rev. Lett. **78**, 3410–3413 (1997). https://doi.org/10.1103/PhysRevLett.78.3410

[LC18] Lai, C.-Y., Chung, K.-M.: Interactive leakage chain rule for quantum min-entropy (2018). arXiv:1809.10694

[LG12] Le Gall, F.: Quantum private information retrieval with sublinear communication complexity. Theory Comput. **8**(16), 369–374 (2012). https://doi.org/10.4086/toc.2012.v008a016

[Lo97] Lo, H.-K.: Insecurity of quantum secure computations. Phys. Rev. A **56**(2), 1154 (1997). https://doi.org/10.1103/PhysRevA.56.1154

[May97] Mayers, D.: Unconditionally secure quantum bit commitment is impossible. Phys. Rev. Lett. **78**, 3414–3417 (1997). https://doi.org/10.1103/PhysRevLett.78.3414

[Moc07] Mochon, C.: Quantum weak coin flipping with arbitrarily small bias (2007). arXiv:0711.4114

[Nay99] Nayak, A.: Optimal lower bounds for quantum automata and random access codes. In: 40th Annual Symposium on Foundations of Computer Science, pp. 369–376 (1999). https://doi.org/10.1109/SFFCS.1999.814608

[ON07] Ogawa, T., Nagaoka, H.: Making good codes for classical-quantum channel coding via quantum hypothesis testing. IEEE Trans. Inf. Theory **53**(6), 2261–2266 (2007). https://doi.org/10.1109/TIT.2007.896874

[vDH03] van Dam, W., Hayden, P.: Universal entanglement transformations without communication. Phys. Rev. A **67**, 060302 (2003). https://doi.org/10.1103/PhysRevA.67.060302

[Wie83] Wiesner, S.: Conjugate coding. SIGACT News **15**(1), 78–88 (1983). https://doi.org/10.1145/1008908.1008920

[Wil13] Wilde, M.M.: Quantum Information Theory. Cambridge University Press, Cambridge (2013). Cambridge Books Online

[Win99] Winter, A.J.: Coding theorem and strong converse for quantum channels. IEEE Trans. Inf. Theory **45**(7), 2481–2485 (1999). https://doi.org/10.1109/18.796385

[YPF14] Yu, L., Pérez-Delgado, C.A., Fitzsimons, J.F.: Limitations on information theoretically secure quantum homomorphic encryption (2014). arXiv:1406.2456

Verifier-on-a-Leash: New Schemes for Verifiable Delegated Quantum Computation, with Quasilinear Resources

Andrea Coladangelo[1,2], Alex B. Grilo[3(✉)], Stacey Jeffery[3], and Thomas Vidick[1,2]

[1] Department of Computing and Mathematical Sciences, California Institute of Technology, Pasadena, USA
[2] CMS, Caltech, Pasadena, USA
{acoladan,vidick}@cms.caltech.edu
[3] QuSoft and CWI, Amsterdam, The Netherlands
{alexg,jeffery}@cwi.nl

Abstract. The problem of reliably certifying the outcome of a computation performed by a quantum device is rapidly gaining relevance. We present two protocols for a classical verifier to verifiably delegate a quantum computation to two non-communicating but entangled quantum provers. Our protocols have near-optimal complexity in terms of the total resources employed by the verifier and the honest provers, with the total number of operations of each party, including the number of entangled pairs of qubits required of the honest provers, scaling as $O(g \log g)$ for delegating a circuit of size g. This is in contrast to previous protocols, whose overhead in terms of resources employed, while polynomial, is far beyond what is feasible in practice. Our first protocol requires a number of rounds that is linear in the depth of the circuit being delegated, and is blind, meaning neither prover can learn the circuit or its input. The second protocol is not blind, but requires only a constant number of rounds of interaction.

Our main technical innovation is an efficient rigidity theorem which allows a verifier to test that two entangled provers perform measurements specified by an arbitrary m-qubit tensor product of single-qubit Clifford observables on their respective halves of m shared EPR pairs, with a robustness that is independent of m. Our two-prover classical-verifier delegation protocols are obtained by combining this rigidity theorem with a single-prover quantum-verifier protocol for the verifiable delegation of a quantum computation, introduced by Broadbent.

1 Introduction

Quantum computers hold the potential to speed up a wide range of computational tasks (see, for example, [Mon16]). Recent progress towards implementing limited quantum devices has added urgency to the already important question of how a classical verifier can test a quantum device. This verifier could be an

Y. Ishai and V. Rijmen (Eds.): EUROCRYPT 2019, LNCS 11478, pp. 247–277, 2019.
https://doi.org/10.1007/978-3-030-17659-4_9

experimentalist running a new experimental setup; a consumer who has purchased a purported quantum device; or a client who wishes to delegate some task to a quantum server. In all cases, the user would like to exert some form of control over the quantum device. For example, the experimentalist may think that she is testing that a particular experiment prepares a certain quantum state by performing a series of measurements, i.e. by state tomography, but this assumes some level of trust in the measurement apparatus being used. For a classical party to truly test a quantum system, that system should be modeled in a device-independent way, having classical inputs (e.g. measurement settings) and classical outputs (e.g. measurement results).

Tests of quantum mechanical properties of a system first appeared in the form of Bell tests [Bel64,CHSH69]. In a Bell test, a verifier asks classical questions to a quantum-device and receives classical answers. These tests make one crucial assumption on the system to be tested: that it consists of two spatially isolated components that are unable to communicate throughout the experiment. One can then upper bound the value of some statistical quantity of interest subject to the constraint that the two devices do not share any entanglement. Such a bound is referred to as a Bell inequality. While the violation of a Bell inequality can be seen as a certificate of entanglement, the area of self-testing, first introduced in [MY04], allows for the certification of much stronger statements, including about which measurements are being performed, and on which state. Informally, a *robust rigidity theorem* is a statement about which kind of apparatus, quantum state and measurements, must be used by a pair of isolated devices in order to succeed in a given statistical test. Following a well-established tradition, we will refer to such tests as *games*, call the devices *players* (or *provers*), and the quantum state and measurements that they implement the *strategy* of the players. A rigidity theorem is a statement about the necessary structure of near-optimal strategies for a game.

In 2012, Reichardt, Unger and Vazirani proved a robust rigidity theorem for playing a sequence of n CHSH games [RUV13]. Aside from its intrinsic interest, this rigidity theorem had two important consequences. One was the first device-independent protocol for quantum key distribution. The second was a protocol whereby a completely classical verifier can test a universal quantum computer consisting of two non-communicating devices. The resulting protocol for delegating quantum computations has received a lot of attention as the first classical-verifier delegation protocol. The task is well-motivated: for the foreseeable future, making use of a quantum computer will likely require delegating the computation to a potentially untrusted cloud service, such as that announced by IBM [Cas17].

Unfortunately, the complexity overhead of the delegation protocol from [RUV13], in terms of both the number of EPR pairs needed for the provers and the overall time complexity of the provers as well as the (classical) verifier, while polynomial, is prohibitively large. Although the authors of [RUV13] do not provide an explicit value for the exponent, in [HPDF15] it is estimated that their protocol requires resources that scale like $\Omega(g^{8192})$, where g is the number of gates in the delegated circuit (notwithstanding the implicit constant, this already makes the approach thoroughly impractical for even a 2-gate circuit!). The large overhead is in part due to a very small (although still inverse polynomial) gap between the completeness and soundness parameters of the rigidity theorem; this requires the verifier to perform many more Bell tests than the actual number of EPR pairs needed to implement the computation, which would scale linearly with the circuit size.

Subsequent work has presented significantly more efficient protocols for achieving the same, or similar, functionality [McK16, GKW15, HPDF15]. We refer to Table 1 for a summary of our estimated lower bounds on the complexity of each of these results (not all papers provide explicit bounds, in which case our estimates, although generally conservative, should be taken with caution). Prior to our work, the best two-prover delegation protocol required resources scaling like g^{2048} for delegating a g-gate circuit. Things improve significantly if we allow for more than two provers, however, the most efficient multi-prover delegation protocols still required resources that scale as at least $\Omega(g^4 \log g)$ for delegating a g-gate circuit on n qubits. Since we expect that in the foreseeable future most quantum computations will be delegated to a third-party server, even such small polynomial overhead is unacceptable, as it already negates the quantum advantage for a number of problems, such as quantum search.

The most efficient classical-verifier delegation protocols known [FH15, NV17], with poly(n) and 7 provers, respectively, require resources that scale as $O(g^3)$, but this efficiency comes at the cost of a technique of "post-hoc" verification. In this technique, the provers must learn the verifier's input even before they are separated, so that they can prepare the history state for the computation.[1] As a result, these protocols are not blind[2]. Moreover, while the method does provide a means for verifying the outcome of an arbitrary quantum computation, in contrast to [RUV13] it does not provide a means for the verifier to test the provers' implementation of the required circuit on a gate-by-gate basis. Other works, such as [HH16], achieve two-prover verifiable delegation with complexity that scales like $O(g^4 \log g)$, but in much weaker models; for example, in [HH16] the provers' private system is assumed a priori to be in tensor product form, with well-defined registers. General techniques are available to remove the strong assumption, but they would lead to similar large overhead as previous results.

In contrast, in the setting where the verifier is allowed to have some limited quantum power, such as the ability to generate single-qubit states and measure them with observables from a small finite set, efficient schemes for blind verifiable delegation do exist [ABE10, FK17, Mor14, Bro18, HM15, MF16, FH17, MTH17] (see also [Fit16] for a recent survey). In this case, only a single prover is needed, and the most efficient *single-prover quantum-verifier* protocols can evaluate a

[1] Using results of Ji [Ji16], this allows the protocol to be single-round. Alternatively, the state can be created by a single prover and teleported to the others with the help of the verifier, resulting in a two-round protocol.

[2] *Blindness* is a property of delegation protocols, which informally states that the prover learns nothing about the verifier's private circuit.

Table 1. Resource requirements of various delegation protocols in the multi-prover model. We use n to denote the number of qubits and g the number of gates in the delegated circuit. "depth" refers to the depth of the delegated circuit. "Total Resources" refers to the gate complexity of the provers, the number of EPR pairs of entanglement needed, and the number of bits of communication in the protocol. To ensure fair comparison, each protocol is required to produce the correct answer with probability 99%. For all protocols except our two new protocols, this requires a polynomial number of sequential repetitions, which is taken into account when computing the total resources.

	Provers	Rounds	Total Resources	Blind
RUV 2012 [RUV13]	2	poly(n)	$\geq g^{8192}$	yes
McKague 2013 [McK16]	poly(n)	poly(n)	$\geq 2^{153}g^{22}$	yes
GKW 2015 [GKW15]	2	poly(n)	$\geq g^{2048}$	yes
HDF 2015 [HPDF15]	poly(n)	poly(n)	$\Theta(g^4 \log g)$	yes
Verifier-on-a-Leash Protocol (Sect. 4)	2	$O(\text{depth})$	$\Theta(g \log g)$	yes
Dog-Walker Protocol (Sect. 5)	2	$O(1)$	$\Theta(g \log g)$	no

quantum circuit with g gates in time $O(g)$. The main reason these protocols are much more efficient than the classical-verifier multi-prover protocols is that they avoid the need for directly testing any of the qubits used by the prover, instead requiring the trusted verifier to directly either prepare or measure the qubits used for the computation.

New Rigidity Results. We overcome the efficiency limitations of multi-prover delegation protocols by introducing a new robust rigidity theorem. Our theorem allows a classical verifier to certify that two non-communicating provers apply a measurement associated with an arbitrary m-qubit tensor product of single-qubit Clifford observables on their respective halves of m shared EPR pairs. This is the first result to achieve self-testing for such a large class of measurements. The majority of previous works in self-testing have been primarily concerned with certifying the state and were limited to simple single-qubit measurements in the X-Z plane. Prior self-testing results for multi-qubit measurements only allow to test for tensor products of σ_X and σ_Z observables. While this is sufficient for verification in the post-hoc model of [FH15], testing for σ_X and σ_Z observables does not directly allow for the verification of a general computation (unless one relies on techniques such as process tomography [RUV13], which introduce substantial additional overhead).

Our first contribution is to extend the "Pauli braiding test" of [NV17], which allows to test tensor products of σ_X and σ_Z observables with constant robustness, to allow for σ_Y observables as well. This is somewhat subtle due to an ambiguity in the complex phase that cannot be detected by any classical two-player test; we formalize the ambiguity and show how it can be effectively accounted for. Our second contribution is to substantially increase the set of elementary gates that can be tested, to include arbitrary m-qubit tensor products of single-qubit Clifford observables. This is achieved by introducing a new "conjugation test", which tests how an observable applied by the provers acts on the Pauli group. The test is inspired by general results of Slofstra [Slo16], but is substantially more direct.

A key feature of our rigidity results is that their robustness scales independently of the number of EPR pairs tested, as in [NV17]. This is crucial for the efficiency of our delegation protocols. The robustness for previous results in parallel self-testing typically had a polynomial dependence on the number of EPR pairs tested. We give an informal statement of our robust rigidity theorem.

Theorem 1 (Informal). *Let* $m \in \mathbb{Z}_{>0}$. *Let* $\mathcal{G}$ *be a fixed, finite set of single-qubit Clifford observables. Then there exists an efficient two-prover test* $\text{RIGID}(\mathcal{G}, m)$ *with* $O(m)$*-bit questions (a constant fraction of which are of the form* $W \in \mathcal{G}^m$*) and answers such that the following properties hold:*

- *(Completeness) There is a strategy for the provers that uses* $m+1$ *EPR pairs and succeeds with probability at least* $1 - e^{-\Omega(m)}$ *in the test.*
- *(Soundness) For any* $\varepsilon > 0$, *any strategy for the provers that succeeds with probability* $1 - \varepsilon$ *in the test must be* $\text{poly}(\varepsilon)$*-close, up to local isometries, to a strategy in which the provers begin with* $(m + 1)$ *EPR pairs and is such that upon receipt of a question of the form* $W \in \mathcal{G}^m$ *the prover measures the "correct" observable* W.

Although we do not strive to obtain the best dependence on ε, we believe it should be possible to obtain a scaling of the form $C\sqrt{\varepsilon}$ for a reasonable constant C. We discuss the test in Sect. 3. The complete analysis can be found in the full version of the paper.

New Delegation Protocols. We employ the new rigidity theorem to obtain two new efficient two-prover classical-verifier protocols in which the complexity of verifiably delegating a g-gate quantum circuit solving a BQP problem scales as $O(g \log g)$.[3]

We achieve our protocols by adapting the efficient single-prover quantum-verifier delegation protocol introduced by Broadbent [Bro18] (we refer to this as the "EPR protocol"), which has the advantage of offering a direct implementation of the delegated circuit, in the circuit model of computation and with very little modification needed to ensure verifiability, as well as a relatively simple and intuitive analysis.

Our first protocol is blind, and requires a number of rounds of interaction that scales linearly with the depth of the circuit being delegated. The second protocol is not blind, but only requires a constant number of rounds of interaction with the provers. Our work is the first to propose verifiable two-prover

[3] The $\log g$ overhead is due to the complexity of sampling from the right distribution in rigidity tests. We leave the possibility of removing this by derandomization for future work. Another source of overhead is in achieving blindness: in order to hide the circuit, we encode it as part of the input to a universal circuit, introducing a factor of $O(\log g)$ overhead.

delegation protocols that overcome the prohibitively large resource requirements of all previous multi-prover protocols, requiring only a quasilinear amount of resources, in terms of number of EPR pairs and time. However, notwithstanding our improvements, a physical implementation of verifiable delegation protocols remains a challenging task for the available technology.

We introduce the protocols in more detail. The protocols provide different methods to delegate the quantum computation performed by the quantum verifier from [Bro18] to a second prover (call him PV for Prover V). The rigidity test is used to verify that the second prover indeed performs the same actions as the honest verifier, which are sequences of single-qubit measurements of Clifford observables from the set $\Sigma = \{X, Y, Z, F, G\}$ (where F and G are defined in (2)).

In the first protocol, one of the provers plays the role of Broadbent's prover (call him PP for Prover P), and the other plays the role of Broadbent's verifier (PV). As PV just performs single-qubit and Bell-basis measurements, universal quantum computational power is not needed for this prover. The protocol is divided into two sub-games; which game is played is chosen by the verifier by flipping a biased coin with appropriately chosen probabilities.

- The first game is a sequential version of the rigidity game RIGID(Σ, m) (from Theorem 1) described in Fig. 9. This aims to enforce that PV performs precisely the right measurements;
- The second game is the delegation game, described in Figs. 6, 7, and 8, and whose structure is summarized in Fig. 4. Here the verifier guides PP through the computation in a similar way as in the EPR Protocol.

We remark that in both sub-games, the questions received by PV are of the form $W \in \Sigma^m$, where $\Sigma = \{X, Y, Z, F, G\}$ is the set of measurements performed by the verifier in Broadbent's EPR protocol. The questions for PV in the two sub-games are sampled from the same distribution. This ensures that the PV is not able to tell which kind of game is being played. Hence, we can use our rigidity result of Theorem 1 to guarantee honest behavior of PV in the delegation sub-game. We call this protocol *Verifier-on-a-Leash Protocol*, or "leash protocol" for short.

The protocol requires $(2d + 1)$ rounds of interaction, where d is the depth of the circuit being delegated (see Sect. 2.3 for a precise definition of how this is computed). The protocol requires $O(n+g)$ EPR pairs to delegate a g-gate circuit on n qubits, and the overall time complexity of the protocol is $O(g \log g)$. The input to the circuit is hidden from the provers, meaning that the protocol can be made blind by encoding the circuit in the input, and delegating a universal circuit. We note that using universal circuits incurs a $\log n$ factor increase in the depth of the circuit [BFGH10].

The completeness of the protocol follows directly from the completeness of [Bro18]. Once we ensure the correct behavior of PV using our rigidity test, soundness follows from [Bro18] as well, since the combined behavior of our verifier and an honest PV is nearly identical to that of Broadbent's verifier.

The second protocol also starts from Broadbent's protocol, but modifies it in a different way to achieve a protocol that only requires a constant number of rounds of interaction. The proof of security is slightly more involved, but the key ideas are the same: we use a combination of our new self-testing results and the techniques of Broadbent's protocol to control the two provers, one of which plays the role of Broadbent's verifier, and the other the role of the prover. Because of the more complicated "leash" structure in this protocol, we call it the *Dog-Walker Protocol*. Like the leash protocol, the Dog-Walker Protocol has overall time complexity $O(g \log g)$. Unlike the leash protocol, the Dog-Walker protocol is not blind. In particular, while PV and PP would have to collude after the protocol is terminated to learn the input in the leash protocol, in the Dog-Walker protocol, PV simply receives the input in clear.

Based on the Dog-Walker Protocol, it is possible to design a classical-verifier two-prover protocol for all languages in QMA. This is achieved along the same lines as the proof that QMIP = MIP* from [RUV13]. The first prover, given the input, creates the QMA witness and teleports it to the second prover with the help of the verifier. The verifier then delegates the verification circuit to the second prover, as in the Dog-Walker Protocol; the first prover can be re-used to verify the operations of the second one.

Subsequent Work. Bowles et al. [BvCA18] have independently re-derived a variant of our rigidity test for multi-qubit σ_X, σ_Y and σ_Z observables in the context of entanglement certification protocols in quantum networks. Their self-test result has a slightly smaller set of questions but significantly weaker robustness bounds.

Recently [Gri17] proposed the first protocol for verifiable delegation of quantum computation by classical clients where such space-like separation can replace the non-communication assumption, but his protocol is not blind.

Open Questions and Directions for Future Work. We have introduced a new rigidity theorem and shown how it can be used to transform a specific quantum-verifier delegation protocol, due to Broadbent, into a classical-verifier protocol with an additional prover, while suffering very little overhead in terms of the efficiency of the protocol. We believe that a similar transformation could be performed starting from delegation protocols based on other models of computation, such as the protocol in the measurement-based model of [FK17] or the protocol based on computation by teleportation considered in [RUV13], and would lead to similar efficiency improvements.

Recently, [HZM+17] provided an experimental demonstration of a two-prover delegation protocol based on [RUV13] for a 3-qubit quantum circuit based on Shor's algorithm to factor the number 15; in order to obtain an actual implementation, necessitating "only" on the order of 6000 CHSH tests, the authors had to make the strong assumption that the devices behave in an i.i.d. manner at each use, and could not use the most general testing results from [RUV13]. We believe that our improved rigidity theorem could lead to an implementation that does not require any additional assumption. We also leave as an open problem investigating whether (a variant of) our protocol can be made fault-tolerant, making it more suitable for future implementation.

We note that our protocols require the verifier to communicate with one prover after at least one round of communication with the other has been completed. Therefore, the requirement that the provers do not communicate throughout the protocol cannot be enforced through space-like separation, and must be taken as an a priori assumption. Since the protocol of [Gri17] is not blind, it is an open question whether there exists a two-prover delegation protocol that consists of a single round of simultaneous communication with each prover, and is blind and verifiable. We also wonder if the fact that blindness is compromised after the provers collude is unavoidable in this model. A different avenue to achieve this is to rely on computational assumptions on the power of the provers to achieve protocols with more properties (non-interactive, blind, verifiable) [DSS16, ADSS17, Mah17, Mah18], albeit not necessarily in a truly efficient manner.

Finally, due to its efficiency and robustness, our ridigity theorem is a potentially useful tool in many other cryptographic protocols. For instance, an interesting direction to explore is the possibility of exploiting our theorem to achieve more efficient protocols for device-independent quantum key distribution, entanglement certification or other cryptographic protocols involving more complex untrusted computation of the users.

Organization. In Sect. 2, we give the necessary preliminaries, including outlining Broadbent's EPR Protocol (Sect. 2.3). In Sect. 3, we introduce our new rigidity theorems. In Sect. 4, we present our first protocol, the leash protocol, and in Sect. 5, we discuss our second protocol, the Dog-Walker Protocol.

2 Preliminaries

2.1 Notation

We often write $\boldsymbol{x} = (x_1, \ldots, x_n) \in \{0,1\}^n$ for a string of bits, and $W = W_1 \cdots W_m \in \Sigma^m$ for a string, where Σ is a finite alphabet. If $S \subseteq \{1, \ldots, m\}$ we write W_S for the sub-string of W indexed by S. For an event E, we use 1_E to denote the indicator variable for that event, so $1_E = 1$ if E is true, and otherwise $1_E = 0$. We write $\text{poly}(\varepsilon)$ for $O(\varepsilon^c)$, where c is a universal constant that may change each time the notation is used.

$\mathcal{H}$ is a finite-dimensional Hilbert space. We denote by $\text{U}(\mathcal{H})$ the set of unitary operators, $\text{Obs}(\mathcal{H})$ the set of binary observables (we omit the term "binary" from here on; in this paper all observables are binary) and $\text{Proj}(\mathcal{H})$ the set of projective measurements on $\mathcal{H}$ respectively. We let $|\text{EPR}\rangle$ denote an EPR pair:

$$|\text{EPR}\rangle = \frac{1}{\sqrt{2}}(|00\rangle + |11\rangle).$$

Observables. We use capital letters $X, Z, W, \ldots$ to denote observables. We use greek letters σ, τ with a subscript σ_W, τ_W, to emphasize that the observable W specified as subscript acts in a particular basis. For example, X is an arbitrary observable but σ_X is specifically the Pauli X matrix defined in (1).

For $a \in \{0,1\}^n$ and commuting observables $\sigma_{W_1}, \ldots, \sigma_{W_n}$, we write $\sigma_W(a) = \prod_{i=1}^n (\sigma_{W_i})^{a_i}$. The associated projective measurements are $\sigma_{W_i} = \sigma_{W_i}^0 - \sigma_{W_i}^1$ and $\sigma_W^u = \mathrm{E}_a(-1)^{u \cdot a} \sigma_W(a)$. Often the σ_{W_i} will be single-qubit observables acting on distinct qubits, in which case each is implicitly tensored with identity outside of the qubit on which it acts.

Pauli and Clifford groups. Let

$$\sigma_I = \begin{pmatrix} 1 & 0 \\ 0 & 1 \end{pmatrix}, \quad \sigma_X = \begin{pmatrix} 0 & 1 \\ 1 & 0 \end{pmatrix}, \quad \sigma_Y = \begin{pmatrix} 0 & -i \\ i & 0 \end{pmatrix} \quad \text{and} \quad \sigma_Z = \begin{pmatrix} 1 & 0 \\ 0 & -1 \end{pmatrix} \tag{1}$$

denote the standard Pauli matrices acting on a qubit. The single-qubit Weyl-Heisenberg group

$$\mathcal{H}^{(1)} = H(\mathbb{Z}_2) = \Big\{ (-1)^c \sigma_X(a) \sigma_Z(b),\ a, b, c \in \{0,1\} \Big\}$$

is the matrix group generated by the Pauli σ_X and σ_Z. We let $\mathcal{H}^{(n)} = H(\mathbb{Z}_2^n)$ be the direct product of n copies of $\mathcal{H}^{(1)}$. The n-qubit Clifford group is the normalizer of $\mathcal{H}^{(n)}$ in the unitary group, up to phase:

$$G_C^{(n)} = \{ G \in \mathrm{U}((\mathbb{C}^2)^{\otimes n}) : G \sigma G^\dagger \in \mathcal{H}^{(n)} \quad \forall \sigma \in \mathcal{H}^{(n)} \}.$$

Some Clifford observables we will use include

$$\sigma_H = \frac{\sigma_X + \sigma_Z}{\sqrt{2}}, \qquad \sigma_{H'} = \frac{\sigma_X - \sigma_Z}{\sqrt{2}}, \qquad \sigma_F = \frac{-\sigma_X + \sigma_Y}{\sqrt{2}}, \qquad \sigma_G = \frac{\sigma_X + \sigma_Y}{\sqrt{2}}. \tag{2}$$

Note that σ_H and $\sigma_{H'}$ are characterized by $\sigma_X \sigma_H \sigma_X = \sigma_{H'}$ and $\sigma_Z \sigma_H \sigma_Z = -\sigma_{H'}$. Similarly, σ_F and σ_G are characterized by $\sigma_X \sigma_F \sigma_X = -\sigma_G$ and $\sigma_Y \sigma_F \sigma_Y = \sigma_G$.

2.2 Quantum Circuits

We use capital letters in sans-serif font to denote gates. We work with the universal quantum gate set $\{\mathsf{CNOT}, \mathsf{H}, \mathsf{T}\}$, where the controlled-not gate is the two-qubit gate with the unitary action

$$\mathsf{CNOT}|b_1, b_2\rangle = |b_1, b_1 \oplus b_2\rangle,$$

and the Hadamard and T gates are single-qubit gates with actions

$$\mathsf{H}|b\rangle = \frac{1}{\sqrt{2}} \left(|0\rangle + (-1)^b |1\rangle \right) \quad \text{and} \quad \mathsf{T}|b\rangle = e^{ib\pi/4} |b\rangle,$$

respectively. We will also use the following gates:

$$\mathsf{X}|b\rangle = |b \oplus 1\rangle, \quad \mathsf{Z}|b\rangle = (-1)^b |b\rangle, \quad \text{and} \quad \mathsf{P}|b\rangle = i^b |b\rangle.$$

Measurements in the Z basis (or computational basis) will be denoted by the standard measurement symbol:

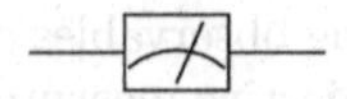

To measure another observable, W, we can perform a unitary change of basis U_W before the measurement in the computational basis.

We assume that every circuit has a specified output wire, which is measured at the end of the computation to obtain the output bit. Without loss of generality, we can assume this is always the first wire. For an n-qubit system, we let Π_b, for $b \in \{0,1\}$, denote the orthogonal projector onto states with $|b\rangle$ in the output wire: $|b\rangle\langle b| \otimes \mathrm{Id}$. For example, the probability that a circuit Q outputs 0 on input $|\boldsymbol{x}\rangle$ is $\|\Pi_0 Q|\boldsymbol{x}\rangle\|^2$.

We can always decompose a quantum circuit into layers such that each layer contains at most one T gate applied to each wire. The minimum number of layers for which this is possible is called the T *depth* of the circuit. We note that throughout this work, we will assume circuits are compiled in a specific form that introduces extra T gates (see the paragraph on the H gadget in Sect. 2.3). The T depth of the resulting circuit is proportional to the depth of the original circuit.

2.3 Broadbent's EPR Protocol

In this section we summarize the main features of a delegation protocol introduced in [Bro18], highlighting the aspects that will be relevant to understanding our subsequent adaptation into two-prover protocols. The "EPR Protocol" from [Bro18] involves the interaction between a verifier V_{EPR} and a prover P. We write P_{EPR} for the "honest" behavior of the prover. The verifier V_{EPR} has limited quantum powers. Her goal is to delegate a BQP computation to the prover P in a verifiable way. Specifically, the verifier has as input a quantum circuit Q on n qubits and an input string $\boldsymbol{x} \in \{0,1\}^n$, and the prover gets as input Q. The verifier and prover interact. At the end of the protocol, the verifier outputs either accept or reject. The protocol is such that there exist values p_{sound} and p_{compl} with $p_{\mathrm{sound}} < p_{\mathrm{compl}}$ such that $p_{\mathrm{compl}} - p_{\mathrm{sound}}$, called the *soundness-completeness gap*, is a constant independent of input size, and moreover:

Completeness: If the prover is honest and $\|\Pi_0 Q|\boldsymbol{x}\rangle\|^2 \geq 2/3$, then the verifier outputs accept with probability at least p_{compl};

Soundness: If $\|\Pi_0 Q|\boldsymbol{x}\rangle\|^2 \leq 1/3$, then the probability the verifier outputs accept is at most p_{sound}.

In the EPR protocol, V_{EPR} and P_{EPR} are assumed to share $(n+t)$ EPR pairs at the start of the protocol, where t is the number of T gates in Q and n the number of input bits. (In [Bro18] the EPR protocol is only considered in the analysis, and it is assumed that the EPR pairs are prepared by the verifier.) The first n EPR pairs correspond to the input to the computation; they are indexed by $N = \{1,\ldots,n\}$. The remaining pairs are indexed by $T = \{n+1,\ldots,n+t\}$; they will be used as ancilla qubits to implement each of the T gates in the delegated circuit.

The behavior of V_{EPR} depends on a *round type* randomly chosen by V_{EPR} after her interaction with P_{EPR}. There are three possible round types:

- Computation round ($r = 0$): the verifier delegates the computation to P_{EPR}, and at the end of the round can recover its output if P_{EPR} behaves honestly;
- X-test round ($r = 1$) and Z-test round ($r = 2$): the verifier tests that P_{EPR} behaves honestly, and rejects if malicious behavior is detected.

For some constant p, V chooses $r = 0$ with probability p, and otherwise chooses $r \in \{1, 2\}$ with equal probability. Since the choice of round type is made after interaction with P_{EPR}, P_{EPR}'s behavior cannot depend on the round type. In particular, any deviating behavior in a computation round is reproduced in both types of test rounds. The analysis amounts to showing that any deviating behavior that affects the outcome of the computation will be detected in at least one of the test rounds.

In slightly more detail, the high-level structure of the protocol is the following. V_{EPR} measures her halves of the n qubits in N in order to prepare the input state on P_{EPR}'s system. As a result the input is quantum one-time padded with keys that depend on V_{EPR}'s measurement results. For example, in a computation round, V_{EPR} measures each input qubit in the Z basis, and gets some result $\boldsymbol{d} \in \{0,1\}^n$, meaning the input on P_{EPR}'s side has been prepared as $\mathsf{X}^{\boldsymbol{d}}|0\rangle^{\otimes n}$. In [Bro18], the input is always considered to be $\mathbf{0}$, but we can also prepare an arbitrary classical input $\boldsymbol{x} \in \{0,1\}^n$ by reinterpreting the one-time pad key as $\boldsymbol{a} = \boldsymbol{d} \oplus \boldsymbol{x}$ so that the input state on P_{EPR}'s side is $\mathsf{X}^{\boldsymbol{a}}|\boldsymbol{x}\rangle$. In a test round, on the other hand, the input is prepared as the one-time pad of either $|0\rangle^{\otimes n}$ or $|+\rangle^{\otimes n}$. Note that as indicated in Fig. 2 this choice of measurements will be made after the interaction with P_{EPR} has taken place.

The honest prover P_{EPR} applies the circuit Q, which we assume is compiled in the universal gate set $\{\mathsf{H}, \mathsf{T}, \mathsf{CNOT}\}$, to his one-time padded input. We will shortly describe gadgets that P_{EPR} can apply in order to implement each of the three gate types. The gadgets are designed in a way that in a test round each gadget amounts to an application of an identity gate; this is what enables V_{EPR} to perform certain tests in those rounds that are meant to identify deviating behavior of a dishonest prover. After each gadget, the one-time padded keys can be updated by V_{EPR}, who is able to keep track of the keys at any point in the circuit using the *update rules* in Table 2.

We now describe the three gadgets, before giving a complete description of the protocol.

CNOT Gadget. To implement a CNOT gate on wires j and j', P_{EPR} simply performs the CNOT gate on those wires of his input qubits. The one-time pad keys are changed by the update rule in Table 2, because $\mathsf{CNOT} \cdot \mathsf{X}^{a_j}\mathsf{Z}^{b_j} \otimes \mathsf{X}^{a_{j'}}\mathsf{Z}^{b_{j'}} = \mathsf{X}^{a_j}\mathsf{Z}^{b_j+b_{j'}} \otimes \mathsf{X}^{a_j+a_{j'}}\mathsf{Z}^{b_{j'}} \cdot \mathsf{CNOT}$. Note that $\mathsf{CNOT}|0\rangle|0\rangle = |0\rangle|0\rangle$ and $\mathsf{CNOT}|+\rangle|+\rangle = |+\rangle|+\rangle$, so in the test runs, P_{EPR} is applying the identity.

H Gadget. To implement an H gate on wire j, P_{EPR} simply performs the H on wire j, and the one-time-pad keys are changed as in Table 2. Unlike CNOT, H

Table 2. Rules for updating the one-time-pad keys after applying each type of gate in the EPR Protocol, in particular: after applying the i-th T gate to the j-th wire; applying an H gate to the j-th wire; or applying a CNOT gate controlled on the j-th wire and targeting the j'-th wire.

		Key Update Rule
T	Computation Round	$(a_j, b_j) \leftarrow (a_j + c_i, b_j + e_i + a_j + c_i + (a_j + c_i)z_i)$
	X-Test, even parity; or Z-test, odd parity	$(a_j, b_j) \leftarrow (e_i, 0)$
	Z-Test, even parity; or X-test, odd parity	$(a_j, b_j) \leftarrow (0, b_j + e_i + z_i)$
H		$(a_j, b_j) \leftarrow (b_j, a_j)$
CNOT		$(a_j, b_j, a_{j'}, b_{j'}) \leftarrow (a_j, b_j + b_{j'}, a_j + a_{j'}, b_{j'})$

does not act as the identity on $|0\rangle$ and $|+\rangle$, so it is not the identity in a test round. To remedy this, assume that Q is compiled so that every H gate appears in a pattern $\mathsf{H}(\mathsf{TTH})^k$, where the maximal such k is odd. This can be accomplished by replacing each H by HTTHTTHTTH, which implements the same unitary. In test rounds, the T gadget, described shortly, implements the identity, and since $\mathsf{H}(\mathsf{Id}\,\mathsf{H})^k$ for odd k implements the identity, $\mathsf{H}(\mathsf{TTH})^k$ will also have no effect in test rounds.

Parity of a T Gate. Within a pattern $\mathsf{H}(\mathsf{TTH})^k$, the H has the effect of switching between an X-test round scenario (the state $|0\rangle$) and a Z-test round scenario (the state $|+\rangle$). In order to consistently talk about the type of a round while evaluating the circuit, we can associate a parity with each T gate in the circuit. The parity of the T gates that are not part of the pattern $\mathsf{H}(\mathsf{TTH})^k$ will be defined to be even. A H will always flip the parity, so that within such a pattern, the first two T gates will be odd, the next two will be even, etc., until the last two T gates will be odd again.

T Gadget. The gadget for implementing the i-th T gate on the j-th wire is performed on P_{EPR}'s j-th input qubit, and his i-th auxiliary qubit (indexed by $n+i$), which we can think of as being prepared in a particular auxiliary state by V_{EPR} measuring her half of the corresponding EPR pair, as shown in Fig. 1. The gadget depends on a random bit z_i that is chosen by V_{EPR} and sent to the prover.

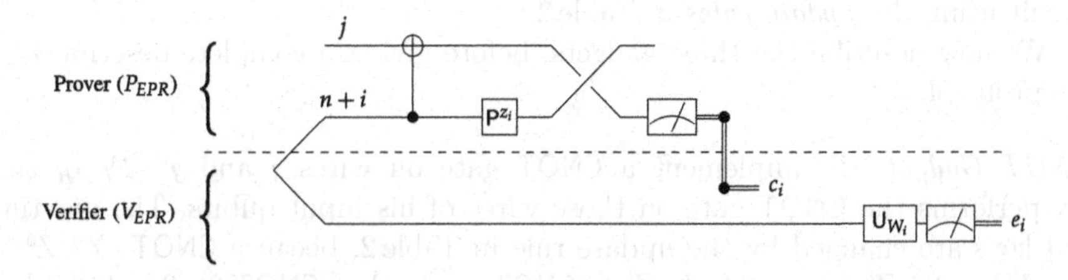

Fig. 1. The gadget for implementing the i-th T gate on the j-th wire. The gate U_{W_i} implementing the change of basis associated with observable W_i is applied as part of the procedure V_{EPR}^r (see Fig. 3b) and is determined by the round type r, the parity of the i-th T gate, z_i, c_i, and a_i' (the X-key going into the i-th T gate), as in Table 3.

Table 3. The choice of U_{W_i} in the T gadget. We also indicate the observable W_i associated with the final measurement $W_i = \mathsf{U}_{W_i}^{\dagger} Z \mathsf{U}_{W_i}$.

			U_{W_i} (observable W_i)
Computation Round	$a_i' \oplus c_i \oplus z_i = 0$		HT (observable G)
	$a_i' \oplus c_i \oplus z_i = 1$		HPT (observable F)
X-Test Round	even T gate		Id (observable Z)
	odd T gate	$z_i = 0$	H (observable X)
		$z_i = 1$	HP (observable Y)
Z-Test Round	odd T gate		Id (observable Z)
	even T gate	$z_i = 0$	H (observable X)
		$z_i = 1$	HP (observable Y)

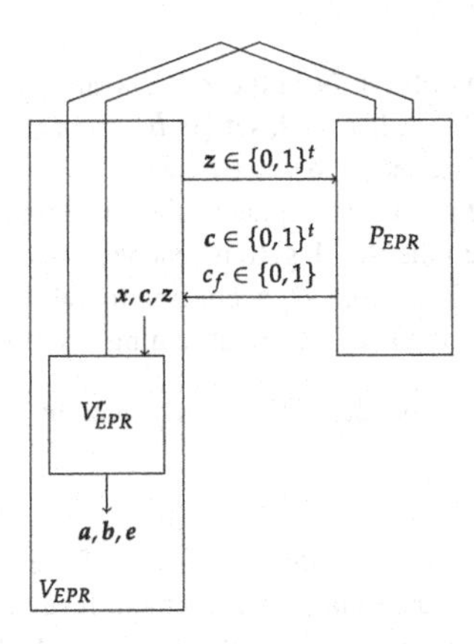

Fig. 2. This figure describes how different pieces of the protocol fit together. V_{EPR} and P_{EPR} share $n+t$ EPR pairs. The honest prover P_{EPR} can be seen as a procedure that acts on $n+t$ qubits—the EPR pair halves—depending on a t-bit string $\boldsymbol{z}$. We have separated the quantum part of V_{EPR} into its own procedure, called V_{EPR}^r, where $r \in \{0,1,2\}$ indicates the *round type*, which V_{EPR} runs on her $n+t$ EPR halves, and the $2t$ bits $\boldsymbol{c}$ and $\boldsymbol{z}$. Aside from running V_{EPR}^r, V_{EPR} is classical.

The EPR Protocol. We show how the gadgets just described are used in the complete protocol. We first describe the protocol for V_{EPR} below. For later convenience we have divided the action of V_{EPR} into classical actions and a single quantum subroutine V_{EPR}^r depending on the round type.

The procedure V_{EPR}^r measures each of the $n+t$ EPR halves according to some observable that depends on r, $\boldsymbol{c}$, and $\boldsymbol{z}$. In the case of a computation round, V_{EPR}^0 measures the qubits in T adaptively. We describe the steps of V_{EPR}, V_{EPR}^r and the honest behaviour of P_{EPR} in Fig. 3.

Completeness and Soundness. We summarize the relevant part of the analysis of the EPR protocol from [Bro18]. First suppose P_{EPR} behaves honestly. If

1. V_{EPR} sends $z \in_R \{0,1\}^t$ to P_{EPR}, and receives back $c \in \{0,1\}^t$ and $c_f \in \{0,1\}$.
2. V_{EPR} chooses a random round type $r \in \{0,1,2\}$ and runs V_{EPR}^r (see Figure 3b) on her EPR halves, x, c and z, to obtain bits $a, b \in \{0,1\}^n$ and $e \in \{0,1\}^t$.
3. V_{EPR} applies the update rules from Table 2 on the initial keys (a, b), gate-by-gate, to obtain, for every $i \in [t]$, the X-key before the i-th T gate is applied, a'_i, and the final X key for the output wire, a'_f. If $r = 1$ (X-test round) and there exists an i such that the i-th T gate is even and $c_i \neq a'_i \oplus e_i$, output reject. If $r = 2$ (Z-test round) and there exists an i such that the i-th T gate is odd and $c_i \neq a'_i \oplus e_i$, output reject. If $r \in \{0,1\}$ (computation or X-test round) and $c_f \oplus a'_f \neq 0$, output reject. Otherwise, output accept.

(a) V_{EPR}'s point of view.

Input: A circuit Q with t T gates, $x \in \{0,1\}^n$, $c, z \in \{0,1\}^t$, an n-qubit system indexed by N, and a t-qubit system indexed by T.

1. If $r \in \{0,1\}$, measure each qubit in N in the Z basis, and otherwise measure in the X basis, to get results $d \in \{0,1\}^n$. If $r = 0$, set $(a, b) = (d \oplus x, 0^n)$; if $r = 1$, set $(a, b) = (d, 0^n)$; and if $r = 2$ set $(a, b) = (0^n, d)$.
2. Going through Q gate-by-gate, use the update rules in Table 2 to update the one-time-pad keys. For every $i \in [t]$, when the i-th T gate is reached, let a'_i be the X key before the i-th T gate is applied. Choose an observable W_i according to Table 3 in which to measure the i-th qubit in T, corresponding to the i-th T gate, obtaining result e_i.

(b) The procedure V_{EPR}^r, employed by V_{EPR}.

1. Receive $z \in \{0,1\}^t$ from V_{EPR}.
2. Evaluate Q gate-by-gate using the appropriate gadget for each gate. In particular, use z_i to implement the i-th T gadget, and obtain measurement result c_i.
3. Measure the output qubit to obtain c_f, and return c and c_f to V_{EPR}.

(c) Honest prover strategy P_{EPR}

Fig. 3. The EPR Protocol.

$\|\Pi_0 Q|0^n\rangle\|^2 = p$, then in a computation round, V_{EPR} outputs accept with probability p, whereas in a test round, V_{EPR} outputs accept with probability 1. This establishes completeness of the protocol:

Theorem 2 (Completeness). *Suppose the verifier executes the EPR Protocol, choosing $r = 0$ with probability p, on an input $(Q, |x\rangle)$ such that $\|\Pi_0 Q|x\rangle\|^2 \geq 1 - \delta$. Then the probability that V_{EPR} accepts when interacting with the honest prover P_{EPR} is at least $(1-p) + p(1-\delta)$.*

The following theorem is implicit in [Bro18, Section 7.6], but we include a brief proof sketch:

Theorem 3 (Soundness). *Suppose the verifier executes the EPR Protocol, choosing $r = 0$ with probability p, on an input $(Q, |x\rangle)$ such that $\|\Pi_0 Q|x\rangle\|^2 \leq \delta$.*

Let P^*_{EPR} be an arbitrary prover such that P^*_{EPR} is accepted by V_{EPR} with probability q_t conditioned on $r \neq 0$, and q_c conditioned on $r = 0$. Then the prover's overall acceptance probability is $pq_c + (1-p)q_t$, and

$$q_c \leq 2\,(q_t\,\delta + (1 - q_t)) - \delta.$$

Proof (Proof sketch). Using the notation of [Bro18], let $A = \sum_k \sum_{Q \in B'_{t,n}} |\alpha_{k,Q}|^2$.[4] For intuition, A should be thought of as the total weight on attacks that could change the outcome of the computation, called non-benign attacks in [Bro18]. By [Bro18], the probability of rejecting in a computation round is $1 - q_c \geq (1-\delta)(1-A)$, whereas the probability of rejecting in a test round is $1 - q_t \geq \frac{1}{2}A$. Combining these gives $q_c \leq 2(q_t\delta + (1 - q_t)) - \delta$.

3 Rigidity

Each of our delegation protocols includes a *rigidity test* that is meant to verify that one of the provers measures his half of shared EPR pairs in a basis specified by the verifier, thereby preparing one of a specific family of post-measurement states on the other prover's space; the post-measurement states will form the basis for the delegated computation. This will be used to certify that one of the provers in our two-prover schemes essentially behaves as the quantum part of V_{EPR} would in the EPR protocol.

In this section we outline the structure of the test, giving the important elements for its use in our delegation protocols. We refer the reader to the full version of the paper for a detailed presentation, including the soundness analysis. The test is parametrized by the number m of EPR pairs to be used. The test consists of a single round of classical interaction between the verifier and the two provers. With constant probability the verifier sends one of the provers a string W chosen uniformly at random from Σ^m where the set $\Sigma = \{X, Y, Z, F, G\}$ contains a label for each single-qubit observable to be tested. With the remaining probability, other queries, requiring the measurement of observables not in Σ^m (such as the measurement of pairs of qubits in the Bell basis), are sent.

In general, an arbitrary strategy for the provers consists of an arbitrary entangled state $|\psi\rangle \in \mathcal{H}_\mathsf{A} \otimes \mathcal{H}_\mathsf{B}$ (which we take to be pure), and measurements (which we take to be projective) for each possible question.[5] This includes an m-bit outcome projective measurement $\{W^u\}_{u \in \{0,1\}^m}$ for each of the queries $W \in \Sigma^m$. Our rigidity result states that any strategy that succeeds with probability $1 - \varepsilon$ in the test is within poly(ε) of the honest strategy, up to local isometries (see

[4] Here, we consider the decomposition of the attack as a sum of tensors of Pauli $A = \sum_k \sum_{Q \in \{I,X,Z,Y\}} \alpha_{k,Q} Q$.

[5] We make the assumption that the players employ a pure-state strategy for convenience, but it is easy to check that all proofs extend to the case of a mixed strategy. Moreover, it is always possible to consider (as we do) projective strategies only by applying Naimark's dilation theorem, and adding an auxiliary local system to each player as necessary, since no bound is assumed on the dimension of their systems.

Theorem 4 for a precise statement). This is almost true, but for an irreconcilable ambiguity in the definition of the complex phase $\sqrt{-1}$. The fact that complex conjugation of observables leaves correlations invariant implies that no classical test can distinguish between the two nontrivial inequivalent irreducible representations of the Pauli group, which are given by the Pauli matrices $\sigma_X, \sigma_Y, \sigma_Z$ and their complex conjugates $\overline{\sigma_X} = \sigma_X$, $\overline{\sigma_Z} = \sigma_Z$, $\overline{\sigma_Y} = -\sigma_Y$ respectively. In particular, the provers may use a strategy that uses a combination of both representations; as long as they do so consistently, no test will be able to detect this behavior.[6] The formulation of our result accommodates this irreducible degree of freedom by forcing the provers to use a single qubit, the $(m+1)$-st, to make their choice of representation (so honest provers require the use of $(m+1)$ EPR pairs to test the operation of m-fold tensor products of observables from Σs).

Theorem 4 below summarizes the guarantees of our main test, which is denoted as RIGID(Σ, m). Informally, Theorem 4 establishes that a strategy that succeeds in RIGID(Σ, m) with probability at least $1-\epsilon$ must be such that (up to local isometries):

- The players' joint state is close to a tensor product of m EPR pairs, together with an arbitrary ancilla register;
- The projective measurements performed by either player upon receipt of a query of the form $W \in \Sigma^m$ are, on average over the uniformly random choice of $W \in \Sigma^m$, close to a measurement that consists in first, measuring the ancilla register to extract a single bit that specifies whether to perform the ideal measurements or their conjugated counterparts, and second, measuring the player's m half-EPR pairs in either the bases indicated by W, or their complex conjugate, depending on the bit obtained from the ancilla register.

For an observable $W \in \Sigma$, let $\sigma_W = \sigma_W^{+1} - \sigma_W^{-1}$ be its eigendecomposition, where σ_W are the "honest" Pauli matrices defined in (1) and (2). For $u \in \{\pm 1\}$ let $\sigma_{W,+}^u = \sigma_W^u$ for $W \in \Sigma$, and

$$\sigma_{X,-}^u = \sigma_X^u, \quad \sigma_{Z,-}^u = \sigma_Z^u, \quad \sigma_{Y,-}^u = \sigma_Y^{-u}, \quad \sigma_{F,-}^u = \sigma_G^{-u}, \quad \sigma_{G,-}^u = \sigma_F^{-u}.$$

(In words, $\sigma_{W,-}^u$ is just the complex conjugate of σ_W^u.) We note that for the purpose of our delegation protocols, we made a particular choice of the set Σ. The result generalizes to any constant-sized set of single-qubit Clifford observables, yielding a test for m-fold tensor products of single-qubit Clifford observables from Σ.

Theorem 4. *Let $\varepsilon > 0$ and m an integer. Suppose a strategy for the players succeeds with probability $1-\varepsilon$ in test* RIGID(Σ, m). *For $W \in \Sigma^m$ and $D \in \{A, B\}$ let $\{W_D^u\}_u$ be the measurement performed by prover D on question W. Let also $|\psi\rangle$ be the state shared by the players. Then for $D \in \{A, B\}$ there exists an isometry*

$$V_D : \mathcal{H}_D \to (\mathbb{C}^2)_{D'}^{\otimes m} \otimes \mathcal{H}_{\hat{D}}$$

[6] See [RUV12, Appendix A] for an extended discussion of this issue, with a similar resolution to ours.

such that

$$\big\|(V_A \otimes V_B)|\psi\rangle_{AB} - |\mathrm{EPR}\rangle^{\otimes m} \otimes |\mathrm{AUX}\rangle_{\widehat{A}\widehat{B}}\big\|^2 = O(\sqrt{\varepsilon}), \tag{3}$$

and positive semidefinite matrices τ_λ *on* $\widehat{A}$ *with orthogonal support, for* $\lambda \in \{+,-\}$*, such that* $\mathrm{Tr}(\tau_+) + \mathrm{Tr}(\tau_-) = 1$ *and*

$$\mathop{\mathrm{E}}_{W\in\Sigma^m} \sum_{u\in\{\pm 1\}^m} \Big\| V_A \mathrm{Tr}_B\big((\mathrm{Id}_A \otimes W_B^u)|\psi\rangle\langle\psi|_{AB}(\mathrm{Id}_A \otimes W_B^u)^\dagger\big)V_A^\dagger - \sum_{\lambda\in\{\pm\}} \Big(\bigotimes_{i=1}^m \frac{\sigma_{W_i,\lambda}^{u_i}}{2}\Big) \otimes \tau_\lambda \Big\|_1 = O(\mathrm{poly}(\varepsilon)).$$

Moreover, players employing the honest strategy succeed with probability $1 - e^{-\Omega(m)}$ *in the test.*

The proof of the theorem is based on standard techniques developed in the literature on "rigidity theorems" for nonlocal games. We highlight two components. The first is a "conjugation test" that allows us to extend the guarantees of elementary tests based on the CHSH game or the Magic Square game, which test for Pauli σ_X and σ_Z observables, to a test for single-qubit Clifford observables—since the latter are characterized by their action on the Pauli group (see full version of the paper for details). The second is an extension of the "Pauli braiding test" from [NV17] to handle tensor products of not only σ_X and σ_Z, but also σ_Y Pauli observables (see full version of the paper for details). As already emphasized in the introduction, the improvements in efficiency of our scheme are partly enabled by the strong guarantees of Theorem 4, and specifically the independence of the final error dependence from the parameter m.

4 The Verifier-on-a-Leash Protocol

4.1 Protocol and Statement of Results

The Verifier-on-a-Leash Protocol (or "Leash Protocol" for short) involves a classical verifier and two quantum provers. The idea behind the Leash Protocol is to have a first prover, nicknamed PV for Prover V, carry out the quantum part of V_{EPR} from Broadbent's EPR Protocol by implementing the procedure V_{EPR}^r. (See Sect. 2.3 for a summary of the protocol and a description of V_{EPR}. Throughout this section we assume that the circuit Q provided as input is compiled in the format described in Sect. 2.3.). A second prover, nicknamed PP for Prover P, will play the part of the prover P_{EPR}. Unlike in the EPR Protocol, the interaction with PV (i.e. running V_{EPR}^r) will take place first, and PV will be asked to perform random measurements from the set $\Sigma = \{X, Y, Z, F, G\}$. The values $\boldsymbol{z}$, rather than being chosen at random, will be chosen based on the corresponding choice of observable. We let n be the number of input bits and t number of T gates in Q.

The protocol is divided into two sub-games; which game is played is chosen by the verifier by flipping a biased coin with probability $(p_r, p_d = 1 - p_r)$.

- The first game is a sequential version of the rigidity game $\text{RIGID}(\Sigma, m)$ described in Fig. 9. This aims to enforce that PV performs precisely the right measurements;
- The second game is the delegation game, described in Figs. 6, 7, and 8, and whose structure is summarized in Fig. 4. Here the verifier guides PP through the computation in a similar way as in the EPR Protocol.

We call the resulting protocol the Leash Protocol with parameters (p_r, p_d). In both sub-games the parameter $m = \Theta(n+t)$ is chosen large enough so that with probability close to 1 each symbol in Σ appears in a random $W \in \Sigma^m$ at least $n+t$ times. It is important that PV is not able to tell which kind of game is being played. Notice also that in order to ensure blindness, we will require that the interaction with PV in the delegation game is sequential (more details on this are found in Sect. 4.4). In order for the two sub-games to be indistinguishable, we also require that the rigidity game $\text{RIGID}(\Sigma, m)$ be played sequentially (i.e. certain subsets of questions and answers are exchanged sequentially, but the acceptance condition in the test is the same). Note, importantly, that the rigidity guarantees of $\text{RIGID}(\Sigma, m)$ hold verbatim when the game is played sequentially, since this only reduces the number of ways that the provers can cheat. The following theorem states the guarantees of the Leash Protocol.

Theorem 5. *There are constants $p_r, p_d = 1 - p_r$, and $\Delta > 0$ such that the following hold of the Verifier-on-a-Leash Protocol with parameters (p_r, p_d), when executed on an input $(Q, |\boldsymbol{x}\rangle)$.*

- (Completeness:) *Suppose that $\|\Pi_0 Q|\boldsymbol{x}\rangle\|^2 \geq 2/3$. Then there is a strategy for* PV *and* PP *that is accepted with probability at least $p_{\text{compl}} = p_r(1 - e^{-\Omega(n+t)}) + 8p_d/9$.*
- (Soundness:) *Suppose that $\|\Pi_0 Q|\boldsymbol{x}\rangle\|^2 \leq 1/3$. Then any strategy for* PV *and* PP *is accepted with probability at most $p_{\text{sound}} = p_{\text{compl}} - \Delta$.*

Further, the protocol leaks no information about $\boldsymbol{x}$ to either prover individually, aside from an upper bound on the length of $\boldsymbol{x}$.

The proof of the completeness property is given in Lemma 1. The soundness property is shown in Lemma 4. Blindness is established in Sect. 4.4. We first give a detailed description of the protocol. We start by describing the delegation game, specified in Figs. 6, 7 and 8, which describe the protocol from the verifier's view, an honest PV's view, and an honest PP's view respectively. This will motivate the need for a sequential version of the game $\text{RIGID}(\Sigma, m)$, described in Fig. 9. As we will show, the rigidity game forces PV to behave honestly. Thus, for the purpose of exposition, we assume for now that PV behaves honestly, which results in the joint behavior of PV and V being similar to that of the verifier V_{EPR} in the EPR Protocol.

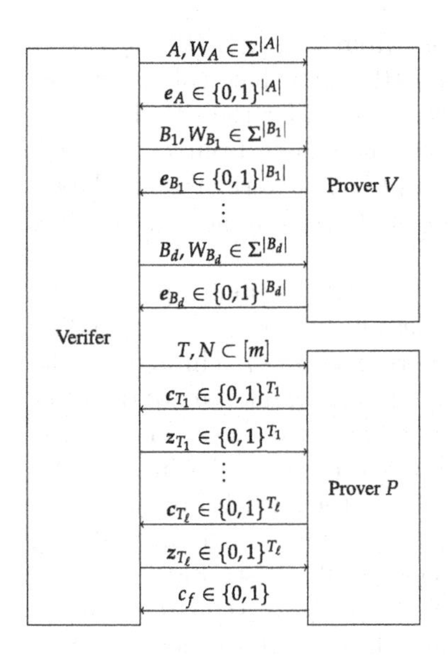

Fig. 4. Structure of the delegation game.

From the rigidity game we may also assume that PV and PP share m EPR pairs, labeled $\{1,\ldots,m\}$, for $m = \Theta(n+t)$. We will assume that the circuit Q is broken into d layers, $Q = Q_1 \ldots Q_d$, such that in every Q_ℓ, each wire has at most one T gate applied to it, after which no other gates are applied to that wire. We will suppose the T gates are indexed from 1 to t, in order of layer.

The protocol begins with an interaction between the verifier and PV. The verifier selects a uniformly random partition $A, B_1, \ldots, B_d$ of $\{1,\ldots,m\}$, with $|A| = \Theta(n)$, and for every $\ell \in \{1,\ldots,d\}$, $|B_\ell| = \Theta(t_\ell)$, where t_ℓ is the number of T gates in Q_ℓ. The verifier also selects a uniformly random $W \in \Sigma^m$, and partitions it into substrings W_A and $W_{B_1}, \ldots, W_{B_d}$, meant to contain observables to initialize the computation qubits and auxiliary qubits for each layer of T gates respectively. The verifier instructs PV to measure his halves of the EPR pairs using the observables W_A first, and then $W_{B_1}, \ldots, W_{B_d}$, sequentially. Upon being instructed to measure a set of observables, PV measures the corresponding half-EPR pairs and returns the results $\boldsymbol{e}$ to the verifier. Breaking this interaction into multiple rounds is meant to enforce that, for example, the results output by PV upon receiving W_{B_ℓ}, which we call $\boldsymbol{e}_{B_\ell}$, cannot depend on the choice of observables $W_{B_{\ell+1}}$. This is required for blindness.

Once the interaction with PV has been completed, as in the EPR Protocol, V selects one of three round types: computation ($r = 0$), X-test ($r = 1$), and Z-test ($r = 2$). The verifier selects a subset $N \subset A$ of size n of qubits to play the role of inputs to the computation. These are chosen from the subset of A corresponding to wires that PV has measured in the appropriate observable for the round type (see Table 4). For example, in an X-test round, PV's EPR halves corresponding to input wires should be measured in the Z basis so that PP is left with a one-time pad of the state $|0\rangle^{\otimes n}$, so in an X-test round, the computation wires are chosen from the set $\{i \in A : W_i = Z\}$. The input wires N are labeled by $\mathcal{X}_1, \ldots, \mathcal{X}_n$.

The verifier also chooses subsets $T_\ell = T_\ell^0 \cup T_\ell^1 \subset B_\ell$ of sizes $t_{\ell,0}$ and $t_{\ell,1} = t_\ell - t_{\ell,0}$ respectively, where $t_{\ell,0}$ is the number of odd T gates in the ℓ-th layer of Q (recall the definition of even and odd T gates from Sect. 2.3). The wires T_ℓ^0 and T_ℓ^1 will play the role of auxiliary states used to perform T gates from the ℓ-th layer. They are chosen from those wires from B_ℓ whose corresponding EPR halves have been measured in a correct basis, depending on the round type. For example, in an X-test round, the auxiliaries corresponding to odd T gates should be prepared by measuring the corresponding EPR half in either the X or Y basis (see Table 3), so in an X-test round, T_ℓ^1 is chosen from $\{i \in B_\ell : W_i \in \{X, Y\}\}$ (see Table 4). We will let $\mathcal{T}_1, \ldots, \mathcal{T}_t$ label those EPR pairs that will be used as auxiliary states. In particular, the system $\mathcal{T}_i$ will be used for the i-th T gate in the circuit, so if the i-th T gate is even, $\mathcal{T}_i$ should be chosen from $T^0 = \cup_\ell T_\ell^0$, and otherwise it should be chosen from $T_1 = \cup_\ell T_\ell^1$. The verifier sends labels $\mathcal{T}_1, \ldots, \mathcal{T}_t$ and $\mathcal{X}_1, \ldots, \mathcal{X}_n$ to PP, who will act as P_{EPR} on the $n + t$ qubits specified by these labels.

Just as in the EPR Protocol, the input on PP's system specified by $\mathcal{X}_1, \ldots, \mathcal{X}_n$ is a quantum one-time pad of either $|\boldsymbol{x}\rangle$, $|0\rangle^{\otimes n}$, or $|+\rangle^{\otimes n}$, depending on the round type, with V holding the keys (determined by $\boldsymbol{e}$). Throughout the interaction, PP always maintains a one-time pad of the current state of the computation, with the verifier in possession of the one-time-pad keys. The verifier updates her keys as the computation is carried out, using the rules in Table 2.

From PP's perspective, the protocol works just as the EPR Protocol, except that he does not receive the bit z_i needed to implement the T gadget until *during* the T gadget, after he has sent V his measurement result c_i (see Fig. 5).

To perform the i-th T gate on the j-th wire, PP performs the circuit shown in Fig. 5. As Fig. 5 shows, PV has already applied the observable specified by V to his half of the EPR pair. The T gadget requires interaction with the verifier, to compute the bit z_i, which depends on the measured c_i, the value W_i, and one-time-pad key a_j, however, this interaction can be done in parallel for T gates in the same layer.

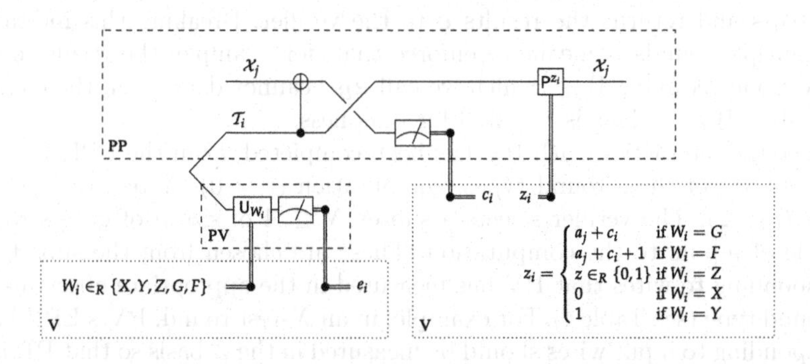

Fig. 5. The gadget for implementing the i-th T gate, on the j-th wire.

It is simple to check that the T gadget in Fig. 5 is the same as the T gadget for the EPR Protocol shown in Fig. 1. In the case of the leash protocol, W is chosen at random, and then $\boldsymbol{z}$ is chosen accordingly, whereas in the case of the EPR Protocol, $\boldsymbol{z}$ is chosen at random and then W is chosen accordingly.

We now give the precise protocols for V (Fig. 6) and honest provers PV (Fig. 7) and PP (Fig. 8).

Table 4. How the verifier chooses index sets $T = T^0 \cup T^1$ and N for each type of round. These sets determine which systems are labeled by $\{T_i\}_{i=1}^t$ and $\{\mathcal{X}_j\}_{j=1}^n$, respectively.

	Computation Round	X-test Round	Z-test Round
N	$\{i \in A : W_i = Z\}$	$\{i \in A : W_i = Z\}$	$\{i \in A : W_i = X\}$
T_ℓ^0	$\{i \in B_\ell : W_i \in \{G, F\}\}$	$\{i \in B_\ell : W_i = Z\}$	$\{i \in B_\ell : W_i \in \{X, Y\}\}$
T_ℓ^1	$\{i \in B_\ell : W_i \in \{G, F\}\}$	$\{i \in B_\ell : W_i \in \{X, Y\}\}$	$\{i \in B_\ell : W_i = Z\}$

Let (Q, x) be the input to the verifier, where Q is compiled in the form described in Section 2.3. Let n be the size of the input to Q. Let d be the T-depth, and for $\ell \in \{1, \ldots, d\}$ let t_ℓ be the number of T gates in the ℓ-th layer.

1. The verifier selects $W \in_R \Sigma^m$. She partitions $\{1, \ldots, m\}$ arbitrarily into non-overlapping subsets A and $B_1, \ldots, B_d$ such that W_A contains at least n copies of each symbol in Σ, and for each $\ell \in \{1, \ldots, d\}$, W_{B_ℓ} contains at least t_ℓ copies of each symbol in Σ.
2. The verifier sends A, W_A to PV, who returns $e_A \in \{0,1\}^A$. The verifier sequentially sends $(B_1, W_{B_1}), \ldots, (B_d, W_{B_d})$ to PV, each time receiving $e_{B_\ell} \in \{0,1\}^{B_\ell}$ as answer.
3. The verifier selects a round type uniformly at random. She selects sets $N \subseteq A$ and $T_\ell \subseteq B_\ell$, for $\ell \in \{1, \ldots, d\}$, of sizes $|N| = n$ and $|T_\ell| = t_\ell$, as follows:
   Computation Round: N is chosen at random from $\{i \in A : W_i = Z\}$. T_ℓ is chosen at random from $\{i \in B_\ell : W_i \in \{G, F\}\}$. She sets $a = e_N + x$ and $b = 0^n$.
   X-test Round: N is chosen at random from $\{i \in A : W_i = Z\}$. $T_\ell = T_\ell^0 \cup T_\ell^1$, where T_ℓ^0 is of size $t_{\ell,0}$ chosen at random from $\{i \in B_\ell : W_i = Z\}$ and T_ℓ^1 is of size $t_{\ell,1}$ chosen at random from $\{i \in B_\ell : W_i \in \{X, Y\}\}$. She sets $a = e_N$ and $b = 0^n$.
   Z-test Round: N is chosen at random from $\{i \in A : W_i = X\}$. $T_\ell = T_\ell^0 \cup T_\ell^1$, where T_ℓ^0 is of size $t_{\ell,0}$ chosen at random from $\{i \in B_\ell : W_i \in \{X, Y\}\}$ and T_ℓ^1 is of size $t_{\ell,1}$ chosen at random from $\{i \in B_\ell : W_i = Z\}$. She sets $a = 0^n$ and $b = e_N$.
   The verifier sends the sets N and T_ℓ^0, T_ℓ^1, for all $\ell \in \{1, \ldots, d\}$, to PP.
4. For $\ell = 1, \ldots, d$,
   (a) For each Clifford gate in the ℓ-th layer, perform the appropriate key update (Table 2).
   (b) The verifier receives $c = \{c_i\}_{i \in T_\ell}$ from PP. If it's an X-test round and $i \in T_\ell^0$, or it's a Z-test round and $i \in T_\ell^1$, reject if $c_i \neq a_j + e_i$, where j is the wire to which the i-th T gate is applied.
   (c) For each $i \in T_\ell$, the verifier computes $z = \{z_i\}_{i \in T_\ell}$ as follows:
   Computation Round $z_i = a_j + 1_{W_i = F} + c_i$;
   X-test Round if $i \in T_\ell^0$, $z_i \in_R \{0,1\}$; else if $i \in T_\ell^1$, $z_i = 1_{W_i = Y}$;
   Z-test Round if $i \in T_\ell^0$, $z_i = 1_{W_i = Y}$; else if $i \in T_\ell^1$, $z_i \in_R \{0,1\}$.
   (d) The verifier sends z to PP and updates keys for each T gate applied (Table 2).
5. The verifier receives a bit c_f from PP. She outputs reject if it's a computation or X-test round and $c_f + a_f \neq 0$, where a_f is the final X-key on the output wire; and accept otherwise.

Fig. 6. The Delegation Game: Verifier's point of view.

Finally, we describe the sequential version of the game RIGID(Σ, m) in Fig. 9. It is no different than RIGID(Σ, m), except for the fact that certain subsets of

1. For $\ell = 0, 1, \ldots, d$,
   (a) PV receives a string $W_S \in \Sigma^S$, for some subset S of $\{1, \ldots, m\}$, from V.
   (b) For $i \in S$, PV measures his half of the i-th EPR pair using the observable indicated by W_i, obtaining an outcome $e_i \in \{0,1\}$.
   (c) PV returns e_S to V.

Fig. 7. Honest strategy for PV

1. PP receives subsets N and T_ℓ^0, T_ℓ^1 of $\{1, \ldots, m\}$, for $\ell \in \{1, \ldots, d\}$, from the verifier.
2. For $\ell = 1, \ldots, d$,
   (a) PP does the Clifford computations in the ℓ-th layer.
   (b) For each $i \in T_\ell = T_\ell^0 \cup T_\ell^1$, PP applies a CNOT from $\mathcal{T}_i$ into the input register corresponding to the wire on which this T gate should be performed, $\mathcal{X}_j$, and measures this wire to get a value c_i. The register $\mathcal{T}_i$ is relabeled $\mathcal{X}_j$. He sends $c_{T_\ell} = \{c_i\}_{i \in T_\ell}$ to V.
   (c) PP receives $z_{T_\ell} = \{z_i\}_{i \in T_\ell}$ from V. For each $i \in T_\ell$, he applies P^{z_i} to the corresponding $\mathcal{X}_j$.
3. PP performs the final computations that occur after the d-th layer of T gates, measures the output qubit, $\mathcal{X}_1$, and sends the resulting bit, c_f, to V.

Fig. 8. Honest strategy for PP

questions and answers are exchanged sequentially, but the acceptance condition is the same. As mentioned earlier, running the game sequentially only reduces the provers' ability to cheat. Hence the guarantees from RIGID(Σ, m) hold verbatim for the sequential version.

Let m, n, and $t_1, \ldots, t_d$ be parameters provided as input, such that $m = \Theta(n + t_1 + \cdots + t_d)$.

1. The verifier selects questions $W, W' \in \Sigma^m$, for the first and second player respectively, according to the distribution of questions in the game RIGID(Σ, m). She partitions $\{1, \ldots, m\}$ at random into subsets A and B_ℓ, for $\ell \in \{1, \ldots, d\}$, of size $|A| = \Theta(n)$ and $|B_\ell| = \Theta(t_\ell)$, exactly as in Step 1 of the Delegation Game.
2. The verifier sends $(A, W_A), (B_1, W_{B_1}), .., (B_d, W_{B_d})$ and $(A, W'_A), (B_1, W'_{B_1}), .., (B_d, W'_{B_d})$ in sequence to the first and second prover respectively. They sequentially return respectively $e_A \in \{0,1\}^{|A|}, e_{B_1} \in \{0,1\}^{|B_1|}, .., e_{B_d} \in \{0,1\}^{|B_d|}$ and $e'_A \in \{0,1\}^{|A|}, e'_{B_1} \in \{0,1\}^{|B_1|}, .., e'_{B_d} \in \{0,1\}^{|B_d|}$.
3. The verifier accepts if and only if e, e' and W, W' satisfy the winning condition of RIGID(Σ, m).

Fig. 9. Sequential version of RIGID(Σ, m).

4.2 Completeness

Lemma 1. *Suppose the verifier executes the rigidity game with probability* p_r *and the delegation game with probability* $p_d = 1 - p_r$*, on an input* $(Q, |\boldsymbol{x}\rangle)$ *such that* $\|\Pi_0 Q|\boldsymbol{x}\rangle\|^2 \geq 2/3$*. Then there is a strategy for the provers which is accepted with probability at least* $p_{\text{compl}} = p_r(1 - e^{-\Omega(n+t)}) + \frac{8}{9}p_d$*.*

Proof. The provers PV and PP play the rigidity game according to the honest strategy, and the delegation game as described in Figs. 7 and 8 respectively. Their success probability in the delegation game is the same as the honest strategy in the EPR Protocol, which is at least $\frac{2}{3} + \frac{2}{3}\frac{1}{3} = \frac{8}{9}$, by Theorem 2 and since in our protocol the verifier chooses each of the three types of rounds uniformly.

4.3 Soundness

We divide the soundness analysis into three parts. First we analyze the case of an honest PV, and a cheating PP (Lemma 2). Then we show that if PV and PP pass the rigidity game with almost optimal probability, then one can construct new provers PV′ and PP′, with PV′ honest, such that the probability that they are accepted in the delegation game is not changed by much (Lemma 3). In Lemma 4, we combine the previous to derive the desired constant soundness-completeness gap, where we exclude that the acceptance probability of the provers in the rigidity game is too low by picking a p_r large enough.

Lemma 2 (Soundness against PP). *Suppose the verifier executes the delegation game on input* $(Q, |\boldsymbol{x}\rangle)$ *such that* $\|\Pi_0 Q|\boldsymbol{x}\rangle\|^2 \leq 1/3$ *with provers* (PV, PP*) *such that* PV *plays the honest strategy. Then the verifier accepts with probability at most 7/9.*

Proof. Let PP* be any prover. Assume that PV behaves honestly and applies the measurements specified by his query W on halves of EPR pairs shared with PP*. As a result the corresponding half-EPR pair at PP* is projected onto the post-measurement state associated with the outcome reported by PV to V.

From PP*, we define another prover, P^*, such that if P^* interacts with V_{EPR}, the honest verifer for the EPR Protocol (Fig. 3a), then V_{EPR} rejects with the same probability that V would reject on interaction with PP*. The main idea of the proof can be seen by looking at Fig. 5, and noticing that: (1) the combined action of V and PV is unchanged if instead of choosing the W_i-values at random and then choosing z_i as a function of these, the z_i are chosen uniformly at random, and then the W_i are chosen as a function of these; and (2) with this transformation, the combined action of V and PV is now the same as the action of V_{EPR} in the EPR Protocol.

We now define P^*. P^* acts on a system that includes $n + t$ qubits that, in an honest run of the EPR Protocol, are halves of EPR pairs shared with V_{EPR}. P^* receives $\{z_i\}_{i=1}^t$ from V_{EPR}. P^* creates $m - (n + t)$ half EPR pairs (i.e. single-qubit maximally mixed states) and randomly permutes these with his $n + t$ unmeasured qubits, n of which correspond to computation qubits on systems $\mathcal{X}_1, \ldots, \mathcal{X}_n$—he sets N to be the indices of these qubits—and t of which correspond to T-auxiliary states—he sets T^0 and T^1 to be the indices of these qubits. P^* simulates PP* on these m qubits in the following way. First, P^* gives PP* the index sets N, T^0, and T^1. In the ℓ-th iteration of the loop (Step 2. in Fig. 8), PP* returns some bits $\{c_i\}_{i \in T_\ell}$, and then expects inputs $\{z_i\}_{i \in T_\ell}$, which P^* provides, using the bits he received from V_{EPR}. Finally, at the end of the computation, PP* returns a bit c_f, and P^* outputs $\{c_i\}_{i \in T}$ and c_f.

This completes the description of P^*. To show the lemma we argue that for any input $(Q, |\boldsymbol{x}\rangle)$ the probability that V outputs accept on interaction with PV and PP* is the same as the probability that V_{EPR} outputs accept on interaction with P^*, which is at most $\frac{2}{3}q_t + \frac{1}{3}q_c$ whenever $\|\Pi_0 Q|\boldsymbol{x}\rangle\|^2 \leq 1/3$, by Theorem 3. Using $\delta = \frac{1}{3}$, Theorem 3 gives $q_c \leq \frac{5}{3} - \frac{4}{3}q_t$, which yields

$$\frac{2}{3}q_t + \frac{1}{3}q_c \leq \frac{5}{9} + \frac{2}{9}q_t \leq \frac{7}{9}.$$

There are two reasons that V_{EPR} might reject: (1) in a computation or X-test round, the output qubit decodes to 1; or (2) in an evaluation of the gadget in Fig. 5 (either an X-test round for an even T gate, or a Z-test round for an odd T gate) the condition $c_i = a_j \oplus e_i$ fails.

We first consider case (1). This occurs exactly when $c_f \oplus a_f = 1$, where a_f is the final X key of the output wire, held by V_{EPR}. We note that a_f is exactly the final X key that V would hold in the Verifier-on-a-Leash Protocol, which follows from the fact that the update rules in both the EPR Protocol and the leash protocol are the same. Thus, the probability that V_{EPR} finds $v_f \oplus a_f = 1$ on interaction with P^* is exactly the probability that V finds $c_f \oplus a_f = 1$ in Step 5 of Fig. 6.

Next, consider case (2). The condition $c_i \neq a_j \oplus e_i$ is exactly the condition in which a verifier interacting with P^* as in Fig. 6 would reject (see Step 4.(b)).

Thus, the probability that V_{EPR} outputs reject upon interaction with P^* is exactly the probability that V outputs reject on interaction with PP*, which, as discussed above, is at most 7/9.

The following lemma shows soundness against cheating PV*.

Lemma 3. *Suppose the verifier executes the leash protocol on input* $(Q, |\boldsymbol{x}\rangle)$ *such that* $\|\Pi_0 Q|\boldsymbol{x}\rangle\|^2 \leq 1/3$ *with provers* (PV*, PP*)*, such that the provers are accepted with probability* $1 - \varepsilon$*, for some* $\varepsilon > 0$*, in the rigidity game, and with probability at least* q *in the delegation game. Then there exist provers* PP$'$ *and* PV$'$ *such that* PV$'$ *applies the honest strategy and* PP$'$ *and* PV$'$ *are accepted with probability at least* $q - \mathrm{poly}(\varepsilon)$ *in the delegation game.*

Proof. By assumption, PP* and PV* are accepted in the rigidity game with probability at least $1-\varepsilon$. Let V_A, V_B be the local isometries guaranteed to exist by Theorem 4, and $\{\tau_\lambda\}$ the sub-normalized densities associated with PP*'s Hilbert space (recall that playing the rigidity game sequentially leaves the guarantees from Theorem 4 unchanged, since it only reduces the provers' ability to cheat).

First define provers PV$''$ and PP$''$ as follows. PP$''$ and PV$''$ initially share the state

$$|\psi'\rangle_{\mathsf{AB}} = \otimes_{i=1}^{m} |\mathrm{EPR}\rangle\langle\mathrm{EPR}|_{\mathsf{AB}} \otimes \sum_{\lambda\in\{\pm\}} |\lambda\rangle\langle\lambda|_{\mathsf{A}'} \otimes |\lambda\rangle\langle\lambda|_{\mathsf{B}'} \otimes (\tau_\lambda)_{\mathsf{A}''},$$

with registers $\mathsf{AA'A''}$ in the possession of PP$''$ and $\mathsf{BB'}$ in the possession of PV$''$. Upon receiving a query $W \in \Sigma^m$, PV$''$ measures B' to obtain a $\lambda \in \{\pm\}$. If $\lambda = +$ he proceeds honestly, measuring his half-EPR pairs exactly as instructed. If $\lambda = -$ he proceeds honestly except that for every honest single-qubit observable specified by W, he instead measures the complex conjugate observable. Note that this strategy can be implemented irrespective of whether W is given at once, as in the game RIGID, or sequentially, as in the Delegation Game. PP$''$ simply acts like PP*, just with the isometry V_A applied.

First note that by Theorem 4, the distribution of answers of PV$''$ to the verifier, as well as the subsequent interaction between the verifier and PP, generate (classical) transcripts that are within statistical distance poly(ε) from those generated by PV* and PP* with the same verifier.

Next we observe that taking the complex conjugate of both provers' actions does not change their acceptance probability in the delegation game, since the interaction with the verifier is completely classical. Define PP$'$ as follows: PP$'$ measures A' to obtain the same λ as PV$''$, and then executes PP$''$ or its complex conjugate depending on the value of λ. Define PV$'$ to execute the honest behavior (he measures to obtain λ, but then discards it and does not take any complex conjugates).

Then PV$'$ applies the honest strategy, and (PV$'$, PP$'$) applies either the same strategy as (PV$''$, PP$''$) (if $\lambda = +$) or its complex conjugate (if $\lambda = -$). Therefore they are accepted in the delegation game with exactly the same probability.

Combining Lemmas 2 and 3 gives us the final soundness guarantee.

Lemma 4. *(Constant soundness-completeness gap) There exist constants* $p_r, p_d = 1 - p_r$ *and* $\Delta > 0$ *such that if the verifier executes the leash protocol with parameters* (p_r, p_d) *on input* $(Q, |\boldsymbol{x}\rangle)$ *such that* $\|\Pi_0 Q|\boldsymbol{x}\rangle\|^2 \leq 1/3$*, any provers* (PV*, PP*) *are accepted with probability at most* $p_{\text{sound}} = p_{\text{compl}} - \Delta$.

Proof. Suppose provers PP* and PV* succeed in the delegation game with probability $\frac{7}{9} + w$ for some $w > 0$, and the testing game with probability $1 - \varepsilon_*(w)$, where $\varepsilon_*(w)$ will be specified below. By Lemma 3, this implies that there exist provers PP$'$ and PV$'$ such that PV$'$ is honest and the provers succeed in the delegation game with probability at least $\frac{7}{9} + w - g(\varepsilon_*(w))$, where $g(\varepsilon) = \text{poly}(\varepsilon)$ is the function from the guarantee of Lemma 3. Let $\varepsilon_*(w)$ be such that $g(\varepsilon_*(w)) \leq \frac{w}{2}$. In particular, $\frac{7}{9} + w - g(\varepsilon_*(w)) \geq \frac{7}{9} + \frac{w}{2} > \frac{7}{9}$. This contradicts Lemma 2.

Thus if provers PP and PV succeed in the delegation game with probability $\frac{7}{9} + w$ they must succeed in the rigidity game with probability less than $1 - \varepsilon_*(w)$. This implies that for any strategy of the provers, on any *no* instance, the probability that they are accepted is at most

$$\max\left\{p_r + (1-p_r)\left(\frac{7}{9}+\frac{1}{18}\right), p_r\left(1-\varepsilon_*\left(\frac{1}{18}\right)\right) + (1-p_r)\cdot 1\right\}. \quad (4)$$

Since $\varepsilon_*(\frac{1}{18})$ is a positive constant, it is clear that one can pick p_r large enough so that

$$p_r\left(1-\varepsilon_*\left(\frac{1}{18}\right)\right) + (1-p_r)\cdot 1 < p_r + (1-p_r)\left(\frac{7}{9}+\frac{1}{18}\right). \quad (5)$$

Select the smallest such p_r. Then the probability that the two provers are accepted is at most

$$p_{\text{sound}} := p_r + (1-p_r)\left(\frac{7}{9}+\frac{1}{18}\right) < p_r\left(1-e^{-\Omega(n+t)}\right) + (1-p_r)\frac{8}{9} = p_{\text{compl}},$$

which gives the desired constant completeness-soundness gap Δ.

4.4 Blindness

We now establish blindness of the Leash Protocol. In Lemma 5, we will prove that the protocol has the property that neither prover can learn anything about the input to the circuit, $\boldsymbol{x}$, aside from its length. Thus, the protocol can be turned into a blind protocol, where Q is also hidden, by modifying any input $(Q,\boldsymbol{x})$ where Q has g gates and acts on n qubits, to an input $(U_{g,n},(Q,\boldsymbol{x}))$, where $U_{g,n}$ is a universal circuit that takes as input a description of a g-gate circuit Q on n qubits, and a string $\boldsymbol{x}$, and outputs $Q|\boldsymbol{x}\rangle$. The universal circuit $U_{g,n}$ can be implemented in $O(g\log n)$ gates. By Lemma 5, running the Leash Protocol on $(U_{g,n},(Q,\boldsymbol{x}))$ reveals nothing about Q or $\boldsymbol{x}$ aside from g and n.

In the form presented in Fig. 6, the verifier V interacts first with PV, sending him random questions that are independent from the input $\boldsymbol{x}$, aside from the input length n. It is thus clear that the protocol is blind with respect to PV.

In contrast, the questions to PP depend on PV's answers and on the input, so it may a priori seem like the questions can leak information to PP. To show that the protocol is also blind with respect to PP, we show that there is an alternative formulation, in which the verifier first interacts with PP, sending him random messages, and then only with PV, with whom the interaction is now adaptive. We argue that, for an arbitrary strategy of the provers, the reduced state of all registers available to either prover, PP or PV, is exactly the same in both formulations of the protocol—the *original* and the *alternative* one. This establishes blindness for both provers. This technique for proving blindness is already used in [RUV13] to establish blindness of a two-prover protocol based on computation by teleportation.

Lemma 5 (Blindness of the Leash Protocol). *For any strategy of* PV* *and* PP*, *the reduced state of* PV* *(resp.* PP*) *at the end of the leash protocol is independent of the input* $\boldsymbol{x}$, *aside from its length.*

Proof. Let PV* and PP* denote two arbitrary strategies for the provers in the leash protocol. Each of these strategies can be modeled as a super-operator

$$\mathcal{T}_{\mathrm{PV}} : \mathrm{L}(\mathcal{H}_{T_{\mathrm{PV}}} \otimes \mathcal{H}_{\mathrm{PV}}) \to \mathrm{L}(\mathcal{H}_{T'_{\mathrm{PV}}} \otimes \mathcal{H}_{\mathrm{PV}}),$$

$$\mathcal{T}_{\mathrm{PP},ad} : \mathrm{L}(\mathcal{H}_{T_{\mathrm{PP}}} \otimes \mathcal{H}_{\mathrm{PP}}) \to \mathrm{L}(\mathcal{H}_{T'_{\mathrm{PP}}} \otimes \mathcal{H}_{\mathrm{PP}}).$$

Here $\mathcal{H}_{T_{\mathrm{PV}}}$ and $\mathcal{H}_{T'_{\mathrm{PV}}}$ (resp. $\mathcal{H}_{T_{\mathrm{PP}}}$ and $\mathcal{H}_{T'_{\mathrm{PP}}}$) are classical registers containing the inputs and outputs to and from PV* (resp. PP*), and $\mathcal{H}_{\mathrm{PV}}$ (resp. $\mathcal{H}_{\mathrm{PP}}$) is the private space of PV* (resp. PP*). Note that the interaction of each prover with the verifier is sequential, and we use $\mathcal{T}_{\mathrm{PV}}$ and $\mathcal{T}_{\mathrm{PP},ad}$ to denote the combined action of the prover and the verifier across all rounds of interaction (formally these are sequences of superoperators).

Consider an alternative protocol, which proceeds as follows. The verifier first interacts with PP. From Fig. 8 we see that the inputs required for PP are subsets N and $T_1, \ldots, T_d$, and values $\{z_i\}_{i \in T_\ell}$ for each $\ell \in \{1, \ldots, d\}$. To select the former, the verifier proceeds as in the first step of the Delegation Game. She selects the latter uniformly at random. The verifier collects values $\{c_i\}_{i \in T_\ell}$ from PP exactly as in the original Delegation Game.

Once the interaction with PP has been completed, the verifier interacts with PV. First, she selects a random string $W_N \in \Sigma^N$, conditioned on the event that W_N contains at least n copies of each symbol in Σ, and sends it to PV, collecting answers $\boldsymbol{e}_N$. The verifier then follows the same update rules as in the delegation game. We describe this explicitly for computation rounds. First, the verifier sets $\boldsymbol{a} = \boldsymbol{e}_N$. Depending on the values $\{c_i\}_{i \in T_1}$ and $\{z_i\}_{i \in T_1}$ obtained in the interaction with PP, using the equation $z_i = a_j + 1_{W_i = F} + c_i$ she deduces a value for $1_{W_i = F}$ for each $i \in T_1 \subseteq B_1$. She then selects a uniformly random $W_{B_1} \in \Sigma^{B_1}$, conditioned on the event that W_{B_1} contains at least t_1 copies of each symbol from Σ, and for $i \in T_1$ it holds that $W_i = F$ if and only if $z_i = a_j + 1 + c_i$. The important observation is that, if T_1 is a uniformly random, unknown subset, the marginal distribution on W_{B_1} induced by the distribution described above is independent of whether $z_i = a_j + 1 + c_i$ or $z_i = a_j + 0 + c_i$: precisely, it is uniform conditioned on the event that W_{B_1} contains at least t_1 copies of each symbol from Σ. The verifier receives outcomes $\boldsymbol{e}_{B_1} \in \{0,1\}^{B_1}$ from PV, and using these outcomes performs the appropriate key update rules; she then proceeds to the second layer of the circuit, until the end of the computation. Finally, the verifier accepts using the same rule as in the last step of the original delegation game.

We claim that both the original and alternative protocols generate the same joint final state:

$$\mathcal{T}_{\mathrm{PP},ad} \circ \mathcal{T}_{\mathrm{PV}}(\rho_{orig}) = \mathcal{T}_{\mathrm{PV},ad} \circ \mathcal{T}_{\mathrm{PP}}(\rho_{alt}) \in \mathcal{H}_{\mathrm{PP}} \otimes \mathcal{H}_{T'_{\mathrm{PP}}} \otimes \mathcal{H}_{\mathrm{V}} \otimes \mathcal{H}_{T'_{\mathrm{PV}}} \otimes \mathcal{H}_{\mathrm{PV}}, \quad (6)$$

where we use ρ_{orig} and ρ_{alt} to denote the joint initial state of the provers, as well as the verifier's initialization of her workspace, in the original and alternative protocols respectively, and $\mathcal{T}_{\mathrm{PV},ad}$ and $\mathcal{T}_{\mathrm{PP}}$ are the equivalent of $\mathcal{T}_{\mathrm{PV}}$ and $\mathcal{T}_{\mathrm{PP},ad}$ for the reversed protocol (in particular they correspond to the same strategies PV* and PP* used to define $\mathcal{T}_{\mathrm{PV}}$ and $\mathcal{T}_{\mathrm{PP},ad}$). Notice that $\mathcal{T}_{\mathrm{PV},ad}$ and $\mathcal{T}_{\mathrm{PP}}$ are well-defined since neither prover can distinguish an execution of the original from the alternative protocol.[7] To see that equality holds in (6), it is possible to re-write the final state of the protocol as the result of the following sequence of operations. First, the verifier initializes the message registers with PP* and PV* using half-EPR pairs, keeping the other halves in her private workspace. This simulates the generation of uniform random messages to both provers. Then, the superoperator $\mathcal{T}_{\mathrm{PV}} \otimes \mathcal{T}_{\mathrm{PP}}$ is executed. Finally, the verifier post-selects by applying a projection operator on $\mathcal{H}_{T_{\mathrm{PV}}} \otimes \mathcal{H}_{T'_{\mathrm{PV}}} \otimes \mathcal{H}_{T_{\mathrm{PP}}} \otimes \mathcal{H}_{T'_{\mathrm{PP}}}$ that projects onto valid transcripts for the original protocol (i.e. transcripts in which the adaptive questions are chosen correctly). This projection can be implemented in two equivalent ways: either the verifier first measures $\mathcal{H}_{T_{\mathrm{PV}}} \otimes \mathcal{H}_{T'_{\mathrm{PV}}}$, and then $\mathcal{H}_{T_{\mathrm{PP}}} \otimes \mathcal{H}_{T'_{\mathrm{PP}}}$; based on the outcomes she accepts a valid transcript for the original protocol or she rejects. Or, she first measures $\mathcal{H}_{T_{\mathrm{PP}}} \otimes \mathcal{H}_{T'_{\mathrm{PP}}}$, and then $\mathcal{H}_{T_{\mathrm{PV}}} \otimes \mathcal{H}_{T'_{\mathrm{PV}}}$; based on the outcomes she accepts a valid transcript for the alternative protocol or she rejects. Using the commutation of the provers' actions, conditioned on the transcript being accepted, the first gives rise to the first final state in (6), and the second to the second final state. The two are equivalent because the acceptance condition for a valid transcript is identical in the two versions of the protocol.

Since in the first case the reduced state on $\mathcal{H}_{T'_{\mathrm{PV}}} \otimes \mathcal{H}_{\mathrm{PV}}$ is independent of the input to the computation, $\boldsymbol{x}$, and in the second the reduced state on $\mathcal{H}_{\mathrm{PP}} \otimes \mathcal{H}_{T'_{\mathrm{PP}}}$ is independent of $\boldsymbol{x}$, we deduce that the protocol hides the input from each of PV* and PP*.

Remark 1. In order to make a fair comparison between previous delegated computation protocols and ours (see Fig. 1), one must analyze their resource requirements under the condition that they produce the correct outcome of the computation with a fixed, let us say 99%, probability. For most protocols, this is achieved by sequentially repeating the original version, in order to amplify the completeness-soundness gap. We refer to the full version of the paper for a sequencial procedure that allows the verifier to obtain the correct output with a fixed probability (or abort whenever the provers are malicious).

5 Dog-Walker Protocol

The Dog-Walker Protocol again involves a classical verifier V and two provers PV and PP. As in the leash protocol presented in Sect. 4, PP and PV take the roles of P_{EPR} and V_{EPR} from [Bro18] respectively. The main difference is that the Dog-Walker Protocol gives up blindness in order to reduce the number of rounds to two (one round of interaction with each prover, played sequentially). After one round of communication with PP, who returns a sequence of measurement outcomes, V communicates all of PP's outcomes, except for the one

[7] One must ensure that a prover does not realize if the alternative protocol is executed instead of the original; this is easily enforced by only interacting with any of the provers at specific, publicly decided times.

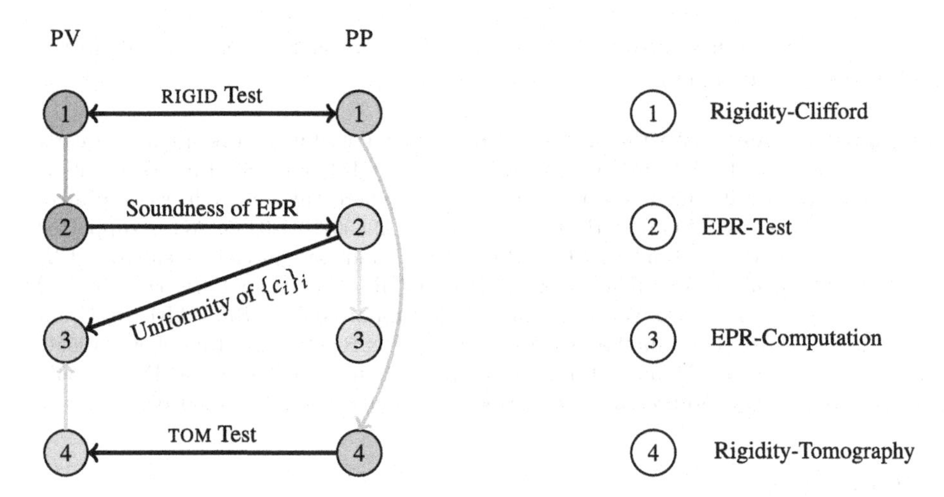

Fig. 10. Overview of the soundness of the Dog-Walker Protocol

corresponding to the output bit of the computation, as well as the input $\boldsymbol{x}$, to PV. With these, PV can perform the required adaptive measurements without the need to interact with V. It may seem risky to communicate bits sent by PP directly to PV—this seems to allow for communication between the two provers! Indeed, blindness is lost. However, if PP is honest, his outcomes $\{c_i\}_i$ in the computation round are the result of measurements he performs on half-EPR pairs, and are uniform random bits. If he is dishonest, and does not return the outcomes obtained by performing the right measurements, he will be caught in the test rounds. It is only in computation rounds that V sends the measurement results $\{c_i\}_i$ to PV.

We notice that PV has a much more important role in this protocol: he decides himself the measurements to perform according to previous measurements' outcomes as well as the input x. For this reason, we must augment the test discussed in Sect. 3 in order to test if PV remains honest with respect to these new tasks. For this reason, we introduce the Tomography test and prove a rigidity theorem that will allow us to prove the soundness of the Dog-walker protocol (see Fig. 10 for a glimpse of the proof structure). Due to space limitations we refer to the full version of the paper for a presentation of the Tomography Test, a formal description of the Dog-walker protocol and the proof for their correctness.

Finally, the Dog-Walker Protocol can be easily extended to a classical-verifier two-prover protocol for all languages in QMA. Along the same lines of the proof that QMIP = MIP* from [RUV13], one of the provers plays the role of PP, running the QMA verification circuit, while the second prover creates and teleports the corresponding QMA witness. In our case, it is not hard to see that the second prover can be re-used as PV in the Dog-Walker Protocol, creating the necessary gadgets for the computation and allowing the Verifier to check the operations performed by the first prover. We describe this approach in more details in the full version of the paper.

Acknowledgments. We thank Anne Broadbent for useful discussions in the early stages of this work. All authors acknowledge the IQIM, an NSF Physics Frontiers Center at the California Institute of Technology, where this research was initiated. AC is supported by AFOSR YIP award number FA9550-16-1-0495. AG is supported by ERC Consolidator Grant 615307-QPROGRESS and was previously supported by ERC QCC when AG was a member of IRIF (CNRS/Université Paris Diderot). SJ is supported by an NWO WISE Grant. TV is supported by NSF CAREER Grant CCF-1553477, MURI Grant FA9550-18-1-0161, AFOSR YIP award number FA9550-16-1-0495, and the IQIM, an NSF Physics Frontiers Center (NSF Grant PHY-1125565) with support of the Gordon and Betty Moore Foundation (GBMF-12500028).

References

[ABE10] Aharonov, D., Ben-Or, M., Eban, E.: Interactive proofs for quantum computations. In: Proceedings of the First Symposium on Innovations in Computer Science (ICS 2010), pp. 453–469 (2010)

[ADSS17] Alagic, G., Dulek, Y., Schaffner, C., Speelman, F.: Quantum fully homomorphic encryption with verification (2017). arXiv preprint arXiv:1708.09156

[Bel64] Bell, J.S.: On the Einstein-Podolsky-Rosen paradox. Physics **1**, 195–200 (1964)

[BFGH10] Bera, D., Fenner, S.A., Green, F., Homer, S.: Efficient universal quantum circuits. Quantum Inf. Comput. **10**(1&2), 16–27 (2010)

[Bro18] Broadbent, A.: How to verify a quantum computation. Theory Comput. **14**(11), 1–37 (2018). arXiv preprint arXiv:1509.09180

[BvCA18] Bowles, J., Šupić, I., Cavalcanti, D., Acín, A.: Self-testing of Pauli observables for device-independent entanglement certification (2018). arXiv:1801.10446

[Cas17] Castelvecchi, D.: IBM's quantum cloud computer goes commercial. Nat. News **543**(7644) (2017)

[CHSH69] Clauser, J.F., Horne, M.A., Shimony, A., Holt, R.A.: Proposed experiment to test local hidden-variable theories. Phys. Rev. Lett. **23**, 880–884 (1969)

[DSS16] Dulek, Y., Schaffner, C., Speelman, F.: Quantum homomorphic encryption for polynomial-sized circuits. In: Robshaw, M., Katz, J. (eds.) CRYPTO 2016. LNCS, vol. 9816, pp. 3–32. Springer, Heidelberg (2016). https://doi.org/10.1007/978-3-662-53015-3_1. arXiv:1603.09717

[FH15] Fitzsimons, J.F., Hajdušek, M.: Post hoc verification of quantum computation (2015). arXiv preprint arXiv:1512.04375

[FH17] Fujii, K., Hayashi, M.: Verifiable fault tolerance in measurement-based quantum computation. Phys. Rev. A **96**, 030301 (2017)

[Fit16] Fitzsimons, J.F.: Private quantum computation: an introduction to blind quantum computing and related protocols (2016). arXiv preprint arXiv:1611.10107

[FK17] Fitzsimons, J.F., Kashefi, E.: Unconditionally verifiable blind quantum computation. Phys. Rev. A **96**(012303) (2017). arXiv preprint arXiv:1203.5217

[GKW15] Gheorghiu, A., Kashefi, E., Wallden, P.: Robustness and device independence of verifiable blind quantum computing. New J. Phys. **17** (2015)

[Gri17] Grilo, A.B.: Relativistic verifiable delegation of quantum computation (2017). arXiv preprint arXiv:1711.09585

[HH16] Hayashi, M., Hajdušek, M.: Self-guaranteed measurement-based quantum computation (2016). arXiv preprint arXiv:1603.02195

[HM15] Hayashi, M., Morimae, T.: Verifiable measurement-only blind quantum computing with stabilizer testing. Phys. Rev. Lett. **115**, 220502 (2015)

[HPDF15] Hajdušek, M., Pérez-Delgado, C.A., Fitzsimons, J.F.: Device-independent verifiable blind quantum computation (2015). arXiv preprint arXiv:1502.02563

[HZM+17] Huang, H.-L., et al.: Experimental blind quantum computing for a classical client. Phys. Rev. Lett. **119**, 050503 (2017)

[Ji16] Ji, Z.: Classical verification of quantum proofs. In: Proceedings of the Forty-eighth Annual ACM SIGACT Symposium on Theory of Computing (STOC 2016), pp. 885–898 (2016)

[Mah17] Mahadev, U.: Classical homomorphic encryption for quantum circuits (2017). arXiv preprint arXiv:1708.02130

[Mah18] Mahadev, U.: Classical verification of quantum computations (2018). arXiv preprint arXiv:1804.01082

[McK16] McKague, M.: Interactive proofs for BQP via self-tested graph states. Theory Comput. **12**(3), 1–42 (2016). arXiv preprint arXiv:1309.5675

[MF16] Morimae, T., Fitzsimons, J.F.: Post hoc verification with a single prover (2016). arXiv preprint arXiv:1603.06046

[Mon16] Montanaro, A.: Quantum algorithms: an overview. npj Quantum Inf. **2**(15023) (2016)

[Mor14] Morimae, T.: Verification for measurement-only blind quantum computing. Phys. Rev. A **89** (2014)

[MTH17] Morimae, T., Takeuchi, Y., Hayashi, M.: Verified measurement-based quantum computing with hypergraph states (2017). arXiv:1701.05688

[MY04] Mayers, D., Yao, A.: Self testing quantum apparatus. Quantum Inf. Comput. **4**, 273–286 (2004)

[NV17] Natarajan, A., Vidick, T.: A quantum linearity test for robustly verifying entanglement. In: Proceedings of the Forty-Ninth Annual ACM SIGACT Symposium on Theory of Computing (STOC 2017), pp. 1003–1015 (2017)

[RUV12] Reichardt, B.W., Unger, F., Vazirani, U.: A classical leash for a quantum system: command of quantum systems via rigidity of CHSH games (2012). arXiv preprint arXiv:1209.0448

[RUV13] Reichardt, B.W., Unger, F., Vazirani, U.: Classical command of quantum systems. Nature **496**, 456–460 (2013). Full version arXiv:1209.0448

[Slo16] Slofstra, W.: Tsirelson's problem and an embedding theorem for groups arising from non-local games (2016). arXiv preprint arXiv:1606.03140

Signatures I

Ring Signatures: Logarithmic-Size, No Setup—from Standard Assumptions

Michael Backes[1(✉)], Nico Döttling[1], Lucjan Hanzlik[1,2], Kamil Kluczniak[1], and Jonas Schneider[1]

[1] CISPA Helmholtz Center for Information Security, Saarland Informatics Campus, Saarbrücken, Germany
{backes,doetling,hanzlik,kamil.kluczniak,jonas.schneider}@cispa.saarland
[2] Stanford University, Stanford, USA
lucjan.hanzlik@stanford.edu

Abstract. Ring signatures allow for creating signatures on behalf of an ad hoc group of signers, hiding the true identity of the signer among the group. A natural goal is to construct a ring signature scheme for which the signature size is short in the number of ring members. Moreover, such a construction should not rely on a trusted setup and be proven secure under falsifiable standard assumptions. Despite many years of research this question is still open.

In this paper, we present the first construction of size-optimal ring signatures which do not rely on a trusted setup or the random oracle heuristic. Specifically, our scheme can be instantiated from standard assumptions and the size of signatures grows only logarithmically in the number of ring members.

We also extend our techniques to the setting of linkable ring signatures, where signatures created using the same signing key can be linked.

Keywords: Ring signatures · Linkable ring signatures · Standard model

1 Introduction

Ring signatures, introduced by Rivest, Shamir and Tauman-Kalai [35] allow a signer to hide in a crowd, or *ring* of potential signers. More specifically, the signing algorithm of a ring signature scheme takes as additional input a list of verification keys R and outputs a signature. Such a signature can be verified given the ring R. The feature of interest of ring signatures is that given such a signature, no one, not even an insider in possession of all the secret keys corresponding to the verification keys in the ring, can tell which key was used to compute this signature. The original motivation for ring signatures was whistleblowing, where the leaking party can hide her identity and at the same time convince outsiders that the leaked information is genuine (by using a ring composed only of people with access to this information). In terms of security two properties are required

Y. Ishai and V. Rijmen (Eds.): EUROCRYPT 2019, LNCS 11478, pp. 281–311, 2019.
https://doi.org/10.1007/978-3-030-17659-4_10

of ring signatures: unforgeability and anonymity. The first property requires that an efficient adversary should not be able to forge a signature on behalf of an honest ring of signers. Anonymity requires that signatures do not give away by which member they were created. This can be cast as an experiment in which the adversary has to guess which one out of two ring members created a signature.

The notion of linkable ring signatures [29] is an extension of the concept of ring signatures such that there is a public way of determining whether two signatures have been produced by the same signer. Linkable ring signatures yield a very elegant approach to e-voting [40]: Every voter is registered with their verification key. To cast a vote, all a voter has to do is to sign his or her vote on behalf of the ring of all registered voters. Linkability prevents voters from casting multiple votes. This can even be turned into an augmentation of the voting functionality by allowing voters to revote, where only the most recently cast votes of a set of votes that link counts.

Recently, linkable ring signature have also drawn attention in the domain of decentralized currencies, where they can be used to implement a mechanism for anonymized transactions. Linkable ring signatures are, for instance, used in a cryptocurrency called *Monero* [32], where they allow payers to hide their identity in an anonymity set composed of identities from previous transactions. Currently Monero uses a setup-free Schnorr based ring signature scheme [37] where the size of signatures scales linearly in the size of the ring. To decrease the size of the transaction by default Monero uses small rings, which provide only a limited amount of anonymity. The anonymity definition for linkable ring signatures needs to be different from the definition for standard ring signatures. We will elaborate further on this topic below. In both of the above applications two aspects are of the essence:

- The ring signature scheme should not rely on a trusted setup. Especially in the e-voting application it is of paramount importance for the acceptance of such a system that there *cannot exist* a trapdoor that enables deanonymization of voters.
- For practical purposes, e.g. for elections with millions of voters, the size of individual signatures should be essentially independent of the size of the ring of signers.

1.1 Our Contributions

In this work, we provide the first construction of ring signatures which simultaneously

- does not rely on a trusted setup or the random oracle heuristic,
- can be proven secure under *falsifiable* standard assumptions, namely the existence of non-interactive witness indistinguishable proofs [4,8,17,25] and additional standard assumptions such as the hardness of the *Decisional Diffie Hellman* problem [18] or the *Learning with Errors* problem [34],

- has signatures of size $\log(\ell) \cdot \mathsf{poly}(\lambda)$, where ℓ is the size of the ring of signers and λ the security parameter.

Our work therefore settles the problem of size-optimal ring signatures in the standard model, which has been a long standing open problem. Furthermore, we extend our techniques to the domain of linkable ring signatures, i.e. we construct linkable ring signatures of size $\log(\ell) \cdot \mathsf{poly}(\lambda)$ without setup and in the standard model. Along the way, we introduce new techniques that enable us to use NIWI proofs instead of NIZK proofs, which may be of independent interest.

As an additional contribution, we propose a stronger security model for linkable ring signatures and prove that our linkable ring signature scheme is secure in this model.

1.2 Our Techniques

To describe our scheme, it is instructive to recall the standard model ring signature scheme of Bender, Katz, and Morselli [6]. In the BKM scheme, a verification key $\mathrm{VK} = (\mathsf{vk}, \mathsf{pk})$ consists of a verification key vk for a standard signature scheme and a public key pk for a public key encryption scheme. To sign a message m given a signing key sk and a ring $\mathsf{R} = (\mathrm{VK}_1, \ldots, \mathrm{VK}_\ell)$, one proceeds as follows. In a first step, locate verification key $\mathrm{VK}_{i^*} = (\mathsf{vk}_{i^*}, \mathsf{pk}_{i^*})$ corresponding to the signing key sk in the ring R. Now compute a signature σ of m using the signing key sk and encrypt σ under pk_{i^*} to obtain a ciphertext ct_{i^*}. Next, for all $i \neq i^*$ compute *filler ciphertexts* ct_i as encryptions of 0^λ under pk_i, where $\mathrm{VK}_i = (\mathsf{vk}_i, \mathsf{pk}_i)$. Finally, use a non-interactive[1] witness-indistinguishable proof π for the statement $(\mathsf{m}, \mathsf{ct}_1, \ldots, \mathsf{ct}_\ell, \mathrm{VK}_1, \ldots, \mathrm{VK}_\ell)$ to show that there exists an index i^* such that ct_{i^*} encrypts a signature σ and that σ verifies for the message m under the verification key vk_{i^*}. The ring signature is now given by $\Sigma = (\mathsf{ct}_1, \ldots, \mathsf{ct}_\ell, \pi)$. To verify a signature Σ for a message m and ring R, use the NIWI verifier to verify that π is a proof for the statement $(\mathsf{m}, \mathsf{ct}_1, \ldots, \mathsf{ct}_\ell, \mathrm{VK}_1, \ldots, \mathrm{VK}_\ell)$.

We also briefly review how unforgeability and anonymity of this scheme are established. To establish unforgeability, note that by the perfect soundness of the NIWI proof π one of the ct_i must actually be an encryption of a signature of m under vk_i. The security reduction can therefore set up all the pk_i such that it knows the corresponding secret keys and decrypt the signature. Establishing anonymity relies on witness indistinguishability of the NIWI proof system. That is, the reduction can set up the signature Σ such that in fact two different ciphertexts ct_{i_0} and ct_{i_1} encrypt a valid signature (each under their corresponding verification key). We can now use witness indistinguishability to switch the witness from index i_0 to i_1. Thus we can establish that signatures computed using sk_{i_0} are computationally indistinguishable from signatures computed using sk_{i_1}. The size of the signature is linear in the ring size ℓ. There are two major obstacles in making the size of the signatures sublinear:

[1] Bender et al. [6] actually use 2-message public-coin witness-indistinguishable proofs (ZAPs) rather than NIWI proofs, which is a slightly weaker primitive than NIWI proofs.

1. The signature contains all the ciphertexts $\mathsf{ct}_1, \ldots, \mathsf{ct}_\ell$.
2. The witness for the validity of statement $(\mathsf{m}, \mathsf{ct}_1, \ldots, \mathsf{ct}_\ell, \mathrm{VK}_1, \ldots, \mathrm{VK}_\ell)$ is also of size linear in ℓ.

Reducing the number of Ciphertexts. Starting from the BKM scheme, our first idea is that if we use an appropriate public key encryption scheme PKE, then we do not need to include all the ciphertexts $\mathsf{ct}_1, \ldots, \mathsf{ct}_\ell$ in the signature, but only two ciphertexts ct and ct'. The additional property we need from PKE is that a ciphertexts ct cannot be linked to the public key pk that was used to compute ct, *unless* one is in the possession of the corresponding secret key sk. This property immediately holds if the public key encryption scheme PKE has pseudorandom ciphertexts. In fact, many constructions of public key encryption have pseudorandom ciphertexts, e.g. the classic ElGamal scheme based on DDH [18] or Regev's scheme based on LWE [34].

Our first modification is thus to compute ct by encrypting the signature σ under pk_{i^*} and choosing ct' uniformly at random. We also compute the proof π differently. Namely, we prove that for a statement of the form $(\mathsf{m}, \mathsf{ct}, \mathsf{ct}', \mathrm{VK}_1, \ldots, \mathrm{VK}_\ell)$ it holds that there exist indices i^* and $i^\dagger$ such that either ct is an encryption of a signature σ^* of m with respect to the verification key vk_{i^*} under the public key pk_{i^*}, *or* ct' is an encryption of a signature $\sigma^\dagger$ of m with respect to the verification key $\mathsf{vk}_{i^\dagger}$ under the public key $\mathsf{pk}_{i^\dagger}$. In this modified scheme, a signature $\Sigma = (\mathsf{ct}, \mathsf{ct}', \pi)$ consists of the two ciphertexts $\mathsf{ct}, \mathsf{ct}'$ and the proof π. Verification checks that π is a proof for the statement $(\mathsf{m}, \mathsf{ct}, \mathsf{ct}', \mathrm{VK}_1, \ldots, \mathrm{VK}_\ell)$. We will briefly argue that this scheme is still unforgeable and anonymous. First observe that if the proof π for the statement $(\mathsf{m}, \mathsf{ct}, \mathsf{ct}', \mathrm{VK}_1, \ldots, \mathrm{VK}_\ell)$ verifies, then by the perfect soundness of the NIWI proof system either ct or ct' must encrypt a signature under a public key pk_{i^*} or $\mathsf{pk}_{i^\dagger}$ respectively. Therefore, we can again construct a reduction which knows all the secret keys corresponding to the pk_i. This way, the reduction will be able to decrypt the signature σ from ct or ct'. To show anonymity, we transform a signature computed with sk_{i_0} into a signature computed with sk_{i_1} via a sequence of hybrids. In the first hybrid step we will make ct' an encryption of a signature σ_1 of m with respect to the key vk_{i_1} under the public key pk_1. This change is possible as the ciphertexts of PKE are pseudorandom. Next, we will use witness-indistinguishability of NIWI to switch the witness for the statement $(\mathsf{m}, \mathsf{ct}, \mathsf{ct}', \mathrm{VK}_1, \ldots, \mathrm{VK}_\ell)$. The new witness shows that ct' encrypts a valid signature of m. This means that we do not need a witness for ct anymore. Thus, in the next hybrid steps, we replace the ciphertext ct by a random string, and then replace this random string by an encryption of the signature σ_1 under the public key pk_{i_1}. In the next steps, we can switch the witness we use to compute the proof π back to using the witness for ct, and in the last hybrid we make ct' uniformly random again. Thus, Σ is now computed using sk_{i_1}.

Compressing the Witness. The bigger challenge, however, is reducing the size of the witness for the membership proofs to linear in $\log(\ell)$. A natural approach would be to prove membership of the verification key VK_i in the ring via a

Merkle-tree accumulator (as e.g. in the ROM-scheme of [15]). In this approach, one first hashes the ring R into a succinct digest h, and can then prove membership of VK_i in the ring via a $\log(\ell)$-sized root-to-leaf path. To sign a message under a ring R, the signer first hashes R into a digest h and computes a NIWI proof π which simultaneously proves membership of his own key VK_i in R via a succinct membership witness *and* that ct encrypts a signature for VK_i. To verify such a signature, the verifier recomputes the root hash h for the ring R and verifies the proof π. While this idea seems to resolve the above issue at first glance, it raises serious issues itself. First and foremost, we will not be able to prove unforgeability as above, as membership proofs for Merkle trees only have computational soundness, but in order to prove unforgeability as above we need perfect soundness. The problem is that an adversary might also produce a proof by finding a collision in the Merkle tree instead of forging a signature. If, in fact, we could use an NIZK proof of knowledge, then this proof strategy can be implemented with routine techniques. NIZK proofs however need a setup, and we only have NIWI proofs at our disposal. Moreover, for a Merkle tree to be binding it is necessary that the hashing key is honestly generated, as unkeyed hash functions are insecure against non-uniform adversaries. Thus, it is also unclear where the hashing key for the Merkle tree should come from. Consequently, the Merkle tree approach seems fundamentally stuck in the standard model.

There is, however a loophole in the above argument. Upon closer inspection, we actually do not need the Merkle tree hash function to be collision resistant. Instead, we need a guarantee that the hash value h binds to at least one specific value in the database, which is under the control of the signer. The key ingredient we use to make the construction work is *somewhere statistically binding* (SSB) hashing [27]. An SSB hash function allows to compress a database into a digest h such that h uniquely binds to a specific database entry. More specifically, the key generator for a SSB hash function takes as an additional input an index i^* and produces a hashing key hk. When a database db is hashed into a digest h using the hashing key hk, the digest h uniquely defines db_{i^*}. In other words, any database db' with $\mathsf{db}'_{i^*} \neq \mathsf{db}_{i^*}$ hashes to a digest $\mathsf{h}' \neq \mathsf{h}$. To enable short membership proofs, we require a SSB hash function with local opening. That is, given a hashing key hk, a digest h of a database db, an index i and a value x, there is witness τ of size linear in $\log(|\mathsf{db}|)$ which demonstrates that $\mathsf{db}_i = x$. Besides the somewhere statistically binding property, we also require that the SSB hash function is index-hiding, i.e. the hashing key hk computationally hides the index i at which it is binding. Finally, as there is no trusted setup which could define the key for the SSB hash function, we must let the signer generate the hashing key hk itself. However, this again introduces an additional problem. The standard notion of SSB hashing requires that the somewhere binding property holds with overwhelming probability over the coins of the key generator, but not with probability 1. However, as we let the signer generate the hashing key, the signer may in fact choose bad random coins for which the hashing key is not binding. We address this problem by using *somewhere perfectly binding* (SPB) hashing instead of SSB hashing. In fact, many constructions of SSB hashing are already SPB, e.g. the LWE-based construction of [27] can be made SPB via standard error-truncation techniques, and the DDH- and DCR-based constructions of [33] are immediately SPB. One additional aspect we require is that generating a hashing key hk for a database db of size ℓ can be performed by a circuit of size linear in $\log(\ell)$, but this is the case for the instantiations above. Equipped with SPB hashing, we can now construct succinct membership proofs with perfect soundness as follows. The signer generates a hashing key hk binding at position i (where VK_i is the signer's verification key) and uses hk to compress R into a digest h. The membership witness shows that hk is binding at position i and that h opens to VK_i at position i. Essentially, a pair $(\mathsf{hk}, \mathsf{h})$ of SPB hashing key hk and digest h form a perfectly binding commitment to VK_i, where we can prove that $(\mathsf{hk}, \mathsf{h})$ opens to VK_i at position i using a witness of size linear in $\log(\ell)$.

Relaxing the requirements on SPB hashing. It turns out that we do not need the opening witnesses for the SPB hashing scheme to be publicly computable. Indeed, we may allow the opening witness to depend on the private coins used by the key generator as we need to prove that hk is binding at position i anyway. We therefore define a slightly weakened notion called *Somewhere Perfectly Binding Hashing with private local Opening*. As observed in [33], this notion can immediately be realized from any *private information retrieval* (PIR) scheme with fully efficient client (i.e. the clients overhead is logarithmic in the database-size). Such a PIR scheme can be immediately constructed from fully homomorphic encryption [12,19,20], avoiding the Merkle tree based approach of [27].

Our Scheme. Armed with these techniques, we can now provide our unlinkable ring signature scheme. Key generation is as described above. To sign a message m with a signing key sk_i, the signer computes a signature σ on m using sk_i and encrypts σ under pk_i obtaining a ciphertext ct. The ciphertext ct' is chosen uniformly random (as in the scheme above). The signer now generates two hashing keys hk and hk' which are binding at position i and computes the hash of $\mathsf{R} = (\mathrm{VK}_1, \ldots, \mathrm{VK}_\ell)$ under both hk and hk', obtaining hash values h and h'. Finally, the signer computes a NIWI proof π which proves that either $(\mathsf{hk}, \mathsf{h})$ bind to a key VK_i and that ct encrypts a signature of m for VK_i *or* $(\mathsf{hk}', \mathsf{h}')$ bind to a key $\mathrm{VK}_{i'}$ and that ct' encrypts a signature of m for $\mathrm{VK}_{i'}$. The signer then outputs the signature $\Sigma = (\mathsf{ct}, \mathsf{ct}', \mathsf{hk}, \mathsf{hk}', \pi)$. To verify a signature $\Sigma = (\mathsf{ct}, \mathsf{ct}', \mathsf{hk}, \mathsf{hk}', \pi)$ for a message m and a ring $\mathsf{R} = (\mathrm{VK}_1, \ldots, \mathrm{VK}_\ell)$, the verifier first computes the hashes h and h' of R using hk and hk' respectively. Now it checks if the NIWI proof π verifies for $(\mathsf{m}, \mathsf{ct}, \mathsf{ct}', \mathsf{hk}, \mathsf{hk}', \mathsf{h}, \mathsf{h}')$, and if so it outputs 1. Unforgeability of this scheme is established in pretty much the same way as described above: If the proof π verifies, then by the somewhere perfectly binding property of SPB and the perfect soundness of the NIWI proof, one of the two ciphertexts $\mathsf{ct}, \mathsf{ct}'$ must in fact encrypt a valid signature. The unforgeability reduction can now recover this signature by setting up the pk_i such that it knows a secret key for each of them and can therefore recover a forge. The idea of establishing anonymity can be outlined as follows. From a high level proof perspective, SPB hashing allows us to *collapse* a ring R of ℓ verification keys into a ring of just two keys. In other words, we only care about the keys to which $(\mathsf{hk}, \mathsf{h})$ and $(\mathsf{hk}', \mathsf{h}')$ bind. With this in mind, we can essentially implement the same proof strategy as before, pretending that our ring just consists of two keys. As before, we will transform a signature computed using a signing key sk_{i_0} into a signature computed using sk_{i_1} via a sequence of hybrids. In the first hybrid, we use the index hiding property of the SPB hash function to move the binding index of hk' from i_0 to i_1. Next, we proceed in a similar way as above, namely compute a signature σ' using sk_{i_1} and encrypt σ' under pk_{i_1} obtaining a ciphertext ct'. Indistinguishability of this hybrid from the previous hybrid can be argued via the pseudorandom ciphertexts property of PKE. In the next step, we switch the witness used to compute the NIWI proof π. That is, instead of proving that ct encrypts a valid signature under pk_{i_0}, we prove that ct' encrypts a valid signature under pk_{i_1}. Both are valid witnesses as we are proving an or-statement. Therefore, witness-indistinguishability of NIWI yields that this hybrid is indistinguishable from the last one. We can now perform the same hybrid modifications to hk and ct and finally switch the witness again. Therefore, in the last hybrid we get a signature Σ computed using sk_{i_1}. For details on this construction, refer to Sect. 4.

Definitions of Linkable Anonymity. The exact definition of linkable anonymity seems to vary between different authors. However, it seems that all these definitions assume that there always remain *unspent* verification keys in an anonymity set. Take for instance the definition of linkable anonymity in [30] (Definition 10 on page 13). Their definition of linkable anonymity is essentially the same as the definition of unlinkable anonymity, with the difference that the adversary is not given access to a signing oracle. We propose a simple definition for linkable anonymity similar in spirit to the blindness definition of blind signatures. The experiment is essentially identical to the anonymity experiment for unlinkable ring signatures, with the following modification:

- The adversary is not allowed to corrupt the challenge keys VK_{i_0} and VK_{i_1}
- In the challenge phase, the adversary submits two message-ring pairs $(\mathsf{m}_0, \mathsf{R}_0)$ and $(\mathsf{m}_1, \mathsf{R}_1)$ such that both R_0 and R_1 contain both VK_{i_0} and VK_{i_1}.
- The experiment flips a bit $b \leftarrow_\$ \{0,1\}$, computes $\Sigma_0 \leftarrow \mathsf{Sign}(\mathrm{SK}_{i_b}, \mathsf{m}_0, \mathsf{R}_0)$ and $\Sigma_1 \leftarrow \mathsf{Sign}(\mathrm{SK}_{i_{1-b}}, \mathsf{m}_1, \mathsf{R}_1)$ and returns (Σ_0, Σ_1) to the adversary
- The adversary must now guess bit b.

Note that the signature Σ_0 is computed exactly as in the experiment for unlinkable anonymity, but now we additionally provide the adversary with a signature Σ_1 computed with the signing key $\mathrm{SK}_{i_{1-b}}$. Consequently, this definition immediately implies e.g. the definition of [30], but does not impose the restriction that no signatures under $\mathrm{VK}_{i_{1-b}}$ can be issued. Like the blindness definition for blind signatures, our definition naturally extends to larger challenge spaces, i.e. considering challenges of size 2 is complete. For details on this new definition refer to Sect. 5.

A linkable Ring Signature Scheme. We will now extend our techniques to the setting of linkable ring signatures. The underlying idea is rather basic. Every verification key VK contains a commitment com to a random tag tag. When a signer signs a message m, he includes tag into the signature Σ and proves that com unveils to tag. This proof can naturally be included in the NIWI proof for the validity of the encrypted signature. Now, whenever a secret key SK is used to sign a message m, its corresponding tag tag is spent. Thus, we can link signatures by checking whether they have the same tag.

While this idea seems to check out at first glance, we run into trouble when trying to prove linkable anonymity. In the linkable anonymity experiment the adversary gets to see the tags of both challenge signatures. This means the reduction must be able to provide witnesses that both the commitment in VK_{i_0} and the commitment in VK_{i_1} open to the respective tags tag_{i_0} and tag_{i_1}. The fact that we need to be able to open both commitments, however, makes it apparently impossible to use the hiding property of the commitments in order to flip the challenge bit in the security proof. Once again, the situation could be resolved easily if we had NIZK proofs at our disposal, yet we can only use witness indistinguishability.

Our way out of this conundrum is based on the following observation. To achieve linkability, we do not actually need that every verification key has a unique tag. Instead, a weaker condition is sufficient. Namely, for a ring of size ℓ it should not be possible to generate $\ell+1$ valid signatures with pairwise distinct tags. We leverage this idea by allowing the commitments in the verification keys to be malformed *in a controlled way.* More specifically, instead of putting only one commitment to a tag tag in the verification key VK, we put 3 commitments to tag in VK.

As before, each signature contains two hashing keys and two hash values. Moreover, in the linkable anonymity proof we will set up things in a way such that for both challenge signatures Σ_0 and Σ_1, one of $(\mathsf{hk}, \mathsf{h})$ and $(\mathsf{hk}', \mathsf{h}')$ will point to VK_{i_0} and the other one to VK_{i_1}. Assume that a signature Σ contains a tag tag and that the SPB hash $(\mathsf{hk}, \mathsf{h})$ points to VK_i, whereas $(\mathsf{hk}', \mathsf{h}')$ points to $\mathrm{VK}_{i'}$. We will make the following consistency requirement: If $i = i'$ we will require that all three commitments in VK_i unveil to the same tag tag. However, if $i \neq i'$, then we only require that out of the six commitments in VK_i and $\mathrm{VK}_{i'}$ that

- at least two unveil to tag,
- at least two unveil to a tag $\mathsf{tag}' \neq \mathsf{tag}$,
- at most one commitment does not unveil correctly.

This relaxed binding condition now allows us to exchange the tags of VK_i and $\mathrm{VK}_{i'}$ even though we are handing out signatures which use these tags! We prove linkable anonymity via a sequence of hybrids. As above, it is instructive to think that SPB hashing collapses a ring R of ℓ keys into a ring of just two verification keys. Call these verification keys VK_0 and VK_1. In the linkable anonymity experiment, there are two signatures, Σ_0 and Σ_1 for m_0 and m_1 respectively. In the first hybrid the challenge bit of the experiment is 0, that is Σ_0 is computed using the signing key SK_0 whereas Σ_1 is computed using SK_1. In the final experiment, Σ_0 will be computed using SK_1 and Σ_1 will be computed using SK_0. The critical part of this proof is to switch the tags. Our proof strategy relies critically on the fact that the tags tag_0 and tag_1 are identically distributed. Namely, we will not switch the tags in the signatures, but switch the tags in the verification keys. More specifically, in the first hybrid VK_{i_0} contains commitments to tag_0 and VK_{i_1} contains commitments to tag tag_1. In the last hybrid, VK_0 will commit to tag_1 and VK_1 will commit to tag_0. But since the tags are identically distributed we can now simply rename them. Therefore, this hybrid is identical to the linkable anonymity experiment with challenge bit 1. In a first step we make both signatures Σ_0 and Σ_1 use both keys VK_0 and VK_1 by modifying the binding indices in hk' appropriately for both signatures. Now, our relaxed binding condition allows us to exchange the tags between VK_0 and VK_1 one by one. That is, the relaxed binding condition allows us to *forget* the unveil information of one of the six commitments in VK_0 and VK_1. Say we forget the unveil information of the first commitment in VK_0. We can then turn this commitment into a commitment of tag_1. Next, we change the first commitment in VK_1 into a commitment of tag_0. We continue like this alternating between VK_0 and VK_1, until we have completely swapped tag_0 and tag_1. Note that in each step the relaxed binding condition holds, thus we can argue via witness indistinguishability and the hiding property of the underlying commitments. Finally, using random tags tag alone does not achieve the strongest notion of non-framemability, where the adversary is allowed to *steal* tags. Thus, we use an idea due to Dolev, Dwork and Naor [16] commonly used to achieve non-malleability[2]: We replace the tag tag by the verification key vk of a signature scheme Sig and additionally sign (m, Σ) with respect to vk. This, however, has the somewhat surprising consequence that we do not need the encrypted signatures anymore, we can rely entirely on the unforgeability of Sig! For details, refer to Sect. 6.

1.3 Related Work

After the initial work of Rivest, Shamir and Tauman [35], a number of works provided constructions in the random oracle model under various computational hardness assumptions [1,9,26]. The scheme of Dodis et al. [15] was the first to achieve sublinear size signatures in the ROM. Libert, Peters, Qian [28] constructed a scheme with logarithmic size ring signatures from DDH in the ROM. Schemes in the CRS model include [10,13,21,23,36,38] achieving varying degrees of compactness but focusing mainly on practical efficiency. Standard model ring signatures were simultaneously proposed by Chow et al. [14] and by Bender, Katz, and Morselli [6]. Malavolta and Schröder [31] build setup free and constant size ring signatures assuming hardness of a variant of the knowledge of exponent assumption. Recently, Backes et al. [3] provided a standard model construction with signatures of size $\sqrt{\ell}$ from a new primitive called *signatures with flexible public key*. Linkable Ring signatures were introduced by Liu et al. [29] as linkable

[2] E.g. in the construction of IND-CCA secure encryption schemes.

spontaneous anonymous group signatures. They propose a notion of linkability which requires that signatures created by the same signer using the same ring must be publicly linkable. In their security model, a scheme achieves a weaker, non-adaptive model of anonymity called signer-ambiguity, if given one signature under signing key SK and ring R as well as a subset of the signing keys corresponding to the keys in the ring which does not include SK, the probability of determining the actual signer as SK is at most negligibly better than guessing one of the remaining keys in the ring uniformly at random. This model is extended by Boyen and Haies [11], introducing signing epochs which allow for forward secure notions of anonymity and unforgeability. Recently, several works described linkable ring signature schemes in post-quantum setting, e.g. [39] based on the hardness of the Ring-SIS problem or [5] based on the Module-SIS and Module-LWE problems. Finally, the idea of replacing NIZK proofs with NIWI proofs in standard model constructions has gained momentum recently, e.g. in the construction of verifiable random functions (VRFs) [7,24].

2 Preliminaries

We will denote by $y \leftarrow \mathcal{A}(x; \mathsf{r})$ the execution of algorithm $\mathcal{A}$ outputting y, on input x and random coins r. We will write $y \leftarrow \mathcal{A}(x)$ if the specific random coins used are not important. By $r \leftarrow_\$ S$ we denote that r is chosen uniformly at random from the set S. We will use $[n]$ to denote the set $\{1, \ldots, n\}$. When defining experiments we implicitly assume the procedures take as input 1^λ as well as some additional parameters which should be clear form the context. We will use the symbol $\emptyset$ to denote an undefined value.

2.1 Signature Schemes

Definition 1. *A signature scheme* Sig *consists of three PPT algorithms* (KeyGen, Sign, Verify) *with the following syntax.*

$\mathsf{KeyGen}(1^\lambda)$**:** *Takes as input the security parameter* 1^λ *and outputs a pair of verification and signing keys* $(\mathsf{vk}, \mathsf{sk})$.

$\mathsf{Sign}(\mathsf{sk}, \mathsf{m})$**:** *Takes as input a signing key* sk *and a message* m *and outputs a signature* σ.

$\mathsf{Verify}(\mathsf{vk}, \mathsf{m}, \sigma)$**:** *Takes as input a verification key* vk*, a message* m *and a signature* σ *and outputs either* 0 *or* 1.

We require the following properties of a signature scheme.

Correctness: *It holds for every security parameter* $\lambda \in \mathbb{N}$ *and every message* m *that given that* $(\mathsf{vk}, \mathsf{sk}) \leftarrow \mathsf{Sig.KeyGen}(1^\lambda)$, $\sigma \leftarrow \mathsf{Sig.Sign}(\mathsf{sk}, \mathsf{m})$, *then it holds that* $\mathsf{Sig.Verify}(\mathsf{vk}, \mathsf{m}, \sigma) = 1$.

Existential Unforgeability under Chosen Message Attacks: *It holds that every PPT adversary* $\mathcal{A}$ *has at most negligible advantage in the following experiment.*

$\mathsf{Exp}_{\mathsf{EUF\text{-}CMA}}(\mathcal{A})$: *1. The experiment generates a pair of verification and signing keys* $(\mathsf{vk}, \mathsf{sk}) \leftarrow \mathsf{Sig.KeyGen}(1^\lambda)$ *and provides* vk *to* $\mathcal{A}$.

2. $\mathcal{A}$ *is allowed to make signing queries of the form* $(\mathtt{sign}, \mathsf{m})$, *upon which the experiment computes* $\sigma \leftarrow \mathsf{Sig.Sign}(\mathsf{sk}, \mathsf{m})$. *Further the experiment keeps a list of all signing queries.*

3. Once $\mathcal{A}$ *outputs a pair* (m^*, σ^*), *the experiment checks if* m^* *was not queried in a signing query and if it holds that* $\mathsf{Sig.Verify}(\mathsf{vk}, \mathsf{m}^*, \sigma^*) = 1$. *If so it outputs* 1, *otherwise* 0.

The advantage of $\mathcal{A}$ *is defined by* $\mathsf{Adv}_{\mathsf{EUF\text{-}CMA}}(\mathcal{A}) = \Pr[\mathsf{Exp}_{\mathsf{EUF\text{-}CMA}}(\mathcal{A}) = 1]$.

2.2 Non-interactive Commitment Schemes

Definition 2. *A commitment scheme* Com *syntactically consists of two PPT algorithms* $(\mathsf{Commit}, \mathsf{Verify})$ *with the following syntax.*

$\mathsf{Commit}(1^\lambda, \mathsf{m})$: *Takes as input a security parameter* 1^λ, *a message* m *and outputs a commitment* com *and unveil information* γ.

$\mathsf{Verify}(\mathsf{com}, \mathsf{m}, \gamma)$: *Takes as input a commitment* com, *a message* m *and unveil information* γ *and outputs either* 0 *or* 1.

We require the following properties of a signature scheme.

Correctness: *It holds for every message* m *that given* $(\mathsf{com}, \gamma) \leftarrow \mathsf{Commit}(\mathsf{m})$, *it holds that* $\mathsf{Verify}(\mathsf{com}, \mathsf{m}, \gamma) = 1$.

Perfect Binding: *It holds that every unbounded adversary* $\mathcal{A}$ *that:*

$$\Pr\left[\begin{array}{l}(\mathsf{com}, \mathsf{m}_0, \gamma_0, \mathsf{m}_1, \gamma_1) \leftarrow \mathcal{A} : \\ \quad \mathsf{m}_0 \neq \mathsf{m}_1 \wedge \mathsf{Verify}(\mathsf{com}, \mathsf{m}_0, \gamma_0) = \mathsf{Verify}(\mathsf{com}, \mathsf{m}_1, \gamma_1) = 1\end{array}\right] = 0.$$

Computational Hiding: *We say that a commitment scheme* $\mathsf{Com} = (\mathsf{Commit}, \mathsf{Verify})$ *is computationally hiding if for every pair of messages* $(\mathsf{m}_0, \mathsf{m}_1)$ *it holds that*

$$\mathsf{com}_0 \approx_c \mathsf{com}_1,$$

where $(\mathsf{com}_0, \gamma_0) \leftarrow \mathsf{Commit}(1^\lambda, \mathsf{m}_0)$ *and* $(\mathsf{com}_1, \gamma_1) \leftarrow \mathsf{Commit}(1^\lambda, \mathsf{m}_1)$. *We denote the advantage of* $\mathcal{A}$ *in distinguishing the commitments as* $\mathsf{Adv}_{\mathsf{Hiding}}(\mathcal{A})$.

Non-interactive commitment schemes can be constructed from any injective one-way function via the Goldreich-Levin hardcore bit [22].

2.3 Public Key Encryption

Definition 3. *A public key encryption scheme* PKE *consists of 3 PPT algorithms* $(\mathsf{KeyGen}, \mathsf{Enc}, \mathsf{Dec})$ *with the following syntax.*

$\mathsf{KeyGen}(1^\lambda)$: *Takes as input a security parameter* 1^λ *and outputs a pair of public and secret keys* $(\mathsf{pk}, \mathsf{sk})$.

$\mathsf{Enc}(\mathsf{pk}, \mathsf{m})$: *Takes as input a public key* pk *and a message* m *and outputs a ciphertext* ct

$\mathsf{Dec}(\mathsf{sk}, \mathsf{ct})$: *Takes as input a secret key* sk *and a ciphertext* ct *and outputs a message* m *or* $\perp$.

We require the following properties of a public key encryption scheme.

Perfect Correctness: *We say a public key encryption scheme* PKE *is perfectly correct, if it holds for all security parameters* $\lambda \in \mathbb{N}$ *and all messages* m *that given that* $(\mathsf{pk}, \mathsf{sk}) \leftarrow \mathsf{PKE.KeyGen}(1^\lambda)$, $\mathsf{ct} \leftarrow \mathsf{PKE.Enc}(\mathsf{pk}, \mathsf{m})$, *then it holds that* $\mathsf{PKE.Dec}(\mathsf{sk}, \mathsf{ct}) = \mathsf{m}$.

Pseudorandom Public Keys: *We require that public keys are computationally indistinguishable from uniform.*

Pseudorandom Ciphertexts: *We require that it holds for every message* m *that*

$$(\mathsf{pk}, u) \approx_c (\mathsf{pk}, \mathsf{Enc}(\mathsf{pk}, \mathsf{m})),$$

where pk *and* u *are chosen uniformly at random.*

We denote the advantages of $\mathcal{A}$ *in breaking pseudorandom public keys and pseudorandom public keys as* $\mathsf{Adv}_{\mathsf{IND\text{-}PK}}(\mathcal{A})$ *and* $\mathsf{Adv}_{\mathsf{IND\text{-}ENC}}(\mathcal{A})$ *respectively. Note that the pseudorandom public keys and pseudorandom ciphertext properties together immediately imply the standard notion of IND-CPA security.*

Such public key encryption schemes can be constructed e.g. from the DDH-problem [18] or the LWE-problem [34].

2.4 Somewhere Perfectly Binding Hashing

Somewhere statistically binding (SSB) hashing [27] allows a negligible fraction of the hashing-keys to be non-binding. For our constructions we actually only require something slightly weaker, a primitive we call *somewhere perfectly binding hashing with private local opening.* This notion relaxes the definition of somewhere perfectly binding hashing in that we allow the Gen algorithm to output a private key shk which the Open algorithm takes as additional input. Below we give our relaxed definition which we use throughout our paper. For completeness we recall the original definition of SSB hashing [27] in the full version of this paper [2] further remark that the LWE-based construction of SBB hashing in [27] can be made somewhere perfectly binding by a noise truncation argument, and the DDH- and DCR-based schemes of [33] are immediately somewhere perfectly binding.

Definition 4. *A somewhere perfectly binding hash family with private local opening* SPB *is given by a tuple of algorithms* $(\mathsf{Gen}, \mathsf{Hash}, \mathsf{Open}, \mathsf{Verify})$ *with the following syntax.*

$\mathsf{Gen}(1^\lambda, n, \mathsf{ind})$: *Takes as input a security parameter* 1^λ, *a database size* n *and an index* ind *and outputs a hashing key* hk *and a private key* shk.

$\mathsf{Hash}(\mathsf{hk}, \mathsf{db})$: *Takes as input a hashing key* hk *and a database* db *and outputs a digest* h.

$\mathsf{Open}(\mathsf{hk}, \mathsf{shk}, \mathsf{db}, \mathsf{ind})$: *Takes as input a hashing key* hk, *a private key* shk *a database* db *and an index* ind *and outputs a witness* τ.

$\mathsf{Verify}(\mathsf{hk}, \mathsf{h}, \mathsf{ind}, x, \tau)$: *Takes as input a hashing key* hk, *a digest* h, *an index* ind, *a value* x *and a witness* τ *and outputs either* 0 *or* 1.

Again, to simplify notation, we will not provide the block size of databases as an input to $\mathsf{SPB.Gen}$ *but rather assume that the block size for the specific application context is hardwired in this function. We require the following properties.*

Correctness: *We say that* $\mathsf{SPB} = (\mathsf{Gen}, \mathsf{Hash}, \mathsf{Open}, \mathsf{Verify})$ *is correct, if it holds for all* $\lambda \in \mathbb{N}$, *all* $n = \mathsf{poly}(\lambda)$, *all databases* db *of size* n *and all indices* $\mathsf{ind} \in [n]$ *that given that* $(\mathsf{hk}, \mathsf{shk}) \leftarrow \mathsf{SPB.Gen}(1^\lambda, n, \mathsf{ind})$, $\mathsf{h} \leftarrow \mathsf{SPB.Hash}(\mathsf{hk}, \mathsf{db})$ *and* $\tau \leftarrow \mathsf{SPB.Open}(\mathsf{hk}, \mathsf{shk}, \mathsf{db}, \mathsf{ind})$, *it holds that*

$$\Pr[\mathsf{SPB.Verify}(\mathsf{hk}, \mathsf{h}, \mathsf{ind}, \mathsf{db}_{\mathsf{ind}}, \tau) = 1] = 1.$$

Efficiency: *The hashing keys* hk *generated by* $\mathsf{Gen}(1^\lambda, n, \mathsf{ind})$ *and the witnesses* τ *generated by* $\mathsf{Open}(\mathsf{hk}, \mathsf{shk}, \mathsf{db}, \mathsf{ind})$ *are of size* $\log(n) \cdot \mathsf{poly}(\lambda)$. *Moreover,* $\mathsf{Verify}(\mathsf{hk}, \mathsf{h}, \mathsf{ind}, x, \tau)$ *can be computed by a circuit of size* $\log(n) \cdot \mathsf{poly}(\lambda)$.

Somewhere Perfectly Binding: *It holds for all* $\lambda \in \mathbb{N}$, *all* $n = \mathsf{poly}(\lambda)$, *all databases* db *of size* n, *all indices* $i \in [n]$, *all database values* x *and all witnesses* τ *that if* $\mathsf{h} = \mathsf{SPB.Hash}(\mathsf{hk}, \mathsf{db})$ *and* $\mathsf{Verify}(\mathsf{hk}, \mathsf{h}, \mathsf{ind}, x, \tau) = 1$, *then it holds that* $x = \mathsf{db}_{\mathsf{ind}}$.

Index Hiding: *Every PPT-adversary* $\mathcal{A}$ *has at most negligible advantage in the following experiment.*

$\mathsf{Exp}_{\mathsf{I\text{-}Hiding}}(\mathcal{A})$:

1. $\mathcal{A}$ *sends* $(n, \mathsf{ind}_0, \mathsf{ind}_1)$ *to the experiment.*
2. *The experiment chooses a random bit* $b \leftarrow_\$ \{0,1\}$, *computes* $(\mathsf{hk}, \mathsf{shk}) \leftarrow \mathsf{SPB.Gen}(1^\lambda, n, \mathsf{ind}_b)$ *and provides* hk *to* $\mathcal{A}$.
3. $\mathcal{A}$ *outputs a guess* b'. *If* $b' = b$, *the experiment outputs* 1, *otherwise* 0.

The advantage of $\mathcal{A}$ *is defined by* $\mathsf{Adv}_{\mathsf{I\text{-}Hiding}}(\mathcal{A}) = \left|\Pr[\mathsf{Exp}_{\mathsf{I\text{-}Hiding}}(\mathcal{A}) = 1] - \frac{1}{2}\right|$

Notice that this definition provides a stronger somewhere perfectly binding guarantee in that we do not have to require that hk has been generated correctly. We can immediately construct a SPB hash family SPB with private local opening from any SPB hash family SPB' with local opening via the following construction.

$\mathsf{SPB.Gen}(1^\lambda, n, \mathsf{ind})$:
Choose random coins $\mathsf{r} \leftarrow \{0,1\}^\lambda$, compute $\mathsf{hk} \leftarrow \mathsf{SPB}'.\mathsf{Gen}(1^\lambda, n, \mathsf{ind}; 1^\lambda; \mathsf{r})$ and output hk and $\mathsf{shk} \leftarrow \mathsf{r}$.

$\mathsf{SPB.Hash}(\mathsf{hk}, \mathsf{db})$:
Output $\mathsf{SPB}'.\mathsf{Hash}(\mathsf{hk}, \mathsf{db})$.

$\mathsf{SPB.Open}(\mathsf{hk}, \mathsf{shk} = \mathsf{r}, \mathsf{db}, \mathsf{ind})$:
Compute $\tau' \leftarrow \mathsf{SPB}'.\mathsf{Open}(\mathsf{hk}, \mathsf{db}, \mathsf{ind})$ and output $\tau \leftarrow (\tau', \mathsf{r})$.

$\mathsf{SPB.Verify}(\mathsf{hk}, \mathsf{h}, \mathsf{ind}, x, \tau = (\tau', \mathsf{r}))$:
If $\mathsf{SPB}'.\mathsf{Gen}(1^\lambda, n, \mathsf{ind}; \mathsf{r}) = \mathsf{hk}$ and $\mathsf{SPB}'.\mathsf{Verify}(\mathsf{hk}, \mathsf{h}, \mathsf{ind}, x, \tau') = 1$ output 1, otherwise 0.

Correctness and index-hiding of SPB follow directly from the corresponding properties of SPB′, the somewhere perfectly binding property follows from the fact that SPB.Verify ensures explicitly that hk is perfectly binding at index ind. Consequently, also this property follows from the corresponding property of SPB′. Moreover, we can also realize a SPB hash family with private local opening from any 2-message private information retrieval scheme with fully efficient verifier and perfect correctness. This was also observed by [33]. The construction is straightforward: A hashing key hk for index i consists of the PIR receiver message, to hash a database db run the PIR sender algorithm on hk and db. The index hiding property follows by PIR receiver privacy, whereas the SPB property follows form perfect correctness. Finally, the receivers private coins serve as succinct private membership witness.

2.5 Non-interactive Witness-Indistinguishable Proof Systems

Let $\mathcal{R}$ be an efficiently computable binary relation, where for $(x, w) \in \mathcal{R}$ we call x a statement and w a witness. Moreover, we denote by $\mathcal{L}_\mathcal{R}$ the language consisting of statements in $\mathcal{R}$, i.e. $\mathcal{L}_\mathcal{R} = \{x | \exists w : (x, w) \in \mathcal{R}\}$.

Definition 5 (Non-interactive Proof System). *Let $\mathcal{R}$ be an efficiently computable witness relation and $\mathcal{L}_\mathcal{R}$ be the language accepted by $\mathcal{R}$. A non-interactive witness-indistinguishable (NIWI) proof system* NIWI *for $\mathcal{L}_\mathcal{R}$ consists of two algorithms* (Prove, Verify) *with the following syntax.*

$\mathsf{Prove}(1^\lambda, x, w)$: *Takes as input a security parameter 1^λ, a statement x and a witness w, output either a proof π or $\perp$.*
$\mathsf{Verify}(x, \pi)$: *Takes as input a statement x, a proof π and outputs either* 0 *or* 1.

We require the following properties.

Perfect Completeness: *It holds for all security parameters $\lambda \in \mathbb{N}$, all statements $x \in \mathcal{L}_\mathcal{R}$ and all witnesses w that if $\mathcal{R}(x, w) = 1$ and $\pi \leftarrow \mathsf{NIWI.Prove}(1^\lambda, x, w)$, then it holds that $\mathsf{NIWI.Verify}(x, \pi) = 1$.*
Perfect Soundness: *It holds for all security parameters $\lambda \in \mathbb{N}$, all statements $x \notin \mathcal{L}_\mathcal{R}$ and all proofs π that $\mathsf{NIWI.Verify}(x, \pi) = 0$.*
Witness-Indistinguishability: *Every PPT adversary $\mathcal{A}$ has at most negligible advantage in the following experiment.*
$\mathsf{Exp}_{\mathsf{WI}}(\mathcal{A})$:
- *$\mathcal{A}$ sends (x, w_0, w_1) with $\mathcal{R}(x, w_0) = 1$ and $\mathcal{R}(x, w_1) = 1$ to the experiment.*
- *The experiment chooses a random bit $b \leftarrow_\$ \{0, 1\}$, computes $\pi^* \leftarrow \mathsf{Prove}(1^\lambda, x, w_b)$ and provides π^* to $\mathcal{A}$.*
- *$\mathcal{A}$ outputs a guess b'. If $b' = b$ the experiment outputs* 1, *otherwise* 0.

The advantage of $\mathcal{A}$ *is defined by* $\mathsf{Adv}_{\mathsf{WI}}(\mathcal{A}) = \left|\Pr[\mathsf{Exp}_{\mathsf{WI}}(\mathcal{A}) = 1] - \frac{1}{2}\right|$.

Proof-Size: *For* $\pi = \mathsf{NIWI.Prove}(1^\lambda, x, w)$ *it holds that* $|\pi| = |C_x| \cdot \mathsf{poly}(\lambda)$, *where* C_x *is a verification circuit for the statement* x, *i.e.* $(x, w) \in \mathcal{R}$ *iff* $C_x(w) = 1$.

Non-interactive witness-indistinguishable proofs can be constructed from NIZK proofs and derandomization assumptions [4,17], from bilinear pairings [25] and indistinguishability obfuscation [8].

3 Ring-Signatures

In this section we provide the definitions related to ring signatures.

Definition 6 (Ring Signatures). *A ring signature scheme* RS *is given by a triple of PPT algorithms* $(\mathsf{KeyGen}, \mathsf{Sign}, \mathsf{Verify})$ *such that*

$\mathsf{KeyGen}(1^\lambda)$: *takes as input the security-parameter* 1^λ *and outputs a pair* $(\mathrm{VK}, \mathrm{SK})$ *of verification and signing keys.*

$\mathsf{Sign}(\mathrm{SK}, \mathsf{m}, \mathsf{R})$: *takes as input a signing key* SK, *a message* $\mathsf{m} \in \mathcal{M}_\lambda$ *and a list of verification keys* $\mathsf{R} = (\mathrm{VK}_1, \ldots, \mathrm{VK}_\ell)$, *and outputs a signature* Σ.

$\mathsf{Verify}(\mathsf{R}, \mathsf{m}, \Sigma)$: *takes as input a ring* $\mathsf{R} = (\mathrm{VK}_1, \ldots, \mathrm{VK}_\ell)$, *a message* $\mathsf{m} \in \mathcal{M}_\lambda$ *and a signature* Σ, *and outputs either* 0 *or* 1.

Correctness: *We say that a ring signature scheme* $\mathsf{RS} = (\mathsf{KeyGen}, \mathsf{Sign}, \mathsf{Verify})$ *is correct, if it holds for all* $\lambda \in \mathbb{N}$, *all* $\ell = \mathsf{poly}(\lambda)$, *all* $i^* \in [\ell]$ *and all messages* $\mathsf{m} \in \mathcal{M}_\lambda$ *that if for* $i \in [\ell]$ $(\mathrm{VK}_i, \mathrm{SK}_i) \leftarrow \mathsf{RS.KeyGen}(1^\lambda)$ *and* $\Sigma \leftarrow \mathsf{RS.Sign}(\mathrm{SK}_i, \mathsf{m}, \mathsf{R})$, *where* $\mathsf{R} = (\mathrm{VK}_1, \ldots, \mathrm{VK}_\ell)$, *then it holds that*

$$\Pr[\mathsf{RS.Verify}(\mathsf{R}, \mathsf{m}, \Sigma) = 1] = 1 - \mathsf{negl}(\lambda),$$

where the probability is taken over the random coins used by $\mathsf{RS.KeyGen}$ *and* $\mathsf{RS.Sign}$.

Anonymity: *We say that a ring signature scheme* $\mathsf{RS} = (\mathsf{KeyGen}, \mathsf{Sign}, \mathsf{Verify})$ *is anonymous against full key exposure, if for every* $q = \mathsf{poly}(\lambda)$ *and every PPT adversary* $\mathcal{A}$ *it holds that* $\mathcal{A}$ *has at most negligible advantage in the following experiment.*

$\mathsf{Exp}_{\mathsf{RS\text{-}Anon}}(\mathcal{A})$:

1. *For all* $i = 1, \ldots, q$ *the experiment generates the keypairs* $(\mathrm{VK}_i, \mathrm{SK}_i) \leftarrow \mathsf{RS.KeyGen}(1^\lambda, \mathsf{r}_i)$ *using random coins* r_i.
2. *The experiment provides* $\mathrm{VK}_1, \ldots, \mathrm{VK}_q$ *and* $\mathsf{r}_1, \ldots, \mathsf{r}_q$ *to* $\mathcal{A}$.
3. *The adversary provides a challenge* $(\mathsf{R}, \mathsf{m}, i_0, i_1)$ *to the experiment, such that* VK_{i_0} *and* VK_{i_1} *are in the ring* R. *The experiment flips a random bit* $b \leftarrow_\$ \{0, 1\}$, *computes* $\Sigma^* \leftarrow \mathsf{RS.Sign}(\mathrm{SK}_{i_b}, \mathsf{m}, \mathsf{R})$ *and outputs* Σ^* *to* $\mathcal{A}$.
4. $\mathcal{A}$ *outputs a guess* b'. *If* $b' = b$, *the experiment outputs* 1, *otherwise* 0

The advantage of $\mathcal{A}$ is defined by $\mathsf{Adv}_{\mathsf{RS\text{-}Anon}}(\mathcal{A}) = \left|\Pr[\mathsf{Exp}_{\mathsf{RS\text{-}Anon}}(\mathcal{A}) = 1] - \frac{1}{2}\right|$.
Note: *We allow that the ring* R *chosen by $\mathcal{A}$ in step 3 may contain maliciously chosen verification keys that were not generated by the challenger.*

Unforgeability: *We say that a ring signature scheme* $\mathsf{RS} = (\mathsf{KeyGen}, \mathsf{Sign}, \mathsf{Verify})$ *is unforgeable with respect to insider corruption, if for every* $q = \mathsf{poly}(\lambda)$ *and every PPT adversary $\mathcal{A}$, it holds that $\mathcal{A}$ has at most negligible advantage in the following experiment.*

$\mathsf{Exp}_{\mathsf{RS\text{-}Unf}}(\mathcal{A})$:

1. *For all* $i = 1, \ldots, q$ *the experiment generates the keypairs* $(\mathrm{VK}_i, \mathrm{SK}_i) \leftarrow \mathsf{RS.KeyGen}(1^\lambda, \mathsf{r}_i)$ *using random coins* r_i. *It sets* $\mathcal{VK} = \{\mathrm{VK}_1, \ldots, \mathrm{VK}_q\}$ *and initializes a set* $\mathcal{C} = \emptyset$.
2. *The experiment provides* $\mathrm{VK}_1, \ldots, \mathrm{VK}_q$ *to $\mathcal{A}$.*
3. *$\mathcal{A}$ is now allowed to make the following queries:*
   $(\mathtt{sign}, i, \mathsf{m}, \mathsf{R})$: *Upon a signing query, the experiment checks if* $\mathrm{VK}_i \in \mathsf{R}$, *and if so computes* $\Sigma \leftarrow \mathsf{RS.Sign}(\mathrm{SK}_i, \mathsf{m}, \mathsf{R})$ *and returns Σ to $\mathcal{A}$. Moreover, the experiment keeps a list of all signing queries.*
   $(\mathtt{corrupt}, i)$: *Upon a corruption query, the experiment adds* VK_i *to $\mathcal{C}$ and returns* r_i *to $\mathcal{A}$.*
4. *In the end, $\mathcal{A}$ outputs a tuple* $(\mathsf{R}^*, \mathsf{m}^*, \Sigma^*)$. *If it holds that* $\mathsf{R}^* \subseteq \mathcal{VK} \backslash \mathcal{C}$ *(i.e. none of the keys in* R^* *were corrupted), $\mathcal{A}$ never made a signing-query of the form* $(\mathtt{sign}, \cdot, \mathsf{m}^*, \mathsf{R}^*)$ *and it holds that*

$$\mathsf{RS.Verify}(\mathsf{R}^*, \mathsf{m}^*, \Sigma^*) = 1,$$

then the experiment outputs 1, *otherwise* 0.

The advantage of $\mathcal{A}$ is defined by $\mathsf{Adv}_{\mathsf{RS\text{-}Unf}}(\mathcal{A}) = \Pr[\mathsf{Exp}_{\mathsf{RS\text{-}Unf}}(\mathcal{A}) = 1]$.

4 Construction of Ring-Signatures

In this section we will provide a construction of a ring signature scheme. Let

- $\mathsf{PKE} = (\mathsf{KeyGen}, \mathsf{Enc}, \mathsf{Dec})$ be a public key encryption scheme with pseudo-random keys and ciphertexts,
- $\mathsf{Sig} = (\mathsf{KeyGen}, \mathsf{Sign}, \mathsf{Verify})$ be a signature scheme,
- $\mathsf{SPB} = (\mathsf{Gen}, \mathsf{Hash}, \mathsf{Open}, \mathsf{Verify})$ be a somewhere perfectly binding hash function with private local opening and,
- $\mathsf{NIWI} = (\mathsf{Prove}, \mathsf{Verify})$ be a NIWI-proof system for the language $\mathcal{L}$ defined as follows. We define a witness-relation $\mathcal{R}$: If $x = (\mathsf{m}, \mathsf{ct}, \mathsf{hk}, \mathsf{h})$ and $w = (\mathrm{VK}, \mathsf{ind}, \tau, \sigma, \mathsf{r}_{\mathsf{ct}})$, where $\mathrm{VK} = (\mathsf{vk}, \mathsf{pk})$, let

$$\begin{aligned}\mathcal{R}(x, w) \Leftrightarrow\ & \mathsf{SPB.Verify}(\mathsf{hk}, \mathsf{h}, \mathsf{ind}, \mathrm{VK}, \tau) = 1\\ & \text{and } \mathsf{PKE.Enc}(\mathsf{pk}, \sigma; \mathsf{r}_{\mathsf{ct}}) = \mathsf{ct}\\ & \text{and } \mathsf{Sig.Verify}(\mathsf{vk}, \mathsf{m}, \sigma) = 1\end{aligned}$$

and let $\mathcal{L}'$ be the language accepted by $\mathcal{R}$. Now, define the language $\mathcal{L}$ by

$$\mathcal{L} = \{(\mathsf{m}, \mathsf{ct}_1, \mathsf{ct}_2, \mathsf{hk}_1, \mathsf{hk}_2, \mathsf{h}_1, \mathsf{h}_2) \mid (\mathsf{m}, \mathsf{ct}_1, \mathsf{hk}_1, \mathsf{h}_1) \in \mathcal{L}' \text{ or } (\mathsf{m}, \mathsf{ct}_2, \mathsf{hk}_2, \mathsf{h}_2) \in \mathcal{L}'\}.$$

Our ring signature scheme $\mathsf{RS} = (\mathsf{KeyGen}, \mathsf{Sign}, \mathsf{Verify})$ is given as follows.

$\mathsf{RS.KeyGen}(1^\lambda; \mathsf{r} = (\mathsf{r_{Sig}}, \mathsf{r_{pk}}))$:
- Compute $(\mathsf{vk}, \mathsf{sk}) \leftarrow \mathsf{Sig.KeyGen}(1^\lambda; \mathsf{r_{Sig}})$
- Compute $\mathsf{pk} \leftarrow \mathsf{r_{pk}}$
- Output $\mathrm{VK} \leftarrow (\mathsf{vk}, \mathsf{pk})$ and $\mathrm{SK} \leftarrow (\mathsf{sk}, \mathrm{VK})$

$\mathsf{RS.Sign}(\mathrm{SK} = (\mathsf{sk}, \mathrm{VK}), \mathsf{m}, \mathsf{R} = (\mathrm{VK}_1, \ldots, \mathrm{VK}_\ell))$:
- Parse $\mathrm{VK} = (\mathsf{vk}, \mathsf{pk})$
- Compute $\sigma \leftarrow \mathsf{Sig.Sign}(\mathsf{sk}, \mathsf{m})$
- Find an index $\mathsf{ind} \in [\ell]$ such that $\mathrm{VK}_{\mathsf{ind}} = \mathrm{VK}$
- Compute $(\mathsf{hk}_1, \mathsf{shk}_1) \leftarrow \mathsf{SPB.Gen}(1^\lambda, |\mathsf{R}|, \mathsf{ind})$
- Compute $(\mathsf{hk}_2, \mathsf{shk}_2) \leftarrow \mathsf{SPB.Gen}(1^\lambda, |\mathsf{R}|, \mathsf{ind})$
- Compute $\mathsf{h}_1 \leftarrow \mathsf{SPB.Hash}(\mathsf{hk}_1, \mathsf{R})$
- Compute $\mathsf{h}_2 \leftarrow \mathsf{SPB.Hash}(\mathsf{hk}_2, \mathsf{R})$
- Compute $\tau \leftarrow \mathsf{SPB.Open}(\mathsf{hk}_1, \mathsf{shk}_1, \mathsf{R}, \mathsf{ind})$
- Compute $\mathsf{ct}_1 \leftarrow \mathsf{PKE.Enc}(\mathsf{pk}, \sigma; \mathsf{r_{ct}})$
- Compute $\mathsf{ct}_2 \leftarrow_\$ \{0,1\}^\lambda$
- Set $x \leftarrow (\mathsf{m}, \mathsf{ct}_1, \mathsf{ct}_2, \mathsf{hk}_1, \mathsf{hk}_2, \mathsf{h}_1, \mathsf{h}_2)$ and $w \leftarrow (\mathrm{VK}, \mathsf{ind}, \tau, \sigma, \mathsf{r_{ct}})$
- Compute $\pi \leftarrow \mathsf{NIWI.Prove}(x, w)$
- Output $\Sigma \leftarrow (\mathsf{ct}_1, \mathsf{ct}_2, \mathsf{hk}_1, \mathsf{hk}_2, \pi)$

$\mathsf{RS.Verify}(\mathsf{R}, \mathsf{m}, \Sigma)$:
- Parse $\Sigma = (\mathsf{ct}_1, \mathsf{ct}_2, \mathsf{hk}_1, \mathsf{hk}_2, \pi)$
- Compute $\mathsf{h}'_1 \leftarrow \mathsf{SPB.Hash}(\mathsf{hk}_1, \mathsf{R})$
- Compute $\mathsf{h}'_2 \leftarrow \mathsf{SPB.Hash}(\mathsf{hk}_2, \mathsf{R})$
- Output $\mathsf{NIWI.Verify}((\mathsf{m}, \mathsf{ct}_1, \mathsf{ct}_2, \mathsf{hk}_1, \mathsf{hk}_2, \mathsf{h}'_1, \mathsf{h}'_2), \pi)$

4.1 Correctness

We will first show that our scheme is correct. Assume that $\mathrm{VK} = (\mathsf{vk}, \mathsf{pk})$ and $\mathrm{SK} = (\mathsf{sk}, \mathrm{VK})$ were generated by $\mathsf{RS.KeyGen}$ and $\Sigma = (\mathsf{ct}_1, \mathsf{ct}_2, \mathsf{hk}_1, \mathsf{hk}_2, \pi)$ is the output of $\mathsf{RS.Sign}(\mathrm{SK}, \mathsf{m}, \mathsf{R})$, where $\mathsf{R} = (\mathrm{VK}_1, \ldots, \mathrm{VK}_\ell)$. We will show that it holds that $\mathsf{RS.Verify}(\mathsf{R}, \mathsf{m}, \sigma) = 1$. First note that since $\mathsf{SPB.Hash}$ is deterministic, it holds that $\mathsf{h}'_1 = \mathsf{h}_1$ and $\mathsf{h}'_2 = \mathsf{h}_2$. Also, it obviously holds that $\mathrm{VK} = \mathrm{VK}_{\mathsf{ind}}$ (where ind is the index of VK in R). Now, notice further that by the correctness of SPB it holds that $\mathsf{SPB.Verify}(\mathsf{hk}_1, \mathsf{h}_1, \mathsf{ind}, \mathrm{VK}_{\mathsf{ind}}, \tau) = 1$. Moreover, by the correctness of Sig it holds that $\mathsf{Sig.Verify}(\mathsf{vk}, \mathsf{m}, \sigma) = 1$. Consequently, $(\mathsf{m}, \mathsf{ct}_1, \mathsf{ct}_2, \mathsf{hk}_1, \mathsf{hk}_2, \mathsf{h}_1, \mathsf{h}_2) \in \mathcal{L}$ and $w = (\mathrm{VK}, \mathsf{ind}, \tau, \sigma, \mathsf{r_{ct}})$ is a witness for membership. Thus, by the correctness of NIWI it holds that

$$\mathsf{NIWI.Verify}((\mathsf{m}, \mathsf{ct}_1, \mathsf{ct}_2, \mathsf{hk}_1, \mathsf{hk}_2, \mathsf{h}_1, \mathsf{h}_2), \pi) = 1$$

and consequently $\mathsf{RS.Verify}(\mathsf{R}, \mathsf{m}, \Sigma)$ outputs 1.

4.2 Signature Size

For a signature $\Sigma = (\mathsf{ct}_1, \mathsf{ct}_2, \mathsf{hk}_1, \mathsf{hk}_2, \pi)$, the size of the ciphertexts $\mathsf{ct}_1, \mathsf{ct}_2$ is $\mathsf{poly}(\lambda)$ and independent of the ring-size ℓ. By the efficiency property of SPB the sizes of the hashing keys $\mathsf{hk}_1, \mathsf{hk}_2$ is bounded by $\log(\ell) \cdot \mathsf{poly}(\lambda)$. Also by the efficiency property of SPB this size of the witness τ is $\log(\ell) \cdot \mathsf{poly}(\lambda)$ and the SPB-verification function Verify can be computed by a circuit of size $\log(\ell) \cdot \mathsf{poly}(\lambda)$.

Consequently, the verification circuit C_x for the language $\mathcal{L}$ and statement $x = (\mathsf{m}, \mathsf{ct}_1, \mathsf{ct}_2, \mathsf{hk}_1, \mathsf{hk}_2, \mathsf{h}_1, \mathsf{h}_2)$ has size $\log(\ell) \cdot \mathsf{poly}(\lambda)$. By the proof-size property of the NIWI proof it holds that $|\pi| = |C_x| \cdot \mathsf{poly}(\lambda) = \log(\ell) \cdot \mathsf{poly}(\lambda)$. All together, the size of signatures Σ is $\log(\ell) \cdot \mathsf{poly}(\lambda)$.

4.3 Unforgeability

We will turn to showing that RS is unforgeable.

Theorem 1. *The ring signature scheme* RS *is unforgeable, given that* NIWI *has perfect soundness,* SPB *is somewhere perfectly binding,* PKE *is perfectly correct,* PKE *has pseudorandom public keys and* Sig *is unforgeable.*

The main idea of the proof is that since the NIWI proof has perfect soundness, it must either hold that $(\mathsf{m}, \mathsf{ct}_1, \mathsf{hk}_1, \mathsf{h}_1) \in \mathcal{L}'$ or $(\mathsf{m}, \mathsf{ct}_2, \mathsf{hk}_2, \mathsf{h}_2) \in \mathcal{L}'$. If the first statement is true, then hk_1 corresponds to an index ind_1 and $\mathcal{A}$ must have produced a forge for a key $\mathrm{VK}_{\mathsf{ind}_1}$ in R. Likewise, if the second statement is true, then $\mathcal{A}$ must have produced a forge for a key $\mathrm{VK}_{\mathsf{ind}_2}$ in R.

Proof. Let $\mathcal{A}$ be a PPT-adversary against the unforgeability experiment of RS and let further $q = \mathsf{poly}(\lambda)$ an upper bound on the number of key queries of $\mathcal{A}$. Consider the following two hybrids.

$\mathcal{H}_0$: This is the real experiment.
$\mathcal{H}_1$: The same as $\mathcal{H}_0$, except that for all $i \in [q]$ the challenger generates the public keys pk_i in VK_i by $(\mathsf{pk}_i, \hat{\mathsf{sk}}_i) \leftarrow \mathsf{PKE.KeyGen}(1^\lambda)$ instead of choosing pk_i uniformly at random. Moreover, the challenger stores all the secret keys $(\hat{\mathsf{sk}}_i)_{i \in [q]}$.

We will first argue that $\mathcal{H}_0$ and $\mathcal{H}_1$ are computationally indistinguishable given that the public keys of PKE are pseudorandom.

Claim. There exists a reduction $\mathcal{R}_1$ such that $\mathsf{Adv}_{\mathsf{IND\text{-}PK}}(\mathcal{R}_1^{\mathcal{A}}) \geq |\Pr[\mathcal{H}_0(\mathcal{A}) = 1] - \Pr[\mathcal{H}_1(\mathcal{A}) = 1]|$

The reduction $\mathcal{R}_1$ is given as follows.

Reduction $\mathcal{R}_1^{\mathcal{A}}(\mathsf{pk}^*)$
- Choose an index $i^* \leftarrow_\$ [q]$ uniformly at random.
- Simulate $\mathcal{H}_0$ with the following modifications. For all indices $i < i^*$ generate $(\mathrm{VK}_i, \mathrm{SK}_i)$ as in $\mathcal{H}_0$. For $i > i^*$ generate $(\mathrm{VK}_i, \mathrm{SK}_i)$ as in $\mathcal{H}_1$.

- Generate $(\mathrm{VK}_{i^*}, \mathrm{SK}_{i^*})$ as follows:
  - Compute $(\mathsf{vk}_{i^*}, \mathsf{sk}_{i^*}) \leftarrow \mathsf{Sig.KeyGen}(1^\lambda; \mathsf{r}_{\mathsf{Sig}})$
  - Set $\mathrm{VK}_{i^*} \leftarrow (\mathsf{vk}_{i^*}, \mathsf{pk}^*)$ and $\mathrm{SK}_{i^*} \leftarrow (\mathsf{sk}_{i^*}, \mathrm{VK}_{i^*})$
- Output whatever the simulated experiment outputs.

Let PK_0 be the uniform distribution and PK_1 be a distribution sampled by computing $(\mathsf{pk}^*, \hat{\mathsf{sk}}^*) \leftarrow \mathsf{PKE.KeyGen}(1^\lambda)$ and outputting pk^*. First observe that when $i^* = q-1$ and pk^* was chosen from PK_0, then $\mathcal{R}_1^{\mathcal{A}}$ perfectly simulates $\mathcal{H}_0(\mathcal{A})$. On the other hand, if $i^* = 0$ and pk^* was chosen from PK_1, then $\mathcal{R}_1^{\mathcal{A}}$ perfectly simulates $\mathcal{H}_1(\mathcal{A})$. Moreover, observe that for $j = 1, \ldots, q-1$ it holds that $\mathcal{R}_1^{\mathcal{A}}(PK_0)|_{i^*=j-1}$ and $\mathcal{R}_1^{\mathcal{A}}(PK_1)|_{i^*=j}$ are identically distributed. Consequently, we get that

$$\begin{aligned}
\mathsf{Adv}_{\mathsf{IND\text{-}PK}}(\mathcal{R}_1^{\mathcal{A}}) &= |\Pr[\mathcal{R}_1^{\mathcal{A}}(PK_0)] - \Pr[\mathcal{R}_1^{\mathcal{A}}(PK_1)]| \\
&= |\sum_{j=0}^{q-1} \Pr[i^* = j] \cdot (\Pr[\mathcal{R}_1^{\mathcal{A}}(PK_0)|i^* = j] - \Pr[\mathcal{R}_1^{\mathcal{A}}(PK_1)|i^* = j])| \\
&= \frac{1}{q} \cdot |(\Pr[\mathcal{R}_1^{\mathcal{A}}(PK_0)|i^* = q-1] - \Pr[\mathcal{R}_1^{\mathcal{A}}(PK_1)|i^* = 0] \\
&\quad + \sum_{j=1}^{q-1} (\Pr[\mathcal{R}_1^{\mathcal{A}}(PK_1)|i^* = j] - \Pr[\mathcal{R}_1^{\mathcal{A}}(PK_0)|i^* = j-1]))| \\
&= \frac{1}{q} \cdot |(\Pr[\mathcal{H}_0(\mathcal{A}) = 1] - \Pr[\mathcal{H}_1(\mathcal{A}) = 1])|.
\end{aligned}$$

Claim. There exists a reduction $\mathcal{R}_2$ such that $\mathcal{R}_2^{\mathcal{A}}$ breaks the EUF-CMA security of Sig with probability $\mathsf{Adv}_{\mathcal{H}_1}(\mathcal{A})/q$.

The reduction $\mathcal{R}_2$ is given as follows.

Reduction $\mathcal{R}_2^{\mathcal{A}}(\mathrm{VK}^*)$
- Guess an index $i^* \leftarrow_\$ [q]$. For all $i \neq i^*$ generate VK_i and SK_i as in $\mathcal{H}_1$. Generate $\mathrm{VK}_{i^*} = \mathrm{VK}^*$ as follows. Generate $(\mathsf{pk}^*, \hat{\mathsf{sk}}^*) \leftarrow \mathsf{KeyGen}(1^\lambda)$ and set $\mathrm{VK}^* \leftarrow (\mathsf{vk}^*, \mathsf{pk}^*)$, where vk^* is the verification key provided by the EUF-CMA experiment. Moreover, store $\hat{\mathsf{sk}}_{i^*} = \hat{\mathsf{sk}}^*$.
- If $\mathcal{A}$ asks to corrupt VK^* abort.
- If $\mathcal{A}$ sends signature query $(\mathsf{m}, \mathrm{VK}^*, \mathsf{R})$, send m to the signing oracle of the EUF-CMA game to obtain a signature σ. Compute the signature Σ by
  - Let ind^* be the index of VK^* in R.
  - Computing $(\mathsf{hk}_1, \mathsf{shk}_1) \leftarrow \mathsf{SPB.Gen}(1^\lambda, |\mathsf{R}|, \mathsf{ind}^*)$
  - Computing $(\mathsf{hk}_2, \mathsf{shk}_2) \leftarrow \mathsf{SPB.Gen}(1^\lambda, |\mathsf{R}|, \mathsf{ind}^*)$
  - Computing $\mathsf{h}_1 \leftarrow \mathsf{SPB.Hash}(\mathsf{hk}_1, R)$
  - Computing $\mathsf{h}_2 \leftarrow \mathsf{SPB.Hash}(\mathsf{hk}_2, R)$
  - Computing $\tau \leftarrow \mathsf{SPB.Open}(\mathsf{hk}_1, \mathsf{shk}_1, R, \mathsf{ind}^*)$
  - Computing $\mathsf{ct}_1 \leftarrow \mathsf{PKE.Enc}(\mathsf{pk}^*, \sigma; \mathsf{r}_{\mathsf{ct}})$

- Computing $\mathsf{ct}_2 \leftarrow_\$ \{0,1\}^\lambda$
- Computing

$$\pi \leftarrow \mathsf{NIWI.Prove}((\mathsf{m}, \mathsf{ct}_1, \mathsf{ct}_2, \mathsf{hk}_1, \mathsf{hk}_2, \mathsf{h}_1, \mathsf{h}_2), (\mathrm{VK}^*, \mathsf{ind}^*, \tau, \sigma, \mathsf{r}_{\mathsf{ct}}))$$

- Output $\Sigma \leftarrow (\mathsf{ct}_1, \mathsf{ct}_2, \mathsf{hk}_1, \mathsf{hk}_2, \pi)$

– Once $\mathcal{A}$ outputs a forge Σ^* for $(\mathsf{m}^*, \mathsf{R}^*)$, check if it is valid, that is in the query phase $\mathcal{A}$ has not requested a signature of m^* for any key in R^*, none of the keys in R^* has been corrupted and it holds that $\mathsf{RS.Verify}(\mathsf{R}, \mathsf{m}^*, \Sigma^*) = 1$. If the forge is valid proceed.
– Parse Σ^* as $\Sigma^* = (\mathsf{ct}_1^*, \mathsf{ct}_2^*, \mathsf{hk}_1^*, \mathsf{hk}_2^*, \pi^*)$.
– Let $|\mathsf{R}^*| = \ell$ and let $i_1, \ldots, i_\ell$ be the indices of the keys in R^*, i.e. $\mathsf{R} = (\mathrm{VK}_{i_1}, \ldots, \mathrm{VK}_{i_\ell})$.
– For $j = 1, \ldots, \ell$:
  * Compute $\check{\sigma}_1 \leftarrow \mathsf{Dec}(\hat{\mathsf{sk}}_{i_j}, \mathsf{ct}_1^*)$ and $\check{\sigma}_2 \leftarrow \mathsf{Dec}(\hat{\mathsf{sk}}_{i_j}, \mathsf{ct}_2^*)$.
  * If $\mathsf{Sig.Verify}(\mathsf{vk}^*, \mathsf{m}^*, \check{\sigma}_1) = 1$ stop and output $\check{\sigma}_1$
  * If $\mathsf{Sig.Verify}(\mathsf{vk}^*, \mathsf{m}^*, \check{\sigma}_2) = 1$ stop and output $\check{\sigma}_2$

First note that the key-pair $(\mathsf{pk}_{i^*}, \hat{\mathsf{sk}}_{i^*})$ is correct for all messages. Notice further that, unless $\mathcal{A}$ asks to corrupt VK^*, $\mathcal{H}_1$ and the simulation of $\mathcal{R}_2$ are identically distributed from the view of $\mathcal{A}$. Observe that with probability at least $1/q$ the adversary $\mathcal{A}$ does not trigger an abort. Thus, conditioned that no abort happened, from the view of $\mathcal{A}$ the index i^* is distributed uniformly random. Assume now that $\mathcal{A}$ outputs a valid forge Σ^* for $(\mathsf{m}^*, \mathsf{R}^*)$ with $\mathsf{R}^* = (\mathrm{VK}_{i_1}, \ldots, \mathrm{VK}_{i_\ell})$. By the perfect soundness of NIWI, it holds that either $(\mathsf{m}^*, \mathsf{ct}_1^*, \mathsf{hk}_1^*, \mathsf{h}_1^*) \in \mathcal{L}'$ or $(\mathsf{m}, \mathsf{ct}_2^*, \mathsf{hk}_2^*, \mathsf{h}_2^*) \in \mathcal{L}'$. Assume w.l.o.g. that $(\mathsf{m}, \mathsf{ct}_1^*, \mathsf{hk}_1^*, \mathsf{h}_1^*) \in \mathcal{L}'$. That is, there exist $(\mathrm{VK}^\dagger, \mathsf{ind}^\dagger, \tau^\dagger, \sigma^\dagger, \mathsf{r}_{\mathsf{ct}})$ with $\mathrm{VK}^\dagger = (\mathsf{vk}^\dagger, \mathsf{pk}^\dagger)$ such that

$$\begin{aligned} \mathsf{SPB.Verify}(\mathsf{hk}_1^*, \mathsf{h}_1^*, \mathsf{ind}^\dagger, \check{\mathrm{VK}}, \check{\tau}) &= 1 \\ \text{and } \mathsf{PKE.Enc}(\mathsf{pk}^\dagger, \sigma^\dagger; \mathsf{r}_{\mathsf{ct}}) &= \mathsf{ct}_1^* \\ \text{and } \mathsf{Sig.Verify}(\mathsf{vk}^\dagger, \mathsf{m}^*, \sigma^\dagger) &= 1 \end{aligned}$$

As $\mathsf{SPB.Verify}(\mathsf{hk}_1^*, \mathsf{h}_1^*, \mathsf{ind}^\dagger, \check{\mathrm{VK}}, \check{\tau}) = 1$ and $\mathsf{h}_1^* = \mathsf{SPB.Hash}(\mathsf{hk}_1^*, \mathsf{R})$ it holds by the somewhere perfectly binding property of SPB that $\mathrm{VK}^\dagger = \mathrm{VK}_{i_{\mathsf{ind}^\dagger}}$, i.e. $\mathsf{vk}^\dagger = \mathsf{vk}_{i_{\mathsf{ind}^\dagger}}^\dagger$ and $\mathsf{pk}^\dagger = \mathsf{pk}_{i_{\mathsf{ind}^\dagger}}$. Moreover, by the above it also holds that $\mathsf{ct}_1^* = \mathsf{PKE.Enc}(\mathsf{pk}_{i_{\mathsf{ind}^\dagger}}, \sigma^\dagger; \mathsf{r}_{\mathsf{ct}})$ and $\mathsf{Sig.Verify}(\mathsf{vk}_{i_{\mathsf{ind}^\dagger}}, \mathsf{m}^*, \sigma^\dagger) = 1$.

Now observe that, as i^* is uniformly random from the view of $\mathcal{A}$, it holds that $i_{\mathsf{ind}^\dagger} = i^*$ with probability at least $1/q$. Assume therefore that $i_{\mathsf{ind}^\dagger} = i^*$. As $(\mathsf{pk}_{i^*}, \hat{\mathsf{sk}}_{i^*})$ are correct for all messages, it holds that $\check{\sigma}_1 = \mathsf{PKE.Dec}(\hat{\mathsf{sk}}_{i^*}, \mathsf{ct}_1^*) = \sigma^\dagger$. Therefore it holds that $\mathsf{Sig.Verify}(\mathsf{vk}_{i_{\mathsf{ind}^\dagger}}, \mathsf{m}^*, \check{\sigma}_1) = 1$ for the signature $\check{\sigma}_1$ decrypted by $\mathcal{R}_2^{\mathcal{A}}$, i.e. $\check{\sigma}_1$ is a valid signature of m^* under vk^*. We conclude that $\mathsf{Adv}_{\mathsf{EUF\text{-}CMA}}(\mathcal{R}_2^{\mathcal{A}}) \geq \frac{1}{q}|\mathsf{Adv}_{\mathcal{H}_1}(\mathcal{A}) - \nu|$.

All together, as $\mathsf{Adv}_{\mathcal{H}_1}(\mathcal{A}) \geq |\mathsf{Adv}_{\mathcal{H}_0}(\mathcal{A}) - q \cdot \mathsf{Adv}_{\mathsf{IND\text{-}PK}}(\mathcal{R}_1^{\mathcal{A}})|$ and $\mathsf{Adv}_{\mathcal{H}_0}(\mathcal{A}) = \mathsf{Adv}_{\mathsf{RS\text{-}Unf}}(\mathcal{A})$, we can conclude that

$$\mathsf{Adv}_{\mathsf{RS\text{-}Unf}}(\mathcal{A}) \leq q \cdot \mathsf{Adv}(\mathcal{R}_1^{\mathcal{A}}) + q \cdot \mathsf{Adv}_{\mathsf{EUF\text{-}CMA}}(\mathcal{R}_2^{\mathcal{A}}) + \nu.$$

This concludes the proof.

On Tightness. Using a public key encryption scheme with tight multi-user security, we can improve the bound on the advantage above to

$$\mathsf{Adv}_{\mathsf{RS\text{-}Unf}}(\mathcal{A}) \leq \mathsf{Adv}(\mathcal{R}_1^{\mathcal{A}}) + q \cdot \mathsf{Adv}_{\mathsf{EUF\text{-}CMA}}(\mathcal{R}_2^{\mathcal{A}}) + \nu.$$

However, getting rid of the q factor for $q \cdot \mathsf{Adv}_{\mathsf{EUF\text{-}CMA}}(\mathcal{R}_2^{\mathcal{A}})$ seems beyond the scope of current techniques.

4.4 Anonymity

We will now turn to establishing anonymity of RS.

Theorem 2. *The ring signature scheme* RS *is anonymous, given that* SPB *is index hiding,* PKE *has pseudorandom ciphertexts and* NIWI *is computationally witness-indistinguishable.*

Our strategy is to first move the index of hk_2 from i_0 to i_1 and argue indistinguishability via the index-hiding property of SPB. Next we switch ct_2 to an encryption of a signature σ' of m for the verification key VK_{i_1}. This modification will not be detected due to the pseudorandom ciphertexts property of the PKE. Now, we can switch the NIWI witness to a witness for $(\mathsf{m}, \mathsf{ct}_2, \mathsf{hk}_2, \mathsf{h}_2) \in \mathcal{L}'$. Next, we perform the first two changes above for hk_1 and ct_1, switch the witness back to the witness for $(\mathsf{m}, \mathsf{ct}_1, \mathsf{hk}_1, \mathsf{h}_1) \in \mathcal{L}'$, and finally replace ct_2 with a random string. The signature in the last experiment is now a real signature of m under VK_{i_1}.

Proof (Sketch). Let in the following ind_0 be the index of VK_{i_0} in R and ind_1 be the index of VK_{i_1} ind R, where $(i_0, i_1, \mathsf{m}^*, \mathsf{R})$ is the challenge query of $\mathcal{A}$. Consider the following hybrids:

$\mathcal{H}_0$: This is the real experiment with challenge-bit $b^* = 0$.

$\mathcal{H}_1$: Same as $\mathcal{H}_0$, except that in Σ^* we compute hk_2^* using $(\mathsf{hk}_2^*, \mathsf{shk}_2^*) \leftarrow \mathsf{SPB.Gen}(1^\lambda, |\mathsf{R}|, \mathsf{ind}_1)$ instead of computing $(\mathsf{hk}_2^*, \mathsf{shk}_2^*) \leftarrow \mathsf{SPB.Gen}(1^\lambda, |\mathsf{R}|, \mathsf{ind}_0)$. Moreover, also compute $\tau' \leftarrow \tau \leftarrow \mathsf{SPB.Open}(\mathsf{hk}_2, \mathsf{shk}_2, \mathsf{R}, \mathsf{ind}_1)$.

$\mathcal{H}_2$: Same as $\mathcal{H}_1$, except that we compute ct_2^* by
- $\sigma' \leftarrow \mathsf{Sig.Sign}(\mathsf{sk}_{i_1}, \mathsf{m}^*)$
- $\mathsf{ct}_2^* \leftarrow \mathsf{PKE.Enc}(\mathsf{pk}_{i_1}, \sigma'; \mathsf{r}_{\mathsf{ct}_2})$

instead of $\mathsf{ct}_2^* \leftarrow_\$ \{0,1\}^\lambda$.

$\mathcal{H}_3$: The same as $\mathcal{H}_2$, except that we use the witness $w' \leftarrow (\mathrm{VK}_{i_1}, \mathsf{ind}_1, \tau', \sigma', \mathsf{r}_{\mathsf{ct}_2})$ instead of $w \leftarrow (\mathrm{VK}_{i_0}, \mathsf{ind}_0, \tau, \sigma, \mathsf{r}_{\mathsf{ct}_1})$ to compute π, i.e. we compute $\pi \leftarrow \mathsf{NIWI.Prove}(x, w')$.

$\mathcal{H}_4$: The same as $\mathcal{H}_3$, except that we compute ct_1^* by $\mathsf{ct}_1^* \leftarrow_\$ \{0,1\}^\lambda$.

$\mathcal{H}_5$: The same as $\mathcal{H}_4$, except that we compute hk_1^* by $(\mathsf{hk}_1^*, \mathsf{shk}_1^*) \leftarrow \mathsf{SPB.Gen}(1^\lambda, |\mathsf{R}|, \mathsf{ind}_1)$ instead of $(\mathsf{hk}_1^*, \mathsf{shk}_1^*) \leftarrow \mathsf{SPB.Gen}(1^\lambda, |\mathsf{R}|, \mathsf{ind}_0)$. Moreover, also compute τ by $\tau \leftarrow \mathsf{SPB.Open}(\mathsf{hk}_1, \mathsf{shk}_1, \mathsf{R}, \mathsf{ind}_1)$.

$\mathcal{H}_6$: The same as $\mathcal{H}_5$, except that we compute ct_1^* by
- $\sigma \leftarrow \mathsf{Sig.Sign}(\mathsf{sk}_{i_1}, \mathsf{m}^*)$

– $\mathsf{ct}_1^* \leftarrow \mathsf{PKE.Enc}(\mathsf{pk}_{i_1}, \sigma; \mathsf{r}_{\mathsf{ct}_1})$ instead of $\mathsf{ct}_1^* \leftarrow_\$ \{0,1\}^\lambda$.

$\mathcal{H}_7$: The same as $\mathcal{H}_6$, except that we use the witness $w'' \leftarrow (\mathrm{VK}_{i_1}, \mathsf{ind}_1, \tau, \sigma, \mathsf{r}_{\mathsf{ct}_1})$ instead of $w' \leftarrow (\mathrm{VK}_{i_1}, \mathsf{ind}_1, \tau', \sigma', \mathsf{r}_{\mathsf{ct}_2})$ to compute π, i.e. we compute $\pi \leftarrow \mathsf{NIWI.Prove}(x, w'')$.

$\mathcal{H}_8$: The same as $\mathcal{H}_7$ except that we compute ct_2^* by $\mathsf{ct}_2^* \leftarrow_\$ \{0,1\}^\lambda$. This is identical to the real experiment with $b^* = 1$.

It follows by inspection that the above hybrids are indistinguishable. The full proof can be found in the full version of this paper [2].

5 Linkable Ring-Signatures

In this section we introduce our new model for linkable ring signatures.

Definition 7 (Linkable Ring Signatures). *Syntactically, a ring signature scheme* LRS *is given by PPT algorithms* (KeyGen, Sign, Verify, Link) *such that*

$\mathsf{KeyGen}(1^\lambda)$: *takes as input the security-parameter* 1^λ *and outputs a pair* $(\mathrm{VK}, \mathrm{SK})$ *of verification and signing keys.*

$\mathsf{Sign}(\mathrm{SK}, \mathsf{m}, \mathsf{R})$: *takes as input a signing key* SK, *a message* $\mathsf{m} \in \mathcal{M}_\lambda$ *and a list of verification keys* $\mathsf{R} = (\mathrm{VK}_1, \ldots, \mathrm{VK}_q)$, *and outputs a signature* Σ.

$\mathsf{Verify}(\mathsf{R}, \mathsf{m}, \Sigma)$: *takes as input a ring* $\mathsf{R} = (\mathrm{VK}_1, \ldots, \mathrm{VK}_q)$, *a message* $\mathsf{m} \in \mathcal{M}$ *and a signature* Σ, *and outputs either* 0 *or* 1.

$\mathsf{Link}(\Sigma_1, \Sigma_2, m_1, m_2)$: *takes as input two signatures and two messages and outputs either* 0 *or* 1.

We say that a linkable ring signature scheme LRS = (KeyGen, Sign, Verify, Link) *is correct, if it holds for all* $\lambda \in \mathbb{N}$, *all* $q = \mathsf{poly}(\lambda)$, *all* $i^* \in [\ell]$ *and all messages* $\mathsf{m} \in \mathcal{M}_\lambda$ *that, if* $(\mathrm{VK}_i, \mathrm{SK}_i) \leftarrow \mathsf{LRS.KeyGen}(1^\lambda)$ *and* $\Sigma \leftarrow \mathsf{LRS.Sign}(\mathrm{SK}_i, \mathsf{m}, \mathsf{R})$, *where* $i \in [q]$ *and* $\mathsf{R} = (\mathrm{VK}_1, \ldots, \mathrm{VK}_q)$, *then*

$$\Pr[\mathsf{LRS.Verify}(\mathsf{R}, \mathsf{m}, \sigma) = 1] = 1 - \mathsf{negl}(\lambda),$$

where the probability is taken over the random coins used by LRS.KeyGen *and* LRS.Sign.

We will now define security properties of linkable ring signatures and begin with the core property called linkability. Informally, we may think of it as the requirement that any two or more uses of a secret key can be publicly linked. We model this property by letting an adversary output q verification keys and signatures, where none of the signatures links with each other. In order to break linkability the adversary has to output one additional signature which does not link with any of the former signatures. Note that producing q signatures which do not link is easy. The adversary only has to use the q different secret keys. But producing the one additional signature without an additional verification key, is required to be infeasible.

Definition 8 (Linkability). *We say that a linkable ring signature scheme* $\mathsf{LRS} = (\mathsf{KeyGen}, \mathsf{Sign}, \mathsf{Verify}, \mathsf{Link})$ *has linkability property, if for every* $q = \mathsf{poly}(\lambda)$ *and every PPT adversary* $\mathcal{A}$*, it holds that* $\mathcal{A}$ *has negligible advantage in the following experiment.*

$\mathsf{Exp}_{\mathsf{LRS\text{-}Link}}(\mathcal{A})$:

1. $\mathcal{A}$ *outputs a set of tuples* $(\mathrm{VK}_i, \Sigma_i, m_i, \mathsf{R}_i)$ *for* $i = 1, \dots, q$ *and another tuple* $(\sigma^*, m^*, \mathsf{R}^*)$. *Denote as* $\mathcal{VK}$ *the set of cardinality* q *such that* $\mathrm{VK}_i \in \mathcal{VK}$ *for* $i = 1, \dots, q$.
2. *The experiment outputs* 1 *if the following conditions hold:*
   - *For all* $i \in [q]$ *we have* $\mathsf{R}_i \subseteq \mathcal{VK}$ *and* $\mathsf{R}^* \subseteq \mathcal{VK}$
   - *For all* $i \in [q]$ *we have* $\mathsf{LRS.Verify}(\mathsf{R}_i, \mathsf{m}_i, \Sigma_i) = 1$ *and* $\mathsf{LRS.Verify}(\mathsf{R}^*, \mathsf{m}^*, \Sigma^*) = 1$
   - *For all* $i, j \in [q]$ *such that* $i \neq j$*, we have* $\mathsf{LRS.Link}(\Sigma_i, \Sigma_j) = 0$ *and* $\mathsf{LRS.Link}(\Sigma_i, \Sigma^*) = 0$

Otherwise, the experiment returns 0.

The advantage of $\mathcal{A}$ *is defined by* $\mathsf{Adv}_{\mathsf{LRS\text{-}Link}}(\mathcal{A}) = \Pr[\mathsf{Exp}_{\mathsf{LRS\text{-}Link}}(\mathcal{A}) = 1]$.

We now turn to anonymity. Since, in linkable ring signatures, there is a public link function, it is easy to tell whether multiple signatures were produced by the same signer or not. However, it should still be infeasible to tell which exact user from a ring produced the signature. We argue that, in contrast to the state-of-the-art definitions, in our definition anonymity is not lost at the moment an adversary obtains the first signature of a user. In reality, even when an adversary obtains multiple signature from the same member, identity of the signer should still be unknown, i.e. it should be infeasible to associate the signatures with a verification key. We model this by letting the adversary choose two users, which need to be always in the same rings, and imposing a permutation on the secret keys. If an adversary would be able to associate a signature of one of this users with its verification key, then the adversary would also be able to guess the permutation.

Definition 9 (Linkable Anonymity). *We say that a linkable ring signature* $\mathsf{LRS} = (\mathsf{KeyGen}, \mathsf{Sign}, \mathsf{Verify}, \mathsf{Link})$ *is linkably anonymous, if for every* $q = \mathsf{poly}(\lambda)$ *and every PPT adversary* $\mathcal{A}$*, it holds that* $\mathcal{A}$ *has negligible advantage in the following experiment.*

$\mathsf{Exp}_{\mathsf{LRS\text{-}Anon}}(\mathcal{A})$:

1. *For all* $i = 1, \dots, q$ *the experiment generates* $(\mathrm{VK}_i, \mathrm{SK}_i) \leftarrow \mathsf{LRS.KeyGen}(1^\lambda, \mathsf{r}_i)$ *using random coins* r_i *and samples* $b \in \{0, 1\}$ *uniformly at random.*
2. *The experiment provides* $\mathcal{VK} = \{\mathrm{VK}_1, \dots, \mathrm{VK}_q\}$ *to* $\mathcal{A}$.
3. $\mathcal{A}$ *outputs a set of verification keys* $\mathcal{U} \subset \mathcal{VK}$ *and two challenge verification keys* $\mathrm{VK}_0^*, \mathrm{VK}_1^* \in \mathcal{VK} \setminus \mathcal{U}$. *We denote the secret keys corresponding to* $\mathrm{VK}_0^*, \mathrm{VK}_1^*$ *as* $\mathrm{SK}_0^*, \mathrm{SK}_1^*$ *respectively. The experiment returns* r_i *for all* $\mathrm{VK}_i \in \mathcal{U}$.

4. *The adversary queries for signatures on input a ring* R *and a verification key* $\mathrm{VK} \in \mathcal{VK} \setminus \mathcal{U}$ *such that* $\mathrm{VK} \in \mathsf{R}$.
   - *If* VK_0^* *or* $\mathrm{VK}_1^* \in \mathsf{R}$ *but* $\{\mathrm{VK}_0^*, \mathrm{VK}_1^*\} \not\subseteq \mathsf{R}$, *then the experiment returns an uniformly random bit and aborts.*
   - *If* $\mathrm{VK} \notin \{\mathrm{VK}_0^*, \mathrm{VK}_1^*\}$, *then the experiment outputs* $\Sigma^* \leftarrow \mathsf{LRS.Sign}(\mathsf{SK}, \mathsf{m}, \mathsf{R})$ *where* SK *corresponds to the queried* VK.
   - *If* $\mathrm{VK} = \mathrm{VK}_0^*$ *the experiment outputs* $\Sigma^* \leftarrow \mathsf{LRS.Sign}(\mathrm{SK}_b^*, \mathsf{m}, \mathsf{R})$.
   - *If* $\mathrm{VK} = \mathrm{VK}_1^*$ *the experiment outputs* $\Sigma^* \leftarrow \mathsf{LRS.Sign}(\mathrm{SK}_{1-b}^*, \mathsf{m}, \mathsf{R})$.
5. $\mathcal{A}$ *submits* $\hat{b} \in \{0, 1\}$ *and the experiment outputs* 1 *if* $\hat{b} = b$, *otherwise it outputs* 0.

The advantage of $\mathcal{A}$ *is defined by* $\mathsf{Adv}_{\mathsf{LRS\text{-}Anon}}(\mathcal{A}) = |\Pr[\mathsf{Exp}_{\mathsf{LRS\text{-}Anon}}(\mathcal{A}, q, \lambda) = 1] - 1/2|$.

Finally, we require that a linkable ring signature is non-frameable. This property guarantees that it is infeasible for an adversary to forge a signature which would link with an honest users' signature, even when the adversary saw a number of his signatures in the past.

Definition 10 (Non-frameability). *We say that a linkable ring signature* $\mathsf{LRS} = (\mathsf{KeyGen}, \mathsf{Sign}, \mathsf{Verify}, \mathsf{Link})$ *is non-frameable, if for every* $q = \mathsf{poly}(\lambda)$ *and every PPT adversary* $\mathcal{A}$, *it holds that* $\mathcal{A}$ *has negligible advantage in the following experiment.*

$\mathsf{Exp}_{\mathsf{LRS\text{-}Frame}}(\mathcal{A})$:

1. *For all* $i = 1, \ldots, q$ *the experiment generates* $(\mathrm{VK}_i, \mathrm{SK}_i) \leftarrow \mathsf{LRS.KeyGen}(1^\lambda, \mathsf{r}_i)$ *using uniformly random coins* r_i. *The experiment sets* $\mathcal{VK} = \{\mathrm{VK}_1, \ldots, \mathrm{VK}_q\}$ *and initializes a set* $\mathcal{C} = \emptyset$.
2. *The experiment provides* $\mathrm{VK}_1, \ldots, \mathrm{VK}_q$ *to* $\mathcal{A}$.
3. $\mathcal{A}$ *is now allowed to make the following queries:*
   $(\mathtt{sign}, \mathrm{VK}_i, \mathsf{m}, \mathsf{R})$: *Upon a signing query, the experiment checks if* $\mathrm{VK}_i \in \mathsf{R}$, *and if so computes* $\Sigma \leftarrow \mathsf{LRS.Sign}(\mathrm{SK}_i, \mathsf{m}, \mathsf{R})$ *and returns* Σ *to* $\mathcal{A}$. *Note that we don't require* $\mathsf{R} \subseteq \mathcal{VK}$, *so the ring* R *may contain verification keys generated by* $\mathcal{A}$.
   $(\mathtt{corrupt}, \mathrm{VK}_i)$: *Upon a corruption query, the experiment adds* VK_i *to* $\mathcal{C}$ *and returns* r_i *to* $\mathcal{A}$.
4. *In the end of Phase-1,* $\mathcal{A}$ *outputs* $(\mathsf{R}^*, \mathsf{m}^*, \Sigma^*)$.
5. *The experiment now provides all random coins* r_i *for all* $i = 1, \ldots, q$ *used to generate the keys to the adversary* $\mathcal{A}$.
6. *The adversary* $\mathcal{A}$ *outputs* $(\mathsf{R}^\dagger, \mathsf{m}^\dagger, \Sigma^\dagger)$ *and the experiment returns* 1 *if the following conditions hold:*
   - $\mathsf{LRS.Verify}(\mathsf{R}^*, \mathsf{m}^*, \Sigma^*) = 1$ *and* $\mathsf{LRS.Verify}(\mathsf{R}^\dagger, \mathsf{m}^\dagger, \Sigma^\dagger) = 1$,
   - $\mathsf{R}^* \subseteq \mathcal{VK}$ *and for all* $\mathrm{VK}_i \in \mathsf{R}^*$ *we have* $\mathrm{VK}_i \notin \mathcal{C}$, *i.e. all verification keys in* R^* *are from honest users,*
   - $\mathcal{A}$ *didn't obtain* Σ^* *from the signing oracle,*
   - $\mathsf{Link}(\Sigma^*, \Sigma^\dagger) = 1$.

Otherwise the experiment returns 0.

The advantage of $\mathcal{A}$ is defined by $\mathsf{Adv}_{\mathsf{LRS\text{-}Frame}}(\mathcal{A}) = \Pr[\mathsf{Exp}_{\mathsf{LRS\text{-}Frame}}(\mathcal{A}) = 1]$.

Remark 1 (Unforgeability). Beside the properties defined above, we also require the standard unforgeability property from ring signatures to hold for linkable ring signatures.

6 Construction of Linkable Ring Signatures

We will now provide a construction of linkable ring signatures from the following primitives. Let

- $\mathsf{Com} = (\mathsf{Commit}, \mathsf{Verify})$ be a non-interactive commitment scheme.
- $\mathsf{Sig} = (\mathsf{KeyGen}, \mathsf{Sign}, \mathsf{Verify})$ be a signature scheme.
- $\mathsf{SPB} = (\mathsf{Gen}, \mathsf{Hash}, \mathsf{Open}, \mathsf{Verify})$ be a somewhere perfectly binding hash function with private local opening.

Before we define the NIWI-proof system NIWI, we will define an algorithm $\mathsf{JointVerify}$. The algorithm takes as input two commitment triples $\mathrm{VK} = (\mathsf{com}_j)_{j\in[3]}$ and $\mathrm{VK}' = (\mathsf{com}'_j)_{j\in[3]}$, two inputs vk and vk' as well as two unveil triples $\boldsymbol{\gamma} = (\gamma_j)_{j\in[3]}$, $\boldsymbol{\gamma}' = (\gamma'_j)_{j\in[3]}$. The algorithm checks that one of the commitments $\mathsf{com}_1, \mathsf{com}_2, \mathsf{com}_3, \mathsf{com}'_1, \mathsf{com}'_2, \mathsf{com}'_3$ at least two open to vk and at least two open to vk' and at least 5 open to either vk or vk'. The last condition can be rephrased as at most one of the 6 commitments does not verify and all the others open to either vk or vk'. As the name suggests, the algorithm verifies if the triples VK and VK′ jointly commit to the values vk and vk', but we allow some leeway which of the 6 commitments actually commit to which value.

$\mathsf{JointVerify}(\mathrm{VK}, \mathrm{VK}', \mathsf{vk}, \mathsf{vk}', \boldsymbol{\gamma}, \boldsymbol{\gamma}')$:
- Parse $\mathrm{VK} = (\mathsf{com}_j)_{j\in[3]}$ and $\mathrm{VK}' = (\mathsf{com}'_j)_{j\in[3]}$
- Parse $\boldsymbol{\gamma} = (\gamma_j)_{j\in[3]}$, $\boldsymbol{\gamma}' = (\gamma'_j)_{j\in[3]}$
- Compute $s \leftarrow \sum_{j=1}^{3}(\mathsf{Com.Verify}(\mathsf{com}_j, \mathsf{vk}, \gamma_j) + \mathsf{Com.Verify}(\mathsf{com}'_j, \mathsf{vk}, \gamma'_j))$
- Compute $s' \leftarrow \sum_{j=1}^{3}(\mathsf{Com.Verify}(\mathsf{com}_j, \mathsf{vk}', \gamma_j) + \mathsf{Com.Verify}(\mathsf{com}'_j, \mathsf{vk}', \gamma'_j))$
- If it holds that $s \geq 2$ and $s' \geq 2$ and $s + s' \geq 5$ output 1, otherwise 0.

We remark that the expression $\mathsf{JointVerify}(\mathrm{VK}, \mathrm{VK}', \mathsf{vk}, \mathsf{vk}', \boldsymbol{\gamma}, \boldsymbol{\gamma}') = 1$ can be unrolled into a short (constant size) sequence of conjunctions and disjunctions over expressions of the form $\mathsf{Com.Verify}(\mathsf{com}_j, \mathsf{vk}, \gamma_j) = 1$, $\mathsf{Com.Verify}(\mathsf{com}'_j, \mathsf{vk}, \gamma'_j) = 1$, $\mathsf{Com.Verify}(\mathsf{com}_j, \mathsf{vk}', \gamma_j) = 1$ and $\mathsf{Com.Verify}(\mathsf{com}'_j, \mathsf{vk}', \gamma'_j) = 1$ for $j = 1, 2, 3$[3].

- $\mathsf{NIWI} = (\mathsf{Prove}, \mathsf{Verify})$ be a NIWI-proof system for the language $\mathcal{L}$ and with witness-relation $\mathcal{R}$ defined as follows. For $x = (\mathsf{vk}, (\mathsf{hk}^{(i)}, \mathsf{h}^{(i)})_{i\in[3]})$ and $w =$

[3] The expression can be unrolled into a disjunction of $6 \cdot \left(\binom{5}{2} + \binom{5}{3}\right) = 480$ clauses, where each clause is a conjunction of 5 $\mathsf{Com.Verify}$ statements.

$((\mathsf{ind}^{(i)}, \mathrm{VK}^{(i)}, \tau^{(i)}, \boldsymbol{\gamma}^{(i)})_{i \in [3]}, \mathsf{vk}')$, where $\mathrm{VK}^{(i)} = (\mathsf{com}_1^{(i)}, \mathsf{com}_2^{(i)}, \mathsf{com}_3^{(i)})$ and $\boldsymbol{\gamma}^{(i)} = (\gamma_1^{(i)}, \gamma_2^{(i)}, \gamma_3^{(i)})$ for $i = 1, \ldots, 3$, let

$$\begin{aligned}
\mathcal{R}(x, w) \Leftrightarrow\ & \mathsf{SPB.Verify}(\mathsf{hk}^{(1)}, \mathsf{h}^{(1)}, \mathsf{ind}^{(1)}, \mathrm{VK}^{(1)}, \tau^{(1)}) = 1 \\
& \text{and } \forall j \in [3] : \mathsf{Com.Verify}(\mathsf{com}_j^{(1)}, \mathsf{vk}, \gamma_j^{(1)}) = 1 \\
& \text{or} \\
& \mathsf{ind}^{(2)} \neq \mathsf{ind}^{(3)} \\
& \text{and } \forall i \in \{2, 3\} : \mathsf{SPB.Verify}(\mathsf{hk}^{(i)}, \mathsf{h}^{(i)}, \mathsf{ind}^{(i)}, \mathrm{VK}^{(i)}, \tau^{(i)}) = 1 \\
& \text{and } \mathsf{JointVerify}(\mathrm{VK}^{(2)}, \mathrm{VK}^{(3)}, \mathsf{vk}, \mathsf{vk}', \boldsymbol{\gamma}^{(2)}, \boldsymbol{\gamma}^{(3)}) = 1.
\end{aligned}$$

Let $\mathcal{L}$ be the language accepted by $\mathcal{R}$.

Our linkable ring signature scheme LRS = (KeyGen, Sign, Verify) is given as follows.

$\mathsf{LRS.KeyGen}(1^\lambda)$:
- Compute $(\mathsf{vk}, \mathsf{sk}) \leftarrow \mathsf{Sig.KeyGen}(1^\lambda)$
- For $i = 1, 2, 3$ compute $(\mathsf{com}_j, \gamma_j) \leftarrow \mathsf{Com.Commit}(1^\lambda, \mathsf{vk})$
- Set $\boldsymbol{\gamma} \leftarrow (\gamma_j)_{j \in [3]}$
- Output $\mathrm{VK} \leftarrow (\mathsf{com}_j)_{j \in [3]}$ and $\mathrm{SK} \leftarrow (\mathsf{sk}, \mathrm{VK}, \mathsf{vk}, \boldsymbol{\gamma})$

$\mathsf{LRS.Sign}(\mathrm{SK}, \mathsf{m}, \mathsf{R} = (\mathrm{VK}_1, \ldots, \mathrm{VK}_\ell))$:
- Parse $\mathrm{SK} = (\mathsf{sk}, \mathrm{VK}, \mathsf{vk}, \boldsymbol{\gamma})$
- Parse $\mathrm{VK} = (\mathsf{com}_j)_{j \in [3]}$
- Find an index $\mathsf{ind} \in [\ell]$ such that $\mathrm{VK}_{\mathsf{ind}} = \mathrm{VK}$
- For $i = 1, 2, 3$ compute $(\mathsf{hk}^{(i)}, \mathsf{shk}^{(i)}) \leftarrow \mathsf{SPB.Gen}(1^\lambda, |\mathsf{R}|, \mathsf{ind})$ and $\mathsf{h}^{(i)} \leftarrow \mathsf{SPB.Hash}(\mathsf{hk}^{(i)}, \mathsf{R})$
- Compute $\tau^{(1)} \leftarrow \mathsf{SPB.Open}(\mathsf{hk}^{(1)}, \mathsf{shk}^{(1)}, \mathsf{R}, \mathsf{ind})$
- Set $x \leftarrow (\mathsf{vk}, (\mathsf{hk}^{(i)}, \mathsf{h}^{(i)})_{i \in [3]})$
- Set $w \leftarrow ((\mathsf{ind}, \mathrm{VK}, \tau^{(1)}, \boldsymbol{\gamma}), \emptyset, \emptyset, \emptyset)$
- Compute $\pi \leftarrow \mathsf{NIWI.Prove}(x, w)$
- Compute $\sigma \leftarrow \mathsf{Sig.Sign}(\mathsf{sk}, (\mathsf{m}, (\mathsf{hk}^{(i)}, \mathsf{h}^{(i)})_{i \in [3]}, \pi))$
- Output $\Sigma \leftarrow (\mathsf{vk}, (\mathsf{hk}^{(i)})_{i \in [3]}, \pi, \sigma)$

$\mathsf{LRS.Verify}(\mathsf{R}, \mathsf{m}, \Sigma)$:
- Parse $\Sigma = (\mathsf{vk}, (\mathsf{hk}^{(i)})_{i \in [3]}, \pi, \sigma)$
- For $i \in [3]$ compute $\tilde{\mathsf{h}}^{(i)} \leftarrow \mathsf{SPB.Hash}(\mathsf{hk}^{(i)}, \mathsf{R})$
- Set $x \leftarrow (\mathsf{vk}, (\mathsf{hk}^{(i)}, \tilde{\mathsf{h}}^{(i)})_{i \in [3]})$
- Check if $\mathsf{NIWI.Verify}(x, \pi) = 1$, if not output 0
- Check if $\mathsf{Sig.Verify}(\mathsf{vk}, (\mathsf{m}, (\mathsf{hk}^{(i)}, \tilde{\mathsf{h}}^{(i)})_{i \in [3]}, \pi), \sigma) = 1$, if not output 0
- Output 1

$\mathsf{LRS.Link}(\Sigma_1, \Sigma_2)$:
- Parse $\Sigma_1 \leftarrow (\mathsf{vk}_1, (\mathsf{hk}_1^{(i)})_{i \in [3]}, \pi_1, \sigma_1)$
- Parse $\Sigma_2 \leftarrow (\mathsf{vk}_2, (\mathsf{hk}_2^{(i)})_{i \in [3]}, \pi_2, \sigma_2)$
- If $\mathsf{vk}_1 = \mathsf{vk}_2$ output 1, otherwise 0

6.1 Correctness

Again, we will first show correctness of our scheme. Assume that $\mathrm{VK} = (\mathsf{com}_1, \mathsf{com}_2, \mathsf{com}_3)$ and $\mathrm{SK} = (\mathsf{sk}, \mathrm{VK}, \mathsf{vk}, \boldsymbol{\gamma})$ with $\boldsymbol{\gamma} = (\gamma_1, \gamma_2, \gamma_3)$ were generated by $\mathsf{LRS.KeyGen}$ and $\Sigma = (\mathsf{vk}, \mathsf{hk}^{(1)}, \mathsf{hk}^{(2)}, \mathsf{hk}^{(3)}, \pi, \sigma)$ is the output of $\mathsf{LRS.Sign}(\mathrm{SK}, \mathsf{m}, \mathsf{R})$, where $\mathsf{R} = (\mathrm{VK}_1, \ldots, \mathrm{VK}_\ell)$. We will show that it holds that $\mathsf{LRS.Verify}(\mathsf{R}, \mathsf{m}, \Sigma) = 1$. As $\mathsf{SPB.Hash}$ is deterministic, it holds for the hashes $\tilde{\mathsf{h}}^{(1)}, \tilde{\mathsf{h}}^{(2)}, \tilde{\mathsf{h}}^{(3)}$ computed by $\mathsf{LRS.Verify}(\mathsf{R}, \mathsf{m}, \Sigma)$ that $\tilde{\mathsf{h}}^{(i)} = \mathsf{h}^{(i)}$ (for $i = 1, 2, 3$), where the $\mathsf{h}^{(i)}$ are the hashes computed by $\mathsf{LRS.Sign}(\mathrm{SK}, \mathsf{m}, \mathsf{R})$. Also, it obviously holds that $\mathrm{VK} = \mathrm{VK}^{(1)}$. Now, notice further that by the correctness of SPB it holds that $\mathsf{SPB.Verify}(\mathsf{hk}^{(1)}, \mathsf{h}^{(1)}, \mathsf{ind}, \mathrm{VK}, \tau^{(1)}) = 1$. By the correctness of the commitment scheme Com, it holds that $\mathsf{Com.Verify}(\mathsf{com}_j, \mathsf{vk}, \gamma_j) = 1$ for $j = 1, 2, 3$. Thus, $w = ((\mathsf{ind}, \mathrm{VK}, \tau^{(1)}, \boldsymbol{\gamma})), \emptyset, \emptyset, \emptyset)$ is a valid witness for the statement $x = (\mathsf{vk}, (\mathsf{hk}^{(i)}, \mathsf{h}^{(i)})_{i \in [3]})$. Consequently, by the correctness of NIWI it holds that $\mathsf{NIWI.Verify}(x, \pi) = 1$. Finally, by the correctness of Sig we get that $\mathsf{Sig.Verify}(\mathsf{vk}, (\mathsf{m}, (\mathsf{hk}^{(i)}, \mathsf{h}^{(i)})_{i \in [3]}, \pi), \sigma) = 1$ and $\mathsf{LRS.Verify}(\mathsf{R}, \mathsf{m}, \Sigma)$ outputs 1.

6.2 Signature Size

For a signature $\Sigma = (\mathsf{vk}, (\mathsf{hk}^{(i)})_{i \in [3]}, \pi, \sigma)$, the size of the signature component σ is $\mathsf{poly}(\lambda)$ and independent of the ring-size ℓ. By the efficiency property of SPB the sizes of the hashing keys $\mathsf{hk}^{(1)}, \mathsf{hk}^{(2)}, \mathsf{hk}^{(3)}$ is bounded by $\log(\ell) \cdot \mathsf{poly}(\lambda)$. Furthermore, for a statement $x = (\mathsf{vk}, (\mathsf{hk}^{(i)}, \tilde{\mathsf{h}}^{(i)})_{i \in [3]})$, the size of the verification circuit C_x is dominated $\mathsf{SPB.Verify}$, which by the efficiency property of SPB can be computed by a circuit of size $\log(\ell) \cdot \mathsf{poly}(\lambda)$. All other algorithms can be computed by circuits of size $\mathsf{poly}(\lambda)$ and independent of ℓ. Consequently, it holds that $|C_x| = \log(\ell) \cdot \mathsf{poly}(\lambda)$. By the efficiency property of the NIWI proof, it holds that $|\pi| = |C_x| \cdot \mathsf{poly}(\lambda) = \log(\ell) \cdot \mathsf{poly}(\lambda)$. All together, we can conclude that $|\Sigma| = \log(\ell) \cdot \mathsf{poly}(\lambda)$.

6.3 Security

Theorem 3. *The ring signature scheme* LRS *is unforgeable, given that* NIWI *has perfect soundness,* SPB *is somewhere perfectly binding,* Com *is perfectly binding and* Sig *is unforgeable.*

Theorem 4. *The ring signature scheme* LRS *is linkably anonymous, given that* SPB *is index hiding,* Com *is computationally hiding and* NIWI *is computationally witness-indistinguishable.*

Theorem 5. *The ring signature scheme* LRS *is perfectly linkable, given* SPB *is somewhere perfectly binding,* Com *is perfectly binding and* NIWI *has perfect soundness.*

Theorem 6. *Given that* Sig *is unforgeable,* Com *is perfectly binding and* NIWI *is perfectly sound, the scheme* LRS *has non-framability.*

Full proofs of these theorems can be found in the full version of this work [2].

7 Conclusions

Ring signatures are a well-studied cryptographic primitive with many applications that include whistleblowing and cryptocurrencies. In this paper we improved the state-of-the-art by introducing a scheme with signature size that is logarithmic in the number of ring members, while at the same time relying on standard assumptions and not requiring a trusted setup. We use novel techniques that combine somewhere statistically binding hashing and NIWI proofs forming a membership proof. An interesting open question is whether one can build structure-preserving SSB hashing that can be combined with pairing based NIWI proofs. Such combination would substantially increase the efficiency of the proposed membership proof and decrease its size.

Acknowledgments. This work has been partially funded/supported by the German Ministry for Education and Research through funding for the project CISPA-Stanford Center for Cybersecurity (Funding numbers: 16KIS0762 and 16KIS0927).

References

1. Abe, M., Ohkubo, M., Suzuki, K.: 1-out-of-n signatures from a variety of keys. In: Zheng, Y. (ed.) ASIACRYPT 2002. LNCS, vol. 2501, pp. 415–432. Springer, Heidelberg (2002). https://doi.org/10.1007/3-540-36178-2_26
2. Backes, M., Döttling, N., Hanzlik, L., Kluczniak, K., Schneider, J.: Ring signatures: logarithmic-size, no setup – from standard assumptions. Cryptology ePrint Archive, Report 2019/196 (2019). http://eprint.iacr.org/2019/196
3. Backes, M., Hanzlik, L., Kluczniak, K., Schneider, J.: Signatures with flexible public key: introducing equivalence classes for public keys. In: Peyrin, T., Galbraith, S. (eds.) ASIACRYPT 2018. LNCS, vol. 11273, pp. 405–434. Springer, Cham (2018). https://doi.org/10.1007/978-3-030-03329-3_14
4. Barak, B., Ong, S.J., Vadhan, S.P.: Derandomization in cryptography. In: Boneh, D. (ed.) CRYPTO 2003. LNCS, vol. 2729, pp. 299–315. Springer, Heidelberg (2003). https://doi.org/10.1007/978-3-540-45146-4_18
5. Baum, C., Lin, H., Oechsner, S.: Towards practical lattice-based one-time linkable ring signatures. Cryptology ePrint Archive, Report 2018/107 (2018). https://eprint.iacr.org/2018/107
6. Bender, A., Katz, J., Morselli, R.: Ring signatures: stronger definitions, and constructions without random oracles. In: Halevi, S., Rabin, T. (eds.) TCC 2006. LNCS, vol. 3876, pp. 60–79. Springer, Heidelberg (2006). https://doi.org/10.1007/11681878_4
7. Bitansky, N.: Verifiable random functions from non-interactive witness-indistinguishable proofs. In: Kalai, Y., Reyzin, L. (eds.) TCC 2017. LNCS, vol. 10678, pp. 567–594. Springer, Cham (2017). https://doi.org/10.1007/978-3-319-70503-3_19
8. Bitansky, N., Paneth, O.: ZAPs and non-interactive witness indistinguishability from indistinguishability obfuscation. In: Dodis, Y., Nielsen, J.B. (eds.) TCC 2015. LNCS, vol. 9015, pp. 401–427. Springer, Heidelberg (2015). https://doi.org/10.1007/978-3-662-46497-7_16

9. Boneh, D., Gentry, C., Lynn, B., Shacham, H.: Aggregate and verifiably encrypted signatures from bilinear maps. In: Biham, E. (ed.) EUROCRYPT 2003. LNCS, vol. 2656, pp. 416–432. Springer, Heidelberg (2003). https://doi.org/10.1007/3-540-39200-9_26
10. Boyen, X.: Mesh signatures. In: Naor, M. (ed.) EUROCRYPT 2007. LNCS, vol. 4515, pp. 210–227. Springer, Heidelberg (2007). https://doi.org/10.1007/978-3-540-72540-4_12
11. Boyen, X., Haines, T.: Forward-secure linkable ring signatures. In: Susilo, W., Yang, G. (eds.) ACISP 2018. LNCS, vol. 10946, pp. 245–264. Springer, Cham (2018). https://doi.org/10.1007/978-3-319-93638-3_15
12. Brakerski, Z., Vaikuntanathan, V.: Efficient fully homomorphic encryption from (standard) LWE. In: Ostrovsky, R. (ed.) 52nd Annual Symposium on Foundations of Computer Science, Palm Springs, CA, USA, 22–25 October 2011, pp. 97–106. IEEE Computer Society Press (2011)
13. Chandran, N., Groth, J., Sahai, A.: Ring signatures of sub-linear size without random oracles. In: Arge, L., Cachin, C., Jurdziński, T., Tarlecki, A. (eds.) ICALP 2007. LNCS, vol. 4596, pp. 423–434. Springer, Heidelberg (2007). https://doi.org/10.1007/978-3-540-73420-8_38
14. Chow, S.S.M., Wei, V.K.-W., Liu, J.K., Yuen, T.H.: Ring signatures without random oracles. In: Lin, F.-C., Lee, D.-T., Lin, B.-S., Shieh, S., Jajodia, S. (eds.) 1st ACM Symposium on Information, Computer and Communications Security, ASIACCS 2006, 21–24 March 2006, Taipei, Taiwan, pp. 297–302. ACM Press (2006)
15. Dodis, Y., Kiayias, A., Nicolosi, A., Shoup, V.: Anonymous identification in *ad hoc* groups. In: Cachin, C., Camenisch, J.L. (eds.) EUROCRYPT 2004. LNCS, vol. 3027, pp. 609–626. Springer, Heidelberg (2004). https://doi.org/10.1007/978-3-540-24676-3_36
16. Dolev, D., Dwork, C., Naor, M.: Non-malleable cryptography (extended abstract). In: 23rd Annual ACM Symposium on Theory of Computing, 6–8 May 1991, New Orleans, LA, USA, pp. 542–552. ACM Press (1991)
17. Dwork, C., Naor, M.: Zaps and their applications. In: 41st Annual Symposium on Foundations of Computer Science, 12–14 November 2000, Redondo Beach, CA, USA, pp. 283–293. IEEE Computer Society Press (2000)
18. ElGamal, T.: A public key cryptosystem and a signature scheme based on discrete logarithms. In: Blakley, G.R., Chaum, D. (eds.) CRYPTO 1984. LNCS, vol. 196, pp. 10–18. Springer, Heidelberg (1985). https://doi.org/10.1007/3-540-39568-7_2
19. Gentry, C.: Fully homomorphic encryption using ideal lattices. In: Mitzenmacher, M. (ed.) 41st Annual ACM Symposium on Theory of Computing, 31 May–2 June 2009, Bethesda, MD, USA, pp. 169–178. ACM Press (2009)
20. Gentry, C., Sahai, A., Waters, B.: Homomorphic encryption from learning with errors: conceptually-simpler, asymptotically-faster, attribute-based. In: Canetti, R., Garay, J.A. (eds.) CRYPTO 2013. LNCS, vol. 8042, pp. 75–92. Springer, Heidelberg (2013). https://doi.org/10.1007/978-3-642-40041-4_5
21. Ghadafi, E.M.: Sub-linear blind ring signatures without random oracles. In: Stam, M. (ed.) IMACC 2013. LNCS, vol. 8308, pp. 304–323. Springer, Heidelberg (2013). https://doi.org/10.1007/978-3-642-45239-0_18
22. Goldreich, O., Levin, L.A.: A hard-core predicate for all one-way functions. In: 21st Annual ACM Symposium on Theory of Computing, 15–17 May 1989, Seattle, WA, USA, pp. 25–32. ACM Press (1989)
23. González, A.: A ring signature of size $O(\sqrt[3]{n})$ without random oracles. Cryptology ePrint Archive, Report 2017/905 (2017). http://eprint.iacr.org/2017/905

24. Goyal, R., Hohenberger, S., Koppula, V., Waters, B.: A generic approach to constructing and proving verifiable random functions. In: Kalai, Y., Reyzin, L. (eds.) TCC 2017. LNCS, vol. 10678, pp. 537–566. Springer, Cham (2017). https://doi.org/10.1007/978-3-319-70503-3_18
25. Groth, J., Ostrovsky, R., Sahai, A.: Non-interactive zaps and new techniques for NIZK. In: Dwork, C. (ed.) CRYPTO 2006. LNCS, vol. 4117, pp. 97–111. Springer, Heidelberg (2006). https://doi.org/10.1007/11818175_6
26. Herranz, J., Sáez, G.: Forking lemmas for ring signature schemes. In: Johansson, T., Maitra, S. (eds.) INDOCRYPT 2003. LNCS, vol. 2904, pp. 266–279. Springer, Heidelberg (2003). https://doi.org/10.1007/978-3-540-24582-7_20
27. Hubacek, P., Wichs, D.: On the communication complexity of secure function evaluation with long output. In: Roughgarden, T. (ed.) 6th Conference on Innovations in Theoretical Computer Science, ITCS 2015, 11–13 January 2015, Rehovot, Israel, pp. 163–172. Association for Computing Machinery (2015)
28. Libert, B., Peters, T., Qian, C.: Logarithmic-size ring signatures with tight security from the DDH assumption. In: Lopez, J., Zhou, J., Soriano, M. (eds.) ESORICS 2018. LNCS, vol. 11099, pp. 288–308. Springer, Cham (2018). https://doi.org/10.1007/978-3-319-98989-1_15
29. Liu, J.K., Wei, V.K., Wong, D.S.: Linkable spontaneous anonymous group signature for ad hoc groups. In: Wang, H., Pieprzyk, J., Varadharajan, V. (eds.) ACISP 2004. LNCS, vol. 3108, pp. 325–335. Springer, Heidelberg (2004). https://doi.org/10.1007/978-3-540-27800-9_28
30. Lu, X., Au, M.H., Zhang, Z.: Raptor: a practical lattice-based (linkable) ring signature. Cryptology ePrint Archive, Report 2018/857 (2018). https://eprint.iacr.org/2018/857
31. Malavolta, G., Schröder, D.: Efficient ring signatures in the standard model. In: Takagi, T., Peyrin, T. (eds.) ASIACRYPT 2017. LNCS, vol. 10625, pp. 128–157. Springer, Cham (2017). https://doi.org/10.1007/978-3-319-70697-9_5
32. Noether, S.: Ring signature confidential transactions for monero. Cryptology ePrint Archive, Report 2015/1098 (2015). http://eprint.iacr.org/2015/1098
33. Okamoto, T., Pietrzak, K., Waters, B., Wichs, D.: New realizations of somewhere statistically binding hashing and positional accumulators. In: Iwata, T., Cheon, J.H. (eds.) ASIACRYPT 2015. LNCS, vol. 9452, pp. 121–145. Springer, Heidelberg (2015). https://doi.org/10.1007/978-3-662-48797-6_6
34. Regev, O.: On lattices, learning with errors, random linear codes, and cryptography. In: Gabow, H.N., Fagin, R. (eds.) 37th Annual ACM Symposium on Theory of Computing, 22–24 May 2005, Baltimore, MA, USA, pp. 84–93. ACM Press (2005)
35. Rivest, R.L., Shamir, A., Tauman, Y.: How to leak a secret. In: Boyd, C. (ed.) ASIACRYPT 2001. LNCS, vol. 2248, pp. 552–565. Springer, Heidelberg (2001). https://doi.org/10.1007/3-540-45682-1_32
36. Schäge, S., Schwenk, J.: A CDH-based ring signature scheme with short signatures and public keys. In: Sion, R. (ed.) FC 2010. LNCS, vol. 6052, pp. 129–142. Springer, Heidelberg (2010). https://doi.org/10.1007/978-3-642-14577-3_12
37. Schnorr, C.P.: Efficient identification and signatures for smart cards. In: Brassard, G. (ed.) CRYPTO 1989. LNCS, vol. 435, pp. 239–252. Springer, New York (1990). https://doi.org/10.1007/0-387-34805-0_22
38. Shacham, H., Waters, B.: Efficient ring signatures without random oracles. In: Okamoto, T., Wang, X. (eds.) PKC 2007. LNCS, vol. 4450, pp. 166–180. Springer, Heidelberg (2007). https://doi.org/10.1007/978-3-540-71677-8_12

39. Torres, W.A.A., et al.: Post-quantum one-time linkable ring signature and application to ring confidential transactions in blockchain (lattice RingCT v1.0). In: Susilo, W., Yang, G. (eds.) ACISP 2018. LNCS, vol. 10946, pp. 558–576. Springer, Cham (2018). https://doi.org/10.1007/978-3-319-93638-3_32
40. Tsang, P.P., Wei, V.K.: Short linkable ring signatures for e-voting, e-cash and attestation. In: Deng, R.H., Bao, F., Pang, H.H., Zhou, J. (eds.) ISPEC 2005. LNCS, vol. 3439, pp. 48–60. Springer, Heidelberg (2005). https://doi.org/10.1007/978-3-540-31979-5_5

Group Signatures Without NIZK: From Lattices in the Standard Model

Shuichi Katsumata[1(✉)] and Shota Yamada[2]

[1] National Institute of Advanced Industrial Science and Technology (AIST), The University of Tokyo, Tokyo, Japan
shuichi_katsumata@it.k.u-tokyo.ac.jp

[2] National Institute of Advanced Industrial Science and Technology (AIST), Tokyo, Japan
yamada-shota@aist.go.jp

Abstract. In a group signature scheme, users can anonymously sign messages on behalf of the group they belong to, yet it is possible to trace the signer when needed. Since the first proposal of lattice-based group signatures in the random oracle model by Gordon, Katz, and Vaikuntanathan (ASIACRYPT 2010), the realization of them in the standard model from lattices has attracted much research interest, however, it has remained unsolved. In this paper, we make progress on this problem by giving the first such construction. Our schemes satisfy CCA-selfless anonymity and full traceability, which are the standard security requirements for group signatures proposed by Bellare, Micciancio, and Warinschi (EUROCRYPT 2003) with a slight relaxation in the anonymity requirement suggested by Camenisch and Groth (SCN 2004). We emphasize that even with this relaxed anonymity requirement, all previous group signature constructions rely on random oracles or NIZKs, where currently NIZKs are not known to be implied from lattice-based assumptions. We propose two constructions that provide tradeoffs regarding the security assumption and efficiency:

- Our first construction is proven secure assuming the standard LWE and the SIS assumption. The sizes of the public parameters and the signatures grow linearly in the number of users in the system.
- Our second construction is proven secure assuming the standard LWE and the subexponential hardness of the SIS problem. The sizes of the public parameters and the signatures are independent of the number of users in the system.

Technically, we obtain the above schemes by combining a secret key encryption scheme with additional properties and a special type of attribute-based signature (ABS) scheme, thus bypassing the utilization of NIZKs. More specifically, we introduce the notion of *indexed* ABS, which is a relaxation of standard ABS. The above two schemes are obtained by instantiating the indexed ABS with different constructions. One is a direct construction we propose and the other is based on previous work.

Y. Ishai and V. Rijmen (Eds.): EUROCRYPT 2019, LNCS 11478, pp. 312–344, 2019.
https://doi.org/10.1007/978-3-030-17659-4_11

1 Introduction

1.1 Background

Group signatures, originally proposed by Chaum and van Heyst [Cv91], allow members of a group to sign on behalf of the group while guaranteeing the properties of authenticity, anonymity, and traceability. The signatures do not reveal the particular identity of the group member who issued it, however, should the need arise, a special entity called the group manager can trace the signature back to the signer using some secret information, thus holding the group members accountable for their signatures. Due to the appealing properties group signatures offer, they have proven to be useful in many real-life applications including privacy-protecting mechanisms, anonymous online communication, e-commerce systems, and trusted hardware attestation such as Intel's SGX.

Since their introduction, numerous constructions of group signatures have been proposed with different flavors: in the random oracle model [BBS04, CL04, GKV10] or standard model [BMW03, BW06, Gro07], supporting static groups [BMW03] or dynamic groups [BSZ05, BCC+16], and constructions based on various number theoretical assumptions such as strong RSA [ACJT00, CL02], paring-based [BW06, Gro07], and lattice-based [GKV10, LLLS13]. Despite the vast amount of research concerning group signatures, in essence all constructions follow the *encrypt-then-prove* paradigm presented by Bellare, Micciancio, and Warinschi [BMW03]. To sign on a message, a group member encrypts its certificate provided by the group manager and then proves in (non-interactive) zero-knowledge of the fact that the ciphertext is an encryption of a valid certificate while also binding the message to the zero-knowledge proof.

Thus far, all group signature schemes have relied on non-interactive zero-knowledge (NIZK) proofs in the proving stage of the encrypt-then-prove paradigm. Since NIZKs for general languages are implied from (certified doubly enhanced) trapdoor permutations [FLS90, BY93] and from bilinear maps [GOS06, GS08], group signatures in the standard model are known to exist from factoring-based and pairing-based assumptions [BMW03, BW06, BW07, Gro07]. In contrast, constructions of lattice-based group signatures in the standard model have shown to be considerably difficult. Since the first lattice-based group signature in the random oracle model (ROM) proposed by Gordon et al. [GKV10], there has been a rich line of subsequent works [LLLS13, NZZ15, LNW15, LLM+16a, LLNW16, LNWX18, PLS18], however, all schemes are only provably secure in the ROM. This situation stems from the notorious fact that lattices are ill-fit with NIZKs. Although more than a decade has passed since the emergence of lattices, there is still only one construction of NIZK known in the standard model [PV08], where the language supported by [PV08] seems unsuitable to devise group signatures. Notably, the open problem of constructing lattice-based group signatures in the standard model, which has explicitly been stated in Laguillaumie et al. [LLLS13] for example, has not made any progress in the past decade or so. Taking prior works on group signatures into consideration, it seems we would require a breakthrough result for lattice-based NIZKs or to come up with a different approach than the encrypt-then-prove paradigm to obtain a lattice-based group signature in the standard model.

1.2 Our Contribution

In this paper, we make progress on this problem and give the first construction of group signatures from lattices in the standard model. Our main result can be stated informally as follows:

Theorem 1 (Informal). *Under the hardness of the LWE and SIS problems with polynomial approximation factors,*[1] *there exists a group signature scheme with full-traceability and CCA-selfless anonymity in the standard model.*

We explain the statement in more details in the following. Here, we basically adopt the syntax and the security notions of the group signatures defined by Bellare, Micciancio, and Warinschi [BMW03], which are presumably one of the most widely accepted definitions. Our construction satisfies the standard notion of full-traceability, which asserts that an adversary cannot forge a valid signature that can be opened to an uncorrupted user or that cannot be traced to anyone. As for anonymity, our construction satisfies CCA-selfless anonymity introduced by Camenisch and Groth [CG05]. The notion of CCA-selfless anonymity is a relaxation of CCA-full anonymity defined by Bellare et al. [BMW03]. Informally, full-anonymity requires that the adversary cannot distinguish signatures from two different members even if *all the signing keys of the members of the system are exposed* and it has access to an open oracle. On the other hand, CCA-selfless anonymity requires anonymity to hold only when *the signing keys of the two members in question are not exposed* and it has access to an open oracle. While the latter definition is weaker, as discussed by Camenisch and Groth [CG05], it is sufficient for some natural situations. For example, consider a situation where an adversary can adaptively corrupt users while the parties cannot erase the data. In this setting, the former security notion does not buy any more security than the latter. We emphasize that even with this relaxed security notion, no group signature from lattices is known in the standard model prior to our work. In particular, regardless of what the security notion we consider for anonymity, all prior lattice-based constructions required random oracles.

One potential drawback of the above construction may be that it has rather large public parameters and signatures, whose sizes grow linearly in the number of users in the system. A natural question would be whether we can make these sizes independent of the number of users. As a side contribution, we answer this question affirmatively under a stronger assumption:

Theorem 2 (Informal). *Under the hardness of the LWE problem with polynomial approximation factors and the subexponential hardness of the SIS problem*

[1] By LWE and SIS problems with polynomial approximation factors, we mean they are problems which are as hard as certain worst case lattice problems with polynomial approximation factor.

with polynomial approximation factors, there exists a group signature scheme with full-traceability and CCA-selfless anonymity whose sizes of the public parameters and signatures are independent of the number of users.

These results are obtained by a generic construction of group signatures from one-time signatures (OTS), secret key encryptions (SKE), and a new primitive which we call *indexed attribute-based signatures* (indexed ABS). We require the standard notion of strong unforgeability for the OTS and it can be instantiated by any existing schemes such as [Moh11]. For the SKE, we require some special properties. Specifically, we require the SKE to be anonymous in addition to standard notions of hiding the message. We also require the SKE to have a decryption circuit with logarithmic depth and the property which we call key-robustness. Intuitively speaking, the key-robustness requires that the ciphertext spaces corresponding to two random secret keys to be disjoint with all but negligible probability. Such an SKE with special properties can be instantiated from the standard LWE assumption. The indexed ABS is a relaxation of the standard notion of ABS, where the setup and key generation algorithms take additional inputs. We require it to satisfy the security notion that we call co-selective unforgeability and (perfect) privacy. We show two ways of instantiating the indexed ABS. As for the first instantiation, we provide a construction of an indexed ABS that is proven to have the required security properties under the standard hardness of the SIS assumption. This instantiation leads us to Theorem 1. As for the second instantiation, we view the constrained signature scheme by Tsabary [Tsa17] as an indexed ABS scheme. Using this we obtain Theorem 2. We note that unlike our first instantiation, since the constrained signature scheme in [Tsa17] does not offer sufficient security properties for our purpose, we need to utilize complexity leveraging that incurs a subexponential reduction loss to when constructing our group signature.

1.3 Overview of Our Technique

Preprocessing NIZKs. The starting point of our work is the recent breakthrough result of *preprocessing* NIZK for **NP** from lattices in the standard model by Kim and Wu [KW18]. In a preprocessing NIZK [DMP88], a trusted third party generates a proving key k_P and a verification key k_V independently of the statement to be proven and provides k_P to the prover and k_V to the verifier. The prover can construct proofs using k_P and the verifier can validate the proofs using k_V. Preprocessing NIZKs can be seen as a general form of NIZKs; if both k_P and k_V need not be secret, then it corresponds to NIZKs in the common reference string (CRS) model; if k_P can be public but k_V needs to be secret, then it corresponds to designated verifier NIZKs [PsV06, DFN06]. The lattice-based preprocessing NIZK of Kim and Wu [KW18] can be viewed as a *designated prover* NIZK (DP-NIZK), where the proving key k_P needs to be kept secret but the verification key k_V can be made public.[2] Here, the zero-knowledge property of DP-NIZKs crucially relies on the fact that the verifier does not know the proving key k_P.

At first glance, DP-NIZKs seem to be all that we require to construct group signatures. The trusted group manager provides the user a (secret) proving key k_P on time of joining the group and publicly publishes the verification key k_V. This meets the criteria of DP-NIZKs since k_P will be kept secret by the group members and the proofs (i.e., signatures) can be publicly verified. Therefore, one might be tempted to substitute NIZKs in the CRS model with lattice-based DP-NIZKs to obtain a lattice-based group signature in the standard model. Unfortunately, this naive approach is trivially insecure. Specifically, the anonymity will be broken the moment a single group member becomes corrupt. If the group manager provides the same proving key k_P to the group members, then in case any of the group members become corrupt, k_P will be in the hands of the adversary. As we mentioned above, the zero-knowledge property of DP-NIZKs will break if the proving key k_P is known. An easy fix may be to instead provide proving keys $(\mathsf{k}_{P_i})_{i \in I}$ respectively to each group members $i \in I$ and publicly publish the corresponding verification keys $(\mathsf{k}_{V_i})_{i \in I}$. In this case, even if some of the group members become corrupt, their proving keys will not affect the zero-knowledge property of the other non-corrupt members using an independent proving key. However, the problem with this approach is that each proof constructed by a proving key k_{P_i} is implicitly associated with a unique verification key k_{V_i}. Since each verification key k_{V_i} is associated to a group member $i \in I$, the adversary can simply check which verification key accepts the proof (i.e., signature) to break anonymity. Therefore, although DP-NIZKs seem to be somewhat useful for constructing group signatures, it itself is not sufficient to be a substitute for NIZKs in the CRS model.

Viewing Attribute-Based Signatures as DP-NIZKs. The problem with the approach using DP-NIZKs is the following: if we give the same proving key k_P to every group member, then the scheme will be insecure against collusion attacks and if we give different proving keys k_{P_i} individually to each group members, then the scheme will lose anonymity. Therefore, the primitive we require for constructing group signatures is something akin to DP-NIZKs that additionally provides us with both collusion resistance and anonymity.

At this point, we would like to draw the attention to attribute-based signatures (ABS) [MPR11]. In ABS, a signer assigned with an attribute $\mathbf{y}$ is provided a signing key $\mathsf{sk}_{\mathbf{y}}$ from the authority and the signer can anonymously sign a message associated with a policy C using $\mathsf{sk}_{\mathbf{y}}$ if and only if $C(\mathbf{y}) = 1$. In addition, using the master public key mpk, anybody can verify the signature regardless of who signed it. The first requirement of an ABS, which captures unforgeability, is that any collusion of signers with attributes $(\mathbf{y}_i)_{i \in I}$ cannot forge a signature on a message associated with a policy C if $C(\mathbf{y}_i) = 0$ for all $i \in I$.

[2] As mentioned in Sect. 4 of [KW18], their scheme is only publicly verifiable when considering a slightly weaker notion of zero-knowledge than the standard notion of zero-knowledge for preprocessing NIZKs. In our work, the weaker notion suffices.

The second requirement, which captures anonymity, is that given a valid signature on a message associated with a policy C, the attribute $\mathbf{y}$ that was used to sign the message must remain anonymous. Namely, signatures generated by $\mathsf{sk}_{\mathbf{y}_0}$ and $\mathsf{sk}_{\mathbf{y}_1}$ are indistinguishable if $C(\mathbf{y}_0) = C(\mathbf{y}_1) = 1$. Looking at the similarity between DP-NIZKs and ABS, it is tempting to view a witness $\mathbf{w}$ as an attribute $\mathbf{y}$ and to set the proving key k_P as the ABS signing key $\mathsf{sk}_{\mathbf{w}}$. To prove that $\mathbf{w}$ is a valid witness to the statement $\mathbf{x}$, i.e., $(\mathbf{x}, \mathbf{w}) \in \mathcal{R}$ for the **NP** relation $\mathcal{R}$, the prover first prepares a circuit $C_{\mathbf{x}}(\mathbf{w}) := \mathcal{R}(\mathbf{x}, \mathbf{w})$ that has the statement $\mathbf{x}$ hard-wired to it. Then the prover signs some message associated with the policy $C_{\mathbf{x}}$ using its proving key $\mathsf{k}_P = \mathsf{sk}_{\mathbf{w}}$ and outputs the signature as the proof π. The verifier can publicly verify the proof π by checking whether or not the signature is valid. At a high level, the soundness of the proof system would follow from the unforgeability of ABS and the zero-knowledge property would follow from the anonymity of ABS. Furthermore, our initial motivation of satisfying collusion resistance and anonymity is met by the properties of ABS; even if the proving keys $(\mathsf{k}_{P_i} = \mathsf{sk}_{\mathbf{w}_i})_{i \in I}$ are compromised, it cannot be used to prove a statement $\mathbf{x}$ such that $\mathcal{R}(\mathbf{x}, \mathbf{w}_i) = 0$ for all $i \in I$ and the proofs constructed by different proving keys are indistinguishable from one another since the single mpk can be used to check the validity of all proofs (unlike the above case where unique verification keys k_{V_i} were assigned to each proving keys k_{P_i}).

Constructing Groups Signatures from ABS. While the idea of viewing ABS as some variant of DP-NIZK seems to be a great step forward, the question of how to use it to construct a group signature remains. Let us come back to the basic but powerful encrypt-then-prove paradigm of Bellare et al. [BMW03]. Recall that with this approach, the group manager issues a certificate to each group member $i \in I$ and publishes a public key for a public-key encryption scheme. To sign, a group member i encrypts its certificate as ct_i under the public key of the group manager and creates a NIZK proof of the fact that ct_i encrypts the certificate. Observe that each group member i implicitly constructs a member-specific statement $\mathbf{x}_i = \mathsf{ct}_i$ when generating the NIZK proof and sets the air of certificate and the randomness used to create ct_i as the witness $\mathbf{w}_i$. Traceability follows since each statement $\mathbf{x}_i$ encrypts the identity of the signer and the group manager who holds the secret key can decrypt them. Anonymity of the group signature is also intact even though the statement $\mathbf{x}_i$ used by each group member is different, due to the semantic security of the underlying public-key encryption scheme. Now, let us look at the above approach through the lens of NIZK-like ABSs: The group manager issues a certificate *and an ABS signing key* $\mathsf{sk}_{\mathbf{w}_i}$ for some witness $\mathbf{w}_i$ to each group member $i \in I$, and to sign, a group member i encrypts its certificate as ct_i under the public key of the group manager and uses the ABS signing key $\mathsf{sk}_{\mathbf{w}_i}$ to *create an ABS signature for some policy* $C_{\mathbf{x}_i}$ which serves as a NIZK proof of the fact that ct_i encrypts the certificate. In order for this approach to work, the witness (i.e., attribute) embedded to the ABS signing key $\mathsf{sk}_{\mathbf{w}_i}$ must be an accepting input to the policy $C_{\mathbf{x}_i}$ which has the statement $\mathbf{x}_i = \mathsf{ct}_i$ hard-wired. Although it may be not obvious at first glance, as a matter of fact, this approach is impossible! Notably, the group manager cannot prepare in advance a witness $\mathbf{w}_i$ to a statement $\mathbf{x}_i$ that will be chosen by the group member at the time of signing. Recall that the witness $\mathbf{w}_i$ to $\mathbf{x}_i = \mathsf{ct}_i$ was the certificate *and* the randomness used to create ct_i. The group manager can embed in the ABS signing key a certificate but not the randomness since there is no way to not know what kind of randomness will be used to generate the ciphertext by the group member beforehand. Therefore, to use the ABS as a type of NIZK proof system, we must devise a mechanism for constructing statements $\mathbf{x}_i$ while keeping the witness $\mathbf{w}_i$ fixed once and for all at the time of preparation of the ABS signing key.

This brings us to our final idea. To overcome the above problem, we embed the group member identifier $i \in I$ and a key K_i of a *secret key encryption* scheme to the ABS signing key $\mathsf{sk}_{i||\mathsf{K}_i}$. We then construct the statements $\mathbf{x}_i$ so that i and K_i can be reused as the fixed witness.[3] The following is the high-level construction of our group signature.

-GS.KeyGen: The group manager provides user $i \in I$ with a key K_i of an SKE scheme and an ABS signing key $\mathsf{sk}_{i||\mathsf{K}_i}$ where the string $i||\mathsf{K}_i$ is interpreted as an attribute.

-GS.Sign: To sign on a message M, the group member $i \in I$ prepares a ciphertext $\mathsf{ct}_i \leftarrow \mathsf{SKE.Enc}(\mathsf{K}_i, i)$, views the statement $\mathbf{x}_i$ as ct_i, and prepares a circuit $C_{\mathbf{x}_i}$ with the statement $\mathbf{x}_i$ hard-wired such that $C_{\mathbf{x}_i}(i||\mathsf{K}_i) := (i \in I) \wedge (i = \mathsf{SKE.Dec}(\mathsf{K}_i, \mathsf{ct}_i))$. Then using $\mathsf{sk}_{i||\mathsf{K}_i}$, it runs the ABS signing algorithm on message M with $C_{\mathbf{x}_i}$ as the policy. The signature is $\Sigma = (\sigma_{\mathsf{ABS}}, \mathsf{ct}_i)$.

-GS.Vrfy: To verify a signature $\Sigma = (\sigma_{\mathsf{ABS}}, \mathsf{ct})$, it prepares the circuit $C_{\mathsf{ct}}(z||y) := (z \in I) \wedge (z = \mathsf{SKE.Dec}(y, \mathsf{ct}))$ and runs the ABS verification algorithm with message M, signature σ_{ABS} and policy C_{ct}.

-GS.Open: To trace a signer from a signature $\Sigma = (\sigma_{\mathsf{ABS}}, \mathsf{ct})$, the group manager uses the secret keys $(\mathsf{K}_i)_{i \in I}$ to extract the group member identifier from the ciphertext ct.

It can be checked that the scheme is correct. If the ciphertext ct_i encrypts $i \in I$, then $\mathsf{sk}_{i||\mathsf{K}_i}$ can be used to construct a signature for the policy $C_{\mathbf{x}_i}$ where $\mathbf{x}_i = \mathsf{ct}_i$. We briefly sketch the traceability and anonymity of our group signature. First, traceability holds from the key robustness of the SKE scheme and the unforgeability of the ABS scheme. The former property states that the ciphertext space of a different set of secret keys must be disjoint. In particular, this implies that the set of statements $\mathbf{x}_i = \mathsf{ct}_i$ (i.e., languages) constructed by each group member will be disjoint. Therefore, since this also implies that the set of policies $C_{\mathbf{x}_i}$ used by each group members will be disjoint, it allows us to reduce the problem of traceability to the unforgeability of the underlying ABS scheme. We note that although key robustness may be a non-standard property to consider

[3] Our core idea of fixing the witness can also be realized by instead embedding $i \in I$ and a (weak) PRF seed into the ABS signing key, and using a *public key* encryption scheme. We provide detailed discussions on our choice of using SKEs in the full version.

for SKE schemes, it is an easy property to satisfy. Second, anonymity holds from the anonymity and semantic security of the SKE scheme and the anonymity of the ABS scheme. Here, anonymity of an SKE scheme informally states that the ciphertext does not leak what secret key was used to construct it. Specifically, if there were two ciphertexts, it must be difficult to tell whether they are an encryption under the same key or two different keys. These two properties allow us to argue that the ciphertext ct_i leaks no information of the group member identity. Furthermore, the anonymity of the ABS scheme ensures that σ_{ABS} does not leak the group member identity as well. Hence the signature $\sigma = (\sigma_{\mathsf{ABS}}, \mathsf{ct}_i)$ remains anonymous.

Interestingly, our construction does not need to explicitly rely on "certificates" anymore as was done in prior constructions. This is because the signing key $\mathsf{sk}_{i||\mathsf{K}_i}$ is not only a proving key for the NIZK proof system, but also implicitly a certificate. In particular, since the ABS can be viewed as a variant of *designated prover* NIZKs, the fact that a signer was able to construct a valid signature implicitly implies that the signer was certified by the group manager. Therefore, there is no need for adding another layer of certificate to our construction as was done in previous group signature constructions. Finally, we point out in advance that our actual construction in Sect. 4 is more complicated than the above high-level structure due to the fact that we additionally capture CCA anonymity rather than only CPA anonymity. In CCA anonymity, the adversary is further provided with an open oracle that opens (i.e., traces) a signature to a signer. Since in the security proof, the reduction algorithm will no longer hold the opening key and must simulate the open oracle on its own, extra complications are incurred compared to the CPA anonymity setting where there is no such open oracle. This situation is analogous to the difference between CPA and CCA-encryption schemes.

To the knowledgeable readers, we remark that the above idea is similar to those of Kim and Wu [KW18] for constructing DP-NIZKs. In particular, the way we embed a key of an SKE scheme, rather than the witness, to the ABS signing key is analogous to the way [KW18] embeds the key of an SKE scheme to a signature of a homomorphic signature scheme [GVW15]. Notably, both schemes crucially rely on the fact that once some private information has been embedded into an ABS signing key (resp. a homomorphic signature), the signing key (resp. signature) can be reused to generate proofs for arbitrary statements.

Constructing ABS with the Desired Properties. We now change the discussion on how to instantiate the above generic construction. Since we can instantiate SKEs through a combination of relatively standard techniques, we focus on how to instantiate ABSs from lattices in this overview. A natural way of instantiating the ABS required in our GS construction would be to use the ABS scheme proposed by Tsabary [Tsa17] proven secure under the SIS assumption, which is the only known ABS construction from lattices.[4] In their paper, two ABS schemes are proposed. The first scheme is constructed from homomorphic

[4] Actually, the paper proposes constructions of constrained signature (CS), which is a slightly different primitive from ABS. However, this primitive readily implies ABS.

signatures and the second is a direct construction. We focus on the second construction here, because the anonymity notion achieved by the first scheme is not sufficient for our purpose.[5] In fact, even the latter scheme does not provide a sufficient security notion that is required for our purpose, namely, for the proof of full-traceability. While Tsabary's ABS scheme achieves selective unforgeability where the adversary is forced to declare its target policy with respect to which it will forge a signature at the beginning of the security game, we require the ABS to be unforgeable even if the adversary is allowed to *adaptively* choose its target policy. The necessity of the adaptiveness of the target policy can be seen by recalling that a forgery in the full-traceability game is of the form $\Sigma^\star = (\sigma^\star_{\mathsf{ABS}}, \mathsf{ct}^\star)$, where $\mathsf{ct}^\star$ is an adaptively chosen ciphertext that specifies the target policy $C_{\mathsf{ct}^\star}$. An easy way to resolve this discrepancy is to assume the subexponential hardness of the SIS problem and prove that Tsabary's scheme is adaptively unforgeable via complexity leveraging [BB04b]. This approach leads us to Theorem 2.

Though the above approach works, it incurs a subexponential security loss, which is not desirable. At first glance, one may think that the underlying ABS must be adaptively unforgeable to be used in our generic GS construction; an adversary can adaptively make arbitrary many key queries and signing queries, and generate a forgery depending on the answers which it gets from these queries. Unfortunately, the only known construction of a lattice-based ABS scheme in the standard model with such a strong security property is provided by complexity leveraging as described above. However, a more careful observation reveals that we do not actually require the full power of adaptive unforgeability. First, the ABS scheme does not have to support an unbounded number of signing keys since the number of members in the group signature is fixed at setup in the static setting. Furthermore, we can relax the syntax of the ABS so that the key generation algorithm takes a user index i as an additional input, since each signing key in the group signature is associated with a user index. Finally, we can relax the unforgeability requirement of the ABS so that the adversary is forced to make all the key queries at the beginning of the security game while the target policy associated with the forgery can be chosen adaptively. We call this security notion *co-selective unforgeability*, since this is somewhat dual to the selective unforgeability notion where the key queries can be adaptive but the target policy is required to be declared at the beginning of the game.

Indeed, co-selective unforgeability is enough for instantiating our generic GS construction, because, in the construction the attributes hardwired to the signing keys of the ABS are $\{i \| \mathsf{K}_i\}$ independent from the public parameter of the ABS and can be chosen at the outset of the security game. With this observation in mind, we define a relaxed version of ABS which we call *indexed* ABS and provide a construction which does not resort to complexity leveraging.

---

[5] More specifically, the first scheme only achieves a so-called weakly-hiding property, where the key attribute is not leaked from a signature, but two signatures that are signed by the same user can be linked. Translated into the setting of group signature, this allows an adversary to link two different signatures by the same user, which trivially breaks anonymity.

Constructing Indexed ABS. Our starting point is the observation made by Tsabary [Tsa17], who showed that a homomorphic signature scheme can be viewed as a very weak form of an ABS scheme. In light of this observation, we can view the fully homomorphic signature scheme by Gorbunov, Vaikuntanathan, and Wichs [GVW15] as a single-user ABS scheme. In the scheme, the master public key is of the form $\mathsf{mpk} = (\mathbf{A}, \vec{\mathbf{B}} = [\mathbf{B}_1 \| \cdots \| \mathbf{B}_k])$ where $\mathbf{A}$ and $\mathbf{B}_i$ are random matrices over $\mathbb{Z}_q^{n\times m}$ and a secret key sk_x for an attribute $x \in \{0,1\}^k$ is a matrix with small entries $\vec{\mathbf{R}} = [\mathbf{R}_1 \| \cdots \| \mathbf{R}_k]$ such that $\vec{\mathbf{B}} = \mathbf{A}\vec{\mathbf{R}} + x \otimes \mathbf{G}$, where $\mathbf{G}$ is the special gadget matrix whose trapdoor is publicly known. To sign on a policy $F : \{0,1\}^k \to \{0,1\}$ and a message M, the signer uses the homomorphic evaluation algorithms [BGG+14, GV15] to compute matrices $\mathbf{R}_F$ and $\mathbf{B}_F$ such that $\mathbf{B}_F = \mathbf{A}\mathbf{R}_F + F(x)\mathbf{G}$ from sk_x, where $\mathbf{R}_F$ is a matrix with small entries and $\mathbf{B}_F$ is a publicly computable matrix. When $F(x) = 1$, the signer can compute the trapdoor for the matrix $[\mathbf{A} \| \mathbf{B}_F]$ from $\mathbf{R}_F$ using the technique of [ABB10a, MP12] and sample a short vector $\mathbf{e}_F$ from a Gaussian distribution such that $[\mathbf{A} \| \mathbf{B}_F]\mathbf{e}_F = \mathbf{0}$ using the trapdoor. The signature on (F, M) is the vector $\mathbf{e}_F$. It can be seen that $\mathbf{e}_F$ does not leak information of x, since the distribution from which it is sampled only depends on the master public key and F. Furthermore, the scheme satisfies a relaxed version of the co-selective unforgeability, where the adversary can corrupt a single user but is *not* allowed to make signing queries. To see this, let us assume that there is an adversary who chooses x at the beginning of the game and generates a forgery $\mathbf{e}_{F^\star}$ for $F^\star$ such that $F^\star(x) = 0$ given $(\mathsf{mpk}, \mathsf{sk}_x)$. Then, we can solve the SIS problem using this adversary. The reduction algorithm is given a matrix $\mathbf{A}$ as the problem instance of SIS and x from the adversary. It then sets $\vec{\mathbf{B}} = \mathbf{A}\vec{\mathbf{R}} + x \otimes \mathbf{G}$ and gives $\mathsf{sk}_x := \vec{\mathbf{R}}$ to the adversary at the beginning of the game. For the forgery $\mathbf{e}_{F^\star}$ output by the adversary, we have $[\mathbf{A} \| \mathbf{B}_{F^\star}]\mathbf{e}_{F^\star} = \mathbf{0}$. Since $\mathbf{B}_{F^\star} = \mathbf{A}\mathbf{R}_{F^\star}$, we can extract a short vector $\mathbf{z} := [\mathbf{I} \| \mathbf{R}_{F^\star}]\mathbf{e}$ such that $\mathbf{A}\mathbf{z} = \mathbf{0}$, which is a solution to the SIS problem.

There are two problems with this scheme. First, the scheme can only support a single user, whereas we need a scheme to support multiple users. It can be seen that the security of the above scheme can be broken in case the adversary obtains the keys of two different users. Second, the unforgeability of the scheme is broken once the adversary is given an access to a signing oracle. Indeed, a valid signature for a policy-message pair (F, M) is also valid for (F, M') with different $\mathsf{M}' \neq \mathsf{M}$, since the above signing and verification algorithms simply ignore the messages M. In other words, the message is not bound to the signature.

We first address the former problem. In order to accommodate multiple users in the system, we change the master public key of the scheme to be $(\mathbf{A}, \{\vec{\mathbf{B}}^{(i)}\}_{i\in[N]})$, where N is the number of users. The secret key for a user i and an attribute $x^{(i)}$ is $\mathbf{R}^{(i)}$ such that $\vec{\mathbf{B}}^{(i)} = \mathbf{A}\vec{\mathbf{R}}^{(i)} + x^{(i)} \otimes \mathbf{G}$. To sign on a message, the user i first computes the trapdoor for $[\mathbf{A} \| \mathbf{B}_F^{(i)}]$ similarly to the above single-user construction. It then extends the trapdoor for the matrix $[\mathbf{A} \| \mathbf{B}_F^{(1)} \| \cdots \| \mathbf{B}_F^{(N)}]$ using the trapdoor extension technique [CHKP10]. Then, it samples a short vector $\mathbf{e}_F$ from a Gaussian distribution such that

$[\mathbf{A}\|\mathbf{B}_F^{(1)}\|\cdots\|\mathbf{B}_F^{(N)}]\mathbf{e}_F = \mathbf{0}$. It can be observed that $\mathbf{e}_F$ does not reveal the attribute x nor the user index i since the distribution from which it is sampled only depends on the master public key and F. Note that the trapdoor extension step is essential for hiding the user index i. We can prove unforgeability for the scheme similarly to the single-user case. A key difference here is that, since there are now N matrices in the master public key, we can embed up to N user attributes $\{x^{(i)}\}_{i\in[N]}$ into the master public key as $\mathbf{B}^{(i)} = \mathbf{A}\mathbf{R}^{(i)} + x^{(i)} \otimes \mathbf{G}$.

Next, we address the latter problem. We apply the classic OR-proof technique [FLS90] and show that a scheme that is unforgeable only when the adversary cannot make signing queries can be generically converted into a scheme that is unforgeable even when the adversary can make signing queries. To do so, we introduce a dummy user that is not used in the real system. In the security proof, the signing queries are answered using the signing key of the dummy user. In order to enable this proof strategy, a naïve approach would be to change the scheme so that in order to sign on (F, M), the signer signs on a modified new policy F', which on input $x \in \{0,1\}^k$ outputs $F(x)$ and outputs 1 on input a special symbol. Then, we associate the attribute of the dummy user with the special symbol. By the privacy property of the original (no signing query) ABS, the fact that the signing queries are answered using the dummy key instead of the key specified by the adversary will be unnoticed. A problem with this approach is that since the reduction algorithm has the secret key associated with the special symbol, it can sign on *any* message and policy. Namely, any forgery output by the adversary will not be useful for the reduction algorithm since it could have constructed it on its own to begin with. To resolve this problem, we partition the space of all possible message-policy pairs into two sets, the challenge set and the controlled set, using an admissible hash [BB04a,FHPS13]. Then, we associate the dummy key with an attribute that can sign on any pair in the controlled set, but not on the challenge set. We then hope that the adversary outputs the pair that falls into the challenge set, which allows us to successfully finish the reduction. By the property of the admissible hash, this happens with noticeable probability and we can prove the security of the resulting scheme.

1.4 Related Works

In the full version, we provide detailed discussions on the different models of group signatures and constructions based on other assumptions.

2 Preliminaries

2.1 Group Signature

Here, we adopt the definition of group signature schemes from the work of Bellare, Micciancio, and Warinschi [BMW03], with the relaxation regarding the anonymity suggested by Camenisch and Groth [CG05].

Syntax. Let $\{\mathcal{M}_\kappa\}_{\kappa\in\mathbb{N}}$ be a family of message spaces. In the following, we occasionally drop the subscript and simply write $\mathcal{M}$ when the meaning is clear. A group signature (GS) scheme is defined by the following algorithms:

$\mathsf{GS.KeyGen}(1^\kappa, 1^N) \to (\mathsf{gpk}, \mathsf{gok}, \{\mathsf{gsk}_i\}_{i\in[N]})$: The key generation algorithm takes as input the security parameter κ and the number of users N both in the unary form and outputs the group public key gpk, the opening key gok, and the set of user secret keys $\{\mathsf{gsk}_i\}_{i\in[N]}$.

$\mathsf{GS.Sign}(\mathsf{gpk}, \mathsf{gsk}_i, \mathsf{M}) \to \Sigma$: The signing algorithm takes as input the group public key gpk, the i-th user's secret key gsk_i (for some $i \in [N]$), and a message $\mathsf{M} \in \mathcal{M}_\kappa$ and outputs a signature Σ.

$\mathsf{GS.Vrfy}(\mathsf{gpk}, \mathsf{M}, \Sigma) \to \top$ or $\bot$: The verification algorithm takes as input the group public key gpk, the message M, and a signature Σ and outputs $\top$ if the signature is deemed valid and $\bot$ otherwise.

$\mathsf{GS.Open}(\mathsf{gpk}, \mathsf{gok}, \mathsf{M}, \Sigma) \to i$ or $\bot$: The opening algorithm takes as input the group public key gpk, the opening key gok, a message M, a signature Σ and outputs an identity i or the symbol $\bot$.

For GS, we require correctness, CCA-selfless anonymity, and full traceability.

Correctness. We require that for all κ, $N \in \mathrm{poly}(\kappa)$, $(\mathsf{gpk}, \mathsf{gok}, \{\mathsf{gsk}_i\}_{i\in[N]}) \in \mathsf{GS.KeyGen}(1^\kappa, 1^N)$, $i \in [N]$, $\mathsf{M} \in \mathcal{M}_\kappa$, and $\Sigma \in \mathsf{GS.Sign}(\mathsf{gpk}, \mathsf{gsk}_i, \mathsf{M})$, $\mathsf{GS.Vrfy}(\mathsf{gpk}, \mathsf{M}, \Sigma) = \top$ holds.

Full Traceability. We now define the full traceability for GS scheme. This security notion is defined by the following game between a challenger and an adversary A. During the game, the challenger maintains lists $\mathcal{Q}$ and $\mathcal{T}$, which are set to be empty at the beginning of the game.

Setup: At the beginning of the game, the challenger runs $\mathsf{GS.KeyGen}(1^\kappa, 1^N) \to (\mathsf{gpk}, \mathsf{gok}, \{\mathsf{gsk}_i\}_{i\in[N]})$ and gives $(1^\kappa, \mathsf{gpk}, \mathsf{gok})$ to A.

Queries: During the game, A can make the following two kinds of queries unbounded polynomially many times.

- **Corrupt Query:** Upon a query $i \in [N]$ from A, the challenger returns gsk_i to A. The challenger also adds i to $\mathcal{T}$.
- **Signing Queries:** Upon a query $(i, \mathsf{M}) \in [N] \times \mathcal{M}_\kappa$ from A, the challenger runs $\mathsf{GS.Sign}(\mathsf{gpk}, \mathsf{gsk}_i, \mathsf{M}) \to \Sigma$ and returns Σ to A. The challenger adds (i, M) to $\mathcal{Q}$.

Forgery: Eventually, A outputs $(\mathsf{M}^\star, \Sigma^\star)$ as the forgery. We say that A wins the game if:

1. $\mathsf{GS.Vrfy}(\mathsf{gpk}, \mathsf{M}^\star, \Sigma^\star) \to \top$, and
2. either of the following conditions (a) or (b) is satisfied:
   (a) $\mathsf{GS.Open}(\mathsf{gpk}, \mathsf{gok}, \mathsf{M}^\star, \Sigma^\star) = \bot$,
   (b) $\mathsf{GS.Open}(\mathsf{gpk}, \mathsf{gok}, \mathsf{M}^\star, \Sigma^\star) = i^\star \notin \mathcal{T} \ \wedge \ (i^\star, \mathsf{M}^\star) \notin \mathcal{Q}$.

We define the advantage of an adversary to be the probability that the adversary A wins, where the probability is taken over the randomness of the challenger and the adversary. A GS scheme is said to satisfy full traceability if the advantage of any PPT adversary A in the above game is negligible for any $N = \text{poly}(\kappa)$.

CCA-Selfless Anonymity. We now define CCA-selfless anonymity for a GS scheme. This security notion is defined by the following game between a challenger and an adversary A.

Setup: At the beginning of the game, the adversary A is given 1^κ as input and sends $i_0^\star, i_1^\star \in [N]$ to the challenger. Then the challenger runs $\mathsf{GS.KeyGen}(1^\kappa, 1^N) \rightarrow (\mathsf{gpk}, \mathsf{gok}, \{\mathsf{gsk}_i\}_{i\in[N]})$ and gives $(\mathsf{gpk}, \{\mathsf{gsk}_i\}_{i\in[N]\setminus\{i_0^\star, i_1^\star\}})$ to A.

Queries: During the game, A can make the following two kinds of queries unbounded polynomially many times.

- **Signing Queries:** Upon a query $(b, \mathsf{M}) \in \{0,1\}\times\mathcal{M}_\kappa$ from A, the challenger runs $\mathsf{GS.Sign}(\mathsf{gpk}, \mathsf{gsk}_{i_b^\star}, \mathsf{M}) \rightarrow \Sigma$ and returns Σ to A.
- **Open Queries:** Upon a query (M, Σ) from A, the challenger runs $\mathsf{GS.Open}(\mathsf{gpk}, \mathsf{gok}, \mathsf{M}, \Sigma)$ and returns the result to A.

Challenge Phase: At some point, A chooses its target message $\mathsf{M}^\star$. The challenger then samples a secret coin $\mathsf{coin} \xleftarrow{\$} \{0,1\}$ and computes $\mathsf{GS.Sign}(\mathsf{gpk}, \mathsf{gsk}_{i^\star_{\mathsf{coin}}}, \mathsf{M}^\star) \rightarrow \Sigma^\star$. Finally, it returns $\Sigma^\star$ to A.

Queries: After the challenge phase, A may continue to make signing and open queries unbounded polynomially many times. Here, we add a restriction that A cannot make an open query for $(\mathsf{M}^\star, \Sigma^\star)$.

Guess: Eventually, A outputs $\widehat{\mathsf{coin}}$ as a guess for coin.

We say that the adversary A wins the game if $\widehat{\mathsf{coin}} = \mathsf{coin}$. We define the advantage of an adversary to be $|\Pr[\mathsf{A} \text{ wins}] - 1/2|$, where the probability is taken over the randomness of the challenger and the adversary. A GS scheme is said to be CCA-selfless anonymous if the advantage of any PPT adversary A is negligible in the above game for any $N = \text{poly}(\kappa)$.

Remark 3. Note that Camenisch and Groth [CG05] defines selfless anonymity slightly differently by allowing the adversary to adaptively choose the targets and corrupt the users other than the targets. However, since the number of users N is polynomially bounded, these two definitions are equivalent w.l.o.g, and we chose the above formalization for simplicity of presentation.

2.2 Secret Key Encryption and Other Primitives

We will use some cryptographic primitives such as secret key encryptions (SKE) and one-time signatures (OTS) to construct a GS scheme. The definitions of these primitives will appear in the full version. Since we require an SKE to have some non-standard properties, we provide a brief explanation here. We require key robustness, which intuitively says that the ciphertext spaces corresponding to two random secret keys are disjoint with all but negligible probability. In addition, we require SKE to satisfy INDr-CCA security, which stipulates that a ciphertext is indistinguishable from a pseudorandom ciphertext that is publicly samplable, even if the distinguisher is equipped with a decryption oracle.

2.3 Admissible Hash Functions

Here, we define the notion of admissible hash, which was first introduced by [BB04a]. We follow the definition of [FHPS13, BV15] with minor changes.

Definition 4. *Let* $\ell := \ell(\kappa)$ *and* $\ell' := \ell'(\kappa)$ *be some polynomials. We define the function* $\mathsf{WldCmp} : \{0,1\}^\ell \times \{0,1\}^\ell \times \{0,1\}^\ell \to \{0,1\}$ *as*

$$\mathsf{WldCmp}(y, z, w) = 0 \Leftrightarrow \forall i \in [\ell] \ \big((y_i = 0) \vee (z_i = w_i)\big)$$

where y_i*,* z_i*, and* w_i *denote the* i*-th bit of* y*,* z*, and* w *respectively. Intuitively,* WldCmp *is a string comparison function with wildcards where it compares* z *and* w *only at those points where* $y_i = 1$*. Let* $\{\mathsf{H}_\kappa : \{0,1\}^{\ell'(\kappa)} \to \{0,1\}^{\ell(\kappa)}\}_{\kappa \in \mathbb{N}}$ *be a family of hash functions. We say that* $\{\mathsf{H}_\kappa\}_\kappa$ *is a family of admissible hash functions if there exists an efficient algorithm* AdmSmp *that takes as input* 1^κ *and* $Q \in \mathbb{N}$ *and outputs* $(y, z) \in \{0,1\}^\ell \times \{0,1\}^\ell$ *such that for every polynomial* $Q(\kappa)$ *and all* $X^\star, X^{(1)}, \ldots X^{(Q)} \in \{0,1\}^{\ell'(\kappa)}$ *with* $X^\star \notin \{X^{(1)}, \ldots, X^{(Q)}\}$*, we have*

$$\Pr_{(y,z)}\left[\mathsf{WldCmp}(y, z, \mathsf{H}(X^\star)) = 0 \wedge \left(\wedge_{j \in [Q]} \mathsf{WldCmp}(y, z, \mathsf{H}(X^{(j)})) = 1\right)\right] \geq \Delta_Q(\kappa),$$

for a noticeable function $\Delta_Q(\kappa)$*, where the probability above is taken over the choice of* $(y, z) \xleftarrow{\$} \mathsf{AdmSmp}(1^\kappa, Q)$*.*

As shown in previous works [Lys02, FHPS13], a family of error correcting codes $\{\mathsf{H}_\kappa : \{0,1\}^{\ell'(\kappa)} \to \{0,1\}^{\ell(\kappa)}\}_{\kappa \in \mathbb{N}}$ with constant relative distance $c \in (0, 1/2)$ is an admissible hash function. Explicit and efficient constructions of such codes are given in [SS96, Zém01, Gol08] to name a few.

3 Indexed Attribute-Based Signatures

In this section, we define the syntax and the security notion of indexed attribute-based signature (indexed ABS). We require indexed ABS to satisfy unforgeability and privacy. For the former, we consider two kinds of security notions that we call co-selective unforgeability and no-signing-query unforgeability. While the latter notion of unforgeability is weaker, we will show that an indexed ABS scheme that only satisfies this weaker security notion can be converted into a scheme with the stronger security notion without loosing privacy.

3.1 Indexed Attribute-Based Signature

Syntax. Let $\{\mathcal{C}_\kappa\}_{\kappa \in \mathbb{N}}$ be a family of circuits, where $\mathcal{C}_\kappa$ is a set of circuits with domain $\{0,1\}^{k(\kappa)}$ and range $\{0,1\}$, and the size of every circuit in $\mathcal{C}_\kappa$ is bounded by $\mathrm{poly}(\kappa)$. Let also $\{\mathcal{M}_\kappa\}_{\kappa \in \mathbb{N}}$ be a family of message spaces. In the following, we occasionally drop the subscript and simply write $\mathcal{C}$ and $\mathcal{M}$ when the meaning is clear. An indexed attribute-based signature (indexed ABS) scheme for the circuit class $\mathcal{C}$ is defined by the following algorithms:

$\mathsf{ABS.Setup}(1^\kappa, 1^N) \to (\mathsf{mpk}, \mathsf{msk})$: The setup algorithm takes as input the security parameter κ and the bound on the number of users N both in the unary form and outputs the master public key mpk and the master secret key msk.

$\mathsf{ABS.KeyGen}(\mathsf{msk}, i, x) \to \mathsf{sk}_x$: The key generation algorithm takes as input the master secret key msk, the user index $i \in [N]$, and the attribute $x \in \{0,1\}^k$ and outputs the user secret key sk_x. We assume that i and x are implicitly included in sk_x.

$\mathsf{ABS.Sign}(\mathsf{mpk}, \mathsf{sk}_x, \mathsf{M}, C) \to \sigma$: The signing algorithm takes as input the master public key mpk, the secret key sk_x associated to x, a message $\mathsf{M} \in \mathcal{M}_\kappa$, and a policy $C \in \mathcal{C}_\kappa$ and outputs the signature σ.

$\mathsf{ABS.Vrfy}(\mathsf{mpk}, \mathsf{M}, C, \sigma) \to \top$ or $\bot$: The verification algorithm takes as input the master public key mpk, a message M, a policy C, and a signature σ. It outputs $\top$ if the signature is deemed valid and $\bot$ otherwise. We assume that the verification algorithm is deterministic.

We require correctness, privacy, and co-selective unforgeability.

Correctness. We require correctness: that is, for all κ, $N \in \mathsf{poly}(\kappa)$, $(\mathsf{mpk}, \mathsf{msk}) \in \mathsf{ABS.Setup}(1^\kappa, 1^N)$, $i \in [N]$, $x \in \{0,1\}^k, C \in \mathcal{C}_\kappa$ such that $C(x) = 1$, $\mathsf{M} \in \mathcal{M}_\kappa$, $\mathsf{sk}_x \in \mathsf{ABS.KeyGen}(\mathsf{msk}, i, x)$, and $\sigma \in \mathsf{ABS.Sign}(\mathsf{mpk}, \mathsf{sk}_x, \mathsf{M}, C)$, $\mathsf{ABS.Vrfy}(\mathsf{mpk}, \mathsf{M}, C, \sigma) = \top$ holds.

Perfect Privacy. We say that the ABS scheme has perfect privacy if for all κ, $N \in \mathsf{poly}(\kappa)$, $(\mathsf{mpk}, \mathsf{msk}) \in \mathsf{ABS.Setup}(1^\kappa, 1^N)$, $x_0, x_1 \in \{0,1\}^k$, $i_0, i_1 \in [N]$, $C \in \mathcal{C}_\kappa$ satisfying $C(x_0) = C(x_1) = 1$, $\mathsf{M} \in \mathcal{M}$, $\mathsf{sk}_{x_0} \in \mathsf{ABS.KeyGen}(\mathsf{msk}, i_0, x_0)$, and $\mathsf{sk}_{x_1} \in \mathsf{ABS.KeyGen}(\mathsf{msk}, i_1, x_1)$, the following distributions are the same:

$$\{\sigma_0 \xleftarrow{\$} \mathsf{ABS.Sign}(\mathsf{mpk}, \mathsf{sk}_{x_0}, \mathsf{M}, C)\} \approx \{\sigma_1 \xleftarrow{\$} \mathsf{ABS.Sign}(\mathsf{mpk}, \mathsf{sk}_{x_1}, \mathsf{M}, C)\}.$$

Co-selective Unforgeability. We now define the co-selective unforgeability for ABS scheme. This security notion is defined by the following game between a challenger and an adversary A. During the game, the challenger maintains a list $\mathcal{Q}$, which is set to be empty at the beginning of the game.

Key Queries: At the beginning of the game, the adversary A is given 1^κ as input. It then sends 1^N, $\{(i, x^{(i)})\}_{i \in [N]}$, and $\mathcal{S} \subseteq [N]$ such that $x^{(i)} \in \{0,1\}^k$ for all $i \in [N]$ to the challenger.

Setup: The challenger runs $\mathsf{ABS.Setup}(1^\kappa, 1^N) \to (\mathsf{mpk}, \mathsf{msk})$ and $\mathsf{ABS.KeyGen}(\mathsf{msk}, i, x^{(i)}) \to \mathsf{sk}_{x^{(i)}}$ for $i \in [N]$. It then gives mpk and $\{\mathsf{sk}_{x^{(i)}}\}_{i \in [\mathcal{S}]}$ to A.

Signing Queries: During the game, A can make signing queries unbounded polynomially many times. When A queries (M, C, i) such that $\mathsf{M} \in \mathcal{M}$, $C \in \mathcal{C}$, $i \in [N]$, and $C(x^{(i)}) = 1$, the challenger runs $\mathsf{ABS.Sign}(\mathsf{mpk}, \mathsf{sk}_{x^{(i)}}, \mathsf{M}, C) \to \sigma$ and returns σ to A. The challenger then adds (M, C) to $\mathcal{Q}$.

Forgery: Eventually, A outputs $(\mathsf{M}^\star, C^\star, \sigma^\star)$ as the forgery. We say that A wins the game if:

1. $C^\star \in \mathcal{C}$,
2. $\mathsf{ABS.Vrfy}(\mathsf{mpk}, \mathsf{M}^\star, C^\star, \sigma^\star) \to \top$,
3. $C^\star(x^{(i)}) = 0$ for $i \in \mathcal{S}$,
4. $(\mathsf{M}^\star, C^\star) \notin \mathcal{Q}$.

We define the advantage of the adversary to be the probability that the adversary A wins in the above game, where the probability is taken over the coin tosses made by A and the challenger. We say that a scheme satisfies co-selective unforgeability if the advantage of any PPT adversary A in the above game is negligible in the security parameter.

No-Signing-Query Unforgeability. We now define a weaker definition of unforgeability. We define the no-signing-query unforgeability game by modifying the co-selective unforgeability game above by adding some more restrictions on A. Namely, we prohibit A from making any signing queries and require $\mathcal{S} \neq \emptyset$. We do not change the winning condition of the game and define the advantage of A as the probability that A wins. Note that Item 4 becomes vacuous because we will always have $\mathcal{Q} = \emptyset$. We say that a scheme satisfies no-signing-query unforgeability if the advantage of any PPT adversary A in the game is negligible.

Remark 5 (Comparing indexed ABS with standard ABS). The syntax of the indexed ABS is a relaxation of the standard ABS [MPR11,OT11,SAH16]: the setup algorithm takes 1^N as an additional input and the key generation algorithm takes an index i as an additional input. It is easy to check that standard ABS can be used as indexed ABS by simply ignoring the additional inputs.

3.2 From No-Signing-Query to Co-selective Unforgeability

Here, we show that an indexed ABS scheme $\mathsf{ABS} = (\mathsf{ABS.Setup}, \mathsf{ABS.KeyGen}, \mathsf{ABS.Sign}, \mathsf{ABS.Vrfy})$ that is no-signing-query unforgeable can be generically converted into a new indexed ABS scheme $\mathsf{ABS}' = (\mathsf{ABS}'.\mathsf{Setup}, \mathsf{ABS}'.\mathsf{KeyGen}, \mathsf{ABS}'.\mathsf{Sign}, \mathsf{ABS}'.\mathsf{Vrfy})$ that is co-selective unforgeable. If ABS is perfectly private, so is ABS'. To enable the resulting scheme ABS' to deal with function class $\mathcal{C} = \{\mathcal{C}_\kappa\}_{\kappa \in \mathbb{N}}$, where $\mathcal{C}_\kappa$ is a set of circuits C such that $C : \{0,1\}^{k(\kappa)} \to \{0,1\}$, we require ABS to be able to deal with a (slightly) more complex function class $\mathcal{F} = \{\mathcal{F}_\kappa\}_{\kappa \in \mathbb{N}}$. We define $\mathcal{F}_\kappa$ as

$$\mathcal{F}_\kappa = \left\{ F[\widetilde{\mathsf{M}}, C] : \{0,1\}^{k(\kappa)+2\ell(\kappa)+1} \to \{0,1\} \;\middle|\; \widetilde{\mathsf{M}} \in \{0,1\}^{\ell(\kappa)},\ C \in \mathcal{C}_\kappa \right\}, \quad (1)$$

where $F[\widetilde{\mathsf{M}}, C]$ is defined in Fig. 1. We assume that the circuit $F[\widetilde{\mathsf{M}}, C]$ is deterministically constructed from $\widetilde{\mathsf{M}}$ and C in a predetermined way. Let $\{\mathcal{H}_\kappa\}_\kappa$ be a family of collision resistant hash functions where an index $h \in \mathcal{H}_\kappa$ specifies a function $h : \{0,1\}^* \to \{0,1\}^{\ell'(\kappa)}$, where $\{0,1\}^{\ell'(\kappa)}$ is the input space of an admissible hash function $\mathsf{H}_\kappa : \{0,1\}^{\ell'(\kappa)} \to \{0,1\}^{\ell(\kappa)}$. We construct ABS' as follows.

$\mathsf{ABS}'.\mathsf{Setup}(1^\kappa, 1^N)$: It runs $\mathsf{ABS}.\mathsf{Setup}(1^\kappa, 1^{N+1}) \to (\mathsf{mpk}, \mathsf{msk})$ and samples a random index of collision resistant hash function $h \xleftarrow{\$} \mathcal{H}_\kappa$. It then outputs the master public key $\mathsf{mpk}' = (\mathsf{mpk}, h)$ and the master secret key $\mathsf{msk}' := \mathsf{msk}$.

$\mathsf{ABS}'.\mathsf{KeyGen}(\mathsf{msk}, i, x)$: It runs $\mathsf{ABS}.\mathsf{KeyGen}(\mathsf{msk}, i, x\|0^{2\ell+1}) \to \mathsf{sk}_{x\|0^{2\ell+1}}$ and returns $\mathsf{sk}'_x := \mathsf{sk}_{x\|0^{2\ell+1}}$.

$\mathsf{ABS}'.\mathsf{Sign}(\mathsf{mpk}', \mathsf{sk}'_x, \mathsf{M}, C)$: It first parses $\mathsf{mpk}' \to (\mathsf{mpk}, h)$ and $\mathsf{sk}'_x \to \mathsf{sk}_{x\|0^{2\ell+1}}$ and computes $\widetilde{\mathsf{M}} = \mathsf{H}(h(\mathsf{M}\|C))$. It then constructs a circuit $F[\widetilde{\mathsf{M}}, C]$ that is defined as in Fig. 1. It finally runs $\mathsf{ABS}.\mathsf{Sign}(\mathsf{mpk}, \mathsf{sk}_{x\|0^{2\ell+1}}, \mathsf{M}, F[\widetilde{\mathsf{M}}, C]) \to \sigma$ and outputs $\sigma' := \sigma$.

$\mathsf{ABS}'.\mathsf{Vrfy}(\mathsf{mpk}', \mathsf{M}, C, \sigma)$: It first parses $\mathsf{mpk}' \to (\mathsf{mpk}, h)$. It then computes $\widetilde{\mathsf{M}} = \mathsf{H}(h(\mathsf{M}\|C))$ and constructs a circuit $F[\widetilde{\mathsf{M}}, C]$ that is defined as in Fig. 1. It then outputs $\mathsf{ABS}.\mathsf{Vrfy}(\mathsf{mpk}, \mathsf{M}, F[\widetilde{\mathsf{M}}, C], \sigma)$.

$F[\widetilde{\mathsf{M}}, C](x\|y\|z\|b)$

Hardwired constants: A bit string $\widetilde{\mathsf{M}} \in \{0,1\}^\ell$ and a circuit $C : \{0,1\}^k \to \{0,1\}$.

1. Parse the input to retrieve $x \in \{0,1\}^k$, $y, z \in \{0,1\}^\ell$, and $b \in \{0,1\}$.
2. If $b = 0$, output $C(x)$.
3. If $b = 1$, output $\mathsf{WldCmp}(y, z, \widetilde{\mathsf{M}}) \in \{0,1\}$.

Fig. 1. Description of the circuit $F[\widetilde{\mathsf{M}}, C]$.

Correctness. We observe that if $C(x) = 1$, we have $F[\widetilde{\mathsf{M}}, C](x\|0^{2\ell+1}) = C(x) = 1$ by the definition of $F[\widetilde{\mathsf{M}}, C]$. The correctness of ABS' therefore follows from that of ABS.

Perfect Privacy. The following addresses the privacy of ABS'.

Theorem 6. *If* ABS *is perfectly private, so is* ABS'.

Proof. If $C(x_0) = C(x_1) = 1$, we have $F[\widetilde{\mathsf{M}}, C](x_0\|0^{2\ell+1}) = C(x_0) = 1$ and $F[\widetilde{\mathsf{M}}, C](x_1\|0^{2\ell+1}) = C(x_1) = 1$ by the definition of $F[\widetilde{\mathsf{M}}, C]$. The theorem therefore follows from the perfect privacy of ABS.

Co-selective Unforgeability. The following theorem addresses the co-selective unforgeability of ABS'. The proof will appear in the full version.

Theorem 7. *If* ABS *is no-signing-query unforgeable and perfectly private,* $\mathcal{H}_\kappa$ *is a family of collision resistant hash functions, and* H_κ *is an admissible hash function, then* ABS' *is co-selective unforgeable.*

4 Generic Construction of Group Signatures

In this section, we give a generic construction of a GS scheme from three building blocks: an indexed ABS, an OTS, and an SKE. As we will show in Sect. 7, by appropriately instantiating the building blocks, we obtain the first lattice-based GS scheme in the standard model.

Ingredients. Here, we give a generic construction of a GS scheme $\mathsf{GS} = (\mathsf{GS.KeyGen}, \mathsf{GS.Sign}, \mathsf{GS.Vrfy}, \mathsf{GS.Open})$ from an indexed ABS scheme $\mathsf{ABS} = (\mathsf{ABS.Setup}, \mathsf{ABS.KeyGen}, \mathsf{ABS.Sign}, \mathsf{ABS.Vrfy})$ with perfect privacy and co-selective unforgeability, an OTS scheme $\mathsf{OTS} = (\mathsf{OTS.KeyGen}, \mathsf{OTS.Sign}, \mathsf{OTS.Vrfy})$ with strong unforgeability, and an SKE scheme $\mathsf{SKE} = (\mathsf{SKE.Gen}, \mathsf{SKE.Enc}, \mathsf{SKE.Dec})$ with key robustness and INDr-CCA security. We require the underlying primitives to satisfy the following constraints:

- $\mathsf{SKE}.\mathcal{M}_\kappa \supseteq [N+1] \times \{0,1\}^{\ell_1(\kappa)}$, where $\mathsf{SKE}.\mathcal{M}_\kappa$ denotes the plaintext space of SKE and $\ell_1(\kappa)$ denotes the upper-bound on the length of ovk that is output by $\mathsf{OTS.Setup}(1^\kappa)$.
- We require the underlying indexed ABS scheme to be able to deal with function class $\mathcal{C} = \{\mathcal{C}_\kappa\}_{\kappa \in \mathbb{N}}$, where $\mathcal{C}_\kappa$ is defined as

$$\mathcal{C}_\kappa = \left\{ C[\mathsf{ovk}, \mathsf{ct}] \;\middle|\; \mathsf{ovk} \in \{0,1\}^{\ell_1(\kappa)}, \mathsf{ct} \in \{0,1\}^{\ell_2(\kappa)} \right\}, \tag{2}$$

  where $C[\mathsf{ovk}, \mathsf{ct}]$ is defined in Fig. 2 and $\ell_2(\kappa)$ is the upper bound on the length of a ciphertext ct output by $\mathsf{SKE.Enc}(\mathsf{K}, \mathsf{M})$ for $\mathsf{K} \in \mathsf{SKE.Gen}(\mathsf{SKE.Setup}(1^\kappa))$ and $\mathsf{M} \in \mathsf{SKE}.\mathcal{M}_\kappa$.
- We require $\mathsf{OTS}.\mathcal{M}_\kappa = \{0,1\}^*$, where $\mathsf{OTS}.\mathcal{M}_\kappa$ denotes the message space of OTS. Note that any OTS scheme with sufficiently large message space can be modified to satisfy this condition by applying a collision resistant hash to a message before signing.

Construction. We construct GS as follows.

$\mathsf{GS.KeyGen}(1^\kappa, 1^N)$: It first samples $\mathsf{pp} \xleftarrow{\$} \mathsf{SKE.Setup}(1^\kappa)$ and $(\mathsf{mpk}, \mathsf{msk}) \xleftarrow{\$} \mathsf{ABS.Setup}(1^\kappa, 1^{N+1})$. It then samples $\mathsf{K}_i \xleftarrow{\$} \mathsf{SKE.Gen}(\mathsf{pp})$ and $\mathsf{sk}_{i\|\mathsf{K}_i} \xleftarrow{\$} \mathsf{ABS.KeyGen}(\mathsf{msk}, i, i\|\mathsf{K}_i)$ for $i \in [N]$. Finally, it outputs

$$\mathsf{gpk} := (\ \mathsf{pp},\ \mathsf{mpk}\ ),\ \mathsf{gok} := \{\ \mathsf{K}_i\ \}_{i\in[N]},\ \left\{\ \mathsf{gsk}_i := \left(\ i,\ \mathsf{K}_i,\ \mathsf{sk}_{i\|\mathsf{K}_i}\ \right)\ \right\}_{i\in[N]}.$$

$\mathsf{GS.Sign}(\mathsf{gsk}_i, \mathsf{M})$: It first samples $(\mathsf{ovk}, \mathsf{osk}) \xleftarrow{\$} \mathsf{OTS.KeyGen}(1^\kappa)$ and computes $\mathsf{ct} \xleftarrow{\$} \mathsf{SKE.Enc}(\mathsf{K}_i, i\|\mathsf{ovk})$. It then runs

$$\mathsf{ABS.Sign}(\mathsf{mpk}, \mathsf{sk}_{i\|\mathsf{K}_i}, C[\mathsf{ovk}, \mathsf{ct}], \mathsf{M}) \to \sigma,$$

where the circuit $C[\mathsf{ovk}, \mathsf{ct}]$ is defined in Fig. 2. It further runs $\mathsf{OTS.Sign}(\mathsf{osk}, \mathsf{M}\|\sigma) \to \tau$. Finally, it outputs $\Sigma := (\mathsf{ovk}, \mathsf{ct}, \sigma, \tau)$.

$\mathsf{GS.Vrfy}(\mathsf{gpk}, \mathsf{M}, \Sigma)$: It first parses $\Sigma \rightarrow (\mathsf{ovk}, \mathsf{ct}, \sigma, \tau)$. It then outputs $\top$ if

$$\mathsf{ABS.Vrfy}(\mathsf{mpk}, \mathsf{M}, C[\mathsf{ovk}, \mathsf{ct}], \sigma) = \top \wedge \mathsf{OTS.Vrfy}(\mathsf{ovk}, \mathsf{M}\|\sigma, \tau) = \top,$$

where $C[\mathsf{ovk}, \mathsf{ct}]$ is defined in Fig. 2. Otherwise, it outputs $\perp$.

$\mathsf{GS.Open}(\mathsf{gpk}, \mathsf{gok}, \mathsf{M}, \Sigma)$: It first runs $\mathsf{GS.Vrfy}(\mathsf{gpk}, \mathsf{M}, \Sigma)$ and returns $\perp$ if the verification result is $\perp$. Otherwise, it parses $\Sigma \rightarrow (\mathsf{ovk}, \mathsf{ct}, \sigma, \tau)$. It then computes $d_i \leftarrow \mathsf{SKE.Dec}(\mathsf{K}_i, \mathsf{ct})$ for $i \in [N]$ and outputs the smallest index i such that $d_i \neq \perp$. If there is not such i, it returns $\perp$.

$C[\mathsf{ovk}, \mathsf{ct}](i\|\mathsf{K})$

Hardwired constants: A verification key ovk of OTS and a ciphertext ct of SKE.

1. **Parse the input to retrieve $i \in [N+1]$ and K. If the input does not conform to the format, output 0.**
2. **If $i = N+1$, output 1.**
3. **Compute $\mathsf{SKE.Dec}(\mathsf{K}, \mathsf{ct}) = i'\|\mathsf{ovk}'$. If $i' = i$ and $\mathsf{ovk}' = \mathsf{ovk}$, output 1. Otherwise, output 0.**

Fig. 2. Description of the circuit $C[\mathsf{ovk}, \mathsf{ct}]$.

Remark 8 (Construction Using Public Key Encryption). We remark that we may be able to obtain an alternative construction using a public key encryption (PKE) instead of an SKE. See full version for further discussion.

Correctness. We show that correctly generated signature $\Sigma = (\mathsf{ovk}, \mathsf{ct}, \sigma, \tau)$ passes the verification. We have $\mathsf{OTS.Vrfy}(\mathsf{ovk}, \mathsf{M}\|\sigma, \tau) = \top$ by the correctness of OTS. Furthermore, we have $\mathsf{ABS.Vrfy}(\mathsf{mpk}, \mathsf{M}, C[\mathsf{ovk}, \mathsf{ct}], \sigma) = \top$ since $C[\mathsf{ovk}, \mathsf{ct}](i\|\mathsf{K}_i) = 1$, which follows from $\mathsf{SKE.Dec}(\mathsf{K}_i, \mathsf{ct}) = i\|\mathsf{ovk}$ by the correctness of SKE.

CCA-Selfless Anonymity. The following theorem addresses the CCA-selfless anonymity of the above GS scheme. The proof will appear in the full version.

Theorem 9. *If* ABS *is perfectly private and co-selective unforgeable,* OTS *is strongly unforgeable, and* SKE *is INDr-CCA-secure and key robust, then* GS *constructed above is CCA-selfless anonymous.*

Traceability. The following addresses the traceability of the above GS scheme.

Theorem 10. *If* ABS *is co-selective unforgeable and* SKE *has key robustness, then* GS *constructed above has full traceability.*

Proof. Let us fix a PPT adversary A and consider the full traceability game played between A and a challenger. Let $(\mathsf{M}^\star, \Sigma^\star)$ be a forgery output by A. We define F_1 to be the event that A wins the game and $\mathsf{GS.Open}(\mathsf{gpk}, \mathsf{gok}, \mathsf{M}^\star, \Sigma^\star) = \bot$ holds, and F_2 be the event that A wins the game and $\mathsf{GS.Open}(\mathsf{gpk}, \mathsf{gok}, \mathsf{M}^\star, \Sigma^\star) = i^\star$ holds for $i^\star$ such that $i^\star \notin \mathcal{T}$. Since both F_1 and F_2 are collectively exhaustive events of a successful forgery, it suffices to prove $\Pr[\mathsf{F}_1] = \mathrm{negl}(\kappa)$ and $\Pr[\mathsf{F}_2] = \mathrm{negl}(\kappa)$.

Lemma 11. *If* ABS *is co-selective unforgeable, we have* $\Pr[\mathsf{F}_1] = \mathrm{negl}(\kappa)$.

Proof. For the sake of the contradiction, let us assume that F_1 happens with non-negligible probability ϵ. We then construct an adversary B that breaks the co-selective unforgeability of ABS with the same probability. The adversary B proceeds as follows.

At the beginning of the game, B is given 1^κ from its challenger. B then samples $\mathsf{pp} \xleftarrow{\$} \mathsf{SKE.Setup}(1^\kappa)$ and $\mathsf{K}_i \xleftarrow{\$} \mathsf{SKE.Gen}(\mathsf{pp})$ for $i \in [N]$ and submits 1^N, $\{(i, i \| \mathsf{K}_i)\}_{i \in [N]}$, and $\mathcal{S} = [N]$ to its challenger. Then, B receives mpk and $\{\mathsf{sk}_{i\|\mathsf{K}_i}\}_{i \in [N]}$ from the challenger. It then gives 1^κ, $\mathsf{gpk} := (\mathsf{pp}, \mathsf{mpk})$, and $\mathsf{gok} := \{\mathsf{K}_i\}_{i \in [N]}$ to A and keeps $\{\mathsf{gsk}_i := (i, \mathsf{K}_i, \mathsf{sk}_{i\|\mathsf{K}_i})\}_{i \in [N]}$ secret. During the game, A makes signing and corrupt queries. These queries are trivial to handle because B has $\{\mathsf{gsk}_i\}_{i \in [N]}$. In particular, B can handle all signing queries from A without making signing query to its challenger. Eventually, A will output a forgery $(\mathsf{M}^\star, \Sigma^\star = (\mathsf{ovk}^\star, \mathsf{ct}^\star, \sigma^\star, \tau^\star))$. If $\mathsf{GS.Vrfy}(\mathsf{gpk}, \mathsf{M}^\star, \Sigma^\star) = \top$ and $\mathsf{GS.Open}(\mathsf{gpk}, \mathsf{gok}, \mathsf{M}^\star, \Sigma^\star) = \bot$ hold, B outputs $(\mathsf{M}^\star, C[\mathsf{ovk}^\star, \mathsf{ct}^\star], \sigma^\star)$ as its forgery. Otherwise, B aborts.

We claim that B wins the game whenever F_1 happens. To prove this, we first observe that $\mathsf{ABS.Vrfy}(\mathsf{mpk}, C[\mathsf{ovk}^\star, \mathsf{ct}^\star], \sigma^\star) = \top$ holds because $\mathsf{GS.Vrfy}(\mathsf{gpk}, \mathsf{M}^\star, \Sigma^\star) = \top$. We then show that B has not made any prohibited key query. Namely, we show $C[\mathsf{ovk}^\star, \mathsf{ct}^\star](i \| \mathsf{K}_i) = 0$ for all $i \in [N]$. This follows since otherwise we have $\mathsf{SKE.Dec}(\mathsf{K}_i, \mathsf{ct}^\star) \neq \bot$ for some i, which contradicts $\mathsf{GS.Open}(\mathsf{gpk}, \mathsf{gok}, \mathsf{M}^\star, \Sigma^\star) = \bot$. We also note that B has not made any signing query. Since B's simulation is perfect, we can conclude that B wins the game with probability ϵ. This concludes the proof of the lemma.

Lemma 12. *If* ABS *is co-selective unforgeable and* SKE *has key robustness, we have* $\Pr[\mathsf{F}_2] = \mathrm{negl}(\kappa)$.

Proof. For the sake of the contradiction, let us assume that F_2 happens with non-negligible probability ϵ. We then construct an adversary B that breaks the co-selective unforgeability of ABS with non-negligible probability. We show this by considering the following sequence of games. In the following, let E_i denote the probability that F_2 occurs and the challenger does not abort in $\mathsf{Game}\ i$.

Game 0: We define Game 0 as the ordinary full traceability game between A and the challenger. By assumption, we have $\Pr[\mathsf{E}_0] = \epsilon$.

Game 1: In this game, the challenger samples $j^\star \xleftarrow{\$} [N]$ at the beginning of the game and aborts if $j^\star \neq i^\star$ at the end of the game. Since the view of A

is independent from $j^\star$ and GS.Open does not output any symbol outside $[N] \cup \{\bot\}$, we have $\Pr[\mathsf{E}_1] = \epsilon/N$.

Game 2: In this game, the challenger aborts the game as soon as $j^\star \neq i^\star$ turns out to be true. Namely, it aborts if A makes a corruption query for $j^\star$, or $i^\star$ defined at the end of the game does not equal to $j^\star$. Since this is only a conceptual change, we have $\Pr[\mathsf{E}_2] = \Pr[\mathsf{E}_3]$.

Game 3: In this game, we change the previous game so that the challenger aborts at the end of the game if $|\{i \in [N] : \mathsf{SKE.Dec}(\mathsf{K}_i, \mathsf{ct}^\star) \neq \bot\}| \neq 1$ for $(\mathsf{M}^\star, \Sigma^\star = (\mathsf{ovk}^\star, \mathsf{ct}^\star, \sigma^\star, \tau^\star))$ output by A as the forgery. We claim that the probability that F_2 and $|\{i \in [N] : \mathsf{SKE.Dec}(\mathsf{K}_i, \mathsf{ct}^\star) \neq \bot\}| \neq 1$ occur at the same time is negligibly small. Note that by the definition of GS.Open, F_2 implies $\mathsf{SKE.Dec}(\mathsf{K}_{i^\star}, \mathsf{ct}^\star) \neq \bot$ for $i^\star \in [N]$. We therefore have $|\{i \in [N] : \mathsf{SKE.Dec}(\mathsf{K}_i, \mathsf{ct}^\star) \neq \bot\}| \geq 2$. However, the probability of this occurring is bounded by

$$\begin{aligned}
&\Pr\left[\, |\{i \in [N] : \mathsf{SKE.Dec}(\mathsf{K}_i, \mathsf{ct}^\star) \neq \bot\}| \geq 2 \,\right] \\
&\leq \Pr\left[\begin{array}{c} \mathsf{pp} \xleftarrow{\$} \mathsf{SKE.Setup}(1^\kappa),\ \mathsf{K}_j \xleftarrow{\$} \mathsf{SKE.Gen}(\mathsf{pp}) \text{ for } j \in [N] : \\ \exists \mathsf{ct}^\star \in \{0,1\}^*,\ \exists i, i^\star \in [N] \\ \text{s.t. } i \neq i^\star \wedge \mathsf{SKE.Dec}(\mathsf{K}_i, \mathsf{ct}^\star) \neq \bot \wedge \mathsf{SKE.Dec}(\mathsf{K}_{i^\star}, \mathsf{ct}^\star) \neq \bot \end{array}\right] \\
&\leq \sum_{i, i^\star \in [N] \text{ s.t. } i \neq i^\star} \Pr\left[\begin{array}{c} \mathsf{pp} \xleftarrow{\$} \mathsf{SKE.Setup}(1^\kappa),\ \mathsf{K}_i, \mathsf{K}_{i^\star} \xleftarrow{\$} \mathsf{SKE.Gen}(\mathsf{pp}) : \\ \exists \mathsf{ct}^\star \in \{0,1\}^* \\ \text{s.t. } \mathsf{SKE.Dec}(\mathsf{K}_i, \mathsf{ct}^\star) \neq \bot \wedge \mathsf{SKE.Dec}(\mathsf{K}_{i^\star}, \mathsf{ct}^\star) \neq \bot \end{array}\right] \\
&\leq N(N-1)/2 \cdot \mathrm{negl}(\kappa) \\
&= \mathrm{negl}(\kappa),
\end{aligned}$$

where the second inequality is by the union bound and the third inequality is by the key robustness of SKE. Therefore, we have $|\Pr[\mathsf{E}_2] - \Pr[\mathsf{E}_3]| = \mathrm{negl}(\kappa)$.

We then replace the challenger in Game 3 with an adversary B against the co-selective unforgeability of ABS with advantage $\Pr[\mathsf{E}_3]$. The adversary B proceeds as follows.

At the beginning of the game, B is given 1^κ from its challenger. Then, B chooses its guess $j^\star \xleftarrow{\$} [N]$ for $i^\star$, samples $\mathsf{pp} \xleftarrow{\$} \mathsf{SKE.Setup}(1^\kappa)$ and $\mathsf{K}_i \xleftarrow{\$} \mathsf{SKE.Gen}(\mathsf{pp})$ for $i \in [N]$, and sends 1^N, $\{(i, i \| \mathsf{K}_i)\}_{i \in [N]}$, and $\mathcal{S} = [N] \backslash \{j^\star\}$ to the challenger. Then, B receives mpk and $\{\mathsf{sk}_{i \| \mathsf{K}_i}\}_{i \in [N] \backslash \{j^\star\}}$ from the challenger. It then sets $\mathsf{gpk} := (\mathsf{pp}, \mathsf{mpk})$, $\mathsf{gsk}_i := (i, \mathsf{K}_i, \mathsf{sk}_{i \| \mathsf{K}_i})$ for $i \in [N] \backslash \{j^\star\}$, and $\mathsf{gok} := \{\mathsf{K}_i\}_{i \in [N]}$ and gives 1^κ, gpk, and gok to A. During the game, A makes two kinds of queries. B answers the queries as follows.

- When A makes a corrupt query for $i \in [N]$, B proceeds as follows. If $i = j^\star$, B aborts. Otherwise, it gives gsk_i to A.
- When A makes a signing query for (i, M), B answers the query using gsk_i if $i \neq j^\star$. If $i = j^\star$, B first samples $(\mathsf{ovk}, \mathsf{osk}) \xleftarrow{\$} \mathsf{OTS.KeyGen}(1^\kappa)$ and computes $\mathsf{ct} \xleftarrow{\$} \mathsf{SKE.Enc}(\mathsf{K}_{j^\star}, j^\star \| \mathsf{ovk})$. It then makes a signing query $(\mathsf{M}, C[\mathsf{ovk}, \mathsf{ct}], j^\star)$ to its challenger, who returns σ to B. Then, it runs $\mathsf{OTS.Sign}(\mathsf{osk}, \mathsf{M} \| \sigma) \to \tau$ and returns $\Sigma := (\mathsf{ovk}, \mathsf{ct}, \sigma, \tau)$ to A.

Eventually, A will output a forgery $(\mathsf{M}^\star, \Sigma^\star = (\mathsf{ovk}^\star, \mathsf{ct}^\star, \sigma^\star, \tau^\star))$. If either of $\mathsf{GS.Vrfy}(\mathsf{gpk}, \mathsf{M}^\star, \Sigma^\star) = \top$ or $i^\star = j^\star$ does not hold, where $i^\star := \mathsf{GS.Open}(\mathsf{gpk}, \mathsf{gok}, \mathsf{M}^\star, \Sigma^\star)$, B aborts. It also aborts if $|\{i \in [N] : \mathsf{SKE.Dec}(\mathsf{K}_i, \mathsf{ct}^\star) \neq \bot\}| \neq 1$. Otherwise, B outputs $(\mathsf{M}^\star, C[\mathsf{ovk}^\star, \mathsf{ct}^\star], \sigma^\star)$ as its forgery.

We claim that B wins the game whenever E_3 occurs. To see this, we first observe that we have $\mathsf{ABS.Vrfy}(\mathsf{mpk}, C[\mathsf{ovk}^\star, \mathsf{ct}^\star], \sigma^\star) = \top$ by $\mathsf{GS.Vrfy}(\mathsf{gpk}, \mathsf{M}^\star, \Sigma^\star) = \top$. We then prove that B has never made prohibited corrupt queries. Namely, we show $C[\mathsf{ovk}^\star, \mathsf{ct}^\star](i \| \mathsf{K}_i) = 0$ for all $i \in [N] \backslash \{i^\star\}$. This follows since we have $|\{i \in [N] : \mathsf{SKE.Dec}(\mathsf{K}_i, \mathsf{ct}^\star) \neq \bot\}| = 1$ and $\mathsf{SKE.Dec}(\mathsf{K}_{i^\star}, \mathsf{ct}^\star) \neq \bot$, where the latter follows from $\mathsf{GS.Open}(\mathsf{gpk}, \mathsf{gok}, \mathsf{M}^\star, \Sigma^\star) = i^\star$. Finally, we show that B has never made prohibited signing queries. Recall that B has only made signing queries of the form $(\mathsf{M}, C[\mathsf{ovk}, \mathsf{ct}], i^\star)$ and all such queries are made in order to answer the signing query $(i^\star, \mathsf{M})$ made by A. Because A has won the game, we have $\mathsf{M}^\star \neq \mathsf{M}$, which implies $(\mathsf{M}^\star, C[\mathsf{ovk}^\star, \mathsf{ct}^\star]) \neq (\mathsf{M}, C[\mathsf{ovk}, \mathsf{ct}])$ as desired. Since B simulates Game 3 perfectly, we have that the winning probability of B is exactly $\Pr[\mathsf{E}_3]$. This concludes the proof of the lemma.

5 Construction of Indexed ABS from Lattices

In this section, we give a new construction of indexed ABS scheme from the SIS assumption. Combined with an appropriate SKE scheme and OTS scheme, we can instantiate the generic construction of GS in Sect. 4 to obtain the first lattice-based GS scheme in the standard model. We refer Sect. 7 to more discussions.

5.1 Preliminaries on Lattices

Here, we recall some facts on lattices that are needed for the exposition of our construction. Throughout this section, n, m, and q are integers such that $n = \mathrm{poly}(\kappa)$ and $m \geq n\lceil \log q \rceil$. In the following, let $\mathsf{SampZ}(\gamma)$ be a sampling algorithm for the truncated discrete Gaussian distribution over $\mathbb{Z}$ with parameter $\gamma > 0$ whose support is restricted to $z \in \mathbb{Z}$ such that $|z| \leq \sqrt{n}\gamma$.[6]

Definition 13 (The SIS Assumption). *Let n, m, q, β be integer parameters. We say that the $\mathsf{SIS}(n, m, q, \beta)$ hardness assumption holds if for any PPT adversaries A we have*

$$\Pr[\mathbf{A} \cdot \mathbf{z} = \mathbf{0} \ \wedge \ 0 < \|\mathbf{z}\|_\infty \leq \beta(\kappa) : \mathbf{A} \xleftarrow{\$} \mathbb{Z}_{q(\kappa)}^{n(\kappa) \times m(\kappa)}, \mathbf{z} \leftarrow \mathsf{A}(1^\kappa, \mathbf{A})] \leq \mathrm{negl}(\kappa).$$

We also say that the $\mathsf{SIS}(n, m, q, \beta)$ problem is subexponentially hard if the above probability is bounded by $2^{-O(n^\epsilon)} \cdot \mathrm{negl}(\kappa)$ for some constant $0 < \epsilon < 1$.

[6] During construction, we fix n and consider this very weak bound for one-dimensional discrete Gaussian samples for simplicity of analysis.

For any $n = \text{poly}(\kappa)$, any $m = \text{poly}(n)$, any $\beta(n) > 0$, and $q \geq \beta\sqrt{n} \cdot \omega(\log n)$, it is known that the $\mathsf{SIS}(n, m, q, \beta)$ problem is as hard as certain worst case lattice problems with approximation factor $\beta(n) \cdot \text{poly}(n)$. We abuse the term and refer to $\mathsf{SIS}(n, m, q, \beta)$ with $\beta \leq \text{poly}(\kappa)$ as the SIS problem with polynomial approximation factor.

Trapdoors. Let $\mathbf{A} \in \mathbb{Z}_q^{n\times m}$. For all $\mathbf{V} \in \mathbb{Z}_q^{n\times m'}$, we let $\mathbf{A}_\gamma^{-1}(\mathbf{V})$ be an output distribution of $\mathsf{SampZ}(\gamma)^{m\times m'}$ conditioned on $\mathbf{A} \cdot \mathbf{A}_\gamma^{-1}(\mathbf{V}) = \mathbf{V}$. A γ-trapdoor for $\mathbf{A}$ is a trapdoor that enables one to sample from the distribution $\mathbf{A}_\gamma^{-1}(\mathbf{V})$ in time $\text{poly}(n, m, m', \log q)$, for any $\mathbf{V}$. We slightly overload notation and denote a γ-trapdoor for $\mathbf{A}$ by $\mathbf{A}_\gamma^{-1}$. We also define the special gadget matrix $\mathbf{G} \in \mathbb{Z}_q^{n\times m}$ as the matrix obtained by padding $\mathbf{I}_n \otimes (1, 2, 4, 8, \ldots, 2^{\lceil \log q\rceil})$ with zero-columns. The following properties had been established in a long sequence of works [GPV08, CHKP10, ABB10a, ABB10b, MP12, BLP+13].

Lemma 14 (Properties of Trapdoors). *Lattice trapdoors exhibit the following properties.*

1. *Given $\mathbf{A}_\gamma^{-1}$, one can obtain $\mathbf{A}_{\gamma'}^{-1}$ for any $\gamma' \geq \gamma$.*
2. *Given $\mathbf{A}_\gamma^{-1}$, one can obtain $[\mathbf{A}\|\mathbf{B}]_\gamma^{-1}$ and $[\mathbf{B}\|\mathbf{A}]_\gamma^{-1}$ for any $\mathbf{B}$.*
3. *For all $\mathbf{A} \in \mathbb{Z}_q^{n\times m}$ and $\mathbf{R} \in \mathbb{Z}^{m\times m}$, one can obtain $[\mathbf{AR} + \mathbf{G}\|\mathbf{A}]_\gamma^{-1}$ for $\gamma = m \cdot \|\mathbf{R}\|_\infty \cdot \omega(\sqrt{\log m})$.*
4. *There exists an efficient procedure $\mathsf{TrapGen}(1^n, 1^m, q)$ that outputs $(\mathbf{A}, \mathbf{A}_{\gamma_0}^{-1})$ where $\mathbf{A} \in \mathbb{Z}_q^{n\times m}$ for some $m = O(n \log q)$ and is 2^{-n}-close to uniform, where $\gamma_0 = \omega(\sqrt{n \log q \log m})$.*

Lemma 15 (Fully Homomorphic Computation [GV15]**).** *There exists a pair of deterministic algorithms* $(\mathsf{PubEval}, \mathsf{TrapEval})$ *with the following properties.*

- $\mathsf{PubEval}(\vec{\mathbf{B}}, F) \to \mathbf{B}_F$. *Here,* $\vec{\mathbf{B}} = [\mathbf{B}_1\|\cdots\|\mathbf{B}_k] \in (\mathbb{Z}_q^{n\times m})^k$ *and* $F : \{0,1\}^k \to \{0,1\}$ *is a circuit.*
- $\mathsf{TrapEval}(\vec{\mathbf{R}}, F, x) \to \mathbf{R}_{F,x}$. *Here,* $\vec{\mathbf{R}} = [\mathbf{R}_1\|\ldots\|\mathbf{R}_k] \in (\mathbb{Z}_q^{n\times m})^k$, $\|\mathbf{R}_i\|_\infty \leq \delta$ *for* $i \in [k]$, $x \in \{0,1\}^k$, *and* $F : \{0,1\}^k \to \{0,1\}$ *is a circuit with depth* d. *We have* $\mathsf{PubEval}(\mathbf{A}\vec{\mathbf{R}} + x \otimes \mathbf{G}) = \mathbf{A}\mathbf{R}_{F,x} + F(x)\mathbf{G}$ *where we denote* $[x_1\mathbf{G}\|\cdots\|x_k\mathbf{G}]$ *by* $x \otimes \mathbf{G}$. *Furthermore, we have* $\|\mathbf{R}_{F,x}\|_\infty \leq \delta \cdot m \cdot 2^{O(d)}$.
- *The running time of* $(\mathsf{PubEval}, \mathsf{TrapEval})$ *is bounded by* $\text{poly}(k, n, m, 2^d, \log q)$.

The above algorithms are taken from [GV15], which is a variant of a similar algorithms proposed by Boneh et al. [BGG+14]. The algorithms in [BGG+14] work for any polynomial-sized circuit F, but $\|\mathbf{R}_{F,x}\|_\infty$ becomes super-polynomial even if the depth of the circuit is shallow (i.e., logarithmic depth). On the other hand, the above algorithm runs in polynomial time only when F is of logarithmic depth, but $\|\mathbf{R}_{F,x}\|_\infty$ can be polynomially bounded. The latter property is useful since our main focus is on the constructions of GS schemes from the SIS assumption with polynomial approximation factors.

5.2 Construction

Here, we show our construction of indexed ABS. The scheme satisfies no-signing-query unforgeability. By applying the conversion in Sect. 3.2 to the scheme, we can obtain a scheme with co-selective unforgeability. Note that the signing and the verification algorithm below ignore the input message M. This is not a problem because the no-signing-query security does not require non-malleability with respect to the message.

We denote the circuit class that is dealt with by the scheme by $\{\mathcal{F}_\kappa\}_\kappa$, where $\mathcal{F}_\kappa$ is a set of circuits F such that $F : \{0,1\}^{k(\kappa)} \to \{0,1\}$ and with depth at most $d_\mathcal{F} = O(\log \kappa)$.

$\mathsf{ABS.Setup}(1^\kappa, 1^N)$: On input 1^κ and 1^N, it sets the parameters n, m, q, γ_0, γ, and β as specified later in this section, where q is a prime number. Then, it picks random matrices $\mathbf{B}_j^{(i)} \xleftarrow{\$} \mathbb{Z}_q^{n\times m}$ for $i \in [N], j \in [k]$. We denote $\vec{\mathbf{B}}^{(i)} = [\mathbf{B}_1^{(i)} \| \cdots \| \mathbf{B}_k^{(i)}]$. It also picks $(\mathbf{A}, \mathbf{A}_{\gamma_0}^{-1}) \xleftarrow{\$} \mathsf{TrapGen}(1^n, 1^m, q)$ such that $\mathbf{A} \in \mathbb{Z}_q^{n\times m}$ and a random vector $\mathbf{r} \xleftarrow{\$} \{0,1\}^m$. It then computes $\mathbf{u} := \mathbf{A}\mathbf{r} \in \mathbb{Z}_q^n$. It finally outputs

$$\mathsf{mpk} = \Big(\mathbf{A},\ \{\vec{\mathbf{B}}^{(i)}\}_{i\in[N]},\ \mathbf{u}, \Big) \quad \text{and} \quad \mathsf{msk} = \Big(\mathbf{A}_{\gamma_0}^{-1},\ \{\vec{\mathbf{B}}^{(i)}\}_{i\in[N]} \Big).$$

$\mathsf{ABS.KeyGen}(\mathsf{msk}, i, x)$: On input $\mathsf{msk} = (\mathbf{A}_{\gamma_0}^{-1}, \{\vec{\mathbf{B}}^{(i)}\}_{i\in[N]})$, $i \in [N]$, and $x \in \{0,1\}^k$, it samples $\vec{\mathbf{R}}^{(i)} \xleftarrow{\$} \mathbf{A}_{\gamma_0}^{-1}(\vec{\mathbf{B}}^{(i)} - x\otimes\mathbf{G})$ where $\vec{\mathbf{R}}^{(i)} \in \mathbb{Z}^{m\times mk}$ using $\mathbf{A}_{\gamma_0}^{-1}$. Note that $\vec{\mathbf{B}}^{(i)} = \mathbf{A}\vec{\mathbf{R}}^{(i)} + x\otimes\mathbf{G}$ and $\|\vec{\mathbf{R}}^{(i)}\|_\infty \le \gamma_0\sqrt{n}$ holds by the definition of the distribution $\mathbf{A}_{\gamma_0}^{-1}(\vec{\mathbf{B}}^{(i)} - x\otimes\mathbf{G})$. It then outputs $\mathsf{sk}_x := (i, \vec{\mathbf{R}}^{(i)})$.

$\mathsf{ABS.Sign}(\mathsf{mpk}, \mathsf{sk}_x, \mathsf{M}, F)$: It outputs $\perp$ if $\mathsf{M} \notin \mathcal{M}_\kappa$, $F \notin \mathcal{F}$, or $F(x) = 0$. Otherwise, it first parses $\mathsf{sk}_x \to (i, \vec{\mathbf{R}}^{(i)})$. It then computes $\mathbf{B}_F^{(i)} := \mathsf{PubEval}(\vec{\mathbf{B}}^{(i)}, F)$ and $\mathbf{R}_{F,x}^{(i)} := \mathsf{TrapEval}(\vec{\mathbf{R}}^{(i)}, F, x)$ such that $\|\mathbf{R}_{F,x}^{(i)}\|_\infty \le \gamma$. By Lemma 15 and since $F(x) = 1$, we have $\mathbf{B}_F^{(i)} = \mathbf{A}\mathbf{R}_{F,x}^{(i)} + \mathbf{G}$. It then computes $[\mathbf{A}\|\mathbf{B}_F^{(i)}]_\beta^{-1}$ from $\mathbf{R}_{F,x}^{(i)}$ (see Item 3 in Lemma 14) and further computes $\Big[\mathbf{A}\|\mathbf{B}_F^{(1)}\| \cdots \|\mathbf{B}_F^{(N)}\Big]_\beta^{-1}$ from $[\mathbf{A}\|\mathbf{B}_F^{(i)}]_\beta^{-1}$ (see Item 2 in Lemma 14). Finally, it samples $\mathbf{e} \xleftarrow{\$} [\mathbf{A}\|\mathbf{B}_F^{(1)}\| \cdots \|\mathbf{B}_F^{(N)}]_\beta^{-1}(\mathbf{u})$ and outputs the signature $\sigma := \mathbf{e} \in \mathbb{Z}^{m(N+1)}$.

$\mathsf{ABS.Vrfy}(\mathsf{mpk}, \mathsf{M}, \sigma, F)$: It outputs $\perp$ if $F \notin \mathcal{F}$ or $\sigma = \mathbf{e} \notin \mathbb{Z}^{m(N+1)}$. Otherwise, it first computes $\mathbf{B}_F^{(i)} = \mathsf{PubEval}(F, \vec{\mathbf{B}}^{(i)})$ for $i \in [N]$. It then checks whether $\|\mathbf{e}\|_\infty \le \sqrt{n}\beta$ and $\Big[\mathbf{A}\|\mathbf{B}_F^{(1)}\| \cdots \|\mathbf{B}_F^{(N)}\Big]\mathbf{e} = \mathbf{u}$. If they hold, it outputs $\top$ and otherwise $\perp$.

Correctness. The correctness of the scheme can be seen by observing that the verification equation and $\|\mathbf{e}\|_\infty \le \sqrt{n}\beta$ follow from the definition of the distribution $[\mathbf{A}\|\mathbf{B}_F^{(1)}\| \cdots \|\mathbf{B}_F^{(N)}]_\beta^{-1}(\mathbf{u})$ from which $\mathbf{e}$ is sampled.

Parameter Selection. As long as the maximum depth of the circuit class $\mathcal{F}_\kappa$ is bounded by $O(\log \kappa)$, we can set all of n, m, γ_0, γ, β, and q to be polynomial in κ. Notably, this allows us to reduce the security of the scheme to $\mathsf{SIS}(n, m, q, \beta_{\mathsf{SIS}})$ with $\beta_{\mathsf{SIS}} = \mathrm{poly}(\kappa)$. We refer to the full version for the precise requirements for these parameters and a concrete selection.

5.3 Security Proofs

Theorem 16. *Our ABS scheme is perfectly private.*

Proof. It can be seen that the signature $\sigma = \mathbf{e}$ for (F, M) is chosen from the distribution $[\mathbf{A}\|\mathbf{B}_F^{(1)}\|\cdots\|\mathbf{B}_F^{(N)}]_\beta^{-1}(\mathbf{u})$, which only depends on mpk and F. The theorem readily follows.

Theorem 17. *Our ABS scheme satisfies no-signing-query unforgeability assuming* $\mathsf{SIS}(n, m, q, \beta_{\mathsf{SIS}})$ *is hard.*

The proof will appear in the full version.

6 Instantiating SKE

Here, we discuss how to instantiate the SKE required for the generic construction of GS in Sect. 4. Since this can be done by a combination of known results and standard techniques, we only give a high level overview here and refer to the full version for the details. We require the SKE to be INDr-CCA secure and to have key robustness and a decryption circuit with $O(\log \kappa)$-depth. The requirement for the depth of the circuit is needed to combine it with our indexed ABS scheme in Sect. 5.2, which can only deal with circuits with logarithmic depth.

To obtain such a scheme, we follow the MAC-then-Encrypt paradigm and show a generic construction of such an SKE from another SKE and a MAC. For the latter SKE, we require INDr-CPA security, key robustness, and a decryption circuit with $O(\log \kappa)$-depth. For the MAC, we require strong unforgeability and a verification circuit with $O(\log \kappa)$-depth. Although an insecure example of the MAC-then-Encrypt approach is known [BN00], we avoid the pitfall by authenticating a part of the ciphertext in addition to the plaintext using the MAC. We also note that the Encrypt-then-MAC approach may not work in our setting, because the MAC part may reveal the information about the user and destroy the INDr-CCA security (in particular, anonymity) of the resulting SKE scheme.

It remains to show how to instantiate the inner SKE and MAC. For the SKE, we use a secret key variant of the Regev encryption scheme [Reg05], where we pad the message with zeroes before encrypting it and the decryption algorithm returns $\perp$ to a ciphertext that does not conform to this format. The padding makes the ciphertext somewhat redundant, and due to this redundancy, we can prove key robustness of the scheme by a standard counting argument. The INDr-CPA security of the scheme is proven from the LWE assumption by a straightforward reduction. The decryption circuit of the scheme can be implemented by an $O(\log \kappa)$-depth circuit, since the decryption algorithm only involves basic algebraic operations such as the computation of an inner-product, modulo reduction, and comparison, all of which are known to be in $\mathbf{NC}^1$. We then discuss how to instantiate the MAC. We need the MAC scheme to have strong unforgeability and a decryption circuit with $O(\log \kappa)$-depth. To obtain such a scheme, we downgrade the (public key) signature scheme proposed by Micciancio and Peikert [MP12] to a MAC scheme. Since the scheme satisfies strong unforgeability as a signature scheme, it is trivial to see that the scheme is strongly unforgeable as a MAC as well. The verification circuit of the scheme can be implemented by an $O(\log \kappa)$-depth circuit, since the verification algorithm only involves basic algebraic operations, similarly to the decryption algorithm of the above SKE.

We finally remark that another way of obtaining the SKE required for the generic construction in Sect. 4 may be to downgrade the CCA-secure public key encryption scheme by Micciancio and Peikert [MP12] to an SKE scheme. However, this approach requires the LWE assumption with larger approximation factor than our approach described above.

7 New Group Signature Constructions

By combining all the results in the previous sections, we obtain the first lattice-based group signatures in the standard model. We show two instantiations, which provide tradeoffs between the security assumption and efficiency. The first instantiation leads to a scheme that is proven secure under the SIS and LWE assumption with polynomial approximation factors, but has long group public key and signatures that are linear in the number of users N. The second instantiation is more efficient and these parameters do not depend on N. However, in order to prove security, we have to assume the subexponential hardness of the SIS problem (with polynomial approximation factors).

First Instantiation. The generic construction of GS schemes in Sect. 4 requires an OTS scheme, an SKE scheme, and an indexed ABS scheme. We instantiate the OTS by the scheme proposed by Mohassel [Moh11], which is strongly unforgeable under the SIS assumption with polynomial approximation factors. We instantiate the SKE by the scheme that is sketched in Sect. 6 (and described in full details in the full version. The scheme satisfies INDr-CCA security under the LWE assumption with polynomial approximation factors, key-robustness, and can have arbitrarily large message space, which are the required properties for the generic construction. Furthermore, the maximum depth of the decryption circuit of the SKE, which is denoted by d_{Dec} hereafter, is $O(\log \kappa)$. We now consider how to instantiate the indexed ABS scheme. In addition to the perfect privacy and co-selective unforgeability, we require the indexed ABS to be capable of dealing with the circuit class $\mathcal{C}_\kappa$ defined in Eq. (2). It is easy to see that we can bound the maximum depth $d_\mathcal{C}$ of circuits in $\mathcal{C}_\kappa$ by $d_\mathcal{C} = O(\log N + \log \ell_1 + d_{\mathsf{Dec}}) = O(\log \kappa)$. To obtain such an indexed ABS scheme, we apply the conversion in Sect. 3.2 to our indexed ABS scheme in Sect. 5.2, whose no-signing-query unforgeability is shown under the SIS assumption with polynomial approximation factors.

Note that the conversion requires a collision resistant hash, which is known to be implied by the same SIS assumption [MR04]. In order to make sure that the ABS scheme obtained through this conversion can deal with the circuit class $\mathcal{C}_\kappa$, we require the original indexed ABS to be capable of dealing with a circuit class $\mathcal{F}_\kappa$ defined in Eq. (1). It is easy to see that the function WldCmp can be implemented by an $O(\log \ell)$-depth circuit and thus we can bound the maximum depth $d_{\mathcal{F}}$ of the circuit class $\mathcal{F}_\kappa$ by $d_{\mathcal{F}} = d_{\mathcal{C}} + O(\log \ell) = O(\log \kappa)$. Since $d_{\mathcal{F}} = O(\log \kappa)$, we can instantiate the latter indexed ABS by the construction in Sect. 5.2. Summing up the above discussion, we have the following theorem:

Theorem 18 (Theorem 1 restated). *Under the hardness of the SIS and LWE with polynomial approximation factors, we have a group signature scheme with CCA-selfless anonymity and full traceability in the standard model whose sizes of the public parameters and signatures are linear in the number of users N.*

Second Instantiation. Here, we show another way of instantiating our generic construction in Sect. 4. We use the same SKE as the first instantiation above, but we instantiate the indexed ABS scheme with the scheme proposed by Tsabary [Tsa17]. To do so, we first state the following theorem.

Theorem 19 (Adapted from Sect. 6 of [Tsa17]**).** *There is an indexed ABS scheme for the circuit class $\mathcal{C}_\kappa$ defined in Eq.* (2) *with perfect privacy and co-selective unforgeability whose master public key and signature sizes are bounded by* poly(κ)*, i.e., independent of the number of users N, assuming the subexponential hardness of the SIS problem with polynomial approximation factors.*

The above theorem is obtained by the result by [Tsa17], but some adaptations are required. We refer to the full version for discussions.

We then combine the ABS scheme given by Theorem 19 with the SKE scheme used in the first instantiation. We then obtain the following theorem.

Theorem 20 (Theorem 2 restated). *Under the hardness of the LWE problem and the subexponential hardness of the SIS problem with polynomial approximation factors, there exists a group signature scheme with full-traceability and CCA-selfless anonymity whose sizes of the public parameters and signatures are* poly(κ)*, i.e., independent of the number of users N.*

Acknowledgement. The authors would like to thank Yusuke Sakai and Ai Ishida for helpful discussions and anonymous reviewers of Eurocrypt 2019 for their valuable comments. The first author was partially supported by JST CREST Grant Number-JPMJCR1302 and JSPS KAKENHI Grant Number 17J05603. The second author was supportedby JST CREST Grant No. JPMJCR1688 and JSPS KAKENHI Grant Number 16K16068.

References

[ABB10a] Agrawal, S., Boneh, D., Boyen, X.: Efficient lattice (H)IBE in the standard model. In: Gilbert, H. (ed.) EUROCRYPT 2010. LNCS, vol. 6110, pp. 553–572. Springer, Heidelberg (2010). https://doi.org/10.1007/978-3-642-13190-5_28

[ABB10b] Agrawal, S., Boneh, D., Boyen, X.: Lattice basis delegation in fixed dimension and shorter-ciphertext hierarchical IBE. In: Rabin, T. (ed.) CRYPTO 2010. LNCS, vol. 6223, pp. 98–115. Springer, Heidelberg (2010). https://doi.org/10.1007/978-3-642-14623-7_6

[ACJT00] Ateniese, G., Camenisch, J., Joye, M., Tsudik, G.: A practical and provably secure coalition-resistant group signature scheme. In: Bellare, M. (ed.) CRYPTO 2000. LNCS, vol. 1880, pp. 255–270. Springer, Heidelberg (2000). https://doi.org/10.1007/3-540-44598-6_16

[BB04a] Boneh, D., Boyen, X.: Secure identity based encryption without random oracles. In: Franklin, M. (ed.) CRYPTO 2004. LNCS, vol. 3152, pp. 443–459. Springer, Heidelberg (2004). https://doi.org/10.1007/978-3-540-28628-8_27

[BB04b] Boneh, D., Boyen, X.: Efficient selective-ID secure identity-based encryption without random oracles. In: Cachin, C., Camenisch, J.L. (eds.) EUROCRYPT 2004. LNCS, vol. 3027, pp. 223–238. Springer, Heidelberg (2004). https://doi.org/10.1007/978-3-540-24676-3_14

[BBS04] Boneh, D., Boyen, X., Shacham, H.: Short group signatures. In: Franklin, M. (ed.) CRYPTO 2004. LNCS, vol. 3152, pp. 41–55. Springer, Heidelberg (2004). https://doi.org/10.1007/978-3-540-28628-8_3

[Boy10] Boyen, X.: Lattice mixing and vanishing trapdoors: a framework for fully secure short signatures and more. In: Nguyen, P.Q., Pointcheval, D. (eds.) PKC 2010. LNCS, vol. 6056, pp. 499–517. Springer, Heidelberg (2010). https://doi.org/10.1007/978-3-642-13013-7_29

[BCC+16] Bootle, J., Cerulli, A., Chaidos, P., Ghadafi, E., Groth, J.: Foundations of fully dynamic group signatures. In: Manulis, M., Sadeghi, A.-R., Schneider, S. (eds.) ACNS 2016. LNCS, vol. 9696, pp. 117–136. Springer, Cham (2016). https://doi.org/10.1007/978-3-319-39555-5_7

[BCH86] Beame, P., Cook, S.A., Hoover, H.J.: Log depth circuits for division and related problems. SIAM J. Comput. **15**(4), 994–1003 (1986)

[BF14] Bellare, M., Fuchsbauer, G.: Policy-based signatures. In: Krawczyk, H. (ed.) PKC 2014. LNCS, vol. 8383, pp. 520–537. Springer, Heidelberg (2014). https://doi.org/10.1007/978-3-642-54631-0_30

[BGG+14] Boneh, D., et al.: Fully key-homomorphic encryption, arithmetic circuit ABE and compact garbled circuits. In: Nguyen, P.Q., Oswald, E. (eds.) EUROCRYPT 2014. LNCS, vol. 8441, pp. 533–556. Springer, Heidelberg (2014). https://doi.org/10.1007/978-3-642-55220-5_30

[BLP+13] Brakerski, Z., Langlois, A., Peikert, C., Regev, O., Stehlé, D.: Classical hardness of learning with errors. In: STOC, pp. 575–584 (2013)

[BMW03] Bellare, M., Micciancio, D., Warinschi, B.: Foundations of group signatures: formal definitions, simplified requirements, and a construction based on general assumptions. In: Biham, E. (ed.) EUROCRYPT 2003. LNCS, vol. 2656, pp. 614–629. Springer, Heidelberg (2003). https://doi.org/10.1007/3-540-39200-9_38

[BN00] Bellare, M., Namprempre, C.: Authenticated encryption: relations among notions and analysis of the generic composition paradigm. In: Okamoto, T. (ed.) ASIACRYPT 2000. LNCS, vol. 1976, pp. 531–545. Springer, Heidelberg (2000). https://doi.org/10.1007/3-540-44448-3_41

[BS04] Boneh, D., Shacham, H.: Group signatures with verifier-local revocation. In: ACM CCS, pp. 168–177 (2004)

[BSZ05] Bellare, M., Shi, H., Zhang, C.: Foundations of group signatures: the case of dynamic groups. In: Menezes, A. (ed.) CT-RSA 2005. LNCS, vol. 3376, pp. 136–153. Springer, Heidelberg (2005). https://doi.org/10.1007/978-3-540-30574-3_11

[BV15] Brakerski, Z., Vaikuntanathan, V.: Constrained key-homomorphic PRFs from standard lattice assumptions - or: how to secretly embed a circuit in your PRF. In: Dodis, Y., Nielsen, J.B. (eds.) TCC 2015, Part II. LNCS, vol. 9015, pp. 1–30. Springer, Heidelberg (2015). https://doi.org/10.1007/978-3-662-46497-7_1

[BW06] Boyen, X., Waters, B.: Compact group signatures without random oracles. In: Vaudenay, S. (ed.) EUROCRYPT 2006. LNCS, vol. 4004, pp. 427–444. Springer, Heidelberg (2006). https://doi.org/10.1007/11761679_26

[BW07] Boyen, X., Waters, B.: Full-domain subgroup hiding and constant-size group signatures. In: Okamoto, T., Wang, X. (eds.) PKC 2007. LNCS, vol. 4450, pp. 1–15. Springer, Heidelberg (2007). https://doi.org/10.1007/978-3-540-71677-8_1

[BY93] Bellare, M., Yung, M.: Certifying cryptographic tools: the case of trapdoor permutations. In: Brickell, E.F. (ed.) CRYPTO 1992. LNCS, vol. 740, pp. 442–460. Springer, Heidelberg (1993). https://doi.org/10.1007/3-540-48071-4_31

[CG05] Camenisch, J., Groth, J.: Group signatures: better efficiency and new theoretical aspects. In: Blundo, C., Cimato, S. (eds.) SCN 2004. LNCS, vol. 3352, pp. 120–133. Springer, Heidelberg (2005). https://doi.org/10.1007/978-3-540-30598-9_9

[CH85] Cook, S.A., Hoover, H.J.: A depth-universal circuit. SIAM J. Comput. **14**(4), 833–839 (1985)

[CHKP10] Cash, D., Hofheinz, D., Kiltz, E., Peikert, C.: Bonsai trees, or how to delegate a lattice basis. In: Gilbert, H. (ed.) EUROCRYPT 2010. LNCS, vol. 6110, pp. 523–552. Springer, Heidelberg (2010). https://doi.org/10.1007/978-3-642-13190-5_27

[CL02] Camenisch, J., Lysyanskaya, A.: Dynamic accumulators and application to efficient revocation of anonymous credentials. In: Yung, M. (ed.) CRYPTO 2002. LNCS, vol. 2442, pp. 61–76. Springer, Heidelberg (2002). https://doi.org/10.1007/3-540-45708-9_5

[CL04] Camenisch, J., Lysyanskaya, A.: Signature schemes and anonymous credentials from bilinear maps. In: Franklin, M. (ed.) CRYPTO 2004. LNCS, vol. 3152, pp. 56–72. Springer, Heidelberg (2004). https://doi.org/10.1007/978-3-540-28628-8_4

[Cv91] Chaum, D., van Heyst, E.: Group signatures. In: Davies, D.W. (ed.) EUROCRYPT 1991. LNCS, vol. 547, pp. 257–265. Springer, Heidelberg (1991). https://doi.org/10.1007/3-540-46416-6_22

[DFN06] Damgård, I., Fazio, N., Nicolosi, A.: Non-interactive zero-knowledge from homomorphic encryption. In: Halevi, S., Rabin, T. (eds.) TCC 2006. LNCS, vol. 3876, pp. 41–59. Springer, Heidelberg (2006). https://doi.org/10.1007/11681878_3

[DMP88] De Santis, A., Micali, S., Persiano, G.: Non-interactive zero-knowledge proof systems. In: Pomerance, C. (ed.) CRYPTO 1987. LNCS, vol. 293, pp. 52–72. Springer, Heidelberg (1988). https://doi.org/10.1007/3-540-48184-2_5

[DRS04] Dodis, Y., Reyzin, L., Smith, A.: Fuzzy extractors: how to generate strong keys from biometrics and other noisy data. In: Cachin, C., Camenisch, J.L. (eds.) EUROCRYPT 2004. LNCS, vol. 3027, pp. 523–540. Springer, Heidelberg (2004). https://doi.org/10.1007/978-3-540-24676-3_31

[FHPS13] Freire, E.S.V., Hofheinz, D., Paterson, K.G., Striecks, C.: Programmable hash functions in the multilinear setting. In: Canetti, R., Garay, J.A. (eds.) CRYPTO 2013, Part I. LNCS, vol. 8042, pp. 513–530. Springer, Heidelberg (2013). https://doi.org/10.1007/978-3-642-40041-4_28

[FLS90] Feige, U., Lapidot, D., Shamir, A.: Multiple non-interactive zero knowledge proofs based on a single random string (extended abstract). In: FOCS, pp. 308–317 (1990)

[FS87] Fiat, A., Shamir, A.: How to prove yourself: practical solutions to identification and signature problems. In: Odlyzko, A.M. (ed.) CRYPTO 1986. LNCS, vol. 263, pp. 186–194. Springer, Heidelberg (1987). https://doi.org/10.1007/3-540-47721-7_12

[GKV10] Gordon, S.D., Katz, J., Vaikuntanathan, V.: A group signature scheme from lattice assumptions. In: Abe, M. (ed.) ASIACRYPT 2010. LNCS, vol. 6477, pp. 395–412. Springer, Heidelberg (2010). https://doi.org/10.1007/978-3-642-17373-8_23

[Gol04] Goldreich, O.: Foundations of Cryptography: Volume 2, Basic Applications. Cambridge University Press, Cambridge (2004)

[Gol08] Goldreich, O.: Computational Complexity - A Conceptual Perspective. Cambridge University Press, Cambridge (2008)

[GOS06] Groth, J., Ostrovsky, R., Sahai, A.: Perfect non-interactive zero knowledge for NP. In: Vaudenay, S. (ed.) EUROCRYPT 2006. LNCS, vol. 4004, pp. 339–358. Springer, Heidelberg (2006). https://doi.org/10.1007/11761679_21

[GPV08] Gentry, C., Peikert, C., Vaikuntanathan, V.: Trapdoors for hard lattices and new cryptographic constructions. In: STOC, pp. 197–206 (2008)

[Gro07] Groth, J.: Fully anonymous group signatures without random oracles. In: Kurosawa, K. (ed.) ASIACRYPT 2007. LNCS, vol. 4833, pp. 164–180. Springer, Heidelberg (2007). https://doi.org/10.1007/978-3-540-76900-2_10

[GS08] Groth, J., Sahai, A.: Efficient non-interactive proof systems for bilinear groups. In: Smart, N. (ed.) EUROCRYPT 2008. LNCS, vol. 4965, pp. 415–432. Springer, Heidelberg (2008). https://doi.org/10.1007/978-3-540-78967-3_24

[GV15] Gorbunov, S., Vinayagamurthy, D.: Riding on asymmetry: efficient ABE for branching programs. In: Iwata, T., Cheon, J.H. (eds.) ASIACRYPT 2015, Part I. LNCS, vol. 9452, pp. 550–574. Springer, Heidelberg (2015). https://doi.org/10.1007/978-3-662-48797-6_23

[GVW15] Gorbunov, S., Vaikuntanathan, V., Wichs, D.: Leveled fully homomorphic signatures from standard lattices. In: STOC, pp. 469–477 (2015)

[HLLR12] Herranz, J., Laguillaumie, F., Libert, B., Ràfols, C.: Short attribute-based signatures for threshold predicates. In: Dunkelman, O. (ed.) CT-RSA 2012. LNCS, vol. 7178, pp. 51–67. Springer, Heidelberg (2012). https://doi.org/10.1007/978-3-642-27954-6_4

[KW18] Kim, S., Wu, D.J.: Multi-theorem preprocessing NIZKs from lattices. In: Shacham, H., Boldyreva, A. (eds.) CRYPTO 2018, Part II. LNCS, vol. 10992, pp. 733–765. Springer, Cham (2018). https://doi.org/10.1007/978-3-319-96881-0_25

[KY06] Kiayias, A., Yung, M.: Secure scalable group signature with dynamic joins and separable authorities. IJSN **1**(1/2), 24–45 (2006)

[LLLS13] Laguillaumie, F., Langlois, A., Libert, B., Stehlé, D.: Lattice-based group signatures with logarithmic signature size. In: Sako, K., Sarkar, P. (eds.) ASIACRYPT 2013, Part II. LNCS, vol. 8270, pp. 41–61. Springer, Heidelberg (2013). https://doi.org/10.1007/978-3-642-42045-0_3

[LLM+16a] Libert, B., Ling, S., Mouhartem, F., Nguyen, K., Wang, H.: Signature schemes with efficient protocols and dynamic group signatures from lattice assumptions. In: Cheon, J.H., Takagi, T. (eds.) ASIACRYPT 2016, Part II. LNCS, vol. 10032, pp. 373–403. Springer, Heidelberg (2016). https://doi.org/10.1007/978-3-662-53890-6_13

[LLM+16b] Libert, B., Ling, S., Mouhartem, F., Nguyen, K., Wang, H.: Zero-knowledge arguments for matrix-vector relations and lattice-based group encryption. In: Cheon, J.H., Takagi, T. (eds.) ASIACRYPT 2016. LNCS, vol. 10032, pp. 101–131. Springer, Heidelberg (2016). https://doi.org/10.1007/978-3-662-53890-6_4

[LLNW14] Langlois, A., Ling, S., Nguyen, K., Wang, H.: Lattice-based group signature scheme with verifier-local revocation. In: Krawczyk, H. (ed.) PKC 2014. LNCS, vol. 8383, pp. 345–361. Springer, Heidelberg (2014). https://doi.org/10.1007/978-3-642-54631-0_20

[LLNW16] Libert, B., Ling, S., Nguyen, K., Wang, H.: Zero-knowledge arguments for lattice-based accumulators: logarithmic-size ring signatures and group signatures without trapdoors. In: Fischlin, M., Coron, J.-S. (eds.) EUROCRYPT 2016, Part II. LNCS, vol. 9666, pp. 1–31. Springer, Heidelberg (2016). https://doi.org/10.1007/978-3-662-49896-5_1

[LNW15] Ling, S., Nguyen, K., Wang, H.: Group signatures from lattices: simpler, tighter, shorter, ring-based. In: Katz, J. (ed.) PKC 2015. LNCS, vol. 9020, pp. 427–449. Springer, Heidelberg (2015). https://doi.org/10.1007/978-3-662-46447-2_19

[LNWX17] Ling, S., Nguyen, K., Wang, H., Xu, Y.: Lattice-based group signatures: achieving full dynamicity with ease. In: Gollmann, D., Miyaji, A., Kikuchi, H. (eds.) ACNS 2017. LNCS, vol. 10355, pp. 293–312. Springer, Cham (2017). https://doi.org/10.1007/978-3-319-61204-1_15

[LNWX18] Ling, S., Nguyen, K., Wang, H., Xu, Y.: Constant-size group signatures from lattices. In: Abdalla, M., Dahab, R. (eds.) PKC 2018, Part II. LNCS, vol. 10770, pp. 58–88. Springer, Cham (2018). https://doi.org/10.1007/978-3-319-76581-5_3

[Lys02] Lysyanskaya, A.: Unique signatures and verifiable random functions from the DH-DDH separation. In: Yung, M. (ed.) CRYPTO 2002. LNCS, vol. 2442, pp. 597–612. Springer, Heidelberg (2002). https://doi.org/10.1007/3-540-45708-9_38

[Moh11] Mohassel, P.: One-time signatures and chameleon hash functions. In: Biryukov, A., Gong, G., Stinson, D.R. (eds.) SAC 2010. LNCS, vol. 6544, pp. 302–319. Springer, Heidelberg (2011). https://doi.org/10.1007/978-3-642-19574-7_21

[MP12] Micciancio, D., Peikert, C.: Trapdoors for lattices: simpler, tighter, faster, smaller. In: Pointcheval, D., Johansson, T. (eds.) EUROCRYPT 2012. LNCS, vol. 7237, pp. 700–718. Springer, Heidelberg (2012). https://doi.org/10.1007/978-3-642-29011-4_41

[MPR11] Maji, H.K., Prabhakaran, M., Rosulek, M.: Attribute-based signatures. In: Kiayias, A. (ed.) CT-RSA 2011. LNCS, vol. 6558, pp. 376–392. Springer, Heidelberg (2011). https://doi.org/10.1007/978-3-642-19074-2_24

[MR04] Micciancio, D., Regev, O.: Worst-case to average-case reductions based on Gaussian measures. In: FOCS, pp. 372–381 (2004)

[Nao91] Naor, M.: Bit commitment using pseudorandomness. J. Cryptol. **4**(2), 151–158 (1991)

[NZZ15] Nguyen, P.Q., Zhang, J., Zhang, Z.: Simpler efficient group signatures from lattices. In: Katz, J. (ed.) PKC 2015. LNCS, vol. 9020, pp. 401–426. Springer, Heidelberg (2015). https://doi.org/10.1007/978-3-662-46447-2_18

[OT11] Okamoto, T., Takashima, K.: Efficient attribute-based signatures for non-monotone predicates in the standard model. In: Catalano, D., Fazio, N., Gennaro, R., Nicolosi, A. (eds.) PKC 2011. LNCS, vol. 6571, pp. 35–52. Springer, Heidelberg (2011). https://doi.org/10.1007/978-3-642-19379-8_3

[Pei09] Peikert, C.: Public-key cryptosystems from the worst-case shortest vector problem: extended abstract. In: STOC, pp. 333–342 (2009)

[PLS18] del Pino, R., Lyubashevsky, V., Seiler, G.: Lattice-based group signatures and zero-knowledge proofs of automorphism stability. ACM-CCS (2018, to appear)

[PsV06] Pass, R., Shelat, A., Vaikuntanathan, V.: Construction of a non-malleable encryption scheme from any semantically secure one. In: Dwork, C. (ed.) CRYPTO 2006. LNCS, vol. 4117, pp. 271–289. Springer, Heidelberg (2006). https://doi.org/10.1007/11818175_16

[PV08] Peikert, C., Vaikuntanathan, V.: Noninteractive statistical zero-knowledge proofs for lattice problems. In: Wagner, D. (ed.) CRYPTO 2008. LNCS, vol. 5157, pp. 536–553. Springer, Heidelberg (2008). https://doi.org/10.1007/978-3-540-85174-5_30

[Reg05] Regev, O.: On lattices, learning with errors, random linear codes, and cryptography. In: STOC, pp. 84–93 (2005)

[Rom90] Rompel, J.: One-way functions are necessary and sufficient for secure signatures. In: STOC, pp. 387–394 (1990)

[SAH16] Sakai, Y., Attrapadung, N., Hanaoka, G.: Attribute-based signatures for circuits from bilinear map. In: Cheng, C.-M., Chung, K.-M., Persiano, G., Yang, B.-Y. (eds.) PKC 2016, Part I. LNCS, vol. 9614, pp. 283–300. Springer, Heidelberg (2016). https://doi.org/10.1007/978-3-662-49384-7_11

[SEH+13] Sakai, Y., Emura, K., Hanaoka, G., Kawai, Y., Matsuda, T., Omote, K.: Group signatures with message-dependent opening. In: Abdalla, M., Lange, T. (eds.) Pairing 2012. LNCS, vol. 7708, pp. 270–294. Springer, Heidelberg (2013). https://doi.org/10.1007/978-3-642-36334-4_18

[SS96] Sipser, M., Spielman, D.A.: Expander codes. IEEE Trans. Inf. Theory **42**(6), 1710–1722 (1996)

[SSE+12] Sakai, Y., Schuldt, J.C.N., Emura, K., Hanaoka, G., Ohta, K.: On the security of dynamic group signatures: preventing signature hijacking. In: Fischlin, M., Buchmann, J., Manulis, M. (eds.) PKC 2012. LNCS, vol. 7293, pp. 715–732. Springer, Heidelberg (2012). https://doi.org/10.1007/978-3-642-30057-8_42

[ST01] Shamir, A., Tauman, Y.: Improved online/offline signature schemes. In: Kilian, J. (ed.) CRYPTO 2001. LNCS, vol. 2139, pp. 355–367. Springer, Heidelberg (2001). https://doi.org/10.1007/3-540-44647-8_21

[Tsa17] Tsabary, R.: An equivalence between attribute-based signatures and homomorphic signatures, and new constructions for both. In: Kalai, Y., Reyzin, L. (eds.) TCC 2017, Part II. LNCS, vol. 10678, pp. 489–518. Springer, Cham (2017). https://doi.org/10.1007/978-3-319-70503-3_16

[Zém01] Zémor, G.: On expander codes. IEEE Trans. Inf. Theory **47**(2), 835–837 (2001)

A Modular Treatment of Blind Signatures from Identification Schemes

Eduard Hauck, Eike Kiltz, and Julian Loss(✉)

Ruhr University Bochum, Bochum, Germany
{eduard.hauck,eike.kiltz,julian.loss}@rub.de

Abstract. We propose a modular security treatment of blind signatures derived from linear identification schemes in the random oracle model. To this end, we present a general framework that captures several well known schemes from the literature and allows to prove their security. Our modular security reduction introduces a new security notion for identification schemes called One-More-Man In the Middle Security which we show equivalent to the classical One-More-Unforgeability notion for blind signatures.

We also propose a generalized version of the Forking Lemma due to Bellare and Neven (CCS 2006) and show how it can be used to greatly improve the understandability of the classical security proofs for blind signatures schemes by Pointcheval and Stern (Journal of Cryptology 2000).

Keyword: Blind signatures

1 Introduction

Blind Signatures are a fundamental cryptographic building block. Informally, a blind signature scheme is an interactive protocol between a signer and an user in which the signer issues signatures on messages chosen by the user. There are two security requirements: *blindness* ensures that the signer cannot link a signature to the run of the protocol in which it was created and *one-more unforgeability* that the user cannot forge a new signature. Originally proposed by Chaum [12] as the basis of his e-cash system, blind signatures have since found numerous applications including e-voting [22] and anonymous credentials [3,5,9–11,13,19]. Despite a flurry of schemes having been published over the past three and a half decades, only a handful of works has considered blind signature schemes which are mutually efficient, instantiable from standard assumptions, and remain secure even when executed in an arbitrarily concurrent fashion. The notoriously difficult task of constructing such schemes was first tackled by Pointcheval and Stern [21]. Their groundbreaking work introduces the well-known *forking lemma* and shows how it can be applied to prove security of the Okamoto-Schnorr blind signature scheme [18] under the discrete logarithm assumption in the random oracle model (ROM) [8]. Their proof technique was subsequently employed to

Y. Ishai and V. Rijmen (Eds.): EUROCRYPT 2019, LNCS 11478, pp. 345–375, 2019.
https://doi.org/10.1007/978-3-030-17659-4_12

Table 1. Examples of linear function families. Group type functions are defined over $\mathbb{G}$ of prime order q with generators g_1, g_2, RSA type functions are defined over an RSA modulus $N = pq$ and $a \in \mathbb{Z}_N^*$ satisfying $\mathrm{ord}(a) > 2\lambda$. Set $\mathcal{S}$ is the challenge set.

Name	Type	Definition of linear function $\mathsf{F} : \mathcal{D} \to \mathcal{R}$	$\mathcal{S}$	Collision resistance
OS	Group	$\mathsf{F} : \mathbb{Z}_q^2 \to \mathbb{G}, \quad (x_1, x_2) \mapsto g_1^{x_1} g_2^{x_2}$	$\mathbb{Z}_q$	DLOG
OGQ	RSA	$\mathsf{F} : \mathbb{Z}_\lambda \times \mathbb{Z}_N^* \to \mathbb{Z}_N^*, \quad (x_1, x_2) \mapsto a^{x_1} x_2^\lambda$	$\mathbb{Z}_\lambda$	RSA
FS	RSA	$\mathsf{F} : (\mathbb{Z}_N^*)^k \to (\mathbb{Z}_N^*)^k, \quad (x_1, \ldots, x_k) \mapsto (x_1^2, \ldots, x_k^2)$	$\mathbb{Z}_2^k$	FACTORING

prove the security of further schemes [4,20,23]. Unfortunately, due to the complexity and subtlety of the argument in [21], these works present either only proof sketches [20] or follow the proof of [21] almost verbatim.

1.1 Our Contribution: A Modular Framework for Blind Signatures

In this work, we propose a general framework which shows how to derive a blind signature scheme from any *linear function family* (with certain properties), as recently introduced by Backendahl et al. [2]. Whereas blindness can be proved directly, one-more unforgeability is proved in two modular steps. In the first step, one builds a linear identification scheme from the linear function family. One-more unforgeability of the blind signature scheme in the random oracle model is shown to be tightly equivalent to a new and natural security notion of the linear identification scheme, which we call *one-more man-in-the-middle* security. In the second, technically involved, step it is shown that the latter is implied by collision resistance of the linear function family. Our framework captures several important schemes from the literature including the Okamoto-Schnorr (OS) [18], the Okamoto-GQ (OGQ) [18], and (a slightly modified version of) the Fiat-Shamir (FS) [20] blind signature schemes and offers, for the first time, a complete and formal proof for some of them. We now provide some details of our contributions.

LINEAR FUNCTION FAMILIES AND IDENTIFICATION SCHEMES. A canonical identification scheme ID [1] is a three-move protocol of a specific form in which a prover P convinces a verifier Ver (holding a public key *pk*) that he knows the corresponding secret key *sk*. ID = ID[LF] is a linear identification scheme [2] if it follows a certain homomorphic structure induced by a linear function LF. For our purpose of building blind signatures, we will also require LF to be perfectly correct, collision resistant, and the kernel to contain a torsion-free element. (Note that this also makes LF many-to-one.) Example instantiations of (collision resistant) linear function families can be derived from OS, OGQ, and FS, cf. Table 1.

We introduce a natural new security notion for (arbitrary, not necessarily linear) canonical identification schemes called *One-More Man-in-the-Middle* (**OMMIM**)-security. Informally, ID is **OMMIM**-secure if it is infeasible to complete $Q_\mathsf{P}+1$ (or more) runs of ID in the role of prover P after completing at most Q_P runs of ID in the role of verifier Ver. Note that **OMMIM** is weaker than standard Man-in-the-Middle security [15] (which we show to be unachievable for linear identification schemes) but stronger than impersonation against active attacks [7,14].

OMMIM SECURITY OF LINEAR IDENTIFICATION SCHEMES. Our first main result can be stated as follows:

Theorem 1 (informal). If LF is collision resistant, then $\mathsf{ID}[\mathsf{LF}]$ is **OMMIM** secure.

Our proof is based on a new Subset Forking Lemma that generalizes the one by Bellare and Neven [6] and contains many technical ingredients from [21] who prove the security of the Okamoto-Schnorr Blind Signature scheme. Unfortunately, the security bound from Theorem 1 is only meaningful if $Q_{\mathsf{V}}^{Q_{\mathsf{P}}+1} \leq |\mathcal{C}| =: q$, where Q_{V} refers to the (potentially large) number of sessions with the verifier and challenge set $\mathcal{C}$ is a parameter of the identification scheme. We next show in Theorem 2 that a natural generalization of Schnorr's ROS-problem [24] to linear functions can be used to break the **OMMIM** of $\mathsf{ID}[\mathsf{LF}]$. The ROS-problem (for the relevant parameters) becomes information theoretically hard when $Q_{\mathsf{V}}^{Q_{\mathsf{P}}+1} \leq q$. For all other cases, it can be solved in sub-exponential time $(Q_{\mathsf{V}}+1)\,2^{\sqrt{\log q}/(1+\log(Q_{\mathsf{V}}+1))}$ using Wagner's k-List algorithm [25]. Our ROS-based attack works whenever $\mathcal{C}$ is a finite field, which is the case for OS and OGQ.

CANONICAL BLIND SIGNATURE SCHEMES. We introduce the notion of *canonical blind signature schemes* (BS), which are three-move blind signature schemes of a specific form. In terms of security we define *blindness* and *one-more unforgeability* (**OMUF**). Intuitively, **OMUF** states that the adversary cannot produce more valid message-signatures pairs, then it has completed successful sessions with the signer. (Note that each such session yields a valid message-signature pair.) Here we consider a natural and strong version of **OMUF** in which abandoned session with the signer (i.e., sessions that are started but never completed) are not counted as a successful sessions with the signer as they do not yield a valid message-signature pair. We propose a general compiler to convert any linear identification scheme $\mathsf{ID}[\mathsf{LF}]$ and a hash function H into a canonical blind signature scheme $\mathsf{BS}[\mathsf{LF},\mathsf{H}]$. Our second main result can be stated as follows:

Theorem 3 (informal). **OMUF** security of $\mathsf{BS}[\mathsf{LF},\mathsf{H}]$ is tightly equivalent to **OMMIM** security of $\mathsf{ID}[\mathsf{LF}]$ in the random oracle model.

Theorem 4 (informal). $\mathsf{BS}[\mathsf{LF},\mathsf{H}]$ is perfectly blind.

Figure 1 summarizes our modular security analysis of $\mathsf{BS}[\mathsf{LF},\mathsf{H}]$. Combining our main theorems, we obtain security proofs for the OS, OGQ, and FS blind signature schemes. Here, the number of random oracle queries Q_{H} corresponds to the number Q_{V} of open sessions with the verifier, whereas the number Q_{S} of signing sessions corresponds to the number of sessions Q_{P} with the prover. Hence, **OMUF** security of $\mathsf{BS}[\mathsf{LF},\mathsf{H}]$ is only guaranteed if $Q_{\mathsf{H}}^{Q_{\mathsf{S}}+1} \ll q$, i.e., for polylogarithmically parallel signing sessions Q_{S}. Our ROS-based attack demonstrates that this restriction is required.

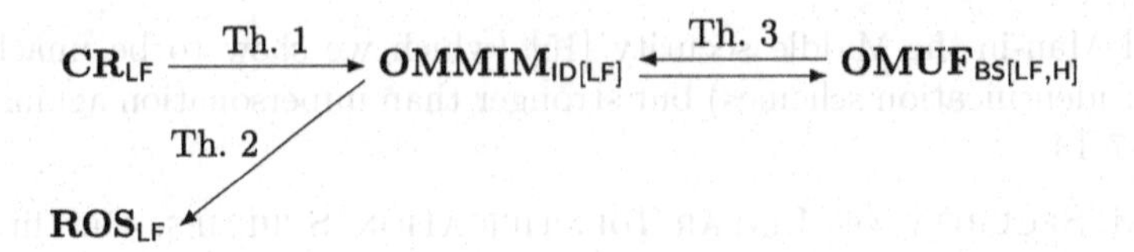

Fig. 1. Overview of our modular security analysis for BS[LF, H]. The arrows denote security implications.

1.2 Technical Details

We now give an intuition for the proof of Theorem 1. Roughly, it states that one can reduce the **OMMIM** security of ID[LF] from the problem of finding a non-trivial collision with respect to the linear function LF. Our proof follows the ideas of Pointcheval and Stern [21], but uses as a key ingredient a novel forking lemma, which enables us to present the proof in [21] in a much more clean and general fashion. The main idea behind our reduction is to run the adversary M against **OMMIM**-security twice, where the instance I and randomness ω in the second run are kept the same, and part of the oracle answers, denoted $\boldsymbol{h}, \boldsymbol{h}'$, are re-sampled uniformly. In this way, we hope to obtain from M two distinct values $\hat{\chi}, \hat{\chi}'$ which yield a collision with respect to LF. The main challenge in our setting is that $\hat{\chi}$ and $\hat{\chi}'$ depend on the internal state of M. To show that $\hat{\chi} \neq \hat{\chi}'$ with high probability, one requires an intricate argument that heavily builds upon a generalized version of Bellare and Neven's Forking Lemma [6]. Our lemma is tailored toward the ideas of the proof in [21] and allows for a more fine-grained replay strategy than the version of [6]. More precisely, our version of the forking lemma considers not only the probability of successfully running an algorithm twice with the same instance I, randomness ω, and (partially distinct) oracle answers $\boldsymbol{h}, \boldsymbol{h}'$, but also allows to analyze in more detail the properties of the triples $(I, \omega, \boldsymbol{h}), (I, \omega, \boldsymbol{h}')$.

1.3 Blind Signatures from Lattices?

We remark that our proof requires linear functions with perfect correctness. This leaves open the question of whether our framework can be extended to cover also the lattice-based identification scheme due to Lyubashevsky [16] and the resulting blind signature scheme due to Rückert [23]. At a technical level, imperfect correctness causes a problem in the proof of Theorem 3 which relates the **OMMIM**-security of ID[LF] to **OMUF**-security of BS[LF, H]. If the adversary manages to abort even a single run of BS[LF, H] in the simulated **OMUF** experiment, our reduction fails at simulating the necessary amount of completed runs of BS[LF, H] to the adversary. This subtlety in the proof arises from the fact that in the **OMMIM** experiment, there is no way of telling whether a run of ID[LF] with the adversary in the role of the verifier was completed. On the other hand, in BS[LF, H], the user can prove to the signer that it obtained an invalid signature for a particular run of the protocol and hence force a restart.

We leave it as an open problem to adapt our framework to linear functions with correctness errors.

2 Preliminaries and Notation

SETS AND ALGORITHMS. We denote as $h \xleftarrow{\$} \mathcal{H}$ the uniform sampling of the variable s from the set $\mathcal{H}$. If ℓ is an integer, then $[\ell]$ is the set $\{1, \ldots, \ell\}$. We write bold lower case letters $\boldsymbol{h}$ to denote a vector of elements and denote the length of $\boldsymbol{h}$ as $|\boldsymbol{h}|$. For $j > 1$, we write $\boldsymbol{h}_{[j]}$ to refer to the first j entries of $\boldsymbol{h}$. For $1 \leq j \leq Q$ and $\boldsymbol{g} \in \mathcal{H}^{j-1}$ we now define the conditional distribution $\boldsymbol{h'} \xleftarrow{\$} \mathcal{H}^Q|\boldsymbol{g}$ which samples $\boldsymbol{h'} \xleftarrow{\$} \mathcal{H}^Q$ conditioned on $\boldsymbol{h'}_{[j-1]} = \boldsymbol{g}$. (This can be implemented by copying vector $\boldsymbol{g}$ into the first $j-1$ entries of $\boldsymbol{h}'$ and next sampling the subvector $\boldsymbol{h}'_j, \ldots, \boldsymbol{h}'_Q \xleftarrow{\$} \mathcal{H}^{Q-j+1}$.)

We write bold upper case letters $\mathbf{A}$ to denote matrices. We denote the i-th row vector of $\mathbf{A}$ as $\mathbf{A}_i$ and the j-th entry of $\mathbf{A}_i$ as $\mathbf{A}_{i,j}$. We use uppercase letters A, B to denote algorithms. Unless otherwise stated, all our algorithms are probabilistic and we write $(y_1, ...) \xleftarrow{\$} \mathsf{A}(x_1, ...)$ to denote that A returns $(y_1, ...)$ when run on input $(x_1, ...)$. We write A^{B} to denote that A has oracle access to B during its execution. Any probabilistic algorithm $\mathsf{A}(x)$, on some input x can be written as a deterministic algorithm $\mathsf{A}(x; \omega)$ on input x and randomness ω. We use standard code-based security games and write $\mathbf{G}^{\mathsf{A}} \Rightarrow 1$ to denote the event that algorithm A is successful in game $\mathbf{G}$.

3 Linear Functions and Identification Schemes

A *module* is specified by two sets $\mathcal{S}$ and $\mathcal{M}$, where $\mathcal{S}$ is a ring with multiplicative identity element $1_{\mathcal{S}}$ and $\langle \mathcal{M}, +, 0 \rangle$ is an additive Abelian group and a mapping $\cdot : \mathcal{S} \times \mathcal{M} \rightarrow \mathcal{M}$, s.t. for all $r, s \in \mathcal{S}$ and $x, y \in \mathcal{M}$ we have (i) $r \cdot (x + y) = r \cdot x + r \cdot y$; (ii) $(r + s) \cdot x = r \cdot x + s \cdot x$; (iii) $(rs) \cdot x = r \cdot (s \cdot x)$; and (iv) $1_{\mathcal{S}} \cdot x = x$.

SYNTAX OF LINEAR FUNCTION FAMILIES. A *linear function family* LF [2] is a tuple of algorithms $(\mathsf{PGen}, \mathsf{F})$. On input the security parameter, the randomized algorithm PGen returns some parameters *par*, which implicitly define the sets $\mathcal{S} = \mathcal{S}(par), \mathcal{D} = \mathcal{D}(par)$ and $\mathcal{R} = \mathcal{R}(par)$. $\mathcal{S}$ is a set of scalars such that $\mathcal{D}$ and $\mathcal{R}$ are modules over $\mathcal{S}$. Further, $\mathsf{F}(par, \cdot)$ implements a mapping from $\mathcal{D}$ to $\mathcal{R}$. To simplify our presentation, we will omit *par* from F's input from now on. $\mathsf{F}(\cdot)$ is required to be a *module homomorphism*, meaning that for any $(x, y) \in (\mathcal{D} \times \mathcal{D})$ and $s \in \mathcal{S}$:

$$\mathsf{F}(s \cdot x + y) = s \cdot \mathsf{F}(x) + \mathsf{F}(y).$$

We say that LF has a *torsion-free element from the kernel* if for all *par* generated with PGen, there exist $z^* \in \mathcal{D} \setminus \{0\}$ such that (i) $\mathsf{F}(z^*) = 0$; and (ii) for all $s \in \mathcal{S}$ satisfying $s \cdot z^* = 0$ we have $s = 0$. Note that this implies that F is a many-to-one mapping.

Security Properties of Linear Function Families. We now define two security properties of a linear function family (collision resistance and ROS security) which will play a significant role in the subsequent sections.

We define the advantage of an adversary A, breaking the *collision resistance* of LF as

$$\mathbf{Adv}^{\mathbf{CR}}_{\mathsf{LF}}(\mathsf{A}) := \Pr_{par \xleftarrow{\$} \mathsf{PGen}, (x_1, x_2) \xleftarrow{\$} \mathsf{A}(par)} [\mathsf{F}(x_1) = \mathsf{F}(x_2) \wedge x_1 \neq x_2]$$

and say that LF is (ε, t)-**CR** if for all adversaries A running in time $\mathbf{Time}(\mathsf{A}) \leq t$ we have $\mathbf{Adv}^{\mathbf{CR}}_{\mathsf{LF}}(\mathsf{A}) \leq \varepsilon$.

The ROS (Random inhomogenities in an Overdetermined, Solvable system of linear equations) problem was introduced by Schnorr [24] (also in the context of blind signatures). Here, we generalize Schnorr's formulation to linear function families. For a linear function family LF we define the advantage of an adversary A as

$$\mathbf{Adv}^{\mathbf{ROS}}_{\mathsf{LF}}(\mathsf{A}) := \Pr[\mathbf{ROS}^{\mathsf{A}}_{\mathsf{LF}} \Rightarrow 1],$$

where game $\mathbf{ROS}_{\mathsf{LF}}$ is defined in Fig. 2. We furthermore say that LF is $(\varepsilon, t, \ell, Q_{\mathsf{H}})$-**ROS** secure if for all adversaries A running in time $\mathbf{Time}(\mathsf{A}) \leq t$ and making at most Q_{H} queries to the random oracle, we have $\mathbf{Adv}^{\mathbf{ROS}}_{\mathsf{LF}}(\mathsf{A}) \leq \varepsilon$.

```
GAME ROS_LF:
00 par <-$ PGen
01 (c ∈ S^{ℓ+1}, A ∈ S^{(ℓ+1)×(ℓ+1)}) <- A^H(par)
02 If (c_{ℓ+1} = −1) ∧ (Ac = 0) ∧ (∀i, j ∈ [ℓ+1] : H(A_{i,1}, ..., A_{i,ℓ}) = A_{i,ℓ+1}) ∧ (A_i ≠ A_j) Then
03     Return 1
04 Return 0
```

Fig. 2. Game $\mathbf{ROS}_{\mathsf{LF}}$, where $\mathsf{H}\colon \{0,1\}^* \to \mathcal{S}$ is a random oracle.

The following Lemma summarizes the known hardness results for the Generalized ROS-Problem for the specific case in which $\mathcal{S}$ is a field of prime order q.

Lemma 1 ([17,24,25]). *Let* LF *be a linear function family for which* $\mathcal{S}$ *is a field of prime order* q*. For every* t*,* LF *is* $(t, \varepsilon = Q_{\mathsf{H}}^{\ell+1}/q, \ell, Q_{\mathsf{H}})$-**ROS** *secure. Conversely,* LF *is not* $(t, 1/4, \ell, Q_{\mathsf{H}})$-**ROS** *secure for* $Q_{\mathsf{H}} = (\ell+1)\, 2^{\log q/(1+\log(\ell+1))}$ *and* $t = O\left((\ell+1)2^{\log q/(1+\log(\ell+1))}\right)$.

Examples of Linear Function Families. We now give three examples of LF with the required properties. We remark that [2] contains more examples of linear functions, but not all of them have a torsion-free element from the kernel.

Okamoto-Schnorr. PGen returns the parameters $par := (\mathbb{G}, g_1, g_2) \xleftarrow{\$} \mathsf{PGen}(1^\lambda)$, where $g_1, g_2 \in \mathbb{G}$, q is prime, and $|\mathbb{G}| = q$. par defines sets $\mathcal{S}, \mathcal{D}, \mathcal{R}$, and the homomorphic evaluation function F as

$$\mathcal{S} := \mathbb{Z}_q; \quad \mathcal{D} := \mathbb{Z}_q^2; \quad \mathcal{R} := \mathbb{G}; \quad \mathsf{F} : \mathbb{Z}_q^2 \to \mathbb{G}, (x_1, x_2) \mapsto g_1^{x_1} g_2^{x_2}.$$

It is not hard to see that F is an homomorphism. It is also not hard to see that collision resistance of LF is equivalent to the discrete logarithm problem over $\mathbb{G}$, i.e., $\mathbf{Adv}_{\mathsf{LF}}^{\mathbf{CR}}(\mathsf{A}) = \mathbf{Adv}_{\mathbb{G}}^{\mathsf{DLOG}}(\mathsf{B})$. For all parameters par and for $w = \log_{g_1}(g_2)$, the element $z^* = (z_1^*, z_2^*) := (w, -1)$, yields a torsion-free in the kernel of LF since $\mathsf{F}(z^*) = g_1^w g_2^{-1} = 1$, where $1 = 0_{\mathbb{G}}$ is the neutral element in $\mathbb{G}$. Furthermore, for all $s \in \mathbb{Z}_q$ satisfying $s \cdot z^* = (s \cdot w, -s) = (0, 0)$ we have $s = 0 \bmod q$ since q is prime.

Okamoto-Guillou-Quisquater. PGen returns the parameters $par := (N = pq, \lambda, a) \xleftarrow{\$} \mathsf{PGen}(1^\lambda)$, where p, q are prime and λ is prime and co-prime with $N, \varphi(N)$ and $a \in \mathbb{Z}_N^*, \mathsf{ord}(a) > 2\lambda$. The parameters par define

$$\mathcal{S} := \mathbb{Z}_\lambda; \quad \mathcal{R} := \mathbb{Z}_N^*; \quad \mathcal{D} = \{(x_1, x_2 = za^{\lfloor \frac{x_1}{\lambda} \rfloor}) \bmod N \mid x_1 \in \mathbb{Z}_\lambda, z \in \mathbb{Z}_N^*\},$$

where $\mathcal{D}$ is an abelian group with the group operation $(x_1, x_2) \circ (y_1, y_2) = (x_1 + y_1 \bmod \lambda, x_2 y_2 a^{\lfloor \frac{x_1+y_1}{\lambda} \rfloor} \bmod N)$. The evaluation function F is defined as

$$\mathsf{F}\colon \mathbb{Z}_\lambda \times \mathbb{Z}_N^* \to \mathbb{Z}_N^*, \mathsf{F}(x_1, x_2) := a^{x_1} x_2^\lambda.$$

F is an homomorphism, since:

$$\begin{aligned}
\mathsf{F}((x_1, x_2) \circ (y_1, y_2)) &= \mathsf{F}(x_1 + y_1 \bmod \lambda, x_2 y_2 a^{\lfloor \frac{x_1+y_1}{\lambda} \rfloor} \bmod N) \\
&= a^{x_1+y_1 \bmod \lambda} \left(x_2 y_2 a^{\lfloor \frac{x_1+y_1}{\lambda} \rfloor} \right)^\lambda \\
&= a^{((x_1+y_1) \bmod \lambda) + \lambda \lfloor \frac{x_1+y_1}{\lambda} \rfloor} (x_2 y_2)^\lambda && (1) \\
&= a^{x_1+y_1} (x_2 y_2)^\lambda && (2) \\
&= \mathsf{F}(x_1, x_2) \mathsf{F}(y_1, y_2),
\end{aligned}$$

where (1) and (2) follow from the identity: $(x \bmod \lambda) = x - \lambda \lfloor \frac{x}{\lambda} \rfloor$.

A collision $(x_1, x_2) \neq (y_1, y_2)$ with $\mathsf{F}(x_1, x_2) = \mathsf{F}(y_1, y_2)$ implies $a^{x_1 - y_1} = (y_2/x_2)^\lambda$ with $\gcd(\lambda, x_1 - x_2) = 1$ from which one can extract the $a^{1/\lambda}$ using the extended Euclidean Algorithm. Hence, collision resistance is implied by the RSA assumption.

For all parameters par, $z^* = (z_1^*, z_2^*) := (-1, a^{1/\lambda})$ is a torsion-free element in the kernel of F since $\mathsf{F}(z^*) = a^{-1 \bmod \lambda} (a^{1/\lambda})^\lambda a^{\lfloor \frac{-1}{\lambda} \rfloor} = a^{(-1 \bmod \lambda) + \lfloor \frac{-1}{\lambda} \rfloor} a = 1$, where $1 = 0_{\mathcal{R}}$ is the neutral element in $\mathcal{R}$. Furthermore, for all $s \in \mathbb{Z}_\lambda$ satisfying $s \cdot z^* = (-s, (a^{1/\lambda})^s a^{\lfloor \frac{-s}{\lambda} \rfloor}) = (0, 1)$ we have $s = 0 \bmod \lambda$.

Fiat-Shamir. PGen returns parameters $par := (N = pq, k)$, where p, q are prime and k is an integer. Parameters par define

$$\begin{aligned}
&\mathcal{S} := \mathbb{Z}_2^k; \quad \mathcal{D} := (\mathbb{Z}_N^*)^k, \mathcal{R} := (\mathbb{Z}_N^*)^k; \\
&\mathsf{F} : (\mathbb{Z}_N^*)^k \to (\mathbb{Z}_N^*)^k, \mathsf{F}(x_1, \ldots, x_k) \mapsto (x_1^2, \ldots, x_k^2).
\end{aligned}$$

Clearly, collision resistance of LF is equivalent to factorization. For all parameters par, $z^* = (z_1^*, \ldots, z_k^*) := (-1, \ldots, -1)$ is a torsion-free element from the kernel of F since $\mathsf{F}(z^*) = (1, \ldots, 1)$, where $(1, \ldots, 1) = 0_{\mathcal{R}}$ is the neutral element in $\mathcal{R}$. Furthermore, for all $\boldsymbol{s} \in \mathbb{Z}_2^k$ satisfying $\boldsymbol{s} \cdot z^* = (-1^{s_1}, \ldots, -1^{s_k}) = (1, \ldots, 1)$ we have $\boldsymbol{s} = \mathbf{0} \bmod 2$.

4 Canonical Identification Schemes

4.1 Syntax and Security

We now recall the definition of define canonical identification schemes [1] and discuss their security notions.

Definition 1 (Canonical Identification Scheme). *A* canonical identification scheme *is a tuple of algorithms* $\mathsf{ID} = (\mathsf{IGen}, \mathsf{P}, \mathsf{Ver})$.

- *The* key generation algorithm IGen *takes as input parameters par and outputs a public/secret key pair* (pk, sk). *We assume that pk implicitly defines a challenge set* $\mathcal{C} = \mathcal{C}(pk)$.
- *The prover algorithm* P *is split into two randomized algorithms* $\mathsf{P}_1, \mathsf{P}_2$, *i.e.,* $\mathsf{P} = (\mathsf{P}_1, \mathsf{P}_2)$. P_1 *takes as input a secret key sk and returns a commitment R and a state st. The deterministic algorithm* P_2 *takes as input a state st, a secret key sk, a commitment R, and a challenge* $c \in \mathcal{C}$. *It returns a response s.*
- *The deterministic verification algorithm* Ver *takes as input a public key pk, a commitment R, a challenge* $c \in \mathcal{C}$, *and a response s. It returns* $b \in \{0, 1\}$.

The diagram below depicts an interaction between prover P *and verifier* V. *For correctness we require that for all* $(pk, sk) \in \mathsf{IGen}(par)$, *all* $(st, R) \in \mathsf{P}_1(sk)$, *all* $c \in \mathcal{C}$, *and all* $s \in \mathsf{P}_2(sk, R, c, st)$, *it holds that* $\mathsf{Ver}(pk, R, c, s) = 1$.

Prover $\mathsf{P}(sk)$		Verifier $\mathsf{V}(pk)$
$(st, R) \xleftarrow{\$} \mathsf{P}_1(sk)$	$\xrightarrow{R}$	
	$\xleftarrow{c}$	$c \xleftarrow{\$} \mathcal{C}$
$s \leftarrow \mathsf{P}_2(sk, R, c, st)$	$\xrightarrow{s}$	$b \leftarrow \mathsf{Ver}(pk, R, c, s)$
Output 1		Output b

Standard security notions for canonical identification schemes include impersonation security against passive and active attacks, and Man-in-the-Middle security [1,7]. We now introduce a new security notion called *One-More Man-in-the-Middle* security. The One-More Man-in-the-Middle (**OMMIM**) security experiment for an identification scheme ID and an adversary A is defined in Fig. 3. Adversary A simultaneously plays against a prover (modeled through oracles $\mathtt{P}_1$ and $\mathtt{P}_2$) and a verifier (modeled through oracles $\mathtt{V}_1$ and $\mathtt{V}_2$). Session identifiers *pSid* and *vSid* are used to model an interaction with the prover and the verifier, respectively. A call to $\mathtt{P}_1$ returns a new prover session identifier *pSid* and sets flag $\mathbf{pSess}_{pSid}$ to $\mathtt{open}$. A call to $\mathtt{P}_2(pSid, \cdot)$ with the same *pSid* sets the flag $\mathbf{pSess}_{pSid}$ to $\mathtt{closed}$. Similarly, a call to $\mathtt{V}_1$ returns a new verifier session identifier *vSid* and sets flag $\mathbf{vSess}_{vSid}$ to $\mathtt{open}$. A call to $\mathtt{V}_2(vSid, \cdot)$ with the same *pSid* sets the flag $\mathbf{vSess}_{vSid}$ to $\mathtt{closed}$. A closed verifier session *vSid* is successful if the oracle $\mathtt{V}_2(vSid, \cdot)$ returns 1. Lines 03–06 define several internal random variables for later references. Variable $Q_{\mathtt{P}_2}(\mathsf{A})$ counts the number of closed prover sessions and $Q_{\mathsf{P}_1}(\mathsf{A})$ counts the number of abandoned sessions (i.e., sessions that were opened but never closed). Most importantly, variable $\ell(\mathsf{A})$ counts the number of successful verifier sessions and variable $Q_{\mathsf{P}_2}(\mathsf{A})$ counts the number of closed sessions with the prover. Adversary A wins the **OMMIM** game, if $\ell(\mathsf{A}) \geq Q_{\mathsf{P}_2}(\mathsf{A}) + 1$, i.e., if A convinces the verifier in at least one more successful verifier sessions than there exist closed sessions with the prover. The **OMMIM** advantage function of an adversary A against ID is defined as $\mathbf{Adv}_{\mathsf{ID}}^{\mathbf{OMMIM}}(\mathsf{A}) := \Pr[\mathbf{OMMIM}_{\mathsf{ID}}^{\mathsf{A}} \Rightarrow 1]$.

We say that ID is $(\varepsilon, t, Q_{\mathsf{V}}, Q_{\mathsf{P}_1}, Q_{\mathsf{P}_2})$-**OMMIM** secure if for all adversaries A satisfying $\mathbf{Time}(\mathsf{A}) \leq t$, $Q_{\mathsf{V}}(\mathsf{A}) \leq Q_{\mathsf{V}}$, $Q_{\mathsf{P}_2}(\mathsf{A}) \leq Q_{\mathsf{P}_2}$, and $Q_{\mathsf{P}_1}(\mathsf{A}) \leq Q_{\mathsf{P}_1}$, we have $\mathbf{Adv}_{\mathsf{ID}}^{\mathbf{OMMIM}}(\mathsf{A}) \leq \varepsilon$.

```
GAME OMMIM_ID^A:
00 (sk, pk) ← IGen
01 pSid ← 0, vSid ← 0                                  //initialize prover/verifier session id
02 A^{P1,P2,V1,V2}(pk)
03 Q_V(A) ← vSid                                        //#total sessions with verifier
04 Q_P1(A) ← #{1 ≤ k ≤ pSid | pSess_k = open}           //#abandoned prover sessions
05 Q_P2(A) ← #{1 ≤ k ≤ pSid | pSess_k = closed}         //#closed prover sessions
06 ℓ(A) ← #{1 ≤ k ≤ vSid | vSess_k = closed ∧ b'_k = 1} //#successful verifier sessions
07 If ℓ(A) ≥ Q_P2(A) + 1 Then                           //A's winning condition
08    Return 1
09 Return 0

Procedure P1                          Procedure V1(R')
10 pSid ← pSid + 1                    19 vSid ← vSid + 1
11 pSess_pSid ← open                  20 vSess_vSid ← open
12 (st_pSid, R_pSid) ←$ P1            21 R'_vSid ← R'; c'_vSid ←$ C
13 Return (pSid, R_pSid)              22 Return (vSid, c'_vSid)

Procedure P2(pSid, c)                 Procedure V2(vSid, s')
14 If pSess_pSid Then                 23 If vSess_vSid ≠ open Then
15    Return ⊥                        24    Return ⊥
16 pSess_pSid ← closed                25 vSess_vSid ← closed
17 s_pSid ← P2(st_pSid, sk, R_pSid, c) 26 b'_vSid ← Ver(pk, R'_vSid, c'_vSid, s')
18 Return s_pSid                      27 Return b'_vSid
```

Fig. 3. The One-More Man-in-the-Middle security game $\mathbf{OMMIM}_{\mathsf{ID}}^{\mathsf{A}}$

We remark that impersonation against active and passive attacks is a weaker notion than **OMMIM** security, whereas Man-in-the-Middle (**MIM**) security is stronger. Concretely, in the standard **MIM** experiment the winning condition is relaxed in the sense that there only has to exist a successful session with the verifier with a transcript that does not result from a closed session with the prover.

4.2 Identification Schemes from Linear Function Families

As showed in [2], a linear function family LF directly implies a canonical identification scheme ID[LF]. The construction is given in Fig. 4, where $par \xleftarrow{\$} \mathsf{PGen}$ are fixed global system parameters. We will prove later that ID[LF] is **OMMIM** secure. This is the best we can hope for since by the linearity of LF, ID[LF] can never be (fully) **MIM** secure. (Concretely, an adversary receiving a commitment R from the prover can send $R' = \mathsf{F}(\hat{r}) + R$ for some $\hat{r} \neq 0$ to the verifier. After forwarding $c' = c$ from verifier to prover, it receives s from the prover and submits $s' = s + \hat{r}$ to the verifier. Since $(R, c, s) \neq (R', c', s')$, A wins the **MIM** experiment with advantage 1.)

Algorithm $\mathsf{IGen}(par)$	Algorithm $\mathsf{P}_1(sk)$
00 $sk \xleftarrow{\$} \mathcal{D}$	07 $r \xleftarrow{\$} \mathcal{D}$; $R \leftarrow \mathsf{F}(r)$
01 $pk \leftarrow \mathsf{F}(sk)$	08 $st_{\mathsf{P}} := r$
02 Return (sk, pk)	09 Return (st_{P}, R)
Algorithm $\mathsf{Ver}(pk, R, c, s)$	Algorithm $\mathsf{P}_2(sk, st_{\mathsf{P}}, c)$
03 $S \leftarrow \mathsf{F}(s)$	10 $r \leftarrow st_{\mathsf{P}}$
04 If $S = c \cdot pk + R$ Then	11 $s \leftarrow c \cdot sk + r$
05 Return 1	12 Return s
06 Return 0	

Fig. 4. Construction of $\mathsf{ID[LF]} := (\mathsf{IGen}, \mathsf{P} := (\mathsf{P}_1, \mathsf{P}_2), \mathsf{Ver})$ with challenge set $\mathcal{C} = \mathcal{S}$.

Theorem 1. *Suppose* LF *is a linear function family with a torsion-free element from the kernel. If* LF *is* (ε', t')-**CR** *secure, then* ID[LF] *is* $(\varepsilon, t, Q_{\mathsf{V}}, Q_{\mathsf{P}_2}, Q_{\mathsf{P}_1})$-**OMMIM** *secure where*

$$t' = 2t, \quad \varepsilon' = O\left(\left(\varepsilon - \frac{(Q_{\mathsf{V}} Q_{\mathsf{P}})^{Q_{\mathsf{P}_2}+1}}{q}\right)^2 \frac{1}{Q_{\mathsf{V}}^2 Q_{\mathsf{P}_2}^3}\right)$$

and $Q_{\mathsf{P}} = Q_{\mathsf{P}_1} + Q_{\mathsf{P}_2}$.

The proof of this theorem will be given in Sect. 6.

Theorem 2. *Let* LF *be a linear function family. If* ID[LF] *is* $(\varepsilon, t, Q_{\mathsf{V}}, Q_{\mathsf{P}_2}, Q_{\mathsf{P}_1} = 0)$-**OMMIM** *secure then* LF *is* $(\varepsilon, t, \ell = Q_{\mathsf{P}_2}, Q_{\mathsf{H}} = Q_{\mathsf{V}})$-**ROS** *secure.*

Proof. Let A be an $(\varepsilon, t, \ell, Q_{\mathsf{H}})$-adversary in game **ROS**. We assume w.l.o.g. that A only makes distinct queries to the random oracle H. In Fig. 5, we show how to construct an $(\varepsilon, t, Q_{\mathsf{V}}, Q_{\mathsf{P}_2}, Q_{\mathsf{P}_1})$-adversary B that is executed in game $\mathbf{OMMIM}_{\mathsf{ID}}$ and uses A as a subroutine. First, B starts Q_{P_2} sessions with the Prover oracle P_1, receiving commitments $\boldsymbol{R}$. Next, A is executed, where B answers a query of the form $\mathsf{H}(\boldsymbol{a})$ from A as $c'_{\boldsymbol{a}}$, where $c'_{\boldsymbol{a}} := \mathsf{V}_1(\sum_{j=1}^{Q_{\mathsf{P}_2}} \boldsymbol{a}_j \boldsymbol{R}_j)$. Note that

```
Adversary B^{P1,P2,V1,V2}(pk):                                  Oracle H(a):
00 For j ∈ [Q_P2] Do:                                           09 R'_a ← Σ_{j=1}^{Q_P2} a_j R_j
01    (pSid_j, R_j) <-$ P1        //start Q_P2 sessions with Prover   10 (vSid_a, c'_a) <-$ V1(R'_a)
02 (c ∈ S^{Q_P2+1}, A ∈ S^{(Q_P2+1)×(Q_P2+1)}) <-$ A^H(par)     11 Return c'_a
03 Parse (Z ∈ S^{(Q_P2+1)×Q_P2}, z ∈ S^{Q_P2+1}) ← A
04 For j ∈ [Q_P2] Do:
05    s_j ← P2(pSid_j, c_j)  //close Q_P2 Prover sessions
06 For i ∈ [Q_P2 + 1] Do:
07    s'_i ← Σ_{j=1}^{Q_P2} A_{i,j} s_j
08    b_i ← V2(vSid_{Z_i}, s'_i)
```

Fig. 5. Adversary B in the $\mathbf{OMMIM}_{\mathsf{ID}}^{\mathsf{B}}$ game

in this manner, each query to H prompts B to open a session with the verifier in $\mathbf{OMMIM}_{\mathsf{ID}}$. Finally, from A's solution to the ROS problem, B successfully closes $Q_{\mathsf{P}_2} + 1$ (out of Q) sessions with the verifier.

If A is successful then $c_{Q_{\mathsf{P}_2}+1} = -1$ and $\wedge$ $\mathbf{A}c = 0$. Furthermore for all $i \in [Q_{\mathsf{P}_2} + 1]$, $\mathsf{H}(\mathbf{Z}_i) = \mathbf{A}_{i,Q_{\mathsf{P}_2}+1}$ and we have

$$\begin{aligned} \mathsf{F}(s'_i) =& \mathsf{F}(\sum_{j=1}^{Q_{\mathsf{P}_2}} \mathbf{A}_{i,j} s_j) = \sum_{j=1}^{Q_{\mathsf{P}_2}} \mathbf{A}_{i,j}(c_j \cdot pk + R_j) = pk \sum_{j=1}^{Q_{\mathsf{P}_2}} \mathbf{A}_{i,j} c_j + R'_{\mathbf{Z}_i} \\ =& pk \cdot c'_{\mathbf{Z}_i} + R'_{\mathbf{Z}_i}, \end{aligned}$$

which is equivalent to $\mathsf{Ver}(pk, R'_{\mathbf{Z}_i}, c'_{\mathbf{Z}_i}, s'_i) = 1$. This shows $b_i = 1$ for all $i \in [Q_{\mathsf{P}_2} + 1]$, which concludes the proof.

5 Canonical Blind Signature Schemes

5.1 Syntax of Canonical Blind Signature Schemes

We now introduce the syntax of a canonical blind signature scheme. We use the term canonical to describe a three-move blind signature protocol in which the signer's and the user's moves consist of picking and sending a random strings of some length, and the user's final signature is a deterministic function of the conversation and the public key. For simplicity, we assume the existence of a public set of parameters *par*.

Definition 2 (Canonical Blind Signature Scheme). *A* canonical blind signature scheme BS *is a tuple of algorithms* BS $= (\mathsf{KG}, \mathsf{S}, \mathsf{U}, \mathsf{Ver})$.

- *The* key generation algorithm KG *outputs a public key/secret key pair* (pk, sk). *We assume that pk implicitly defines a challenge set* $\mathcal{C} = \mathcal{C}(pk)$.

- *The Signer algorithm* S *is split into two algorithms* $\mathsf{S} = (\mathsf{S}_1, \mathsf{S}_2)$. S_1 *returns the first message of the transcript, commitment* R *and the Signers's state* st_S. *Deterministic algorithm* S_2 *takes as input the Signer's state* st_S, *a secret key* sk, *a commitment* R, *and a challenge* $c \in C$. *It returns with the last message of the transcript, the answer* s.
- *The User algorithm* U *is split into two algorithms* $\mathsf{U} = (\mathsf{U}_1, \mathsf{U}_2)$. U_1 *takes as input the public key* pk, *a commitment* R, *a message* m *and returns the Users' state* st_U *and the second message of the transcript, a challenge* $c \in C$. *Deterministic algorithm* U_2 *takes as input the public key* pk, *the transcript* (R, c, s), *a message* m, *the Users' state* st_U *and outputs a signature* σ.
- *The deterministic verification algorithm* Ver *takes as input a message* m, *a signature* σ, *a public key* pk *and outputs a bit* b *indicating accept* ($b = 1$) *or reject* ($b = 0$).

The diagram below depicts an interaction between signer S *and user* U. *For perfect correctness we require that for all* $(pk, sk) \xleftarrow{\$} \mathsf{KG}(par)$, $m \in \{0,1\}^*$, σ *being the output of the interaction of* $\mathsf{S}(sk)$ *and* $\mathsf{U}(pk, m)$ *we have* $\mathsf{Ver}(pk, \sigma, m) = 1$.

Signer $\mathsf{S}(sk)$		User $\mathsf{U}(pk, m)$
$(st_\mathsf{S}, R) \xleftarrow{\$} \mathsf{S}_1(sk)$	$\xrightarrow{R}$	
	$\xleftarrow{c}$	$(st_\mathsf{U}, c) \xleftarrow{\$} \mathsf{U}_1(pk, R, m)$
$s \leftarrow \mathsf{S}_2(sk, R, c, st_\mathsf{S})$	$\xrightarrow{s}$	$\sigma \leftarrow \mathsf{U}_2(pk, R, c, s, m, st_\mathsf{U})$
Output 1		Output σ

We remark that modeling S_2 and U_2 as deterministic algorithms is w.l.o.g. since randomness can be transmitted through the states.

5.2 Security of Canonical Blind Signature Schemes

Security of a Canonical Blind Signature Scheme BS is captured by two security notions: *blindness* and *one more unforgability.*

BLINDNESS. Intuitively, blindness ensures that a signer S that issues signatures on two messages $(\boldsymbol{m}_0, \boldsymbol{m}_1)$ of its own choice to a user U, can not tell in what order it issues them. In particular, S is given both resulting signatures $\boldsymbol{\sigma}_0, \boldsymbol{\sigma}_1$, and gets to keep the transcripts of both interactions with U. Let A be an adversary in the $\mathbf{Blind}^{\mathsf{A}}_{\mathsf{BS}}$ experiment. In BS, the experiment takes the role of an User and A takes the role of the signer. First, the experiment selects a random bit b which will decide the order of adversarially chosen messages in both transcripts. Then A is given access to all three oracles $\mathtt{Init}, \mathtt{U}_1$ and $\mathtt{U}_2$. By convention, A first has to query oracle $\mathtt{Init}$. Then, by the definition of the experiment, A may query at most two sessions. During these two sessions A learns two sets of transcripts $\boldsymbol{T}_0 = (\boldsymbol{R}_0, \boldsymbol{c}_0, \boldsymbol{s}_0)$ and $\boldsymbol{T}_1 = (\boldsymbol{R}_1, \boldsymbol{c}_1, \boldsymbol{s}_1)$. In transcripts $\boldsymbol{T}_0$ and $\boldsymbol{T}_1$, the experiment embeds messages $\boldsymbol{m}_b$ and $\boldsymbol{m}_{1-b}$, respectively. If A behaves honestly, A learns signatures $\boldsymbol{\sigma}_b$ and $\boldsymbol{\sigma}_{1-b}$ on messages $\boldsymbol{m}_b$ and $\boldsymbol{m}_{1-b}$, else nothing at all. At the

```
GAME Blind^A_BS:                                   Oracle U2(sid, s)
00 b <-$ {0,1}; b1 <- b; b2 <- 1 - b               11 If sess_sid ≠ open Then
01 b' <-$ A^{Init,U1,U2}()                         12   Return ⊥
02 Return b = b'                                   13 sess_sid <- closed
                                                   14 s_sid <- s
Oracle Init(pk, m0, m1)    //one, first query      15 σ_{b_sid} <-$ U2(pk, st_sid, R_sid, c_sid, s_sid)
03 Absorb pk as public key                         16 If sess1 = sess2 = closed Then
04 sess1 <- sess2 <- init                          17   If σ0 = ⊥ ∨ σ1 = ⊥ Then
                                                   18     (σ0, σ1) := (⊥, ⊥)
Oracle U1(sid, R)                                  19   Return (σ0, σ1)   //return both signatures
05 If sid ∉ {1,2} ∨ sess_sid ≠ init Then           20 Else
06   Return ⊥             //max. two sessions      21   Return ε
07 sess_sid <- open
08 R_sid <- R
09 (st_sid, c_sid) <-$ U1(pk, R_sid, m_{b_sid})
10 Return (sid, c_sid)
```

Fig. 6. Games defining $\mathbf{Blind}^{\mathsf{A}}_{\mathsf{BS}}$ for a canonical blind signature scheme BS, with the convention that A makes exactly one query to `Init` at the beginning of its execution.

end of the experiment, for A to win, A has to guess the bit b. In Fig. 6 we formally define the $\mathbf{Blind}^{\mathsf{A}}_{\mathsf{BS}}$ experiment. Formally, the advantage function of an adversary A in attacking the blindness of BS is defined as $\mathbf{Adv}^{\mathbf{Blind}}_{\mathsf{BS}}(\mathsf{A}) := \Pr[\mathbf{Blind}^{\mathsf{A}}_{\mathsf{BS}} \Rightarrow 1] - \frac{1}{2}$. We say BS is *perfectly blind* if $\mathbf{Adv}^{\mathbf{Blind}}_{\mathsf{BS}}(\mathsf{A}) = 0$.

OMUF-Security of Blind Signature Schemes. We now define the standard unforgeability notion for blind signatures, namely *one-more unforgeability.* Intuitively, One-More Unforgeability ensures that a user U can not produce a single signature more than it should be able to learn from interactions with the signer S. Let A be an adversary in the $\mathbf{OMUF}^{\mathsf{A}}_{\mathsf{BS}}$ experiment, which takes the role of the User. Let $Q_{\mathsf{S}} \leftarrow Q_{\mathsf{S}_1} + Q_{\mathsf{S}_2}$. Session identifier $sid \in [Q_{\mathsf{S}}]$ is used to model one interaction with the signer. A call to $\mathtt{S}_1$ returns a new session identifier $sid \in [Q_{\mathsf{S}}]$ and sets flag $\mathbf{sess}_{sid}$ to $\mathtt{open}$. A call to $\mathtt{S}_2(sid, \cdot)$ with the same sid sets the flag $\mathbf{sess}_{sid}$ to $\mathtt{closed}$. The closed sessions result in Q_{S_2} different transcripts $(\boldsymbol{R}_k, \boldsymbol{c}_k, \boldsymbol{s}_k)$, where each challenge $\boldsymbol{c}_i$ is adversarially chosen. (The remaining Q_{S_1} abandoned sessions are of the form $(\boldsymbol{R}_k, \perp, \perp)$ and hence do not contain a complete transcript.) A wins the experiment, if it is able to produce $\ell(\mathsf{A}) \geq Q_{\mathsf{S}_2}(\mathsf{A}) + 1$ signatures (on distinct messages) after having interacted with $Q_{\mathsf{S}_2}(\mathsf{A}) \leq Q_{\mathsf{S}_2}$ closed signer sessions (from which he should be able to compute ℓ signatures). In Fig. 7 we formally define the $\mathbf{OMUF}^{\mathsf{A}}_{\mathsf{BS}}$ experiment. Formally, the advantage function of an adversary A in attacking the One-More Unforgeability of BS is defined as $\mathbf{Adv}^{\mathbf{OMUF}}_{\mathsf{BS}}(\mathsf{A}) := \Pr[\mathbf{OMUF}^{\mathsf{A}}_{\mathsf{BS}} \Rightarrow 1]$.

We say that BS is $(\varepsilon, t, Q_{\mathsf{S}_1}, Q_{\mathsf{S}_2})$-**OMUF** secure if for all adversaries A satisfying $\mathbf{Time}(\mathsf{A}) \leq t$, $Q_{\mathsf{S}_2}(\mathsf{A}) \leq Q_{\mathsf{S}_2}$, and $Q_{\mathsf{S}_1}(\mathsf{A}) \leq Q_{\mathsf{S}_1}$, we have $\mathbf{Adv}^{\mathbf{OMUF}}_{\mathsf{BS}}(\mathsf{A}) \leq \varepsilon$. In the random oracle model we say BS is $(\varepsilon, t, Q_{\mathsf{S}_1}, Q_{\mathsf{S}_2}, Q_{\mathsf{H}})$-**OMUF** secure if for all adversaries A variables $\varepsilon, t, Q_{\mathsf{S}_1}$ and Q_{S_2} satisfy the latter conditions and Q_{H} is the number of queries to H.

GAME $\mathrm{OMUF}_{\mathsf{BS}}^{\mathsf{A}}$:

00 $(sk, pk) \leftarrow \mathsf{KG}(par)$
01 $sid \leftarrow 0$ //initialize signer session id
02 $((\boldsymbol{m}_1, \boldsymbol{\sigma}_1), ..., (\boldsymbol{m}_{\ell(\mathsf{A})}, \boldsymbol{\sigma}_{\ell(\mathsf{A})})) \leftarrow \mathsf{A}^{\mathsf{S}_1,\mathsf{S}_2}(pk)$
03 If $\exists i \neq j : \boldsymbol{m}_i = \boldsymbol{m}_j$ Then //all messages have to be distinct
04 Return 0
05 If $\exists i \in [\ell(\mathsf{A})] : \mathsf{Ver}(pk, \boldsymbol{m}_i, \boldsymbol{\sigma}_i) = 0$ Then //all signatures have to be valid
06 Return 0
07 $Q_{\mathsf{S}_1}(\mathsf{A}) \leftarrow \#\{k \mid \mathbf{sess}_k = \mathbf{open}\}$ //#abandoned signer sessions
08 $Q_{\mathsf{S}_2}(\mathsf{A}) \leftarrow \#\{k \mid \mathbf{sess}_k = \mathtt{closed}\}$ //#closed signer sessions
09 If $\ell(\mathsf{A}) \geq Q_{\mathsf{S}_2}(\mathsf{A}) + 1$ Then
10 Return 1
11 Return 0

Oracle S_1
12 $sid \leftarrow sid + 1$
13 $\mathbf{sess}_{sid} \leftarrow \mathbf{open}$
14 $(\boldsymbol{st}_{sid}, \boldsymbol{R}_{sid}) \xleftarrow{\$} \mathsf{S}_1(sk)$
15 Return $(sid, \boldsymbol{R}_{sid})$

Oracle $\mathsf{S}_2(sid, c)$
16 If $\mathbf{sess}_{sid} \neq \mathbf{open}$ Then
17 Return $\perp$
18 $\mathbf{sess}_{sid} = \mathtt{closed}$
19 $\boldsymbol{s}_{sid} \leftarrow \mathsf{S}_2(sk, \boldsymbol{st}_{sid}, \boldsymbol{R}_{sid}, c)$
20 Return $\boldsymbol{s}_{sid}$

Fig. 7. $\mathbf{OMUF}_{\mathsf{BS}}^{\mathsf{A}}$ Game

5.3 Linear Blind Signature Schemes

Let LF be a linear function family and H a random oracle. Figure 8 shows how to construct a blind signature scheme $\mathsf{BS}[\mathsf{LF}, \mathsf{H}]$.

Algorithm $\mathsf{KG}(par)$
00 $sk \xleftarrow{\$} \mathcal{D}$
01 $pk \leftarrow \mathsf{F}(sk)$
02 Return (sk, pk)

Algorithm $\mathsf{S}_1(sk)$
03 $r \xleftarrow{\$} \mathcal{D}$; $R \leftarrow \mathsf{F}(r)$
04 $st_{\mathsf{S}} := r$
05 Return (st_{S}, R)

Algorithm $\mathsf{S}_2(sk, st_{\mathsf{S}}, c)$
06 $r \leftarrow st_{\mathsf{S}}$
07 $s \leftarrow c \cdot sk + r$
08 Return s

Algorithm $\mathsf{U}_1(pk, R, m)$
09 $\alpha \xleftarrow{\$} \mathcal{D}, \beta \xleftarrow{\$} \mathcal{S}$
10 $R' \leftarrow R + \mathsf{F}(\alpha) + \beta \cdot pk$
11 $c' \leftarrow \mathsf{H}(R', m)$
12 $c \leftarrow c' + \beta$
13 $st_{\mathsf{U}} \leftarrow (\alpha, \beta)$
14 Return (c, st_{U})

Algorithm $\mathsf{U}_2(pk, R, c, s, m, st_{\mathsf{U}})$
15 $(\alpha, \beta) \leftarrow st_{\mathsf{U}}$
16 $R' \leftarrow R + \mathsf{F}(\alpha) + \beta \cdot pk$
17 $c' \leftarrow \mathsf{H}(R', m)$
18 $s' \leftarrow s + \alpha$
19 Return $\sigma \leftarrow (c', s')$

Algorithm $\mathsf{Ver}(pk, m, \sigma)$
20 $(c', s') \leftarrow \sigma$
21 $R' \leftarrow \mathsf{F}(s') - c' \cdot pk$
22 If $c' \neq \mathsf{H}(R', m)$ Then
23 Return 0
24 Return 1

Fig. 8. Let LF be a linear function and $\mathsf{H} : \{0,1\}^* \rightarrow \mathcal{C}$ be a hash function. This figure shows the construction of the canonical blind signature scheme $\mathsf{BS}[\mathsf{LF}, \mathsf{H}] = (\mathsf{KG}, \mathsf{S} = (\mathsf{S}_1, \mathsf{S}_2), \mathsf{U} = (\mathsf{U}_1, \mathsf{U}_2), \mathsf{Ver})$.

Theorem 3. *Let* LF *be a linear function family and* H *be a random oracle.* ID[LF] *is* $(\varepsilon', t', Q_{\mathsf{V}}, Q_{\mathsf{P}_1}, Q_{\mathsf{P}_2})$-**OMMIM** *secure if and only if* BS[LF, H] *is* $(\varepsilon, t, Q_{\mathsf{S}_1}, Q_{\mathsf{S}_2}, Q_{\mathsf{H}})$-**OMUF** *secure, where*

$$t' = t, \quad \varepsilon' = \varepsilon, \quad Q_{\mathsf{V}} = Q_{\mathsf{H}} + Q_{\mathsf{S}_2} + 1, \quad Q_{\mathsf{P}_1} = Q_{\mathsf{S}_1}, \quad Q_{\mathsf{P}_2} = Q_{\mathsf{S}_2}.$$

Proof. Let A be an $(\varepsilon, t, Q_{\mathsf{S}_1}, Q_{\mathsf{S}_2}, Q_{\mathsf{H}})$-**OMUF** adversary in the $\mathbf{OMUF}_{\mathsf{BS}}$ experiment. In Fig. 9 we construct an $(\varepsilon', t', Q_{\mathsf{V}}, Q_{\mathsf{P}_1}, Q_{\mathsf{P}_2})$-**OMMIM** adversary B that is executed in the $\mathbf{OMMIM}_{\mathsf{ID}}$ experiment that perfectly simulates A's oracles S_1, S_2 and H via its own oracles P_1, P_2, and V_1, respectively. Suppose that A is successful, i.e., it outputs $Q_{\mathsf{P}_2} + 1$ valid signatures on distinct messages and the number of successfully sessions with the signer is at most Q_{P_2}. Since $\boldsymbol{\sigma}_i$ is a valid signature on $\boldsymbol{m}_i$, B can make a successful query to oracle $\mathsf{V}_2(vSid, \boldsymbol{s}_i')$ in line 06 resulting in $\boldsymbol{b}_i = 1$. Overall, B makes $Q_{\mathsf{P}_2} + 1$ successful queries to V_2 such that the internal counter $\ell(\mathsf{A})$ is set to $Q_{\mathsf{P}_2} + 1$ and B wins. This proves $\varepsilon' \geq \varepsilon$. Moreover, the number of abandoned sessions (denoted as Q_{S_1}) in the $\mathbf{OMUF}_{\mathsf{BS}}$ experiment equals the number of abandoned sessions (denoted as Q_{P_1}) in the $\mathbf{OMMIM}_{\mathsf{ID}}$ experiment and the number of calls to oracle V_1 is bounded by Q_{H} plus additional $Q_{\mathsf{P}} + 1$ implicit calls in Line 04.

```
Adversary B^{P1,P2,V1,V2}(pk):
00 ((m_1, σ_1), ..., (m_{Q_P2+1}, σ_{Q_P2+1})) ← A^{S1,S2,H}(pk)
01 For i ∈ [Q_P2 + 1] do
02    (c'_i, s'_i) ← σ_i
03    R'_i ← F(s'_i) − c'_i · pk
04    h_i ← H(R'_i, m_i)
05    vSid ← SID_(R'_i, m_i)
06    b_i ← V2(vSid, s'_i)

Oracle S1
07 (pSid, R_pSid) ←$ P1
08 Return (pSid, R_pSid)

Oracle S2(pSid, c)
09 s_pSid ← P2(pSid, c)
10 Return s_pSid

Oracle H(R', m)
11 if H_(R',m) ≠ ⊥ Then
12    Return H_(R',m)
13 (vSid, h) ←$ V1(R')
14 SID_(R',m) ← vSid
15 Return H_(R',m) ← h
```

Fig. 9. Reduction from $\mathbf{OMMIM}_{\mathsf{ID}}^{\mathsf{B}}$ to $\mathbf{OMUF}_{\mathsf{BS}}^{\mathsf{A}}$

Let B be an $(\varepsilon, t, Q_{\mathsf{V}}, Q_{\mathsf{P}_1}, Q_{\mathsf{P}_2})$-**OMMIM** adversary in the $\mathbf{OMMIM}_{\mathsf{ID}}$ experiment. In Fig. 10 we construct an $(\varepsilon', t', Q_{\mathsf{S}_1}, Q_{\mathsf{S}_2}, Q_{\mathsf{H}})$-**OMUF** adversary A that is executed in the $\mathbf{OMUF}_{\mathsf{BS}}$ experiment that perfectly simulates B's oracles $\mathsf{P}_1, \mathsf{P}_2$ and V_1 via its own oracles $\mathsf{S}_1, \mathsf{S}_2$ and H, respectively. To simulate oracle V_2, A executes the same code as specified in the $\mathbf{OMMIM}_{\mathsf{ID}}$ experiment, with the only difference being line 20. This additional line does not change the behavior of V_2 and is thus not detectable by B. Suppose that B is successful, i.e., it completes Q_{P_2} sessions as a verifier and $Q_{\mathsf{P}_2} + 1$ sessions as a prover (denoted as $\ell(\mathsf{B})$ in the $\mathbf{OMMIM}_{\mathsf{ID}}$ experiment). From the $Q_{\mathsf{P}_2} + 1$ successful calls of B to V_2, it follows that A learns $Q_{\mathsf{P}_2} + 1$ transcripts $(\boldsymbol{R}, \boldsymbol{c}, \boldsymbol{s})$ from the view of an honest User in BS. Since messages $\boldsymbol{m}$ are constructed by calling U_1, A creates $Q_{\mathsf{P}_2} + 1$ signatures after learning values $\boldsymbol{s}$ by simply following the protocol specification of U_2. This proves $\varepsilon' \geq \varepsilon$. Moreover the number of abandoned sessions (denoted as $Q_{\mathsf{P}_1}(\mathsf{B})$) in the $\mathbf{OMMIM}_{\mathsf{ID}}$ experiment equals the number of abandoned sessions (denoted as $Q_{\mathsf{S}_1}(\mathsf{A})$) in the $\mathbf{OMUF}_{\mathsf{BS}}$ experiment.

```
Adversary A^{S1,S2,H}(pk):
00 vSid ← 0
01 B^{P1,P2,V1,V2}(pk)
02 i ← 1
03 For all k where vSess_k = closed :
04     m_i ← k, σ_i ← (c'_k := c_k − β_k, s'_k := s_k + α_k)
05     i ← i + 1
06 Return (m_1, σ_1), ..., (m_{ℓ+1}, σ_{ℓ+1})

Oracle P1
07 (pSid, R_pSid) ←$ S1
08 Return (pSid, R_pSid)

Oracle P2(pSid, c)
14 s_pSid ← S2(pSid, c)
15 Return s_pSid

Oracle V1(R)
09 vSid ← vSid + 1
10 vSess_vSid ← open
11 (c_vSid, st_vSid) ← U1(pk, R, m := vSid)
12 (α_vSid, β_vSid) ← st_vSid
13 Return (vSid, c_vSid)

Oracle V2(vSid, s)
16 If vSess_vSid ≠ open Then
17     Return ⊥
18 b_vSid ← Ver(pk, R_vSid, c_vSid, s)
19 vSess_vSid ← closed
20 s_vSid ← s
21 Return b_vSid
```

Fig. 10. Reduction from $\mathbf{OMUF}^{\mathsf{A}}_{\mathsf{BS}}$ to $\mathbf{OMMIM}^{\mathsf{B}}_{\mathsf{ID}}$

Theorem 4. *If* LF *is a linear function, then* $\mathsf{BS}[\mathsf{LF},\mathsf{H}]$ *is perfectly blind.*

Proof. Let A be an adversary playing in game $\mathbf{Blind}^{\mathsf{A}}_{\mathsf{BS}[\mathsf{LF},\mathsf{H}]}$. After its execution, A holds $(\boldsymbol{m}_0, \boldsymbol{\sigma}_0)$, $(\boldsymbol{m}_1, \boldsymbol{\sigma}_1)$ where $\boldsymbol{\sigma}_0$ is a signature on $\boldsymbol{m}_0$ and $\boldsymbol{\sigma}_1$ is a signature on $\boldsymbol{m}_1$. (Here we assume without loss of generality that both signatures are valid as otherwise A obtains $\boldsymbol{\sigma}_0 = \boldsymbol{\sigma}_1 = \perp$ and thus $\mathbf{Adv}^{\mathsf{A}}_{\mathbf{Blind},\mathsf{BS}[\mathsf{LF},\mathsf{H}]} = 0$.) Adversary A furthermore learns two transcripts $\boldsymbol{T}_1 = (\boldsymbol{R}_1, c_1, \boldsymbol{s}_1)$ and $\boldsymbol{T}_2 = (\boldsymbol{R}_2, c_2, \boldsymbol{s}_2)$ from its interaction with the first and the second signer session, respectively. The goal of A is to match the message/signature pairs with the two transcripts.

We show that there exists no adversary which is able to distinguish, whether the message $\boldsymbol{m}_0$ was used by the experiment to create Transcript $\boldsymbol{T}_1$ or $\boldsymbol{T}_2$. We argue that for all sessions $1 \leq i \leq 2$ and indexes $0 \leq j \leq 1$, the tuple $(\boldsymbol{T}_i, \boldsymbol{m}_j, \boldsymbol{\sigma}_j)$ completely determines $\boldsymbol{st}_j = (\boldsymbol{\alpha}_{(i,j)}, \boldsymbol{\beta}_{(i,j)})$. This implies that given A's view, it is equally likely that the experiment was executed with $b = 0$ or $b = 1$ since for both choices $b \in \{0, 1\}$ there exists properly distributed states $(\boldsymbol{st}_0, \boldsymbol{st}_1)$ that would have resulted in A's view.

It remains to argue that $\boldsymbol{T}_i = (\boldsymbol{R}_i, c_i, \boldsymbol{s}_i)$, $\boldsymbol{m}_j$, and $\boldsymbol{\sigma}_j = (c_j', \boldsymbol{s}_j')$ determine values $\boldsymbol{\alpha}_{(i,j)}, \boldsymbol{\beta}_{(i,j)}$ such that $c_j' = \mathsf{H}(\boldsymbol{R}_i + \boldsymbol{\beta}_{(i,j)} \cdot pk + \mathsf{F}(\boldsymbol{\alpha}_{(i,j)}), \boldsymbol{m}_j)$ and $\boldsymbol{\alpha}_{(i,j)} =$

$\boldsymbol{s}_j' - \boldsymbol{s}_i, \boldsymbol{\beta}_{(i,j)} = \boldsymbol{c}_i - \boldsymbol{c}_j'$. Uniformity of $(\boldsymbol{\alpha}_{(i,j)}, \boldsymbol{\beta}_{(i,j)})$ is implied by uniformity of $(\boldsymbol{s}_j', \boldsymbol{c}_j')$, which come from the experiment.

Since $\boldsymbol{T}_i$ is a valid transcript, we have $\mathsf{F}(\boldsymbol{s}_i) = \boldsymbol{R}_i + \boldsymbol{c}_i \cdot pk$. Therefore

$$\begin{aligned} \boldsymbol{R}_i + \boldsymbol{\beta}_{(i,j)} \cdot pk + \mathsf{F}(\boldsymbol{\alpha}_{(i,j)}) &= \boldsymbol{R}_i + (\boldsymbol{c}_i - \boldsymbol{c}_j') \cdot pk + \mathsf{F}(\boldsymbol{s}_j' - \boldsymbol{s}_i) \\ &= \boldsymbol{R}_i + \boldsymbol{c}_i \cdot pk - \mathsf{F}(\boldsymbol{s}_i) + \mathsf{F}(\boldsymbol{s}_j') - \boldsymbol{c}_j' \cdot pk \\ &= \mathsf{F}(\boldsymbol{s}_j') - \boldsymbol{c}_j' \cdot pk. \end{aligned}$$

Since $\boldsymbol{\sigma}_j$ is a valid signature on $\boldsymbol{m}_j$ we have $\mathsf{H}(\mathsf{F}(\boldsymbol{s}_j') - \boldsymbol{c}_j' \cdot pk, \boldsymbol{m}_j) = \boldsymbol{c}_j'$ which concludes the proof.

Corollary 1. *Let* LF *be a linear function family with a torsion-free element from the kernel. If* LF *is* (ε', t')-**CR** *secure, then* $\mathsf{BS}[\mathsf{LF}, \mathsf{H}]$ *is* $(\varepsilon, t, Q_{\mathsf{S}_1}, Q_{\mathsf{S}_2}, Q_{\mathsf{H}})$-**OMUF** *secure where*

$$t' = 2t, \quad \varepsilon' = O\left(\left(\varepsilon - \frac{(Q + Q_{\mathsf{S}})^{Q_{\mathsf{S}_2}+1}}{q}\right)^2 \frac{1}{Q^2 Q_{\mathsf{S}_2}^3}\right),$$

$Q_{\mathsf{S}} = Q_{\mathsf{S}_2} + Q_{\mathsf{S}_1}$ *and* $Q = Q_{\mathsf{H}} + Q_{\mathsf{S}_2} + 1$. *Moreover,* $\mathsf{BS}[\mathsf{LF}, \mathsf{H}]$ *is perfectly blind.*

Proof. The proof of the one-more unforgability security follows from combining Theorems 1 and 3. Perfect blindness follows directly from Theorem 4.

6 Proof of Theorem 1

Before we give the proof of Theorem 1, we provide some intuition about the difficulty that arises in the context of proving the **OMMIM**-security of $\mathsf{ID}[\mathsf{LF}]$ and how our proof overcomes it. The main issue is that the adversary M in **OMMIM** can interleave sessions between the oracles $\mathsf{P}_1, \mathsf{P}_2$ and $\mathtt{V}_1, \mathtt{V}_2$. This gives M strong adaptive capabilities which lead to the ROS-attack described in Sect. 4.2. The ROS-attack is reflected in Corollary 2, which can be translated into an upper bound on M's success probability of providing our reduction with two identical values $\hat{\chi}, \hat{\chi}'$ that result from running the adversary twice with fixed public key pk and randomness ω, but (partially) different replies $\boldsymbol{h}, \boldsymbol{h}'$ to $\mathtt{V}_1$. If the adversary succeeds in setting $\hat{\chi} = \hat{\chi}'$, the reduction fails in recovering a collision with respect to LF, i.e., values $\hat{\chi} \neq \hat{\chi}'$ s.t. $\mathsf{LF}(\hat{\chi}) = \mathsf{LF}(\hat{\chi}')$.

To prove the bound in Corollary 2, our proof follows the ideas of [21], but takes into account also the abandoned sessions with P_1, which [21] does not consider. The intuitive idea behind ensuring $\hat{\chi} \neq \hat{\chi}'$ is to run M on an instance $I = pk$ that could be the result of applying F to either sk or $\widehat{sk} = sk + z^*$ from the domain $\mathcal{D}$ of F. One can show that from M's perspective, the resulting view is identical in both cases (Lemma 7). On the other hand, since $\hat{\chi}$ depends non-trivially on sk (or $\widehat{sk}$, respectively), it should take (with high probability) different values from the reduction's point of view, depending on whether the reduction used sk or $sk + z^*$ as a preimage to pk. Indeed, this intuition is supported by Corollary 2.

However, Corollary 2 can only be translated into an upper bound on the probability that $\hat{\chi}$ takes the same *particular* value $C(sk, \omega, \boldsymbol{h})$, regardless of whether sk or $\hat{sk}$ was used by the reduction. Intuitively, $C(sk, \omega, \boldsymbol{h})$ is the value that is most likely taken by the random variable $\hat{\chi}'$, which occurs as the result of rewinding M with the same sk, ω, but a partially different set of V_1-replies $\boldsymbol{h}'$ (i.e., the probability is over the fresh values in $\boldsymbol{h}'$). To ensure that $\hat{\chi} \neq \hat{\chi}'$, the analysis first defines the set $\mathcal{B}$ of tuples $(sk, \omega, \boldsymbol{h})$ which yield a successful run of M, but for which $\hat{\chi}(sk, \omega, \boldsymbol{h}) \neq C(sk, \omega, \boldsymbol{h})$. It then estimates the probability that both tuples $(sk, \omega, \boldsymbol{h}), (sk, \omega, \boldsymbol{h}')$ that are used to run M, are tuples from the set $\mathcal{B}$. The final step of the proof is to leverage this fact to obtain a lower bound on the success probability of the reduction, i.e., to ensure that $\hat{\chi} \neq \hat{\chi}'$ (Lemma 2). To argue that not only both runs of M are successful, but yield tuples in $\mathcal{B}$, we present a more general version of the forking lemma by Bellare and Neven [6].

6.1 The Reduction Algorithm

Let M be an $(\varepsilon, t, Q_{\mathsf{V}}, Q_{\mathsf{P}_1}, Q_{\mathsf{P}_2})$-**OMMIM** adversary that plays in game $\mathbf{OMMIM}_{\mathsf{ID[LF]}}$. Without loss of generality, we will assume throughout the proof that $Q_{\mathsf{P}_1}(\mathsf{M}) = Q_{\mathsf{P}_1}, Q_{\mathsf{P}_2}(\mathsf{M}) = Q_{\mathsf{P}_2}, Q_{\mathsf{V}}(\mathsf{M}) = Q_{\mathsf{V}}, \ell(\mathsf{M}) = Q_{\mathsf{P}_2} + 1$, as well as $Q_{\mathsf{P}_1} \geq Q_{\mathsf{P}_2}$.

For $1 \leq i \leq Q_{\mathsf{P}_2}+1$, we define an auxiliary algorithm A_i which 'sandboxes' M and that will be used later by another adversary B to break collision resistance of LF. More concretely, A_i obtains as input an instance $I = sk$, runs M on random tape ω and uses vector $\boldsymbol{h} \in \mathcal{C}^{Q_{\mathsf{V}}}$ to answer M's Q_{V} queries to V_1. The description of algorithm A_i is given in Fig. 11. Note that A_i is deterministic for fixed randomness ω.

ANALYSIS OF A_i. To analyze A_i, we now introduce some notation. First, consider the variables $\hat{\boldsymbol{J}}_i, \hat{\boldsymbol{\chi}}_i, \hat{\boldsymbol{s}}'$, and $\hat{\boldsymbol{h}}_i$ defined on Lines 32 through 35 of Fig. 11. These variables are introduced to simplify the referencing of values associated with successful calls to the verification oracle $\mathsf{V}_2(vSid, \cdot)$ over the course of the proof. Concretely, the variable

$$\hat{\boldsymbol{\chi}}_i = \hat{\boldsymbol{s}}_i' - \hat{\boldsymbol{h}}_i \cdot sk$$

results from the *i-th successful call* to the verification oracle $\mathsf{V}_2(vSid, \cdot)$, whereas the index $\hat{\boldsymbol{J}}_i$ indicates which session identity $vSid$ corresponds to this call.

We will fix an execution of A_i via the tuples $I = sk$, $\boldsymbol{h}$, and A_i's randomness ω. We define the set $\mathcal{W}$ of *successful inputs of* A_i as the set of all such tuples $(I, \omega, \boldsymbol{h})$ which lead to a successful run of A_i, i.e.,

$$\mathcal{W} := \{(I, \omega, \boldsymbol{h}) \mid \hat{\boldsymbol{J}}_i \neq 0; (\hat{\boldsymbol{J}}_i, \hat{\boldsymbol{\chi}}_i) \leftarrow \mathsf{A}_i(I, \boldsymbol{h}; \omega)\}$$

Note that $\mathcal{W}$ is independent of i and, by construction of A_i,

$$\Pr_{(I,\omega,\boldsymbol{h}) \xleftarrow{\$} (\mathcal{I} \times \Omega \times \mathcal{C}^{Q_{\mathsf{V}}})} [(I, \omega, \boldsymbol{h}) \in \mathcal{W}] = \mathbf{Adv}_{\mathsf{ID[LF]}}^{\mathbf{OMMIM}}(\mathsf{M}) = \varepsilon.$$

We can view $\hat{\boldsymbol{J}}_i, \hat{\boldsymbol{\chi}}_i, \hat{\boldsymbol{s}}'$, and $\hat{\boldsymbol{h}}_i$ as random variables whose distribution is induced by the the uniform distribution on $(\mathcal{I} \times \Omega \times \mathcal{C}^{Q_V})$. Furthermore, their outcome is uniquely determined given $(I, \omega, \boldsymbol{h}) \in \mathcal{W}$, so let us write in this case

$$\left(\hat{\boldsymbol{J}}_i(I,\omega,\boldsymbol{h}), \hat{\boldsymbol{\chi}}_i(I,\omega,\boldsymbol{h})\right) \leftarrow \mathsf{A}_i(I,\boldsymbol{h};\omega).$$

```
Adversary A_i(I = sk, h; ω):
00 Parse (ω_M, r) ← ω
01 R ← F(r)
02 pk ← F(sk)
03 ctr ← 0; pSid ← 0; vSid ← 0
04 M^{P1,P2,V1,V2}(pk)
05 ℓ(M) ← #{k | vSess_k = closed ∧ b_k = 1}
06 Q_P2(M) ← #{k | pSess_k = closed}
07 Q_P1(M) ← #{k | pSess_k = open}
08 Q_V(M) ← vSid
09 If (ℓ(M) ≥ Q_P2(M) + 1) Then
10     Return (Ĵ_i, χ̂_i)
11 Return (Ĵ_i, χ̂_i) ← (0,0)

Procedure P1
12 pSid ← pSid + 1
13 pSess_pSid ← open
14 c_pSid ← ⊥
15 Return (pSid, R_pSid)

Procedure P2(pSid, c)
16 If pSess_pSid ≠ open Then
17     Return ⊥
18 pSess_pSid ← closed
19 s_pSid ← c · sk + r_pSid
20 c_pSid ← c
21 Return s_pSid

Procedure V1(R')
22 vSid ← vSid + 1
23 R'_vSid ← R'
24 vSess_pSid ← open
25 Return (vSid, h_vSid)

Procedure V2(vSid, s')
26 If vSess_vSid ≠ open Then
27     Return ⊥
28 S'_vSid ← F(s')
29 vSess_vSid ← closed
30 If S'_vSid = h_vSid · pk + R'_vSid Then
31     ctr ← ctr + 1
32     ŝ'_ctr ← s'
33     ĥ_ctr ← h_vSid
34     χ̂_ctr ← ŝ'_ctr − ĥ_ctr · sk
35     Ĵ_ctr ← vSid
36     b'_vSid ← 1
37 Else
38     b'_vSid ← 0
39 Return b'_vSid
```

Fig. 11. Wrapping adversaries A_i for $1 \leq i \leq Q_{\mathsf{P}_2} + 1$

In the following, when stating probability distributions over I, ω, and $\boldsymbol{h}$, unless specified differently, we will always refer to the uniform distributions. That is, $(I, \omega, \boldsymbol{h}) \xleftarrow{\$} (\mathcal{I} \times \Omega \times \mathcal{C}^{Q_V})$.

We consider the following probability for fixed $(I, \omega, \boldsymbol{h}), j, c$ and i:

$$\Pr_{\boldsymbol{h}' \xleftarrow{\$} \mathcal{C}^{Q_V} | \boldsymbol{h}_{[j-1]}} [\hat{\boldsymbol{J}}_i(I,\omega,\boldsymbol{h}') = j \wedge \hat{\boldsymbol{\chi}}_i(I,\omega,\boldsymbol{h}') = c], \tag{3}$$

where the conditional probability $\boldsymbol{h}' \xleftarrow{\$} \mathcal{C}^{Q_V} | \boldsymbol{h}_{[j-1]}$ was introduced in Sect. 2.

We denote by $c_{i,j}(I, \omega, \boldsymbol{h})$ the lexicographically first value c s.t. the probability in (3) is maximized when $(I, \omega, \boldsymbol{h}), j, i$ are fixed. We then write $C_i(I, \omega, \boldsymbol{h}) = c_{i,\hat{\boldsymbol{J}}_i(I,\omega,\boldsymbol{h})}(I, \omega, \boldsymbol{h})$. For fixed i, j, let us define $\mathcal{B}_{i,j} \subset \mathcal{W}$ as

```
Adversary B(par):
00 i* <-$ [Q_P2 + 1]
01 h <-$ C^Qv
02 ω <-$ Ω
03 sk <-$ D
04 (Ĵ_i*, X̂_i*) <- A_i*(I = sk, h; ω)          //First execution of A_i*
05 If Ĵ_i* = 0
06     Return ⊥
07 h' <-$ C^Qv | h_[Ĵ_i* - 1]                   //Conditionally resample h'
08 (Ĵ'_i*, X̂'_i*) <- A_i*(I = sk, h'; ω)        //Second execution of A_i*
09 If (Ĵ'_i* = Ĵ_i*) ∧ (X̂_i* ≠ X̂'_i*) Then
10     return (X̂_i*, X̂'_i*)
11 Return ⊥
```

Fig. 12. Adversary B against CR of LF.

$$\mathcal{B}_{i,j} := \{(I, \omega, \boldsymbol{h}) \in \mathcal{W} \mid \hat{\boldsymbol{J}}_i(I, \omega, \boldsymbol{h}) = j \wedge \hat{\mathcal{X}}_i(I, \omega, \boldsymbol{h}) \neq C_i(I, \omega, \boldsymbol{h})\}.$$

and

$$\beta_{i,j} = \Pr_{(I,\omega,\boldsymbol{h}) \xleftarrow{\$} (\mathcal{I} \times \Omega \times \mathcal{C}^{Q_\mathsf{V}})} [(I, \omega, \boldsymbol{h}) \in \mathcal{B}_{i,j}]$$

$$\delta_{i,j} = \Pr_{(I,\omega,\boldsymbol{h}) \xleftarrow{\$} (\mathcal{I} \times \Omega \times \mathcal{C}^{Q_\mathsf{V}}), \boldsymbol{h}' \xleftarrow{\$} \mathcal{C}^{Q_\mathsf{V}} | \boldsymbol{h}_{[j-1]}} \begin{bmatrix} \hat{\mathcal{X}}_i(I, \omega, \boldsymbol{h}') \neq \hat{\mathcal{X}}_i(I, \omega, \boldsymbol{h}) \\ \wedge \hat{\boldsymbol{J}}_i(I, \omega, \boldsymbol{h}) = \hat{\boldsymbol{J}}_i(I, \omega, \boldsymbol{h}') = j \end{bmatrix}.$$

Lemma 2. *For all* i, j*:* $\delta_{i,j} \geq \beta_{i,j} \left(\frac{\beta_{i,j}}{8} - \frac{1}{2q} \right)$.

The proof of this lemma is postponed to Sect. 6.3.

Lemma 3. *There exist* $i \in [Q_{\mathsf{P}_2} + 1], j \in [Q_\mathsf{V}]$ *such that* $\beta_{i,j} > \left(\varepsilon - \frac{Q_\mathsf{V}^{Q_{\mathsf{P}_2}+1} \cdot \binom{Q_{\mathsf{P}_2} + Q_{\mathsf{P}_1}}{Q_{\mathsf{P}_1}}}{q} \right) \cdot \frac{1}{2 Q_\mathsf{V} (Q_{\mathsf{P}_2} + 1)}$.

The proof of this lemma is postponed to Sect. 6.4.

Adversary B against CR of LF. We are now ready to describe our (ε', t')-adversary B depicted in Fig. 12, which plays in the collision resistance game of LF. B works roughly as follows. It first samples randomness $\omega \xleftarrow{\$} \Omega$, a secret key $sk \xleftarrow{\$} \mathcal{D}$, a vector $\boldsymbol{h} \xleftarrow{\$} \mathcal{C}^{Q_\mathsf{V}}$, and an index $i^* \xleftarrow{\$} [Q_{\mathsf{P}_2} + 1]$ and runs A_{i^*} on input $(I = sk, \boldsymbol{h}; \omega)$. It samples a second random vector $\boldsymbol{h}'$ as $\boldsymbol{h}' \xleftarrow{\$} \mathcal{C}^{Q_\mathsf{V}} | \boldsymbol{h}_{[\hat{\boldsymbol{J}}_{i^*} - 1]}$ and runs A_{i^*} a second time with the same randomness ω and the same instance I, but replacing $\boldsymbol{h}$ by $\boldsymbol{h}'$. In the case that B does not abort, note that by definition of A_{i^*},

$$\begin{aligned} \mathsf{F}(\hat{\mathcal{X}}_{i^*}) &= \mathsf{F}(\hat{\boldsymbol{s}}'_{i^*} - \hat{\boldsymbol{h}}_{i^*} \cdot sk) \\ &= \boldsymbol{S}'_{\hat{\boldsymbol{J}}_{i^*}} - \boldsymbol{h}_{\hat{\boldsymbol{J}}_{i^*}} \cdot pk = \boldsymbol{R}'_{\hat{\boldsymbol{J}}_{i^*}} \end{aligned}$$

Because A_{i^*} sees identical answers for the first $\hat{\boldsymbol{J}}_{i^*} - 1$ queries to V_1, it behaves identically in both runs until it receives the answer to the $\hat{\boldsymbol{J}}_{i^*}$-th query to V_1. In particular, A_{i^*} poses the same $\hat{\boldsymbol{J}}_{i^*}$-th query to V_1 which means that $\mathsf{F}(\hat{\chi}'_{i^*}) = \boldsymbol{R}'_{\hat{\boldsymbol{J}}_{i^*}}$ and therefore also $\mathsf{F}(\hat{\chi}_{i^*}) = \mathsf{F}(\hat{\chi}'_{i^*})$. We now consider

$$\begin{aligned}
\varepsilon' = \mathbf{Adv}^{\mathbf{CR}}_{\mathsf{LF}}(\mathsf{B}) &= \Pr_{par \xleftarrow{\$} \mathsf{PGen}, (\hat{\chi}_{i^*}, \hat{\chi}'_{i^*}) \xleftarrow{\$} \mathsf{B}(par)} [\hat{\chi}_{i^*} \neq \hat{\chi}'_{i^*} \wedge \mathsf{F}(\hat{\chi}_{i^*}) = \mathsf{F}(\hat{\chi}'_{i^*})] \\
&= \sum_{j=1}^{Q_{\mathsf{V}}} \Pr[\hat{\chi}_{i^*} \neq \hat{\chi}'_{i^*} \wedge \mathsf{F}(\hat{\chi}_{i^*}) = \mathsf{F}(\hat{\chi}'_{i^*}) \wedge \hat{\boldsymbol{J}}_{i^*} = \hat{\boldsymbol{J}}'_{i^*} = j] \\
&= \sum_{j=1}^{Q_{\mathsf{V}}} \Pr[\hat{\chi}_{i^*} \neq \hat{\chi}'_{i^*} \wedge \hat{\boldsymbol{J}}_{i^*} = \hat{\boldsymbol{J}}'_{i^*} = j] = \sum_{j=1}^{Q_{\mathsf{V}}} \delta_{i^*,j} \\
&\geq \frac{1}{Q_{\mathsf{P}_2} + 1} \cdot \max_{i \in [Q_{\mathsf{P}_2}+1]} \sum_{j=1}^{Q_{\mathsf{V}}} \delta_{i,j} \\
&\geq \max_{i,j} \frac{\beta_{i,j}}{2(Q_{\mathsf{P}_2} + 1)} \left(\frac{\beta_{i,j}}{4} - \frac{1}{q} \right),
\end{aligned}$$

where for the first inequality we used that $\sum \delta_{i^*,j} = \max_i \sum \delta_{i,j}$ with probability at least $1/(Q_{\mathsf{P}_2} + 1)$ and in the last step we applied Lemma 2. By Lemma 3 we finally obtain

$$\begin{aligned}
\varepsilon' &\geq \frac{\varepsilon - \frac{Q_{\mathsf{V}}^{Q_{\mathsf{P}_2}+1} \cdot \binom{Q_{\mathsf{P}_2}+Q_{\mathsf{P}_1}}{Q_{\mathsf{P}_1}}}{q}}{32 Q_{\mathsf{V}}^2 (Q_{\mathsf{P}_2} + 1)^3} \cdot \left(\varepsilon - \frac{Q_{\mathsf{V}}^{Q_{\mathsf{P}_2}+1} \cdot \binom{Q_{\mathsf{P}_2}+Q_{\mathsf{P}_1}}{Q_{\mathsf{P}_1}}}{q} - \frac{16 Q_{\mathsf{V}}^2 (Q_{\mathsf{P}_2} + 1)^2}{q} \right) \\
&= O\left(\left(\varepsilon - \frac{(Q_{\mathsf{V}} Q_{\mathsf{P}_1})^{Q_{\mathsf{P}_2}+1}}{q} \right)^2 \frac{1}{Q_{\mathsf{V}}^2 Q_{\mathsf{P}_2}^3} \right),
\end{aligned}$$

where the last equality holds for $Q_{\mathsf{P}_1} \geq Q_{\mathsf{P}_2}$.

6.2 A Generalized Forking Lemma

In this section we will introduce our *Subset Forking Lemma*, a generalization of the forking lemma that will be useful for proving Lemma 2.

Lemma 4 (Subset Splitting Lemma). *Let $\mathcal{B} \subset \mathcal{X} \times \mathcal{Y}$ be such that*

$$\Pr_{(x,y) \xleftarrow{\$} \mathcal{X} \times \mathcal{Y}} [(x, y) \in \mathcal{B}] \geq \varepsilon.$$

For any $\alpha \leq \varepsilon$, define

$$\mathcal{B}_\alpha = \left\{ (x, y) \in \mathcal{X} \times \mathcal{Y} \mid \Pr_{y' \xleftarrow{\$} \mathcal{Y}} [(x, y') \in \mathcal{B}] \geq \varepsilon - \alpha \right\}.$$

Then

$$\Pr_{y, y' \xleftarrow{\$} \mathcal{Y}, x \xleftarrow{\$} \mathcal{X}} [(x, y') \in \mathcal{B} \wedge (x, y) \in \mathcal{B}] \geq (\varepsilon - \alpha) \cdot \alpha.$$

Proof. The standard splitting lemma [21] states that

$$\forall (x,y) \in \mathcal{B}_\alpha \colon \Pr_{y' \xleftarrow{\$} \mathcal{Y}}[(x,y') \in \mathcal{B}] \geq \varepsilon - \alpha \tag{4}$$

$$\Pr_{(x,y) \xleftarrow{\$} \mathcal{B}}[(x,y) \in \mathcal{B}_\alpha] \geq \alpha/\varepsilon \tag{5}$$

For the conditional probability, we have that

$$\begin{aligned}
&\Pr_{y,y' \xleftarrow{\$} \mathcal{Y}, x \xleftarrow{\$} \mathcal{X}}[(x,y') \in \mathcal{B} \mid (x,y) \in \mathcal{B}] \\
\geq\; &\Pr_{y,y' \xleftarrow{\$} \mathcal{Y}, x \xleftarrow{\$} \mathcal{X}}[(x,y') \in \mathcal{B} \wedge (x,y) \in \mathcal{B}_\alpha \mid (x,y) \in \mathcal{B}] \\
=\; &\Pr_{y,y' \xleftarrow{\$} \mathcal{Y}, x \xleftarrow{\$} \mathcal{X}}[(x,y') \in \mathcal{B} \mid (x,y) \in \mathcal{B}_\alpha \cap \mathcal{B}] \cdot \Pr_{(x,y) \xleftarrow{\$} \mathcal{X} \times \mathcal{Y}}[(x,y) \in \mathcal{B}_\alpha \mid (x,y) \in \mathcal{B}] \\
=\; &\Pr_{y,y' \xleftarrow{\$} \mathcal{Y}, x \xleftarrow{\$} \mathcal{X}}[(x,y') \in \mathcal{B} \mid (x,y) \in \mathcal{B}_\alpha] \cdot \Pr_{(x,y) \xleftarrow{\$} \mathcal{X} \times \mathcal{Y}}[(x,y) \in \mathcal{B}_\alpha \mid (x,y) \in \mathcal{B}] \\
=\; &\Pr_{y,y' \xleftarrow{\$} \mathcal{Y}, x \xleftarrow{\$} \mathcal{X}}[(x,y') \in \mathcal{B} \mid (x,y) \in \mathcal{B}_\alpha] \cdot \Pr_{(x,y) \xleftarrow{\$} \mathcal{B}}[(x,y) \in \mathcal{B}_\alpha] \\
\geq\; &(\varepsilon - \alpha) \cdot \frac{\alpha}{\varepsilon},
\end{aligned}$$

where the inequalities follow from (4) and (5), respectively. We conclude the proof by

$$\begin{aligned}
&\Pr_{y,y' \xleftarrow{\$} \mathcal{Y}, x \xleftarrow{\$} \mathcal{X}}[(x,y') \in \mathcal{B} \wedge (x,y) \in \mathcal{B}] \\
&= \Pr_{y,y' \xleftarrow{\$} \mathcal{Y}, x \xleftarrow{\$} \mathcal{X}}[(x,y') \in \mathcal{B} \mid (x,y) \in \mathcal{B}] \cdot \Pr_{(x,y) \xleftarrow{\$} \mathcal{X} \times \mathcal{Y}}[(x,y) \in \mathcal{B}] \\
&\geq (\varepsilon - \alpha) \cdot \frac{\alpha}{\varepsilon} \cdot \varepsilon = (\varepsilon - \alpha) \cdot \alpha.
\end{aligned}$$

Lemma 5 (Subset Forking Lemma). *Fix any integer $Q \geq 1$ and a set $\mathcal{H}$ of size > 2 as well as a set of side outputs Σ, instances $\mathcal{I}$, and a randomness space Ω. Let C be an algorithm that on input $(I, \boldsymbol{h}) \in \mathcal{I} \times \mathcal{H}^Q$ and randomness $\omega \in \Omega$ returns a tuple (j, σ), where $1 \leq j \leq Q$ and $\sigma \in \Sigma$. We partition its input space $\mathcal{I} \times \Omega \times \mathcal{H}^Q$ into sets $\mathcal{W}_1, \ldots, \mathcal{W}_Q$ where for fixed $1 \leq j \leq Q$, $\mathcal{W}_j$ is the set of all $(I, \omega, \boldsymbol{h})$ that result in $(j, \sigma) \leftarrow \mathsf{C}(\boldsymbol{h}, I; \omega)$ for some arbitrary side output σ.*

For any $1 \leq j \leq Q$ and $\mathcal{B} \subseteq \mathcal{W}_j$ define

$$\mathtt{acc}(\mathcal{B}) := \Pr_{(I,\omega,\boldsymbol{h}) \xleftarrow{\$} \mathcal{I} \times \Omega \times \mathcal{H}^Q}[(I, \omega, \boldsymbol{h}) \in \mathcal{B}]$$

$$\mathtt{frk}(\mathcal{B}, j) := \Pr_{(I,\omega,\boldsymbol{h}) \xleftarrow{\$} \mathcal{I} \times \Omega \times \mathcal{H}^Q, \boldsymbol{h}' \xleftarrow{\$} \mathcal{C}^Q \vee | \boldsymbol{h}_{[j-1]}} \begin{bmatrix} \boldsymbol{h}_j \neq \boldsymbol{h}'_j \\ (I, \omega, \boldsymbol{h}) \in \mathcal{B} \wedge (I, \omega, \boldsymbol{h}') \in \mathcal{B} \end{bmatrix}.$$

Then

$$\mathtt{frk}(\mathcal{B}, j) \geq \mathtt{acc}(\mathcal{B}) \cdot \left(\frac{\mathtt{acc}(\mathcal{B})}{4} - \frac{1}{|\mathcal{H}|} \right).$$

Proof. By applying Lemma 4 to $\varepsilon = \mathtt{acc}(B)$, $\alpha := \varepsilon/2$, and to the two sets $\mathcal{X} = \mathcal{I} \times \Omega \times \mathcal{H}^{j-1}$ and $\mathcal{Y} = \mathcal{H}^{Q-j+1}$, we obtain

$$\Pr_{(I,\omega,\boldsymbol{h}) \xleftarrow{\$} \mathcal{I}\times\Omega\times\mathcal{H}^Q, \boldsymbol{h}' \xleftarrow{\$} \mathcal{C}^Q \vee |\boldsymbol{h}_{[j-1]}} [(I,\omega,\boldsymbol{h}) \in \mathcal{B} \wedge (I,\omega,\boldsymbol{h}') \in \mathcal{B}] \geq \frac{\mathtt{acc}^2(\mathcal{B})}{4}.$$

Next, we observe that

$$\begin{aligned}
&\mathtt{frk}(\mathcal{B}, j) = \Pr[(I,\omega,\boldsymbol{h}) \in \mathcal{B} \wedge (I,\omega,\boldsymbol{h}') \in \mathcal{B} \wedge \boldsymbol{h}_j \neq \boldsymbol{h}'_j] \\
&= \Pr[(I,\omega,\boldsymbol{h}) \in \mathcal{B} \wedge (I,\omega,\boldsymbol{h}') \in \mathcal{B}] - \Pr[(I,\omega,\boldsymbol{h}) \in \mathcal{B} \wedge (I,\omega,\boldsymbol{h}') \in \mathcal{B} \wedge \boldsymbol{h}_j = \boldsymbol{h}'_j] \\
&\geq \Pr[(I,\omega,\boldsymbol{h}) \in \mathcal{B} \wedge (I,\omega,\boldsymbol{h}') \in \mathcal{B}] - \Pr[(I,\omega,\boldsymbol{h}) \in \mathcal{B} \wedge \boldsymbol{h}_j = \boldsymbol{h}'_j] \\
&= \Pr[(I,\omega,\boldsymbol{h}) \in \mathcal{B} \wedge (I,\omega,\boldsymbol{h}') \in \mathcal{B}] - \frac{\Pr[(I,\omega,\boldsymbol{h}) \in \mathcal{B}]}{|\mathcal{H}|},
\end{aligned}$$

where the last equation follows from independence and uniformity of $\boldsymbol{h}_j$ and $\boldsymbol{h}'_j$. We continue with

$$\begin{aligned}
&= \Pr[(I,\omega,\boldsymbol{h}) \in \mathcal{B} \wedge (I,\omega,\boldsymbol{h}') \in \mathcal{B}] - \frac{\Pr[(I,\omega,\boldsymbol{h}) \in \mathcal{B}]}{|\mathcal{H}|} \\
&\geq \frac{\mathtt{acc}^2(\mathcal{B})}{4} - \frac{\Pr[(I,\omega,\boldsymbol{h}) \in \mathcal{B}]}{|\mathcal{H}|} = \frac{\mathtt{acc}^2(\mathcal{B})}{4} - \frac{\mathtt{acc}(\mathcal{B})}{|\mathcal{H}|} \\
&= \mathtt{acc}(\mathcal{B}) \cdot \left(\frac{\mathtt{acc}(\mathcal{B})}{4} - \frac{1}{|\mathcal{H}|} \right),
\end{aligned}$$

which completes the proof.

Note that Lemma 5 implies the version of the Forking Lemma in [6]. Namely, by, defining the set $\mathcal{W} = \bigcup_j \mathcal{W}_j$, $\mathtt{acc}(\mathcal{W}) = \Pr_{(I,\omega,\boldsymbol{h}) \xleftarrow{\$} \mathcal{I}\times\Omega\times\mathcal{H}^Q, (j,\sigma) \leftarrow \mathsf{C}(I,\boldsymbol{h};\omega)} [j \geq 1]$ and $\mathtt{frk} := \sum_{j=1}^{Q} \mathtt{frk}(\mathcal{W}_j, j)$, we obtain

$$\begin{aligned}
\mathtt{frk} &= \sum_{j=1}^{Q} \mathtt{frk}(\mathcal{W}_j, j) = \sum_{j=1}^{Q} \mathtt{acc}(\mathcal{W}_j) \cdot \left(\frac{\mathtt{acc}(\mathcal{W}_j)}{4} - \frac{1}{|\mathcal{H}|} \right) \\
&= \left(\sum_{j=1}^{Q} \frac{\mathtt{acc}^2(\mathcal{W}_j)}{4} \right) - \frac{\mathtt{acc}(\mathcal{W})}{|\mathcal{H}|} \geq \frac{1}{4Q} \left(\sum_{j=1}^{Q} \mathtt{acc}(\mathcal{W}_j) \right)^2 - \frac{\mathtt{acc}(\mathcal{W})}{|\mathcal{H}|} \\
&= \frac{1}{4Q} \mathtt{acc}^2(\mathcal{W}) - \frac{\mathtt{acc}(\mathcal{W})}{|\mathcal{H}|} = \mathtt{acc}(\mathcal{W}) \cdot \left(\frac{\mathtt{acc}(\mathcal{W})}{4Q} - \frac{1}{|\mathcal{H}|} \right),
\end{aligned}$$

where the inequality follows from Jensen's inequality (Lemma 3 in [6]).

6.3 Proof of Lemma 2

We will show in the following that for all $(I, \omega, \boldsymbol{h}) \xleftarrow{\$} (\mathcal{I} \times \Omega \times \mathcal{C}^{Q_V}), d \in \mathcal{D}$:

$$\alpha_{i,j}(I, \omega, \boldsymbol{h}, d) := \Pr_{\boldsymbol{h}' \xleftarrow{\$} \mathcal{C}^{Q_V}|\boldsymbol{h}_{[j-1]}} [\hat{\boldsymbol{\chi}}_i(I, \omega, \boldsymbol{h}') \neq d \wedge \hat{\boldsymbol{J}}_i(I, \omega, \boldsymbol{h}') = j] \geq \mu_{i,j}(I, \omega, \boldsymbol{h})/2, \tag{6}$$

where

$$\mu_{i,j}(I, \omega, \boldsymbol{h}) := \Pr_{\boldsymbol{h}' \xleftarrow{\$} \mathcal{C}^{Q_V}|\boldsymbol{h}_{[j-1]}} [(I, \omega, \boldsymbol{h}') \in \mathcal{B}_{i,j} \wedge \boldsymbol{h}_j \neq \boldsymbol{h}'_j].$$

For a true/false statement s, define $B(s)$ as 1 if s is true and 0 otherwise. It is easy to see that (6) implies the theorem statement since

$$\begin{aligned}
\delta_{i,j} &= \Pr_{(I,\omega,\boldsymbol{h}) \xleftarrow{\$} (\mathcal{I} \times \Omega \times \mathcal{C}^{Q_V}), \boldsymbol{h}' \xleftarrow{\$} \mathcal{C}^{Q_V}|\boldsymbol{h}_{[j-1]}} \begin{bmatrix} \hat{\boldsymbol{\chi}}_i(I, \omega, \boldsymbol{h}') \neq \hat{\boldsymbol{\chi}}_i(I, \omega, \boldsymbol{h}) \\ \wedge \hat{\boldsymbol{J}}_i(I, \omega, \boldsymbol{h}) = \hat{\boldsymbol{J}}_i(I, \omega, \boldsymbol{h}') = j \end{bmatrix} \\
&= \sum_d \Pr_{(I,\omega,\boldsymbol{h}) \xleftarrow{\$} (\mathcal{I} \times \Omega \times \mathcal{C}^{Q_V}), \boldsymbol{h}' \xleftarrow{\$} \mathcal{C}^{Q_V}|\boldsymbol{h}_{[j-1]}} \begin{bmatrix} \hat{\boldsymbol{\chi}}_i(I, \omega, \boldsymbol{h}') \neq d \wedge \hat{\boldsymbol{\chi}}_i(I, \omega, \boldsymbol{h}) = d \\ \wedge \hat{\boldsymbol{J}}_i(I, \omega, \boldsymbol{h}) = \hat{\boldsymbol{J}}_i(I, \omega, \boldsymbol{h}') = j \end{bmatrix} \\
&= \sum_d \mathbf{E}_{I,\omega,\boldsymbol{h}} [B(\hat{\boldsymbol{\chi}}_i(I, \omega, \boldsymbol{h}) = d \wedge \hat{\boldsymbol{J}}_i(I, \omega, \boldsymbol{h}) = j) \cdot \alpha_{i,j}(I, \omega, \boldsymbol{h}, d)] \\
&\geq \frac{1}{2} \sum_d \mathbf{E}_{I,\omega,\boldsymbol{h}} [B(\hat{\boldsymbol{\chi}}_i(I, \omega, \boldsymbol{h}) = d \wedge \hat{\boldsymbol{J}}_i(I, \omega, \boldsymbol{h}) = j) \cdot \mu_{i,j}(I, \omega, \boldsymbol{h})],
\end{aligned}$$

where in the last step, we have applied linearity and monotonicity of the expectation and the fact that due to (6), for all $I, \omega, \boldsymbol{h} \in \mathcal{C}^{Q_V}, d$, we have $\alpha_{i,j}(I, \omega, \boldsymbol{h}, d) \geq \mu_{i,j}(I, \omega, \boldsymbol{h})/2$. We continue with

$$\begin{aligned}
&\frac{1}{2} \sum_d \mathbf{E}_{I,\omega,\boldsymbol{h}} [B(\hat{\boldsymbol{\chi}}_i(I, \omega, \boldsymbol{h}) = d \wedge \hat{\boldsymbol{J}}_i(I, \omega, \boldsymbol{h}) = j) \cdot \mu_{i,j}(I, \omega, \boldsymbol{h})] \\
&= \frac{1}{2} \cdot \sum_d \Pr_{(I,\omega,\boldsymbol{h}) \xleftarrow{\$} (\mathcal{I} \times \Omega \times \mathcal{C}^{Q_V}), \boldsymbol{h}' \xleftarrow{\$} \mathcal{C}^{Q_V}|\boldsymbol{h}_{[j-1]}} \begin{bmatrix} \hat{\boldsymbol{\chi}}_i(I, \omega, \boldsymbol{h}) = d \wedge \hat{\boldsymbol{J}}_i(I, \omega, \boldsymbol{h}) = j \\ \wedge (I, \omega, \boldsymbol{h}') \in \mathcal{B}_{i,j} \wedge \boldsymbol{h}_j \neq \boldsymbol{h}'_j \end{bmatrix} \\
&= \frac{1}{2} \cdot \Pr_{(I,\omega,\boldsymbol{h}) \xleftarrow{\$} (\mathcal{I} \times \Omega \times \mathcal{C}^{Q_V}), \boldsymbol{h}' \xleftarrow{\$} \mathcal{C}^{Q_V}|\boldsymbol{h}_{[j-1]}} \begin{bmatrix} \hat{\boldsymbol{J}}_i(I, \omega, \boldsymbol{h}) = j \\ \wedge (I, \omega, \boldsymbol{h}') \in \mathcal{B}_{i,j} \wedge \boldsymbol{h}_j \neq \boldsymbol{h}'_j \end{bmatrix} &(7) \\
&\geq \frac{1}{2} \cdot \Pr_{(I,\omega,\boldsymbol{h}) \xleftarrow{\$} (\mathcal{I} \times \Omega \times \mathcal{C}^{Q_V}), \boldsymbol{h}' \xleftarrow{\$} \mathcal{C}^{Q_V}|\boldsymbol{h}_{[j-1]}} [(I, \omega, \boldsymbol{h}) \in \mathcal{B}_{i,j} \wedge (I, \omega, \boldsymbol{h}') \in \mathcal{B}_{i,j} \wedge \boldsymbol{h}_j \neq \boldsymbol{h}'_j] &(8) \\
&= \frac{1}{2} \cdot \mathtt{frk}(\mathcal{B}_{i,j}, j) &(9) \\
&\geq \beta_{i,j} \left(\beta_{i,j}/8 - \frac{1}{2q} \right), &(10)
\end{aligned}$$

where from (7) to (8), we have used the fact that $(I, \omega, \boldsymbol{h}') \in \mathcal{B}_{i,j}$ implies $\hat{\boldsymbol{J}}_i(I, \omega, \boldsymbol{h}') = j$. The inequality from (9) to (10) follows directly from Lemma 5.

We prove (6) by analyzing two cases. For all $I, \omega, \boldsymbol{h}, d$, we define

$$\gamma_{i,j}(I, \omega, \boldsymbol{h}, d) := \Pr_{\boldsymbol{h}' \xleftarrow{\$} \mathcal{C}^{Q_V}|\boldsymbol{h}_{[j-1]}}[\hat{\boldsymbol{\chi}}_i(I, \omega, \boldsymbol{h}') = d \wedge (I, \omega, \boldsymbol{h}') \in \mathcal{B}_{i,j} \wedge h_j \neq h'_j].$$

Case 1: $\gamma_{i,j}(I, \omega, \boldsymbol{h}, d) \geq \mu_{i,j}(I, \omega, \boldsymbol{h})/2$.

Note that in this case we can assume $d \neq C_i(I, \omega, \boldsymbol{h})$. (This is because if $d = C_i(I, \omega, \boldsymbol{h})$, then $\gamma_{i,j}(I, \omega, \boldsymbol{h}, d) \leq \Pr[\hat{\boldsymbol{\chi}}_i(I, \omega, \boldsymbol{h}') = C_i(I, \omega, \boldsymbol{h}) \wedge (I, \omega, \boldsymbol{h}') \in \mathcal{B}_{i,j}] = 0$ which would trivialize the claim.)

$$\begin{aligned}
\alpha_{i,j}(I, \omega, \boldsymbol{h}, d) &= \Pr_{\boldsymbol{h}' \xleftarrow{\$} \mathcal{C}^{Q_V}|\boldsymbol{h}_{[j-1]}}[\hat{\boldsymbol{\chi}}_i(I, \omega, \boldsymbol{h}') \neq d \wedge \hat{\boldsymbol{J}}_i(I, \omega, \boldsymbol{h}') = j] \\
&\geq \Pr[\hat{\boldsymbol{\chi}}_i(I, \omega, \boldsymbol{h}') = C_i(I, \omega, \boldsymbol{h}) \wedge \hat{\boldsymbol{J}}_i(I, \omega, \boldsymbol{h}') = j] \\
&\geq \Pr[\hat{\boldsymbol{\chi}}_i(I, \omega, \boldsymbol{h}') = d \wedge \hat{\boldsymbol{J}}_i(I, \omega, \boldsymbol{h}') = j]
\end{aligned}$$

Using again that $(I, \omega, \boldsymbol{h}') \in \mathcal{B}_{i,j}$ implies $\hat{\boldsymbol{J}}_i(I, \omega, \boldsymbol{h}') = j$, we obtain

$$\begin{aligned}
&\Pr[\hat{\boldsymbol{\chi}}_i(I, \omega, \boldsymbol{h}') = d \wedge \hat{\boldsymbol{J}}_i(I, \omega, \boldsymbol{h}') = j] \geq \Pr[\hat{\boldsymbol{\chi}}_i(I, \omega, \boldsymbol{h}') = d \wedge (I, \omega, \boldsymbol{h}') \in \mathcal{B}_{i,j}] \\
&\geq \gamma_{i,j}(I, \omega, \boldsymbol{h}, d) \geq \mu_{i,j}(I, \omega, \boldsymbol{h})/2.
\end{aligned}$$

Case 2: $\gamma_{i,j}(I, \omega, \boldsymbol{h}, d) < \mu_{i,j}(I, \omega, \boldsymbol{h})/2$. Now,

$$\begin{aligned}
\alpha_{i,j}(I, \omega, \boldsymbol{h}, d) &= \Pr_{\boldsymbol{h}' \xleftarrow{\$} \mathcal{C}^{Q_V}|\boldsymbol{h}_{[j-1]}}[\hat{\boldsymbol{\chi}}_i(I, \omega, \boldsymbol{h}') \neq d \wedge \hat{\boldsymbol{J}}_i(I, \omega, \boldsymbol{h}') = j] \\
&\geq \Pr[\hat{\boldsymbol{\chi}}_i(I, \omega, \boldsymbol{h}') \neq d \wedge (I, \omega, \boldsymbol{h}') \in \mathcal{B}_{i,j} \wedge \boldsymbol{h}_j \neq \boldsymbol{h}'_j] \\
&= \Pr[(I, \omega, \boldsymbol{h}') \in \mathcal{B}_{i,j} \wedge \boldsymbol{h}_j \neq \boldsymbol{h}'_j] \\
&\quad - \Pr[\hat{\boldsymbol{\chi}}_i(I, \omega, \boldsymbol{h}') = d \wedge (I, \omega, \boldsymbol{h}') \in \mathcal{B}_{i,j} \wedge \boldsymbol{h}_j \neq \boldsymbol{h}'_j] \\
&= \mu_{i,j}(I, \omega, \boldsymbol{h}) - \gamma_{i,j}(I, \omega, \boldsymbol{h}, d) > \mu_{i,j}(I, \omega, \boldsymbol{h})/2.
\end{aligned}$$

This proves (6) and hence the lemma.

6.4 Proof of Lemma 3

Consider again the algorithm A_i in Fig. 11 and its internal variables. On input $(I = sk, \omega = (\omega_\mathsf{M}, \boldsymbol{r}), \boldsymbol{h})$, A_i invokes M on $pk = \mathsf{F}(sk)$ and randomness ω_M and answers its queries using the values in $\boldsymbol{r}, \boldsymbol{h}$. Similarly as before, this allows us to fix an execution of M (within A_i) via a tuple of the form $(I, \omega, \boldsymbol{h}) = (I, (\omega_\mathsf{M}, \boldsymbol{r}), \boldsymbol{h})$. Let $\boldsymbol{c}(I, \omega, \boldsymbol{h})$ denote the vector of challenge values as defined in Line 20 of Fig. 11.

Recall that we have assumed that $\mathsf{F} : \mathcal{D} \longrightarrow \mathcal{R}$ and the existence of a torsion-free element $z^* \in \mathcal{D} \setminus \{0\}$ such that (i) $\mathsf{F}(z^*) = 0$; and (ii) $\forall s \in \mathcal{C} : s \cdot z^* = 0 \implies s = 0$.

Lemma 6. *Consider the mapping*

$$\Phi : \mathcal{W} \longrightarrow (\mathcal{I} \times \Omega \times \mathcal{C}^{Q_V}), \quad (sk, (\omega_M, \boldsymbol{r}), \boldsymbol{h}) \mapsto (sk + z^*, (\omega_M, \boldsymbol{r} - z^* \cdot \boldsymbol{c}(I, \omega, \boldsymbol{h})), \boldsymbol{h}),$$

where we make the convention that for $v \in \mathcal{D} \cup \mathcal{C} \cup \mathcal{R}, v \cdot \bot := 0$. Then Φ is a permutation on $\mathcal{W}$.

For the proof we require the following lemma.

Lemma 7. *Let $(I, \omega, \boldsymbol{h}) \in \mathcal{W}$. Then the tuples $(I, \omega, \boldsymbol{h})$ and $\Phi(I, \omega, \boldsymbol{h})$ fix the same execution of* M.

Proof. We show that M sees identical values in both executions corresponding to $(I, \omega, \boldsymbol{h})$ and $\Phi(I, \omega, \boldsymbol{h})$. To this end we consider all values in the view of M.

- **Initial input to** M**.** Since Φ does not alter the values of ω_M, we only need to verify that M obtains the same public key in both executions. This is ensured via $F(sk + z^*) = F(sk) + F(z^*) = F(sk) = pk$.
- **Outputs of oracle** P_1. Oracle P_1 consecutively returns the values from $\boldsymbol{R} = F(\boldsymbol{r})$, as defined in Line 01 of Fig. 11. They remain the same in both executions since $F(\boldsymbol{r}) = \boldsymbol{R} = \boldsymbol{R} - 0 \cdot \boldsymbol{c}(I, \omega, \boldsymbol{h}) = F(\boldsymbol{r}) - F(z^*) \cdot \boldsymbol{c}(I, \omega, \boldsymbol{h}) = F(\boldsymbol{r} - z^* \cdot \boldsymbol{c}(I, \omega, \boldsymbol{h}))$.
- **Outputs of oracle** P_2. Oracle P_2 consecutively returns the values from $\boldsymbol{s} = \boldsymbol{c}sk + \boldsymbol{r}$, as defined in Line 01 of Fig. 11. They remain the same in both executions since $\boldsymbol{r} + sk \cdot \boldsymbol{c}(I, \omega, \boldsymbol{h}) = \boldsymbol{s} = \boldsymbol{r} - z^* \cdot \boldsymbol{c}(I, \omega, \boldsymbol{h}) + z^* \cdot \boldsymbol{c}(I, \omega, \boldsymbol{h}) + sk \cdot \boldsymbol{c}(I, \omega, \boldsymbol{h}) = (\boldsymbol{r} - z^* \cdot \boldsymbol{c}(I, \omega, \boldsymbol{h})) + (sk + z^*) \cdot \boldsymbol{c}(I, \omega, \boldsymbol{h})$.
- **Outputs of oracle** V_2. Oracle P_2 consecutively returns the values from $\boldsymbol{b}$. They remain the same in both executions since they depend on $\boldsymbol{R}$, $\boldsymbol{h}$, and the randomness ω_M.

Thus, $(I, \omega, \boldsymbol{h})$ and $\Phi(I, \omega, \boldsymbol{h})$ fix the same executions of M.

Proof (Proof of Lemma 6). First note that Lemma 7 implies that Φ maps to $\mathcal{W}$. It remains to prove that Φ is also a bijection. Suppose Φ is not injective. Thus, for distinct tuples $(I, (\omega_M, \boldsymbol{r}), \boldsymbol{h}) \neq (I', (\omega'_M, \boldsymbol{r}'), \boldsymbol{h}')$, $\Phi(I, (\omega_M, \boldsymbol{r}), \boldsymbol{h}) = \Phi(I', (\omega'_M, \boldsymbol{r}'), \boldsymbol{h}')$. This implies $\omega_M = \omega'_M$ and $\boldsymbol{h} = \boldsymbol{h}'$. Similarly, $sk + z^* = sk' + z^*$, which implies that $sk = sk'$. Lastly, $\boldsymbol{r} - z^* \cdot \boldsymbol{c}(I, (\omega_M, \boldsymbol{r}), \boldsymbol{h}) = \boldsymbol{r}' - z^* \cdot \boldsymbol{c}(I', \omega'_M, \boldsymbol{r}', \boldsymbol{h}')$. Since $\Phi(I, (\omega_M, \boldsymbol{r}), \boldsymbol{h}) = \Phi(I', (\omega'_M, \boldsymbol{r}'), \boldsymbol{h}')$, by Claim 7, $(I, (\omega_M, \boldsymbol{r}), \boldsymbol{h})$ and $(I', (\omega'_M, \boldsymbol{r}'), \boldsymbol{h}')$ fix the same execution and therefore also $\boldsymbol{c}(I, (\omega_M, \boldsymbol{r}), \boldsymbol{h}) = \boldsymbol{c}(I', (\omega'_M, \boldsymbol{r}'), \boldsymbol{h}')$. This implies $\boldsymbol{r} = \boldsymbol{r}'$, leading to the contradiction $(I, (\omega_M, \boldsymbol{r}), \boldsymbol{h}) = (I', (\omega'_M, \boldsymbol{r}'), \boldsymbol{h}')$.

To prove that Φ is surjective, we consider the function $\Phi^{-1} : (\mathcal{I} \times \Omega \times \mathcal{C}^{Q_V}) \longrightarrow (\mathcal{I} \times \Omega \times \mathcal{C}^{Q_V})$, defined as $\Phi^{-1}(sk, (\omega_M, \boldsymbol{r}), \boldsymbol{h}) = (sk - z^*, (\omega_M, \boldsymbol{r} + z^* \cdot \boldsymbol{c}(I, \omega, \boldsymbol{h})), \boldsymbol{h})$, which is the inverse of Φ. With the same argument as above, one can also prove that Φ^{-1} is injective which implies the surjectivity of Φ.

We now introduce the following notation. Let $\mathcal{B} = \bigcup_{i,j} \mathcal{B}_{i,j}$ and let $\mathcal{G} = \mathcal{W} \setminus \mathcal{B}$. That is, for all $(I, \omega, \boldsymbol{h}) \in \mathcal{G}$, we have $\forall k \in [Q_{P_2} + 1] : \hat{\chi}_k(I, \omega, \boldsymbol{h}) = C_k(I, \omega, \boldsymbol{h})$.

The following combinatorial lemma lower bounds the probability that $\hat{\chi}$ takes different values (i.e., differs in at least one component) as a result of distinct instances $I = sk, I' = sk + z^*$.

Lemma 8. *For any fixed* $(I, (\omega_M, \boldsymbol{r})) \in \mathcal{I} \times \Omega$,

$$\Pr_{\boldsymbol{h} \xleftarrow{\$} \mathcal{C}^{Q_V}} [(I, (\omega_M, \boldsymbol{r}), \boldsymbol{h}) \in \mathcal{G} \wedge \Phi(I, (\omega_M, \boldsymbol{r}), \boldsymbol{h}) \in \mathcal{G}] \leq \frac{Q_V^{Q_{P_2}+1} \cdot \binom{Q_{P_2}+Q_{P_1}}{Q_{P_1}}}{q}.$$

Proof. We argue by contradiction. Thus, assume that for some $(I, (\omega_M, \boldsymbol{r})) \in \mathcal{I} \times \Omega$,

$$\Pr_{\boldsymbol{h} \xleftarrow{\$} \mathcal{C}^{Q_V}} [(I, (\omega_M, \boldsymbol{r}), \boldsymbol{h}) \in \mathcal{G} \wedge \Phi(I, (\omega_M, \boldsymbol{r}), \boldsymbol{h}) \in \mathcal{G}] > \frac{Q_V^{Q_{P_2}+1} \cdot \binom{Q_{P_2}+Q_{P_1}}{Q_{P_1}}}{q}.$$

Then there exist a set $\{u_1, ..., u_{Q_{P_2}+1}\}$ of $Q_{P_2}+1$ distinct indices from $[Q_V]$ such that

$$\Pr_{\boldsymbol{h} \xleftarrow{\$} \mathcal{C}^{Q_V}} \begin{bmatrix} ((I, (\omega_M, \boldsymbol{r}), \boldsymbol{h}) \in \mathcal{G}) \wedge \ (\Phi(I, (\omega_M, \boldsymbol{r}), \boldsymbol{h}) \in \mathcal{G}) \\ \wedge \forall j : \hat{\boldsymbol{J}}_j (I, (\omega_M, \boldsymbol{r}), \boldsymbol{h}) = u_j \end{bmatrix} > \frac{\binom{Q_{P_2}+Q_{P_1}}{Q_{P_1}}}{q}.$$

Similarly, there exists a vector $\boldsymbol{d} \in (\mathcal{C} \cup \{\bot\})^{Q_{P_2}+Q_{P_1}}$ of challenges such that $\boldsymbol{d}$ has exactly Q_{P_1} entries which are $\bot$ and furthermore has the property that

$$\Pr_{\boldsymbol{h} \xleftarrow{\$} \mathcal{C}^{Q_V}} \begin{bmatrix} ((I, (\omega_M, \boldsymbol{r}), \boldsymbol{h}) \in \mathcal{G}) \wedge \ (\Phi(I, (\omega_M, \boldsymbol{r}), \boldsymbol{h}) \in \mathcal{G}) \\ \wedge (\boldsymbol{c}(I, (\omega_M, \boldsymbol{r}), \boldsymbol{h}) = \boldsymbol{d}) \wedge \left(\forall j : \hat{\boldsymbol{J}}_j (I, (\omega_M, \boldsymbol{r}), \boldsymbol{h}) = u_j\right) \end{bmatrix} > \frac{1}{q^{Q_{P_2}+1}}.$$

Lastly, there exists a set $\{v_1, ..., v_{Q_V - Q_{P_2} - 1}\}$ of $Q_V - Q_{P_2} - 1$ distinct indices from $[Q_V] \setminus \{u_1, ..., u_{Q_{P_2}+1}\}$ and a vector $(\tilde{\boldsymbol{h}}_{v_1}, ..., \tilde{\boldsymbol{h}}_{v_{Q_V - Q_{P_2} - 1}}) \in \mathcal{C}^{Q_V - Q_{P_2} - 1}$ such that

$$\Pr_{\boldsymbol{h} \xleftarrow{\$} \mathcal{C}^{Q_V}} \begin{bmatrix} ((I, (\omega_M, \boldsymbol{r}), \boldsymbol{h}) \in \mathcal{G}) \wedge (\Phi(I, (\omega_M, \boldsymbol{r}), \boldsymbol{h}) \in \mathcal{G}) \wedge \ (\boldsymbol{c}(I, (\omega_M, \boldsymbol{r}), \boldsymbol{h}) = \boldsymbol{d}) \\ \wedge \left(\forall j : \hat{\boldsymbol{J}}_j (I, (\omega_M, \boldsymbol{r}), \boldsymbol{h}) = u_j\right) \wedge \left(\forall j : \boldsymbol{h}_{v_j} = \tilde{\boldsymbol{h}}_{v_j}\right) \end{bmatrix}$$
$$> \frac{1}{q^{Q_{P_2}+1} q^{Q_V - Q_{P_2} - 1}} = \frac{1}{q^{Q_V}}.$$

Since the random variable $\boldsymbol{h}$ takes a particular value $\boldsymbol{k} \in \mathcal{C}^{Q_V}$ with probability exactly q^{-Q_V}, the statement inside the probability term above must be true for at least two distinct vectors $\boldsymbol{k}, \boldsymbol{k}' \in \mathcal{C}^{Q_V}$. Furthermore, since the condition in the probability term above fixes all but the $Q_{P_2}+1$ components $\{u_1, ..., u_{Q_{P_2}+1}\}$ of $\boldsymbol{k}$ and $\boldsymbol{k}'$, there exists an index $i \in [Q_{P_2}+1]$ s.t. $\boldsymbol{k}_{u_i} \neq \boldsymbol{k}'_{u_i}$.

W.l.o.g., let i be the smallest such index. This implies that $\forall j < u_i : \boldsymbol{k}_j = \boldsymbol{k}'_j$ and $\boldsymbol{k}_{u_i} \neq \boldsymbol{k}'_{u_i}$. Therefore,

$$C_i(I, (\omega_M, \boldsymbol{r}), \boldsymbol{k}) = C_i(I, (\omega_M, \boldsymbol{r}), \boldsymbol{k}'). \tag{11}$$

Furthermore, by Lemma 7,

$$\begin{aligned}
C_i(I,(\omega_{\mathsf{M}},\boldsymbol{r}),\boldsymbol{k}) &= \hat{\boldsymbol{s}}'_i(I,(\omega_{\mathsf{M}},\boldsymbol{r}),\boldsymbol{k}) - sk\cdot \boldsymbol{k}_{u_i}\\
&= \hat{\boldsymbol{s}}'_i(\Phi(I,(\omega_{\mathsf{M}},\boldsymbol{r}),\boldsymbol{k})) - sk\cdot \boldsymbol{k}_{u_i}\\
&= \hat{\boldsymbol{s}}'_i(\Phi(I,(\omega_{\mathsf{M}},\boldsymbol{r}),\boldsymbol{k})) - sk\cdot \boldsymbol{k}_{u_i} + z^*\cdot \boldsymbol{k}_{u_i} - z^*\cdot \boldsymbol{k}_{u_i}\\
&= \hat{\boldsymbol{s}}'_i(\Phi(I,(\omega_{\mathsf{M}},\boldsymbol{r}),\boldsymbol{k})) - (sk+z^*)\cdot \boldsymbol{k}_{u_i} + z^*\cdot \boldsymbol{k}_{u_i}\\
&= C_i(\Phi(I,(\omega_{\mathsf{M}},\boldsymbol{r}),\boldsymbol{k})) + z^*\cdot \boldsymbol{k}_{u_i}\\
&= C_i(I,\omega_{\mathsf{M}},\boldsymbol{r} - z^*\cdot \boldsymbol{c}(I,(\omega_{\mathsf{M}},\boldsymbol{r}),\boldsymbol{k}),\boldsymbol{k}) + z^*\cdot \boldsymbol{k}_{u_i}. \qquad (12)
\end{aligned}$$

Analogously, we infer

$$\begin{aligned}
C_i(I,(\omega_{\mathsf{M}},\boldsymbol{r}),\boldsymbol{k}') &= \hat{\boldsymbol{s}}'_i(I,(\omega_{\mathsf{M}},\boldsymbol{r}),\boldsymbol{k}') - sk\cdot \boldsymbol{k}'_{u_i}\\
&= C_i(I,\omega_{\mathsf{M}},\boldsymbol{r} - z^*\cdot \boldsymbol{c}(I,(\omega_{\mathsf{M}},\boldsymbol{r}),\boldsymbol{k}'),\boldsymbol{k}') + z^*\cdot \boldsymbol{k}'_{u_i}. \qquad (13)
\end{aligned}$$

Combining (in this order) Eqs. 12, 11, and 13, we obtain:

$$\begin{aligned}
&C_i(I,\omega_{\mathsf{M}},\boldsymbol{r} - z^*\cdot \boldsymbol{c}(I,(\omega_{\mathsf{M}},\boldsymbol{r}),\boldsymbol{k}),\boldsymbol{k}) + z^*\cdot \boldsymbol{k}_{u_i}\\
&= C_i(I,(\omega_{\mathsf{M}},\boldsymbol{r}),\boldsymbol{k}) = C_i(I,(\omega_{\mathsf{M}},\boldsymbol{r}),\boldsymbol{k}')\\
&= C_i(I,\omega_{\mathsf{M}},\boldsymbol{r} - z^*\cdot \boldsymbol{c}(I,(\omega_{\mathsf{M}},\boldsymbol{r}),\boldsymbol{k}'),\boldsymbol{k}') + z^*\cdot \boldsymbol{k}'_{u_i}. \qquad (14)
\end{aligned}$$

Since above we have fixed $\boldsymbol{c}(I,(\omega_{\mathsf{M}},\boldsymbol{r}),\boldsymbol{k}) = \boldsymbol{c}(I,(\omega_{\mathsf{M}},\boldsymbol{r}),\boldsymbol{k}') = \boldsymbol{d}$, we also know that

$$\begin{aligned}
&C_i(I,\omega_{\mathsf{M}},\boldsymbol{r} - z^*\cdot \boldsymbol{c}(I,(\omega_{\mathsf{M}},\boldsymbol{r}),\boldsymbol{k}),\boldsymbol{k})\\
&= C_i(I,\omega_{\mathsf{M}},\boldsymbol{r} - z^*\cdot \boldsymbol{d},\boldsymbol{k})\\
&= C_i(I,\omega_{\mathsf{M}},\boldsymbol{r} - z^*\cdot \boldsymbol{d},\boldsymbol{k}') \qquad (15)\\
&= C_i(I,\omega_{\mathsf{M}},\boldsymbol{r} - z^*\cdot \boldsymbol{c}(I,(\omega_{\mathsf{M}},\boldsymbol{r}),\boldsymbol{k}'),\boldsymbol{k}'), \qquad (16)
\end{aligned}$$

where 15 follows again from the fact that $\forall j < u_i : \boldsymbol{k}_j = \boldsymbol{k}'_j$. By combining 14 and 16, it now follows that $z^*\cdot \boldsymbol{k}_{u_i} = z^*\cdot \boldsymbol{k}'_{u_i}$ or, equivalently, $z^*\cdot(\boldsymbol{k}_{u_i} - \boldsymbol{k}'_{u_i}) = 0$. Thus, torsion-freeness of z^* implies that $\boldsymbol{k}_{u_i} = \boldsymbol{k}'_{u_i}$ which contradicts the assumption that $\boldsymbol{k}_{u_i} \neq \boldsymbol{k}'_{u_i}$. This completes the proof.

Corollary 2. $\Pr_{(I,\omega,\boldsymbol{h})\xleftarrow{\$}(\mathcal{I}\times\Omega\times\mathcal{C}^{Q_{\mathsf{V}}})}[(I,\omega,\boldsymbol{h})\in\mathcal{G} \wedge \Phi(I,\omega,\boldsymbol{h})\in\mathcal{G}] \leq \frac{Q_{\mathsf{V}}^{Q_{\mathsf{P}_2}+1}\cdot\binom{Q_{\mathsf{P}_2}+Q_{\mathsf{P}_1}}{Q_{\mathsf{P}_1}}}{q}$.

Discussion. The lower bound in Corollary 2 exponentially depreciates with the number Q_{P_2} of parallel sessions allowed in the **OMMIM** experiment. Unfortunately, the ROS-attack in Sect. 4.2 shows that the bound in Corollary 2 can not be improved beyond a factor of $\binom{Q_{\mathsf{P}_2}+Q_{\mathsf{P}_1}}{Q_{\mathsf{P}_1}}$. The reason for this is that our attacker computes $\hat{\chi}$ in a manner that does not depend on $\boldsymbol{h}$, but only on ω, I (more precisely, any contribution of $\boldsymbol{h}$ 'cancels out' in the values returned by the attacker). Therefore, $\hat{\chi}$ *always* takes the 'most likely' value according to 3 in the sense that, regardless of $\boldsymbol{h}$, the attacker can force $(\omega, I, \boldsymbol{h}) \in \mathcal{G}$ *and* $\Phi(\omega, I, \boldsymbol{h}) \in \mathcal{G}$.

Lemma 9. $\Pr_{(I,\omega,\boldsymbol{h}) \xleftarrow{\$} (\mathcal{I}\times\Omega\times\mathcal{C}^{Q_\mathsf{V}})}[(I,\omega,\boldsymbol{h}) \in \mathcal{B}] \geq \frac{1}{2}\left(\varepsilon - \frac{Q_\mathsf{V}^{Q_{\mathsf{P}_2}+1}\cdot\binom{Q_{\mathsf{P}_2}+Q_{\mathsf{P}_1}}{Q_{\mathsf{P}_1}}}{q}\right).$

Proof. We partition $\mathcal{G}$ into subsets $\mathcal{G}_g, \mathcal{G}_b$ such that all elements in $\mathcal{G}_g$ are mapped into $\mathcal{G}$ via Φ and all elements in $\mathcal{G}_b$ are mapped into $\mathcal{B}$ via Φ. It follows that

$$\begin{aligned}&\Pr_{(I,\omega,\boldsymbol{h}) \xleftarrow{\$} (\mathcal{I}\times\Omega\times\mathcal{C}^{Q_\mathsf{V}})}[(I,\omega,\boldsymbol{h}) \in \mathcal{G}] \\ &= \Pr_{(I,\omega,\boldsymbol{h}) \xleftarrow{\$} (\mathcal{I}\times\Omega\times\mathcal{C}^{Q_\mathsf{V}})}[(I,\omega,\boldsymbol{h}) \in \mathcal{G}_g] + \Pr_{(I,\omega,\boldsymbol{h}) \xleftarrow{\$} (\mathcal{I}\times\Omega\times\mathcal{C}^{Q_\mathsf{V}})}[(I,\omega,\boldsymbol{h}) \in \mathcal{G}_b]. \quad (17)\end{aligned}$$

By Corollary 2 and because Φ is a bijection, we can infer that

$$\Pr_{(I,\omega,\boldsymbol{h}) \xleftarrow{\$} (\mathcal{I}\times\Omega\times\mathcal{C}^{Q_\mathsf{V}})}[(I,\omega,\boldsymbol{h}) \in \mathcal{G}_g] \leq \frac{Q_\mathsf{V}^{Q_{\mathsf{P}_2}+1}\cdot\binom{Q_{\mathsf{P}_2}+Q_{\mathsf{P}_1}}{Q_{\mathsf{P}_1}}}{q}, \quad (18)$$

$$\Pr_{(I,\omega,\boldsymbol{h}) \xleftarrow{\$} (\mathcal{I}\times\Omega\times\mathcal{C}^{Q_\mathsf{V}})}[(I,\omega,\boldsymbol{h}) \in \mathcal{G}_b] \leq \Pr_{(I,\omega,\boldsymbol{h}) \xleftarrow{\$} (\mathcal{I}\times\Omega\times\mathcal{C}^{Q_\mathsf{V}})}[(I,\omega,\boldsymbol{h}) \in \mathcal{B}]. \quad (19)$$

It follows from 17, 18, 19 that

$$\Pr[(I,\omega,\boldsymbol{h}) \in \mathcal{G}] \leq \frac{Q_\mathsf{V}^{Q_{\mathsf{P}_2}+1}\cdot\binom{Q_{\mathsf{P}_2}+Q_{\mathsf{P}_1}}{Q_{\mathsf{P}_1}}}{q} + \Pr[(I,\omega,\boldsymbol{h}) \in \mathcal{B}]. \quad (20)$$

From 20, we can bound $\Pr[(I,\omega,\boldsymbol{h}) \in \mathcal{B}]$ as

$$\begin{aligned}\Pr[(I,\omega,\boldsymbol{h}) \in \mathcal{B}] &= \Pr[(I,\omega,\boldsymbol{h}) \in \mathcal{W}] - \Pr[(I,\omega,\boldsymbol{h}) \in \mathcal{G}] \\ &\geq \Pr[(I,\omega,\boldsymbol{h}) \in \mathcal{W}] - \Pr[(I,\omega,\boldsymbol{h}) \in \mathcal{B}] - \frac{Q_\mathsf{V}^{Q_{\mathsf{P}_2}+1}\cdot\binom{Q_{\mathsf{P}_2}+Q_{\mathsf{P}_1}}{Q_{\mathsf{P}_1}}}{q}.\end{aligned}$$

Since $\varepsilon = \Pr_{(I,\omega,\boldsymbol{h}) \xleftarrow{\$} (\mathcal{I}\times\Omega\times\mathcal{C}^{Q_\mathsf{V}})}[(I,\omega,\boldsymbol{h}) \in \mathcal{W}]$, we finally obtain

$$\Pr_{(I,\omega,\boldsymbol{h}) \xleftarrow{\$} (\mathcal{I}\times\Omega\times\mathcal{C}^{Q_\mathsf{V}})}[(I,\omega,\boldsymbol{h}) \in \mathcal{B}] \geq \frac{1}{2}\left(\varepsilon - \frac{Q_\mathsf{V}^{Q_{\mathsf{P}_2}+1}\cdot\binom{Q_{\mathsf{P}_2}+Q_{\mathsf{P}_1}}{Q_{\mathsf{P}_1}}}{q}\right).$$

We are now ready to prove Lemma 3, i.e., we show that there exist $i \in [Q_{\mathsf{P}_2}+1], j \in [Q_\mathsf{V}]$ such that $\beta_{i,j} > \left(\varepsilon - \frac{Q_\mathsf{V}^{Q_{\mathsf{P}_2}+1}\cdot\binom{Q_{\mathsf{P}_2}+Q_{\mathsf{P}_1}}{Q_{\mathsf{P}_1}}}{q}\right)\cdot\frac{1}{2Q_\mathsf{V}(Q_{\mathsf{P}_2}+1)}$. Toward a contradiction, suppose instead that for all $i \in [Q_{\mathsf{P}_2}+1], j \in [Q_\mathsf{V}]$, we have that

$$\Pr_{(I,\omega,\boldsymbol{h}) \xleftarrow{\$} (\mathcal{I}\times\Omega\times\mathcal{C}^{Q_\mathsf{V}})}[(I,\omega,\boldsymbol{h}) \in \mathcal{B}_{i,j}] < \left(\varepsilon - \frac{Q_\mathsf{V}^{Q_{\mathsf{P}_2}+1}\cdot\binom{Q_{\mathsf{P}_2}+Q_{\mathsf{P}_1}}{Q_{\mathsf{P}_1}}}{q}\right)\cdot\frac{1}{2Q_\mathsf{V}(Q_{\mathsf{P}_2}+1)}.$$

By Lemma 9,

$$\frac{1}{2}\left(\varepsilon - \frac{Q_{\mathsf{V}}^{Q_{\mathsf{P}_2}+1} \cdot \binom{Q_{\mathsf{P}_2}+Q_{\mathsf{P}_1}}{Q_{\mathsf{P}_1}}}{q}\right) \leq \Pr[(I,\omega,\boldsymbol{h}) \in \mathcal{B}] = \Pr[(I,\omega,\boldsymbol{h}) \in \bigcup_{i,j} \mathcal{B}_{i,j}]$$

$$\leq \sum_{i,j} \Pr[(I,\omega,\boldsymbol{h}) \in \mathcal{B}_{i,j}] < \frac{1}{2}\left(\varepsilon - \frac{Q_{\mathsf{V}}^{Q_{\mathsf{P}_2}+1} \cdot \binom{Q_{\mathsf{P}_2}+Q_{\mathsf{P}_1}}{Q_{\mathsf{P}_1}}}{q}\right).$$

This is a contradiction.

Acknowledgments. We would like to thank David Pointcheval for helpful discussions and for answering many of our questions.

References

1. Abdalla, M., An, J.H., Bellare, M., Namprempre, C.: From identification to signatures via the Fiat-Shamir transform: minimizing assumptions for security and forward-security. In: Knudsen, L.R. (ed.) EUROCRYPT 2002. LNCS, vol. 2332, pp. 418–433. Springer, Heidelberg (2002). https://doi.org/10.1007/3-540-46035-7_28
2. Backendal, M., Bellare, M., Sorrell, J., Sun, J.: The Fiat-Shamir zoo: relating the security of different signature variants. Cryptology ePrint Archive, Report 2018/775 (2018). https://eprint.iacr.org/2018/775
3. Baldimtsi, F., Lysyanskaya, A.: Anonymous credentials light. In: Sadeghi, A.-R., Gligor, V.D., Yung, M. (eds.) ACM CCS 2013, pp. 1087–1098. ACM Press, November 2013
4. Baldimtsi, F., Lysyanskaya, A.: On the security of one-witness blind signature schemes. In: Sako, K., Sarkar, P. (eds.) ASIACRYPT 2013. LNCS, vol. 8270, pp. 82–99. Springer, Heidelberg (2013). https://doi.org/10.1007/978-3-642-42045-0_5
5. Belenkiy, M., Camenisch, J., Chase, M., Kohlweiss, M., Lysyanskaya, A., Shacham, H.: Randomizable proofs and delegatable anonymous credentials. In: Halevi, S. (ed.) CRYPTO 2009. LNCS, vol. 5677, pp. 108–125. Springer, Heidelberg (2009). https://doi.org/10.1007/978-3-642-03356-8_7
6. Bellare, M., Neven, G.: Multi-signatures in the plain public-key model and a general forking lemma. In: Juels, A., Wright, R.N., Vimercati, S. (eds.) ACM CCS 2006, pp. 390–399. ACM Press, October/November 2006
7. Bellare, M., Palacio, A.: GQ and Schnorr identification schemes: proofs of security against impersonation under active and concurrent attacks. In: Yung, M. (ed.) CRYPTO 2002. LNCS, vol. 2442, pp. 162–177. Springer, Heidelberg (2002). https://doi.org/10.1007/3-540-45708-9_11
8. Bellare, M., Rogaway, P.: Random oracles are practical: a paradigm for designing efficient protocols. In: Ashby, V. (ed.) ACM CCS 1993, pp. 62–73. ACM Press, November 1993
9. Brands, S.: Untraceable off-line cash in wallet with observers (extended abstract). In: Stinson, D.R. (ed.) CRYPTO 1993. LNCS, vol. 773, pp. 302–318. Springer, Heidelberg (1994). https://doi.org/10.1007/3-540-48329-2_26

10. Camenisch, J., Hohenberger, S., Lysyanskaya, A.: Compact e-cash. In: Cramer, R. (ed.) EUROCRYPT 2005. LNCS, vol. 3494, pp. 302–321. Springer, Heidelberg (2005). https://doi.org/10.1007/11426639_18
11. Camenisch, J., Lysyanskaya, A.: An efficient system for non-transferable anonymous credentials with optional anonymity revocation. In: Pfitzmann, B. (ed.) EUROCRYPT 2001. LNCS, vol. 2045, pp. 93–118. Springer, Heidelberg (2001). https://doi.org/10.1007/3-540-44987-6_7
12. Chaum, D.: Blind signatures for untraceable payments. In: Chaum, D., Rivest, R.L., Sherman, A.T. (eds.) CRYPTO 1982, pp. 199–203. Plenum Press, New York (1982)
13. Chaum, D., Fiat, A., Naor, M.: Untraceable electronic cash. In: Goldwasser, S. (ed.) CRYPTO 1988. LNCS, vol. 403, pp. 319–327. Springer, New York (1990). https://doi.org/10.1007/0-387-34799-2_25
14. Feige, U., Fiat, A., Shamir, A.: Zero-knowledge proofs of identity. J. Cryptol. **1**(2), 77–94 (1988)
15. Gennaro, R.: Multi-trapdoor commitments and their applications to proofs of knowledge secure under concurrent man-in-the-middle attacks. In: Franklin, M. (ed.) CRYPTO 2004. LNCS, vol. 3152, pp. 220–236. Springer, Heidelberg (2004). https://doi.org/10.1007/978-3-540-28628-8_14
16. Lyubashevsky, V.: Lattice-based identification schemes secure under active attacks. In: Cramer, R. (ed.) PKC 2008. LNCS, vol. 4939, pp. 162–179. Springer, Heidelberg (2008). https://doi.org/10.1007/978-3-540-78440-1_10
17. Minder, L., Sinclair, A.: The extended k-tree algorithm. In: Mathieu, C. (ed.) 20th SODA, pp. 586–595. ACM-SIAM, January 2009
18. Okamoto, T.: Provably secure and practical identification schemes and corresponding signature schemes. In: Brickell, E.F. (ed.) CRYPTO 1992. LNCS, vol. 740, pp. 31–53. Springer, Heidelberg (1993). https://doi.org/10.1007/3-540-48071-4_3
19. Okamoto, T., Ohta, K.: Universal electronic cash. In: Feigenbaum, J. (ed.) CRYPTO 1991. LNCS, vol. 576, pp. 324–337. Springer, Heidelberg (1992). https://doi.org/10.1007/3-540-46766-1_27
20. Pointcheval, D., Stern, J.: New blind signatures equivalent to factorization (extended abstract). In: ACM CCS 1997, pp. 92–99. ACM Press, April 1997
21. Pointcheval, D., Stern, J.: Security arguments for digital signatures and blind signatures. J. Cryptol. **13**(3), 361–396 (2000)
22. Rodriuguez-Henriquez, F., Ortiz-Arroyo, D., Garcia-Zamora, C.: Yet another improvement over the Mu-Varadharajan e-voting protocol. Comput. Stand. Interfaces **29**(4), 471–480 (2007)
23. Rückert, M.: Lattice-based blind signatures. In: Abe, M. (ed.) ASIACRYPT 2010. LNCS, vol. 6477, pp. 413–430. Springer, Heidelberg (2010). https://doi.org/10.1007/978-3-642-17373-8_24
24. Schnorr, C.P.: Security of blind discrete log signatures against interactive attacks. In: Qing, S., Okamoto, T., Zhou, J. (eds.) ICICS 2001. LNCS, vol. 2229, pp. 1–12. Springer, Heidelberg (2001). https://doi.org/10.1007/3-540-45600-7_1
25. Wagner, D.: A generalized birthday problem. In: Yung, M. (ed.) CRYPTO 2002. LNCS, vol. 2442, pp. 288–304. Springer, Heidelberg (2002). https://doi.org/10.1007/3-540-45708-9_19

Best Paper Awards

Efficient Verifiable Delay Functions

Benjamin Wesolowski[1,2(✉)]

[1] EPFL IC LACAL, Station 14, 1015 Lausanne, Switzerland
[2] Cryptology Group, CWI, Amsterdam, The Netherlands
benjamin.wesolowski@cwi.nl

Abstract. We construct a verifiable delay function (VDF). A VDF is a function whose evaluation requires running a given number of sequential steps, yet the result can be efficiently verified. They have applications in decentralised systems, such as the generation of trustworthy public randomness in a trustless environment, or resource-efficient blockchains. To construct our VDF, we actually build a *trapdoor* VDF. A trapdoor VDF is essentially a VDF which can be evaluated efficiently by parties who know a secret (the trapdoor). By setting up this scheme in a way that the trapdoor is unknown (not even by the party running the setup, so that there is no need for a trusted setup environment), we obtain a simple VDF. Our construction is based on groups of unknown order such as an RSA group, or the class group of an imaginary quadratic field. The output of our construction is very short (the result and the proof of correctness are each a single element of the group), and the verification of correctness is very efficient.

1 Introduction

We describe a function that is slow to compute and easy to verify: a *verifiable delay function* (henceforth, VDF) in the sense of [4][1]. These functions should be computable in a prescribed amount of time Δ, but not faster (the *time* measures an amount of sequential work, that is work that cannot be performed faster by running on a large number of parallel cores), and the result should be easy to verify (i.e., for a cost polylog(Δ)). These special functions are used in [15] (under the name of *slow-timed hash functions*) to construct a trustworthy randomness beacon: a service producing publicly verifiable random numbers, which are guaranteed to be unbiased and unpredictable. These randomness beacons, introduced by Rabin in [17], are a valuable tool in a public, decentralised setting, as it is not trivial for someone to flip a coin and convince their peers that

B. Wesolowski—The author is currently affiliated to CWI[2], yet this work was completed at EPFL[1].

[1] The paper [4] was developed independently of the present work, yet we adopt their terminology for verifiable delay functions, for the sake of uniformity.

Y. Ishai and V. Rijmen (Eds.): EUROCRYPT 2019, LNCS 11478, pp. 379–407, 2019.
https://doi.org/10.1007/978-3-030-17659-4_13

the outcome was not rigged. A number of interesting applications of VDFs have recently emerged—see [4] for an overview. Most notably, they can be used to design resource-efficient blockchains, eliminating the need for massively power-consuming mining farms. VDFs play a key role in the Chia blockchain design (`chia.net`), and the Ethereum Foundation (`ethereum.org`) and Protocol Labs (`protocol.ai`) are teaming up to investigate the technology of VDFs which promise to play a key role in their respective platforms.

There is thereby a well-motivated need for an efficient construction. This problem was left open in [4], and we address it here with a new, simple, and efficient VDF.

1.1 Contribution

An efficient construction. The starting point of our construction is the time-lock puzzle of Rivest, Shamir and Wagner [18]: given as input an RSA group $(\mathbf{Z}/N\mathbf{Z})^\times$, where N is a product of two large, secret primes, a random element $x \in (\mathbf{Z}/N\mathbf{Z})^\times$, and a timing parameter t, compute x^{2^t}. Without the factorisation of N, this task requires t sequential squarings in the group. More generally, one could work with any group G of unknown order. This construction is only a time-lock puzzle and not a VDF, because given an output y, there is no efficient way to verify that $y = x^{2^t}$.

The new VDF construction consists in solving an instance of the time-lock puzzle of [18], and computing a proof of correctness, which allows anyone to efficiently verify the result. Fix a timing parameter Δ, a security level k (say, 128, 192, or 256), and a group G. Our construction has the following properties:

1. It is Δ-sequential (meaning that it requires Δ sequential steps to evaluate) assuming the classic time-lock assumption of [18] in the group G.
2. It is sound (meaning that one cannot produce a valid proof for an incorrect output) under some group theoretic assumptions on G, believed to be true for RSA groups and class groups of quadratic imaginary number fields.
3. The output and the proof of correctness are each a single element of the group G (also, the output can be recovered from the proof and a $2k$-bit integer; so it is possible to transmit a single group element and a small integer instead of 2 group elements).
4. The verification of correctness requires essentially two exponentiations in the group G, with exponents of bit-length $2k$.
5. The proof can be produced in $O(\Delta/\log(\Delta))$ group operations.

For applications where a lot of these proofs need to be stored, widely distributed, and repeatedly verified, having very short and efficiently verifiable proofs is invaluable.

Following discussions about the present work at the August 2018 workshop at Stanford hosted by the Ethereum Foundation and the Stanford Center for Blockchain Research, we note that our construction features two other useful properties: the proofs can be *aggregated* and *watermarked*. Aggregating consists in producing a single short proof that simultaneously proves the correctness of several VDF evaluations. Watermarking consists in tying a proof to the evaluator's identity; in a blockchain setting, this allows to give credit (and a reward) to the party who spent time and resources evaluating the VDF. These properties are discussed in Sect. 7.

Note that the method we describe to compute the proof requires an amount $O(\Delta/\log(\Delta))$ group operations. Hence, there is an interval between the guaranteed sequential work Δ and the total work $(1+\varepsilon)\Delta$, where $\varepsilon = O(1/\log(\Delta))$. For practical parameters, this ε is in the order of 0.05, and this small part of the computation is easily parallelizable, so that the total evaluation time with s cores is around $(1 + 1/(20s))\Delta$. This gap should be of no importance since anyways, computational models do not capture well small constant factors with respect to real-world running time. Precise timing is unlikely to be achievable without resorting to trusted hardware, thus applications of VDFs are designed not to be too sensitive to these small factors.

If despite these facts it is still problematic in some application to know the output of the VDF slightly before having the proof, it is possible to eliminate this gap by artificially considering the proof as part of the output (the output is now a pair of group elements, and the proof is empty). The resulting protocol is still Δ-sequential (trivially), and as noted in Remark 5, it is also sound. We also propose a second method in Sect. 4.3 which allows to exponentially reduce the overhead of the proof computation at the cost of lengthening the resulting proof by a few group elements.

Trapdoor verifiable delay function. The construction proposed is actually a *trapdoor* VDF, from which we can derive an actual VDF. A party, Alice, holds a secret key sk (the trapdoor), and an associated public key pk. Given a piece of data x, a trapdoor VDF allows to compute an output y from x such that anyone can easily verify that either y has been computed by Alice (i.e., she used her secret trapdoor), or the computation of y required an amount of time at least Δ (where, again, time is measured as an amount of sequential work). The verification that y is the correct output of the VDF for input x should be efficient, with a cost $\text{polylog}(\Delta)$.

Deriving a verifiable delay function. Suppose that a public key pk for a trapdoor VDF is given without any known associated secret key. This results in a simple VDF, where the evaluation requires a prescribed amount of time Δ for everyone (because there is no known trapdoor).

Now, how to publicly generate a public key without any known associated private key? In the construction we propose, this amounts to the public generation of a group of unknown order. A standard choice for such groups are RSA groups, but it is hard to generate an RSA number (a product of two large primes) with a strong guarantee that nobody knows the factorisation. It is possible to generate a random number large enough that with high probability it is divisible by two large primes (as done in [19]), but this approach severely damages the efficiency of the construction, and leaves more room for parallel optimisation of the arithmetic modulo a large integer, or for specialised hardware acceleration. It is also possible to generate a modulus by a secure multiparty execution of the RSA key generation procedure among independent parties, each contributing some secret random seeds (as done in [6]). However, in this scenario, a third party would have to assume that the parties involved in this computation did not collude to retrieve the secret. We propose to use the class group of an imaginary quadratic order. One can easily generate an imaginary quadratic order by choosing a random discriminant, and when the discriminant is large enough, the order of the class group cannot be computed. These class groups were introduced in cryptography by Buchmann and Williams in [9], exploiting the difficulty of computing their orders (and the fact that this order problem is closely related to the discrete logarithm and the root problems in this group). To this day, the best known algorithms for computing the order of the class group of an imaginary quadratic field of discriminant d are still of complexity $L_{|d|}(1/2)$ under the generalised Riemann hypothesis, for the usual function $L_t(s) = \exp\left(O\left(\log(t)^s \log\log(t)^{1-s}\right)\right)$, as shown in [14] and [20].

Circumventing classic impossibility results. Finally, we further motivate the notion of *trapdoor* VDF by showing that it constitutes an original tool to circumvent classic impossibility results. We illustrate this in Sect. 8 with a simple and efficient identification protocol with surprising zero-knowledge and deniability properties.

1.2 Time-Sensitive Cryptography and Related Work

Rivest, Shamir and Wagner [18] introduced in 1996 the use of *time-locks* for encrypting data that can be decrypted only in a predetermined time in the future. This was the first time-sensitive cryptographic primitive taking into account the parallel power of possible attackers. Other timed primitives appeared in different contexts: Bellare and Goldwasser [1,2] suggested *time capsules* for key escrowing in order to counter the problem of early recovery. Boneh and Naor [7] introduced *timed commitments*: a hiding and binding commitment scheme, which can be *forced open* by a procedure of determined running time. More recently, and as already mentioned, the notion of slow-timed hash function was introduced in [15] as a tool to provide trust to the generation of public random numbers.

Verifiable delay functions. These slow-timed hash functions were recently revisited and formalised by Boneh *et al.* in [4] under the name of verifiable delay functions. The function proposed in [15], *sloth*, is not asymptotically efficiently verifiable: the verification procedure (given x and y, verify that $\mathsf{sloth}(x) = y$) is faster than the evaluation procedure (given x, compute the value $\mathsf{sloth}(x)$) only by a constant factor. The authors of [4] proposed practical constructions that achieve an exponential gap between evaluation and verification, but do not strictly achieve the requirements of a VDF. For one of them, the evaluation requires an amount polylog(Δ) of parallelism to run in parallel time Δ. The other one is insecure against an adversary that can run a large (but feasible) pre-computation, so the setup must be regularly updated. The new construction we propose does not suffer these disadvantages.

Pietrzak's verifiable delay function. Independently from the present work, another efficient VDF was proposed in [16]. The author describes an elegant construction, provably secure under the classic time-lock assumption of [18] when implemented over an RSA group $(\mathbf{Z}/N\mathbf{Z})^\times$ where N is a product of two safe primes. The philosophy of [16] is close to our construction: it consists in solving the puzzle of [18] (for a timing parameter Δ), and computing a proof of correctness. Their proofs can be computed with $O(\sqrt{\Delta}\log(\Delta))$ group multiplications. However, the proofs obtained are much longer (they consist of $O(\log(\Delta))$ group elements, versus a single group element in our construction), and the verification procedure is less efficient (it requires $O(\log(\Delta))$ group exponentiations, versus essentially two group exponentiations in our construction—for exponents of bit-length the security level k in both cases).

In the example given in [18], the group G is an RSA group for a 2048 bit modulus, and the time Δ is set to 2^{40} sequential squarings in the group, so the proofs are $10KB$ long. In comparison, in the same setting, our proofs are $0.25KB$ long.

1.3 Notation

Throughout this paper, the integer k denotes a security level (typically 128, 192, or 256), and the map $H : \{\mathtt{0},\mathtt{1}\}^* \rightarrow \{\mathtt{0},\mathtt{1}\}^{2k}$ denotes a secure cryptographic hash function. For simplicity of exposition, the function H is regarded as a map from $\mathcal{A}^*$ to $\{\mathtt{0},\mathtt{1}\}^{2k}$, where $\mathcal{A}^*$ is the set of strings over some alphabet $\mathcal{A}$ such that $\{\mathtt{0},\mathtt{1}\} \subset \mathcal{A}$. The alphabet $\mathcal{A}$ contains at least all nine digits and twenty-six letters, and a special character $\star$. Given two strings $s_1, s_2 \in \mathcal{A}^*$, denote by $s_1||s_2$ their concatenation, and by $s_1|||s_2$ their concatenation separated by $\star$. The function $\mathtt{int} : \{\mathtt{0},\mathtt{1}\}^* \rightarrow \mathbf{Z}_{\geq 0}$ maps $x \in \{\mathtt{0},\mathtt{1}\}^*$ in the canonical manner to the non-negative integer with binary representation x. The function $\mathtt{bin} : \mathbf{Z}_{\geq 0} \rightarrow \{\mathtt{0},\mathtt{1}\}^*$ maps any non-zero integer to its binary representation with no leading $\mathtt{0}$-characters, and $\mathtt{bin}(0) = \mathtt{0}$.

2 Trapdoor Verifiable Delay Functions

Let $\Delta : \mathbf{Z}_{>0} \rightarrow \mathbf{R}_{>0}$ be a function of the (implicit) security parameter k. This Δ is meant to represent a time duration, and what is precisely meant by *time* is explained in Sect. 3 (essentially, it measures an amount of sequential work). A party, Alice, has a public key pk and a secret key sk. Let x be a piece of data. Alice, thanks to her secret key sk, is able to quickly evaluate a function $\mathsf{trapdoor}_{\mathsf{sk}}$ on x. On the other hand, other parties knowing only pk can compute $\mathsf{eval}_{\mathsf{pk}}(x)$ in time Δ, but not faster (and important parallel computing power does not give a substantial advantage in going faster; remember that Δ measures the sequential work), such that the resulting value $\mathsf{eval}_{\mathsf{pk}}(x)$ is the same as $\mathsf{trapdoor}_{\mathsf{sk}}(x)$.

More formally, a trapdoor VDF consists of the following components (very close to the classic VDF defined in [4]):

$\mathsf{keygen} \rightarrow (\mathsf{pk}, \mathsf{sk})$ is a key generation procedure, which outputs Alice's public key pk and secret key sk. As usual, the public key should be publicly available, and the secret key is meant to be kept secret.

$\mathsf{trapdoor}_{\mathsf{sk}}(x, \Delta) \rightarrow (y, \pi)$ takes as input the data $x \in \mathcal{X}$ (for some input space $\mathcal{X}$), and uses the secret key sk to produce the output y from x, and a (possibly empty) proof π. The parameter Δ is the amount of sequential work required to compute the same output y without knowledge of the secret key.

$\mathsf{eval}_{\mathsf{pk}}(x, \Delta) \rightarrow (y, \pi)$ is a procedure to evaluate the function on x using only the public key pk, for a targeted amount of sequential work Δ. It produces the output y from x, and a (possibly empty) proof π. This procedure is meant to be infeasible in time less than Δ (this will be expressed precisely in the security requirements).

$\mathsf{verify}_{\mathsf{pk}}(x, y, \pi, \Delta) \rightarrow$ true or false is a procedure to check if y is indeed the correct output for x, associated to the public key pk and the evaluation time Δ, possibly with the help of the proof π.

Note that the security parameter k is implicitly an input to each of these procedures. Given any key pair (pk, sk) generated by the keygen procedure, the functionality of the scheme is the following. Given any input x and time parameter Δ, let $(y, \pi) \leftarrow \mathsf{eval}_{\mathsf{pk}}(x, \Delta)$ and $(y', \pi') \leftarrow \mathsf{trapdoor}_{\mathsf{sk}}(x, \Delta)$. Then, $y = y'$ and the procedures $\mathsf{verify}_{\mathsf{pk}}(x, y, \pi, \Delta)$ and $\mathsf{verify}_{\mathsf{pk}}(x, y', \pi', \Delta)$ both output true.

We also require the protocol to be *sound*, as in [4]. Intuitively, we want that if y' is not the correct output of $\mathsf{eval}_{\mathsf{pk}}(x, \Delta)$ then $\mathsf{verify}_{\mathsf{pk}}(x, y', \Delta)$ outputs false. We however allow the holder of the trapdoor to generate misleading values y'.

Definition 1 (Soundness). *A trapdoor VDF is* sound *if any polynomially bounded algorithm solves the following* soundness-breaking *game with negligible probability (in k): given as input the public key* pk*, output a message x, a value y' and a proof π' such that* $y' \neq \mathsf{eval}_{\mathsf{pk}}(x, \Delta)$, *and* $\mathsf{verify}_{\mathsf{pk}}(x, y', \pi', \Delta) = \mathsf{true}$.

The second security property is that the correct output cannot be produced in time less than Δ without knowledge of the secret key sk. This is formalised in the next section via the Δ-*evaluation race* game. A trapdoor VDF is Δ-*sequential* if any polynomially bounded adversary wins the Δ-evaluation race game with negligible probability.

3 Wall-Clock Time and Computational Assumptions

Primitives such as verifiable delay functions or time-lock puzzles wish to deal with the delicate notion of real-world time. This section discusses how to formally handle this concept, and how it translates in practice.

3.1 Theoretical Model

A precise notion of wall-clock time is difficult to capture formally. However, we can get a first approximation by choosing a model of computation, and defining *time* as an amount of sequential work in this model. A model of computation is a set of allowable operations, together with their respective costs. For instance, working with circuits with gates $\vee$, $\wedge$ and $\neg$ which each have cost 1, the notion of time complexity of a circuit $\mathcal{C}$ can be captured by its depth $d(\mathcal{C})$, i.e., the length of the longest path in $\mathcal{C}$. The time-complexity of a boolean function f is then the minimal depth of a circuit implementing f, but this does not reflect the time it might take to actually compute f in the real world where one is not bound to using circuits. A random access machine might perform better, or maybe a quantum circuit.

A good model of computation for analysing the actual time it takes to solve a problem should contain all the operations that one could use in practice (in particular the adversary). From now on, we suppose the adversary works in a model of computation $\mathcal{M}$. We do not define exactly $\mathcal{M}$, but only assume that it allows all operations a potential adversary could perform, and that it comes with a cost function c and a time-cost function t. For any algorithm $\mathcal{A}$ and input x, the cost $C(\mathcal{A}, x)$ measures the overall cost of computing $\mathcal{A}(x)$ (i.e., the sum of the costs of all the elementary operations that are executed), while the time-cost $T(\mathcal{A}, x)$ abstracts the notion of time it takes to run $\mathcal{A}(x)$ in the model $\mathcal{M}$. For the model of circuits, one could define the cost as the size of the circuit and the time-cost as its depth. For concreteness, one can think of the model $\mathcal{M}$ as the model of parallel random-access machines.

All forthcoming security claims are (implicitly) made with respect to the model $\mathcal{M}$. The time-lock assumption of Rivest, Shamir and Wagner [18] can be expressed as Assumption 1 below.

Definition 2 ((δ, t)-time-lock game). *Let $k \in \mathbf{Z}_{>0}$ be a difficulty parameter, and $\mathcal{A}$ be an algorithm playing the game. The parameter t is a positive integer, and $\delta : \mathbf{Z}_{>0} \to \mathbf{R}_{>0}$ is a function. The* (δ, t)-time-lock game *goes as follows:*

1. *An RSA modulus N is generated at random by an RSA key-generation procedure, for the security parameter k;*
2. *$\mathcal{A}(N)$ outputs an algorithm $\mathcal{B}$;*
3. *An element $g \in \mathbf{Z}/N\mathbf{Z}$ is generated uniformly at random;*
4. *$\mathcal{B}(g)$ outputs $h \in \mathbf{Z}/N\mathbf{Z}$.*

Then, $\mathcal{A}$ wins the game if $h = g^{2^t} \mod N$ and $T(\mathcal{B}, g) < t\delta(k)$.

Assumption 1 (Time-lock assumption). *There is a cost function $\delta : \mathbf{Z}_{>0} \to \mathbf{R}_{>0}$ such that the following two statements hold:*

1. *There is an algorithm $\mathcal{S}$ such that for any modulus N generated by an RSA key-generation procedure with security parameter k, and any element $g \in \mathbf{Z}/N\mathbf{Z}$, the output of $\mathcal{S}(N, g)$ is the square of g, and $T(\mathcal{S}, (N, g)) < \delta(k)$;*

2. *For any $t \in \mathbf{Z}_{>0}$, no algorithm $\mathcal{A}$ of polynomial cost*[2] *wins the (δ, t)-time-lock game with non-negligible probability (with respect to the difficulty parameter k).*

The function δ encodes the time-cost of computing a single modular squaring, and Assumption 1 expresses that without knowledge of the factorisation of N, there is no faster way to compute $g^{2^t} \bmod N$ than performing t sequential squarings.

With this formalism, we can finally express the security notion of a trapdoor VDF.

Definition 3 (Δ-evaluation race game). *Let $\mathcal{A}$ be a party playing the game. The parameter $\Delta : \mathbf{Z}_{>0} \to \mathbf{R}_{>0}$ is a function of the (implicit) security parameter k. The* Δ-evaluation race game *goes as follows:*

1. *The random procedure* keygen *is run and it outputs a public key* pk*;*
2. *$\mathcal{A}(\mathsf{pk})$ outputs an algorithm $\mathcal{B}$;*
3. *Some data $x \in \mathcal{X}$ is generated according to some random distribution of min-entropy at least k;*
4. *$\mathcal{B}^{\mathcal{O}}(x)$ outputs a value y, where $\mathcal{O}$ is an oracle that outputs the evaluation* $\mathsf{trapdoor}_{\mathsf{sk}}(x', \Delta)$ *on any input $x' \neq x$.*

Then, $\mathcal{A}$ wins the game if $T(\mathcal{B}, x) < \Delta$ and $\mathsf{eval}_{\mathsf{pk}}(x, \Delta)$ *outputs y.*

Definition 4 (Δ-sequential). *A trapdoor VDF is* Δ-sequential *if any polynomially bounded player (with respect to the implicit security parameter) wins the above Δ-evaluation race game with negligible probability.*

Observe that it is useless to allow $\mathcal{A}$ to adaptively ask for oracle evaluations of the VDF during the execution of $\mathcal{A}(\mathsf{pk})$: for any data x', the procedure $\mathsf{eval}_{\mathsf{pk}}(x', \Delta)$ produces the same output as $\mathsf{trapdoor}_{\mathsf{sk}}(x', \Delta)$, so any such request can be computed by the adversary in time $O(\Delta)$.

Remark 1. Suppose that the input x is hashed as $H(x)$ (by a secure cryptographic hash function) before being evaluated (as is the case in the construction we present in the next section), i.e.

$$\mathsf{trapdoor}_{\mathsf{sk}}(x, \Delta) = t_{\mathsf{sk}}(H(x), \Delta),$$

for some procedure t, and similarly for eval and verify. Then, it becomes unnecessary to give to $\mathcal{B}$ access to the oracle $\mathcal{O}$. We give a proof in Appendix A when H is modeled as a random oracle.

Remark 2. At the third step of the game, the bound on the min-entropy is fixed to k. The exact value of this bound is arbitrary, but forbidding low entropy is important: if x has a good chance of falling in a small subset of $\mathcal{X}$, the adversary can simply precompute the VDF for all the elements of this subset.

[2] i.e., $C(\mathcal{A}, g) = O(f(\mathrm{len}(g)))$ for a polynomial f, with $\mathrm{len}(g)$ the binary length of g.

3.2 Timing Assumptions in the Real World

Given an algorithm, or even an implementation of this algorithm, its actual running time will depend on the hardware on which it is run. If the algorithm is executed independently on several single-core general purpose CPUs, the variations in running time between them will be reasonably small as overclocking records on clock-speeds barely achieve 9 GHz (cf. [10]), only a small factor higher than a common personal computer. Assuming the computation is not parallelisable, using multiple CPUs would not allow to go faster. However, specialized hardware could be built to perform a certain computation much more efficiently than on any general purpose hardware.

For these reasons, the theoretical model developed in Sect. 3.1 has a limited accuracy. To resolve this issue, and evaluate precisely the security of a timing assumption like Assumption 1, one must estimate the speed at which the current state of technology allows to perform a certain task, given a possibly astronomical budget. To this end, the Ethereum Foundation and Protocol Labs [13] are currently investigating extremely fast hardware implementations of RSA multiplication, and hope to construct a piece of hardware close enough to today's technological limits, with the goal of using the present construction in their future platforms. Similarly, the Chia Network has opened a competition in the near future for very fast multiplication in the class group of a quadratic imaginary field.

4 Construction of the Verifiable Delay Function

Let $x \in \mathcal{A}^*$ be the input at which the VDF is to be evaluated. Alice's secret key sk is the order of a finite group G, and her public key is a description of G allowing to compute the group multiplication efficiently. We also assume that any element g of G can efficiently be represented in a canonical way as binary strings $\mathtt{bin}(g)$. Also part of Alice's public key is a hash function $H_G : \mathcal{A}^* \to G$.

Example 1 (RSA setup). A natural choice of setup is the following: the group G is $(\mathbf{Z}/N\mathbf{Z})^\times$ where $N = pq$ for a pair of distinct prime numbers p and q, where the secret key is $(p-1)(q-1)$ and the public key is N, and the hash function $H_G(x) = \mathtt{int}(H(\text{"}\mathtt{residue}\text{"}||x)) \mod N$ (where H is a secure cryptographic hash function). For a technical reason explained later in Remark 4, we actually need to work in $(\mathbf{Z}/N\mathbf{Z})^\times/\{\pm 1\}$, and we call this the *RSA setup*.

Example 2 (Class group setup). For a public setup where we do not want the private key to be known by anyone, one could choose G to be the class group of an imaginary quadratic field. The construction is simple. Choose a random, negative, square-free integer d, of large absolute value, and such that $d \equiv 1 \mod 4$. Then, let $G = \mathrm{Cl}(d)$ be the class group of the imaginary quadratic field $\mathbf{Q}(\sqrt{d})$. Just as we wish, there is no known algorithm to efficiently compute the order of this group. The multiplication can be performed efficiently, and each class can be represented canonically by its reduced ideal. Note that the even part of $|\text{Cl}(d)|$ can be computed if the factorisation of d is known. Therefore one should choose d to be a negative prime, which ensures that $|\text{Cl}(d)|$ is odd. See [8] for a review of the arithmetic in class groups of imaginary quadratic orders, and a discussion on the choice of cryptographic parameters.

Consider a targeted evaluation time given by $\Delta = t\delta$ for a timing parameter t, where δ is the time-cost (i.e., the amount of sequential work) of computing a single squaring in the group G (as done in Assumption 1 for the RSA setup).

To evaluate the VDF on input x, first let $g = H_G(x)$. The basic idea (which finds its origins in [18]) is that for any $t \in \mathbf{Z}_{>0}$, Alice can efficiently compute g^{2^t} with two exponentiations, by first computing $e = 2^t \mod |G|$, followed by g^e. The running time is logarithmic in t. Any other party who does not know $|G|$ can also compute g^{2^t} by performing t sequential squarings, with a running time $t\delta$. Therefore anyone can compute $y = g^{2^t}$ but only Alice can do it fast, and any other party has to spend a time linear in t. However, verifying that the published value y is indeed g^{2^t} is long: there is no shortcut to the obvious strategy consisting in recomputing g^{2^t} and checking if it matches. To solve this issue, we propose the following public-coin succinct argument, for proving that $y = g^{2^t}$. The input of the interaction is (G, g, y, t). Let $\text{Primes}(2k)$ denote the set containing the 2^{2k} first prime numbers.

1. The verifier samples a prime ℓ uniformly at random from $\text{Primes}(2k)$.
2. The prover computes $\pi = g^{\lfloor 2^t/\ell \rfloor}$ and sends it to the verifier.
3. The verifier computes $r = 2^t \mod \ell$, (the least positive residue of 2^t modulo ℓ), and accepts if $g, y, \pi \in G$ and $\pi^\ell g^r = y$.

Now, it might not be clear how Alice or a third party should compute $\pi = g^{\lfloor 2^t/\ell \rfloor}$. For Alice, it is simple: she can compute $r = 2^t \mod \ell$. Then we have $\lfloor 2^t/\ell \rfloor = (2^t - r)/\ell$, and since she knows the order of the group, she can compute $q = (2^t - r)/\ell \mod |G|$ and $\pi = g^q$. We explain in Sect. 4.1 how anyone else can compute π without knowing $|G|$, with a total of $O(t/\log(t))$ group multiplications.

This protocol is made non-interactive using the Fiat-Shamir transformation, by letting $\ell = H_{\texttt{prime}}(\texttt{bin}(g)|||\texttt{bin}(y))$, where $H_{\texttt{prime}}$ is a hash function which sends any string s to an element of $\text{Primes}(2k)$. We assume in the security analysis below that this function is a uniformly distributed random oracle. The procedures trapdoor, verify and eval are fully described in Algorithms 1, 2 and 3 respectively.

Remark 3. Instead of hashing the input x into the group G as $g = H_G(x)$, one could simply consider $x \in G$. However, the function $x \mapsto x^{2^t}$ being a group homomorphism, bypassing the hashing step has undesirable consequences. For instance, given x^{2^t}, one can compute $(x^\alpha)^{2^t}$ for any integer α at the cost of only an exponentiation by α.

Verification. It is straightforward to check that the verification condition $\pi^\ell g^r = y$ holds if the evaluator is honest. Now, what can a dishonest evaluator do? That question is answered formally in Sect. 6, but the intuitive idea is easy to

Data: a public key $\mathsf{pk} = (G, H_G)$ and a secret key $\mathsf{sk} = |G|$, some input $x \in \mathcal{A}^*$, a targeted evaluation time $\Delta = t\delta$.
Result: the output y, and the proof π.
$g \leftarrow H_G(x) \in G$;
$e \leftarrow 2^t \mod |G|$;
$y \leftarrow g^e$;
$\ell \leftarrow H_{\texttt{prime}}(\texttt{bin}(g)|||\texttt{bin}(y))$;
$r \leftarrow$ least residue of 2^t modulo ℓ;
$q \leftarrow (2^t - r)\ell^{-1} \mod |G|$;
$\pi \leftarrow g^q$;
return (y, π);

Algorithm 1: $\mathsf{trapdoor}_{\mathsf{sk}}(x, t) \to (y, \pi)$

understand. We will show that given x, finding a pair (y, π) different from the honest one amounts to solve a root-finding problem in the underlying group G (supposedly hard for anyone who does not know the secret order of the group).. As a result, only Alice can produce misleading proofs.

Consider the above interactive succinct argument, and suppose that the verifier accepts, i.e., $\pi^\ell g^r = y$, where r is the least residue of 2^t modulo ℓ. Since $r = 2^t - \ell\lfloor 2^t/\ell \rfloor$, the verification condition is equivalent to

$$yg^{-2^t} = \left(\pi g^{-\lfloor 2^t/\ell \rfloor}\right)^\ell.$$

Before the generation of ℓ, the left-hand side $\alpha = yg^{-2^t}$ is already determined. Once ℓ is revealed, the evaluator is able to compute $\beta = \pi g^{-\lfloor 2^t/\ell \rfloor}$, which is an ℓ-th root of α. For a prover to succeed with good probability, he must be able to extract ℓ-th roots of α for arbitrary values of ℓ. This is hard in our groups of interest, unless $\alpha = \beta = 1_G$, in which case (y, π) is the honest output.

Remark 4. Observe that in the RSA setup, this task is easy if $\alpha = \pm 1$, i.e. $y = \pm g^{2^t}$. It is however a difficult problem, given an RSA modulus N, to find an element $\alpha \mod N$ other than ± 1 from which ℓ-th roots can be extracted for any ℓ. This explains why we need to work in the group $G = (\mathbf{Z}/N\mathbf{Z})^\times/\{\pm 1\}$ instead of $(\mathbf{Z}/N\mathbf{Z})^\times$ in the RSA setup. This problem is formalized (and generalised to other groups) in Definition 6.

4.1 Computing the Proof π in $O(t/\log(t))$ Group Operations

In this section, we describe how to compute the proof $\pi = g^{\lfloor 2^t/\ell \rfloor}$ with a total of $O(t/\log(t))$ group multiplications. First, we mention a very simple algorithm to compute π, which simply computes the long division $\lfloor 2^t/\ell \rfloor$ on the fly, as pointed out by Boneh, Bünz and Fisch [5], but requires between t and $2t$ group operations. It is given in Algorithm 4.

Data: a public key $\mathsf{pk} = (G, H_G)$, some input $x \in \mathcal{A}^*$, a targeted evaluation time $\Delta = t\delta$, a VDF output y and a proof π.
Result: true if y is the correct evaluation of the VDF at x, false otherwise.

```
g ← H_G(x);
ℓ ← H_prime(bin(g)|||bin(y));
r ← least residue of 2^t modulo ℓ;
if π^ℓ g^r = y then
  | return true;
else
  | return false;
end
```

Algorithm 2: $\mathsf{verify}_{\mathsf{pk}}(x, y, \pi, t) \to$ true or false

Data: a public key $\mathsf{pk} = (G, H_G)$, some input $x \in \mathcal{A}^*$, a targeted evaluation time $\Delta = t\delta$.
Result: the output value y and a proof π.

```
g ← H_G(x) ∈ G;
y ← g^(2^t);                         // via t sequential squarings
ℓ ← H_prime(bin(g)|||bin(y));
π ← g^⌊2^t/ℓ⌋;    // following the simple Algorithm 4, or the faster
Algorithm 5
return (y, π);
```

Algorithm 3: $\mathsf{eval}_{\mathsf{pk}}(x, t) \to (y, \pi)$

We now describe how to perform the same computation with only $O(t/\log(t))$ group operations. Fix a parameter κ. The idea is to express $\lfloor 2^t/\ell \rfloor$ in base 2^κ as

$$\lfloor 2^t/\ell \rfloor = \sum_i b_i 2^{\kappa i} = \sum_{b=0}^{2^\kappa - 1} b \left(\sum_{i \text{ such that } b_i = b} 2^{\kappa i} \right).$$

Similarly to Algorithm 4, each coefficient b_i can be computed as

$$b_i = \left\lfloor \frac{2^\kappa (2^{t-\kappa(i+1)} \mod \ell)}{\ell} \right\rfloor,$$

where $2^{t-\kappa(i+1)} \mod \ell$ denotes the least residue of $2^{t-\kappa(i+1)}$ modulo ℓ. For each κ-bits integer $b \in \{0, \dots, 2^\kappa - 1\}$, let $I_b = \{i \mid b_i = b\}$. We get

$$g^{\lfloor 2^t/\ell \rfloor} = \prod_{b=0}^{2^\kappa - 1} \left(\prod_{i \in I_b} g^{2^{\kappa i}} \right)^b. \tag{1}$$

Suppose first that all the values $g^{2^{\kappa i}}$ have been memorised (from the sequential computation of the value $y = g^{2^t}$). Then, each product $\prod_{i \in I_b} g^{2^{\kappa i}}$ can be computed in $|I_b|$ group multiplications (for a total of $\sum_b |I_b| = t/\kappa$ multiplications), and the full product (1) can be deduced with about $\kappa 2^\kappa$ additional group

Data: an element g in a group G (with identity 1_G), a prime number ℓ and a positive integer t.
Result: $g^{\lfloor 2^t/\ell \rfloor}$.
$x \leftarrow 1_G \in G$;
$r \leftarrow 1 \in \mathbf{Z}$;
for $i \leftarrow 0$ **to** $T - 1$ **do**
  $b \leftarrow \lfloor 2r/\ell \rfloor \in \{0, 1\} \in \mathbf{Z}$;
  $r \leftarrow$ least residue of $2r$ modulo ℓ;
  $x \leftarrow x^2 g^b$;
end
return x;

Algorithm 4: Simple algorithm to compute $g^{\lfloor 2^t/\ell \rfloor}$, with an on-the-fly long division [5].

operations. In total, this strategy requires about $t/\kappa + \kappa 2^\kappa$ group operations. Choosing, for instance, $\kappa = \log(t)/2$, we get about $t \cdot 2/\log(t)$ group operations. Of course, this would require the storage of t/κ group elements.

We now show that the memory requirement can easily be reduced to, for instance, $O(\sqrt{t})$ group elements, for essentially the same speedup. Instead of memorising each κ element of the sequence g^{2^i}, only memorise every $\kappa\gamma$ element (i.e., the elements $g^{2^0}, g^{2^{\kappa\gamma}}, g^{2^{2\kappa\gamma}}, \dots$), for some parameter γ (we will show that $\gamma = O(\sqrt{t})$ is sufficient). For each integer j, let $I_{b,j} = \{i \in I_b \mid i \equiv j \mod \gamma\}$. Now,

$$g^{\lfloor 2^t/\ell \rfloor} = \prod_{b=0}^{2^\kappa - 1} \left(\prod_{j=0}^{\gamma-1} \prod_{i \in I_{b,j}} g^{2^{\kappa i}} \right)^b = \prod_{j=0}^{\gamma-1} \left(\prod_{b=0}^{2^\kappa - 1} \left(\prod_{i \in I_{b,j}} g^{2^{\kappa(i-j)}} \right)^b \right)^{2^{\kappa j}}.$$

In each factor of the final product, $i - j$ is divisible by γ, so $g^{2^{\kappa(i-j)}}$ is one of the memorised values. A straightforward approach allows to compute this product with a total amount of group operations about $t/\kappa + \gamma\kappa 2^\kappa$, yet one can still do better. Write $y_{b,j} = \prod_{i \in I_{b,j}} g^{2^{\kappa(i-j)}}$, and split κ into two halves, as $\kappa_1 = \lfloor \kappa/2 \rfloor$ and $\kappa_0 = \kappa - \kappa_1$. Now, observe that for each index j,

$$\prod_{b=0}^{2^\kappa - 1} y_{b,j}^b = \prod_{b_1=0}^{2^{\kappa_1}-1} \left(\prod_{b_0=0}^{2^{\kappa_0}-1} y_{b_1 2^{\kappa_0} + b_0, j} \right)^{b_1 2^{\kappa_0}} \cdot \prod_{b_0=0}^{2^{\kappa_0}-1} \left(\prod_{b_1=0}^{2^{\kappa_1}-1} y_{b_1 2^{\kappa_0} + b_0, j} \right)^{b_0}$$

The right-hand side provides a way to compute the product with a total of about $2(2^\kappa + \kappa 2^{\kappa/2})$ (instead of $\kappa 2^\kappa$ as in the more obvious strategy). The full method is summarised in Algorithm 5 (on page 29), and requires about $t/\kappa + \gamma 2^{\kappa+1}$ group multiplications.

The algorithm requires the storage of about $t/(\kappa\gamma) + 2^\kappa$ group elements. Choosing, for instance, $\kappa = \log(t)/3$ and $\gamma = \sqrt{t}$, we get about $t \cdot 3/\log(t)$ group operations, with the storage of about $\sqrt{t}$ group elements. This algorithm can also be parallelised.

Data: an element g in a group G (with identity 1_G), a prime number ℓ, a positive integer t, two parameters $\kappa, \gamma > 0$, and a table of precomputed values $C_i = g^{2^{i\kappa\gamma}}$, for $i = 0, \ldots, \lceil t/(\kappa\gamma) \rceil$.
Result: $g^{\lfloor 2^t/\ell \rfloor}$.

```
// define a function get_block such that ⌊2^t/ℓ⌋ = Σ_i get_block(i)2^{κi}
get_block ← the function that on input i returns ⌊2^κ(2^{t−κ(i+1)} mod ℓ)/ℓ⌋;
// split κ into to halves
κ1 ← ⌊κ/2⌋;
κ0 ← κ − κ1;
x ← 1_G ∈ G;
for j ← γ − 1 to 0 (descending order) do
    x ← x^{2^κ};
    for b ∈ {0, ..., 2^κ − 1} do
        y_b ← 1_G ∈ G;
    end
    for i ← 0, ..., ⌈t/(κγ)⌉ − 1 do
        b ← get_block(iγ + j);   // this could easily be optimised by
        computing the blocks iteratively as in Algorithm 4 (but
        computing blocks of κ bits and taking steps of κγ bits),
        instead of computing them one by one.
        y_b ← y_b · C_i;
    end
    for b1 ∈ {0, ..., 2^{κ1} − 1} do
        z ← 1_G ∈ G;
        for b0 ∈ {0, ..., 2^{κ0} − 1} do
            z ← z · y_{b1 2^{κ0} + b0};
        end
        x ← x · z^{b1 2^{κ0}};
    end
    for b0 ∈ {0, ..., 2^{κ0} − 1} do
        z ← 1_G ∈ G;
        for b1 ∈ {0, ..., 2^{κ1} − 1} do
            z ← z · y_{b1 2^{κ0} + b0};
        end
        x ← x · z^{b0};
    end
end
return x;
```

Algorithm 5: Faster algorithm to compute $g^{\lfloor 2^t/\ell \rfloor}$, given some precomputations.

4.2 A Practical Bandwidth and Storage Improvement

Typically, an *evaluator* would compute the output y and the proof π, and send the pair (y, π) to the *verifiers*. Each verifier would compute the Fiat-Shamir challenge

$$\ell \leftarrow H_{\texttt{prime}}(\texttt{bin}(g)|||\texttt{bin}(y)),$$

then check $y = \pi^\ell g^{2^t \bmod \ell}$. Instead, it is possible for the evaluator to transmit (ℓ, π), which has almost half the size (typically, ℓ is in the order of hundreds of bits while group elements are in the order of thousands of bits). The verifiers would recover

$$y \leftarrow \pi^\ell g^{2^t \bmod \ell},$$

and then verify that $\ell = H_{\texttt{prime}}(\texttt{bin}(g)|||\texttt{bin}(y))$. The two strategies are equivalent, but the second divides almost by 2 the bandwidth and storage footprint.

4.3 A Trade-Off Between Proof Shortness and Prover Efficiency

The evaluation of the VDF, i.e., the computation of $y = g^{2^t}$, takes time $T = \delta t$, where δ is the time of one squaring in the underlying group. As demonstrated in Sect. 4.1, the proof π can be computed in $O(t/\log(t))$ group operations. Say that the total time of computing the proof is a fraction T/ω; considering Algorithm 5, one can think of $\omega = 20$, a reasonable value for practical parameters. One potential issue with the proposed VDF is that the computation of π can only start after the evaluation of the VDF output g^{2^t} is completed. So after the completion of the VDF evaluation, there still remains a total amount T/ω of work to compute the proof. We call *overhead* these computations that must be done after the evaluation of $y = g^{2^t}$. Even though this part of the computation can be parallelised, it might be advantageous for some applications to reduce the overhead to a negligible amount of work.

We show in the following that using only two parallel threads, the overhead can be reduces to a total cost of about T/ω^n, at the cost of lengthening the proofs to n group elements (instead of a single one), and $n-1$ small prime numbers. Note that the value of ω varies with T, yet for simplicity of exposition, we assume that it is constant in the following analysis (a reasonable approximation for practical purposes).

The idea is to start proving some intermediate results before the full evaluation is over. For instance, consider $t_1 = t\frac{\omega}{\omega+1}$. Run the evaluator, and when the intermediate value $g_1 = g^{2^{t_1}}$ is reached, store it (but keep the evaluator running in parallel), and compute the proof π_1 for the statement $g_1 = g^{2^{t_1}}$. The computation of this proof takes time about $\delta t_1/\omega = T/(\omega+1)$, which is the time it takes to finish the full evaluation (i.e., going from g_1 to $y = g^{2^t} = g_1^{2^{t/(\omega+1)}}$). Therefore, the evaluation of y and the first proof π_1 finish at the same time. It only remains to produce a proof π_2 for the statement $y = g_1^{2^{t/(\omega+1)}}$, which can be done in total time $\frac{\delta t}{\omega(\omega+1)} \leq T/\omega^2$. Therefore the overhead is at most T/ω^2. At first glance, it seems the verification requires the triple (g_1, π_1, π_2), but in fact, the value g_1 can be recovered from π_1 and the prime number $\ell_1 = H_{\texttt{prime}}(\texttt{bin}(g)|||\texttt{bin}(g_1))$ via $g_1 = \pi_1^\ell g^{t_1 \bmod \ell}$, as done is Sect. 4.2. Therefore, the proof can be compressed to (ℓ_1, π_1, π_2).

More generally, one could split the computation into n segments of length $t_i = t\omega^{n-i}\frac{\omega-1}{\omega^n-1}$, for $i = 1, \ldots, n$. We have that $t = \sum_{i=1}^n t_i$, and $t_i = t_{i-1}/\omega$, so during the evaluation of each segment (apart from the first), one can compute the proof corresponding to the previous segment. The overhead is only the proof of the last segment, which takes time $T\frac{\omega-1}{\omega(\omega^n-1)} \leq T/\omega^n$. The proof consists of the n intermediate proofs and the $n-1$ intermediate prime challenges.

5 Analysis of the Sequentiality

In this section, the proposed construction is proven to be $(t\delta)$-sequential, meaning that no polynomially bounded player can win the associated $(t\delta)$-evaluation race game with non-negligible probability (in other words, the VDF cannot be evaluated in time less than $t\delta$). For the RSA setup, it is proved under the classic time-lock assumption of Rivest, Shamir and Wagner [18] (formalised in Assumption 1), and more generally, it is secure for groups where a generalisation of this assumption holds (Assumption 2).

5.1 Generalised Time-Lock Assumptions

The following game generalises the classic time-lock assumption to arbitrary families of groups of unknown orders.

Definition 5 (Generalised (δ, t)-time-lock game). *Consider a sequence $(\mathcal{G}_k)_{k\in\mathbf{Z}_{>0}}$, where each $\mathcal{G}_k$ is a set of finite groups (supposedly of unknown orders), associated to a "difficulty parameter" k. Let* keygen *be a procedure to generate a random group from $\mathcal{G}_k$, according to the difficulty k.*

Fix the difficulty parameter $k \in \mathbf{Z}_{>0}$, and let $\mathcal{A}$ be an algorithm playing the game. The parameter t is a positive integer, and $\delta : \mathbf{Z}_{>0} \to \mathbf{R}_{>0}$ is a function. The (δ, t)-time-lock game *goes as follows:*

1. *A group G is generated by* keygen*;*
2. *$\mathcal{A}(G)$ outputs an algorithm $\mathcal{B}$;*
3. *An element $g \in G$ is generated uniformly at random;*
4. *$\mathcal{B}(g)$ outputs $h \in G$.*

Then, $\mathcal{A}$ wins the game if $h = g^{2^t}$ and $T(\mathcal{B}, g) < t\delta(k)$.

Assumption 2 (Generalised time-lock assumption). *The generalised time-lock assumption for a given family of groups $(\mathcal{G}_k)_{k\in\mathbf{Z}_{>0}}$ is the following. There is a cost function $\delta : \mathbf{Z}_{>0} \to \mathbf{R}_{>0}$ such that the following two statements hold:*

1. *There is an algorithm $\mathcal{S}$ such that for any group $G \in \mathcal{G}_k$ (for the difficulty parameter k), and any element $g \in G$, the output of $\mathcal{S}(G, g)$ is the square of g, and $T(\mathcal{S}, (G, g)) < \delta(k)$;*
2. *For any $t \in \mathbf{Z}_{>0}$, no algorithm $\mathcal{A}$ of polynomial cost wins the (δ, t)-time-lock game with non-negligible probability (with respect to the difficulty parameter k).*

The function δ encodes the time-cost of computing a single squaring in a group of $\mathcal{G}_k$, and Assumption 2 expresses that without more specific knowledge about these groups (such as their orders), there is no faster way to compute g^{2^t} than performing t sequential squarings.

5.2 Sequentiality in the Random Oracle Model

Proposition 1 ($t\delta$-sequentiality of the trapdoor VDF in the random oracle model). *Let $\mathcal{A}$ be a player winning with probability p_{win} the $(t\delta)$-evaluation race game associated to the proposed construction, assuming H_G and $H_{\mathtt{prime}}$ are random oracles and $\mathcal{A}$ is limited to q oracle queries*[3]. *Then, there is a player $\mathcal{C}$ for the (generalised) (δ, t)-time-lock game, with winning probability $p \geq (1 - q/2^k)p_{\mathsf{win}}$, and with same running time as $\mathcal{A}$ (up to a constant factor*[4]*).*

Proof. Build $\mathcal{C}$ as follows. Upon receiving the group G, $\mathcal{C}$ starts running $\mathcal{A}$ on input G. The random oracles H_G and $H_{\mathtt{prime}}$ are simulated in a straightforward manner, maintaining a table of values, and generating a random outcome for any new request (with distribution uniform in G and in Primes$(2k)$ respectively). When $\mathcal{A}(G)$ outputs an algorithm $\mathcal{B}$, $\mathcal{C}$ generates a random $x \in \mathcal{X}$ (according to the same distribution as the $(t\delta)$-evaluation race game). If x has been queried by the oracle already, $\mathcal{C}$ aborts; this happens with probability at most $q/2^k$, since the min-entropy of the distribution of messages in the $(t\delta)$-evaluation race game is at least k. Otherwise, $\mathcal{C}$ outputs the following algorithm $\mathcal{B}'$. When receiving as input the challenge g, $\mathcal{B}'$ adds g to the table of oracle H_G, for the input x (i.e., $H_G(x) = g$). As discussed in Remark 1, we can assume that the algorithm $\mathcal{B}$ does not call the oracle $\mathsf{trapdoor}_{\mathsf{sk}}(-, y, \Delta)$. Then $\mathcal{B}'$ can invoke $\mathcal{B}$ on input x while simulating the oracles H_G and $H_{\mathtt{prime}}$. Whenever $\mathcal{B}$ outputs y, $\mathcal{B}'$ outputs y, which equals g^{2^t} whenever y is the correct evaluation of the VDF at x. We assume that simulating the oracle has a negligible cost, so $\mathcal{B}'(g)$ has essentially the same time-cost as $\mathcal{B}(x)$. Then, $\mathcal{C}$ wins the (δ, t)-time-lock game with probability $p \geq p_{\mathsf{win}}(1 - q/2^k)$. □

6 Analysis of the Soundness

In this section, the proposed construction is proven to be sound, meaning that no polynomially bounded player can produce a misleading proof for an invalid output of the VDF. For the RSA setup, it is proved under a new number theoretic assumption expressing that it is hard to find an integer $u \neq 0, \pm 1$ for which ℓ-th roots modulo an RSA modulus N can be extracted for arbitrary ℓ-values sampled uniformly at random from Primes$(2k)$, when the factorisation of N is unknown. More generally, the construction is sound if a generalisation of this assumptions holds in the group of interest.

---

[3] In this game, the output of $\mathcal{A}$ is another algorithm $\mathcal{B}$. When we say that $\mathcal{A}$ is limited to q queries, we limit the total number of queries by $\mathcal{A}$ and $\mathcal{B}$ combined. In other words, if $\mathcal{A}$ did $x \leq q$ queries, then its output $\mathcal{B}$ is limited to $q - x$ queries.

[4] Note that this constant factor does not affect the chances of $\mathcal{C}$ to win the (δ, t)-time-lock game, since it concerns only the running time of $\mathcal{C}$ itself and not of the algorithm output by $\mathcal{C}(G)$.

6.1 The Root Finding Problem

The following game formalises the root finding problem.

Definition 6 (The root finding game $\mathcal{G}^{\text{root}}$). *Let $\mathcal{A}$ be a party playing the game. The* root finding game *$\mathcal{G}^{\text{root}}(\mathcal{A})$ goes as follows: first, the* keygen *procedure is run, resulting in a group G which is given to $\mathcal{A}$ (G is supposedly of unknown order). The player $\mathcal{A}$ then outputs an element u of G. An integer ℓ is sampled uniformly from* Primes$(2k)$ *and given to $\mathcal{A}$. The player $\mathcal{A}$ outputs an integer v and wins the game if $v^{\ell} = u \neq 1_G$.*

In the RSA setup, the group G is the quotient $(\mathbf{Z}/N\mathbf{Z})^{\times}/\{\pm 1\}$, where N is a product of two random large prime numbers. It is not known if this problem can easily be reduced to a standard assumption such as the difficulty of factoring N or the RSA problem, for which the best known algorithms have complexity $L_N(1/3)$.

Similarly, in the class group setting, this problem is not known to reduce to a standard assumption, but it is closely related to the order problem and the root problem (which are tightly related to each other, see [3, Theorem 3]), for which the best known algorithms have complexity $L_{|d|}(1/2)$ where d is the discriminant.

We now prove that to win this game $\mathcal{G}^{\text{root}}$, it is sufficient to win the following game $\mathcal{G}_X^{\text{root}}$, which is more convenient for our analysis.

Definition 7 (The oracle root finding game $\mathcal{G}_X^{\text{root}}$). *Let $\mathcal{A}$ be a party playing the game. Let X be a function that takes as input a group G and a string s in $\mathcal{A}^*$, and outputs an element $X(G,s) \in G$. Let $\mathcal{O} : \mathcal{A}^* \rightarrow$* Primes$(2k)$ *be a random oracle with the uniform distribution. The player has access to the random oracle $\mathcal{O}$. The* oracle root finding game *$\mathcal{G}_X^{\text{root}}(\mathcal{A}, \mathcal{O})$ goes as follows: first, the* keygen *procedure is run and the resulting group G is given to $\mathcal{A}$. The player $\mathcal{A}$ then outputs a string $s \in \mathcal{A}^*$, and an element v of G. The game is won if $v^{\mathcal{O}(s)} = X(G,s) \neq 1_G$.*

Lemma 1. *If there is a function X and an algorithm $\mathcal{A}$ limited to q queries to the oracle $\mathcal{O}$ winning the game $\mathcal{G}_X^{\text{root}}(\mathcal{A}, \mathcal{O})$ with probability p_{win}, there is an algorithm $\mathcal{B}$ winning the game $\mathcal{G}^{\text{root}}(\mathcal{B})$ with probability at least $p_{\text{win}}/(q+1)$, and same running time, up to a small constant factor.*

Proof. Let $\mathcal{A}$ be an algorithm limited to q oracle queries, and winning the game with probability p_{win}. Build an algorithm $\mathcal{A}'$ which does exactly the same thing as $\mathcal{A}$, but with possibly additional oracle queries at the end to make sure the output string s' is always queried to the oracle, and the algorithm always does exactly $q+1$ (distinct) oracle queries.

Build an algorithm $\mathcal{B}$ playing the game $\mathcal{G}^{\text{root}}$, using $\mathcal{A}'$ as follows. Upon receiving $\mathsf{pk} = G$, $\mathcal{B}$ starts running $\mathcal{A}'$ on input pk. The oracle $\mathcal{O}$ is simulated as follows. First, an integer $i \in \{1, 2, ..., q+1\}$ is chosen uniformly at random. For the first $i-1$ (distinct) queries from $\mathcal{A}'$ to $\mathcal{O}$, the oracle value is chosen uniformly at random from Primes$(2k)$. When the ith string $s \in \mathcal{A}^*$ is queried to the oracle, the algorithm $\mathcal{B}$ outputs $u = X(G, s)$, concluding the first round of the game $\mathcal{G}^{\mathsf{root}}$. The game continues as the integer ℓ is received (uniform in Primes$(2k)$). This ℓ is then used as the value for the ith oracle query $\mathcal{O}(s)$, and the algorithm $\mathcal{A}'$ can continue running. The subsequent oracle queries are handled like the first $i-1$ queries, by picking random primes in Primes$(2k)$. Finally, $\mathcal{A}'$ outputs a string $s' \in \mathcal{A}^*$ and an element v of G. To conclude the game $\mathcal{G}^{\mathsf{root}}(\mathcal{B})$, $\mathcal{B}$ returns v.

Since $\mathcal{O}$ simulates a random oracle with uniform outputs in Primes$(2k)$, $\mathcal{A}'$ outputs with probability p_{win} a pair (s', v) such that $v^{\mathcal{O}(s')} = X(G, s') \neq 1_G$; denote this event $\mathsf{win}_{\mathcal{A}'}$. If $s = s'$, this condition is exactly $v^\ell = u \neq 1_G$, where $u = X(G, s)$ is the output for the first round of $\mathcal{G}^{\mathsf{root}}$, and $\mathcal{O}(s) = \ell$ is the input for the second round. If these conditions are met, the game $\mathcal{G}^{\mathsf{root}}(\mathcal{B})$ is won. Therefore

$$\Pr[\mathcal{B} \text{ wins } \mathcal{G}^{\mathsf{root}}] \geq p_{\mathsf{win}} \cdot \Pr\left[s = s' | \mathsf{win}_{\mathcal{A}'}\right].$$

Let $\mathcal{Q} = \{s_1, s_2, ..., s_{q+1}\}$ be the $q+1$ (distinct) strings queried to $\mathcal{O}$ by $\mathcal{A}'$, indexed in chronological order. By construction, we have $s = s_i$. Let j be such that $s' = s_j$ (recall that $\mathcal{A}'$ makes sure that $s' \in \mathcal{Q}$). Then,

$$\Pr\left[s = s' | \mathsf{win}_{\mathcal{A}'}\right] = \Pr\left[i = j | \mathsf{win}_{\mathcal{A}'}\right]$$

The integer i is chosen uniformly at random in $\{1, 2, ..., q+1\}$, and the values given to $\mathcal{A}'$ are independent from i (the oracle values are all independent random variables). So $\Pr\left[i = j | \mathsf{win}_{\mathcal{A}'}\right] = 1/(q+1)$. Therefore $\Pr[\mathcal{B} \text{ wins } \mathcal{G}^{\mathsf{root}}] \geq p_{\mathsf{win}}/(q+1)$. Since $\mathcal{B}$ mostly consists in running $\mathcal{A}$ and simulating the random oracle, it is clear than both have the same running time, up to a small constant factor. □

6.2 Soundness in the Random Oracle Model

Proposition 2 (Soundness of the trapdoor VDF in the random oracle model). *Let $\mathcal{A}$ be a player winning with probability p_{win} the soundness-breaking game associated to the proposed scheme, assuming H_G and $H_{\mathtt{prime}}$ are random oracles and $\mathcal{A}$ is limited to q oracle queries[5]. Then, there is a player $\mathcal{D}$ for the root finding game $\mathcal{G}^{\mathsf{root}}$ with winning probability $p \geq p_{\mathsf{win}}/(q+1)$, and with same running time as $\mathcal{A}$ (up to a constant factor).*

Proof. Instead of directly building $\mathcal{D}$, we build an algorithm $\mathcal{D}'$ playing the game $\mathcal{G}_X^{\mathsf{root}}(\mathcal{D}', \mathcal{O})$, and invoke Lemma 1. Define the function X as follows. Recall that for any group G that we consider in the construction, each element $g \in G$ admits a canonical binary representation $\mathtt{bin}(g)$. For any such group G, any elements $g, h \in G$, let

$$X(G, \mathtt{bin}(g)|||\mathtt{bin}(h)) = h/g^{2^t},$$

[5] In this game, the output of $\mathcal{A}$ is another algorithm $\mathcal{B}$. When we say that $\mathcal{A}$ is limited to q queries, we limit the total number of queries by $\mathcal{A}$ and $\mathcal{B}$ combined. In other words, if $\mathcal{A}$ did $x \leq q$ queries, then its output $\mathcal{B}$ is limited to $q - x$ queries.

and let $X(G, s) = 1_G$ for any other string s. When receiving pk, $\mathcal{D}'$ starts running $\mathcal{A}$ with input pk. The oracle H_G is simulated by generating random values in the straightforward way, and $H_{\mathtt{prime}}$ is set to be exactly the oracle $\mathcal{O}$. The algorithm $\mathcal{A}$ outputs a message x, and pair $(y, \pi) \in G \times G$ (if it is not of this form, abort). Output $s = \mathtt{bin}(H_G(x))|||\mathtt{bin}(y)$ and $v = \pi / H_G(x)^{\lfloor 2^t/\mathcal{O}(s) \rfloor}$. If $\mathcal{A}$ won the simulated soundness-breaking game, the procedure did not abort, and $v^{\mathcal{O}(s)} = X(G, s) \neq 1_G$, so $\mathcal{D}'$ wins the game. Hence $\mathcal{D}'$ has winning probability p_{win}. Since $\mathcal{A}$ was limited to q oracle queries, $\mathcal{D}'$ also does not do more than q queries. Applying Lemma 1, there is an algorithm $\mathcal{D}$ winning the game $\mathcal{G}^{\mathsf{root}}(\mathcal{B})$ with probability $p \geq p_{\mathsf{win}}(1 - \varepsilon)/(q + 1)$. □

Remark 5. The construction remains sound if instead of considering the output y and the proof π, we consider the output to be the pair (y, π), with an empty proof. The winning probability of $\mathcal{D}$ in Proposition 2 becomes $p \geq p_{\mathsf{win}}(1 - \varepsilon)/(q + 1)$, where $\varepsilon = \operatorname{negl}\left(\frac{k}{\log\log(|G|)\log(q)}\right)$, by accounting for the unlikely event that the large random prime $\mathcal{O}(s)$ is a divisor of $|G|$.

7 Aggregating and Watermarking Proofs

In this section, we present two useful properties of the VDF: the proofs can be aggregated, and watermarked. The methods of this section follow from discussions at the August 2018 workshop at Stanford hosted by the Ethereum Foundation and the Stanford Center for Blockchain Research. The author wishes to thank the participants for their contribution.

7.1 Aggregation

If the VDF is evaluated at multiple inputs, it is possible to produce a single proof $\tilde{\pi} \in G$ that simultaneously proves the validity of all the outputs. Suppose that n inputs are given, $x_1, \ldots, x_n$. For each index i, let $g_i = H_G(x_i)$. The following public-coin interactive succinct argument allows to prove that a given list $(y_i)_{i=1}^n$ satisfies $y_i = g_i^{2^t}$:

1. The verifier samples a prime ℓ uniformly at random from Primes$(2k)$, and n uniformly random integers $(\alpha_i)_{i=1}^n$ of k bits.
2. The prover computes
$$\tilde{\pi} = \left(\prod_{i=1}^{n} g_i^{\alpha_i}\right)^{\lfloor 2^t/\ell \rfloor}$$
and sends it to the verifier.
3. The verifier computes $r = 2^t \mod \ell$, (the least positive residue of 2^t modulo ℓ), and accepts if
$$\tilde{\pi}^{\ell}\left(\prod_{i=1}^{n} g_i^{\alpha_i}\right)^{r} = \prod_{i=1}^{n} y_i^{\alpha_i}.$$

The single group element $\widetilde{\pi}$ serves as proof for the whole list of n statements $y_i = g_i^{2^t}$: it is an aggregated proof. The protocol can be made non-interactive by a Fiat-Shamir transformation: let

$$s = \mathtt{bin}(g_1)|||\mathtt{bin}(g_2)|||\dots|||\mathtt{bin}(g_n)|||\mathtt{bin}(y_1)|||\mathtt{bin}(y_2)|||\dots|||\mathtt{bin}(y_n),$$

and let $\ell = H_{\mathtt{prime}}(s)$, and for each index i, let $\alpha_i = \mathtt{int}(H(\mathtt{bin}(i)|||s))$ (where H is a secure cryptographic hash function). For simplicity, we prove the soundness in the interactive setup (the non-interactive soundness then follows from the Fiat-Shamir heuristic).

Remark 6. One could harmlessly fix $\alpha_1 = 1$, leaving only α_i to be chosen at random for $i > 1$. We present the protocol as above for simplicity, to avoid dealing with $i = 1$ as a special case in the proof below.

Theorem 1. *If there is a malicious prover $\mathcal{P}$ breaking the soundness of the above interactive succinct argument with probability p, then there is a player $\mathcal{B}$ winning the root finding game $\mathcal{G}^{\mathsf{root}}$ with probability at least $(p^2 - 2^{-k})/3$, with essentially the same running time as $\mathcal{P}$.*

Proof. Let $\mathcal{I} = \{0, 1, \dots, 2^k - 1\}$, and let $\mathcal{Z} = \mathcal{I}^{n-1} \times \text{Primes}(2k)$. Let $Z = (\alpha_2, \dots, \alpha_n, \ell)$ be a uniformly distributed random variable in $\mathcal{Z}$, and let α_1 and α'_1 be two independent, uniformly distributed random variables in $\mathcal{I}$. Let win and win' be the events that $\mathcal{P}$ breaks soundness when given $(\alpha_1, \alpha_2, \dots, \alpha_n, \ell)$ and $(\alpha'_1, \alpha_2, \dots, \alpha_n, \ell)$ respectively. We wish to estimate the probability of the event $\mathsf{double\_win} = \mathsf{win} \wedge \mathsf{win}' \wedge (\alpha_1 \neq \alpha'_1)$. Observe that conditioning over $Z = z$ for an arbitrary, fixed $z \in \mathcal{Z}$, the events win and win' are independent and have same probability, so

$$\Pr[\mathsf{win} \wedge \mathsf{win}'] = \frac{1}{|Z|} \sum_{z \in \mathcal{Z}} \Pr[\mathsf{win} \wedge \mathsf{win}' \mid Z = z] = \frac{1}{|Z|} \sum_{z \in \mathcal{Z}} \Pr[\mathsf{win} \mid Z = z]^2.$$

From the Cauchy-Schwarz inequality, we get

$$\frac{1}{|Z|} \sum_{z \in \mathcal{Z}} \Pr[\mathsf{win} \mid Z = z]^2 \geq \left(\frac{1}{|Z|} \sum_{z \in \mathcal{Z}} \Pr[\mathsf{win} \mid Z = z] \right)^2 = \Pr[\mathsf{win}]^2 = p^2.$$

We conclude that $\Pr[\mathsf{win} \wedge \mathsf{win}'] \geq p^2$, and therefore, $\Pr[\mathsf{double\_win}] \geq p^2 - 2^{-k}$.

With these probabilities at hand, we can now construct the player $\mathcal{B}$ for the root finding game $\mathcal{G}^{\mathsf{root}}$. Run $\mathcal{P}$, which outputs values g_i and y_i. If $y_i = g_i^{2^t}$ for all i, abort. Up to some reindexing, we can now assume $y_1 \neq g_1^{2^t}$. Draw $\alpha_1, \alpha'_1, \alpha_2, \dots, \alpha_n$ uniformly at random from $\mathcal{I}$. Define

$$x_0 = y_1/g_1^{2^t},\ x_1 = \prod_{i=1}^{n} (y_i^{\alpha_i}/g_i^{2^t})^{\alpha_i},\ x_2 = (y_1/g_1^{2^t})^{\alpha'_1} \prod_{i=2}^{n} (y_i^{\alpha_i}/g_i^{2^t})^{\alpha_i}.$$

Let b be a uniformly random element of $\{0, 1, 2\}$. The algorithm $\mathcal{B}$ outputs x_b. We get back a challenge ℓ. Run the prover $\mathcal{P}$ twice, independently, for the challenges

$(\alpha_1, \alpha_2, \ldots, \alpha_n, \ell)$ and $(\alpha_1', \alpha_2, \ldots, \alpha_n, \ell)$, and suppose that both responses break soundness, and $\alpha_1 \neq \alpha_1'$ (i.e., the event double_win occurs). If $x_1 \neq 1_G$ or $x_2 \neq 1_G$, the winning responses from $\mathcal{P}$ allow to extract an ℓ-th root of either x_1 or x_2 respectively. Otherwise, we have $x_1 = x_2$, which implies that $x_0^{\alpha_1 - \alpha_1'} = 1_G$, so x_0 is an element of order dividing $\alpha_1 - \alpha_1'$, and one can easily extract any ℓ-th root of x_0. In conclusion, under the event double_win, one can always extract an ℓ-th root of either x_0, x_1 or x_2, so the total winning probability of algorithm $\mathcal{B}$ is at least $(p^2 - 2^{-k})/3$. □

7.2 Watermarking

When using a VDF to build a decentralised randomness beacon (e.g., as a backbone for an energy-efficient blockchain design), people who spent time and energy evaluating the VDF should be rewarded for their effort. Since the output of the VDF is supposed to be unique, it is hard to reliably identify the person who computed it. A naive attempt of the evaluator to sign the output would not prevent theft: since the output is public, a dishonest party could as easily sign it and claim it their own.

Let the evaluator's identity be given as a string id. One proposed method (see [12]) essentially consists in computing the VDF twice: once on the actual input, and once on a combination of the input with the evaluator's identity id. Implemented carefully, this method could allow to reliably reward the evaluators for their work, but it also doubles the required effort. In the following, we sketch two cost-effective solutions to this problem.

The first cost-effective approach consists in having the evaluator prove that he knows some hard-to-recover intermediate value of the computation of the VDF. Since the evaluation of our proposed construction requires computing in sequence the elements $g_i = g^{2^i}$ for $i = 1, \ldots, t$, and only the final value $y = g_t$ of the sequence is supposed to be revealed, one can prove that they performed the computation by proving that they know g_{t-1} (it is a square root of y, hence the fastest way for someone else to recover it would be to recompute the full sequence). A simple way to do so would be for the evaluator to reveal the value $c_{\mathsf{id}} = g_{t-1}^{p_{\mathsf{id}}}$ (a *certificate*), where $p_{\mathsf{id}} = H_{\mathtt{prime}}(\mathsf{id})$. The validity of the certificate can be checked via the equation $y^{p_{\mathsf{id}}} = c_{\mathsf{id}}^2$. The security claim is the following: given the input x, the output y, the proof π, and the certificate c_{id}, the cost for an adversary with identifier id′ (distinct from id) to produce a valid certificate $c_{\mathsf{id}'}$ is as large as actually recomputing the output of the VDF by themself.

The above method is cost-effective as it does not require the evaluator to perform much more work than evaluating the VDF. However, it makes the output longer by adding an extra group element: the certificate. Another approach consists in producing a single group element that plays simultaneously the role of the proof and the certificate. This element is a *watermarked proof*, tied to the evaluator's identity. This can be done easily with our construction. In the evaluation procedure (Algorithm 3), replace the definition of the prime ℓ by $H_{\mathtt{prime}}(\mathsf{id}|||\mathtt{bin}(g)|||\mathtt{bin}(y))$ (and the corresponding change must be made in the verification procedure). The resulting proof π_{id} is now inextricably tied to id. Informally, the security claim is the following: given the input x, the output y, and the watermarked proof π_{id}, the cost for an adversary with identifier id' (distinct from id) to produce a valid proof $\pi_{\mathsf{id}'}$ is about as large as reevaluating the VDF altogether. Indeed, a honest prover, after having computed the output y, can compute π_{id} at a reduced cost thanks to some precomputed intermediate values. But an adversary does not have these intermediate values, so they would have to compute $\pi_{\mathsf{id}'}$ from scratch. This is an exponentiation in G, with exponent of bit-length close to t; without any intermediate values, it requires in the order of t sequential group operations, which is the cost of evaluating the VDF.

8 Circumventing Impossibility Results with Timing Assumptions

In addition to the applications mentioned in the introduction, we conclude this paper by showing that a *trapdoor* VDF also constitutes a new tool for circumventing classic impossibility results. We illustrate this through a simple identification protocol constructed from a trapdoor VDF, where a party, Alice, wishes to identify herself to Bob by proving that she knows the trapdoor. Thanks to the VDF timing properties, this protocol features surprising zero-knowledge and deniability properties challenging known impossibility results.

As this discussion slightly deviates from the crux of the article (the construction of a trapdoor VDF), most of the details are deferred to Appendices B and C, and this section only introduces the main ideas. As in the rest of the paper, the parameter k is the security level. The identification protocol goes as follows:

1. Bob chooses a challenge $c \in \{\mathtt{0}, \mathtt{1}\}^k$ uniformly at random. He sends it to Alice, along with a time limit Δ, and starts a timer.
2. Alice responds by sending the evaluation of the VDF on input c (with time parameter Δ), together with the proof. She can respond fast using her trapdoor.
3. Upon receiving the response, Bob stops the timer. He accepts if the verification of the VDF succeeds and the elapsed time is smaller than Δ.

Remark 7. We present here only an identification protocol, but it is easy to turn it into an authentication protocol for a message m by having Alice use the concatenation $c||m$ as input to the VDF.

Since only Alice can respond immediately thanks to her secret, Bob is convinced of her identity. Since anyone else can compute the response to the challenge in time Δ, the exchange is perfectly simulatable, hence perfectly zero-knowledge. It is well-known (and in fact clear from the definition) that a classic interactive zero-knowledge proof cannot have only one round (this would be a challenge-response exchange, and the simulator would allow to respond to the challenge in polynomial time, violating soundness). The above protocol avoids this impossibility thanks to a modified notion of soundness, ensuring that only Alice can respond *fast enough.* This is made formal in Appendix B, via the notion of zero-knowledge timed challenge-response protocol.

Remark 8. Note that this very simple protocol is also efficient: the "time-lock" evaluation of the VDF does not impact any of the honest participants, it is only meant to be used by the simulator. Only the trapdoor evaluation and the verification are actually executed.

Finally, this protocol has strong deniability properties. Indeed, since anyone can produce in time Δ a response to any challenge, any transcript of a conversation that is older than time Δ could have been generated by anyone. In fact the protocol is *on-line deniable* against any judge that suffers a communication delay larger than $\Delta/2$. Choosing Δ to be as short as possible (while retaining soundness) yields a strongly deniable protocol. Full on-line deniability is known to be impossible in a PKI (see [11]), and this delay assumption provides a new way to circumvent this impossibility. This is discussed in more detail in Appendix C.

Acknowledgements. The author wishes to thank a number of people with whom interesting discussions helped improve the present work, in alphabetical order, Dan Boneh, Justin Drake, Alexandre Gélin, Novak Kaluđerović, Arjen K. Lenstra and Serge Vaudenay.

A Proof of Remark 1

Model H as a random oracle. Suppose that

$$\begin{aligned}\mathsf{trapdoor}^H_{\mathsf{sk}}(x,\Delta) &= t_{\mathsf{sk}}(H(x),\Delta),\\ \mathsf{eval}^H_{\mathsf{pk}}(x,\Delta) &= e_{\mathsf{pk}}(H(x),\Delta), \textit{and}\\ \mathsf{verify}_{\mathsf{pk}}(x,y,\Delta) &= v_{\mathsf{pk}}(H(x),y,\Delta),\end{aligned}$$

for procedures t, e and v that do not have access to H.

Let $\mathcal{A}$ be a player of the Δ-evaluation race game. Assume that the output $\mathcal{B}$ of $\mathcal{A}$ is limited to a number q of queries to $\mathcal{O}$ and H. We are going to build an algorithm $\mathcal{A}'$ that wins with same probability as $\mathcal{A}$ when its output $\mathcal{B}'$ is not given access to $\mathcal{O}$.

Let $(Y_i)_{i=1}^q$ be a sequence of random hash values (i.e., uniformly distributed random values in $\{0,1\}^{2k}$). First observe that $\mathcal{A}$ wins the Δ-evaluation race game with the same probability if the last step runs the algorithm $\mathcal{B}^{\mathcal{O}',H'}$ instead of $\mathcal{B}^{\mathcal{O},H}$, where

1. H' is the following procedure: for any new requested input x, if x has previously been requested by $\mathcal{A}$ to H then output $H'(x) = H(x)$; otherwise set $H'(x)$ to be the next unassigned value in the sequence (Y_i);
2. $\mathcal{O}'$ is an oracle that on input x outputs $t_{\mathsf{sk}}(H'(x),\Delta)$.

With this observation in mind, we build $\mathcal{A}'$ as follows. On input pk, $\mathcal{A}'$ first runs $\mathcal{A}^H$ which outputs $\mathcal{A}^H(\mathsf{pk}) = \mathcal{B}$. Let X be the set of inputs of the requests that $\mathcal{A}$ made to H. For any $x \in X$, $\mathcal{A}'$ computes and stores the pair $(H(x), \mathsf{eval}_{\mathsf{pk}}(x, \Delta))$ in a list L. In addition, it computes and stores $(Y_i, e_{\mathsf{pk}}(Y_i, \Delta))$ for each $i = 1, \ldots, q$, and adds them to L.

Consider the following procedure $\mathcal{O}'$: on input x, look for the pair of the form $(H'(x), \sigma)$ in the list L, and output σ. The output of $\mathcal{A}'$ is the algorithm $\mathcal{B}' = \mathcal{B}^{\mathcal{O}', H'}$. It does not require access to the oracle $\mathcal{O}$ anymore: all the potential requests are available in the list of precomputed values. Each call to $\mathcal{O}$ is replaced by a lookup in the list L, so $\mathcal{B}'$ has essentially the same running time as $\mathcal{B}$. Therefore $\mathcal{A}'$ wins the Δ-evaluation race game with same probability as $\mathcal{A}$ even when its output $\mathcal{B}'$ is not given access to a evaluation oracle.

B Timed Challenge-Response Identification Protocols

A timed challenge-response identification protocol has four procedures:

$\mathsf{keygen} \to (\mathsf{pk}, \mathsf{sk})$ is a key generation procedure, which outputs a prover's public key pk and secret key sk.
$\mathsf{challenge} \to c$ which outputs a random challenge.
$\mathsf{respond}_{\mathsf{sk}}(c, \Delta) \to r$ is a procedure that uses the prover's secret key to respond to the challenge c, for the time parameter Δ.
$\mathsf{verify}_{\mathsf{pk}}(c, r, \Delta) \to \mathsf{true}$ or false is a procedure to check if r is a valid response to c, for the public key pk and the time parameter Δ.

The security level k is implicitly an input to each of these procedures. The keygen procedure is used the generate Alice's public and secret keys, then the identification protocol is as follows:

1. Bob generates a random c with the procedure $\mathsf{challenge}$. He sends it to Alice, along with a time limit Δ, and starts a timer.
2. Alice responds $r = \mathsf{respond}_{\mathsf{sk}}(c, \Delta)$.
3. Bob stops the timer. He accepts if $\mathsf{verify}_{\mathsf{pk}}(c, r, \Delta) = \mathsf{true}$ and the elapsed time is smaller than Δ.

Given a time parameter Δ, a *Δ-response race game* and an associated notion of *Δ-soundness* can be defined in a straightforward manner as follows.

Definition 8 (Δ-response race game). *Let $\mathcal{A}$ be a party playing the game. The parameter $\Delta : \mathbf{Z}_{>0} \to \mathbf{R}_{>0}$ is a function of the (implicit) security parameter k. The* Δ-response race game *goes as follows:*

1. *The random procedure* keygen *is run and it outputs a public key* pk*;*
2. *$\mathcal{A}(\mathsf{pk})$ outputs an algorithm $\mathcal{B}$;*
3. *A random challenge c is generated according to the procedure* $\mathsf{challenge}$*;*
4. *$\mathcal{B}^{\mathcal{O}}(c)$ outputs a value r, where $\mathcal{O}$ is an oracle that outputs the evaluation* $\mathsf{respond}_{\mathsf{sk}}(c', \Delta)$ *on any input $c' \neq c$.*

Then, $\mathcal{A}$ wins the game if $T(\mathcal{B}, c) < \Delta$ and $\mathsf{verify}_{\mathsf{pk}}(c, r, \Delta) = \mathsf{true}$.

Definition 9 (Δ-soundness). *A timed challenge-response identification protocol is Δ-sound if any polynomially bounded player (with respect to the implicit security parameter) wins the above Δ-response race game with negligible probability.*

It is as immediate to verify that a sound and Δ-sequential VDF gives rise to a Δ-sound identification protocol (via the construction of Sect. 8). Similarly, the *completeness* of the identification protocol (that a honest run of the protocol terminates with a successful verification) is straightforward to derive from the fact that the verification of a valid VDF output always outputs true. There simply is one additional requirement: if the procedure $\mathsf{respond}_{\mathsf{sk}}(c, \Delta)$ requires computation time at least ϵ_1, and the channel of communication has a transmission delay at least ϵ_2, we must have $\epsilon_1 + 2\epsilon_2 < \Delta$. Finally the *zero-knowledge* property is defined as follows.

Definition 10 (Zero-knowledge). *A timed challenge-response identification protocol is (perfectly, computationally, or statistically) zero-knowledge if there is an algorithm $\mathcal{S}$ that on input k, Δ,* pk *and a random* $\mathsf{challenge}(k, \Delta)$ *produces an output (perfectly, computationally, or statistically) indistinguishable from* $\mathsf{respond}_{\mathsf{sk}}(c, k, \Delta)$, *and the running time of $\mathcal{S}$ is polynomial in k.*

In a classical cryptographic line of though, this zero-knowledge property is too strong to provide any soundness, since an adversary can respond to the challenge with a running time polynomial in the security parameter of Alice's secret key. This notion starts making sense when the complexity of the algorithm $\mathcal{S}$ is governed by another parameter, here Δ, independent from Alice's secret.

For the protocol derived from a VDF, the zero-knowledge property is ensured by the fact that anyone can compute Alice's response to the challenge in time polynomial in k, with the procedure eval.

C Local Identification

The challenge-response identification protocol derived from a VDF in Sect. 8 is totally deniable against a judge, Judy, observing the communication from a long distance. The precise definition of on-line deniability is discussed in [11]. We refer the reader there for the details, but the high level idea is as follows. Alice is presumably trying to authenticate her identity to Bob. Judy will rule whether or not the identification was attempted. Judy interacts with an informant who is witnessing the identification and who wants to convince Judy that it happened. This informant could also be a misinformant, who is not witnessing any identification, but tries to deceive Judy into believing it happened. The protocol is on-line deniable if no efficient judge can distinguish whether she is talking to an informant or a misinformant. The (mis)informant is allowed to corrupt Alice or Bob, at which point he learns their secret keys and controls their future actions. When some party is corrupted, Judy learns about it.

It is shown in [11] that this strong deniability property is impossible to achieve in a PKI. To mitigate this issue, they propose a secure protocol in a relaxed setting, allowing incriminating aborts. We propose an alternative relaxation of the setting, where Judy is assumed to be far away from Alice and Bob (more precisely: the travel time of a message between Alice and Bob is shorter than between Alice (or Bob) and Judy[6]). For example, consider a building whose access is restricted to authorised card holders. Suppose the card holders do not want anyone other than the card reader to get convincing evidence that they are accessing the building (even if the card reader is corrupted, it cannot convince anyone else). Furthermore, Alice herself cannot convince anyone that the card reader ever acknowledged her identification attempt. In this context, the card and the card reader benefit from very efficient communications, while a judge farther away would communicate with an additional delay. An identification protocol can exploit this delay to become deniable, and this is achieved by the timed challenge-response identification protocol derived from a VDF.

The idea is the following. Suppose that the distance between Alice and Judy is long enough to ensure that the travel time of a message from Alice to Judy is larger than $\Delta/2$. Then, Judy cannot distinguish a legitimate response of Alice that took some time to reach her from a response forged by a misinformant that is physically close to Judy.

More precisely, considering an informant I who established a strategy with Judy, we can show that there is a misinformant M that Judy cannot distinguish from I. First of all, Bob cannot be incriminated since he is not using a secret key. It all boils down to tracking the messages that depend on Alice's secret key. Consider a run of the protocol with the informant I. Let t_0 be the point in time where Alice computed $s = \mathsf{trapdoor}_{\mathsf{sk}}(c, \Delta)$. The delay implies two things:

1. The challenge c is independent of anything Judy sent after point in time $t_0 - \Delta/2$.
2. The first message Judy receives that can depend on s (and therefore the first message that depends on Alice's secret) arrives after $t_0 + \Delta/2$.

From Point 1, at time $t_0 - \Delta/2$, the misinformant (who is close to Judy) can already generate c (following whichever procedure I and Judy agreed on), and start evaluating $\mathsf{eval}_{\mathsf{pk}}(c, \Delta)$. The output is ready at time $t_0 + \Delta/2$, so from Point 2, the misinformant is on time to send to Judy messages that should depend on the signature s.

In practice. The protocol is deniable against a judge at a certain distance away from Alice and Bob, and the minimal value of this distance depends on Δ. An

[6] A message does not travel directly from Alice (or Bob) to Judy, since Judy is only communicating with the (mis)informant. What is measured here is the sum of the delay between Alice and the (mis)informant and the delay between the (mis)informant and Judy. There is no constraint on the location of the (mis)informant, but we assume a triangular inequality: he could be close to Alice and Bob, in which case his communications with Judy suffer a delay, or he could be close to Judy, in which case his interactions with Alice and Bob are delayed.

accurate estimation of this distance would require in the first place an equally accurate estimation of the real time Δ (in seconds) a near-optimal adversary would need to forge the response. This non-trivial task relates to the discussion of Sect. 3.2.

Assuming reasonable bounds for Δ have been established, one can relate the distance and the communication delay in a very conservative way through the speed of light. We want Judy to stand at a sufficient distance to ensure that any message takes at least $\Delta/2$ s to travel between them, so Judy should be at least $c\Delta/2$ m away, where $c \approx 3.00 \times 10^8$ m/s is the speed of light. For security against a judge standing 100 m away, one would require $\Delta \approx 0.66$ µs. Alice should be able to respond to Bob's challenge in less time than that. At this point, it seems unreasonable to assume that such levels of precision can be achieved (although in principle, distance bounding protocols do deal with such constraints), yet it remains interesting that such a simple and efficient protocol provides full deniability against a judge that suffers more serious communication delays.

References

1. Bellare, M., Goldwasser, S.: Encapsulated key escrow. Technical report (1996)
2. Bellare, M., Goldwasser, S.: Verifiable partial key escrow. In: Proceedings of the 4th ACM Conference on Computer and Communications Security, CCS 1997, pp. 78–91. ACM, New York, NY, USA (1997)
3. Biehl, I., Buchmann, J., Hamdy, S., Meyer, A.: A signature scheme based on the intractability of computing roots. Des. Codes Crypt. **25**(3), 223–236 (2002)
4. Boneh, D., Bonneau, J., Bünz, B., Fisch, B.: Verifiable delay functions. In: Shacham, H., Boldyreva, A. (eds.) CRYPTO 2018. LNCS, vol. 10991, pp. 757–788. Springer, Cham (2018). https://doi.org/10.1007/978-3-319-96884-1_25
5. Boneh, D., Bünz, B., Fisch, B.: A survey of two verifiable delay functions. Cryptology ePrint Archive, report 2018/712 (2018). https://eprint.iacr.org/2018/712
6. Boneh, D., Franklin, M.: Efficient generation of shared RSA keys. In: Kaliski, B.S. (ed.) CRYPTO 1997. LNCS, vol. 1294, pp. 425–439. Springer, Heidelberg (1997). https://doi.org/10.1007/BFb0052253
7. Boneh, D., Naor, M.: Timed commitments. In: Bellare, M. (ed.) CRYPTO 2000. LNCS, vol. 1880, pp. 236–254. Springer, Heidelberg (2000). https://doi.org/10.1007/3-540-44598-6_15
8. Buchmann, J., Hamdy, S.: A survey on IQ cryptography. In: Proceedings of Public Key Cryptography and Computational Number Theory, pp. 1–15 (2001)
9. Buchmann, J., Williams, H.C.: A key-exchange system based on imaginary quadratic fields. J. Cryptol. **1**(2), 107–118 (1988)
10. CPU-Z OC world records (2018). http://valid.canardpc.com/records.php
11. Dodis, Y., Katz, J., Smith, A., Walfish, S.: Composability and on-line deniability of authentication. In: Reingold, O. (ed.) TCC 2009. LNCS, vol. 5444, pp. 146–162. Springer, Heidelberg (2009). https://doi.org/10.1007/978-3-642-00457-5_10
12. Drake, J.: Ethereum 2.0 randomness. August 2018 Workshop at Stanford Hosted by the Ethereum Foundation and the Stanford Center for Blockchain Research (2018)

13. Drake, J.: Minimal VDF randomness beacon. Ethereum Research Post (2018). https://ethresear.ch/t/minimal-vdf-randomness-beacon/3566
14. Hafner, J.L., McCurley, K.S.: A rigorous subexponential algorithm for computation of class groups. J. Am. Math. Soc. **2**(4), 837–850 (1989)
15. Lenstra, A.K., Wesolowski, B.: Trustworthy public randomness with sloth, unicorn and trx. Int. J. Appl. Cryptol. **3**, 330–343 (2016)
16. Pietrzak, K.: Simple verifiable delay functions. In: Blum, A. (ed.), 10th Innovations in Theoretical Computer Science Conference, ITCS 2019, San Diego, California, USA, 10–12 January 2019, pp. 60:1–60:15 (2019)
17. Rabin, M.O.: Transaction protection by beacons. J. Comput. Syst. Sci. **27**(2), 256–267 (1983)
18. Rivest, R.L., Shamir, A., Wagner, D.A.: Time-lock puzzles and timed-release crypto. Technical report (1996)
19. Sander, T.: Efficient accumulators without trapdoor extended abstract. In: Varadharajan, V., Mu, Y. (eds.) ICICS 1999. LNCS, vol. 1726, pp. 252–262. Springer, Heidelberg (1999). https://doi.org/10.1007/978-3-540-47942-0_21
20. Vollmer, U.: Asymptotically fast discrete logarithms in quadratic number fields. In: Bosma, W. (ed.) ANTS 2000. LNCS, vol. 1838, pp. 581–594. Springer, Heidelberg (2000). https://doi.org/10.1007/10722028_39

Quantum Lightning Never Strikes the Same State Twice

Mark Zhandry(✉)

Princeton University, Princeton, USA
mzhandry@princeton.edu

Abstract. Public key quantum money can be seen as a version of the quantum no-cloning theorem that holds even when the quantum states can be verified by the adversary. In this work, we investigate *quantum lightning* where no-cloning holds *even when the adversary herself generates the quantum state to be cloned*. We then study quantum money and quantum lightning, showing the following results:

- We demonstrate the usefulness of quantum lightning beyond quantum money by showing several potential applications, such as generating random strings with a proof of entropy, to completely decentralized cryptocurrency without a block-chain, where transactions is instant and local.
- We give Either/Or results for quantum money/lightning, showing that either signatures/hash functions/commitment schemes meet very strong recently proposed notions of security, or they yield quantum money or lightning. Given the difficulty in constructing public key quantum money, this suggests that natural schemes do attain strong security guarantees.
- We show that instantiating the quantum money scheme of Aaronson and Christiano [STOC'12] with *indistinguishability obfuscation* that is secure against quantum computers yields a secure quantum money scheme. This construction can be seen as an instance of our Either/Or result for signatures, giving the first separation between two security notions for signatures from the literature.
- Finally, we give a plausible construction for quantum lightning, which we prove secure under an assumption related to the multi-collision resistance of degree-2 hash functions. Our construction is inspired by our Either/Or result for hash functions, and yields the first plausible standard model instantiation of a *non-collapsing* collision resistant hash function. This improves on a result of Unruh [Eurocrypt'16] which is relative to a quantum oracle.

1 Introduction

Unlike classical bits, which can be copied ad nauseum, quantum bits—called qubits—cannot in general be copied, as a result of the Quantum No-Cloning Theorem. No-cloning has various negative implications to the handling of quantum information; for example it implies that classical error correction cannot be

Y. Ishai and V. Rijmen (Eds.): EUROCRYPT 2019, LNCS 11478, pp. 408–438, 2019.
https://doi.org/10.1007/978-3-030-17659-4_14

applied to quantum states, and that it is impossible to transmit a quantum state over a classical channel. On the flip side, no-cloning has tremendous potential for cryptographic purposes, where the adversary is prevented from various strategies that involve copying. For example, Wiesner [36] shows that if a quantum state is used as a banknote, no-cloning means that an adversary cannot duplicate the note. This is clearly impossible with classical bits. Wiesner's idea can also be seen as the starting point for quantum key distribution [9], which can be used to securely exchange keys over a public channel, even against computationally unbounded eavesdropping adversaries.

In this work, we investigate no-cloning in the presence of computationally bounded adversaries, and it's implications to cryptography. To motivate this discussion, consider the following two important applications:

- A public key quantum money scheme allows anyone to verify banknotes. This remedies a key limitation of Wiesner's scheme, which requires sending the banknote back to the mint for verification. The mint has a *secret* classical description of the banknote which it can use to verify; if this description is made public, then the scheme is completely broken. Requiring the mint for verification represents an obvious logistical hurdle. In contrast, a public key quantum money scheme can be verified locally without the mint's involvement. Yet, even with the ability to verify a banknote, it is impossible for anyone (save the mint) to create new notes.
- Many cryptographic settings such as multiparty computation require a random string to be created by a trusted party during a set up phase. But what if the randomness creator is not trusted? One would still hope for some way to verify that the strings it produces are still random, or at least have some amount of (min-)entropy. At a minimum, one would hope for a guarantee that their string is different from any previous or future string that will be generated for anyone else. Classically, these goals are impossible. But quantumly, one may hope to create proofs that are unclonable, so that only a single user can possibly ever receive a valid proof for a particular string.

The settings above are subtly different from those usually studied in quantum cryptography. Notice that in both settings above, a computationally unbounded adversary can always break the scheme. For public key quantum money, the following attack produces a valid banknote from scratch in exponential-time: generate a random candidate quantum money state and apply the verification procedure. If it accepts, output the state; otherwise try again. Similarly, in the verifiable randomness setting, an exponential-time adversary can always run the randomness generating procedure until it gets two copies of the same random string, along with two valid proofs for that string. Then it can give the same string (but different valid proofs) to two different users. With the current state of knowledge of complexity theory, achieving security against a computationally bounded adversary means computational assumptions are required; in particular, both scenarios imply at a minimum one-way functions.

Unfortunately, most of the techniques developed in quantum cryptography are inherently information theoretic, and porting these techniques over to the computational setting can be tricky task. For example, whereas information-theoretic security can be often proved directly, computational security must always be proved by a reduction to the underlying hard computational problem.

We stress that the underlying problem should still be a *classical* problem (that is, the inputs and outputs of the problem are classical), rather than some quantum problem that talks about manipulating quantum states. For one, we want a violation of the assumption to lead to a mathematically interesting result, and this seems much more likely for classical problems. Furthermore, it is much harder for the research community to study and analyze a quantum assumption, since it will be hard to isolate the features of the problem that make it hard. For this work, we want to:

Combine no-cloning and computational assumptions about classical problems to obtain no-cloning-with-verification.

In addition to the underlying assumption being classical, it would ideally also be one that has been previously studied by cryptographers, and ideally used in other cryptographic contexts. This would give the strongest possible evidence that the assumption, and hence application, are secure.

For now, we focus on the setting of public key quantum money. Constructing such quantum money from a classical hardness assumption is a surprisingly difficult task. One barrier is the following. Security would be proved by reduction, an algorithm that interacts with a supposed quantum money adversary and acts as an adversary for the underlying *classical* computational assumption. Note that the adversary expects as input a valid banknote, which the reduction must supply. Then it appears the reduction should somehow use the adversary's forgery to break the computational assumption. But if the reduction can generate a single valid banknote, there is nothing preventing it from generating a second—recall that the underlying assumption is classical, so we cannot rely on the assumption to provide us with an un-clonable state. Therefore, if the reduction works, it would appear that the reduction can create two banknotes for itself, in which case it can break the underlying assumption without the aid of the adversary. This would imply that the underlying assumption is in fact false.

The above difficulties become even more apparent when considering the known public key quantum money schemes. The first proposed scheme by Aaronson [2] had no security proof, and was subsequently broken by Lutomirski et al. [28]. The next proposed scheme by Farhi et al. [20] also has no security proof, though this scheme still remains unbroken. However, the scheme is complicated, and it is unclear which quantum states are accepted by the verification procedure; it might be that there are dishonest banknotes that are both easy to construct, but are still accepted by the verification procedure.

Finally, the third candidate by Aaronson and Christiano [3] actually *does* prove security using a classical computational problem. However, in order to circumvent the barrier discussed above, the classical problem has a highly non-standard format. They observe that a polynomial-time algorithm can, by random guessing, produce a valid banknote with some exponentially-small probability p, while random guessing can only produce n valid banknotes with probability p^n. Therefore, their reduction first generates a valid banknote with probability p, runs the adversary on the banknote, and then uses the adversary's forgery to increase its success probability for some task. This reduction strategy requires a very carefully crafted assumption, where it is assumed hard to solve a particular problem in polynomial time with exponentially-small probability p, even though it can easily be solved with probability p^2.

In contrast, typical assumptions in cryptography involve polynomial-time algorithms and *inverse-polynomial* success probabilities, rather than exponential. (Sub)exponential hardness assumptions are sometimes made, but even then the assumptions are usually closed under polynomial changes in adversary running times or success probabilities, and therefore make no distinction between p and p^2. In addition to the flavor of assumption being highly non-standard, Aaronson and Christiano's assumption—as well as their scheme—have been subsequently broken [1,31].

Turning to the verifiable randomness setting, things appear even more difficult. Indeed, our requirements for verifiable randomness imply an even stronger version of computational no-cloning: an adversary should not be able to copy a state, even if it can verify the state, and moreover *even if it devised the original state itself.* Indeed, without such a restriction, an adversary may be able to come up with a dishonest proof of randomness, perhaps by deviating from the proper proof generating procedure, that it *can* clone arbitrarily many times. Therefore, a fascinating objective is to

Obtain a no-cloning theorem, even for settings where the adversary controls the entire process for generating the original state.

1.1 This Work: Strong Variants of No-Cloning and Connections to Post-quantum Cryptography

In this work, we study strong computational variants of quantum no-cloning, in particular public key quantum money, and uncover interesting relationships between no-cloning and various cryptographic applications.

Quantum Lightning Never Strikes the Same State Twice. The old adage about lightning is of course false, but the idea nonetheless captures some of the features we would like for the verified randomness setting discussed above. Suppose a magical randomness generator could go out into a thunderstorm, and "freeze" and "capture" lightning bolts as they strike. Every lightning bolt will be different. The randomness generator then somehow extracts a fingerprint or serial number from the frozen lightning bolt (say, hashing the image of the bolt from a particular direction). The serial number will serve as the random string, and the frozen lightning bolt will be the proof of randomness; since every bolt is different, this ensures that the bolts, and hence serial numbers, have some amount of entropy.

Of course, it may be that there are other ways to create lightning other than walking out into a thunderstorm (Tesla coils come to mind). We therefore would like that, no matter how the lightning is generated, be it from thunderstorms or in a carefully controlled laboratory environment, every bolt has a unique fingerprint/serial number.

We seek a complexity-theoretic version of this magical frozen lightning object, namely a phenomenon which guarantees different outcomes every time, no matter how the phenomenon is generated. We will necessarily rely on quantum no-cloning—since in principle a classical phenomenon can be replicated by starting with the same initial conditions—and hence we call our notion *quantum lightning*. Quantum lightning, roughly, is a strengthening of public key quantum money where the procedure to generate new banknotes itself is public, allowing anyone to generate banknotes. Nevertheless, it is impossible for an adversary to construct two notes with the same serial number. This is a surprising and counter-intuitive property, as the adversary knows how to generate banknotes, and moreover has full control over how it does so; in particular it can deviate from the generation procedure any way it wants, as long as it is computationally efficient. Nonetheless, it cannot devise a malicious note generation procedure that allows it to construct the same note twice. This concept of quantum money can be seen as a formalization of the concept of "collision-free" public key quantum money due to Lutomirski et al. [28].

Slightly more precisely, a quantum lightning protocol consists of two efficient (quantum) algorithms. The first is a bolt generation procedure, or storm, ⛈, which generates a quantum state $|⚡\rangle$ on each invocation. The second algorithm, Ver, meanwhile verifies bolts as valid and also extracts a fingerprint/serial number of the bolt. For correctness, we require that (1) Ver always accepts bolts produced by ⛈, (2) it does not perturb valid bolts, and (3) that it will always output the same serial number on a given bolt.

For security, we require the following: it is computationally infeasible to produce two bolts $|⚡_0\rangle$ and $|⚡_1\rangle$ such that Ver accepts both and outputs identical serial numbers. This is true for even for *adversarial* storms 🌩—those that depart from ⛈ or produce entangled bolts—so long as 🌩 is efficient.

Applications. Quantum lightning as described has several interesting applications:

- **Quantum money.** Quantum lightning easily gives quantum money. A banknote is just a bolt, with the associated serial number signed by the bank using an arbitrary classical signature scheme. Any banknote forgery must either forge the bank's signature, or must produce two bolts with the same serial number, violating quantum lightning security.
- **Verifiable min-entropy.** Quantum lightning also gives a way to generate random strings along with a proof that the string is random, or at least has min-entropy. Indeed, consider an adversarial bolt generation procedure that produces bolts such that the associated serial number has low min-entropy. Then by running this procedure several times, one will eventually obtain in polynomial time two bolts with the same serial number, violating security.

Therefore, to generate a verifiable random string, generate a new bolt using ⛈. The string is the bolt's serial number, and ⛈ serves as a proof of min-entropy, which is verified using Ver.

- **Decentralized Currency.** Finally, quantum lightning yields a simple new construction of totally decentralized digital currency. Coins are just bolts, except the serial number must hash to a string that begins with a certain number of 0's. Anyone can produce coins by generating bolts until the hash begins with enough 0's. Moreover, verification is just Ver, and does not require any interaction or coordination with other users of the system. This is an advantage over classical cryptocurrencies such as BitCoin, which require a large public and dynamic ledger, and requires a pool of miners to verify transactions. Our protocol does have significant limitations relative to classical cryptocurrencies, which likely make it only a toy object. We hope that further developments will yield a scheme that overcomes these limitations.

Connections to Post-quantum Security. One simple folklore way to construct a state that can only be constructed once but never a second time is to use a collision-resistant hash function H. First, generate a uniform superposition of inputs. Then apply the H in superposition, and measure the result y. The state collapses to the superposition $|\psi_y\rangle$ of all pre-images x of y.

Notice that, while it is easy to sample states $|\psi_y\rangle$, it is impossible to sample two copies of the same $|\psi_y\rangle$. Indeed, given two copies of $|\psi_y\rangle$, simply measure both copies. Since these are superpositions over many inputs, each state will likely yield a different x. The two x's obtained are both pre-images of the same y, and therefore constitute a collision for H.

The above idea does not yet yield quantum lightning. For verification, one can hash the state to get the serial number y, but this alone is insufficient. For example, an adversarial storm can simply choose a random string x, and output $|x\rangle$ twice as its two copies of the same state. Of course, $|x\rangle$ is not equal to $|\psi_y\rangle$ for any y. However, the verification procedure just described does not distinguish between these two states.

What one needs therefore is mechanism to distinguish a random $|x\rangle$ from a random $|\psi_y\rangle$. Interestingly, as observed by Unruh [34], this is exactly *the opposite* what one would normally want from a hash function. Consider the usual way of building a computationally binding commitment from a collision resistant hash function: to commit to a message m, choose a random r and output $H(m, r)$. Classically, this is computationally binding by the collision resistance of H: if an adversary can open the commitment to two different values, this immediately yields a collision for H. Unruh [34] shows in the quantum setting, collision resistance—even against quantum adversaries—is not enough. Indeed, he shows that for certain hash functions H it may be possible for the adversary to produce a commitment, and only afterward decide on the committed value. Essentially, the adversary constructs a superposition of pre-images $|\psi_y\rangle$ as above, and then uses particular properties of H to perturb $|\psi_y\rangle$ so that it becomes a different superposition of pre-images of y. Then one simply de-commits to any message by first modifying the superposition and then measuring. This does not violate the collision-resistance of H: since the adversary cannot copy $|\psi_y\rangle$, the adversary can only ever perform this procedure once and obtain only a single de-commitment.

To overcome this potential limitation, Unruh defines a notion of *collapsing* hash functions. Roughly, these are hash functions for which $|x\rangle$ and $|\psi_y\rangle$ are *indistinguishable*. Using such hash functions to build commitments, one obtains *collapse-binding* commitments, for which the attack above is impossible. Finally, he shows that a random oracle is collapse binding.

More generally, an implicit assumption in many classical settings is that, if an adversary can modify one value into another, then it can produce both the original and modified value simultaneously. For example, in a commitment scheme, if a classical adversary can de-commit to both 0 or 1, it can then also simultaneously de-commit to both 0 and 1 by first de-committing to 0, and then re-winding and de-committing to 1. Thus it is natural classically to require that it is impossible to simultaneously produce de-commitments to both 0 and 1. Similarly, for signatures, if an adversary can modify a signed message m_0 into a signed message m_1, then it can simultaneously produce two signed messages m_0, m_1. This inspires the Boneh-Zhandry [10,11] definition of security against quantum adversaries, which says that after seeing a (superposition of) signed messages, the adversary cannot produce two signed messages.

However, a true quantum adversary may be able, for some schemes, to set things up so that it can modify a (superposition) of values into one of many possibilities, but still only be able to ever produce a single value. For example, it may be that an adversary sees a superposition of signed messages that always begin with 0, but somehow modifies the superposition to obtain a signed message that begins with a 1. This limitation for signatures was observed by Garg, Yuen, and Zhandry [23], who then give a much stronger notion to fix this issue[1].

Inspired by the above, we formulate a series Either/Or results for quantum lightning and quantum money. In particular, in Sect. 4, we show, roughly,

Theorem 1 (informal). *If H is a hash function that is collision resistant against quantum adversaries, then either (1) H is collapsing or (2) it can be used to build quantum lightning without any additional computational assumptions.*[2]

The construction of quantum lightning is inspired by the outline above. One difficulty is that above we needed a perfect distinguisher, whereas a collapsing adversary may only have a non-negligible advantage. To obtain an actual quantum lightning scheme, we need to repeat the scheme in parallel many times to

[1] Garg et al. only actually discuss message authentication codes, but the same idea applies to signatures.

[2] Technically, there is a slight gap due to the difference between *non-negligible* and *inverse polynomial.* Essentially what we show is that the theorem holds for fixed values of the security parameter, but whether (1) or (2) happens may vary across different security parameters.

boost the distinguish advantage to essentially perfect. Still, defining verification so that we can prove security is a non-trivial task. Indeed, it is much harder to analyze what sorts of invalid bolts might be accepted by the verification procedure, especially since we know virtually nothing about the types of states the given adversary for collapsing accepts.

For example, in order to base security on collision resistance, we would like to say that if a bolt passes verification, we can measure it and obtain a collision. But then we need that the classical test (namely evaluating $H(x)$) *and* the quantum test (namely, that it is superposition) both succeed *simultaneously*. Unfortunately, these two tests are non-commuting operations, so it is impossible to test both with certainty simultaneously. If we perform the classical test before the quantum test, it could be that the second test perturbs the quantum state so that it is in superposition, but no longer a superposition of pre-images. Similarly, if we perform the quantum test first, it could be that running the classical test collapses the state to a singleton. In this case, measuring two accepting bolts could give us the same pre-image, so we do not get a collision.

Using a careful argument, we show nonetheless how to verify and prove security. The intuition is to only perform a single test, and which test is performed is chosen at random independent of the input. We demonstrate that if a state had a reasonably high probability of passing, then it must have *simultaneously* had a noticeable probability of passing each of the two tests. This is enough to get a collision. Next, we just repeat the scheme many times in parallel; now if a bolt even has a non-negligible chance of passing, one of the components must have a high chance of passing, which in turn gives a collision.

Next, we move on to other Either/Or results. We show that:

Theorem 2 (informal). *Any* non-interactive *commitment scheme that is computationally binding against quantum adversaries is either collapse-binding, or it can be used to build quantum lightning without any additional computational assumptions.*

The above theorem crucially relies on the commitment scheme being non-interactive: the serial number of the bolt is the sender's single message, along with his private quantum state. If the commitment scheme is not collapse-binding, the sender's private state can be verified to be in superposition. If an adversary produces two identical bolts, these bolts can be measured to obtain two openings, violating computational binding. In contrast, in the case of interactive commitments, the bolt should be expanded to the transcript of the interaction between the sender and receiver. Unfortunately, for quantum lightning security, the transcript is generated by an adversary, who can deviate from the honest receiver's protocol. Since the commitment scheme is only binding when the receiver is run honestly, we cannot prove security in this setting.

Instead, we consider the weaker goal of constructing public key quantum money. Here, since the mint produces bolts, the original bolt is honestly generated. The mint then signs the transcript using a standard signature scheme (which can be built from one-way functions, and hence implied by commitments).

If the adversary duplicates this banknote, it is duplicating an honest commitment transcript, but the note can be measured to obtain two different openings, breaking computational binding. This gives us the following:

Theorem 3 (informal). *Any interactive commitment scheme that is computationally binding against quantum adversaries is either collapse-binding, or it can be used to build public key quantum money without any additional computational assumptions.*

Finally, we extend these ideas to quantum money and digital signatures:

Theorem 4 (informal). *Any one-time signature scheme that is Boneh-Zhandry secure is either Garg-Yuen-Zhandry secure, or it can be used to build public key quantum money without any additional computational assumptions.*

Given the difficulty of constructing public key quantum money (let alone quantum lightning), the above results suggest that most natural constructions of collision resistant hash functions, including all of those used in practice, are likely already collapsing, with analogous statements for commitment schemes and signatures. If they surprisingly turn out to not meet the stronger quantum notions, then we would immediately obtain a construction of public key quantum money from simple tools.

Notice that using our Either/Or results give a potential route toward proving the security of quantum money/lightning in a way that avoids the barrier discussed above. Consider building quantum money from quantum lightning, and in turn building quantum lightning from a collision-resistant non-collapsing hash function. Recall that a banknote is a bolt, together with the mint's signature on the bolt's serial number. A quantum money adversary either (1) duplicates a bolt to yield two bolts with the same serial number (and hence same signature), or (2) produces a second bolt with a different serial number, as well as a forged signature on that serial number. Notice that (2) is impossible simply by the unforgeability of the mint's signature. Meanwhile, in proving that (1) is impossible, our reduction actually *can* produce arbitrary quantum money states (for this step, we assume the reduction is given the signing key). The key is that the reduction on its own *cannot* produce the same quantum money state twice, but it *can* do so using the adversary's cloning abilities, allowing it to break the underlying hard problem.

Quantum Money from Obfuscation. We now consider the task of constructing public key quantum money. One possibility is based on Aaronson and Christiano's broken scheme [3]. In their scheme, a quantum banknote $|\$\rangle$ is a uniform superposition over some subspace S, that is known only to the bank. The quantum Fourier transform of such a state is the uniform superposition over the dual subspace $S^\perp$. This gives a simple way to check the banknote: test if $|\$\rangle$ lies in S, and whether it's Fourier transform lies in $S^\perp$. Aaronson and Christiano show that the only state which can pass verification is $|\$\rangle$.

To make this scheme public key, one gives out a mechanism to test for membership in S and $S^{\perp}$, without actually revealing $S, S^{\perp}$. This essentially means obfuscating the functions that decide membership. Aaronson and Christiano's scheme can be seen as a candidate obfuscator for subspaces. While unfortunately their obfuscator has since been broken, one may hope to instantiate their scheme using recent advances in general-purpose program obfuscation, specifically indistinguishability obfuscation (iO) [7,22].

On the positive side, Aaronson and Christiano show that their scheme is secure if the subspaces are provided as quantum-accessible black boxes, giving hope that some obfuscation of the subspaces will work. Unfortunately, proving security relative to iO appears a difficult task. One limitation is the barrier discussed above, that any reduction must be able to produce a valid banknote, which means it can also produce two banknotes. Yet at the same time, it somehow has to use the adversary's forgery (a second banknote) to break the iO scheme. Note that this situation is different from the quantum lightning setting, where there were many valid states, and no process could generate the same state twice. Here, there is a single valid state (the state $|\$\rangle$), and it would appear the reduction must be able to construct this precise state exactly once, but not twice. Such a reduction would clearly be impossible. As discussed above Aaronson and Christiano circumvent this issue by using a non-standard type of assumption; their technique is not relevant for standard definitions of iO.

In Sect. 5, we prove the security of Aaronson and Christiano's scheme using iO. Our solution is to separate the proof into two phases. In the first, we change the spaces obfuscated from $S, S^{\perp}$ to T_0, T_1, where T_0 is a random unknown subspace containing S, and T_1 is a unknown random subspace containing $S^{\perp}$. This modification can be proved undetectable using a weak form of obfuscation we define, called subspace-hiding obfuscation, which in turn is implied by iO. Note that in this step, we even allow the reduction to know S (but not T_0, T_1), so it can produce as many copies of $|\$\rangle$ as it would like to feed to the adversary. The reduction does not care about the adversary's forgery directly, only whether or not the adversary successfully forges. If the adversary forges when given obfuscations of $S, S^{\perp}$, it must also forge under T_0, T_1, else it can distinguish the two cases and hence break the obfuscation. By using the adversary in this way, we avoid the apparent difficulties above.

In the next step, we notice that, conditioned on T_0, T_1, the space S is a random subspace between $T_1^{\perp}$ and T_0. Thus conditioned on T_0, T_1, the adversary clones a state $|\$\rangle$ defined by a random subspace S between $T_1^{\perp}$ and T_0. The number of possible S is much larger than the dimension of the state $|\$\rangle$, so in particular the states cannot be orthogonal. Thus, by no-cloning, duplication is impossible. We need to be careful however, since we want to rule out adversaries that forge with even very low success probabilities. To do so, we need to precisely quantify the no-cloning theorem, which we do. We believe our new no-cloning theorem may be of independent interest. We note that when applying no-cloning, we do not rely on the secrecy of T_0, T_1, but only that S is hidden. Intuitively, there are exponentially many more S's between T_0, T_1 than the dimension of the space $|\$\rangle$ belongs to, so no-cloning implies that a forger has negligible success probability. Thus we reach a contradiction, showing that the original adversary could not exist.

We also show how to view Aaronson and Christiano's scheme as a signature scheme; we show that the signature scheme satisfies the Boneh-Zhandry definition, but not the strong Garg-Yuen-Zhandry notion. Thus, we can view Aaronson and Christiano's scheme as an instance of our Either/Or results, and moreover provide the first separation between the two security notions for signatures.

We note that our result potentially relies on a much weaker notion of obfuscation that full iO, giving hope that security can be based on weaker assumptions. For example, an intriguing open question is whether or not recent constructions of obfuscation for certain evasive functions [26,35] based on LWE can be used to instantiate our notion of subspace hiding obfuscation. This gives another route toward building quantum money from hard lattice problems. This is particularly important at the present time, where the security of iO in the quantum setting is somewhat uncertain (see below for a discussion).

Constructing Quantum Lightning. In Sect. 6, we finally turn to actually building quantum lightning, and hence giving another route to quantum money. Following our Either/Or results, we would like a *non-collapsing* collision-resistant hash function. Unfortunately, Unruh's counterexample does not yield an explicit construction. Instead, he builds on techniques of [5] to give a hash function relative to a *quantum* oracle[3]. As it is currently unknown how to obfuscate quantum oracles with a meaningful notion of security, this does not give even a candidate construction of quantum lightning. Instead, we focus on specific standard-model constructions of hash functions. Finding suitable hash functions is surprisingly challenging; we were only able to find a single family of candidates, and leave finding additional candidates as a challenging open problem.

To motivate our construction, we consider the following approach to building quantum lightning from the short integer solution (SIS) problem. In SIS, an underdetermined system of homogeneous linear equations is given, specified by a wide matrix $\mathbf{A}$, and the goal is to find a solution consisting of "small" entries; that is, a "short" vector $\mathbf{x}$ such that $\mathbf{A}.\mathbf{x} = 0$. For random linear constraints, SIS is conjectured to be computationally difficult, which is backed up by reductions from the hardness of worst-case lattice problems [29]. SIS gives a simple collision resistant hash function $f_{\mathbf{A}}(\mathbf{x}) = \mathbf{A} \cdot \mathbf{x}$, where the domain is constrained to be small; given a collision $\mathbf{x}, \mathbf{x}'$, one obtains a SIS solution as $\mathbf{x} - \mathbf{x}'$.

One may hope that SIS is also non-collapsing, in which case we would obtain quantum lightning. One (failed) attempt to obtaining a collapsing distinguisher is the following. Start with superposition of "short" vectors $\mathbf{x}$, weighted by a Gaussian function. When $f_{\mathbf{A}}$ is applied, the superposition collapses to a superposition over short vectors $\mathbf{x}$ that all have the same value of $\mathbf{A} \cdot \mathbf{x}$. This will be a bolt in the scheme, and the serial number will be the common hash. To

[3] That is, the oracle itself performs quantum operations.

verify the bolt, we first check the hash. Then, to verify the bolt is in superposition, we apply the quantum Fourier transform. Note that if $\mathbf{x}$ were a uniform superposition over *all* vectors, the QFT would give a uniform superposition over all vectors in the row-span of $\mathbf{A}$ (with some phase terms). Instead, since $\mathbf{x}$ is a superposition over "short" vectors, using the rules of Fourier transforms is possible to show that the QFT gives a superposition over vectors of the form $\mathbf{r} \cdot \mathbf{A} + \mathbf{e}$, where $\mathbf{r}$ is a random row vector, and $\mathbf{e}$ is a Gaussian-weighted random short row vector.

Intuitively, we just need to distinguish these types of vectors from random vectors. Unfortunately, distinguishing $\mathbf{r} \cdot \mathbf{A} + \mathbf{e}$ from random for a random matrix $\mathbf{A}$ is an instance of the Learning With Errors (LWE) problem, which is widely believed to be comptuationally intractable, as evidenced by quantum reductions from worst-case lattice problems [32].

We therefore need to "break" LWE by given some trapdoor information. The usual way to break LWE is to provide a short vector $\mathbf{t}$ in the kernel of $\mathbf{A}$. Then, to distinguish an input $\mathbf{u}$, simply compute $\mathbf{u} \cdot \mathbf{t}$, and check if the result is small. In the case $\mathbf{u} = \mathbf{r} \cdot \mathbf{A} + \mathbf{e}$, then $\mathbf{u} \cdot \mathbf{t} = \mathbf{e} \cdot \mathbf{t}$, which will be small. In contrast, if $\mathbf{u}$ is random, $\mathbf{t} \cdot \mathbf{u}$ will be large with overwhelming probability.

Unfortunately, the trapdoor $\mathbf{t}$ is a SIS solution! In particular, in order for the distinguisher to work, one can show that $\mathbf{t}$ needs to be somewhat smaller than the size bound on the domain of $f_{\mathbf{A}}$. With such a trapdoor, it is therefore easy to manufacture collisions for $f_{\mathbf{A}}$, so $f_{\mathbf{A}}$ is no longer collision-resistant. Worse yet, it is straightforward to use the trapdoor to come up with a superposition of inputs that fools the distinguisher.

We do not know how to make the above approach work, as all ways we are aware of for breaking LWE involve handing out a SIS solution. One possible approach would be to obfuscate an LWE distinguisher that has the trapdoor hardcoded. This allows for distinguishing LWE samples without *explicitly* handing out a SIS solution. However, it might be possible to construct a SIS solution from any such distinguishing program.

We now turn to our actual construction. Our idea is to use linear equations over restricted domains as in SIS, but will restrict the domain in different ways. In particular, we will view vectors as specifying symmetric matrices (that is, an $(n+1)n/2$-dimensional vector will correspond to an $n \times n$ symmetric matrix, with the vector entries specifying the upper-triangular part of the matrix). Instead of restricting the size of entries, will instead restrict the *rank* of the symmetric matrix. Our construction then follows the rough outline of the SIS-based approach above, intuitively using rank as a stand-in for vector norm. By switching from vector norm to matrix rank, we are able to arrive at a construction whose security follows from a plausible computational assumption.

A bolt is then a superposition over rank-bounded matrices satisfying the linear constraints. Analogous to the SIS approach, we are able to show that applying the Quantum Fourier transform on such bolts results in a state whose support consists of matrices $\mathbf{A}$ that can be written as $\mathbf{A} = \mathbf{B} + \mathbf{C}$, where $\mathbf{B}$ is a sum of a few known matrices (based on the precise linear functions), whereas

C is an arbitrary low-rank matrix. We show how to generate the constraints along with a public "trapdoor" which allows for such matrices can be identified. Our trapdoor is simply a row rank matrix in the kernel of the linear constraints, analogous to how the LWE trapdoor is a short vector in the kernel.

One may be rightfully concerned at this point, as our trapdoor has the same form as domain elements for our hash function. Indeed, if the rank of the trapdoor was smaller than the rank of the domain, the trapdoor would completely break the construction. Importantly for our construction, we show that this matrix can have higher rank than the allowed inputs to the hash function; as such, it does not appear useful for generating collisions.

Our scheme can easily be proved secure under the assumed collision-resistance of our hash function. Unfortunately, this assumption is false. Indeed, the family of matrices $\mathbf{B}^T\mathbf{B}$ for wide and short matrices $\mathbf{A}$ is low rank. By evaluating our hash function on such matrices, we turn it into a degree-2 polynomial over the $\mathbf{B}$ matrices. Unfortunately, Ding and Yang [19] and Applebaum et al. [6] show that such hash functions are not collision resistant[4].

However, we will apply a simple trick in order to get our scheme to work. Namely, we show how to use the attacks above to actually generate *superpositions* over k colliding inputs for some parameter k that depends on the various parameters of the scheme. At the same time, the attacks do not seem capable of generating collisions beyond k. We will therefore set our bolt to be this superposition over several colliding inputs. Now, we can apply our testing procedure to each of the inputs separately to verify the bolt. If an adversary creates two bolts with the same serial number, we can measure to obtain $2k$ colliding inputs. By assuming the plausible *$2k$-multi-collision resistance* of our hash functions, we obtain security.

Our construction requires a common reference string, namely the sequence of linear constraints and the trapdoor. We show that we can convert our scheme into the common *random* string (crs) model by using the common random string to generate the trapdoor and linear constraints.

1.2 Related Works

Quantum Money. Lutomirski [27] shows another weakness of Wiesner's scheme: a merchant, who is allowed to interact with the mint for verification, can use the verification oracle to break the scheme and forge new currency. Public key quantum money is necessarily secure against adversaries with a verification oracle, since the adversary can implement the verification oracle for itself. Several alternative solutions to the limitations of Wiesner's scheme have been proposed [24,30], though the "ideal" solution still remains public key quantum money.

[4] Technically, they only show this is true if the degree-2 polynomials are random, whereas ours are more structured, but we show that their analysis extends to our setting as well.

Randomness Expansion and Certifiable Randomness. Colbeck [16] proposed the idea of a classical experimenter, interacting with several potentially untrustworthy quantum devices, can expand a small random seed into a certifiably random longer seed. Subsequent to our work, Brakerski et al. [12] consider certifiable randomness in the computational setting. Of of these results are related, but entirely different from, our version of verifiable randomness. In particular, their protocols are interactive and privately verifiable, but allows for a classical verifier. In contrast, our protocol is non-interactive (in the crs model) and publicly verifiable, but requires a quantum verifier.

Obfuscation and Multilinear Maps. There is a vast body of literature on strong notions of obfuscation, starting with the definitional work of Barak et al. [7]. Garg et al. [22] propose the first obfuscator plausibly meeting the strong notion of iO, based on cryptographic multilinear maps [17,21,25]. Unfortunately, there have been numerous attacks on multilinear maps, which we do not fully elaborate on here. There have been several quantum attacks [4,14,15,18] on obfuscators, but there are still schemes that remain unbroken. Moreover, there has been some success in transforming applications of obfuscation to be secure under assumptions on lattices [13,26,35], which are widely believed to be quantum hard. We therefore think it plausible that subspace-hiding obfuscation, which is all we need for this work, can be based on similar lattice problems. Nonetheless, obfuscation is a very active area of research, and we believe that one of the current obfuscators so some future variant will likely be secure quantum resistant.

Computational No-Cloning. We note that computational assumptions and no-cloning have been combined in other contexts, such as Unruh's revocable time-released encryption [33]. We note however, that these settings do not involve verification, the central theme of this work.

2 Preliminaries

Throughout this paper, we will let λ be a security parameter. When inputted into an algorithm, λ will be represented in unary. A function $\epsilon(\lambda)$ is *negligible* if for any inverse polynomial $1/p(\lambda)$, $\epsilon(\lambda) < 1/p(\lambda)$ for sufficiently large λ. A function is *non-negligible* if it is not negligible, that is there exists an inverse polynomial $1/p(\lambda)$ such that $\epsilon(\lambda) \geq 1/p(\lambda)$ infinitely often.

2.1 Quantum Computation

Here, we very briefly recall some basics of quantum computation. A quantum system Q is defined over a finite set B of classical states. A **pure** state over Q is an L_2-normalized vector in $\mathbb{C}^{|B|}$, which assigns a (complex) weight to each element in B. We will think of pure states as column vectors. The pure state that assigns weight 1 to x and weight 0 to each $y \neq x$ is denoted $|x\rangle$.

A pure state $|\phi\rangle$ can be manipulated by performing a unitary transformation U to the state $|\phi\rangle$. We will denote the resulting state as $|\phi'\rangle = U|\phi\rangle$. A unitary is *quantum polynomial time* (QPT) if it can be represented as a polynomial-sized circuit of gates from a finite gate set. $|\phi\rangle$ can also be measured; the measurement outputs the value x with probability $|\langle x|\phi\rangle|^2$. The normalization of $|\phi\rangle$ ensures that the distribution over x is indeed a probability distribution. After measurement, the state "collapses" to the state $|x\rangle$. Notice that subsequent measurements will always output x, and the state will always stay $|x\rangle$.

We define the Euclidean distance $\||\phi\rangle - |\psi\rangle\|$ between two states as the value $(\sum_x |\alpha_x - \beta_x|^2)^{\frac{1}{2}}$ where $|\phi\rangle = \sum_x \alpha_x|x\rangle$ and $|\psi\rangle = \sum_x \beta_x|x\rangle$.

We will be using the following lemma:

Lemma 1 ([8]). *Let $|\varphi\rangle$ and $|\psi\rangle$ be quantum states with Euclidean distance at most ϵ. Then, performing the same measurement on $|\varphi\rangle$ and $|\psi\rangle$ yields distributions with statistical distance at most 4ϵ.*

2.2 Public Key Quantum Money

Here, we define public key quantum money. We will slightly modify the usual definition [2], though the definition will be equivalent under simple transformations.

- We only will consider what Aaronson and Christiano [3] call a quantum money *mini-scheme*, where there is just a single valid banknote. It is straightforward to extend to general quantum money using a signatures
- We will change the syntax to more closely resemble our eventual quantum lightning definition, in order to clearly compare the two objects.

Quantum money consists of two quantum polynomial time algorithms $\mathsf{Gen}, \mathsf{Ver}$.

- Gen takes as input the security parameter, and samples a banknote $|\$\rangle$
- Ver verifies a banknote, and if the verification is successful, produces a serial number for the note.

For correctness, we require that verification always accepts money produced by Gen. We also require that verification does not perturb the money. Finally, since Ver is a quantum algorithm, we must ensure that multiple runs of Ver on the same money will always produce the same serial number. This is captured by the following two of requirements:

- For a money state $|\$\rangle$, let $H_\infty(|\$\rangle) = -\log_2 \min_s \Pr[\mathsf{Ver}(|\$\rangle) = s]$ be the min-entropy of s produced by applying Ver to $|\$\rangle$, were we do not count the rejecting output $\bot$ as contributing to the min-entropy. We insist that $\mathbb{E}[H_\infty(|\$\rangle)]$ is negligible, the expectation over $|\$\rangle \leftarrow \mathsf{Gen}(1^\lambda)$. This ensures the serial number is essentially a deterministic function of the money.
- For a money state $|\$\rangle$, let $|\psi\rangle$ be the state left over after running $\mathsf{Ver}(|\$\rangle)$. We insist that $\mathbb{E}[|\langle\psi|\$\rangle|^2] \geq 1 - \mathsf{negl}(\lambda)$, where the expectation is over $|\$\rangle \leftarrow \mathsf{Gen}(1^\lambda)$, and any affect Ver has on $|\psi\rangle$. This ensures that verification does not perturb the money.

For security, consider the following game between an adversary A and a challenger

- The challenger runs $\mathsf{Gen}(1^\lambda)$ to get a banknote $|\$\rangle$. It runs Ver on the banknote to extract a serial number s.
- The challenger sends $|\$\rangle$ to A.
- A produces two candidate quantum money states $|\$_0\rangle, |\$_1\rangle$, which are potentially entangled.
- The challenger runs Ver on both states, to get two serial numbers s_0, s_1.
- The challenger accepts if and only if both runs of Ver pass, and the serial numbers satisfy $s_0 = s_1 = s$.

Definition 1. *A quantum money scheme* $(\mathsf{Gen}, \mathsf{Ver})$ *is secure if, for all QPT adversaries A, the probability the challenger accepts in the above experiment is negligible.*

3 Quantum Lightning

3.1 Definitions

The central object in a quantum lightning system is a lightning bolt, a quantum state that we will denote as $|\lightning\rangle$. Bolts are produced by a storm, ⛈, a polynomial time quantum algorithm which takes as input a security parameter λ and samples new bolts. Additionally, there is a quantum polynomial-time bolt verification procedure, Ver, which serves two purposes. First, it verifies that a supposed bolt is actually a valid bolt; if not it rejects and outputs $\perp$. Second, if the bolt is valid, it extracts a fingerprint/serial number of the bolt, denoted s.

Rather than having a single storm ⛈ and single verifier Ver, we will actually have a family $\mathcal{F}_\lambda$ of (⛈, Ver) pairs for each security parameter. We will have a setup procedure $\mathsf{SetupQL}(1^\lambda)$ which samples a (⛈, Ver) pair from some distribution over $\mathcal{F}_\lambda$.

For correctness, we have essentially the same requirements as quantum money. We require that verification always accepts bolts produced by ⛈. We also require that verification does not perturb the bolt. Finally, since Ver is a quantum algorithm, we must ensure that multiple runs of Ver on the same bolt will always produce the same fingerprint. This is captured by the following two of requirements:

- For a bolt $|\lightning\rangle$, let

$$H_\infty(|\lightning\rangle, \mathsf{Ver}) = -\log_2 \min_s \Pr[\mathsf{Ver}(|\lightning\rangle) = s]$$

  be the min-entropy of s produced by applying Ver to $|\lightning\rangle$, were we do not count the rejecting output $\perp$ as contributing to the min-entropy. We insist that $\mathbb{E}[H_\infty(|\lightning\rangle, \mathsf{Ver})]$ is negligible, where the expectation is over (⛈, Ver) $\leftarrow \mathsf{SetupQL}(\lambda)$ and $|\lightning\rangle \leftarrow$ ⛈. This ensures the serial number is essentially a deterministic function of the bolt.

- For a bolt $|\text{⚡}\rangle$, let $|\psi\rangle$ be the state left over after running $\mathsf{Ver}(|\text{⚡}\rangle)$. We insist that $\mathbb{E}[|\langle\psi|\text{⚡}\rangle|^2] \geq 1 - \mathsf{negl}(\lambda)$, where the expectation is over $(\text{☁}, \mathsf{Ver}) \leftarrow \mathsf{SetupQL}(\lambda), |\text{⚡}\rangle \leftarrow \text{☁}$, and any affect Ver has on $|\psi\rangle$. This ensures that verification does not perturb the bolt.

Remark 1. We note that it suffices to only consider the first requirement, since the serial number is essentially a deterministic function of the bolt. Indeed, by un-computing the Ver computation after obtaining the serial number, a straightforward calculation shows the result will be negligibly close to the original state.

For security, informally, we ask that no adversarial storm ⛈ can produce two bolts with the same serial number. More precisely, consider the following experiment between a challenger and a malicious bolt generation procedure ⛈:

- The challenger runs $(\text{☁}, \mathsf{Ver}) \leftarrow \mathsf{SetupQL}(1^\lambda)$, and sends $(\text{☁}, \mathsf{Ver})$ to ⛈.
- ⛈ produces two (potentially entangled) quantum states $|\text{⚡}_0\rangle, |\text{⚡}_1\rangle$, which it sends to the challenger.
- The challenger runs Ver on each state, obtaining two fingerprints s_0, s_1. The challenger accepts if and only if $s_0 = s_1 \neq \bot$.

Definition 2. *A quantum lightning scheme has* uniqueness *if, for all QPT adversarial storms* ⛈*, the probability the challenger accepts in the game above is negligible in* λ.

4 Either/Or Results

4.1 Infinity-Often Security

Before describing our Either/Or results, we need to introduce a non-standard notion of security. Typically, a security statement says that no polynomial-time adversary can win some game, except with negligible probability. A violation of the security statement is a polynomial-time adversary that can win with *non*-negligible probability; that is, some probability ϵ that is lower bounded by an inverse-polynomial *infinitely often.* In our proofs below, we use such an adversary to devise a scheme for another problem. But to actually get an efficient scheme, we need the adversary's success probability to actually be inverse-polynomial, not non-negligible. This motivates the notion of *infinitely often security.* A scheme has infinitely-often security if security holds for an infinite number of security parameters, but not necessarily all security parameters. It is straightforward to modify all security notions in this work to infinitely-often variants.

4.2 Collision Resistant Hashing

A hash function is a function H that maps large inputs to small inputs. We will considered keyed functions, meaning it takes two inputs: a key $k \in \{0,1\}^\lambda$, and the actual input to be compressed, $x \in \{0,1\}^{m(\lambda)}$. The output of H is $n(\lambda)$ bits. For the hash function to be useful, we will require $m(\lambda) \gg n(\lambda)$.

The usual security property for a hash function is collision resistance, meaning it is computationally infeasible to find two inputs that map to the same output.

Definition 3. *H is collision resistant if, for any QPT adversary A,* $\Pr[H(x_0) = H(x_1) \wedge x_0 \neq x_1 : (x_0, x_1) \leftarrow A(k), k \leftarrow \{0,1\}^\lambda] < \mathsf{negl}(\lambda)$.

Unruh [34] points out weaknesses in the usual collision resistance definition, and instead defines a stronger notion called *collapsing*. Intuitively, it is easy for an adversary to obtain a superposition of pre-images of some output, by running H on a uniform superposition and then measuring the output. Collapsing requires, however, that this state is computationally indistinguishable from a random input x. More precisely, for an adversary A, consider the following experiment between A and a challenger

- The challenger has an input bit b.
- The challenger chooses a random key k, which it gives to A.
- A creates a superposition $|\psi\rangle = \sum_x \alpha_x |x\rangle$ of elements in $\{0,1\}^{m(\lambda)}$.
- In superposition, the challenger evaluates $H(k, \cdot)$ to get the state $|\psi'\rangle = \sum_x \alpha_x |x, H(k,x)\rangle$.
- Then, the challenger either:
  - If $b = 0$, measures the $H(k,x)$ register, to get a string y. The state $|\psi'\rangle$ collapses to $|\psi_y\rangle \propto \sum_{x:H(k,x)=y} \alpha_x |x, y\rangle$
  - If $b = 1$, measures the entire state, to get a string $x, H(k,x)$. The state $|\psi'\rangle$ collapses to $|x, H(k,x)\rangle$
- The challenger returns whatever state remains of $|\psi'\rangle$ (namely $|\psi_y\rangle$ or $|x, H(k,x)\rangle$) to A.
- A outputs a guess b' for b. Define $\texttt{Collapse-Exp}_b(A, \lambda)$ as b'.

Definition 4. *H is collapsing if, for all QPT adversaries A,*
$|\Pr[\texttt{Collapse-Exp}_0(A,\lambda) = 1] - \Pr[\texttt{Collapse-Exp}_1(A,\lambda) = 1]| < \mathsf{negl}(\lambda)$.

Theorem 5. *Suppose H is collision resistant. Then both of the following are true:*

- *Either H is collapsing, or H can be used to build a quantum lightning scheme that is infinitely often secure.*
- *Either H is infinitely often collapsing, or H can be used to build a quantum lightning scheme that is secure.*

Proof. Let A be a collapsing adversary; the only difference between the two cases above are whether A's advantage is non-negligible or actually inverse polynomial. The two cases are nearly identical, but the inverse polynomial case will simplify notation. We therefore assume that H is not infinitely-often collapsing, and will design a quantum lightning scheme that is secure.

Let A_0 be the first phase of A: it receives a hash key k as input, and produces a superposition of pre-images, as well as it's own internal state. Let A_1 be the second phase of A: it receives the internal state from A_0, plus the superposition of input/output pairs returned by the challenger. It outputs 0 or 1.

Define $q_b(\lambda) = \Pr[\texttt{Collapse-Exp}_b(A, \lambda) = 1]$. By assumption, we have that $|q_0(\lambda) - q_1(\lambda)| \geq 1/p(\lambda)$ for some polynomial p. We will assume $q_0 < q_1$, the other case handled analogously.

For an integer r, consider the function $H^{\otimes r}(k, \cdot)$ which takes as input a string $x \in (\{0,1\}^{m(\lambda)})^r$, and outputs the vector $(H(k, x_1), \ldots, H(k, x_r))$. The collision resistance of H easily implies the collision resistance of $H^{\otimes r}$, for any polynomial r. Moreover, we will use A to derive a collapsing adversary $A^{\otimes r}$ for $H^{\otimes r}$ which has near-perfect distinguishing advantage. $A^{\otimes r}$ works as follows.

- First, it runs A_0 in parallel r times to get r independent states $|\psi_i\rangle$, where each $|\psi_i\rangle$ contains a superposition of internal state values, as well as inputs to the hash function.
- It assembles the r superpositions of inputs into a superposition of inputs for $H^{\otimes r}$, which it then sends to the challenger.
- The challenger responds with a potential superposition over input/output pairs (through the output value in $(\{0,1\}^{n(\lambda)})^r$ is fixed).
- $A^{\otimes r}$ disassembles the input/output pairs into r input/output pairs for H.
- It then runs A_1 in parallel r times, on each of the corresponding state/input/output superpositions, to get bits $b'_1, \ldots, b'_r$.
- $A^{\otimes r}$ then computes $f = (\sum_i b'_i)/r$, the fraction of b'_i that are 1.
- If $f > (q_0 + q_1)/2$ (in other words, f is closer to q_1 than it is to q_0), A outputs 1; otherwise it outputs 0.

Notice that if $A^{\otimes r}$'s challenger uses $b = 0$ (so it only measures the output registers), this corresponds to each A seeing a challenger with $b = 0$. In this case, each b'_i with be 1 with probability q_0. This means that f will be a (normalized) Binomial distribution with expected value q_0. Analogously, if $b = 1$, each b'_i will be 1 with probability q_1, so f will be a normalized Binomial distribution with expected value q_1. Since $q_1 - q_0 \geq 1/p(\lambda)$, we can use Hoeffding's inequality to choose r large enough so that in the $b = 0$ case, $f < (q_0 + q_1)/2 = q_0 + 1/2p(\lambda)$ except with probability $2^{-\lambda}$. Similarly, in the $b = 1$ case, $f > (q_0 + q_1)/2 = q_1 - 1/2p(\lambda)$ except with probably $2^{-\lambda}$. This means $A^{\otimes r}$ outputs the correct answer except with probability $2^{-\lambda}$.

We now describe a first attempt at a quantum lightning scheme:

- $\mathsf{SetupQL}_0$ simply samples and outputs a random hash key k. This key will determine ⛈$_0$, Ver_0 as defined below.
- ⛈$_0$ runs $A_0^{\otimes r}(k)$, where r is as chosen above and $A_0^{\otimes r}$ represents the first phase of $A^{\otimes r}$.
When $A_0^{\otimes r}$ produces a superposition $|\psi\rangle$ over inputs $x \in \{0,1\}^{rm}$ for $H^{\otimes r}(k, \cdot)$ as well as some private state, ⛈$_0$ applies $H^{\otimes r}$ in superposition, and measures the result to get $y \in \{0,1\}^{rn}$.
Finally, ⛈$_0$ outputs the resulting state $|\text{ϟ}\rangle = |\psi_y\rangle$.

- Ver_0 on input a supposed bolt $|\lightning\rangle$, first applies $H^{\otimes r}(k, \cdot)$ in superposition to the input registers to obtain y, which it measures. It saves y, which will be the serial number for the bolt.
  Next, consider two possible tests Test_0 and Test_1. In Test_0, run $A_1^{\otimes r}$—the second phase of $A^{\otimes r}$—on the $|\lightning\rangle$ and measure the result. If the result is 1 (meaning $A^{\otimes r}$ guesses that the challenger measured the entire input/output registers), then abort and reject. Otherwise if the result is 0 (meaning $A^{\otimes r}$ guess that the challenger only measured the output), then it un-computes $A_1^{\otimes r}$. Note that since we measured the output of $A_1^{\otimes r}$, un-computing does not necessarily return the bolt to its original state.
  Test_1 is similar to Test_0, except that the input registers x are measured before running $A_1^{\otimes r}$. This measurement is not a true measurement, but is instead performed by copying x into some private registers. Moreover, the abort condition is flipped: if the result of applying $A_1^{\otimes r}$ is 0, then abort and reject. Otherwise un-compute $A_1^{\otimes r}$, and similarly "un-measure" x by un-computing x from the private registers.
  Ver_0 chooses a random c, and applies Test_c. If the test accepts, then it outputs the serial number y, indicated that it accepts the bolt.

Correctness. For a valid bolt, Test_0 corresponds to the $b = 0$ challenger, in which case we know $A_1^{\otimes r}$ outputs 0 with near certainty. This means Ver continues, and when it un-computes, the result will be negligibly close to the original bolt. Similarly, Test_1 corresponds to the $b = 1$ challenger, in which case $A_1^{\otimes r}$ outputs 1 with near certainty. Un-computing returns the bolt to (negligibly close to) its original state. For a valid bolt, the serial number is always the same. Thus, ⛈, Ver satisfy the necessary correctness requirements.

Security. Security is more tricky. Suppose instead of applying a random Test_c, Ver_0 applied both tests. The intuition is that if Ver accepts, it means that the two possible runs of $A_1^{\otimes r}$ would output different results, which in turn means that $A_1^{\otimes r}$ detected whether or not the input registers were measured. For such detection to even be possible, it must be the case that the input registers are in superposition. Then suppose an adversarial storm ⛈ generates two bolts $|\lightning_0\rangle, |\lightning_1\rangle$ that are potentially entangled such that both pass verification with the same serial number. Then we can measure both states, and the result will (with reasonable probability) be two distinct pre-images of the same y, representing a collision. By the assumed collision-resistance of H (and hence $H^{\otimes r}$), this will means a contradiction.

The problem with the above informal argument is that we do not know how $A_1^{\otimes r}$ will behave on non-valid bolts that did not come from $A_0^{\otimes r}$. In particular, maybe it passes verification with some small, but non-negligible success probability. It could be that after passing Test_0, the superposition has changed significantly, and maybe is no longer a superposition over pre-images of y, but instead a single pre-image. Nonetheless, if the auxiliary state registers are not those generated by $A_0^{\otimes r}$, it may be that the second test still accepts—for example, it may be that if $A^{\otimes r}$'s private state contains a particular string, it will always accept; normally this string would not be present, but the bolt that remains after performing one of Test_c may contain this string. We have to be careful to show that this case cannot happen, or if it does there is still nonetheless a way to extract a collision.

Toward that end, we only choose a single test at random. We will first show a weaker form of security, namely that an adversary cannot produce two bolts that are both accepted with probability close to 1 and have the same serial number. Then we will show how to modify the scheme so that it is impossible to produce bolts that are even accepted with small probability.

Consider a bolt where, after measuring $H(k, \cdot)$, the inputs registers are *not* in superposition at all. In this case, the measurement in Test_1 is redundant, and we therefore know that both runs of Test_c are the same, except the acceptance conditions are flipped. Since the choice of test is random, this means that such a bolt can only pass verification with probability at most $1/2$.

More generally, suppose the bolt was in superposition, but most of the weight was on a single input x_0. Precisely, suppose that when measuring the x registers, x_0 is obtained with probability $1 - \alpha$ for some relatively small α. We prove:

Claim. Consider a quantum state $|\phi\rangle$ and a projective partial measurement on some of the registers. Let $|\phi_x\rangle$ be the state left after performing the measurement and obtaining x. Suppose that some outcome of the measurement x_0 occurs with probability $1 - \alpha$. Then $\||\phi_{x_0}\rangle - |\phi\rangle\| < \sqrt{2\alpha}$.

Proof. First, the $|\phi_x\rangle$ are all orthogonal since the measurement was projective. Let $\Pr[x]$ be the probability that the partial measurement obtains x. It is straightforward to show that $|\phi\rangle = \sum_y \sqrt{\Pr[x]}\beta_x|\phi_x\rangle$ for some β_x of unit norm. The overall phase can be taken to be arbitrary, so we can set $\beta_{x_0} = 1$. Then we have $\langle\phi_{x_0}|\phi\rangle = \sqrt{1-\alpha}$. This means $\||\phi_{x_0}\rangle - |\phi\rangle\|^2 = 2 - 2(\langle\phi_{x_0}|\phi\rangle) = 2 - 2\sqrt{1-\alpha} \leq 2\alpha$ for $\alpha \in [0, 1]$. □

Now, suppose for the bolt that Test_0 passes with probability t. Suppose $\alpha \leq 1/200$. Then Test_1 can only pass with probability at most $3/2 - t$. This is because with probability at least $199/200$, the measurement in Test_1 yields x_0. Applying Claim, the result in this case is at most a distance $\sqrt{2/200} = \frac{1}{10}$ from the original bolt. In this case, since the acceptance criteria for Test_1 is the opposite of Test_0, the probability Test_1 passes is at most $1 - t + \frac{4}{10}$ by Lemma 1. Over all then, Test_1 passes with probability at most $(199/200)\left(1 - t + \frac{4}{10}\right) + (1/200) \leq \frac{3}{2} - t$.

Therefore, since the test is chosen at random, the probability of passing the test is the average of the two cases, which is at most $\frac{3}{4}$ regardless of t. Therefore, for any candidate pair of bolts $|\not{z}_0\rangle|\not{z}_1\rangle$, either:

(1) If the bolts are measured, two different pre-images of the same y, and hence a collision for $H^{\otimes r}$, will be obtained with probability at least $1/200$
(2) The probability that both bolts accept and have the same serial number is at most $\frac{3}{4}$.

Notice that if $|ϟ_0\rangle, |ϟ_1\rangle$ are produced by an adversarial storm ⛈, then event (1) can only happen with negligible probability, else we obtain a collision-finding adversary. Therefore, we have that for any efficient ⛈, except with negligible probability, the probability that both bolts produced by ⛈ accept and have the same serial number is at most $\frac{3}{4}$.

In the full scheme, a bolt is simply a tuple of λ bolts produced by 🌧$_0$, and the serial number is the concatenation of the serial numbers from each constituent bolt. The above analysis show that for any efficient adversarial storm ⛈ that produces two bolt sequences $|ϟ_b\rangle = (|ϟ_{b,1}\rangle, \ldots, |ϟ_{b,\lambda}\rangle)$, the probability that both sequences completely accept and agree on the serial numbers is, except with negligible probability, at most $\left(\frac{3}{4}\right)^\lambda$, which is negligible. Thus we obtain a valid quantum lightning scheme. $\square$

5 Quantum Money from Obfuscation

In this section, we show that, assuming injective one-way functions exist, applying indistinguishability obfuscation to Aaronson and Christiano's abstract scheme [3] yields a secure quantum money scheme.

5.1 Obfuscation

Definition 5. *A* subspace hiding obfuscator *(shO) for a field* $\mathbb{F}$ *and dimensions* d_0, d_1 *is a PPT algorithm* shO *such that:*

- **Input.** shO *takes as input the description of a linear subspace* $S \subseteq \mathbb{F}^n$ *of dimension* $d \in \{d_0, d_1\}$*. For concreteness, we will assume* S *is given as a matrix whose rows form a basis for* S*.*
- **Output.** shO *outputs a circuit* $\hat{S}$ *that computes membership in* S*. Precisely, let* $S(x)$ *be the function that decides membership in* S*. Then*

$$\Pr[\hat{S}(x) = S(x) \forall x : \hat{S} \leftarrow \mathsf{shO}(S)] \geq 1 - \mathsf{negl}(n)$$

- **Security.** *For security, consider the following game between an adversary and a challenger, indexed by a bit* b*.*
  - *The adversary submits to the challenger a subspace* S_0 *of dimension* d_0
  - *The challenger chooses a random subspace* $S_1 \subseteq \mathbb{F}^n$ *of dimension* d_1 *such that* $S_0 \subseteq S_1$*. It then runs* $\hat{S} \leftarrow \mathsf{shO}(S_b)$*, and gives* $\hat{S}$ *to the adversary*
  - *The adversary makes a guess* b' *for* b*.*

  The adversary's advantage is the probability $b' = b$*, minus* $1/2$*.* shO *is secure if, all PPT adversaries have negligible advantage.*

In the full version [37], we show the following theorem, which demonstrates that *indistinguishability obfuscation* can be used to build subspace-hiding obfuscation:

Theorem 6. *If injective one-way functions exist, then any indistinguishability obfuscator, appropriately padded, is also a subspace hiding obfuscator for field* $\mathbb{F}$ *and dimensions* d_0, d_1*, as long as* $|\mathbb{F}|^{n-d_1}$ *is exponential.*

5.2 Quantum Money from Obfuscation

Here, we recall Aaronson and Christiano's [3] construction, when instantiated with a subspace-hiding obfuscator.

Generating Banknotes. Let $\mathbb{F} = \mathbb{Z}_q$ for some prime q. Let λ be the security parameter. To generate a banknote, choose n a random even integer that is sufficiently large; we will choose n later, but it will depend on q and λ. Choose a random subspace $S \subseteq \mathbb{F}^n$ of dimension $n/2$. Let $S^\perp = \{x : x \cdot y = 0 \forall y \in S\}$ be the dual space to S.

Let $|\$_S\rangle = \frac{1}{|\mathbb{F}|^{n/4}} \sum_{x \in S} |x\rangle$. Let $P_0 = \mathsf{shO}(S)$ and $P_1 = \mathsf{shO}(S^\perp)$. Output $|\$_S\rangle, P_0, P_1$ as the quantum money state.

Verifying Banknotes. Given a banknote state, first measure the program registers, obtaining P_0, P_1. These will be the serial number. Let $|\$\rangle$ be the remaining registers. First run P_0 in superposition, and measure the output. If P_0 outputs 0, reject. Otherwise continue. Notice that if $|\$\rangle$ is the honest banknote state $|\$_S\rangle$ and P_0 is the obfuscation of S, then P_0 will output 1 with certainty.

Next, perform the quantum Fourier transform (QFT) to $|\$\rangle$. Notice that if $|\$\rangle = |\$_S\rangle$, now the state is $|\$_{S^\perp}\rangle$. Next, apply P_1 in superposition and measure the result. In the case of an honest banknote, the result is 1 with certainty. Finally, perform the inverse QFT to return the state. In the case of an honest banknote, the state goes back to being exactly $|\$_S\rangle$. The above shows that the scheme is correct. Next, we argue security:

Theorem 7. *If* shO *is a secure subspace-hiding obfuscator for* $d_0 = n/2$ *and some* d_1 *such that both* $|\mathbb{F}|^{n-d_1}$ *and* $|\mathbb{F}|^{d_1 - n/2}$ *are exponentially-large, then the construction above is a secure quantum money scheme.*

Corollary 1. *If injective one-way functions and iO exist, then quantum money exists.*

Proof. We now prove Theorem 7 through a sequence of hybrids

- H_0 is the normal security experiment for quantum money. Suppose the adversary, given a valid banknote, is able to produce two banknotes that pass verification with probability ϵ.
- H_1: here, we recall that Aaronson and Christiano's scheme is *projective*, so verification is equivalent to projecting onto the valid banknote state. Verifying two states is equivalent to projecting onto the product of two banknote states. Therefore, in H_1, instead of running verification, the challenger measures in the basis containing $|\$_S\rangle \times |\$_S\rangle$, and accepts if and only if the output is $|\$_S\rangle \times |\$_S\rangle$. The adversary's success probability is still ϵ.
- H_2: Here we invoke the security of shO to move P_0 to a higher-dimensional space. P_0 is moved to a random d_1 dimensional space containing S_0.
  We prove that the adversary's success probability in H_2 is negligibly close to

ϵ. Suppose not. Then we construct an adversary B that does the following. B chooses a random $d_0 = n/2$-dimensional space S_0. It queries the challenger on S_0, to obtain a program P_0. It then obfuscates $S_0^\perp$ to obtain P_1. B constructs the quantum state $|\$_{S_0}\rangle$, and gives $P_0, P_1, |\$_{S_0}\rangle$ to A. A produces two (potentially entangled) quantum states $|\$_0\rangle|\$_1\rangle$. B measures in a basis containing $|\$_{S_0}\rangle \otimes |\$_{S_0}\rangle$, and outputs 1 if and only if $|\$_{S_0}\rangle \otimes |\$_{S_0}\rangle$.
If B is given P_0 which obfuscates S_0, then A outputs 1 with probability ϵ, since it perfectly simulates A's view in H_1. If P_0 obfuscates a random space containing S_0, then B simulates H_2. By the security of shO, we must have that B outputs 1 with probability at least $\epsilon - \mathsf{negl}$. Therefore, in H_2, A succeeds with probability $\epsilon - \mathsf{negl}$.
- H_3: Here we invoke security of shO to move P_1 to a random d_1-dimensional space containing $S_0^\perp$. By an almost identical analysis to he above, we have that A still succeeds with probability at least $\epsilon - \mathsf{negl}$.
- H_4. Here, we change how the subspaces are constructed. First, a random space T_0 of dimension d_1 is constructed. Then a random space T_1 of dimension d_1 is constructed, subject to $T_0^\perp \subseteq T_1$. These spaces are obfuscated using shO to get programs P_0, P_1. A random $n/2$-dimensional space S_0 is chosen such that $T_1^\perp \subseteq S_0 \subseteq T_0$. S_0 is used to construct the state $|\$_{S_0}\rangle$, which is given to A along with P_0, P_1. Then during verification, the space S_0 is used again.
The distribution on spaces is identical to that in H_3, to A succeeds in H_4 with probability $\epsilon - \mathsf{negl}$.

Since on average over T_0, T_1, A succeeds with probability $\epsilon - \mathsf{negl}$, there exist fixed $T_0, T_1, T_0^\perp \subseteq T_1$, such that the adversary succeeds for these T_0, T_1 with probability at least $\epsilon - \mathsf{negl}$.

We now construct a no-cloning adversary C. C is given a state $|\$_{S_0}\rangle$ for a random S_0 such that $T_1^\perp \subseteq S_0 \subseteq T_0$, and it tries to clone $|\$_{S_0}\rangle$. To do so, it constructs obfuscations P_0, P_1 of T_0, T_1 using shO, and gives them along with $|\$_{S_0}\rangle$ to A. C then outputs whatever A outputs. C's probability of cloning is exactly the probability A succeeds in H_4, which is $\epsilon - \mathsf{negl}$. This gives an instance of the no-cloning problem. In the full version [37], we prove that the probability of cloning in this instance is at most $2|\mathbb{F}|^{-n'/2} = 2|\mathbb{F}|^{d_1 - n/2}$, which is exponentially small by the assumptions of the theorem. □

6 Constructing Quantum Lightning

6.1 Background

Degree-2 Polynomials over $\mathbb{Z}_q$. Consider a set $\mathcal{A}$ of n degree-2 polynomials over m variables in $\mathbb{Z}_q$ for some large prime q. Let $f_\mathcal{A} : \mathbb{Z}_q^m \to \mathbb{Z}_q^n$ be the function that evaluates each of the polynomials in $\mathcal{A}$ on its input. We will be interested in the compressing case, where $n < m$.

As shown by [6,19], the function $f_\mathcal{A}$ is *not* collision resistant when the coefficients of the polynomials are random. Here, we recreate the proof, and also discuss the multi-collision resistance of the function.

To find a collision for $f_{\mathcal{A}}$, choose a random $\Delta \in \{0,1\}^m$. We will find a collision of the form $\mathbf{x}, \mathbf{x}+\Delta$. The condition that $\mathbf{x}, \mathbf{x}+\Delta$ collide means $P(\mathbf{x}+\Delta)-P(\mathbf{x}) = 0$ for all polynomials in $\mathcal{A}$. Now, since P has degree 2, all the order-2 terms in $\mathbf{x}$ in this difference will cancel out, leaving only terms that are linear in $\mathbf{x}$ (and potentially quadratic in Δ). This gives us a system of linear equations over $\mathbf{x}$, which we can solve provided the equations are consistent. As shown in [6], these equations are consistent with overwhelming probability provided $n \leq m$.

This attack can be generalized to find $k+1$ colliding inputs. Choose random $\Delta_1, \ldots, \Delta_k$. We will compute an $\mathbf{x}$ such that $\mathbf{x}, \mathbf{x}+\Delta_1, \ldots, \mathbf{x}+\Delta_k$ form $k+1$ colliding points. Each Δ_j generates a system of n equations for $\mathbf{x}$ as described above. Let $B = B_{\Delta_1,\ldots,\Delta_k}$ be the matrix consisting of all the rows of B_{Δ_j} as j varies. As long as B is full rank, a solution for $\mathbf{x}$ is guaranteed. Again, B will be full rank with overwhelming probability, provided $m \geq kn$. However, if $m \ll kn$, this procedure will fail, and it therefore appears reasonable to assume the multi-collision resistance of such functions.

Using the above, we can even generate superpositions over $k+1$ inputs such that all the inputs map to the same output. Consider the following procedure:

- Generate the uniform superposition $|\phi_0\rangle \propto \sum_{\Delta_1,\ldots,\Delta_k} |\Delta_1, \ldots, \Delta_k\rangle$
- Write $\mathbf{\Delta} = (\Delta_1, \ldots, \Delta_k)$ In superposition, run the computation above that maps $\mathbf{\Delta}$ to the affine space $S_{\mathbf{\Delta}}$ such that, for all $\mathbf{x} \in S$, $f_{\mathcal{A}}(\mathbf{x}) = f_{\mathcal{A}}(\mathbf{x}+\Delta_j)$ for all j. This will be an affine space of dimension $m - nk$ with overwhelming probability. Then construct a uniform superposition of elements in $S_{\mathbf{\Delta}}$. The resulting state is then: $|\phi_1\rangle \propto \sum_{\mathbf{\Delta}} \frac{1}{\sqrt{|S_{\mathbf{\Delta}}|}} \sum_{\mathbf{x}\in S_{\mathbf{\Delta}}} |\Delta, \mathbf{x}\rangle$
- Next, in superposition, compute $f_{\mathcal{A}}(\mathbf{x})$, and measure the result to get a string $\mathbf{y}$. The resulting state is $|\phi_y\rangle \propto \sum_{\mathbf{\Delta},\mathbf{x}\in S_{\mathbf{\Delta}}:f_{\mathcal{A}}(\mathbf{x})=\mathbf{y}} \frac{1}{\sqrt{|S_{\mathbf{\Delta}}|}} |\mathbf{x}, \mathbf{\Delta}\rangle$
- Finally, in superposition, map $(\mathbf{x}, \Delta_1, \ldots, \Delta_k)$ to $(\mathbf{x}, \mathbf{x}+\Delta_1, \ldots, \mathbf{x}+\Delta_k)$. The resulting state is $|\psi_y\rangle \propto \sum_{\mathbf{\Delta},\mathbf{x}\in S_{\mathbf{\Delta}}:f_{\mathcal{A}}(\mathbf{x})=\mathbf{y}} \frac{1}{|S_{\mathbf{\Delta}}|} |\mathbf{x}, \mathbf{x}+\Delta_1, \ldots\rangle$

  We note that the support of this state is *all* vectors $(\mathbf{x}_0, \ldots, \mathbf{x}_k)$ such that $f_{\mathcal{A}}(\mathbf{x}_i) = \mathbf{y}$ for all $i \in [0, k]$. Moreover, for all but a negligible fraction, the weight $|S_{\mathbf{\Delta}}|$ is the same, and so the weights for these components are the same. Even more, the total weight of the other points is negligible. Therefore, the this state is negligibly close to the state $\sum_{\mathbf{x}_0,\ldots,\mathbf{x}_k:f_{\mathcal{A}}(\mathbf{x}_i)=\mathbf{y}\forall i} |\mathbf{x}_0, \ldots, \mathbf{x}_k\rangle = \left(\sum_{\mathbf{x}:f_{\mathcal{A}}(\mathbf{x})=\mathbf{y}} |\mathbf{x}\rangle\right)^{\otimes(k+1)} \propto |\psi'_{\mathbf{y}}\rangle^{\otimes(k+1)}$, where $|\psi'_{\mathbf{y}}\rangle \propto \sum_{\mathbf{x}:f_{\mathcal{A}}(\mathbf{x})=\mathbf{y}} |\mathbf{x}\rangle$.

Linear Functions over Rank-Constrained Matrices. Here, we consider a related problem. Consider a set of n linear functions $\mathcal{A}$ over rank-d matrices in $\mathbb{Z}_q^{m\times m}$. Since q is large, a random rank-d matrix in $\mathbb{Z}_q^{m\times m}$ will have it's first d columns span the entire column space. Therefore, most rank constrained matrices can be written as $(\mathbf{A}\ \ \mathbf{A}\cdot\mathbf{B})$ for a $m \times d$ matrix $\mathbf{A}$ and a $d \times (m-d)$ matrix $\mathbf{B}$.

Let $f_{\mathcal{A}} : \mathbb{Z}_q^{m^2} \to \mathbb{Z}_q^n$ be the function that evaluates each of the functions in $\mathcal{A}$ on its input. We can therefore think of $f_{\mathcal{A}}$ as a degree-2 polynomial over $\mathbf{A}, \mathbf{B}$. Note, however, that in this case, the function is bipartite: it can be divided into two sets of variables ($\mathbf{A}$ and $\mathbf{B}$) such that it is linear in each set. This means we can easily invert the function by choosing an arbitrary selection for one of the sets of variables, and then solving for the other.

Linear Functions over Rank-Constrained Symmetric Matrices. By instead considering only symmetric matrices, we essentially become equivalent to degree-2 polynomials. In particular, $\mathbf{A} \cdot \mathbf{A}^T$ for $\mathbf{A} \in \mathbb{Z}_q^{m \times d}$ is a symmetric rank-d matrix. Moreover, any degree-2 polynomial over $\mathbb{Z}_q^m$ can be represented as a linear polynomial over rank-1 symmetric matrices by tensoring the input with itself.

Note, however, that since $\mathbb{Z}_q$ is not a closed field, in general we cannot decompose any symmetric rank-d matrix into $\mathbf{A} \cdot \mathbf{A}^T$ (though we can over the closure). Therefore, linear functions over rank-constrained symmetric matrices can be seen as a slightly relaxed version of the degree-2 polynomials discussed above. In particular it is straightforward to generalize the algorithm for generating superpositions of colliding inputs to generate uniform superpositions of low-rank matrices that collide.

6.2 Hardness Assumption

Our assumption will have parameters n, m, q, d, e, k, to be described in the following discussion. Let $\mathcal{D}$ be the set of symmetric $m \times m$ matrices over $\mathbb{Z}_q$ whose rank is at most d. We will alternately think of $\mathcal{D}$ as matrices, as well as vectors by writing out all of the $\binom{m+1}{2}$ entries on and above the diagonal.

Let $\mathcal{A}$ be a set of n linear functions over $\mathcal{D}$, which we will think of as being n linear functions over $\binom{m+1}{2}$ variables. Consider the function $f_{\mathcal{A}} : \mathcal{D} \to \mathbb{Z}_q^n$ given by evaluating each linear function in $\mathcal{A}$.

As discussed above, we could imagine assuming that $f_{\mathcal{A}}$ is multi-collision resistant for a random set of linear functions $\mathcal{A}$. However, in order for our ultimate bolt verification procedure to work, we will need $\mathcal{A}$ to have a special form.

$\mathcal{A}$ is sampled as follows. Let $\mathbf{R} \in \mathbb{Z}_q^{e \times m}$ be chosen at random. Consider the set of symmetric matrices $\mathbf{A} \in \mathbb{Z}_q^{m \times m}$ such that $\mathbf{R} \cdot \mathbf{A} \cdot \mathbf{R}^T = 0$. This is a linear subspace of dimension $\binom{m+1}{2} - \binom{e+1}{2}$ (note that since $\mathbf{B}$ is symmetric, $\mathbf{R} \cdot \mathbf{A} \cdot \mathbf{R}^T$ is guaranteed to be symmetric, so we only get $\binom{e+1}{2}$ equations). We can think of each $\mathbf{A}$ as represented by its $\binom{m+1}{2}$ upper-triangular entries, which gives us an equation over $\binom{m+1}{2}$ variables. Let $\mathcal{A}$ be a basis for this space of linear functions. Thus, we set $n = \binom{m+1}{2} - \binom{e+1}{2}$. Note that we will *not* keep $\mathbf{R}$ secret. Rank d symmetric matrices in $\mathbb{Z}_q^{m \times m}$ have $\binom{d+1}{2} + d \times (m - d) = d \times m - \binom{d}{2}$ degrees of freedom. Therefore, the function $f_{\mathcal{A}}$ will be compressing provided that $d \times m - \binom{d}{2} > n = \binom{m+1}{2} - \binom{e+1}{2}$.

By choosing $f_{\mathcal{A}}$ in this way, we provide a "trapdoor" $\mathbf{R}$ that will be used for verifying bolts. This trapdoor is a rank-e matrix in the kernel of $f_{\mathcal{A}}$. If $e < d$, this would allow us to compute many colliding inputs, as, for any rank-1 $\mathbf{S}$, the whole affine space $\mathbf{S} + \alpha\mathbf{R}$ has rank at most $e + 1 \leq d$ and maps to the same value. However, if we choose $e > d$, $\mathbf{R}$ does not appear to let us find collisions.

Our Assumption. We now make the following hardness assumption. We say a hash function f is k-multi-collision resistant (k-MCR) if it is computationally infeasible to find k colliding inputs.

Assumption 8. *There exists some functions n, d, e, k in m such that $n = \binom{m+1}{2} - \binom{e+1}{2} < d \times m - \binom{d}{2}$, $kn \leq d \times m - \binom{d}{2} < (2k+1)n$, and $e > d$, such that $f_{\mathcal{A}}$ as sampled above is $(2k+2)$-MCR, even if $\mathbf{R}$ is given to the adversary.*

For example, we can choose e, d such that $m = e + d - 1$, in which case $n = d \times m - \binom{d}{2} - e$. By choosing $e \approx d$, we have $d \times m - \binom{d}{2} \ll 3n$, so we can set k to be 1. We therefore assume that it is computationally infeasible to find 4 colliding inputs to $f_{\mathcal{A}}$.

We stress that this assumption is highly speculative and untested. In the full version [37], we discuss in more depth possible attacks on the assumption, as well as weakened versions that are still sufficient for our purposes.

6.3 Quantum Lightning

We now describe our quantum lightning construction.

Parameters. Our scheme will be parameterized by integers n, m, q, d, e, k.

Setup. To set up the quantum lightning scheme, simply sample $\mathcal{A}, \mathbf{R}$ as above, and output $\mathcal{A}, \mathbf{R}$.

Bolt Generation. We generate a bolt $|\lightning_{\mathbf{y}}\rangle$ as a superposition of $k+1$ colliding inputs, following the procedure described above. The result is statistically close to $|\lightning'_{\mathbf{y}}\rangle^{\otimes(k+1)}$ where $|\lightning'_{\mathbf{y}}\rangle$ is the equally-weighted superposition over rank-d symmetric matrices such that applying $f_{\mathcal{A}}$ gives $\mathbf{y}$. We will call $|\lightning'_{\mathbf{y}}\rangle$ a mini-bolt.

Verifying a Bolt. Full verification of a bolt will run a mini verification on each of the $k+1$ mini-bolts. Each mini verification will output an element in $\mathbb{Z}_q^n \cup \{\bot\}$. Full verification will accept and output y only if each mini verification accepts and outputs the same string y. We now describe the mini verification.

Roughly, our goal is to be able to distinguish $|\lightning'_{\mathbf{y}}\rangle$ for some $\mathbf{y}$ from any singleton state. We will output $\mathbf{y}$ in this case, and for any other state, reject.

Mini verification on a state $|\phi\rangle$ will proceed in two steps. Recall that superposition is over the upper triangular portion of $m \times m$ matrices. We first apply, in superposition, the procedure that flips some external bit if the input does not correspond to a matrix of rank at most d. The bit is initially set to 0. Then we measure this bit, and abort if it is 1. Notice that for the honest $|\lightning'_{\mathbf{y}}\rangle$ state, this will pass with certainty and not affect the state.

In the next step, we apply the procedure that evaluates $f_{\mathcal{A}}$ in superposition, and flips some external bit if the result is not $\mathbf{y}$. The bit is initially set to 0. Then we measure this bit, and abort if it is 1. Notice that for the honest $|\lightning'_{\mathbf{y}}\rangle$ state, this will pass with certainty and not affect the state.

At this point, if we have not aborted, our state is a superposition of pre-images of $f_{\mathcal{A}}$ which correspond to symmetric rank-d matrices. If our input was $|\lightning'_{\mathbf{y}}\rangle$, the state is the uniform superposition over such pre-images.

Next, we verify that the state is in superposition and not a singleton state. To do so, we perform the quantum Fourier transform (QFT) to the state. We now analyze what the QFT does to $|\lightning_{\mathbf{y}}\rangle$.

The support of $|\lightning'_{\mathbf{y}}\rangle$ is the intersection of sets S, T where S is the set of all pre-images of $\mathbf{y}$ (not necessarily rank constrained) and T is the set of all rank-d matrices. We analyze the Fourier transform applied to each set separately.

Recall that the Fourier transform takes the uniform superposition over the kernel of a matrix to the uniform superposition over its row-span. Therefore, the superposition over pre-images of 0 is just the uniform superposition of symmetric matrices $\mathbf{A}$ such that $\mathbf{R} \cdot \mathbf{A} \cdot \mathbf{R}^T = 0$ (or technically, just the upper triangular part). The fact that the superposition lies in a coset of the kernel simply introduces a phase term to each element in the superposition.

In the full version [37], we prove the following claim:

Claim. The Fourier transform of the uniform superposition over upper-triangular parts of rank d symmetric matrices is negligibly close to the uniform superposition over upper triangular parts of rank $m - d$ symmetric matrices.

Putting this together, since multiplication in the primal domain becomes convolution in the Fourier domain, after we apply the Fourier transform to our mini bolt state, the result is the superposition of upper triangular parts of matrices $\mathbf{A} + \mathbf{B}$ where $\mathbf{B}$ is symmetric and rank $m - d$ and $\mathbf{A}$ is symmetric such that $\mathbf{R} \cdot \mathbf{A} \cdot \mathbf{R}^T = 0$. The superposition is uniform, though there will be a phase factor associated with each element.

We therefore compute $\mathbf{R} \cdot (\mathbf{A} + \mathbf{B}) \cdot \mathbf{R}^T = \mathbf{R} \cdot \mathbf{B} \cdot \mathbf{R}^T$ and compute the rank. Notice that the rank is at most $m - d$ for honest bolt states. Therefore, if the rank is indeed at most $m - d$ we will accept, otherwise we will reject. Next, we un-compute $\mathbf{R} \cdot (\mathbf{A} + \mathbf{B}) \cdot \mathbf{R}^T$, and undo the Fourier transform. The analysis above shows that for an honest state $|\lightning_{\mathbf{y}}\rangle$, we will accept with overwhelming probability, and the final both state will be negligibly close to the original bolt.

Note that, in contrast, if the bolt state is a singleton state, then the Fourier transform will result in a uniform superposition over all symmetric matrices; when we apply$\mathbf{R} \cdot (\cdot) \cdot \mathbf{R}^T$, the result will have rank e with overwhelming probability. So we set $m - d < e$ to have an almost perfect distinguishing advantage.

Security. We now prove security. Consider a quantum adversary A that is given $\mathcal{A}$ and tries to construct two (possibly entangled) bolts $|\lightning_0\rangle, |\lightning_1\rangle$. Assume toward contradiction that with non-negligible probability, verification accepts on both bolts, and outputs the same serial number $\mathbf{y}$.

Conditioned on acceptance, by the above arguments the resulting mini bolts must all be far from singletons when we trace out the other bolts. This means that if we measure the mini-bolts, the resulting superpositions will have high min-entropy. Therefore, we measure all $2k + 2$ mini bolts, and we obtain $2k + 2$ colliding inputs that are distinct except with negligible probability. This violates our hardness assumption.

Theorem 9. *If Assumption 8 holds, then the scheme above is a secure quantum lightning scheme.*

In the full version [37], we show how to modify our construction to get a collapse-non-binding hash function.

References

1. Aaronson, S.: http://www.scottaaronson.com/blog/?p=2854
2. Aaronson, S.: Quantum copy-protection and quantum money. In: Proceedings of the 2009 24th Annual IEEE Conference on Computational Complexity, CCC 2009, Washington, DC, USA, pp. 229–242. IEEE Computer Society (2009)
3. Aaronson, S., Christiano, P.: Quantum money from hidden subspaces. In: Karloff, H.J., Pitassi, T. (eds.) 44th ACM STOC, pp. 41–60. ACM Press, May 2012
4. Albrecht, M.R., Bai, S., Ducas, L.: A subfield lattice attack on overstretched NTRU assumptions - cryptanalysis of some FHE and graded encoding schemes. In: Robshaw, M., Katz, J. (eds.) CRYPTO 2016, Part I. LNCS, vol. 9814, pp. 153–178. Springer, Heidelberg (2016). https://doi.org/10.1007/978-3-662-53018-4_6
5. Ambainis, A., Rosmanis, A., Unruh, D.: Quantum attacks on classical proof systems: the hardness of quantum rewinding. In: 55th FOCS, pp. 474–483. IEEE Computer Society Press, October 2014
6. Applebaum, B., Haramaty, N., Ishai, Y., Kushilevitz, E., Vaikuntanathan, V.: Low-complexity cryptographic hash functions. In: Papadimitriou, C.H. (ed.) ITCS 2017. vol. 4266, pp. 7:1–7:31, 67. LIPIcs, January 2017
7. Barak, B., et al.: On the (im)possibility of obfuscating programs. In: Kilian, J. (ed.) CRYPTO 2001. LNCS, vol. 2139, pp. 1–18. Springer, Heidelberg (2001). https://doi.org/10.1007/3-540-44647-8_1
8. Bennett, C.H., Bernstein, E., Brassard, G., Vazirani, U.: Strengths and weaknesses of quantum computing. SIAM J. Comput. **26**(5), 1510–1523 (1997)
9. Bennett, C.H., Brassard, G.: Quantum public key distribution reinvented. SIGACT News **18**(4), 51–53 (1987)
10. Boneh, D., Zhandry, M.: Quantum-secure message authentication codes. In: Johansson, T., Nguyen, P.Q. (eds.) EUROCRYPT 2013. LNCS, vol. 7881, pp. 592–608. Springer, Heidelberg (2013). https://doi.org/10.1007/978-3-642-38348-9_35
11. Boneh, D., Zhandry, M.: Secure signatures and chosen ciphertext security in a quantum computing world. In: Canetti, R., Garay, J.A. (eds.) CRYPTO 2013, Part II. LNCS, vol. 8043, pp. 361–379. Springer, Heidelberg (2013). https://doi.org/10.1007/978-3-642-40084-1_21
12. Brakerski, Z., Christiano, P., Mahadev, U., Vazirani, U.V., Vidick, T.: A cryptographic test of quantumness and certifiable randomness from a single quantum device. In: Thorup, M. (ed.) 59th FOCS, pp. 320–331. IEEE Computer Society Press, October 2018
13. Brakerski, Z., Vaikuntanathan, V., Wee, H., Wichs, D.: Obfuscating conjunctions under entropic ring LWE. In: Sudan, M. (ed.) ITCS 2016, pp. 147–156. ACM, January 2016
14. Chen, Y., Gentry, C., Halevi, S.: Cryptanalyses of candidate branching program obfuscators. In: Coron, J.-S., Nielsen, J.B. (eds.) EUROCRYPT 2017, Part III. LNCS, vol. 10212, pp. 278–307. Springer, Cham (2017). https://doi.org/10.1007/978-3-319-56617-7_10

15. Cheon, J.H., Jeong, J., Lee, C.: An algorithm for CSPR problems and cryptanalysis of the GGH multilinear map without an encoding of zero. Technical report, Cryptology ePrint Archive, Report 2016/139 (2016)
16. Colbeck, R.: Quantum and relativistic protocols for secure multi-party computation (2009)
17. Coron, J.-S., Lepoint, T., Tibouchi, M.: Practical multilinear maps over the integers. In: Canetti, R., Garay, J.A. (eds.) CRYPTO 2013, Part I. LNCS, vol. 8042, pp. 476–493. Springer, Heidelberg (2013). https://doi.org/10.1007/978-3-642-40041-4_26
18. Cramer, R., Ducas, L., Peikert, C., Regev, O.: Recovering short generators of principal ideals in cyclotomic rings. In: Fischlin, M., Coron, J.-S. (eds.) EUROCRYPT 2016, Part II. LNCS, vol. 9666, pp. 559–585. Springer, Heidelberg (2016). https://doi.org/10.1007/978-3-662-49896-5_20
19. Ding, J., Yang, B.-Y.: Multivariates polynomials for hashing. In: Pei, D., Yung, M., Lin, D., Wu, C. (eds.) Inscrypt 2007. LNCS, vol. 4990, pp. 358–371. Springer, Heidelberg (2008). https://doi.org/10.1007/978-3-540-79499-8_28
20. Farhi, E., Gosset, D., Hassidim, A., Lutomirski, A., Shor, P.W.: Quantum money from knots. In: Goldwasser, S. (ed.) ITCS 2012, pp. 276–289. ACM, January 2012
21. Garg, S., Gentry, C., Halevi, S.: Candidate multilinear maps from ideal lattices. In: Johansson, T., Nguyen, P.Q. (eds.) EUROCRYPT 2013. LNCS, vol. 7881, pp. 1–17. Springer, Heidelberg (2013). https://doi.org/10.1007/978-3-642-38348-9_1
22. Garg, S., Gentry, C., Halevi, S., Raykova, M., Sahai, A., Waters, B.: Candidate indistinguishability obfuscation and functional encryption for all circuits. In: 54th FOCS, pp. 40–49. IEEE Computer Society Press, October 2013
23. Garg, S., Yuen, H., Zhandry, M.: New security notions and feasibility results for authentication of quantum data. In: Katz, J., Shacham, H. (eds.) CRYPTO 2017, Part II. LNCS, vol. 10402, pp. 342–371. Springer, Cham (2017). https://doi.org/10.1007/978-3-319-63715-0_12
24. Gavinsky, D.: Quantum money with classical verification (2011)
25. Gentry, C., Gorbunov, S., Halevi, S.: Graph-induced multilinear maps from lattices. In: Dodis, Y., Nielsen, J.B. (eds.) TCC 2015, Part II. LNCS, vol. 9015, pp. 498–527. Springer, Heidelberg (2015). https://doi.org/10.1007/978-3-662-46497-7_20
26. Goyal, R., Koppula, V., Waters, B.: Lockable obfuscation. In: Umans, C. (ed.) 58th FOCS, pp. 612–621. IEEE Computer Society Press, October 2017
27. Lutomirski, A.: An online attack against Wiesner's quantum money (2010)
28. Lutomirski, A., et al.: Breaking and making quantum money: toward a new quantum cryptographic protocol. In: Yao, A.C.-C. (ed.) ICS 2010, pp. 20–31. Tsinghua University Press, January 2010
29. Micciancio, D., Regev, O.: Worst-case to average-case reductions based on Gaussian measures. SIAM J. Comput. **37**(1), 267–302 (2007)
30. Mosca, M., Stebila, D.: Quantum coins. In: Error-Correcting Codes, Finite Geometries and Cryptography, vol. 523, pp. 35–47 (2010)
31. Pena, M.C., Faugère, J.-C., Perret, L.: Algebraic cryptanalysis of a quantum money scheme: the noise-free case. In: Katz, J. (ed.) PKC 2015. LNCS, vol. 9020, pp. 194–213. Springer, Heidelberg (2015). https://doi.org/10.1007/978-3-662-46447-2_9
32. Regev, O.: On lattices, learning with errors, random linear codes, and cryptography. In: Gabow, H.N., Fagin, R. (eds.) 37th ACM STOC, pp. 84–93. ACM Press, May 2005
33. Unruh, D.: Revocable quantum timed-release encryption. In: Nguyen, P.Q., Oswald, E. (eds.) EUROCRYPT 2014. LNCS, vol. 8441, pp. 129–146. Springer, Heidelberg (2014). https://doi.org/10.1007/978-3-642-55220-5_8

34. Unruh, D.: Computationally binding quantum commitments. In: Fischlin, M., Coron, J.-S. (eds.) EUROCRYPT 2016, Part II. LNCS, vol. 9666, pp. 497–527. Springer, Heidelberg (2016). https://doi.org/10.1007/978-3-662-49896-5_18
35. Wichs, D., Zirdelis, G.: Obfuscating compute-and-compare programs under LWE. In: Umans, C. (ed.) 58th FOCS, pp. 600–611. IEEE Computer Society Press, October 2017
36. Wiesner, S.: Conjugate coding. SIGACT News **15**(1), 78–88 (1983)
37. Zhandry, M.: Quantum lightning never strikes the same state twice. Cryptology ePrint Archive, Report 2017/1080 (2017). https://eprint.iacr.org/2017/1080

Information-Theoretic Cryptography

Secret-Sharing Schemes for General and Uniform Access Structures

Benny Applebaum[1(✉)], Amos Beimel[2], Oriol Farràs[3], Oded Nir[1], and Naty Peter[2]

[1] Tel Aviv University, Tel Aviv, Israel
benny.applebaum@gmail.com, odednir123@gmail.com
[2] Ben-Gurion University of the Negev, Be'er-Sheva, Israel
amos.beimel@gmail.com, naty@post.bgu.ac.il
[3] Universitat Rovira i Virgili, Tarragona, Catalonia, Spain
oriol.farras@urv.cat

Abstract. A secret-sharing scheme allows some authorized sets of parties to reconstruct a secret; the collection of authorized sets is called the access structure. For over 30 years, it was known that any (monotone) collection of authorized sets can be realized by a secret-sharing scheme whose shares are of size $2^{n-o(n)}$ and until recently no better scheme was known. In a recent breakthrough, Liu and Vaikuntanathan (STOC 2018) have reduced the share size to $O(2^{0.994n})$. Our first contribution is improving the exponent of secret sharing down to 0.892. For the special case of linear secret-sharing schemes, we get an exponent of 0.942 (compared to 0.999 of Liu and Vaikuntanathan).

Motivated by the construction of Liu and Vaikuntanathan, we study secret-sharing schemes for uniform access structures. An access structure is k-uniform if all sets of size larger than k are authorized, all sets of size smaller than k are unauthorized, and each set of size k can be either authorized or unauthorized. The construction of Liu and Vaikuntanathan starts from protocols for conditional disclosure of secrets, constructs secret-sharing schemes for uniform access structures from them, and combines these schemes in order to obtain secret-sharing schemes for general access structures. Our second contribution in this paper is constructions of secret-sharing schemes for uniform access structures. We achieve the following results:

- A secret-sharing scheme for k-uniform access structures for large secrets in which the share size is $O(k^2)$ times the size of the secret.
- A linear secret-sharing scheme for k-uniform access structures for a binary secret in which the share size is $\tilde{O}(2^{h(k/n)n/2})$ (where h is the binary entropy function). By counting arguments, this construction is optimal (up to polynomial factors).
- A secret-sharing scheme for k-uniform access structures for a binary secret in which the share size is $2^{\tilde{O}(\sqrt{k \log n})}$.

Our third contribution is a construction of ad-hoc PSM protocols, i.e., PSM protocols in which only a subset of the parties will compute a function on their inputs. This result is based on ideas we used in the construction of secret-sharing schemes for k-uniform access structures for a binary secret.

Y. Ishai and V. Rijmen (Eds.): EUROCRYPT 2019, LNCS 11478, pp. 441–471, 2019.
https://doi.org/10.1007/978-3-030-17659-4_15

1 Introduction

A secret-sharing scheme is a method in which a dealer that holds a secret information (e.g., a password or private medical data) can store it on a distributed system, i.e., among a set of parties, such that only some predefined authorized sets of parties can reconstruct the secret. The process of storing the secret information is called secret sharing and the collections of authorized sets of parties are called access structures. Interestingly, secret-sharing schemes are nowadays used in numerous applications (in addition to their obvious usage for secure storage), e.g., they are used for secure multiparty computation [15,19], threshold cryptography [24], access control [37], attribute-based encryption [29,43], and oblivious transfer [39,42]. The original and most important secret-sharing schemes, introduced by Blakley [18] and Shamir [38], are threshold secret-sharing schemes, in which the authorized sets are all the sets whose size is larger than some threshold.

Secret-sharing schemes for general access structures were introduced in [32] more than 30 years ago. However, we are far from understanding constraints on the share size of these schemes. In the original constructions of secret-sharing schemes in [32], the share size of each party is $2^{n-O(\log n)}$. New constructions of secret-sharing schemes followed, e.g., [16,17,33]; however, the share size of each party in these schemes remains $2^{n-O(\log n)}$. In a recent breakthrough, Liu and Vaikuntanathan [34] (building on [36]) showed, for the first time, that it is possible to construct secret-sharing schemes in which the share size of each party is $O(2^{cn})$ with an exponent c strictly smaller than 1. In particular, they showed that every access structure can be realized with an exponent of $\mathbf{S}_{\text{LV}} = 0.994$. Moreover, they showed that every access structure can be realized by a linear secret-sharing scheme with an exponent of 0.994 (a scheme is linear if each share can be written as a linear combination of the secret and some global random field elements; see Sect. 2 for a formal definition). On the negative size, the best lower bound on the total share size required for sharing a secret for some access structure is $\Omega(n^2/\log n)$ [22,23]. Thus, there is a huge gap between the known upper and lower bounds.

1.1 Our Results

Our first result is an improvement of the secret-sharing exponent of general access structure. In Sect. 3, we prove the following theorem.

Theorem 1.1. *Every access structure over n parties can be realized by a secret-sharing scheme with a total share size of* $2^{0.8916n+o(n)}$ *and by a linear secret-sharing scheme with a total share size of* $2^{0.942n+o(n)}$.

In a nutshell, the construction of [34] together with combinatorial covering designs are being used to establish a recursive construction, which eventually leads to the improved bounds.

We next construct secret-sharing schemes for uniform access structures. An access structure is k-uniform if all sets of size larger than k are authorized, all sets of size smaller than k are unauthorized, and every set of size k can be either authorized or unauthorized. Our second contribution is on the construction of secret-sharing schemes for uniform access structures. The motivation for studying uniform access structures is twofold. First, they are related to protocols for conditional disclosure of secrets (CDS), a primitive introduced by Gertner et al. [28]. By various transformations [2,11–13,34], CDS protocols imply secret-sharing schemes for uniform access structures. Furthermore, as shown in [34], CDS protocols and secret-sharing schemes for uniform access structures are a powerful primitives to construct secret-sharing schemes for general access structures. Thus, improvements on secret-sharing schemes for uniform access structures can lead to better constructions of secret-sharing schemes for general access structures. Second, as advocated in [2], uniform access structures should be studied as they are a useful scaled-down version of general access structures. Studying uniform access structures can shed light on the share size required for general access structures, which is a major open problem.

Three regimes of secret-sharing schemes for uniform access structures have been studied. The first regime is the obvious one of secret-sharing schemes with short secrets (e.g., a binary secret). The second regime is secret-sharing schemes with long secrets. Surprisingly, there are secret-sharing for this regime that are much more efficient than schemes with short secrets [2]. The third regime is *linear* secret-sharing schemes with short secrets. Linear secret-sharing schemes are schemes where the sharing of the secret is done using a linear transformation; such schemes are interesting since in many applications linearity is required, e.g., in the construction of secure multiparty computation protocols in [21] and in the constructions of Attrapadung [7] and Wee [44] of public-key (multi-user) attribute-based encryption.

In this paper we improve the constructions of secret-sharing schemes for uniform access structures in these three regimes. We describe our results according to the order that they appear in the paper.

Long secrets. In Sect. 4, we construct secret-sharing schemes for n-party k-uniform access structures for large secrets, i.e., secrets of size at least 2^{n^k}. Previously, the share size in the best constructions for such schemes was either e^k times the length of the secret [2] or n times the length of the secret (implied by the CDS protocol of [2] and a transformation of [12]). We show a construction in which the share size is $O(k^2)$ times the size of the secret. For this construction, we use the CDS protocol of [2] with k^2 parties (in contrast to [2], which uses it with k parties) with an appropriate k^2-input function. Combined with the results of [12], we get a share size which is at most $\min(k^2, n)$-times larger than the secret size.

Linear schemes. In Sect. 5, we design a linear secret-sharing scheme for k-uniform access structures for a binary secret in which the share size is $\tilde{O}(2^{h(k/n)n/2})$ (where h is the binary entropy function). By counting arguments, our construction is optimal (up to polynomial factors). Previously, the best construction was implied by the CDS protocols of [13,36] and had share size $\tilde{O}(2^{n/2})$.

Our construction is inspired by a linear 2-party CDS protocol of [27] and the linear k-party CDS protocols of [13]. We use the ideas of these CDS protocols to design a linear secret-sharing schemes for balanced k-uniform access structures (where there is a set B of $n/2$ parties such that any minimal authorized set of size k in the access structure contains exactly $k/2$ parties from B). Using a probabilistic argument, we show that every k-uniform access structure can be written as a union of $O(n^{3/2})$ balanced access structures, thus, we can share the secret independently for each balanced access structure in the union.

Short secrets. In Sect. 6, we describe a secret-sharing scheme for k-uniform access structures for a binary secret in which the share size is $2^{\tilde{O}(\sqrt{k \log n})}$. Previously, the best share size in a secret-sharing scheme realizing such access structure was $\min\left\{2^{O(k)+\tilde{O}(\sqrt{k \log n})}, 2^{\tilde{O}(\sqrt{n})}\right\}$ (by combining results of [2,12,36]). To achieve this result we define a new transformation from a k-party CDS protocol to secret-sharing schemes for k-uniform access structures. The idea of this transformation is that the shares of the parties contain the messages in a CDS protocol of an appropriate function. The difficulty is how to ensure that parties of an unauthorized set of size k cannot obtain two messages of the same party in the CDS protocol (otherwise, the privacy of the CDS protocol can be violated). We achieve this goal by appropriately sharing the CDS messages among the parties.

Ad-hoc PSM. We also study private simultaneous messages (PSM) protocols, which is a minimal model of secure multiparty computation protocols. In a PSM protocol there are k parties and a referee; each party holds a private input x_i and sends one message to the referee without seeing the messages of the other parties. The referee should learn the output of a pre-defined function $f(x_1, \ldots, x_n)$ without learning any additional information on the inputs. We use the ideas of the last transformation to construct ad-hoc PSM protocols (a primitive introduced in [14]), i.e., PSM protocols in which only a subset of the parties will compute a function on their inputs. We show that if a function f has a k-party PSM protocol with complexity C, then it has a k-out-of-n ad-hoc PSM protocol with complexity $O(knC)$.

1.2 Related Work

Constructions of secret-sharing schemes. Shamir [38] and Blakley [18] showed that threshold access structures can be realized by linear secret-sharing schemes, in which the size of every share is the maximum between the $\log n$ and the secret size. Ito, Saito, and Nishizeki constructed secret-sharing schemes for general access structures in which the share size is proportional either to the DNF or CNF representation of the access structure. Benaloh and Leichter [16] showed that access structures that can be described by small monotone formulas can be realized by efficient secret-sharing schemes. Later, Karchmer and Wigderson [33] showed that access structures that can be described by small monotone span programs can also be realized by efficient secret-sharing schemes. Bertilsson and

Ingemarsson [17] presented multi-linear secret-sharing schemes for general access structures. All the above schemes have share size $2^{n-O(\log n)}$. This was recently improved in [34] (as we have already explained).

Secret-sharing schemes for uniform access structures. Secret-sharing schemes for 2-uniform access structures were first introduced by Sun and Shieh [41]. Such schemes are called schemes for prohibited or forbidden graphs. 2-uniform access structures were studied in many papers, such as [2,3,9–11,13,27,35,36]. Beimel et al. [11] proved that every 2-uniform access structure can be realized by a (non-linear) secret-sharing scheme in which the share size of every party is $O(n^{1/2})$. Later, Gay et al. [27] presented linear secret-sharing schemes for such access structures with the same share size. Liu et al. [35] constructed non-linear secret-sharing scheme for 2-uniform access structures in which the share size of every party is $2^{O(\sqrt{\log n \log \log n})} = n^{o(1)}$. The notion of k-uniform access structures was explicitly introduced by [2,12] and was implicit in the work of [34]. By combining the CDS protocol of [36] and transformations of [2,12], we obtain that every k-uniform access structure can be realized by a secret-sharing scheme in which the share size of every party is $\min\left\{2^{O(k)+\tilde{O}(\sqrt{k \log n})}, 2^{\tilde{O}(\sqrt{n})}\right\}$. Applebaum and Arkis [2] (extending the work of Applebaum et al. [3]) showed a secret-sharing scheme for k-uniform access structures for long secrets, in which the share size of every party is $O(e^k)$ times the secret size (for long secrets). Recently, Beimel and Peter [13] proved that every k-uniform access structure can be realized by a linear secret-sharing scheme in which the share of every party is $\min\left\{(O(n/k))^{(k-1)/2}, O(n \cdot 2^{n/2})\right\}$.

Conditional disclosure of secrets (CDS) Protocols. Our constructions, described in Sect. 1.1, start from CDS protocols and transform them to secret-sharing schemes. In a conditional disclosure of secrets protocol, there are k parties and a referee; each party holds a private input, a common secret, and a common random string. The referee holds all private inputs but, prior to the protocol, it does not know neither the secret nor the random string. The goal of the protocol is that the referee will learn the secret if and only if the inputs of the parties satisfy some pre-defined condition (e.g., all inputs are equal). The challenge is that the communication model is minimal – each party sends one message to the referee, without seeing neither the inputs of the other parties nor their inputs.

CDS protocols were introduced by Gertner et al. [28], who presented a linear k-party CDS protocol for k-input functions $f : [N]^k \rightarrow \{0,1\}$ with message size $O(N^k)$. CDS protocols are used in the constructions of many cryptographic protocols, for example, symmetrically-private information retrieval protocols [28], attribute based encryption [7,27,44], and priced oblivious transfer [1].

CDS protocols have been studied in many papers [2,3,9,10,12,13,27,31,35, 36]. In the last few years there were dramatic improvements in the message size of CDS protocols. For a function $f : [N]^k \rightarrow \{0,1\}$, the message size in the best known CDS protocols is as follows: (1) For a binary secret, the message size is $2^{\tilde{O}(\sqrt{k \log N})}$ [36]. (2) For long secrets (of size at least 2^{N^k-1}), the message size is

4 times the size of the secret [2]. (3) For a binary secret, there is a linear CDS protocol with message size $O(N^{(k-1)/2})$ [13,36]. The best known lower-bound for general CDS protocol is $\Omega(\log N)$ [3,5,6].

Private simultaneous messages (PSM) Protocols. The model of k-party PSM protocols for k-input functions $f:[N]^k \rightarrow \{0,1\}$ was first introduced by Feige et al. [26], for $k=2$, and was generalized to any k in [26,30]. In [26], it was shown that every 2-input function has a 2-party PSM protocol with message size $O(N)$. Beimel et al. [11] improved this result by presenting a 2-party PSM protocol with messages size $O(N^{1/2})$. The best known lower bound for such 2-party PSM protocol is $3\log N - O(\log\log N)$ [5,26]. It was shown by Beimel et al. [12] that there exists a k-party PSM protocol with message size $O(k^3 \cdot N^{k/2})$.

Ad-hoc PSM protocols were presented by Beimel et al. in [14]. They showed that if there is a k-party PSM protocol for a symmetric function f with message size C, then there is a k-out-of-n ad-hoc PSM protocol for f with message size $O(k^3 \cdot e^k \cdot \log n \cdot C)$. Thus, by the PSM protocol of [12], there is a k-out-of-n ad-hoc PSM protocol for every symmetric function with message size $O(k^6 \cdot e^k N^{k/2} \cdot \log n)$. In [14], they also showed that if there is a n-party PSM protocol for a function f' related to f, with message size C, then there is a k-out-of-n ad-hoc PSM protocol for f with message size $n \cdot C$. This construction implies, in particular, that ad-hoc PSM protocols with poly(n)-communication exist for NC1 and different classes of log-space computation.

2 Preliminaries

Secret-Sharing Schemes. We present the definition of secret-sharing schemes, similar to [8,20].

Definition 2.1 (Access Structures). *Let $P=\{P_1,\dots,P_n\}$ be a set of parties. A collection $\Gamma \subseteq 2^P$ is* monotone *if $B \in \Gamma$ and $B \subseteq C$ imply that $C \in \Gamma$.* An access structure *is a monotone collection $\Gamma \subseteq 2^P$ of non-empty subsets of P. Sets in Γ are called* authorized, *and sets not in Γ are called* unauthorized. *The family of minimal authorized subsets is denoted by $\min\Gamma$. We represent a subset of parties $A \subseteq P$ by its characteristic string $x_A=(x_1,\dots,x_k) \in \{0,1\}^n$, where for every $j \in [n]$ it holds that $x_j=1$ if and only if $P_j \in A$. For an access structure Γ, we define the function $f_\Gamma:\{0,1\}^n \rightarrow \{0,1\}$, where for every subset of parties $A \subseteq P$, it holds that $f_\Gamma(x_A)=1$ if and only if $A \in \Gamma$.*

Definition 2.2 (Secret-Sharing Schemes). *A* secret-sharing scheme *with domain of secrets S is a pair $\Sigma=\langle \Pi,\mu\rangle$, where μ is a probability distribution on some finite set R called the set of random strings and Π is a mapping from $S \times R$ to a set of n-tuples $S_1 \times S_2 \times \cdots \times S_n$, where S_j is called the* domain of shares *of P_j. A dealer distributes a secret $s \in S$ according to Σ by first sampling a random string $r \in R$ according to μ, computing a vector of* shares *$\Pi(s,r)=(s_1,\dots,s_n)$, and privately communicating each share s_j to party P_j. For a set $A \subseteq P$, we* *denote* $\Pi_A(s,r)$ *as the restriction of* $\Pi(s,r)$ *to its* A*-entries (i.e., the shares of the parties in* A*).*

Given a secret-sharing scheme Σ*, define the* size *of the secret as* $\log|S|$*, the* share size *of party* P_j *as* $\log|S_j|$*, the* max share size *as* $\max_{1\leq j\leq n}\{\log|S_j|\}$*, and the* total share size *as* $\sum_{j=1}^{n}\log|S_j|$*.*

Let S *be a finite set of secrets, where* $|S|\geq 2$*. A secret-sharing scheme* $\Sigma = \langle\Pi,\mu\rangle$ *with domain of secrets* S *realizes an access structure* Γ *if the following two requirements hold:*

Correctness. *The secret* s *can be reconstructed by any authorized set of parties. That is, for any set* $B=\{P_{i_1},\ldots,P_{i_{|B|}}\}\in\Gamma$ *there exists a* reconstruction function $\mathrm{Recon}_B : S_{i_1}\times\cdots\times S_{i_{|B|}}\rightarrow S$ *such that for every secret* $s\in S$ *and every random string* $r\in R$*,* $\mathrm{Recon}_B(\Pi_B(s,r))=s$*.*

Privacy. *Every unauthorized set cannot learn anything about the secret from its shares. Formally, there exists a randomized function* Sim*, called the* simulator*, such that for any set* $T=\{P_{i_1},\ldots,P_{i_{|T|}}\}\notin\Gamma$*, every secret* $s\in S$*, and every vector of shares* $(s_{i_1},\ldots,s_{i_{|T|}})\in S_{i_1}\times\cdots\times S_{i_{|T|}}$*,*

$$\Pr[\,\textsc{Sim}(T)=(s_{i_1},\ldots,s_{i_{|T|}})\,]=\Pr[\,\Pi_T(s,r)=(s_{i_1},\ldots,s_{i_{|T|}})\,],$$

where the first probability is over the randomness of the simulator Sim *and the second probability is over the choice of* r *from* R *at random according to* μ*.*

A scheme is linear if the mapping that the dealer uses to generate the shares that are given to the parties is linear, as we formalize at the following definition.

Definition 2.3 (Linear Secret-Sharing Schemes). *Let* $\Sigma=\langle\Pi,\mu\rangle$ *be a secret-sharing scheme with domain of secrets* S*, where* μ *is a probability distribution on a set* R *and* Π *is a mapping from* $S\times R$ *to* $S_1\times S_2\times\cdots\times S_n$*. We say that* Σ *is a* linear secret-sharing scheme over a finite field $\mathbb{F}$ *if* $S=\mathbb{F}$*, the sets* $R,S_1,\ldots,S_n$ *are vector spaces over* $\mathbb{F}$*,* Π *is an* $\mathbb{F}$*-linear mapping, and* μ *is the uniform probability distribution over* R*.*

Definition 2.4 (Uniform Access Structures). *Let* $P=\{P_1,\ldots,P_n\}$ *be a set of parties. An access structure* $\Gamma\subseteq 2^P$ *is a* k*-uniform access structure, where* $1\leq k\leq n$*, if all sets of size less than* k *are unauthorized, all sets of size greater than* k *are authorized, and each set of size exactly* k *can be either authorized or unauthorized.*

Definition 2.5 (Threshold Secret-Sharing Schemes). *Let* Σ *be a secret-sharing scheme on a set of* n *parties* P*. We say that* Σ *is a* t-out-of-n secret-sharing scheme *if it realizes the access structure* $\Gamma_{t,n}=\{A\subseteq P:|A|\geq t\}$*.*

Claim 2.6 ([38])**.** *For every set of* n *parties* P *and for every* $t\in[n]$*, there is a linear* t*-out-of-*n *secret-sharing scheme realizing* $\Gamma_{t,n}\subseteq 2^P$ *for secrets of size* ℓ *in which the share size of every party is* $\max\{\ell,\log n\}$*.*

Fact 2.7 ([16]). *Let $\Gamma_1, \ldots, \Gamma_t$ be access structures over the same set of n parties, and let $\Gamma = \Gamma_1 \cup \cdots \cup \Gamma_t$ and $\Gamma' = \Gamma_1 \cap \cdots \cap \Gamma_t$. If there exist secret-sharing schemes with share size at most k realizing $\Gamma_1, \ldots, \Gamma_t$, then there exist secret-sharing schemes realizing Γ and Γ' with share size at most kt. If the former schemes are linear over a finite field $\mathbb{F}$, then there exist linear secret-sharing schemes over $\mathbb{F}$ realizing Γ and Γ' with share size at most kt.*

Conditional Disclosure of Secrets Protocols. We next define k-party conditional disclosure of secrets (CDS) protocols, first presented in [28]. For more details, see [4].

Definition 2.8 (Conditional Disclosure of Secrets Protocols – Syntax and Correctness). *Let $f : X_1 \times \cdots \times X_k \to \{0,1\}$ be some k-input function. A k-party* CDS *protocol $\mathcal{P}$ for f with domain of secrets S consists of: (1) A finite domain of common random strings R, and k finite message domains $M_1, \ldots, M_k$, (2) Deterministic message computation functions $\text{Enc}_1, \ldots, \text{Enc}_k$, where $\text{Enc}_i : X_i \times S \times R \to M_i$ for every $i \in [k]$, and (3) A deterministic reconstruction function $\text{Dec} : X_1 \times \cdots \times X_k \times M_1 \times \cdots \times M_k \to \{0,1\}$. We say that a CDS protocol $\mathcal{P}$ is correct (with respect to f) if for every inputs $(x_1, \ldots, x_k) \in X_1 \times \cdots \times X_k$ for which $f(x_1, \ldots, x_k) = 1$, every secret $s \in S$, and every common random string $r \in R$, $\text{Dec}(x_1, \ldots, x_k, \text{Enc}_1(x_1, s, r), \ldots, \text{Enc}_k(x_k, s, r)) = s$.*

The message size of a CDS protocol $\mathcal{P}$ is defined as the size of largest message sent by the parties, i.e., $\max_{1 \leq i \leq k} \{\log |M_i|\}$.

Definition 2.9 (Conditional Disclosure of Secrets Protocols – Privacy). *We say that a CDS protocol $\mathcal{P}$ is private (with respect to f) if there exists a randomized function Sim, called the* simulator, *such that for every inputs $(x_1, \ldots, x_k) \in X_1 \times \cdots \times X_k$ for which $f(x_1, \ldots, x_k) = 0$, every secret $s \in S$, and every k messages $(m_1, \ldots, m_k) \in M_1 \times \cdots \times M_k$, the probability that $\text{Sim}(x_1, \ldots, x_k) = (m_1, \ldots, m_k)$ is equal to the probability that $(\text{Enc}_1(x_1, s, r), \ldots, \text{Enc}_k(x_k, s, r)) = (m_1, \ldots, m_k)$ where the first probability is over the randomness of the simulator Sim and the second probability is over the choice of r from R with uniform distribution (the same r is chosen for all encryptions).*

Private Simultaneous Messages Protocols. We next define k-party ad-hoc private simultaneous messages (PSM) protocols, as presented in [14]. For more details, see [4].

Definition 2.10 (Ad-hoc Private Simultaneous Messages Protocols – Syntax and Correctness). *Let $P = \{P_1, \ldots, P_n\}$ be a set of parties and let $f : X^k \to Y$ be some k-input function. A k-*out-of-*n* ad-hoc PSM *protocol $\mathcal{P}$ for f consists of: (1) A finite domain of common random strings R, and a finite message domain M, (2) Deterministic message computation functions $\text{Enc}_1, \ldots, \text{Enc}_n$, where $\text{Enc}_i : X \times R \to M$ for every $i \in [n]$, and (3) A deterministic reconstruction function $\text{Dec} : \binom{P}{k} \times M^k \to Y$. We say that an ad-hoc*

PSM protocol $\mathcal{P}$ is correct (with respect to f) if for any set $A = \{P_{i_1}, \ldots, P_{i_k}\} \in \binom{P}{k}$, every inputs $(x_{i_1}, \ldots, x_{i_k}) \in X^k$, and every common random string $r \in R$, $\textsc{Dec}(A, \textsc{Enc}_{i_1}(x_{i_1}, r), \ldots, \textsc{Enc}_{i_k}(x_{i_k}, r)) = f(x_{i_1}, \ldots, x_{i_k})$.

The message size of an ad-hoc PSM protocol $\mathcal{P}$ is the size of the messages sent by each of the parties, i.e., $\log |M|$.

Definition 2.11 (Ad-hoc Private Simultaneous Messages Protocols – Privacy). *We say that an ad-hoc PSM protocol $\mathcal{P}$ is private (with respect to f) if:*

- *There exists a randomized function* $\textsc{Sim}$*, called a* simulator*, such that for every $A = \{P_{i_1}, \ldots, P_{i_k}\} \in \binom{P}{k}$, every inputs $(x_{i_1}, \ldots, x_{i_k}) \in X^k$, and every k messages $(m_{i_1}, \ldots, m_{i_k}) \in M^k$,*
$$\begin{aligned}&\Pr[\,\textsc{Sim}(A, f(x_{i_1}, \ldots, x_{i_k})) = (m_{i_1}, \ldots, m_{i_k})\,]\\&= \Pr[\,(\textsc{Enc}_{i_1}(x_{i_1}, r), \ldots, \textsc{Enc}_{i_k}(x_{i_k}, r)) = (m_{i_1}, \ldots, m_{i_k})\,],\end{aligned}$$
where the first probability is over the randomness of the simulator $\textsc{Sim}$ *and the second probability is over the choice of r from R with uniform distribution (the same r is chosen for all encryptions).*
- *There exists a randomized function* $\textsc{Sim}'$*, called a* simulator*, such that for every $k' < k$, every $A' = \{P_{i_1}, \ldots, P_{i_{k'}}\} \in \binom{P}{k'}$, every inputs $(x_{i_1}, \ldots, x_{i_{k'}}) \in X^{k'}$, and every k' messages $(m_{i_1}, \ldots, m_{i_{k'}}) \in M^{k'}$,*
$$\begin{aligned}&\Pr[\,\textsc{Sim}'(A') = (m_{i_1}, \ldots, m_{i_{k'}})\,]\\&= \Pr[\,(\textsc{Enc}_{i_1}(x_{i_1}, r), \ldots, \textsc{Enc}_{i_{k'}}(x_{i_{k'}}, r)) = (m_{i_1}, \ldots, m_{i_{k'}})\,],\end{aligned}$$
where the first probability is over the randomness of the simulator $\textsc{Sim}'$ *and the second probability is over the choice of r from R with uniform distribution (the same r is chosen for all encryptions).*

A PSM protocol is a k-out-k ad-hoc PSM protocol, where the privacy requirement only holds for sets of size k (we do not require that a referee that gets messages from less than k parties will not learn any information).

Notation. We denote the logarithmic function with base 2 and base e by log and ln, respectively. Additionally, we use the notation $[n]$ to denote the set $\{1, \ldots, n\}$. For $0 \leq \alpha \leq 1$, we denote the binary entropy of α by $h(\alpha) \stackrel{\text{def}}{=} -\alpha \log \alpha - (1-\alpha)\log(1-\alpha)$. Next, we present an approximation of the binomial coefficients.

Fact 2.12. *For every k and every n such that $k \in [n]$, it holds that $\binom{n}{k} = \Theta(k^{-1/2} \cdot 2^{h(k/n)n})$.*

3 Secret-Sharing Schemes Realizing General Access Structures from CDS Protocols

In this section we present a construction of secret-sharing schemes for a general access structure. The starting point of our results is a work by Liu and Vaikuntanathan [34], in which they presented the first general construction with share size $O(2^{cn})$ with a constant c smaller than 1. In the first part of the section, we give an outline of the construction in [34], presenting their results in terms of access structures. Our main result, is the following theorem.

Theorem 3.1. *Every access structure over n parties can be realized by a secret-sharing scheme with a total share size of $2^{0.892n+o(n)}$ and by a linear secret-sharing scheme with a total share size of $2^{0.942n+o(n)}$.*

We say that an access structure Γ can be *realized with an exponent of S* (resp., *linearly realized with an exponent of S*) if Γ can be realized by a secret-sharing scheme (resp., linear secret-sharing scheme) with shares of size at most $2^{Sn+o(n)}$ where n is the number of participants.[1]

3.1 Our Construction

Following Liu and Vaikuntanathan [34], we decompose an access structure Γ to three parts: a bottom part (that handles small sets), middle part (that handles medium-size sets) and a top part (that handles large sets). Formally, we have the following proposition.

Proposition 3.2 ([34]). *For every access structure Γ over a set of n participants, and every slice $\delta \in (0, \frac{1}{2})$, define the following access structures over the same set of participants.*

$$\Gamma_{\text{bot}} : A \in \Gamma_{\text{bot}} \;\; \textit{iff} \;\; \exists A' \in \Gamma \textit{ s.t. } A' \subseteq A \textit{ and } |A'| \leq (\frac{1}{2} - \delta)n,$$

$$\Gamma_{\text{mid}} : A \in \Gamma_{\text{mid}} \;\; \textit{iff} \;\; A \in \Gamma \textit{ and } (\frac{1}{2} - \delta)n \leq |A| \leq (\frac{1}{2} + \delta)n, \textit{ or } |A| \geq (\frac{1}{2} + \delta)n$$

$$\Gamma_{\text{top}} : A \notin \Gamma_{\text{top}} \;\; \textit{iff} \;\; \exists A' \notin \Gamma \textit{ s.t. } A \subseteq A' \textit{ and } |A'| \geq (\frac{1}{2} + \delta)n.$$

Then $\Gamma = \Gamma_{\text{top}} \cap (\Gamma_{\text{mid}} \cup \Gamma_{\text{bot}})$. Consequently, if $\Gamma_{\text{top}}, \Gamma_{\text{mid}}$, and Γ_{bot} can be realized (resp., linearly realized) with exponent of S then so is Γ.

The "consequently" part follows from standard closure properties of secret-sharing schemes (see Fact 2.7). Thus realizing Γ reduces to realizing Γ_{top}, Γ_{bot}, and Γ_{mid}. The main work in [34] is devoted to realizing the access structure Γ_{mid}. Their main construction can be summarized as follows.

Lemma 3.3 ([34]). *For every access structure Γ and every slice parameter $\delta \in (0, \frac{1}{2})$, the access structure Γ_{mid} can be realized with an exponent of $\mathbf{M}(\delta) = h(0.5 - \delta) + 0.2h(10\delta) + 10\delta - 0.2\log(10)$, and can be linearly realized with an exponent of $\mathbf{M}_\ell(\delta) = h(0.5 - \delta) + 0.2h(10\delta) + 10\delta - 0.1\log(10)$.*[2]

[1] Formally, such a statement implicitly refers to an infinite sequence of (collections of) access structures that is parameterized by the number of participants n.

[2] The notation $\mathbf{M}$ stands for "middle".

The extreme slices. Liu and Vaikuntanathan [34] realized Γ_{top} and Γ_{bot} with an exponent of $h(\frac{1}{2}+\delta)$ by exploiting the fact that the number of authorized (or non-authorized) sets is exponential in $h(\frac{1}{2}+\delta)$. (The actual implementation is based on the classical schemes of [32].) We show that the nice structure of these access structures can be further exploited.

In particular, for a *covering parameter* α, the minimal authorized sets of Γ_{bot} can be covered by exponentially-many αn-subsets of n. (A dual statement applies to the maximal unauthorized sets of Γ_{top}.) This property allows us to realize Γ_{bot} and Γ_{top} by decomposing each of them into (exponentially) many access structures over αn parties and realizing each access structure via a general secret-sharing scheme. Overall, we get a tradeoff between the size of the decomposition (i.e., number of sub-access structures) and the number of players αn in each part. Formally, in Sect. 3.2 we prove the following statement.

Lemma 3.4. *Suppose that every access structure can be realized (resp., linearly realized) with an exponent of S. Then, for every covering parameter $\alpha \in (\frac{1}{2}, 1)$, every access structure Γ and every slice parameter $\delta \in (0, \frac{1}{2})$, the access structures Γ_{top} and Γ_{bot} can be realized (resp., linearly realized) with an exponent of* $\mathbf{X}(S, \delta, \alpha) \stackrel{def}{=} \alpha S + h(0.5-\delta) - h\left((0.5-\delta)/\alpha\right)\alpha$.[3]

By combining Lemmas 3.3 and 3.4 with Proposition 3.2, we derive the following Theorem.

Theorem 3.5. *Suppose that every access structure can be realized (resp., linearly realized) with an exponent of S (resp., S_ℓ). Then, for every covering parameter $\alpha \in (\frac{1}{2}, 1)$ and slice parameter $\delta \in (0, \frac{1}{2})$, every access structure can be realized with an exponent of* $\max\left(\mathbf{M}(\delta), \mathbf{X}(S, \delta, \alpha)\right)$, *and can be linearly realized with an exponent of* $\max\left(\mathbf{M}_\ell(\delta), \mathbf{X}(S_\ell, \delta, \alpha)\right)$.

We can improve the secret sharing exponent by applying Theorem 3.5 recursively as follows. Start with the Liu-Vaikuntanathan bound $\mathbf{S}_{\text{LV}} = 0.994$ as an initial value, and iterate with carefully chosen values for δ and α.

Example 3.6. Consider a single application of Theorem 3.5 starting with $\mathbf{S}_{\text{LV}} = 0.994$ and taking $\delta = 0.037$ and $\alpha = 0.99$. In this case, $\mathbf{M}(\delta) < 0.897$ and $\mathbf{X}(\mathbf{S}_{\text{LV}}, \delta, \alpha) < 0.9931$, thus we get an exponent smaller than 0.9931.

Since each step of the recursion is parameterized by both δ and α, the problem of finding the best choice of parameters in every step of the recursion becomes a non-trivial optimization problem. In Sect. 3.3, we analyze the recursive process and derive an analytic expression for the infimum of the process (over all sequences of (δ_i, α_i)). This leads to a general scheme with an exponent of 0.897 and a linear scheme with an exponent of 0.955. Finally, an additional (minor) improvement is obtained by analyzing a low-level optimization to the middle slice that was suggested by [34] (see Sect. 3.4). This leads to Theorem 3.1.

[3] The notation **X** stands for eXternal slices.

3.2 Realizing Γ_{bot} and Γ_{top} (Proof of Lemma 3.4)

We start by introducing a fact about *combinatorial covering designs* by Erdos and Spenser:

Fact 3.7 ([25]). *Let P be a set of size n. For every positive integers $c \leq a \leq n$, there exists a family $\mathcal{G} = \{G_i\}_{i=1}^{L}$ of a-subsets of P, such that every c-subset of P is contained in at least one member of $\mathcal{G}$, and $L = L(n, a, c) = O((\binom{n}{c}\log\binom{a}{c})/\binom{a}{c})$.*

We next prove Lemma 3.4.

Proof (of Lemma 3.4). Let $a = \alpha n$ and $c = (0.5-\delta)n$ and let $\mathcal{G} = \{G_i\}_{i\in[L]}$ be the family of a-subsets of $P = \{P_1, \ldots, P_n\}$ promised by Fact 3.7. Using Fact 2.12, the number of sets L satisfies $\log L \leq n(h(0.5-\delta) - h((0.5-\delta)/\alpha)\alpha + o(1))$. Hence, to prove the lemma it suffices to realize Γ_{bot} and Γ_{top} with share size of $L \cdot 2^{S\alpha n + o(n)}$. Towards this end, we decompose Γ_{bot} and Γ_{top} according to $\mathcal{G}$ as follows.

Γ_{bot}: Let $\mathcal{T}$ be the set of minimal authorized sets of Γ_{bot}. Recall that all these sets are of size of at most c. For every $i \in [L]$, let $\mathcal{T}_i = \{T \in \mathcal{T} : T \subseteq G_i\}$, and let Γ_i be the access structure whose minimal authorized sets are the sets in $\mathcal{T}_i$. By Fact 3.7, $\mathcal{T} = \bigcup \mathcal{T}_i$ and therefore $\Gamma_{\text{bot}} = \bigcup_{i\in[L]} \Gamma_i$. Indeed, both in the RHS and in the LHS, A is an authorized set iff there exists some T in $\mathcal{T} = \bigcup \mathcal{T}_i$ such that $T \subseteq A$. We further note that every minimal authorized set in Γ_i is a subset of G_i and therefore Γ_i can be implemented as an access structure over αn parties with share size of $2^{S\alpha n + o(n)}$. To share a secret s according to $\Gamma_{\text{bot}} = \bigcup_{i\in[L]} \Gamma_i$, for every $i \in [L]$ independently share s via the scheme of Γ_i. The share size of the resulting scheme realizing Γ_{bot} is $L \cdot 2^{S\alpha n + o(n)}$, as required.

Γ_{top}: We use a dual construction for Γ_{top}. Let $\mathcal{T}'$ be the set of maximal unauthorized sets of Γ_{top}. Recall that all these sets are of size at least $n - c$. For every $i \in [L]$, let $\mathcal{T}'_i = \{T \in \mathcal{T}' : \overline{G_i} \subseteq T\} = \{T \in \mathcal{T}' : \overline{T} \subseteq G_i\}$ and let Γ'_i be the access structure whose maximal unauthorized sets are the sets in $\mathcal{T}_i$. By Fact 3.7, $\mathcal{T}' = \bigcup \mathcal{T}'_i$ and therefore $\Gamma_{\text{top}} = \bigcap_{i\in[L]} \Gamma'_i$. Indeed, both in the RHS and in the LHS, A is an unauthorized set iff $A \subseteq T$ for some T in $\mathcal{T}' = \bigcup \mathcal{T}'_i$. We further note that all minimal authorized sets of Γ'_i are subsets of $\bigcap_{T\in\mathcal{T}'_i} \overline{T} \subseteq G_i$, and therefore Γ_i can be implemented as an access structure over αn parties with share size of $2^{S\alpha n + o(n)}$. To share a secret s according to Γ_{top}, sample L random elements $s_1, \ldots, s_L$ in the domain of s satisfying $s = s_1 + \ldots + s_L$, and share s_i via the scheme for Γ_i. A set A can reconstruct the secret iff it can reconstruct each s_i iff $A \in \Gamma'_i$ for every i iff $A \in \bigcap_{i\in[L]} \Gamma'_i = \Gamma_{\text{top}}$. Thus we can realize Γ_{top} with share size of $L \cdot 2^{(\alpha S + o(1))n} = 2^{(\alpha S + h(0.5-\delta) - h((0.5-\delta)/\alpha)\alpha + o(1))n}$, as required. $\square$

3.3 Analyzing the Recursion

In this section, we analyze the exponent achievable by repeated applications of Theorem 3.5 by considering the following single-player game.

The exponent game. The goal of the player is to minimize a positive number S. The value of S is initialized to the LV-exponent 0.994, and can be updated by making an arbitrary number of moves. In each move the player can choose $\delta \in (0, \frac{1}{2})$ and $\alpha \in (\frac{1}{2}, 1)$, if $S < \max(\mathbf{X}(S, \delta, \alpha), \mathbf{M}(\delta))$, update S to $\max(\mathbf{X}(S, \delta, \alpha), \mathbf{M}(\delta))$; otherwise, S remains unchanged.

Recall that the function $\mathbf{X}(S, \delta, \alpha)$ represents the exponent of the external slices and the function $\mathbf{M}(\delta)$ represents the exponent of middle slice. We denote by **opt** the infimum of S over all finite sequences of (δ_i, α_i). Our goal is to determine **opt**. A (δ, α)-move improves S if and only if (1) $\mathbf{X}(S, \delta, \alpha) < S$ and (2) $\mathbf{M}(\delta) < S$. If the first condition holds we say that S is X-improved by (δ, α). We begin by showing that the question of whether a given S can be X-improved by a (δ, α)-move depends only on δ and S (and is independent of α and n).

Lemma 3.8. *Fix a parameter $\delta \in (0, \frac{1}{2})$ and let $\mathbf{X}'(\delta) \stackrel{def}{=} h(0.5 - \delta) - (0.5 - \delta) \cdot \log((0.5 + \delta)/(0.5 - \delta))$.*

- *If $S \leq \mathbf{X}'(\delta)$, then there does not exist any α for which $S > \mathbf{X}(S, \delta, \alpha)$.*
- *For every $S' > \mathbf{X}'(\delta)$ there exists an $\alpha < 1$ such that $\mathbf{X}(S, \delta, \alpha) \leq \alpha S + (1 - \alpha)S'$ for every $S > \mathbf{X}'(\delta)$.*

Proof. Fix some S. The exponent S is X-improved by (δ, α) if and only if

$$\frac{h(0.5 - \delta) - h\left(\frac{0.5-\delta}{\alpha}\right)\alpha}{1 - \alpha} < S. \tag{1}$$

Denote the LHS by $\mathbf{X}'(\delta, \alpha)$. Clearly, S can be X-improved by (δ, α) if and only if S is larger than $\inf_\alpha(\mathbf{X}'(\delta, \alpha))$ (assuming that the infimum exists). We next show that $\inf_\alpha(\mathbf{X}'(\delta, \alpha)) = \mathbf{X}'(\delta)$. Indeed, for any fixed δ, the function $\mathbf{X}'(\delta, \alpha)$ is monotonically decreasing with α, and since $\alpha < 1$, we get that $\inf_\alpha(\mathbf{X}'(\delta, \alpha)) = \lim_{\alpha \to 1} \frac{h(0.5-\delta) - h((0.5-\delta)/\alpha)\alpha}{1-\alpha}$, which by l'Hôpitals Rule, simplifies to $\mathbf{X}'(\delta)$. The first item of the claim follows.

For the second item, take any $\alpha \in (0, 1)$ such that $\frac{h(0.5-\delta) - h\left(\frac{0.5-\delta}{\alpha}\right)\alpha}{1-\alpha} \leq S'$ (by the definition of the limit and since $S' > \mathbf{X}'(\delta)$, such α exists). Thus, $\mathbf{X}(S, \delta, \alpha) = \alpha S + h(0.5 - \delta) - h((0.5 - \delta)/\alpha)\alpha \leq \alpha S + (1 - \alpha)S'$. Note that the choice of α is independent of S (as long as $S > \mathbf{X}'(\delta)$). □

Lemma 3.8 takes into account only the effect of the outer slices, Γ_{top} and Γ_{bot}. Recall, however, that the cost of the medium slice Γ_{mid} prevents us from going below $\mathbf{M}(\delta)$. Let $\delta^\star \in (0, 0.5)$ denote the positive value that satisfies $\mathbf{X}'(\delta^\star) = \mathbf{M}(\delta^\star)$. Let us denote by $S^\star$ the value of $\mathbf{X}'(\delta^\star) = \mathbf{M}(\delta^\star)$. The curves of $\mathbf{M}(\delta)$ and $\mathbf{X}(\delta)$ are depicted in Fig. 1, and $\delta^\star \approx 0.037, S^\star \approx 0.897$. The following two claims show that the infimum of the game, **opt**, equals to $S^\star$. Overall, we get that **opt** $= S^\star$, which is about 0.897, and Theorem 3.11. The proof of Claim 3.10 is deferred to [4].

Claim 3.9. *For every constant $S'' > S^\star$ there exists an $\alpha < 1$ and an integer i (where α and i are independent of n) such that a sequence of i (δ, α)-moves improve the exponent to S''.*

Proof. Choose any constant S' such that $S^\star < S' < S''$ and let α be a constant guaranteed by Lemma 3.8 for $\delta^\star$ and S', that is for every $S > S'$, the exponent S can be improved to $\alpha S + (1-\alpha)S'$. Furthermore, let $S_0 = 0.994$ be the exponent of the secret-sharing scheme of [34] and define $S_j = \alpha S_{j-1} + (1-\alpha)S'$ for every $j > 0$. By Lemma 3.8, the exponent S_j can be achieved after j (δ, α)-moves. By induction, $S_j = \alpha^j S_0 + (1-\alpha^j)S' < \alpha^j S_0 + S'$. Taking an integer i such that $\alpha^i \leq (S'' - S')/S_0$ completes the proof. □

Claim 3.10. *There is no (δ, α)-move that takes a value $A > S^\star$ to a value $B < S^\star$. Consequently, any finite number of steps ends in a value $S > S^\star$ and* $\mathbf{opt} \geq S^\star$.

Theorem 3.11. *Every access structure can be realized with share size* $2^{(0.897+o(1))n}$.

Remark 3.12. We note that our analysis holds even if the function $\mathbf{M}(\delta)$ is replaced with a different function that represents the exponent of the middle slice. That is, for any choice of $\mathbf{M}(\delta)$ the value of $\mathbf{opt}$ equals to $\inf_\delta \max(\mathbf{X}'(\delta), \mathbf{M}(\delta))$ (assuming that the initial starting point is over the $\mathbf{M}(\delta)$ curve).

In particular, the following theorem is obtained by replacing $\mathbf{M}(\delta)$ with the exponent $\mathbf{M}_\ell(\delta)$ for a linear realization of the middle layer (from Lemma 3.3).

Theorem 3.13. *Every access structure can be linearly realized with share size of* $2^{(0.955+o(1))n}$.

3.4 Minor Improvement of Share Size for Γ_{mid}

In this section we give a tighter analysis for the constructions of $\mathbf{M}(\delta)$ and $\mathbf{M}_\ell(\delta)$ from [34]. These ideas were suggested in [34], but were not implemented.

Lemma 3.14. *For every access structure Γ and every slice parameter $\delta \in (0, \frac{1}{2})$, the access structure Γ_{mid} can be realized with an exponent of* $\mathbf{M}(\delta) = h(0.5-\delta) + 0.2h(10\delta) + 2\log(26)\delta - 0.2\log(10)$, *and can be linearly realized with an exponent of* $\mathbf{M}_\ell(\delta) = h(0.5-\delta) + 0.2h(10\delta) + 2\log(26)\delta - 0.1\log(10)$.

The above expressions slightly improves over the ones obtained in Lemma 3.3. In particular, the third summand in both $\mathbf{M}(\delta)$ and $\mathbf{M}_\ell(\delta)$ is reduced from 10δ to $2\log(26)\delta$.

Proof. We assume familiarity with the construction of [34]. In the original analysis of reduction 4 in [34, Section 3.4], the expression 10δ is added to the exponent (of both $\mathbf{M}(\delta)$ and $\mathbf{M}_\ell(\delta)$) due to an enumeration over all possible subsets that are taken from a universe of size $10\delta n$. It is noted there that it actually suffices to enumerate only over subsets T that satisfy the following condition. For a given (fixed) partition of the universe to $2\delta n$ bins of size 5 each, the set T must contain at least 2 elements from each bin. The number of such sets is $\left(2^5 - \binom{5}{0} - \binom{5}{1}\right)^{2\delta n} = 2^{2\log(26)\delta n}$, and so the lemma follows. □

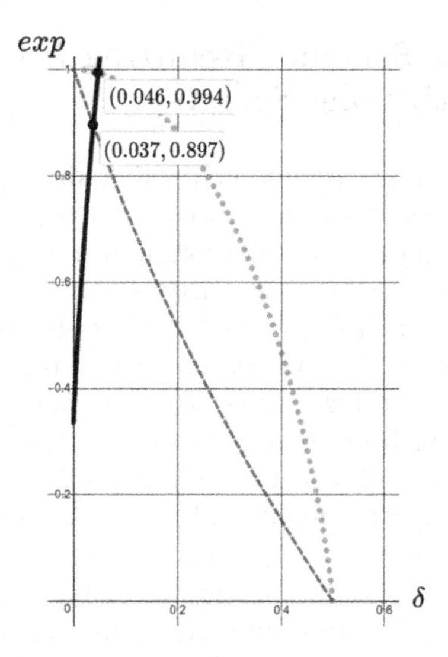

Fig. 1. A description of the functions $\mathbf{M}(\delta)$ and $\mathbf{X}'(\delta)$. The horizontal axis represents the value of δ and the vertical axis represents the resulting exponents. The solid black curve corresponds to the exponent $\mathbf{M}(\delta)$ of the middle slice Γ_{mid}, as defined in Lemma 3.3 (the minor improvements of Sect. 3.4 do not appear here). The function $\mathbf{X}'(\delta)$ appears as the dashed blue line. For comparison, we plot in the dotted red line the exponent that is achieved for Γ_{top} and Γ_{bot} via the simple (non-recursive) construction from [34]. Our exponent appears as the y-coordinate of the intersection of the black and blue curves, and the exponent of [34] appears at the y-coordinate of the intersection of the red and black curves. (Color figure online)

We can further improve the exponent of the linear scheme by reducing the last summand as follows.

Lemma 3.15. *For every access structure Γ and every slice parameter $\delta \in (0, \frac{1}{2})$, the access structure Γ_{mid} can be linearly realized with an exponent of* $\mathbf{M}_\ell(\delta) = h(0.5 - \delta) + 0.2h(10\delta) + 2\log(26)\delta - (0.1 + \delta)\log(10)$.

Proof. Again, we assume familiarity with the construction of [34]. The last reduction of [34, Section 3.5] utilizes a protocol for *conditional disclosure of secrets* (CDS) with an input size of $\binom{n/k}{a/k}^k$ for $a = \frac{1}{2} - \delta$ and $k = n/5$. As the authors note, for this choice of parameters, the input size of the CDS is actually $\binom{n/k}{a/k}^{k-2\delta}$. (In general, this holds whenever $\binom{n/k}{\lfloor a/k \rfloor} = \binom{n/k}{\lceil a/k \rceil}$.) This improvement becomes useful in the linear case (which employs linear CDS), and eventually it leads to the improvement stated in the lemma. □

Proof (proof of Theorem 3.1). As stated in Remark 3.12, the analysis of the recursive process holds when $\mathbf{M}(\delta)$ is updated to some $\mathbf{M}'(\delta)$, and the new game becomes $\mathbf{X}'(\delta^\star)$, where $\delta^\star$ satisfies $\mathbf{X}'(\delta^\star) = \mathbf{M}'(\delta^\star)$. By using the bounds obtained in Lemmas 3.14 and 3.15, we derive Theorem 3.1. □

4 Secret-Sharing Schemes Realizing k-Uniform Access Structures with Long Secrets

In this section, we present a construction of secret-sharing schemes for k-uniform access structures on n parties using k^2-party CDS protocols. Using the CDS protocols of [2] with long secrets, we obtain secret-sharing schemes for long secrets in which the share size of every party is only $O(k^2)$ times the secret size.

In [2], it was shown how to construct a secret-sharing scheme realizing any k-uniform access structure Γ in which the share size of every party is $O(e^k)$ times the message size, with big secrets. To construct this scheme, they used a family of perfect hash functions from $[n]$ to $[k]$, where each such function defines a k-uniform access structure, and use a k-party CDS protocol to realize every such access structure. Since the number of perfect hash functions for sets of size k and range of size k is bigger than e^k, each share in the secret-sharing scheme of [2] contains $O(e^k)$ messages of the CDS protocol. We improve their construction by taking a family of perfect hash functions from $[n]$ to $[k^2]$, and construct a secret-sharing scheme using a k^2-party CDS protocol for every function in this family, such that the resulting scheme realizes Γ.

The definition of a family of perfect hash functions is presented next.

Definition 4.1 (Families of Perfect Hash Functions). *A set of functions $H_{n,k,t} = \{h_i : [n] \to [t] : i \in [\ell]\}$ is a family of perfect hash functions if for every set $A \subseteq [n]$ such that $|A| = k$ there exists at least one index $i \in [\ell]$ such that $|h_i(A)| = |\{h_i(a) : a \in A\}| = k$, i.e., h_i restricted to A is one-to-one.*

To construct a secret-sharing scheme that realizes the k-uniform access structure Γ, we construct a scheme, using a CDS protocol for the function f (defined in Definition 4.2), that realizes the access structure Γ_h (defined in Definition 4.3), for every function h among a family of perfect hash functions.

Definition 4.2 (The Function f). *Let Γ be a k-uniform access structure with n parties. The k^2-input function $f : \{0, 1, \ldots, n\}^{k^2} \to \{0, 1\}$ is the function that satisfies $f(x_1, \ldots, x_{k^2}) = 1$ if and only if $\{P_{x_i} : i \in [k^2], x_i \neq 0\} \in \Gamma$.*

For example, if $k = 2$, $n = 5$, and the authorized sets of size $k = 2$ are exactly $\{P_1, P_2\}, \{P_3, P_5\}$, then $f(1, 3, 4, 0) = f(1, 2, 0, 0) = f(0, 2, 0, 1) = f(3, 0, 0, 5) = 1$ and $f(0, 0, 0, 0) = f(0, 2, 0, 0) = f(0, 2, 3, 0) = f(2, 0, 0, 5) = 0$.

Definition 4.3 (The Access Structure Γ_h). *Let Γ be a k-uniform access structure with n parties and let $h : [n] \to [k^2]$ be a function. The k-uniform access structure Γ_h is the access structure that contains all the subsets of parties of size greater than k, and all authorized subsets from Γ of size k such that h restricted to the indices of the parties of such subset is one-to-one. That is, $\Gamma_h = \{A \subseteq P : |A| > k\} \cup \{A \subseteq P : A \in \Gamma, |A| = k, \text{ and } |\{h(j) : P_j \in A\}| = k\}$.*

Using a simple probabilistic proof, we show that there exists a family of perfect hash function $H_{n,k,k^2} = \{h_1, \ldots, h_\ell\}$ with $\ell = \Theta(k \cdot \log n)$ functions. Moreover, if H_{n,k,k^2} is a family of perfect hash functions, then $\Gamma = \cup_{h \in H_{n,k,k^2}} \Gamma_h$.

Thus, constructing secret-sharing schemes realizing Γ_h for every $h \in H_{n,k,k^2}$, we get a secret-sharing scheme realizing Γ.

We start with a scheme that realizes the k-uniform access structure Γ_h, as defined in Definition 4.3; this scheme uses a CDS protocol for f, as defined in Definition 4.2. The scheme is described in Fig. 2; we next give an informal description. For every $j \in [n]$, we give the message of the $h(j)$th party in the CDS protocol when holding the input j, and for every $i \in [k^2]$, we share the message of the ith party in the CDS protocol when holding the input 0 using a k-out-of-k^2 scheme among the parties P_j for which $h(j) \neq i$.

Every authorized set $A \in \Gamma_h$ can reconstruct the secret, since every $P_j \in A$ gets the message of the $h(j)$th party in the CDS protocol when holding the input j, and the parties in A can reconstruct the messages of the other parties in the CDS protocol from the k-out-of-k^2 scheme, because $|\{h(j) : P_j \in A\}| = k$. Thus, the parties in A can reconstruct the secret since they hold messages for a 1-input of f.

Every unauthorized set $A \notin \Gamma$ that does not collides on h (that is, for every two different parties $P_j, P_{j'} \in A$ it must hold that $h(j) \neq h(j')$), the parties in A cannot learn any other messages except for the above mentioned messages. Thus, if $|A| = k$ then the parties in A hold messages for a 0-input of f, so by the privacy of the CDS protocol for f they cannot learn any information about the secret.

However, if A collides on h, then the parties in A hold two different messages of the same party in the CDS protocol for f, and CDS protocols cannot ensure any privacy in this scenario. To overcome this problem, we choose two random elements s_1, s_2 from the domain of secrets such that $s = s_1 + s_2$. We share s_1 using a k-out-of-k^2 scheme and give the $h(j)$th share to party P_j, and apply the above scheme using a CDS protocol for f with the secret s_2. Now, if A collides on h, the parties in A may learn information about s_2, but they cannot learn s_1, so the privacy of the scheme holds.

Lemma 4.4. *Let Γ be a k-uniform access structure with n parties, and let $h : [n] \to [k^2]$ be a function. Assume that for every k-input function $f : [N]^k \to \{0,1\}$ there is a k-party CDS protocol for f, for secrets of size t, in which the message size is $c(k, N, t)$. Then, the scheme Σ_h described in Fig. 2, is a secret-sharing scheme for secrets of size t realizing Γ_h in which the share size of every party is $O(\log n + k^2 \cdot c(k^2, n+1, t))$.*

Using a simple probabilistic argument, we show in Lemma 4.5 the existence of a family of perfect hash functions with a small number of functions (satisfying a stronger requirement than in Definition 4.1). The proofs of Lemmas 4.4 and 4.5 are deferred to the full version [4].

Lemma 4.5. *There exists a family of perfect hash functions $H_{n,k,k^2} = \{h_i : [n] \to [k^2] : i \in [\ell]\}$, where $\ell = \Theta(k \cdot \log n)$, such that for every subset $A \subseteq [n]$ there are at least $\ell/4$ functions $h \in H_{n,k,k^2}$ for which $|h(A)| = k$.*

Using a family of perfect hash function H_{n,k,k^2} as in Lemma 4.5 and the scheme of Lemma 4.4 for every function in H_{n,k,k^2}, we get the a scheme in which

The secret: An element $s \in S$.
The scheme: Let $f : \{0,1,\dots,n\}^{k^2} \to \{0,1\}$ be a k^2-input function as in Definition 4.2, and let $\mathcal{P}$ be a k^2-party CDS protocol for f.

1. Share the secret s using a $(k+1)$-out-of-n secret-sharing scheme among the parties in P, that is, for every $j \in [n]$, give the jth share from this scheme to party P_j.
2. Choose a random element $s_1 \in S$, and let $s_2 = (s_1 + s) \mod |S|$.
3. Share the element s_1 using a k-out-of-k^2 secret-sharing scheme. For every $j \in [n]$, we give the $h(j)$th share from this scheme to party P_j.
4. Apply the k^2-party CDS protocol $\mathcal{P}$ to the k^2-input function f with the secret s_2 and a common random string r that is chosen at random. For every $i \in [k^2]$, let $m_{i,x}$ be the message of the ith party in the CDS protocol $\mathcal{P}$ when holding the input $x \in \{0,1,\dots,n\}$.
5. For every $j \in [n]$, give the message $m_{h(j),j}$ to party P_j.
6. For every $i \in [k^2]$, share the message $m_{i,0}$ using a k-out-of-k^2 secret-sharing scheme. For every $j \in [n]$ such that $h(j) \neq i$, give the $h(j)$th share from this scheme to party P_j.

Fig. 2. A secret-sharing scheme Σ_h realizing the k-uniform access structure Γ_h.

the share size is $O(k^3 \cdot \log n)$ times the message size in the CDS protocol. Using Stinson's decomposition [40], we reduce the overhead.

Theorem 4.6. *Let Γ be a k-uniform access structure with n parties. Assume that for every k-input function $f : [N]^k \to \{0,1\}$ there is a k-party CDS protocol for f, for secrets of size t, in which the message size is $c(k, N, t)$. Then, for every $t' > \log n$, there is a secret-sharing scheme realizing Γ, for secrets of size $t = t' \cdot \Theta(k \cdot \log n)$, in which the share size of every party is $O(k^3 \cdot \log n \cdot c(k^2, n+1, t'))$.*

Proof. By Lemma 4.5, there exists a family of perfect hash functions $H_{n,k,k^2} = \{h_i : [n] \to [k^2] : i \in [\ell]\}$ with $\ell = \Theta(k \cdot \log n)$ functions, such that for every subset $A \subseteq [n]$ there is at least $\ell/4$ functions $h \in H_{n,k,k^2}$ for which $|h(A)| = k$. By Lemma 4.4, for every $i \in [\ell]$ there is a secret-sharing scheme Σ_{h_i} realizing the k-uniform access structure Γ_{h_i}, for secrets of size t, in which the share size of every party is $O(k^2 \cdot c(k^2, n+1, t))$. Also, by the definition of a family of perfect hash functions it holds that $\Gamma = \cup_{h \in H_{n,k,k^2}} \Gamma_h$.

To construct the desired secret-sharing scheme that realizes Γ, we use the Stinson's decomposition technique [40]. Let $\mathbb{F}$ be a finite field that contains at least $\max\{n, \ell\}$ elements. By an abuse of notation, we will assume that $\mathbb{F}$ is a prime field. Let $s = (s_1, \dots, s_{\ell/4}) \in \mathbb{F}^{\ell/4}$ be the secret. We use a $(0, \ell/4)$-ramp secret-sharing scheme (that is, a scheme in which every set of size $\ell/4$ can reconstruct the secret, while there are no requirements on smaller sets) to generate shares $s_1, \dots, s_\ell \in \mathbb{F}$ of s (that is, we choose a polynomial Q of degree $\ell/4 - 1$ such that $Q(i) = s_i$ for every $i \in [\ell/4]$ and define $s_i = Q(i)$ for every $i \in \{\ell/4+1, \dots, \ell\}$). Then, for every $1 \leq i \leq \ell$, we independently generate shares of s_i using the scheme Σ_{h_i} that realizes the k-uniform access structure

Γ_{h_i}, and give the shares to the parties in P. Since every set $A \subseteq P$ such that $|A| = k$ satisfies $|h_i(A)| = k$ for at least $\ell/4$ values of $i \in [\ell]$, every authorized set $A \in \Gamma$ such that $|A| = k$ can reconstruct at least $\ell/4$ values from $s_1, \dots, s_\ell$. Thus, by the property of the ramp scheme, the parties in A can reconstruct $s = (s_1, \dots, s_{\ell/4})$.

Finally, let $t' = \log |\mathbb{F}|$. The combined scheme is a secret-sharing scheme that realizes the access structure Γ, in which the share size of every party is $\ell \cdot O(k^2 \cdot c(k^2, n+1, t')) = O(k^3 \cdot \log n \cdot c(k^2, n+1, t'))$. □

Remark 4.7. In the secret-sharing scheme of Theorem 4.6, if we start with a linear or multi-linear CDS protocol, then we result with a multi-linear secret-sharing scheme (i.e., a scheme in which the secret is a vector over a finite field $\mathbb{F}$, the random string is a vector over $\mathbb{F}$ chosen with uniform distribution, and each share is a vector over $\mathbb{F}$, where every element in the vector is a linear combination of the secret and the random elements).

Using the multi-linear CDS protocol of [2] for long secrets, in which the message size is $O(t)$, for secrets of size t (for big enough t), we get the following result.

Corollary 4.8. *Let Γ be a k-uniform access structure with n parties. Then, there is a multi-linear secret-sharing scheme realizing Γ, for secrets of size $t = \Omega(k \cdot \log n \cdot 2^{(n+1)^{k^2}})$, in which the share size of every party is $O(k^2 \cdot t)$.*

Remark 4.9. We can apply the transformation of Theorem 4.6 also to CDS protocols with short secrets. However, the best known k-party CDS protocol for such secrets of [36] (for k-input functions $f : [N]^k \to \{0,1\}$) have message size $2^{\tilde{O}(\sqrt{k \log N})}$, thus, using a k^2-party CDS would result in an inefficient secret-sharing scheme.

5 Optimal Linear Secret-Sharing Schemes Realizing k-Uniform Access Structures

In this section, we show how to construct a *linear* secret-sharing scheme realizing n-party k-uniform access structures in which the share size of every party is $O(n \cdot 2^{h(k/n)n/2})$. Using a result of [9], we prove a matching lower bound, which shows that our construction is optimal (up to a small polynomial factor).

We start by giving some high-level ideas of our linear secret-sharing scheme. We are inspired by the linear CDS protocols of [13], where for every Boolean n-input function they construct a linear CDS protocol with message size $O(2^{n/2})$ (a similar protocol with the same message size was independently constructed in [36]). By a transformation of [12], this implies that for every uniform access structure, there is a linear secret-sharing scheme with share size $O(n \cdot 2^{n/2})$. We want to optimize this construction for k-uniform access structures for $k < n/2$.

As a first step, we define balanced k-uniform access structures, where a k-uniform access structure is balanced if there exists a set of parties B of size

$n/2$ such that every authorized set A of size k contains exactly $k/2$ parties in B (that is, $|A \cap B| = k/2$ and $|A \setminus B| = k/2$). We construct an optimized secret-sharing scheme for balanced k-uniform access structures. We then show (using a probabilistic argument) that every k-uniform access structure Γ is a union of $O(k^{1/2} \cdot n)$ balanced k-uniform access structures. Thus, to realize Γ, we independently share the secret for each of the balanced access structures.

Definition 5.1 (The Access Structure Γ_B). *Let Γ be a k-uniform access structure with n parties for some even k and let $B \subseteq P$ be a subset of parties. The k-uniform access structure Γ_B is the access structure that contains all the subsets of parties of size greater than k and all authorized subsets from Γ of size k that contain exactly $k/2$ parties from the subset B. That is, $\Gamma_B = \{A \subseteq P : |A| > k\} \cup \{A \subseteq P : A \in \Gamma, |A| = k, \text{ and } |A \cap B| = k/2\}$.*

Next, we show in Lemma 5.2 our basic linear scheme, which realizes the access structure Γ_B. The proof is deferred to the full version [4].

Lemma 5.2. *Let $\mathbb{F}$ be a finite field and Γ be a k-uniform access structure with n parties for some even k and some even n, and let B be a subset of parties such that $|B| = n/2$. Then, the scheme Σ_B, described in Fig. 3, is a linear secret-*

The secret: An element $s \in \mathbb{F}$, where $\mathbb{F}$ is a finite field.
The scheme: Assume without loss of generality that $B = \{P_1, \ldots, P_{n/2}\}$, and let $\mathcal{U} = \{U \subseteq B : |U| = k/2\}$ and $\mathcal{V} = \{V \subseteq \overline{B} : |V| = k/2\}$.

1. Share the secret s using a $(k+1)$-out-of-n secret-sharing scheme among the parties in P, that is, for every $j \in [n]$, give the jth share from this scheme to party P_j.
2. Choose two random elements $s_1, s_2 \in \mathbb{F}$, and let $s_3 = s_1 + s_2 + s$.
3. Share the element s_1 using a $k/2$-out-of-$n/2$ secret-sharing scheme among the parties in B, that is, for every $j \in [n/2]$, give the jth share from this scheme to party P_j.
4. Share the element s_2 using a $k/2$-out-of-$n/2$ secret-sharing scheme among the parties in $\overline{B}$, that is, for every $j \in [n/2]$, give the jth share from this scheme to party $P_{n/2+j}$.
5. For every $U \in \mathcal{U}$, choose a random element $r_U \in \mathbb{F}$.
6. For every $V = \left\{P_{j_1}, \ldots, P_{j_{k/2}}\right\} \in \mathcal{V}$, choose $k-1$ random elements $q_V^{j_1}, \ldots, q_V^{j_{k/2-1}} \in \mathbb{F}$.
7. Let $q_V = s_3 + \sum_{U \in \mathcal{U}: U \cup V \notin \Gamma} r_U$ and $q_V^{j_{k/2}} = q_V - (q_V^{j_1} + \cdots + q_V^{j_{k/2-1}})$. [a]
8. For every $j \in \{1, \ldots, n/2\}$, give the elements $(r_U)_{U \in \mathcal{U}: P_j \notin U}$ to P_j.
9. For every $j \in \{n/2+1, \ldots, n\}$, give the elements $(q_V^j)_{V \in \mathcal{V}: P_j \in V}$ to P_j.

[a] We assume that for every $V \in \mathcal{V}$ there exists a $U \in \mathcal{U}$ such that $U \cup V \notin \Gamma$; for example, this can be achieved by adding a "dummy" $U_0 \in \mathcal{U}$.

Fig. 3. A linear secret-sharing scheme Σ_B realizing the k-uniform access structure Γ_B.

sharing scheme over $\mathbb{F}$ *realizing* Γ_B *in which the share size of every party is* $O(k^{-1/2} \cdot 2^{h(k/n)n/2} \cdot \log|\mathbb{F}|)$.

In the following we show that a k-uniform access structure can be decomposed to $\ell = O(k^{1/2} \cdot n)$ balanced access structures. Its proof is also deferred to the full version [4].

Lemma 5.3. *Let* P *be a set of* n *parties for some even* n*, and let* k *be an even integer. Then, there are* $\ell = \Theta(k^{1/2} \cdot n)$ *subsets* $B_1, \dots, B_\ell \subseteq P$*, each of them of size* $n/2$*, such that for every subset* $A \subseteq P$ *of size* k *it holds that* $|A \cap B_i| = k/2$*, for at least one* $i \in [\ell]$.

Now, we are ready to present our final linear scheme, which realizes every k-uniform access structure.

Theorem 5.4. *Let* Γ *be a* k*-uniform access structure with* n *parties. Then, for every finite field* $\mathbb{F}$*, there is a linear secret-sharing scheme realizing* Γ*, for secrets from* $\mathbb{F}$*, in which the share size of every party is* $O(n \cdot 2^{h(k/n)n/2} \cdot \log|\mathbb{F}|)$.

Proof. Let $s \in \mathbb{F}$ be the secret. By adding dummy parties (which either belong to all authorized sets or belong to none of them), we can assume without loss of generality that k and n are even. By Lemma 5.3, there exist $\ell = \Theta(k^{1/2} \cdot n)$ subsets $B_1, \dots, B_\ell \subseteq P$, where $|B_i| = n/2$ for every $i \in [\ell]$, such that for every subset $A \subseteq P$ of size k it holds that $|A \cap B_i| = k/2$ for at least one $i \in [\ell]$. Thus, we get that $\Gamma = \cup_{i=1}^{\ell} \Gamma_{B_i}$. By Lemma 5.2, for every $i \in [\ell]$ there is a linear secret-sharing scheme Σ_{B_i} realizing the k-uniform access structure Γ_{B_i}, in which the share size of every party is $O(k^{-1/2} \cdot 2^{h(k/n)n/2} \cdot \log|\mathbb{F}|)$. We independently realize every access structure Γ_{B_i} using the linear scheme Σ_{B_i} with secret s; the combined scheme is a linear secret-sharing scheme realizing the access structure Γ in which the share size of every party is $\ell \cdot O(k^{-1/2} \cdot 2^{h(k/n)n/2} \cdot \log|\mathbb{F}|) = O(n \cdot 2^{h(k/n)n/2} \cdot \log|\mathbb{F}|)$. □

5.1 A Lower Bound for Linear Schemes Realizing k-Uniform Access Structures

Using a result of [9] we prove a lower bound of $\tilde{O}(2^{h(k/n)n/2})$ on the share size of at least one party in every linear secret-sharing scheme that realizes k-uniform access structures, for one-bit secrets. As we have shown above for one-bit secrets (that is, $\mathbb{F} = \{0, 1\}$), this bound is tight up to a poly(n) factor.

Theorem 5.5. *For most* k*-uniform access structures* Γ *with* n *parties, the share size of at least one party for sharing a one-bit secret in every linear secret-sharing scheme realizing* Γ *is* $\Omega(k^{-3/4} \cdot n^{-1/2} \cdot 2^{h(k/n)n/2})$.

Proof. If we share a one-bit secret using a linear secret-sharing scheme over $\mathbb{F}$ in which the largest share contains s field elements, then the size of the share of at least one party is $s \cdot \log|\mathbb{F}|$. For the share size of every party to be less than

$k^{-3/4} \cdot n^{-1/2} \cdot 2^{h(k/n)n/2}$, it must be that $|\mathbb{F}| \le 2^{k^{-3/4} \cdot n^{-1/2} \cdot 2^{h(k/n)n/2}}$ (otherwise, each share contains at least $k^{-3/4} \cdot n^{-1/2} \cdot 2^{h(k/n)n/2}$ bits), and, obviously, $s \cdot \log|\mathbb{F}| \le k^{-3/4} \cdot n^{-1/2} \cdot 2^{h(k/n)n/2}$.

We say that the rank of an access structure Γ is r if the size of every minimal authorized set in Γ is at most r, so the rank of k-uniform access structures is $k+1$. By [9], for every finite field $\mathbb{F}$ and integers s, r, n such that $s > \log n$, there are at most $2^{2rns^2 \cdot \log|\mathbb{F}|}$ access structures Γ with n parties and rank r such that there exists a linear secret-sharing scheme over $\mathbb{F}$ realizing Γ in which each share contains at most s field elements. Let $\theta = s \cdot \log|\mathbb{F}|$. Thus, there are at most $2^{2(k+1)n(\theta/\log|\mathbb{F}|)^2 \cdot \log|\mathbb{F}|} < 2^{2(k+1)n\theta^2}$ k-uniform access structures Γ with n parties such that there exists a linear secret-sharing scheme over $\mathbb{F}$ realizing Γ in which the share size of each party is at most θ.

Next, we count the number of linear schemes that realize k-uniform access structures in which the share size of each party is at most $\theta < k^{-3/4} \cdot n^{-1/2} \cdot 2^{h(k/n)n/2}$. Since we are counting linear schemes, we need to sum the number of linear schemes that realizes k-uniform access structures for every possible finite field (there are at most $2^{k^{-3/4} \cdot n^{-1/2} \cdot 2^{h(k/n)n/2}}$ such fields, because $|\mathbb{F}| \le 2^{k^{-3/4} \cdot n^{-1/2} \cdot 2^{h(k/n)n/2}}$). From all the above, the number of such linear schemes is at most $2^{k^{-3/4} \cdot n^{-1/2} \cdot 2^{h(k/n)n/2} + 2(k+1)n\theta^2}$.

By Fact 2.12, the number of k-uniform access structures is $2^{\binom{n}{k}} = 2^{\Theta(k^{-1/2} \cdot 2^{h(k/n)n})}$. Thus, if half of the k-uniform access structures Γ with n parties have linear secret-sharing schemes in which the share size of every party is at most θ, then $2^{k^{-3/4} \cdot n^{-1/2} \cdot 2^{h(k/n)n/2} + 2(k+1)n\theta^2} \ge \frac{1}{2} \cdot 2^{\Theta(k^{-1/2} \cdot 2^{h(k/n)n})}$, i.e., $k^{-3/4} \cdot n^{-1/2} \cdot 2^{h(k/n)n/2} + 2(k+1)n\theta^2 \ge \Theta(k^{-1/2} \cdot 2^{h(k/n)n})$, so $\theta = \Omega(k^{-3/4} \cdot n^{-1/2} \cdot 2^{h(k/n)n/2})$. □

6 Transformation from CDS to Secret-Sharing and Implications to Ad-Hoc PSM

In this section, we describe a new transformation from a k-party CDS protocol to a secret-sharing scheme for k-uniform access structure. This construction improves the secret-sharing schemes for k-uniform access structures, for short secrets, compared to the scheme implied by the construction of [34]. We also show how to use the ideas of our transformation to construct a k-out-of-n ad-hoc PSM protocol from k-party PSM protocol.

6.1 The Transformation for Uniform Access Structures

We show how to realize any k-uniform access structure Γ with n parties using a k-party CDS protocol for the function g, defined in Definition 6.1.

Definition 6.1 (The Function g). *Let Γ be a k-uniform access structure with n parties. The k-input function $g : [n]^k \to \{0,1\}$ is the function that satisfies $g(x_1, \dots, x_k) = 1$ if and only if $x_1 < \dots < x_k$ and $A = \{P_{x_1}, \dots, P_{x_k}\}$ is an authorized set, that is, $A \in \Gamma$.*

We say that a party P_{x_i} is the ith party in A if and only if there are $i-1$ parties before it and there are $k-i$ parties after it, when the indices of the parties are sorted. The idea of our scheme is that if party P_x is the ith party in a set A of size k, then its share will contain the message of the ith party in the CDS protocol for g (with the shared secret) when holding the input x. The problem with this idea is that the dealer does not know which set of parties will try to reconstruct the secret and it does not know if P_x is the ith party. If the dealer gives to two parties in the set the message of the ith party in the CDS protocol, then these parties get two different messages of the same party in the CDS protocol with different input, so we cannot ensure the privacy of the CDS protocol. Hence, some unauthorized sets may learn information about the secret. To solve this problem, the message of the ith party in the CDS protocol that party P_x gets will be masked by two random elements, such that only if P_x is the ith party in A, then the parties in A can learn this message. For this, the dealer shares one of the above mentioned random elements using a $(i-1)$-out-of-$(x-1)$ secret-sharing scheme and gives the shares to all parties before P_x, and shares the second random element using a $(k-i)$-out-of-$(n-x)$ secret-sharing scheme and gives the shares to all parties after P_x.

Theorem 6.2. *Let Γ be a k-uniform access structure with n parties, and assume that for every k-input function $f : [N]^k \to \{0,1\}$ there is a k-party CDS protocol for f for a one-bit secret, in which the message size is $c(k, N, 1)$. Then, the scheme Σ_g, described in Fig. 4, is a secret-sharing scheme realizing Γ, for a one-bit secret, in which the share size of every party is $O(k \cdot n \cdot c(k, n, 1))$.*

Proof. We prove that the secret-sharing scheme Σ_g is a scheme that realizes Γ with share size as in the theorem. Let $s \in \{0,1\}$ be the secret and $\mathcal{P}$ be a k-party CDS protocol for $g : [n]^k \to \{0,1\}$ (defined in Definition 6.1), for a one-bit secret, in which the message size is $c(k, n, 1)$. We prove that every subset of parties A of size k can learn only the messages corresponding to the parties in A of the CDS protocol for the function g (that is, party $P_j \in A$ can learn only the message $m_{i,j}$, where P_j is the ith party in A), so A can reconstruct the secret using these messages if and only if it is an authorized set. Additionally, we show that subsets of parties of size less than k cannot learn any messages of the CDS protocol for the function g, so such subsets cannot learn any information about the secret.

Correctness. An authorized set of size greater than k can reconstruct the secret using the $(k+1)$-out-of-n secret-sharing scheme.

Let $A = \{P_{x_1}, \dots, P_{x_k}\}$ be an authorized set of size k such that $x_1 < \cdots < x_k$. For every $i \in [k]$, party P_{x_i} gets the string $m_{i,x_i} \oplus r_{i,x_i} \oplus q_{i,x_i}$. Additionally, the parties $P_{x_1}, \dots, P_{x_{i-1}}$ get $i-1$ shares from the $(i-1)$-out-of-(x_i-1) scheme for the string q_{i,x_i}, so they can reconstruct q_{i,x_i}, and the parties $P_{x_{i+1}}, \dots, P_{x_k}$ get $k-i$ shares from the $(k-i)$-out-of-$(n-x_i)$ scheme for the string r_{i,x_i}, so they can reconstruct r_{i,x_i}. Overall, for every $i \in [k]$, the parties $P_{x_1}, \dots, P_{x_k}$ learn the strings $m_{i,x_i} \oplus r_{i,x_i} \oplus q_{i,x_i}$, r_{i,x_i}, and q_{i,x_i}, so they can reconstruct the message m_{i,x_i} of the CDS protocol for g. Since $g(x_1, \dots, x_k) = 1$, and the parties

The secret: An element $s \in \{0,1\}$.
The scheme: Let $g : [n]^k \to \{0,1\}$ be the k input function defined in Definition 6.1 and let $\mathcal{P}$ be a k-party CDS protocol for g.

1. Share the secret s using a $(k+1)$-out-of-n secret-sharing scheme among the parties in P, that is, for every $j \in [n]$, give the jth share from this scheme to party P_j.
2. Apply the k-party CDS protocol $\mathcal{P}$ to the k-input function g with the secret s and a common random string r that is chosen at random. For every $i \in [k]$, let $m_{i,x}$ be the message of the ith party in the CDS protocol $\mathcal{P}$ when holding the input $x \in [n]$.
3. For every $i \in \{2,\dots,k\}$ and every $j \in \{i,\dots,n-k+i\}$, choose a random string $q_{i,j}$ (of the same size as $m_{i,j}$), and for every $j \in \{1,\dots,n-k+1\}$, let $q_{1,j} = \mathbf{0}$ (i.e., a string of zeroes).
4. For every $i \in \{1,\dots,k-1\}$ and every $j \in \{i,\dots,n-k+i\}$, choose a random string $r_{i,j}$ (of the same size as $m_{i,j}$), and for every $j \in \{k,\dots,n\}$, let $r_{k,j} = \mathbf{0}$.
5. For every $i \in \{1,\dots,k\}$ and every $j \in \{i,\dots,n-k+i\}$, give the string $m_{i,j} \oplus r_{i,j} \oplus q_{i,j}$ to party P_j.
6. For every $i \in \{2,\dots,k\}$ and every $j \in \{i,\dots,n-k+i\}$, share the string $q_{i,j}$ using a $(i-1)$-out-of-$(j-1)$ secret-sharing scheme among the first $j-1$ parties (i.e., the parties $P_1,\dots,P_{j-1}$), that is, for every $w \in [j-1]$, give the wth share from this scheme to party P_w.
7. For every $i \in \{1,\dots,k-1\}$ and every $j \in \{i,\dots,n-k+i\}$, share the string $r_{i,j}$ using a $(k-i)$-out-of-$(n-j)$ secret-sharing scheme among the last $n-j$ parties (i.e., the parties $P_{j+1},\dots,P_n$), that is, for every $w \in [n-j]$, give the wth share from this scheme to party P_{j+w}.

Fig. 4. A secret-sharing scheme Σ_g realizing a k-uniform access structure Γ.

in A hold the messages $m_{1,x_1},\dots,m_{k,x_k}$, they can reconstructs the secret s from those messages of the CDS protocol for g.

Privacy. Let $A = \{P_{x_1},\dots,P_{x_k}\}$ be an unauthorized set of size k such that $x_1 < \cdots < x_k$. As claimed above, the parties in A can learn the messages $m_{1,x_1},\dots,m_{k,x_k}$, but since $g(x_1,\dots,x_k) = 0$, the parties in A cannot learn the secret from the CDS protocol for g (by the privacy of the CDS protocol).

We show that the parties in A cannot learn any other messages from the CDS protocol for g. For $x \in [n]$ such that $P_x \notin A$, the parties in A cannot learn $m_{i,x}$ for every $i \in [k]$, since they do not get this message (even masked by random strings). Consider an $x \in [n]$ such that $P_x \in A$ and $x \neq x_i$ for some $i \in [k]$. If $x < x_i$ (that is, P_x is smaller than the ith party in A) then the parties in A cannot learn the string $q_{i,x}$, since they hold less than $i-1$ shares from the $(i-1)$-out-of-$(x-1)$ for the string $q_{i,x}$, so the parties in A cannot learn the message $m_{i,x}$. Otherwise, if $x > x_i$ (that is, P_x is bigger than the ith party in A) then the parties in A cannot learn the string $r_{i,x}$, since they hold less than $k-i$ shares from the $(k-i)$-out-of-$(n-x)$ for the string $r_{i,x}$, so the parties in A cannot learn the message $m_{i,x}$. Thus, the parties in A cannot learn any information about the secret s.

The last argument holds for unauthorized sets of size less than k, so such sets cannot learn any messages from the CDS protocol for g, and cannot learn any information about the secret s. The formal privacy proof is deferred to [4].

SHARE SIZE. The share size of every party in the scheme Σ_h is $O(k \cdot n \cdot c(k,n,1) + \log n) = O(k \cdot n \cdot c(k,n,1))$. □

Using the CDS protocol of [36], in which the message size is $2^{O(\sqrt{k \log n} \log(k \log n))}$, for a one-bit secret, we get the following result.

Corollary 6.3. *Let Γ be a k-uniform access structure with n parties. Then, there is a secret-sharing scheme realizing Γ, for a one-bit secret, in which the share size of every party is $k \cdot n \cdot 2^{O(\sqrt{k \log n} \log(k \log n))}$.*

6.2 The Transformation for Ad-Hoc PSM Protocols

We use the same ideas as in the above transformation to construct a k-out-of-n ad-hoc PSM protocol for a function $f : [N]^k \to Y$ using a k-party PSM protocol for f. Recall that some k parties $P_{i_1}, \ldots, P_{i_k}$, holding inputs $x_{i_1}, \ldots, x_{i_k} \in [N]$ respectively, participate in the protocol, and they want to compute $f(x_{i_1}, \ldots, x_{i_k})$. However, the participating parties do not know which k parties among the n parties participate in the protocol. In Fig. 5, we describe our ad-hoc PSM protocol; in the protocol there is an offline stage, which contains computation that only depends on the common string, and an online stage in which each participating party sends its message.

Theorem 6.4. *Let $f : [N]^k \to Y$ be a k-input function, for some integer k, and assume that there is a k-party PSM protocol for f with message size $c_f(k,N)$. Then, the protocol $\mathcal{P}_f$, described in Fig. 5, is a k-out-of-n ad-hoc PSM protocol for f with message size $O(k \cdot n \cdot c_f(k,N))$.*

Proof. The correctness of the protocol follows from the fact that given k parties $P_{i_1}, \ldots, P_{i_k}$, the referee learns the messages $m_{1,x_{i_1}}, \ldots, m_{k,x_{i_k}}$, as explained in the proof of Theorem 6.2, and thus, by the correctness of the PSM protocol for f, the referee can learn $f(x_{i_1}, \ldots, x_{i_k})$. The privacy of the protocol follows from the privacy of the PSM protocol and the fact that the referee learns only the messages $m_{1,x_{i_1}}, \ldots, m_{k,x_{i_k}}$, as proved in Theorem 6.2. Note that for less than k parties, the referee cannot learn any message of the PSM protocol, again like in Theorem 6.2. □

By the PSM protocol of [12], in which the message size is $O(k^3 \cdot N^{k/2})$, we get the following result.

Offline stage of the protocol: Let $\mathcal{P}$ be a k-party PSM protocol for f.

1. Apply the k-party PSM protocol $\mathcal{P}$ for the k-input function f with a common random string r that is chosen at random. For every $i \in [k]$ let $m_{i,x}$ be the message of the ith party in the PSM protocol $\mathcal{P}$ when holding the input $x \in [N]$.
2. For every $i \in \{2, \ldots, k\}$ and every $j \in \{i, \ldots, n-k+i\}$, choose a random string $q_{i,j}$ (of the same size as $m_{i,j}$), and for every $j \in \{1, \ldots, n-k+1\}$, let $q_{1,j} = \mathbf{0}$ (i.e., a string of zeroes).
3. For every $i \in \{1, \ldots, k-1\}$ and every $j \in \{i, \ldots, n-k+i\}$, choose a random string $r_{i,j}$ (of the same size as $m_{i,j}$), and for every $j \in \{k, \ldots, n\}$, let $r_{k,j} = \mathbf{0}$.
4. For every $i \in \{2, \ldots, k\}$ and every $j \in \{i, \ldots, n-k+i\}$, share the string $q_{i,j}$ using a $(i-1)$-out-of-$(j-1)$ secret-sharing scheme among the first $j-1$ parties (i.e., the parties $P_1, \ldots, P_{j-1}$). For every $w \in \{1, \ldots, j-1\}$, let $q_{i,j}^w$ be the share of party P_w.
5. For every $i \in \{1, \ldots, k\}$ and every $j \in \{i, \ldots, n-k+i\}$, share the string $r_{i,j}$ using a $(k-i)$-out-of-$(n-j)$ secret-sharing scheme among the last $n-j$ parties (i.e., the parties $P_{j+1}, \ldots, P_n$). For every $w \in \{j+1, \ldots, n\}$, let $r_{i,j}^w$ be the share of party P_w.

Online stage of the protocol for a set A of k parties: Each party $P_j \in A$ holds an input $x_j \in [N]$.

1. Every party $P_j \in A$ sends to the referee the string $m_{i,x_j} \oplus r_{i,j} \oplus q_{i,j}$ for every $i \in \{1, \ldots, k\}$.
2. Every party $P_w \in A$ sends to the referee the string $q_{i,j}^w$ for every $i \in \{1, \ldots, k\}$ and every $j > w$.
3. Every party $P_w \in A$ sends to the referee the string $r_{i,j}^w$ for every $i \in \{1, \ldots, k\}$ and every $j < w$.

Fig. 5. A k-out-of-n ad-hoc PSM protocol $\mathcal{P}_f$ for a k-input function $f : [N]^k \to Y$.

Corollary 6.5. *Let $f : [N]^k \to Y$ be a k-input function, for some integer k. Then, there is a k-out-of-n ad-hoc PSM protocol for f with message size $O(k^4 \cdot n \cdot N^{k/2})$.*

6.3 Improving the Ad-Hoc PSM Protocol for Symmetric Functions

We combine the protocol of Sect. 6.2 with the ideas of Sect. 4, and construct a better k-out-of-n ad-hoc PSM protocol for symmetric functions $f : [N]^k \to Y$, where a function f is symmetric if for a given input $x = (x_1, \ldots, x_k)$, the output of f on the input x is the same as the output of f on any permutation on the order of the x_i's, that is, for every $x = (x_1, \ldots, x_k)$ and every permutation $\pi : [k] \to [k]$, it holds that $f(x_1, \ldots, x_k) = f(x_{\pi(1)}, \ldots, x_{\pi(k)})$.

Our construction consists of two steps. First, we show that we can construct a k-out-of-n ad-hoc PSM protocol for f using $\Theta(k \cdot \log n)$ invocations of a k-out-of-k^2 ad-hoc PSM protocol. Then, we use the protocol of Theorem 6.4 with

k^2 parties, and get a k-out-of-n ad-hoc PSM protocol for f with message size $O(k^4 \cdot \log n \cdot c_f(k, N))$, where $c_f(k, N)$ is the message size of a k-party PSM protocol for f.

For the first step, we show a general transformation from k-out-of-t ad-hoc PSM protocols to k-out-of-n ad-hoc PSM protocols, for every $k \leq t \leq n$. This transformation generalizes and improves the construction of [11], which only works when $t = k$. As mentioned above, we use this transformation for $t = k^2$. For our transformation, we take a family of perfect hash functions $H_{n,k,t}$, and construct a k-out-of-n ad-hoc PSM protocol for f using independent copies of a k-out-of-t ad-hoc PSM protocol for f, one copy for each hash function $h \in H_{n,k,t}$.

In the k-out-of-t ad-hoc PSM protocol for h, denoted by $\mathcal{P}_h$, party P_j simulates the $h(j)$th party in $\mathcal{P}_h$. If h is one-to-one on a set of k parties, that is, the set does not collide on h, then the referee gets k messages of k different parties of the protocol $\mathcal{P}_h$, so it can compute the output of f on the inputs of the parties.

If a set collides on h, then the referee gets at least two messages of the same party of the protocol $\mathcal{P}_h$, so the privacy is not guaranteed. To solve this problem, every party encrypts its message of the protocol $\mathcal{P}_h$ using an information-theoretic encryption system that is secure as long as the adversary sees at most k encryptions. We also share the encryption key using a k-out-of-t secret-sharing scheme, and party P_j sends to the referee the $h(j)$th share from this scheme. For sets of size less than k and sets of size k that collide on h, the referee cannot reconstruct the key and sees at most k encrypted messages, thus cannot learn any information on the messages of the protocol $\mathcal{P}_h$. For the encryption system, we use a polynomial of degree k as the encryption key; to encrypt a message each party masks it by a unique point of the polynomial.

Observe that the referee might learn the output of f from more than one protocol, for several functions from $H_{n,k,t}$, so the requirement for symmetric functions is necessary, since the order of the parties in a set of size k can change according to the different hash functions.

Lemma 6.6. *Let $f : [N]^k \to Y$ be a k-input symmetric function, for some integer k, and assume that there is a k-out-of-t ad-hoc PSM protocol $\mathcal{P}$ for f with message size $c_f(k, t, N)$, and that there is a family of perfect hash function $H_{n,k,t} = \{h_i : [n] \to [t] : i \in [\ell]\}$. Then, there is a k-out-of-n ad-hoc PSM protocol for f with message size $O(\ell \cdot k \cdot \max\{c_f(k, t, N), \log n\})$.*

The proof of Lemma 6.6 is deferred to the full version [4]. By taking $t = k^2$ and using our ad-hoc PSM protocol from Theorem 6.4 and the family of perfect hash functions from Lemma 4.5, we get the following result (Fig. 6).

Offline stage of the protocol: Let $\mathcal{P}$ be a k-out-of-t ad-hoc PSM protocol for $f : [N]^k \to Y$ and let $h : [n] \to [t]$ be a hash function.

1. Apply the k-out-of-t ad-hoc PSM protocol $\mathcal{P}$ for the k-input function f with a common random string r chosen at random with uniform distribution. For every $i \in [t]$ let $m_{i,x}$ be the message of the ith party in the ad-hoc PSM protocol $\mathcal{P}$ when holding the input $x \in [N]$.
2. Choose a random polynomial Q of degree k over a finite field $\mathbb{F}$ such that $\log |\mathbb{F}| > \max \{\log n, c_f(k, t, N)\}$.
3. Share the polynomial Q (i.e., its coefficients) using a k-out-of-t secret-sharing scheme. For every $i \in \{1, \ldots, t\}$, let q^i be the ith share from this scheme.

Online stage of the protocol for a set A of k parties: Each party $P_j \in A$, who holds an input $x_j \in [N]$, sends $m_{h(j),x_j} \oplus Q(j)$ and $q^{h(j)}$ to the referee.

Fig. 6. A k-out-of-n ad-hoc PSM protocol $\mathcal{P}_h$ for a symmetric k-input function $f : [N]^k \to Y$.

Theorem 6.7. *Let $f : [N]^k \to Y$ be a k-input symmetric function, for some integer k, and assume that there is a k-party PSM protocol for f with message size $c_f(k, N)$. Then, there is a k-out-of-n ad-hoc PSM protocol for f with message size $O(k^5 \cdot \log n \cdot c_f(k, N))$.*

Proof. By Theorem 6.4, there is a k-out-of-k^2 ad-hoc PSM protocol for f with message size $O(k \cdot k^2 \cdot c_f(k, N)) = O(k^3 \cdot c_f(k, N))$, and by Lemma 4.5, there is a family of perfect hash functions H_{n,k,k^2} with $\ell = \Theta(k \cdot \log n)$ functions.

Thus, by Lemma 6.6, there is a k-out-of-n ad-hoc PSM protocol for f with message size $O(\ell \cdot k \cdot k^3 \cdot c_f(k, N)) = O(k^5 \cdot \log n \cdot c_f(k, N))$. □

Finally, again by the PSM protocol of [12], we obtain the next result.

Corollary 6.8. *Let $f : [N]^k \to Y$ be a k-input symmetric function, for some integer k. Then, there is a k-out-of-n ad-hoc PSM protocol for f with message size $O(k^8 \cdot \log n \cdot N^{k/2})$.*

Acknowledgement. The first and fourth authors are supported by the European Union's Horizon 2020 Programme (ERC-StG-2014-2020) under grant agreement no. 639813 ERC-CLC, and the Check Point Institute for Information Security. Part of this work was done while the second author was visiting Georgetown university, supported by NSF grant no. 1565387, TWC: Large: Collaborative: Computing Over Distributed Sensitive Data. The second author is also supported by ISF grant 152/17 and by a grant from the Cyber Security Research Center at Ben-Gurion University of the Negev. The third author is supported by the Spanish Government through TIN2014-57364-C2-1-R and by the Government of Catalonia through Grant 2017 SGR 705. The fifth author is supported by ISF grant 152/17, by a grant from the Cyber Security Research Center at Ben-Gurion University of the Negev, and by the Frankel center for computer science.

References

1. Aiello, B., Ishai, Y., Reingold, O.: Priced oblivious transfer: how to sell digital goods. In: Pfitzmann, B. (ed.) EUROCRYPT 2001. LNCS, vol. 2045, pp. 119–135. Springer, Heidelberg (2001). https://doi.org/10.1007/3-540-44987-6_8
2. Applebaum, B., Arkis, B.: On the power of amortization in secret sharing: d-uniform secret sharing and CDS with constant information rate. In: Beimel, A., Dziembowski, S. (eds.) TCC 2018. LNCS, vol. 11239, pp. 317–344. Springer, Cham (2018). https://doi.org/10.1007/978-3-030-03807-6_12
3. Applebaum, B., Arkis, B., Raykov, P., Vasudevan, P.N.: Conditional disclosure of secrets: amplification, closure, amortization, lower-bounds, and separations. In: Katz, J., Shacham, H. (eds.) CRYPTO 2017. LNCS, vol. 10401, pp. 727–757. Springer, Cham (2017). https://doi.org/10.1007/978-3-319-63688-7_24
4. Applebaum, B., Beimel, A., Farràs, O., Nir, O., Peter, N.: Secret-sharing schemes for general and uniform access structures. Technical report 2019/231, IACR Cryptology ePrint Archive (2019)
5. Applebaum, B., Holenstein, T., Mishra, M., Shayevitz, O.: The communication complexity of private simultaneous messages, revisited. In: Nielsen, J.B., Rijmen, V. (eds.) EUROCRYPT 2018. LNCS, vol. 10821, pp. 261–286. Springer, Cham (2018). https://doi.org/10.1007/978-3-319-78375-8_9
6. Applebaum, B., Vasudevan, P.: Placing conditional disclosure of secrets in the communication complexity universe. In: 10th Innovations in Theoretical Computer Science Conference, ITCS. LIPIcs, vol. 124, pp. 4:1–4:14 (2019)
7. Attrapadung, N.: Dual system encryption via doubly selective security: framework, fully secure functional encryption for regular languages, and more. In: Nguyen, P.Q., Oswald, E. (eds.) EUROCRYPT 2014. LNCS, vol. 8441, pp. 557–577. Springer, Heidelberg (2014). https://doi.org/10.1007/978-3-642-55220-5_31
8. Beimel, A., Chor, B.: Universally ideal secret-sharing schemes. IEEE Trans. Inf. Theory **40**(3), 786–794 (1994)
9. Beimel, A., Farràs, O., Mintz, Y., Peter, N.: Linear secret-sharing schemes for forbidden graph access structures. In: Kalai, Y., Reyzin, L. (eds.) TCC 2017. LNCS, vol. 10678, pp. 394–423. Springer, Cham (2017). https://doi.org/10.1007/978-3-319-70503-3_13
10. Beimel, A., Farràs, O., Peter, N.: Secret sharing schemes for dense forbidden graphs. In: Zikas, V., De Prisco, R. (eds.) SCN 2016. LNCS, vol. 9841, pp. 509–528. Springer, Cham (2016). https://doi.org/10.1007/978-3-319-44618-9_27
11. Beimel, A., Ishai, Y., Kumaresan, R., Kushilevitz, E.: On the cryptographic complexity of the worst functions. In: Lindell, Y. (ed.) TCC 2014. LNCS, vol. 8349, pp. 317–342. Springer, Heidelberg (2014). https://doi.org/10.1007/978-3-642-54242-8_14
12. Beimel, A., Kushilevitz, E., Nissim, P.: The complexity of multiparty PSM protocols and related models. In: Nielsen, J.B., Rijmen, V. (eds.) EUROCRYPT 2018. LNCS, vol. 10821, pp. 287–318. Springer, Cham (2018). https://doi.org/10.1007/978-3-319-78375-8_10
13. Beimel, A., Peter, N.: Optimal linear multiparty conditional disclosure of secrets protocols. In: Peyrin, T., Galbraith, S. (eds.) ASIACRYPT 2018. LNCS, vol. 11274, pp. 332–362. Springer, Cham (2018). https://doi.org/10.1007/978-3-030-03332-3_13
14. Beimel, A., Ishai, Y., Kushilevitz, E.: Ad hoc PSM protocols: secure computation without coordination. In: Coron, J.-S., Nielsen, J.B. (eds.) EUROCRYPT 2017. LNCS, vol. 10212, pp. 580–608. Springer, Cham (2017). https://doi.org/10.1007/978-3-319-56617-7_20

15. Ben-Or, M., Goldwasser, S., Wigderson, A.: Completeness theorems for noncryptographic fault-tolerant distributed computations. In: Proceedings of the 20th ACM Symposium on the Theory of Computing, pp. 1–10 (1988)
16. Benaloh, J., Leichter, J.: Generalized secret sharing and monotone functions. In: Goldwasser, S. (ed.) CRYPTO 1988. LNCS, vol. 403, pp. 27–35. Springer, New York (1990). https://doi.org/10.1007/0-387-34799-2_3
17. Bertilsson, M., Ingemarsson, I.: A construction of practical secret sharing schemes using linear block codes. In: Seberry, J., Zheng, Y. (eds.) AUSCRYPT 1992. LNCS, vol. 718, pp. 67–79. Springer, Heidelberg (1993). https://doi.org/10.1007/3-540-57220-1_53
18. Blakley, G.R.: Safeguarding cryptographic keys. In: Proceedings of the 1979 AFIPS National Computer Conference, vol. 48, pp. 313–317 (1979)
19. Chaum, D., Crépeau, C., Damgård, I.: Multiparty unconditionally secure protocols. In: Proceedings of the 20th ACM Symposium on the Theory of Computing, pp. 11–19 (1988)
20. Chor, B., Kushilevitz, E.: Secret sharing over infinite domains. J. Cryptol. **6**(2), 87–96 (1993)
21. Cramer, R., Damgård, I., Maurer, U.: General secure multi-party computation from any linear secret-sharing scheme. In: Preneel, B. (ed.) EUROCRYPT 2000. LNCS, vol. 1807, pp. 316–334. Springer, Heidelberg (2000). https://doi.org/10.1007/3-540-45539-6_22
22. Csirmaz, L.: The size of a share must be large. In: De Santis, A. (ed.) EUROCRYPT 1994. LNCS, vol. 950, pp. 13–22. Springer, Heidelberg (1995). https://doi.org/10.1007/BFb0053420. Journal Version in: J. Cryptol. **10**(4), 223–231 (1997)
23. Csirmaz, L.: The dealer's random bits in perfect secret sharing schemes. Studia Sci. Math. Hungar. **32**(3–4), 429–437 (1996)
24. Desmedt, Y., Frankel, Y.: Shared generation of authenticators and signatures. In: Feigenbaum, J. (ed.) CRYPTO 1991. LNCS, vol. 576, pp. 457–469. Springer, Heidelberg (1992). https://doi.org/10.1007/3-540-46766-1_37
25. Erdös, P., Spencer, J.: Probabilistic Methods in Combinatorics. Academic Press, Cambridge (1974)
26. Feige, U., Kilian, J., Naor, M.: A minimal model for secure computation. In: 26th STOC 1994, pp. 554–563 (1994)
27. Gay, R., Kerenidis, I., Wee, H.: Communication complexity of conditional disclosure of secrets and attribute-based encryption. In: Gennaro, R., Robshaw, M. (eds.) CRYPTO 2015. LNCS, vol. 9216, pp. 485–502. Springer, Heidelberg (2015). https://doi.org/10.1007/978-3-662-48000-7_24
28. Gertner, Y., Ishai, Y., Kushilevitz, E., Malkin, T.: Protecting data privacy in private information retrieval schemes. J. Comput. Syst. Sci. **60**(3), 592–629 (2000)
29. Goyal, V., Pandey, O., Sahai, A., Waters, B.: Attribute-based encryption for fine-grained access control of encrypted data. In: Proceedings of the 13th ACM Conference on Computer and Communications Security, pp. 89–98 (2006)
30. Ishai, Y., Kushilevitz, E.: Private simultaneous messages protocols with applications. In: 5th Israel Symposium on Theory of Computing and Systems, pp. 174–183 (1997)
31. Ishai, Y., Wee, H.: Partial garbling schemes and their applications. In: Esparza, J., Fraigniaud, P., Husfeldt, T., Koutsoupias, E. (eds.) ICALP 2014. LNCS, vol. 8572, pp. 650–662. Springer, Heidelberg (2014). https://doi.org/10.1007/978-3-662-43948-7_54

32. Ito, M., Saito, A., Nishizeki, T.: Secret sharing schemes realizing general access structure. In: Globecom 1987, pp. 99–102 (1987). Journal Version: Multiple assignment scheme for sharing secret. J. Cryptol. **6**(1), 15–20 (1993)
33. Karchmer, M., Wigderson, A.: On span programs. In: 8th Structure in Complexity Theory, pp. 102–111 (1993)
34. Liu, T., Vaikuntanathan, V.: Breaking the circuit-size barrier in secret sharing. In: 50th STOC 2018, pp. 699–708 (2018)
35. Liu, T., Vaikuntanathan, V., Wee, H.: Conditional disclosure of secrets via non-linear reconstruction. In: Katz, J., Shacham, H. (eds.) CRYPTO 2017. LNCS, vol. 10401, pp. 758–790. Springer, Cham (2017). https://doi.org/10.1007/978-3-319-63688-7_25
36. Liu, T., Vaikuntanathan, V., Wee, H.: Towards breaking the exponential barrier for general secret sharing. In: Nielsen, J.B., Rijmen, V. (eds.) EUROCRYPT 2018. LNCS, vol. 10820, pp. 567–596. Springer, Cham (2018). https://doi.org/10.1007/978-3-319-78381-9_21
37. Naor, M., Wool, A.: Access control and signatures via quorum secret sharing. In: 3rd ACM Conference on Computer and Communications Security, pp. 157–167 (1996)
38. Shamir, A.: How to share a secret. Commun. ACM **22**, 612–613 (1979)
39. Shankar, B., Srinathan, K., Rangan, C.P.: Alternative protocols for generalized oblivious transfer. In: Rao, S., Chatterjee, M., Jayanti, P., Murthy, C.S.R., Saha, S.K. (eds.) ICDCN 2008. LNCS, vol. 4904, pp. 304–309. Springer, Heidelberg (2007). https://doi.org/10.1007/978-3-540-77444-0_31
40. Stinson, D.R.: Decomposition construction for secret sharing schemes. IEEE Trans. Inf. Theory **40**(1), 118–125 (1994)
41. Sun, H., Shieh, S.: Secret sharing in graph-based prohibited structures. In: INFOCOM 1997, pp. 718–724 (1997)
42. Tassa, T.: Generalized oblivious transfer by secret sharing. Des. Codes Crypt. **58**(1), 11–21 (2011)
43. Waters, B.: Ciphertext-policy attribute-based encryption: an expressive, efficient, and provably secure realization. In: Catalano, D., Fazio, N., Gennaro, R., Nicolosi, A. (eds.) PKC 2011. LNCS, vol. 6571, pp. 53–70. Springer, Heidelberg (2011). https://doi.org/10.1007/978-3-642-19379-8_4
44. Wee, H.: Dual system encryption via predicate encodings. In: Lindell, Y. (ed.) TCC 2014. LNCS, vol. 8349, pp. 616–637. Springer, Heidelberg (2014). https://doi.org/10.1007/978-3-642-54242-8_26

Towards Optimal Robust Secret Sharing with Security Against a Rushing Adversary

Serge Fehr[1,2] and Chen Yuan[1(✉)]

[1] CWI, Amsterdam, The Netherlands
{serge.fehr,Chen.Yuan}@cwi.nl

[2] Mathematical Institute, Leiden University, Leiden, The Netherlands

Abstract. Robust secret sharing enables the reconstruction of a secret-shared message in the presence of up to t (out of n) *incorrect* shares. The most challenging case is when $n = 2t + 1$, which is the largest t for which the task is still possible, up to a small error probability $2^{-\kappa}$ and with some overhead in the share size.

Recently, Bishop, Pastro, Rajaraman and Wichs [3] proposed a scheme with an (almost) optimal overhead of $\widetilde{O}(\kappa)$. This seems to answer the open question posed by Cevallos et al. [6] who proposed a scheme with overhead of $\widetilde{O}(n+\kappa)$ and asked whether the linear dependency on n was necessary or not. However, a subtle issue with Bishop et al.'s solution is that it (implicitly) assumes a *non-rushing* adversary, and thus it satisfies a *weaker* notion of security compared to the scheme by Cevallos et al. [6], or to the classical scheme by Rabin and BenOr [13].

In this work, we almost close this gap. We propose a new robust secret sharing scheme that offers full security against a rushing adversary, and that has an overhead of $O(\kappa n^{\varepsilon})$, where $\varepsilon > 0$ is arbitrary but fixed. This n^{ε}-factor is obviously worse than the polylog(n)-factor hidden in the $\widetilde{O}$ notation of the scheme of Bishop et al. [3], but it greatly improves on the linear dependency on n of the best known scheme that features security against a rushing adversary (when κ is substantially smaller than n).

A small variation of our scheme has the same $\widetilde{O}(\kappa)$ overhead as the scheme of Bishop et al. *and* achieves security against a rushing adversary, but suffers from a (slightly) superpolynomial reconstruction complexity.

1 Introduction

Background. Robust secret sharing is an extended version of secret sharing as originally introduced by Shamir [14] and Blakley [4], where the reconstruction is required to work even if some of the shares are incorrect (rather than missing, as in the standard notion). Concretely, a robust secret sharing scheme needs to satisfy t-privacy: any t shares reveal no information on the secret, as well as t-robust-reconstructability: as long as no more than t shares are incorrect the

Y. Ishai and V. Rijmen (Eds.): EUROCRYPT 2019, LNCS 11478, pp. 472–499, 2019.
https://doi.org/10.1007/978-3-030-17659-4_16

secret can be reconstructed from the full set of n (partly correct, partly incorrect) shares. For $t < n/3$ this can easily be achieved by means of error correct techniques, whereas for $t \geq n/2$ the task is impossible. Thus, the interesting region is $n/3 \leq t < n/2$, respectively $n = 2t + 1$ if we want t maximal, where robust secret sharing is possible but only up to a small error probability $2^{-\kappa}$ (which can be controlled by a statistical security parameter κ) and only with some overhead in the share size (beyond the size of the "ordinary", e.g., Shamir share). There are many works [2,3,5–9,11,13] in this direction.[1]

The classical scheme proposed by Rabin and BenOr [13] has an overhead in share size of $O(\kappa n)$, i.e., next to the actual share of the secret, each player has to hold an additional $O(\kappa n)$ bits of information as part of his share. This additional information is in the form of $n - 1$ authentication tags and keys. Concretely, every player P_i holds $n - 1$ authentication keys $key_{i,j}$ that allow him to verify the (Shamir) shares s_j of all parties P_j, plus $n - 1$ authentication tags $\sigma_{i,j}$ that allow the other parties to verify his share s_i by means of their keys. By this way, the honest parties can recognize all incorrect shares—and, in case the reconstructor is not a share holder, he would keep those shares that are correctly verified by at least $t + 1$ other parties and dismiss the others (note that a dishonest share holder may also lie about his authentication key, and thus make look a correct share incorrect).

Cevallos, Fehr, Ostrovsky and Rabani [6] proposed an improvement, which results in an overhead in share size of $\widetilde{O}(n + \kappa)$ instead. The core insight is that in the Rabin-BenOr scheme, one can reduce the size of the authentication keys and tags (and thus weaken the security of the authentication) at the expense of a slightly more involved reconstruction procedure—and a significantly more involved analysis. Since the linear dependency of the overhead on κ is unavoidable, they posed the question of whether the linear dependency on n is necessary, or whether an overhead of $\widetilde{O}(\kappa)$ is possible.

Bishop, Pastro, Rajaraman and Wichs [3] gave a positive answer to this question by proposing a scheme that indeed has an overhead in share size of $\widetilde{O}(\kappa)$. At first glance, this seems to settle the case. However, a subtle issue is that their scheme is proven secure only against a weaker attacker than what is considered in the above works. Concretely, the security of their scheme relies on the (implicit) assumption that the attacker is *non-rushing*, whereas the above discussed schemes remain secure in the presence of a *rushing* attacker. As such, the open question of Cevallos et al. [6] is not fully answered.

Recall that for the attacker to be rushing, it means that during the reconstruction procedure, when the parties announce their shares, he can decide on the incorrect shares of the corrupt parties *depending* on the shares that the honest parties announce (rather than on the shares of the corrupt parties alone). This is in particular meaningful and desirable if it is the parties themselves that do the reconstruction—in this case there is little one can do to prevent the corrupt parties from waiting and receiving the honest parties's shares, and then "rush" and

[1] In particular, [3,7,9,11] use partly similar tools than we do but achieve weaker or incomparable results.

announce their own shares before the end of the communication round. Even if the shares (which may include authentication tags and keys etc.) are announced gradually, in multiple rounds, in each round the attacker can still rush in that sense.

To the best of our knowledge, it has not been pointed out before that the scheme of Bishop et al. does not (necessarily) offer security against a rushing attacker. It is also not explicitly discussed in [3], but becomes clear when inspecting the considered security definition carefully.

The Scheme in [3]**.** We briefly discuss some of the features of the scheme by Bishop, Pastro, Rajaraman and Wichs [3], and why it is not secure against a rushing adversary. Like the schemes above, their scheme is also based on verification of shares by means of pairwise authentication using a message authentication code (MAC). However, in order to reduce the number of keys and tags so as to obtain an overhead that is independent of n, every party can now verify only a *subset* of the shares of the other parties, where the subset is randomly chosen (during the sharing phase) and of constant size. However, this makes the reconstruction procedure much more delicate, and Bishop et al. [3] need to pair this basic idea with various additional clever tricks in order to get the reconstruction working. One of these enhancements is that they need to avoid that a dishonest party can make an honest party look dishonest by announcing an incorrect authentication key without being identified as a cheater by other honest parties. This is done by authenticating not only the Shamir share of the party under consideration, but also that party's authentication keys. Concretely, if P_j is chosen to be one of the parties that P_i can verify, then this verification is enabled by means of an authentication tag $\sigma_{i,j}$ that is computed as

$$\sigma_{i,j} = MAC_{key_{i,j}}(s_j, key_j)$$

where $key_{i,j}$ is P_i's verification key, and key_j is the collection of keys that P_j holds for verification of the shares (and keys) that he can verify.[2]

It is now not hard to see that this construction design is inherently insecure against a rushing adversary. Even if the reconstruction is done in multiple rounds where first the Shamir shares and authentication tags are announced and only then the keys (which is what one has to do to make the Rabin-BenOr scheme and the scheme by Cevallos et al. secure in the rushing setting), given that a rushing adversary can choose an incorrect key_j *depending* on the authentication key $key_{i,j}$, the MAC offers no security. Worse, this cannot be fixed by, say, enhancing the MAC: either the adversary has some freedom in choosing incorrect key_j once given $key_{i,j}$, or then it is uniquely determined by $key_{i,j}$ and so P_i knows it—and so it cannot serve the purpose of an authentication key.[3]

[2] One might feel uncomfortable about that there seems to be some circularity there; but it turns out that this is no issue.

[3] The actual scheme is significantly more involved than the simplifies exposition given here, e.g., the identities of the parties that P_j can verify are authenticated as well, and the authentication tags are not stored "locally" but in a "robust and distributed" manner, but the issue pointed out here remains.

We emphasize that we do not claim an explicit rushing attack against the scheme of Bishop et al. [3]. What the above shows is the existence of an attack that prevents a certain property on the consistency graph to hold upon which the reconstruction procedure appears to crucially rely—certainly the proof does. Thus, our claim is that we see no reason why the scheme of Bishop et al. should offer security against a rushing adversary.

Our Result. In this work, we propose a new robust secret sharing scheme. Our new scheme is secure against a rushing adversary *and* close to optimal in terms of overhead. By "close to optimal" we mean that our new scheme has an overhead of $O(\kappa n^{\varepsilon})$ and runs in polynomial time for any arbitrary fixed constant $\varepsilon > 0$. This is obviously slightly worse than the scheme of Bishop et al. [3], which has an overhead of $O(\kappa \cdot \text{polylog}(n))$, but it greatly improves over the best known scheme that features security against a rushing adversary when n is significantly larger than κ.

Our approach recycles some of the ideas from Bishop et al. [3] (e.g. to use "small random subsets" for the verifying parties, and to store the authentication tags in a "robust and distributed" manner) but our scheme also differs in many aspects. The crucial difference is that we do not require the authentication keys to be authenticated; this is what enables us to obtain security against a rushing adversary. Also, how the reconstruction actually works—and *why* it works—is very different. In our approach, we mainly exploit the expander property of the "verification graph" given by the randomly chosen set of neighbors for each P_i, i.e., the set of parties whose share P_i can verify.

For instance, in a first step, our reconstruction procedure checks if the number of incorrect Shamir shares is almost t (i.e., maximal) or whether there is a small *linear* gap. It does so by checking if there are $t+1$ parties that accept sufficiently more than half of the shares they verify. This works because by the random choice of the neighbors, the local view of each honest party provides a good estimate of the global picture (except with small probability). e.g., if almost half of all the shares are incorrect, then for each honest party roughly half of his neighbors has an incorrect share.

If the outcome is that there is a (small but positive) linear gap between the number of incorrect shares and t then we can employ list-decoding to obtain a poly-size list of possible candidates for the secret. In order to find the right secret from the list we need to further inspect the "consistency graph", given by who accepts whom. Concretely, for every secret on the list (and the corresponding *error-corrected* list of shares) it is checked if there exist $t+1$ parties whose shares are deemed correct *and* who accept a party whose share is deemed incorrect. It is clear that this cannot happen if the secret in question is the right one, because no one of the $t+1$ honest parties would accept an incorrect share. And, vice versa, if the secret in question is incorrect then, because of the promised redundancy in the correct shares, there must be a certain number of correct shares that are deemed incorrect and, with the parameters suitably chosen, each honest party has one of them as neighbor (except with small probability) and so will accept that one.

If, on the other hand, the outcome of the initial check is that there are almost t incorrect shares, then the reconstruction procedure uses a very different approach to find the correct shares. Explaining the strategy in detail is beyond the scope of this high-level sketch, but the idea is to start with a set that consists of a single party and then recursively pull those parties into the set that are accepted by the current parties in the set. The hope is that when we start with an honest party then we keep mostly including other honest parties. Of course, we cannot expect to end up with honest parties only, because an honest party may pull a party into the set that has announced a correct share (and thus looks honest) but which is actually dishonest and accepts incorrect shares of other dishonest parties, which then get pulled into the set as well. But note, given that we are in the case of almost t incorrect shares, there are not many such "passively dishonest" parties, and we can indeed control this and show that if we stop at the right moment, the set will consist of mainly honest parties and only a few dishonest parties with incorrect shares. By further inspection of the "consistency graph", trying to identify the missing honest parties and removing dishonest ones, we are eventually able to obtain a set of parties that consists of all honest parties, plus where the number of "actively dishonest" parties is at most half the number of "passively dishonest" parties (except with small probability), so that we have sufficient redundancy in the shares to recover the secret (using Reed-Solomon error correction).

By choosing the out-degree of the "verification graph" (i.e., the number of parties each party can verify) appropriately, so that the above informal reasoning can be rigorously proven (to a large extent by exploiting the randomness of each party's neighborhood and applying the Chernoff-Hoeffding bound), we obtain the claimed overhead $O(\kappa n^{\varepsilon})$ for an arbitrary choice of $\varepsilon > 0$.

As a simple variation of our approach, by choosing the out-degree of the "verification graph" to be polylog(n), we obtain the same (asymptotic) $\widetilde{O}(\kappa)$ overhead as Bishop et al. [6] and still have security against a rushing adversary, but then the reconstruction becomes (slightly) superpolynomial (because the size of the list produced by the list-decoder becomes superpolynomial).

2 Preliminaries

2.1 Graph Notation

Let $G = (V, E)$ be a graph with vertex set V and edge set E. By convention, $(v, w) \in E$ is the edge directed from v to w. For $S \subseteq V$, we let $G|_S$ be the restriction of G to S, i.e., $G|_S = (S, E|_S)$ with $E|_S = \{(u, v) \in E : u, v \in S\}$. Furthermore, we introduce the following notation.

For $v \in V$, we set

$$N^{\mathsf{out}}(v) = \{w \in V : (v, w) \in E\} \quad \text{and} \quad N^{\mathsf{in}}(v) = \{w \in V : (w, v) \in E\}.$$

We often write E_v as a short hand for $N^{\mathsf{out}}(v)$, and call it the *neighborhood* of v. For $S \subseteq V$, we set

$$N_S^{\mathsf{out}}(v) = N^{\mathsf{out}}(v) \cap S \quad \text{and} \quad N_S^{\mathsf{in}}(v) = N^{\mathsf{out}}(v) \cap S.$$

We extend this notation to a *labeled* graph, i.e., when G comes with a function $L : E \to \{\mathsf{good}, \mathsf{bad}\}$ that labels each edge. Namely, for $v \in V$ we set

$$N^{\mathsf{out}}(v, \mathsf{good}) = \{w \in N^{\mathsf{out}}(v) : L(v, w) = \mathsf{good}\},$$
$$N^{\mathsf{in}}(v, \mathsf{good}) = \{w \in N^{\mathsf{in}}(v) : L(w, v) = \mathsf{good}\},$$

and similarly $N^{\mathsf{out}}(v, \mathsf{bad})$ and $N^{\mathsf{in}}(v, \mathsf{bad})$. Also, $N_S^{\mathsf{out}}(v, \mathsf{good})$, $N_S^{\mathsf{in}}(v, \mathsf{good})$, $N_S^{\mathsf{out}}(v, \mathsf{bad})$ and $N_S^{\mathsf{in}}(v, \mathsf{bad})$ are defined accordingly for $S \subseteq V$. Finally, we set

$$n^{\mathsf{out}}(v) = |N^{\mathsf{out}}(v)| \quad \text{and} \quad n_S^{\mathsf{in}}(v, \mathsf{bad}) = |N_S^{\mathsf{in}}(v, \mathsf{bad})|$$

and similarly for all other variations.

We refer to a graph $G = (V, E)$ as a *randomized* graph if the edges E are chosen in a randomized manner, i.e., if E is actually a random variable. We are particularly interested in randomized graphs where (some or all of) the E_v's are uniformly random and independent subsets $E_v \subset V \setminus \{v\}$ of a given size d. For easier terminology, we refer to such neighborhoods E_v as being *random and independent*.

2.2 Chernoff Bound

Like for [3], much of our analysis relies on the Chernoff-Hoeffding bound, and its variation to "sampling without replacement". Here and throughout, $[n]$ is a short hand for $\{1, 2, \dots, n\}$.

Definition 1 (Negative Correlation [1]). *Let $X_1, \dots, X_n$ be binary random variables. We say that they are negatively correlated if for all $I \subset [n]$:*

$$\Pr[X_i = 1 \, \forall i \in I] \le \prod_{i \in I} \Pr[X_i = 1], \quad \Pr[X_i = 0 \, \forall i \in I] \le \prod_{i \in I} \Pr[X_i = 0].$$

Theorem 1 (Chernoff-Hoeffding Bound). *Let $X_1, \dots, X_n$ be random variables that are independent and in the range $0 \le X_i \le 1$, or binary and negatively correlated, and let $u = E\left[\sum_{i=1}^{n} X_i\right]$. Then, for any $0 < \delta < 1$:*

$$\Pr\left[\sum_{i=1}^{n} X_i \le (1 - \delta)u\right] \le e^{-\delta^2 u/2} \quad \textit{and} \quad \Pr[\left[\sum_{i=1}^{n} X_i \ge (1 + \delta)u\right] \le e^{-\delta^2 u/3}.$$

As immediate consequence, we obtain the following two bounds. The first follows from Chernoff-Hoeffding with independent random variables, and the latter from Chernoff-Hoeffding with negatively correlated random variables. We refer to [1] for more details, e.g., for showing that the random variables $X_j = 1$ if $j \in E_v$ and 0 otherwise are negatively correlated for E_v as in Corollary 1.

Corollary 1. *Let G be a randomized graph with the property that, for some fixed $v \in V$, the neighborhood E_v is a random subset of $V \setminus \{v\}$ of size d. Then, for any fixed subset $T \subset V$, we have*

$$\Pr\left[n_T^{\mathsf{out}}(v) \ge (1+\epsilon)\tfrac{|T|d}{|V|}\right] \le e^{-\epsilon^2 \frac{|T|d}{2|V|}} \quad \textit{and} \quad \Pr\left[n_T^{\mathsf{out}}(v) \le (1-\epsilon)\tfrac{|T|d}{|V|}\right] \le e^{-\epsilon^2 \frac{|T|d}{3|V|}}.$$

Corollary 2. *Let G be a randomized graph with the property that, for some fixed $T \subset V$, the neighborhoods E_v for $v \in T$ are random and independent of size d (in the sense as explained in Sect. 2.1). Then, for any $v \notin T$, we have*

$$\Pr\left[n_T^{\mathsf{in}}(v) \geq (1+\epsilon)\frac{|T|d}{|V|}\right] \leq e^{-\epsilon^2 \frac{|T|d}{2|V|}} \text{ and } \Pr\left[n_T^{\mathsf{in}}(v) \leq (1-\epsilon)\frac{|T|d}{|V|}\right] \leq e^{-\epsilon^2 \frac{|T|d}{3|V|}}.$$

We emphasize than when we apply these corollaries, we consider a graph G where *a priori* all E_v's are random and independent. However, our reasoning typically is applied *a posteriori*, given some additional information on G, like the adversaries view. As such, we have to be careful each time that the considered neighborhoods are still random *conditioned* on this additional information on G.

2.3 Robust Secret Sharing

A robust secret sharing scheme consists of two interactive protocols: the *sharing* protocol **Share** and the *reconstruction* protocol **Rec**. The sharing protocol is executed by a *dealer* D and n parties $1, \ldots, n$: the dealer takes as input a message **msg**, and each party $i \in \{1, \ldots, n\}$ obtains as output a so-called *share*. Typically, these shares are locally computed by the dealer and then individually sent to the parties. The reconstruction protocol is executed by a *receiver* R and the n parties: each party is supposed to use its share as input, and the goal is that R obtains **msg** as output. Here, the protocol is typically so that the parties send their shares to R (possibly "piece-wise", distributed over multiple rounds of communication), and R then performs some local computation.

Such a robust secret sharing should be "secure" in the presence of an adversary that can adaptively corrupt up to t of the parties $1, \ldots, n$. Once a party is corrupted, the adversary is able to see the share of this party, and he can choose the next corruption based on the shares of the currently corrupt parties. Furthermore, in the reconstruction protocol, the corrupt parties can arbitrarily deviated from the protocol and, e.g., use incorrect shares. The following captures the security of a robust secret sharing in the list of such an adversary.

Definition 2 (Robust Secret Sharing). *Such a pair* (**Share**, **Rec**) *of protocols is called a* (t, δ)-robust secret sharing scheme *if it satisfies the following properties hold for any distribution of* **msg** *(from a given domain).*

- Privacy: *Before* **Rec** *is started, the adversary has no more information on the shared secret* **msg** *than he had before the execution of* **Share**.
- Robust reconstructability: *At the end of* **Rec**, *the reconstructor R output* **msg**$'$ = **msg** *except with probability at most δ.*

2.4 On the Power of Rushing

As defined above, there is still some ambiguity in the security notion, given that we have not specified yet the adversary's (dis)ability of *eavesdropping* on the communication of the sharing and the reconstruction protocols. Obviously, for the *privacy* condition to make sense, it has to be assumed that during the execution of the sharing protocol the adversary has no access to the communication between D and the uncorrupt parties. On the other hand, it is commonly assumed that the adversary *has* access to the communication between the parties and R during the execution of the reconstruction protocol. This in particular means that the adversary can choose the incorrect shares, which the corrupt parties send to R, *depending* on the honest parties' (correct) shares.[4] Such an adversary is referred to as a *rushing* adversary. In contrast, if it is assumed that the adversary has to choose the incorrect shares depending on the shares of the corrupt parties only, one speaks of a *non-rushing* adversary. Thus, Definition 2 above comes in two flavors, depending on whether one considers a rushing or a non-rushing adversary. Obviously, considering a rushing adversary gives rise to a *stronger* notion of security. In order to deal with a rushing adversary, it is useful to reveal the shares "in one go" but piece-by-piece, so as to limit the dependence between incorrect and correct shares.

In this work, in order to be in-par with [13] and [6], we require security against a *rushing* adversary. On the other hand, the scheme by Bishop, Pastro, Rajaraman and Wichs [3] offers security against a non-rushing adversary only—and, as explained in the introduction, there are inherent reasons why it cannot handle a rushing adversary.

3 Overview of Scheme

Our Approach. As in [3], the sharing phase is set up in such a manner that every party i can verify (by means of a MAC) the Shamir shares of the parties j of a *randomly sampled* subset $E_i \subset [n]\backslash\{i\}$ of parties. However, in contrast to [3], in our scheme *only* the Shamir share is authenticated; in particular, we do not authenticate the authentication keys (nor the set E_j).

If the reconstruction is then set up in such a way that first the Shamir shares are announced, and only afterwards the authentication keys, it is ensured (even in the presence of a rushing adversary) that the consistency graph, which labels an edge from i to $j \in E_i$ as "good" if and only if i correctly verifies j's Shamir share, satisfies the following:

- All edges from honest parties to passive or honest parties are labeled good.
- All edges form honest parties to active parties are labeled bad.

Here, and in the remainder, a corrupt party is called *active* if it announced an incorrect Shamir share in the reconstruction, and it is called *passive* if it

[4] This may look artificial at first glance, but one motivation comes from the fact that in some applications one might want to do the reconstruction *among the parties*, where then each party individually plays the role of R (and performs the local computation that the reconstruction protocol prescribes). In this case, every party sends his share to every other party, and thus the corrupt parties unavoidably get to see the shares of the honest parties and can decide on the incorrect shares depending on those.

announced a correct Shamir share, but may still lie about other parts, like the authentication keys. This is a significant difference to [3], where it is also ensured that corrupt parties that lie about their authentication keys are recognized as well.

Divide the Discussion. Similarly to [3], the reconstructor first tries to distinguish between the (non-exclusive) cases $|P| \leq \epsilon n$ and $|P| \geq \frac{\epsilon n}{4}$, where P denotes the set of passive parties (as defined above). In order to do so, we observe that the honest party is expected to have $(|P| + |H|)\frac{d}{n}$ good outgoing edges, where H is the set of honest parties. Thus, if there exist $t+1$ parties with more than $(1+\epsilon)d/2$ good outgoing edges, we are likely to be in the case $|P| \geq \frac{\epsilon n}{4}$, and otherwise $|P| \leq \epsilon n$.

Based on this distinction, the reconstructor will then refer to either of the following two algorithms to recover the secret.

Code Based Algorithm. This algorithm is used to handle the case $|P| \geq \frac{\epsilon n}{4}$. Here, one can use the redundancy provided by the correct shares of the parties in P to do *list-decoding.* This works given that the Shamir sharing is done by means of a *folded* Reed-Solomon code. Since those are maximum distance separable (MDS) codes, the corresponding secret sharing scheme is still threshold; moreover, it enjoys the nice feature that we can apply list decoding to correct up to $t - \epsilon n/4$ corruptions for any small constant ϵ. Finding the right entry in the list can then be done by a further inspection of the consistency graph.

Graph Algorithm. This graph algorithm is used in case $|P| \leq \epsilon n$. The basic algorithm starts off with a particular party, and produces the correct secret (with high probability) if that party happens to be honest. Hence, applying this algorithm to all choices for that party and taking a majority enables to reconstruct the secret.

The algorithm consists of three steps.

- The first step is to find a big subset V that contains many honest parties and very small proportion of dishonest parties. We do so by starting off with $V = \{i\}$ for a particular party i (which we assume to be honest for the discussion) and recursively include all parties into V that are correctly verified by the parties in V. A simple argument shows that in each step, we expect to include $d/2$ honest parties and at most ϵd passive parties.
  By the expander property of the consistency graph restricted to the honest parties ensures that the set V will soon be expanded to a set containing many honest parties. On the other hand, we can limit the "damage" done by passive parties by only including parties that have at most $\frac{d}{2}(1+3\epsilon)$ good outgoing edges. Given that there are only very few passive parties and that we limit the number of active parties they can pull into V, we can show that V can be expanded to a set of size $\Omega(\epsilon n)$ such that at most a $O(\sqrt{\epsilon})$-fraction of the parties in V are corrupt.
- The next step is to rely on the authentication of parties in V to include all honest parties and few dishonest parties where the majority is passive. We first expand V to contain *all* honest parties and at most $O(\sqrt{\epsilon} n)$ dishonest

parties. Then, we remove all active parties from V (but possibly also some honest and passive ones). Let W be the set of all parties removed from V. We show that the resulting set V still contains almost all honest parties and few passive parties, and W is of size $O(\sqrt{\epsilon}n)$ and contains the rest of honest parties.
A subtle issue now is that the sets V and W above depend on the E_i's of the honest parties; this then means that now given these two sets, we cannot rely anymore on the randomness of the E_i's. In order to circumvent this, we resort to another layer of authentication that is done in parallel to the former, with fresh E_i''s, and by means of this, we can then eventually identify a subset $S \subseteq [n]$ that contains all $t+1$ honest parties, as well as some number h of passive parties and at most $\frac{h}{2}$ active parties (with high probability).
- Given that we have "sufficiently more redundancy that errors", the secret can now be recovered by means of Reed-Solomon error correction (noting that a codeword of a folded Reed-Solomon code is also a codeword of some classic Reed-Solomon code).

4 Building Blocks

We present three building blocks here which are used in our construction.

4.1 Shamir Secret Sharing with List Decoding

It is well known that the share-vector in Shamir's secret sharing scheme is nothing else than a codeword of a Reed-Solomon code. Thus, Reed-Solomon decoding techniques can be applied when we are in the regime of unique decoding. Furthermore, if we use a *folded* Reed-Solomon code, then we still get a Shamir-like threshold secret sharing scheme, but in addition we can employ *list-decoding* when we are in a regime were decoding is not unique anymore.

In summary, we have the following (see Appendix A.1 and [12] for the details).

Proposition 1. *Let γ be any small constant. There exists $2t+1$-party threshold secret sharing scheme over $\mathbb{F}_q$ with $q = t^{O(\frac{1}{\gamma^2})}$ such that:*

- *This scheme enjoys t-privacy and $t+1$-reconstruction.*
- *There is a randomized list decoding algorithm that corrects up to $t - \gamma(2t+1)$ incorrect shares and outputs a list of candidates containing the correct secret with probability at least $1 - 2^{-\Omega(t)}$. The list size is $\lambda = (\frac{1}{\gamma})^{\frac{1}{\gamma}\log\frac{1}{\gamma}}$ and this list decoding algorithm runs in time poly(t, λ).*
- *There exists an efficient decoding algorithm that reconstructs the secret from any $t+1+2a$ shares of which at most a are incorrect.*

4.2 MAC Construction

Similarly to [3], our construction requires a message authentication code (MAC) with some additional features. There is some overlap with the features needed in [3], but also some differences.

Definition 3. *A message authentication code (MAC) for a finite message space $\mathcal{M}$ consists of a family of functions $\{MAC_{key} : \mathcal{M} \times \mathcal{R} \to \mathcal{T}\}_{key \in \mathcal{K}}$. This MAC is said to be (ℓ, ϵ)-secure if the following three conditions hold.*

1. Authentication security: *For all $(m, r) \neq (m', r') \in \mathcal{M} \times \mathcal{R}$ and all $\sigma, \sigma' \in \mathcal{T}$,*

$$\Pr_{key \leftarrow \mathcal{K}}[MAC_{key}(m', r') = \sigma' | MAC_{key}(m, r) = \sigma] \leq \epsilon.$$

2. Privacy over Randomness: *For all $m \in \mathcal{M}$ and $key_1, \ldots, key_\ell \in \mathcal{K}$, the distribution of ℓ values $\sigma_i = MAC_{key_i}(m, r)$ is independent of m over the choice of random string $r \in \mathcal{R}$, i.e.,*

$$\Pr_{r \leftarrow \mathcal{R}}[(\sigma_1, \ldots, \sigma_\ell) = \mathbf{c} | m] = \Pr_{r \leftarrow \mathcal{R}}[(\sigma_1, \ldots, \sigma_\ell) = \mathbf{c}]$$

for any $\mathbf{c} \in \mathcal{T}^\ell$.

3. Uniformity: *For all $(m, r) \in \mathcal{M} \times \mathcal{R}$, the distribution of $\sigma = MAC_{key}(m, r)$ is uniform at random over the random element $key \in \mathcal{K}$.*

The above privacy condition will be necessary for the privacy of the robust secret sharing scheme, since the Shamir shares will be authenticated by means of such a MAC but the corresponding tags will not be hidden from the adversary.

The uniformity property will be crucial in a *lazy sampling* argument, were we need to "simulate" certain tags *before* we know which messages they actually authenticate. With the uniformity property, this can obviously be done by picking σ uniformly at random from $\mathcal{T}$. When m and r become available, we can then sample a uniformly random key *key* subject to $MAC_{key}(m, r) = \sigma$. This has the same distribution as when *key* is chosen uniformly at random and σ is computed as $\sigma = MAC_{key}(m, r)$.

The following variation of the standard polynomial-evaluation MAC construction meets all the requirements.

Theorem 2 (Polynomial Evaluation). *Let $\mathbb{F}$ be a finite field. Let $\mathcal{M} = \mathbb{F}^a$, $\mathcal{R} = \mathbb{F}^\ell$ and $\mathcal{T} = \mathbb{F}$ such that $\frac{a+\ell}{|\mathbb{F}|} \leq \epsilon$. Define the family of MAC functions $\{MAC_{(x,y)} : \mathbb{F}^a \times \mathbb{F}^\ell \to \mathbb{F}\}_{(x,y) \in \mathbb{F}^2}$ such that*

$$MAC_{(x,y)}(\mathbf{m}, \mathbf{r}) = \sum_{i=1}^{a} m_i x^{i+\ell} + \sum_{i=1}^{\ell} r_i x^i + y$$

for all $\mathbf{m} = (m_1, \ldots, m_a) \in \mathbb{F}^a$, $\mathbf{r} = (r_1, \ldots, r_\ell) \in \mathbb{F}^\ell$ and $(x, y) \in \mathbb{F}^2$. Then, this family of MAC functions is (ℓ, ϵ)-secure.

4.3 Robust Distributed Storage

A *robust distributed storage scheme* is a robust secret sharing scheme as in Definition 2 but without the privacy requirement. This was used in [3] in order to ensure that dishonest parties cannot lie about the tags that authenticate their shares (so as to, say, provoke disagreement among honest parties about the correctness of the share).[5] Also our construction uses a robust distributed storage scheme for storing the authentication tags. It does not play such a crucial role here as in [3], but it makes certain things simpler.

A robust distributed storage scheme can easily be obtained by encoding the message by means of a list-decodable code and distribute the components of the code word among the parties, and to give each party additionally a random key and the hash of the message under the party's key (using almost-universal hashing). In order to reconstruct, each party runs the list-decoding algorithm and uses his key and hash to find the correct message in the list. The exact parameters then follow from list-decoding parameters.

Therefore, each party i holds two components, the i-th share of list-decodable code, p_i, and a hash-key and the hash of the message, jointly referred to as q_i. While [3] did not consider a rushing adversary, it is easy to see that security of this robust distributed storage scheme against a rushing adversary can be obtained by having the parties reveal p_i and q_i in two different communication rounds (so that the adversary has to decide on an incorrect p_i *before* he knows the keys that the honest parties will use). Therefore, the following result from [3], which is obtained by using suitable parameters for the list decoding an hashing, is also applicable in rushing-adversary model.

Theorem 3 ([3]). *For any $n = 2t + 1$ and $u \geq \log n$, there exists a robust distributed storage with messages of length $m = \Omega(nu)$ and shares of length $O(u)$ that can recover the message with probability $1 - O(\frac{n^2}{2^u})$ up to t corruptions.*

In our application, the length of shares is $O(u) = O(n^{\sqrt{\epsilon}})$ and the length of messages is $m = \Omega(nu)$. If we apply this theorem directly, the size of $\mathbb{F}_q$ in their construction is 2^u which is unnecessarily big. Instead, We pick $\mathbb{F}_q$ with $q = \Omega(n^5)$. Then, we obtain following.

Theorem 4. *For any $n = 2t + 1$ and $u = O(n^{\sqrt{\epsilon}})$ for small constant ϵ, there exists a robust distributed storage against rushing adversary with messages of length $m = \Omega(nu)$, shares of length $O(u)$ that can recover the message with probability $1 - O(\frac{1}{n^2})$ up to t corruptions. In this robust distributed storage, party i holds two components, p_i and q_i, revealed in two rounds.*

[5] On the other hand, this is why the additional privacy property of the MAC is necessary, since the robust distributed storage does not offer privacy, and thus the tags are (potentially) known.

5 The Robust Secret Sharing Scheme

5.1 Sharing Protocol

Let t be an arbitrary positive integer and $n = 2t + 1$. Let $\epsilon > 0$ be a small constant and $d = n^{\sqrt{\epsilon}}$. Let $(\mathbf{Sh}, \mathbf{Lis})$ be the sharing and list decoding algorithm of the threshold secret sharing scheme in Proposition 1 with $\gamma = \frac{\epsilon}{4}$. Also, we use the MAC construction from Theorem 2 with $\ell = 4d$, and with the remaining parameters to be determined later (but chosen so that a share produced by **Sh** can be authenticated).

On input $\mathbf{msg} \in \mathbb{F}_q$, our sharing procedure $\mathbf{Share}(\mathbf{msg})$ proceeds as follows.

1. Let $(s_1, \ldots, s_n) \leftarrow \mathbf{Sh}(\mathbf{msg})$ to be a non-robust secret sharing of **msg**.
2. For each $i \in [n]$, sample MAC randomness $r_i \leftarrow \mathcal{T}^{4d}$ and do the following operation twice.
   (a) For each $i \in [n]$, choose a random set $E_i \subseteq [n] \backslash \{i\}$ of size d. If there exists $j \in [n]$ with in-degree more than $2d$, do it again.[6]
   (b) For each $i \in [n]$, sample the d random MAC keys $key_{i,j} \in \mathcal{T}^2$ for $j \in E_i$. Define $\mathcal{K}_i = \{key_{i,j} : j \in E_i\}$ to be the collection of these d random keys.
   (c) Compute the MAC
   $$\sigma_{i \to j} = MAC_{key_{i,j}}(s_j, r_j) \in \mathcal{T} \quad \forall j \in E_i.$$
   Let E_i, $key_{i,j}$ and $\sigma_{i \to j}$ be the output of the first round and E'_i, $key'_{i,j}$ and $\sigma'_{i \to j}$ be the output of the second round.
3. For each $i \in [n]$, define $\mathbf{tag}_i = \{\sigma_{i \to j} : j \in E_i\} \in \mathcal{T}^d$ and $\mathbf{tag}'_i = \{\sigma'_{i \to j} : j \in E'_i\} \in \mathcal{T}^d$. Let $\mathbf{tag} = (\mathbf{tag}_1, \mathbf{tag}'_1, \ldots, \mathbf{tag}_n, \mathbf{tag}'_n) \in \mathcal{T}^{2nd}$. Use the robust distributed storage scheme to store **tag**. Party i holds p_i and q_i.
4. For $i \in [n]$, define $\mathbf{s}_i = (s_i, E_i, E'_i, \mathcal{K}_i, \mathcal{K}'_i, r_i, p_i, q_i)$ to be the share of party i. Output $(\mathbf{s}_1, \ldots, \mathbf{s}_n)$.

5.2 Reconstruction Protocol

1. The first round: Every party i sends (s_i, r_i, p_i) to the reconstructor R.
2. The second round: Every party i sends $(q_i, E_i, \mathcal{K}_i)$ to the reconstructor R.
3. The third round: Every party i sends $(E'_i, \mathcal{K}'_i)$ to the reconstructor R.

Remark 1. We emphasize that since the keys for the authentication tags are announced *after* the Shamir shares, it is ensured that the MAC does its job also in the case of a rushing adversary. Furthermore, it will be crucial that also the E_i's are revealed in the second round only, so as to ensure that once the (correct and incorrect) Shamir shares are "one the table", the E_i's for the honest parties are still random and independent. Similarly for the E'_i's in the third round.

On receiving the shares of n parties, our reconstruction scheme $\mathbf{Rec}(\mathbf{s_1}, \ldots, \mathbf{s_n})$ goes as follows:

[6] This is for the privacy purpose.

1. R collects the share of robust distributed storage: $(p_i, q_i)_{i \in [n]}$.
2. Reconstruct the $\mathbf{tag} = (\mathbf{tag}_1, \mathbf{tag}'_1, \dots, \mathbf{tag}_n, \mathbf{tag}'_n)$ and parse $\mathbf{tag}_i = \{\sigma_{i \to j} : j \in E_i\}$ and $\mathbf{tag}'_i = \{\sigma'_{i \to j} : j \in E'_i\}$.
3. Define two graphs $G = ([n], E)$ and $G' = ([n], E')$ such that $E = \{(i,j) : i \in [n], j \in E_i\}$ and $E' = \{(i,j) : i \in [n], j \in E'_i\}$.
4. Assign a label $L(e) \in \{\mathtt{good}, \mathtt{bad}\}$ to each edge $e = (i,j) \in E$ such that $L(e) = \mathtt{good}$ if

$$\sigma_{i \to j} = MAC_{key_{i,j}}(s_j, r_j)$$

   and $\mathtt{bad}$ otherwise. Do the same thing to the edge $e \in E'$.
5. Run the Check(G, L, ϵ),
   (a) If the output is Yes, Let $\mathbf{s} = (s_1, \dots, s_n)$ and $\mathbf{c} = \text{List}(G, \mathbf{s}, \epsilon/4)$.
   (b) Otherwise, for each $i \in [n]$, let $\mathbf{c}_i = \text{Graph}(G, G', \epsilon, i)$. If there exists a codeword $\mathbf{c}_i$ repeating at least $t+1$ times, let $\mathbf{c} = \mathbf{c}_i$. Otherwise, $\mathbf{c} = \perp$.
6. Output $\mathbf{c}$.

Note that step 5 in the reconstruction refers to subroutines: Check, List and Graph, which we specify only later.

5.3 The Privacy Property

Theorem 5. *The scheme* (**Share**, **Rec**) *satisfies perfect privacy.*

Proof. Let $C \subset [n]$ be of size t. We let $\mathbf{msg} \in \mathcal{M}$ be arbitrarily distributed and consider $\mathbf{Share}(\mathbf{msg}) = (\mathbf{s}_1, \dots, \mathbf{s}_n)$. Our goal is to show that the distribution of $(\mathbf{s}_i)_{i \in C}$ is independent of $\mathbf{msg}$. Note that $\mathbf{s}_i = (s_i, r_i, p_i, q_i, E_i, E'_i, \mathcal{K}_i, \mathcal{K}'_i)$. Since our threshold secret sharing scheme has t-privacy, the collection of shares s_i for $i \in C$ is independent of $\mathbf{msg}$. By construction, $r_i, E_i, E'_i, \mathcal{K}_i, \mathcal{K}'_i$ are independently chosen as well. Since the (p_i, q_i)'s are computed from $\mathbf{tag}$ (using independent randomness), it suffices to show that $\mathbf{tag}$ reveals no information on $\mathbf{msg}$. Recall that $\mathbf{tag}$ is used to verify the integrity of (s_j, r_j) for all j. For any $j \notin C$, there are at most $4d$ tags $\sigma_{i \to j} = MAC_{key_{i,j}}(s_j, r_j)$ corresponding to the total degree of vertex i in two graphs. By the "privacy over randomness" of the MAC, $\mathbf{tag}$ is independent of these shares s_j, and hence the privacy of $\mathbf{msg}$ is ensured.

6 The Robustness Property

6.1 Preliminary Observations

From the security properties of the robust distributed storage and of the MAC, we immediately obtain the following results.

Lemma 1. $\mathbf{tag}$ *is correctly reconstructed expect with probability* $\epsilon_{tag} = O(1/n^2)$.

This is not negligible; this will be dealt with later by parallel repetition.

Here and in the remainder of the analysis of the robustness property, H denotes the set of honest parties, and C denotes the set of dishonest parties. Furthermore, we decompose C into $C = A \cup P$, where A is the set of dishonest parties that announced an *incorrect* (s_i, r_i) in the first communication round and P denotes the (complementary) set of dishonest parties that announced a *correct* (s_i, r_i). The parties in A are called *active* parties, and the parties in P are referred to as *passive* parties.

Proposition 2. *If* tag *was correctly reconstructed then the labelling of the graph G satisfies the following, except with probability $\epsilon_{mac} \le 4(t+4d)dt/|\mathcal{T}|$. For every $h \in H$ and for every edge $e = (h, j) \in E_h = N^{\text{out}}(h)$, it holds that*

$$L(e) = \begin{cases} \text{bad} & if j \in A \\ \text{good} & if j \in H \cup P \end{cases}.$$

I.e., all the edges from honest parties to active parties are labeled bad and all edges from honest parties to honest parties or passive parties are labeled good. The same holds for G'.

Proof. By the definition of passive parties and the construction of MAC, all edges from honest parties to passive parties are labeled good. It remains to prove the first half of the claim. Let us fix an active party i. According to the definition of active party, he claims $(s'_j, r'_j) \neq (s_j, r_j) \in \mathcal{T}^a \times \mathcal{T}^{4d}$. For any honest party i with $j \in E_i$ or $j \in E'_i$, the edge (i, j) is label good if $\sigma_{i \to j} = MAC_{key_{i,j}}(s'_j, r'_j)$. This event happens with

$$\Pr_{key_{i,j} \leftarrow \mathcal{T}^2}[\sigma_{i \to j} = MAC_{key_{i,j}}(s'_j, r'_j)] \le \frac{a + 4d}{|\mathcal{T}|}$$

due to the authentication of MAC. Note that each vertex has at most $4d$ incoming edges. Taking a union bound over all these edges and the active parties, the desired result follows.

6.2 On the Randomness of the Graph

Much of our analysis relies on the randomness of the graph G (and G'). A subtle point is that even though *a priori* all of the E_i are chosen to be random and independent (in the sense as explained in Sect. 2.1), we have to be careful about the *a posteriori* randomness of the E_i *given the adversary's view*. In particular, since the adversary can corrupt parties *adaptively* (i.e., depending on what he has seen so far), if we consider a particular dishonest party j then the mere fact that this party is dishonest may affect the *a posteriori* distribution of G.

However, and this is what will be crucial for us is that, conditioned on the adversary's view, the E_i's *of the honest parties* i remain random and independent.

Proposition 3. *Up to right before the second communication round of the reconstruction protocol, conditioned on the adversary's view of the protocol, the graph G is such that the E_i for $i \in H$ are random and independent.*
The corresponding holds for G' up to right before the third communication round.

Proof. The claim follows from a straightforward lazy-sampling argument. It follows by inspection of the protocol that one can delay the random choice of each E_i (and E_i') to the point where party i gets corrupted, or is announced in the corresponding round in the reconstruction protocol. The only subtle issue is that, at first glance it seems that the computation of the tags $\sigma_{i \to j} = MAC_{key_{i,j}}(s_j, r_j)$ for $j \in E_i$ requires knowledge of E_i. However, by the uniformity property of MAC, these tags can instead be "computed" by sampling d tags uniformly at random from $\mathcal{T}$, and once party i gets corrupted and E_i is sampled, one can choose the keys $key_{i,j}$ appropriately for $j \in E_i$.

Remark 2. In the sequel, when making probabilistic statements, they should be understood as being *conditioned* on an arbitrary but fixed choice of the adversary's view. This in particular means that H, P and A are *fixed* sets then (since they are determined by the view of the adversary), and we can quantify over, say, all honest parties. The randomness in the statements then stems from the randomness of the E_i's (and E_i's) of the honest parties $i \in H$, as guaranteed by Proposition 3 above.

Remark 3. The remaining analysis below is done under the implicit assumption that **tag** is correctly reconstructed and that the labelling of G and of G' is as specified in Proposition 2. We will incorporate the respective error probabilities then in the end.

6.3 The Check Subroutine

Roughly speaking, the following subroutine allows the reconstructor to find out if $|P|$, the number of passive parties, is linear in n or not.

Check(G, L, ϵ)

- Input: $G = ([n], E, L)$ and ϵ.
- If $|\{i \in [n] : n^{\mathsf{out}}(i, \mathsf{good}) \geq \frac{d}{2}(1+\epsilon)\}| \geq t+1$, then output "Yes".
- Otherwise, output "No".

Theorem 6. *Except with probability $\epsilon_{check} \leq 2^{-\Omega(\epsilon d)}$,* Check$(G, L, \epsilon)$ *outputs "Yes" if $|P| \geq \epsilon n$ and "No" if $|P| \leq \epsilon n/4$ (and either of the two otherwise).*

Proof. We only analyze the case that $|P| \leq \frac{\epsilon n}{4}$. The same kind of reasoning can be applied to the case $|P| \geq \epsilon n$. By Proposition 3 (and Remark 2), we know that for given P and H, the E_i's for $i \in H$ are random and independent. It follows that $n^{\text{out}}(i, \textsf{good}) = n^{\text{out}}_{P\cup H}(i)$ is expected to be $\frac{|P|+|H|}{n} \le \frac{d}{2}(1+\frac{\epsilon}{2})$, and thus, by Corollary 1,

$$\Pr\left[n^{\text{out}}(i, \textsf{good}) \ge \tfrac{d}{2}(1+\epsilon)\right] \le 2^{-\Omega(\epsilon d)}.$$

Taking a union bound over all honest parties, we have

$$\Pr\left[\exists i \in H : n^{\text{out}}(i, \textsf{good}) \ge \tfrac{d}{2}(1+\epsilon)\right] \le (t+1)2^{-\Omega(\epsilon d)} = 2^{-\Omega(\epsilon d)}.$$

Thus, except with probability $(t+1)2^{-\Omega(\epsilon d)}$, $n^{\text{out}}(i, \textsf{good}) \ge \frac{d}{2}(1+\epsilon)$ can only hold for dishonest parties i, and in this case Check(G, L, ϵ) outputs "No". The desired result follows.

Remark 4. The subroutine Check(G, L, ϵ) allows us to find out if $|P| \ge \frac{\epsilon n}{4}$ or $|P| \le \epsilon n$. If Check(G, L, ϵ) tells us that $|P| \ge \frac{\epsilon n}{4}$ (by outputting "Yes") then we use the redundancy provided by the shares of the parties in P to recover the secret by means of the code based algorithm List$(G, \mathbf{s}, \epsilon/4)$. If Check$(G, L, \epsilon)$ tells us that $|P| \le \epsilon n$ (by outputting "No") then we run the graph algorithm Graph(G, G', ϵ, v) for every choice of party v. For honest v, it is ensure to output the correct secret (with high probability), and so we can do a majority decision. We leave the description and the analysis of the code based algorithm and the graph algorithm to the respective next two sections.

6.4 Code Based Algorithm

Recall that $\gamma = \frac{\epsilon}{4}$. H is the set of honest parties, P is the set of passive parties and A is the set of active parties. In this section, we present an algorithm List$(G, \mathbf{s}, \gamma)$ based on the list decoding algorithm of secret sharing scheme in Proposition 1 up to $t - \gamma n$ errors.

Code Based Algorithm, List$(G, \mathbf{s}, \gamma)$

- Input $G = ([n], E, L), \mathbf{s}, \gamma$.
- Run the list decoding algorithm on $\mathbf{s}$ to correct up to $t - \gamma n$ errors and output the list of candidates $(\mathbf{c}_1, \ldots, \mathbf{c}_\ell)$.
- Let S_i (T_i) be the set of parties whose shares agree (do not agree) with $\mathbf{c}_i$.
- For $1 \le i \le \ell$, run Cand(G, S_i, T_i). If the output is "succeed" then output $\mathbf{c}_i$.
- Output "fail".

We proceed to the analysis of the algorithm. The input of the list decoding algorithm is n shares of $\mathbf{s}$. Since $|P| \ge \gamma n$, there are at most $t - \gamma n = \frac{n}{2}(1 - 2\gamma)$ shares that are corrupted. Thus, by Proposition 1, the output of this list decoding algorithm will contain the correct codeword with probability at least $1 - 2^{-\Omega(n)}$. Moreover, the list size of this algorithm is at most $(\frac{1}{\gamma})^{O(1/\gamma \log 1/\gamma)}$. We may assume that this list only include codewords that are at most $t - \gamma n$ away from the n shares. Let $\mathbf{c}_1, \ldots, \mathbf{c}_\ell$ be the candidates on this list. To find the correct one, we resort to the labelled graph $G = ([n], E, L)$. Note that $\mathbf{c}_i$ for $1 \le i \le \ell$ are determined right after the first communication round. Thus, by Proposition 3, conditioned on $\mathbf{c}_1, \ldots, \mathbf{c}_\ell$ (and the entire view of the adversary at this point), the E_i's for $i \in H$ are random and independent. For each candidate $\mathbf{c}_i$, we run the algorithm $\text{Cand}(G, S_i, T_i, \gamma)$ to check if it is the correct codeword. For the correct codeword $\mathbf{c}_r$, we claim that this algorithm $\text{Cand}(G, S_r, T_r, \gamma)$ will always output succeed. To see this, we notice that S_r must contain the set of $t+1$ honest parties H. Meanwhile, T_r is a subset of active parties A. By our assumption, there does not exist good edge from H to T_r. The desired results follows as these $t+1$ honest parties will remain in S_r after calling $\text{Cand}(G, S_r, T_r, \gamma)$.

Verify the candidate, Cand(G, S, T)

- Input: $G = ([n], E, L)$, S, T.
- Remove all i from S if $n_T^{\mathsf{out}}(i, \mathsf{good}) \geq 1$.
- If $|S| \geq t+1$, output "succeed". Otherwise, output "fail".

It remains to show that with high probability this algorithm will output fail for all of the incorrect candidates.

Lemma 2. *If $\mathbf{c}_i$ is not a correct codeword then the algorithm* $\text{Cand}(G, S_i, T_i)$ *will output fail except with probability at most $2^{-\Omega(\gamma d)}$.*

Proof. By the guarantee of the list decoding algorithm, it is ensured that $|T_i| \leq t - \gamma n$ and thus $|S_i| \geq t+1+\gamma n$. Let W_i be the set of passive and honest parties in T_i. We observe that $|W_i| \geq \gamma n$; otherwise, $\mathbf{c}_i$ and the correct codeword would have $t+1$ shares in common, which would imply that they are the same codeword. Furthermore, for every honest party j in S_i, we have that

$$n_T^{\mathsf{out}}(j, \mathsf{good}) \geq n_{W_i}^{\mathsf{out}}(j, \mathsf{good}) = n_{W_i}^{\mathsf{out}}(j)$$

and Corollary 1 ensures that this is 0 with probability at most $2^{-\Omega(\gamma d)}$. Taking union bound over all honest parties in S_i, with probability at least $1 - 2^{-\Omega(\gamma d)}$, all the honest parties will be removed from S_i. The desired result follows as S_i has size at most t when all honest parties are removed.

Taking a union bound over all these ℓ candidates, we obtain the following.

Theorem 7. *Assume $|P| \geq \gamma n$. With probability ϵ_{code} at least $1 - 2^{-\Omega(\gamma d)}$, the algorithm List$(G, \mathbf{s}, \gamma)$ will output the correct codeword in time poly$\big(m, n, (\frac{1}{\epsilon})^{\widetilde{O}(\frac{1}{\epsilon})}\big)$.*

6.5 Graph Algorithm

In this section, we assume that our graph algorithm $\text{Graph}(G, G', \epsilon, v)$ starts with an honest party v. Under this assumption and $|P| \leq \epsilon n$, we show that this algorithm will output the correct secret with high probability. Recall that the out-degree of vertices in G and G' is $d = n^{\sqrt{\epsilon}}$ for some small constant ϵ and that by assumption (justified by Proposition 2) the edges from honest parties to active parties are labeled bad, and the edges from honest parties to honest or passive parties are labeled good.

We also recall that, by definition, whether a corrupt party $i \in C$ is *passive* or *active*, i.e., in P or in A, only depends on s_i and r_i announced in the first communication round in the reconstruction protocol; a passive party may well lie about, say, his neighborhood E_i. Our reasoning only relies on the neighborhoods of the honest parties, which are random and independent conditioned on the adversary's view, as explained in Proposition 3 and Remark 2.

The graph algorithm Graph(G, G', ϵ, v) goes as follows. Note that n'^{out}_W refers to n^{out}_W but for the graph G' rather than G, and similarly for n'^{in}_V.

The algorithm Graph(G, G', ϵ, v)

i. Input $G = ([n], E, L), G' = ([n], E', L'), d, \epsilon$ and $v \in [n]$.

ii. Expand set $V = \{v\}$ to include more honest parties:

$$\text{While } |V| \le \frac{\epsilon t}{d} \text{ do } V := \text{Expan}(G, V, \epsilon).$$

iii. Include all honest parties into V:

$$V := V \cup \{v \notin V : n^{\mathsf{in}}_V(v, \mathsf{good}) \ge \tfrac{d|V|}{2n}\}.$$

iv. Remove all active parties from V (and maybe few honest parties as well):

$$W := \{v \in V : n^{\mathsf{in}}_V(v, \mathsf{bad}) \ge \tfrac{d}{4}\} \quad \text{and} \quad V := V \setminus W.$$

v. 1. Bound the degree of parties in V:

$$V := V \setminus \{v \in V : n'^{\mathsf{out}}_W(v) \ge \tfrac{d}{8}\}.$$

2. Include the honest parties from W (and perhaps few active parties):

$$V := V \cup \{v \in W : n'^{\mathsf{in}}_V(v, \mathsf{good}) \ge \tfrac{d}{4}\}.$$

3. Error correction: run the unique decoding algorithm on the shares of parties in V and output the result.

Remark 5. Each time we call Expan(G, V, ϵ), the size of V increases. After Step *ii*, we hope that the V has size $\Omega(\epsilon n)$ instead of barely bigger than $\frac{\epsilon t}{d}$. To achieve this, we require that the input V of the last loop to be of size $\Omega(\frac{\epsilon t}{d})$. It can be achieved as follows. Assume that $|V| > \frac{\epsilon t}{d}$. Take out each party in V with same probability such that the expectation of resulting V is less than $\frac{\epsilon t}{2d}$. Then, the proportion of honest party and dishonest party stays almost the same but the size of V is below the threshold $\frac{\epsilon t}{d}$ with probability at least $1 - 2^{-\Omega(\epsilon n/d)}$. It will not affect our randomness argument since we treat each party equally. We skip this step for simplicity. In our following analysis, we assume that V has size $\Omega(\epsilon n)$ at the end of Step *ii*.

Graph expansion algorithm Expan(G, V, ϵ)

- Input: $G = ([n], E, L)$, V and ϵ.
- Set $V' = \emptyset$. For each vertex $v \in V$ do the following:

$$\text{if } n^{\mathsf{out}}(v, \mathsf{good}) \leq \tfrac{d}{2}(1 + 3\epsilon) \text{ then } V' := V' \cup N^{\mathsf{out}}(v, \mathsf{good}).$$

- Output $V' \cup V$.

Theorem 8. *Under the assumption in Remark 3, and assuming that the graph algorithm takes an honest party v as input and that $|P| \leq \epsilon n$, the following holds. Except with failure probability $\epsilon_{graph} \leq 2^{-\Omega(\epsilon^2 d)}$, the algorithm will output a correct secret. Moreover, it runs in time poly$(n, m, \frac{1}{\epsilon})$.*

We will prove this theorem in the following subsections.

6.6 Graph Expansion

We start by analyzing the expansion property of $G|_H$, the subgraph of G restricted to the set of honest parties H.

Lemma 3 (Expansion property of $G|_H$). *If $H' \subset H$ is so that $|H'| \leq \frac{\epsilon|H|}{d}$ and the E_v's for $v \in H'$ are still random and independent in G when given H' and H, then*

$$n_H^{\mathsf{out}}(H') := \left| \bigcup_{v \in H'} N_H^{\mathsf{out}}(v) \right| \geq \frac{d}{2}(1 - 2\epsilon)|H'|$$

except with probability $2^{-\Omega(\epsilon^2 d|H'|)}$.

Informally, this ensures that, as long as H' is still reasonably small, including all the honest "neighbours" increases the set essentially by a factor $d/2$, as is to be expected: each party in H' is expected to pull in $d/2$ new honest parties. The formal proof is almost the same as the proof for a random expander graph except that we require a different parameter setting for our own purpose.

Proof. By assumption on the E_i's and by Corollary 1, the probability for any vertex $v \in H'$ to have $n_H^{\mathsf{out}}(v) < \frac{1}{2}(1-\epsilon)d$ is at most $\leq e^{-\epsilon^2 d/4} = 2^{-\Omega(\epsilon^2 d)}$. Taking the union bound, this hold for all $v \in H'$. In the remainder of the proof, we may thus assume that $N_H^{\mathsf{out}}(v)$ consist of $d' := \frac{1}{2}(1 - \epsilon)d$ random outgoing edges. Let $N := |H|$, $N' := |H'|$, and let $v_1, \ldots, v_{d'N'}$ denote the list of neighbours of all $v \in H'$, with repetition. To prove the conclusion, it suffices to bound the probability p_f that more than $\frac{d}{2}\epsilon N'$ of these $d'N'$ vertices are repeated.

The probability that a vertex v_i is equal to one of $v_1, \ldots, v_{i-1}$ is at most

$$\frac{i}{N-1} \leq \frac{d'N'}{N-1} = \frac{1}{2}(1 - \epsilon)d \cdot \frac{\epsilon|H|}{d} \cdot \frac{1}{N-1} \leq \frac{\epsilon}{2}.$$

Taking over all vertex sets of size $\frac{d}{2}\epsilon N'$ in these $d'N'$ neighbours, the union bound shows that p_f is at most

$$\begin{aligned}\binom{d'N'}{\frac{d}{2}\epsilon N'}\left(\frac{\epsilon}{2}\right)^{\frac{d}{2}\epsilon N'} &\leq 2^{d'N'H(\frac{\epsilon}{1-\epsilon})+\frac{d}{2}\epsilon N'(\log\epsilon-1)}\\ &\leq 2^{\frac{d(1-\epsilon)}{2}N'(-\frac{\epsilon}{1-\epsilon}\log\epsilon+\frac{\epsilon}{\ln 2}+O(\epsilon^2))+\frac{d}{2}N'\epsilon(\log\epsilon-1)}\\ &\leq 2^{\frac{d}{2}N'\epsilon(\frac{1}{\ln 2}-1+O(\epsilon))}\\ &\leq 2^{-\Omega(dN'\epsilon)}\end{aligned}$$

The first inequality is due to that $\binom{n}{k} \leq 2^{nH(\frac{k}{n})}$ and the second due to

$$\begin{aligned}H\Big(\frac{\epsilon}{1-\epsilon}\Big) &= -\frac{\epsilon}{1-\epsilon}\log\frac{\epsilon}{1-\epsilon} - \frac{1-2\epsilon}{1-\epsilon}\log\frac{1-2\epsilon}{1-\epsilon}\\ &\leq -\frac{\epsilon}{1-\epsilon}\log\epsilon - \log\Big(1-\frac{\epsilon}{1-\epsilon}\Big)\\ &= -\frac{\epsilon}{1-\epsilon}\log\epsilon + \frac{1}{\ln 2}\Big(\frac{\epsilon}{1-\epsilon}+O(\epsilon^2)\Big)\\ &\leq -\frac{\epsilon}{1-\epsilon}\log\epsilon + \frac{\epsilon}{\ln 2}+O(\epsilon^2)\end{aligned}$$

for small ϵ and the Taylor series $\ln(1-x) = \sum_{i\geq 1}\frac{x^i}{i}$.

6.7 Analysis of Step ii

The following shows that after Step ii, at most an $O(\sqrt{\epsilon})$-fraction of the parties in V is dishonest. This is pretty much a consequence of Lemma 3.

Proposition 4. *At the end of Step ii, with probability at least $1 - 2^{-\Omega(\epsilon^2 d)}$, V is a set of size $\Omega(\epsilon n)$ with $|H \cap V| \geq (1 - O(\sqrt{\epsilon}))|V|$ and $|C \cap V| \leq O(\sqrt{\epsilon})|V|$.*

Proof. Let V_i be the set V after Expan has been called i times, i.e., $V_0 = \{v\}$, $V_1 = \text{Expan}(G, V_0, \epsilon)$ etc., and let $H_0 = \{v\}$ and $H_1 = \text{Expan}(G, H_0, \epsilon) \cap H$, $H_2 = \text{Expan}(G, H_2, \epsilon) \cap H$ etc. be the corresponding sets when we include only honest parties into the sets.

Using a similar lazy-sampling argument as for Proposition 3, it follows that conditioned on $H_0, H_1, \ldots, H_i$, the E_j's for $j \in H_i \setminus H_{i-1}$ are random and independent for any i.[7] Therefore, we can apply Lemma 3 to $H'_i = H_i \setminus H_{i-1}$ to obtain that $|H_{i+1}| \geq |H'_i|\frac{d}{2}(1-2\epsilon)$. It follows[8] that $|H_i| \geq (\frac{d}{2}(1-2\epsilon))^i$ except with probability $2^{-\Omega(\epsilon^2 d)}$. According to Remark 5, our algorithm jumps out of Step *ii* when V is of size $\Omega(\epsilon n)$. We bound the number of rounds in this step. For $i = \frac{2}{\sqrt{\epsilon}}$, noting that $d = n^{\sqrt{\epsilon}}$, it thus follows that

$$|V_i| \geq |H_i| \geq \Big(\frac{d}{2}(1-2\epsilon)\Big)^i = \frac{n^2}{2^{\frac{2}{\sqrt{\epsilon}}}}(1-2\epsilon)^{\frac{2}{\sqrt{\epsilon}}} \geq \Omega(n^2).$$

[7] The crucial point here is that H_i is determined by the E_j's with $j \in H_{i-1}$ only.

[8] The size of H_{i-1} is negligible compared to H_i; indeed, $|H_i| = \Omega(d|H_{i-1}|)$ and thus $|H_i \setminus H_{i-1}| = (1 - o(1))|H_i|$. So, we may ignore the difference between H_i and H'_i.

That means $\text{Expan}(G, V, \epsilon)$ is called $r \leq \frac{2}{\sqrt{\epsilon}}$ times assuming n is large enough.

On the other hand, we trivially have $|V_r| \leq (\frac{d}{2}(1+3\epsilon))^r$ by specification of Expan. Thus,

$$|V_r| - |H_r| \leq \left(\frac{d}{2}(1+3\epsilon)\right)^r - \left(\frac{d}{2}(1-2\epsilon)\right)^r$$
$$= \frac{5\epsilon d}{2}\left(\sum_{i=0}^{r-1}\left(\frac{d}{2}(1+3\epsilon)\right)^i\left(\frac{d}{2}(1-2\epsilon)\right)^{r-1-i}\right) \leq \frac{5\epsilon d}{2} r\left(\frac{d}{2}(1+3\epsilon)\right)^{r-1}$$
$$\leq 5r\epsilon\left(\frac{d}{2}(1+3\epsilon)\right)^r \leq 10\sqrt{\epsilon}|V_r|.$$

The first equality is due to $a^n - b^n = (a-b)(\sum_{i=0}^{n-1} a^i b^{n-1-i})$ and the last one is due to $r \leq \frac{2}{\sqrt{\epsilon}}$.

This upper bound implies that there are at least $|V_r|(1-10\sqrt{\epsilon})$ honest parties in V_r while the number of dishonest parties is at most $10\sqrt{\epsilon}|V_r|$.

6.8 Analysis of Step iii

The intuition for the next observation is simply that because V consists almost entirely of honest parties, every honest party v not yet in V will get sufficient support in Step iii from the parties in V to be included as well; indeed, any such v is expected have close to $\frac{d}{n}|V|$ good incoming edges from the parties in V.

Proposition 5. *At the end of Step iii, with probability at least* $1 - 2^{-\Omega(\epsilon d)}$, V *contains* all *honest parties and* $O(\sqrt{\epsilon} n)$ *dishonest parties.*

Proof. Recall the notation from the proof in the previous section, and the observation that conditioned on H_r, the E_i's for $i \in H_r \setminus H_{r-1}$ are random and independent.

Setting $\tilde{H} := H_r \setminus H_{r-1}$ and $d_1 := \frac{|V|d}{n} = \Omega(\epsilon d)$, and using Corollary 2 for the final bound, it follows that for a given honest party $v \notin \tilde{H}$,

$$\Pr\left[n_V^{\mathsf{in}}(v, \mathsf{good}) < \frac{d_1}{2}\right] \leq \Pr\left[n_{\tilde{H}}^{\mathsf{in}}(v, \mathsf{good}) < \frac{d_1}{2}\right] = \Pr\left[n_{\tilde{H}}^{\mathsf{in}}(v) < \frac{d_1}{2}\right] \leq 2^{-\Omega(\epsilon d)}.$$

By union bound over all honest parties outside $\tilde{H}$, all these honest parties are added to V with probability at least $1 - 2^{-\Omega(\epsilon d)}$.

On the other hand, any active party w outside V needs at least $\frac{d_1}{2}$ good incoming edges to be admitted. These edges must come from dishonest parties in V. Since there are at most $O(\sqrt{\epsilon})|V|$ of them in V and each of them contributes to at most d good incoming edges, the number of active parties admitted to V is at most $\frac{O(\sqrt{\epsilon})|V|d}{d_1/2} = O(\sqrt{\epsilon} n)$.

6.9 Analysis of Step iv

By construction, after Step iv, V and W together obviously still contain all honest parties. Furthermore, as we show below, there is now no active party left in V and only few honest parties ended up in V. The idea here is that the active parties in V will be recognized as being dishonest by many honest parties in V.

Proposition 6. *At the end of Step iv, with probability at least* $1 - 2^{-\Omega(\epsilon d)}$, V *consists of* $t+1-O(\sqrt{\epsilon}n)$ *honest parties and no active parties, and* W *consists of the rest of honest parties and* $O(\sqrt{\epsilon}n)$ *dishonest parties.*

Proof. Observe that $\frac{|H|d}{n} \geq \frac{d}{2}$. It follows, again using Corollary 2, that for an active party w in V, we have

$$\Pr\left[n_V^{\mathsf{in}}(w,\mathsf{bad}) < \frac{d}{4}\right] \leq \Pr\left[n_H^{\mathsf{in}}(w,\mathsf{bad}) < \frac{d}{4}\right] = \Pr\left[n_H^{\mathsf{in}}(w) < \frac{d}{4}\right] \leq 2^{-\Omega(d)}.$$

By union bound over all active parties in V, all of them are removed from V with probability at least $1 - t2^{-\Omega(d)} = 1 - 2^{-\Omega(d)}$.

On the other hand, if the honest party v is removed from V, he must receive at least $\frac{d}{4}$ bad incoming edges from dishonest parties in V. Since the number of dishonest parties is at most $a := O(\sqrt{\epsilon}n)$, there are at most $\frac{ad}{d/4} = O(\sqrt{\epsilon}n)$ honest parties removed from V in Step 2.

In order to analyze the last step (see next section), we introduce the following notation. We partition V into the set of honest parties V_H and the set of passive parties V_P. We also partition W into the set of honest parties W_H and the set of dishonest parties W_C. From above, we know that $|W| = |W_H|+|W_C| = O(\sqrt{\epsilon}n)$, $V_H \cup W_H = H$ and $|V_H| = t+1-O(\sqrt{\epsilon}n)$.

6.10 Analysis of Step v

Proposition 7. *Except with probability* $2^{-\Omega(d)}$, *after Step v the set* V *will contain all honest parties and at least twice as many passive parties as active ones. Therefore, Step v will output the correct secret with probability at least* $1-2^{-\Omega(d)}$.

Note that, given the adversary's strategy, all the previous steps of the graph algorithm are determined by the graph G. Therefore, by Proposition 3, at this point in the algorithm the E_i''s for $i \in H$ are still random and independent given V_H, V_P, W_C, W_H.

Proof. **Step v.1.** For any $i \in V_H$, $n_W'^{\mathsf{out}}(i)$ is expected to be $\frac{|W|d}{n} = O(\sqrt{\epsilon}d)$. By Corollary 1 we thus have

$$\Pr\left[n_W'^{\mathsf{out}}(i) \geq \frac{d}{8}\right] \leq 2^{-\Omega(d)}.$$

Hence, by union bound, all honest parties in V remain in V except with probability $2^{-\Omega(d)}$.

Let V'_P be the set of passive parties left in V after this step, and set $p := |V'_P|$. Note that $n_W^{\mathsf{out}}(v) \leq d/8$ for every $v \in V$.

Step v.2. Observe that $\frac{d|V_H|}{n} = (\frac{1}{2} - O(\sqrt{\epsilon}))d$. It follows from Corollary 2 that for any honest party $i \in W_H$,

$$\Pr\Big[n'^{\mathsf{in}}_V(i, \mathsf{good}) \leq \frac{d}{4}\Big] \leq \Pr\Big[n'^{\mathsf{in}}_{V_H}(i, \mathsf{good}) \leq \frac{d}{4}\Big] = \Pr\Big[n'^{\mathsf{in}}_{V_H}(i) \leq \frac{d}{4}\Big] \leq 2^{-\Omega(d)}.$$

Thus, all honest parties in W are added to V, except with probability $2^{-\Omega(d)}$.

On the other hand, the active parties only receive good incoming edges from passive parties in V'_P. Observe that each party in V is allowed to have at most $\frac{d}{8}$ outgoing neighbours in W. This implies there are at most $\frac{pd/8}{d/4} = \frac{p}{2}$ active parties admitted to V in this step, proving the first part of the statement.

Step v.3. Observe that the shares of the parties in S form a code with length $|V|$ and dimension $(t+1)$. Since the fraction of errors is at most $\frac{p}{2|V|} < \frac{(|V|-t-1)}{2|V|}$, by Proposition 1, the unique decoding algorithm will output a correct secret.

7 Parameters of Construction and Parallel Repetition

We first determine the parameters in our algorithm and then show how to reach the security parameter κ by parallel repeating this algorithm for $O(\kappa)$ times. This parallel repetition idea comes from [3]. Assume that there are n parties and m-bit secret **msg** to share among these n parties. Note that we have already set $d = n^{\sqrt{\epsilon}}$ with ϵ a small constant. Let $\log q = O(\frac{m+\log n}{\epsilon^2})$. We choose $\log|\mathcal{T}| = \log m + 5\log n$ and then the random string r_i has length $4d\log|\mathcal{T}|$. The key $key_{i,j}$ is defined over $\mathcal{T}^2$ and thus has length $2\log|\mathcal{T}|$. It follows that $|\mathcal{K}_i| = |\mathcal{K}'_i| = 2d\log|\mathcal{T}|$ and **tag** has length $4nd\log|\mathcal{T}|$. By Theorem 4, (p_i, q_i) has length $\widetilde{O}(d)$. By Theorem 2 and plug $a = O(\frac{m+\log n}{\epsilon^2})$, the error probability of the MAC ϵ_{mac} is at most $\frac{4(a+4d)dt}{|\mathcal{T}|} = O(\frac{mnd}{\epsilon^2 mn^5}) \leq O(\frac{1}{n^3})$. The failure probability of our reconstruction scheme consists of the error probability $\epsilon_{mac} = O(1/n^3)$ of the MAC authentication, error probability $\epsilon_{tag} = O(1/n^2)$ of reconstructing **tag**, error probability $\epsilon_{check} = 2^{-O(\epsilon d)}$ of determining the situation whether $|P| \geq \epsilon n/4$ or $|P| \leq \epsilon n$, error probability $\epsilon_{code} = 2^{-\Omega(\epsilon d)}$ of code based algorithm and error probability $\epsilon_{graph} = 2^{-\Omega(\epsilon^2 d)}$ of graph algorithm. The total error probability of our algorithm is

$$\delta = \epsilon_{mac} + \epsilon_{tag} + \epsilon_{check} + (t+1)\epsilon_{graph} + \epsilon_{code} = O(\frac{1}{n^2}).$$

We summarize our result as follows.

Theorem 9. *The scheme* (**Share**, **Rec**) *is a* $2t+1$*-party* $(t, O(\frac{1}{n^2}))$*-robust secret sharing scheme with running time* $poly(m, n, (\frac{1}{\epsilon})^{\widetilde{O}(\frac{1}{\epsilon})})$ *and share size* $\widetilde{O}(m+n^{\sqrt{\epsilon}})$.

Next, we describe how to achieve the security parameter $\delta = 2^{-\kappa}$ based on this "weakly-robust" secret sharing scheme (**Share**, **Rec**) with $\delta = O(\frac{1}{n^2})$.

Given a secret **msg**, we run **Share(msg)** $q = O(\kappa)$ times to produce q robust sharings **msg**, except that the first step of **Share(msg)** is executed only *once*, i.e., only one set of non-robust shares $(s_1, \ldots, s_n) \leftarrow \mathbf{Sh}(\mathbf{msg})$ is produced, and then re-used in the otherwise independent q executions of **Share(msg)**. This is exactly the same idea as that in [3]. The resulting share size is then $\widetilde{O}(m+\kappa n^{\sqrt{\epsilon}})$.

The analysis is almost the same as that in [3], so we omit it here.

Theorem 10. *The scheme* (**Share′**, **Rec′**) *is a* $2t+1$*-party* $(t, 2^{-\kappa})$*-robust secret sharing scheme against rushing adversary with share size* $\widetilde{O}(m + \kappa n^{\sqrt{\epsilon}})$ *and running time poly*$(\kappa, m, n, (\frac{1}{\epsilon})^{\widetilde{O}(\frac{1}{\epsilon})})$.

8 Further Improvement and Existence Result

If we do not consider the efficiency of our algorithm, by setting proper parameters, our algorithm can also achieve the optimal $\widetilde{O}(m+\kappa)$ share size. The algorithm is exactly the same as we described above except that we set $\epsilon = \frac{1}{\log^2 n}$, $d = \Omega(\log^5 n)$ and $\gamma = \frac{\epsilon}{4} = O(\frac{1}{\log^2 n})$. This parameter setting will affect the efficiency and error probability of our algorithm. We briefly review this improvement by pointing out the differences. Since $d = \Omega(\log^5 n)$, we use the tag construction in Theorem 3 with $u = \Omega(\log^5 n)$. The error probability ϵ_{tag} of reconstructing tags now becomes $2^{-\Omega(u)} = 2^{-\Omega(\log^5 n)}$. The error probability ϵ_{Check} becomes $2^{-\Omega(\epsilon d)} = 2^{-\Omega(\log^3 n)}$. In the code based algorithm, the list size $(\frac{1}{\gamma})^{\widetilde{O}(\frac{1}{\gamma})}$ now becomes $2^{\widetilde{O}(\log^2 n)}$.[9] By taking union bound over candidates on this new list, we get

$$\epsilon_{code} = 2^{\widetilde{O}(\log^2 n)} 2^{-\Omega(\gamma d)} = 2^{\widetilde{O}(\log^2 n)} 2^{-\Omega(\log^4 n)} = 2^{-\Omega(\log^4 n)}.$$

In the graph algorithm, we bound the size of V_r and H_r. First, we notice that $d^{\log n} = n^{\log \log n}$. This implies that $r < \log n = \sqrt{\frac{1}{\epsilon}}$. In worst case scenario, we assume that $|V_r| = (\frac{d}{2}(1+3\epsilon))^r$ and $|H_r| = (\frac{d}{2}(1-2\epsilon))^r$. It follows that the number of dishonest parties is at most

$$|V_r| - |H_r| = \left(\frac{d}{2}(1+3\epsilon)\right)^r - \left(\frac{d}{2}(1-2\epsilon)\right)^r$$
$$= \frac{5\epsilon d}{2}\left(\sum_{i=0}^{r-1} \left(\frac{d}{2}(1+3\epsilon)\right)^i \left(\frac{d}{2}(1-2\epsilon)\right)^{r-1-i}\right) \leq 5r\epsilon\left(\frac{d}{2}(1+3\epsilon)\right)^r \leq 5\sqrt{\epsilon}|V_r|.$$

The rest of the algorithm are the same. Therefore, the error probability of our graph algorithm now becomes $\epsilon_{graph} = 2^{-\Omega(\epsilon^2 d)} = 2^{-\Omega(\log n)}$. It follows that the total error probability of our algorithm is $2^{-\Omega(\log n)}$. The overhead of share size is $O(d) = O(\log^5 n)$. By the parallel repetition technique, we can reduce it to $2^{-\kappa}$. The share size then becomes $\widetilde{O}(m+\kappa)$. As a trade-off, the running time of our algorithm now becomes $2^{\widetilde{O}(\log^2 n)}$ which is super-polynomial in n.

[9] Here, we hide the poly$(\log \log n)$ in $\widetilde{O}(\cdot)$.

Theorem 11. *There exists $2t+1$-party $(t, 2^{-\kappa})$-robust secret sharing scheme against rushing adversary with share size $\widetilde{O}(m+\kappa)$ and running time $2^{\widetilde{O}(\log^2 n)}$.*

Acknowledgements. CY was partially supported by the European Union Horizon 2020 research and innovation programme under grant agreement No. 74079 (ALGSTRONGCRYPTO) and the National Research Foundation, Prime Minister's Office, Singapore, under its Strategic Capability Research Centres Funding Initiative.

A Appendix

A.1 Folded Reed-Solomon Codes

Instead of using the Reed-Solomon codes to share our secret, our robust secret sharing scheme is encoded by the folded Reed-Solomon codes. Since the folded Reed-Solomon code is a class of MDS codes, it is an eligible candidate for threshold secret sharing scheme. Moreover, the folded Reed-Solomon codes first introduced by Guruswami and Rudra [10] can be list decoded up to $1-R-\gamma$ fraction of errors for any constant γ. This extra nice property allows us to divide our reconstruction scheme into two scenarios, one with small number of passive parties and another with big one. Let us first introduce the formal definition of fold Reed-Solomon codes.

Let q be a prime power, $n+1 \le \frac{q-1}{s}$ and β be a primitive element of $\mathbb{F}_q$. The folded Reed-Solomon code $\mathsf{FRS}_{q,s}(n+1, d)$ is a code over $\mathbb{F}_q^s$. To every polynomial $P(X) \in \mathbb{F}_q[X]$ of degree at most d, the encoding algorithm goes as follows:

$$P(X) \mapsto \mathbf{c}_P = \left(\begin{bmatrix} P(\beta) \\ P(\beta^2) \\ \vdots \\ P(\beta^{s-1}) \end{bmatrix}, \begin{bmatrix} P(\beta^s) \\ P(\beta^{s+1}) \\ \vdots \\ P(\beta^{2s-1}) \end{bmatrix}, \dots, \begin{bmatrix} P(\beta^{ns}) \\ P(\beta^{ns+1}) \\ \vdots \\ P(\beta^{(n+1)s-1}) \end{bmatrix} \right).$$

It is easy to verify that $\mathsf{FRS}_{q,s}(n+1, d)$ is an $\mathbb{F}_q$-linear code with code length $n+1$, rate $\frac{d+1}{(n+1)s}$ and distance at least $(n+1) - \lfloor \frac{d}{s} \rfloor$. The folded Reed-Solomon code is a class of MDS code when $d+1$ is divisible by s. In our robust secret sharing scheme, we set $n = 2t+1$ and $d+1 = (t+1)s$. For every secret $\mathbf{s} \in \mathbb{F}_q^s$, we find the $P(X)$ of degree at most d uniform at random such that $\mathbf{s} = (P(\beta), P(\beta^2), \dots, P(\beta^{s-1}))$. The party i receives the $i+1$-th component of $\mathbf{c}_P$. It is easy to verify that this scheme is a threshold secret sharing scheme with t-privacy and $t+1$-reconstruction. Moreover, if we write the n shares as

$$(P(\beta^s), P(\beta^{s+1}), \dots, P(\beta^{(n+1)s-1})) \in \mathbb{F}_q^{ns}.$$

Then, it becomes a classic Reed-Solomon codes with length ns, dimension $(t+1)s$ and distance $(n-(t+1))s+1$. We will use this fact in our robust secret sharing scheme.

Besides the MDS property, the folded Reed-Solomon codes enjoy a large list decoding radius up to the Singleton bound while the list size is bounded by a polynomial in q. There are many works aimed at reducing the list size of the folded Reed-Solomon codes. Recently, Kopparty et al. [12] proved that the list size of the folded Reed-Solomon codes is at most a constant in γ.

Theorem 12 (Theorem 3.1 [12]). *Let $\gamma > 0$ such that $\frac{16}{\gamma^2} \leq s$. The folded Reed-Solomon code $\mathsf{FRS}_{q,s}(n,d)$ can be list decoded up to $1 - \frac{d}{sn} - \gamma$ with list size at most $(\frac{1}{\gamma})^{\frac{1}{\gamma}\log\frac{1}{\gamma}}$. Moreover, there exists a randomized algorithm that list decodes this code with above parameters in time poly$(\log q, s, d, n, (\frac{1}{\gamma})^{\frac{1}{\gamma}\log\frac{1}{\gamma}})$.*

Remark 6. By running this polynomial list decoding algorithm n times and taking the union of all its output, with probability at least $1 - 2^{-\Omega(n)}$, we will find all the codewords within distance $1 - \frac{d}{sn} - \gamma$ to the corrupted vector. This error probability is good enough for our robust secret sharing scheme. Compared with the approach in [10], the new algorithm runs faster and ensures a significantly small list of candidates.

A.2 Proof of Theorem 2

Proof. We need to verify three conditions in Definition 3.

Privacy over Randomness: It suffices to consider that all the ℓ keys are distinct. Otherwise, we keep one key for each value and apply the argument to these distinct keys. Let $(x_1, y_1), \dots, (x_\ell, y_\ell) \in \mathbb{F}^2$ be the ℓ distinct keys. Let $\sigma_i = MAC_{(x_i,y_i)}(\mathbf{m}, \mathbf{r})$. For any $\mathbf{m} \in \mathbb{F}^a$, we will show that $(\sigma_1, \dots, \sigma_\ell) \in \mathbb{F}^\ell$ are distributed uniformly at random. To see this, we write

$$MAC_{x,y}(\mathbf{m}, \mathbf{r}) = f_{\mathbf{m}}(x) + g_{\mathbf{r}}(x) + y$$

where $f_{\mathbf{m}}(x) = \sum_{i=1}^{a} m_i x^{i+\ell}$ and $g_{\mathbf{r}}(x) = \sum_{i=1}^{\ell} r_i x^i$. For any ℓ-tuple $(\sigma_1, \dots, \sigma_\ell) \in \mathbb{F}^\ell$, we obtain the evaluation of $g_{\mathbf{r}}(x)$ at ℓ points, i.e., $g_{\mathbf{r}}(x_i) = \sigma_i - f_{\mathbf{m}}(x_i) - y_i$. Since $g_{\mathbf{r}}$ is a polynomial of degree $\ell - 1$, the polynomial interpolation yields an unique $g_{\mathbf{r}}(x)$. This implies that for any $\mathbf{m} \in \mathbb{F}^a$, the distribution of $(\sigma_1, \dots, \sigma_\ell)$ is uniform at random over $\mathbf{r} \in \mathbb{F}^\ell$.

Authentication: For $(\mathbf{m}, \mathbf{r}) \neq (\mathbf{m}', \mathbf{r}') \in \mathbb{F}^a \times \mathbb{F}^\ell$, $MAC_{(x,y)}(\mathbf{m}, \mathbf{r})$-$MAC_{(x,y)}(\mathbf{m}', \mathbf{r}')$ is a nonzero polynomial in x of degree at most $t + \ell$ over $\mathbb{F}$. Thus, for any $b \in \mathbb{F}$, the equation

$$MAC_{(x,y)}(\mathbf{m}, \mathbf{r}) - MAC_{(x,y)}(\mathbf{m}', \mathbf{r}') = b$$

has at most $(a + \ell)|\mathbb{F}|$ pairs (x, y) as its solutions. The desired result follows as $\frac{(a+\ell)(|\mathbb{F}|)}{|\mathbb{F}|^2} \leq \epsilon$.

Uniformity: We need to show that given any $(\mathbf{m}, \mathbf{r}) \in \mathbb{F}^a \times \mathbb{F}^\ell$, the tag $\sigma = MAC_{(x,y)}(\mathbf{m}, \mathbf{r})$ is uniform at random over the random key $(x, y) \in \mathbb{F}^2$. Let us fix $(\mathbf{m}, \mathbf{r})$. By the definition of MAC, we have

$$\sigma = MAC_{(x,y)}(\mathbf{m}, \mathbf{r}) = f_{\mathbf{m}}(x) + g_{\mathbf{r}}(x) + y.$$

For each $\sigma \in \mathbb{F}$, there exists exactly q distinct keys (x, y) to satisfy this MAC. Thus, the tag σ is uniform at random over the random key. The desired result follows.

References

1. Auger, A., Doerr, B.: Theory of Randomized Search Heuristics. World Scientific, Singapore (2011)
2. Bishop, A., Pastro, V.: Robust secret sharing schemes against local adversaries. In: Cheng, C.-M., Chung, K.-M., Persiano, G., Yang, B.-Y. (eds.) PKC 2016. LNCS, vol. 9615, pp. 327–356. Springer, Heidelberg (2016). https://doi.org/10.1007/978-3-662-49387-8_13
3. Bishop, A., Pastro, V., Rajaraman, R., Wichs, D.: Essentially optimal robust secret sharing with maximal corruptions. In: Fischlin, M., Coron, J.-S. (eds.) EUROCRYPT 2016. LNCS, vol. 9665, pp. 58–86. Springer, Heidelberg (2016). https://doi.org/10.1007/978-3-662-49890-3_3
4. Blakley, G.R.: Safeguarding cryptographic keys. In: International Workshop on Managing Requirements Knowledge, AFIPS, pp. 313–317, November 1979
5. Carpentieri, M., De Santis, A., Vaccaro, U.: Size of shares and probability of cheating in threshold schemes. In: Helleseth, T. (ed.) EUROCRYPT 1993. LNCS, vol. 765, pp. 118–125. Springer, Heidelberg (1994). https://doi.org/10.1007/3-540-48285-7_10
6. Cevallos, A., Fehr, S., Ostrovsky, R., Rabani, Y.: Unconditionally-secure robust secret sharing with compact shares. In: Pointcheval, D., Johansson, T. (eds.) EUROCRYPT 2012. LNCS, vol. 7237, pp. 195–208. Springer, Heidelberg (2012). https://doi.org/10.1007/978-3-642-29011-4_13
7. Cheraghchi, M.: Nearly optimal robust secret sharing. In: 2016 IEEE International Symposium on Information Theory, ISIT, pp. 2509–2513, July 2016
8. Cramer, R., Damgård, I., Fehr, S.: On the cost of reconstructing a secret, or VSS with optimal reconstruction phase. In: Kilian, J. (ed.) CRYPTO 2001. LNCS, vol. 2139, pp. 503–523. Springer, Heidelberg (2001). https://doi.org/10.1007/3-540-44647-8_30
9. Cramer, R., Damgård, I.B., Döttling, N., Fehr, S., Spini, G.: Linear secret sharing schemes from error correcting codes and universal hash functions. In: Oswald, E., Fischlin, M. (eds.) EUROCRYPT 2015. LNCS, vol. 9057, pp. 313–336. Springer, Heidelberg (2015). https://doi.org/10.1007/978-3-662-46803-6_11
10. Guruswami, V., Rudra, A.: Explicit codes achieving list decoding capacity: error-correction with optimal redundancy. IEEE Trans. Inf. Theory **54**(1), 135–150 (2008)
11. Hemenway, B., Ostrovsky, R.: Efficient robust secret sharing from expander graphs. Cryptogr. Commun. **10**(1), 79–99 (2018)
12. Kopparty, S., Ron-Zewi, N., Saraf, S., Wootters, M.: Improved decoding of folded Reed-Solomon and multiplicity codes. Electron. Colloq. Comput. Complex. (ECCC) **25**, 91 (2018)
13. Rabin, T., Ben-Or, M.: Verifiable secret sharing and multiparty protocols with honest majority (extended abstract). In: Proceedings of the 21st Annual ACM Symposium on Theory of Computing, Seattle, Washington, USA, 14–17 May 1989, pp. 73–85 (1989)
14. Shamir, A.: How to share a secret. Commun. ACM **22**(11), 612–613 (1979)

Simple Schemes in the Bounded Storage Model

Jiaxin Guan(✉) and Mark Zhandary(✉)

Princeton University, Princeton, NJ 08544, USA
jiaxin@guan.io, mzhandry@cs.princeton.edu

Abstract. The bounded storage model promises unconditional security proofs against computationally unbounded adversaries, so long as the adversary's space is bounded. In this work, we develop simple new constructions of two-party key agreement, bit commitment, and oblivious transfer in this model. In addition to simplicity, our constructions have several advantages over prior work, including an improved number of rounds and enhanced correctness. Our schemes are based on Raz's lower bound for learning parities.

1 Introduction

For the vast majority of cryptographic applications, security relies on the assumed hardness of certain computational problems, such as factoring large integers or inverting certain hash functions. Unfortunately, with the current state of complexity theory the hardness of these problems can only be conjectured. This means that the security of such schemes is always *conditional* on such conjectures being true.

Maurer proposes the Bounded Storage Model [Mau92] as an alternate model for constraining the adversary; here, instead of constraining the adversary's time, the adversary's memory is bounded. Amazingly, it is actually possible to give *unconditional* proofs of security for schemes in this model. The core idea is that the honest parties exchange so much information that the adversary cannot possibly store it all. Then, schemes are cleverly devised to exploit the adversary's lack of knowledge about the scheme.

Moreover, the space bounds are only necessary when the protocol is run, and even if the adversary later gains more space the protocol remains secure. This means schemes only need to be designed with current storage capacities in mind. This is fundamentally different than the usual approach of time-bounding adversaries, where an adversary can later break the protocol if its computational abilities increase. Hence, traditional schemes must be designed with future computational abilities in mind. This is especially important in light of recent developments in quantum computing, as Grover's algorithm [Gro96] and Shor's algorithm [Sho94] can speed up attacks on many current cryptographic protocols. Hence, much of the communication taking place today will be revealed once quantum computers become reality.

Y. Ishai and V. Rijmen (Eds.): EUROCRYPT 2019, LNCS 11478, pp. 500–524, 2019.
https://doi.org/10.1007/978-3-030-17659-4_17

This Work. In this work, we devise very simple round-optimal protocols for bit-commitment and oblivious transfer (namely, 1 round and 2 rounds, respectively) in the Bounded Storage Model, improving 5 rounds needed in prior works. We additionally develop a new key agreement protocol with several advantages over prior works. Our results rely on Raz's recent space lower bound for learning parities [Raz17], and in particular the simple encryption scheme based on this lower bound. Our key observation is that Raz's encryption scheme has several useful properties—including additive homomorphism and leakage resilience—that can be useful for building higher-level protocols. Our core technical contribution is a new "encrypt zero" protocol for Raz's encryption scheme, which may be of independent interest.

Our schemes are based on entirely different techniques than most of the prior literature—most of which is based on the birthday paradox—and we believe our work will therefore be a useful starting point for future work in the bounded storage model.

1.1 Prior Work in the Bounded Storage Model

Prior work in the Bounded Storage Model [Mau92, CM97, CCM98, Lu02, AR99, Din01, DHRS04] typically uses something akin to the birthday paradox to achieve security against space-bounded adversaries.

In slightly more detail, the key agreement scheme of Maurer [Mau92] works as follows. One party sends a stream of roughly n^2 random bits to the other party[1]. Each party records a random secret subset of n bits of the stream. By the birthday paradox, the two parties will have recorded one bit position in common with constant probability. They therefore share the bit positions they recorded with each other, and set their secret key to be the bit of the stream at the shared position.

An eavesdropper first sees n^2 random bits. If the eavesdropper's storage is somewhat lower than n^2, he cannot possibly remember the entire sequence of random bits. In particular, it can be shown that the adversary has little information about the bit shared by the two honest parties. This remains true even after the parties share their bit positions. Notice that the honest parties require space n, and security holds even for adversaries with space Cn^2 for some constant C. Therefore, by tuning n so that n storage is feasible, but Cn^2 is not, one obtains the desired security.

Much of the literature on the Bounded Storage Model relies on this sort of birthday attack property. Unfortunately, this leads to several difficulties:

- The two honest parties only achieve success with constant probability. In order to achieve success with high probability, the protocol either needs to be repeated many times (thus requiring more than n^2 communication) or requires the honest users to store more than n positions (thus requiring more

[1] In most works in the Bounded Storage Model, the random bit stream is assumed to come from a trusted third party. In this work we will insist on there being no trusted third party, and instead the bit stream comes from the parties themselves.

than n space, and making the gap between the honest users and adversaries less than quadratic).

- Remembering n random positions out of n^2 requires $O(n \log n)$ space just to record the indices. To compress the space requirements of the honest parties, the positions are actually chosen by a pairwise independent function, complicating the scheme slightly.
- The adversary has a $1/n^2$ chance of guessing the bit position shared by the two users. As such, the adversary has a non-negligible advantage in guessing the bit. To get statistical security, a randomness extraction step is applied, adding slightly to the complexity of the protocol.
- More importantly, there is very little structure to exploit with the birthday approach. For more advanced applications such as oblivious transfer or bit commitment, the protocols end up being somewhat complicated and require several rounds.

1.2 Space Lower Bounds for Learning Parities

In this work, we exploit recent space lower bounds due to Raz [Raz17]. Raz considers a setting where one party holds a secret key $\mathbf{k} \in \{0,1\}^n$, and streams random tuples $(\mathbf{r}_i, \mathbf{r}_i \cdot \mathbf{k})$, where $\mathbf{r}_i$ is random in $\{0,1\}^n$ and the inner product is taken mod 2. Raz asks: given these random tuples, and only limited storage (namely Cn^2 for some constant C), how hard is it to recover $\mathbf{k}$? Clearly, if $C \approx 1$, then one can store n tuples, and then recover $\mathbf{k}$ using linear algebra. But if $C \ll 1$, then the adversary has no hope of storing enough tuples to perform linear algebra.

Raz proves that, for some constant C (roughly $1/20$), then either the adversary needs an exponential (in n) number of samples, or the adversary's probability of correctly guessing $\mathbf{k}$ is exponentially small.

Raz observes that his lower bound easily leads to a secret key encryption scheme in the bounded storage model. The key will be an n-bit string $\mathbf{k}$. To encrypt a message bit b, choose a random $\mathbf{r}$, and produce the ciphertext $(\mathbf{r}, \mathbf{r} \cdot \mathbf{k} \oplus b)$. Raz's lower bound shows that after seeing fewer than exponentially many encrypted messages, an adversary with Cn^2 space has an exponentially small probability of guessing $\mathbf{k}$. This means $\mathbf{k}$ always has some min-entropy conditioned on the adversaries' view. Then using the fact that the inner product is a good extractor, we have that for any new ciphertext $\mathbf{r} \cdot \mathbf{k}$ is statistically close to random, and hence masks the message b.

1.3 This Work

In this work, we use Raz's scheme in order to develop simple new constructions in the Bounded Storage Model that have several advantages over prior work.

Our main observation is that Raz's encryption scheme has several attractive properties. First, it is leakage resilient: since inner products are strong extractors, the scheme remains secure even if the adversary has partial knowledge of the key, as long as the conditional min-entropy of the key is large.

Next, we note that Raz's scheme is additively homomorphic: given encryptions $(\mathbf{r_0}, \mathbf{r_0} \cdot \mathbf{k} \oplus m_0)$ and $(\mathbf{r_1}, \mathbf{r_1} \cdot \mathbf{k} \oplus m_1)$ of m_0, m_1, we can compute an encryption of $m_0 \oplus m_1$ by simply taking the componentwise XOR of the two ciphertexts, yielding $(\mathbf{r_0} \oplus \mathbf{r_1}, (\mathbf{r_0} \oplus \mathbf{r_1}) \cdot \mathbf{k} \oplus (m_0 \oplus m_1))$. This additive homomorphism will prove very useful. We can also toggle the bit being encrypted by toggling the last bit of a ciphertext.

For example, Rothblum [Rot11] shows that any additively homomorphic secret key encryption scheme can be converted into a public key (additively homomorphic) encryption scheme. The rough idea is that the public key consists of many encryptions of zero. Then, to devise an encryption of a bit m, simply add a random subset sum of the public key ciphertexts to get a "fresh" encryption of zero, and then toggle the encrypted bit as necessary to achieve an encryption of m.

Key Agreement. In the case of Raz's scheme, the public key will end up containing $O(n)$ ciphertexts, meaning the public key is too large for the honest users to even write down. However, we can re-interpret this protocol as a *key-agreement* protocol. Here, the public key is streamed from user A to user B, who applies the additive homomorphism to construct the fresh encryption on the fly. Now one party knows the secret key, and the other has a fresh ciphertext with a known plaintext. So the second party just sends the ciphertext back to the first party, who decrypts. The shared key is the plaintext value.

Bit Commitment. Next, we observe that the public key encryption scheme obtained above is *committing*: for any public key there is a unique secret key. Therefore, we can use the scheme to get a bit commitment scheme as follows: to commit to a bit b, the Committer simply chooses a random secret key, streams the public key to the receiver, and then sends an encryption of b. To open the commitment, the Committer simply sends the secret decryption key. The Verifier, on the other hand, constructs several fresh encryptions of 0 by reading the Committer's stream, as user B did in our key agreement protocol. Upon receiving a supposed secret key, the Verifier checks that all the encryptions do in fact decrypt to 0. If so, then it decrypts the commitment to get the committed value.

Oblivious Transfer. We can also turn this commitment scheme into an oblivious transfer protocol: the Receiver, on input b, commits to the bit b. Then the Sender, on input x_0, x_1, using the homomorphic properties of the encryption scheme, turns the encryption of b in the commitment into encryptions of $(1 - b)x_0$ and bx_1. To maintain privacy of x_{1-b}, the Sender will *re-randomize* the encryptions, again using the homomorphic properties. To re-randomize, the Sender will construct some fresh encryptions of zero, again just as user B did in our key agreement protocol. The Receiver can then decrypt these ciphertexts, which yield 0 and x_b.

Malicious Security. The commitment scheme and the oblivious transfer protocol are secure as long as the public key is generated correctly. This occurs, for example, if the randomness for the encryptions of 0 is generated and streamed by a trusted third party. This is the setting considered in much of the prior work in the bounded storage model.

On the other hand, if we do not wish to rely on a trusted third party to generate the encryption randomness, a malicious Committer can choose a public key with bad randomness, which will allow him to break the commitment, as explained below. This also would let the Receiver break the security of the oblivious transfer protocol. We therefore additionally show how to modify the constructions above to obtain security for malicious parties without relying on a trusted third party. The result is round-optimal protocols for bit-commitment and oblivious transfer without a trusted third party.

1.4 Additional Technical Details

The Encrypt Zero Protocol. Notice that all of our schemes have a common feature: one user has a secret key, and the other user obtains encryptions of 0. Importantly for security, these encryptions of 0 should be independent of the view of the first user.

In order to unify our schemes, we abstract the common features required with an *Encrypt Zero* protocol for Raz's encryption scheme. The goal of the protocol is to give one party, the Keeper, a random key $\mathbf{s}$, and another party, the Recorder, λ random encryptions $\{c_1, \ldots, c_\lambda\}$ of 0. Here, λ is a parameter that will be chosen based on application. Recorder security dictates that the Keeper learns nothing about the λ encryptions stored by the Recorder (aside from the fact that they encrypt 0). Keeper security requires that the min-entropy of the key $\mathbf{s}$ conditioned on the Recorder's view is $\Omega(n)$. We additionally require that the Keeper's space is $O(n)$ (which is optimal since the Keeper must store a secret key of $O(n)$ bits), and the Recorder's space is $O(\lambda n)$ (which is also optimal, since the Recorder must store λ encryptions of $O(n)$ bits each).

Our basic protocol for Raz's scheme works as follows:

- The Keeper chooses a random key $\mathbf{k} \in \{0,1\}^n$. Let $m = O(n)$ be a parameter. The Recorder chooses a secret matrix $\boldsymbol{\Sigma} \in \{0,1\}^{\lambda \times m}$.
- The Keeper streams m encryptions $(\mathbf{r}_i, a_i = \mathbf{r}_i \cdot \mathbf{k} + 0)$ to the Recorder, for random $\mathbf{r}_i \in \{0,1\}^n$ and $i = 1, 2, \ldots, m$. From now on, we use the convention that "+" and "·" are carried out mod 2.
- The Recorder maintains matrix $\boldsymbol{\Psi} \in \{0,1\}^{\lambda \times n}$ and column vector $\boldsymbol{\kappa} \in \{0,1\}^\lambda$. Each row of $(\boldsymbol{\Psi}|\boldsymbol{\kappa})$ will be a random subset-sum of the encryptions sent by the Keeper, with each subset-sum chosen according to $\boldsymbol{\Sigma}$. The matrices will be computed on the fly. So when $(\mathbf{r}_i, a_i)$ comes in, the Recorder will map $\boldsymbol{\Psi} \to \boldsymbol{\Psi} + \boldsymbol{\sigma}_i \cdot \mathbf{r}_i$, $\boldsymbol{\kappa} \to \boldsymbol{\kappa} + \boldsymbol{\sigma}_i a_i$. Here, $\boldsymbol{\sigma}_i$ is the i-th column of $\boldsymbol{\Sigma}$, and $\mathbf{r}_i$ is interpreted as a row vector.
- At the end of the protocol, the Keeper outputs its key $\mathbf{s} = \mathbf{k}$, and the Recorder outputs $(\boldsymbol{\Psi}|\boldsymbol{\kappa})$, whose rows are the ciphertexts $c_1, \ldots, c_\lambda$.

Let $\mathbf{R}$ be the matrix whose rows are the $\mathbf{r}_i$'s, and let $\mathbf{a}$ be the column vector of the a_i's. Then we have that $\mathbf{a} = \mathbf{R} \cdot \mathbf{k}$, $\boldsymbol{\Psi} = \boldsymbol{\Sigma} \cdot \mathbf{R}$, and $\boldsymbol{\kappa} = \boldsymbol{\Sigma} \cdot \mathbf{a} = \boldsymbol{\Psi} \cdot \mathbf{k}$. Hence, the rows of $(\boldsymbol{\Psi}|\boldsymbol{\kappa})$ are encryptions of zero, as desired.

For Keeper security, Raz's theorem directly shows that $\mathbf{k}$ has min-entropy relative to the Recorder's view. For Recorder security, notice that $\boldsymbol{\Sigma}$ is independent of the Keeper's view. Therefore, if the Keeper follows the protocol and m is slightly larger than n so that $\mathbf{R}$ is full rank with high probability, then $\boldsymbol{\Psi}$ is a random matrix independent of the adversary's view. Therefore the ciphertexts c_i are actually *random* encryptions of 0. Thus we get security for honest-but-curious Keepers.

Key Agreement. This protocol gives a simple key-agreement scheme. Basically, one party acts as the Keeper, and one as the Recorder. We set $\lambda = 1$. The result of the Encrypt Zero protocol is that the Recorder contains a uniformly random encryption of 0. The Recorder simply flips the bit encrypted with probability 1/2 to get a random encryption of a random bit b, and sends the resulting ciphertext to the Keeper. The Keeper decrypts, and the shared secret key is just the resulting plaintext b.

Security of the protocol follows from the fact that after the Encrypt Zero protocol, the Keeper's key has min-entropy relative to any eavesdropper (since the eavesdropper learns no more than the Recorder). Moreover, the Keeper acts honestly, so the final ciphertext is always a fresh encryption. Finally, the encryption scheme is leakage resilient so it hides the bit b even though the adversary may have some knowledge of the key.

Notice that this scheme has *perfect* correctness, in that the two parties always arrive at a secret key. This is in contrast to the existing schemes based on the birthday paradox, where security is only statistical, and moreover this holds only if the adversary's space bounds are asymptotically smaller than n^2. In contrast, we get perfect correctness and statistical security for adversarial space bounds that are $O(n^2)$. The honest users only require $O(n)$ space.

Bit Commitment. We now describe a simple bit-commitment protocol using the above Encrypt Zero protocol. Recall that in a bit-commitment scheme, there are two phases: a commit phase where the Committer commits to a bit b, and a reveal or de-commit phase where the Committer reveals b and proves that b was the value committed to. After the commit phase, we want that the bit b is hidden. On the other hand, we want the commit phase to be binding, in that the Committer cannot later change the committed bit to something else.

The Committer and the Verifier will run the Encrypt Zero protocol, with Committer playing the role of Keeper and Verifier the role of Recorder. The protocol works as follows:

- Run the Encrypt Zero protocol, giving the Committer a random key $\mathbf{s}$ and the Verifier λ random encryptions c_i of 0.
- The Committer then sends an encryption of b relative to the key $\mathbf{s}$.

- To open the commitment, the Committer sends $\mathbf{s}$. The Verifier checks that $\mathbf{s}$ correctly decrypts all the c_i to 0. If so, it decrypts the final ciphertext to get b.

The security of the Encrypt Zero protocol and the leakage resilience of the encryption scheme show that this scheme is hiding. For binding, we note that an honest Committer will have no idea what encryptions c_i the Verifier has. As such, if the Committer later tries to change its committed bit by sending a malicious key $\mathbf{s}'$, $\mathbf{s}'$ will cause each ciphertext c_i to decrypt to 1 with probability 1/2. Therefore, the Committer will get caught with probability $1 - 2^{-\lambda}$.

Already, this gives a very simple protocol for bit commitment that is *non-interactive*; in contrast, the prior work of Ding et al. [DHRS04] required five rounds. One limitation is that we require the Committer to behave honestly during the commit phase. For example, if the Committer chooses $\mathbf{R}$ to be low rank, then the encryptions obtained by the Verifier will not be independent of the Committer's view, and hence the Committer may be able to cheat during the de-commit phase.

To get around this, we tweak the Encrypt Zero protocol slightly to get security even against malicious Keepers. Our Enhanced Encrypt Zero protocol is as follows:

- The Keeper chooses a random key $\mathbf{k} \in \{0,1\}^n$ and an independent random secret $\mathbf{s} \in \{0,1\}^m$. We will let $m = 2n$. The Recorder chooses a secret matrix $\boldsymbol{\Sigma} \in \{0,1\}^{\lambda \times m}$.
- The Keeper streams random encryptions of the bits of s_i. We will write this in matrix form as $(\mathbf{R}, \mathbf{a} = \mathbf{R} \cdot \mathbf{k} + \mathbf{s})$.
- The Recorder computes $\boldsymbol{\Psi} = \boldsymbol{\Sigma} \cdot \mathbf{R}$ and $\boldsymbol{\kappa} = \boldsymbol{\Sigma} \cdot \mathbf{a}$.
- The Keeper then sends its key $\mathbf{k}$ *in the clear*.
- The Keeper outputs its secret $\mathbf{s}$ as the key, and the Recorder outputs $(\boldsymbol{\Sigma}, \boldsymbol{\kappa} - \boldsymbol{\Psi} \cdot \mathbf{k})$.

Notice that $\boldsymbol{\kappa} - \boldsymbol{\Psi} \cdot \mathbf{k} = \boldsymbol{\Sigma} \cdot \mathbf{s}$, a list of λ encryptions of 0 relative to the key $\mathbf{s}$, as desired. Moreover, these encryptions are random encryptions, even if $\mathbf{R}$ is chosen adversarially by the Keeper, since the Keeper has no knowledge or control over $\boldsymbol{\Sigma}$.

To prove the min-entropy of $\mathbf{s}$ relative to a malicious Recorder, we note that the real-or-random CPA security of the encryption scheme shows that just prior to receiving $\mathbf{k}$, the Recorder has essentially no information about $\mathbf{s}$. Then, since $\mathbf{k}$ is n bits, revealing it can only reveal n bits of $\mathbf{s}$. But $\mathbf{s}$ is a uniformly random $m = 2n$ bit string, meaning it has roughly n bits of min-entropy remaining, as desired. Thus we get both our security properties, even for malicious parties.

Our Enhanced Encrypt Zero protocol roughly doubles the communication, but otherwise maintains all the attractive properties of the original scheme: it is non-interactive and has perfect correctness.
Putting it all together, our bit commitment protocol is the following:

- To commit to a bit b, the Committer streams $\mathbf{R}$, $\mathbf{a} = \mathbf{R} \cdot \mathbf{k} + \mathbf{s}$ followed by $\mathbf{k}, \boldsymbol{\gamma}, c = \boldsymbol{\gamma} \cdot \mathbf{s} + b$ for random $\mathbf{R}, \mathbf{k}, \mathbf{s}, \boldsymbol{\gamma}$.

- The Verifier records $\boldsymbol{\Sigma}, \boldsymbol{\Psi} = \boldsymbol{\Sigma} \cdot \mathbf{R}, \boldsymbol{\kappa} = \boldsymbol{\Sigma} \cdot \mathbf{a}$ for a random choice of $\boldsymbol{\Sigma}$, and then once $\mathbf{k}$ comes in it computes $\boldsymbol{\phi} = \boldsymbol{\kappa} - \boldsymbol{\Psi} \cdot \mathbf{k} = \boldsymbol{\Sigma} \cdot \mathbf{s}$.
- To reveal the bit b, the Committer just sends $\mathbf{x} = \mathbf{s}$.
- The Verifier checks that $\boldsymbol{\phi} = \boldsymbol{\Sigma} \cdot \mathbf{x}$. If so, it computes $b' = c - \boldsymbol{\gamma} \cdot \mathbf{x}$.

Oblivious Transfer. We now turn to constructing an oblivious transfer (OT) protocol. In an OT protocol, one party, the Sender, has two input bits x_0, x_1. Another party, the Receiver, has a bit b. The Receiver would like to learn x_b without revealing b, and the Sender would like to ensure that the Receiver learns nothing about x_{1-b}.

In our protocol, the Receiver will play the role of Committer in our commitment scheme, committing to its input b. The Sender will play the role of Recorder in the Encrypt Zero protocol, setting $\lambda = 2$. The hiding property of the commitment scheme ensures that the space-bounded Sender learns nothing about the Receiver's bit b.

At the end of the Receiver's message, the Sender has an encryption $(\boldsymbol{\gamma}, c^* = \boldsymbol{\gamma} \cdot \mathbf{s} + b)$ of b with secret key $\mathbf{s}$. Additionally, it also has two encryptions of 0, namely $(\boldsymbol{\sigma}_0, c_0 = \boldsymbol{\sigma}_0 \cdot \mathbf{s})$ and $(\boldsymbol{\sigma}_1, c_1 = \boldsymbol{\sigma}_1 \cdot \mathbf{s})$ for random vectors $\boldsymbol{\sigma}_0, \boldsymbol{\sigma}_1$. Importantly, $\boldsymbol{\sigma}_0, \boldsymbol{\sigma}_1$ are independent of the Receiver's view, as they were chosen by the Sender.

The Sender will now exploit the additive homomorphism of the encryption scheme once more. In particular, it will compute encryptions of $(1-b)x_0$ and bx_1, which it will then send back to the Receiver. To compute an encryption of bx_1, it simply multiplies the ciphertext $(\boldsymbol{\gamma}, c^*)$ by x_1. Similarly, to compute an encryption of $(1-b)x_0$, it toggles c^* (to get an encryption of $1-b$) and then multiplies the entire ciphertext by x_0.

Now clearly these two ciphertexts reveal both x_0 and x_1, so the Sender cannot send them directly to the Receiver. Instead, it will *re-randomize* them by adding the two encryptions of 0. Now it obtains *fresh* encryptions of $(1-b)x_0$ and bx_1:

$$\begin{aligned}\boldsymbol{\sigma}_0 + x_0\boldsymbol{\gamma},\ c_0 + x_0(1-c^*) &= (\boldsymbol{\sigma}_0 + x_0\boldsymbol{\gamma}) \cdot \mathbf{s} + \big((1-b)x_0\big)\\ \boldsymbol{\sigma}_1 + x_1\boldsymbol{\gamma},\ c_1 + x_1 c^* &= (\boldsymbol{\sigma}_1 + x_1\boldsymbol{\gamma}) \cdot \mathbf{s} + \big(bx_1\big)\end{aligned}$$

It sends these ciphertexts to the Receiver, who then decrypts. All the Receiver learns then is $(1-b)x_0$ and bx_1. One of these plaintexts will be x_b as desired, and the other will be 0. Thus, the Receiver learns nothing about x_{1-b}.

Our protocol is round-optimal, since it involves only a single message in each direction. This improves on the best prior work of Ding et al. [DHRS04] requiring 5 rounds. Additionally, our protocol is much simpler than the prior work.

1.5 Discussion

Just as homomorphic encryption has been an extremely useful tool in traditional cryptography, our work demonstrates that the homomorphic properties of Raz's encryption scheme are also fruitful for the Bounded Storage model. We believe our work will be a useful starting point for much future work in this area.

1.6 Other Related Work

A recent work by Ball et al. [BDKM18] shows another application of Raz's encryption scheme, where they use it to construct unconditional non-malleable codes against streaming, space-bounded tempering.

2 Preliminaries

Here, we recall some basic cryptographic notions, translated into the setting of the bounded storage model. In the following definitions, n will be a security parameter.

A symmetric encryption scheme is a pair of algorithms $\Pi = (\mathtt{Enc}, \mathtt{Dec})$ with an associated key space $\mathcal{K}_n$, message space $\mathcal{M}$, and ciphertext space $\mathcal{C}_n$. Notice that the key space and ciphertext space depend on n; the message space will not depend on n. We require that:

- $\mathtt{Enc} : \mathcal{K}_n \times \mathcal{M} \rightarrow \mathcal{C}_n$ is a probabilistic polynomial time (PPT) algorithm
- $\mathtt{Dec} : \mathcal{K}_n \times \mathcal{C}_n \rightarrow \mathcal{M}$ is a deterministic polynomial time algorithm.
- Correctness: for any $k \in \mathcal{K}_n$ and any message $m \in \mathcal{M}$,

$$\Pr[\mathtt{Dec}(k, \mathtt{Enc}(k, m)) = m] = 1.$$

Additionally, we will require a security notion. In this work, we will focus on the following notion.

Definition 1 (Real-or-Random-Ciphertext (RoRC) Security). *Let $\mathcal{A}$ be an adversary. $\mathcal{A}$ plays the following game $\mathsf{RoRC}_{\mathcal{A},\Pi,b}(n, q)$:*

- *The challenger's input is a bit $b \in \{0, 1\}$.*
- *The challenger chooses a random key $k \in \mathcal{K}_n$.*
- *$\mathcal{A}$ makes q adaptive queries on messages $m_1, \ldots, m_q \in \mathcal{M}$.*
- *In response to each query, the challenger does the following:*
  - *If $b = 0$, the challenger responds with $c_i \leftarrow \mathtt{Enc}(k, m_i)$.*
  - *If $b = 1$, the challenger responds with a random ciphertext $c_i \in \mathcal{C}_n$.*
- *Finally, $\mathcal{A}$ outputs a guess b' for b.*

We say that Π is $(S(n), Q(n), \epsilon)$-secure if for all adversaries that use at most $S(n)$ memory bits and $Q(n)$ queries (i.e. $q \leq Q(n)$),

$$|\Pr[\mathsf{RoRC}_{\mathcal{A},\Pi,0}(n, q) = 1] - \Pr[\mathsf{RoRC}_{\mathcal{A},\Pi,1}(n, q) = 1]| \leq \epsilon.$$

In this work, a lot of the proofs are based on the Leftover Hash Lemma for Conditional Min-Entropy due to Impagliazzo, Levin, and Luby [ILL89].

For random distributions X and Y, let $H_\infty(X|Y)$ denote the min-entropy of X conditioned on Y. Let $X \approx_\epsilon Y$ denote that the two distributions are ϵ-close, i.e. the statistical distance between these two distributions $\Delta(X, Y) \leq \epsilon$. Furthermore, let U_m denote a uniformly distributed random variable of m bits for some positive integer m.

Lemma 1 (Leftover Hash Lemma for Conditional Min-Entropy [ILL89]). *Let X, E be a joint distribution. If $H_\infty(X|E) \geq k$, and $m = k - 2\log(1/\epsilon)$, then*

$$(H(X), H, E) \approx_{\epsilon/2} (U_m, U_d, E),$$

where m is the output length of a universal hash function H, and d is the length of the description of H.

3 Raz's Encryption Scheme

Our constructions of the commitment scheme and the oblivious transfer scheme are largely based on the bit encryption scheme from parity learning proposed by Raz [Raz17]. Raz sketches how his lower bound for learning implies the security of his encryption scheme. Below we reproduce the construction of the encryption scheme, and formalize the security proof.

Construction 1 (Bit Encryption Scheme from Parity Learning). For a given security parameter n, the encryption scheme consists of a message space $\mathcal{M} = \{0,1\}$, a ciphertext space $\mathcal{C}_n = \{0,1\}^n \times \{0,1\}$, a key space $\mathcal{K}_n = \{0,1\}^n$, and a pair of algorithms $\Pi = (\mathtt{Enc}, \mathtt{Dec})$ as specified below:

- $\mathtt{Enc}(\mathbf{k}, m \in \mathcal{M})$: Samples a random row vector $\mathbf{r} \leftarrow \{0,1\}^n$, computes $a = \mathbf{r} \cdot \mathbf{k} + m$, and outputs the ciphertext $c = (\mathbf{r}, a)$ as a pair.
- $\mathtt{Dec}(\mathbf{k}, c = (\mathbf{r}, a) \in \mathcal{C}_n)$: Computes and outputs $m' = \mathbf{r} \cdot \mathbf{k} + a$.

To prove Real-or-Random-Ciphertext security of the above scheme, we rely on a result from Raz [Raz17], reproduced below.

Lemma 2 ([Raz17]). *For any $C < \frac{1}{20}$, there exists $\alpha > 0$, such that: for uniform $\mathbf{k} \in \{0,1\}^n$, $m \leq 2^{\alpha n}$, and algorithm $\mathcal{A}$ that takes a stream of $(\mathbf{x}_1, y_1)$, $(\mathbf{x}_2, y_2)$, ..., $(\mathbf{x}_m, y_m)$, where $\mathbf{x}_i$ is a uniform distribution over $\{0,1\}^n$ and $y_i = \mathbf{x}_i \cdot \mathbf{k}$ for every i, under the condition that $\mathcal{A}$ uses at most Cn^2 memory bits and outputs $\tilde{\mathbf{k}} \in \{0,1\}^n$, then $Pr[\tilde{\mathbf{k}} = \mathbf{k}] \leq O(2^{-\alpha n})$.*

We also rely on the Goldreich-Levin Algorithm, reproduced below.

Lemma 3 (Goldreich-Levin Algorithm [GL89]). *Assume that there exists a function $f : \{0,1\}^n \to \{0,1\}$ s.t. for some unknown $\mathbf{x} \in \{0,1\}^n$, we have*

$$\Pr_{\mathbf{r} \in \{0,1\}^n} [f(\mathbf{r}) = \langle \mathbf{x}, \mathbf{r} \rangle] \geq \frac{1}{2} + \epsilon$$

for $\epsilon > 0$.

Then there exists an algorithm $\mathcal{GL}$ that runs in time $O(n^2\epsilon^{-4}\log n)$, makes $O(n\epsilon^{-4}\log n)$ oracle queries into f, and outputs $\mathbf{x}$ with probability $\Omega(\epsilon^2)$.

Instead of directly proving RoRC security of the encryption scheme, we prove Modified Real-or-Random-Ciphertext (RoRC') security, which differs from RoRC security in that for all but the last query, the challenger always responds with the valid encryption of the message; for the last query, the challenger responds either with a valid encryption or a random ciphertext, each with probability 1/2. A detailed definition is given below.

Definition 2 (Modified Real-or-Random-Ciphertext (RoRC') Security). *Let $\mathcal{A}$ be an adversary. $\mathcal{A}$ plays the following game* $\mathsf{RoRC'}_{\mathcal{A},\Pi,b}(n,q)$*:*

- *The challenger's input is a bit $b \in \{0,1\}$.*
- *The challenger chooses a random key $k \in \mathcal{K}_n$.*
- *$\mathcal{A}$ makes q adaptive queries on messages $m_1, \ldots, m_q \in \mathcal{M}$.*
- *In response to query m_i with $1 \le i \le q-1$, the challenger responds with $c_i \leftarrow \mathtt{Enc}(k, m_i)$.*
- *In response to query m_q, the challenger does the following:*
  - *If $b = 0$, the challenger responds with $c_q \leftarrow \mathtt{Enc}(k, m_q)$.*
  - *If $b = 1$, the challenger responds with a random ciphertext $c_q \in \mathcal{C}_n$.*
- *Finally, $\mathcal{A}$ outputs a guess b' for b.*

We say that Π is $(S(n), Q(n), \epsilon)$-secure if for all adversaries that use at most $S(n)$ memory bits and $Q(n)$ queries (i.e. $q \le Q(n)$),

$$|\Pr[\mathsf{RoRC'}_{\mathcal{A},\Pi,0}(n,q) = 1] - \Pr[\mathsf{RoRC'}_{\mathcal{A},\Pi,1}(n,q) = 1]| \le \epsilon.$$

We now show that RoRC' security implies RoRC security.

Lemma 4. *An encryption scheme that is $(S(n), Q(n), \epsilon)$-secure under the RoRC' setting is $(S(n), Q(n), Q(n)\epsilon)$-secure under the RoRC setting.*

Proof. We prove this using a hybrid argument. For any $q \le Q(n)$, consider the hybrid security games H_0, H_1, ..., H_q, where H_j describes the following hybrid game:

- The challenger chooses a random key $k \in \mathcal{K}_n$.
- $\mathcal{A}$ makes q adaptive queries on messages $m_1, \ldots, m_q \in \mathcal{M}$.
- In response to query m_i with $1 \le i \le j$, the challenger responds with $c_i \leftarrow \mathtt{Enc}(k, m_i)$.
- In response to query m_i with $j+1 \le i \le q$, the challenger responds with a random ciphertext $c_i \in \mathcal{C}_n$.

Particularly, notice that H_0 corresponds to a game where the challenger always responds with random ciphertexts, and that H_q corresponds to a game where the challenger always responds with valid encryptions of the messages. In that way, the $\mathsf{RoRC}_{\mathcal{A},\Pi,b}(n,q)$ game is equivalent to distinguishing H_q from H_0.

To put this formally, let D be an arbitrary distinguisher, and $h \leftarrow H_j$ denote a randomly sampled instance of the game H_j, we have

$$\begin{aligned} &|\Pr[\mathsf{RoRC}_{\mathcal{A},\Pi,0}(n,q) = 1] - \Pr[\mathsf{RoRC}_{\mathcal{A},\Pi,1}(n,q) = 1]| \\ &= \left| \Pr_{h \leftarrow H_q}[D(h) = 1] - \Pr_{h \leftarrow H_0}[D(h) = 1] \right|. \end{aligned}$$

By the hybrid argument, there exists j, s.t. $0 \le j < q$ and

$$\left| \Pr_{h \leftarrow H_q}[D(h)=1] - \Pr_{h \leftarrow H_0}[D(h)=1] \right| \le q \left| \Pr_{h \leftarrow H_{j+1}}[D(h)=1] - \Pr_{h \leftarrow H_j}[D(h)=1] \right|.$$

To distinguish between H_{j+1} and H_j, consider the following security game $\mathsf{Dist}_{\mathcal{A},\Pi,b}(n,q,j)$:

- The challenger's input is a bit $b \in \{0,1\}$.
- The challenger chooses a random key $k \in \mathcal{K}_n$.
- $\mathcal{A}$ makes q adaptive queries on messages $m_1, \ldots, m_q \in \mathcal{M}$.
- In response to query m_i with $1 \le i \le j$, the challenger responds with $c_i \leftarrow \mathtt{Enc}(k, m_i)$.
- In response to query m_{j+1}, the challenger does the following:
  - If $b = 0$, the challenger responds with $c_{j+1} \leftarrow \mathtt{Enc}(k, m_{j+1})$.
  - If $b = 1$, the challenger responds with a random ciphertext $c_{j+1} \in \mathcal{C}_n$.
- In response to query m_i with $j+1 < i \le q$, the challenger responds with a random ciphertext $c_i \in \mathcal{C}_n$.
- Finally, $\mathcal{A}$ outputs a guess b' for b.

This directly gives us

$$\begin{aligned} &\left| \Pr_{h \leftarrow H_{j+1}}[D(h)=1] - \Pr_{h \leftarrow H_j}[D(h)=1] \right| \\ &= \left| \Pr[\mathsf{Dist}_{\mathcal{A},\Pi,0}(n,q,j)=1] - \Pr[\mathsf{Dist}_{\mathcal{A},\Pi,1}(n,q,j)=1] \right|. \end{aligned}$$

Next, we show that we can use an adversary $\mathcal{A}$ for the $\mathsf{Dist}_{\mathcal{A},\Pi,b}(n,q,j)$ game to construct an adversary $\mathcal{A}'$ for the $\mathsf{RoRC'}_{\mathcal{A}',\Pi,b}(n,j+1)$ game. Notice that the only difference between $\mathsf{RoRC'}_{\mathcal{A}',\Pi,b}(n,j+1)$ and $\mathsf{Dist}_{\mathcal{A},\Pi,b}(n,q,j)$ is that $\mathsf{Dist}_{\mathcal{A},\Pi,b}(n,q,j)$ has $(q-j-1)$ extra queries at the end. An adversary $\mathcal{A}'$ for $\mathsf{RoRC'}_{\mathcal{A}',\Pi,b}(n,j+1)$ can simulate $\mathsf{Dist}_{\mathcal{A},\Pi,b}(n,q,j)$ for adversary $\mathcal{A}$ by forwarding each of $\mathcal{A}$'s first $(j+1)$ queries to the challenger in $\mathsf{RoRC'}_{\mathcal{A}',\Pi,b}(n,j+1)$, and similarly forward the responses from the challenger back to $\mathcal{A}$. For the additional $(q-j-1)$ queries in the end, $\mathcal{A}'$ can simply respond by drawing random ciphertexts from $\mathcal{C}_n$. $\mathcal{A}'$ will output whatever is output by $\mathcal{A}$.

Notice that adversary $\mathcal{A}'$ does *not* require any additional memory space besides the space used by adversary $\mathcal{A}$. All that $\mathcal{A}'$ needs to do is to forward $\mathcal{A}$'s queries and the challenger's responses, and to sample random ciphertexts from $\mathcal{C}_n$. These operations do not require $\mathcal{A}'$ to store any persistent states.

Therefore, we have

$$\begin{aligned} &\left| \Pr[\mathsf{Dist}_{\mathcal{A},\Pi,0}(n,q,j)=1] - \Pr[\mathsf{Dist}_{\mathcal{A},\Pi,1}(n,q,j)=1] \right| \\ &\le \left| \Pr[\mathsf{RoRC'}_{\mathcal{A},\Pi,0}(n,j+1)=1] - \Pr[\mathsf{RoRC'}_{\mathcal{A},\Pi,1}(n,j+1)=1] \right|. \end{aligned}$$

Bringing all these parts together, assuming that the encryption scheme Π is $(S(n), Q(n), \epsilon)$-secure yields

$$\begin{aligned}
&|\Pr[\mathsf{RoRC}_{\mathcal{A},\Pi,0}(n,q)=1]-\Pr[\mathsf{RoRC}_{\mathcal{A},\Pi,1}(n,q)=1]|\\
=&\left|\Pr_{h\leftarrow H_q}[D(h)=1]-\Pr_{h\leftarrow H_0}[D(h)=1]\right|\\
\leq &q\left|\Pr_{h\leftarrow H_{j+1}}[D(h)=1]-\Pr_{h\leftarrow H_j}[D(h)=1]\right|\\
=&q\,|\Pr[\mathsf{Dist}_{\mathcal{A},\Pi,0}(n,q,j)=1]-\Pr[\mathsf{Dist}_{\mathcal{A},\Pi,1}(n,q,j)=1]|\\
\leq &q\,|\Pr[\mathsf{RoRC'}_{\mathcal{A},\Pi,0}(n,j+1)=1]-\Pr[\mathsf{RoRC'}_{\mathcal{A},\Pi,1}(n,j+1)=1]|\\
\leq &q\epsilon\leq Q(n)\epsilon.
\end{aligned}$$

Therefore, Π is $(S(n)), Q(n), Q(n)\epsilon)$-secure under the RoRC setting. □

Theorem 1. *For any $C<\frac{1}{20}$, there exists $\alpha>0$, s.t. the bit encryption scheme from parity learning is $(Cn^2, 2^{\alpha n}, O(2^{-\alpha n/2}))$-secure under the RoRC' setting.*

Proof. We prove this result by reducing a parity learning game to an RoRC' game.

To start off, we consider a weaker variant of the parity learning game described in Lemma 2, denoted as $\mathsf{PL}_{\mathcal{A},b}(n,q)$:

- The challenger's input is a bit $b\in\{0,1\}$.
- The challenger chooses a random $\mathbf{k}\in\{0,1\}^n$.
- The challenger streams $(\mathbf{x}_1,y_1)$, $(\mathbf{x}_2,y_2)$, ..., $(\mathbf{x}_{q-1},y_{q-1})$, where $\mathbf{x}_i$ is uniformly distributed over $\{0,1\}^n$ and $y_i=\mathbf{x}_i\cdot\mathbf{k}$ for all i.
- The challenger sends $(\mathbf{x}_q,y_q)$, where $\mathbf{x}_q$ is uniformly distributed over $\{0,1\}^n$ and:
  - If $b=0$, $y_q=\mathbf{x}_q\cdot\mathbf{k}$.
  - If $b=1$, y_q is a random bit.
- Finally, $\mathcal{A}$ outputs a guess b' for b.

We now show how we can use an adversary $\mathcal{A}$ for $\mathsf{RoRC'}_{\mathcal{A},\Pi,b}(n,q)$ to build an adversary $\mathcal{A}'$ for $\mathsf{PL}_{\mathcal{A}',b}(n,q)$. The adversary $\mathcal{A}'$ works as follows:

- Simulate for $\mathcal{A}$ an $\mathsf{RoRC'}_{\mathcal{A},\Pi,b}(n,q)$ game.
- For every query m_i submitted by $\mathcal{A}$, respond with $(\mathbf{x}_i,y_i+m_i)$ where $\mathbf{x}_i$ and y_i come from the i-th pair of the $\mathsf{PL}_{\mathcal{A}',b}(n,q)$ game.
- If the adversary $\mathcal{A}$ outputs 0, output 0. Otherwise, output 1.

This should be easily verifiable. First, notice that $\mathcal{A}'$ faithfully simulates $\mathsf{RoRC'}_{\mathcal{A},\Pi,b}(n,q)$. For $1\leq i\leq q-1$, $\mathcal{A}$ receives $(\mathbf{x}_i,y_i+m_i)=(\mathbf{x}_i,\mathbf{x}_i\cdot\mathbf{k}+m_i)$, which is a valid encryption of m_i. Also, for the last query m_q, $\mathcal{A}$ receives either $(\mathbf{x}_q,y_q+m_q)=(\mathbf{x}_q,\mathbf{x}_q\cdot\mathbf{k}+m_q)$, i.e. a valid encryption, or $(\mathbf{x}_q,y_q+m_q)$ for a random bit y_q, i.e. a random ciphertext. Secondly, if $\mathcal{A}$ outputs 0, that implies $(\mathbf{x}_q,y_q+m_q)=\mathsf{Enc}(\mathbf{k},m_q)=(\mathbf{x}_q,\mathbf{x}_q\cdot\mathbf{k}+m_q)$, and hence $y_q=\mathbf{x}_q\cdot\mathbf{k}$ and $\mathcal{A}'$

should output 0. Lastly, if $\mathcal{A}$ outputs 1, we have $y_q + m_q$ being a random bit. Since m_q is fixed, we have y_q a random bit and hence $\mathcal{A}'$ should output 1.

This yields

$$|\Pr[\mathsf{RoRC'}_{\mathcal{A},\Pi,0}(n,q) = 1] - \Pr[\mathsf{RoRC'}_{\mathcal{A},\Pi,1}(n,q) = 1]|$$
$$\leq |\Pr[\mathsf{PL}_{\mathcal{A},0}(n,q) = 1] - \Pr[\mathsf{PL}_{\mathcal{A},1}(n,q) = 1]| .$$

Let $\beta = |\Pr[\mathsf{PL}_{\mathcal{A},0}(n,q) = 1] - \Pr[\mathsf{PL}_{\mathcal{A},1}(n,q) = 1]|$. Then we have an algorithm that distinguishes between $(\mathbf{x}_q, y_q = \mathbf{x}_q \cdot \mathbf{k})$ and $(\mathbf{x}_q, y_q \leftarrow \{0,1\})$ with probability $(1+\beta)/2$, i.e. it outputs 0 if y_q is a valid inner product and 1 if it is random. This can be easily converted into an algorithm that given $\mathbf{x}_q$, outputs $\mathbf{x}_q \cdot \mathbf{k}$ with probability $(1+\beta)/2$ (simply XOR the output of the previous algorithm with y_q). Let f be the function computed by this algorithm. Then for given $\mathbf{x}_q \in \{0,1\}^n$ and unknown $\mathbf{k} \in \{0,1\}^n$, $f(\mathbf{x}_q) = \langle \mathbf{k}, \mathbf{x}_q \rangle$ with probability $(1+\beta)/2$. By applying Lemma 3, there is an algorithm that runs in time $O(n^2\beta^{-4}\log n)$ and outputs $\mathbf{k}$ with probability at least $\Omega(\beta^2)$.

Recall from Lemma 2 that for any $C < 1/20$, there is a positive α such that any potentially *computationally unbounded* algorithm that uses up to Cn^2 memory bits and has access to at most $2^{\alpha n}$ $(\mathbf{x}_i, y_i)$ pairs can output $\mathbf{k}$ with probability at most $O(2^{-\alpha n})$. Therefore, for adversaries that are space-bounded by Cn^2 bits and submit at most $2^{\alpha n}$ queries, $\Omega(\beta^2) \leq O(2^{-\alpha n})$. And hence $\beta = O(2^{-\alpha n/2})$.

Therefore, for any $C < 1/20$, there is a positive α such that for all adversaries that use at most Cn^2 memory bits and at most $2^{\alpha n}$ queries ($q \leq 2^{\alpha n}$), we have

$$|\Pr[\mathsf{RoRC'}_{\mathcal{A},\Pi,0}(n,q) = 1] - \Pr[\mathsf{RoRC'}_{\mathcal{A},\Pi,1}(n,q) = 1]| \leq \beta = O(2^{-\alpha n/2}),$$

i.e. the scheme is $(Cn^2, 2^{\alpha n}, O(2^{-\alpha n/2}))$-secure under the RoRC' setting as desired. □

Corollary 1 (RoRC Security of the Bit Encryption Scheme from Parity Learning). *For any $C < \frac{1}{20}$, there exists $\alpha > 0$, s.t. the bit encryption scheme from parity learning is $(Cn^2, 2^{\alpha n/4}, O(2^{-\alpha n/2}))$-secure under the RoRC' setting (here we further bound the number of queries to $\alpha n/4$ instead of αn). By Lemma 4, this scheme is also $(Cn^2, 2^{\alpha n/4}, 2^{\alpha n/4} \cdot O(2^{-\alpha n/2}) = O(2^{-\alpha n/4}))$-secure under the RoRC setting. Put another way, for any $C < \frac{1}{20}$, there exists $\alpha'(= \alpha/4) > 0$, s.t. the bit encryption scheme from parity learning is $(Cn^2, 2^{\alpha' n}, O(2^{-\alpha' n}))$-secure under the RoRC setting.*

4 Encrypt Zero Protocols

In this section, we introduce two constructions of the Encrypt Zero Protocol. They both have the same goal: to give one party, the *Keeper*, a random key $\mathbf{s}$, and the other party, known as the *Recorder*, several encryptions of 0 under the key $\mathbf{s}$. They differ in that the simple construction is only secure against honest-but-curious Keepers, while the enhanced construction is secure even against malicious Keepers.

Before we jump into the constructions, we first define an Encrypt Zero Protocol and its security properties.

An Encrypt Zero Protocol Π involves two parties, a Keeper $\mathcal{K}$ and a Recorder $\mathcal{R}$. The protocol takes three parameters $n, m = O(n)$ and λ, and produces $(\mathsf{s}, \{c_1, c_2, \ldots, c_\lambda\}, \mathsf{trans})$, where s is a random key output by $\mathcal{K}$, $\{c_1, c_2, \ldots, c_\lambda\}$ is a set of ciphertexts output by $\mathcal{R}$, and trans is the transcript of their communication.

The correctness of an Encrypt Zero Protocol requires that the set of ciphertexts output by $\mathcal{R}$ are encryptions of zero under the key s output by $\mathcal{K}$. Put formally, we require that $\mathtt{Dec}(\mathsf{s}, c_i) = 0$ for all i.

Now, we define two desired security properties for the Encrypt Zero Protocol, namely Keeper security and Recorder security.

The security of the Keeper ensures that the Keeper's key s has enough min-entropy conditioned on the Recorder's view $\mathsf{view}_{\mathcal{R}}$.

Definition 3 (Keeper Security). *Let the view of the Recorder be* $\mathsf{view}_{\mathcal{R}}$, *we say that a protocol* Π *is* $(S(n), h)$*-secure for the Keeper if for all Recorders* $\mathcal{R}$ *that use up to* $S(n)$ *memory bits,*

$$H_\infty(\mathsf{s}|\mathsf{view}_{\mathcal{R}}) \geq h.$$

The security of the Recorder ensures that the Keeper learns nothing about $c_1, c_2, \ldots, c_\lambda$ (except that they are encryptions of zero).

For an honest-but-curious Keeper $\mathcal{K}$, this means that given all the Keeper's randomness and the transcript produced by the protocol, it is hard to distinguish the output ciphertexts $(c_1, c_2, \ldots, c_\lambda)$ from some random ciphertexts that encrypt zero.

Definition 4 (Recorder Security with Honest-but-Curious Keeper). *Let* $C = \{c_1, c_2, \ldots, c_\lambda\}$ *be the ciphertexts output by* $\mathcal{R}$ *at the end of the protocol, and* $C' = \{c'_1, c'_2, \ldots, c'_\lambda\}$ *where* $c'_i \leftarrow \mathtt{Enc}(\mathsf{s}, 0)$ *be fresh encryptions of zero under the key* s. *Let* $\mathsf{state}_{\mathcal{K}}$ *consist of all the random coins used by* $\mathcal{K}$ *together with* trans. *Given the Keeper's state* $\mathsf{state}_{\mathcal{K}}$, *the key* s, *the protocol* Π *is* ϵ*-secure for the Recorder if for any distinguisher* D,

$$\left| \Pr_{c \leftarrow C}[D_{\mathsf{state}_{\mathcal{K}}, \mathsf{s}}(c) = 1] - \Pr_{c \leftarrow C'}[D_{\mathsf{state}_{\mathcal{K}}, \mathsf{s}}(c) = 1] \right| \leq \epsilon.$$

In the case of a malicious Keeper $\mathcal{K}^*$ who can have arbitrary behavior, we let $\mathsf{state}_{\mathcal{K}^*}$ be the state of $\mathcal{K}^*$ at the end of the protocol. Notice that regardless of the possible behaviors that $\mathcal{K}^*$ could have, it is constrained to the state that it has stored at the end of the protocol. It has no additional information besides what it has stored in $\mathsf{state}_{\mathcal{K}^*}$.

Definition 5 (Recorder Security with Malicious Keeper). *Let* $C = \{c_1, c_2, \ldots, c_\lambda\}$ *be the ciphertexts output by* $\mathcal{R}$ *at the end of the protocol, and* $C' = \{c'_1, c'_2, \ldots, c'_\lambda\}$ *where* $c'_i \leftarrow \mathtt{Enc}(\mathsf{s}, 0)$ *be fresh encryptions of zero under the* *key* $\mathbf{s}$. *Given the malicious Keeper's state* $\mathsf{state}_{\mathcal{K}^*}$, *the key* $\mathbf{s}$, *the protocol* Π *is* ϵ*-secure for the Recorder if for any distinguisher* D,

$$\left|\Pr_{c\leftarrow C}[D_{\mathsf{state}_{\mathcal{K}^*},\mathbf{s}}(c)=1]-\Pr_{c\leftarrow C'}[D_{\mathsf{state}_{\mathcal{K}^*},\mathbf{s}}(c)=1]\right|\leq\epsilon.$$

4.1 Simple Encrypt Zero Protocol

Here we present the Simple Encrypt Zero Protocol, which achieves Keeper Security and Recorder security against honest-but-curious Keeper. The main idea here is simple: the Keeper will stream a sequence of ciphertexts which are encryptions of zero, and Recorder will obtain fresh encryptions of zero by taking random subset-sums of the ciphertexts received.

Construction 2 (Simple Encrypt Zero Protocol). A Simple Encrypt Zero Protocol instance $\mathtt{EZ}(n,m,\lambda)$ for the Keeper $\mathcal{K}$ and the Recorder $\mathcal{R}$ proceeds as follows:

- $\mathcal{K}$ chooses a random key $\mathbf{k}\in\{0,1\}^n$, and $\mathcal{R}$ chooses a random secret matrix $\boldsymbol{\Sigma}\in\{0,1\}^{\lambda\times m}$.
- $\mathcal{K}$ streams encryptions $(\mathbf{r}_i, a_i=\mathbf{r}_i\cdot\mathbf{k}+0)$ to $\mathcal{R}$, for $i=1,2,\ldots,m$ and random $\mathbf{r}_i\in\{0,1\}^n$.
- $\mathcal{R}$ maintains matrix $\boldsymbol{\Psi}\in\{0,1\}^{\lambda\times n}$ and column vector $\boldsymbol{\kappa}\in\{0,1\}^{\lambda}$. Each row of $(\boldsymbol{\Psi}|\boldsymbol{\kappa})$ will be a random subset-sum of the encryptions sent by $\mathcal{K}$, with each subset-sum chosen according to $\boldsymbol{\Sigma}$. $\boldsymbol{\Psi}$ and $\boldsymbol{\kappa}$ will be computed on the fly. Specifically, when encryption $(\mathbf{r}_i, a_i)$ comes in, $\mathcal{R}$ will update $\boldsymbol{\Psi}$ to be $\boldsymbol{\Psi}+\boldsymbol{\sigma}_i\cdot\mathbf{r}_i$ and $\boldsymbol{\kappa}$ to be $\boldsymbol{\kappa}+\boldsymbol{\sigma}_i a_i$. Here, $\boldsymbol{\sigma}_i$ is the i-th column of $\boldsymbol{\Sigma}$, and $\mathbf{r}_i$ is interpreted as a row vector.
- At the end of the protocol, $\mathcal{K}$ outputs its key $\mathbf{s}=\mathbf{k}$, and $\mathcal{R}$ outputs $(\boldsymbol{\Psi}|\boldsymbol{\kappa})$, whose rows are the ciphertexts $c_1, c_2, \ldots, c_\lambda$.

Remark 1. For the ease of analysis, we combine all the encryptions sent together, and denote $\mathbf{R}=\begin{bmatrix}\mathbf{r}_1\\ \mathbf{r}_2\\ \cdots\\ \mathbf{r}_m\end{bmatrix}\in\{0,1\}^{m\times n}$, and $\mathbf{a}=\begin{bmatrix}a_1\\ a_2\\ \cdots\\ a_m\end{bmatrix}\in\{0,1\}^m$. This gives us

$$\mathbf{a}=\mathbf{R}\cdot\mathbf{k}.$$

Correspondingly, notice that $\mathcal{R}$ is essentially recording $\boldsymbol{\Sigma}$, $\boldsymbol{\Psi}=\boldsymbol{\Sigma}\cdot\mathbf{R}$ and $\boldsymbol{\kappa}=\boldsymbol{\Sigma}\cdot\mathbf{a}=\boldsymbol{\Sigma}\cdot\mathbf{R}\cdot\mathbf{k}=\boldsymbol{\Psi}\cdot\mathbf{k}$.

It is easy to verify that the rows of $(\boldsymbol{\Psi}|\boldsymbol{\kappa})$ are encryptions of 0 under the key $\mathbf{s}=\mathbf{k}$, as they are simply sums of encryptions of 0 under $\mathbf{s}$ and by the additive homomorphism of Raz's encryption scheme they also must encrypt 0. Therefore, this construction meets the correctness requirement for an Encrypt Zero Protocol.

Next, we show that this construction achieves Keeper security and Recorder security against honest-but-curious Keepers.

Theorem 2 (Keeper Security of EZ). *The Simple Encrypt Zero Protocol is* $(Cn^2, \Omega(\alpha n))$*-secure for the Keeper, for some* $C < \frac{1}{20}$ *and* α *dependent on* C.

Proof. This follows directly from Lemma 2. Here $\mathsf{view}_{\mathcal{R}}$ essentially contains m pairs of $(\mathbf{r}_i, a_i)$, where $a_i = \mathbf{r}_i \cdot \mathbf{s}$ for $i = 1, 2, \ldots, m$ and random $\mathbf{r}_i \leftarrow \{0,1\}^n$. For adversaries space-bounded to Cn^2 memory bits for some $C < \frac{1}{20}$ and α dependent on C, by applying Lemma 2, we get that the probability of an adversary outputting $\mathbf{s}$ is no more than $O(2^{-\alpha n})$. Hence, the average min-entropy of $\mathbf{s}$ conditioned on $\mathsf{view}_{\mathcal{R}}$ is $\Omega(\alpha n)$. □

Theorem 3 (Recorder Security of EZ). *The Simple Encrypt Zero Protocol with parameter* $m = 2n$ *and an honest-but-curious Keeper is* $O(2^{-n})$*-secure for the Recorder.*

Proof. Since the Keeper is honest and follows the protocol, $\mathbf{R}$ is a random $m \times n$ matrix. For $m = 2n$, we have $\mathbf{R}$ being a random $2n \times n$ matrix, which is full rank with probability $1 - O(2^{-n})$. Notice that if $\mathbf{R}$ is full rank, given that $\boldsymbol{\Sigma}$ is a random matrix conditioned on the Keeper's state $\mathsf{state}_{\mathcal{K}}$ and $\mathbf{s}$, $\boldsymbol{\Psi} = \boldsymbol{\Sigma} \cdot \mathbf{R}$ is also a random matrix conditioned on $\mathsf{state}_{\mathcal{K}}$ and $\mathbf{s}$.

In this way, conditioned on $\mathsf{state}_{\mathcal{K}}$ and $\mathbf{s}$, $(\boldsymbol{\Psi}|\boldsymbol{\kappa})$ contains *random* encryptions of 0. Therefore, by definition, these encryptions $\{c_1, \ldots, c_\lambda\}$ cannot be distinguished from $\{c'_1, \ldots, c'_\lambda\}$ where c'_i is a random encryption of 0. Hence, the probability of distinguishing C from C' is bounded by the probability that $\mathbf{R}$ is *not* full rank, which is $O(2^{-n})$. Thus we have

$$\left| \Pr_{c \leftarrow C}[D_{\mathsf{trans},\mathbf{s}}(c) = 1] - \Pr_{c \leftarrow C'}[D_{\mathsf{trans},\mathbf{s}}(c) = 1] \right| \leq 2O(2^{-n}) = O(2^{-n})$$

as desired. □

Kindly notice that this simple construction of an Encrypt Zero protocol is only secure for the Recorder if the Keeper is honest. For malicious Keepers, they could, for example, generate the matrix $\mathbf{R}$ with bad randomness so that it is very likely to be low rank.

One way to tackle this is to have the random matrix $\mathbf{R}$ generated and streamed by a trusted third party, which is a common practice in much of the prior work in the bounded storage model. However, if we do not wish to rely on a trusted third party (notice that the model without a trusted third party is stronger than one with a trusted third party), we show in the following subsection how we can tweak our simple construction to have Recorder security even against malicious Keepers.

4.2 Enhanced Encrypt Zero Protocol

In the Enhanced Encrypt Zero Protocol construction, we tweak the simple construction slightly to account for malicious Keepers.

Construction 3 (Enhanced Encrypt Zero Protocol). An Enhanced Encrypt Zero Protocol instance $\mathtt{EZ}^+(n, m, \lambda)$ with the Keeper $\mathcal{K}$ and the Recorder $\mathcal{R}$ proceeds as follows:

- $\mathcal{K}$ chooses a random key $\mathbf{k} \in \{0,1\}^n$ and an independent random secret $\mathbf{s} \in \{0,1\}^m$. $\mathcal{R}$ chooses a random secret matrix $\boldsymbol{\Sigma} \in \{0,1\}^{\lambda \times m}$.
- $\mathcal{K}$ streams random encryptions of the bits in $\mathbf{s}$. Namely, in matrix form, $\mathcal{K}$ sends $(\mathbf{R}, \mathbf{a} = \mathbf{R} \cdot \mathbf{k} + \mathbf{s})$ for random $\mathbf{R} \in \{0,1\}^{m \times n}$.
- $\mathcal{R}$ maintains matrix $\boldsymbol{\Psi} = \boldsymbol{\Sigma} \cdot \mathbf{R}$ and column vector $\boldsymbol{\kappa} = \boldsymbol{\Sigma} \cdot \mathbf{a}$.
- $\mathcal{K}$ sends its key $\mathbf{k}$ *in the clear*, and $\mathcal{R}$ uses that to compute $\boldsymbol{\phi} = \boldsymbol{\kappa} - \boldsymbol{\Psi} \cdot \mathbf{k}$.
- $\mathcal{K}$ outputs $\mathbf{s}$ as its key, and $\mathcal{R}$ outputs $(\boldsymbol{\Sigma}|\boldsymbol{\phi})$, whose rows are the ciphertexts $c_1, c_2, \ldots, c_\lambda$.

Notice that $\boldsymbol{\phi} = \boldsymbol{\kappa} - \boldsymbol{\Psi} \cdot \mathbf{k} = \boldsymbol{\Sigma} \cdot \mathbf{s}$, and hence the rows of $(\boldsymbol{\Sigma}|\boldsymbol{\phi})$ are indeed encryptions of 0 using key $\mathbf{s}$, as desired in the correctness property.

Theorem 4 (Keeper Security of $\mathtt{EZ}^+$). *The Simple Encrypt Zero Protocol is $(Cn^2, \Omega(n))$-secure for the Keeper, for some $C < \frac{1}{20}$ and α dependent on C.*

Proof. First, notice that before the Keeper sends over $\mathbf{k}$, the two distributions $(\mathbf{s}, \mathbf{R}, \mathbf{R} \cdot \mathbf{k} + \mathbf{s})$ and $(\mathbf{s}, \mathbf{R}, \mathbf{R} \cdot \mathbf{k} + \mathbf{s}')$ for random $\mathbf{s}' \in \{0,1\}^m$ are statistically indistinguishable, due to the RoRC security of Raz's encryption scheme.

Now, notice that in the second distribution, the probability the Recorder can guess $\mathbf{s}$ is 2^{-m}. In this case, if it later receives $\mathbf{k}$, the probability it guesses $\mathbf{s}$ is still at most 2^{n-m}, which is 2^{-n}.

Now, we use the following simple fact: suppose two distributions X, Y are ϵ-close. Then there is a procedure P which first samples $x \leftarrow X$, and then based which x it samples, it may replace x with a different sample x'. P satisfies the property that (1) its output distribution is identical to Y, and (2) the probability it re-samples is ϵ.

We use this simple fact by assigning X to $(\mathbf{s}, \mathbf{R}, \mathbf{R} \cdot \mathbf{k} + \mathbf{s}')$ for random $\mathbf{s}' \in \{0,1\}^m$ and Y to $(\mathbf{s}, \mathbf{R}, \mathbf{R} \cdot \mathbf{k} + \mathbf{s})$.

Now consider the probability of guessing $\mathbf{s}$. In the case X, we know it is 2^{-n}. So if we consider Y sampled from P, we know that the probability of guessing $\mathbf{s}$ in the non-replacing case is 2^{-n}. But the replacing case only happens with probability ϵ, meaning overall the probability of outputting $\mathbf{s}$ is at most $\epsilon + 2^{-n}$. □

Theorem 5 (Recorder Security of $\mathtt{EZ}^+$). *The Enhanced Encrypt Zero Protocol with parameter $m = 2n$ and any possibly malicious Keeper $\mathcal{K}^*$ is perfectly secure for the Recorder.*

Proof. Notice that regardless of the Keeper's state $\mathsf{state}_{\mathcal{K}^*}$ (even if one of a malicious Keeper), $\boldsymbol{\Sigma}$ is always random conditioned on $\mathsf{state}_{\mathcal{K}^*}$ and $\mathbf{s}$, since it is solely sampled by the Recorder. Therefore, $(\boldsymbol{\Sigma}|\boldsymbol{\phi})$ is already random encryptions of 0 conditioned on $\mathsf{state}_{\mathcal{K}^*}$ and $\mathbf{s}$. Hence, to distinguish it from other random encryptions of 0, one can do no better than a random guess. Thus, the advantage that any distinguisher D could have in distinguishing C and C' is 0 as desired. □

5 Two-Party Key-Agreement Protocol

Consider a pair of interactive PPT algorithms $\Pi = (\mathsf{A}, \mathsf{B})$. Each of A, B take n as input. We will let $(a, b, \mathsf{trans}) \leftarrow \Pi(n)$ denote the result of running the protocol on input n. Here, a is the output of A, b the output of B, and trans is the transcript of their communication.

A two-party key-agreement protocol is a protocol $\Pi = (\mathsf{A}, \mathsf{B})$ with the correctness property that $\Pr[a = b] = 1$. In this case, we will define $\hat{k} = a = b$ and write $(\hat{k}, \mathsf{trans}) \leftarrow \Pi(n)$. Additionally, we will require eavesdropping security:

Definition 6 (Eavesdropping Security of Two-Party Key-Agreement Protocol). *We say that Π is $(S(n), \epsilon)$-secure if for all adversaries $\mathcal{A}$ that use at most $S(n)$ memory bits,*

$$|\Pr[\mathcal{A}(\hat{k}, \mathsf{trans}) = 1 : (\hat{k}, \mathsf{trans}) \leftarrow \Pi(n)]$$
$$- \Pr[\mathcal{A}(k', \mathsf{trans}) = 1 : k' \leftarrow \mathcal{K}_n, (k, \mathsf{trans}) \leftarrow \Pi(n)]| \leq \epsilon.$$

In this section we demonstrate how we can use the Simple Encrypt Zero Protocol to implement a two-party key-agreement protocol. For simplicity, we consider a key space of one single bit.

Construction 4 (Two-Party Key-Agreement Protocol). For two parties $\mathcal{P}$ and $\mathcal{Q}$ trying to derive a shared key $\hat{k} \in \{0, 1\}$, they will first run a Simple Encrypt Zero Protocol $\mathsf{EZ}(n, m, \lambda = 1)$ with $\mathcal{P}$ as the Keeper and $\mathcal{Q}$ as the Recorder. At the end of the EZ protocol, $\mathcal{P}$ gets a key $\mathbf{s}$, and $\mathcal{Q}$ gets an encryption of 0 using $\mathbf{s}$, namely $(\boldsymbol{\Psi}|\boldsymbol{\kappa})$ (notice that $\boldsymbol{\kappa}$ is of dimension $\lambda \times 1$, and hence is a single bit here). To derive a shared key, $\mathcal{Q}$ sends $\boldsymbol{\Psi}$ to $\mathcal{P}$. The shared key is thus $\boldsymbol{\kappa}$, which is known to $\mathcal{Q}$, and is computable by $\mathcal{P}$ as $\boldsymbol{\kappa} = \boldsymbol{\Psi} \cdot \mathbf{s}$.

Remark 2. For key spaces $\{0, 1\}^d$, we can simply tune the protocol to use $\lambda = d$, and that will yield a shared key $\hat{\mathbf{k}} \in \{0, 1\}^d$.

Theorem 6. *The two-party key-agreement protocol presented above is $(Cn^2, O(2^{-\alpha n/2}))$-secure against eavesdropping adversaries.*

Proof. First, by the Keeper security of the EZ protocol, for adversaries with up to Cn^2 memory bits for some $C < \frac{1}{20}$, $H_\infty(\mathsf{s}|\mathsf{view}_\mathcal{R}) \geq \Omega(\alpha n)$. Subsequently, $H_\infty(\boldsymbol{\Psi}, \mathsf{s}|\mathsf{view}_\mathcal{R}) \geq \Omega(\alpha n)$. Let $H : \{0, 1\}^n \times \{0, 1\}^n \to \{0, 1\}$ compute the inner product. Using the fact that the inner product is a universal hash function and applying Lemma 1, we have

$$(H(\boldsymbol{\Psi}, \mathsf{s}), H, \mathsf{view}_\mathcal{R}) \approx_{\epsilon/2} (U_1, U_d, \mathsf{view}_\mathcal{R}),$$

where $1 + 2\log(1/\epsilon) = \Omega(\alpha n)$. Solving for ϵ yields that $\epsilon = O(2^{-\alpha n/2})$, i.e. an adversary has advantage at most $O(2^{-\alpha n/2})$ in distinguishing $H(\boldsymbol{\Psi}, \mathsf{s})$ and U_1. Recall that in the eavesdropping security game for Two-Party Key-Agreement Protocols, the adversary need to distinguish between actual derived keys

$\hat{k} = \boldsymbol{\Psi} \cdot \mathbf{s}$ from random k' sampled directly from the key space $\{0,1\}$. Observe that $H(\boldsymbol{\Psi}, \mathbf{s}) = \boldsymbol{\Psi} \cdot \mathbf{s} = \hat{k}$, and k' is drawn from U_1. Therefore, we have

$$\begin{aligned}&|\Pr[\mathcal{A}(\hat{k}, \mathsf{trans}) = 1 : (\hat{k}, \mathsf{trans}) \leftarrow \Pi(n)] \\ &- \Pr[\mathcal{A}(k', \mathsf{trans}) = 1 : k' \leftarrow \mathcal{K}_n, (k, \mathsf{trans}) \leftarrow \Pi(n)]| \leq \epsilon = O(2^{-\alpha n/2})\end{aligned}$$

as desired. □

6 Bit Commitment Scheme

Let n and λ be security parameters. A bit commitment scheme Π consists of a tuple of algorithm (`Commit`, `Reveal`, `Verify`) for a committer $\mathcal{C}$ and a verifier $\mathcal{V}$.

- The `Commit` algorithm is run by the committer, and it takes as input the security parameter n and a bit b to be committed to. A transcript of the communication, a committer state, and a verifier state $(\mathsf{trans}, \mathsf{state}_\mathcal{C}, \mathsf{state}_\mathcal{V}) \leftarrow \mathtt{Commit}(n, \lambda, b)$ is output by the `Commit` algorithm.
- The `Reveal` algorithm is also run by the committer, and it takes as input a committer state $\mathsf{state}_\mathcal{C}$ and a bit b'. It outputs a revealing, denoted as x, together with the committed bit b'.
- The `Verify` algorithm is run by the Verifier and takes input a verifier state $\mathsf{state}_\mathcal{V}$ and outputs of a `Reveal` algorithm, (x, b'). It outputs a bit u.

There are two desired security properties for a bit commitment scheme, namely hiding and binding. We will give out formal definitions below.

The hiding property of a bit commitment scheme essentially states that the committed bit b should be hidden from the Verifier given the Verifier's view after the `Commit` algorithm. Notice that the Verifier's view after the `Commit` algorithm consists of exactly trans and $\mathsf{state}_\mathcal{V}$. Put formally:

Definition 7 (Hiding Property of a Bit Commitment Scheme). *For some given security parameters n, λ and a bit b, let $(\mathsf{trans}, \mathsf{state}_\mathcal{C}, \mathsf{state}_\mathcal{V}) \leftarrow \mathtt{Commit}(n, \lambda, b)$, we say that the bit commitment scheme is $(S(n), \epsilon)$-hiding if for all Verifiers $\mathcal{V}$ with up to $S(n)$ memory bits,*

$$(b, \mathsf{trans}, \mathsf{state}_\mathcal{V}) \approx_\epsilon (r, \mathsf{trans}, \mathsf{state}_\mathcal{V})$$

for random r uniformly sampled from $\{0,1\}$.

The binding property of a bit commitment scheme essentially requires that a committer is not able to open a commitment to both 0 and 1. Notice that this applies to all committers, who can be potentially malicious. A malicious committer $\mathcal{A}$ can run an arbitrary Commit^* procedure, which has no guarantees except that it produces some $(\mathsf{trans}, \mathsf{state}_\mathcal{A}, \mathsf{state}_\mathcal{V})$. Note that this Commit^* procedure does not necessarily commit to a bit b, so it does not take b as a parameter.

Definition 8 (Binding Property of a Bit Commitment Scheme). *Let $\mathcal{A}$ be an adversary. $\mathcal{A}$ plays the following game* $\mathsf{Binding}_{\mathcal{A},\Pi}(n,\lambda)$ *for some given security parameters n and λ:*

- *The adversary $\mathcal{A}$ runs an arbitrary commit procedure (potentially malicious)* $\mathtt{Commit}^*(n,\lambda)$ *with an honest Verifier $\mathcal{V}$ and produces* $(\mathsf{trans}, \mathsf{state}_{\mathcal{A}}, \mathsf{state}_{\mathcal{V}})$.
- *The adversary produces $(x_0, 0)$ and $(x_1, 1)$.*
- *The game outputs 1 if both* $\mathtt{Verify}(\mathsf{state}_{\mathcal{V}}, (x_0, 0))$ *and* $\mathtt{Verify}(\mathsf{state}_{\mathcal{V}}, (x_1, 1))$ *output 1, and 0 otherwise.*

We say that Π is ϵ-binding if for all adversary $\mathcal{A}$

$$\Pr[\mathsf{Binding}_{\mathcal{A},\Pi}(n,\lambda) = 1] \leq \epsilon.$$

Now we present the construction for a bit commitment scheme using the Enhanced Encrypt Zero Protocol.

Construction 5 (Bit Commitment Scheme from Parity Learning). For security parameters n, λ and committer input bit b, we construct the bit commitment scheme by specifying each of the $(\mathtt{Commit}, \mathtt{Reveal}, \mathtt{Verify})$ algorithms.

- $\mathtt{Commit}(n, b)$: Runs the Enhanced Encrypt Zero Protocol $\mathtt{EZ}^+(n, 2n, \lambda)$ with $\mathcal{C}$ as the Keeper and $\mathcal{V}$ as the Recorder. Set trans to be the transcript of the $\mathtt{EZ}^+$ protocol, $\mathsf{state}_{\mathcal{C}}$ to be the output of $\mathcal{C}$ after the $\mathtt{EZ}^+$ protocol, i.e. a secret key $\mathbf{s}$, and $\mathsf{state}_{\mathcal{V}}$ to be the output of $\mathcal{V}$ after the $\mathtt{EZ}^+$ protocol, namely $(\boldsymbol{\Sigma}|\boldsymbol{\phi})$, which contains multiple encryptions of 0 under the key $\mathbf{s}$. Additionally, samples random $\boldsymbol{\gamma} \in \{0,1\}^{2n}$, and sends $(\boldsymbol{\gamma}, c = \boldsymbol{\gamma} \cdot \mathbf{s} + b)$ to the Verifier (notice that this also gets appended to trans).
- $\mathtt{Reveal}(\mathsf{state}_{\mathcal{C}}, b')$: Outputs $(\mathbf{x}, b') = (\mathbf{s}, b')$.
- $\mathtt{Verify}(\mathsf{state}_{\mathcal{V}}, \mathbf{x}, b')$: Checks that $\boldsymbol{\phi} = \boldsymbol{\Sigma} \cdot \mathbf{x}$, and that $c = \boldsymbol{\gamma} \cdot \mathbf{x} + b'$. If any of the checks fail, output 0; otherwise, output 1.

Theorem 7. *The bit commitment construction above is $(Cn^2, O(2^{-n/2}))$-hiding for some $C < 1/20$.*

Proof. First, by the Keeper security of the $\mathtt{EZ}^+$ protocol, for adversaries with up to Cn^2 memory bits for some $C < \frac{1}{20}$, $H_\infty(\mathbf{s}|\mathsf{view}_{\mathcal{V}}) \geq \Omega(n)$. Recall that $\mathsf{view}_{\mathcal{V}}$ is exactly $(\mathsf{trans}, \mathsf{state}_{\mathcal{V}})$. Subsequently, $H_\infty(\boldsymbol{\gamma}, \mathbf{s}|\mathsf{trans}, \mathsf{state}_{\mathcal{V}}) \geq \Omega(n)$. Let $H : \{0,1\}^n \times \{0,1\}^n \to \{0,1\}$ compute the inner product. Using the fact that the inner product is a universal hash function and applying Lemma 1, we have

$$(H(\boldsymbol{\gamma}, \mathbf{s}), H, \mathsf{trans}, \mathsf{state}_{\mathcal{V}}) \approx_{\epsilon/2} (U_1, U_d, \mathsf{trans}, \mathsf{state}_{\mathcal{V}}),$$

where $1 + 2\log(1/\epsilon) = \Omega(n)$. Furthermore, we have

$$(H(\boldsymbol{\gamma}, \mathbf{s}) + c, H, \mathsf{trans}, \mathsf{state}_{\mathcal{V}}) \approx_{\epsilon/2} (U_1 + c, U_d, \mathsf{trans}, \mathsf{state}_{\mathcal{V}}),$$

Solving for ϵ yields that $\epsilon = O(2^{-n/2})$, i.e. an adversary has advantage at most $O(2^{-n/2})$ in distinguishing $H(\boldsymbol{\gamma}, \mathbf{s}) + c$ and $U_1 + c$. Notice that $H(\boldsymbol{\gamma}, \mathbf{s}) + c =$

$\gamma \cdot \mathbf{s} + c = b$, and that $U_1 + c$ is yet another uniformly random bit $r \leftarrow \{0,1\}$. Therefore, we have

$$(b, H, \mathsf{trans}, \mathsf{state}_{\mathcal{V}}) \approx_{\epsilon/2} (r, U_d, \mathsf{trans}, \mathsf{state}_{\mathcal{V}})$$

for $\epsilon = O(2^{-n/2})$ and r a uniformly random bit. Thus, by

$$(b, \mathsf{trans}, \mathsf{state}_{\mathcal{V}}) \approx_{\epsilon'} (r, \mathsf{trans}, \mathsf{state}_{\mathcal{V}})$$

for $\epsilon' = \frac{1}{2}O(2^{-n/2}) = O(2^{-n/2})$ and r a uniformly random bit, we have shown that the bit commitment scheme presented above is $(Cn^2, O(2^{-n/2}))$-hiding as desired. □

Theorem 8. *The bit commitment scheme presented above is* $(2^{-\lambda})$*-binding.*

Proof. We show that the scheme is statistically binding by arguing that the probability that an adversary can win the Binding game is no more than $\frac{1}{2^\lambda}$.

Notice that in order for the adversary to win the game, the adversary need to output $(\mathbf{x}_0, 0)$ and $(\mathbf{x}_1, 1)$ that both pass the Verify algorithm. Recall that the Verify Algorithm checks for two things:

- $c = \gamma \cdot \mathbf{x}_0 + 0$ and $c = \gamma \cdot \mathbf{x}_1 + 1$ where c and γ are part of the transcript trans and are stored in the Verifier's state $\mathsf{state}_{\mathcal{V}}$. This leads to that $\gamma \cdot \mathbf{x}_0 \neq \gamma \cdot \mathbf{x}_1$ and hence $\mathbf{x}_0 \neq \mathbf{x}_1$.
- $\phi = \boldsymbol{\Sigma} \cdot \mathbf{x}_0 = \boldsymbol{\Sigma} \cdot \mathbf{x}_1$ where $\boldsymbol{\Sigma}$ and ϕ are sampled and computed by the Verifier and stored in $\mathsf{state}_{\mathcal{V}}$. Notice this leads to $\boldsymbol{\Sigma} \cdot (\mathbf{x}_0 - \mathbf{x}_1) = 0$.

Now let $\mathbf{x}' = \mathbf{x}_0 - \mathbf{x}_1$. From $\mathbf{x}_0 \neq \mathbf{x}_1$, we know that $\mathbf{x}' \neq \mathbf{0}$. Therefore, we need to find a non-trivial root for the equation $\boldsymbol{\Sigma} \cdot \mathbf{x}' = \mathbf{0}$. Recall that by the Recorder's perfect security of the $\mathtt{EZ}^+$ protocol, the matrix $\boldsymbol{\Sigma}$ stored in $\mathsf{state}_{\mathcal{V}}$ is random conditioned on the Committer's view. For each row of $\boldsymbol{\Sigma}$, denoted as $\boldsymbol{\Sigma}_i$ for the i-th row, the probability that $\boldsymbol{\Sigma}_i \cdot \mathbf{x}' = 0$ is no more than a random guess, i.e. $\frac{1}{2}$. Since to pass the $\mathtt{Verify}$ algorithm requires $\boldsymbol{\Sigma} \cdot \mathbf{x}' = \mathbf{0}$, i.e. $\boldsymbol{\Sigma}_i \cdot \mathbf{x}' = 0$ for all $i = 1, 2, \ldots, \lambda$, and recall that the rows of $\boldsymbol{\Sigma}$ are independent, the probability that the adversary can find such a $\mathbf{x}'$ is no more than $(\frac{1}{2})^\lambda = \frac{1}{2^\lambda}$. □

7 Oblivious Transfer Protocol

In an oblivious transfer (OT) protocol, one party, the Sender $\mathcal{S}$, has two input bits x_0, x_1, and the other party, known as the Receiver $\mathcal{R}'$ (not to be confused with the Recorder $\mathcal{R}$ in the Encrypt Zero Protocols), has an input bit b. After some communication between the two parties, $\mathcal{R}'$ outputs x_b. The OT protocol requires two security properties, namely Sender security and Receiver security. Sender security dictates that $\mathcal{R}'$ should have no information about x_{1-b}, and Receiver security requires that $\mathcal{S}$ has no information about b.

Before we proceed to our construction of an OT protocol, we first formally define these two security properties.

The security of the Sender ensures that an adversarial Receiver can learn about at most one of x_0 and x_1. In other words, there always exists a b' s.t. the Receiver has no information about $x_{b'}$. Put formally:

Definition 9 (Sender Security). *An OT protocol is said to be ϵ-secure for the Sender if there exists some b' s.t. for any arbitrary distinguisher D and Receiver's view* $\mathsf{view}_{\mathcal{R}'}$,

$$\left|\Pr[D_{\mathsf{view}_{\mathcal{R}'}}(x_{b'}) = 1] - \Pr[D_{\mathsf{view}_{\mathcal{R}'}}(r) = 1]\right| \leq \epsilon$$

for a uniformly random bit r.

The security of the Receiver requires that the sender $\mathcal{S}$ has no information about b. In other words, given the view of the Sender, one should not be able to distinguish between b and a random bit r. Put formally:

Definition 10 (Receiver Security). *Let* $\mathsf{view}_{\mathcal{S}}$ *denote the view of the sender, the OT protocol Π is said to be $(S(n), \epsilon)$-secure for the Receiver if for all possible Senders that use up to $S(n)$ memory bits,*

$$(b, \mathsf{view}_{\mathcal{S}}) \approx_{\epsilon} (r, \mathsf{view}_{\mathcal{S}}),$$

where r is a uniformly random bit.

Now we give out our construction of the OT protocol.

The key idea is that the Receiver will send a commitment of its bit b to the Sender. And the Sender therefore uses the additive homomorphism of Raz's encryption scheme to compute the encryptions of $(1-b)x_0$ and bx_1. The Sender further re-randomizes these two ciphertexts by adding fresh encryptions of zero before sending them to the Receiver. The Receiver decrypts these two ciphertexts and obtains 0 and x_b as desired.

Construction 6 (Oblivious Transfer Protocol from Parity Learning). For given security parameter n, a Sender $\mathcal{S}$ and a receiver $\mathcal{R}'$:

- Run an Enhanced Encrypt Zero Protocol $\mathtt{EZ}^{+}(n, 2n, \lambda = 2)$ with $\mathcal{R}'$ as the Keeper and $\mathcal{S}$ as the Recorder. At the end of the protocol, $\mathcal{R}'$ has as output a secret key $\mathbf{s}$, and $\mathcal{S}$ has output $(\boldsymbol{\Sigma}|\boldsymbol{\phi})$, which consists of two encryptions of 0 under the key $\mathbf{s}$. Additionally, $\mathcal{R}'$ samples random $\gamma \in \{0,1\}^{2n}$, and sends $(\gamma, c = \gamma \cdot \mathbf{s} + b)$ to the Sender. Kindly notice that in this step the Receiver $\mathcal{R}'$ is actually just executing $\mathtt{Commit}(n, b)$.
- For Sender $\mathcal{S}$, let $\boldsymbol{\sigma}_0$, $\boldsymbol{\sigma}_1$ be the first and second row of $\boldsymbol{\Sigma}$, and ϕ_0, ϕ_1 be the two elements in $\boldsymbol{\phi}$. Notice that $\phi_0 = \boldsymbol{\sigma}_0 \cdot \mathbf{s}$ and $\phi_1 = \boldsymbol{\sigma}_1 \cdot \mathbf{s}$. The Sender then sends to the Receiver two ciphertexts:

$$\begin{aligned} \boldsymbol{\sigma}_0 + x_0\gamma,\ \phi_0 + x_0(1-c) &= (\boldsymbol{\sigma}_0 + x_0\gamma) \cdot \mathbf{s} + ((1-b)x_0) \\ \boldsymbol{\sigma}_1 + x_1\gamma,\ \phi_1 + x_1 c &= (\boldsymbol{\sigma}_1 + x_1\gamma) \cdot \mathbf{s} + (bx_1). \end{aligned}$$

- $\mathcal{R}'$ decrypts both ciphertexts that it has received using the key $\mathbf{s}$, and learns $(1-b)x_0$ and bx_1. Notice that one of these two values will be x_b as desired and gets output by $\mathcal{R}'$.

We then proceed to prove desired security properties for the above construction of the OT protocol.

Theorem 9. *The OT protocol described above is perfectly secure for the Sender.*

Proof. We show that right after the first part of the protocol where $\mathcal{R}'$ executes $\texttt{Commit}(n, b)$, there is a fixed $b' = c + \boldsymbol{\gamma} \cdot \mathbf{s} + 1$ such that the Receiver will have no information about $x_{b'}$. Notice that this does not break Receiver security, since although b' is fixed, $\mathcal{S}$ has no way to compute b' as $\mathbf{s}$ is only known to the Receiver $\mathcal{R}'$.

If $b' = c + \boldsymbol{\gamma} \cdot \mathbf{s} + 1 = 0$, we show that the Receiver has no information about x_0, i.e. x_0 is random given the Receiver's view. Notice that we have $1 - c = \boldsymbol{\gamma} \cdot \mathbf{s}$. And hence the two ciphertext that the Receiver receives are

$$\begin{aligned} \boldsymbol{\sigma}_0 + x_0\boldsymbol{\gamma},\ \phi_0 + x_0(1-c) &= (\boldsymbol{\sigma}_0 + x_0\boldsymbol{\gamma}) \cdot \mathbf{s} \\ \boldsymbol{\sigma}_1 + x_1\boldsymbol{\gamma},\ \phi_1 + x_1 c \qquad &= (\boldsymbol{\sigma}_1 + x_1\boldsymbol{\gamma}) \cdot \mathbf{s} + x_1. \end{aligned}$$

The only source that the Receiver might be able to gather information about x_0 is from the first ciphertext. However, since $\boldsymbol{\sigma}_0$ is uniformly random given the Receiver's view, $\boldsymbol{\sigma}_0 + x_0\boldsymbol{\gamma}$ is also uniformly random given the Receiver's view, i.e., it does not give any additional information to the Receiver. The Receiver also gets no information from $(\boldsymbol{\sigma}_0 + x_0\boldsymbol{\gamma}) \cdot \mathbf{s}$, as this value can be easily simulated by the Receiver since it knows both $\boldsymbol{\sigma}_0 + x_0\boldsymbol{\gamma}$ and $\mathbf{s}$. Therefore, x_0 is random given the Receiver's view.

If $b' = c + \boldsymbol{\gamma} \cdot \mathbf{s} + 1 = 1$, by a similar argument, we have that x_1 is random given the Receiver's view. Bringing these parts together, we have shown that for $b' = c + \boldsymbol{\gamma} \cdot \mathbf{s} + 1$, $x_{b'}$ *is* random conditioned on the Receiver's view, i.e.

$$\left| \Pr[D_{\mathsf{view}_{\mathcal{R}'}}(x_{b'}) = 1] - \Pr[D_{\mathsf{view}_{\mathcal{R}'}}(r) = 1] \right| = 0.$$

Thus, the OT protocol above is perfectly secure for the Sender as desired. □

Theorem 10. *The OT protocol described above is $(Cn^2, O(2^{-n/2}))$-secure for the Receiver, for some $C < \frac{1}{20}$.*

Proof. The proof for this is extremely straightforward. As observed above, the receiver $\mathcal{R}'$ is exactly executing $\texttt{Commit}(n, b)$, i.e. it is committing the bit b to the Sender, who is playing the role of the Verifier in the commitment scheme. Hence, by the $(Cn^2, O(2^{-n/2}))$-hiding property of the commitment scheme, we have that for all possible Sender $\mathcal{S}$ that uses at most Cn^2 memory bits,

$$(b, \mathsf{trans}, \mathsf{state}_{\mathcal{S}}) \approx_{\epsilon} (r, \mathsf{trans}, \mathsf{state}_{\mathcal{S}})$$

for $\epsilon = O(2^{-n/2})$) and a uniformly random bit r. Notice that $\mathsf{view}_{\mathcal{S}}$ is actually just $(\mathsf{trans}, \mathsf{state}_{\mathcal{S}})$. Therefore, the above equation can be rewritten as

$$(b, \mathsf{view}_{\mathcal{S}}) \approx_{\epsilon} (r, \mathsf{view}_{\mathcal{S}}).$$

This is the exact definition for (Cn^2, ϵ)-Receiver-security. Therefore, the OT protocol above is $(Cn^2, O(2^{-n/2}))$-secure for the Receiver as desired. □

References

[AR99] Aumann, Y., Rabin, M.O.: Information theoretically secure communication in the limited storage space model. In: Wiener, M. (ed.) CRYPTO 1999. LNCS, vol. 1666, pp. 65–79. Springer, Heidelberg (1999). https://doi.org/10.1007/3-540-48405-1_5

[BDKM18] Ball, M., Dachman-Soled, D., Kulkarni, M., Malkin, T.: Non-malleable codes from average-case hardness: AC^0, decision trees, and streaming space-bounded tampering. In: Nielsen, J.B., Rijmen, V. (eds.) EUROCRYPT 2018. LNCS, vol. 10822, pp. 618–650. Springer, Cham (2018). https://doi.org/10.1007/978-3-319-78372-7_20

[CCM98] Cachin, C., Crépeau, C., Marcil, J.: Oblivious transfer with a memory-bounded receiver. In: 39th FOCS, pp. 493–502. IEEE Computer Society Press, November 1998

[CM97] Cachin, C., Maurer, U.: Unconditional security against memory-bounded adversaries. In: Kaliski Jr., B.S. (ed.) CRYPTO 1997. LNCS, vol. 1294, pp. 292–306. Springer, Heidelberg (1997). https://doi.org/10.1007/BFb0052243

[DHRS04] Ding, Y.Z., Harnik, D., Rosen, A., Shaltiel, R.: Constant-round oblivious transfer in the bounded storage model. In: Naor, M. (ed.) TCC 2004. LNCS, vol. 2951, pp. 446–472. Springer, Heidelberg (2004). https://doi.org/10.1007/978-3-540-24638-1_25

[Din01] Ding, Y.Z.: Oblivious transfer in the bounded storage model. In: Kilian, J. (ed.) CRYPTO 2001. LNCS, vol. 2139, pp. 155–170. Springer, Heidelberg (2001). https://doi.org/10.1007/3-540-44647-8_9

[GL89] Goldreich, O., Levin, L.A.: A hard-core predicate for all one-way functions. In: 21st ACM STOC, pp. 25–32. ACM Press, May 1989

[Gro96] Grover, L.K.: A fast quantum mechanical algorithm for database search. In: 28th ACM STOC, pp. 212–219. ACM Press, May 1996

[ILL89] Impagliazzo, R., Levin, L.A., Luby, M.: Pseudo-random generation from one-way functions (extended abstracts). In: 21st ACM STOC, pp. 12–24. ACM Press, May 1989

[Lu02] Lu, C.-J.: Hyper-encryption against space-bounded adversaries from on-line strong extractors. In: Yung, M. (ed.) CRYPTO 2002. LNCS, vol. 2442, pp. 257–271. Springer, Heidelberg (2002). https://doi.org/10.1007/3-540-45708-9_17

[Mau92] Maurer, U.M.: Conditionally-perfect secrecy and a provably-secure randomized cipher. J. Cryptol. **5**(1), 53–66 (1992)

[Raz17] Raz, R.: A time-space lower bound for a large class of learning problems. In: 58th FOCS, pp. 732–742. IEEE Computer Society Press (2017)

[Rot11] Rothblum, R.: Homomorphic encryption: from private-key to public-key. In: Ishai, Y. (ed.) TCC 2011. LNCS, vol. 6597, pp. 219–234. Springer, Heidelberg (2011). https://doi.org/10.1007/978-3-642-19571-6_14

[Sho94] Shor, P.W.: Algorithms for quantum computation: discrete logarithms and factoring. In: 35th FOCS, pp. 124–134. IEEE Computer Society Press, November 1994

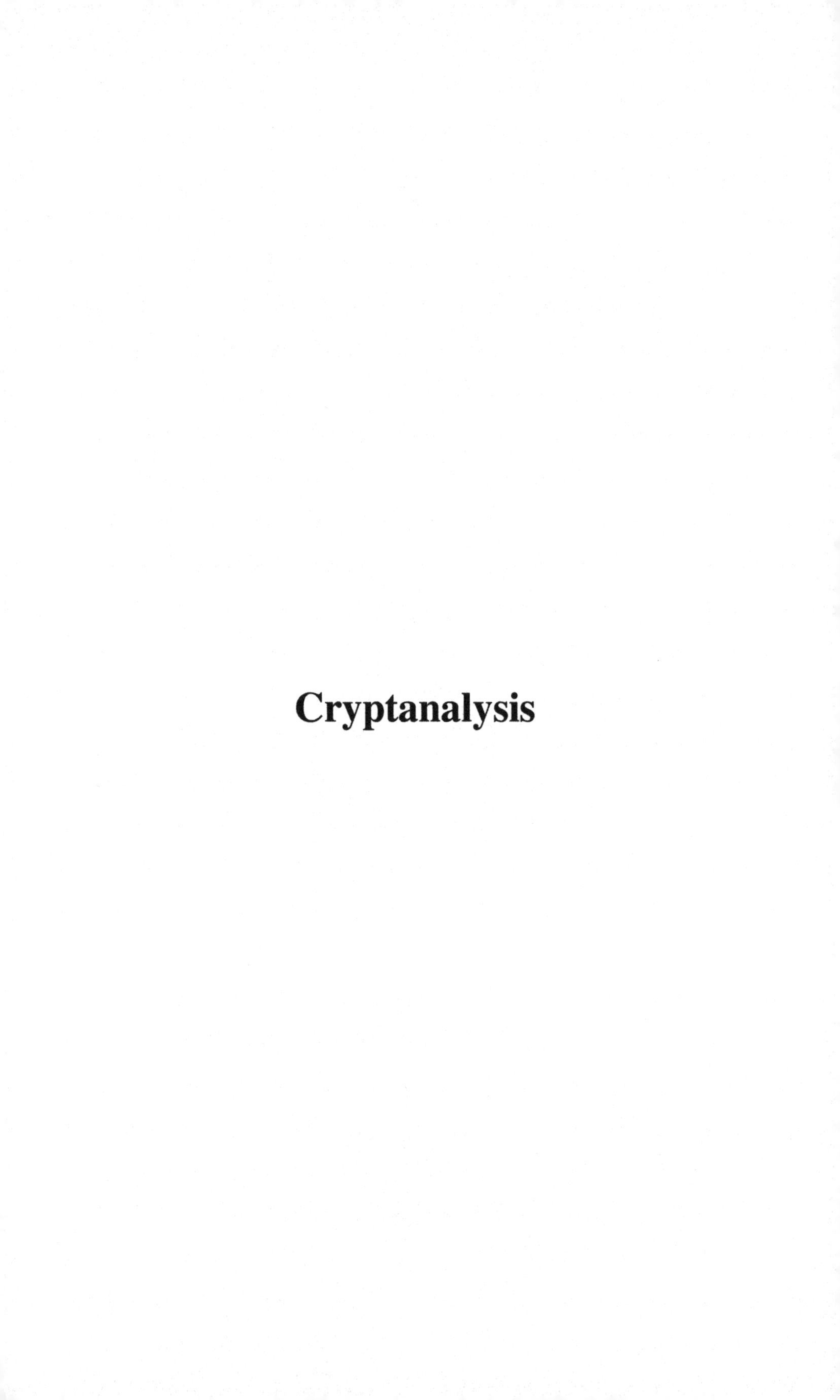

Cryptanalysis

From Collisions to Chosen-Prefix Collisions Application to Full SHA-1

Gaëtan Leurent[1(✉)] and Thomas Peyrin[2,3(✉)]

[1] Inria, Paris, France
gaetan.leurent@inria.fr
[2] Nanyang Technological University, Singapore, Singapore
thomas.peyrin@ntu.edu.sg
[3] Temasek Laboratories, Singapore, Singapore

Abstract. A chosen-prefix collision attack is a stronger variant of a collision attack, where an arbitrary pair of challenge prefixes are turned into a collision. Chosen-prefix collisions are usually significantly harder to produce than (identical-prefix) collisions, but the practical impact of such an attack is much larger. While many cryptographic constructions rely on collision-resistance for their security proofs, collision attacks are hard to turn into break of concrete protocols, because the adversary has a limited control over the colliding messages. On the other hand, chosen-prefix collisions have been shown to break certificates (by creating a rogue CA) and many internet protocols (TLS, SSH, IPsec).

In this article, we propose new techniques to turn collision attacks into chosen-prefix collision attacks. Our strategy is composed of two phases: first a birthday search that aims at taking the random chaining variable difference (due to the chosen-prefix model) to a set of pre-defined target differences. Then, using a multi-block approach, carefully analysing the clustering effect, we map this new chaining variable difference to a colliding pair of states using techniques developed for collision attacks.

We apply those techniques to MD5 and SHA-1, and obtain improved attacks. In particular, we have a chosen-prefix collision attack against SHA-1 with complexity between $2^{66.9}$ and $2^{69.4}$ (depending on assumptions about the cost of finding near-collision blocks), while the best-known attack has complexity $2^{77.1}$. This is within a small factor of the complexity of the classical collision attack on SHA-1 (estimated as $2^{64.7}$). This represents yet another warning that industries and users have to move away from using SHA-1 as soon as possible.

Keywords: Hash function · Cryptanalysis · Chosen-prefix collision · SHA-1 · MD5

1 Introduction

Cryptographic hash functions are crucial components in many information security systems, used for various purposes such as building digital signature schemes,

Y. Ishai and V. Rijmen (Eds.): EUROCRYPT 2019, LNCS 11478, pp. 527–555, 2019.
https://doi.org/10.1007/978-3-030-17659-4_18

message authentication codes or password hashing functions. Informally, a cryptographic hash function H is a function that maps an arbitrarily long message M to a fixed-length hash value of size n bits. Hash functions are classically defined as an iterative process, such as the Merkle-Damgård design strategy [7,16]. The message M is first divided into blocks m_i of fixed size (after appropriate padding) that will successively update an internal state (also named chaining variable), initialised with a public initial value (IV), using a so-called compression function h. The security of the hash function is closely related to the security of the compression function.

The main security property expected from such functions is *collision resistance*: it should be hard for an adversary to compute two distinct messages M and M' that map to the same hash value $H(M) = H(M')$, where "hard" means not faster than the generic birthday attack that can find a collision for any hash function with about $2^{n/2}$ computations. A stronger variant of the collision attack, the so-called *chosen-prefix* collision attack is particularly important. The attacker is first challenged with two message prefixes P and P', and its goal is to compute two messages M and M' such that $H(P\|M) = H(P'\|M')$, where $\|$ denotes concatenation. Such collisions are much more dangerous than simple collisions in practice, as they indicate the ability of an attacker to obtain a collision even though random differences (thus potentially some meaningful information) were inserted as message prefix. In particular, this is an important threat in the key application of digital signatures: chosen-prefix collisions for MD5 were demonstrated in [28], eventually leading to the creation of colliding X.509 certificates, and later of a rogue certificate authority [29]. Chosen-prefix collisions have also been shown to break important internet protocols, including TLS, IKE, and SSH [1], because they allow forgeries of the handshake messages.

SHA-1 is one the most famous cryptographic hash functions in the world, having been the NIST and de-facto worldwide hash function standard for nearly two decades until very recently. Largely inspired by MD4 [22] and then MD5 [23], the American National Security Agency (NSA) first designed a 160-bit hash function SHA-0 [17] in 1993, but very slightly tweaked one part of the design two years later to create a corrected version SHA-1 [18]. It remained a standard until its deprecation by the NIST in 2011 (and disallowed to be used for digital signatures at the end of 2013). Meanwhile, hash functions with larger output sizes were standardized as SHA-2 [19] and due to impressive advances in hash function cryptanalysis in 2004, in particular against hash functions of the MD-SHA family [31–34], the NIST decided to launch a hash design competition that eventually led to the standardization in 2015 of SHA-3 [20].

There has been a lot of cryptanalysis done on SHA-1. After several first advances on SHA-0 and SHA-1 [3,5], researchers managed to find for the first time in 2004 a theoretical collision attack on the whole hash function, with an estimated cost of 2^{69} hash function calls [32]. This attack was extremely complex and involved many details, so the community worked on better evaluating and further improving the actual cost of finding a collision on SHA-1 with these new techniques. Collisions on reduced-step versions of SHA-1 were computed [8,11], or even collisions on the whole SHA-1 compression function [27], which eventually led to the announcement in 2017 of the first SHA-1 collision [26].

Even though SHA-1 has been broken in 2004, it is still deployed in many security systems, because collision attacks do not seem to directly threaten most protocols, and migration is expensive. Web browsers have recently started to reject certificates with SHA-1 signatures, but there are still many users with older browsers, and many protocols and softwares that allow SHA-1 signatures. Concretely, it is still possible to buy a SHA-1 certificate from a trusted CA, and many email clients accept a SHA-1 certificate when opening a TLS connection. SHA-1 is also widely supported to authenticate TLS and IKE handshake messages.

Main SHA-1 cryptanalysis techniques. Attacks against SHA-1 are based on differential cryptanalysis, where an attacker manages to somewhat control the output of the compression function. Several important ideas were used to turn differential cryptanalysis into an effective tool against hash functions:

Linearization [5]. In order to build differential trails with high probability, a linearized version of the step function is used. Differential trails with a low-weight output difference δ_O can be used to find near-collisions in the compression function (i.e. two outputs that are close to a collision, the distance metric being for example the Hamming distance).

Message modification [2,32]. In a differential attack against a hash function, the attacker can choose messages that directly satisfy some of the constraints of the path, because there is no secret key. While the conditions in the first steps are easy to satisfy, more advanced techniques have been introduced to extend the usage of these degree of freedom to later rounds in order to speed-up collision search: neutral bits (firstly introduced for cryptanalysis SHA-0 [2,3]), message modifications [32] and boomerangs/tunnels [10,12].

Non-linear trails [32]. In order to get more flexibility on the differential trails, the first few steps can use non-linearity instead of following the constraints of the linearized step function. This does not affect the complexity of the search for conforming messages (thanks to messages modification techniques), but it allows to build trails from an arbitrary input difference to a good fixed output difference δ_O (or its opposite).

Multi-block technique [5,32]. The multi-block technique takes advantage of the Davies-Meyer construction used inside the compression function. Indeed, it can be written as $h(x, m) = x + E_m(x)$ where E is a block cipher, and $+$ denotes some group operation. Because of this feed-forward, an attacker can use several differential trails in E, and several near-collisions blocks, to iteratively affect the internal state. In particular, using non-linearity in the first steps, he can derive two related trails $0 \overset{\delta_M}{\rightsquigarrow} \delta_O$ and $\delta_O \overset{-\delta_M}{\rightsquigarrow} -\delta_O$ in E from a single linear trail, by swapping the message pair. When conforming messages are found for each block, this leads to a collision because the internal state differences cancel out (see Fig. 1).

Birthday phase for chosen-prefix collisions [28]. Differential techniques have also been used for chosen-prefix collision attacks. An attacker can relax the last steps of the differential trail to allow a set $\mathcal{D}$ of output differences rather than a single δ_O. He can also use several differential trails, and use the union of the corresponding sets. Starting from two different prefixes P, P', the chosen-prefix collision attack has two stages (see Fig. 2):

- In the *birthday stage*, the attacker uses a generic collision search with message blocks m_0, m'_0 to reach a difference $\delta = H(P' \| m'_0) - H(P \| m_0)$ in $\mathcal{D}$ with complexity roughly $\sqrt{\pi \cdot 2^n / |\mathcal{D}|}$.
- In the *near-collision stage*, he builds a differential trail $\delta \rightsquigarrow -\delta$ using non-linearity in the first steps, and searches a conforming message to build the chosen-prefix collision.

Multi-block for chosen-prefix collisions [28]**.** If a collection of differential trails affecting separate parts of the internal state is available, chosen-prefix collision attacks can be greatly improved. In particular, if an arbitrary input difference δ_R can be decomposed as $\delta_R = -(\delta_O^{(1)} + \delta_O^{(2)} + \cdots + \delta_O^{(r)})$, where each $\delta_O^{(i)}$ can be reached as the output of a differential trail, the attacker just has to find near-collision blocks with output differences $\delta_O^{(1)}, \ldots, \delta_O^{(r)}$ (see Fig. 3). Alternatively, if this only covers a subset of input differences, the multi-block technique is combined with a birthday stage.

Our contributions. In this work, we describe new chosen-prefix collision attacks against the MD-SHA family, using several improvements to reduce the complexity.

1. The main improvement comes from the use of several near-collision blocks, while Stevens uses a single near-collision block [25]. For instance, using two blocks we can start from any difference in the set $\mathcal{S} := \{\delta_1 + \delta_2 \mid \delta_1, \delta_2 \in \mathcal{D}\}$, and cancel it iteratively with a first block following a trail $\delta_1 + \delta_2 \rightsquigarrow -\delta_1$ and a second block following a trail $\delta_2 \rightsquigarrow -\delta_2$ (see Fig. 4). The set $\mathcal{S}$ grows with the number of blocks: this reduces the cost of the birthday search in exchange for a more expensive near-collision stage.
While there are previous chosen-prefix collision attacks using several near-collision blocks [13,15,21,28,29], these attacks use a collection of differential trails to impact different parts of the state (each block uses a different trail). On the opposite, our technique can be used with a single differential trail, or a collection of trails without any special property. In particular, previous chosen-prefix collision attacks based on a single trail (against SHA-1 [25] and MD5 [29, Sect. 6]) used only one near-collision block.
2. In addition, we use a clustering technique to optimize the near-collision stage, taking advantage of multiple ways to cancel a given difference. For instance, let's assume that we have to cancel a difference δ in the internal state that can be written in two different ways: $\delta = \delta_1 + \delta_2 = \delta'_1 + \delta'_2$, with $\delta_1, \delta'_1, \delta_2, \delta'_2 \in \mathcal{D}$, knowing trails $\delta \rightsquigarrow -\delta_1$ and $\delta \rightsquigarrow -\delta'_1$ with the same message constraints. Then, an attacker can target simultaneously $-\delta_1$ and $-\delta'_1$ for the first near-collision block (and use either a trail $\delta_2 \rightsquigarrow -\delta_2$ or $\delta'_2 \rightsquigarrow -\delta'_2$ for the second block, depending on the first block found). This can reduce the cost of finding the first block by a factor two.

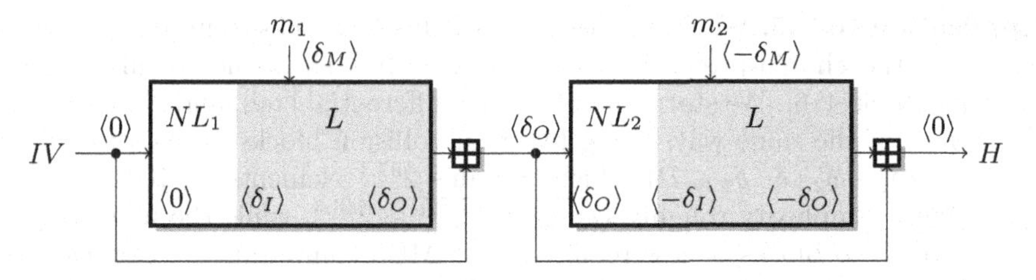

Fig. 1. 2-block collision attack using a linear trail $\delta_I \overset{\delta_M}{\rightsquigarrow} \delta_O$ and two non-linear trails $0 \rightsquigarrow \delta_I$ and $\delta_O \rightsquigarrow -\delta_I$. Green values between bracket represent differences in the state. (Color figure online)

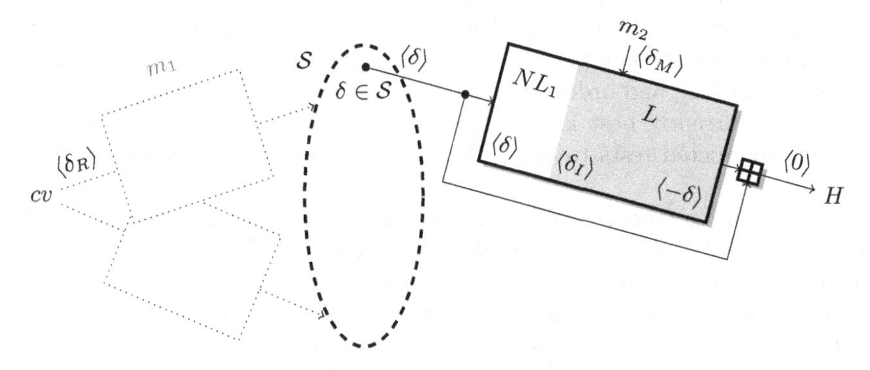

Fig. 2. Single-block chosen-prefix collision attack with a birthday stage. The linear trail $\delta_I \rightsquigarrow \delta_O$ is relaxed to reach a set $\mathcal{S}$ of feasible differences.

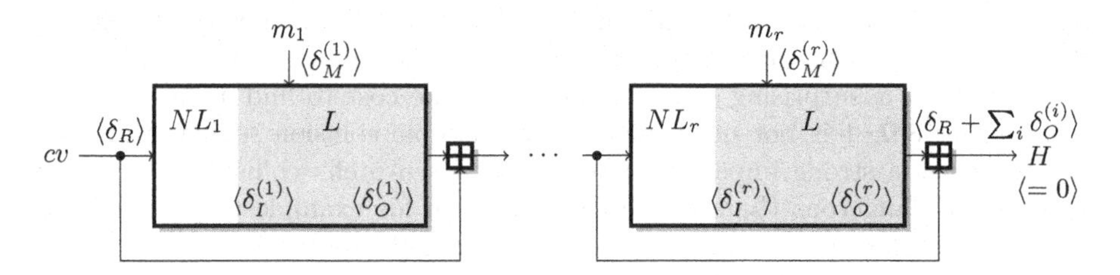

Fig. 3. Multi-block chosen-prefix collision attack. We assume that an arbitrary difference δ_R can be decomposed as $\delta_R = -\left(\delta_O^{(1)} + \delta_O^{(2)} + \cdots + \delta_O^{(r)}\right)$, where each $\delta_O^{(i)}$ can be reached as the output of a differential trail.

This technique can be seen as a generalization of an optimisation used for collision attacks with two blocks, where the first is less constrained and several output differences are allowed (for instance the SHA-1 collision attack of [25] allows 6 output differences, so that the first block is 6 times less expensive than the second).

Using these techniques, we obtain significant improvements to chosen-prefix collision attacks against MD5 and SHA-1.

Application to MD5. We use multiple near-collision blocks to improve the complexity of the chosen-prefix collision attack with a single near-collision block given in [29, Sect. 6]. We start with the same differential trail, and a set $\mathcal{D}$ of size $2^{25.2}$, built in the same way. Using two near-collision blocks, we can target the set $\mathcal{S} := \{\delta_1 + \delta_2 \mid \delta_1, \delta_2 \in \mathcal{D}\}$ which contains $2^{37.1}$ elements. This leads to an attack with complexity roughly $\sqrt{\pi \cdot 2^{128}/2^{37.1}} \approx 2^{46.3}$, while the best previous attack with two blocks or less required $2^{53.2}$ MD5 computations. However, the best chosen-prefix collision attack against MD5 is still the attack from [29] with complexity $2^{39.1}$ using 9 near-collision blocks.

Application to SHA-1. For SHA-1, we start with the attack of Stevens [25], and after using several improvements we obtain a chosen-prefix collision attack with estimated complexity between $2^{66.9}$ and $2^{69.4}$ SHA-1 computations. This is within a small factor of the complexity of a plain collision attack, estimated at $2^{64.7}$ on average [26,32], and orders of magnitude better than the $2^{77.1}$ computations cost of the currently best known chosen-prefix collision attack [25] on SHA-1. We have conducted tests to check that our assumptions are indeed verified in practice.

First, we use a more relaxed version of the differential trail than used in [25], so that we have a set $\mathcal{S}$ of size 8768 rather than 192. This directly reduce the attack cost by a factor 6.75, down to $2^{74.3}$. Next, we use the multi-block technique to build a large set $\mathcal{S}$ and to reduce further the cost of the birthday stage. Using a set $\mathcal{S}$ of size $2^{29.4}$ with a near-collision cost at most $12 \times 2^{64.7}$, this reduces the cost of the attack down to $2^{68.6}$ (with an optimistic estimate). Finally, we use the clustering technique to reduce the near-collision cost. After optimization, we have a set $\mathcal{S}$ of $2^{32.67}$ differences that can be reached with a maximum cost of $3.5 \times 2^{64.7}$ (with an optimistic estimate), leading to a full attack with complexity $2^{66.9}$—about five time more expensive than the collision attack (Table 1).

Our result is surprising since we show that the cost to find a chosen-prefix collision for SHA-1 is not much more than a simple collision search. Moreover our work has a strong impact in practice as chosen-prefix collision attacks are much more dangerous than simple collisions (see for example the case of MD5 [29]). This is yet another warning that SHA-1 should be totally removed from any security product as soon as possible. The thinking "a collision attack is not directly exploitable, thus we are fine" is clearly wrong for SHA-1, and we give a proof here.

Our method is in essence quite generic, even though a lot of details have to be taken care of in order to make it work. Since most collision attacks on members of the MD-SHA family are built on the same principles as SHA-1 attacks, we believe similar ideas would apply and a collision attack can probably be transformed into a chosen-prefix collision attack for a reasonable extra cost factor. We do not foresee any reason why this technique would not apply to non MD-SHA hash functions as well (except wide-pipe hash functions which would make the birthday part too costly).

Table 1. Comparison of previous and new cryptanalysis results on MD5 and SHA-1. A free-start collision is a collision or the compression function only, where the attacker has full control on all the primitive's inputs. Complexities in the table are given in terms of SHA-1 equivalents on a GTX-970 GPU (when possible)

Function	Collision type	Complexity	Ref.
SHA-1	free-start collision	$2^{57.5}$	[28]
	collision	2^{69}	[33]
		$2^{64.7}$	[27]
	chosen-prefix collision	$2^{77.1}$	[26]
		$2^{66.9}$ — $2^{69.4}$	New
MD5	collision	2^{40}	[34]
		2^{16}	[30]
	chosen-prefix collision (9 blocks)	$2^{39.1}$	[30]
	(3 blocks)	2^{49}	[30]
	(1 block)	$2^{53.2}$	[30]
	(2 blocks)	$2^{46.3}$	New

Outline. We first consider the impact of this result and give some recommendations in Sect. 2. Then, we describe SHA-1 and previous cryptanalysis works on this hash function in Sect. 3. The generic high-level description of our attack is given in Sect. 4, while the details regarding its application to MD5 and SHA-1 are provided in Sects. 5 and 6, respectively. Eventually, we conclude and propose future works in Sect. 7.

2 Implications and Recommendations

Our work shows that finding a chosen-prefix collision is much easier than previously expected, and potentially not much harder than a normal collision search for SHA-1. As a real collision has already been computed for this hash function, one can now assume that chosen-prefix collisions are reachable even by medium funded organisations. Since a chosen-prefix collision attack can break many widely used protocols, we strongly urge users to migrate to SHA-2 or SHA-3, and to disable SHA-1 to avoid downgrade attacks.

Cost Estimation. We use the same estimation process as in [26]. With an optimistic spot-price scenario on `g2.8xlarge` instances of Amazon EC2, the authors estimated that the workload spent to find the SHA-1 collision was equivalent to a cost of about US$ 110 K, with $2^{63.4}$ SHA-1 equivalent calls on GTX-970 GPUs. We recall that they found the collision with less computations than expected. Using expected computational cost, the average workload required to find a SHA-1 collision is equivalent to a cost of about US$ 275 K, or $2^{64.7}$ SHA-1 calls.

An optimistic analysis of our attack leads to a complexity of $2^{66.9}$ SHA-1 equivalent calls on GTX-970 GPUs, corresponding to a cost of US$ 1.2M, while a more conservative analysis yields a complexity of $2^{69.4}$, or a cost of US$ 7M.

Hardware will improve as well as cryptanalysis and we can expect that collision together with chosen-prefix collision attacks will get cheaper over the years. Migration from SHA-1 to the secure SHA-2 or SHA-3 hash algorithms should now be done as soon as possible, if not already.

Impact of Chosen-prefix Collisions. Chosen-prefix collision attacks have been demonstrated already for MD5, and they are much more dangerous that identical-prefix collision attacks, with a strong impact in practice. For example, they have been shown to break important internet protocols, including TLS, IKE, and SSH by allowing the forgery of handshake messages. The SLOTH attacks [1] can break various security properties of these protocols using MD5 chosen-prefix collisions, such as client impersonation in TLS 1.2. It was also shown [28] that one can generate colliding X.509 certificates and later a rogue certificate authority [29] from a chosen-prefix collision attack on MD5, undermining the security of websites. MD5 has been removed from most security applications, but the very same threats are now a reality for SHA-1.

The SLOTH attacks with SHA-1 chosen-prefix collisions allow client impersonation in TLS 1.2 and peer impersonation in IKEv2. In particular, IKEv2 can be broken with an *offline* chosen-prefix collision, which is now practical for a powerful adversary. On the other hand, the creation of a rogue CA requires to predict in advance all the fields of the signed certificate. Hopefully, this is not possible with current certificate authorities, because they should randomize the serial number field.

Usage of SHA-1. Even if practically broken only very recently, SHA-1 has been theoretically broken since 2004. It is therefore surprising that SHA-1 remains deployed in many security systems. In particular, as long as SHA-1 is *allowed*, even if it is not used in normal operation, an attacker can use weaknesses in SHA-1 to forge a signature, and the signature will be accepted.

First, SHA-1 is still widely used to authenticate handshake messages in secure protocols such as TLS, SSH or IKE. As shown with the SLOTH attack [1], this allows various attacks using chosen-prefix collision, such as breaking authentication. These protocols have removed support for MD5 after the SLOTH attack, but SHA-1 is still widely supported. Actually, more than[1] 5% of the web servers from Alexa's top 1M (including skype.com) *prefer* to use SHA-1 to authenticate TLS handshake messages.

An important effort is underway to remove SHA-1 certificates from the Web, and major browsers are now refusing to connect to servers still using SHA-1-based certificates. Yet SHA-1-based certificates remains present: according to scans of the top 1 million websites from Alexa by censys.io, there are still about 35

[1] https://censys.io/domain?q=tags:https+and+443.https.tls.signature.hash_algorithm:sha1.

thousand[2] servers with SHA-1 certificates out of 1.2 million servers with HTTPS support. SHA-1-based certificates are also used with other protocols: for instance 700 thousand[3] out of 4.4 million mail servers (with IMAPS) use a SHA-1 certificate. Actually, it is still possible to buy a SHA-1 certificate from a trusted root[4]! Even though recent web browsers reject those certificates, they are accepted by older browsers and by many clients for other protocols. For instance, the "Mail" application included in Windows 10 still accepts SHA-1 certificates without warnings when opening an IMAPS connection.

Unfortunately, many industry players did not consider moving away from SHA-1 a priority, due to important costs and possible compatibility and bug issues induced by this move. An often-heard argument is that a simple collision attacks against a hash function is not very useful for an attacker, because he doesn't have much control over the colliding messages. Therefore, there seemed to be a long way to go before really useful collision attacks would be found for SHA-1, if ever. Indeed, the current best chosen-prefix collision attack against SHA-1 requires $2^{77.1}$ hash calls [25], thus orders of magnitude harder than the cost of finding a simple collision. Similarly, in the case of MD5, the cost goes from 2^{16} to 2^{39} for the currently best known collision and chosen-prefix collision attacks. However, this is a dangerous game to play as the history showed that cryptanalysis only keep improving, and attackers will eventually come up with ways to further improve their cryptanalysis techniques. For example, in the case of MD5, collisions for the compression function were found [9] in 1993, collisions for the whole hash function were found [33] in 2004, colliding X.509 MD5-based certificates were computed [28] in 2007, and rogue Certificate Authority certificate [29] was eventually created in 2009.

3 Preliminaries

3.1 Description of SHA-1

We describe here the SHA-1 hash function, but we refer to [18] for all the complete details.

SHA-1 is a 160-bit hash function based on the well-known Merkle-Damgård paradigm [6,16]. The message input is first padded (with message length encoded) to a multiple of 512 bits, and divided into blocks m_i of 512 bits each. Then, each block is processed via the SHA-1 compression function h to update a 160-bit chaining variable cv_i that is initialised to a constant and public initial value (IV): $cv_0 = IV$. More precisely, we have $cv_{i+1} = h(cv_i, m_{i+1})$. When all blocks have been processed, the hash output is the last chaining variable.

[2] https://censys.io/domain?q=tags:https+and+443.https.tls.certificate.parsed.signature_algorithm.name:SHA1*.

[3] https://censys.io/ipv4?q=tags:imaps+and+993.imaps.tls.tls.certificate.parsed.signature_algorithm.name:SHA1*.

[4] https://www.secure128.com/online-security-solutions/products/ssl-certificate/symantec/sha-1-private-ssl/.

The compression function is similar to other members of the MD-SHA family of hash functions. It is based on the Davies-Meyer construction, that turns a block cipher E into a compression function: $cv_{i+1} = E_{m_{i+1}}(cv_i) + cv_i$, where $E_k(y)$ is the encryption of the plaintext y with the key k, and $+$ is a word-wise modular addition.

The internal block cipher is composed of 80 steps (4 rounds of 20 steps each) processing a generalised Feistel network. More precisely, the state is divided into five registers $(A_i, B_i, C_i, D_i, E_i)$ of 32-bit each. At each step, an extended message word W_i updates the registers as follows:

$$\begin{cases} A_{i+1} = (A_i \lll 5) + f_i(B_i, C_i, D_i) + E_i + K_i + W_i \\ B_{i+1} = A_i \\ C_{i+1} = B_i \ggg 2 \\ D_{i+1} = C_i \\ E_{i+1} = D_i \end{cases}$$

where K_i are predetermined constants and f_i are boolean functions (in short: IF function for the first round, XOR for the second and fourth round, MAJ for the third round, see Table 2). Since only a single register value is updated (A_{i+1}), the other registers being only rotated copies, we can express the SHA-1 step function using a single variable:

$$\begin{aligned} A_{i+1} = {} & (A_i \lll 5) + f_i(A_{i-1}, A_{i-2} \ggg 2, A_{i-3} \ggg 2) \\ & + (A_{i-4} \ggg 2) + K_i + W_i. \end{aligned}$$

For this reason, the differential trails figures in this article will only represent A_i, the other register values at a certain point of time can be deduced directly.

Table 2. Boolean functions and constants of SHA-1

Step i	$f_i(B, C, D)$	K_i
$0 \leq i < 20$	$f_{IF} = (B \wedge C) \oplus (\overline{B} \wedge D)$	0x5a827999
$20 \leq i < 40$	$f_{XOR} = B \oplus C \oplus D$	0x6ed6eba1
$40 \leq i < 60$	$f_{MAJ} = (B \wedge C) \oplus (B \wedge D) \oplus (C \wedge D)$	0x8fabbcdc
$60 \leq i < 80$	$f_{XOR} = B \oplus C \oplus D$	0xca62c1d6

The extended message words W_i are computed linearly from the incoming 512-bit message block m, the process being called message extension. One first splits m into 16 32-bit words $M_0, \ldots, M_{15}$, and then the W_i's are computed as follows:

$$W_i = \begin{cases} M_i, & \text{for } 0 \leq i \leq 15 \\ (W_{i-3} \oplus W_{i-8} \oplus W_{i-14} \oplus W_{i-16}) \lll 1, & \text{for } 16 \leq i \leq 79 \end{cases}$$

3.2 Previous Works

Collision attacks on SHA-1. We quickly present here without details the previous advances on SHA-1 collision search. First results on SHA-0 and SHA-1 were obtained by linearizing the compression function and constructing differential trails based on the probabilistic event that difference spreads will indeed happen linearly. These linear trails are generated with a succession of so-called local collisions (small message disturbances whose influence is corrected with other message differences inserted in the subsequent SHA-1 steps) that follows the SHA-1 message expansion. However, with this linear construction, impossibilities might appear in the first 20 steps of SHA-1 (for example due to the f_{IF} boolean function that never behaves linearly in some specific situations) and the cheapest linear trail candidates might not be the ones that start and end with the same difference (which is a property required to obtain a collision after the compression function feed-forward). Thus, since [32], collision attacks on SHA-1 are performed using two blocks containing differences. The idea is to simply pick the cheapest linear trail from roughly step 20 to 80, without paying any attention to the f_{IF} constraints or to the input/output differences. Then, the attacker will generate a non-linear differential trail for the first 20 steps in order to connect the actual incoming input difference to the linear part difference at step 20. With two successive blocks using the same linear trail (just ensuring that the output difference of the two blocks have opposite signs), one can see in Fig. 1 that a collision is obtained at the end of the second block.

Once the differential trail is set, the attacker can concentrate on finding a pair of messages that follows it for each successive block. For this, he will construct a large number of messages that follow the trail up to some predetermined step, and compute the remaining steps to test whether the output difference is the required one. The computational cost is minimized by using a simple early-abort strategy for the 16 first steps, but also more advanced amortization methods such as neutral bits [3], boomerangs [10] or message modification [32] for a few more steps. Usually, the first 20 or so steps do not contribute the complexity of the attack.

Chosen-prefix collision attacks. The first concrete application of a chosen-prefix collision attack was proposed in [28] for MD5. This work was also the first to introduce a birthday search phase in order to find such collisions. The idea is to process random message blocks after the challenged prefixes, until the chaining variable difference δ belongs to a large predetermined set $\mathcal{S}$. Since the message blocks after each prefix are chosen independently, this can be done with birthday complexity $\sqrt{\pi \cdot 2^n/|\mathcal{S}|}$. Then, from that difference δ, the authors eventually manage to reach a collision by slowly erasing the unwanted difference bits by successfully applying some near-collision blocks. We note that the starting difference set $\mathcal{S}$ during the birthday phase must not be too small, otherwise this phase would be too costly. Moreover, the near-collisions blocks must not be too expensive either, and this will of course depend on the cryptanalysis advancements of the compression function being studied.

Finally, using also this two-phase strategy, in [25] is described a chosen-prefix collision attack against the full SHA-1, for a cost of $2^{77.1}$ hash calls. The improvement compared to the generic 2^{80} attack is not very large, due to the difficulty for an attacker to generate enough allowable differences that can later be erased efficiently with a near-collisions block. Indeed, the compression function of SHA-1 being much stronger than that of MD5, few potential candidates are found. Actually, Stevens only considers one type of near collision block, following the best differential trail used for the collision attack. By varying the signs of the message and output differences, and by letting some uncontrolled differences spread during the very last steps of the compression function, a set $\mathcal{S}$ of 192 allowable differences is obtained. However, having such a small set makes the birthday part by far the most expensive phase of the attack.

In this article, we will use essentially the same strategy: a birthday phase to reach a set $\mathcal{S}$ of allowable differences and a near-collision phase to remove the remaining differences. We improve over previous works on several points. First, we further generalise for SHA-1 the set of possible differences that can be obtained for a cheap cost with a single message block. Secondly, we propose a multi-block strategy for SHA-1 that allows to greatly increase the size of the set $\mathcal{S}$. Finally, we study the clustering effect that appears when using a multi-block strategy. This can be utilised by the attacker to select dynamically the allowable differences at the output of each successive blocks, to further reduce the attack complexity. Notably, and in contrary to previous works, our set $\mathcal{S}$ is not the direct sum of independent subspaces corresponding to distinct trails. On the opposite, our applications use the same core differential trail for all the near-collision blocks. Overall, we improve substantially previous attack [25] from $2^{77.1}$ SHA-1 calls to only $2^{66.9}$. Surprisingly, our attack is very close to some sort of optimal since its cost is not much greater than that of finding a simple collision. Our attack being rather generic, we believe that this might be the case for many hash functions, which contradicts the idea that chosen-prefix collisions are much harder to obtain than simple collisions.

One can mention other parallel researches conducted on finding chosen-prefix collision attacks for various hash functions. For example, in [21], the author explains how to compute collisions with random incoming differences in the internal state for the GRINDAHL hash function, the strategy being to slowly remove these differences thanks to the many degrees of freedom available every step. Such a divide-and-conquer technique is not applicable at all to SHA-1 as the degrees of freedom are much fewer and only available at the beginning of the compression function. In [15], inspired by the second-preimage attack against SMASH [13], the authors proposed a chosen-prefix collision attack on a reduced version of the GROSTL hash function. However, this attack strongly relies on the ability of the attacker to perform a rebound attack, which seems really hard to achieve in the case of SHA-1.

4 From Collision to Chosen-Prefix Collision

4.1 The Chosen-Prefix Collision Attack

We assume that the hash function considered is an n-bit narrow pipe primitive, based on a Merkle-Damgård-like operating mode. In addition, we assume that the compression function is built upon a block cipher in a Davies-Meyer mode.

Preparing the attack. The attacker first builds a set $\mathcal{S}$ and a graph $\mathcal{G}$; $\mathcal{S}$ corresponds to a set of differences that can be cancelled with near-collision blocks, and $\mathcal{G}$ is used to find the sequence of blocks needed to cancel a difference in $\mathcal{S}$. We first explain how to execute the attack when $\mathcal{S}$ and $\mathcal{G}$ are given, and we will explain how to build them in Sect. 4.2.

The prefixes (stage 1 of Fig. 4). The attacker receives the two challenge prefixes and pads them to two prefixes of the same length, to avoid differences in the final length padding. After processing the two padded prefixes starting from the IV, he reaches states cv/cv', and we denote the corresponding difference as δ_R.

The birthday search (stage 2 of Fig. 4). The goal of the attacker is now to find one message block pair (u, u') to reach a chaining variable pair with a difference δ that belongs to $\mathcal{S}$, the set of acceptable chaining variable differences.

For this stage, we use the parallel collision search algorithm of van Oorschot and Wiener [30]. When a memory $M \gg C$ is available, this algorithm can find C collisions in a function $f : \{0,1\}^k \rightarrow \{0,1\}^k$ with complexity $\sqrt{\pi/2 \cdot 2^k \cdot C}$, and is efficiently parallelizable. It computes chains of iterates of the function f, and stops when the end point is a *distinguished point*, *i.e.* it satisfies some easy to detect property. In practice, we stop a chain when $x < 2^k \cdot \theta$, with $\theta \gg \sqrt{C/2^k}$, and we store on average the starting points and end points of $\theta \cdot \sqrt{\pi/2 \cdot 2^k \cdot C}$ chains (the expected length of a chain is $1/\theta$). When colliding end points are detected, we restart the corresponding chains to locate the collision point, with an expected cost of $2C/\theta$, which is small compared to the total complexity if $\theta \gg \sqrt{C/2^k}$.

In our case, we are looking for message blocks (u, u') such that $h(cv, u) - h(cv', u') \in \mathcal{S}$. Therefore, we need a function f such that a collision in f corresponds to good (u, u') with high probability. First, we consider a truncation function $\tau : \{0,1\}^n \rightarrow \{0,1\}^k$, so that pairs (x, x') with $x - x' \in \mathcal{S}$ have $\tau(x) = \tau(x')$ with high probability:

$$p = \Pr_{x,x'}\left[\tau(x) = \tau(x') \mid x - x' \in \mathcal{S}\right] \approx 1.$$

For functions of the `MD-SHA` family, the group operation $+$ is a word-wise modular addition, and we build τ by removing bits that are directly affected when adding a value in $\mathcal{S}$, and bits that are affected by a carry with a relatively high probability. This typically leads to p close to one (as seen in previous attacks [25,29], and the new attacks in this paper). Then, we build f as:

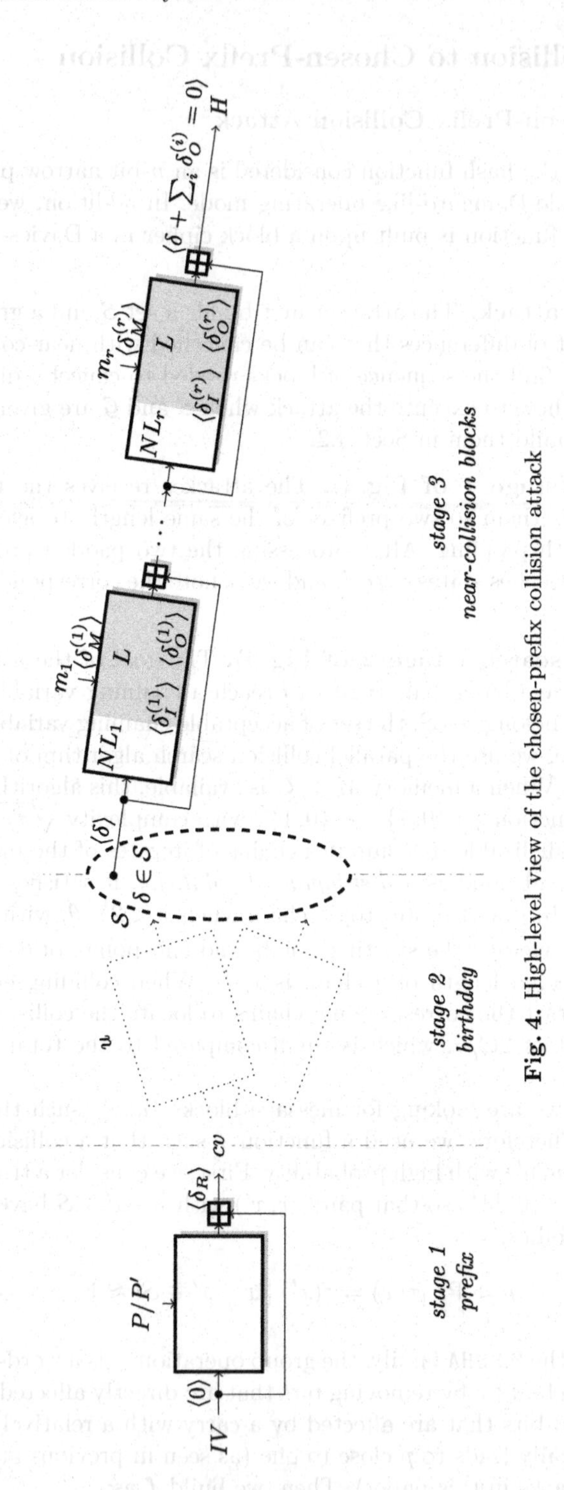

Fig. 4. High-level view of the chosen-prefix collision attack

$$f(u) := \begin{cases} \tau(h(cv\ , \text{pad}(u))) & \text{if } u[0] = 1; \\ \tau(h(cv', \text{pad}(u))) & \text{else.} \end{cases}$$

The probability that a collision in f corresponds to a pair (u, u') with $h(cv, u) - h(cv', u') \in \mathcal{S}$ can be estimated as:

$$p_f = \frac{1}{2} \cdot \Pr_{x,x'} [x - x' \in \mathcal{S} \mid \tau(x) = \tau(x')] = \frac{p}{2} \cdot \frac{|\mathcal{S}|/2^n}{2^{-k}}$$

Finally, we need $1/p_f$ collisions in f, and the total complexity of the birthday stage is on average:

$$\sqrt{\frac{\pi}{2} \cdot \frac{2^k}{p_f}} = \sqrt{\frac{\pi \cdot 2^n}{p \cdot |\mathcal{S}|}} \approx \sqrt{\frac{\pi \cdot 2^n}{|\mathcal{S}|}}.$$

The multi-block collision search (stage 3 of Fig. 4). The attacker now uses the graph $\mathcal{G}$ to build a sequence of near-collision blocks that ends up with a collision. Each node of the graph represents one chaining variable difference in the set $\mathcal{S}$. To each node i of the graph is associated a cost value w_i that represents the cost an attacker will expect to pay from this particular chaining variable difference i in order to reach a colliding state (with one or multiple message blocks). Of course, a null cost will be associated with the zero difference ($w_0 = 0$). A directed edge from node i to node j represents a way for an attacker to reach chaining variable difference j from difference i with a single message block. Note that the graph is acyclic, as we will ensure that the edges will always go to strictly lower costs (i.e. an edge from i to j is only possible if $w_j < w_i$). To each edge is attached the details of the differential trail and message difference to use for that transition to happen. However, a *very* important point is that all edges going out of a node i will share the same core differential trail (by core differential trail, we mean the entire differential trail except the last steps for which one can usually accept a few divergences in the propagation of the differences). For example, during the attack, from a chaining difference i, an attacker can potentially reach difference j or difference k using the same core differential trail (and thus without having to commit in advance which of the two differences he would like to reach). Thus, in essence, the details of the differential trail and message difference to use can be directly attached to the source node.

Once the attacker hits a starting difference $\delta \in \mathcal{S}$ in the birthday phase, he will pick the corresponding node in $\mathcal{G}$, and use the differential trail and message difference attached to this node. He will use this differential trail until he reaches one of the target nodes (which has necessarily a lower expected cost attached to it). As explained in the next section, targeting several nodes simultaneously reduces the cost of the attack, because it is easier to hit one node out of many than a fixed one. We call this the clustering effect, because we use a cluster of paths in the graph. When a new node is reached, the attacker repeats this process until he eventually reaches the colliding state (i.e. null difference). Overall, the expected computational cost for this phase is the cost attached to the node δ (in practice, when actually computing one collision, he might pay a slightly lower or higher computational cost as the w_i's are expected values).

We note that any suffix message blocks that do not contain differences can of course be added after this colliding state, as the Merkle-Damgård-like mode will maintain the collision throughout the subsequent compression function calls.

4.2 Building the Set $\mathcal{S}$ and the Graph $\mathcal{G}$

We now describe how the set $\mathcal{S}$ and the graph $\mathcal{G}$ can be built during the preparation of the attack. For that, we first need to describe what an adversary can do when he tries to attack the compression function. We consider that the attacker knows some good core differential trails for the internal block cipher E, that is differential trails that go from early steps to late steps of E. For each core differential trail $\mathtt{CDT}_i$ there are several possible output differences δ^i_j for E. This is typically what happens in the chosen-prefix collision attack on SHA-1 [25] where some differences in the very last steps can be allowed to spread differently than planned, thus generating new output differences. We denote the set of all possible output differences as $\mathcal{D}$ (in particular, we have $0 \in \mathcal{D}$, and $\delta \in \mathcal{D} \iff -\delta \in \mathcal{D}$ because of symmetries).

We finally assume that any input difference for E can be mapped to any of the core differential trails inside the primitive. In the case of a SHA-1 attack, this is achieved with the non-linear part of the trail in the first steps of the function. As shown in previous works, it allows to map any input difference to any internal difference. The non-linear part has a low probability, but during the near-collision search this is solved using the many degrees of freedom available in the first steps of the function.

Building the graph $\mathcal{G}'$. The attacker will first build a graph $\mathcal{G}'$ and filter it later to create $\mathcal{G}$. The graph $\mathcal{G}'$ is similar to the graph $\mathcal{G}$: each node will represent a chaining variable difference. A directed edge from node i to node j represents again a way for an attacker to reach chaining variable difference j from difference i with a single message block, stored with the details of the differential trail attached to it, and the cost to find the corresponding block. The differences with $\mathcal{G}$ are that (see Figs. 5 and 6):

- $\mathcal{G}'$ can potentially be cyclic, as we do not ensure that an edge goes from a higher to a strictly smaller cost;
- all outgoing edges from a node i will not necessarily share the same core differential trail;
- there can be several edges from i to j, with different core differential trails.

In order to build the graph $\mathcal{G}'$, starting from the colliding state $\delta = 0$, we will simply proceed backward. We go through all possible core differential trails for E and their possible output differences δ^i_j. Due to the feed-forward of the

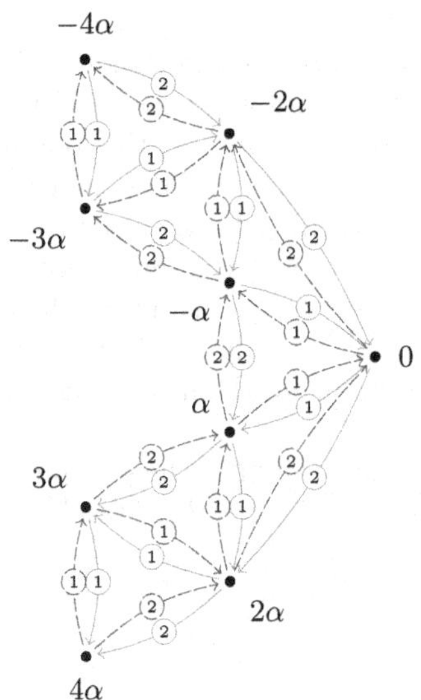

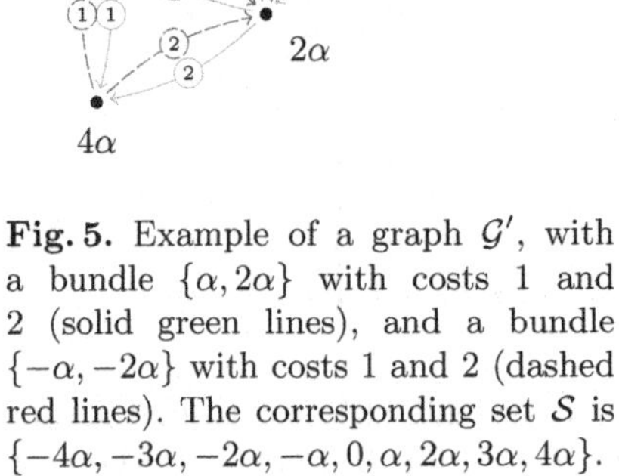

Fig. 5. Example of a graph $\mathcal{G}'$, with a bundle $\{\alpha, 2\alpha\}$ with costs 1 and 2 (solid green lines), and a bundle $\{-\alpha, -2\alpha\}$ with costs 1 and 2 (dashed red lines). The corresponding set $\mathcal{S}$ is $\{-4\alpha, -3\alpha, -2\alpha, -\alpha, 0, \alpha, 2\alpha, 3\alpha, 4\alpha\}$. (Color figure online)

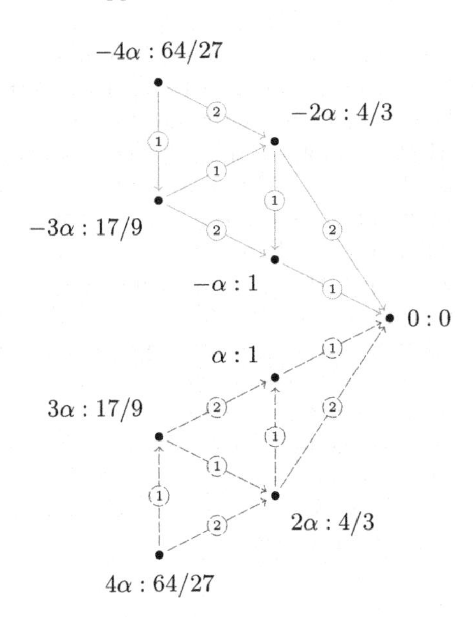

Fig. 6. The graph $\mathcal{G}$ corresponding to $\mathcal{G}'$. We show the cost of each edge and each node. In particular, note that use of clustering reduces the cost of node 4α from 4 to $64/27 \approx 2.37$ (Color figure online)

Davies-Meyer construction, all these differences can be used to reach the colliding state by simply forcing their respective opposite $-\delta^i_j$ as input difference of the cipher (since we assumed that any input difference for E can be mapped to any of the core differential trails inside the primitive, this is always possible). Thus, for each such difference δ^i_j coming from a core differential trail $\texttt{CDT}_i$, we will add a node $-\delta^i_j$ in the graph $\mathcal{G}'$, and an edge going from this new node to the colliding state. If a node with that difference already exists in the graph, we add the edge between this node and the colliding state. This means that nodes can have several incoming and outgoing edges.

We iteratively repeat this process again with all the newly created nodes as starting points (instead of the collision state). This will create a bigger and bigger graph as we keep iterating, and the attacker can simply stop when he believes that he has enough nodes in the graph (alternatively, he can set an upper limit on the cost of the nodes to consider, or on the depth of the search, which will naturally limit the size of the graph).

The clustering effect. A simple way to build a graph $\mathcal{G}$ for the attack is to keep only the minimal cost paths in the graph $\mathcal{G}'$ (the corresponding edges form a tree), and to set the cost of the nodes to cost of the minimal path. However, the attack cost can be greatly improved with the *clustering effect*: during the last phase of the attack, when the attacker is currently located at a node N, he does not necessarily need to choose in advance which outgoing edge of N he will try to follow. Indeed, the only commitment he needs to make at this point is which core differential trail he will use to go to the next node. Thus, he can simultaneously target several output differences corresponding to the same core differential trail, and the cost to reach one difference out of many is significantly lower than the cost of reaching any given output difference. For instance, when computing the first block of a SHA-1 collision, Stevens [25] allows six output differences with a similar cost, so that the cost to reach one of them is one sixth of the cost to reach a predetermined one.

For a given node, we call *bundle* of a core differential trail $\mathtt{CDT}_i$ the grouping of all outgoing edges of that node that use $\mathtt{CDT}_i$ (see Fig. 5). Let $\mathcal{B}_N$ stand for the set of all bundles of a node N, where each bundle $\beta \in \mathcal{B}_N$ corresponds to a distinct core differential trail $\mathtt{CDT}_i$. Then, for each node of $\mathcal{G}'$, we compute its assigned cost as follows[5]:

$$w_N = \min_{\beta \in \mathcal{B}_N} \left\{ \frac{1 + \sum\limits_{(N,j) \in \beta | w_j < w_N} \left(w_j / c_j^\beta \right)}{\sum\limits_{(N,j) \in \beta | w_j < w_N} \left(1/c_j^\beta \right)} \right\}, \tag{1}$$

where for an edge (N, j) of the bundle β, c_j^β represents the cost to find a conforming message pair with difference output $j - N$ for E, and w_j is the cost assigned to the node j.

We initialize the costs of the nodes in $\mathcal{G}'$ using the shortest path in the graph, and update them iteratively until we can't find any more opportunity for improvement.

Building $\mathcal{S}$ and $\mathcal{G}$. The graph $\mathcal{G}$ is obtained from $\mathcal{G}'$ by only keeping the edges that goes from a higher to a strictly lower cost (in order to render the graph acyclic), and by only keeping for each node the outgoing edges for the bundle that minimizes the cost w_N in (1).

---

[5] In order to explain this formula, we consider that when the adversary uses a bundle β, he has to perform C_β operations to find a pair conforming to the core differential trail up to some intermediate step, and those pairs lead to an output difference $j - N$ (*i.e.* to node j) with probability p_j^β (with $p_j^\beta = C_\beta / c_j^\beta$). If none of the predetermined output differences is reached (or if the target node reached has a cost $w_j \geq w_N$), then he stays at node N and will have to still pay w_N to reach the colliding state. Thus, we have that:

$$w_N = C_\beta + \sum_{j \in \beta | w_j < w_N} \left(p_j^\beta \cdot w_j \right) + \left(1 - \sum_{j \in \beta | w_j < w_N} p_j^\beta \right) \cdot w_N$$

which leads to (1) with $c_j^\beta = C_\beta / p_j^\beta$.

The set $\mathcal{S}$ is then finally deduced by harnessing all the differences corresponding to every node in $\mathcal{G}$ (one node in $\mathcal{G}$ will correspond to one difference in $\mathcal{S}$). In particular, if $\mathcal{G}'$ includes all nodes at depth at most r, then $\mathcal{S} = \{\delta_1 + \delta_2 + \cdots + \delta_r \mid \delta_1, \delta_2, \ldots, \delta_r \in \mathcal{D}\}$.

5 Application to MD5

Our techniques can easily be applied to MD5, to build an alternative chosen-prefix collision attack. We can't reach an attack complexity as low as 2^{39} (the best attack from [29]), because this would require to build a set $\mathcal{S}$ and graph $\mathcal{G}$ of size roughly 2^{50}, which is impractical. However, when the number of blocks available for the chosen-prefix collision is limited, the complexity of the best-known attack grows; for instance, the chosen-prefix collision used to create a rogue certificate was limited to 3 blocks, and this increased the complexity to 2^{49}. In this scenario we can improve the currently best-known attack with our multi-block technique using a single differential trail.

We start from the single-block chosen-prefix collision attack given in [29, Sect. 6]: this attack uses a high probability trail for MD5 collisions, where the last steps are relaxed to allow a set $\mathcal{D}$ of size $2^{23.3}$. Therefore the birthday stage has complexity roughly $\sqrt{\pi \cdot 2^{128}/2^{23.3}} \approx 2^{53.2}$, and the near-collision block is found with a complexity of $2^{40.8}$. In our analysis, we recomputed the set $\mathcal{D}$ used by Stevens *et al.*, but we actually found a set of size $2^{24.2}$ using the same trail, with a maximum cost of $2^{26} \cdot 2^{14.8}$ (following [29], we only consider output differences with $\delta a = -2^5, \delta d = -2^5 + 2^{25}, \delta c = -2^5 \bmod 2^{20}$). Then, we extend $\mathcal{D}$ by adding the zero value and the opposite of each value, to generate $\mathcal{D}' := \mathcal{D} \cup -\mathcal{D} \cup \{0\}$. Finally, we build the set $\mathcal{S}$ and the graph $\mathcal{G}'$ corresponding to an attack with at most 2 blocks, with $\mathcal{S} := \{\delta_1 + \delta_2 \mid \delta_1, \delta_2 \in \mathcal{D}'\}$. Since the cost of the near-collision stage is negligible (at most $2 \cdot 2^{40.8}$), we do not need to use clustering, and we can just use the minimal cost paths of $\mathcal{G}'$ as the graph $\mathcal{G}$.

We find that the set $\mathcal{S}$ contains $2^{37.1}$ elements, so that the birthday stage has a complexity of roughly $\sqrt{\pi \cdot 2^{128}/2^{37.1}} \approx 2^{46.3}$. Therefore, we have a simple chosen-prefix collision attack with two near-collision blocks with complexity $2^{46.3}$, while the best previous attack with two blocks or less requires $2^{53.2}$ MD5 computations, and even the best attack with three blocks requires 2^{49} MD5 computations.

6 Application to SHA-1

For the attack on SHA-1, we directly recycle the details of the collision attack from [26]: we will use the same linear part for our successive near-collision blocks (even though the very last steps might behave slightly differently as we will explain in this section). We assume that the attacker can generate non-linear parts on the fly and can apply amortization methods just like in [26]. In order to validate this assumption, we have tried to generate a non-linear part with several random input differences from $\mathcal{S}$ and random input chaining values.

In our experiments, we have successfully generated such non-linear part, and we could even make it limited to the very first SHA-1 steps. We discuss this assumption and our experiments in more details in Sect. 6.3.

We now explain how to apply the framework of Sect. 4 to a chosen-prefix collision against SHA-1. As mentioned, our attack uses the best core trail known for attacks against SHA-1, as used in previous attacks [24–26]. This allows us to have a relatively good complexity estimation for the attack, because this trail has been well studied, and a full collision attack with this trail was recently implemented. In the following we denote the complexity to find a block conforming to the trail (with an optimal output difference) as C_{block}. In the case of the recent collision attack, this cost was estimated as $C_{\text{block}} = 2^{64.7}$ SHA-1 evaluations on a GTX-970 GPU [26, Sect. 5.7]. In this work, we consider several hypothesis for the cost of finding near-collision blocks: an optimistic hypothesis with $C_{\text{block}} = 2^{64.7}$ (following [26]) and a conservative hypothesis with $C_{\text{block}} = 2^{67.7}$. This parameter depends on the difficulty of linking an arbitrary input difference to the core trail, and will be discussed in more detail below.

As in the previous chosen-prefix collision attack on SHA-1 [25], we consider several variants of the core trail by changing some of the message constraints in the last steps (in particular, we flip the sign of some message bits), and we relax the last steps to reach a larger set of output difference. However, we do this more exhaustively than Stevens: he only describes a set $\mathcal{D}$ of size 192 with cost at most $1.15 \cdot C_{\text{block}}$, but we found a set of size 8768 with cost at most $8 \cdot C_{\text{block}}$, including 576 values with cost at most $1.15 \cdot C_{\text{block}}$. In particular, this directly leads to an improvement of the single-block chosen-prefix collision from [25], with complexity roughly $\sqrt{\pi \cdot 2^{160}/8768} \approx 2^{74.3}$, rather than $\sqrt{\pi \cdot 2^{160}/192} \approx 2^{77}$ (ignoring some technical details of the birthday step).

More precisely, we allow the signs of the message differences to not necessarily follow local collision patterns in the last steps. Instead, we consider variants of the trail where each of those constraints is either followed or not. In addition, we fix the sign of some additional state bits to reduce the cost to reach a given output difference. Table 3 compares our message constraints with those used for the second-block of the attack from [26]. This leads to 288 differences with optimal probability ($2^{-19.17}$ in steps 61 to 79), and 288 with almost optimal probability ($2^{-19.36}$ in steps 61 to 79), as listed in Table 4. Moreover, we consider output differences whose cost is higher than the optimal cost C_{block}, up to roughly $8 \cdot C_{\text{block}}$ (we allow a probability up to 2^{-22} in steps 61 to 79).

Instead of building the corresponding set of output differences and their probability analytically, we used a heuristic approach. For each choice of the message constraints z_i in Table 3 (up to some symmetries), we generated 2^{30} intermediate states at step 60, and we computed the corresponding output differences in order to identify high probability ones. We also keep track of the differences reached with the same constraints, to build the corresponding bundles of differences. Next, we used symmetries in the set of differences to verify the consistency of the results, and to increase the precision of the heuristic probabilities. This strategy leads to a set of 8768 possible output differences, grouped in 2304 bun-

Table 3. Message constraints in the final steps. The z_i are fixed to 0 or 1 to define variant of the trail with distinct output differences. We use three more constraints than [26].

Stevens *et al.* constraints [27]	Our constraints
$W_{68}^{[5]} = W_{67}^{[0]} \oplus 1$	$W_{68}^{[5]} = W_{67}^{[0]} \oplus 1$
$W_{72}^{[30]} = W_{67}^{[0]} \oplus 1$	$W_{72}^{[30]} = W_{67}^{[0]} \oplus 1$
$W_{71}^{[6]} = W_{70}^{[1]} \oplus 1$	$W_{71}^{[6]} = W_{70}^{[1]} \oplus 1$
$W_{72}^{[5]} = W_{71}^{[0]} \oplus 1$	$W_{72}^{[5]} = W_{71}^{[0]} \oplus 1$
$W_{76}^{[30]} = W_{71}^{[0]} \oplus 1$	$W_{76}^{[30]} = W_{71}^{[0]} \oplus 1$
$W_{74}^{[7]} = W_{73}^{[2]} \oplus 1$	$W_{74}^{[7]} = W_{73}^{[2]} \oplus 1, \ W_{73}^{[2]} = z_1$
$W_{75}^{[6]} = W_{74}^{[1]} \oplus 1$	$W_{75}^{[6]} = W_{74}^{[1]} \oplus 1$
$W_{76}^{[6]} = W_{75}^{[1]} \oplus 1$	$W_{76}^{[6]} = W_{75}^{[1]} \oplus 1$
$W_{76}^{[1]} = W_{76}^{[0]} \oplus 1$	$W_{76}^{[1]} = z_2, \quad W_{76}^{[0]} = z_3$
$W_{77}^{[1]} = W_{77}^{[0]} \oplus z_1$	$W_{77}^{[1]} = z_4, \quad W_{77}^{[0]} = z_5$
$W_{77}^{[2]} = W_{77}^{[1]} \oplus 1$	$W_{77}^{[2]} = z_6$
$W_{77}^{[8]} = W_{76}^{[3]} \oplus 1, \ W_{76}^{[3]} = z_2$	$W_{76}^{[3]} = z_7, \quad W_{77}^{[8]} = z_8$
$W_{78}^{[0]} = W_{74}^{[7]}$	$W_{78}^{[0]} = z_9$
$W_{78}^{[3]} = z_3$	$W_{78}^{[3]} = z_{10}$
$W_{78}^{[7]} = z_4$	$W_{78}^{[7]} = z_{11}$
$W_{79}^{[2]} = z_5$	$W_{79}^{[2]} = z_{12}$
$W_{79}^{[4]} = z_6$	$W_{79}^{[4]} = z_{13}$

dles. We list the output differences with (almost) optimal probability that we have identified in Table 4, with the corresponding bundles (we do not give the set with all considered differences due its large size).

Next, we build the set $\mathcal{S}$ of acceptable differences, and the graph $\mathcal{G}$ that indicates the sequences of near-collision blocks to use to cancel the differences in $\mathcal{S}$. We first build the graph $\mathcal{G}'$ as explained in Sect. 4.2. We use a limit on the cost of the nodes added to graph: we only consider nodes that have a path with cost at most $18 \cdot C_{\text{block}}$ in the graph $\mathcal{G}'$ (where this cost is computed with a single path, without using clustering). This yield a set of $2^{33.78}$ unique differences. After optimizing the cost with clustering, most of the nodes have a cost at most $4.5 \cdot C_{\text{block}}$, and we use a subset of the graph by bounding the cost of the near-collision stage. We describe various trade-offs in Table 5: a larger set reduces the cost of the birthday stage, but increase the cost of the near-collision stage.

We note that the memory requirements of our attack are rather limited: one just has to store the graph, and the chains for the birthday phase. With the parameters we propose, this represents less than 1 TB.

Table 4. Bundles of trails with (near) optimal cost and the corresponding probability for steps 61–79. For each bundle $\mathcal{B}_i$ in the table, there are 32 related bundles where we flip some of the messages bits, that can be constructed as:
$B_0 = \{\mathcal{B}_i\}$
$B_1 = \Big\{\{\beta + (2^5, 0, 0, 0, 0) \mid \beta \in \mathcal{B}\} \;\Big|\; \mathcal{B} \in B_0\Big\} \cup B_0$
$B_2 = \Big\{\{\beta + (2^3, 0, 0, 0, 0) \mid \beta \in \mathcal{B}\} \;\Big|\; \mathcal{B} \in B_1\Big\} \cup B_1$
$B_3 = \Big\{\{\beta + (2^{13}, 2^8, 0, 0, 0) \mid \beta \in \mathcal{B}\} \;\Big|\; \mathcal{B} \in B_2\Big\} \cup B_2$
$B_4 = \Big\{\{\beta + (2^9, 2^4, 0, 0, 0) \mid \beta \in \mathcal{B}\} \;\Big|\; \mathcal{B} \in B_3\Big\} \cup B_3$
$B_5 = \Big\{\{\beta + (2^6, 2, 0, 0, 0) \mid \beta \in \mathcal{B}\} \;\Big|\; \mathcal{B} \in B_4\Big\} \cup B_4$.
The set used in [25] corresponds to bundles $\mathcal{B}_1$ to $\mathcal{B}_4$, with extension rules B_1 to B_4. Note that most output differences appears in several bundles.

Bundle	Output difference					Proba (− log)
$\mathcal{B}_1$	0xffffedea	0xffffff70	0x00000000	0x00000002	0x80000000	19.17
	0xffffedee	0xffffff70	0x00000000	0x00000002	0x80000000	19.17
	0xffffefea	0xffffff80	0x00000000	0x00000002	0x80000000	19.17
	0xffffefee	0xffffff80	0x00000000	0x00000002	0x80000000	19.17
	0xffffe5ec	0xffffff30	0x80000000	0x00000002	0x80000000	19.36
	0xffffe7ec	0xffffff40	0x80000000	0x00000002	0x80000000	19.36
$\mathcal{B}_2$	0xffffedea	0xffffff70	0x00000000	0xfffffffe	0x80000000	19.17
	0xffffedee	0xffffff70	0x00000000	0xfffffffe	0x80000000	19.17
	0xffffefea	0xffffff80	0x00000000	0xfffffffe	0x80000000	19.17
	0xffffefee	0xffffff80	0x00000000	0xfffffffe	0x80000000	19.17
	0xffffe5ec	0xffffff30	0x80000000	0xfffffffe	0x80000000	19.36
	0xffffe7ec	0xffffff40	0x80000000	0xfffffffe	0x80000000	19.36
$\mathcal{B}_3$	0xffffedea	0xffffff70	0x00000000	0x00000002	0x80000000	19.17
	0xffffedee	0xffffff70	0x00000000	0x00000002	0x80000000	19.17
	0xffffefea	0xffffff80	0x00000000	0x00000002	0x80000000	19.17
	0xffffefee	0xffffff80	0x00000000	0x00000002	0x80000000	19.17
	0xfffff5ec	0xffffffb0	0x80000000	0x00000002	0x80000000	19.36
	0xfffff7ec	0xffffffc0	0x80000000	0x00000002	0x80000000	19.36
$\mathcal{B}_4$	0xffffedea	0xffffff70	0x00000000	0xfffffffe	0x80000000	19.17
	0xffffedee	0xffffff70	0x00000000	0xfffffffe	0x80000000	19.17
	0xffffefea	0xffffff80	0x00000000	0xfffffffe	0x80000000	19.17
	0xffffefee	0xffffff80	0x00000000	0xfffffffe	0x80000000	19.17
	0xfffff5ec	0xffffffb0	0x80000000	0xfffffffe	0x80000000	19.36
	0xfffff7ec	0xffffffc0	0x80000000	0xfffffffe	0x80000000	19.36
$\mathcal{B}_5$	0xffffedaa	0xffffff6e	0x00000000	0x00000002	0x80000000	19.17
	0xffffedae	0xffffff6e	0x00000000	0x00000002	0x80000000	19.17
	0xffffefaa	0xffffff7e	0x00000000	0x00000002	0x80000000	19.17
	0xffffefae	0xffffff7e	0x00000000	0x00000002	0x80000000	19.17
	0xffffe5ac	0xffffff2e	0x80000000	0x00000002	0x80000000	19.36
	0xffffe7ac	0xffffff3e	0x80000000	0x00000002	0x80000000	19.36
$\mathcal{B}_6$	0xffffedaa	0xffffff6e	0x00000000	0xfffffffe	0x80000000	19.17
	0xffffedae	0xffffff6e	0x00000000	0xfffffffe	0x80000000	19.17
	0xffffefaa	0xffffff7e	0x00000000	0xfffffffe	0x80000000	19.17
	0xffffefae	0xffffff7e	0x00000000	0xfffffffe	0x80000000	19.17
	0xffffe5ac	0xffffff2e	0x80000000	0xfffffffe	0x80000000	19.36
	0xffffe7ac	0xffffff3e	0x80000000	0xfffffffe	0x80000000	19.36
$\mathcal{B}_7$	0xffffedaa	0xffffff6e	0x00000000	0x00000002	0x80000000	19.17
	0xffffedae	0xffffff6e	0x00000000	0x00000002	0x80000000	19.17
	0xffffefaa	0xffffff7e	0x00000000	0x00000002	0x80000000	19.17
	0xffffefae	0xffffff7e	0x00000000	0x00000002	0x80000000	19.17
	0xfffff5ac	0xffffffae	0x80000000	0x00000002	0x80000000	19.36
	0xfffff7ac	0xffffffbe	0x80000000	0x00000002	0x80000000	19.36
$\mathcal{B}_8$	0xffffedaa	0xffffff6e	0x00000000	0xfffffffe	0x80000000	19.17
	0xffffedae	0xffffff6e	0x00000000	0xfffffffe	0x80000000	19.17
	0xffffefaa	0xffffff7e	0x00000000	0xfffffffe	0x80000000	19.17
	0xffffefae	0xffffff7e	0x00000000	0xfffffffe	0x80000000	19.17
	0xfffff5ac	0xffffffae	0x80000000	0xfffffffe	0x80000000	19.36
	0xfffff7ac	0xffffffbe	0x80000000	0xfffffffe	0x80000000	19.36

6.1 Limiting the Number of Near-Collision Blocks

The attack above is optimized to minimize the time complexity of the attack, but this can result in long paths in the graph. For instance, when starting from a random difference with cost at most $3.0 \cdot C_{\text{block}}$, a random path has on average 15.7 near-collision blocks, but the maximal length is 26 near-collision blocks. This might be impractical for some applications of chosen-prefix collision attacks, and the work needed to generate all the differential trails for the near-collision blocks might also be an issue.

Therefore, we propose an alternative attack where we limit the length of the paths in the graph $\mathcal{G}$. This result in a slightly higher complexity, but might be better in practice. More precisely, we first construct a graph with only paths of length 1, and we iteratively build graphs by increasing the length of allowed paths. Note that a given difference can often be reached by many paths of varying length, and the cost of a node decreases when allowing longer paths.

We have constructed exactly the graph with all paths of length at most 4, and all paths of length at most 8 and cost at most $3.5 \cdot C_{\text{block}}$; for larger parameters, we cannot build the full graph, but we can build an approximation by limiting the set of values as in the previous construction. We give the size of the corresponding sub-graphs in Table 6. As we can see, with 8 near-collision blocks we already have a set $\mathcal{S}$ almost as large as the set corresponding to the previous attack (cf. Table 5), so that limiting the attack to 8 blocks has a small impact on the complexity. We can even find chosen-prefix collisions with just 4 near-collision blocks with a small cost increase, using a larger threshold on the maximum cost per block. We evaluate the complexity of such attacks in detail in Table 7.

We can also study the sparseness of the values in $\mathcal{S}$ to better understand the difficulty of building the differential trails for the near collision blocks. Using the set of size $2^{29.71}$ with a limit of 8 near-collision blocks and a maximum cost of $3.0 \cdot C_{\text{block}}$, the maximum weight in the differences is 26, and the average is 15.4 (using the non-adjacent form—NAF).

6.2 Birthday Stage

For the birthday stage of the attack, we follow the approach given in [25]: we consider a truncation of the SHA-1 state by keeping bits which are likely to contain a difference, and we use the distinguished points technique of [30]. Parameters for the birthday step with various choice of $\mathcal{G}$ are given in Table 5; we now explain in detail the case where the maximum cost of the near-collision stage is set to $3.0 \cdot C_{\text{block}}$. First, we truncate the state to 98 bits[6] so that for a random pair of values with their difference in $\mathcal{S}$, there is a probability 0.78 that the values collide on 98 bits (this probability has been computed with the tools from [14]). Reciprocally, if two truncated SHA-1 outputs are equal, then their difference is

[6] Given by mask 0x7f800000, 0xfffc0001, 0x7ffff800, 0x7fffff80, 0x7fffffff.

in the set $\mathcal{S}$ with probability $2^{-31.97}$. Therefore, the birthday stage will require on average $2 \cdot 2^{31.97}$ collisions in the following function:

$$f(r) := \begin{cases} \tau(h(cv\,, \mathrm{pad}(u))) & \text{if } u[0] = 1; \\ \tau(h(cv', \mathrm{pad}(u))) & \text{else.} \end{cases}$$

In order to keep the cost of rerunning the trail low, we use chains of average length 2^{31} (*i.e.* a point u is distinguished when $u < 2^{98-31}$). Therefore, the expected complexity of the birthday stage is[7]:

$$T = \sqrt{\pi/2 \cdot 2^{98} \cdot 2^{32.97}} \approx 2^{65.81} \text{ SHA-1 computations}$$
$$M = 2^{65.81}/2^{31} \cdot 19\,\text{bytes} \approx 570\,\text{GB},$$

and the cost to re-run the chains to locate collisions is only $2^{32.97} \cdot 2 \cdot 2^{31} \approx 2^{64.97}$. Finally, we can evaluate the complexity of the full attacks as: $2^{65.81} + 2^{64.97} + 3.0 \cdot C_{\text{block}}$.

Table 5. Trade-offs between the cost of birthday phase and the near-collision phase.

Set $\mathcal{S}$			Birthday parameters				
Max cost	Size	Mask	Proba	# coll.	Chain len.	# chain	Attack cost
$2.0 \cdot C_{\text{block}}$	$2^{24.66}$	106 bits	0.71	$2^{30.83}$	2^{34}	$2^{34.74}$	$2^{68.74} + 2^{65.83} + 2.0 \cdot C_{\text{block}}$
$2.5 \cdot C_{\text{block}}$	$2^{28.59}$	102 bits	0.65	$2^{31.03}$	2^{32}	$2^{34.84}$	$2^{66.84} + 2^{64.03} + 2.5 \cdot C_{\text{block}}$
$3.0 \cdot C_{\text{block}}$	$2^{30.95}$	98 bits	0.76	$2^{32.44}$	2^{31}	$2^{34.55}$	$2^{65.55} + 2^{64.44} + 3.0 \cdot C_{\text{block}}$
$3.5 \cdot C_{\text{block}}$	$2^{32.70}$	98 bits	0.76	$2^{30.70}$	2^{30}	$2^{34.68}$	$2^{64.68} + 2^{61.70} + 3.5 \cdot C_{\text{block}}$
$4.0 \cdot C_{\text{block}}$	$2^{33.48}$	98 bits	0.74	$2^{29.95}$	2^{30}	$2^{34.30}$	$2^{64.30} + 2^{60.95} + 4.0 \cdot C_{\text{block}}$
$4.5 \cdot C_{\text{block}}$	$2^{33.66}$	98 bits	0.74	$2^{29.77}$	2^{30}	$2^{34.21}$	$2^{64.21} + 2^{60.77} + 4.5 \cdot C_{\text{block}}$

6.3 Near-Collision Stage

An important parameter to evaluate the cost of the attack is C_{block}, the complexity to find near-collision blocks. An optimistic hypothesis is that we can find them with same complexity as in the attack of [26], *i.e.* $C_{\text{block}} = 2^{64.7}$. As mentioned earlier, we have conducted tests to verify that one can easily find short non-linear differential paths, regardless of the input chaining difference and value, to allow for a good use of neutral bits (one path example is given in Table 8).

We note that our trails are somewhat more constrained than the trails used in the collision attack, because we have denser chaining value differences and we have a few more conditions in the last round, as seen in Table 3. This could lead to fewer degrees of freedom than in the collision attack of Stevens *et al.*, and increase the cost of finding a conforming block. In particular, this can affect the

[7] To store a chain, we use 40 bits for the starting point, 40 bits for the length, and $98 - 31 = 67$ bits for the output, *i.e.* 19 bytes in total.

Table 6. Size of the set $\mathcal{S}$ with various limits on the maximum cost and on the number of near-collision blocks. We give a lower bound when we couldn't compute the full set.

Max Cost	1 bl.	2 bl.	3 bl.	4 bl.	5 bl.	6 bl.	7 bl.	8 bl.
$2.0 \cdot C_{\text{block}}$	$2^{9.17}$	$2^{16.30}$	$2^{19.92}$	$2^{22.05}$	$2^{23.13}$	$2^{23.95}$	$2^{24.44}$	$2^{24.55}$
$2.5 \cdot C_{\text{block}}$	$2^{10.17}$	$2^{16.62}$	$2^{21.04}$	$2^{23.76}$	$2^{25.50}$	$2^{26.58}$	$2^{27.38}$	$2^{27.92}$
$3.0 \cdot C_{\text{block}}$	$2^{10.17}$	$2^{17.10}$	$2^{21.76}$	$2^{24.66}$	$2^{26.58}$	$2^{27.95}$	$2^{28.96}$	$2^{29.71}$
$3.5 \cdot C_{\text{block}}$	$2^{12.53}$	$2^{17.89}$	$2^{22.47}$	$2^{25.62}$	$2^{27.70}$	$2^{29.18}$	$2^{30.29}$	$2^{31.22}$
$4.0 \cdot C_{\text{block}}$	$2^{12.53}$	$2^{18.60}$	$2^{22.97}$	$2^{26.34}$	$\geq 2^{28.68}$	$\geq 2^{30.35}$	$\geq 2^{31.55}$	$\geq 2^{32.15}$
$5.0 \cdot C_{\text{block}}$	$2^{12.53}$	$2^{19.65}$	$2^{24.18}$	$2^{27.44}$	$\geq 2^{29.83}$	$\geq 2^{31.64}$	$\geq 2^{32.95}$	$\geq 2^{33.04}$
$6.0 \cdot C_{\text{block}}$	$2^{12.53}$	$2^{19.79}$	$2^{24.81}$	$2^{28.26}$	$\geq 2^{30.74}$	$\geq 2^{32.55}$	$\geq 2^{33.59}$	$\geq 2^{33.59}$
$7.0 \cdot C_{\text{block}}$	$2^{13.09}$	$2^{20.37}$	$2^{25.30}$	$2^{28.82}$	$\geq 2^{31.33}$	$\geq 2^{32.93}$	$\geq 2^{33.77}$	$\geq 2^{33.77}$
$8.0 \cdot C_{\text{block}}$	$2^{13.09}$	$2^{20.62}$	$2^{25.72}$	$2^{29.27}$	$\geq 2^{31.72}$	$\geq 2^{33.09}$	$\geq 2^{33.81}$	$\geq 2^{33.81}$

Table 7. Trade-offs between the cost of birthday phase and the near-collision phase with a limited number of near-collision blocks (4 or 8).

Set $\mathcal{S}$			Birthday parameters					
Max bl.	Max cost	Size	Mask	Proba	# coll.	Chain len.	# chain	Attack cost
4	$4.0 \cdot C_{\text{block}}$	$2^{26.34}$	106 bits	0.48	$2^{29.70}$	2^{33}	$2^{35.18}$	$2^{68.18} + 2^{63.70} + 4.0 \cdot C_{\text{block}}$
4	$5.0 \cdot C_{\text{block}}$	$2^{27.44}$	102 bits	0.67	$2^{32.14}$	2^{32}	$2^{35.40}$	$2^{67.40} + 2^{65.14} + 5.0 \cdot C_{\text{block}}$
4	$6.0 \cdot C_{\text{block}}$	$2^{28.26}$	102 bits	0.65	$2^{31.35}$	2^{32}	$2^{35.00}$	$2^{67.00} + 2^{64.35} + 6.0 \cdot C_{\text{block}}$
4	$7.0 \cdot C_{\text{block}}$	$2^{28.82}$	102 bits	0.64	$2^{30.82}$	2^{32}	$2^{34.74}$	$2^{66.74} + 2^{63.82} + 7.0 \cdot C_{\text{block}}$
4	$8.0 \cdot C_{\text{block}}$	$2^{29.26}$	102 bits	0.63	$2^{30.39}$	2^{32}	$2^{34.52}$	$2^{66.52} + 2^{63.39} + 8.0 \cdot C_{\text{block}}$
8	$2.0 \cdot C_{\text{block}}$	$2^{24.55}$	106 bits	0.71	$2^{30.94}$	2^{34}	$2^{34.80}$	$2^{68.80} + 2^{65.94} + 2.0 \cdot C_{\text{block}}$
8	$2.5 \cdot C_{\text{block}}$	$2^{27.92}$	102 bits	0.63	$2^{31.75}$	2^{32}	$2^{35.20}$	$2^{67.20} + 2^{64.75} + 2.5 \cdot C_{\text{block}}$
8	$3.0 \cdot C_{\text{block}}$	$2^{29.71}$	98 bits	0.73	$2^{33.73}$	2^{31}	$2^{35.19}$	$2^{66.19} + 2^{65.73} + 3.0 \cdot C_{\text{block}}$
8	$3.5 \cdot C_{\text{block}}$	$2^{31.22}$	98 bits	0.72	$2^{32.23}$	2^{30}	$2^{35.44}$	$2^{65.44} + 2^{63.23} + 3.5 \cdot C_{\text{block}}$

use of accelerating techniques such as neutral bits and boomerangs; boomerangs are the most powerful technique, but they require significant degrees of freedom in the path construction. Therefore, we also consider a conservative complexity estimate, where we assume that boomerangs are no longer available. Since there are three boomerangs in the trail of [26], this would give $C_{\text{block}} = 2^{67.7}$.

Our experiments show that that those assumptions are reasonable. The path given in Table 8 is about as constrained as the path used for the second block of the collision attack [26] in the first round. In particular, most condition are in the first 6 steps, and don't affect the use of neutral bits, and the same three boomerang are available. In general we expect similar results with a few boomerangs, but this might of course vary depending on the exact chaining input difference/value.

Finally, with the optimistic hypothesis, the best trade-off is to use a limit of $3.5 \cdot C_{\text{block}}$, for a total complexity of

$$2^{64.68} + 2^{61.70} + 3.5 \cdot C_{\text{block}} \approx 2^{66.9} \qquad (\text{using } C_{\text{block}} = 2^{64.7})$$

Table 8. Example of a SHA-1 non-linear differential path generated for one of the differences in $\mathcal{S}$. Notations follow [8]. $\delta = [-2^{17} - 2^{15} + 2^{10} - 2^{8} + 2^{5} + 2^{6} - 2^{2} + 2^{0}, -2^{13} + 2^{11} + 2^{10} + 2^{5} - 2^{3}, -2^{5} + 2^{0}, -2^{4} - 2^{0}, 0]$, with cost $2.954 \cdot C_{\text{block}}$.

i	A_i	W_i
−4:	10110011001011000101111011010101	
−3:	110110001100001000100111un01un01	
−2:	001010100011010111011unnn1011n11	
−1:	00101010111000011un1nn0010n0u111	
00:	1010111000111un1u0110n1u0nn0un1n	1011un0101000101010111110-10-u0
01:	10100u101u0u10101n1nu0111n-u1-1u	nu1110011101100011110-00-00n0110
02:	1u1u0nn010nuunnu11011uuuu0001uu1	u1nn0u000100010001010111---unn00
03:	u1un01uunn1u1010u0u101101nu11uu1	00uuun1111010111100111010000-u1-
04:	n0110unnnnnnnnnnnnn11nu1000u1n1	n0nunu00----001----0---1000uu0u1
05:	un0n011100--11001--111-1u1uu1u11	10u-1--101110-000-1100-0110n-00-
06:	1101-0-101101011110101-10nun01uu	--u--u1----------------0101uun--
07:	0nuu-00----------------0100100uu	xun-nu------------1-1----11u0u--
08:	---u01----------------0--0n010-u	----un------------------------u0
09:	0n--------------------0--1-1--0u	xn----------------------0--n-0--
10:	1--1-1------------------0---1---	x-nx-x--------------1------uxx--
11:	-1n-----------------------0-----	--u0nn-------------------1-1-u--
12:	----0---------------------1-----	n-nxxu---------------------un---
13:	n---1---------------------------	x-uu-0---------------------u----
14:	--n-----------------------------	--------------------------1-un--
15:	u-1-1---------------------------	x-nxn----------------------n----
16:	un0-0---------------------------	----u----------------------nu---

With the conservative hypothesis, the best trade-off is to set the limit at $2.5 \cdot C_{\text{block}}$, for a total complexity of

$$2^{66.84} + 2^{64.03} + 2.5 \cdot C_{\text{block}} \approx 2^{69.35} \qquad (\text{using } C_{\text{block}} = 2^{67.7})$$

There are other trade-offs possible between the various parameters of attack. For instance, we discussed attacks with a limited number of near-collision blocks in Sect. 6.1; we can now evaluate the complexity of the resulting attacks. If we limit the attack to 8 near-collision blocks, the best trade-offs give the following complexities for the optimistic and conservative hypothesis respectively:

$$2^{65.44} + 2^{63.23} + 3.5 \cdot C_{\text{block}} \approx 2^{67.2} \qquad (\text{using } C_{\text{block}} = 2^{64.7})$$
$$2^{67.20} + 2^{64.75} + 2.5 \cdot C_{\text{block}} \approx 2^{69.5} \qquad (\text{using } C_{\text{block}} = 2^{67.7})$$

Even with a limit of only 4 near-collision blocks, we have a relatively small increase of the complexity, with the following trade-offs:

$$2^{66.74} + 2^{63.82} + 7.0 \cdot C_{\text{block}} \approx 2^{68.3} \qquad (\text{using } C_{\text{block}} = 2^{64.7})$$
$$2^{68.18} + 2^{63.70} + 4.0 \cdot C_{\text{block}} \approx 2^{70.2} \qquad (\text{using } C_{\text{block}} = 2^{67.7})$$

7 Conclusion and Future Works

This work puts another nail in the SHA-1 coffin, with almost practical chosen-prefix collisions, between five and twenty-six times more expensive than the identical-prefix collisions recently demonstrated. This shows that continued usage of SHA-1 for certificates or for authentication of handshake messages in TLS, SSH or IKE is dangerous, and could already be abused today by a well-motivated adversary. SHA-1 has been broken since 2004, but it is still used in many security systems; we strongly advise users to remove SHA-1 support to avoid downgrade attacks.

More generally, our results show that, for some hash functions, chosen-prefix collision attacks are much easier than previously expected, and potentially not much harder than a normal collision search.

Our research opens several new directions. Obviously, future work will have to implement this attack to demonstrate a real chosen-prefix collision for SHA-1. While the computation cost of our attack is somewhat practical, SHA-1 attacks still require a huge computation power (thousands of GPUs in order to obtain the chosen-prefix collision in a reasonable time) and a large implementation effort. For a concrete demonstration, a good target would be to break a protocol such as TLS or IKE, or to build a rogue certificate authority.

Another research direction is to study how one can improve SHA-1 collision attacks, not only for minimising the cost of finding a simple collision, but to improve our chosen-prefix collision search complexity. In particular, our attack requires the ability to reach many distinct output differences for the compression function. In this paper, to simplify our analysis, we only considered the differential trail from [26] because a real collision was found with this trail, and a precise complexity evaluation was conducted. However, it should be possible to increase the pool of available differences, and further reduce the total complexity, by using other (slightly more costly) differential trails.

Finally, a last direction is to evaluate how our strategy actually applies to other hash functions, such as RIPEMD, (reduced-round) SHA-2, or even others. Again, this will require a deep knowledge of the functions studied, as many details might impact the overall complexity. We can however expect that our attack strategy will be applicable mostly on classical Davies-Meier constructions inside a single-pipe Merkle-Damgård operating mode.

In order to make our easier to verify, we are publishing some additional data and code online at: https://github.com/Cryptosaurus/sha1-cp.

Acknowledgments. The authors would like to thank the anonymous referees for their helpful comments. The second author is supported by Temasek Laboratories, Singapore.

References

1. Bhargavan, K., Leurent, G.: Transcript collision attacks: breaking authentication in TLS, IKE and SSH. In: NDSS 2016. The Internet Society, February 2016
2. Biham, E., Chen, R.: Near-collisions of SHA-0. In: Franklin, M. (ed.) CRYPTO 2004. LNCS, vol. 3152, pp. 290–305. Springer, Heidelberg (2004). https://doi.org/10.1007/978-3-540-28628-8_18
3. Biham, E., Chen, R., Joux, A., Carribault, P., Lemuet, C., Jalby, W.: Collisions of SHA-0 and reduced SHA-1. In: Cramer, R. (ed.) EUROCRYPT 2005. LNCS, vol. 3494, pp. 36–57. Springer, Heidelberg (2005). https://doi.org/10.1007/11426639_3
4. Brassard, G. (ed.): CRYPTO 1989. LNCS, vol. 435. Springer, New York (1990). https://doi.org/10.1007/0-387-34805-0
5. Chabaud, F., Joux, A.: Differential collisions in SHA-0. In: Krawczyk, H. (ed.) CRYPTO 1998. LNCS, vol. 1462, pp. 56–71. Springer, Heidelberg (1998). https://doi.org/10.1007/BFb0055720
6. Damgård, I.: A design principle for hash functions. In: [4], pp. 416–427
7. Damgård, I.B.: A design principle for hash functions. In: Brassard, G. (ed.) CRYPTO 1989. LNCS, vol. 435, pp. 416–427. Springer, New York (1990). https://doi.org/10.1007/0-387-34805-0_39
8. De Cannière, C., Mendel, F., Rechberger, C.: Collisions for 70-step SHA-1: on the full cost of collision search. In: Adams, C.M., Miri, A., Wiener, M.J. (eds.) SAC 2007. LNCS, vol. 4876, pp. 56–73. Springer, Heidelberg (2007). https://doi.org/10.1007/978-3-540-77360-3_4
9. den Boer, B., Bosselaers, A.: Collisions for the compression function of MD5. In: Helleseth, T. (ed.) EUROCRYPT 1993. LNCS, vol. 765, pp. 293–304. Springer, Heidelberg (1994). https://doi.org/10.1007/3-540-48285-7_26
10. Joux, A., Peyrin, T.: Hash functions and the (amplified) boomerang attack. In: Menezes, A. (ed.) CRYPTO 2007. LNCS, vol. 4622, pp. 244–263. Springer, Heidelberg (2007). https://doi.org/10.1007/978-3-540-74143-5_14
11. Karpman, P., Peyrin, T., Stevens, M.: Practical free-start collision attacks on 76-step SHA-1. In: Gennaro, R., Robshaw, M.J.B. (eds.) CRYPTO 2015. LNCS, vol. 9215, pp. 623–642. Springer, Heidelberg (2015). https://doi.org/10.1007/978-3-662-47989-6_30
12. Klima, V.: Tunnels in hash functions: MD5 collisions within a minute. Cryptology ePrint Archive, Report 2006/105 (2006). http://eprint.iacr.org/2006/105
13. Lamberger, M., Pramstaller, N., Rechberger, C., Rijmen, V.: Second preimages for SMASH. In: Abe, M. (ed.) CT-RSA 2007. LNCS, vol. 4377, pp. 101–111. Springer, Heidelberg (2006). https://doi.org/10.1007/11967668_7
14. Leurent, G.: Analysis of differential attacks in ARX constructions. In: Wang, X., Sako, K. (eds.) ASIACRYPT 2012. LNCS, vol. 7658, pp. 226–243. Springer, Heidelberg (2012). https://doi.org/10.1007/978-3-642-34961-4_15
15. Mendel, F., Rijmen, V., Schläffer, M.: Collision attack on 5 rounds of Grøstl. In: Cid, C., Rechberger, C. (eds.) FSE 2014. LNCS, vol. 8540, pp. 509–521. Springer, Heidelberg (2015). https://doi.org/10.1007/978-3-662-46706-0_26
16. Merkle, R.C.: One way hash functions and DES. [4] 428–446
17. National Institute of Standards and Technology: FIPS 180: secure hash standard, May 1993
18. National Institute of Standards and Technology: FIPS 180–1: secure hash standard, April 1995

19. National Institute of Standards and Technology: FIPS 180–2: secure hash standard, August 2002
20. National Institute of Standards and Technology: FIPS 202: SHA-3 standard: permutation-based hash and extendable-output functions, August 2015
21. Peyrin, T.: Cryptanalysis of Grindahl. In: Kurosawa, K. (ed.) ASIACRYPT 2007. LNCS, vol. 4833, pp. 551–567. Springer, Heidelberg (2007). https://doi.org/10.1007/978-3-540-76900-2_34
22. Rivest, R.L.: The MD4 message digest algorithm. In: Menezes, A.J., Vanstone, S.A. (eds.) CRYPTO 1990. LNCS, vol. 537, pp. 303–311. Springer, Heidelberg (1991). https://doi.org/10.1007/3-540-38424-3_22
23. Rivest, R.L.: RFC 1321: the MD5 message-digest algorithm. Internet Activities Board, April 1992
24. Stevens, M.: Attacks on hash functions and applications. Ph.D. thesis, Leiden University, June 2012
25. Stevens, M.: New collision attacks on SHA-1 based on optimal joint local-collision analysis. In: Johansson, T., Nguyen, P.Q. (eds.) EUROCRYPT 2013. LNCS, vol. 7881, pp. 245–261. Springer, Heidelberg (2013). https://doi.org/10.1007/978-3-642-38348-9_15
26. Stevens, M., Bursztein, E., Karpman, P., Albertini, A., Markov, Y.: The first collision for full SHA-1. In: Katz, J., Shacham, H. (eds.) CRYPTO 2017. LNCS, vol. 10401, pp. 570–596. Springer, Cham (2017). https://doi.org/10.1007/978-3-319-63688-7_19
27. Stevens, M., Karpman, P., Peyrin, T.: Freestart collision for full SHA-1. In: Fischlin, M., Coron, J.S. (eds.) EUROCRYPT 2016. LNCS, vol. 9665, pp. 459–483. Springer, Heidelberg (2016). https://doi.org/10.1007/978-3-662-49890-3_18
28. Stevens, M., Lenstra, A.K., de Weger, B.: Chosen-prefix collisions for MD5 and colliding X.509 certificates for different identities. In: Naor, M. (ed.) EUROCRYPT 2007. LNCS, vol. 4515, pp. 1–22. Springer, Heidelberg (2007). https://doi.org/10.1007/978-3-540-72540-4_1
29. Stevens, M., Sotirov, A., Appelbaum, J., Lenstra, A., Molnar, D., Osvik, D.A., de Weger, B.: Short chosen-prefix collisions for MD5 and the creation of a rogue CA certificate. In: Halevi, S. (ed.) CRYPTO 2009. LNCS, vol. 5677, pp. 55–69. Springer, Heidelberg (2009). https://doi.org/10.1007/978-3-642-03356-8_4
30. van Oorschot, P.C., Wiener, M.J.: Parallel collision search with cryptanalytic applications. J. Cryptol. **12**(1), 1–28 (1999)
31. Wang, X., Lai, X., Feng, D., Chen, H., Yu, X.: Cryptanalysis of the hash functions MD4 and RIPEMD. In: Cramer, R. (ed.) EUROCRYPT 2005. LNCS, vol. 3494, pp. 1–18. Springer, Heidelberg (2005). https://doi.org/10.1007/11426639_1
32. Wang, X., Yin, Y.L., Yu, H.: Finding collisions in the full SHA-1. In: Shoup, V. (ed.) CRYPTO 2005. LNCS, vol. 3621, pp. 17–36. Springer, Heidelberg (2005). https://doi.org/10.1007/11535218_2
33. Wang, X., Yu, H.: How to break MD5 and other hash functions. In: Cramer, R. (ed.) EUROCRYPT 2005. LNCS, vol. 3494, pp. 19–35. Springer, Heidelberg (2005). https://doi.org/10.1007/11426639_2
34. Wang, X., Yu, H., Yin, Y.L.: Efficient collision search attacks on SHA-0. In: Shoup, V. (ed.) CRYPTO 2005. LNCS, vol. 3621, pp. 1–16. Springer, Heidelberg (2005). https://doi.org/10.1007/11535218_1

Preimage Attacks on Round-Reduced KECCAK-224/256 via an Allocating Approach

Ting Li[1,2] and Yao Sun[1,2](✉)

[1] State Key Laboratory of Information Security,
Institute of Information Engineering, Chinese Academy of Sciences, Beijing, China
[2] School of Cyber Security, University of Chinese Academy of Sciences,
Beijing, China
sunyao@iie.ac.cn

Abstract. We present new preimage attacks on standard KECCAK-224 and KECCAK-256 that are reduced to 3 and 4 rounds. An allocating approach is used in the attacks, and the whole complexity is allocated to two stages, such that fewer constraints are considered and the complexity is lowered in each stage. Specifically, we are trying to find a 2-block preimage, instead of a 1-block one, for a given hash value, and the first and second message blocks are found in two stages, respectively. Both the message blocks are constrained by a set of newly proposed conditions on the middle state, which are weaker than those brought by the initial values and the hash values. Thus, the complexities in the two stages are both lower than that of finding a 1-block preimage directly. Together with the basic allocating approach, an improved method is given to balance the complexities of two stages, and hence, obtains the optimal attacks. As a result, we present the best theoretical preimage attacks on KECCAK-224 and KECCAK-256 that are reduced to 3 and 4 rounds. Moreover, we practically found a (second) preimage for 3-round KECCAK-224 with a complexity of $2^{39.39}$.

Keywords: Cryptanalysis · KECCAK · SHA-3 · Preimage attack

1 Introduction

The KECCAK sponge function family [3], which was designed by Bertoni et al., became a candidate for the SHA-3 competition in 2008 [20]. It won this competition in 2012, and the U.S. National Institute of Standards and Technology (NIST) standardized KECCAK as Secure Hash Algorithm-3 (SHA-3) in 2015 [26]. KECCAK has received numerous security analysis since it was publicly available in 2008.

This work was supported by National Natural Science Foundation of China under Grant No. 61877058, and Strategy Cooperation Project AQ-1701.

Y. Ishai and V. Rijmen (Eds.): EUROCRYPT 2019, LNCS 11478, pp. 556–584, 2019.
https://doi.org/10.1007/978-3-030-17659-4_19

On practical collision attacks, Dinur et al. presented the first actual collision attack on 3-round KECCAK-384 based on a generalized internal differential attack [8]. Besides, they obtained practical complexities up to 4 out of 24 rounds of KECCAK-224/256 [7,9]. Then, Qiao et al. extended Dinur et al.'s framework and achieved the first practical collision attack against 5-round SHAKE128 [22]. By improving Qiao et al.'s method, Song et al. proposed a practical collision attack on KECCAK-224 reduced to 5 rounds in 2017 [23]. Most of these collision attacks depend on the differential trails. Daemen et al. analyzed the differential propagation of KECCAK in 2012 [6]. After that, Kölbl et al. went a step further to study the differential properties of KECCAK-f[800] and KECCAK-f[1600], and presented collision attacks with practical complexity on KECCAK when the permutation is reduced to 4 rounds [14].

For preimage attacks, Naya-Plasencia et al. [19] and Morawiecki et al. [18] presented practical attacks up to 2 rounds. Guo et al. developed the technique of *linear structures*, and presented a practical attack on 3-round SHAKE128 [12]. Besides, the analysis of theoretical preimage attacking results on 3-round and 4-round instances of KECCAK are also given in their paper. Li et al. constructed a new kind of structures, called *cross-linear structures*, and improved the theoretical preimage complexities on 3-round KECCAK-256/SHA3-256/SHAKE256 [15]. Theoretical preimage attacks up to 7/8/9 rounds on KECCAK-224/256/512 are considered in [2,5,17]. In addition, Aumasson and Meier presented a new type of distinguisher and applied it to reduced versions of the KECCAK-f permutation in 2009 [1].

The KECCAK permutation is also used for authenticated encryption, and many researches have been made in this field. In 2014, Dinur et al. gave the first cube attacks on round-reduced KECCAK sponge function and applied it to attack MAC and stream cipher mode [11]. Then they analyzed the problems of key recovery, MAC forgery and other types of attacks on the keyed mode of KECCAK as well as the security margin of Keyak—a KECCAK-based authenticated encryption scheme [10]. After that, Huang et al. developed the conditional cube tester to analyze KECCAK in keyed modes and improved the previous distinguishing attacks in 2017 [13]. Since then, the keyed modes of KECCAK have attracted more intensive cryptanalysis [4,16,24].

In this paper, we present an allocating approach to make preimage attacks on KECCAK-224 and KECCAK-256 that are reduced to 3 and 4 rounds. Generally, to find a 1-block preimage for a given hash value, a system is constructed by this hash value and hash algorithm. This system contains two kinds of constraints. The first kind comes from the initial value. For example, the last $224 \times 2 = 448$ bits in the initial state of KECCAK-224 must be 0's and these bits will not XOR messages. Constraints of the second kind are produced by the hash value. That is, the first l output bits of the hash algorithm must equal the given l-bit hash value. The unknowns of the system are usually the bit values of the messages. Constrained by these two kinds of constraints, the system is often with high nonlinearity and hard to be solved. The motivation of the allocating approach is to allocate these two kinds of constraints to *two* different sets of unknowns. Specifically, we prefer to find a 2-block preimage instead of a 1-block one. The unknowns from the first message block are constrained by the first kind of constraints, while those from the second message block should make the second kind of constraints hold. Thus, the whole attacking complexity is allocated to two stages, and we expect the complexity of each stage to be lower than that of finding a 1-block preimage. Our motivation is shown in Fig. 1.

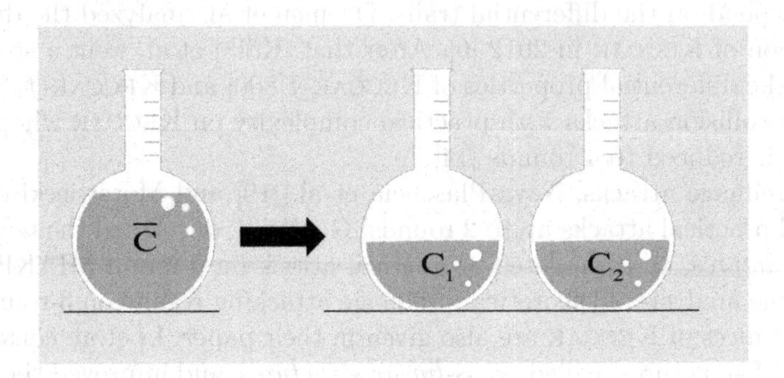

Fig. 1. The motivation of the allocating approach. The volume of water in a flask stands for the complexity of solving a system. Let $\bar{C}$ be the complexity of finding a 1-block preimage, C_1 and C_2 be the complexities of the two stages used to find a 2-block preimage by the allocating approach. We expect $\bar{C} > C_1 + C_2$.

The key step of applying this allocating approach is to find *suitable* constraints on the *middle state*, i.e. the output state of the first block as well as the initial state of the second block. Since more constraints usually cost more operations for solving the systems, to make the complexities of the two stages both lower than that of finding a 1-block preimage, the constraints on the middle state must be weaker than both kinds of constraints mentioned in the last paragraph. We improve the structure proposed by Li et al. [15], and obtain a set of *suitable* constraints on the middle state. For example, the number of constraints on the middle state of KECCAK-224 is 129, which is smaller than 448 (the number of constraints from the initial value) and 224 (the number of constraints from the hash value). For KECCAK-256, the number of constraints on the middle state is 193, which is also good enough to improve the preimage attacks on KECCAK-256.

The contributions of this paper are summarized in three aspects.

1. We present an *allocating approach* for preimage attacks on round-reduced KECCAK. This approach allocates the whole attack complexity to two stages, called *Precomputation Stage* and *Online Stage* for convenience. The complexity of each stage is lower than that of finding a 1-block preimage directly. To the best of our knowledge, this is the first two-block attack on standard KECCAK, although multi-block methods have been successfully applied to MD5 [27] and SHA-1 [25].

Table 1. Summary of preimage attacks on 3/4-round KECCAK-224/256

Rounds	Digest length	Instances	Complexity	Reference
3	224	KECCAK-224	2^{97}	[12]
			$\mathbf{2^{38}}$	Sect. 4.2
		SHA3-224	$\mathbf{2^{41}}$	Sect. 4.2
	256	KECCAK-256	2^{192}	[12]
			2^{150}	[15]
			$\mathbf{2^{81}}$	Sect. 4.2
		SHA3-256	2^{151}	[15]
			$\mathbf{2^{84}}$	Sect. 4.2
		SHAKE256	2^{153}	[15]
			$\mathbf{2^{86}}$	Sect. 4.2
4	224	KECCAK-224	2^{213}	[12]
			$\mathbf{2^{207}}$	Sect. 4.3
		SHA3-224	$\mathbf{2^{207}}$	Sect. 4.3
	256	KECCAK-256	2^{251}	[12]
			$\mathbf{2^{239}}$	Sect. 4.3
		SHA3-256	$\mathbf{2^{239}}$	Sect. 4.3
		SHAKE256	$\mathbf{2^{239}}$	Sect. 4.3

2. We propose a new set of constraints on the middle state by improving Li et al.'s structure [15]. The improved structure could linearize the generated system at a low cost, such that we obtain more degrees of freedom to solve for the second message block.
3. We improve theoretical complexities of preimage attacks on 3/4-round KECCAK-224 and KECCAK-256, as well as SHA3-224/256 and SHAKE256. Particularly, we give the first practical preimage attack on 3-round KECCAK-224 with about $2^{39.39}$ operations. The theoretical results of preimage attacks in this paper, as well as the previous best ones, are summarized in Table 1. Detailed theoretical complexities of the two stages of attacking 3/4-round KECCAK-224/256 are given in Table 2.

This paper is organized as follows. Some preliminaries and notations are given in Sect. 2. The allocating approach is proposed in Sect. 3. Theoretical analyses are presented in Sect. 4, and a practical preimage attack on 3-round KECCAK-224 comes in Sect. 5. At last, we conclude this paper in Sect. 6.

Table 2. Detailed theoretical complexities of the two stages. C_1 and C_2 represent the complexities of Precomputation Stage and Online Stage. Basic and improved allocating approaches are used in Sects. 4.1 and 4.2, respectively.

Rounds	Instances	C_1 (Pre. Stage)	C_2 (Onl. Stage)	Overall Complexity	Reference
3	KECCAK-224	2^{66}	2^{31}	2^{66}	Sect. 4.1
		$2^{35.62}$	2^{38}	2^{38}	Sect. 4.2
	KECCAK-256	2^{162}	2^{62}	2^{162}	Sect. 4.1
		$2^{80.06}$	2^{81}	2^{81}	Sect. 4.2
4	KECCAK-224	2^{129}	2^{207}	2^{207}	Sect. 4.3
	KECCAK-256	2^{193}	2^{239}	2^{239}	Sect. 4.3

2 Preliminaries

2.1 The Sponge Construction

The sponge construction is used in KECCAK algorithm. As shown in Fig. 2, it processes messages in two phases—absorbing phase and squeezing phase. With these two phases, a sponge construction receives an input stream of any length and produces an output bit stream of any desired length.

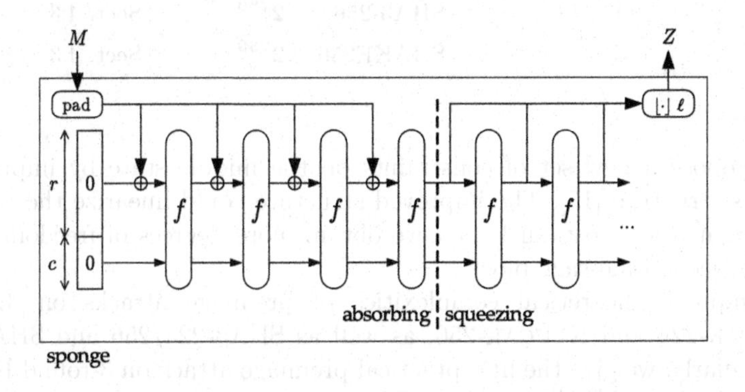

Fig. 2. The sponge construction.

At the beginning, the internal state of b-bits is initialized to be all 0's, which is the **initial value (IV)**. The message is padded and split into blocks of r-bits. In the absorbing phase, the first r bits of b-bits state are XORed with the message block, followed by the application of permutation f. This procedure is repeated until all the message blocks are processed. Then in the squeezing phase, the first l bits are output. With an additional application of f, another l output bits are obtained. The algorithm iterates this step until the required length of a digest is reached.

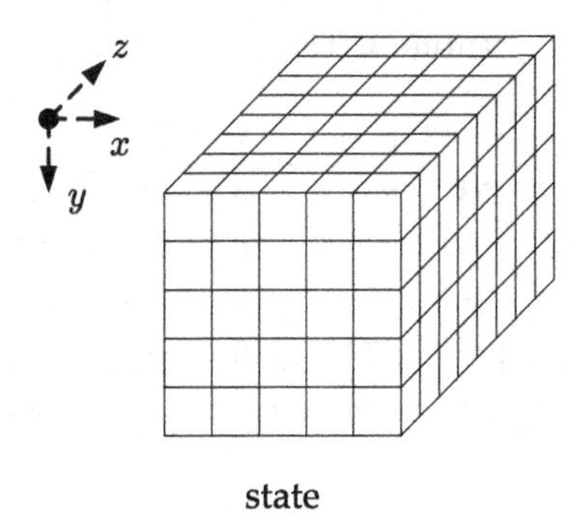

Fig. 3. The KECCAK state.

2.2 The KECCAK-f Permutations

According to the KECCAK reference [3], there are 7 KECCAK-f permutations, indicated by KECCAK-$f[b]$, where $b \in \{25, 50, 100, 200, 400, 800, 1600\}$. We call b the **width** of the permutation. In this paper, we only focus on the case $b = 1600$, since KECCAK-$f[1600]$ is used widely in practice, which can be described as a 5×5 **64-bits lanes** as depicted in Fig. 3. In this paper, we use L to denote the number of bits in a lane. In KECCAK-$f[1600]$, we have $L = 64$. Each bit is denoted as $A_{x,y,z}$. The integer triples (x, y, z) are the indices of bits, where x, y come from the set $\{0, 1, 2, 3, 4\}$ and $0 \leq z \leq L - 1$. The values of x and y are taken modulo 5 and we take z's values modulo L in the rest of this paper. The axis z is omitted sometimes for simplification.

The function KECCAK-$f[1600]$ consists of 24 rounds permutation R. Each round R consists of five steps:

$$R = \iota \circ \chi \circ \pi \circ \rho \circ \theta,$$

where

$$\theta : A_{x,y,z} = A_{x,y,z} \oplus \bigoplus_{j=0}^{4} (A_{x-1,j,z} \oplus A_{x+1,j,z-1}),$$

$$\rho : A_{x,y,z} = A_{x,y,(z+r_{x,y})},$$

$$\pi : A_{y,2x+3y,z} = A_{x,y,z},$$

$$\chi : A_{x,y,z} = A_{x,y,z} \oplus (A_{x+1,y,z} \oplus 1) \cdot A_{x+2,y,z},$$

$$\iota : A_{0,0,z} = A_{0,0,z} \oplus RC_z.$$

In the above definitions, bit-wise XOR is denoted by "$\oplus$" and bit-wise logic AND by "$\cdot$". Besides, "$r_{x,y}$" refers to a lane-dependent rotation constant which equals the corresponding value in Table 3 taken modulo the lane length L. And "RC_z" is a round-dependent constant. The details about RC are omitted since they do not affect our attacks. For further details about KECCAK, please refer to [3].

Table 3. The offsets of ρ.

	$x = 3$	$x = 4$	$x = 0$	$x = 1$	$x = 2$
$y = 2$	153	231	3	10	171
$y = 1$	55	276	36	300	6
$y = 0$	28	91	0	1	190
$y = 4$	120	78	210	66	253
$y = 3$	21	136	105	45	15

2.3 Instances of Keccak

The hash function Keccak $[r, c, l]$ means an instance of Keccak sponge function family with **capacity** c, **bitrate** r, and **output length** l. The official versions of Keccak-l have $r = 1600 - c$ and $c = 2 \cdot l$, where $l \in \{128, 224, 256, 384, 512\}$. Their padding rules are identical. The message is padded by appending a bit string of "10*1", where "0*" means the shortest string of 0's such that the padded message is of multiple of r bits.

The digests of the standard SHA-3 have lengths of 224, 256, 384, and 512 bits. SHA-3 is similar to Keccak except for the padding rule. SHA-3 pads the message with another two bits "01" before applying the Keccak padding rule, *i.e.*, the padded string becomes "0110*1".

The SHA-3 family also includes two SHAKE instances (SHAKE128 and SHAKE256), which are called extendable-output functions (XOF's). Specifically, SHAKE128(M, l) and SHAKE256(M, l) are defined as Keccak $[r = 1344, c = 256]$ and Keccak $[r = 1088, c = 512]$. And the message M is padded with a suffix "1111".

In this paper, our attacks on instances of Keccak, SHA-3 and SHAKE use the same parameters, and we focus on the instances with $l = 224/256$.

2.4 Notations

To find preimages for given hash values, we need to construct algebraic systems during the attacks. Some bits of the internal state will be set as **unknowns** and some are set as constants, where the unknowns are the bits that we need to solve. When some bits are set as unknowns in some state, bits in the consequent states can be represented as polynomials of these unknowns. For convenience, if a bit is represented as a linear polynomial of unknowns, we say **this bit is linear**; similarly, we say **a bit is quadratic** if it is represented as a quadratic polynomial. Please note that the polynomial representation of a bit is unique in our attacks. Similarly, each column of a specific state contains 5 bits. We say **a column is linear** if all bits in this column are linear bits or constants.

We also give names to some states for sake of convenience. As the message is split into several **message blocks**, there are many hash blocks in the absorbing phase. We call the starting state of each hash block **the initial state** of this block, while the ending state of each block is called **the output state** or **outputs** of this block. Note that bits in the initial state of the first block are all 0's, and the output state of the i-th block is just the initial state of the $(i+1)$-th block. Particularly, the output state of the first block is also called **the middle state** in our attacks, since there are only two blocks in consideration. We call the state after XORing the initial state with the message **the messaged state**, and the state after the operation θ is called **the θ-diffused state**.

For notations, x, y, z refer to the axises of bits in states, and $a_{x,y,z}$, $d_{x,y,z}$, and $e_{x,y,z}$ are always used to represent bits in the messaged state, θ-diffused state, and output state of a hash block, respectively. We use i_j and o_j to denote input and output bits of the operation χ. The notation $s_{x,y}$ is used to represent the sum of a column in some state.

3 The Allocating Approach

With the motivation introduced in Sect. 1, we present the constraints on the middle state firstly by improving Li et al.'s structure [15]. Next, we show the details of the allocating approach.

3.1 Constraints on the Middle State

In [15], Li et al. proposed a structure by setting some bits of the messaged state as unknowns and constants, such that the output bits after 2 forward rounds are *almost* all linear. However, there are still some nonlinear bits due to the constraints brought by the initial value. In this section, we improve this structure, and make *all* output bits linear after 2 forward rounds.

We start by studying the properties of the only nonlinear operator χ of each forward permutation. The operator χ can be regarded as a small S-box with 5 input and output bits, and the algebraic normal form of χ is

$$o_j = i_j \oplus (i_{j+1} \oplus 1) \cdot i_{j+2},\ j = 0, 1, 2, 3, 4,$$

where i_j and o_j are the j-th input and output bits. When building the algebraic systems, the input and output bits of χ are all represented as polynomials of unknowns and constants. Our goal is to find an input pattern such that: (1) the inputs contain as many linear bits as possible, i.e. the degrees of freedom need to be high; (2) the outputs contain as few non-constant bits as possible, i.e. the unknowns are not diffused much; (3) the outputs do not contain nonlinear bits. By the requirement (1) and (3), we known that there are at most 2 linear bits in the inputs and these two linear bits must not be consecutive as well. Let x and c stand for the linear bit and constant bit, respectively. Then the input pattern is 'xcxcc', while other patterns satisfying (1) and (3) are all rotations of this one. Since each constant c could be 1 or 0, there are 8 possible cases of this input pattern. We list them and their corresponding outputs in Table 4.

Table 4. Input and output bits of χ for the 'xcxcc' input pattern.

inputs	x0x01	x0x00	x0x11	x1x01	x0x10	x1x00	x1x11	x1x10
outputs	x0x01	x0xx0	xxx11	x1x0x	xxxx0	x1xxx	xxx1x	xxxxx
#(linear output bits)	2	3	3	3	4	4	4	5

As shown in Table 4, only the input pattern 'x0x01' meets the requirement (2).

Based on the above study, we improve Li et al.'s structure in Lemma 1.

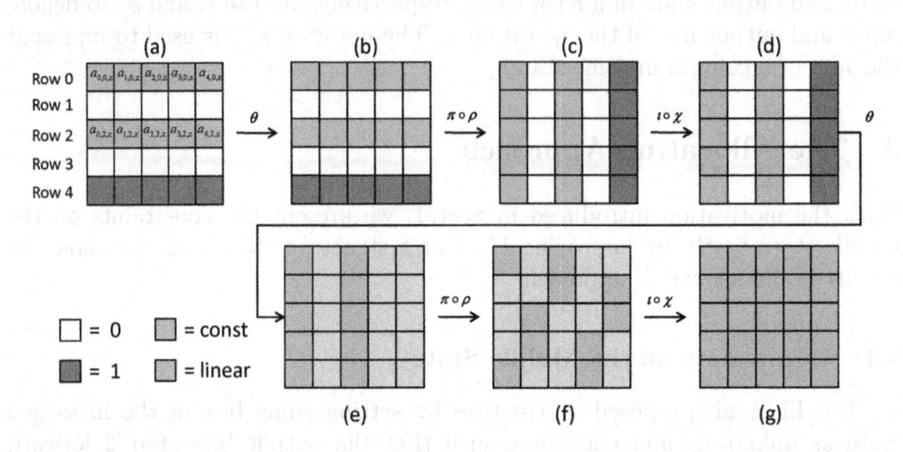

Fig. 4. The improved linear structure. Only one slice is shown, while the structures of other slices are the same.

Lemma 1. *Let the messaged state be (a) in Fig. 4, i.e. bits in Row 0, 2 are unknowns, bits in Row 1, 3 are 0's, and bits in Row 4 are 1's. Then the* Keccak-*f[1600] permutation can be linearized up to 2 rounds with 194 degrees of freedom left.*

Proof. To avoid the propagation of unknowns after the θ operation, we assume that the bitewise sum of two columns is 0, i.e., $\bigoplus_{j=0}^{4} A_{x-1,j,z} + \bigoplus_{j=0}^{4} A_{x+1,j,z-1} = 0$ in state (a). In this way, after the operation θ, constant bits in state (a) are unchanged in state (b), but the linear bits in state (b) may be different from those in state (a) by some constants. Initially, there are $10 \times 64 = 640$ unknowns, say $a_{x,0,z}$ and $a_{x,2,z}$. The sum assumption generates $5 \times 64 = 320$ linear constraints:

$$a_{x-1,0,z} \oplus a_{x-1,2,z} \oplus a_{x+1,0,z-1} \oplus a_{x+1,2,z-1} = 0, \text{ where } 0 \leq x < 5, \text{ and } 0 \leq z < 64.$$

And there is 1 constraint linear dependent on the other $320-1 = 319$ constraints. To show the linear dependence, we denote $p_{x,z} := a_{x-1,0,z} \oplus a_{x-1,2,z} \oplus a_{x+1,0,z-1} \oplus a_{x+1,2,z-1}$. Then we have

$$\bigoplus_{x,z} p_{x,z} = \bigoplus_{x,z} a_{x-1,0,z} \oplus \bigoplus_{x,z} a_{x-1,2,z} \oplus \bigoplus_{x,z} a_{x+1,0,z-1} \oplus \bigoplus_{x,z} a_{x+1,2,z-1}.$$

Since $\bigoplus_{x,z} a_{x-1,0,z} = \bigoplus_{x,z} a_{x+1,0,z-1}$ and $\bigoplus_{x,z} a_{x-1,2,z} = \bigoplus_{x,z} a_{x+1,2,z-1}$, we have $\bigoplus_{x,z} p_{x,z} = 0$, which means each $p_{x,z}$ equals the sum of the others. So after θ, there are $640 - 319 = 321$ degrees of freedom left. After ρ and π, each row of the state (c) has the pattern 'x0x01', and it is just the optimal case we studied above. So nonlinear bits are not generated and the unknowns are not diffused after the χ operation.

To keep the outputs of the second forward round linear, we also need to assume the sums of bits in Column 0 and 2 are constants in state (d), which produces $2 \times 64 = 128$ linear constraints and 1 of them is linear dependent on the other $128 - 1 = 127$ as well, while the proof is similar. Then there are $321 - 127 = 194$ degrees of freedom left. The consequent operations will not cost degrees of freedom, and all the bits in state (g) are linear.

In general, the layout of state (a) in Fig. 4 is hard to meet, as it has rigid requirements on the values of constants. So we consider a more general case in Theorem 1.

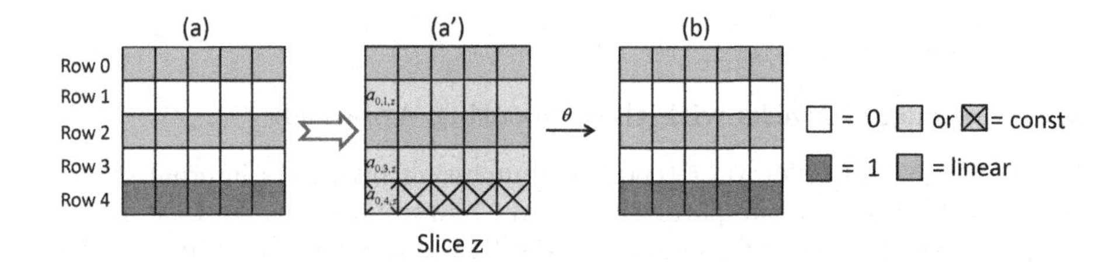

Fig. 5. Transforming state (a) to a more general case (a').

Theorem 1. *Let the messaged state be (a') in Fig. 5, i.e. bits in Row 0, 2 are unknowns, and bits in Row 1, 3, 4 are constants such that*

(i) $a_{x,1,z} = a_{x,3,z} = a_{x,4,z} \oplus 1$, *and*
(ii) $\bigoplus_{x,z} a_{x,4,z} = 0$,

where $a_{x,y,z}$ *stands for the linear or constant bit at the position* (x, y, z), $0 \leq x, y < 5$, *and* $0 \leq z < 64$. *Then there exist constants* $s_{x,z}$*'s with* $0 \leq x < 5$ *and* $0 \leq z < 64$, *such that if assuming* $\bigoplus_{j=0}^{4} a_{x,j,z} = s_{x,z}$, *then the state (b) in Fig. 5 can be obtained by operating* θ *on (a'). And hence, the* KECCAK-$f[1600]$ *permutation can be linearized up to 2 rounds with 194 degrees of freedom left.*

Proof. As introduced in Sect. 2.2, the operation θ is defined as:

$$\theta : d_{x,y,z} = a_{x,y,z} \oplus \bigoplus_{j=0}^{4} (a_{x-1,j,z} \oplus a_{x+1,j,z-1}) = a_{x,y,z} \oplus \bigoplus_{j=0}^{4} a_{x-1,j,z} \oplus \bigoplus_{j=0}^{4} a_{x+1,j,z-1},$$

where $d_{x,y,z}$ is a bit in the state (b) that is diffused from $a_{x,y,z}$'s. Let $s_{x,z} := \bigoplus_{j=0}^{4} a_{x,j,z}$. Then we have

$$d_{x,y,z} = a_{x,y,z} \oplus s_{x-1,z} \oplus s_{x+1,z-1}. \tag{1}$$

To ensure that the state (b) can be obtained after the operation θ, i.e. $d_{x,0,z}$ and $d_{x,2,z}$ are linear, $d_{x,1,z} = d_{x,3,z} = 0$, and $d_{x,4,z} = 1$, we only need to make the following equations hold by the condition $a_{x,1,z} = a_{x,3,z} = a_{x,4,z} \oplus 1$:

$$a_{x,4,z} \oplus s_{x-1,z} \oplus s_{x+1,z-1} = 1, \text{ where } 0 \le x < 5, 0 \le z < 64. \tag{2}$$

There are $5\times64 = 320$ equations in Eq. (2). All $a_{x,4,z}$ are constants. Regarding $s_{x,z}$ as symbols, the reduced Gröbner basis of the ideal $\langle a_{x,4,z} \oplus s_{x-1,z} \oplus s_{x+1,z-1} \oplus 1 \mid 0 \le x < 5, 0 \le z < 64\rangle$ over $(GF(2)[a_{x,4,z}])[s_{x,z}]$ w.r.t. some lexicographic ordering on $\{s_{x,z}\}$ contains 320 polynomials. Among these polynomials, the only one that does not involve $\{s_{x,z}\}$ is $\bigoplus_{x,z} a_{x,4,z}$. By the properties of Gröbner bases, Eq. (2) have solutions for $\{s_{x,z}\}$, if and only if $\bigoplus_{x,z} a_{x,4,z} = 0$. This is just the condition (ii), and the theorem is proved.

In fact, the above proof implies that the condition (i) and (ii) in Theorem 1 are also necessary conditions for the existence of $s_{x,z}$'s, if the messaged state is set as (a').

3.2 Preimage Attacks with the Allocating Approach

For an instance of KECCAK-f[1600], the number of bits in its internal state is 1600, which consists of two parts with r and c bits respectively. All 1600 bits are set as 0 initially. The first r bits need to XOR the message, and the last c bits remain 0's. The number c is the capacity of this instance.

By Theorem 1, if the capacity c of an instance of KECCAK-f[1600] is smaller than $5\times64 = 320$, e.g. SHAKE128 whose capacity is 256, we can set the messaged state like (a') in Fig. 5 by choosing the message carefully. Unfortunately, the capacities of most KECCAK instances are bigger than 320. This means the results in Sect. 3.1 cannot be used directly, because the number of 0's in the tail of the messaged state is more than 320 and the condition (i) does not hold.

Fortunately, the state (a') could serve as a good internal state in the allocating approach, where 2-block messages are considered. The outputs of the first block are not all 0's generally. We can adjust the values of the first r bits by choosing the second message block carefully. To make the messaged state of the second block meet the conditions in Theorem 1, we also need some constraints on the last c bits in the output state of the first block.

Specifically, as shown in Fig. 6, we consider the initial state (A) and messaged state (B) of the second block. The state (A) is the middle state in our attack, and it is also the output state of the first block.

For KECCAK-224, its capacity is 448, which means the last 7 lanes of bits can not be changed after the second message block being XORed. So the last 7 lanes in the state (A) and (B) are identical. Since the values of the first 18 lanes

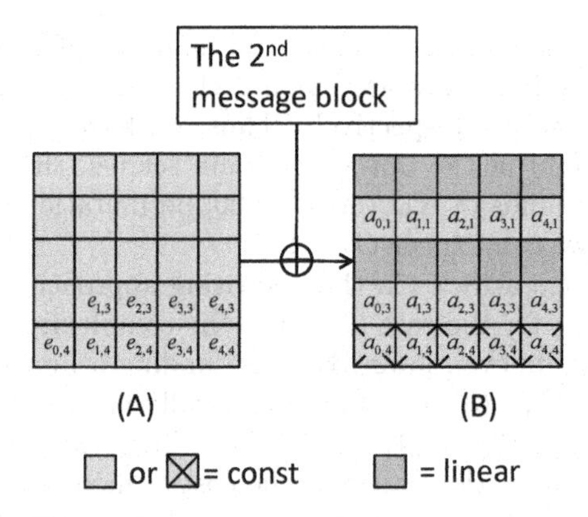

Fig. 6. The initial and messaged states of the second block of KECCAK-224/256. The axis z is omitted for simplification.

in state (B) can be adjusted by the second message block[1], to make bits in the state (B) meet condition (i) and (ii) in Theorem 1, it suffices to ensure

$$e_{3,3,z} \oplus 1 = e_{3,4,z}, e_{4,3,z} \oplus 1 = e_{4,4,z}, \text{ and } \bigoplus_{x,z} e_{x,4,z} = 0, \tag{3}$$

which consists of $64 + 64 + 1 = 129$ equations.

The case for KECCAK-256 is similar, except that the last 8 lanes in the state (A) and (B) are identical. To make bits in the state (B) meet conditions in Theorem 1, we need the following $64 + 64 + 64 + 1 = 193$ equations hold:

$$e_{2,3,z} \oplus 1 = e_{2,4,z}, e_{3,3,z} \oplus 1 = e_{3,4,z}, e_{4,3,z} \oplus 1 = e_{4,4,z}, \text{ and } \bigoplus_{x,z} e_{x,4,z} = 0. \tag{4}$$

So in all, attacks on KECCAK-224/256 via the allocating approach consist of two stages:

1. Precomputation Stage: Find a first message block, such that Eqs. (3) or (4) hold for the output bits of the first block. Let C_1 be the complexity of finding this message block.
2. Online Stage: Construct an algebraic system using the structure in Theorem 1 for a given hash value, and solve this system for a second message block. The complexity of this stage is denoted as C_2.

We call the first stage "Precomputation Stage", because it does not need to be re-executed for different hash values, if a good first message block has been found.

Consequently, the complexity of the whole preimage attack is $C_1 + C_2$. Let $\bar{C}$ denote the complexity of finding a 1-block preimage, then we have

[1] The paddings will be dealt with sooner.

$\max\{C_1, C_2\} < \bar{C}$. On one hand, since the numbers of equations in Eqs. (3) and (4) are 129 and 193, they are smaller than 224 and 256, which are the lengths of the hash values, respectively. Thus, we have $C_1 < \bar{C}$. On the other hand, if the last c bits in the initial state are set 0's, there is no way to linearize the first two rounds of KECCAK-f[1600] permutation of with 194 degrees of freedom left. So we can expect $C_2 < \bar{C}$.

The basic preimage attack via the allocating approach will be described in Sect. 4.1. Particularly, for the case $C_1 > C_2$, the complexity of the whole preimage attack can be made even lower. Because finding a first message block such that all equations in Eqs. (3) or (4) hold is usually harder than that if we allow some equations in Eqs. (3) or (4) *not to hold.* This means we can decrease the complexity of Precomputation Stage at the cost of increasing the complexity of Online Stage. Thus, the balanced complexity will be smaller than $\max\{C_1, C_2\}$. This balanced method will be given in Sect. 4.2, and it also leads to a practical preimage attack on 3-round KECCAK-224 in Sect. 5.

4 Theoretical Results on Round-Reduced KECCAK-224/256

In this section, we use the allocating approach to attack KECCAK-224/256 that are reduced to 3 and 4 rounds. Theoretical complexities of these instances, as well as instances of SHA-3 and SHAKE, are given. The complexity is measured by the number of times for solving systems of linear equations, i.e., the complexity of solving a linear system is assumed to be a constant, which is the same as done in [12].

4.1 Attacks on 3-Round KECCAK-224/256

In this section, we give detailed preimage attacks on 3-round KECCAK-224. Attacks on 3-round KECCAK-256 and instances of SHA-3 and SHAKE are similar.

Attacks on 3-round KECCAK-224. The attack consists of three parts. First, we find a first message block such that Eq. (3) holds in Precomputation Stage. Second, we find a second message block such that the state (B) meets conditions in Theorem 1 and the outputs of the second block equal the given hash value in Online Stage. At last, we show how to deal with the paddings.

Part 1: finding a first message block

To find the first message block satisfying Eq. (3), we use the structure presented by Guo et al. [12] in Fig. 7, which keeps 2.5 rounds linear with 128 degrees of freedom left.

In the messaged state of the first round, bits of 8 lanes are set as unknowns (shown in yellow boxes), which means there are $8 \times 64 = 512$ unknowns. White boxes and dark gray boxes in this state mean constant 0's and 1's. In the state (c) of the 3rd round, all bits are linear, and all of them become quadratic after

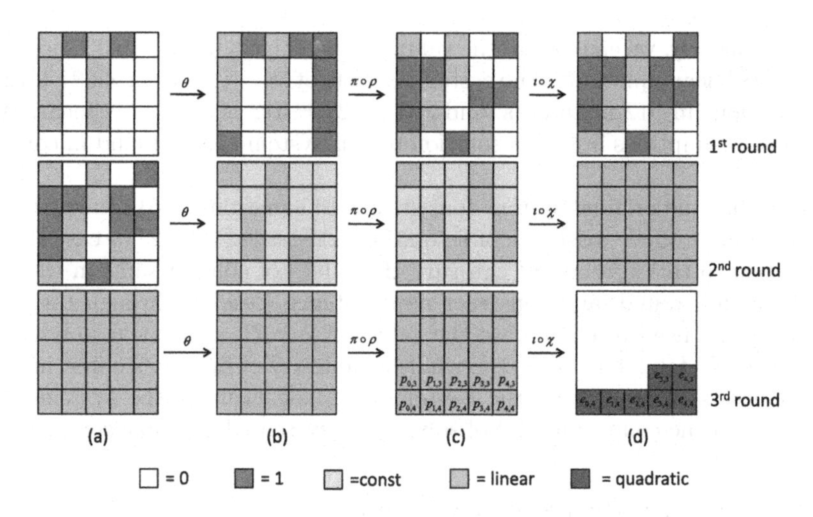

Fig. 7. The 3 forward rounds of the first block of KECCAK-224. The axis z is omitted for simplification. (Color figure online)

the operation χ. During this procedure, to avoid the propagation by θ in the first and second rounds, $2 \times 64 + 4 \times 64 = 384$ linear constraints are added to the system by assuming sums of linear columns as constants. Here, a linear column refers to a column whose bits are linear or constant. By Eq. (3), we obtain another $2 \times 64 + 1 = 129$ quadratic equations.

In all, we construct a system with $384 + 129 = 513$ equations in 512 unknowns. Although 384 equations are linear, this system is still not easy to solve in general. Fortunately, after noticing that there is only *one* quadratic term in each polynomial representation of $e_{i,j}$, we can enumerate 2 values of linear polynomials and obtain 4 linear equations like done in [15].

Specifically, let $p_{i,j}$ be the (linear) polynomial representation of bits in the state (c) of the 3rd round in Fig. 7. By the χ operation, we have:

$$e_{3,4} = p_{3,4} \oplus (p_{4,4} \oplus 1) \cdot p_{0,4},\ e_{4,4} = p_{4,4} \oplus (p_{0,4} \oplus 1) \cdot p_{1,4},$$

$$e_{3,3} = p_{3,3} \oplus (p_{4,3} \oplus 1) \cdot p_{0,3},\ e_{4,3} = p_{4,3} \oplus (p_{0,3} \oplus 1) \cdot p_{1,3},$$

where the axis z is omitted for simplification. By Eq. (3), we have the following equations:

$$p_{3,4} \oplus (p_{4,4} \oplus 1) \cdot p_{0,4} \oplus p_{3,3} \oplus (p_{4,3} \oplus 1) \cdot p_{0,3} = 1, \tag{5}$$

$$p_{4,4} \oplus (p_{0,4} \oplus 1) \cdot p_{1,4} \oplus p_{4,3} \oplus (p_{0,3} \oplus 1) \cdot p_{1,3} = 1. \tag{6}$$

If the values of the pair $(p_{0,3}, p_{0,4})$ are enumerated, then both Eqs. (5) and (6) are linearized in each slice. Together with equations from the enumeration, we obtain 4 linear equations totally.

Consequently, we enumerate the values of the pair $(p_{0,3}, p_{0,4})$ in 32 slices, and obtain 128 linear equations. The system consists of $384 + (129 - 2 \times 32) + 128 = 577$ equations in 512 unknowns, and $384 + 128 = 512$ of them are linear. With the same assumptions in [12], a solution to this system can be found in constant time.

Note that the original system consists of 513 equations in 512 unknowns, so the probability of the existence of a solution is regarded as 1/2. In case there is no solution to this system, we can vary the values of column sums in the state (a) of the 2nd round, and construct new systems. Besides, through the above procedure, we need to enumerate the values of $2 \times 32 = 64$ linear polynomials to solve the system. Therefore, the whole complexity of finding the first message block consists of two parts, the complexity 2^1 of ensuring the system has a solution, and the complexity 2^{64} of solving the system. The whole complexity of is $2^{1+64} = 2^{65}$.

Part 2: finding a second message block
By part 1, we obtain an initial state of the second block satisfying Eq. (3). After setting bits in the second message block carefully, the messaged state, depicted in (a) of the 1st round in Fig. 8, meets the conditions in Theorem 1. So there are 194 degrees of freedom left in the end of the 2nd round.

Bits in the initial state of the 3rd round are all linear, and after the linear operation θ, π, and ρ, bits in the state (c) of the 3rd round remain linear. On the other hand, KECCAK-224 generates a 224-bit hash value, which is supposed to be known for preimage attacks. Next, we construct relations between the bits before and after the operation χ. The relations are first studied in [12], and the operation ι is omitted here for simplification.

Let i_j and o_j be the input and output bit of χ. We have

$$o_j = i_j \oplus (i_{j+1} \oplus 1) \cdot i_{j+2},\ j = 0, 1, 2, 3, 4,$$

by definition. Next, we can deduce

$$\begin{aligned} o_j &= i_j \oplus ((o_{j+1} \oplus (i_{j+2} \oplus 1) \cdot i_{j+3}) \oplus 1) \cdot i_{j+2} \\ &= i_j \oplus (o_{j+1} \oplus 1) \cdot i_{j+2} \oplus (i_{j+2} \oplus 1) \cdot i_{j+3} \cdot i_{j+2} \\ &= i_j \oplus (o_{j+1} \oplus 1) \cdot i_{j+2}. \end{aligned} \tag{7}$$

Assume the i_j's are linear. If the values of 4 consecutive output bits are known, e. g. $o_0, \ldots, o_3$ are constants, then we have 3 linear equations

$$o_0 = i_0 \oplus (o_1 \oplus 1) \cdot i_2, o_1 = i_1 \oplus (o_2 \oplus 1) \cdot i_3, o_2 = i_2 \oplus (o_3 \oplus 1) \cdot i_4,$$

and 1 quadratic equation

$$o_3 = i_3 \oplus (i_4 \oplus 1) \cdot i_0. \tag{8}$$

Fortunately, Eq. (8) can be simplified to linear as below.

$$i_0 = o_0 \oplus (o_1 \oplus 1) \cdot i_2 = o_0 \oplus (o_1 \oplus 1) \cdot (o_2 \oplus (o_3 \oplus 1) \cdot i_4) = A \oplus B \cdot i_4,$$

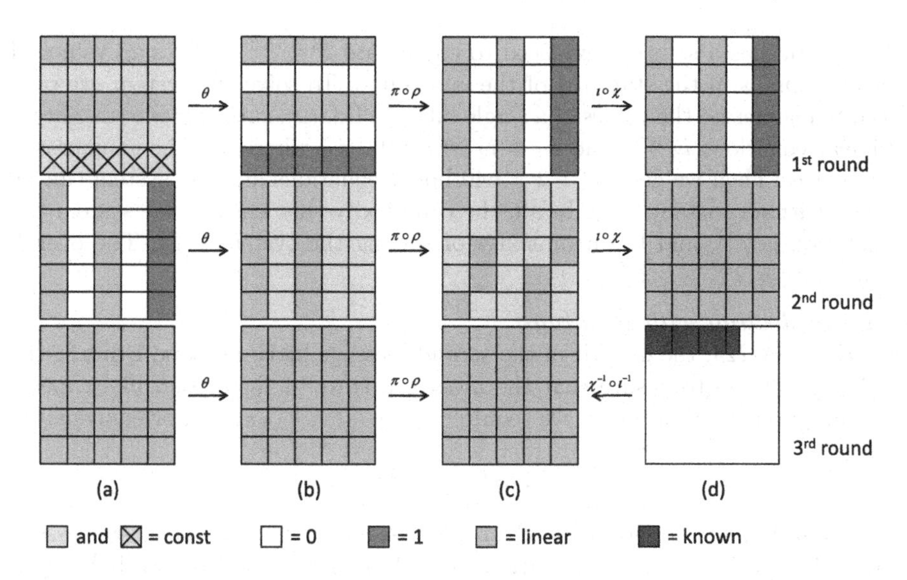

Fig. 8. The forward 3 rounds of the second block of KECCAK-224. The axis z is omitted for simplification.

where $A = o_0 \oplus (o_1 \oplus 1) \cdot o_2$ and $B = (o_1 \oplus 1) \cdot (o_3 \oplus 1)$. Thus,

$$o_3 = i_3 \oplus (i_4 \oplus 1) \cdot i_0 = i_3 \oplus (i_4 \oplus 1) \cdot (A \oplus B \cdot i_4) = i_3 \oplus (i_4 \oplus 1) \cdot A.$$

Similarly, if the values of 3 consecutive output bits are known, say o_0, o_1, o_2, then we get 2 linear equations $o_0 = i_0 \oplus (o_1 \oplus 1) \cdot i_2$, $o_1 = i_1 \oplus (o_2 \oplus 1) \cdot i_3$, and a quadratic one

$$o_2 = i_2 \oplus (i_3 \oplus 1) \cdot i_4. \tag{9}$$

But this quadratic equation cannot be simplified.

In a digest of KECCAK-224, we have 4 consecutive bits in 32 slices, and 3 consecutive bits in the other 32 slices. Since the bits in the state (c) of the 3rd round are linear, we can set up $(4 + 2) \times 32 = 192$ linear equations, and $1 \times 32 = 32$ quadratic ones.

To sum up, bits of 10 lanes in the messaged state are unknowns. The number of unknowns is $10 \times 64 = 640$. To avoid the propagation by θ in the first and second rounds, $5 \times 64 + 2 \times 64 = 448$ linear constraints are added to the system by assuming that the bitewise sums of two linear columns are constants. Please note that there are 2 linear equations linear dependent on others, and the reason is shown in the proof of Lemma 1. We should also pay attention to that the values of these bitewise sums in the state (a) of the 1st round in Fig. 8 must equal the values of $s_{x,z}$'s which are obtained from the proof of Theorem 1. But the sums of linear columns in the state (a) of the 2nd round could be set randomly.

Together with equations constructed by the hash value, the system has $448 - 2 + 192 + 32 = 670$ equations in 640 unknowns, and among these equations, $448 - 2 + 192 = 638$ are linear and linear independent on each other. To ensure this system has a solution, we need to enumerate $2^{670-640} = 2^{30}$ sum values of linear columns in the state (a) of the 2nd round. To solve the system, we only need to enumerate the values of a single bit i_3 in Eq. (9), such that we can obtain 2 linear equations $i_3 = c$ and $o_2 = i_2 \oplus (c \oplus 1) \cdot i_4$, where c is the enumerated value of i_3. Then we get $638 + 2 = 640$ linear equations and the system can be solved within constant time. In all, the complexity that ensures the system has a solution is 2^{30}, and the complexity of solving the system is 2^1. The overall complexity is $2^{30+1} = 2^{31}$.

Part 3: dealing with paddings
For KECCAK-224, the last bit of the second message block must be 1 due to the padding rule. So to ensure that the messaged state of the second block meets the conditions in Theorem 1, we require $e_{2,3,63} \oplus 1 = e_{2,4,63} \oplus 1$, or equivalently

$$e_{2,3,63} = e_{2,4,63}. \tag{10}$$

This equation should be included in the system for finding the first message block. That is, we have 514 equations in 512 unknowns, and the probability of the existence of a solution is 1/4. The complexity for solving the first message block becomes 2^{66}.

The overall complexities of attacking 3-round* KECCAK*-224/SHA3-224
Summing up the analyses in the above three parts, the theoretical preimage attack on 3-round KECCAK-224 costs about 2^{66} operations. With similar analyses, the complexity of attacking 3-round SHA3-224 is 2^{69}.

Attacks on 3-round KECCAK-256. To find a first message block for 3-round KECCAK-256, the messaged state is set as (a) in Fig. 9. And Eq. (4) are considered. Following a similar procedure, bits in the output of the 2nd round have algebraic degree 1 at most.

To solve a first message block, we set $2 \times 64 + 3 \times 64 = 320$ linear equations by assuming the sums of linear columns in the state (a) and (d) are constants. Besides, there are $3 \times 64 + 1 = 193$ quadratic equations in Eq. (4). The number of unknowns is $6 \times 64 = 384$. So this system consists of $320 + 193 = 513$ equations in 384 unknowns, and the probability of the existence of a solution is $1/2^{513-384} = 1/2^{129}$. That is, we need to enumerate 2^{129} sum values of linear columns in the state (d) to ensure the system has a solution. To solve this system, similar to the case of KECCAK-224, we need to enumerate the values of the pair $(p_{0,3}, p_{0,4})$ in 16 slices, and obtain 64 linear equations. Then we obtain $320 + 64 = 384$ linear equations, and the system can be solved with a constant time complexity. Thus, the complexity of finding a first message block is $2^{129+32} = 2^{161}$.

To solve a second message block, the procedure is the same as that of KECCAK-224, except that we obtain $4 \times 64 = 256$ linear equations from the hash value. The system of this stage consists of $5 \times 64 + 2 \times 64 - 2 + 256 = 702$ linear equations in 640 knowns. We need to try $2^{702-640} = 2^{62}$ sum values of the linear columns in the state (a) of the 2nd round, to ensure there is a solution to this system. So the complexity of this stage is 2^{62}.

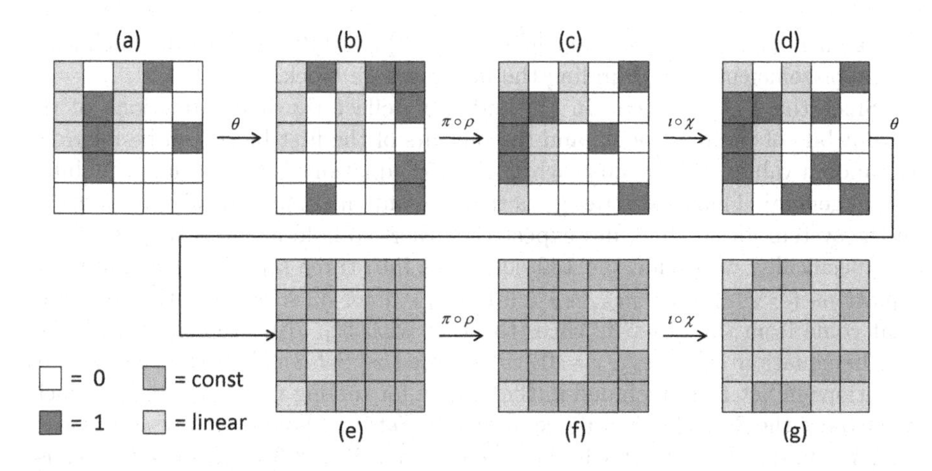

Fig. 9. The first message block of 3-round KECCAK-256. The axis z is omitted for simplification.

The overall complexities of attacking 3-round KECCAK-***256/SHA3-256/SHAKE256***
After dealing with paddings for the instances of KECCAK-256, SHA3-256, and SHAKE256, their attack complexities are 2^{162}, 2^{165}, and 2^{167}, respectively. Note that these results are not as good as those in [15].

4.2 Improved Attacks on 3-Round KECCAK-224/256

In the attacks of Sect. 4.1, the complexity of Precomputation Stage is much higher than that of Online Stage. One reason is that we require the state (B) in Fig. 6 meets all conditions in Theorem 1. In fact, to obtain better attacks on KECCAK-224/256, we can give up some constraints in Eqs. (3) or (4). In such a way, the complexity of finding the first message block will decrease at the cost of increasing the difficulty of finding the second message block. Then the complexities of the two stages will be balanced, and hence, the overall complexities of attacks will be improved.

Improved attacks on 3-round KECCAK**-224.** The improved attack contains two parts.

Part 1: finding a first message block
As discussed in Sect. 4.1, it costs 2^{66} operations to find a first message block such that all the $129+1$ equations in (3) and (10) hold. Consequently, the messaged state (B) in Fig. 6 meets all conditions in Theorem 1, and hence, we can obtain the best complexity 2^{31} for the second block. Note that the complexity in the second stage is much lower than that in the first one. To improve the overall complexity of attacking KECCAK-224, we can balance the complexities of the two stages by allowing the messaged state (B) *to not* meet all conditions in

Theorem 1. This means, not all equations in (3) and (10) must hold, which leads to a more efficient way of finding the first message block.

Since the 130 equations in (3) and (10) reflect the linear relations of the output bits of the first block, and the outputs of the first block can be regarded as random values, this means each of the 130 equations holds at the probability 1/2 in general. Thus, our strategy is that, we only include some of the equations in the system for solving, and expect the others to hold as many as possible.

Specifically, we divide the 130 equations into three sets. Set 1 contains the equations $\{e_{3,3,z} \oplus 1 = e_{3,4,z}, e_{4,3,z} \oplus 1 = e_{4,4,z}\}$ for 32 slices, and the equations that come from the other 32 slices together with Eq. (10) are contained in Set 2. The equation $\{\bigoplus_{x,z} e_{x,4,z} = 0\}$ is regarded as Set 3. Then the $2 \times 32 = 64$ equations in Set 1 are included in the system for solving the first message block. We expect the $2 \times 32 + 1 = 65$ equations in Set 2 to hold as many as possible. And we do not care about whether the equation in Set 3 holds or not, because it does not affect the complexity of Online Stage.

Next, we estimate the complexity of finding a first message block. First, we have $2 \times 64 + 4 \times 64 = 384$ linear constraints by assigning the values of sums of linear columns. Second, we have $2 \times 32 = 64$ quadratic equations from Set 1. So there are $384 + 64 = 448$ equations in $8 \times 64 = 512$ unknowns. We have $512 - 448 = 64$ degrees of freedom left to solve this system. For the equations from Set 1, they are selected from 32 slices. Similar as discussed in Sect. 4.1, we can obtain 32 linear equations by guessing the values of 16 pairs $(p_{0,3}, p_{0,4})$, and get another 32 linear equations after the linearization of Eqs. (5) and (6). Then we get $384 + 64 + 64 = 512$ linear equations in 512 unknowns, which means we can obtain a solution to this system with a constant complexity. Note that when the values of $p_{0,3}$ and $p_{0,4}$ varies, solutions to the system change as well.

At last, we estimate how many operations are necessary to make as many equations in Set 2 hold as possible. In theoretical aspects, since each equation holds with a probability 1/2, for any first message block, it makes n of the 65 equations hold with probability $\frac{C_{65}^{n}}{2^{65}}$. So the theoretical complexity of making n equations hold is $\frac{2^{65}}{C_{65}^{n}}$, and the complexities for different n's are shown in Table 5. In experimental aspects, we can obtain a lot of solutions of the systems by varying the values of $p_{0,3}$ and $p_{0,4}$. Then the number of messages that make n equations hold can be counted, and the practical probabilities are obtained as well. Complexities of practical attacks are reported in Sect. 5.1.

Part 2: finding a second message block

From Part 1, equations in Set 1 always hold, but some in Set 2 are not. Besides, we do not care about whether $\bigoplus_{x,z} e_{x,4,z} = 0$ holds or not. In this section, we deal with the troubles brought by these unsatisfied equations.

There are two types of unsatisfied equations for the first message block found in Part 1:

I. $e_{2,3,63} = e_{2,4,63}$, $e_{x,3,z} \oplus 1 = e_{x,4,z}$ where $x = 3$ or 4 for some z,
II. $\bigoplus_{x,z} e_{x,4,z} = 0$.

Our strategy is as follows. First, we construct an adjusted state by flipping the values of some bits in the unsatisfied equations and keeping others unchanged, such that all equations hold for the bits in the adjusted state. Second, like what we have done in the proof of Theorem 1, we can solve for the values of $s_{x,z}$'s such that the adjusted state transforms to the state (b) in Fig. 5 after the operation θ. At last, we apply θ to the real outputs of the first block by assuming that the sums of columns are $s_{x,z}$'s. The obtained state will be different from the state (b) in Fig. 5 only in a few bits, and hence, we only need to deal with these different bits afterwards.

Constructing the adjusted state and solving for $s_{x,z}$'s. Assume there are n_I unsatisfied equations of Type I and $n_{II} \in \{0, 1\}$ unsatisfied one of Type II. We do not consider the case $n_I = 0$ since it does not happen in general cases. Our adjusting method only needs to flip n_I values of bits from the unsatisfied equations of Type I. Note that in each equation of Type I, there is one bit in Row 3 and one in Row 4. We totally flip n_{II} bit in Row 4 to ensure the sum of all bits in Row 4 of the adjusted state is 0, and flip the other $n_I - n_{II}$ bits in Row 3. Specifically, let $e'_{x,4,z} := e_{x,4,z} \oplus 1$ for n_{II} equation out of the n_I unsatisfied Type I equations, and let $e'_{x,3,z} := e_{x,3,z} \oplus 1$ from the other $n_I - n_{II}$ unsatisfied equations of Type I. Other bits are unchanged.

We illustrate the above method using a toy example with $n_I = 3$ and $n_{II} = 1$. Let $e_{x,y,z}$'s be the bits output by the first block for a given first message block, and we assume the 3 unsatisfied equations of Type I and the unsatisfied one of Type II are

$$e_{2,3,63} = e_{2,4,63} \oplus 1, e_{3,3,0} = e_{3,4,0}, e_{4,3,1} = e_{4,4,1}, \text{ and } \bigoplus_{x,z} e_{x,4,z} = 1.$$

Since $n_{II} = 1$, we let $e'_{3,4,0} := e_{3,4,0} \oplus 1$. For the other $n_I - n_{II} = 2$ bits, we set $e'_{2,3,63} := e_{2,3,63} \oplus 1$, $e'_{4,3,1} := e_{4,3,1} \oplus 1$. For the rest of bits, we have $e'_{x,y,z} := e_{x,y,z}$. In this way, the bits $e'_{x,y,z}$'s construct the adjusted state and the equations in (3) and (10) all hold for $e'_{x,y,z}$'s. Then Eq. (2) has solutions to $s_{x,z}$'s by the proof of Theorem 1. Thus, we find the desired values of $s_{x,z}$'s.

Dealing with the different bits in the θ-defused states. For the original state consisting of $e_{x,y,z}$'s, let us see what happens to the state after the operation θ by assuming that the sums of columns equal the precomputed $s_{x,z}$'s. Let $d_{x,y,z}$'s be the bits in the state after the operation θ. By Eq. (1), we have $d_{x,y,z} = e_{x,y,z} \oplus s_{x-1,z} \oplus s_{x+1,z-1}$. Since $s_{x,z}$'s are precomputed constants, the value of $d_{x,y,z}$ is determined by $e_{x,y,z}$. Note that there are only n_I $e_{x,y,z}$'s different from $e'_{x,y,z}$'s, so only n_I bits of the θ-defused states of the original and adjusted states are not identical, and the differences only lie in Row 3 or Row 4. Based on the row that the different bit appears, we consider two cases as shown in Fig. 10.

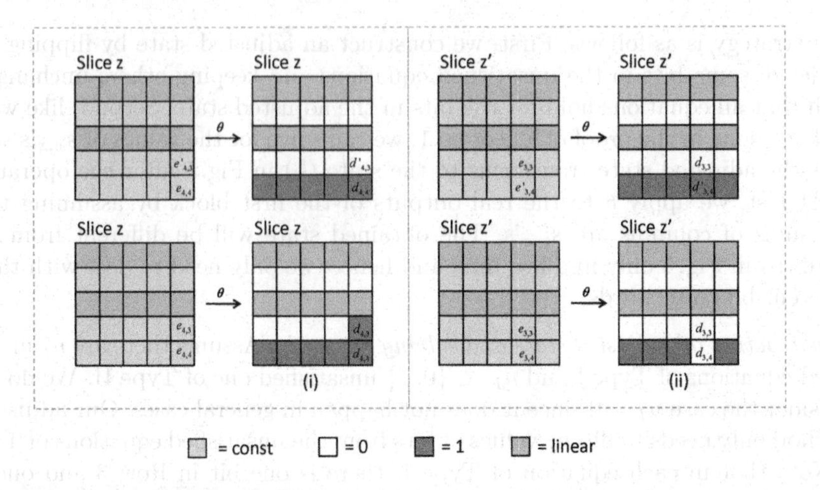

Fig. 10. The top row shows the adjusted states before and after the operation θ, and the bottom row shows the original states before and after θ. The state (i) and (ii) show two types of troubles.

Figure 11 shows how the trouble generated by the state (i) in Fig. 10 is handled. The bit at (Row 3, Column 4, Slice z) of the θ-diffused state is 1 instead of 0, and it will produce two quadratic bits after two rounds. The method of eliminating this effect is that, we can enumerate the values of the bit in the orange box in the end of the first round, such that this bit becomes a constant and no quadratic bits are generated after two forward rounds. This enumeration costs 1 degree of freedom. Similarly, we can also handle the state (ii) at the cost of 1 degree of freedom in Fig. 12.

To sum up, if n_I equations of Type I do not hold, the complexity of finding a second message block is 2^{31+n_I}.

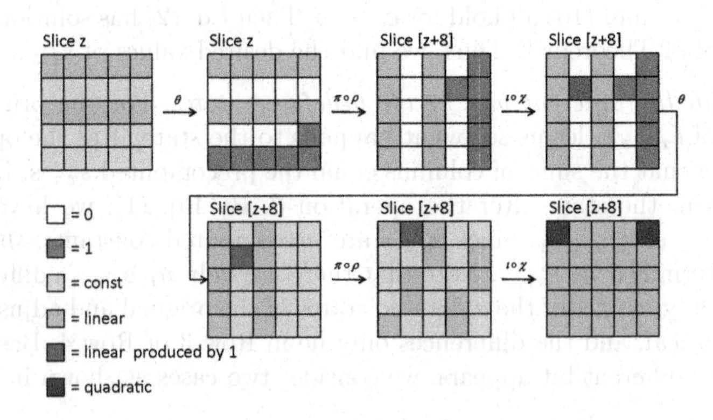

Fig. 11. The effects caused by the state (i) of Fig. 10.

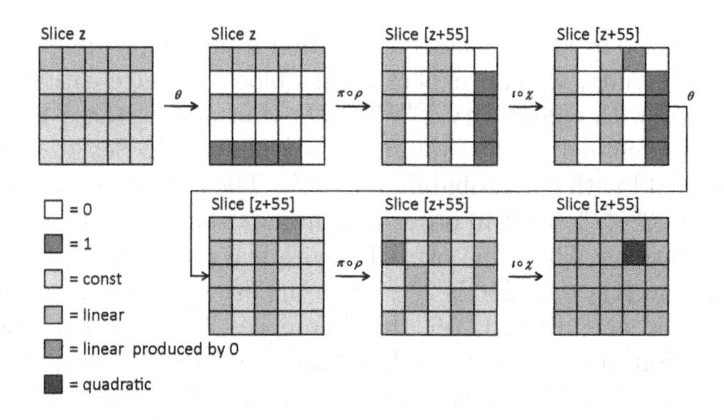

Fig. 12. The effects caused by the state (ii) of Fig. 10.

The overall complexity of the improved preimage attacks on* KECCAK*-224/SHA3-224

With the improved attacks, we estimate the theoretical complexities in Table 5 for the cases that $n = 56, \ldots, 60$ equations from Set 2 hold. In this table, $n_I = 65-n$ is the number of equations that do not hold. C_1 and C_2 are the complexities of finding the first and second message blocks. The overall complexity is $C = C_1+C_2$. Please note that C_1 is estimated by probability, so its values are rounded. In Online Stage, the complexity C_2 is obtained by calculating degrees of freedom, so the values of C_2 are accurate integers.

Table 5. Theoretical complexities of preimage attacks with $n = 56, \ldots, 60$, where n is the number of holding equations in Set 2.

n	n_I	C_1	C_2	$C = C_1 + C_2$
56	9	$2^{30.10}$	2^{40}	2^{40}
57	8	$2^{32.77}$	2^{39}	2^{39}
58	7	$2^{35.62}$	2^{38}	2^{38}
59	6	$2^{38.70}$	2^{37}	$2^{38.70}$
60	5	$2^{42.02}$	2^{36}	$2^{42.02}$

Table 5 shows that we can obtain the best theoretical attack on 3-round KECCAK-224 with complexity 2^{38} when $n = 58$. Since this complexity is low enough, we perform a practical attack on 3-round KECCAK-224 in Sect. 5.

Similarly, the complexity of attacking SHA3-224 is at most 2^{41} considering padding bits.

Improved attacks on 3-round KECCAK-256. The improved preimage attack works on 3-round KECCAK-256 as well. There are 193 equations in Eq. (4). The last 1 equation is in Set 3 and is not considered, so we hope 192 of these equations to hold, and we also need to consider 1 equation from the padding. Based on the theoretical probability, we can expect a first block message satisfying 174 out of 193 equations with complexity $2^{80.06}$. Note that among these 174 equations, 32 of them are included in the system for solving, i.e. in Set 1, and the other 142 equations hold with the probability $2^{-80.06}$. The complexity of Online Stage also increases to $2^{62+19} = 2^{81}$ in order to eliminate the impact of $193 - 174 = 19$ unsatisfied equations. Thus, the overall theoretical complexity of the preimage attack on 3-round KECCAK-256 is 2^{81}, while the previous best result is 2^{150} [15].

For SHA3-256 and SHAKE256, the differences lie in the padding rules. So extra computations are needed to find first message blocks. Using the same approach, the complexities of attacking 3-round SHA3-256/SHAKE256 are $2^{84}/2^{86}$.

4.3 Attacks on 4-Round KECCAK-224/256

Attacks on 4-round KECCAK-224. The first message block for KECCAK-224 can be found by probability. Since a hash function outputs bits in a 'random' manner, the probability of finding a preimage by the random preimage attack is $1/2^l$, where l is the number of bits in digests [21]. The reason is that each output bit could be 0 or 1 with probability 1/2. The first message block are constrained by Eq. (3). For the pair $(e_{3,3,z}, e_{3,4,z})$ in each slice, it has four possible values, and two of them make the equation $e_{3,3,z} \oplus e_{3,4,z} \oplus 1 = 0$ hold, which means this equation holds with a probability 1/2. The case is similar for the pair $(e_{4,3,z}, e_{4,4,z})$. As the value of $e_{x,4,z}$ is supposed to be random, the probability that $\bigoplus_{x,z} e_{x,4,z} = 0$ holds is also 1/2. Thus, the complexity of finding the desired first message block by random preimage attack is $2^{64+64+1} = 2^{129}$.

For the second block, as the messaged state meets conditions in Theorem 1, there are $(5+2) \times 64 - 2 = 446$ linear constraints in $10 \times 64 = 640$ unknowns after two forward rounds. The bits in the output state of the 3rd round are all quadratic and remain quadratic before the χ operation in the 4th round. Similarly to the analysis in Sect. 4.1, 224-bit digest leads to 192 quadratic equations and 32 quartic equations. Note that 160 out of the 192 quadratic equations are constructed by Eq. (7). That is, $o_j = i_j \oplus (o_{j+1} \oplus 1) \cdot i_{j+2}$. So if one bit o_{j+1} of the hash value is 1, then the quadratic equation becomes $o_j = i_j$ which has at most 11 quadratic terms as well. These equations are hard to solve, so the following analysis follows from the attacks on 4-round KECCAK-224 in [12]. By Guo et al.'s study, this quadratic equation can be linearized by guessing 10 values of linear polynomials. Figure 13 illustrates how to linearize this quadratic equation.

For any quadratic bit in the state before the χ operation, it corresponds to a bit in the θ-defused state of the same round, since ρ and π are both permutations. Next, taking one quadratic bit $Q_{1,1,z}$ in the θ-defused state in the 4th round for example, we explain how to linearize $Q_{1,1,z}$ by guessing the values of 10 linear polynomials. By the definition of θ, we have $Q_{1,1,z} = q_{1,1,z} \oplus \bigoplus_{j=0}^{4}(q_{0,j,z} \oplus q_{2,j,z-1})$, where $q_{x,y,z}$ represents the output bit of the 3rd round and it is quadratic. Next, let us study how $q_{x,y,z}$ is generated by linear bits. From the states in Fig. 8, the bits in the state before χ in the 3rd round are all linear,

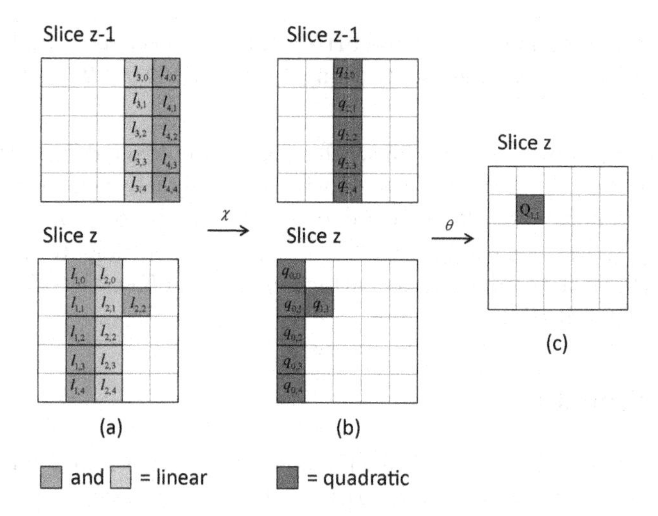

Fig. 13. Linearizing a quadratic bit by guessing the values of 10 linear polynomials. (Color figure online)

and we donate them as $l_{x,y,z}$'s. So we have $q_{x,y,z} = l_{x,y,z} \oplus (l_{x+1,y,z} \oplus 1) \cdot l_{x+2,y,z}$. Note that $q_{x,y,z}$ consists of only 1 quadratic term which is produced by two linear polynomials, $l_{x+1,y,z}$ and $l_{x+2,y,z}$. By guessing the values of either one, $q_{x,y,z}$ can be linearized. Another observation in Fig. 13 is that, $q_{0,1,z}$ and $q_{1,1,z}$ share a common linear factor $l_{2,1,z}$. Thus, we can guess the values of 10 bits in the state (a), i.e. the light green bits, to linearize 11 blue quadratic bits in the state (b). Since $Q_{1,1,z}$ is represented by these blue bits, $Q_{1,1,z}$ is linearized as well. Consequently, we obtain 11 linear equations by enumerating the values of 10 linear polynomials.

As the values of bits in the digest can be regarded as random values, half of the 224 bits are supposed to be 1's. This means, we can expect 80 out of the 160 quadratic equations to have at most 11 quadratic terms. Next, we only consider $\lfloor \frac{640-446}{11} \rfloor = 17$ quadratic equations from the above 80 ones, and leave the other $224 - 17 = 207$ equations hold by probabilities. By enumerating the values of $17 \times 10 = 170$ linear polynomials, we obtain $17 \times (10+1) = 187$ linear equations. To solve the system, we guess the values of another $640 - 446 - 187 = 7$ linear polynomials. Thus, the system consists of $446 + 187 + 7 = 640$ linear equations with 640 unknowns, so it has a solution and we can find it within constant time. Such a solution makes all the other $224 - 17 = 207$ equations hold at a probability 2^{-207}. So the complexity of the second block is 2^{207}.

Compared to Guo et al.'s attacks on 4-round KECCAK-224, our improvement is that, we obtain $640 - 446 = 194$ degrees of freedom after two forward rounds by using the allocating approach, while Guo et al. only get 127 degrees of freedom after the same rounds. So they only match $\lfloor \frac{127}{11} \rfloor = 11$ hash bits, and they need $2^{224-11} = 2^{213}$ operations to ensure all the other $224 - 11 = 213$ equations hold.

The paddings of KECCAK-224 and SHA3-224 only affect the complexity of the first block. So the complexities of attacking 4-round KECCAK-224 and SHA3-224 are both 2^{207}.

Attacks on 4-round KECCAK-256. The attacks on 4-round KECCAK-256 are quite similar. For the first message block, constraints become Eq. (4). With the random preimage attack, the complexity of finding a first message block is $2^{3\times 64+1} = 2^{193}$. And we need $2^{256-\lfloor\frac{640-446}{11}\rfloor} = 2^{239}$ operations to find a second message block. So the complexities of attacking 4-round KECCAK-256, SHA3-256 and SHAKE256 are all 2^{239}.

5 Experiments

In this section, we give a practical preimage attack on 3-round KECCAK-224. Related codes, including those for verifying the found messages and for solving the systems on GPU, are available at https://github.com/ysun0102/keccak224.

5.1 Results of Precomputation Stage

We found about $2^{43.41}$ solutions with more than 10 NVIDIA GTX 1080 Ti cards in weeks. The numbers of solutions $\#(sol.)$ that make $n = 50, \ldots, 60$ equations hold are reported in Table 6, together with the practical and theoretical probabilities.

Table 6. Comparisons of practical and theoretical probabilities

n	$\#(sol.)$	Practical probability	Theoretical probability
50	65 469 825	$2^{-17.44}$	$2^{-17.44}$
51	19 262 179	$2^{-19.21}$	$2^{-19.21}$
52	5 185 994	$2^{-21.10}$	$2^{-21.10}$
53	1 271 108	$2^{-23.13}$	$2^{-23.13}$
54	281 252	$2^{-25.30}$	$2^{-25.30}$
55	56 771	$2^{-27.61}$	$2^{-27.62}$
56	9 986	$2^{-30.12}$	$2^{-30.10}$
57	1 591	$2^{-32.77}$	$2^{-32.77}$
58	227	$2^{-35.58}$	$2^{-35.63}$
59	26	$2^{-38.70}$	$2^{-38.70}$
60	2	$2^{-42.41}$	$2^{-42.02}$

From the table, we can see that the theoretical and practical probabilities match very well when the number of solutions is large enough, and there is only a bit of difference between the theoretical and practical probabilities of $n = 58, 60$.

This is mainly because the samples are not adequate. The results in hexadecimal format are given below.

First message block with $n = 58$:

```
|3867ED3B88A48506|FFFFFFFFFFFFFFFF|DD2D9BE5549AE517|FFFFFFFFFFFFFFFF|
|0000000000000000|97CBA3B4524267F6|0000000000000000|F607605E0D17724B|
|0000000000000000|0000000000000000|59E591785BB04788|0000000000000000|
|87A44FB877A61A6E|0000000000000000|0000000000000000|F649DFF78156A578|
|0000000000000000|AC8EB4032E2B8D32|
```

First message block with $n = 59$:

```
|35EF68DC35F1E5EB|FFFFFFFFFFFFFFFF|A1D249A40996BB5F|FFFFFFFFFFFFFFFF|
|0000000000000000|42F30B16705F6ECA|0000000000000000|26A9E432AE324F66|
|0000000000000000|0000000000000000|BBB37F56A6F28967|0000000000000000|
|A9590D7698444C80|0000000000000000|0000000000000000|CCAF1C9CE35C0246|
|0000000000000000|2E22A0E03FE0B8B9|
```

First message block with $n = 60$:

```
|CBB53657E0A66871|FFFFFFFFFFFFFFFF|7537C0597B751AA7|FFFFFFFFFFFFFFFF|
|0000000000000000|A27C4639BB60DFF0|0000000000000000|561C6D11D6A8DE58|
|0000000000000000|0000000000000000|22DF18C837CF65DB|0000000000000000|
|37C8309A24DD20E7|0000000000000000|0000000000000000|4B1668A66C09D25A|
|0000000000000000|14E39DD28900E418|
```

5.2 Results of Preimage Attacks on 3-Round KECCAK-224

Example 1. Let the hash algorithm be 3-round KECCAK-224. Find a second preimage for the message '1' with $length = 1$.

The padded message M and its digest H are given below.

$M(length = 1152)$:

```
|0000000000000003|0000000000000000|0000000000000000|0000000000000000|
|0000000000000000|0000000000000000|0000000000000000|0000000000000000|
|0000000000000000|0000000000000000|0000000000000000|0000000000000000|
|0000000000000000|0000000000000000|0000000000000000|0000000000000000|
|0000000000000000|8000000000000000|
```

$H(length = 224)$:

```
|F4FE7CCEA5D8B144|60F6C316572983A8|A2564CA289E5F897|        CA30DB85|
```

Using the methods in Sect. 4.2, we find three second preimages M_{58}, M_{59}, M_{60} of H based on the first message blocks computed in the last subsection. When $n = 58$, finding the second message block takes a week on 6 GPU cards with approximate $2^{39.39}$ operations, while the second message block of M_{59} costs 3 days on 6 GPU cards by solving about $2^{38.20}$ linear systems. The second message block of M_{60} is obtained in 2 days using 4 GPU cards. Note that these practical complexities of Online Stage are all larger than those estimated in Table 5. We think this is because we only calculate one second message block for each of them. Interested readers can generate more preimages with the published codes, and we believe the complexities will be reasonable then.

From Table 6, the averaged complexities of Precomputation Stage for finding the first message blocks of $n = 58, 59, 60$ are $2^{35.58}$, $2^{38.70}$, and $2^{42.41}$, respectively. So the overall practical complexities of finding M_{58}, M_{59}, and M_{60}, are $2^{39.39}$, $2^{39.47}$, and $2^{42.41}$. However, since the first message blocks need to be computed only once, we suggest using the first message block of $n = 60$, if we want to produce more preimages of H or other digests. The values of M_{58}, M_{59}, M_{60} are listed below.

- $M_{58}(length = 2301)$:
```
|3867ED3B88A48506|FFFFFFFFFFFFFFFF|DD2D9BE5549AE517|FFFFFFFFFFFFFFFF|
|0000000000000000|97CBA3B4524267F6|0000000000000000|F607605E0D17724B|
|0000000000000000|0000000000000000|59E591785BB04788|0000000000000000|
|87A44FB877A61A6E|0000000000000000|0000000000000000|F649DFF78156A578|
|0000000000000000|AC8EB4032E2B8D32|
|C84C8045515BF0C7|61FD4B2BBE00140E|00B252887E479E1D|4CA8454ECB4032EC|
|0980778FEAFC137D|1B4109C0E732BD96|820D1264F56CED03|E3A15B12575B72A2|
|1A068D85C2B37FE0|5DCA726A8F294970|D41129BE08A68BD4|301DD29F5E9BE657|
|98A7904810694A48|B3E8566CE50EA6AA|48C3E4DEB3ADD02B|853EF9C96DC6F02D|
|A72B40AD1F31A630|AAD47114F4750BFC|
```
- $M_{59}(length = 2302)$:
```
|35EF68DC35F1E5EB|FFFFFFFFFFFFFFFF|A1D249A40996BB5F|FFFFFFFFFFFFFFFF|
|0000000000000000|42F30B16705F6ECA|0000000000000000|26A9E432AE324F66|
|0000000000000000|0000000000000000|BBB37F56A6F28967|0000000000000000|
|A9590D7698444C80|0000000000000000|0000000000000000|CCAF1C9CE35C0246|
|0000000000000000|2E22A0E03FE0B8B9|
|CCADB05484618913|CB72585A10CF1D24|5142B6082D69F648|55FF802052E9AFA7|
|5002434225118309|4673F9FF53CF4651|422091CBEE6ED26C|2CED676FB523B95D|
|AF5FD173FA98BE32|1BB7489625D2A58A|1B58D9FB91AD563D|D2F304B902CD182E|
|9F519823A0C16E4D|A54F438AFE22755C|8C39E80475FCDBB0|B908F9B8CD448A94|
|63EF7F66EA21A245|D0A64F63C7333027|
```
- $M_{60}(length = 2302)$:
```
|CBB53657E0A66871|FFFFFFFFFFFFFFFF|7537C0597B751AA7|FFFFFFFFFFFFFFFF|
|0000000000000000|A27C4639BB60DFF0|0000000000000000|561C6D11D6A8DE58|
|0000000000000000|0000000000000000|22DF18C837CF65DB|0000000000000000|
|37C8309A24DD20E7|0000000000000000|0000000000000000|4B1668A66C09D25A|
|0000000000000000|14E39DD28900E418|
|B5B27127B16157CE|1D9CDF75F80E635D|D2024BC09980F06E|43E0D61A974E2162|
|D3E8E4C133283C19|291ADC10C38952D3|0D79C02584D59EB5|5B6EDBF95FBBD637|
|FDF01822DC1C43A3|516EB953B657C03F|8C83A4CFE46AFA61|8EF91ECCAD2D5731|
|3510F4267D8A4D55|13A2BACDCE43348D|0A22C2B955093C72|8836257614188A4E|
|AFBB582F7829B0EB|6CF33EA53BEC3299|
```

6 Conclusion

In this paper, we propose preimage attacks with an allocating approach. We improved the attacks on KECCAK-224/SHA3-224 and KECCAK-256/SHA3-256 /SHAKE256 that are reduced to 3 and 4 rounds. The main idea is to divide the attacking procedure into two stages and to find a 2-block preimage, such that the complexity of each stage is lower than that of finding a 1-block preimage directly. The key step is that, the conditions in Theorem 1 have fewer constraints on the middle state, such that we obtain more degrees of freedom to solve for the first and second message blocks. This is why we obtain better results, compared with previous attacks.

References

1. Aumasson, J., Meier, W.: Zero-sum distinguishers for reduced Keccak-f and for the core functions of Luffa and Hamsi (2009). https://131002.net/data/papers/AM09.pdf
2. Bernstein, D.: Second preimages for 6(7?(8??)) rounds of Keccak. In: NIST Mailing List (2010)
3. Bertoni, G., Daemen, J., Peeters, M., Assche, G.V.: The Keccak reference, version 3.0 (2011). https://keccak.team/keccak.html
4. Chaigneau, C., et al.: Key-recovery attacks on full kravatte. IACR Trans. Symmetric Cryptol. **2018**, 5–28 (2018). https://doi.org/10.13154/tosc.v2018.i1.5-28. https://tosc.iacr.org/index.php/ToSC/article/view/842
5. Chang, D., Kumar, A., Morawiecki, P., Sanadhya, S.: 1st and 2nd preimage attacks on 7, 8 and 9 rounds of Keccak-224,256,384,512. In: SHA-3 Workshop (2014)
6. Daemen, J., Van Assche, G.: Differential propagation analysis of Keccak. In: Canteaut, A. (ed.) FSE 2012. LNCS, vol. 7549, pp. 422–441. Springer, Heidelberg (2012). https://doi.org/10.1007/978-3-642-34047-5_24
7. Dinur, I., Dunkelman, O., Shamir, A.: New attacks on keccak-224 and Keccak-256. In: Canteaut, A. (ed.) FSE 2012. LNCS, vol. 7549, pp. 442–461. Springer, Heidelberg (2012). https://doi.org/10.1007/978-3-642-34047-5_25. Revised Selected Papers
8. Dinur, I., Dunkelman, O., Shamir, A.: Collision attacks on up to 5 rounds of SHA-3 using generalized internal differentials. In: Moriai, S. (ed.) FSE 2013. LNCS, vol. 8424, pp. 219–240. Springer, Heidelberg (2014). https://doi.org/10.1007/978-3-662-43933-3_12. Revised Selected Papers
9. Dinur, I., Dunkelman, O., Shamir, A.: Improved practical attacks on round-reduced Keccak. J. Cryptol. **27**(2), 183–209 (2014)
10. Dinur, I., Morawiecki, P., Pieprzyk, J., Srebrny, M., Straus, M.: Cube attacks and cube-attack-like cryptanalysis on the round-reduced Keccak sponge function. In: Oswald, E., Fischlin, M. (eds.) EUROCRYPT 2015. LNCS, vol. 9056, pp. 733–761. Springer, Heidelberg (2015). https://doi.org/10.1007/978-3-662-46800-5_28
11. Dinur, I., Morawiecki, P.L., Pieprzyk, J., Srebrny, M., Straus, M.L.: Practical complexity cube attacks on round-reduced Keccak sponge function. IACR Cryptology ePrint Archive **2014**, 259 (2014)
12. Guo, J., Liu, M., Song, L.: Linear structures: applications to cryptanalysis of round-reduced KECCAK. In: Cheon, J.H., Takagi, T. (eds.) ASIACRYPT 2016. LNCS, vol. 10031, pp. 249–274. Springer, Heidelberg (2016). https://doi.org/10.1007/978-3-662-53887-6_9
13. Huang, S., Wang, X., Xu, G., Wang, M., Zhao, J.: Conditional cube attack on reduced-round Keccak sponge function. In: Coron, J.-S., Nielsen, J.B. (eds.) EUROCRYPT 2017. LNCS, vol. 10211, pp. 259–288. Springer, Cham (2017). https://doi.org/10.1007/978-3-319-56614-6_9

14. Kölbl, S., Mendel, F., Nad, T., Schläffer, M.: Differential cryptanalysis of Keccak variants. In: Stam, M. (ed.) IMACC 2013. LNCS, vol. 8308, pp. 141–157. Springer, Heidelberg (2013). https://doi.org/10.1007/978-3-642-45239-0_9
15. Li, T., Sun, Y., Liao, M., Wang, D.: Preimage attacks on the round-reduced Keccak with cross-linear structures. IACR Trans. Symmetric Cryptol. **2017**, 39–57 (2017)
16. Li, Z., Bi, W., Dong, X., Wang, X.: Improved conditional cube attacks on Keccak keyed modes with MILP method. In: Takagi, T., Peyrin, T. (eds.) ASIACRYPT 2017. LNCS, vol. 10624, pp. 99–127. Springer, Cham (2017). https://doi.org/10.1007/978-3-319-70694-8_4
17. Morawiecki, P., Pieprzyk, J., Srebrny, M.: Rotational cryptanalysis of round-reduced Keccak. In: Moriai, S. (ed.) FSE 2013. LNCS, vol. 8424, pp. 241–262. Springer, Heidelberg (2014). https://doi.org/10.1007/978-3-662-43933-3_13. Revised Selected Papers
18. Morawiecki, P., Srebrny, M.: A sat-based preimage analysis of reduced Keccak hash functions. Inf. Process. Lett. **113**(10–11), 392–397 (2013)
19. Naya-Plasencia, M., Röck, A., Meier, W.: Practical analysis of reduced-round Keccak. In: Bernstein, D.J., Chatterjee, S. (eds.) INDOCRYPT 2011. LNCS, vol. 7107, pp. 236–254. Springer, Heidelberg (2011). https://doi.org/10.1007/978-3-642-25578-6_18
20. NIST: SHA-3 competition (2007-2012). http://csrc.nist.gov/groups/ST/hash/sha-3/index.html
21. Preneel, B.: The state of cryptographic hash functions. In: Damgård, I.B. (ed.) EEF School 1998. LNCS, vol. 1561, pp. 158–182. Springer, Heidelberg (1999). https://doi.org/10.1007/3-540-48969-X_8
22. Qiao, K., Song, L., Liu, M., Guo, J.: New collision attacks on round-reduced Keccak. In: Coron, J.-S., Nielsen, J.B. (eds.) EUROCRYPT 2017. LNCS, vol. 10212, pp. 216–243. Springer, Cham (2017). https://doi.org/10.1007/978-3-319-56617-7_8
23. Song, L., Liao, G., Guo, J.: Non-full Sbox linearization: applications to collision attacks on round-reduced Keccak. In: Katz, J., Shacham, H. (eds.) CRYPTO 2017. LNCS, vol. 10402, pp. 428–451. Springer, Cham (2017). https://doi.org/10.1007/978-3-319-63715-0_15
24. Song, L., Guo, J., Shi, D.: New MILP modeling: improved conditional cube attacks to Keccak-based constructions. IACR Cryptology ePrint Archive **2017**, 1030 (2017)
25. Stevens, M., Bursztein, E., Karpman, P., Albertini, A., Markov, Y.: The first collision for full SHA-1. In: Katz, J., Shacham, H. (eds.) CRYPTO 2017. LNCS, vol. 10401, pp. 570–596. Springer, Cham (2017). https://doi.org/10.1007/978-3-319-63688-7_19
26. The U.S. National Institute of Standards and Technology Technology: SHA-3 standard: Permutation-based hash and extendable-output functions. In: Federal Information Processing Standard, FIPS 202 (2015). http://nvlpubs.nist.gov/nistpubs/FIPS/NIST.FIPS.202.pdf
27. Wang, X., Yu, H.: How to break MD5 and other hash functions. In: Cramer, R. (ed.) EUROCRYPT 2005. LNCS, vol. 3494, pp. 19–35. Springer, Heidelberg (2005). https://doi.org/10.1007/11426639_2

BISON
Instantiating the Whitened Swap-Or-Not Construction

Anne Canteaut[1], Virginie Lallemand[2], Gregor Leander[2], Patrick Neumann[2], and Friedrich Wiemer[2](✉)

[1] Inria, Paris, France
anne.canteaut@inria.fr
[2] Horst Görtz Institute for IT-Security, Ruhr University Bochum, Bochum, Germany
{virginie.lallemand,gregor.leander,patrick.neumann, friedrich.wiemer}@rub.de

Abstract. We give the first practical instance – BISON – of the Whitened Swap-Or-Not construction. After clarifying inherent limitations of the construction, we point out that this way of building block ciphers allows easy and very strong arguments against differential attacks.

Keywords: Whitened Swap-Or-Not · Instantiating provable security · Block cipher design · Differential cryptanalysis

1 Introduction

Block ciphers are among the most important cryptographic primitives as they are at the core responsible for a large fraction of all our data that is encrypted. Depending on the mode of operation (or used construction), a block cipher can be turned into an encryption function, a hash-function, a message authentication code or an authenticated encryption function.

Due to their importance, it is not surprising that block ciphers are also among the best understood primitives. In particular the Advanced Encryption Standard (AES) [2] has been scrutinized by cryptanalysts ever since its development in 1998 [19] without any significant security threat discovered for the full cipher (see e.g. [6,7,23,26–29]).

The overall structure of AES, being built on several (round)-permutations interleaved with a (binary) addition of round keys is often referred to as key-alternating cipher and is depicted in Fig. 1.

The first cipher following this approach was, to the best of our knowledge, the cipher MMB [17], while the name key-alternating cipher first appears in [20] and in the book describing the design of the AES [21]. Nowadays many secure ciphers follow this construction.

Y. Ishai and V. Rijmen (Eds.): EUROCRYPT 2019, LNCS 11478, pp. 585–616, 2019.
https://doi.org/10.1007/978-3-030-17659-4_20

Interestingly, besides its overwhelming use in practice and the intense cryptanalytic efforts spent to understand its practical security, the generic (or idealized) security of key-alternating ciphers has not been investigated until 2012. Here, generic or idealized security refers to the setting where the round functions R_i are modeled as random permutations. An (computational unbounded) attacker is given access to those round functions via oracle queries and additional oracle access to either the block cipher or a random permutation. The goal of the attacker is to tell apart those two cases. As any attack in this setting is obviously independent of any particular structure of the round function, those attacks are generic for all key-alternating ciphers. In this setting, the construction behind key-alternating ciphers is referred to as the iterated Even-Mansour construction. Indeed, the Even-Mansour cipher [25] can be seen as a one-round version of the key-alternating cipher where the round function is a random permutation.

The first result on the iterated Even-Mansour construction (basically focusing on the two-round version) was given in [10]. Since then, quite a lot of follow-up papers, e.g. [3,30,32,38], managed to improve and generalize this initial result significantly. In particular, [15] managed to give a tight security bound for any number of rounds. Informally, for breaking the r-round Even-Mansour construction, any attacker needs to make roughly $2^{\frac{r}{r+1}n}$ oracle queries.

While this bound can be proven tight for the iterated Even-Mansour construction, it is unsatisfactory for two reasons. First, one might hope to get better security bounds with different constructions and second one might hope to lower the requirement of relying on r random permutations.

Motivated by this theoretical defect and the importance of encrypting small domains with full security (see e.g. [42]), researchers started to investigate alternative ways to construct block ciphers with the highest possible security level under minimal assumptions in ideal models. The most interesting result along those lines is the construction by Tessaro [48]. His construction is based on the Swap-or-Not construction by [31], which was designed for the setting where the component functions are secret. Instead of being based on random permutations, this construction requires only a set of random (Boolean) functions. Tessaro's construction, coined Whitened Swap-Or-Not (WSN for short), requires only two public random (Boolean) functions f_i with n-bit input, and can be proven to achieve full security, see Sect. 2 for more details.

However, and this is the main motivation for our work, *no instance of this construction is known.* This situation is in sharp contrast to the case of the

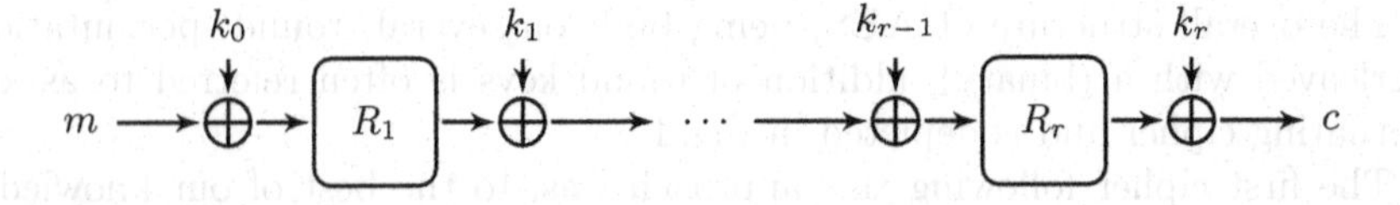

Fig. 1. Key-alternating construction for r rounds, using unkeyed round permutations R_1 to R_r. In practical instantiations, the round keys k_i are typically derived from a master key by some key schedule.

iterated Even-Mansour construction, where many secure instances are known for a long time already, as discussed above.

Without such a concrete instance, the framework of [48] remains of no avail. As soon as one wants to use the framework in any way, one fundamentally has to instantiate the Boolean functions modeled as ideal functionalities by efficiently computable functions. Clearly, the above mentioned bound in the ideal model does not say anything about any concrete instance. Tessaro phrases this situation as follows:

> Heuristically, however, one hopes for even more: Namely, that under a careful implementation of the underlying component, the construction retains the promised security level [48].

There has actually been one instance of the previous construction [31], but this has been broken almost instantaneously and completely, as parts of the encryption function were actually linear, see [52]. This failure to securely instantiate the construction points to an important hurdle. Namely, proving the generic bounds and analyzing the security of an instance are technically very different tasks. The security of any block cipher is, with the current state of knowledge, always the security against known attacks. In particular, when designing any concrete block cipher, one has to argue why linear and differential attacks do not threaten the construction.

Our Contribution

Consequently, in this paper we investigate the important, but so far overlooked, aspect of instantiating the WSN construction with a practical secure instance. Practical secure meaning, just like in the case of AES, that the block cipher resists all known attacks. We denote this instance as **BISON** (for Bent whItened Swap Or Not). Our insights presented here are twofold.

First, we derive some inherent restrictions on the choice of the round function f_i. In a nutshell, we show that f_i has to be rather strong, in the sense that its output bit has to basically depend on all input bits. Moreover, we show that using less than n rounds will always result in an insecure construction. Those, from a cryptanalytic perspective rather obvious, results are presented in Sect. 3. Again, but from a different angle, this situation is in sharp contrast to key-alternating ciphers. In the case of key-alternating ciphers, even with a rather small number of rounds (e.g. ten in the case of AES-128) and rather weak round functions (in case of the AES round function any output bit depends on 32 input bits only and the whole round function decomposes into four parallel functions on 32 bits each) we get ciphers that provide, to the best of our knowledge today and after a significant amount of cryptanalysis, full security.

Second, despite those restrictions of the WSN construction, that have significant impact on the performance of any instance, there are very positive aspects of the WSN construction as well. In Sect. 4, we first define a family of WSN instances which fulfill our initial restrictions.

As we will show in detail, this allows to argue very convincingly that our instance is secure against differential attacks. Indeed, under standard assumptions, we can show that the probability of any (non-trivial) *differential* is upper bounded by 2^{-n+1} where n is the block size, a value that is close to the ideal case. This significantly improves upon what is the state of the art for key-alternating ciphers. Deriving useful bounds on differentials is notoriously hard and normally one therefore has to restrict to bounding the probability of *differential characteristics* only. Our results for differential cryptanalysis can be of independent interest in the analysis of maximally unbalanced Feistel networks or nonlinear feedback shift registers.

We specify our concrete instance as a family of block ciphers for varying input length in Sect. 5. In our instance, we attach importance to simplicity and mathematical clarity. It is making use of bent functions, i.e. maximally non-linear Boolean functions, for instantiating f and linear feedback shift registers (LFSRs) for generating the round keys. Another advantage of BISON is that it defines a whole family of block ciphers, one for any odd block size. In particular it allows the straightforward definition of small scale variants to be used for experiments.

Finally we discuss various other attacks and argue why they do not pose a threat for BISON in Sect. 6. Particularly the discussion on algebraic attacks might be of independent interest. For this we analyse the growth of the algebraic degree over the rounds. In contrast to what we intuitively expect – an exponential growth (until a certain threshold) as in the case for SPNs [11] – the degree of the WSN construction grows linearly in the degree of the round function f_i. This result can also be applied in the analysis of maximally unbalanced Feistel networks or nonlinear feedback shift registers.

Related Work

The first cipher, a Feistel structure, that allowed similarly strong arguments against differential attacks was presented by Nyberg and Knudsen [45], see also [44] for a nice survey on the topic. This cipher was named CRADIC, as Cipher Resistant Against DIfferential Cryptanalysis but is often simply referenced as the KN cipher. However, the cipher has been broken quickly afterwards, with the invention of interpolation attacks [34]. Another, technically very different approach to get strong results on resistance against attacks we would like to mention is the decorrelation theory [51]. Interestingly, both previous approaches rely rather on one strong component, i.e. round function, to ensure security, while the WSN approach, and in particular BISON, gains its resistance against differential attacks step by step.

Regarding the analysis of differentials, extensive efforts have been expended to evaluate the MEDP/MELP of SPN ciphers, and in particular of the AES. Some remarkable results were published by [46] and then subsequently improved by [35] with a sophisticated pruning algorithm. Interestingly, further work by [22] and later by [13] revealed that such bounds are not invariant under affine transformations – an equivalence notion often exploited for classification of S-boxes when studying their strength against differential cryptanalysis. All these works

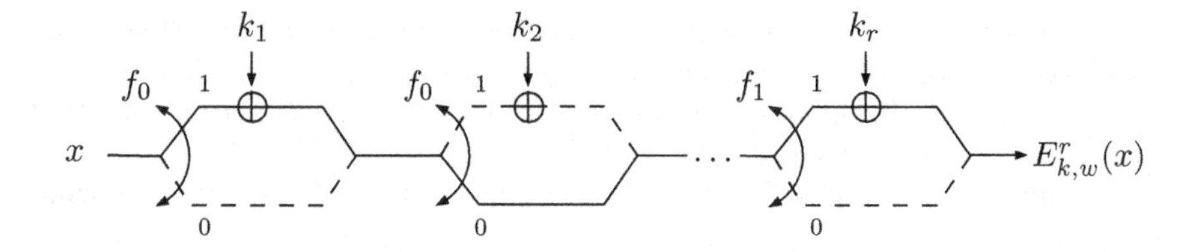

Fig. 2. Schematic view of the WSN construction.

stress out how difficult it is to evaluate the MEDP/MELP of SPNs, even for a small number of rounds. On the contrary, and as we are going to elaborate in the remaining of this paper, computing the MEDP of BISON is rather straightforward and independent of the exact details of the components. This can be compared to the wide trail strategy that, making use of the branch number and the superbox argument, allows bounding the probability of any differential characteristic for a large class of SPNs. Our arguments allow to bound the differential probability for a large class of WSN instances.

2 Preliminaries

We briefly recall the Whitened Swap-or-Not construction, recapitulate properties of Boolean functions and shortly cover differential and linear cryptanalysis. We denote by $\mathbb{F}_2$ the finite field with two elements and by $\mathbb{F}_2^n$ the n-dimensional vector space over $\mathbb{F}_2$, i.e. the set of all n-bit vectors with a bitwise xor as the addition.

2.1 Whitened Swap-or-Not

The WSN is defined as follows. Given two round keys k_i, w_i, the ith round R_{k_i,w_i} computes

$$\begin{aligned} R_{k_i,w_i} &: \mathbb{F}_2^n \to \mathbb{F}_2^n \\ R_{k_i,w_i}(x) &:= x + f_{b(i)}(w_i + \max\{x, x + k_i\}) \cdot k_i \end{aligned}$$

where $f_{0,1} : \mathbb{F}_2^n \to \mathbb{F}_2$ are modeled as two ideal random functions, the max function returns the lexicographic biggest value in the input set, and $+$ denotes the addition in $\mathbb{F}_2$ (the bitwise xor). The index $b(i)$ equals zero for the first half of the rounds and one for the second half (see Fig. 2 for a graphical overview of the encryption process).

In the remainder of the paper, we denote by $E^r_{k,w}(x)$ the application of r rounds of the construction to the input x with round keys k_i and w_i derived from the master key (k, w). Every round is involutory, thus for decryption one only has to reverse the order of the round keys.

Note that the usage of the maximum function is not decisive but that it can be replaced by any function Φ_k that returns a unique representative of the set $\{x, x+k\}$, see [48]. In other words it can be replaced by any function such that $\Phi_k(x) = \Phi_k(y)$ if and only if $y \in \{x, x+k\}$.

The main result given by Tessaro on the security of the WSN is the following:

Proposition 1 (Security of the WSN (Informal) [48]). *The WSN construction is* $(2^{n-\mathcal{O}(\log n)}, 2^{n-\mathcal{O}(1)})$*-secure for* $\mathcal{O}(n)$ *rounds.*

Thus, any adversary trying to distinguish the WSN construction from a random permutation and making at most $2^{n-\mathcal{O}(\log n)}$ queries to the block cipher and $2^{n-\mathcal{O}(1)}$ queries to the underlying function has negligible advantage. Here, the round keys are modeled as independent and uniformly distributed random variables.

2.2 Boolean Functions

A *Boolean function* is defined as a function f mapping n bits to one bit. Any Boolean function

$$f : \mathbb{F}_2^n \rightarrow \mathbb{F}_2$$

can be uniquely expressed by its algebraic normal form (ANF), i.e. as a (reduced) multivariate polynomial with n variables $x_0, \ldots, x_{n-1}$. For $u \in \mathbb{F}_2^n$ we denote

$$x^u = \prod_{i=0}^{n-1} x_i^{u_i}.$$

The ANF of f can be expressed as

$$f(x) = \sum_{u \in \mathbb{F}_2^n} \lambda_u x^u$$

for suitable choices of $\lambda_u \in \mathbb{F}_2$. The degree of f, denoted by $\deg(f)$ is defined as the maximal weight of a monomial present in the ANF of f. That is

$$\deg(f) = \max\{\mathrm{wt}(u) \mid u \in \mathbb{F}_2^n \text{ such that } \lambda_u \neq 0\}.$$

In the context of symmetric cryptography, the differential and linear behavior of a Boolean function play an important role.

The *derivative* of a function f *in direction* α is defined as $\Delta_\alpha(f)(x) := f(x) + f(x+\alpha)$. Informally, studying the behavior of this derivative is at the core of differential cryptanalysis. If we generalize to the derivative of a vectorial Boolean function $F : \mathbb{F}_2^n \rightarrow \mathbb{F}_2^n$, we can additionally specify an output difference β. The differential distribution table (DDT) captures the distribution of all possible derivatives; its entries are

$$\mathrm{DDT}_F[\alpha, \beta] := |\{x \in \mathbb{F}_2^n \mid \Delta_\alpha(F)(x) = \beta\}|,$$

where we leave out the subscript, if it is clear from the context. Note that α is usually referred to as the input difference and β as the output difference.

In a similar way, we can approach the linear behavior of a Boolean function, that is its similarity to any linear function. The *Fourier coefficient* of a function $f : \mathbb{F}_2^n \to \mathbb{F}_2$, which is defined as

$$\widehat{f}(\alpha) := \sum_{x \in \mathbb{F}_2^n} (-1)^{\langle \alpha, x \rangle + f(x)},$$

is a very useful way to measure this similarity. Here, the notation $\langle a, b \rangle$ denotes the inner product, defined as $\langle a, b \rangle := \sum_{i=1}^n a_i b_i$. Recall that any affine Boolean function can be written as $x \mapsto \langle \alpha, x \rangle + c$ for some fixed $\alpha \in \mathbb{F}_2^n$ and a constant $c \in \mathbb{F}_2$. In particular, it follows that any such affine function has one Fourier coefficient equal to $\pm 2^n$. More generally, the *nonlinearity* of f, defined as $\mathrm{NL}(f) := 2^n - \max_\alpha |\widehat{f}(\alpha)|$, measures the minimal Hamming-distance of f to the set of all affine functions.

Analogously to the DDT, for a vectorial Boolean function $F : \mathbb{F}_2^n \to \mathbb{F}_2^n$, we define

$$\widehat{F}(\alpha, \beta) = \sum_{x \in \mathbb{F}_2^n} (-1)^{\langle \alpha, x \rangle + \langle \beta, F(x) \rangle},$$

and the linear approximation table (LAT) contains the Fourier coefficients

$$\mathrm{LAT}_F[\alpha, \beta] := \widehat{F}(\alpha, \beta).$$

Again we leave out the subscript, if it is clear from the context. Here α is usually referred to as the input mask and β as the output mask. Another representation that is sometimes preferred is the *correlation matrix* that in a similar way contains the correlation values for all possible linear approximations, see [18]. The correlation values are simply scaled versions of the Fourier coefficients, i.e.

$$\Pr\left[\langle \alpha, x \rangle + \langle \beta, F(x) \rangle = 0\right] = \frac{1}{2} + \frac{\mathbf{cor}_F(\alpha, \beta)}{2} = \frac{1}{2} + \frac{\widehat{F}(\alpha, \beta)}{2^{n+1}}.$$

The advantage of the correlation matrix notation is that the correlation matrix of a composition of functions is nothing but the product of the corresponding matrices. For the linear approximation table, additional scaling is required.

Bent Functions. As they will play an important role in our design of BISON, we recall the basic facts of bent functions. Boolean functions on an even number n of input bits that achieve the highest possible nonlinearity of $2^n - 2^{\frac{n}{2}}$ are called *bent*. Bent functions have been introduced by Rothaus [47] and studied ever since, see also [14, Section 8.6]. Even so bent functions achieve the highest possible nonlinearity, their direct use in symmetric cryptography is so far very limited. This is mainly due to the fact that bent functions are not balanced, i.e. the distribution of zeros and ones is (slightly) biased.

Using Parseval's equality, it is easy to see that a function is bent if and only if all its Fourier coefficients are $\pm 2^{\frac{n}{2}}$. Moreover, an alternative classification that will be of importance for BISON, is that a function is bent if and only if all (non-trivial) derivatives $\Delta_\alpha(f)$ are balanced Boolean functions [41].

While there are many different primary and secondary constructions[1] of bent functions known, for simplicity and for the sake of ease of implementation, we decided to focus on the simplest known bent functions which we recall next, see also [14, Section 6.2].

Lemma 1 ([24]). *Let $n = 2m$ be an even integer. The function*

$$f : \mathbb{F}_2^m \times \mathbb{F}_2^m \to \mathbb{F}_2$$
$$f(x, y) := \langle x, y \rangle$$

is a quadratic bent function. Moreover, any quadratic bent function is affine equivalent to f.

2.3 Differential and Linear Cryptanalysis

The two most important attacks on symmetric primitives are *differential* and *linear* cryptanalysis, respectively developed by Biham and Shamir [5] and by Matsui [40] for the analysis of the Data Encryption Standard. The general idea for both is to find a non-random characteristic in the differential, resp/linear, behavior of the scheme under inspection. Such a property can then be used as a distinguisher between the cipher and a random permutation and in many cases leads to key-recovery attacks.

It is inherently hard to compute these properties for the whole function, thus one typically exploits the special structure of the cipher. For round-based block ciphers one usually makes use of linear and differential characteristics that specify not only the input and output masks (resp/differences) but also all intermediate masks after the single rounds.

In the case of differential cryptanalysis, an r-round characteristic δ is defined by $(r + 1)$ differences

$$\delta = (\delta_0, \ldots, \delta_r) \in \mathbb{F}_2^{(r+1)n}.$$

For so-called Markov ciphers and assuming the *independence of round keys*, we can compute the probability of a characteristic *averaged over all round-key sequences*:

$$\mathrm{EP}(\delta) = \prod_{i=0}^{r-1} \Pr\left[F(x) + F(x + \delta_i) = \delta_{i+1}\right] = \prod_{i=0}^{r-1} \frac{\mathrm{DDT}_F[\delta_i, \delta_{i+1}]}{2^n},$$

[1] Primary constructions give bent functions from scratch, while secondary constructions build new bent functions from previously defined ones.

where the encryption iterates the round function F for r rounds. Moreover we usually assume the *hypothesis of stochastic equivalence* introduced by Lai *et al.* [37], stating that the actual probability for any fixed round key equals the average.

In contrast to the normal *characteristic* that defines the exact differences before and after each round, a *differential* takes every possible intermediate differences into account and fixes only the overall input and output differences (which are the two values an attacker can typically control).

However, while bounding the average probability of a differential characteristic is easily possible for many ciphers (using in particular the wide-trail strategy introduced in [16]), bounding the average probability of a differential, which is denoted as the expected differential probability (EDP), is not. Nevertheless, some effort was spent to prove bounds on the maximum EDP (MEDP) for two rounds of some key-alternating ciphers [13,21,33,46].

Similarly, for linear cryptanalysis, an r-round characteristic (also called trail or path) for a round function F is defined by $(r+1)$ masks

$$\theta = (\theta_0, \ldots, \theta_r) \in \mathbb{F}_2^{(r+1)n}$$

and its correlation is defined as

$$\mathbf{cor}_F(\theta) := \prod_{i=0}^{r-1} \mathbf{cor}_F(\theta_i, \theta_{i+1}) = \prod_{i=0}^{r-1} \frac{\widehat{F}(\theta_i, \theta_{i+1})}{2^n}$$

and it can be shown that the correlation of a composition can be computed as the sum over the trail correlations. More precisely,

$$\mathbf{cor}_{E_k^r}(\alpha, \beta) = \sum_{\substack{\theta \\ \theta_0=\alpha, \theta_r=\beta}} \mathbf{cor}_F(\theta), \tag{1}$$

where the encryption E_k^r iterates the round function F for r rounds.

This is referred to as the linear hull (see [43]). While not visible in order to simplify notation, the terms in Eq. (1) are actually key dependent and thus for some keys they either could cancel out or amplify the overall correlation. For more background, we refer to e.g. [9] and [36]. For a key-alternating cipher with independent round keys, the average over all round-key sequences of the correlation $\mathbf{cor}_{E_k^r}(\alpha, \beta)$ is zero for any pair of nonzero masks (α, β) (see e.g. [21, Section 7.9]). Then, the most relevant parameter of the distribution is its variance, which corresponds to the average square correlation, and is called the *expected linear potential*. Again, bounding the ELP is out of reach for virtually any practical cipher, while for bounding the correlation of a single trail, one can again use the wide-trail strategy mentioned above. Upper bounds for the MELP of two rounds of AES are also given in [13,33,46].

Finally we would like to note that the round keys in an actual block cipher instance are basically never independent and identically distributed over the full key space, but instead derived from a key schedule, rendering the above assumption plain wrong. While the influence of key schedules is a crucially understudied topic and for specific instances strange effects can occur, see [1,36], the above assumption are seen as valid heuristics for most block ciphers.

3 Inherent Restrictions

In this section we point out two inherent restrictions on any practical secure instance, i.e. generic for the WSN construction. Those restrictions result in general conditions on both the minimal number of rounds to be used and general properties of the round functions $f_{b(i)}$. In particular, those insights are taken into account for BISON. While these restrictions are rather obvious from a cryptanalytic point of view, they have a severe impact on the performance of any concrete instance. We discuss performance in more detail in the full version [12, Section 7].

3.1 Number of Rounds

As in every round of the cipher, we simply add (or not) the current round key k_i, the ciphertext can always be expressed as the addition of the plaintext and a (message dependent) linear combination of all round keys k_i. The simple but important observation to be made here is that, as long as the round keys do not span the full space, the block cipher is easily attackable.

Phrased in terms of linear cryptanalysis we start with the following lemma.

Lemma 2. *For any number of rounds $r < n$ there exists an element $u \in \mathbb{F}_2^n \setminus \{0\}$ such that*

$$\widehat{E_{k,w}^r}(u, u) = 2^n,$$

that is the equation

$$\langle u, x \rangle = \langle u, E_{k,w}^r(x) \rangle$$

holds for all $x \in \mathbb{F}_2^n$.

Proof. Let $k_1, \ldots, k_r$ be the round keys derived from k and denote by

$$U = \text{span}\{k_1, \ldots, k_r\}^\perp$$

the dual space of the space spanned by the round keys, i.e.

$$\forall u \in U, \forall 1 \leqslant i \leqslant r \text{ it holds that } \langle u, k_i \rangle = 0.$$

As $r < n$ by assumption, the dimension of span $\{k_1, \ldots, k_r\}$ is smaller than n and thus $U \neq \{0\}$. Therefore, U contains a non-zero element

$$u \in \text{span}\{k_1, \ldots, k_r\}^\perp$$

and it holds that

$$\langle u, E^r_{k,w}(x)\rangle = \langle u, x + \sum_{i=1}^{r} \lambda_i k_i\rangle = \langle u, x\rangle + \langle u, \sum_{i=1}^{r} \lambda_i k_i\rangle = \langle u, x\rangle. \quad (2)$$

Even more importantly, this observation leads directly to a known plaintext attack with very low data-complexity. Given a set of t plaintext/ciphertext (p_i, c_i) pairs, an attacker simply computes

$$V = \text{span}\{p_i + c_i \mid 1 \leqslant i \leqslant t\} \subseteq \text{span}\{k_j \mid 1 \leqslant j \leqslant r\}.$$

Given $t > r$ slightly more pairs than rounds, and assuming that $p_i + c_i$ is uniformly distributed in span $\{k_j\}$ (otherwise the attack only gets even stronger)[2] implies that

$$V = \text{span}\{k_j\}$$

with high probability and V can be efficiently computed. Furthermore, as above dim(span $\{k_j\}$) is at most r, we have $V^\perp \neq \{0\}$. Given any $u \neq 0$ in $V^\perp$ allows to compute one bit of information on the plaintext given only the ciphertext and particularly distinguish the cipher from a random permutation in a chosen-plaintext setting efficiently.

A similar argument shows the following:

Lemma 3. *For any number of rounds r smaller than $2n-3$ there exist nonzero α and β, such that*

$$\widehat{E^r_{k,w}}(\alpha, \beta) = 0$$

Proof. We restrict to the case $r \geqslant n$ as otherwise the statement follows directly from the lemma above. Indeed, from Parseval equality, the fact that $\widehat{E^r_{k,w}}(\alpha, \alpha) = 2^n$ implies that $\widehat{E^r_{k,w}}(\alpha, \beta) = 0$ for all $\beta \neq \alpha$. Let $k_1, \ldots, k_r$ be the round keys derived from k and choose non-zero elements $\alpha \neq \beta$ such that

$$\alpha \in \text{span}\{k_1, \ldots, k_{n-2}\}^\perp \qquad \text{and} \qquad \beta \in \text{span}\{k_{n-1}, \ldots, k_r\}^\perp.$$

Note that, as $r \leq 2n-3$ by assumption such elements always exist. Next, we split the encryption function in two parts, the first $n-2$ rounds E_1 and the remaining $r-(n-2) < n$ rounds E_2, i.e.

$$E^r_{k,w} = E_2 \circ E_1.$$

[2] E.g. if, with high probability, the $p_i + c_i$ do not depend on one or more k_j's, the described attack can be extended to one or more rounds with high probability.

We can compute the Fourier coefficient of $E^r_{k,w}$ as

$$\widehat{E^r_{k,w}}(\alpha, \beta) = \sum_{\gamma \in \mathbb{F}_2^n} \frac{\widehat{E_1}(\alpha, \gamma)}{2^n} \cdot \frac{\widehat{E_2}(\gamma, \beta)}{2^n}.$$

Now, the above lemma and the choices of α and β imply that $\widehat{E_1}(\alpha, \gamma) = 0$ for $\gamma \neq \alpha$ and $\widehat{E_2}(\gamma, \beta) = 0$ for $\gamma \neq \beta$. Recalling that $\alpha \neq \beta$ by construction concludes the proof. ♠

However, as the masks α and β depend on the key, and unlike above there does not seem to be an efficient way to compute those, we do not see a direct way to use this observation for an attack.

Summarizing the observations above, we get the following conclusion:

Rationale 1. *Any practical instance must iterate at least n rounds. Furthermore, it is beneficial if any set of n consecutive round keys are linearly independent.*[3]

After having derived basic bounds on the number of rounds for any secure instance, we move on to criteria on the round function itself.

3.2 Round Function

Here, we investigate a very basic criterion on the round function, namely dependency on all input bits. Given the Boolean functions $f_{b(i)} : \mathbb{F}_2^n \to \mathbb{F}_2$ used in the round function of $E^r_{k,w}$, an important question is, if it is necessary that the output bit of $f_{b(i)}$ has to depend on all input bits. It turns out that this is indeed strictly necessary for any secure instance, as summarized in the next rational.

Rationale 2. *For a practical instance, the functions $f_{b(i)}$ has to depend on all bits. Even more, for any $\delta \in \mathbb{F}_2^n$ the probability of*

$$f_{b(i)}(x) = f_{b(i)}(x + \delta)$$

should be close to $\frac{1}{2}$.

Due to page constraints, we refer to [12, Lemma 4] for more details. It is worth noticing that the analysis leading to Rationale 2 applies to the original round function. However, as pointed out in [49, Section 3.1], in the definition of the round function, we can replace the function

$$x \mapsto \max\{x, x + k\}$$

[3] If (some) round keys are linearly dependent, Lemma 3 can easily be extended to more rounds.

by any function Φ_k such that $\Phi_k(x) = \Phi_k(x + k)$ for all x. While the following sections will focus on the case when Φ_k is linear, we will prove that Rationale 2 is also valid in this other setting.

Again, this should be compared to key-alternating ciphers, where usually not all output bits of a single round function depend on all input bits. For example for AES any output bit after one round depends only on 32 input bits and for PRESENT any output bit only depends on 4 input bits. However, while for key-alternating ciphers this does not seem to be problematic, and indeed allows rather weak round functions to result in a secure scheme, for the WSN construction the situation is very different.

Before specifying our exact instance, we want to discuss differential cryptanalysis of a broader family of instances.

4 Differential Cryptanalysis of BISON-Like Instances

We coin an instance of the WSN construction "BISON-like", if it iterates at least n rounds with linearly independent round keys $k_1, \ldots, k_n$ and applies Boolean functions $f_{b(i)}$. As explained in [49, Section 3.1], in order to enable decryption it is required that the Boolean functions $f_{b(i)}$ return the same result for both x and $x + k$. In the original proposition by Tessaro, this is achieved by using the max function in the definition of the round function. Using this technique reduces the number of possible inputs for the $f_{b(i)}$ to 2^{n-1}. To simplify the analysis and to ease notation, we replace the max function with a *linear function* $\Phi_k : \mathbb{F}_2^n \to \mathbb{F}_2^{n-1}$ with $\ker(\Phi_k) = \{0, k\}$. From now on, we assume that any BISON-like instance uses such a Φ_k instead of the max function. The corresponding round function has then the following form

$$R_{k_i,w_i}(x) := x + f_{b(i)}(w_i + \Phi_{k_i}(x))k_i. \tag{3}$$

With the above conditions, any BISON-like instance of the WSN construction is resistant to differential cryptanalysis, as we show in the remainder of this section.

For our analysis, we make two standard assumptions in symmetric cryptanalysis as mentioned above: the *independence of whitening round keys* w_i and the *hypothesis of stochastic equivalence* with respect to these round keys. That is, we assume round keys w_i to be independently uniformly drawn and the resulting EDP to equal the differential probabilities averaged over all w. Thus, the keys w_i act very much like the round key for a key-alternating cipher with respect to the probabilities of characteristics. We further back up this intuition by practical experiments (see Sect. 6.3 and [12, Appendix B]). For the round keys k_i we do not have to make such assumptions.

We first discuss the simple case of differential behaviour for one round only and then move up to an arbitrary number of rounds and devise the number of possible output differences and their probabilities.

4.1 From One-Round Differential Characteristics

Looking only at one round, we can compute the DDT explicitly:

Proposition 2. *Let $R_{k_i,w_i} : \mathbb{F}_2^n \to \mathbb{F}_2^n$ be the WSN round function as in Eq. (3). Then its* DDT *consists of the entries*

$$\text{DDT}_R[\alpha,\beta] = \begin{cases} 2^{n-1} + \widehat{\Delta_{\Phi_k(\alpha)}(f)}(0) & \text{if } \beta = \alpha \\ 2^{n-1} - \widehat{\Delta_{\Phi_k(\alpha)}(f)}(0) & \text{if } \beta = \alpha + k \cdot \\ 0 & \text{otherwise.} \end{cases} \tag{4}$$

Most notably, if f is bent, we have

$$\text{DDT}_R[\alpha,\beta] = \begin{cases} 2^{n} & \text{if } \alpha = \beta = k \text{ or } \alpha = \beta = 0 \\ 2^{n-1} & \text{if } \beta \in \{\alpha, \alpha + k\} \text{ and } \alpha \notin \{0,k\} \ . \\ 0 & \text{otherwise.} \end{cases}$$

Proof. We have to count the number of solutions of $R(x) + R(x+\alpha) = \beta$:

$$\begin{aligned} \text{DDT}_R[\alpha,\beta] &= |\{x \in \mathbb{F}_2^n \mid R(x) + R(x+\alpha) = \beta\}| \\ &= |\{x \in \mathbb{F}_2^n \mid \alpha + [f(w + \Phi_k(x)) + f(w + \Phi_k(x+\alpha))] \cdot k = \beta\}| \end{aligned}$$

Since f takes its values in $\mathbb{F}_2$, the only output differences β such that $\text{DDT}_R[\alpha,\beta]$ may differ from 0 are $\beta = \alpha$ and $\beta = \alpha + k$. When $\beta = \alpha$, we have

$$\begin{aligned} \text{DDT}_R[\alpha,\alpha] &= |\{x \in \mathbb{F}_2^n \mid f(w + \Phi_k(x)) + f(w + \Phi_k(x+\alpha)) = 0\}| \\ &= |\{x \in \mathbb{F}_2^n \mid f(w + \Phi_k(x)) + f(w + \Phi_k(x) + \Phi_k(\alpha)) = 0\}| \\ &= 2 \cdot \left|\{x' \in \mathbb{F}_2^{n-1} \mid f(x') + f(x' + \Phi_k(\alpha)) = 0\}\right| \\ &= 2\left(2^{n-2} + \frac{1}{2}\widehat{\Delta_{\Phi_k(\alpha)}(f)}(0)\right). \end{aligned}$$

Similarly,

$$\begin{aligned} \text{DDT}_R[\alpha,\alpha+k] &= |\{x \in \mathbb{F}_2^n \mid f(w + \Phi_k(x)) + f(w + \Phi_k(x+\alpha)) = 1\}| \\ &= 2\left(2^{n-2} - \frac{1}{2}\widehat{\Delta_{\Phi_k(\alpha)}(f)}(0)\right). \end{aligned}$$

Most notably, when $\alpha \in \{0,k\}$, $\widehat{\Delta_{\Phi_k(\alpha)}(f)}(0) = 2^{n-1}$. Moreover, when f is bent, $\widehat{\Delta_{\Phi_k(\alpha)}(f)}(0) = 2^{n-2}$ for all other values of α. ∎

4.2 To Differentials over More Rounds

As previously explained, it is possible to estimate the probability of a differential characteristic over several rounds, averaged over the round keys, when the cipher is a Markov cipher. We now show that this assumption holds for any BISON-like instance of the WSN construction.

Lemma 4. *Let $R_{k,w} : \mathbb{F}_2^n \to \mathbb{F}_2^n$ be the WSN round function as in Eq. (3). For any fixed $k \in \mathbb{F}_2^n$ and any differential $(\alpha, \beta) \in \mathbb{F}_2^n \times \mathbb{F}_2^n$, we have that*

$$\Pr_w\left[R_{k,w}(x+\alpha) + R_{k,w}(x) = \beta\right]$$

is independent of x. More precisely

$$\Pr_w\left[R_{k,w}(x+\alpha) + R_{k,w}(x) = \beta\right] = \Pr_x\left[R_{k,w}(x+\alpha) + R_{k,w}(x) = \beta\right] .$$

Proof. We have

$$\begin{aligned}
&\left\{w \in \mathbb{F}_2^{n-1} \mid \Delta_\alpha(R_{k,w})(x) = \beta\right\} \\
&= \left\{w \in \mathbb{F}_2^{n-1} \mid \left(\Delta_{\Phi_k(\alpha)}(f)(w + \Phi_k(x))\right) \cdot k = \alpha + \beta\right\} \\
&= \begin{cases} \emptyset & \text{if } \beta \notin \{\alpha, \alpha + k\} \\ \Phi_k(x) + \mathsf{Supp}\left(\Delta_{\Phi_k(\alpha)}(f)\right) & \text{if } \beta = \alpha + k \\ \Phi_k(x) + \left(\mathbb{F}_2^{n-1} \setminus \mathsf{Supp}\left(\Delta_{\Phi_k(\alpha)}(f)\right)\right) & \text{if } \beta = \alpha, \end{cases}
\end{aligned}$$

where $\mathsf{Supp}(g)$ denotes the support of a Boolean function g, i.e., the values x for which $g(x) = 1$. Clearly, the cardinality of this set does not depend on x. Moreover, this cardinality, divided by 2^{n-1}, corresponds to the value of

$$\Pr_x\left[R_{k,w}(x+\alpha) + R_{k,w}(x) = \beta\right]$$

computed in the previous proposition. ∎

By induction on the number of rounds, we then directly deduce that any BISON-like instance of the WSN construction is a Markov cipher in the sense of the following corollary.

Corollary 1. *Let $E_{k,w}^i$ denote i rounds of a BISON-like instance of the WSN construction with round function R_{k_i,w_i}. For any number of rounds r and any round keys $(k_1, \ldots, k_r)$, the probability of an r-round characteristic δ satisfies*

$$\Pr_w\left[E_{k,w}^i(x) + E_{k,w}^i(x + \delta_0) = \delta_i, \forall 1 \leqslant i \leqslant r\right] = \prod_{i=1}^{r} \Pr_x\left[R_{k_i,w_i}(x) + R_{k_i,w_i}(x + \delta_{i-1}) = \delta_i\right].$$

For many ciphers several differential characteristics can cluster in a *differential* over more rounds. This is the main reason why bounding the probability of differentials is usually very difficult if possible at all. For BISON-like instances the situation is much nicer; we can actually compute the EDP, i.e., the probabilities of the differentials averaged over all whitening key sequences $(w_1, \ldots, w_r)$. This comes from the fact that any differential for less than n rounds contains at most one differential characteristic with non-zero probability. To understand this behavior, let us start by analyzing the EDP (averaged over the w_i) and by determining the number of possible output differences.

In the following, we assume that the input difference α is fixed, and we calculate the number of possible output differences. We show that this quantity depends on the relation between α and the k_i.

Lemma 5. *Let us consider r rounds of a* BISON*-like instance of the WSN construction with round function involving Boolean functions $f_{b(i)}$ having no (non-trivial) constant derivative. Assume that the first n round keys $k_1, \ldots, k_n$ are linearly independent, and that $k_{n+1} = k_1 + \sum_{i=2}^{n} \gamma_i k_i$ for $\gamma_i \in \mathbb{F}_2$. For any non-zero input difference α, the number of possible output differences β such that*

$$\Pr_{w,x}\left[E^r_{k,w}(x+\alpha) + E^r_{k,w}(x) = \beta\right] \neq 0$$

is

$$\begin{cases} 2^r & \text{if } \alpha \notin \operatorname{span}\{k_i\} \text{ and } r < n, \\ 2^r - 2^{r-\ell} & \text{if } \alpha = k_\ell + \sum_{i=1}^{\ell-1} \lambda_i^\alpha k_i \text{ and } r \leqslant n, \\ 2^n - 1 & \text{if } r > n. \end{cases}$$

Proof. By combining Corollary 1 and Proposition 2, we obtain that the average probability of a characteristic $(\delta_0, \delta_1, \ldots, \delta_{r-1}, \delta_r)$ can be non-zero only if $\delta_i \in \{\delta_{i-1}, \delta_{i-1} + k_i\}$ for all $1 \leqslant i \leqslant r$. Therefore, the output difference δ_r must be of the form $\delta_r = \delta_0 + \sum_{i=1}^{r} \lambda_i k_i$ with $\lambda_i \in \mathbb{F}_2$. Moreover, for those characteristics, the average probability is non-zero unless there exists $1 \leqslant i < r$ such that $|\widehat{\Delta_{\Phi_{k_i}(\delta_i)}(f_{b(i)})}(0)| = 2^{n-1}$, i.e. $\Delta_{\Phi_{k_i}(\delta_i)}(f_{b(i)})$ is constant. By hypothesis, this only occurs when $\delta_i \in \{0, k_i\}$, and the impossible characteristics correspond to the case when either $\delta_i = 0$ or $\delta_{i+1} = 0$. It follows that the valid characteristics are exactly the characteristics of the form

$$\delta_i = \delta_0 + \sum_{j=1}^{i} \lambda_j k_j$$

where none of the δ_i vanishes.

- When the input difference $\alpha \notin \operatorname{span}\{k_i\}$, for any given output difference $\beta = \alpha + \sum_{i=1}^{r} \lambda_i k_i$, the r-round characteristic

$$(\alpha, \alpha + \lambda_1 k_1, \alpha + \lambda_1 k_1 + \lambda_2 k_2, \ldots, \alpha + \sum_{i=1}^{r} \lambda_i k_i)$$

  is valid since none of the intermediate differences vanishes.

– When $r \leqslant n$ and $\alpha = k_\ell + \sum_{i=1}^{\ell-1} \lambda_i^\alpha k_i$, the only possible characteristic from α to $\beta = \alpha + \sum_{i=1}^{r} \lambda_i k_i$ satisfies

$$\delta_j = \begin{cases} \sum_{i=1}^{j}(\lambda_i + \lambda_i^\alpha)k_i + \sum_{i=j+1}^{\ell} \lambda_i^\alpha k_i & \text{if } j \leqslant \ell \\ \sum_{i=1}^{\ell}(\lambda_i + \lambda_i^\alpha)k_i + \sum_{i=\ell+1}^{j} \lambda_i k_i & \text{if } j > \ell\,. \end{cases}$$

Since the involved round keys are linearly independent, we deduce that $\delta_j = 0$ only when $j = \ell$ and $\lambda_i = \lambda_i^\alpha$ for all $1 \leqslant i \leqslant \ell$. It then follows that there exists a valid characteristic from α to β unless $\lambda_i = \lambda_i^\alpha$ for all $1 \leqslant i \leqslant \ell$. The number of possible outputs β is then

$$(2^\ell - 1)2^{r-\ell} = 2^r - 2^{r-\ell}.$$

– If we increase the number of rounds to more than n, we have $\alpha = k_\ell + \sum_{i=1}^{\ell-1} \lambda_i^\alpha k_i$ for some $\ell \leqslant n$. If $\beta = \alpha + \sum_{i=1}^{n} \lambda_i k_i$ with $\sum_{i=1}^{\ell} \lambda_i k_i \neq \alpha$, then we can obviously extend the previous n-round characteristic to

$$(\alpha, \alpha + \lambda_1 k_1, \ldots, \alpha + \sum_{i=1}^{n-1} \lambda_i k_i, \beta, \beta, \ldots, \beta).$$

If $\sum_{i=1}^{\ell} \lambda_i k_i = \alpha$, β cannot be the output difference of an n-round characteristic. However, the following $(n+1)$-round characteristic starting from $\delta_0 = \alpha$ is valid:

$$\delta_j = \begin{cases} k_1 + \sum_{i=2}^{j} \gamma_i k_i + \sum_{i=j+1}^{\ell} \lambda_i^\alpha k_i & \text{if } j \leqslant \ell \\ k_1 + \sum_{i=2}^{j} \gamma_i k_i + \sum_{i=\ell+1}^{j} \lambda_i k_i & \text{if } \ell < j \leqslant n \\ \beta & \text{if } j = n+1 \end{cases}$$

Indeed, $\delta_n = \beta + k_n$ implying that the last transition is valid. Moreover, it can be easily checked that none of these δ_j vanishes, unless $\beta = 0$. This implies that all non-zero output differences β are valid.

The last case in the above lemma is remarkable, as it states any output difference is possible after $n + 1$ rounds. To highlight this, we restate it as the following corollary.

Corollary 2. *For* BISON*-like instances with more than n rounds whose round keys $k_1, \ldots, k_{n+1}$ satisfy the hypothesis of Lemma 5, and for any non-zero input difference, every non-zero output difference is possible.*

We now focus on a reduced version of the cipher limited to exactly n rounds and look at the probabilities for every possible output difference. Most notably, we exhibit in the following lemma an upper-bound on the MEDP which is minimized when n is odd and the involved Boolean functions $f_{b(i)}$ are bent. In other words, Rationale 2 which was initially motivated by the analysis in Sect. 3 for the original round function based on $x \mapsto \max(x, x+k)$ [48] is also valid when a linear function Φ_k is used.

Lemma 6. *Let us consider n rounds of a* BISON*-like instance of the WSN construction with round function involving Boolean functions $f_{b(i)}$. Let $k_1, \ldots, k_n$ be any linearly independent round keys. Then, for any input difference $\alpha \neq 0$ and any β, we have*

$$\begin{aligned}\mathrm{EDP}(\alpha, \beta) &= \Pr_{w,x}\left[E_{k,w}(x+\alpha) + E_{k,w}(x) = \beta\right] \\ &\leqslant \left(\frac{1}{2} + 2^{-n} \max_{1 \leqslant i \leqslant n} \max_{\delta \neq 0} \left|\widehat{\Delta_\delta(f_{b(i)})}(0)\right|\right)^{n-1}.\end{aligned}$$

More precisely, if all $f_{b(i)}$ are bent*,*

$$\mathrm{EDP}(\alpha, \beta) = \begin{cases} 0 & \text{if } \beta = \sum_{i=\ell+1}^{n} \lambda_i k_i, \quad (5) \\ 2^{-n+1} & \text{if } \beta = k_\ell + \sum_{i=\ell+1}^{n} \lambda_i k_i, \quad (6) \\ 2^{-n} & \text{otherwise}, \quad (7) \end{cases}$$

where ℓ denotes as previously the latest round key that appears in the decomposition of α into the basis $(k_1, \ldots, k_n)$, that is $\alpha = k_\ell + \sum_{i=1}^{\ell-1} \lambda_i k_i$.

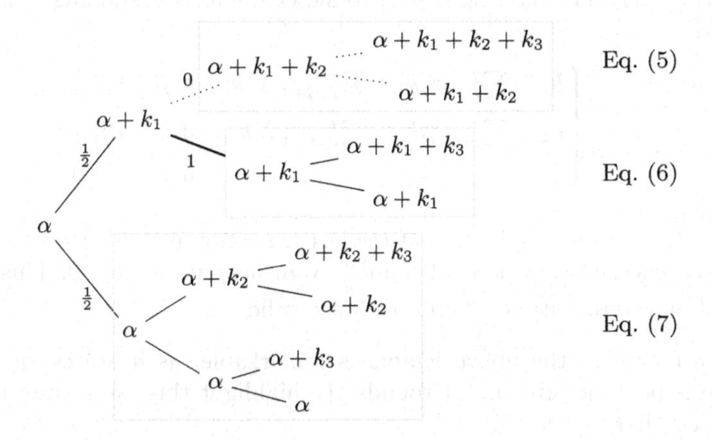

Fig. 3. Probabilities of output differences for three rounds and the cases of the input difference $\alpha = k_1 + k_2$, thus $\ell = 2$. Dotted transitions are impossible.

The case of bent functions is visualized in Fig. 3, where we give an example of the three possibilities for three rounds.

Proof. As proved in Lemma 5, (α, β) is an impossible differential if and only if $\beta = \sum_{i=\ell+1}^{n} \lambda_i k_i$. For all other values of $\beta = \alpha + \sum_{i=1}^{n} \lambda_i k_i$, we have

$$\mathrm{EDP}(\alpha, \beta) = \prod_{i=1}^{n} \left(\frac{1}{2} + (-1)^{\lambda_i} 2^{-n} \widehat{\Delta_{\Phi_{k_i}(\delta_i)}(f_{b(i)})}(0)\right)$$

where $\delta_i = \alpha + \sum_{j=1}^{i} \lambda_j k_j$. The i-th term in the product is upper-bounded by

$$\frac{1}{2} + 2^{-n} \max_{1 \leqslant i \leqslant n} \max_{\delta \neq 0} \left| \widehat{\Delta_\delta(f_{b(i)})}(0) \right|$$

except if $\Phi_{k_i}(\delta_i) = 0$, i.e., $\delta_i \in \{0, k_i\}$. As seen in Lemma 5, the case $\delta_i = 0$ cannot occur in a valid characteristic. The case $\delta_i = k_i$ occurs if and only if $i = \ell$ and $\beta = k_\ell + \sum_{j=\ell+1}^{n} \lambda_j k_j$. In this situation, the ℓ-th term in the product equals 1. In the tree of differences this is visible as the collapsing of the two branches from two possible succeeding differences into only one, which then of course occurs with probability one, see upper branch of Fig. 3.

Most notably, all $f_{b(i)}$ are bent if and only if

$$\max_{1 \leqslant i \leqslant n} \max_{\delta \neq 0} \left| \widehat{\Delta_\delta(f_{b(i)})}(0) \right| = 0,$$

leading to the result.

This can be seen on Fig. 3: the $2^{n-\ell}$ possible differences coming from the collapsed branch have a transition of probability one in that round, resulting in an overall probability of 2^{-n+1}, see Eq. (6). For the lower part of Fig. 3, all the other differences are not affected by this effect and have a probability of 2^{-n}, see Eq. (7).

Because they allow us to minimize the MEDP, we now concentrate on the case of bent functions for the sake of simplicity, which implies that the block size is odd. However, if an even block size is more appropriate for implementation reasons, we could also define BISON-like instances based on maximally nonlinear functions.

It would be convenient to assume in differential cryptanalysis that the EDP of a differential does not increase when adding more rounds, while this does not hold in general. However, this argument can easily be justified for BISON-like instances using bent functions, when averaging over the whitening keys w.

Proposition 3. *Let us consider $r \geqslant n$ rounds of a BISON-like instance of the WSN construction with bent functions $f_{b(i)}$. Let $k_1, \dots, k_n$ be any linearly independent round keys. Then the probability of any non-trivial differential, averaged over all whitening key sequences w is upper bounded by 2^{-n+1}.*

In other words, the MEDP of BISON-like instances with bent $f_{b(i)}$ for $r \geqslant n$ rounds is 2^{-n+1}.

Proof. By induction over r. The base case for $r = n$ rounds comes from Lemma 6. In the induction step, we first consider the case when the output difference β after r rounds differs from k_r. Then the output difference $\delta_r = \beta$ can be reached if and only if the output difference after $(r-1)$ rounds δ_{r-1} belongs to $\{\beta, \beta + k_r\}$.

Then,

$$\begin{aligned}\mathrm{EDP}^r(\alpha,\beta) &= \Pr_{w_r}\left[R_{k_r,w_r}(x_r) + R_{k_r,w_r}(x_r+\beta) = \beta\right]\mathrm{EDP}^{r-1}(\alpha,\beta)\\ &\quad + \Pr_{w_r}\left[R_{k_r,w_r}(x_r) + R_{k_r,w_r}(x_r+\beta+k_r) = \beta\right]\mathrm{EDP}^{r-1}(\alpha,\beta+k_r)\\ &= \frac{1}{2}\left(\mathrm{EDP}^{r-1}(\alpha,\beta) + \mathrm{EDP}^{r-1}(\alpha,\beta+k_r)\right) \leqslant 2^{-n+1}\,.\end{aligned}$$

When the output difference β after r rounds equals k_r, it results from $\delta_{r-1} = k_r$ with probability 1. In this case

$$\mathrm{EDP}^r(\alpha,\beta) = \mathrm{EDP}^{r-1}(\alpha,\beta) \leqslant 2^{-n+1}\,.$$

This bound is close to the ideal case, in which each differential has probability $1/(2^n-1)$.

We now give a detailed description of our instance BISON.

5 Specification of BISON

As BISON-like instances should obviously generalise BISON, this concrete instance inherits the already specified parts. Thus BISON uses two bent functions $f_{b(i)}$, replaces the max function by Φ_k, and uses a key schedule that generates round keys, where all n consecutive round keys are linearly independent. The resulting instance for n bits iterates the WSN round function as defined below over $3 \cdot n$ rounds. The chosen number of rounds mainly stems from the analysis of the algebraic degree that we discuss in Sect. 6.

Security Claim. *We claim n-bit security for* BISON *in the single-key model. We emphasize that we do not claim any security in the related-key, chosen-key or known-key model.*

5.1 Round Function

For any nonzero round key k, we define $\Phi_k : \mathbb{F}_2^n \to \mathbb{F}_2^{n-1}$ as

$$\Phi_k(x) := (x_{i(k)} \cdot k + x)[1,\ldots,i(k)-1, i(k)+1,\ldots,n], \tag{8}$$

where $i(k)$ denotes the index of the lowest bit set to 1 in k, and the notation $x[1,\ldots,j-1,j+1,\ldots,n]$ returns the $(n-1)$-bit vector, consisting of the bits of x except the jth bit.

Lemma 7. *The function* $\Phi_k : \mathbb{F}_2^n \to \mathbb{F}_2^{n-1}$ *is linear and satisfies*

$$\ker(\Phi_k) = \{0, k\}.$$

The proof can be done by simply computing both outputs for x and $x + k$. For the preimage of $y \in \mathbb{F}_2^{n-1}$ and $j = i(k)$ we have

$$\Phi_k^{-1}(y) \in \left\{ \begin{array}{l} (y[1:j-1], 0, y[j:n-1]) + k[1:n], \\ (y[1:j-1], 0, y[j:n-1]) \end{array} \right\}. \tag{9}$$

Due to the requirement for the $f_{b(i)}$ being bent, we are limited to functions taking an even number of bits as input. The simplest example of a bent function is the inner product.

Eventually we end up with the following instance of the WSN round.

BISON's Round Function

For round keys $k_i \in \mathbb{F}_2^n$ and $w_i \in \mathbb{F}_2^{n-1}$ the round function computes

$$R_{k_i,w_i}(x) := x + f_{b(i)}(w_i + \Phi_{k_i}(x))k_i. \tag{10}$$

where

- Φ_{k_i} is defined as in Eq. (8),
- $f_{b(i)}$ is defined as

$$\begin{aligned} f_{b(i)} &: \mathbb{F}_2^{n-1} \to \mathbb{F}_2 \\ f_{b(i)}(x) &:= \langle x[1:(n-1)/2], x[(n+1)/2:n] \rangle + b(i), \end{aligned}$$

- and $b(i)$ is 0 if $i \leqslant \frac{r}{2}$ and 1 otherwise.

5.2 Key Schedule

In the ith round, the key schedule has to compute two round keys: $k_i \in \mathbb{F}_2^n$ and $w_i \in \mathbb{F}_2^{n-1}$. We compute those round keys as the states of LFSRs after i clocks, where the initial states are given by a master key K. The master key consists of two parts of n and $n-1$ bits, i.e.

$$K = (k, w) \in \mathbb{F}_2^n \times \mathbb{F}_2^{n-1}.$$

As the all-zero state is a fixed point for any LFSR, we *exclude the zero key* for both k and w. In particular $k = 0$ is obviously a weak key that would result in a ciphertext equal to the plaintext $p = E_{0,w}^r(p)$ for all p, independently of w or of the number of rounds r.

It is well-known that choosing a feedback polynomial of an LFSR to be primitive results in an LFSR of maximal period. Clocking the LFSR then corresponds to multiplication of its state with the companion matrix of this polynomial. Interpreted as elements from the finite field, this is the same as multiplying with a primitive element.

In order to avoid structural attacks, e.g. invariant attacks [28,39,50], as well as the propagation of low-weight inputs, we add round constants c_i to the round key w_i.

These round constants are also derived from the state of an LFSR with the same feedback polynomial as the w_i LFSR, initialized to the unit vector with the least significant bit set. To avoid synchronization with the w_i LFSR, the c_i LFSR clocks backwards.

BISON's Key Schedule

For two primitive polynomials $p_w(x), p_k(x) \in \mathbb{F}_2[x]$ with degrees $\deg(p_w) = n-1$ and $\deg(p_k) = n$ and the master key $K = (k,w) \in \mathbb{F}_2^n \times \mathbb{F}_2^{n-1}$, $k, w \neq 0$ the key schedule computes the ith round keys as

$$\begin{aligned}\mathrm{KS}_i : \mathbb{F}_2^n \times \mathbb{F}_2^{n-1} &\to \mathbb{F}_2^n \times \mathbb{F}_2^{n-1}\\ \mathrm{KS}_i(k,w) &:= \left(C(p_k)^i k, C(p_w)^{-i} e_1 + C(p_w)^i w\right) = (k_i, c_i + w_i)\end{aligned}$$

where $C(\cdot)$ is the companion matrix of the corresponding polynomial, and $0 \leqslant i < r$.
In [12, Appendix A] we give concrete polynomials for $5 \leqslant n \leqslant 129$-bit block sizes.

As discussed above, this key schedule has the following property, see also Rationale 1.

Lemma 8. *For* BISON*'s key schedule, the following property holds: Any set of n consecutive round keys k_i are linearly independent. Moreover there exist coefficients λ_i such that*

$$k_{n+i} = k_i + \sum_{j=i+1}^{n+i-1} \lambda_j k_j.$$

Proof. To prove this, we start by showing that the above holds for the first n round keys, the general case then follows from a similar argumentation. We need to show that there exists no non-trivial $c_i \in \mathbb{F}_2$ so that $\sum_{i=1}^{n} c_i C(p_k)^i k = 0$, which is equivalent to showing that there exists no non-trivial $c_i \in \mathbb{F}_2$ so that $\sum_{i=0}^{n-1} c_i C(p_k)^i k = 0$. In this regard, we recall the notion of *minimal polynomial of k with respect to $C(p_k)$*, defined as the monic polynomial of smallest degree $Q_L(k)(x) = \sum_{i=0}^{d} q_i x^i \in \mathbb{F}_2[x]$ such that $\sum_{i=0}^{d} q_i C(p_k)^i k = 0$. Referring to a discussion that has been done for instance in [4], we know that the minimal polynomial of k is a divisor of the minimal polynomial of $C(p_k)$. Since in our case our construction has been made so that this later is equal to p_k which is a primitive polynomial, we deduce that the minimal polynomial of $k \neq 0$ is p_k itself. Since the degree of p_k is equal to n, this prove that the first n keys are linearly independent.
The equation holds, since $p_k(0) = 1$. ∎

6 Security Analysis

As we have already seen, BISON is resistant to differential cryptanalysis. In this section, we argue why BISON is also resistant to other known attacks.

6.1 Linear Cryptanalysis

For linear cryptanalysis, given the fact that BISON is based on a bent function, i.e. a maximally non-linear function, arguing that no linear characteristic with high correlation exist is rather easy. Again, we start by looking at the Fourier coefficients for one round.

From One Round. Using the properties of f being bent, we get the following.

Proposition 4. *Let $R_{k,w} : \mathbb{F}_2^n \to \mathbb{F}_2^n$ be the round function as defined in Eq. (10). Then, its* LAT *consists of the entries*

$$\widehat{R_{k,w}}(\alpha, \beta) = \begin{cases} 2^n & \text{if } \alpha = \beta \text{ and } \langle \beta, k \rangle = 0 \\ \pm 2^{\frac{n+1}{2}} & \text{if } \langle \alpha, k \rangle = 1 \text{ and } \langle \beta, k \rangle = 1 \\ 0 & \text{if } \langle \alpha + \beta, k \rangle = 1 \text{ or } (\alpha \neq \beta \text{ and } \langle \beta, k \rangle = 0) \end{cases} . \quad (11)$$

We prove the proposition in [12, Section 6.1.1, Proposition 4].

To More Rounds. When we look at more than one round, we try to approximate the linear hull by looking at the strongest linear trail. As already discussed in Lemma 2, for $r < n$ there are trails with probability one. We now show that any trail's correlation for $r \geqslant n$ rounds is actually upper bounded by $2^{-\frac{n+1}{2}}$:

Proposition 5. *For $r \geqslant n$ rounds, the correlation of any non-trivial linear trail for* BISON *is upper bounded by $2^{-\frac{n+1}{2}}$.*

Proof. It is enough to show the above for any n-round trail. By contradiction, assume there exists a non-trivial trail $\theta = (\theta_0, \ldots, \theta_n)$ with correlation one. Following Proposition 4, for every round i the intermediate mask θ_i has to fulfill $\langle \theta_i, k_i \rangle = 0$. Further $\theta_i = \theta_{i+1}$ for $0 \leqslant i < n$. Because all n round keys are linearly independent, this implies that $\theta_i = 0$, which contradicts our assumption. Thus, in at least one round the second or third case of Eq. (11) has to apply. ∎

It would be nice to be able to say more about the linear hull, analogously to the differential case. However, for the linear cryptanalysis this looks much harder, due to the denser LAT. In our opinion developing a framework where bounding linear hulls is similarly easy as it is for BISON with respect to differentials is a fruitful future research topic.

6.2 Higher-Order Differentials and Algebraic Attacks

High-order differential attacks, cube attacks, algebraic attacks and integral attacks all make use of non-random behaviour of the ANF of parts of the encryption function. In all these attacks the algebraic degree of (parts of) the encryption function is of particular interest. In this section, we argue that those attacks do not pose a threat to BISON.

We next elaborate in more detail on the algebraic degree of the WSN construction. In particular, we are going to show that the algebraic degree increases at most linearly with the number of rounds. More precisely, if the round function is of degree d, the algebraic degree after r rounds is upper bounded by $r(d-1)+1$.

Actually, we are going to consider a slight generalization of the WSN construction and prove the above statement for this generalization.

General Setting. Consider an initial state of n bits given as $x = (x_0, \dots, x_{n-1})$ and a sequence of Boolean functions

$$f_i : \mathbb{F}_2^{n+i} \to \mathbb{F}_2$$

for $0 \leqslant i < r$. We define a sequence of values y_i by setting $y_0 = f_0(x)$ and

$$y_i = f_i(x_0, \dots, x_{n-1}, y_0, \dots, y_{i-1}),$$

for $1 \leqslant i < r$. Independently of the exact choice of f_i the degree of any y_ℓ, as a function of x can be upper bounded as stated in the next proposition.

Proposition 6. *Let f_i be a sequence of functions as defined above, such that $\deg(f_i) \leqslant d$. The degree of y_ℓ at step ℓ seen as a function of the bits of the initial state $x_0, \dots, x_{n-1}$ satisfies*

$$\deg(y_\ell) \leqslant (d-1)(\ell+1)+1.$$

Moreover, for any $I \subseteq \{0, \dots, \ell\}$,

$$\deg(\prod_{i \in I} y_i) \leqslant (d-1)(\ell+1)+|I|.$$

Proof. The first assertion is of course a special case of the second one, but we add it for the sake of clarity. We prove the second, more general, statement by induction on ℓ.

Starting with $\ell = 0$, we have to prove that $\deg(y_0) \leqslant d$, which is obvious, as

$$y_0 = f_0(x_0, \dots, x_{n-1})$$

and $\deg(f_0) \leq d$.

Now, we consider some $I \subseteq \{0, \dots, \ell\}$ and show that

$$\deg(\prod_{i \in I} y_i) \leqslant (d-1)(\ell+1)+|I|.$$

We assume that $\ell \in I$, otherwise the result directly follows the induction hypothesis.

Since f_ℓ depends both on $y_0, \dots, y_{\ell-1}$ and x, we decompose it as follows:

$$y_\ell = f_\ell(y_0, \dots, y_{\ell-1}, x) = \sum_{\substack{J \subseteq \{0,\dots,\ell-1\} \\ 0 \leqslant |J| \leqslant \min(d,\ell)}} \left(\prod_{j \in J} y_j \right) g_J(x)$$

with $\deg(g_J) \leqslant d - |J|$ for all J since $\deg(f_\ell) \leqslant d$.

Then, for $I = \{\ell\} \cup I'$, we look at

$$y_\ell \left(\prod_{i \in I'} y_i \right) = \sum_{\substack{J \subseteq \{0,\dots,\ell-1\} \\ 0 \leqslant |J| \leqslant \min(d,\ell)}} \left(\prod_{j \in J \cup I'} y_j \right) g_J(x) \, .$$

From the induction hypothesis, the term of index J in the sum has degree at most

$$\begin{aligned} (d-1)\ell + |J \cup I'| + \deg(g_J) &= (d-1)\ell + |J \cup I'| + d - |J| \\ &\leqslant (d-1)(\ell+1) + |J \cup I'| - |J| + 1 \\ &\leqslant (d-1)(\ell+1) + |J| + |I'| - |J| + 1 \\ &\leqslant (d-1)(\ell+1) + |I| \, . \end{aligned}$$

Special Case of BISON. In the case of BISON, we make use of quadratic functions, and thus Proposition 6 implies that after r rounds the degree is upper bounded by $r+1$ only. Thus, it will take at least $n-2$ rounds before the degree reaches the maximal possible degree of $n-1$. Moreover, due to the construction of WSN, if all component functions of $E^r_{k,w}$ are of degree at most d, there will be at least one component function of $E^{r+n-1}_{k,w}$ of degree at most d. That is, there exist a vector $\beta \in \mathbb{F}_2^n$ such that

$$\langle \beta, E^{r+n-1}_{k,w}(x) \rangle$$

has degree at most d. Namely, for

$$\beta \in \operatorname{span}\{k_r, \dots, k_{r+s}\}^\perp$$

it holds that

$$\deg\left(\langle \beta, E^{r+s}_{k,w}(x) \rangle\right) = \deg\left(\langle \beta, E^r_{k,w}(x) \rangle + \sum_{i=r}^{r+s} \lambda_i \langle \beta, k_i \rangle \right) = \deg\left(\langle \beta, E^r_{k,w}(x) \rangle\right) .$$

We conclude there exists a component function of $E^{r+s}_{k,w}$ of non-maximal degree, as long as $0 \leqslant r \leqslant n-2$ and $0 \leqslant s \leqslant n-1$. Thus for BISON there will be at least one component function of degree less than $n-1$ for any number of rounds $0 \leqslant r \leqslant 2n-3$. However, similarly to the case of zero-correlation properties as described in Lemma 3, the vector β is key dependent and thus this property does not directly lead to an attack.

Finally, so far we only discussed upper bounds on the degree, while for arguing security, lower bounds on the degree are more relevant. As it seems very hard (just like for any cipher) to prove such lower bounds, we investigated experimentally how the degree increases in concrete cases. As can be seen in [12, Figure 4] the maximum degree is reached for almost any instance for $n+5$ rounds. Most importantly, the fraction of instances where it takes more than $n+2$ rounds decreases with increasing block length n. Moreover, the round function in BISON experimentally behaves with this respect as a random function, as can be seen in [12, Figure 5]. Thus, as the number of rounds is $3n$, we are confident that attacks exploiting the algebraic degree do not pose a threat for BISON.

Besides the WSN construction, a special case of the above proposition worth mentioning is a non linear feedback generator (NLFSR).

Degree of NLFSRs. One well-known special case of the above general setting is an NLFSR or, equivalently a maximally unbalanced Feistel cipher, depicted below.

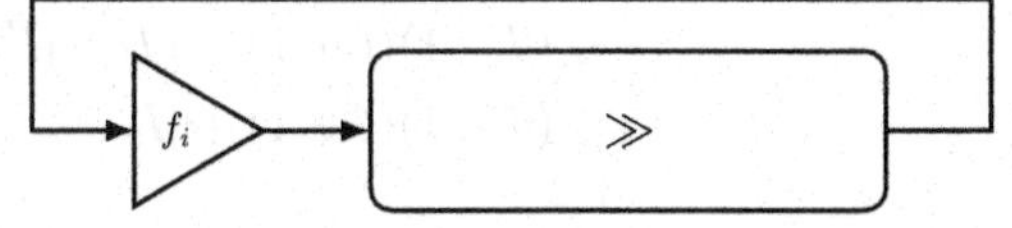

Proposition 6 implies that the degree of any NLFSR increases linearly with the number of rounds. To the best of our knowledge, this is the first time this have been observed in this generality. We like to add that this is in sharp contrast to how the degree increases for SPN ciphers. For SPN ciphers the degree usually increases exponentially until a certain threshold is reached [11].

6.3 Other Attacks

We briefly discuss other cryptanalytic attacks.

Impossible Differentials. In Lemma 5 and Corollary 2, we discuss that every output difference is possible after more than n rounds. Consequently, there are no impossible differentials for BISON.

Truncated Differentials. Due to our strong bounds on differentials it seems very unlikely that any strong truncated differential exists.

Zero Correlation Linear Cryptanalysis. In Lemma 3 we already discussed generic zero correlation linear hulls for the WSN construction. Depending on the actual key used, this technique may be used to construct a one-round-longer zero-correlation trail. For this, we need two *distinct* elements $\alpha \in \langle k_1, \dots, k_{n-1} \rangle^{\perp}$, $\beta \in \langle k_n, \dots, k_{2n-2} \rangle^{\perp}$, and construct the trail analogously to Lemma 3 (which may not exist, due to the key dependency).

Invariant Attacks. For an invariant attack, we need a Boolean function g, s.t. $g(x) + g(E^r_{k,w}(x))$ is constant *for all* x and some *weak keys* (k, w). As the encryption of any message is basically this message with some of the round keys added, key addition is the only operation which is performed. It has been shown in [4, Proposition 1] that any g which is invariant for a linear layer followed by the addition of the round key k_i as well as for the same up to addition of a different k_j, has a linear space containing $k_i + k_j$. In the case of the linear layer being the identity, the linear space actually contains also the k_i and k_j (by definition).

Thus, the linear space of any invariant for our construction has to contain span $\{k_1, \dots, k_{3n}\}$ which is obviously the full space $\mathbb{F}_2^n$. Following the results of [4], there are thus no invariant subspace or nonlinear invariant attack on BISON.

Related-Key Attacks. In generic related-key attacks, the attacker is also allowed to exploit encryptions under a related, that is $k' = f(k)$, key – in the following, we restrict our analysis to the case where f is the addition with a constant. That is, the attacker cannot only request $E_{k,w}(x)$, and $E_{k,w}(x + \alpha)$, but also $E_{k+\beta,w+\beta'}(x)$ or $E_{k+\beta,w+\beta'}(x + \alpha)$, for α (difference in the input x), β (difference in the key k) and β' (difference in the key w) of her choice. As $\beta = \beta' = 0$ would result in the standard differential scenario, we exclude it for the remainder of this discussion. Similar, $\beta = k$ results in $\Phi_{k+\beta} = \Phi_0$, which we did not define, thus we also skip this case and refer to the fact that if an attacker chooses $\beta = k$, she basically already has guessed the secret key correctly.

For BISON, the following proposition holds.

Proposition 7. *For r rounds, the probability of any related-key differential characteristic for* BISON*, averaged over all whieting key sequences $(w_1, \dots, w_r)$, is upper bounded by $\left(\frac{3}{4}\right)^r$.*

For more details and a proof of the proposition, see [12, Section 6.3.5, Proposition 7].

Further Observations. During the design process, we observed the following interesting point: For sparse master keys k and w and message m, e.g. $k = w = m = 1$, in the first few rounds, nothing happens. This is mainly due to the choice of sparse key schedule polynomials p_w and p_k and the fact that f_0 outputs 0 if only one bit in its input is set (as $\langle 0, x \rangle = 0$ for any x).

To the best of our knowledge, this observation cannot be exploited in an attack.

Experimental Results. We conducted experiments on small-scale versions of BISON with $n = 5$. The DDTs and LATs, depicted using the "Jackson Pollock representation" [8], for one to ten rounds are listed in [12, Appendix B]. In [12, Appendix B.1] one can see that the two cases of averaging over all possible w_i and choosing a fixed w_i results in very similar differential behaviors. Additionally, after $5 = n$ rounds, the plots do not change much.

The results in the linear case, see [12, Appendix B.2], are quite similar. The major difference here, is the comparable bigger entries for a fixed w_i. Nonetheless, most important is that there are no high entries in the average LAT which would imply a strong linear approximation for many keys. Additionally one also expects for a random permutation not too small LAT entries. Note that one can well observe the probability-one approximation for $4 = n - 1$ rounds (lower right corner of the corresponding plot).

7 Conclusion

Efficiency of symmetric ciphers have been significantly improved further and further, in particular within the trend of lightweight cryptography. However, when it comes to arguing about the security of ciphers, the progress is rather limited and the arguments basically did not get easier nor stronger since the development of the AES. In our opinion it might be worth shifting the focus to improving security arguments for new designs rather than (incrementally) improving efficiency. We see BISON as a first step in this direction.

With our instance for the WSN construction and its strong resistance to differential cryptanalysis, this framework emerges as an interesting possibility to design block ciphers. Unfortunately, we are not able to give better then normal arguments for the resistance to linear cryptanalysis. It is thus an interesting question, if one can find a similar instance of the WSN construction for which comparable strong arguments for the later type of cryptanalysis exist.

Alternative designs might also be worth looking at. For example many constructions for bent functions are known and could thus be examined as alternatives for the scalar product used in BISON. One might also look for a less algebraic design – but we do not yet see how this would improve or ease the analysis or implementation of an instance.

Finally, for an initial discussion of implementation figures, see [12, Section 7]. Another line of future work in this direction is the in-depth analysis of implementation optimizations and side channel-resistance of BISON.

Acknowledgments. We would like to thank the anonymous reviewers and Christof Beierle for their helpful comments, and Lucas Hartmann for the artistic design of BISON.

References

1. Abdelraheem, M.A., Ågren, M., Beelen, P., Leander, G.: On the distribution of linear biases: three instructive examples. In: Safavi-Naini, R., Canetti, R. (eds.) CRYPTO 2012. LNCS, vol. 7417, pp. 50–67. Springer, Heidelberg (2012). https://doi.org/10.1007/978-3-642-32009-5_4
2. Advanced Encryption Standard (AES), November 2001
3. Andreeva, E., Bogdanov, A., Dodis, Y., Mennink, B., Steinberger, J.P.: On the indifferentiability of key-alternating ciphers. In: Canetti, R., Garay, J.A. (eds.) CRYPTO 2013. LNCS, vol. 8042, pp. 531–550. Springer, Heidelberg (2013). https://doi.org/10.1007/978-3-642-40041-4_29
4. Beierle, C., Canteaut, A., Leander, G., Rotella, Y.: Proving resistance against invariant attacks: how to choose the round constants. In: Katz, J., Shacham, H. (eds.) CRYPTO 2017. LNCS, vol. 10402, pp. 647–678. Springer, Cham (2017). https://doi.org/10.1007/978-3-319-63715-0_22
5. Biham, E., Shamir, A.: Differential cryptanalysis of DES-like cryptosystems. In: Menezes, A.J., Vanstone, S.A. (eds.) CRYPTO 1990. LNCS, vol. 537, pp. 2–21. Springer, Heidelberg (1991). https://doi.org/10.1007/3-540-38424-3_1
6. Biryukov, A., Khovratovich, D.: Related-key cryptanalysis of the full AES-192 and AES-256. In: Matsui, M. (ed.) ASIACRYPT 2009. LNCS, vol. 5912, pp. 1–18. Springer, Heidelberg (2009). https://doi.org/10.1007/978-3-642-10366-7_1
7. Biryukov, A., Khovratovich, D., Nikolić, I.: Distinguisher and related-key attack on the full AES-256. In: Halevi, S. (ed.) CRYPTO 2009. LNCS, vol. 5677, pp. 231–249. Springer, Heidelberg (2009). https://doi.org/10.1007/978-3-642-03356-8_14
8. Biryukov, A., Perrin, L.: On reverse-engineering S-boxes with hidden design criteria or structure. In: Gennaro, R., Robshaw, M. (eds.) CRYPTO 2015. LNCS, vol. 9215, pp. 116–140. Springer, Heidelberg (2015). https://doi.org/10.1007/978-3-662-47989-6_6
9. Blondeau, C., Nyberg, K.: Improved parameter estimates for correlation and capacity deviates in linear cryptanalysis. IACR Trans. Symm. Cryptol. **2016**(2), 162–191 (2016). http://tosc.iacr.org/index.php/ToSC/article/view/570
10. Bogdanov, A., Knudsen, L.R., Leander, G., Standaert, F.-X., Steinberger, J., Tischhauser, E.: Key-alternating ciphers in a provable setting: encryption using a small number of public permutations (extended abstract). In: Pointcheval, D., Johansson, T. (eds.) EUROCRYPT 2012. LNCS, vol. 7237, pp. 45–62. Springer, Heidelberg (2012). https://doi.org/10.1007/978-3-642-29011-4_5
11. Boura, C., Canteaut, A., De Cannière, C.: Higher-order differential properties of KECCAK and *Luffa*. In: Joux, A. (ed.) FSE 2011. LNCS, vol. 6733, pp. 252–269. Springer, Heidelberg (2011). https://doi.org/10.1007/978-3-642-21702-9_15
12. Canteaut, A., Lallemand, V., Leander, G., Neumann, P., Wiemer, F.: BISON - instantiating the whitened swap-or-not construction. Cryptology ePrint Archive, Report 2018/1011 (2018)
13. Canteaut, A., Roué, J.: On the behaviors of affine equivalent Sboxes regarding differential and linear attacks. In: Oswald, E., Fischlin, M. (eds.) EUROCRYPT 2015. LNCS, vol. 9056, pp. 45–74. Springer, Heidelberg (2015). https://doi.org/10.1007/978-3-662-46800-5_3
14. Carlet, C.: Boolean functions for cryptography and error correcting codes. In: Crama, Y., Hammer, P. (eds.) Boolean Methods and Models. Cambridge University Press (2007)

15. Chen, S., Steinberger, J.: Tight security bounds for key-alternating ciphers. In: Nguyen, P.Q., Oswald, E. (eds.) EUROCRYPT 2014. LNCS, vol. 8441, pp. 327–350. Springer, Heidelberg (2014). https://doi.org/10.1007/978-3-642-55220-5_19
16. Daemen, J.: Cipher and hash function design, strategies based on linear and differential cryptanalysis. Ph.D. thesis. K.U.Leuven (1995). http://jda.noekeon.org/
17. Daemen, J., Govaerts, R., Vandewalle, J.: Block ciphers based on modular arithmetic. In: Wolfowicz, W. (ed.) State and Progress in the Research of Cryptography, pp. 80–89. Fondazione Ugo Bordoni (1993)
18. Daemen, J., Govaerts, R., Vandewalle, J.: Correlation matrices. In: Preneel, B. (ed.) FSE 1994. LNCS, vol. 1008, pp. 275–285. Springer, Heidelberg (1995). https://doi.org/10.1007/3-540-60590-8_21
19. Daemen, J., Rijmen, V.: The block cipher Rijndael. In: Quisquater, J.-J., Schneier, B. (eds.) CARDIS 1998. LNCS, vol. 1820, pp. 277–284. Springer, Heidelberg (2000). https://doi.org/10.1007/10721064_26
20. Daemen, J., Rijmen, V.: The wide trail design strategy. In: Honary, B. (ed.) Cryptography and Coding 2001. LNCS, vol. 2260, pp. 222–238. Springer, Heidelberg (2001). https://doi.org/10.1007/3-540-45325-3_20
21. Daemen, J., Rijmen, V.: The Design of Rijndael: AES - The Advanced Encryption Standard. Information Security and Cryptography. Springer, Heidelberg (2002). https://doi.org/10.1007/978-3-662-04722-4
22. Daemen, J., Rijmen, V.: Understanding two-round differentials in AES. In: De Prisco, R., Yung, M. (eds.) SCN 2006. LNCS, vol. 4116, pp. 78–94. Springer, Heidelberg (2006). https://doi.org/10.1007/11832072_6
23. Derbez, P., Fouque, P.-A., Jean, J.: Improved key recovery attacks on reduced-round AES in the single-key setting. In: Johansson, T., Nguyen, P.Q. (eds.) EUROCRYPT 2013. LNCS, vol. 7881, pp. 371–387. Springer, Heidelberg (2013). https://doi.org/10.1007/978-3-642-38348-9_23
24. Dillon, J.F.: A survey of bent functions. NSA Tech. J. **191**, 215 (1972)
25. Even, S., Mansour, Y.: A construction of a cipher from a single pseudorandom permutation. J. Cryptol. **10**(3), 151–162 (1997)
26. Ferguson, N., et al.: Improved cryptanalysis of Rijndael. In: Goos, G., Hartmanis, J., van Leeuwen, J., Schneier, B. (eds.) FSE 2000. LNCS, vol. 1978, pp. 213–230. Springer, Heidelberg (2001). https://doi.org/10.1007/3-540-44706-7_15
27. Gilbert, H., Minier, M.: A collision attack on 7 rounds of Rijndael. In: AES Candidate Conference, vol. 230, p. 241 (2000)
28. Grassi, L., Rechberger, C., Rønjom, S.: Subspace trail cryptanalysis and its applications to AES. IACR Trans. Symm. Cryptol. **2016**(2), 192–225 (2016). http://tosc.iacr.org/index.php/ToSC/article/view/571
29. Grassi, L., Rechberger, C., Rønjom, S.: A new structural-differential property of 5-round AES. In: Coron, J.-S., Nielsen, J.B. (eds.) EUROCRYPT 2017. LNCS, vol. 10211, pp. 289–317. Springer, Cham (2017). https://doi.org/10.1007/978-3-319-56614-6_10
30. Guo, C., Lin, D.: On the indifferentiability of key-alternating feistel ciphers with no key derivation. In: Dodis, Y., Nielsen, J.B. (eds.) TCC 2015. LNCS, vol. 9014, pp. 110–133. Springer, Heidelberg (2015). https://doi.org/10.1007/978-3-662-46494-6_6
31. Hoang, V.T., Morris, B., Rogaway, P.: An enciphering scheme based on a card shuffle. In: Safavi-Naini, R., Canetti, R. (eds.) CRYPTO 2012. LNCS, vol. 7417, pp. 1–13. Springer, Heidelberg (2012). https://doi.org/10.1007/978-3-642-32009-5_1

32. Hoang, V.T., Tessaro, S.: Key-alternating ciphers and key-length extension: exact bounds and multi-user security. In: Robshaw, M., Katz, J. (eds.) CRYPTO 2016. LNCS, vol. 9814, pp. 3–32. Springer, Heidelberg (2016). https://doi.org/10.1007/978-3-662-53018-4_1
33. Hong, S., Lee, S., Lim, J., Sung, J., Cheon, D., Cho, I.: Provable security against differential and linear cryptanalysis for the SPN structure. In: Goos, G., Hartmanis, J., van Leeuwen, J., Schneier, B. (eds.) FSE 2000. LNCS, vol. 1978, pp. 273–283. Springer, Heidelberg (2001). https://doi.org/10.1007/3-540-44706-7_19
34. Jakobsen, T., Knudsen, L.R.: The interpolation attack on block ciphers. In: Biham, E. (ed.) FSE 1997. LNCS, vol. 1267, pp. 28–40. Springer, Heidelberg (1997). https://doi.org/10.1007/BFb0052332
35. Keliher, L., Sui, J.: Exact maximum expected differential and linear probability for two-round advanced encryption standard. IET Inf. Secur. **1**(2), 53–57 (2007)
36. Kranz, T., Leander, G., Wiemer, F.: Linear cryptanalysis: key schedules and tweakable block ciphers. IACR Trans. Symm. Cryptol. **2017**(1), 474–505 (2017)
37. Lai, X., Massey, J.L., Murphy, S.: Markov ciphers and differential cryptanalysis. In: Davies, D.W. (ed.) EUROCRYPT 1991. LNCS, vol. 547, pp. 17–38. Springer, Heidelberg (1991). https://doi.org/10.1007/3-540-46416-6_2
38. Lampe, R., Seurin, Y.: Security analysis of key-alternating feistel ciphers. In: Cid, C., Rechberger, C. (eds.) FSE 2014. LNCS, vol. 8540, pp. 243–264. Springer, Heidelberg (2015). https://doi.org/10.1007/978-3-662-46706-0_13
39. Leander, G., Abdelraheem, M.A., AlKhzaimi, H., Zenner, E.: A cryptanalysis of PRINTCIPHER: the invariant subspace attack. In: Rogaway, P. (ed.) CRYPTO 2011. LNCS, vol. 6841, pp. 206–221. Springer, Heidelberg (2011). https://doi.org/10.1007/978-3-642-22792-9_12
40. Matsui, M.: On correlation between the order of S-boxes and the strength of DES. In: De Santis, A. (ed.) EUROCRYPT 1994. LNCS, vol. 950, pp. 366–375. Springer, Heidelberg (1995). https://doi.org/10.1007/BFb0053451
41. Meier, W., Staffelbach, O.: Nonlinearity criteria for cryptographic functions. In: Quisquater, J.-J., Vandewalle, J. (eds.) EUROCRYPT 1989. LNCS, vol. 434, pp. 549–562. Springer, Heidelberg (1990). https://doi.org/10.1007/3-540-46885-4_53
42. Miracle, S., Yilek, S.: Cycle slicer: an algorithm for building permutations on special domains. Cryptology ePrint Archive, Report 2017/873 (2017). http://eprint.iacr.org/2017/873
43. Nyberg, K.: Linear approximation of block ciphers (rump session). In: De Santis, A. (ed.) EUROCRYPT 1994. LNCS, vol. 950, pp. 439–444. Springer, Heidelberg (1995). https://doi.org/10.1007/BFb0053460
44. Nyberg, K.: "Provable" security against differential and linear cryptanalysis (invited talk). In: Canteaut, A. (ed.) FSE 2012. LNCS, vol. 7549, pp. 1–8. Springer, Heidelberg (2012). https://doi.org/10.1007/978-3-642-34047-5_1
45. Nyberg, K., Knudsen, L.R.: Provable security against a differential attack. J. Cryptol. **8**(1), 27–37 (1995)
46. Park, S., Sung, S.H., Lee, S., Lim, J.: Improving the upper bound on the maximum differential and the maximum linear hull probability for SPN structures and AES. In: Johansson, T. (ed.) FSE 2003. LNCS, vol. 2887, pp. 247–260. Springer, Heidelberg (2003). https://doi.org/10.1007/978-3-540-39887-5_19
47. Rothaus, O.S.: On 'bent' functions. J. Comb. Theory Ser. A **20**(3), 300–305 (1976)
48. Tessaro, S.: Optimally secure block ciphers from ideal primitives. In: Iwata, T., Cheon, J.H. (eds.) ASIACRYPT 2015. LNCS, vol. 9453, pp. 437–462. Springer, Heidelberg (2015). https://doi.org/10.1007/978-3-662-48800-3_18

49. Tessaro, S.: Optimally secure block ciphers from ideal primitives. Cryptology ePrint Archive, Report 2015/868 (2015). http://eprint.iacr.org/2015/868
50. Todo, Y., Leander, G., Sasaki, Y.: Nonlinear invariant attack. In: Cheon, J.H., Takagi, T. (eds.) ASIACRYPT 2016. LNCS, vol. 10032, pp. 3–33. Springer, Heidelberg (2016). https://doi.org/10.1007/978-3-662-53890-6_1
51. Vaudenay, S.: Provable security for block ciphers by decorrelation. In: Morvan, M., Meinel, C., Krob, D. (eds.) STACS 1998. LNCS, vol. 1373, pp. 249–275. Springer, Heidelberg (1998). https://doi.org/10.1007/BFb0028566
52. Vaudenay, S.: The end of encryption based on card shuffling. CRYPTO 2012 Rump Session (2012). crypto.2012.rump.cr.yp.to

Foundations II

Worst-Case Hardness for LPN and Cryptographic Hashing via Code Smoothing

Zvika Brakerski[1(✉)], Vadim Lyubashevsky[2], Vinod Vaikuntanathan[3], and Daniel Wichs[4]

[1] Weizmann Institute of Science, Rehovot, Israel
zvika.brakerski@weizmann.ac.il
[2] IBM Research – Zurich, Rüschlikon, Switzerland
[3] Massachusetts Institute of Technology, Cambridge, USA
[4] Northeastern University, Boston, USA

Abstract. We present a worst case decoding problem whose hardness reduces to that of solving the Learning Parity with Noise (LPN) problem, in some parameter regime. Prior to this work, no worst case hardness result was known for LPN (as opposed to syntactically similar problems such as Learning with Errors). The caveat is that this worst case problem is only mildly hard and in particular admits a quasi-polynomial time algorithm, whereas the LPN variant used in the reduction requires extremely high noise rate of $1/2 - 1/\text{poly}(n)$. Thus we can only show that "very hard" LPN is harder than some "very mildly hard" worst case problem. We note that LPN with noise $1/2 - 1/\text{poly}(n)$ already implies symmetric cryptography.

Specifically, we consider the (n, m, w)-nearest codeword problem ((n, m, w)-NCP) which takes as input a generating matrix for a binary linear code in m dimensions and rank n, and a target vector which is very close to the code (Hamming distance at most w), and asks to find the codeword nearest to the target vector. We show that for balanced (unbiased) codes and for relative error $w/m \approx \log^2 n/n$, (n, m, w)-NCP can be solved given oracle access to an LPN distinguisher with noise ratio $1/2 - 1/\text{poly}(n)$.

Our proof relies on a smoothing lemma for codes which we show to have further implications: We show that (n, m, w)-NCP with the aforementioned parameters lies in the complexity class Search-$\mathcal{BPP}^{\mathcal{SZK}}$ (i.e. reducible to a problem that has a statistical zero knowledge protocol) implying that it is unlikely to be $\mathcal{NP}$-hard. We then show that the hardness of LPN with very low noise rate $\log^2(n)/n$ implies the existence of

Z. Brakerski—Supported by the Israel Science Foundation (Grant No. 468/14), Binational Science Foundation (Grants No. 2016726, 2014276), and by the European Union Horizon 2020 Research and Innovation Program via ERC Project REACT (Grant 756482) and via Project PROMETHEUS (Grant 780701).
V. Lyubashevsky—Supported by the SNSF ERC Transfer Grant CRETP2-166734 – FELICITY.
D. Wichs—Research supported by NSF grants CNS1314722, CNS-1413964.

Y. Ishai and V. Rijmen (Eds.): EUROCRYPT 2019, LNCS 11478, pp. 619–635, 2019.
https://doi.org/10.1007/978-3-030-17659-4_21

collision resistant hash functions (our aforementioned result implies that in this parameter regime LPN is also in $\mathcal{BPP}^{\mathcal{SZK}}$).

1 Introduction

The hardness of noisy learning problems such as learning parity with noise (LPN) [BFKL93,BKW03] and learning with errors (LWE) [Reg05] have proved to be a goldmine in modern cryptography. The hardness of LWE has been instrumental in solving long-standing problems such as fully homomorphic encryption [Gen09,BV11]. Both LPN and LWE have given us efficient and plausibly quantum-proof cryptographic constructions [KPC+11,BCD+16,ADPS16]. However, while we know several structural results about LWE, relatively little is known about the 25-year old LPN problem.

Before we proceed, let us define the LPN and LWE problems. In the (search version of the) LPN problem, the algorithm is given access to an oracle that produces samples $(\mathbf{a}_i, \mathbf{s}^T\mathbf{a}_i + e_i)$ where $\mathbf{s} \in \mathbb{Z}_2^n$ is the "secret" vector, $\mathbf{a}_i \in \mathbb{Z}_2^n$ are uniformly distributed and $e_i \in \mathbb{Z}_2$ come from the Bernoulli distribution (that is, it is 1 with probability ϵ and 0 otherwise). The goal is to recover $\mathbf{s}$. The (search version of the) LWE problem is the same but for two key changes: first, the vectors $\mathbf{a}_i \in \mathbb{Z}_q^n$ are uniformly random with entries from some large enough finite field $\mathbb{Z}_q$ and second, each error term e_i is chosen from the discrete Gaussian distribution over the integers. The exact choice of the error distribution does not matter much: what is important is that in LWE, each sample has an error with bounded absolute value (at least with high probability). These seemingly minor differences seem to matter a great deal: we know worst-case to average-case reductions for LWE [Reg05,Pei09,BLP+13] while no such result is known for LPN;[1] we know that (a decisional version of) LWE is in the complexity class $\mathcal{SZK}$ [MV03] (statistical zero-knowledge) while no such result is known for LPN; and we can build a dizzying array of cryptographic primitives assuming the hardness of LWE (e.g. attribute based encryption and homomorphic encryption to name the more exotic examples) while the repertoire of LPN is essentially limited to one-way functions and public-key encryption (and primitives that can be constructed generically from it). In particular, we do not know how to construct even simple, seemingly "unstructured", primitives such as a collision-resistant hash function from the hardness of LPN, even with extreme parameter choices. Can we bridge this puzzling gap between LWE and LPN?

In a nutshell, the goal of this paper is to solve all three of these problems. Our main tool is a *smoothing lemma for binary linear codes.* We proceed to describe our results and techniques in more detail.

1.1 Overview of Our Results and Techniques

Worst-Case to Average-Case Reduction. We consider the promise nearest codeword problem (NCP), a worst-case analog of the learning parity with noise

[1] Feldman et al. [FGKP09] showed a worst-case to average-case reduction with respect to the noise distribution, but not with respect to the samples themselves.

problem. Roughly speaking, in the search version of the (n, m, w)-promise nearest codeword problem, one is given the generator matrix $\mathbf{C} \in \mathbb{Z}_2^{n \times m}$ of a linear code, along with a vector $\mathbf{t} \in \mathbb{Z}_2^m$ such that $\mathbf{t} = \mathbf{s}^T\mathbf{C} + \mathbf{x}^T$ for some $\mathbf{s} \in \mathbb{Z}_2^n$ and $\mathbf{x} \in \mathbb{Z}_2^m$ *with the promise* that $\mathsf{wt}(\mathbf{x}) = w$. The problem is to find $\mathbf{s}$. The non-promise version of this problem (which is commonly called the nearest codeword problem) is known to be $\mathcal{NP}$-hard, even to approximately solve [ABSS93] and the promise problem is similarly $\mathcal{NP}$-hard in the large-error regime (that is, when the Hamming weight of $\mathbf{x}$ exceeds $(1/2 + \epsilon)d$ where d is the minimum distance of the code and $\epsilon > 0$ is an arbitrarily small constant) [DMS99].

In terms of algorithms, Berman and Karpinski [BK02] show how to find an $O(n/\log n)$-approximate nearest codeword in polynomial time. In particular, this means that if the Hamming weight of $\mathbf{x}$ in the promise version is at most $O(d \cdot \log n/n)$, their algorithm finds the unique nearest codeword's $\mathbf{s}$ efficiently. To the best of our knowledge, this result is the current limit of polynomial-time solvability of the promise nearest codeword problem. Alon, Panigrahy and Yekhanin [APY09] show a deterministic nearly-polynomial time algorithm with the same parameters. In this work, we consider the promise NCP for *balanced codes*, where all nonzero codewords have Hamming weight between $(1/2 - \beta)m$ and $(1/2+\beta)m$ for some balance parameter $\beta > 0$. We are not aware of improved NCP algorithms that apply to balanced codes.

Our first result (in Sect. 4) shows a reduction from the *worst-case* promise NCP for balanced codes where $w/m \approx \frac{\log^2 n}{n}$ to the *average-case* hardness of LPN_ϵ^n with very high error-rate $\epsilon = 1/2 - 1/O(n^4)$. We note that a random linear code is β-balanced with overwhelming probability when $\beta \geq 3\sqrt{n/m}$ so for a sufficiently large m the restriction on β is satisfied by most codes. Thus, qualitatively speaking, our result shows that solving LPN with very high error *on the average* implies solving NCP with very low error *for most codes.* While the parameters we achieve are extreme, we emphasize that no worst-case to average-case reduction for LPN was known prior to our work.

The worst-case to average-case reduction is a simple consequence of a smoothing lemma for codes that we define and prove in Sect. 3. In a nutshell, our smoothing lemma shows a simple randomized procedure that maps a worst-case linear code $\mathcal{C}$ and a vector $\mathbf{t}$ to a random linear code $\mathcal{C}'$ and a vector $\mathbf{t}'$ such that if $\mathbf{t}$ is super-close to $\mathcal{C}$, then $\mathbf{t}'$ is somewhat close to $\mathcal{C}'$. Our worst-case to average-case reduction then follows simply by applying the smoothing lemma to the worst-case code and vector. We show a simple Fourier-analytic proof of the smoothing lemma, in a way that is conceptually similar to analogous statements in the context of lattices [MR04] (see more details in the end of Sect. 3). Similar statements have been shown before in the list-decoding high-error regime [KS10], whereas our setting for NCP is in the unique decoding (low error) regime.

Statistical Zero-Knowledge. Another consequence of our smoothing lemma is a statistical zero-knowledge proof for the NCP problem for balanced codes with low noise, namely where $w/m \approx \frac{\log^2 n}{n}$. In particular, we show that the search problem is in $\mathcal{BPP}^{\mathcal{SZK}}$. Membership in $\mathcal{BPP}^{\mathcal{SZK}}$ should be viewed as an easiness result: a consequence of this result and a theorem of Mahmoody and Xiao [MX10] is that NCP with low noise cannot be $\mathcal{NP}$-hard unless the polynomial hierarchy collapses. Our result is the first non-$\mathcal{NP}$-hardness result we know for NCP, complementing the $\mathcal{NP}$-hardness result of Dumer, Micciancio and Sudan [DMS99] for noise slightly larger than half the minimum distance, namely where $w/m \approx 1/2$ (but leaves a large gap in between). This is the LPN/codes analog of a result for LWE/lattices that we have known for over a decade [MV03]. We refer the reader to Sect. 5 for this result.

Collision-Resistant Hashing. Finally, we show a new cryptographic consequence of the hardness of LPN with low noise, namely a construction of a collision-resistant hash (CRH) function. Again, collision-resistant hashing from LWE/lattices has been known for over two decades [Ajt96, GGH96] and we view this result as an LPN/codes analog. The construction is extremely simple: the family of hash functions is parameterized by a matrix $\mathbf{A} \in \mathbb{Z}_2^{n \times n^{1+c}}$ for some $c > 0$, its domain is the set of vectors $\mathbf{x} \in \mathbb{Z}_2^{n^{1+c}}$ with Hamming weight $2n/(c \log n)$ and the output is simply $\mathbf{Ax} \pmod 2$. This is similar to a CRH construction from the recent work of Applebaum et al. [AHI+17] modulo the setting of parameters; what is new in our work is a reduction from the LPN problem with error rate $O(\log^2 n/n)$ to breaking this CRH function.

Related Work. Our LPN-based collision-resistant hash function was used in [BLSV17] as a basis for constructing an identity based encryption scheme based on LPN with very low noise. Concurrently with, and independently from, our work, Yu et al. [YZW+17] constructed a family of collision-resistant hash functions based on the hardness of LPN using the same main idea as in Sect. 6 of the present work. While the core ideas of the construction in the two works is identical, [YZW+17] further discusses different parameter settings and some heuristics upon whose reliance one can obtain a tighter connection between the hardness of the CRH and the LPN problem.

2 Preliminaries

2.1 Notation

Throughout the paper, we will be working with elements in the additive group $\mathbb{Z}_2$ with the usual addition operation. We will denote by bold lower-case letters vectors over $\mathbb{Z}_2^n$ for $n > 1$, and by bold upper-case letters matrices over $\mathbb{Z}_2^{m \times n}$ for $m, n > 1$. We will make the assumption that all vectors are column vectors and write $\mathbf{a}^T$ to denote the row vector which is the transpose of $\mathbf{a}$. The Hamming weight of $\mathbf{a} \in \mathbb{Z}_2^n$, written as $\mathsf{wt}(\mathbf{a})$, denotes the number of 1's in $\mathbf{a}$. For a set S, we write $s \leftarrow S$ to denote that s is chosen uniformly at random from S. When D is some probability distribution, then $s \leftarrow D$ means that s is chosen according to D.

The Ber_ϵ distribution over $\mathbb{Z}_2$ is the Bernoulli distribution that outputs 1 with probability ϵ and 0 with probability $1 - \epsilon$. Let S_k^m be the set of all the elements $\mathbf{s} \in \mathbb{Z}_2^m$ such that $\mathsf{wt}(\mathbf{s}) = k$.

A negligible function $\text{negl}(n)$ is any function that grows slower than inverse polynomial in n. In particular, for every polynomial p there is an $n_0 \in \mathbb{N}$ such that for every $n > n_0$, $\text{negl}(n) < 1/p(n)$.

2.2 The Learning Parity with Noise (LPN) Problem

For an $\mathbf{s} \in \mathbb{Z}_2^n$, and an $\epsilon \in [0, .5]$ let $\mathcal{O}_{\mathbf{s},\epsilon}^n$ be an algorithm that, when invoked, chooses a random $\mathbf{a} \leftarrow \mathbb{Z}_2^n$ and $e \leftarrow \mathsf{Ber}_\epsilon$ and outputs $(\mathbf{a}, \mathbf{s}^T\mathbf{a} + e)$. An algorithm A is said to solve the search LPN_ϵ^n problem with probability δ if

$$\Pr[\mathsf{A}^{\mathcal{O}_{\mathbf{s},\epsilon}^n} \Rightarrow \mathbf{s}\ ;\ \mathbf{s} \leftarrow \mathbb{Z}_2^n] \geq \delta.$$

Let $\mathcal{U}^n$ be an algorithm that, when invoked, chooses random $\mathbf{a} \leftarrow \mathbb{Z}_2^n$ and $b \leftarrow \mathbb{Z}_2$ and outputs $(\mathbf{a}, b)$. We say that an algorithm A has advantage δ in solving the decisional LPN_ϵ^n problem if

$$\left|\Pr[\mathsf{A}^{\mathcal{O}_{\mathbf{s},\epsilon}^n} \Rightarrow 0;\ \mathbf{s} \leftarrow \mathbb{Z}_2^n] - \Pr[\mathsf{A}^{\mathcal{U}^n} \Rightarrow 0]\right| \geq \delta.$$

The LPN problem has a search to decision reduction (c.f. [KS06]). Namely, if there is an algorithm that runs in time t and has advantage δ in solving the decisional LPN_ϵ^n problem, then there is an algorithm that runs in time $O(nt/\delta)$ that solves the search LPN_ϵ^n problem with probability ≈ 1.

The following fact is known in some contexts as The Piling-Up Lemma [Mat93].

Lemma 2.1. *For all* $\epsilon \in [0, \frac{1}{2}]$ *it holds that* $\Pr[e_1 + \ldots + e_k = 0;\ e_i \leftarrow \mathsf{Ber}_\epsilon] = \frac{1}{2} + \frac{1}{2} \cdot (1 - 2\epsilon)^k$.

2.3 The Nearest Codeword Problem

An (binary) (n, m, d)-code $\mathcal{C}$ is a subset of $\{0,1\}^m$ such that $|\mathcal{C}| = 2^n$ and for any two codewords $\mathbf{x}, \mathbf{y} \in \mathcal{C}$, $\mathsf{wt}(\mathbf{x} \oplus \mathbf{y}) \leq d$. The code is linear (denoted $[n, m, d]$-code) if $\mathcal{C}$ is the row span of some matrix $\mathbf{C} \in \{0,1\}^{n \times m}$.

Definition 2.1 (Nearest Codeword Problem (NCP)). *The nearest codeword problem* $\mathsf{NCP}_{n,m,w}$ *is characterized by* $n, m, w \in \mathbb{Z}$ *and is defined as follows. The input consists of a matrix* $\mathbf{C} \in \mathbb{Z}^{n \times m}$ *which is the generator of a code, along with a vector* $\mathbf{t} \in \mathbb{Z}^m$ *such that* $\mathbf{t} = \mathbf{s}^T\mathbf{C} + \mathbf{x}^T$ *for some* $\mathbf{s} \in \mathbb{Z}_2^n, \mathbf{x} \in \mathbb{Z}_2^m$ *with* $\mathsf{wt}(\mathbf{x}) = w$. *The problem is to find* $\mathbf{s}$.

Note that our definition requires $\mathsf{wt}(\mathbf{x}) = w$, as opposed to the more relaxed requirement $\mathsf{wt}(\mathbf{x}) \leq w$. However since w comes from a polynomial domain $\{0, \ldots, m\}$ the difference is not very substantial (in particular, to solve the relaxed version one can go over all polynomially-many relevant values of w and try solving the exact version).

In this work, we consider a variant of the problem which is restricted to *balanced* codes, which are codes where all non-zero codewords have hamming weight close to 1/2. We start by defining balanced codes and then present balanced NCP.

Definition 2.2. *A code $\mathcal{C} \subseteq \{0,1\}^m$ is β-balanced if its minimum distance is at least $\frac{1}{2}(1-\beta)m$ and maximum distance is at most $\frac{1}{2}(1+\beta)m$.*

Definition 2.3 (balanced NCP (balNCP)). *The balanced nearest codeword problem $\mathsf{balNCP}_{n,m,w,\beta}$ is characterized by $n, m, w \in \mathbb{Z}$ and $\beta \in (0,1)$, and is defined as follows. The input consists of a matrix $\mathbf{C} \in \mathbb{Z}^{n\times m}$ which is the generator of a β-balanced code, along with a vector $\mathbf{t} \in \mathbb{Z}^m$ such that $\mathbf{t}^T = \mathbf{s}^T\mathbf{C} + \mathbf{x}^T$ for some $\mathbf{s} \in \mathbb{Z}_2^n, \mathbf{x} \in \mathbb{Z}_2^m$ with $\mathsf{wt}(\mathbf{x}) = w$. The problem is to find $\mathbf{s}$.*

The $\mathsf{balNCP}_{n,m,w,\beta}$ problem has a unique solution when $w \leq \frac{1}{4}(1-\beta)m$.

Standard decoding algorithms allow to solve NCP in polynomial time with success probability $(1-\frac{w}{m})^n$ [BK02] or even deterministically in time $(1-\frac{w}{m})^{-n} \cdot \mathrm{poly}(n,m)$ [APY09]. We are not aware of improved methods that apply to balanced codes.

To conclude this section we show via a straightforward probabilistic argument that most sparse linear codes are indeed balanced (this is essentially the Gilbert-Varshamov Bound). This is to serve as sanity check that the definition is not vacuous and will also be useful when we apply our $\mathcal{SZK}$ results to the LPN_ϵ^n problem which naturally induces random codes.

Lemma 2.2. *A random linear code $\mathcal{C} \subseteq \mathbb{Z}_2^m$ of dimension n is β-balanced with probability at least $1 - 2^{n-\beta^2 m/4+1}$. In particular, when $\beta \geq 3\sqrt{n/m}$ a random linear code is β-balanced with probability* $1 - \mathrm{negl}(n)$.

Proof. Let $\mathbf{C} \leftarrow \mathbb{Z}_2^{n\times m}$ be a randomly chosen generator matrix. Then the associated code $\mathcal{C}$ fails to be β-balanced if and only if there exists some $\mathbf{s} \neq \mathbf{0} \in \mathbb{Z}_2^n$ such that $|\mathsf{wt}(\mathbf{s}^T\mathbf{C}) - \frac{m}{2}| > \frac{\beta}{2}m$. For any fixed $\mathbf{s} \neq \mathbf{0}$ the vector $\mathbf{s}^T\mathbf{C}$ is uniformly random in $\mathbb{Z}_2^m$ and therefore by the Chernoff bound:

$$\Pr\left[\left|\mathsf{wt}(\mathbf{s}^T\mathbf{C}) - \frac{m}{2}\right| > \frac{\beta m}{2}\right] \leq 2\exp\left(-\frac{\beta^2 m}{4}\right)$$

By the union bound, the probability that the code is not β-balanced is at most

$$2^{n+1}\exp\left(-\frac{\beta^2 m}{4}\right) \leq 2^{n-\frac{\beta^2 m}{4}+1}.$$

This is negligible in n when $\beta \geq 3\sqrt{n/m}$.

2.4 Statistical Zero Knowledge

Statistical zero-knowledge ($\mathcal{SZK}$) is the class of all problems that admit a zero-knowledge proof [GMR89] with a statistically sound simulation. Sahai and Vadhan [SV03] showed that the following problem is complete for $\mathcal{SZK}$.

Definition 2.4. *The promise problem Statistical Distance (SD) is defined by the following YES and NO instances. For a circuit $C : \{0,1\}^n \to \{0,1\}^m$, we let $C(U_n)$ denote the probability distribution on m-bit strings obtained by running C*

on a uniformly random input. Let $\mathsf{SD}(D_0, D_1)$ *denote the statistical (variation) distance between the distributions* D_0 *and* D_1.

$$\Pi_{YES} := \{(C_0, C_1): \ C_0, C_1 : \{0,1\}^n \to \{0,1\}^m \text{ and } \mathsf{SD}(C_0(U_n), C_1(U_n)) \geq 2/3\}$$
$$\Pi_{NO} := \{(C_0, C_1): \ C_0, C_1 : \{0,1\}^n \to \{0,1\}^m \text{ and } \mathsf{SD}(C_0(U_n), C_1(U_n)) \leq 1/3\}$$

By $\mathcal{BPP}^{\mathcal{SZK}}$, we mean decision problems that can be reduced to the statistical distance problem using randomized reductions. While in general such reduction could query the SD oracle on inputs that violate the promise (namely, a pair of circuits/distributions whose statistical distance lies strictly between 1/3 and 2/3), the reductions we present in this paper will respect the SD promise. Search-$\mathcal{BPP}^{\mathcal{SZK}}$ is defined analogously.

3 A Smoothing Lemma for Noisy Codewords

Let $\mathcal{C} \subseteq \mathbb{Z}_2^m$ be a binary linear code with generating matrix $\mathbf{C} \in \mathbb{Z}_2^{n \times m}$. We say that a distribution $\mathcal{R}$ over $\mathbb{Z}_2^m$ smooths $\mathcal{C}$ if the random variable $\mathbf{Cr}$ for $\mathbf{r} \leftarrow \mathcal{R}$ is statistically close to uniform over $\mathbb{Z}_2^n$. We say that $\mathcal{R}$ also smooths noisy codewords if for every vector $\mathbf{x}$ of sufficiently low Hamming weight, it holds that $(\mathbf{Cr}, \mathbf{x}^T\mathbf{r})$ is statistically close to the distribution $\mathcal{U}_{\mathbb{Z}_2^n} \times \mathsf{Ber}_\epsilon$ for some ϵ.

The notion of smoothing will play an important role in our reductions in this work. In particular, we would like to characterize families of codes that are smoothed by distributions supported over low Hamming weight vectors. To this end, we show that for balanced codes, there exist such smoothing distributions. (Similar statements have been shown before in the high-error regime, e.g., by Kopparty and Saraf [KS10].)

We note that while our proof uses harmonic analysis, it is also possible to prove it using the Vazirani XOR Lemma [Vaz86, Gol95]. However, we find that our method of using harmonic analysis demonstrates more straightforwardly the analogy of our lemma to smoothing in the lattice world (which is most often proved using harmonic analysis), see comparison in the end of this section. Furthermore, this suggests an approach if one wants to analyze the non-binary setting.

We start by defining our family of smoothing distributions $\mathcal{R}_{d,m}$.

Definition 3.1. *Let* $d, m \in \mathbb{N}$. *The distribution* $\mathcal{R}_{d,m}$ *over* $\mathbb{Z}_2^m$ *is defined as follows. Sample (with replacement)* d *elements* $t_1, \ldots, t_d$ *uniformly and independently from* $[m]$. *Output* $\mathbf{x} = \oplus_{i=1}^d \mathbf{u}_{t_i}$, *where* $\mathbf{u}_j$ *is the* j*-th standard basis vector. One can easily verify that* $\mathcal{R}_{d,m}$ *is supported only over vectors of Hamming weight at most* d.

We can now state and prove our smoothing lemma for noisy codewords.

Lemma 3.1. *Let* $\beta \in (0,1)$ *and let* $\mathbf{C} \in \mathbb{Z}_2^{n \times m}$ *be a generating matrix for a* β*-balanced binary linear code* $\mathcal{C} \subseteq \mathbb{Z}_2^m$. *Let* $\mathbf{c} \in \mathbb{Z}_2^m$ *be a word of distance* w *from* $\mathcal{C}$. *Let* $\mathbf{s}, \mathbf{x}$ *be s.t.* $\mathbf{c}^T = \mathbf{s}^T\mathbf{C} + \mathbf{x}^T$ *and* $\mathsf{wt}(\mathbf{x}) = w$.

Consider the distribution $(\mathbf{a}, b)$ *generated as follows. Sample* $\mathbf{r} \leftarrow \mathcal{R}_{d,m}$ *and set* $\mathbf{a} = \mathbf{C}\mathbf{r}$, $b = \mathbf{c}^T\mathbf{r}$. *Then it holds that the joint distribution of* $(\mathbf{a}, b - \mathbf{s}^T\mathbf{a})$ *is within statistical distance* δ *from the product distribution* $\mathcal{U}_{\mathbb{Z}_2^n} \times \mathsf{Ber}_\epsilon$, *where*

$$\delta \leq 2^{(n+1)/2} \cdot (\beta + \tfrac{2w}{m})^d \text{ and}$$
$$\epsilon = \tfrac{1}{2} - \tfrac{1}{2}(1 - \tfrac{2w}{m})^d.$$

Proof. Let e denote the value $b - \mathbf{s}^T\mathbf{a}$. We bound the distance of $\left[\begin{smallmatrix}\mathbf{a}\\ e\end{smallmatrix}\right] = \left[\begin{smallmatrix}\mathbf{C}\\ \mathbf{x}^T\end{smallmatrix}\right]\mathbf{r}$ from $\mathcal{U}_{\{0,1\}^n} \times \mathsf{Ber}_\epsilon$ using simple harmonic analysis. Let f be the probability density function of $\left[\begin{smallmatrix}\mathbf{a}\\ e\end{smallmatrix}\right]$, and consider its (binary) Fourier Transform:

$$\hat{f}(\mathbf{y}, z) = \mathop{\mathbb{E}}_{\mathbf{a},e}\left[(-1)^{\mathbf{y}^T\mathbf{a}+ze}\right] = \mathop{\mathbb{E}}_{\mathbf{r}}[(-1)^{(\mathbf{y}^T\mathbf{C}+z\mathbf{x}^T)\mathbf{r}}], \tag{1}$$

It immediately follows that $\hat{f}(\mathbf{0}, 0) = 1$. Moreover

$$\hat{f}(\mathbf{0}, 1) = \mathop{\mathbb{E}}_{\mathbf{r}}[(-1)^{\mathbf{x}^T\mathbf{r}}]. \tag{2}$$

Recalling that $\mathbf{r} = \oplus_{i=1}^d \mathbf{u}_{t_i}$ we have

$$\mathop{\mathbb{E}}_{\mathbf{r}}[(-1)^{\mathbf{x}^T\mathbf{r}}] = \prod_{i=1}^{d} \mathop{\mathbb{E}}_{t_i}[(-1)^{\mathbf{x}^T\mathbf{u}_{t_i}}] = (1 - \tfrac{2w}{m})^d,$$

since each t_i is sampled uniformly and independently in $[m]$ and thus has a $\frac{w}{m}$ probability to hit a coordinate where $\mathbf{x}$ is one. Recalling the definition of ϵ, we have $\hat{f}(\mathbf{0}, 1) = 1 - 2\epsilon$.

Now let us consider the setting where $\mathbf{y} \neq \mathbf{0}$. In that case, let us denote $\mathbf{v} = \mathbf{y}^T\mathbf{C}$, a nonzero codeword in $\mathcal{C}$. Since $\mathcal{C}$ is balanced it follows that $\mathsf{wt}(\mathbf{v}) \in [\frac{1}{2}(1-\beta)m, \frac{1}{2}(1+\beta)m]$. Let us further denote $(\mathbf{v}')^T = \mathbf{y}^T\mathbf{C}+z\mathbf{x}^T$, since $\mathsf{wt}(\mathbf{x}) \leq w$ it follows that $\mathsf{wt}(\mathbf{v}') \in \frac{1}{2}(1 \pm \beta')m$ for $\beta' = \beta + \frac{2w}{m}$. For $\mathbf{y} \neq \mathbf{0}$ we thus get

$$\hat{f}(\mathbf{y}, z) = \mathop{\mathbb{E}}_{\mathbf{r}}[(-1)^{(\mathbf{v}')^T\mathbf{r}}] = \prod_{i=1}^{d} \mathop{\mathbb{E}}_{t_i}[(-1)^{(\mathbf{v}')^T\mathbf{u}_{t_i}}]. \tag{3}$$

Since each t_i is sampled uniformly from $[m]$, it follows that $\mathbf{v}'\mathbf{u}_{t_i} \pmod 2 = 0$ with probability $\epsilon_i \in \frac{1}{2}(1 \pm \beta')$. Therefore for all $i \in [d]$ it holds that

$$\left|\mathop{\mathbb{E}}_{t_i}[(-1)^{\mathbf{v}'\mathbf{u}_{t_i}}]\right| = |1 - 2\epsilon_i| \leq \beta'. \tag{4}$$

We conclude that

$$\left|\hat{f}(\mathbf{y}, z)\right| \leq (\beta')^d. \tag{5}$$

Now we are ready to compare with $\mathcal{U}_{\mathbb{Z}_2^n} \times \mathsf{Ber}_\epsilon$. Let g be the probability density function of $\mathcal{U}_{\mathbb{Z}_2^n} \times \mathsf{Ber}_\epsilon$, and let $\hat{g}$ be its Fourier Transform. Then $\hat{g}(\mathbf{0}, 0) = 1$, $\hat{g}(\mathbf{0}, 1) = 1 - 2\epsilon$ and $\hat{g}(\mathbf{y}, z) = 0$ for all $\mathbf{y} \neq 0$. Therefore

$$\left\|\hat{f} - \hat{g}\right\|_2^2 = \sum_{\mathbf{y},z} \left|\hat{f}(\mathbf{y}) - \hat{g}(\mathbf{y})\right|^2 \leq \sum_{\substack{\mathbf{y}\in\mathbb{Z}_2^n\setminus\{0\}\\ z\in\mathbb{Z}_2}} (\beta')^{2d} \leq 2^{n+1}(\beta')^{2d}. \tag{6}$$

By Parseval's theorem, we have that

$$\|f-g\|_2^2 = \frac{1}{2^{n+1}} \left\|\hat{f}-\hat{g}\right\|_2^2 \leq (\beta')^{2d}, \tag{7}$$

and going to ℓ_1 norm we have

$$\|f-g\|_1 \leq 2^{(n+1)/2} \cdot \left\|\hat{f}-\hat{g}\right\|_2 \leq 2^{(n+1)/2} \cdot (\beta')^d, \tag{8}$$

which completes the proof. □

Relation to Lattice Smoothing. To conclude this section, let us briefly explain the analogy to smoothing lemmas for lattices [MR04]. Our explanation is intended mostly for readers who are familiar with lattice smoothing and the notion of q-ary lattices, and wish to better understand the connection to our notion of smoothing. Other readers may safely skip this paragraph. We recall that the goal of smoothing in the lattice world is to find a distribution $\mathcal{D}$ (a Gaussian in the lattice case), such that reducing it modulo a lattice $\mathcal{L}$ results in an almost uniform distribution over the cosets of the lattice. Let us restrict our attention to integer lattices, integer distributions and integer cosets. Formally, $\mathcal{D} \pmod{\mathcal{L}}$ is uniform over $\mathbb{Z}/\mathcal{L}$. Now let $\mathbf{C}$ be a generating matrix for a binary code, and consider the so called "perp lattice" $\mathcal{L} = \Lambda_2^{\perp}(\mathbf{C}) = \{\mathbf{x} \in \mathbb{Z}^m : \mathbf{Cx} = 0 \pmod{2}\} = \mathcal{C}^{\perp} + 2\mathbb{Z}^m$. That is, the lattice corresponding to the dual code of $\mathcal{C}$ plus all even vectors. Each integer cosets of the lattice $\mathcal{L}$ corresponds to a vector $\mathbf{y}$ where the respective coset is $\mathcal{L}_{\mathbf{y}} = \{\mathbf{x} \in \mathbb{Z}^m : \mathbf{Cx} = \mathbf{y} \pmod{2}\}$. Thus a smoothing distribution $\mathcal{D}$ is one where drawing $\mathbf{r}$ from $\mathcal{D}$ and computing $\mathbf{Cr} \pmod{2}$ is close to uniform. Therefore, our smoothing lemma above shows that for 2-ary lattices one can devise non-trivial (and useful) smoothing distributions, and these distributions are not discrete Gaussians as usually considered in the lattice literature. Finally, the fact that we can smooth the code together with a noisy codeword is somewhat analogous to Gaussian leftover hash lemmas in the context of lattices.

4 A Worst Case Balanced NCP to Average Case LPN Reduction

Theorem 4.1. *Assume there is an algorithm that solves the search* LPN_ϵ^n *problem with success probability* α *in the average case by running in time* T *and making* q *queries. Then, for every* $d \leq m \in \mathbb{Z}$ *there is an algorithm that solves search* $\mathsf{balNCP}_{n,m,w,\beta}$ *in the worst case in time* $T \cdot \mathrm{poly}(n,m)$ *with success probability at least* $\alpha - q \cdot \delta$ *where*

$$\delta \leq 2^{(n+1)/2} \cdot (\beta + \tfrac{2w}{m})^d$$
$$\epsilon = \tfrac{1}{2} - \tfrac{1}{2}(1 - \tfrac{2w}{m})^d.$$

Proof. Assume $\mathcal{A}$ is an algorithm for the LPN problem as in the theorem. Define $\mathcal{B}$ as follows:

– Input: $\mathbf{C} \in \mathbb{Z}_2^{n\times m}$, $\mathbf{t} \in \{0,1\}^m$. By assumption $\mathbf{C}$ is the generator of a β-balanced code and $\mathbf{t}^T = \mathbf{s}^T\mathbf{C} + \mathbf{x}^T$ for some $\mathbf{s} \in \mathbb{Z}_2^n, \mathbf{x} \in \mathbb{Z}_2^m$ with $\mathsf{wt}(\mathbf{x}) \le w$.

1. Sample $\mathbf{s}' \leftarrow \mathbb{Z}_2^n$ and set $\mathbf{c}^T = \mathbf{t}^T + (\mathbf{s}')^T\mathbf{C} = (\mathbf{s}+\mathbf{s}')^T\mathbf{C} + \mathbf{x}^T$.
2. Run the algorithm $\mathcal{A}$. Every time $\mathcal{A}$ request a new LPN sample, choose $\mathbf{r} \leftarrow \mathcal{R}_{d,m}$ and set $\mathbf{a} = \mathbf{Cr}$, $b = \mathbf{c}^T\mathbf{r}$ and give $\mathbf{a}, b$ to $\mathcal{A}$.
3. If at some point $\mathcal{A}$ outputs $\mathbf{s}^* \in \mathbb{Z}_2^n$ then output $\mathbf{s}^* - \mathbf{s}'$.

By Lemma 3.1 each of the values $(\mathbf{a}, b)$ given to $\mathcal{A}$ during step 2 is δ-close to a fresh sample from $\mathcal{O}^n_{\mathbf{s}^*,\epsilon}$ where $\mathbf{s}^* = \mathbf{s}+\mathbf{s}'$ is uniformly random over $\mathbb{Z}_2^n$. By assumption, if $\mathcal{A}$ were actually given samples from $\mathcal{O}^n_{\mathbf{s}^*,\epsilon}$ is step 2 it would output $\mathbf{s}^*$ in step 3 with probability α. Therefore if $\mathcal{A}$ makes q queries in step 2, the probability that it outputs $\mathbf{s}^*$ in step 3 is at least $\alpha - q\delta$. This proves the theorem. □

Corollary 4.1. *Let $m = n^c$ for some constant $c > 1$, $\beta = \frac{1}{\sqrt{n}}, w = \lceil m\frac{\log^2 n}{n}\rceil$. Assume that search $\mathsf{balNCP}_{n,m,w,\beta}$ is hard in the worst case, meaning that for every polynomial time algorithm its success probability on the worst case instance is at most* negl(n). *Then for some $\epsilon < \frac{1}{2} - \frac{1}{O(n^4)}$ search LPN^n_ϵ is hard in the average case, meaning that for every polynomial time algorithm its success probability on a random instance is at most* negl(n).

Proof. Follows directly from the theorem by setting $d = \lceil 2n/\log n\rceil$ and noting that:

$$\delta \le 2^{(n+1)/2}\cdot(\beta + \tfrac{2w}{m})^d \le 2^{(n+1)/2-(d/2)\log n + O(1)} \le 2^{-n/2+O(1)} = \mathrm{negl}(n)$$

$$\epsilon = \tfrac{1}{2} - \tfrac{1}{2}(1-\tfrac{2w}{m})^d \le \frac{1}{2} - 2^{-(4\frac{w}{m}d+1)} \le \frac{1}{2} - 1/O(n^4)$$

□

The above says that the wost-case hardness of balNCP with very low error-rate $w/m \approx \frac{\log^2 n}{n}$ implies the average-case hardness of LPN^n_ϵ with very high error-rate $\epsilon = 1/2 - 1/O(n^4)$. Note that a random linear code is β-balanced with overwhelming probability when $\beta \ge 3\sqrt{n/m}$ so for a sufficiently large m the restriction on β is satisfied by most codes.

Other choices of parameters may also be interesting. For example, we can set the error-rate to be $w/m \approx 1/\sqrt{n}$ and $d = 2n/\log n$ while keeping $m = n^c$ for some $c > 1$, $\beta = 1/\sqrt{n}$ the same as before. Then if we assume that $\mathsf{balNCP}_{n,m,w,\beta}$ is $(T(n), \delta(n))$ hard in the worst case (meaning that for every $T(n)$ time algorithm the success probability on the worst case instance is at most $\delta(n)$) this implies LPN^n_ϵ is $(T'(n), \delta'(n))$ hard in the average where $\epsilon(n) = 1/2 - 2^{-\sqrt{n}/\log n}$, $T'(n) = T(n)/\mathrm{poly}(n)$ and $\delta'(n) = \delta(n) + T'(n)2^{-(n-1)/2}$. Note that, as far as we know, the $\mathsf{balNCP}_{n,m,w,\beta}$ with noise rate $w/m = 1\sqrt{n}$ may be $(T(n), \delta(n))$ hard for some $T(n) = 2^{\Omega(\sqrt{n})}$, $\delta(n) = 2^{-\Omega(\sqrt{n})}$, which would imply the same asymptotic hardness for LPN^n_ϵ. Although the error-rate $\epsilon = 1/2 - 2^{-\sqrt{n}/\log n}$ is extremely high, it is not high enough for the conclusion to hold statistically and therefore this connection may also be of interest.

5 Statistical Zero Knowledge for Balanced NCP and LPN

In this section, we show that for certain parameter regimes, balNCP $\in$ Search-$\mathcal{BPP}^{\mathcal{SZK}}$ and is thus unlikely to be $\mathcal{NP}$-hard [MX10]. Towards this end, we use a decision to search reduction analogous to the canonical one known for the LPN problem. We consider the following randomized samplers (with an additional implicit parameter d):

- Randomized sampler $\mathsf{Samp}_0(\mathbf{C}, \mathbf{t})$ takes as input a matrix $\mathbf{C} \in \{0,1\}^{n\times m}$ and a word $\mathbf{t} \in \{0,1\}^m$. It samples $\mathbf{r} \xleftarrow{\$} \mathcal{R}_{d,m}$ and outputs $(\mathbf{Cr}, \mathbf{t}^T\mathbf{r})$.
- Randomized sampler $\mathsf{Samp}_{i,\sigma}(\mathbf{C}, \mathbf{t})$ is parameterized by $i \in [n]$, $\sigma \in \{0,1\}$, takes as input a matrix $\mathbf{C} \in \{0,1\}^{n\times m}$ and a word $\mathbf{t} \in \{0,1\}^m$. It samples $\mathbf{r} \xleftarrow{\$} \mathcal{R}_{d,m}$ and $\rho \in \{0,1\}$ and outputs $(\mathbf{Cr} + \rho\mathbf{u}_i, \mathbf{t}^T\mathbf{r} + \rho\sigma)$.

Lemma 5.1. *Consider a generating matrix $\mathbf{C} \in \{0,1\}^{n\times m}$ for a β-balanced code, and let $\mathbf{t} = \mathbf{s}^T\mathbf{C} + \mathbf{x}^T$ for some $\mathbf{s} \in \{0,1\}^n$ and $\mathbf{x}$ with hamming weight w. Then the following hold:*

1. *The sampler $\mathsf{Samp}_0(\mathbf{C}, \mathbf{t})$ samples from a distribution that is δ-close to $\mathcal{U}_{\{0,1\}^n} \times \mathsf{Ber}_\epsilon$.*
2. *If $\mathbf{s}_i = \sigma$ then $\mathsf{Samp}_{i,\sigma}(\mathbf{C}, \mathbf{t})$ samples from a distribution that is δ-close to $\mathcal{U}_{\{0,1\}^n} \times \mathsf{Ber}_\epsilon$.*
3. *If $\mathbf{s}_i \neq \sigma$ then $\mathsf{Samp}_{i,\sigma}(\mathbf{C}, \mathbf{t})$ samples from a distribution that is δ-close to $\mathcal{U}_{\{0,1\}^n} \times \mathcal{U}_{\{0,1\}}$.*

Here, $\epsilon = \frac{1}{2} - \frac{1}{2}(1 - \frac{2w}{m})^d$, $\delta = 2^{(n+1)/2} \cdot (\beta + \frac{2w}{m})^d$.

Proof. Assertion 1 follows directly from Lemma 3.1.

For Assertion 2 we note that if $\mathbf{s}_i = \sigma$ then

$$(\mathbf{Cr} + \rho\mathbf{u}_i, \mathbf{t}^T\mathbf{r} + \rho\sigma) = (\mathbf{Cr}, \mathbf{t}^T\mathbf{r}) + \rho(\mathbf{u}_i, \sigma) = (\mathbf{Cr}, \mathbf{t}^T\mathbf{r}) + (\rho\mathbf{u}_i, \mathbf{s}^T(\rho\mathbf{u}_i)).$$

By Lemma 3.1 this distribution is within δ statistical distance to

$$(\mathbf{a}, \mathbf{s}^T\mathbf{a} + e) + (\rho\mathbf{u}_i, \mathbf{s}^T(\rho\mathbf{u}_i)),$$

with $(\mathbf{a}, e)$ distributed $\mathcal{U}_{\{0,1\}^n} \times \mathsf{Ber}_\epsilon$. Finally, we can write

$$(\mathbf{a}, \mathbf{s}^T\mathbf{a} + e) + (\rho\mathbf{u}_i, \mathbf{s}^T(\rho\mathbf{u}_i)) = ((\mathbf{a} + \rho\mathbf{u}_i), \mathbf{s}^T(\mathbf{a} + \rho\mathbf{u}_i) + e),$$

and since $\mathbf{a}' = \mathbf{a} + \rho\mathbf{u}_i$ is also uniformly distributed, the assertion follows.

For Assertion 3 we note that when $\mathbf{s}_i \neq \sigma$, i.e. $\sigma = \mathbf{s}_i + 1$ then

$$(\mathbf{Cr} + \rho\mathbf{u}_i, \mathbf{t}^T\mathbf{r} + \rho\sigma) = (\mathbf{Cr}, \mathbf{t}^T\mathbf{r}) + \rho(\mathbf{u}_i, \sigma) = (\mathbf{Cr}, \mathbf{t}^T\mathbf{r}) + (\rho\mathbf{u}_i, \mathbf{s}^T(\rho\mathbf{u}_i)) + (\mathbf{0}, \rho).$$

As above, by Lemma 3.1, this distribution is within δ statistical distance to

$$(\mathbf{a}, \mathbf{s}^T\mathbf{a} + e) + (\rho\mathbf{u}_i, \mathbf{s}^T(\rho\mathbf{u}_i)) = (\underbrace{(\mathbf{a} + \rho\mathbf{u}_i)}_{\mathbf{a}'}, \mathbf{s}^T(\mathbf{a} + \rho\mathbf{u}_i) + e) + (\mathbf{0}, \rho) = (\mathbf{a}', \mathbf{s}^T\mathbf{a}' + e + \rho),$$

with $(\mathbf{a}, e)$ distributed $\mathcal{U}_{\{0,1\}^n} \times \mathsf{Ber}_\epsilon$, and thus also $(\mathbf{a}', e)$ distributed $\mathcal{U}_{\{0,1\}^n} \times \mathsf{Ber}_\epsilon$ and independent of ρ. Since ρ is uniform and independent of $(\mathbf{a}', e)$ it follows that $(\mathbf{a}', \mathbf{s}^T\mathbf{a}' + e + \rho)$ is distributed $\mathcal{U}_{\{0,1\}^n} \times \mathcal{U}_{\{0,1\}}$. □

The following is an immediate corollary of Lemma 5.1.

Corollary 5.1. *If* $\mathbf{s}_i = \sigma$ *then the distributions generated by* $\mathsf{Samp}_{i,\sigma}(\mathbf{C}, \mathbf{t})$ *and* $\mathsf{Samp}_0(\mathbf{C}, \mathbf{t})$ *are within statistical distance* at most 2δ.

If $\mathbf{s}_i \neq \sigma$ *then the distributions generated by* $\mathsf{Samp}_{i,\sigma}(\mathbf{C}, \mathbf{t})$ *and* $\mathsf{Samp}_0(\mathbf{C}, \mathbf{t})$ *are within statistical distance* at least $(1 - 2\epsilon) - 2\delta$.

Proof. A direct calculation shows that the statistical distance between Ber_ϵ and $\mathcal{U}_{\{0,1\}}$ is $1 - 2\epsilon$. Plugging in Lemma 5.1, the result follows. □

We define the notion of a direct product sampler. This is just a sampler that outputs multiple samplers.

Definition 5.1. *Let* $\mathcal{D}$ *be a distribution and let* $k \in \mathbb{N}$, *then* $\mathcal{D}^{(k)}$ *is the distribution defined by* k *independent samples from* $\mathcal{D}$.

Lemma 5.2. *Consider distributions* $\mathcal{D}_1, \mathcal{D}_2$ *and values* $0 \leq \delta_1 \leq \delta_2 \leq 1$ *s.t.* $\mathsf{dist}(\mathcal{D}_1, \mathcal{D}_2) \in (\delta_1, \delta_2)$. *Let* $k \in \mathbb{N}$ *then* $\mathsf{dist}(\mathcal{D}_1^{(k)}, \mathcal{D}_2^{(k)}) \in (1 - c_1 e^{-c_2 \delta_1^2 k}, k\delta_2)$. *For some positive constants* c_1, c_2.

Proof. The upper bound follows by union bound and the lower bound from the Chernoff bound. □

Theorem 5.1. *There exists a* Search-$\mathcal{BPP}^{\mathcal{SZK}}$ *algorithm for solving* balNCP *on instances of the following form. Letting* $\mathbf{C} \in \{0,1\}^{n \times m}$, $\mathbf{t} \in \{0,1\}^m$, $w \in [m]$ *denote the* balNCP *input, we require that the code generated by* $\mathbf{C}$ *is* β*-balanced and that* n, m, w, β *are such that there exist* $d \in [m]$ *and* $k \leq \mathrm{poly}(n, m)$ *for which*

$$2\delta k < 1/3 \tag{9}$$

for $\delta = 2^{(n+1)/2} \cdot (\beta + \frac{2w}{m})^d$, *and*

$$c_1 e^{-c_2(1-2\epsilon-2\delta)^2 k} < 1/3 \tag{10}$$

for δ *as above,* $\epsilon = \frac{1}{2} - \frac{1}{2}(1 - \frac{2w}{m})^d$, *and* c_1, c_2 *are the constants from Lemma 5.2.*

Proof. We recall the problem Statistical Distance (SD) which is in $\mathcal{SZK}$. This problem takes as input two sampler circuits and outputs 0 if the inputs sample distributions that are within statistical distance $< 1/3$ and 1 if the distributions are within statistical distance $> 2/3$. We will show how to solve balNCP for the above parameters using an oracle to SD.

Specifically, for all $i = 1, \ldots, n$ and $\sigma \in \{0,1\}$, the algorithm will call the SD oracle on input $(\mathsf{Samp}_0^{(k)}(\mathbf{C}, \mathbf{t}), \mathsf{Samp}_{i,\sigma}^{(k)}(\mathbf{C}, \mathbf{t}))$, where $\mathsf{Samp}_{(\cdot)}^{(k)}$ is the algorithm that runs the respective Samp k times and outputs all k generated samples.

Let $\alpha_{i,\sigma}$ denote the oracle response on the (i, σ) call. Then if for any i it holds that $\alpha_{i,0} = \alpha_{i,1}$, then return $\perp$. Otherwise set $\mathbf{s}_i$ to the value σ for which $\alpha_{i,\sigma} = 0$. Return $\mathbf{s}$.

By definition of our samplers, they run in polynomial time, so if k is polynomial then our inputs to SD are indeed valid. Combining Corollary 5.1 and Lemma 5.2, it holds that $\alpha_{i,\sigma} = 0$ if and only if $\mathbf{s}^*_i = \sigma$, where $\mathbf{s}^*$ is the vector for which $\mathbf{t}^T = (\mathbf{s}^*)^T\mathbf{C} + \mathbf{x}^T$ and $\mathsf{wt}(\mathbf{x}) = w$. The correctness of the algorithm follows. □

Corollary 5.2. *Let $m = n^c$ for some constant $c > 1$, $\beta = \frac{1}{\sqrt{n}}$, $w = \lceil m\frac{\log^2 n}{n}\rceil$. Then search $\mathsf{balNCP}_{n,m,w,\beta} \in \text{Search-}\mathcal{BPP}^{\mathcal{SZK}}$.*

Proof. In Theorem 5.1 set $d = \lceil 2n/\log n\rceil$ and $k = n^9$. By the same calculation as in Corollary 4.1 we have $\delta = \text{negl}(n)$ and $\epsilon \leq \frac{1}{2} - 1/O(n^4)$. Therefore for large enough n we have $2\delta k < 1/3$ and $c_1 e^{-c_2(1-2\epsilon-2\delta)^2 k} = e^{-\Omega(n)} < 1/3$ as required by the theorem. □

On Statistical Zero Knowledge and LPN. We notice that since sparse random codes are balanced with overwhelming probability (Lemma 2.2), our results in this section also imply that the LPN problem is in Search-$\mathcal{BPP}^{\mathcal{SZK}}$ for error value $\frac{\log^2 n}{n}$. We note that even though in LPN the weight of the noise vector (the distance from the code) is not fixed as in our definition of balNCP, the domain of possible weights is polynomial and thus the exact weight can be guessed with polynomial success probability. Once a successful guess had been made, it can be verified once a solution had been found.

6 Collision-Resistant Hashing

In this section, we describe a collision-resistant hash function family whose security is based on the hardness of the (decisional) $\mathsf{LPN}^n_{O(\log^2 n/n)}$ problem. For any positive constant $c \in \mathbb{R}^+$ and a matrix $\mathbf{A} \in \mathbb{Z}_2^{n\times n^{1+c}}$, define the function

$$h_{\mathbf{A}} : \mathcal{S}^{n^{1+c}}_{2n/(c\log n)} \to \mathbb{Z}_2^n \quad \text{as} \quad h_{\mathbf{A}}(\mathbf{r}) := \mathbf{A}\mathbf{r}. \tag{11}$$

Notice that because

$$\left|\mathcal{S}^{n^{1+c}}_{2n/(c\log n)}\right| = \binom{n^{1+c}}{2n/(c\log n)} > \left(\frac{n^{1+c}}{2n/(c\log n)}\right)^{2n/(c\log n)} > 2^{2n}$$

and the size of $\mathbb{Z}_2^n$ is exactly 2^n, the function $h_{\mathbf{A}}$ is compressing.

We now relate the hardness of finding collisions in the function $h_{\mathbf{A}}$, for a random $\mathbf{A}$, to the hardness of the decisional LPN^n_ϵ problem.

Theorem 6.1. *For any constant $c > 0$, if there exists an algorithm A_1 running in time t such that*

$$\Pr\left[\mathsf{A}_1(h_{\mathbf{A}}) \Rightarrow (\mathbf{r}_1, \mathbf{r}_2) \in \mathcal{S}^{n^{1+c}}_{2n/(c\log n)} \text{ s.t. } \mathbf{r}_1 \neq \mathbf{r}_2 \text{ and } h_{\mathbf{A}}(\mathbf{r}_1) = h_{\mathbf{A}}(\mathbf{r}_2);\ \mathbf{A} \leftarrow \mathbb{Z}_2^{n\times n^{1+c}}\right] \geq \delta,$$

then there exists an algorithm A_2 *that runs in time* $\approx t$ *and solves the decisional* LPN^n_ϵ *problem for any* $\epsilon \leq \frac{1}{4}$ *with advantage at least* $\delta \cdot 2^{-16n\epsilon/(c\log n)-1}$*.*

In particular, for $\epsilon = O(\log^2 n/n)$ *and any* $\delta = 1/\text{poly}(n)$*, the advantage is* $1/\text{poly}(n)$*.*

Proof. The algorithm A_2 has access to an oracle that is either $\mathcal{O}^n_{\mathbf{s},\epsilon}$ or $\mathcal{U}^n$. He calls the oracle n^{1+c} times to obtain samples of the form $(\mathbf{a}_i, b_i)$. He arranges the $\mathbf{a}_i$ and b_i into a matrix $\mathbf{A}$ and vector $\mathbf{b}$ as

$$\mathbf{A} = \left[\ \mathbf{a}_1 \mid \cdots \mid \mathbf{a}_{n^{1+c}}\ \right] \in \mathbb{Z}_2^{n \times n^{1+c}}, \quad \mathbf{b} = \begin{bmatrix} b_1 \\ \cdots \\ b_{n^{1+c}} \end{bmatrix} \in \mathbb{Z}_2^{n^{1+c}}$$

and sends $\mathbf{A}$ to A_1. If A_1 fails to return a valid answer, then A_2 outputs $\mathsf{ans} \leftarrow \{0,1\}$. If A_1 does return valid distinct $\mathbf{r}_1$ and $\mathbf{r}_2$ such that $h_{\mathbf{A}}(\mathbf{r}_1) = h_{\mathbf{A}}(\mathbf{r}_2)$, then A_2 returns $\mathsf{ans} = \mathbf{b}^T(\mathbf{r}_1 - \mathbf{r}_2)$.

We first look at the distribution of ans when the oracle that A_2 has access to is $\mathcal{U}^n$. In this case it's easy to see that regardless of whether A_1 returns a valid answer, we'll have $\Pr[\mathsf{ans} = 0] = \frac{1}{2}$ because $\mathbf{b}$ is completely uniform in $\mathbb{Z}_2^{n^{1+c}}$.

On the other hand, if the oracle is $\mathcal{O}^n_{\mathbf{s},\epsilon}$, then we know that for all i, $b_i = \mathbf{s}^T\mathbf{a}_i + e_i$, where $e_i \leftarrow \mathsf{Ber}_\epsilon$. This can be rewritten as $\mathbf{s}^T\mathbf{A} + \mathbf{e}^T = \mathbf{b}^T$ where $\mathbf{e} = \begin{bmatrix} e_1 \\ \cdots \\ e_{n^{1+c}} \end{bmatrix}$. Therefore

$$\mathbf{b}^T(\mathbf{r}_1 - \mathbf{r}_2) = \mathbf{A}(\mathbf{r}_1 - \mathbf{r}_2) + \mathbf{e}^T(\mathbf{r}_1 - \mathbf{r}_2) = \mathbf{e}^T(\mathbf{r}_1 - \mathbf{r}_2).$$

Since $\mathsf{wt}(\mathbf{r}_i) = 2n/(c\log n)$, we know that $\mathsf{wt}(\mathbf{r}_1 - \mathbf{r}_2) \leq 4n/(c\log n)$. Since the $\mathbf{A}$ that is sent to A_1 is independent of $\mathbf{e}$, we have that

$$\Pr[\mathbf{e}^T \cdot (\mathbf{r}_1 - \mathbf{r}_2) = 0\ ;\ e_i \leftarrow \mathsf{Ber}_\epsilon] \geq \frac{1}{2} + \frac{1}{2}(1-2\epsilon)^{4n/(c\log n)} \geq \frac{1}{2} + 2^{-16n\epsilon/(c\log n)-1}, \tag{12}$$

where the first inequality follows from Lemma 2.1 and the second inequality is due to the assumption that $\epsilon \leq \frac{1}{4}$ and the fact that $1 - x \geq 2^{-2x}$ for $x \leq 1/2$.

Thus when the oracle is $\mathcal{O}^n_{\mathbf{s},\epsilon}$, we have

$$\Pr[\mathsf{ans} = 0] \geq \frac{1}{2} \cdot (1 - \delta) + \left(\frac{1}{2} + 2^{-16n\epsilon/(c\log n)-1}\right) \cdot \delta = \frac{1}{2} + \delta \cdot 2^{-16n\epsilon/(c\log n)-1}.$$

□

6.1 Observations and Other Parameter Regimes

As far as we know, the best attack against the hash function in (11) with $c = 1$ requires $2^{\Omega(n)}$ time, whereas the $\mathsf{LPN}^n_{\log^2 n/n}$ problem, from which we can show a polynomial-time reduction, can be solved in time $2^{O(\log^2 n)}$. Thus there is possibly a noticeable loss in the reduction for this parameter setting. It was observed in [YZW+17, Theorem 2, Theorem 3] that there are other ways to set the parameters in Theorem 6.1 which achieve different connections between the hash function and the underlying LPN problem. For example, defining $n = \log^2 m$ and $c = \log m / \log\log m - 1$ implies that there exists a hash function defined by the matrix $\mathbf{A} \in \mathbb{Z}_2^{\log^2 m \times 2m}$ such that succeeding with probability δ in finding collisions in this hash function is at least as hard as solving $\mathsf{LPN}_\epsilon^{\log^2 m}$ problem with advantage $\delta \cdot m^{-O(\kappa\epsilon)}$ for a constant κ. This is exactly the parameter setting in [YZW+17, Theorem 3].[2]

Based on the state of the art of today's algorithms, it's clear that using a hash function defined by an $n \times n^2$ matrix $\mathbf{A}$ is more secure than one defined by a $\log^2 n \times 2n$ matrix (since one can trivially find collisions in the latter in time $2^{O(\log^2 n)}$). There is, however, no connection that we're aware of between the LPN problems on which they are based via Theorem 6.1. In particular, we do not know of any polynomial-time (in n) reductions that relate the hardness of the $\mathsf{LPN}^n_{\log^2 n/n}$ problem to the $\mathsf{LPN}_\epsilon^{\log^2 n}$ problem for a constant ϵ.

Acknowledgments. The first author wishes to thank Ben Berger and Noga Ron-Zewi for discussions on the hardness of decoding problems. We also thank Yu Yu and anonymous Eurocrypt reviewers for their helpful feedback.

References

[ABSS93] Arora, S., Babai, L., Stern, J., Sweedyk, Z.: The hardness of approximate optimia in lattices, codes, and systems of linear equations. In: 34th Annual Symposium on Foundations of Computer Science, Palo Alto, CA, USA, 3–5 November 1993, pp. 724–733. IEEE Computer Society (1993)

[ADPS16] Alkim, E., Ducas, L., Pöppelmann, T., Schwabe, P.: Post-quantum key exchange - a new hope. In: 25th USENIX Security Symposium, USENIX Security 16, Austin, TX, USA, 10–12 August 2016, pp. 327–343 (2016)

[AHI+17] Applebaum, B., Haramaty, N., Ishai, Y., Kushilevitz, E., Vaikuntanathan, V.: Low-complexity cryptographic hash functions. In: 8th Innovations in Theoretical Computer Science Conference, ITCS 2017, 9-11 January 2017, Berkeley, CA, USA, pp. 7:1–7:31 (2017)

[Ajt96] Ajtai, M.: Generating hard instances of lattice problems (extended abstract). In: Proceedings of the Twenty-Eighth Annual ACM Symposium on the Theory of Computing, Philadelphia, PA, USA, 22–24 May 1996, pp. 99–108 (1996)

[APY09] Alon, N., Panigrahy, R., Yekhanin, S.: Deterministic approximation algorithms for the nearest codeword problem. In: Dinur, I., Jansen, K., Naor, J., Rolim, J.D.P. (eds.) APPROX/RANDOM 2009. LNCS, vol. 5687, pp. 339–351. Springer, Heidelberg (2009). https://doi.org/10.1007/978-3-642-03685-9_26

[2] The error ϵ in [YZW+17] can be any constant $< \frac{1}{2}$, whereas our Theorem 6.1 restricts it to $< \frac{1}{4}$. Our restriction is made simply for obtaining a "clean" inequality in Eq. (12) since we were only interested in LPN instances with much lower noise.

[BCD+16] Bos, J.W., et al.: Frodo: take off the ring! Practical, quantum-secure key exchange from LWE. In: Proceedings of the 2016 ACM SIGSAC Conference on Computer and Communications Security, Vienna, Austria, 24–28 October 2016, pages 1006–1018 (2016)

[BFKL93] Blum, A., Furst, M.L., Kearns, M.J., Lipton, R.J.: Cryptographic primitives based on hard learning problems. In: Stinson, D.R. (ed.) CRYPTO 1993. LNCS, vol. 773, pp. 278–291. Springer, Heidelberg (1994). https://doi.org/10.1007/3-540-48329-2_24

[BK02] Berman, P., Karpinski, M.: Approximating minimum unsatisfiability of linear equations. In: Eppstein, D. (ed.) Proceedings of the Thirteenth Annual ACM-SIAM Symposium on Discrete Algorithms, 6–8 January 2002, San Francisco, CA, USA, pp. 514–516. ACM/SIAM (2002)

[BKW03] Blum, A., Kalai, A., Wasserman, H.: Noise-tolerant learning, the parity problem, and the statistical query model. J. ACM **50**(4), 506–519 (2003)

[BLP+13] Brakerski, Z., Langlois, A., Peikert, C., Regev, O., Stehlé, D.: Classical hardness of learning with errors. In: Symposium on Theory of Computing Conference, STOC 2013, Palo Alto, CA, USA, 1–4 June 2013, pp. 575–584 (2013)

[BLSV17] Brakerski, Z., Lombardi, A., Segev, G., Vaikuntanathan, V.: Anonymous IBE, leakage resilience and circular security from new assumptions. Cryptology ePrint Archive, Report 2017/967 (2017). https://eprint.iacr.org/2017/967. EUROCRYPT 2018 [BLSV18]

[BLSV18] Brakerski, Z., Lombardi, A., Segev, G., Vaikuntanathan, V.: Anonymous IBE, leakage resilience and circular security from new assumptions. In: Nielsen, J.B., Rijmen, V. (eds.) EUROCRYPT 2018. LNCS, vol. 10820, pp. 535–564. Springer, Cham (2018). https://doi.org/10.1007/978-3-319-78381-9_20

[BV11] Brakerski, Z., Vaikuntanathan, V.: Efficient fully homomorphic encryption from (standard) LWE. In: IEEE 52nd Annual Symposium on Foundations of Computer Science, FOCS 2011, Palm Springs, CA, USA, 22–25 October 2011, pp. 97–106 (2011)

[DMS99] Dumer, I., Micciancio, D., Sudan, M.: Hardness of approximating the minimum distance of a linear code. In: 40th Annual Symposium on Foundations of Computer Science, FOCS 1999, 17–18 October 1999, New York, NY, USA, pp. 475–485. IEEE Computer Society (1999)

[FGKP09] Feldman, V., Gopalan, P., Khot, S., Ponnuswami, A.K.: On agnostic learning of parities, monomials, and halfspaces. SIAM J. Comput. **39**(2), 606–645 (2009)

[Gen09] Gentry, C.: Fully homomorphic encryption using ideal lattices. In: Proceedings of the 41st Annual ACM Symposium on Theory of Computing, STOC 2009, Bethesda, MD, USA, 31 May–2 June 2009, pp. 169–178 (2009)

[GGH96] Goldreich, O., Goldwasser, S., Halevi, S.: Collision-free hashing from lattice problems. In: Electronic Colloquium on Computational Complexity (ECCC), vol. 3, no. 42 (1996)

[GMR89] Goldwasser, S., Micali, S., Rackoff, C.: The knowledge complexity of interactive proof systems. SIAM J. Comput. **18**(1), 186–208 (1989)

[Gol95] Goldreich, O.: Three XOR-lemmas - an exposition. In: Electronic Colloquium on Computational Complexity (ECCC), vol. 2, no. 56 (1995)

[KPC+11] Kiltz, E., Pietrzak, K., Cash, D., Jain, A., Venturi, D.: Efficient authentication from hard learning problems. In: Paterson, K.G. (ed.) EUROCRYPT 2011. LNCS, vol. 6632, pp. 7–26. Springer, Heidelberg (2011). https://doi.org/10.1007/978-3-642-20465-4_3

[KS06] Katz, J., Shin, J.S.: Parallel and concurrent security of the HB and hb^+ protocols. In: Vaudenay, S. (ed.) EUROCRYPT 2006. LNCS, vol. 4004, pp. 73–87. Springer, Heidelberg (2006). https://doi.org/10.1007/11761679_6

[KS10] Kopparty, S., Saraf, S.: Local list-decoding and testing of random linear codes from high error. In: Schulman, L.J. (ed.) Proceedings of the 42nd ACM Symposium on Theory of Computing, STOC 2010, Cambridge, MA, USA, 5–8 June 2010, pp. 417–426. ACM (2010)

[Mat93] Matsui, M.: Linear cryptanalysis method for DES cipher. In: Helleseth, T. (ed.) EUROCRYPT 1993. LNCS, vol. 765, pp. 386–397. Springer, Heidelberg (1994). https://doi.org/10.1007/3-540-48285-7_33

[MR04] Micciancio, D., Regev, O.: Worst-case to average-case reductions based on Gaussian measures. In: Proceedings of 45th Symposium on Foundations of Computer Science (FOCS 2004), 17–19 October 2004, Rome, Italy, pp. 372–381 (2004)

[MV03] Micciancio, D., Vadhan, S.P.: Statistical zero-knowledge proofs with efficient provers: lattice problems and more. In: Boneh, D. (ed.) CRYPTO 2003. LNCS, vol. 2729, pp. 282–298. Springer, Heidelberg (2003). https://doi.org/10.1007/978-3-540-45146-4_17

[MX10] Mahmoody, M., Xiao, D.: On the power of randomized reductions and the checkability of SAT. In: Proceedings of the 25th Annual IEEE Conference on Computational Complexity, CCC 2010, Cambridge, MA, 9–12 June 2010, pp. 64–75 (2010)

[Pei09] Peikert, C.: Public-key cryptosystems from the worst-case shortest vector problem (extended abstract). In: Proceedings of the 41st Annual ACM Symposium on Theory of Computing, STOC 2009, Bethesda, MD, USA, 31 May–2 June 2009, pp. 333–342 (2009)

[Reg05] Regev, O.: On lattices, learning with errors, random linear codes, and cryptography. In: Proceedings of the 37th Annual ACM Symposium on Theory of Computing, Baltimore, MD, USA, 22–24 May 2005, pp. 84–93 (2005)

[SV03] Sahai, A., Vadhan, S.P.: A complete problem for statistical zero knowledge. J. ACM **50**(2), 196–249 (2003)

[Vaz86] Vazirani, U.V.: Randomness, adversaries and computation (random polynomial time). Ph.D. thesis (1986)

[YZW+17] Yu, Y., Zhang, J., Weng, J., Guo, C., Li, X.: Learning parity with noise implies collision resistant hashing. Cryptology ePrint Archive, Report 2017/1260 (2017). https://eprint.iacr.org/2017/1260

New Techniques for Obfuscating Conjunctions

James Bartusek[1(✉)], Tancrède Lepoint[2], Fermi Ma[1], and Mark Zhandry[1]

[1] Princeton University, Princeton, USA
bartusek.james@gmail.com, fermima1@gmail.com, mzhandry@princeton.edu
[2] SRI International, New York, NY, USA
tancrede@google.com

Abstract. A conjunction is a function $f(x_1, \ldots, x_n) = \bigwedge_{i \in S} l_i$ where $S \subseteq [n]$ and each l_i is x_i or $\neg x_i$. Bishop et al. (CRYPTO 2018) recently proposed obfuscating conjunctions by embedding them in the error positions of a noisy Reed-Solomon codeword and placing the codeword in a group exponent. They prove distributional virtual black box (VBB) security in the generic group model for random conjunctions where $|S| \geq 0.226n$. While conjunction obfuscation is known from LWE [31,47], these constructions rely on substantial technical machinery.

In this work, we conduct an extensive study of *simple* conjunction obfuscation techniques.

- We abstract the Bishop et al. scheme to obtain an equivalent yet more efficient "dual" scheme that can handle conjunctions over exponential size alphabets. This scheme admits a straightforward proof of generic group security, which we combine with a novel combinatorial argument to obtain distributional VBB security for $|S|$ of *any size*.
- If we replace the Reed-Solomon code with a *random binary linear code*, we can prove security from standard LPN and avoid encoding in a group. This addresses an open problem posed by Bishop et al. to prove security of this simple approach in the standard model.
- We give a new construction that achieves *information theoretic* distributional VBB security and weak functionality preservation for $|S| \geq n - n^{\delta}$ and $\delta < 1$. Assuming discrete log and $\delta < 1/2$, we satisfy a stronger notion of functionality preservation for computationally bounded adversaries while still achieving information theoretic security.

1 Introduction

Program obfuscation [7] scrambles a program in order to hide its implementation details, while still preserving its functionality. Much of the recent

J. Bartusek and F. Ma—This work was done while the author was an intern at SRI International.

T. Lepoint—Now at Google.

Y. Ishai and V. Rijmen (Eds.): EUROCRYPT 2019, LNCS 11478, pp. 636–666, 2019.
https://doi.org/10.1007/978-3-030-17659-4_22

attention on obfuscation focuses on obfuscating *general* programs. Such obfuscation is naturally the most useful [15,43], but currently the only known constructions are extremely inefficient and rely on new uncertain complexity assumptions about cryptographic multilinear maps [23,27,29]. Despite recent advances [1,2,8,13,28,35–37,39], obfuscating general programs remains out of reach.

For some specific functionalities, it is possible to avoid multilinear maps. A series of works have shown how to obfuscate point functions (i.e., boolean functions that output 1 on a single input) and hyperplanes [10,21,24,38,46,48]. Brakerski, Vaikuntanathan, Wee, and Wichs [19] showed how to obfuscate conjunctions under a variant of the Learning with Errors (LWE) assumption. More recently it has been shown how to obfuscate a very general class of evasive functions including conjunctions under LWE [31,47].

1.1 This Work: Conjunction Obfuscation

In this work, we primarily consider obfuscating conjunctions. This class of programs has also been called pattern matching with wildcards [12], and in related contexts is known as bit-fixing [14].

A conjunction is any boolean function $f(x_1, \ldots, x_n) = \bigwedge_{i \in S} l_i$ for some $S \subseteq [n]$, where each literal l_i is either x_i or $\neg x_i$. This functionality can be viewed as pattern-matching for a pattern $\mathsf{pat} \in \{0, 1, *\}^n$, where the $*$ character denotes a wildcard. An input string $x \in \{0, 1\}^n$ matches a pattern pat if and only if x matches pat at all non-wildcard positions. So for example $x = \mathtt{0100}$ matches $\mathsf{pat} = \mathtt{*10*}$ but not $\mathsf{pat} = \mathtt{1**0}$.

We are interested in obfuscating the boolean functions $f_{\mathsf{pat}} \colon \{0, 1\}^n \to \{0, 1\}$ which output 1 if and only if x matches pat. We additionally give obfuscation constructions for two generalizations of the pattern matching functionality: multi-bit conjunction programs $f_{\mathsf{pat},m}$ which output a secret message $m \in \{0, 1\}^\ell$ on an accepting input, and conjunctions over arbitrary size alphabets.[1]

In particular, we consider the notion of *distributional virtual black-box obfuscation* (VBB), which guarantees that the obfuscation of a pattern drawn from some distribution can be simulated efficiently, given only oracle access to the truth table of the function defined by the pattern. We consider this notion of obfuscation in the *evasive* setting, where given oracle access to a pattern drawn from the distribution, the polynomial time simulator cannot find an accepting input except with negligible probability. Thus our goal will be to produce obfuscations that are easily simulatable without any information about the sampled pattern other than its distribution.

Recently, Bishop, Kowalczyk, Malkin, Pastro, Raykova, and Shi [12] gave a simple and elegant obfuscation scheme for conjunctions, which they prove secure

---

[1] Conjunctions over boolean/binary inputs naturally generalize to alphabets $[\ell]$ for $\ell \geq 2$. In this setting, each $x_i \in [\ell]$, and ℓ_i specifies the setting on the ith character. Positions not fixed by the ℓ_i are the wildcards.

in the generic group model [44]. Unfortunately, they did not prove security relative to any concrete (efficiently falsifiable [30,41]) assumption on cryptographic groups. Before their work, obfuscation for conjunctions was already known from LWE as a consequence of lockable obfuscation (also known as obfuscation for compute-and-compare programs) [31,47]. However, for the restricted setting of conjunctions, the Bishop et al. [12] construction is significantly simpler and more efficient.

Our Results. In this work, we show how to alter the Bishop et al. construction in various ways, obtaining the following results.

- *A New Generic Group Construction.* We give a new group-based construction that can be viewed as "dual" to the construction of Bishop et al. [12]. Our construction offers significant efficiency improvements by removing the dependence on alphabet size from the construction.
  We also improve upon the generic group security analysis of Bishop et al. [12] by simplifying the proof steps and extending the argument to handle a larger class of distributions.
- *Security from LPN.* We show that a few modifications to the group-based construction allows us to remove groups from the scheme entirely. We prove security of the resulting construction under the (constant-rate) Learning Parity with Noise (LPN) assumption. Along the way, we give a reduction from standard LPN to a specific, structured-error version of LPN, which we believe may be of independent interest.
- *Information-Theoretic Security.* Finally, we show how to extend our techniques to the information-theoretic setting if the number of wildcards is sublinear. We stress that this requires considering a weaker notion of functionality preservation. We also give an alternative information theoretic scheme that achieves an intermediate "computational" notion of functionality preservation assuming discrete log.

In Table 1, we compare our results with prior works on conjunction obfuscation achieving distributional-VBB security (we omit the [19] and [17] constructions from entropic-ring-LWE and multilinear maps).

1.2 Technical Overview

Review of the Bishop et al. Construction [12]. We first recall the Bishop et al. scheme for obfuscating a pattern $\mathsf{pat} \in \{0,1,*\}^n$. Begin by fixing a prime q exponential in n. Then sample uniformly random $s_1, \ldots, s_{n-1} \leftarrow \mathbb{Z}_q$ and define the polynomial $s(t) := \sum_{k=1}^{n-1} s_k t^k \in \mathbb{Z}_q[t]$. Note that $s(t)$ is a uniformly random degree $n-1$ polynomial conditioned on $s(0) = 0$.

Now visualize a $2 \times n$ grid with columns indexed as $i = 1, \ldots, n$ and rows indexed as $j = 0, 1$. To obfuscate $\mathsf{pat} \in \{0,1,*\}^n$, for each (i,j) such that $\mathsf{pat}_i \in \{j,*\}$, we place $s(2i+j)$ in grid cell (i,j) and otherwise, we place $r_{2i+j} \leftarrow \mathbb{Z}_q$. For example, if the pattern is $\mathsf{pat} = \mathtt{11*0}$, we write

Table 1. A comparison between our constructions and prior work. Let $\mathcal{U}_{n,w}$ be the uniform distribution over all patterns in $\{0,1,*\}^n$ with exactly w wildcards. For any pattern $\mathsf{pat} \in \{0,1,*\}^n$, define $\mathsf{pat}^{-1}(*) := \{j \mid \mathsf{pat}_j = *\}$ the *positions* of the wildcards and let $\mathbf{b} \in \{0,1\}^{n-w}$ denote the fixed bits of pat. When we say the alphabet is exponential, we mean any alphabet with size at most exponential in the security parameter. FP refers to functionality preservation.

	Assumption	Alphabet	Distribution	FP
[31,47]	LWE	Exponential	$H_\infty(\mathbf{b} \vert \mathsf{pat}^{-1}(*)) \geq \log(n)$	Strong
Bishop et al.	GGM	Binary	$\mathcal{U}_{w,n}$ for $w < .774n$	Strong
This work	GGM	Exponential	$\mathcal{U}_{w,n}$ for $w < n - \omega(\log(n))$[a]	Strong
This work	LPN	Binary	$\mathcal{U}_{w,n}$ for $w = cn, c < 1$	Weak
This work	None	Binary	$H_\infty(\mathbf{b} \vert \mathsf{pat}^{-1}(*)) \geq n^{1-\gamma}$[b]	Weak

[a]In a concurrent work [11], Beullens and Wee achieved the same improvement in parameters and show how to base security on a new knowledge assumption secure in the generic group model. In the full version [9], we also obtain security for more general distributions that satisfy a certain min-entropy requirement.
[b]For patterns with n^δ wildcards, and $\gamma < 1 - \delta$.

r_2	r_4	$s(6)$	$s(8)$
$s(3)$	$s(5)$	$s(7)$	r_9

Bishop et al. [12] observe that these $2n$ field elements are essentially a noisy Reed-Solomon codeword with the white grid cells representing error positions. If the number of error positions is small enough, an attacker can run any Reed-Solomon error correction algorithm to recover $s(t)$ and learn pat. However, all known error-correction algorithms for Reed-Solomon codes are *non-linear*. Thus, the final step is to place the $2n$ field elements in the exponent of a group $\mathbb{G} = \langle g \rangle$ of order q. The crucial observation in [12] is that we can perform honest evaluation on an input $x \in \{0,1\}^n$ with linear operations in the exponent. For example, to evaluate on input $x = \mathtt{1110}$, we generate Lagrange reconstruction coefficients L_3, L_5, L_7, L_8 for the cells corresponding to x and reconstruct

$$g^{L_3 s(3) + L_5 s(5) + L_7 s(7) + L_8 s(8)} = g^{s(0)} = g^0.$$

Evaluation accepts if and only if the result is g^0. Notice that if a single element from a white cell is included in the reconstruction, the evaluator fails to recover g^0 with overwhelming probability $(q-1)/q$. For security, they prove the following:

Theorem **([12]).** *Let $\mathcal{U}_{n,w}$ be the uniform distribution over all patterns in $\{0,1,*\}^n$ with exactly w wildcards. For any $w < 0.774n$, this construction attains distributional virtual black box security in the generic group model.*

Bishop et al. [12] do not address whether the scheme becomes insecure for $0.774n < w < n - \omega(\log n)$, or if the bound is a limitation of their analysis.[2]

This Work. We provide several new interpretations of the [12] scheme. Through these interpretations, we are able to obtain improved security, efficiency, and generality, as well as novel constructions secure under standard cryptographic assumptions. We summarize the properties of these new constructions in Table 1.

Interpretation 1: The Primal. Our first observation is that the $2n$ field elements generated by the [12] construction can be rewritten as a product of a transposed Vandermonde matrix A and a random vector s, plus a certain "error vector" e. So if the pattern is $\mathsf{pat} = \mathtt{11{*}0}$, instead of writing the elements in grid form as above, we can stack them in a column as

$$\begin{pmatrix} r_2 \\ s(3) \\ r_4 \\ s(5) \\ s(6) \\ s(7) \\ s(8) \\ r_9 \end{pmatrix} = \begin{pmatrix} 2^1 & 2^2 & 2^3 \\ 3^1 & 3^2 & 3^3 \\ 4^1 & 4^2 & 4^3 \\ 5^1 & 5^2 & 5^3 \\ 6^1 & 6^2 & 6^3 \\ 7^1 & 7^2 & 7^3 \\ 8^1 & 8^2 & 8^3 \\ 9^1 & 9^2 & 9^3 \end{pmatrix} \cdot \begin{pmatrix} s_1 \\ s_2 \\ s_3 \end{pmatrix} + \begin{pmatrix} r_2' \\ 0 \\ r_4' \\ 0 \\ 0 \\ 0 \\ 0 \\ r_9' \end{pmatrix}.$$

So far, nothing has changed—the Bishop et al. [12] obfuscation scheme is precisely $g^{A \cdot s + e}$. But if we revisit the evaluation procedure in the $A \cdot s + e$ format, a possible improvement to the construction becomes apparent. Recall that evaluation is simply polynomial interpolation: on input $x \in \{0,1\}^n$, the evaluator generates a vector $v \in \mathbb{Z}_q^{2n}$ where $v_{2i+x_i-1} = 0$ for all $i \in [n]$, and the n non-zero elements of v are Lagrange coefficients. For any input x (even ones not corresponding to accepting inputs), the Lagrange coefficients ensure v satisfies $v^\top \cdot A = 0 \in \mathbb{Z}_q^{2n}$ and the corresponding scalar equation $v^\top \cdot A \cdot s = 0$. This means an input x is only accepted if $v^\top \cdot (A \cdot s + e) = v^\top \cdot e = 0$. Indeed, we can verify that if there exists a position $i \in [n]$ where $x_i \neq \mathsf{pat}_i$ (note that if $\mathsf{pat}_i = *$ we take this to mean $x_i = \mathsf{pat}_i$), this corresponds to an entry where v is non-zero and e is uniformly random, making $v^\top \cdot e$ non-zero with overwhelming probability.

Interpretation 2: The Dual. Observe that evaluation only required the A matrix and e vector. The random degree $n-1$ polynomial $s(t)$ generated in the [12] scheme, whose coefficients form the random s vector, does not play a role in functionality. This suggests performing the following "dual" transformation to the $A \cdot s + e$ scheme. Let B be an $(n+1) \times 2n$ dimensional matrix whose

[2] If $w = n - O(\log n)$, the distributional virtual black box security notion is vacuous since an attacker can guess an accepting input and recover pat entirely.

rows span the left kernel of A. Since $B \cdot A = \mathbf{0} \in \mathbb{Z}_q^{(n+1)\times(n-1)}$, multiplying $B \cdot (A \cdot s + e)$ yields the $n+1$ dimensional vector $B \cdot e$. We claim this dual $g^{B \cdot e}$ scheme captures all the information needed for secure generic group obfuscation, but with $n+1$ group elements rather than $2n$.

Evaluation in the Dual. A similar evaluation procedure works for the dual scheme. On input x, the evaluator solves for a vector $k \in \mathbb{F}_q^{n+1}$ so that the $2n$-dimensional vector $k^\top \cdot B$ is 0 at position $2i + x_i - 1$ for each $i \in [n]$. Note that such a k exists since we only place n constraints on $n+1$ variables. $k^\top \cdot B$ will play exactly the same role as the $v^\top$ vector from the $A \cdot s + e$ evaluation. On accepting input, $k^\top \cdot B$ will be 0 in all the positions where e is non-zero, so $k^\top \cdot B \cdot e = 0$. On rejecting inputs, $(k^\top \cdot B)$ will have a non-zero entry where the corresponding entry of e is uniformly random, so $k^\top \cdot B \cdot e \neq 0$ with overwhelming probability.

Proving Generic Group Security. The Bishop et al. [12] proof of distributional VBB security uses over 10 pages of generic group and combinatorial analysis. They derive their bound of $0.774n$ on the number of wildcards by numerically solving a non-linear equation arising from their analysis of a certain combinatorial problem.

Our first contribution is to show that by analyzing our dual scheme, we can give a short and extremely intuitive proof of generic group model security from a linear independence argument (Sect. 3). Part of the simplification arises from the fact that our dual scheme completely removes the random polynomial from the construction. Our generic model proof steps end with the same combinatorial problem Bishop et al. [12] consider, but instead of deferring to their analysis, we give a simple combinatorial argument from a Chernoff bound. This allows us to improve their $0.774n$ wildcard bound to $n - \omega(\log n)$. This bound is optimal; for $n - O(\log n)$ wildcards, a polynomial time adversary can guess an accepting input and learn the pattern entirely. We remark that our new combinatorial analysis implies the original Bishop et al. [12] scheme is also secure up to $n - \omega(\log n)$ wildcards. In the full version [9], we show how to generalize this analysis to certain distributions with sufficient entropy.

Conjunctions over Large Alphabets. If we go beyond binary alphabets, the dual scheme actually reduces the obfuscation size by far more than a factor of 2. Suppose the alphabet is $[\ell]$ for some integer ℓ, so a conjunction is specified by a length n pattern $\mathsf{pat} \in \{[\ell] \cup \{*\}\}^n$. $f_{\mathsf{pat}}(x) = 1$ only if $x_i = \mathsf{pat}_i$ on all non-wildcard positions.

We can give a natural generalization of the $A \cdot s + e$/Bishop et al. [12] scheme to handle larger alphabets. For an alphabet of size ℓ, we use an error vector $e \in \mathbb{Z}_q^{n\ell}$, which we imagine partitioning into n blocks of length ℓ. The ith block of e corresponds with the ith pattern position. As in the binary case, if $\mathsf{pat}_i = *$, we set every entry of e in the ith block to 0. If $\mathsf{pat}_i = j$ for $j \in [\ell]$, we set the jth position in the ith block of e to a uniformly random value in $\mathbb{F}_q$, and set the remaining $\ell - 1$ entries in the ith block to 0. A is now a transposed

Vandermonde matrix of dimension $n\ell \times n\ell - n - 1$, and s is drawn as a random vector from $\mathbb{Z}_p^{n\ell-n-1}$. To evaluate on $x \in [\ell]^n$, we solve for $v^\top \cdot A = 0$ where v is restricted to be zero only at $v_{(i-1)\ell+x_i}$ for each $i \in [n]$.[3] However, this scheme is fundamentally stuck at polynomial-size alphabets, since $A \cdot s + e$ contains $n\ell$ elements.

If we switch to the dual view, this same scheme can be implemented as $g^{B \cdot e}$ where $B \in \mathbb{Z}_q^{(n+1)\times n\ell}, e \in \mathbb{Z}_q^{n\ell}$. But the number of group elements in $g^{B \cdot e}$ is simply $n+1$, which has *no* dependence on the alphabet size. Of course B will have dimension $(n+1) \times n\ell$, but if we choose B to be a Vandermonde matrix, we can demonstrate that neither the evaluator nor the obfuscator ever have to store B or e in their entirety, since e is sparse for large ℓ. In particular, we set the (i, j)th entry of B to j^i. We simply need q to grow with $\log \ell$ to ensure this implicit B satisfies the certain linear independence conditions that arise from our security analysis.

Moving Out of the Exponent. Returning to the $A \cdot s + e$ view of the scheme for a moment, we see that its form begs an interesting question:

Can the (transposed) Vandermonde matrix A be replaced with other matrices?

In [12], the transposed Vandermonde matrix A plays at least two crucial roles: it allows for evaluation by polynomial interpolation and at the same time is vital for their security analysis. However, the structure of the transposed Vandermonde matrix is what leads to Reed-Solomon decoding attacks on the plain scheme, necessitating encoding the values in a cryptographic group. Furthermore, observe that our abstract evaluation procedure described for our primal interpretation made no reference to the specific *structure* of A; in particular, it works for *any* public matrix A. In the case of the transposed Vandermonde matrix, applying this abstract procedure results in the Lagrange coefficients used in [12], but we can easily perform evaluation for other matrices.

Furthermore, the matrix form of the scheme is strongly reminiscent of the Learning Parity with Noise (LPN) problem and in particular its extension to $\mathbb{F}_q$, known as the Random Linear Codes (RLC) problem [33].

We recall the form of the RLC problem over $\mathbb{F}_q$ for noise rate ρ and n^c samples. Here, we have a uniformly random matrix $A \leftarrow \mathbb{F}_q^{n^c \times n}$, a uniformly random column vector $s \in \mathbb{F}_q^n$, and an error vector $e \in \mathbb{F}_q^{n^c}$ generated as follows. For each $i \in [n^c]$, set $e_i = 0$ with independent probability $1 - \rho$, and otherwise draw $e_i \leftarrow \mathbb{F}_q$ uniformly at random. The search version of this problem is to recover the secret vector s given $(A, A \cdot s + e)$, and the decision version is to is to distinguish $(A, A \cdot s + e)$ from (A, v) for uniformly random $v \leftarrow \mathbb{F}_q^{n^c}$. The standard search RLC and decisional RLC assumptions are that these problems are intractable for any computationally bounded adversary for constant noise rate $0 < \rho < 1$.

[3] We note that if we set $\ell = 2$, this generalization flips the role of 0 and 1, but is functionally equivalent.

This suggests the following approach to obtaining a secure obfuscation scheme from the original scheme: simply replace A with a *random matrix* over $\mathbb{F}_q$. A would be publicly output along with $A \cdot s + e$. The hope would be that we could invoke the RLC assumption to show that even given A, the obfuscation $A \cdot s + e$ is computationally indistinguishable from a vector v of $2n$ random elements. This would allow us to simultaneously avoid encoding in a group exponent *and* obtain security under a standard assumption.

Structured Error Distributions. However, we cannot invoke security of RLC right away. The main problem lies in the fact that the error vector in our setting is *structured*: for any pair of positions e_{2i-1}, e_{2i} for $i \in [n]$, the construction ensures that at least one of e_{2i-1} or e_{2i} is 0. Recall that if the ith bit of the pattern is b, then $e_{2i-b} = 0$ while $e_{2i-(1-b)}$ is drawn randomly from $\mathbb{F}_q$. If the ith bit of the pattern is $*$, then $e_{2i-1} = e_{2i} = 0$. But if both e_{2i-1} and e_{2i} are random elements from $\mathbb{F}_q$, this corresponds to a position where the input string can be neither 0 nor 1, which can never arise in the obfuscation construction.

To the best of our knowledge, the only work that considers this particular structured error distribution is the work of Arora and Ge [5], which shows that this problem is actually *insecure* in the binary case (corresponding to a structured error version of LPN). Their attack uses re-linearization and it is easy to see that it extends to break the problem we would like to assume hard as long as A has $\Omega(n^2)$ rows.

This leaves some hope for security, as our construction only requires that A have $2n$ rows. Thus, we give a reduction that proves hardness of the structured error RLC assumption with $2n$ samples assuming the hardness of the standard RLC assumption for polynomially many samples. We note that our reductions handle both the search and decision variants, and both LPN and RLC. We give a high-level overview of our reduction below.

The Reduction to Structured Error. For our reduction, we return to the $B \cdot e$ view of the scheme and consider the equivalent "dual" version of the decisional RLC problem,[4] where the goal is to distinguish $(B, B \cdot e)$ from (B, u) for $B \leftarrow \mathbb{F}_q^{(n^c-n)\times n^c}$, $u \leftarrow \mathbb{F}_q^{n^c-n}$, and e as drawn previously. The advantage of considering the dual version is that the resulting technical steps of the reduction are slightly easier to see, and we stress that our proof implies the hardness of structured error RLC in its primal $A \cdot s + e$ form.

Note that the problem of distinguishing between $(B, B \cdot e)$ and (B, u) for $n^c - n$ samples and error vector e of dimension n^c is equivalent to the setting where the number of samples is $n - n^{1/c}$ and the error vector is of dimension n. Since the standard RLC problem is conjectured hard for any constant c, we set $\epsilon = 1/c$ and assume hardness for any $0 < \epsilon < 1$.

We show how to turn an instance of this problem into a structured error RLC instance, where the challenge is to distinguish between $(B, B \cdot e)$ and (B, u) for

[4] In the context of LWE this duality/transformation has been observed a number of times, see e.g. [40]. For RLC, this is essentially syndrome decoding.

uniformly random $B \leftarrow \mathbb{F}_q^{(n+1)\times 2n}$, a *structured* error vector $e \in \mathbb{F}_q^{2n}$ with noise rate ρ, and uniformly random $u \in \mathbb{F}_q^{n+1}$.

To perform this transformation, we need to somehow inject the necessary structure into the standard RLC error vector e, which means introducing a zero element in each pair. The most natural way to do this given the regular RLC instance $(B, B \cdot e)$ is to draw n new uniformly random columns and insert them into B in random locations to produce the structured matrix B'. Now $B \cdot e = B' \cdot e'$, where e' is a structured error vector with a 0 element in every $(2i-1, 2i)$ index pair. This immediately gives us a structured error RLC instance with a matrix B' of dimension $(n - n^{\epsilon}) \times 2n$. However, we require B to have $n+1$ rows to enable evaluation of the corresponding obfuscation. We would like to simply extend the $(n-n^{\epsilon}) \times 2n$-dimensional B' to an $(n+1) \times 2n$-dimensional B'' by appending $n^{\epsilon} + 1$ newly generated uniformly random rows, but this appears impossible since we will be unable to fill in the $n^{\epsilon} + 1$ additional entries of $B'' \cdot e'$ without knowledge of e'.

As a first attempt, we can try to extend B' to B'' by appending random linear combinations of the $n - n^{\epsilon}$ rows of B'. This would allow us to properly generate $B'' \cdot e'$ by extending $B' \cdot e'$ with the corresponding linear combinations. Unfortunately, this is not quite sufficient since the matrix B'' is distinguishable from random, since its bottom $n^{\epsilon} + 1$ rows are in the row span of the first $n - n^{\epsilon}$.

We now appeal to the fact that the reduction algorithm itself chose the locations of the newly generated columns in B', and thus it knows the location of n elements of e' set to 0. The reduction can therefore introduce randomness into the last $n^{\epsilon} + 1$ rows of B'' by modifying only the entries in these n columns, since any changes it makes will not affect the dot product with e'. After this process, the last $n^{\epsilon} + 1$ rows of B'' are no longer restricted to being in the row span of the top $n - n^{\epsilon}$ rows. By appealing to leftover hash lemma arguments, we can prove the resulting $(n+1) \times 2n$ dimensional B'' matrix is statistically close uniform and that $B'' \cdot e'$ is correctly distributed.

A Note on Functionality Preservation. Some previous works on conjunction obfuscation [12,17] explicitly prove a weak notion of functionality preservation, where on any given input the obfuscation is required to be correct with overwhelming probability. This is in contrast to strong functionality preservation, which requires simultaneous correctness on all inputs with overwhelming probability. Both [12] and [17] remark that if desired, their constructions can be boosted to achieve the stronger notion by scaling parameters until the error probability on any given input can be union bounded over all inputs.[5]

A notable weakness of our analysis is that the above argument used for proving the B'' matrix is statistically close to uniform does not work for q as large as 2^n. Further complications arise when we attempt to equip a search-to-decision reduction with a predicate (for more detail, see Sect. 4), and thus we limit $q = 2$ for our formal obfuscation construction.[6] Our reduction allows us to

[5] This holds for our generic group model constructions as well.

[6] RLC for field size $q = 2$ is equivalent to LPN.

add slightly more than $n^\epsilon + 1$ additional rows, and it turns out these rows can be used to boost correctness—to a point. On any input, our final construction has an error probability of $1/2^{n^\delta}$ (for any $\delta < 1/2$), and therefore settles for weak functionality preservation.

Information Theoretic Security. Our third and final contribution is a new *statistically* secure conjunction obfuscator. As a starting point, we recall a simple proposal for distributional VBB secure point obfuscation informally discussed by Bishop et al. [12]. The idea of their proposal (modified slightly for our setting) is roughly the following. To obfuscate a point $\mathsf{p} \in \{0,1\}^n$, output n uniformly random elements $a^{(1)}, \ldots, a^{(n)}$ from $\mathbb{F}_q$ conditioned on $\sum_{i|\mathsf{p}_i=1} a^{(i)} = 0$. Equality checking on an input $x \in \{0,1\}^n$ would be done by checking whether $\sum_{i|x_i=1} a^{(i)} = 0$.[7]

While this idea seems like a plausible starting point for point obfuscation, there is no room to support conjunctions. Any wildcard element must be set to 0 to preserve functionality, and thus the obfuscation trivially leaks information on the underlying pattern. This barrier appears inherent if we are limited to summing a set of elements in $\mathbb{F}_q$ and checking if the result is 0. But what if we use matrices in $\mathbb{F}_q$ instead of scalar elements? Evaluation could now involve checking the *rank* of the resulting matrix sum.

We prove security of this scheme by applying the leftover hash lemma (LHL), which shows that as long as the non-wildcard bits of pat have sufficient min-entropy, the matrix F is statistically close to a uniformly random matrix. Then the rank deficient matrix B is statistically hidden from view, so if there are fewer than k wildcards, all of the $A^{(i)}$ matrices are distributed as uniformly random $k \times k$ rank 1 matrices. The number of wildcards this scheme can handle is $k-1$, but we cannot make the matrices arbitrarily large. The limitation arises from our statistical security arguments which only work for k as large as n^δ (for any $\delta < 1$), so we obtain statistical distributional VBB security for patterns with a sublinear number of wildcards.

Computational Functionality Preservation. Although we obtain weak functionality preservation with the above construction, it *necessarily* falls short of strong functionality preservation. Without relaxing correctness, statistical VBB security is impossible since a computationally unbounded adversary can recover pat from the truth table of the obfuscated function.

A Motivating Scenario from [47]. A natural question to ask is when weak functionality preservation is "good enough." To shed light on this, we take a step back and recall a motivating example for general evasive circuit obfuscation. Even this might not be immediately obvious: *what good is an obfuscated circuit if a user*

[7] To the best of our knowledge, this scheme had not appeared in the literature before [12]. However, most prior work on point obfuscation considers stronger correctness, security, and functionality requirements (such as multi-bit output) that this scheme falls short of, which may preclude its use in certain settings.

can never find an accepting input? Wichs and Zirdelis [47] address precisely this question with the following scenario. Suppose we have a set of users where a subset of them has access to additional privileged information. If we publicly give out an obfuscated circuit containing this privileged information, then security assures us that the un-privileged users cannot find accepting inputs. For them, functionality preservation is unimportant since the circuit may as well be the all 0's circuit.[8] However, it does matter for the privileged users who may actually find accepting inputs (for these users, security does not hold).

In this example, a secure obfuscation that only achieves weak functionality preservation is good enough to ensure the un-privileged users never learn anything about the hidden circuit. However, it might not be enough for certain applications. Weak functionality preservation does not explicitly rule out the possibility that a user with privileged information can detect that the obfuscated circuit functionality differs from the intended circuit functionality. In addition, it does not rule out the possibility that a user (privileged or not) can find an input that causes the obfuscated circuit to wrongly accept. For example, in many cases the hash of a password can be viewed as an "obfuscation" of a point function for that password; simply accept if the input hashes to the stored hash [38]. Even if we guarantee that a computationally unbounded adversary cannot learn any information about the original password just given the hash, this does not rule out the possibility that an attacker can find a different string that causes the obfuscated password checker to accept.

An Intermediate Definition. To address this gap, we use a notion (between weak and strong) we refer to as *computational functionality preservation.* In the context of point obfuscation, this notion is essentially equivalent to the correctness definition for oracle hashing[9] considered by Canetti [21] (also achieved by Canetti, Micciancio, and Reingold [22] and Dodis and Smith [25]), as observed by Wee [46]. It is also roughly the same definition considered by Brakerski and Vaikuntanathan [18] for constrained PRFs. For us, computational functionality preservation guarantees that even a user who knows the real circuit (in this work, "real circuit" means the obfuscated pattern) cannot find a point x on which the obfuscated circuit and the real circuit differ, provided they are computationally bounded.

In Sect. 5.3, we describe a simple modification to our basic sum-of-matrices scheme that allows us to achieve computational functionality preservation from discrete log. We note that the resulting construction is still information theoretically secure. Mapping this to the above example, this means even computationally unbounded un-privileged users cannot learn any predicate on the hidden pattern. This is only possible because our obfuscated circuit computes the *wrong output* on exponentially many inputs. Despite this, a computationally bounded user (who might even know the hidden pattern) cannot even find one of these incorrect inputs, assuming discrete log.

[8] This is slightly informal, since it requires a notion of input-hiding obfuscation [6].
[9] This was re-named to "perfectly one-way functions" in [22].

1.3 Related Work

Conjunction Obfuscation. Previously, Brakerski and Rothblum had shown how to obfuscate conjunctions using multilinear maps [16]. This was followed by a work of Brakerski et al. which showed how to obfuscate conjunctions under entropic ring LWE [19]. More recently, Wichs and Zirdelis showed how to obfuscate compute-and-compare programs under LWE [47]. Goyal, Koppula, and Waters concurrently and independently introduced lockable obfuscation and proved security under LWE [31]. Both of these works easily imply secure obfuscation of conjunctions under LWE, though with a complicated construction that encodes branching programs in a manner reminiscent of the GGH15 multilinear map [29]. The main contribution of [12] then was the simplicity and efficiency of their conjunction obfuscation scheme. In this work, we provide constructions and proofs that maintain these strengths while addressing the major weaknesses of the [12] construction—lack of generality (to more wildcards, more distributions, and more alphabet sizes) and lack of security based on a falsifiable assumption.

2 Preliminaries

Notation. Let $\mathbb{Z}, \mathbb{N}$ be the set of integers and positive integers. For $n \in \mathbb{N}$, we let $[n]$ denote the set $\{1, \ldots, n\}$. For $q \in \mathbb{N}$, denote $\mathbb{Z}/q\mathbb{Z}$ by $\mathbb{Z}_q$, and denote the finite field of order q by $\mathbb{F}_q$. A vector v in $\mathbb{F}_q$ (represented in column form by default) is written as a lower-case letter and its coefficients $v_i \in \mathbb{F}_q$ are indexed by i; a matrix A is written as a capital letter and its columns $(A)_j$ are indexed by j. We denote by $0^{n \times m}$ the (n, m)-dimensional matrix filled with zeros. For any matrix M, let $\text{colspan}(M)$ denote the column span of M.

We use the usual Landau notations. A function $f(n)$ is said to be negligible if it is $n^{-\omega(1)}$ and we denote it by $f(n) := \mathsf{negl}(n)$. A probability $p(n)$ is said to be overwhelming if it is $1 - n^{-\omega(1)}$.

If D is a distribution, we denote $\mathsf{Supp}(D) = \{x : D(x) \neq 0\}$ its support. For a set S of finite weight, we let $U(S)$ denote the uniform distribution on S. The statistical distance between two distributions D_1 and D_2 over a countable support S is $\Delta(D_1, D_2) := \frac{1}{2} \sum_{x \in S} |D_1(x) - D_2(x)|$. We naturally extend those definitions to random variables. Let $\epsilon > 0$. We say that two distributions D_1 and D_2 are ϵ-statistically close if $\Delta(D_1, D_2) \leq \epsilon$. We say that D_1 and D_2 are statistically close, and denote $D_1 \approx_s D_2$, if there exists a negligible function ϵ such that D_1 and D_2 are $\epsilon(n)$-statistically close.

The distinguishing advantage of an algorithm $\mathcal{A}$ between two distributions D_0 and D_1 is defined as $\mathsf{Adv}_{\mathcal{A}}(D_0, D_1) := |\Pr_{x \leftarrow D_0}[\mathcal{A}(x) = 1] - \Pr_{x \leftarrow D_1}[\mathcal{A}(x) = 1]|$, where the probabilities are taken over the randomness of the input x and the internal randomness of $\mathcal{A}$. We say that D_1 and D_2 are computationally indistinguishable, and denote $D_1 \approx_c D_2$, if for any non-uniform probabilistic polynomial-time (PPT) algorithm $\mathcal{A}$, there exists a negligible function ϵ such that $\mathsf{Adv}_{\mathcal{A}} = \epsilon(n)$.

Finally, we let $x \leftarrow X$ denote drawing x *uniformly at random* from the space X, and define $\mathcal{U}_{n,w}$ to be the uniform distribution over $\{0,1,*\}^n$ with a fixed w number of $*$ (wildcard) characters.

The min-entropy of a random variable X is $H_\infty(X) := -\log(\max_x \Pr[X = x])$. The (average) conditional min-entropy of a random variable X conditioned on a correlated variable Y, denoted as $H_\infty(X|Y)$, is defined by

$$H_\infty(X|Y) := -\log\left(\mathbb{E}_{y \leftarrow Y}\left[\max_x \Pr[X = x|Y = y]\right]\right).$$

We recall the leftover hash lemma below.

Lemma 1 (Leftover hash lemma). *Let $\mathcal{H} = \{h\colon \mathcal{X} \to \mathcal{Y}\}$ be a 2-universal hash function family. For any random variable $X \in \mathcal{X}$ and Z, for $\epsilon > 0$ such that $\log(|\mathcal{Y}|) \leq H_\infty(X|Z) - 2\log(1/\epsilon)$, the distributions $(h, h(X), Z)$ and $(h, U(\mathcal{Y}), Z)$ are ϵ-statistically close.*

2.1 Security Notions for Evasive Circuit Obfuscation

We recall the definition of a distributional virtual black-box (VBB) obfuscator. We roughly follow the definition of Brakerski and Rothblum [16], but we include a computational functionality preservation definition.

Definition 1 (Distributional VBB Obfuscation). *Let $\mathcal{C} = \{\mathcal{C}_n\}_{n\in\mathbb{N}}$ be a family of polynomial-size circuits, where $\mathcal{C}_n$ is a set of boolean circuits operating on inputs of length n, and let* Obf *be a PPT algorithm which takes as input an input length $n \in \mathbb{N}$ and a circuit $C \in \mathcal{C}_n$ and outputs a boolean circuit $\mathsf{Obf}(C)$ (not necessarily in $\mathcal{C}$). Let $\mathcal{D} = \{\mathcal{D}_n\}_{n\in\mathbb{N}}$ be an ensemble of distribution families $\mathcal{D}_n$ where each $D \in \mathcal{D}_n$ is a distribution over $\mathcal{C}_n$.*

Obf *is a distributional VBB obfuscator for the distribution class $\mathcal{D}$ over the circuit family $\mathcal{C}$ if it has the following properties:*

1. *Functionality Preservation: We give three variants:*
   - *(Weak) Functionality Preservation: For every $n \in \mathbb{N}$, $C \in \mathcal{C}_n$, and $x \in \{0,1\}^n$, there exists a negligible function μ such that*

     $$\Pr[\mathsf{Obf}(C, 1^n)(x) = C(x)] = 1 - \mu(n).$$

   - *(Computational) Functionality Preservation: For every PPT adversary $\mathcal{A}$, $n \in \mathbb{N}$, and $C \in \mathcal{C}_n$, there exists a negligible function μ such that*

     $$\Pr[x \leftarrow \mathcal{A}(C, \mathsf{Obf}(C, 1^n)) : C(x) \neq \mathsf{Obf}(C, 1^n)(x)] = \mu(n).$$

   - *(Strong) Functionality Preservation: For every $n \in \mathbb{N}$, $C \in \mathcal{C}_n$, there exists a negligible function μ such that*

     $$\Pr[\mathsf{Obf}(C, 1^n)(x) = C(x)\ \forall x \in \{0,1\}^n] = 1 - \mu(n).$$

2. *Polynomial Slowdown: For every $n \in \mathbb{N}$ and $C \in \mathcal{C}_n$, the evaluation of* $\mathsf{Obf}(C, 1^n)$ *can be performed in time poly*$(|C|, n)$.
3. *Distributional Virtual Black-Box: For every PPT adversary $\mathcal{A}$, there exists a (non-uniform) polynomial size simulator $\mathcal{S}$ such that for every $n \in \mathbb{N}$, every distribution $D \in \mathcal{D}_n$ (a distribution over $\mathcal{C}_n$), and every predicate $\mathcal{P}\colon \mathcal{C}_n \to \{0,1\}$, there exists a negligible function μ such that*

$$\left| \Pr_{C \leftarrow \mathcal{D}_n} [\mathcal{A}(\mathsf{Obf}(C, 1^n)) = \mathcal{P}(C)] - \Pr_{C \leftarrow \mathcal{D}_n} [\mathcal{S}^C(1^{|C|}, 1^n) = \mathcal{P}(C)] \right| = \mu(n).$$

We note that computational functionality preservation has appeared before in the obfuscation literature [25,46], and our definition is also the same as the functionality preservation notion considered in Definition 3.1 of [18] in the context of constrained PRFs. We motivate and discuss this definition in Sect. 1.2, and demonstrate an obfuscation scheme achieving it in Sect. 5.3.

We now extend the above definition to give the notion of *statistical* security in the context of average-case obfuscation.

Definition 2 ($\epsilon(n)$-Statistical Distributional VBB Obfuscation). *Let $\mathcal{C}$,* Obf*, and $\mathcal{D}$, be as in Definition 1.* Obf *is a $\epsilon(n)$-statistical distributional VBB obfuscator if it satisfies the notions of Functionality Preservation and Polynomial Slowdown and a modified notion of Distributional Virtual Black-Box where for any* unbounded *adversary $\mathcal{A}$, the distinguishing advantage is bounded by $\epsilon(n)$.*

We recall the definition of *perfect-circuit hiding*, introduced by Barak, Bitansky, Canetti, Kalai, Paneth, and Sahai [6].

Definition 3 (Perfect Circuit-Hiding [6]). *Let $\mathcal{C}$ be a collection of circuits. An obfuscator* Obf *for a circuit collection $\mathcal{C}$ is perfect circuit-hiding if for every PPT adversary $\mathcal{A}$ there exists a negligible function μ such that for every balanced predicate $\mathcal{P}$, every $n \in \mathbb{N}$ and every auxiliary input $z \in \{0,1\}^{\mathsf{poly}(n)}$ to $\mathcal{A}$:*

$$\Pr_{C \leftarrow \mathcal{C}_n} [\mathcal{A}(z, \mathsf{Obf}(C)) = \mathcal{P}(C)] \leq \frac{1}{2} + \mu(n),$$

where the probability is also over the randomness of Obf.

Barak et al. [6] prove that perfect-circuit hiding security is equivalent to distributional virtual black-box security, i.e. property 3 in Definition 1 is equivalent to Definition 3. We rely on this equivalence to simplify the proof of Theorem 3.

2.2 The Generic Group Model

Part of our analysis occurs in the generic group model [44], which assumes that an adversary interacts with group elements in a *generic* way. To model this, it is common to associate each group element with an independent and uniformly random string (drawn from a sufficiently large space) with we refer to as a "handle." The adversary has access to a generic group oracle which maintains the mapping between group elements and handles. The adversary is initialized with the handles corresponding to the group elements that comprise the scheme in question. It can query its generic group oracle with two handles, after which the oracle performs the group operation on the associated group elements and returns the handle associated with the resulting group element.

It will be convenient to associate each of these group operation queries performed by the adversary to a linear combination over the initial handles that it receives. The adversary can also request a "ZeroTest" operation on a handle, to which the oracle replies with a bit indicating whether or not that handle is associated with the identity element of the group.

There is a natural extension of the notion of distributional VBB security to the generic group model. In Definition 1, we simply give the obfuscation Obf and adversary $\mathcal{A}$ access to the generic group oracle $\mathcal{G}$. We refer to this definition as *Distributional VBB Obfuscation in the Generic Group Model.*

2.3 Learning Parity with Noise

We give the precise definition of the Learning Parity with Noise (LPN) problem in its dual formulation. Let $\rho \in (0,1)$ and m be an integer. Let $\mathcal{B}_\rho^m$ denote the distribution on $\mathbb{F}_2^m$ for which each component of the output independently takes the value 1 with probability ρ and 0 with probability $1-\rho$.

Definition 4. *Let n, m be integers and $\rho \in (0,1)$. The Decisional Learning Parity with Noise (DLPN) problem with parameters n, m, ρ, denoted $\mathsf{DLPN}(n, m, \rho)$, is hard if for every probabilistic polynomial-time (in n) algorithm $\mathcal{A}$, there exists a negligible function μ such that*

$$\left| \Pr_{B,e}[\mathcal{A}(B, B \cdot e) = 1] - \Pr_{B,u}[\mathcal{A}(B, B \cdot u) = 1 \right| \leq \mu(n),$$

where $B \leftarrow \mathbb{F}_2^{(m-n)\times m}$, $e \leftarrow \mathcal{B}_\rho^m$, and $u \leftarrow \mathbb{F}_2^{m-n}$.

Remark 1. The primal version of the above problem is, for $A \leftarrow \mathbb{F}_2^{m\times n}, s \leftarrow \mathbb{F}_2^n, e \leftarrow \mathcal{B}_\rho^m$, and $v \leftarrow \mathbb{F}_2^m$, to distinguish between $(A, As+e)$ and (A, v). These problems are equivalent for *any* error distribution when $m = n + \omega(\log n)$, as discussed for example in [40, Sect. 4.2].

3 Obfuscating Conjunctions in the Generic Group Model

In this section, we present our generalized dual scheme for obfuscating conjunctions in the generic group model. We then show a simple proof of security that applies to the uniform distribution over binary patterns with *any* fixed number of wildcards. In particular, our distributional VBB security result holds for up to $n - \omega(\log n)$ wildcards, but distributional VBB security is vacuously satisfied for $w > n - O(\log n)$ wildcards. This extends the generic model analysis of [12] that proved security up to $w < .774n$. We note that the combinatorial argument we give can be used to show that the original [12] construction achieves security for all values of w as well.

In the full version [9], we show how to extend these generic group model results in a number of ways. In particular, we prove security for general distributions with sufficient min-entropy (over a fixed number of wildcards). We then give a formal description of how to extend our construction to large alphabets, though we stress the construction is essentially the one sketched in Sect. 1.2. We also prove that our min-entropy results extend to the large alphabet setting.

Here and throughout the remainder of paper, the length n of the pattern will double as the security parameter.

3.1 Generic Group Construction

Throughout this section, we will refer to a fixed matrix B.

Definition 5. *Let $B_{n+1,k,q} \in \mathbb{Z}_q^{(n+1)\times k}$ be the matrix whose (i,j)th entry is j^i:*

$$B_{n+1,k,q} = \begin{pmatrix} 1 & 2 & \dots & k \\ 1 & 2^2 & \dots & k^2 \\ \vdots & \vdots & \vdots & \vdots \\ 1 & 2^{n+1} & \dots & k^{n+1} \end{pmatrix}.$$

Construction.

- $\mathsf{Setup}(n)$. Let $\mathbb{G}$ be a group of prime order $q > 2^n$ with generator g. We let $B := B_{n+1,2n,q}$ where $B_{n+1,2n,q}$ is as in Definition 5.
- $\mathsf{Obf}(\mathsf{pat} \in \{0,1,*\}^n)$. Set $e \in \mathbb{Z}_q^{2n\times 1}$ as follows. For each $i \in [n]$:
  - If $\mathsf{pat}_i = *$, set $e_{2i-1} = e_{2i} = 0$.
  - If $\mathsf{pat}_i = b$, sample $e_{2i-b} \leftarrow \mathbb{Z}_q$ and set $e_{2i-(1-b)} = 0$.

  Output
  $$g^{B\cdot e} \in \mathbb{G}^{n+1}.$$
- $\mathsf{Eval}(v \in \mathbb{G}^{n+1}, x \in \{0,1\}^n)$. Define B_x to be the $(n+1)\times n$ matrix where column j is set as $(B_x)_j := (B)_{2j-x_j}$. Solve[10] $tB_x = 0$ for a non-zero $t \in \mathbb{Z}_q^{1\times(n+1)}$. Compute
  $$\prod_{i=1}^{n+1} v_i^{t_i}$$
  and accept if and only if the result is g^0.

Alternative Setup. For concreteness (and efficiency), we define Obf and Eval to use the matrix $B_{n+1,2n,q}$. However, Setup can be modified to output any $B \in \mathbb{Z}_q^{(n+1)\times 2n}$ with the property that any $n+1$ columns of B form a full rank matrix (with overwhelming probability), and Obf and Eval will work as above with the matrix B. We note that if B is viewed as the generator matrix for a linear code (of length $2n$ and rank $n+1$), then this property is equivalent to the code having distance n. This requirement on B is sufficient to prove Theorem 1 below.

[10] See the full version [9] for a description of how to do this in $O(n\log^2(n))$ time.

Functionality Preservation. We first state a useful lemma.

Lemma 2. *If $k < q$, any set of $n+1$ columns of $B_{n+1,k,q}$ are linearly independent over $\mathbb{Z}_q$.*

Proof. This follows from inspecting the form of the determinant of the Vandermonde matrix, and noting that none of the factors of the determinant will divide q as long as $k < q$. □

Fix an x which matches pat and let t be the row vector computed in the Eval procedure. By construction, the vector tB is zero in all of the positions for which e is non-zero and thus

$$\prod_{i=1}^{n+1} v_i^{t_i} = g^{tBe} = g^0.$$

On the other hand, for an x which does not match pat, by construction there is at least one index $i \in [2n]$ such that $(B)_i$ is not part of B_x and e_i is a uniformly random field element. Then appealing to Lemma 2, $t(B)_i \neq 0$ since otherwise the $n+1$ columns B_x and $(B)_i$ would be linearly dependent. Then the product $t(B)_i e_i$ is distributed as a uniformly random field element, which means that tBe is as well. Thus x is only accepted with probability $1/q = \mathsf{negl}(n)$.[11]

Security. We prove the distributional VBB security of our construction.

Theorem 1. *Fix any function $w(n) \leq n$. The above construction is a distributional VBB obfuscator in the generic group model for the distribution $\mathcal{U}_{n,w(n)}$ over strings $\{0,1,*\}^n$.*

Proof. First we consider the case where $w(n) = n - \omega(\log(n))$. Let $c(n) = n - w(n) = \omega(\log(n))$. Let $\mathcal{H}$ be the space of handles used in the generic group instantiation of the obfuscation and let $|\mathcal{H}| > 2^n$ so that two uniformly drawn handles collide with negligible probability. For any adversary $\mathcal{A}$, we consider the simulator $\mathcal{S}$ that acts as the generic group model oracle and initializes $\mathcal{A}$ with $n+1$ uniformly random handles. On a group operation query by $\mathcal{A}$, $\mathcal{S}$ responds with a uniformly random handle unless $\mathcal{A}$ had previously requested the same linear combination of initial elements, in which case $\mathcal{S}$ responds with the same handle as before. $\mathcal{S}$ can easily implement this with a lookup table. We assume without loss of generality that $\mathcal{A}$ only submits linear combinations over initial elements that are not identically zero. On any ZeroTest query by $\mathcal{A}$, $\mathcal{S}$ will return "not zero". Finally, $\mathcal{S}$ will output whatever $\mathcal{A}$ outputs after it has finished interacting with the generic group model simulation.

We show that with all but negligible probability, $\mathcal{A}$'s view of the generic group model oracle that is honestly implementing the obfuscation is *identical* to its view of the simulated oracle, which completes the proof of security. Observe that the only way that $\mathcal{A}$'s view diverges is if when interacting with the honest oracle,

[11] As noted in [12], we can boost this to strong functionality preservation by setting $q > 2^{2n}$.

$\mathcal{A}$ either gets a successful ZeroTest, or receives the same handle on two group operation queries corresponding to different linear combinations of the initial handles. If we subtract these two linear combinations, we see that in both cases $\mathcal{A}$ has formed a non-trivial linear combination of the initial $n+1$ group elements that evaluates to zero. Consider the first time that this occurs and denote the vector of coefficients as $k = (k_1, \ldots, k_{n+1}) \in \mathbb{Z}_q^{1\times(n+1)}$. Let $e \in \mathbb{Z}_q^{2n\times 1}$ be the vector drawn in the Obf procedure on input a pattern pat drawn from $\mathcal{U}_{n,n-c(n)}$, so the resulting evaluation is equal to kBe. We show that the probability that $kBe = 0$ over the randomness of the pattern and of the obfuscation is negligible.

Since the coefficients of k are specified by $\mathcal{A}$ before its view has diverged from the simulated view, we can treat k as completely independent of e. Now by Lemma 2, any $n+1$ columns of B form a full rank matrix, so the vector $kB \in \mathbb{Z}_q^{1\times 2n}$ is 0 in at most n positions. Now if there exists $i \in [2n]$ for which $(kB)_i$ is non-zero and e_i is uniformly random, then with overwhelming probability $kBe \neq 0$ over the randomness of the obfuscation.

Partition e into the n pairs $\{e_{2j-1}, e_{2j}\}_{j\in[n]}$. Sampling pat from $\mathcal{U}_{n,n-c(n)}$ corresponds to uniformly randomly picking $c(n)$ of the pairs to have one uniformly random e component, and then within each of these $c(n)$ sets, picking either e_{2j-1} or e_{2j} with probability $1/2$ to be the uniformly random component.

Let $S \subset [2n]$ be any fixed set of n indices. At least $n/2$ of these pairs must contain at least one e_i such that $i \in S$, and among them, an expected $c(n)/2$ number of them have a uniformly random e component. This random variable is an instance of a *hypergeometric* random variable, and in Lemma 3 we use a Chernoff bound to show that it is greater than $c(n)/8$ except with negligible probability. Now for each of these $n/2$ pairs that contains a uniformly random component e_i, we have that $i \in S$ with probability $1/2$. Then the probability that there does not exist any $i \in S$ such that e_i is uniformly random is at most $(1/2)^{c(n)/8} + \mathsf{negl}(n)$ which is $\mathsf{negl}(n)$ for $c(n) = \omega(\log n)$.

Now we handle the case where $w(n) = n - O(\log(n))$. In this parameter regime, distributional VBB security is a vacuous security notion since a random input will satisfy the pattern with $1/\mathsf{poly}(n)$ probability. Thus a polynomial time simulator $\mathcal{S}$ can find an accepting input with overwhelming probability. Then it simply varies the accepting input one bit at a time in queries to the function oracle, and recovers the pattern in full. At this point it can run the obfuscation itself and simulate $\mathcal{A}$ on the honest obfuscation. □

We now state Lemma 3. While tail bounds are known for hypergeometric random variables, we were unable to find bounds strong enough for our parameter settings. In particular, plugging in the bounds summarized by Skala [45] into the proof of Theorem 1 imply security when $c(n)$ is as small as $1/n^{\epsilon}$ for $\epsilon < 1/2$. Using Lemma 3, we obtain $c(n) = \omega(\log n)$. We note that our bound is specifically tailored for our application and should not be misinterpreted as a strengthening of known bounds on hypergeometric random variables.

Lemma 3. *A bag initially contains n balls, of which $c(n)$ are black and $n - c(n)$ are white. If $n/2$ balls are randomly drawn without replacement, then*

$$\Pr\left[\#\,\text{black balls drawn} \geq \frac{c(n)}{8}\right] \geq 1 - e^{-c(n)/12}.$$

This claim follows from Chernoff bounds; a detailed proof is given in the full version [9].

4 Obfuscating Conjunctions from Constant-Noise LPN

In this section, we present our second obfuscation construction. As described in Sect. 1.2, this is our "dual" construction instantiated with a random matrix B over $\mathbb{F}_2$ and taken out of the group exponent. Security will be based on the standard constant-noise LPN assumption.

LPN vs. RLC. We note that under the Random Linear Codes (RLC) assumption (i.e., a generalization of LPN to $\mathbb{F}_q$ for $q \geq 2$—see the full version [9] or [33]), we could use the techniques from this section to prove that our construction over large fields is indistinguishable from random. However, indistinguishability from random *does not* imply distributional VBB security.[12] The problem arises from the fact that distributional VBB security requires indistinguishability from a simulated obfuscation even if the adversary knows a one bit predicate on the circuit (the pattern in our case). This requires us to prove that the decisional "structured error" LPN/RLC problem is indistinguishable from random even if the adversary knows a predicate on the positions of the non-zero error vector entries, which encode the pattern, which can be accomplished by modifying an appropriate search-to-decision reduction. Unfortunately, no search-to-decision reductions are known for RLC with super-polynomial modulus q, preventing our approach from extending beyond polynomial size q [3]. Since no (asymptotic) improvements to our construction result from considering polynomial size q, we restrict to $q = 2$ for concreteness and prove security from LPN.

In Sect. 4.1, we define the relevant LPN variants we consider for our construction, which we formally describe in Sect. 4.2. We then observe in Sect. 4.3 that prior work implies hardness of our structured error LPN notion still holds even if an arbitrary predicate on the error vector is known.

In the full version [9], we give a core technical reduction from standard RLC to structured error RLC that works for q up to size $2^{n^{\gamma}}$. Plugging in $q = 2$ suffices for our constructions, but we state our result for maximal generality as the reduction may be of independent interest.

[12] Consider for example the distributional point obfuscator that simply outputs the single accepting point in the clear as the "obfuscation." To evaluate, we simply compare the input point with the accepting point. Notice this trivially insecure obfuscation is perfectly indistinguishable from random for point functions drawn from the uniform distribution. However, we note that in the generic group model, indistinguishability from random *does* imply distributional VBB.

Strong Functionality Preservation. We note that simply plugging our reduction into our obfuscation scheme only gives us weak functionality preservation (Definition 1). Other works such as [12] address this issue by increasing the size of the field, but this will not work here since LPN restricts us to $q = 2$. We can still boost our scheme and satisfy strong functionality preservation by making use of additional *regular* (as opposed to structured) LPN samples (as we describe in the full version [9]). However, this modification has one caveat: the evaluation is polynomial-time *in expectation*, requiring a relaxation of the polynomial slowdown requirement in Definition 1.

Multi-bit Output. As a consequence of the reduction from constant noise LPN, our scheme can handle random conjunctions where a constant fraction ρ of the bits are wildcards, but it cannot handle a sub-constant fraction of wildcards. This is surprising, since obfuscation for evasive functionalies should intuitively get *easier* as we reduce the number of accepting inputs. However, our construction is completely broken if there are no wildcards, and in fact there is an easy brute force attack on our scheme for any $\rho = 1 - O(\log n/n)$.

In the full version [9], we show how to adapt our construction to support multi-bit output. In this setting, the obfuscator can embed a fixed message into the obfuscation, which an evaluator recovers upon finding an accepting input. This allows us to handle conjunctions with a sub-constant (or even zero) fraction of wildcards. The idea is to set some of the non-wildcard bits to be wildcards, and then use the multi-bit output to specify the true settings of those bits.

4.1 Exact Structured Learning Parity with Noise

We begin by recalling the decisional Exact Learning Parity with Noise (DxLPN) problem considered by Jain et al. [34]. The word "exact" modifies the standard decisional Learning Parity with Noise (DLPN) problem by changing the sampling procedure for the error vector. Instead of setting each component of $e \in \mathbb{F}_q^m$ to be 1 with independent probability ρ, we sample e uniformly from the set of error vectors with exactly $\lfloor \rho m \rfloor$ entries set to 1 (we refer to these as vectors of weight $\lfloor \rho m \rfloor$). DLPN is polynomially equivalent to the exact version following the search to decision reduction given in [4], as noted in [26,34]. We give the precise definition in its dual formulation.

Let $\rho \in [0, 1]$ and $m > 0$ be an integer. Let χ_ρ^m denote the distribution on $\mathbb{F}_2^m$ which outputs uniformly random vectors in $\mathbb{F}_2^m$ of weight $\lfloor \rho m \rfloor$.

Definition 6 (Exact Learning Parity with Noise). *Let n, m be integers and $\rho \in (0, 1)$. The (dual) Decisional Exact Learning Parity with Noise (DxLPN) problem with parameters n, m, ρ, denoted DxLPN(n, m, ρ), is hard if, for every probabilistic polynomial-time (in n) algorithm $\mathcal{A}$, there exists a negligible function μ such that*

$$\left| \Pr_{B,e}[\mathcal{A}(B, B \cdot e) = 1] - \Pr_{B,u}[\mathcal{A}(B, u) = 1] \right| \leq \mu(n)$$

where $B \leftarrow \mathbb{F}_2^{(m-n)\times m}$, $e \leftarrow \chi_\rho^m$, *and* $u \leftarrow \mathbb{F}_2^{m-n}$.

Exact Structured LPN. We now introduce a modification of the Exact Learning Parity with Noise (DxLPN) problem where we enforce that the error vector is *structured.* Concretely, the error vector e is now $2m$-dimensional, and we enforce that in any of the pairs $(2i-1, 2i)$ for $i \in [m]$, at least one of e_{2i-1} and e_{2i} is 0. As we are considering the exact version of the problem, we enforce that $\lfloor \rho m \rfloor$ components of e are non-zero. Note that while the error vector has doubled in size, the number of non-zero components is unchanged.

We first introduce some notation. For a distribution $\mathcal{D}$ on $\mathbb{F}_2^m$, we define

$$\sigma(\mathcal{D}) = \left\{ \begin{pmatrix} s_1 \\ \vdots \\ s_{2m} \end{pmatrix} \middle| \begin{array}{l} x \leftarrow \{0,1\}^m \\ e' \leftarrow \mathcal{D} \\ \text{for all } i \in [m], \begin{cases} s_{2i-x_i} = e'_i \\ s_{2i-(1-x_i)} = 0 \end{cases} \end{array} \right\}.$$

Definition 7 (Exact Structured LPN). *Let n, m be integers and $\rho \in (0,1)$. The (dual) Decisional Exact Structured Learning Parity with Noise (DxSLPN) problem with parameters $n, 2m, \rho$, denoted DxSLPN$(n, 2m, \rho)$, is hard if, for every probabilistic polynomial-time (in n) algorithm $\mathcal{A}$, there exists a negligible function μ such that*

$$\left| \Pr_{B,e}[\mathcal{A}(B, B \cdot e) = 1] - \Pr_{B,u}[\mathcal{A}(B, u) = 1] \right| \leq \mu(n)$$

where $B \leftarrow \mathbb{F}_2^{(2m-n)\times 2m}$, $e \leftarrow \sigma(\chi_\rho^m)$, *and* $u \leftarrow \mathbb{F}_2^{2m-n}$.

In other words, the error vector $e \in \mathbb{F}_2^{2m}$ in the DxSLPN problem can be derived from the error vector $e' \in \mathbb{F}_2^m$ of the DxLPN problem; for each $i \in [m]$, randomly set one of e_{2i-1} or e_{2i} to e'_i and the other to 0.

We prove the following theorem in the full version [9].

Theorem 2. *Fix constants $\epsilon, \delta, \in [0, 1/2)$ and constant $\rho \in (0,1)$. If DxLPN(n^ϵ, n, ρ) is hard, then DxSLPN$(n - n^\delta, 2n, \rho)$ is hard.*

4.2 Construction

The following is parameterized by a pattern length n and a constant $\delta \in [0, 1/2)$.

- Obf(pat $\in \{0,1,*\}^n$): Draw $B \leftarrow \mathbb{F}_2^{(n+n^\delta)\times 2n}$ and $e \in \mathbb{F}_2^{2n}$ as follows. For each $i \in [n]$
  - If pat$_i = *$, $e_{2i-1} = e_{2i} = 0$
  - If pat$_i = b$, $e_{2i-b} = 1$, $e_{2i-(1-b)} = 0$

  Output (B, Be).
- Eval$((B, v), x)$: Define B_x to be the $(n+n^\delta) \times n$ matrix where column j is set as $(B_x)_j := (B)_{2j-x_j}$. Solve for a full rank matrix $T \in \mathbb{F}_2^{n^\delta \times (n+n^\delta)}$ such that $T \cdot B_x = 0$. Compute $T \cdot v$ and if the result is $0^{n^\delta \times 1}$ output 1 and otherwise output 0.

Weak Functionality Preservation. We show that for all $\mathsf{pat} \in \{0,1,*\}^n$ and $x \in \{0,1\}^n$, it holds that

$$\Pr[\mathsf{Eval}(\mathsf{Obf}(\mathsf{pat}), x) = f_{\mathsf{pat}}(x)] = 1 - \mathsf{negl}(n),$$

over the randomness of the Obf procedure. Let B, e be drawn as in the Obf procedure. Let T, B_x be as defined in the Eval procedure and $B_{\overline{x}}$ be the n columns of B not in B_x. Let $e_{\overline{x}}$ be defined analogously. First, if $f_{\mathsf{pat}}(x) = 1$, then $e_{\overline{x}} = 0$ by construction. Then $T \cdot v = T \cdot B \cdot e = (T \cdot B_{\overline{x}}) \cdot e_{\overline{x}} = 0$. Hence, $\mathsf{Eval}(\mathsf{Obf}(\mathsf{pat}), x) = 1$ with probability 1. Now if $f_{\mathsf{pat}}(x) = 0$, then $e_{\overline{x}} \neq 0$ by construction. Since $T \cdot B_{\overline{x}}$ is a uniformly random rank n^δ matrix independent of $e_{\overline{x}}$, it holds that

$$\Pr[T \cdot v = 0] = \frac{1}{2^{n^\delta}} = \mathsf{negl}(n).$$

4.3 Security

Lemma 4. *Fix any predicate $\mathcal{P}\colon \{0,1,*\}^n \to \{0,1\}$. Assuming the hardness of $\mathsf{DxSLPN}(n, 2m, \rho)$ implies that for all probabilistic polynomial-time $\mathcal{A}$,*

$$\left| \Pr_{B,e}[\mathcal{A}(B, Be, \mathcal{P}(e)) = 1] - \Pr_{B,u}[\mathcal{A}(B, u, \mathcal{P}(e)) = 1] \right| = \mathsf{negl}(n)$$

where $B \leftarrow \mathbb{F}_2^{(2m-n)\times 2m}$, $e \leftarrow \sigma(\chi_\rho^m)$, and $u \leftarrow \mathbb{F}_2^{2m-n}$.

Proof. The hardness of $\mathsf{DxSLPN}(n, 2m, \rho)$ immediately implies that for all probabilistic polynomial-time $\mathcal{A}'$,

$$\Pr_{B,e}[\mathcal{A}'(B, Be, \mathcal{P}(e)) = e] = \mathsf{negl}(n),$$

where B, e are drawn as in the lemma statement. This follows since the reduction can simply guess the value of $\mathcal{P}(e)$ and be correct with probability at least $1/2$. Thus we just need to show a search to decision reduction for structured LPN with a one bit predicate. This follows from the proof of Lemma 5 in [26] (equivalence of search and decision "leaky LPN"), which is a slight tweak of the search-to-decision reduction presented in [4]. We can easily adapt the proof to our case by letting the underlying problem be structured LPN rather than regular LPN and considering the special case of leakage functions corresponding to one bit predicates. This proof is presented for the $As + e$ version of LPN, but the same technique works for the dual Be version, as shown for example in the proof of Lemma 2.3 in [32]. □

Theorem 3. *Fix any constant $\rho \in (0,1)$. Assuming the hardness of $\mathsf{DLPN}(n^\epsilon, n, \rho)$ for some $\epsilon < 1/2$, the above obfuscation with parameters (n, δ) for $\delta < 1/2$ is Distributional-VBB secure for patterns $\mathsf{pat} \leftarrow \mathcal{U}_{n,n-\rho n}$.*

Proof. We show that the above obfuscator satisfies the definition of Perfect Circuit-Hiding (Definition 3), which implies Distributional VBB security [6]. We want to show that for any probabilistic polynomial-time adversary $\mathcal{A}$ and any *balanced* predicate $\mathcal{P}\colon \{0,1,*\}^n \to \{0,1\}$ (that is, $\mathcal{P}$ takes the values 0 and 1 with probability 1/2 over the randomness of $\mathsf{pat} \leftarrow \mathcal{U}_{n,n-\rho n}$),

$$\Pr_{\mathsf{pat} \leftarrow \mathcal{U}_{n,n-\rho n}}[\mathcal{A}(\mathsf{Obf}(\mathsf{pat})) = \mathcal{P}(\mathsf{pat})] = \frac{1}{2} + \mathsf{negl}(n).$$

We know by assumption and from Theorem 2 and Lemma 4 that, for any predicate $\mathcal{P}\colon \{0,1,*\} \to \{0,1\}$ and for all probabilistic polynomial-time $\mathcal{B}$,

$$\left| \Pr[\mathcal{B}(\mathsf{Obf}(\mathsf{pat}), \mathcal{P}(\mathsf{pat})) = 1] - \Pr[\mathcal{B}((B, u), \mathcal{P}(\mathsf{pat})) = 1] \right| = \mathsf{negl}(n),$$

where $\mathsf{pat} \leftarrow \mathcal{U}_{n,n-\rho n}$, $B \leftarrow \mathbb{F}_2^{(n+n^\delta)\times 2n}$, and $u \leftarrow \mathbb{F}_2^{n+n^\delta}$.

Now assume that there exists a balanced predicate $\mathcal{P}$ such that there exists a probabilistic polynomial-time adversary $\mathcal{A}$ with non-negligible advantage $\mu(n)$ in the above Perfect Circuit-Hiding definition. Consider an adversary $\mathcal{B}$ that receives $((B, u), \mathcal{P}(\mathsf{pat}))$, runs $\mathcal{A}$ on (B, u) and outputs 1 if $\mathcal{A}(B, u) = \mathcal{P}(\mathsf{pat})$ and 0 otherwise. If (B, u) was an honest obfuscation, then $\mathcal{B}$ outputs 1 with probability $\frac{1}{2} + \mu(n)$. If (B, u) was uniformly random, then $\mathcal{A}(B, u)$ is independent of $\mathcal{P}(\mathsf{pat})$, so since $\mathcal{P}$ is balanced, $\mathcal{B}$ outputs 1 with probability exactly 1/2. Thus, $\mathcal{B}$'s distinguishing advantage is $\mu(n)$, which is non-negligible. □

5 Information-Theoretic Security

In this section, we consider a third construction, which relies on subset sums of random rank one matrices. We prove this construction attains a notion of statistical distributional VBB security, as well as weak functionality preservation. In order to achieve statistical security, however, we must limit the number of wildcards to at most n^δ for any $\delta < 1$. In Sect. 5.3, we show how to modify this base construction to achieve an intermediate notion of *computational functionality preservation*, assuming the discrete log assumption. The resulting scheme has the curious property of being distributional-VBB secure against computationally unbounded adversaries, but functionality preserving in the view of any computationally bounded adversary (even those who know pat).

5.1 Construction

We begin by drawing a $k \times k$ matrix B by choosing its first $k-1$ rows at random, and then picking its last row to be in the row span of the first $k-1$. We could also have drawn B as a uniformly random rank $k-1$ matrix; however, "pushing" the rank deficiency to the last row of B will simplify both the security analysis and the modified construction in Sect. 5.3.

Notation. We will frequently write a matrix M as $\begin{pmatrix} \overline{M} \\ \underline{M} \end{pmatrix}$ where $\overline{M}$ is the submatrix of M consisting of every row but the last, and $\underline{M}$ denotes the last row.

Construction. The following is parameterized by a pattern length n and field size $q = 2^{n^\gamma}$ for a $\gamma > 0$. We let $\mathbb{F}_q$ denote a field of size q.

- $\mathsf{Obf}(\mathsf{pat} \in \{0,1,*\}^n)$. Partition $[n]$ into $S_0 \cup S_1 \cup S_*$ so that $S_0 = \{i \mid \mathsf{pat}_i = 0\}$, $S_1 = \{i \mid \mathsf{pat}_i = 1\}$, and $S_* = \{i \mid \mathsf{pat}_i = *\}$, and let $k = |S_*| + 1$.
  - Draw $\overline{B} \leftarrow \mathbb{F}_q^{(k-1)\times k}$, $r \leftarrow \mathbb{F}_q^{1\times(k-1)}$ and let $B := \begin{pmatrix} \overline{B} \\ r \cdot \overline{B} \end{pmatrix}$
  - For all $i \in S_0 \cup S_1$, sample a uniformly random rank 1 $A^{(i)} \in \mathbb{F}_q^{k\times k}$.
  - For all $i \in S_*$, sample a uniformly random rank 1 $\overline{A}^{(i)} \in \mathbb{F}_q^{(k-1)\times k}$. Let
  $$A^{(i)} := \begin{pmatrix} \overline{A}^{(i)} \\ r \cdot \overline{A}^{(i)} \end{pmatrix}.$$
  - Define $F := B - \sum_{i\in S_1} A^{(i)}$, and output $(F, A^{(1)}, \ldots, A^{(n)})$.
- $\mathsf{Eval}((F, A^{(1)}, \ldots, A^{(n)}), x \in \{0,1\}^n)$. Output 1 if $\det\left(F + \sum_{i|x_i=1} A^{(i)}\right) = 0$ and 0 otherwise.

Weak Functionality Preservation. By construction, for an x that matches pat, we have that

$$\mathrm{colspan}\left(F + \sum_{i|x_i=1} A^{(i)}\right) = \mathrm{colspan}\left(B + \sum_{i|x_i=1\wedge \mathsf{pat}_i=*} A^{(i)}\right) \subseteq \mathrm{colspan}(B).$$

It then follows that $\det(F + \sum_{i|x_i=1} A^{(i)}) = 0$ since B has rank at most $k-1$. For an x that does not match pat, consider the matrix

$$F + \sum_{i|x_i=1} A^{(i)} = \underbrace{B + \sum_{i|x_i=1\wedge \mathsf{pat}_i=*} A^{(i)}}_{B'} + \underbrace{\sum_{i|x_i=1\wedge \mathsf{pat}_i=0} A^{(i)} - \sum_{i|x_i=0\wedge \mathsf{pat}_i=1} A^{(i)}}_{A'}.$$

Since the first $k-1$ rows of B are all uniformly random, the same is true of first $k-1$ rows of B', denoted as $\overline{B}'$. Furthermore, we know by construction that there exists at least one i such that $\mathsf{pat}_i \neq x_i$ and $\mathsf{pat}_i \in \{0,1\}$, so A' contains at least one of these $A^{(i)}$ matrices. Note that the last row of $A^{(i)}$ (and hence $\underline{A}'$) is uniformly random and independent of $\overline{B}'$. Thus $F + \sum_{i|x_i=1} A^{(i)}$ is distributed as a uniformly random matrix, so its determinant is non-zero with overwhelming probability $1 - k/q = 1 - \mathsf{negl}(n)$ by the Schwartz–Zippel lemma.

5.2 Security

We prove our construction attains statistical distributional VBB security, defined in Definition 1.

For any pattern $\mathsf{pat} \in \{0,1,*\}^n$, define $\mathsf{pat}^{-1}(*) := \{j \mid \mathsf{pat}_j = *\}$ the *positions* of the wildcards and let $\mathbf{b} \in \{0,1\}^{n-w}$ denote the fixed bits of pat.

Theorem 4. *The above construction with field size q is a $\epsilon(n)$-Statistical Distributional VBB obfuscator for any distribution over patterns with $w \leq n$ wildcards such that $H_\infty(\mathbf{b}|\mathsf{pat}^{-1}(*)) \geq (w+1)\log(q) + 2\log(1/\epsilon(n)) + 1$.*

Corollary 1. *Fix any $\delta \in [0,1)$. The above construction can be used to satisfy $\epsilon(n)$-Statistical Distributional VBB security for a negligible function $\epsilon(n)$, for any distribution over patterns with $w = n^\delta$ wildcards such that $H_\infty(\mathbf{b}|\mathsf{pat}^{-1}(*)) \geq n^{1-\gamma}$ for some $\gamma < 1-\delta$.*

The proof of Theorem 4 follows from standard applications of the leftover hash lemma. We show that as long as there is sufficient entropy on the fixed bits, the leftover hash lemma will imply the matrix F is statistically close to a uniformly random matrix. Then the low rank matrix B is hidden from view, and the $k-1$ random wildcard matrices $A^{(i)}$ drawn from the column space of B are distributed as uniformly random rank 1 matrices, just like all the other $A^{(i)}$ matrices. The formal proof is done in the full version [9].

5.3 Computational Functionality Preservation

We now consider the notion of computational functionality preservation from Definition 1, which is strictly weaker than strong functionality preservation, and strictly stronger than weak functionality preservation.[13] Refer to Sect. 1.2 for general discussion motivating this definition.

Remark 2. For the setting of conjunction obfuscation, computational functionality preservation combined with distributional VBB security imply that a computationally bounded adversary can never find an accepting input to the obfuscated program.[14] If the adversary can find an accepting input to the program that actually matches the hidden pattern pat, the adversary can learn a predicate on pat, violating distributional VBB. If they find an accepting input to the program that does not match the hidden pattern, they violate computational functionality preservation.

[13] To see this informally, consider any obfuscation scheme for an evasive functionality given by $(\mathsf{Obf}, \mathsf{Eval})$ that achieves weak functionality preservation. Now define $(\mathsf{Obf}', \mathsf{Eval}')$ where $\mathsf{Obf}'(C)$ samples a random y from the input space and then outputs $\mathsf{Obf}(C), y$. Then $\mathsf{Eval}(\mathsf{Obf}', x)$ returns $\mathsf{Eval}(\mathsf{Obf}, x)$ if $x \neq y$, but returns 1 if $x = y$. It is not hard to see that this scheme still satisfies weak functionality preservation, but now an adversary can easily tell that functionality preservation is violated at y, so computational functionality preservation is violated.

[14] This is reminiscent of the notion of input-hiding obfuscation [6], but different in that we require that the adversary cannot find an accepting input for the *obfuscated* circuit rather than the original circuit.

We show that the following simple tweaks to our scheme allow us to base computational functionality preservation on the hardness of solving discrete log.

- **Modification 1:** All of the matrices $F, A^{(1)}, \ldots, A^{(n)}$ have their last row encoded in the exponent of the group.
- **Modification 2:** On evaluation, we first check if $\text{rank}(\overline{F} + \sum_{i|x_i=1} \overline{A}^{(i)}) = k-1$, and if not, immediately reject.

Our functionality proof will use a reduction from the *representation problem*, introduced by Brands [20], which we denote as FIND-REP following [42].

Instance: A group $\mathbb{G}$ of order q, and random $g^{s_1}, \ldots, g^{s_n} \leftarrow \mathbb{G}$.

Problem: Find non-trivial $d_1, \ldots, d_n \in \mathbb{Z}_q$ such that $g^{\sum_{i=1}^{n} d_i s_i} = g^0$.

Brands [20] proves that solving FIND-REP in $\mathbb{G}$ is as hard as solving discrete log in $\mathbb{G}$. Now we prove a theorem similar to Theorem 4, but with different parameters than Corollary 1.

Theorem 5. *Fix any $\delta \in [0, \frac{1}{2})$. Assuming discrete log, this construction satisfies computational functionality preservation for any distribution over patterns with $w = n^\delta$ wildcards such that $H_\infty(\mathbf{b}|\mathsf{pat}^{-1}(*)) \geq n^{1-\epsilon}$ for some $\epsilon < 1 - 2\delta$.*

Proof. We prove that a PPT adversary that can find some point x for which $f_{\mathsf{pat}}(x) \neq \mathsf{Obf}(f_{\mathsf{pat}})(x)$, even given $\mathsf{Obf}(f_{\mathsf{pat}})$, can solve discrete log in $\mathbb{G}$. We break up the analysis into two cases: we denote inputs x for which $f_{\mathsf{pat}}(x) = 1$ and $\mathsf{Obf}(f_{\mathsf{pat}})(x) = 0$ as false negatives, and denote inputs for which $f_{\mathsf{pat}}(x) = 0$ and $\mathsf{Obf}(f_{\mathsf{pat}})(x) = 1$ as false positives.

For $\delta \in [0, 1/2)$, pick $\delta' > \delta$ and set the field size q to $2^{n^{\delta'}}$.

Lemma 5. *For $q = 2^{n^{\delta'}}$ and $w = n^\delta$ where $\delta' > \delta$, with overwhelming probability our construction has no false negatives.*

Proof. For any x where $f_{\mathsf{pat}}(x) = 1$, $\mathsf{Obf}(f_{\mathsf{pat}})(x)$ can only evaluate to 0 if

$$\text{rank}\left(\overline{B} + \sum_{i|x_i=1, \mathsf{pat}_i=*} \overline{A}^{(i)}\right) < k-1.$$

Recall from the construcion that $\overline{B}$ is sampled as a uniformly random matrix, and for i where $\mathsf{pat}_i = *$, $\overline{A}^{(i)}$ is sampled as a uniformly random rank 1 matrix. Thus, each of the 2^{n^δ} possible $(k-1) \times k$ subset sums is distributed as a uniformly random $(k-1) \times k$ matrix, and is thus rank deficient with probability at most $\frac{k-1}{q^2}$. Since we set q to be at least $2^{n^{\delta'}}$ for $\delta' > \delta$, the probability that any of these subset sum matrices is rank deficient is at most $\frac{(k-1) \cdot 2^{n^\delta}}{q^2} = \mathsf{negl}(n)$. □

Thus with overwhelming probability, an adversary that finds an x where $f_{\mathsf{pat}}(x) \neq \mathsf{Obf}(f_{\mathsf{pat}})(x)$ must return a false positive. We show that finding a false positive is as hard as solving FIND-REP.

Lemma 6. *If there exists an algorithm $\mathcal{A}$ that finds a false positive with non-negligible probability, there exists an algorithm $\mathcal{A}'$ that solves* FIND-REP *with non-negligible probability.*

Proof. On input $g^{s_1}, \ldots, g^{s_n}$, $\mathcal{A}'$ constructs an obfuscation for a pattern pat with $w = n^\delta$ wildcards drawn from an arbitrary distribution. Given pat, define the same sets S_0, S_1, and S_* and as before, let $k = w + 1$. Note that throughout this proof, when we add/subtract matrices that include group elements, we multiply/divide the group element components of the matrices. Likewise, when we multiply a vector of group elements by a scalar, we actually raise each group element to the appropriate power. $\mathcal{A}$ constructs the obfuscation as follows.

- Let $r \in \mathbb{G}^{1\times(k-1)} = [\ldots g^{s_j} \ldots]$ for $j \in S_*$, draw $\overline{B} \leftarrow \mathbb{Z}_q^{(k-1)\times k}$, and let

$$B := \begin{pmatrix} \overline{B} \\ r \cdot \overline{B} \end{pmatrix}$$

- For each $i \in S_0 \cup S_1$, sample a uniformly random rank 1 matrix $\overline{A}^{(i)} \in \mathbb{F}_q^{(k-1)\times k}$, and let $A^{(i)} := \begin{pmatrix} \overline{A}^{(i)} \\ g^{s_i} \cdot \overline{A}_1^{(i)} \end{pmatrix}$
- For each $i \in S_*$, sample $c_i \leftarrow \mathbb{F}_q^{k-1}$ and $d_i \leftarrow \mathbb{F}_q^{1\times k}$, and let $A^{(i)} := \begin{pmatrix} c_i \\ r \cdot c_i \end{pmatrix} \cdot d_i$.
- Define $F := B - \sum_{i\in S_1} A^{(i)}$ and output $(F, A^{(1)}, \ldots, A^{(n)})$.

Then $\mathcal{A}'$ sends $(F, A^{(1)}, \ldots, A^{(n)}, \mathsf{pat})$ to $\mathcal{A}$ and if $\mathcal{A}$ is successful, $\mathcal{A}'$ receives back a set T with the following properties:

- $\det(F + \sum_{i\in T} A^{(i)}) = 0$;
- $\det(\overline{F} + \sum_{i\in T} \overline{A}^{(i)}) \neq 0$;
- $T \setminus S_* \neq S_1$.

The determinant polynomial reduces to a linear combination of the elements in the last row of $F + \sum_{i\in T} A^{(i)}$. By the second property above, this linear combination is not identically zero. Now $\mathcal{A}'$ will plug in the random values it chose in constructing the obfuscation to recover a linear combination over $s_1, \ldots, s_n$ that evaluates to zero, by the first property above. It then submits this linear combination to the FIND-REP challenger.

So it just remains to show that this final linear combination is not identically zero. As in our weak functionality preservation proof, we can re-write the summation as

$$F + \sum_{i\in T} A^{(i)} = \underbrace{B + \sum_{i\in T\cap S_*} A^{(i)}}_{B'} + \underbrace{\sum_{i\in T\cap S_0} A^{(i)} - \sum_{i\in([n]\setminus T)\cap S_1} A^{(i)}}_{A'}.$$

By the third property above, there exists some i such that A' includes the matrix $A^{(i)}$. We show that with overwhelming probability, this implies that there is some setting of $s_1, \ldots, s_n$ that produces a non-zero evaluation, which shows that the final linear combination must not be identically zero.

We condition on the fact that with overwhelming probability, for each of the 2^{n^δ} possible sets $T \cap S_*$, and each $i \notin S_*$, the row span of $A^{(i)}$ is outside of the row span of $\overline{B}'$. Indeed, this fails to happen with probability at most $n2^{n^\delta}/q = \mathsf{negl}(n)$.

Thus since we can assume $\overline{B}'$ has rank $k-1$ for each $T \cap S_*$ (by the arguments from the proof of Lemma 5), and since $\underline{A}'$ must include a row from some $A^{(i)}$, we conclude that the row $\underline{B}' + \underline{A}'$ could be anything in the entire k dimensional space, depending on the values of $s_1, \ldots, s_n$. In particular it could be outside of the $k-1$ dimensional space spanned by $\overline{A}' + \overline{B}'$, in which case the determinant polynomial would evaluate to non-zero. □

Together, Lemmas 5 and 6 imply that any adversary that breaks computational functionality preservation can solve discrete log in $\mathbb{G}$. □

References

1. Ananth, P., Jain, A.: Indistinguishability obfuscation from compact functional encryption. In: Gennaro, R., Robshaw, M. (eds.) CRYPTO 2015. LNCS, vol. 9215, pp. 308–326. Springer, Heidelberg (2015). https://doi.org/10.1007/978-3-662-47989-6_15
2. Ananth, P., Sahai, A.: Projective arithmetic functional encryption and indistinguishability obfuscation from degree-5 multilinear maps. In: Coron, J.-S., Nielsen, J.B. (eds.) EUROCRYPT 2017. LNCS, vol. 10210, pp. 152–181. Springer, Cham (2017). https://doi.org/10.1007/978-3-319-56620-7_6
3. Applebaum, B., Avron, J., Brzuska, C.: Arithmetic cryptography: extended abstract. In: Roughgarden, T. (ed.) ITCS 2015, pp. 143–151. ACM (2015)
4. Applebaum, B., Ishai, Y., Kushilevitz, E.: Cryptography with constant input locality. J. Cryptol. **22**(4), 429–469 (2009)
5. Arora, S., Ge, R.: New algorithms for learning in presence of errors. In: Aceto, L., Henzinger, M., Sgall, J. (eds.) ICALP 2011. LNCS, vol. 6755, pp. 403–415. Springer, Heidelberg (2011). https://doi.org/10.1007/978-3-642-22006-7_34
6. Barak, B., Bitansky, N., Canetti, R., Kalai, Y.T., Paneth, O., Sahai, A.: Obfuscation for evasive functions. In: Lindell, Y. (ed.) TCC 2014. LNCS, vol. 8349, pp. 26–51. Springer, Heidelberg (2014). https://doi.org/10.1007/978-3-642-54242-8_2
7. Barak, B., et al.: On the (im)possibility of obfuscating programs. In: Kilian, J. (ed.) CRYPTO 2001. LNCS, vol. 2139, pp. 1–18. Springer, Heidelberg (2001). https://doi.org/10.1007/3-540-44647-8_1
8. Bartusek, J., Guan, J., Ma, F., Zhandry, M.: Return of GGH15: provable security against zeroizing attacks. In: Beimel, A., Dziembowski, S. (eds.) TCC 2018. LNCS, vol. 11240, pp. 544–574. Springer, Cham (2018). https://doi.org/10.1007/978-3-030-03810-6_20
9. Bartusek, J., Ma, F., Lepoint, T., Zhandry, M.: New techniques for obfuscating conjunctions. Cryptology ePrint Archive, Report 2018/936 (2018). https://eprint.iacr.org/2018/936

10. Bellare, M., Stepanovs, I.: Point-function obfuscation: a framework and generic constructions. In: Kushilevitz, E., Malkin, T. (eds.) TCC 2016. LNCS, vol. 9563, pp. 565–594. Springer, Heidelberg (2016). https://doi.org/10.1007/978-3-662-49099-0_21
11. Beullens, W., Wee, H.: Obfuscating simple functionalities from knowledge assumptions. In: PKC. LNCS. Springer (2019)
12. Bishop, A., Kowalczyk, L., Malkin, T., Pastro, V., Raykova, M., Shi, K.: A simple obfuscation scheme for pattern-matching with wildcards. In: Shacham, H., Boldyreva, A. (eds.) CRYPTO 2018. LNCS, vol. 10993, pp. 731–752. Springer, Cham (2018). https://doi.org/10.1007/978-3-319-96878-0_25
13. Boneh, D., Ishai, Y., Sahai, A., Wu, D.J.: Lattice-based SNARGs and their application to more efficient obfuscation. In: Coron, J.-S., Nielsen, J.B. (eds.) EUROCRYPT 2017. LNCS, vol. 10212, pp. 247–277. Springer, Cham (2017). https://doi.org/10.1007/978-3-319-56617-7_9
14. Boneh, D., Waters, B.: Constrained pseudorandom functions and their applications. In: Sako, K., Sarkar, P. (eds.) ASIACRYPT 2013. LNCS, vol. 8270, pp. 280–300. Springer, Heidelberg (2013). https://doi.org/10.1007/978-3-642-42045-0_15
15. Boneh, D., Zhandry, M.: Multiparty key exchange, efficient traitor tracing, and more from indistinguishability obfuscation. In: Garay, J.A., Gennaro, R. (eds.) CRYPTO 2014. LNCS, vol. 8616, pp. 480–499. Springer, Heidelberg (2014). https://doi.org/10.1007/978-3-662-44371-2_27
16. Brakerski, Z., Rothblum, G.N.: Obfuscating conjunctions. In: Canetti, R., Garay, J.A. (eds.) CRYPTO 2013. LNCS, vol. 8043, pp. 416–434. Springer, Heidelberg (2013). https://doi.org/10.1007/978-3-642-40084-1_24
17. Brakerski, Z., Rothblum, G.N.: Obfuscating conjunctions. J. Cryptol. **30**(1), 289–320 (2017)
18. Brakerski, Z., Vaikuntanathan, V.: Constrained key-homomorphic PRFs from standard lattice assumptions. In: Dodis, Y., Nielsen, J.B. (eds.) TCC 2015. LNCS, vol. 9015, pp. 1–30. Springer, Heidelberg (2015). https://doi.org/10.1007/978-3-662-46497-7_1
19. Brakerski, Z., Vaikuntanathan, V., Wee, H., Wichs, D.: Obfuscating conjunctions under entropic ring LWE. In: Sudan, M. (ed.) ITCS 2016, pp. 147–156. ACM (2016)
20. Brands, S.: Untraceable off-line cash in wallet with observers. In: Stinson, D.R. (ed.) CRYPTO 1993. LNCS, vol. 773, pp. 302–318. Springer, Heidelberg (1994). https://doi.org/10.1007/3-540-48329-2_26
21. Canetti, R.: Towards realizing random oracles: Hash functions that hide all partial information. In: Kaliski, B.S. (ed.) CRYPTO 1997. LNCS, vol. 1294, pp. 455–469. Springer, Heidelberg (1997). https://doi.org/10.1007/BFb0052255
22. Canetti, R., Micciancio, D., Reingold, O.: Perfectly one-way probabilistic hash functions (preliminary version). In: 30th ACM STOC, pp. 131–140. ACM Press (1998)
23. Coron, J.-S., Lepoint, T., Tibouchi, M.: Practical multilinear maps over the integers. In: Canetti, R., Garay, J.A. (eds.) CRYPTO 2013. LNCS, vol. 8042, pp. 476–493. Springer, Heidelberg (2013). https://doi.org/10.1007/978-3-642-40041-4_26
24. Dodis, Y., Kalai, Y.T., Lovett, S.: On cryptography with auxiliary input. In: Mitzenmacher, M. (ed.) 41st ACM STOC, pp. 621–630. ACM Press (2009)
25. Dodis, Y., Smith, A.: Correcting errors without leaking partial information. In: Gabow, H.N., Fagin, R. (eds.) 37th ACM STOC, pp. 654–663. ACM Press (2005)

26. Döttling, N.: Low noise LPN: key dependent message secure public key encryption an sample amplification. IET Inf. Secur. **10**(6), 372–385 (2016)
27. Garg, S., Gentry, C., Halevi, S.: Candidate multilinear maps from ideal lattices. In: Johansson, T., Nguyen, P.Q. (eds.) EUROCRYPT 2013. LNCS, vol. 7881, pp. 1–17. Springer, Heidelberg (2013). https://doi.org/10.1007/978-3-642-38348-9_1
28. Garg, S., Miles, E., Mukherjee, P., Sahai, A., Srinivasan, A., Zhandry, M.: Secure obfuscation in a weak multilinear map model. In: Hirt, M., Smith, A. (eds.) TCC 2016. LNCS, vol. 9986, pp. 241–268. Springer, Heidelberg (2016). https://doi.org/10.1007/978-3-662-53644-5_10
29. Gentry, C., Gorbunov, S., Halevi, S.: Graph-induced multilinear maps from lattices. In: Dodis, Y., Nielsen, J.B. (eds.) TCC 2015. LNCS, vol. 9015, pp. 498–527. Springer, Heidelberg (2015). https://doi.org/10.1007/978-3-662-46497-7_20
30. Gentry, C., Wichs, D.: Separating succinct non-interactive arguments from all falsifiable assumptions. In: Fortnow, L., Vadhan, S.P. (eds.) 43rd ACM STOC, pp. 99–108. ACM Press (2011)
31. Goyal, R., Koppula, V., Waters, B.: Lockable obfuscation. In: 58th FOCS, pp. 612–621. IEEE Computer Society Press (2017)
32. Hazay, C., Orsini, E., Scholl, P., Soria-Vazquez, E.: TinyKeys: a new approach to efficient multi-party computation. In: Shacham, H., Boldyreva, A. (eds.) CRYPTO 2018. LNCS, vol. 10993, pp. 3–33. Springer, Cham (2018). https://doi.org/10.1007/978-3-319-96878-0_1
33. Ishai, Y., Prabhakaran, M., Sahai, A.: Secure arithmetic computation with no honest majority. In: Reingold, O. (ed.) TCC 2009. LNCS, vol. 5444, pp. 294–314. Springer, Heidelberg (2009). https://doi.org/10.1007/978-3-642-00457-5_18
34. Jain, A., Krenn, S., Pietrzak, K., Tentes, A.: Commitments and efficient zero-knowledge proofs from learning parity with noise. In: Wang, X., Sako, K. (eds.) ASIACRYPT 2012. LNCS, vol. 7658, pp. 663–680. Springer, Heidelberg (2012). https://doi.org/10.1007/978-3-642-34961-4_40
35. Lin, H.: Indistinguishability obfuscation from constant-degree graded encoding schemes. In: Fischlin, M., Coron, J.-S. (eds.) EUROCRYPT 2016. LNCS, vol. 9665, pp. 28–57. Springer, Heidelberg (2016). https://doi.org/10.1007/978-3-662-49890-3_2
36. Lin, H., Tessaro, S.: Indistinguishability obfuscation from trilinear maps and block-wise local PRGs. In: Katz, J., Shacham, H. (eds.) CRYPTO 2017. LNCS, vol. 10401, pp. 630–660. Springer, Cham (2017). https://doi.org/10.1007/978-3-319-63688-7_21
37. Lin, H., Vaikuntanathan, V.: Indistinguishability obfuscation from DDH-like assumptions on constant-degree graded encodings. In: Dinur, I. (ed.) 57th FOCS, pp. 11–20. IEEE Computer Society Press (2016)
38. Lynn, B., Prabhakaran, M., Sahai, A.: Positive results and techniques for obfuscation. In: Cachin, C., Camenisch, J.L. (eds.) EUROCRYPT 2004. LNCS, vol. 3027, pp. 20–39. Springer, Heidelberg (2004). https://doi.org/10.1007/978-3-540-24676-3_2
39. Ma, F., Zhandry, M.: The MMap strikes back: obfuscation and new multilinear maps immune to CLT13 zeroizing attacks. In: Beimel, A., Dziembowski, S. (eds.) TCC 2018. LNCS, vol. 11240, pp. 513–543. Springer, Cham (2018). https://doi.org/10.1007/978-3-030-03810-6_19
40. Micciancio, D., Mol, P.: Pseudorandom Knapsacks and the sample complexity of LWE search-to-decision reductions. In: Rogaway, P. (ed.) CRYPTO 2011. LNCS, vol. 6841, pp. 465–484. Springer, Heidelberg (2011). https://doi.org/10.1007/978-3-642-22792-9_26

41. Naor, M.: On cryptographic assumptions and challenges. In: Boneh, D. (ed.) CRYPTO 2003. LNCS, vol. 2729, pp. 96–109. Springer, Heidelberg (2003). https://doi.org/10.1007/978-3-540-45146-4_6
42. Peikert, C.: On error correction in the exponent. In: Halevi, S., Rabin, T. (eds.) TCC 2006. LNCS, vol. 3876, pp. 167–183. Springer, Heidelberg (2006). https://doi.org/10.1007/11681878_9
43. Sahai, A., Waters, B.: How to use indistinguishability obfuscation: deniable encryption, and more. In: Shmoys, D.B. (ed.) 46th ACM STOC, pp. 475–484. ACM Press (2014)
44. Shoup, V.: Lower bounds for discrete logarithms and related problems. In: Fumy, W. (ed.) EUROCRYPT 1997. LNCS, vol. 1233, pp. 256–266. Springer, Heidelberg (1997). https://doi.org/10.1007/3-540-69053-0_18
45. Skala, M.: Hypergeometric tail inequalities: ending the insanity. arXiv preprint arXiv:1311.5939 (2013)
46. Wee, H.: On obfuscating point functions. In: Gabow, H.N., Fagin, R. (eds.) 37th ACM STOC, pp. 523–532. ACM Press (2005)
47. Wichs, D., Zirdelis, G.: Obfuscating compute-and-compare programs under LWE. In: 58th FOCS, pp. 600–611. IEEE Computer Society Press (2017)
48. Yu, Y., Zhang, J.: Cryptography with auxiliary input and trapdoor from constant-noise LPN. In: Robshaw, M., Katz, J. (eds.) CRYPTO 2016. LNCS, vol. 9814, pp. 214–243. Springer, Heidelberg (2016). https://doi.org/10.1007/978-3-662-53018-4_9

Distributional Collision Resistance Beyond One-Way Functions

Nir Bitansky[1], Iftach Haitner[1], Ilan Komargodski[2], and Eylon Yogev[3](✉)

[1] School of Computer Science, Tel Aviv University, Tel Aviv, Israel
{nirbitan,iftachh}@tau.ac.il
[2] Cornell Tech, New York, NY, USA
komargodski@cornell.edu
[3] Technion, Haifa, Israel
eylony@gmail.com

Abstract. Distributional collision resistance is a relaxation of collision resistance that only requires that it is hard to sample a collision (x, y) where x is uniformly random and y is uniformly random conditioned on colliding with x. The notion lies between one-wayness and collision resistance, but its exact power is still not well-understood. On one hand, distributional collision resistant hash functions cannot be built from one-way functions in a black-box way, which may suggest that they are stronger. On the other hand, so far, they have not yielded any applications beyond one-way functions.

Assuming distributional collision resistant hash functions, we construct *constant-round* statistically hiding commitment scheme. Such commitments are not known based on one-way functions, and are impossible to obtain from one-way functions in a black-box way. Our construction relies on the reduction from inaccessible entropy generators to statistically hiding commitments by Haitner et al. (STOC '09). In the converse direction, we show that two-message statistically hiding commitments imply distributional collision resistance, thereby establishing a loose equivalence between the two notions.

A corollary of the first result is that constant-round statistically hiding commitments are implied by average-case hardness in the class SZK (which is known to imply distributional collision resistance). This implication seems to be folklore, but to the best of our knowledge has not been proven explicitly. We provide yet another proof of this implication, which is arguably more direct than the one going through distributional collision resistance.

1 Introduction

Distributional collision resistant hashing (dCRH), introduced by Dubrov and Ishai [9], is a relaxation of the notion of collision resistance. In (plain) collision resistance, it is guaranteed that no efficient adversary can find *any* collision given a random hash function in the family. In dCRH, it is only guaranteed that no efficient adversary can sample *a random* collision given a random hash function

Y. Ishai and V. Rijmen (Eds.): EUROCRYPT 2019, LNCS 11478, pp. 667–695, 2019.
https://doi.org/10.1007/978-3-030-17659-4_23

in the family. More precisely, given a random hash function h from the family, it is computationally hard to sample a pair (x, y) such that x is uniform and y is uniform in the preimage set $h^{-1}(x) = \{z: h(x) = h(z)\}$. This hardness is captured by requiring that the adversary cannot get statistically-close to this distribution over collisions.[1]

The power of dCRH. Intuitively, the notion of dCRH seems quite weak. The adversary may even be able to sample collisions from the set of *all* collisions, but only from a skewed distribution, far from the random one. Komargodski and Yogev [26] show that dCRH can be constructed assuming average-case hardness in the complexity class *statistical zero-knowledge* (SZK), whereas a similar implication is not known for multi-collision resistance.[2] (let alone plain collision resistance). This can be seen as evidence suggesting that dCRH may be weaker than collision resistance, or even multi-collision resistance [4,6,24,25].

Furthermore, dCRH has not led to the same cryptographic applications as collision resistance, or even multi-collision resistance. In fact, dCRH has no known applications beyond those implied by one-way functions.

At the same time, dCRH is not known to follow from one-way functions, and actually, cannot follow based on black-box reductions [34]. In fact, it can even be separated from indistinguishability obfuscation (and one-way functions) [2]. Overall, we are left with a significant gap in our understanding of the power of dCRH:

Does the power of dCRH go beyond one-way functions?

1.1 Our Results

We present the first application of dCRH that is not known from one-way functions and is provably unachievable from one-way functions in a black-box way.

Theorem 1. *dCRH implies* constant-round *statistically hiding commitment scheme.*

Such commitment schemes cannot be constructed from one-way functions (or even permutations) in a black-box way due to a result of Haitner, Hoch, Reingold and Segev [15]. They show that the number of rounds in such commitments must grow quasi-linearly in the security parameter.

The heart of Theorem 1 is a construction of an inaccessible-entropy generator [17,18] from dCRH.

[1] There are some subtleties in defining this precisely. The definition we use differs from previous ones [9,21,26]. We elaborate on the exact definition and the difference in the technical overview below and in Sect. 3.4.

[2] Multi-collision resistance is another relaxation of collision resistance, where it is only hard to find multiple elements that all map to the same image. Multi-collision resistance does not imply dCRH in a black-box way [25], but Komargodski and Yogev [26] give a non-black-box construction.

An implication of the above result is that constant-round statistically hiding commitments can be constructed from average-case hardness in SZK. Indeed, it is known that such hardness implies the existence of a dCRH [26].

Corollary 1. *A Hard-on-average problem in SZK implies a* constant-round *statistically hiding commitment scheme.*

The statement of Corollary 1 has been treated as known in several previous works (c.f. [5,10,18]), but a proof of this statement has so far not been published or (to the best of our knowledge) been publicly available. We also provide an alternative proof of this statement (and in particular, a different commitment scheme) that does not go through a construction of a dCRH, and is arguably more direct.

A limit on the power of dCRH. We also show a converse connection between dCRH and statistically hiding commitments. Specifically, we show that *any* two-message statistically hiding commitment implies a dCRH function family.

Theorem 2. *Any two-message statistically hiding commitment scheme implies dCRH.*

This establishes a loose equivalence between dCRH and statistically hiding commitments. Indeed, the commitments we construct from dCRH require more than two messages. Interestingly, we can even show that such commitments imply a stronger notion of dCRH where the adversary's output distribution is not only noticeably far from the random collision distribution, but is $(1 - \mathsf{negl}(n))$-far.

1.2 Related Work on Statistically Hiding Commitments

Commitment schemes, the digital analog of sealed envelopes, are central to cryptography. More precisely, a commitment scheme is a two-stage interactive protocol between a sender S and a receiver R. After the commit stage, S is bound to (at most) one value, which stays hidden from R, and in the reveal stage R learns this value. The immediate question arising is what it means to be "bound to" and to be "hidden". Each of these security properties can come in two main flavors, either *computational security*, where a polynomial-time adversary cannot violate the property except with negligible probability, or the stronger notion of *statistical security*, where even an unbounded adversary cannot violate the property except with negligible probability. However, it is known that there do *not* exist commitment schemes that are simultaneously statistically hiding and statistically binding.

There exists a one-message (i.e., non-interactive) statistically binding commitment schemes assuming one-way permutations (Blum [7]). From one-way functions, such commitments can be achieved by a two-message protocol (Naor [28] and Håstad, Impagliazzo, Levin and Luby [22]).

Statistically hiding commitments schemes have proven to be somewhat more difficult to construct. Naor, Ostrovsky, Venkatesan and Yung [29] gave a statistically hiding commitment scheme protocol based on one-way permutations, whose linear number of rounds matched the lower bound of [15] mentioned above. After many years, this result was improved by Haitner, Nguyen, Ong, Reingold and Vadhan [16] constructing such commitment based on the minimal hardness assumption that one-way functions exist. The reduction of [16] was later simplified and made more efficient by Haitner, Reingold, Vadhan and Wee [17,18] to match, in some settings, the round complexity lower bound of [15]. Constant-round statistically hiding commitment protocols are known to exist based on families of collision resistant hash functions [8,20,30]. Recently, Berman, Degwekar, Rothblum and Vasudevan [4] and Komargodski, Naor and Yogev [25] constructed constant-round statistically hiding commitment protocols assuming the existence of *multi*-collision resistant hash functions.

Constant-round statistically hiding commitments are a basic building block in many fundamental applications. Two prominent examples are constructions of *constant-round* zero-knowledge proofs for all NP (Goldreich and Kahan [12]) and *constant-round* public-coin statistical zero-knowledge arguments for NP (Barak [3], Pass and Rosen [33]).

Statistically hiding commitment are also known to be tightly related to the hardness of the class of problems that posses a statistical zero-knowledge protocol, i.e., the class SZK. Ong and Vadhan [31] showed that a language in NP has a zero-knowledge protocol if and only if the language has an "instance-dependent" commitment scheme. An instance-dependent commitment scheme for a given language is a commitment scheme that can depend on an instance of the language, and where the hiding and binding properties are required to hold only on the YES and NO instances of the language, respectively.

1.3 Directions for Future Work

The security notions of variants of collision resistance, including plain collision resistance and multi-collision resistance, can be phrased in the language of entropy. For example, plain collision resistance requires that once a hash value y is fixed the (max) entropy of preimages that any efficient adversary can find is zero. In multi-collision resistance, it may be larger than zero, even for every y, but still bounded by the size of allowed multi collisions. In distributional collision resistance, the (Shannon) entropy is close to maximal.

Yet, the range of applications of collision resistance (or even multi-collision resistance) is significantly larger than those of distributional collision resistance. Perhaps the most basic such application is *succinct* commitment protocols which are known from plain/multi-collision resistance but not from distributional collision resistance (by *succinct* we mean that the total communication is shorter than the string being committed to). Thus, with the above entropy perspective in mind, a natural question is to characterize the full range or parameters between distributional and plain collision resistance and understand for each of them what are the applications implied. A more concrete question is to find the minimal notion of security for collision resistance that implies succinct commitments.

A different line of questions concerns understanding better the notion of distributional collision resistance and constructing it from more assumptions. Komargodski and Yogev constructed it from multi-collision resistance and from the average-case hardness of SZK. Can we construct it, for example, from the multivariate quadratic (MQ) assumption [27] or can we show an attack for random degree 2 mappings? Indeed, we know that random degree 2 mappings cannot be used for plain collision resistant hashing [1, Theorem 5.3].

2 Technical Overview

In this section, we give an overview of our techniques. We start with a more precise statement of the definition of dCRH and a comparison with previous versions of its definition.

A dCRH is a family of functions $\mathcal{H}_n = \{h\colon \{0,1\}^n \to \{0,1\}^m\}$. (The functions are not necessarily compressing.) The security guarantee is that there exists a universal polynomial $p(\cdot)$ such that for every efficient adversary A it holds that

$$\mathbf{\Delta}\left((h, \mathsf{A}(1^n, h)), (h, \mathsf{Col}(h))\right) \geq \frac{1}{p(n)},$$

where $\mathbf{\Delta}$ denotes statistical distance, $h \leftarrow \mathcal{H}_n$ is chosen uniformly at random, and Col is a random variable that is sampled in the following way: Given h, first sample $x_1 \leftarrow \{0,1\}^n$ uniformly at random and then sample x_2 uniformly at random from the set of all preimages of x_1 relative to h (namely, from the set $\{x\colon h(x) = h(x_1)\}$). Note that Col may not be efficiently samplable and intuitively, the hardness of dCRH says that there is no efficient way to sample from Col, even approximately.

Our definition is stronger than previous definitions of dCRH [9,21,26] by that we require the existence of a universal polynomial $p(\cdot)$, whereas previous definitions allow a different polynomial per adversary. Our modification seems necessary to get non-trivial applications of dCRH, as the previous definitions are not known to imply one-way functions. In contrast, our notion of dCRH implies distributional one-way functions which, in turn, imply one-way functions [23] (indeed, the definition of distributional one-way functions requires a universal polynomial rather than one per adversary).[3] We note that previous constructions of dCRH (from multi-collision resistance and SZK-hardness) [26] apply to our stronger notion as well.

2.1 Commitments from dCRH and Back

We now describe our construction of constant-round statistically hiding commitments from dCRH. To understand the difficulty, let us recall the standard

[3] The previous definition is known to imply a weaker notion of distributional one-way functions (with a different polynomial bound per each adversary) [21], which is not known to imply one-way functions.

approach to constructing statistically hiding commitments from (fully) collision resistant hash functions [8,20,30]. Here to commit to a bit b, we hash a random string x, and output $(h(x), s, b \oplus Ext_s(x))$, where s is a seed for a strong randomness extractor Ext and b is padded with a (close to) random bit extracted from x. When h is collision resistant, x is computationally fixed and thus so is the bit b. However, for a dCRH h, this is far from being the case: for any y, the sender might potentially be able to sample preimages from the set of all preimages.

The hash $h(x)$, however, does yield a weak binding guarantee. For simplicity of exposition, let us assume that any $y \in \{0,1\}^m$ has exactly 2^k preimages under h in $\{0,1\}^n$. Then, for a noticeable fraction of commitments y, the adversary cannot open y to a uniform x in the preimage set $h^{-1}(y)$. In particular, the adversary must choose between two types of *entropy losses*: it either outputs a commitment y of entropy m' noticeably smaller than m, or after the commitment, it can only open to a value x of entropy k' noticeably smaller than k. One way or the other, in total $m' + k'$ must be noticeably smaller than $n = m + k$. This naturally leads us to the notion of *inaccessible entropy* defined by Haitner, Reingold, Vadhan and Wee [17,18].

Let us briefly recall what inaccessible entropy is (see Sect. 4.1 for a precise definition). The entropy of a random variable X is a measure of "the amount of randomness" that X contains. The notion of (in)accessible entropy measures the feasibility of sampling high-entropy strings that are *consistent* with a given random process. Consider the two-block generator (algorithm) G that samples $x \leftarrow \{0,1\}^n$, and then outputs $y = h(x)$ and x. The *real entropy* of G is defined as the entropy of the generator's (total) output in a random execution, and is clearly equal to n, the length of x. The *accessible entropy* of G measures the entropy of these output blocks from the point of view of an efficient G-*consistent* generator, which might act arbitrarily, but still outputs a value in the support of G.

Assume for instance that h had been (fully) collision resistant. Then from the point of view of any efficient G-consistent generator $\widetilde{G}$, conditioned on its first block y, and its internal randomness, its second output block is fixed (otherwise, G can be used for finding a collision). In other words, while the value of x given y may have entropy $k = n - m$, this entropy is completely *inaccessible* for an efficient G-consistent generator. (Note that we do not measure here the entropy of the output blocks of $\widetilde{G}$, which clearly can be as high as the real entropy of G by taking $\widetilde{G} = G$. Rather, we measure the entropy of the block from *$\widetilde{G}$'s point of view*, and in particular, the entropy of its second block given the randomness used for generating the first block.) Haitner et al. show that any noticeable gap between the real entropy and the inaccessible entropy of such an efficient generator can be leveraged for constructing statistically hiding commitments, with a number of rounds that is linear in the number of blocks.

Going back to dCRH, we have already argued that in the simple case that h is regular and onto $\{0,1\}^m$, we get a noticeable gap between the real entropy $n = m + k$ and the accessible entropy $m' + k' \leq m + k - 1/\mathsf{poly}(n)$. We prove that this is, in fact, true for any dCRH:

Lemma 1. *dCRH implies a two-block inaccessible entropy generator.*

The block generator itself is the simple generator described above:

$$\text{output } h(x) \text{ and then } x, \text{ for } x \leftarrow \{0,1\}^n.$$

The proof, however, is more involved than in the case of collision resistance. In particular, it is sensitive to the exact notion of entropy used. Collision resistant hash functions satisfy a very clean and simple guarantee—the *maximum entropy*, capturing the support size, is always at most $m < n$. In contrast, for dCRH (compressing or not), the maximum entropy could be as large as n, which goes back to the fact that the adversary may be able to sample from the set of *all* collisions (albeit from a skewed distribution). Still, we show a gap with respect to average (a.k.a Shannon) accessible entropy, which suffices for constructing statistically hiding commitments [17].

From commitments back to dCRH. We show that any two-message statistically hiding commitment implies a dCRH function family. Let $(\mathcal{S}, \mathcal{R})$ be the sender and receiver of a statistically hiding bit commitment. The first message sent by the receiver is the description of the hash function: $h \leftarrow \mathcal{R}(1^n)$. The sender's commitment to a bit b, using randomness r, is the hash of $x = (b, r)$. That is, $h(x) = \mathcal{S}(h, b; r)$.

To argue that this is a dCRH, we show that any attacker that can sample collisions that are close to the random collision distribution Col can also break the binding of the commitment scheme. For this, it suffices to show that a collision $(b, r), (b', r')$ sampled from Col, translates to equivocation—the corresponding commitment can be opened to two distinct bits $b \neq b'$. Roughly speaking, this is because statistical hiding implies that a random collision to a random bit b (corresponding to a random hash value) is statistically independent of the underlying committed bit. In particular, a random preimage of such a commitment will consist of a different bit b' with probability roughly $1/2$. See details in Sect. 4.3.

2.2 Commitments from SZK Hardness

We now give an overview of our construction of statistically hiding commitments directly from average-case hardness in SZK. Our starting point is a result of Ong and Vadhan [31] showing that any promise problem in SZK has an *instance-dependent commitment*. These are commitments that are also parameterized by an instance x, such that if x is a *yes instance*, they are statistically hiding and if x is a *no instance*, they are statistically binding. We construct statistically hiding commitments from instance-dependent commitments for a hard-on-average problem $\Pi = (\Pi_N, \Pi_Y)$ in SZK.

A first attempt: using zero-knowledge proofs. To convey the basic idea behind the construction, let us first assume that Π satisfies a strong form of average-case hardness where we can efficiently sample no-instances from Π_N and yes-instances from Π_Y so that the two distributions are computationally indistinguishable. Then a natural protocol for committing to a message m is the following: The receiver $\mathcal{R}$ would sample a yes-instance $x \leftarrow \Pi_Y$, and send it to the sender $\mathcal{S}$ along with zero-knowledge proof [14] that x is indeed a yes-instance. The sender $\mathcal{S}$ would then commit to m using an x-dependent commitment.

To see that the scheme is statistically hiding, we rely on the soundness of the proof which guarantees that x is indeed a yes-instance, and then on the hiding of the instance-dependent scheme. To prove (computational) binding, we rely on zero knowledge property and the hardness of Π. Specifically, by zero knowledge, instead of sampling x from Π_Y, we can sample it from any computationally indistinguishable distribution, without changing the probability that an efficient malicious sender breaks binding. In particular, by the assumed hardness of Π, we can sample x from Π_N. Now, however, the instance-dependent commitment guarantees binding, implying that the malicious sender will not be able to equivocate.

The main problem with this construction is that constant-round zero-knowledge proofs (with a negligible soundness error) are only known assuming constant-round statistically hiding commitments [12], which is exactly what we are trying to construct.

A second attempt: using witness-indistinguishable proofs. Instead of relying on zero-knowledge proofs, we rely on the weaker notion of witness-indistinguishable proofs and use the *independent-witnesses paradigm* of Feige and Shamir [11]. (Indeed such proofs are known for all of NP, based average-case hardness in SZK [13,28,32], see Sect. 5 for details.) We change the previous scheme as follows: the receiver $\mathcal{R}$ will now sample *two* instances x_0 and x_1 and provide a witness-indistinguishable proof that at least one of them is a yes-instance. The sender, will secret share the message m into two random messages m_0, m_1 such that $m = m_0 \oplus m_1$, and return two instance-dependent commitments to m_0 and m_1 relative to x_0 and x_1, respectively.

Statistical hiding follows quite similarly to the previous protocol—by the soundness of the proof one of the instances x_b is a yes-instance, and by the hiding of the x_b-dependent commitment, the corresponding share m_b is statistically hidden, and thus so is m. To prove binding, we first note that by witness indistinguishability, to prove its statement, the receiver could use x_b for either $b \in \{0,1\}$. Then, relying on the hardness of Π, we can sample x_{1-b} to be a no-instance instead of a yes-instance. If b is chosen at random, the sender cannot predict b better than guessing. At the same time, in order to break binding, the sender must equivocate with respect to at least one of the instance-dependent commitments, and since it cannot equivocate with respect to the no-instance x_{1-b}, it cannot break binding unless it can get an advantage in predicting b.

Our actual scheme. The only gap remaining between the scheme just described and our actual scheme is our assumption regarding the strong form of average-case hardness of Π. In contrast, the standard form of average-case hardness only implies a single samplable distribution D, such that given a sample x from D it is hard to tell whether x is a yes-instance or a no-instance better than guessing.

This requires the following changes to the protocol. First, lacking a samplable distribution on yes-instances, we consider instead the product distribution D^n, as a way to sample *weak yes instances*—n-tuples of instances where at least one is a yes-instance in Π_Y. Unlike before, where everything in the support of the yes-instance sampler was guaranteed to be a yes-instance, now we are only guaranteed that a random tuple is a weak yes instance with overwhelming probability. To deal with this weak guarantee, we add a *coin-tossing into the well* phase [13], where the randomness for sampling an instance from D^n is chosen together by the receiver and sender. We refer the reader to Sect. 5 for more details.

3 Preliminaries

Unless stated otherwise, the logarithms in this paper are base 2. For a distribution $\mathcal{D}$ we denote by $x \leftarrow \mathcal{D}$ an element chosen from $\mathcal{D}$ uniformly at random. For an integer $n \in \mathbb{N}$ we denote by $[n]$ the set $\{1, \ldots, n\}$. We denote by U_n the uniform distribution over n-bit strings. We denote by $\circ$ the string concatenation operation. A function $\mathsf{negl}\colon \mathbb{N} \to \mathbb{R}^+$ is *negligible* if for every constant $c > 0$, there exists an integer N_c such that $\mathsf{negl}(n) < n^{-c}$ for all $n > N_c$.

3.1 Cryptographic Primitives

A function f, with input length $m_1(n)$ and outputs length $m_2(n)$, specifies for every $n \in \mathbb{N}$ a function $f_n\colon \{0,1\}^{m_1(n)} \to \{0,1\}^{m_2(n)}$. We only consider functions with polynomial input lengths (in n) and occasionally abuse notation and write $f(x)$ rather than $f_n(x)$ for simplicity. The function f is computable in polynomial time (efficiently computable) if there exists a probabilistic machine that for any $x \in \{0,1\}^{m_1(n)}$ outputs $f_n(x)$ and runs in time polynomial in n.

A function family ensemble is an infinite set of function families, whose elements (families) are indexed by the set of integers. Let $\mathcal{F} = \{\mathcal{F}_n\colon \mathcal{D}_n \to \mathcal{R}_n\}_{n\in\mathbb{N}}$ stand for an ensemble of function families, where each $f \in \mathcal{F}_n$ has domain $\mathcal{D}_n$ and range $\mathcal{R}_n$. An efficient function family ensemble is one that has an efficient sampling and evaluation algorithms.

Definition 1 (Efficient function family ensemble). *A function family ensemble $\mathcal{F} = \{\mathcal{F}_n\colon \mathcal{D}_n \to \mathcal{R}_n\}_{n\in\mathbb{N}}$ is efficient if:*

- *$\mathcal{F}$ is samplable in polynomial time: there exists a probabilistic polynomial-time machine that given 1^n, outputs (the description of) a uniform element in $\mathcal{F}_n$.*
- *There exists a deterministic algorithm that given $x \in \mathcal{D}_n$ and (a description of) $f \in \mathcal{F}_n$, runs in time $\mathsf{poly}(n, |x|)$ and outputs $f(x)$.*

3.2 Distance and Entropy Measures

Definition 2 (Statistical distance). *The* statistical distance *between two random variables X, Y over a finite domain Ω, is defined by*

$$\boldsymbol{\Delta}(X,Y) \triangleq \frac{1}{2} \cdot \sum_{x \in \Omega} |\mathbf{Pr}[X = x] - \mathbf{Pr}[Y = x]| .$$

We say that X and Y are δ-close (resp. -far) if $\boldsymbol{\Delta}(X,Y) \leq \delta$ (resp. $\boldsymbol{\Delta}(X,Y) \geq \delta$).

Entropy. Let X be a random variable. For any $x \in \mathsf{supp}(X)$, the sample-entropy of x with respect to X is

$$\mathsf{H}_X(x) = \log\left(\frac{1}{\mathbf{Pr}[X = x]}\right).$$

The Shannon entropy of X is defined as:

$$\mathsf{H}(X) = \mathop{\mathbf{E}}_{x \leftarrow X}[\mathsf{H}_X(x)] .$$

Conditional entropy. Let (X, Y) be a jointly distributed random variable.

- For any $(x, y) \in \mathsf{supp}(X, Y)$, the conditional sample-entropy to be

$$\mathsf{H}_{X|Y}(x \mid y) = \log\left(\frac{1}{\mathbf{Pr}[X = x \mid Y = y]}\right).$$

- The conditional Shannon entropy is

$$\mathsf{H}(X \mid Y) = \mathop{\mathbf{E}}_{(x,y)\leftarrow(X,Y)}[\mathsf{H}_{X|Y}(x \mid y)] = \mathop{\mathbf{E}}_{y \leftarrow Y}[\mathsf{H}(X|_{Y=y})] = \mathsf{H}(X,Y) - \mathsf{H}(Y).$$

Relative entropy. We also use basic facts about relative entropy (also known as, Kullback-Leibler divergence).

Definition 3 (Relative entropy). *Let X and Y be two random variables over a finite domain Ω. The* relative entropy *is*

$$\mathbf{D}_{\mathsf{KL}}(X\|Y) = \sum_{x \in \Omega} \mathbf{Pr}[X = x] \cdot \log\left(\frac{\mathbf{Pr}[X = x]}{\mathbf{Pr}[Y = x]}\right).$$

Proposition 1 (Chain rule). *Let (X_1, X_2) and (Y_1, Y_2) be random variables. It holds that*

$$\mathbf{D}_{\mathsf{KL}}((X_1, X_2)\|(Y_1, Y_2)) = \mathbf{D}_{\mathsf{KL}}(X_1\|Y_1) + \mathop{\mathbf{E}}_{x \leftarrow X_1}[\mathbf{D}_{\mathsf{KL}}(X_2|_{X_1=x}\|Y_2|_{Y_1=x})] .$$

A well-known relation between statistical distance and relative entropy is given by Pinsker's inequality.

Proposition 2 (Pinsker's inequality). *For any two random variables X and Y over a finite domain it holds that*

$$\mathbf{\Delta}(X,Y) \leq \sqrt{\frac{\ln 2}{2} \cdot \mathbf{D}_{\mathsf{KL}}(X\|Y)}.$$

Another useful inequality is Jensen's inequality.

Proposition 3 (Jensen's inequality). *If X is a random variable and f is concave, then*

$$\mathbf{E}[f(X)] \leq f(\mathbf{E}[X]).$$

3.3 Commitment Schemes

A commitment scheme is a two-stage interactive protocol between a sender $\mathcal{S}$ and a receiver $\mathcal{R}$. The goal of such a scheme is that after the first stage of the protocol, called the commit protocol, the sender is bound to at most one value. In the second stage, called the opening protocol, the sender opens its committed value to the receiver. Here, we are interested in statistically hiding and computationally binding commitments. Also, for simplicity, we restrict our attention to protocols that can be used to commit to bits (i.e., strings of length 1).

In more detail, a commitment scheme is defined via a pair of probabilistic polynomial-time algorithms $(\mathcal{S}, \mathcal{R}, \mathcal{V})$ such that:

- The commit protocol: $\mathcal{S}$ receives as input the security parameter 1^n and a bit $b \in \{0,1\}$. $\mathcal{R}$ receives as input the security parameter 1^n. At the end of this stage, $\mathcal{S}$ outputs decom (the decommitment) and $\mathcal{R}$ outputs com (the commitment).
- The verification: $\mathcal{V}$ receives as input the security parameter 1^n, a commitment com, a decommitment decom, and outputs either a bit b or $\perp$.

A commitment scheme is *public coin* if all messages sent by the receiver are independent random coins.

Denote by $(\mathsf{decom}, \mathsf{com}) \leftarrow \langle \mathcal{S}(1^n, b), \mathcal{R} \rangle$ the experiment in which $\mathcal{S}$ and $\mathcal{R}$ interact with the given inputs and uniformly random coins, and eventually $\mathcal{S}$ outputs a decommitment string and $\mathcal{R}$ outputs a commitment. The completeness of the protocol says that for all $n \in \mathbb{N}$, every $b \in \{0,1\}$, and every tuple $(\mathsf{decom}, \mathsf{com})$ in the support of $\langle \mathcal{S}(1^n, b), \mathcal{R} \rangle$, it holds that $\mathcal{V}(\mathsf{decom}, \mathsf{com}) = b$. Unless otherwise stated, $\mathcal{V}$ is the canonical verifier that receives the sender's coins as part of the decommitment and checks their consistency with the transcript.

Below we define two security properties one can require from a commitment scheme. The properties we list are *statistical-hiding* and *computational-binding*. These roughly say that after the commit stage, the sender is *bound* to a specific value but the receiver cannot know this value.

Definition 4 (binding). *A commitment scheme* $(\mathcal{S}, \mathcal{R}, \mathcal{V})$ *is binding if for every probabilistic polynomial-time adversary* $\mathcal{S}^*$ *there exits a negligible function* $\mathsf{negl}(n)$ *such that*

$$\mathbf{Pr}\left[\begin{array}{c}\mathcal{V}(\mathsf{decom}, \mathsf{com}) = 0 \text{ and} \\ \mathcal{V}(\mathsf{decom}', \mathsf{com}) = 1\end{array} : (\mathsf{decom}, \mathsf{decom}', \mathsf{com}) \leftarrow \langle \mathcal{S}^*(1^n), \mathcal{R}\rangle\right] \leq \mathsf{negl}(n)$$

for all $n \in \mathbb{N}$*, where the probability is taken over the random coins of both* $\mathcal{S}^*$ *and* $\mathcal{R}$*.*

Given a commitment scheme $(\mathcal{S}, \mathcal{R}, \mathcal{V})$ and an adversary $\mathcal{R}^*$, we denote by $\mathsf{view}_{\langle \mathcal{S}(b), \mathcal{R}^*\rangle}(n)$ the distribution on the view of $\mathcal{R}^*$ when interacting with $\mathcal{S}(1^n, b)$. The view consists of $\mathcal{R}^*$'s random coins and the sequence of messages it received from $\mathcal{S}$. The distribution is taken over the random coins of both $\mathcal{S}$ and $\mathcal{R}$. Without loss of generality, whenever $\mathcal{R}^*$ has no computational restrictions, we can assume it is deterministic.

Definition 5 (hiding). *A commitment scheme* $(\mathcal{S}, \mathcal{R}, \mathcal{V})$ *is statistically hiding if there exists a negligible function* $\mathsf{negl}(n)$ *such that for every (deterministic) adversary* $\mathcal{R}^*$ *it holds that*

$$\mathbf{\Delta}\left(\{\mathsf{view}_{\langle \mathcal{S}(0), \mathcal{R}^*\rangle}(n)\}, \{\mathsf{view}_{\langle \mathcal{S}(1), \mathcal{R}^*\rangle}(n)\}\right) \leq \mathsf{negl}(n)$$

for all $n \in \mathbb{N}$*.*

3.4 Distributional Collision Resistant Hash Functions

Roughly speaking, a distributional collision resistant hash function [9] guarantees that no efficient adversary can sample a uniformly random collision. We start by defining more precisely what we mean by a random collision throughout the paper, and then move to the actual definition.

Definition 6 (Ideal collision finder). *Let* Col *be the random function that given a (description) of a function* $h\colon \{0,1\}^n \to \{0,1\}^m$ *as input, returns a collision* (x_1, x_2) *with respect to* h *as follows: it samples a uniformly random element,* $x_1 \leftarrow \{0,1\}^n$*, and then samples a uniformly random element that collides with* x_1 *under* h*,* $x_2 \leftarrow \{x \in \{0,1\}^n : h(x) = h(x_1)\}$*. (Note that possibly,* $x_1 = x_2$*).*

Definition 7 (Distributional collision resistant hashing). *Let* $\mathcal{H} = \{\mathcal{H}_n : \{0,1\}^n \to \{0,1\}^{m(n)}\}_{n\in\mathbb{N}}$ *be an efficient function family ensemble. We say that* $\mathcal{H}$ *is a secure* distributional collision resistant hash *(dCRH) function family if there exists a polynomial* $p(\cdot)$ *such that for any probabilistic polynomial-time algorithm* A*, it holds that*

$$\mathbf{\Delta}\left((h, \mathsf{A}(1^n, h)), (h, \mathsf{Col}(h))\right) \geq \frac{1}{p(n)},$$

for $h \leftarrow \mathcal{H}_n$ *and large enough* $n \in \mathbb{N}$*.*

Comparison with the previous definition. Our definition deviates from the previous definition of distributional collision resistance considered in [9,21,26]. The definition in the above-mentioned works is equivalent to requiring that for any efficient adversary A, there exists a polynomial p_{A}, such that the collision output by A is $\frac{1}{p_{\mathsf{A}}(n)}$-far from a random collision on average (over h). Our definition switches the order of quantifiers, requiring that there is one such polynomial $p(\cdot)$ for all adversaries A.

We note that the previous definition is, in fact, not even known to imply one-way functions. In contrast, the definition presented here strengthens that of *distributional one-way functions*, which in turn implies one-way functions [23]. Additionally, note that both constructions of distributional collision resistance in [26] (from multi-collision resistance and from SZK hardness) satisfy our stronger notion of security (with a similar proof).

On compression. As opposed to classical notions of collision resistance (such as plain collision resistance or multi-collision resistance), it makes sense to require distributional collision resistance even for *non-compressing* functions. So we do not put a restriction on the order between n and $m(n)$. As a matter of fact, by padding, the input, arbitrary polynomial compression can be assumed without loss of generality.

4 From dCRH to Statistically Hiding Commitments and Back

We show distributional collision resistant hash functions imply constant-round statistically hiding commitments.

Theorem 3. *Assume the existence of a distributional collision resistant hash function family. Then, there exists a constant-round statistically hiding and computationally binding commitment scheme.*

Our proof relies on the transformation of Haitner et al. [17,18], translating inaccessible-entropy generators to statistically hiding commitments. Concretely, we construct appropriate inaccessible-entropy generators from distributional collision resistant hash functions. In Sect. 4.1, we recall the necessary definitions and the result of [17], and then in Sect. 4.2, we prove Theorem 3.

We complement the above result by showing a loose converse to Theorem 3, namely that two message statistically hiding commitments (with possibly large communication) imply the existence of distributional collision resistance hashing.

Theorem 4. *Assume the existence of a binding and statistically hiding two-message commitment scheme. Then, there exists a dCRH function family.*

This proof of Theorem 4 appears in Sect. 4.3.

4.1 Preliminaries on Inaccessible Entropy Generators

The following definitions of real and accessible entropy of protocols are taken from [17].

Definition 8 (Block generators). *Let n be a security parameter, and let $c = c(n)$, $s = s(n)$ and $m = m(n)$. An m-*block generator *is a function $G\colon \{0,1\}^c \times \{0,1\}^s \mapsto (\{0,1\}^*)^m$. It is* efficient *if its running time on input of length $c(n) + s(n)$ is polynomial in n.*

We call parameter n the security parameter, *c the* public parameter length, *s the* seed length, *m the* number of blocks, *and $\ell(n) = \max_{(z,x)\in\{0,1\}^{c(n)}\times\{0,1\}^{s(n)},i\in[m(n)]} |G(z,x)_i|$ the* maximal block length *of G.*

Definition 9 (Real sample-entropy). *Let G be an m-block generator over $\{0,1\}^c \times \{0,1\}^s$, let $n \in \mathbb{N}$, let Z_n and X_n be uniformly distributed over $\{0,1\}^{c(n)}$ and $\{0,1\}^{s(n)}$, respectively, and let $\mathbf{Y}_n = (Y_1, \ldots, Y_m) = G(Z_n, X_n)$. For $n \in \mathbb{N}$ and $i \in [m(n)]$, define the* real sample-entropy of $\mathbf{y} \in \operatorname{Supp}(Y_1, \ldots, Y_i)$ given *$z \in \operatorname{Supp}(Z_n)$ as*

$$\operatorname{RealH}_{G,n}(\mathbf{y}|z) = \sum_{j=1}^{i} \mathsf{H}_{Y_j|Z_n,Y_{<j}}(\mathbf{y}_j|z, \mathbf{y}_{<j}).$$

We omit the security parameter from the above notation when clear from the context.

Definition 10 (Real entropy). *Let G be an m-block generator, and let Z_n and $\mathbf{Y}_n$ be as in Definition 9. Generator G has* real entropy at least $k = k(n)$, *if*

$$\mathop{\mathbf{E}}_{(z,\mathbf{y})\leftarrow(Z_n,\mathbf{Y}_n)}[\operatorname{RealH}_{G,n}(\mathbf{y}|z)] \geq k(n)$$

for every $n \in \mathbb{N}$.

The generator G has real min-entropy at least $k(n)$ in its i'th block *for some $i = i(n) \in [m(n)]$, if*

$$\Pr_{(z,\mathbf{y})\leftarrow(Z_n,\mathbf{Y}_n)}\left[\mathsf{H}_{Y_i|Z_n,Y_{<i}}(\mathbf{y}_i|z, \mathbf{y}_{<i}) < k(n)\right] = \mathsf{negl}(n).$$

We say the above bounds are invariant to the public parameter *if they hold for any fixing of the public parameter Z_n.*[4]

It is known that the real Shannon entropy amounts to measuring the standard conditional Shannon entropy of G's output blocks.

Lemma 2 ([17, Lemma 3.4]). *Let G, Z_n and $\mathbf{Y}_n$ be as in Definition 9 for some $n \in \mathbb{N}$, then*

$$\mathop{\mathbf{E}}_{(z,\mathbf{y})\leftarrow(Z_n,\mathbf{Y}_n)}[\operatorname{RealH}_{G,n}(\mathbf{y}|z)] = \mathsf{H}(\mathbf{Y}_n|Z_n).$$

[4] In particular, this is the case when there is no public parameter, i.e., $c = 0$.

Toward the definition of *inaccessible entropy*, we first define *online block-generators* which are a special type of block generators that toss fresh random coins before outputting each new block.

Definition 11 (Online block generator). *Let n be a security parameter, and let $c = c(n)$ and $m = m(n)$. An m-block* online *generator is a function $\widetilde{G}\colon \{0,1\}^c \times (\{0,1\}^v)^m \mapsto (\{0,1\}^*)^m$ for some $v = v(n)$, such that the i'th output block of $\widetilde{G}$ is a function of (only) its first i input blocks. We denote the* transcript *of $\widetilde{G}$ over random input by $T_{\widetilde{G}}(1^n) = (Z, R_1, Y_1, \ldots, R_m, Y_m)$, for $Z \leftarrow \{0,1\}^c$, $(R_1, \ldots, R_m) \leftarrow (\{0,1\}^v)^m$ and $(Y_1, \ldots, Y_m) = \widetilde{G}(Z, R_1, \ldots, R_i)$.*

That is, an online block generator is a special type of block generator that tosses fresh random coins before outputting each new block. In the following, we let $\widetilde{G}(z, r_1, \ldots, r_i)_i$ stand for $\widetilde{G}(z, r_1, \ldots, r_i, x^*)_i$ for arbitrary $x^* \in (\{0,1\}^v)^{m-i}$ (note that the choice of x^* has no effect on the value of $\widetilde{G}(z, r_1, \ldots, r_i, x^*)_i$).

Definition 12 (Accessible sample-entropy). *Let n be a security parameter, and let $\widetilde{G}$ be an online $m = m(n)$-block online generator. The* accessible sample-entropy of $\mathbf{t} = (z, r_1, y_1, \ldots, r_m, y_m) \in \mathrm{Supp}(Z, R_1, Y_1 \ldots, R_m, Y_m) = T_{\widetilde{G}}(1^n)$ *is defined by*

$$\mathrm{AccH}_{\widetilde{G},n}(\mathbf{t}) = \sum_{i=1}^{m} \mathsf{H}_{Y_i|Z,R_{<i}}(y_i|z, r_{<i}).$$

Again, we omit the security parameter from the above notation when clear from the context.

As in the case of real entropy, the expected accessible entropy of a random transcript can be expressed in terms of the standard conditional Shannon entropy.

Lemma 3 ([17, Lemma 3.7])**.** *Let $\widetilde{G}$ be an online m-block generator and let $(Z, R_1, Y_1, \ldots, R_m, Y_m) = T_{\widetilde{G}}(1^n)$ be its transcript. Then,*

$$\mathop{\mathbf{E}}_{\mathbf{t} \leftarrow T_{\widetilde{G}}(Z,1^n)}\left[\mathrm{AccH}_{\widetilde{G}}(\mathbf{t})\right] = \sum_{i \in [m]} \mathsf{H}(Y_i|Z, R_{<i}).$$

We focus on efficient generators that are consistent with respect to G. That is, the support of their output is contained in that of G.

Definition 13 (Consistent generators). *Let G be a block generator over $\{0,1\}^{c(n)} \times \{0,1\}^{s(n)}$. A block (possibly online) generator G' over $\{0,1\}^{c(n)} \times \{0,1\}^{s'(n)}$ is G* consistent *if, for every $n \in \mathbb{N}$, it holds that $\mathrm{Supp}(G'(U_{c(n)}, U_{s'(n)})) \subseteq \mathrm{Supp}(G(U_{c(n)}, U_{s(n)}))$.*

Definition 14 (Accessible entropy). *A block generator G has* accessible entropy at most *$k = k(n)$ if, for every efficient G-consistent, online generator $\widetilde{G}$ and all large enough n,*

$$\mathop{\mathbf{E}}_{\mathbf{t} \leftarrow T_{\widetilde{G}}(1^n)}\left[\mathrm{AccH}_{\widetilde{G}}(\mathbf{t})\right] \le k.$$

We call a generator whose real entropy is noticeably higher than it accessible entropy an inaccessible entropy generator.

We use the following reduction from inaccessible entropy generators to constant round statistically hiding commitment.

Theorem 5 ([17, Theorem 6.24]). *Let G be an efficient block generator with constant number of blocks. Assume G's real Shannon entropy is at least $k(n)$ for some efficiently computable function k, and that its accessible entropy is bounded by $k(n) - 1/p(n)$ for some $p \in \mathsf{poly}$. Then there exists a constant-round statistically hiding and computationally binding commitment scheme. Furthermore, if the bound on the real entropy is invariant to the public parameter, then the commitment is receiver public-coin.*

Remark 1 (Inaccessible max/average entropy). Our result relies on the reduction from inaccessible *Shannon* entropy generators to statistically hiding commitments, given in [17]. The proof of this reduction follows closely the proof in previous versions [18,19], where the reduction was from inaccessible *max* entropy generators. The extension to Shannon entropy generators is essential for our result.

4.2 From dCRH to Inaccessible Entropy Generators – Proof of Theorem 3

In this section we show that there is a block generator with two blocks in which there is a gap between the real entropy and the accessible entropy. Let $\mathcal{H} = \{\mathcal{H}_n : \{0,1\}^n \to \{0,1\}^m\}_{n\in\mathbb{N}}$ be a dCRH for $m = m(n)$ and assume that each $h \in \mathcal{H}_n$ requires $c = c(n)$ bits to describe. By Definition 7, there exists a polynomial $p(\cdot)$ such that for any probabilistic polynomial-time algorithm A, it holds that

$$\boldsymbol{\Delta}\left((h, \mathsf{A}(1^n, h)), (h, \mathsf{Col}(h))\right) = \mathop{\mathbf{E}}_{h \leftarrow \mathcal{H}_n}\left[\boldsymbol{\Delta}\left(\mathsf{A}(1^n, h), \mathsf{Col}(h)\right)\right] \geq \frac{1}{p(n)}$$

for large enough $n \in \mathbb{N}$, where $h \leftarrow \mathcal{H}_n$.

The generator $G\colon \{0,1\}^c \times \{0,1\}^n \to \{0,1\}^m \times \{0,1\}^n$ is defined by

$$G(h, x) = (h(x),\ x).$$

The public parameter length is c (this is the description size of h), the generator consists of two blocks, and the maximal block length is $\max\{n, m\}$. Since the random coins of G define x and x is completely revealed, the real Shannon entropy of G is n. That is,

$$\mathop{\mathbf{E}}_{y \leftarrow G(U_c, U_n)}\left[\mathsf{RealH}_G(y)\right] = n.$$

Our goal in the remaining of this section is to show a non-trivial upper bound on the accessible entropy of G. We prove the following lemma.

Lemma 4. *There exists a polynomial $q(\cdot)$ such that for every G-consistent online generator $\widetilde{G}$, it holds that*

$$\mathop{\mathbf{E}}_{t \leftarrow T_{\widetilde{G}}(Z,1^n)} \left[\mathsf{AccH}_{\widetilde{G}}(t) \right] \leq n - \frac{1}{q(n)}$$

for all large enough $n \in \mathbb{N}$.

Proof. Fix a G-consistent online generator $\widetilde{G}$. Let us denote by Y a random variable that corresponds to the first part of G's output (i.e., the first m bits) and by X the second part (i.e., the last n bits). Denote by R the randomness used by the adversary to sample Y. Denote by Z the random variable that corresponds to the description of the hash function h. Fix $q(n) \triangleq 4 \cdot p(n)^2$ Assume towards contradiction that for infinitely many n's it holds that

$$\mathop{\mathbf{E}}_{t \leftarrow T_{\widetilde{G}}(Z,1^n)} \left[\mathsf{AccH}_{\widetilde{G}}(t) \right] > n - \frac{1}{q(n)}.$$

By Lemma 3, this means that

$$\mathsf{H}(Y \mid Z) + \mathsf{H}(X \mid Y, Z, R) > n - \frac{1}{q(n)} \tag{1}$$

We show how to construct an adversary A that can break the security of the dCRH. The algorithm A, given a hash function $h \leftarrow \mathcal{H}$, does the following:

1. Sample r and let $y = \widetilde{G}(h, r)_1$
2. Sample r_1, r_2 and output $x_1 = \widetilde{G}(h, r, r_1)_2$ and $x_2 = \widetilde{G}(h, r, r_2)_2$.

In other words, A tries to create a collision by running G to get the first block, y, and then running it twice (by rewinding) to get two inputs x_1, x_2 that are mapped to y. Indeed, A runs in polynomial-time and if $\widetilde{G}$ is G-consistent, then x_1 and x_2 collide relative to h. Denote by Y^{A}, X_1^{A}, and X_2^{A} be random variables that correspond to the output of the emulated $\widetilde{G}$. Furthermore, denote by $(X_1^{\mathsf{Col}}, X_2^{\mathsf{Col}})$ a random collision that $\mathsf{Col}(h)$ samples. To finish the proof it remains to show that

$$\mathop{\mathbf{E}}_{h \leftarrow \mathcal{H}_n} \left[\mathbf{\Delta}((X_1^{\mathsf{A}}, X_2^{\mathsf{A}}), (X_1^{\mathsf{Col}}, X_2^{\mathsf{Col}})) \right] \leq \frac{1}{p(n)}$$

which is a contradiction.

By Pinsker's inequality (Proposition 2) and the chain rule from Proposition 1, it holds that

$$\begin{aligned}
\mathbf{\Delta}\left((X_1^{\mathsf{A}}, X_2^{\mathsf{A}}), (X_1^{\mathsf{Col}}, X_2^{\mathsf{Col}})\right) &\leq \sqrt{\frac{\ln(2)}{2} \cdot \mathbf{D}_{\mathsf{KL}}(X_1^{\mathsf{A}}, X_2^{\mathsf{A}} \| X_1^{\mathsf{Col}}, X_2^{\mathsf{Col}})} \\
&= \sqrt{\mathbf{D}_{\mathsf{KL}}\left(X_1^{\mathsf{A}} \| X_1^{\mathsf{Col}}\right) + \mathop{\mathbf{E}}_{x_1 \leftarrow X_1^{\mathsf{A}}} \left[\mathbf{D}_{\mathsf{KL}}(X_2^{\mathsf{A}}|_{X_1^{\mathsf{A}}=x_1} \| X_2^{\mathsf{Col}}|_{X_1^{\mathsf{Col}}=x_1}) \right]} \\
&\leq \sqrt{\mathbf{D}_{\mathsf{KL}}\left(X_1^{\mathsf{A}} \| X_1^{\mathsf{Col}}\right)} + \sqrt{\mathop{\mathbf{E}}_{x_1 \leftarrow X_1^{\mathsf{A}}} \left[\mathbf{D}_{\mathsf{KL}}(X_2^{\mathsf{A}}|_{X_1^{\mathsf{A}}=x_1} \| X_2^{\mathsf{Col}}|_{X_1^{\mathsf{Col}}=x_1}) \right]}.
\end{aligned}$$

Hence, by Jensen's inequality (Proposition 3), it holds that

$$\mathop{\mathbf{E}}_{h\leftarrow\mathcal{H}_n}\left[\boldsymbol{\Delta}((X_1^{\mathsf{A}},X_2^{\mathsf{A}}),(X_1^{\mathsf{Col}},X_2^{\mathsf{Col}}))\right] \leq \sqrt{\mathop{\mathbf{E}}_{h\leftarrow\mathcal{H}_n}\left[\mathbf{D}_{\mathsf{KL}}(X_1^{\mathsf{A}}\|X_1^{\mathsf{Col}})\right]+} \\ \sqrt{\mathop{\mathbf{E}}_{\substack{h\leftarrow\mathcal{H}_n\\ x_1\leftarrow X_1^{\mathsf{A}}}}\left[\mathbf{D}_{\mathsf{KL}}(X_2^{\mathsf{A}}|_{X_1^{\mathsf{A}}=x_1}\|X_2^{\mathsf{Col}}|_{X_1^{\mathsf{Col}}=x_1})\right]}.$$

We complete the proof using the following claims.

Claim 1. *It holds that*

$$\mathop{\mathbf{E}}_{h\leftarrow\mathcal{H}_n}\left[\mathbf{D}_{\mathsf{KL}}(X_1^{\mathsf{A}}\|X_1^{\mathsf{Col}})\right] \leq \frac{1}{p(n)^2}.$$

Claim 2. *It holds that*

$$\mathop{\mathbf{E}}_{\substack{h\leftarrow\mathcal{H}_n\\ x_1\leftarrow X_1^{\mathsf{A}}}}\left[\mathbf{D}_{\mathsf{KL}}(X_2^{\mathsf{A}}|_{X_1^{\mathsf{A}}=x_1}\|X_2^{\mathsf{Col}}|_{X_1^{\mathsf{Col}}=x_1})\right] \leq \frac{1}{p(n)^2}.$$

Proof (Proof of Claim 1). Recall that X_1^{Col} is the *uniform* distribution over the inputs of the hash function and thus

$$\mathbf{D}_{\mathsf{KL}}(X_1^{\mathsf{A}}\|X_1^{\mathsf{Col}}) = \sum_x \mathbf{Pr}[X_1^{\mathsf{A}}=x]\cdot\log\frac{\mathbf{Pr}[X_1^{\mathsf{A}}=x]}{2^{-n}} = n - \mathsf{H}(X_1^{\mathsf{A}}).$$

To sample X_1^{A}, the algorithm A first runs $\widetilde{G}(r)_1$ to get y and then runs $G(r,r_1)$ to get x_1. Thus, by Eq. (1), it holds that

$$\mathop{\mathbf{E}}_{h\leftarrow\mathcal{H}_n}\left[\mathsf{H}(X_1^{\mathsf{A}})\right] = \mathop{\mathbf{E}}_{h\leftarrow\mathcal{H}_n}[\mathsf{H}(X)] = \mathsf{H}(X,Y\mid Z) = \mathsf{H}(Y\mid Z) + \mathsf{H}(X\mid Y,Z,R) \geq n - \frac{1}{q(n)},$$

where the second equality follows since $\widetilde{G}$ is G-consistent and thus X fully determines Y. This implies that

$$\mathop{\mathbf{E}}_{h\leftarrow\mathcal{H}_n}\left[\mathbf{D}_{\mathsf{KL}}(X_1^{\mathsf{A}}\|X_1^{\mathsf{Col}})\right] \leq \frac{1}{q(n)} = \frac{1}{p(n)^2},$$

as required.

Proof (Proof of Claim 2).
For $x_1 \in \mathsf{supp}(X_1^{\mathsf{A}})$, it holds that

$$\mathbf{D}_{\mathsf{KL}}(X_2^{\mathsf{A}}|_{X_1^{\mathsf{A}}=x_1}\|X_2^{\mathsf{Col}}|_{X_1^{\mathsf{Col}}=x_1}) = \sum_x \mathbf{Pr}\left[X_2^{\mathsf{A}}=x|_{X_1^{\mathsf{A}}=x_1}\right]\cdot\log\frac{\mathbf{Pr}\left[X_2^{\mathsf{A}}=x|_{X_1^{\mathsf{A}}=x_1}\right]}{|h^{-1}(h(x_1))|^{-1}} \\ = \log|h^{-1}(h(x_1))| - \mathsf{H}(X_2^{\mathsf{A}}|_{X_1^{\mathsf{A}}=x_1}).$$

Hence,

$$\mathop{\mathbf{E}}_{\substack{h \leftarrow \mathcal{H}_n \\ x_1 \leftarrow X_1^{\mathsf{A}}}} \left[\mathbf{D}_{\mathsf{KL}}(X_2^{\mathsf{A}}|_{X_1^{\mathsf{A}}=x_1} \| X_2^{\mathsf{Col}}|_{X_1^{\mathsf{Col}}=x_1}) \right] = \mathop{\mathbf{E}}_{\substack{h \leftarrow \mathcal{H}_n \\ x_1 \leftarrow X_1^{\mathsf{A}}}} \left[\log |h^{-1}(h(x_1))| - \mathsf{H}(X_2^{\mathsf{A}}|_{X_1^{\mathsf{A}}=x_1}) \right].$$

Notice that the distribution of X_2^{A} only depends on $y = h(x_1)$, that is, $X_2^{\mathsf{A}}|_{X_1^{\mathsf{A}}=x_1}$ is distributed exactly as $X_2^{\mathsf{A}}|_{X_1^{\mathsf{A}}=x_1'}$ for every x_1 and x_1' that such that $y = h(x_1) = h(x_1')$. Thus, we have that $X_2^{\mathsf{A}}|_{X_1^{\mathsf{A}}=x_1}$ is distributed exactly as $X|_{Y=y}$ and the distribution of $h(X_1)$ is distributed as Y. Namely,

$$\begin{aligned}
\mathop{\mathbf{E}}_{\substack{h \leftarrow \mathcal{H}_n \\ x_1 \leftarrow X_1^{\mathsf{A}}}} \left[\mathbf{D}_{\mathsf{KL}}(X_2^{\mathsf{A}}|_{X_1^{\mathsf{A}}=x_1} \| X_2^{\mathsf{Col}}|_{X_1^{\mathsf{Col}}=x_1}) \right] &= \mathop{\mathbf{E}}_{\substack{h \leftarrow \mathcal{H}_n \\ x_1 \leftarrow X_1^{\mathsf{A}}}} \left[\log |h^{-1}(y)| \right] - \mathop{\mathbf{E}}_{h \leftarrow \mathcal{H}_n} [\mathsf{H}(X \mid Y, R)] \\
&= \mathop{\mathbf{E}}_{\substack{h \leftarrow \mathcal{H}_n \\ x_1 \leftarrow X_1^{\mathsf{A}}}} \left[\log |h^{-1}(y)| \right] - \mathsf{H}(X \mid Y, Z, R) \\
&\leq \mathop{\mathbf{E}}_{\substack{h \leftarrow \mathcal{H}_n \\ x_1 \leftarrow X_1^{\mathsf{A}}}} \left[\log |h^{-1}(y)| \right] + \mathsf{H}(Y \mid Z) - n + \frac{1}{q(n)} \\
&= \frac{1}{q(n)},
\end{aligned}$$

where the first inequality follows by Eq. (1) and the second follows since

$$\begin{aligned}
\mathop{\mathbf{E}}_{\substack{h \leftarrow \mathcal{H}_n \\ y \leftarrow Y}} \left[\log |h^{-1}(y)| \right] + \mathsf{H}(Y \mid Z) &= \mathop{\mathbf{E}}_{\substack{h \leftarrow \mathcal{H}_n \\ y \leftarrow Y}} \left[\log |h^{-1}(y)| + \mathsf{H}_Y(y) \right] \\
&= \mathop{\mathbf{E}}_{\substack{h \leftarrow \mathcal{H}_n \\ y \leftarrow Y}} \left[\log \frac{|h^{-1}(y)|}{\mathbf{Pr}[Y = y]} \right] \\
&\leq \log \mathop{\mathbf{E}}_{\substack{h \leftarrow \mathcal{H}_n \\ y \leftarrow Y}} \left[\frac{|h^{-1}(y)|}{\mathbf{Pr}[Y = y]} \right] = n,
\end{aligned}$$

where the inequality is by Jensen's inequality (Proposition 3). Thus, overall

$$\mathop{\mathbf{E}}_{\substack{h \leftarrow \mathcal{H}_n \\ x_1 \leftarrow X_1^{\mathsf{A}}}} \left[\mathbf{D}_{\mathsf{KL}}(X_2^{\mathsf{A}}|_{X_1^{\mathsf{A}}=x_1} \| X_2^{\mathsf{Col}}|_{X_1^{\mathsf{Col}}=x_1}) \right] \leq \frac{1}{q(n)} = \frac{1}{p(n)^2},$$

as required.

4.3 From Statistically Hiding Commitments to dCRH– Proof of Theorem 4

Let $\pi = (\mathcal{S}, \mathcal{R}, \mathcal{V})$ be a binding and statistically hiding two-message commitment scheme. We show that there exists a dCRH family $\mathcal{H}$.

To sample a hash function in the family with security parameter n, we use the receiver's first message of the protocol. Namely, we set the hash function as

$h \leftarrow \mathcal{R}(1^n)$. Then, to evaluate h on input x we first parse x as $x = (b, r)$, where b is a bit, and output a commitment to the bit b using randomness r, with respect to the receiver message h. That is, we set

$$h(x) = \mathcal{S}(h, b; r).$$

Since π is efficient, then sampling and evaluating h are polynomial-time procedures. This concludes the definition of our family $\mathcal{H}$ of hash functions. (Note that the functions in the family are not necessarily compressing).

We next argue security. Suppose toward contradiction that $\mathcal{H}$ is not a dCRH according to Definition 7. Then, for any $\delta(n) = n^{-O(1)}$ there exists an adversary A, such that

$$\mathbf{\Delta}\left((h, \mathsf{A}(1^n, h)), (h, \mathsf{Col}(h))\right) \leq \delta, \tag{2}$$

for infinitely many n's. From hereon, we fix δ to be any function such that $n^{-O(1)} < \delta < \frac{1}{2} - n^{-O(1)}$.

We show how to use A to break the binding property of the commitment scheme. Our cheating receiver $\mathcal{R}^*$ is defined as follows: On input h, $\mathcal{R}^*$ runs $\mathsf{A}(h)$ to get x and x', interprets $x = (b, r)$ and $x' = (b', r')$ and outputs b and b' along with their openings r and r', respectively. Our goal is to show that $x = (b, r)$ and $x' = (b', r')$ are two valid distinct openings to the commitment scheme.

By Eq. (2), it suffices to analyze the success probability when the pair (x, x') is sampled according to the distribution Col_h, and show that it is at least $1/2 - \mathsf{negl}(n)$. From the definition of Col_h, we have that $h(x) = h(x')$ and thus $\mathcal{S}(h, b; r) = \mathcal{S}(h, b'; r') := y$. In other words, the second message of the protocol for b with randomness r and b' with randomness r' are the same, and thus both pass as valid openings in the reveal stage of the protocol: $\mathcal{V}(h, y, b, r) = 1$ and $\mathcal{V}(h, y, b', r') = 1$.

We are left to show that these are two *distinct* openings for the commitment, namely, $b \neq b'$. To show this, we use the statistically hiding property of the commitment scheme. The following claim concludes the proof.

Claim. Fix any h. Then for $((b, r), (b', r')) \leftarrow \mathsf{Col}(h)$ it holds that $\mathbf{Pr}[b \neq b'] \geq 1/2 - \mathsf{negl}(n)$.

Proof. Let B be the uniform distribution on bits and R the uniform distribution on commitment randomness. For every commitment c, let B_c be the distribution on bits given by sampling $(b, r) \leftarrow (B, R)$ conditioned on $\mathcal{S}(h, b; r) = c$. Let C be the distribution on random commitments to a random bit.

By the statistical hiding property of the commitment scheme,

$$\Delta((\mathcal{S}(h, B, R), B), (\mathcal{S}(h, B', R), B)) \leq \varepsilon,$$

where B' is an independent copy of B, and $\epsilon = \mathsf{negl}(n)$ is a negligible function. Furthermore,

$$\Delta((\mathcal{S}(h, B, R), B), (\mathcal{S}(h, B', R), B)) = \Delta((C, B_C), (C, B)) = \mathop{\mathbf{E}}_{c \leftarrow C}\left[\Delta(B_c, B)\right].$$

By Markov's inequality, it holds that

$$\Pr_{c \leftarrow C}\left[\Delta(B_c, B) \geq \sqrt{\varepsilon}\right] \leq \sqrt{\varepsilon}.$$

To conclude the proof note that

$$\mathbf{Pr}[b = b' : (b,r),(b',r') \leftarrow \mathsf{Col}_h] = \mathbf{Pr}\left[b = b' : \begin{array}{l}(b,r) \leftarrow (B,R)\\ c = \mathcal{S}(h,b;r)\\ b' \leftarrow B_c\end{array}\right] \leq$$

$$\mathbf{Pr}\left[b = b' : \begin{array}{l}(b,r) \leftarrow (B,R)\\ c = \mathcal{S}(h,b;r)\\ b' \leftarrow B_c\\ \Delta(B_c,B) \leq \sqrt{\varepsilon}\end{array}\right] + \Pr_{c \leftarrow C}\left[\Delta(B_c, B) \geq \sqrt{\varepsilon}\right] \leq$$

$$\left(\frac{1}{2} + \sqrt{\varepsilon}\right) + \sqrt{\varepsilon} = \frac{1}{2} + \mathsf{negl}(n).$$

Overall, the success probability of A is at least $1/2 - \mathsf{negl}(n) - \delta \geq n^{-O(1)}$.

Using string commitments. The above proof constructs dCRH from statistically hiding *bit* commitment schemes. For schemes that support commitments to *strings*, following the above proof gives a stronger notion of dCRH, where the adversary's output distribution is $(1 - \mathsf{negl}(n))$-far from a random collision distribution.

Technically, the change in the proof is to interpret b in $x = (b, r)$ as a string of length n, rather than as a single bit. The proof remains the same except that the probability that $b = b'$ is (negligibly close to) 2^{-n} instead of $1/2$. Thus, overall the success probability of A is at least $1 - \mathsf{negl}(n) - \delta$. To ensure a polynomial success probability we can allow any $\delta = 1 - n^{-O(1)}$.

5 From SZK-Hardness to Statistically Hiding Commitments

In this section, we give a direct construction of a constant-round statistically hiding commitment from average-case hardness in SZK. This gives an alternative proof to Corollary 1.

5.1 Hard on Average Promise Problems

Definition 15. *A promise problem (Π_Y, Π_N) consists of two disjoint sets of* yes instances *Π_Y and* no instances *Π_N.*

Definition 16. *A promise problem (Π_Y, Π_N) is hard on average if there exists a probabilistic polynomial-time sampler Π with support $\Pi_Y \cup \Pi_N$, such that for any probabilistic polynomial-time decider D, there exists a negligible function $\mathsf{negl}(n)$, such that*

$$\Pr_{r \leftarrow \{0,1\}^n}\left[x \in \Pi_{D(x)} \mid x \leftarrow \Pi(r)\right] \leq \frac{1}{2} + \mathsf{negl}(n).$$

5.2 Instance-Dependent Commitments

Definition 17 ([31]). *An instance-dependent commitment scheme $\mathcal{IDC}$ for a promise problem (Π_Y, Π_N) is a commitment scheme where all algorithms get as auxiliary input an instance $x \in \{0,1\}^*$. The induced family of schemes $\{\mathcal{IDC}_x\}_{x\in\{0,1\}^*}$ is*

- *statistically binding when $x \in \Pi_N$,*
- *statistically hiding when $x \in \Pi_Y$.*

Theorem 6 ([31]). *Any promise problem $(\Pi_Y, \Pi_N) \in$ SZK has a constant-round instance-dependent commitment.*

5.3 Witness-Indistinguishable Proofs

Definition 18. *A proof system $\mathcal{WI}$ for an NP relation R is witness indistinguishable if for any x, w_0, w_1 such that $(x, w_0), (x, w_1) \in R$, the verifier's view given a proof using w_0 is computationally indistinguishable from its view given a proof using w_1.*

Constant-round $\mathcal{WI}$ proofs systems are known from any constant-round statistically-binding commitments [13]. Statistically-binding commitments can be constructed from one-way functions [28], and thus can also be obtained from average-case hardness in SZK [32].

Theorem 7 ([13,28,32]). *Assuming hard-on-average problems in SZK, there exist constant-round witness-indistinguishable proof systems.*

5.4 The Commitment Protocol

Here, we give the details of our protocol. Our protocol uses the following ingredients and notation:

- A $\mathcal{WI}$ proof for NP.
- A hard-on average SZK problem (Π_Y, Π_N) with sampler Π.
- An instance-dependent commitment scheme $\mathcal{IDC}$ for Π.

We describe the commitment scheme in Fig. 1.

5.5 Analysis

Proposition 4. *Protocol 1 is computationally binding.*

Proof. Let $\mathcal{S}^*$ be any probabilistic polynomial-time sender that breaks binding in Protocol 1 with probability ε. We use $\mathcal{S}^*$ to construct a probabilistic polynomial-time decider D for the SZK problem Π with advantage $\varepsilon/4n - \mathsf{negl}(n)$.

Protocol 1

Sender input: a bit $m \in \{0,1\}$.
Common input: security parameter 1^n.

Coin tossing into the well

- $\mathcal{R}$ samples $2n$ independent random strings $\rho_{i,b} \leftarrow \{0,1\}^n$, for $i \in [n], b \in \{0,1\}$.
- The parties then execute (in parallel) $2n$ statistically-binding commitment protocols $\mathcal{SBC}$ in which $\mathcal{R}$ commits to each of the strings $\rho_{i,b}$. We denote the transcript of each such commitment by $C_{i,b}$.
- $\mathcal{S}$ samples $2n$ independent random strings $\sigma_{i,b} \leftarrow \{0,1\}^n$, and sends them to $\mathcal{R}$.
- $\mathcal{R}$ sets $r_{i,b} = \rho_{i,b} \oplus \sigma_{i,b}$.

Generating hard instances

- $\mathcal{R}$ generates $2n$ instances $x_{i,b} \leftarrow \Pi(r_{i,b})$, using the strings $r_{i,b}$ as randomness, and sends the instances to $\mathcal{S}$.
- The parties then execute a $\mathcal{WI}$ protocol in which $\mathcal{R}$ proves to $\mathcal{S}$ that there exists a $b \in \{0,1\}$ such that for all $i \in [n]$, $x_{i,b}$ was generated consistently. That is, there exist strings $\{\rho_{i,b}\}_{i\in[n]}$ that are consistent with the receiver's commitments $\{C_{i,b}\}_{i\in[n]}$, and $x_{i,b} = \Pi(\rho_{i,b} \oplus \sigma_{i,b})$.
  As the witness, $\mathcal{R}$ uses $b = 0$ and the strings $\{\rho_{i,0}\}_{i\in[n]}$ sampled earlier in the protocol.

Instance-binding commitment

- The sender samples $2n$ random bits $m_{i,b}$ subject to $m = \bigoplus_{i,b} m_{i,b}$.
- The parties then execute (in parallel) $2n$ instance-dependent commitment protocols $\mathcal{IDC}_{x_{i,b}}$ in which $\mathcal{S}$ commits to each bit $m_{i,b}$ using the instance $x_{i,b}$.

Fig. 1. A constant round statistically hiding commitment from SZK hardness.

Given an instance $x \leftarrow \Pi$, the decider D proceeds as follows:

- It samples at random $i^* \in [n]$ and $b^* \in \{0,1\}$.
- It executes the protocol $(\mathcal{S}^*, \mathcal{R})$ with the following exceptions:
  - The instance x_{i^*,b^*}, generated by $\mathcal{R}$, is replaced with the instance x, given to D as input.
  - In the $\mathcal{WI}$ protocol, as the witness we use $1 \oplus b^*$ and the strings $\{\rho_{i,1\oplus b^*}\}_{i\in[n]}$ (instead of 0 and the strings $\{\rho_{i,0}\}_{i\in[n]}$).
- Then, at the opening phase, if $\mathcal{S}^*$ equivocally opens the (i^*, b^*)-th instance-dependent commitment, D declares that $x \in \Pi_Y$. Otherwise, it declares that $x \in \Pi_\beta$ for a random $\beta \in \{Y, N\}$.

Analyzing D's advantage. Denote by E the event that in the above experiment $\mathcal{S}^*$ equivocally opens the (i^*, b^*)-th instance-dependent commitment. We first observe that the advantage of D in deciding Π is at least as large as the probability that E occurs.

Claim 3. $\mathbf{Pr}\left[x \in \Pi_{D(x)}\right] \geq \frac{1+\mathbf{Pr}[E]}{2} - \mathsf{negl}(n)$.

Proof. By the definition of D,

$$\mathbf{Pr}\left[x \in \Pi_{D(x)} \mid E\right] = \mathbf{Pr}[x \in \Pi_Y \mid E] = 1 - \mathbf{Pr}[x \in \Pi_N \mid E] \geq 1 - \frac{\mathbf{Pr}[E \mid x \in \Pi_N]}{\mathbf{Pr}[E]},$$

$$\mathbf{Pr}\left[x \in \Pi_{D(x)} \mid \overline{E}\right] = \frac{1}{2}.$$

Furthermore, if $x \in \Pi_N$ (namely, it is a no instance), then $\mathcal{IDC}_x$ is binding, and thus

$$\mathbf{Pr}[E \mid x \in \Pi_N] = \mathsf{negl}(n).$$

Claim 3 now follows by the law of total probability.

From hereon, we focus on showing that E occurs with high probability.

Claim 4. $\mathbf{Pr}[E] \geq \frac{\varepsilon}{2n} - \mathsf{negl}(n)$.

Proof. To prove the claim, we consider hybrid experiments $\mathcal{H}_0, \ldots, \mathcal{H}_4$, and show that the view of the sender $\mathcal{S}^*$ changes in a computationally indistinguishable manner throughout the hybrids. We then bound the probability that E occurs in the last hybrid experiment.

$\mathcal{H}_0$: In this experiment, we consider an execution of $D(x)$ as specified above.

$\mathcal{H}_1$: Here x is not sampled ahead of time, but rather first the value σ_{i^*,b^*} is obtained from $\mathcal{S}^*$, then a random value $\rho' \leftarrow \{0,1\}^n$ is sampled, and x is sampled using randomness $r_{i^*,b^*} = \sigma_{i^*,b^*} \oplus \rho'$. Since ρ' is sampled independently of the rest of the experiment, the sender's view in $\mathcal{H}_1$ is identically distributed to its view in $\mathcal{H}_0$.

$\mathcal{H}_2$: Here the (i^*, b^*)-th commitment to ρ_{i^*,b^*} is replaced with a commitment to ρ'. By the (computational) hiding of the commitment $\mathcal{SBC}$, the sender's view in $\mathcal{H}_2$ is computationally indistinguishable from its view in $\mathcal{H}_1$.

$\mathcal{H}_3$: Here, in the $\mathcal{WI}$ protocol, instead of using as the witness $1 \oplus b^*$ and the strings $\{\rho_{i,1\oplus b^*}\}_i$, we use 0 and the strings $\{\rho_{i,0}\}_i$. By the (computational) witness-indistinguishability of the protocol, the sender's view in $\mathcal{H}_3$ is computationally indistinguishable from its view in $\mathcal{H}_2$.

$\mathcal{H}_4$: In this experiment, we consider a standard execution of the protocol between $\mathcal{S}^*$ and $\mathcal{R}$ (without any exceptions). The sender's view in this hybrid is identical to its view in $\mathcal{H}_3$ (by renaming $\rho' = \rho_{i^*,b^*}$ and $x = x_{i^*,b^*}$).

It is left to bound from below the probability that E occurs in $\mathcal{H}_4$. That is, when we consider a standard execution of $(\mathcal{S}^*, \mathcal{R})$ and sample (i^*, b^*) independently at random.

Indeed, note that since the plaintext bit m is uniquely determined by the bits $\{m_{i,b}\}_{i,b}$. Whenever $\mathcal{S}^*$ equivocally opens the commitment to two distinct bits, there exists (at least one) (i, b) such that $\mathcal{S}^*$ equivocally opens the (i, b)-th instance-dependent commitment. Since in a standard execution $\mathcal{S}^*$ equivocally opens the commitment with probability at least ε, and (i^*, b^*) is sampled independently, E occurs in this experiment with probability at least $\frac{\varepsilon}{2n}$.

Claim 4 follows.

This completes the proof that the scheme is binding.

Proposition 5. *Protocol 1 is statistically hiding.*

Proof. Let $\mathcal{R}^*$ be any (computationally unbounded) receiver. We show that the view of $\mathcal{R}^*$ given a commitment to $m = 0$ is statistically indistinguishable from its view given a commitment to $m = 1$.

For this purpose, consider the view of the receiver $\mathcal{R}^*$ after the coin tossing and instance-generation phase (and before the instance-dependent commitment phase). We shall refer to this as the *preamble view*. We say that the preamble view is *admissible*, if either of the following occurs:

- Let $\{x_{i,b}\}_{i,b}$ be the instances sent by $\mathcal{R}^*$. Then there exists i^*, b^* such that $x_{i^*,b^*} \in \Pi_Y$.
- The sender $\mathcal{S}$ rejects the $\mathcal{WI}$ proof that $\{x_{i,b}\}_{i,b}$ were properly generated.

To complete the proof, we show that the preamble view is admissible with overwhelming probability, and that conditioned on any admissible preamble view, the view of $\mathcal{R}^*$ given a commitment to $m = 0$ is statistically indistinguishable from its view given a commitment to $m = 1$. Since the preamble view is completely independent of m, the above two conditions are sufficient to establish statistical indistinguishability of the total views.

Claim 5. *The probability that the preamble view is not admissible is negligible.*

Proof. Let A be the event that the $\mathcal{WI}$ proof is accepted and let Y be the event that for some (i, b), $x_{i,b}$ is a yes instance. To show that the preamble view is not admissible with negligible probability, we would like to prove that

$$\mathbf{Pr}\left[A \wedge \overline{Y}\right] \leq \mathsf{negl}(n).$$

Let T be the event that the statement proven by $\mathcal{R}^*$ in the $\mathcal{WI}$ protocol is true. Namely, there exists $b \in \{0, 1\}$ such that all $\{x_{i,b}\}_i$ are generated consistently with the coin-tossing phase (and in particular where the coin-tossing phase consists of valid commitments $\{C_{i,b}\}_i$).

First, note that by the soundness of the $\mathcal{WI}$ system, the probability that the preamble is admissible, and in particular the proof is accepted, when the statement is false, is negligible:

$$\mathbf{Pr}\left[A \wedge \overline{T}\right] \leq \mathsf{negl}(n).$$

We now show:

$$\mathbf{Pr}\left[\overline{Y} \wedge T\right] \leq \mathsf{negl}(n).$$

For this purpose, fix any $\mathcal{SBC}$ commitments $\{C_{i,b}\}_{i,b}$. Let $F = F[\{C_{i,b}\}_{i,b}]$ be the event, over the sender randomness $\{\sigma_{i,b}\}_{i,b}$, that there exists $\beta \in \{0,1\}$ such that $\{C_{i,\beta}\}_i$ are valid commitments to strings $\{\rho_{i,\beta}\}_i$ and for all i, $\Pi(\rho_{i,\beta} \oplus \sigma_{i,\beta}) = x_{i,\beta} \in \Pi_N$. We show

$$\mathbf{Pr}[F] \leq 2^{-\Omega(n)}.$$

This is sufficient since

$$\mathbf{Pr}\left[\overline{Y} \wedge T\right] \leq \max_{\substack{C_{1,0}\ldots C_{n,0} \\ C_{1,1}\ldots C_{n,1}}} \mathbf{Pr}[F] \leq 2^{-\Omega(n)}.$$

To bound the probability that F occurs, fix any β and commitments $\{C_{i,\beta}\}_i$ to strings $\{\rho_{i,\beta}\}_i$. Then the strings $\rho_{i,\beta} \oplus \sigma_{i,\beta}$ are distributed uniformly and independently at random. Since $\Pi \in \Pi_Y$ with probability at least 0.49, and taking a union bound over both $\beta \in \{0,1\}$, the bound follows.

This concludes the proof of Claim 5.

Claim 6. *Fix any admissible preamble view V. Then, conditioned on V the view of $\mathcal{R}^*$ when given a commitment to $m = 0$ is statistically indistinguishable from its view when given a commitment to $m = 1$.*

Proof. If V is such that the $\mathcal{WI}$ proof is rejected then $\mathcal{S}$ aborts and the view of $\mathcal{R}^*$ remains independent of m. Thus, from hereon, we assume that the instances corresponding to V include an instance $x_{i^*,b^*} \in \Pi_Y$. In particular, the corresponding instance-dependent commitment $\mathcal{IDC}_{x_{i^*,b^*}}$ is statistically hiding.

It is left to note that in any execution $(\mathcal{S}, \mathcal{R}^*)$, with either $m \in \{0,1\}$, the bits $M_{-i} := \{m_{i,b}\}_{(i,b)\neq(i^*,b^*)}$ are distributed uniformly and independently at random. Conditioned on V and M_{-i}, only the bit

$$m_{i^*,b^*} = m \bigoplus_{m' \in M_{-i}} m'$$

depends on m. By the statistical hiding of $\mathcal{IDC}_{x_{i^*,b^*}}$ a commitment to $0 \bigoplus_{m' \in M_{-i}} m'$ is statistically indistinguishable from a commitment to $1 \bigoplus_{m' \in M_{-i}} m'$.

This concludes the proof of Claim 6.

Acknowledgments. Nir Bitansky is a member of the Check Point Institute of Information Security. Supported by ISF grant 18/484, the Alon Young Faculty Fellowship, and by Len Blavatnik and the Blavatnik Family foundation. Iftach Haitner is a member of the Check Point Institute for Information Security. Research supported by ERC starting grant 638121. Ilan Komargodski is supported in part by an AFOSR grant FA9550-15-1-0262. Eylon Yogev is supported by the European Union's Horizon 2020 research and innovation program under grant agreement No. 742754.

References

1. Applebaum, B., Haramaty, N., Ishai, Y., Kushilevitz, E., Vaikuntanathan, V.: Low-complexity cryptographic hash functions. In: 8th Innovations in Theoretical Computer Science Conference, ITCS, pp. 7:1–7:31 (2017)
2. Asharov, G., Segev, G.: Limits on the power of indistinguishability obfuscation and functional encryption. SIAM J. Comput. **45**(6), 2117–2176 (2016)
3. Barak, B.: How to go beyond the black-box simulation barrier. In: 42nd IEEE Annual Symposium on Foundations of Computer Science, FOCS, pp. 106–115 (2001)
4. Berman, I., Degwekar, A., Rothblum, R.D., Vasudevan, P.N.: Multi-collision resistant hash functions and their applications. In: Nielsen, J.B., Rijmen, V. (eds.) EUROCRYPT 2018. LNCS, vol. 10821, pp. 133–161. Springer, Cham (2018). https://doi.org/10.1007/978-3-319-78375-8_5
5. Bitansky, N., Degwekar, A., Vaikuntanathan, V.: Structure vs. hardness through the obfuscation lens. In: Katz, J., Shacham, H. (eds.) CRYPTO 2017. LNCS, vol. 10401, pp. 696–723. Springer, Cham (2017). https://doi.org/10.1007/978-3-319-63688-7_23
6. Bitansky, N., Kalai, Y.T., Paneth, O.: Multi-collision resistance: a paradigm for keyless hash functions. In: Proceedings of the 50th Annual ACM SIGACT Symposium on Theory of Computing, STOC, pp. 671–684 (2018)
7. Blum, M.: Coin flipping by telephone. In: Advances in Cryptology - CRYPTO, pp. 11–15 (1981)
8. Damgård, I.B., Pedersen, T.P., Pfitzmann, B.: On the existence of statistically hiding bit commitment schemes and fail-stop signatures. In: Stinson, D.R. (ed.) CRYPTO 1993. LNCS, vol. 773, pp. 250–265. Springer, Heidelberg (1994). https://doi.org/10.1007/3-540-48329-2_22
9. Dubrov, B., Ishai, Y.: On the randomness complexity of efficient sampling. In: Proceedings of the 38th Annual ACM Symposium on Theory of Computing, pp. 711–720 (2006)
10. Dvir, Z., Gutfreund, D., Rothblum, G.N., Vadhan, S.P.: On approximating the entropy of polynomial mappings. In: Innovations in Computer Science - ICS, pp. 460–475 (2011)
11. Feige, U., Shamir, A.: Witness indistinguishable and witness hiding protocols. In: Proceedings of the 22nd Annual ACM Symposium on Theory of Computing, STOC, pp. 416–426 (1990)
12. Goldreich, O., Kahan, A.: How to construct constant-round zero-knowledge proof systems for NP. J. Cryptology **9**(3), 167–190 (1996)
13. Goldreich, O., Micali, S., Wigderson, A.: How to play any mental game or a completeness theorem for protocols with honest majority. In: Proceedings of the 19th Annual ACM Symposium on Theory of Computing, STOC, pp. 218–229 (1987)
14. Goldwasser, S., Micali, S., Rackoff, C.: The knowledge complexity of interactive proof systems. SIAM J. Comput. **18**(1), 186–208 (1989)
15. Haitner, I., Hoch, J.J., Reingold, O., Segev, G.: Finding collisions in interactive protocols - tight lower bounds on the round and communication complexities of statistically hiding commitments. SIAM J. Comput. **44**(1), 193–242 (2015)
16. Haitner, I., Nguyen, M., Ong, S.J., Reingold, O., Vadhan, S.P.: Statistically hiding commitments and statistical zero-knowledge arguments from any one-way function. SIAM J. Comput. **39**(3), 1153–1218 (2009)

17. Haitner, I., Reingold, O., Vadhan, S., Wee, H.: Inaccessible entropy I: inaccessible entropy generators and statistically hiding commitments from one-way functions (2018). www.cs.tau.ac.il/~iftachh/papers/AccessibleEntropy/IE1.pdf. Prelimanry version, named Inaccessible Entropy, appeared in STOC 2009
18. Haitner, I., Reingold, O., Vadhan, S.P., Wee, H.: Inaccessible entropy. In: Mitzenmacher, M. (ed.) Proceedings of the 41st Annual ACM Symposium on Theory of Computing, STOC, pp. 611–620 (2009)
19. Haitner, I., Vadhan, S.: The many entropies in one-way functions. Tutorials on the Foundations of Cryptography. ISC, pp. 159–217. Springer, Cham (2017). https://doi.org/10.1007/978-3-319-57048-8_4
20. Halevi, S., Micali, S.: Practical and provably-secure commitment schemes from collision-free hashing. In: Koblitz, N. (ed.) CRYPTO 1996. LNCS, vol. 1109, pp. 201–215. Springer, Heidelberg (1996). https://doi.org/10.1007/3-540-68697-5_16
21. Harnik, D., Naor, M.: On the compressibility of NP instances and cryptographic applications. SIAM J. Comput. **39**(5), 1667–1713 (2010)
22. Håstad, J., Impagliazzo, R., Levin, L.A., Luby, M.: A pseudorandom generator from any one-way function. SIAM J. Comput. **28**(4), 1364–1396 (1999)
23. Impagliazzo, R., Luby, M.: One-way functions are essential for complexity based cryptography (extended abstract). In: 30th IEEE Annual Symposium on Foundations of Computer Science, FOCS, pp. 230–235 (1989)
24. Komargodski, I., Naor, M., Yogev, E.: White-box vs. black-box complexity of search problems: Ramsey and graph property testing. In: 58th IEEE Annual Symposium on Foundations of Computer Science, FOCS, pp. 622–632 (2017)
25. Komargodski, I., Naor, M., Yogev, E.: Collision resistant hashing for paranoids: dealing with multiple collisions. In: Nielsen, J.B., Rijmen, V. (eds.) EUROCRYPT 2018. LNCS, vol. 10821, pp. 162–194. Springer, Cham (2018). https://doi.org/10.1007/978-3-319-78375-8_6
26. Komargodski, I., Yogev, E.: On distributional collision resistant hashing. In: Shacham, H., Boldyreva, A. (eds.) CRYPTO 2018. LNCS, vol. 10992, pp. 303–327. Springer, Cham (2018). https://doi.org/10.1007/978-3-319-96881-0_11
27. Matsumoto, T., Imai, H.: Public quadratic polynomial-tuples for efficient signature-verification and message-encryption. In: Barstow, D., et al. (eds.) EUROCRYPT 1988. LNCS, vol. 330, pp. 419–453. Springer, Heidelberg (1988). https://doi.org/10.1007/3-540-45961-8_39
28. Naor, M.: Bit commitment using pseudorandomness. J. Cryptology **4**(2), 151–158 (1991)
29. Naor, M., Ostrovsky, R., Venkatesan, R., Yung, M.: Perfect zero-knowledge arguments for *NP* can be based on general complexity assumptions. In: Brickell, E.F. (ed.) CRYPTO 1992. LNCS, vol. 740, pp. 196–214. Springer, Heidelberg (1993). https://doi.org/10.1007/3-540-48071-4_14
30. Naor, M., Yung, M.: Universal one-way hash functions and their cryptographic applications. In: Proceedings of the 21st Annual ACM Symposium on Theory of Computing, pp. 33–43. ACM (1989)
31. Ong, S.J., Vadhan, S.P.: An equivalence between zero knowledge and commitments. In: Canetti, R. (ed.) TCC 2008. LNCS, vol. 4948, pp. 482–500. Springer, Heidelberg (2008). https://doi.org/10.1007/978-3-540-78524-8_27
32. Ostrovsky, R., Wigderson, A.: One-way functions are essential for non-trivial zero-knowledge. In: Second Israel Symposium on Theory of Computing Systems, ISTCS, pp. 3–17. IEEE Computer Society (1993)

33. Pass, R., Rosen, A.: Concurrent nonmalleable commitments. SIAM J. Comput. **37**(6), 1891–1925 (2008). https://doi.org/10.1137/060661880
34. Simon, D.R.: Finding collisions on a one-way street: can secure hash functions be based on general assumptions? In: Nyberg, K. (ed.) EUROCRYPT 1998. LNCS, vol. 1403, pp. 334–345. Springer, Berlin (1998). https://doi.org/10.1007/BFb0054137

Signatures II

Multi-target Attacks on the Picnic Signature Scheme and Related Protocols

Itai Dinur[1(✉)] and Niv Nadler[2]

[1] Department of Computer Science, Ben-Gurion University, Beersheba, Israel
dinuri@cs.bgu.ac.il
[2] Beersheba, Israel

Abstract. Picnic is a signature scheme that was presented at ACM CCS 2017 by Chase et al. and submitted to NIST's post-quantum standardization project. Among all submissions to NIST's project, Picnic is one of the most innovative, making use of recent progress in construction of practically efficient zero-knowledge (ZK) protocols for general circuits.

In this paper, we devise multi-target attacks on Picnic and its underlying ZK protocol, ZKB++. Given access to S signatures, produced by a single or by several users, our attack can (information theoretically) recover the κ-bit signing key of a user in complexity of about $2^{\kappa-7}/S$. This is faster than Picnic's claimed 2^{κ} security against classical (non-quantum) attacks by a factor of $2^7 \cdot S$ (as each signature contains about 2^7 attack targets).

Whereas in most multi-target attacks, the attacker can easily sort and match the available targets, this is not the case in our attack on Picnic, as different bits of information are available for each target. Consequently, it is challenging to reach the information theoretic complexity in a computational model, and we had to perform cryptanalytic optimizations by carefully analyzing ZKB++ and its underlying circuit. Our best attack for $\kappa = 128$ has time complexity of $T = 2^{77}$ for $S = 2^{64}$. Alternatively, we can reach the information theoretic complexity of $T = 2^{64}$ for $S = 2^{57}$, given that all signatures are produced with the same signing key.

Our attack exploits a weakness in the way that the Picnic signing algorithm uses a pseudo-random generator. The weakness is fixed in the recent Picnic 2.0 version.

In addition to our attack on Picnic, we show that a recently proposed improvement of the ZKB++ protocol (due to Katz, Kolesnikov and Wang) is vulnerable to a similar multi-target attack.

Keywords: Cryptanalysis · Multi-target attack · Picnic · Signature scheme · Zero-knowledge protocol · ZKB++ · MPC · Block cipher · LowMC

1 Introduction

Multi-target attacks are among the most basic attacks against cryptosystems that are built using symmetric-key primitives. In a typical example, the attacker

N. Nadler—Independent

Y. Ishai and V. Rijmen (Eds.): EUROCRYPT 2019, LNCS 11478, pp. 699–727, 2019.
https://doi.org/10.1007/978-3-030-17659-4_24

first obtains G possible *targets*, which correspond to outputs of the cryptosystem, evaluated with different secret keys (or secret inputs, in general). Then, the attacker guesses a key, evaluates the cryptosystem, and compares the result with all targets. Based on a standard birthday paradox argument, the expected workload of the attacker for hitting one of the targets is reduced by a factor of (at least[1]) G, compared to the workload of hitting a single target.[2]

In our multi-target attack model, we deal with a cryptosystem with U users, each with a long-term key. For each user $i \in [1, U]$, the attacker obtains D_i data points created by this user and we denote $D = \sum_{i=1}^{U} D_i$. Each data point may be additionally associated with a short-term key. The goal of the attacker is to recover one of the keys for the cryptosystem (either a short or a long-term key). For example, in a signature scheme, each user has a long-term signing key, a data point may be a signature and a short-term key is (secret) randomness used in creating the signature. We note that in many cases the recovery of a short-term key allows recovering the corresponding user's long-term key, but this possibility is not directly captured by our simple model.

We distinguish between three types of multi-target attacks according to the number of targets G they present to an attacker.[3]

1. *Multi-user single-target attack*: G is determined by the number of users U, i.e., $G = U$. Typically, this occurs if the long-term user keys are vulnerable to a multi-target attack.
2. *Single-user multi-target attack*: G is determined separately for each user as $G_i = D_i$. Hence, the best attack uses $G = \text{argmax}_i\{D_i\}$. In this case, the short-term keys of each user are vulnerable to a multi-target attack.
3. *Multi-user multi-target attack* (or generic multi-target attack): G is determined by the total number of available data points D, i.e., $G = D$. Here, all short-term keys are vulnerable to a multi-target attack. In principle, this is the most powerful type of multi-target attack, as all data points can simultaneously be used by the attacker as targets.

A standard way to mitigate multi-target attacks is to add a public random input to the cryptosystem (i.e., a salt), thus creating a different tweaked variant of it per salt. Since one has to choose a particular salt in order to evaluate the cryptosystem with a secret key, salting forces the attacker to focus on only one target per secret key guess.

In this paper, we are mainly interested in public key cryptosystems that are based on symmetric-key primitives. These cryptosystems have received significant attention recently due to their alleged post-quantum security. The most well-known category within this class consists of hash-based signatures, which originate from Lamport's one-time signatures [15]. In recent years, these signatures have been subject to many optimizations and improvements until the

[1] If the keys are not generated uniformly, the workload of the attack could be lower.
[2] Throughout this paper, we focus on attacks run on classical computers, but our analysis can be extended to deal with attacks on quantum computers.
[3] Our model is related to the one of [12], but our classification is at a higher level.

recent development of practical stateless hash-based signatures [3]. As all cryptosystems built with symmetric-key primitives, hash-based signature are potentially vulnerable to multi-target attacks and substantial effort has been put into their efficient mitigation (cf. [12]).

Another public key cryptosystem that is based on symmetric-key components is the Picnic signature scheme. It was presented at ACM CCS 2017 [6] by Chase et al. and submitted [5] to NIST's post-quantum standardization project [20].[4] Picnic's design is solely based on symmetric-key primitives, yet is completely different from the design of hash-based signatures. Our main goal in this paper is to investigate the resistance of Picnic against multi-target attacks. As we demonstrate, this requires dedicated analysis due to Picnic's novel design. We note that our description of Picnic and its analysis applies to Picnic 1.0 and not to the recent Picnic 2.0 version [5].

Picnic. The Picnic signature scheme uses the ZKB++ zero-knowledge (ZK) protocol (that improves upon the original ZKBoo protocol [11] in terms of efficiency), which allows to non-interactively prove knowledge of a preimage x to a public value y under a one-way function f. In Picnic, y is part of the public key, whereas x is the secret signing key. In order to sign a message, the signer uses ZKB++ to prove knowledge of x, where the message is embedded in the signing process to generate (pseudo) random bits.

The ZKB++ protocol employs the "MPC-in-the-head" paradigm due to Ishai et al. [13]. In order to prove knowledge of x, the prover (signer), simulates a multi-party computation (MPC) protocol between several players (whose number is 3 in ZKB++) that receive shares of x and compute $f(x) = y$. The prover then commits to the different internal states (views) of each of the players, and the verifier challenges the prover by asking to open the commitments of a subset of the players, revealing their views.

The *correctness* of the MPC protocol guarantees that if the prover does not know x and tries to cheat, then the joint views of some of the players are inconsistent. Hence, the verifier can catch a cheating prover with some probability, which is amplified by repeating the process. The *privacy* guarantee of the MPC protocol ensures that opening the views of a (sufficiently small) subset of players does not reveal any information about x, hence the secret signing key is not leaked. The proof is made non-interactive using the Fiat-Shamir transform [10]. More specifically, the prover computes the challenge by hashing the commitments, where in Picnic, the message to be signed is hashed as well (making the signature depend on the message).

[4] The ACM CCS 2017 paper [6] introduced two signature scheme variants: Fish (which uses the Fiat-Shamir transform [10]), with claimed security against classical computer attacks, and Picnic (which uses Unruh's transform [22]), with claimed security against quantum computer attacks. In the NIST submission [5], these variants were renamed to Picnic-FS and Picnic-UR, respectively. Our analysis applies to both variants, but we focus on Picnic-FS for simplicity.

A Picnic signature thus comprises of partial transcripts of several independent runs of the MPC protocol, where for each run, the views of two out of three participating (virtual) players are opened. As noted above, a view contains the player's internal states computed during the MPC protocol. The signature also includes the player's sampled random bits, so that the view's consistency can be checked by a verifier of the signature. However, having the signature include all the random bits sampled by the "opened players" blows up its size. Hence, Picnic uses a standard optimization, where each player only samples a short seed of size κ bits (where κ is the security level against classical attacks), and produces the random bits required by the protocol using a deterministic pseudo-random generator (PRG), initialized with the seed. Thus, the short seeds of the opened players are included in the signature for each run and the verifier uses them to compute the required pseudo-random bits. Obviously, the random seed (and view) of the remaining "unopened player" in each run must not be included in the signature, as it may expose the secret key x.

Multi-target Attacks on Picnic. Our main result is a multi-target attack on Picnic. The first step of the attack involves collecting signatures (produced by one or several users) containing (partial) transcripts of various runs of the MPC protocol. Then, by independently guessing a value of the κ-bit seed and evaluating the PRG, the attacker can match and detect that the seed is used by the unopened player in a particular run. Once the seed of the unopened player in a run is revealed, the secret signing key of the corresponding user can be computed easily. Thus, given a total of D runs, the attacker needs to test an average of $2^{\kappa}/D$ seeds until a match with a run is detected. The attack is thus a generic multi-target attack (i.e., a multi-user multi-target attack) and it violates Picnic's claims of κ-bit security (against attacks by classical computers).

A crucial detail missing from the attack's outline above is how to detect a match between a guessed seed and the seed used by an unopened player in an available run. In fact, this may seem impossible, as the privacy of the MPC protocol should presumably prevent the pseudo-random bits used by the unopened player from leaking. This issue is related to a subtlety about MPC protocols: their privacy guarantees apply to the input of each player and not (necessarily) to the (pseudo) random bits that each player uses. In other words, MPC protocols are allowed to (and mostly do) expose some (pseudo) random bits used by each player and still remain private, i.e., protect the players' inputs. On the other hand, it is generally important that not all of a player's randomness is exposed, as this leaks the player's input. In the context of Picnic, in each run, some output bits of the PRG used by the unopened player can be easily computed by the attacker, which makes it possible to detect that the unopened player uses a certain seed once it is correctly guessed (and then compute the secret key).

We note that the Picnic designers attempted to protect it against multi-target attacks. For example, the public key of each owner i defines a different one-way function f_i,[5] rather than having all owners prove knowledge of a preimage under the same function f. Indeed, a global choice of f allows the attacker to mount a multi-user single-target attack by computing a preimage to one out of many images available in the different public keys. Yet, Picnic was not protected against our generic (and more powerful) multi-user multi-target attack against the seeds, presumably because it is not obvious that such an attack is possible (as previously mentioned). Internally, the security proof of Picnic (published in its design document [5]) simply does not consider attacker queries with arbitrary seed values to the PRG and hence does not cover our attack.

Randomness Extraction. According to the birthday paradox, the expected complexity of our attack is $T = 2^{\kappa}/D$ (for $D \leq 2^{\kappa/2}$). However, this information theoretic analysis assumes that the attacker wins once the PRG is evaluated with a seed that is used by the unopened player in one of the available runs (as enough information is available to recover the key). In practice, achieving the information theoretic complexity is challenging, since the PRG output bits of the unopened player that can be computed in each run, vary according to the run. Therefore, a standard matching algorithm which sorts the runs according to the available PRG output bits does not work, while its naive extension has very high complexity (e.g., at least 2^{102} for $\kappa = 128$). Consequently, we carefully analyze Picnic (and its underlying block cipher LowMC that implements f [1]) in order to extract the maximal amount of PRG output data from each run. We then utilize this data by devising a dedicated attack algorithm that recovers the signing key and outperforms the naive algorithm by a factor of up to 2^{25} for $\kappa = 128$ and by more than 2^{30} for larger κ values.

The techniques we use for extracting the maximal amount of PRG output data mainly involve exploiting dependencies among private values computed by a player and masked with PRG output bits. As a simple example, assume that a player outputs 3 bits z_1, z_2, z_3 such that $z_1 = v_1 \cdot v_2 \oplus r_1$, $z_2 = v_2 \cdot v_3 \oplus r_2$ and $z_3 = v_1 \cdot v_3 \oplus r_3$ where v_1, v_2, v_3 are internal private bit values and r_1, r_2, r_3 are PRG output bits. Observe that the triplet $v_1 \cdot v_2, v_2 \cdot v_3, v_1 \cdot v_3$ can only attain 5 values (as the values 011, 101, 110 as impossible). Hence, given z_1, z_2, z_3, the triplet of bits r_1, r_2, r_3 can only attain 5 out of 8 possible values, revealing information about them.

Although our techniques are tailored to Picnic, they can be easily adapted and applied to other MPC protocol implementations in order to extract information about the random bits that are used by the players. Such extraction techniques may be relevant to attackers is scenarios that extend beyond multi-target attacks. For example, the attacker's ability to exploit a weak PRG in a cryptographic protocol (e.g., by predicting its output) may depend on the number of PRG output bits available. This was demonstrated in [7] by Checkoway

[5] Internally, Picnic uses a block cipher encryption $f_i(x) = \text{Enc}_x(p(i))$, where a different plaintext $p = p(i)$ is used for each public key owner (defining a different encryption function).

et al.[6] which investigated the exploitability of the backdoored Dual EC PRG in TLS implementations. Additionally, in case protocol implementations generate seeds with low entropy, extraction techniques may allow an attacker to efficiently detect that two protocol executions use the same seed and to violate their security.

Concrete Complexity of the Main Attack. In terms of concrete complexity, we are interested in attacks that utilize at most 2^{64} signatures. This is the limit set in NIST's Call for Proposals document [20] on the number of signatures produced per signing key.[7] For Picnic, each signature contains $R \in \{219, 324, 438\}$ runs depending on the desired security level against classical attacks, $\kappa \in \{128, 192, 256\}$, respectively. The complexities of our main attack for each desired security level are summarized below.

- For $\kappa = 128$, we can reach the information theoretic complexity of $T = 2^{\kappa}/D$ up to $D = 2^{42}$ (using about 2^{35} signatures), and obtain $T \approx 2^{128-42} = 2^{86}$. When 2^{64} signatures are available, we can recover a secret signing key with complexity of about 2^{77}.
- For $\kappa = 192$, we achieve the information theoretic complexity $T = 2^{\kappa}/D$ for almost all $D \leq 324 \cdot 2^{64} \approx 2^{72}$. The best complexity is $T = 2^{124}$, obtained for $D = 2^{72}$.
- For $\kappa = 256$, we achieve the information theoretic complexity $T = 2^{\kappa}/D$ for all $D \leq 438 \cdot 2^{64}$.

Seed Collision Attack. Interestingly, for $\kappa = 128$, we can reach the information theoretic complexity for the specific case of $D = 2^{64}$ (i.e. utilizing about 2^{57} signatures) using another attack, given that all the available signatures are produced with the same signing key.[8] While the attack resembles a single-user multi-target attack, it is not a classical multi-target attack in the sense that the attacker does not guess any key material (such as PRG seeds). Instead, the attacker waits for a specific *seed collision* event (in which two different runs use the same PRG seed for the unopened player) to occur on the observed data. Once the event is detected, the user's signing key can be efficiently recovered using the known PRG output bits of the unopened player in both runs. The attack can be extended (with a limited range of parameters) to recover the signing key of one out of many users, e.g., if the attacker collects 2^{64} signatures ($D \approx 2^{71}$), produced with up to 2^{14} private keys.

[6] We thank an anonymous reviewer for pointing out the link between [7] and our paper.

[7] As the attacker may acquire signatures produced with various signing keys, our model is somewhat more restrictive than NIST's.

[8] The attack can also be applied to $\kappa \in \{192, 256\}$, but it requires significantly more than 2^{64} signatures.

Multi-target Attacks on Additional Cryptosystems. Our multi-target attack is, in fact, an attack on the ZKB++ protocol, as well as the previous ZKBoo protocol. Therefore, the attack also carries over to additional cryptosystems that were built using these protocols. This includes the ring-signature and additional constructions of [4,8], whose implementations are based in ZKB++.

We further analyze in this paper a recently proposed protocol due to Katz, Kolesnikov and Wang [14] (KKW), which was presented at ACM CCS 2018. The KKW protocol describes a new way to instantiate the MPC-in-the-head approach, yielding shorter proofs compared to ZKB++. Interestingly, the KKW protocol is vulnerable to a multi-target attack which is similar to our main attack on Picnic (and ZKB++).

In the penultimate section of the paper, we describe multi-target attacks on additional cryptosystems, which are made possible due to several design optimizations (mostly for MPC protocols). Unlike the case of Picnic, these multi-target attacks are standard and their descriptions only requires a very high-level understanding of the cryptosystems. Yet, the aim of this section is to show that some common optimizations do not come without a cost, which needs to be considered in cryptosystems that are designed for practical use.

Picnic 2.0. We notified the Picnic designers about the attack and they confirmed our findings. The weakness is addressed in the Picnic 2.0 version [5] by appending an additional salt to each signature. The salt is carefully used in generating the pseudo-random bits of each player in each run, such that multi-target attacks are mitigated. We further note that Picnic 2.0 added additional instances that use the KKW protocol, while in this paper we describe and analyze Picnic 1.0 which only uses ZKB++.

Paper Organization. The rest of this paper is organized as follows. In Sect. 2 we describe Picnic and its building blocks, while in Sect. 3 we partially summarize the KKW protocol. Next, in Sect. 4, we outline the main steps of our multi-target attacks on Picnic and the KKW protocol. In Sect. 5, we elaborate on our main multi-target attack on Picnic, while our seed collision attack is described in Sect. 6. Finally, we describe multi-target attacks on additional cryptosystems in Sect. 7 and conclude in Sect. 8.

2 ZKBoo, ZKB++ and Picnic

ZKBoo is a ZK protocol described in [11]. An optimized variant of ZKBoo (called ZKB++) was later described in [6], which used it to construct the Picnic signature scheme. In this section, we give a brief overview of these constructions.

2.1 Overview of ZKBoo

The goal ZKBoo is to prove knowledge of a witness for a relation $Re := \{(x,y), f(x) = y\}$, where y is public and x is kept private. For example, given a

256-bit string y, we aim to prove knowledge of a preimage of y under SHA-256, namely, a string x such that $y = \text{SHA-256}(x)$.

ZKBoo employs the MPC-in-the-head paradigm of Ishai et al. [13], that we now outline very briefly. It uses some MPC protocol that implements f on input shares of the secret witness x. The prover simulates the MPC protocol "in the head" and commits to the state and transcripts of all players. The verifier then "corrupts" a random subset of the simulated players by requesting to see their complete states. The verifier checks that the computation was done correctly from the perspective of the corrupted players, obtaining some assurance that the output is correct and the prover knows x. Iterating this procedure many times gives the verifier high assurance.

ZKBoo improves upon the practical efficiency of the MPC-in-the-head approach by replacing the MPC with a *circuit decomposition*, which does not necessarily need to satisfy classical MPC protocol properties. The circuit decompositions in ZKBoo involves 3 players. Given a circuit ϕ that computes f, it defines the following functions.

- Share: splits the input x into 3 shares.
- $\text{Output}_{i\in\{1,2,3\}}$: takes as input all of the input shares and some randomness and produces an output share for each of the players.
- Reconstruct: takes as input the three output shares and reconstructs the circuit's final output.

The circuit decomposition should satisfy the correctness property which means that its execution on input x must yield $f(x)$. It must further satisfy the 2-privacy property which requires that revealing the views (i.e., the values of the intermediate computation states) of any two players does not leak information about the witness x.

Given a circuit decomposition for ϕ, the ZKBoo protocol is a Σ-protocol for languages of the form $L := \{y \mid \exists x : y = \phi(x)\}$. As outlined below, it gives a non-interactive ZK proof of knowledge system for the relation using the Fiat-Shamir transform [10].

The computation $\phi(x)$ using the decomposition is a randomized algorithm called a *run*. As indicated above, in each run, each player $P_{i\in\{1,2,3\}}$ uses some (pseudo) random bits, generated by random seeds k_1, k_2, k_3, respectively. For a parameter R that determines the total number of runs, a proof is constructed as below.

1. For each run $i \in [1, R]$:
   (a) Sample $k_1^{(i)}, k_2^{(i)}, k_3^{(i)}$ and compute run i using the circuit decomposition outlined above.
   (b) For each player $P_1^{(i)}, P_2^{(i)}, P_3^{(i)}$, compute a commitment to its view during the run. The commitment for each player is computed by applying a hash function (modeled as a random oracle) to the player's view and additional randomness.

2. Using the Fiat-Shamir transform, send the $3R$ commitments and output shares of each player in all runs to a random oracle (implemented as a hash function).
3. Interpret the output of the random oracle as a challenge $\{e^{(i)}\}_{i=1}^{R}$. For each run $i \in [1, R]$, the challenge element $e^{(i)} \in \{1,2,3\}$ specifies to open the views of the two players $P_{e^{(i)}}^{(i)}, P_{e^{(i)}+1}^{(i)}$ (where $3+1=1$).
4. The proof contains for each run $i \in [1, R]$:
   - The commitments and output shares of all 3 players.
   - The two views and commitment openings (i.e., additional randomness) of the players $P_{e^{(i)}}^{(i)}, P_{e^{(i)}+1}^{(i)}$, indicated by the challenge.
   - The values $k_{e^{(i)}}^{(i)}, k_{e^{(i)}+1}^{(i)}$. Namely, the random seeds used by the two players whose views are opened.

Due to the 2-privacy property, opening two views for each run does not leak information about the witness. The number of runs, R, is chosen to achieve negligible soundness error, i.e., it should be infeasible for the prover to cheat without getting caught in at least one of the runs. More specifically, in order to achieve soundness error of $2^{-\kappa}$, we set $R = \lceil \kappa(\log_2 3 - 1)^{-1} \rceil$.

The verifier checks that: (1) for each run, the output shares of the three views reconstruct to y, (2) for each run, each of the two open views was computed correctly and their commitment openings are valid, and (3) the challenge was computed correctly,

In the following, we describe Step 1(a) in the above ZKBoo protocol in more detail.

2.2 (2, 3)-Function Decomposition

ZKBoo uses the following circuit decomposition.

Definition 1. *Let f be a function that is computed by an N-gate circuit ϕ such that $f(x) = \phi(x) = y$, and let κ be the security parameter. Let k_1, k_2, k_3 be seeds chosen uniformly at random from $\{0,1\}^\kappa$, corresponding to players P_1, P_2, P_3, respectively. A $(2,3)$-decomposition of ϕ is a tuple of algorithms $\mathcal{D} = (\text{Share}, \text{Update}, \text{Output}, \text{Reconstruct})$:*

- $(\text{view}_1^{(0)}, \text{view}_2^{(0)}, \text{view}_3^{(0)}) \leftarrow \text{Share}(x, k_1, k_2, k_3)$
  On input of the secret value x and random seeds, outputs the initial views for each player containing the secret share x_i of x.
- $\text{view}_i^{(j+1)} \leftarrow \text{Update}(\text{view}_i^{(j)}, \text{view}_{i+1}^{(j)}, k_i, k_{i+1})$
  On input of the views $\text{view}_i^{(j)}, \text{view}_{i+1}^{(j)}$ and random seeds k_i, k_{i+1}, computes wire values for the next gate and returns the updated view $\text{view}_i^{(j+1)}$.
- $y_i \leftarrow \text{Output}(\text{view}_i^{(N)})$
  On input of the final view $\text{view}_i^{(N)}$, returns the output share y_i.
- $y \leftarrow \text{Reconstruct}(y_1, y_2, y_3)$
  On input of output shares y_i, reconstructs and returns y.

In order to compute a run for the computation $\phi(x)$ using the decomposition $\mathcal{D}$ defined above, the prover executes the steps detailed below.

1. Choose the seeds k_1, k_2, k_3 uniformly at random from $\{0,1\}^\kappa$.
2. $(\text{view}_1^{(0)}, \text{view}_2^{(0)}, \text{view}_3^{(0)}) \leftarrow \text{Share}(x, k_1, k_2, k_3)$
3. For each of the three views, call the Update function successively for every gate in the circuit:
$$\text{view}_i^{(j+1)} \leftarrow \text{Update}(\text{view}_i^{(j)}, \text{view}_{i+1}^{(j)}, k_i, k_{i+1}),$$
for $i \in \{1,2,3\}$, $j \in [1,N]$.
4. From the final views, compute the output share of each view:
$$y_i \leftarrow \text{Output}(\text{view}_i^{(N)}),$$
for $i \in \{1,2,3\}$.
5. $y \leftarrow \text{Reconstruct}(y_1, y_2, y_3)$

The correctness property requires that the output y above satisfies $y = \phi(x)$. The 2-privacy property requires that revealing the views of any two players reveals nothing about x.

2.3 The ZKBoo (2, 3)-Function Decomposition

The ZKBoo protocol works over some finite ring $\mathbb{R}$. Let $f : \mathbb{R}^m \rightarrow \mathbb{R}^\ell$ be a function and ϕ an arithmetic circuit realizing f with N gates that include addition by constant, multiplication by constant, binary addition and binary multiplication gates. The $(2,3)$-decomposition of ϕ in ZKBoo is a *linear decomposition*: denote by w_k the value of the k'th wire of ϕ. Then, each party P_i has a corresponding wire value $w_k^{(i)}$. The linear decomposition maintains the invariant that for all wires, $w_k = w_k^{(1)} + w_k^{(2)} + w_k^{(3)}$. In detail, the $(2,3)$-decomposition is defined using the following tuple of algorithms:

- $\text{Share}(x, k_1, k_2, k_3)$: Samples uniform $x_1, x_2 \in \mathbb{R}^m$ and computes x_3 such that $x_1 + x_2 + x_3 = x$ (or $x_3 = x - x_1 - x_2$). Returns views containing x_1, x_2, x_3.
- $\text{Update}(\text{view}_i^{(j)}, \text{view}_{i+1}^{(j)}, k_i, k_{i+1})$: Computes P_i's view of the output wire of gate g_j and appends it to the view. For the k'th wire w_k (where $w_k^{(i)}$ denotes P_i's view for the wire), the update operation is defined as follows:
  Addition by constant: $(w_b = w_a + d)$: $w_b^{(i)} = w_a^{(i)} + d$ if $i = 1$ and $w_b^{(i)} = w_a^{(i)}$, otherwise.
  Multiplication by constant: $(w_b = w_a \cdot d)$: $w_b^{(i)} = w_a^{(i)} \cdot d$.
  Binary addition: $(w_c = w_a + w_b)$: $w_c^{(i)} = w_a^{(i)} + w_b^{(i)}$.
  Binary multiplication: $(w_c = w_a \cdot w_b)$:
$$w_c^{(i)} = (w_a^{(i)} \cdot w_b^{(i)}) + (w_a^{(i+1)} \cdot w_b^{(i)}) + (w_a^{(i)} \cdot w_b^{(i+1)}) + R_i(c) - R_{i+1}(c),$$

where $R_i(c)$ is the c'th output of a pseudorandom generator (PRG) seeded with k_i.
- Output($\text{view}_i^{(N)}$): Returns the output wires of view, $\text{view}_i^{(N)}$.
- Reconstruct(y_1, y_2, y_3): Returns $y = y_1 + y_2 + y_3$.

It is easy to verify that the decomposition maintains the invariant $w_k = w_k^{(1)} + w_k^{(2)} + w_k^{(3)}$ for all wires, which implies that it is correct. Note that P_i can compute all gate types locally with the exception of binary multiplication gates which require inputs from P_{i+1}.

Serializing the Views. It is sufficient for the prover to include in the proof only the wire values of the gates that require non-local computations (namely, the binary multiplication gates). The verifier can recompute these omitted parts of the view by local computations (i.e., they do not need to be serialized). In ZKBoo, a serialized view includes: (1) the input share, (2) output wire values for binary multiplication gates, and (3) the output share.

2.4 ZKB++

ZKB++ is an improved version of ZKBoo, obtained using several optimizations which reduce the proof size to less than a half. In general, these optimizations mainly show that some values included in the ZKBoo proof (as outlined above) can be directly computed by the verifier and hence can be omitted in the proof of ZKB++. In our context, most of these optimization are not very relevant as the attacker (verifier) has access to all data included in the original ZKBoo proof (since it is either directly included in the shorter ZKB++ proof, or can be easily computed from it).

The only optimization that is directly exploited in our attack involves the Share function: instead of uniformly sampling the input shares x_1, x_2, the Share function of ZKB++ uses pseudo-random shares for the first 2 players, generated by PRG invocations seeded with the corresponding player's random seed (k_1 or k_2). Since the random seeds of two players are revealed in the proof, the verifier can compute some of the shares (the ones of the first two players whose seeds are revealed) using the known seeds and they do not have to be included in the proof.

2.5 The Picnic Signature Scheme

The Picnic signature scheme is based on the ZKB++ protocol, where the input to the hash function that computes the challenge also includes the message m to be signed (in addition to the $3R$ commitments and output shares of each player, which are input to the hash function in ZKB++).[9]

In order to define the statement to be proved by the signer, Picnic uses a block cipher, Enc. In the classical setting (on which we focus in this paper),

[9] In the NIST submission, the public key is hashed as well.

the block size of the block cipher and its key size in bits are both equal to the security parameter κ.

During key generation, the signer chooses a plaintext p and a key x for the block cipher uniformly at random from $\{0,1\}^{\kappa}$, encrypts the plaintext using the key and obtains the ciphertext y (of length κ bits). The public key is the plaintext-ciphertext pair (p, y) and the private signing key is the pair (x, p) (i.e., the chosen block cipher key and plaintext).

During signing, the signer proves knowledge of the key x, which encrypts p to y. Namely, Picnic uses ZKB++ in order to prove knowledge of a witness for the relation $Re := \{((p, y), x), \text{Enc}_x(p) = y\}$, where $\text{Enc}_x(p)$ is the block cipher encryption of plaintext p with of the key x.

The specific block cipher used by Picnic is LowMC [1], implemented using a Boolean circuit. LowMC is an iterative block cipher that employs a certain number of encryption rounds to its input. The most relevant components of LowMC for this paper are its identical 3×3 Sboxes (all the other operations are linear over $GF(2)$). Each LowMC round applies a certain number of Sboxes in parallel to the encryption state. In all LowMC variants used in Picnic, 10 parallel Sboxes are applied in a round. The algebraic normal form of an Sbox is given as

$$S(w_{a_1}, w_{a_2}, w_{a_3}) = \\ (w_{a_1} \oplus (w_{a_2} \cdot w_{a_3}), w_{a_1} \oplus w_{a_2} \oplus (w_{a_1} \cdot w_{a_3}), w_{a_1} \oplus w_{a_2} \oplus w_{a_3} \oplus (w_{a_1} \cdot w_{a_2})). \tag{1}$$

In particular, the Sbox employs 3 non-linear AND operations

$$w_{a_2} \cdot w_{a_3}, w_{a_1} \cdot w_{a_3}, w_{a_1} \cdot w_{a_2}$$

in computing the 3 output bits, respectively.

Picnic defines a total of 6 instances depending on a desired security level and on whether they are intended to resist attacks by quantum computers. We focus on the instances that are deemed secure (only) against attacks by classical computers, whose parameters are given in Table 1. However, our attacks are applicable to all instances. Note that all LowMC instances have at least 200 Sboxes, where each Sbox employs 3 AND operations. Since evaluating an AND operation in Picnic requires a PRG output bit from each player, then each player computes at least $200 \cdot 3 = 600$ PRG output bits during a run.

Table 1. Picnic instances (for classical security)

Instance	κ	LowMC rounds	Sboxes\round	PRG	R
picnic-L1-FS	128	20	10	SHAKE128	219
picnic-L3-FS	192	30	10	SHAKE256	324
picnic-L5-FS	256	38	10	SHAKE256	438

3 The KKW Protocol [14]

In this section we give a very brief overview of the KKW protocol [14], focusing on details relevant for the paper.

The KKW protocol describes a new way to instantiate the MPC-in-the-head approach which leads to shorter proofs compared to ZKB++. The main idea is to instantiate the MPC protocol in the preprocessing model, which makes it possible to use protocols designed for a large number of players with small communication complexity (which translates to small proofs in the ZK proof protocol) and low soundness error per protocol execution (i.e., run). In the following, we only partially summarize the details of KKW's MPC protocol and refer the reader to the original paper [14] for more details about the full protocol.

The KKW MPC protocol involves n players that compute a Boolean circuit on the secret input x. The privacy property of the protocol assures that revealing the states and randomness of $n-1$ (all-but-one) players reveals nothing about the secret input x. The protocol maintains the invariant that, for each wire in the circuit α, the players hold an n-out-of-n XOR-based secret sharing of a random mask λ_α, denoted by $[\lambda_\alpha]$, along with the public masked value of the wire $\hat{z}_\alpha = z_\alpha \oplus \lambda_\alpha$ on the input x.

During the preprocessing phase, shares are distributed among the players as follows. For each wire α that is either an input wire of the circuit or the output wire of an AND gate, the players are given $[\lambda_\alpha]$, where $\lambda_\alpha \in \{0,1\}$ is uniform. For an XOR gate with input wires α, β and output wire γ, let $\lambda_\gamma = \lambda_\alpha \oplus \lambda_\beta$ (the players can compute $[\lambda_\gamma]$ locally). Finally, for each AND gate with input wires α, β, the players are given $[\lambda_{a,b}]$, where $\lambda_{\alpha,\beta} = \lambda_\alpha \cdot \lambda_\beta$.

The uniform shares of $\{\lambda_\alpha\}$ are generated by each player P_i by applying a PRG to its short input seed $k_i \in \{0,1\}^\kappa$ (where κ is the claimed security level). Then, each $\{\lambda_\alpha\}$ is defined implicitly by these shares. The shares of each $\{\lambda_{a,b}\}$ are also generated this way, but the final shares of P_n are constrained by the values of $\{\lambda_\alpha\}$. Therefore, P_n is given additional $|C|$ "correction bits" (where $|C|$ is the number of AND gates in the circuit) that determine its share of $\{\lambda_\alpha\}$ for each AND gate.

In the online phase, the players are given a masked value $\hat{z}_\alpha$ for each input wire α. The players inductively compute $\hat{z}_\alpha$ for all wires in the circuit. The full details of the online protocol are given in [14]. We remark that when used to instantiate MPC-in-the-head, an unopened player $i \in [1,n]$ is selected, while the views of the remaining $n-1$ players are opened. For each player this involves revealing its secret seed, while for P_n this additionally involves revealing the auxiliary $|C|$ correction bits.

4 Multi-target Attacks on Zero-Knowledge Protocols

4.1 Outline of the Attacks

In this section we give a general overview of our multi-target attacks on the KKW protocol and Picnic. We assume that the attacker has access to $D = 2^d$ runs of the underlying MPC-in-the-head protocol (generated by a single of by multiple users). In each run, the views of all-but-one player are opened, along with their randomness. We refer to the player whose view is not opened as the unopened player. The randomness used by each player is generated by a PRG initialized with a seed of length κ bits, where κ is the claimed security level against classical attack algorithms. A crucial assumption required for the multi-target attacks is that for each run, the attacker can extract a string of bits output by the PRG of the unopened player.

Below, we provide a very rough outline of the steps of the multi-user multi-target attack and its analysis in the setting described above.

1. For each run $r \in [1, 2^d]$, extract a string of bits b_r that are output by the PRG of the unopened player, and store b_r along with run r.
2. For each PRG seed $k \in [1, 2^{\kappa-d}]$,[a] derive a corresponding PRG output string b'_k using the seed k, and compare with the 2^d stored strings b_r.
3. For each matching pair r, k such that $b_r = b'_k$, compute and output the corresponding secret witness x.

[a] The seed values can be selected arbitrarily.

After trying $2^{\kappa-d}$ random seeds to Step 2, according to the birthday paradox, the attacker will test a seed used in one of the 2^d runs with high probability (assuming that the players' seeds are selected uniformly at random). Given that in Step 1 the attacker can extract sufficiently many PRG output bits from each run,[10] then the expected number of matches will be (a small) constant. Finally, assuming that Step 3 can indeed be performed, the attacker will recover the secret witness for the corresponding run. In the information theoretic model assumed in the security analysis of Picnic and KKW, the complexity of the attack is $2^{\kappa-d}$ invocations of the PRG (as long as $d \leq \kappa/2$). However, in practice the computational complexity could be higher, depending on how efficiently the matching in Step 2 is performed.

Next, we describe each one of these steps for the KKW protocol. The dedicated attack on Picnic (detailed in Sect. 5) uses a variant of the attack above in order to optimize its complexity. In particular, for a range of parameter values, it filters out some of the 2^d runs in the first step and keeps only those that satisfy a certain condition which allows more efficient matching in the second step.

4.2 A Multi-target Attack on the KKW Protocol

We describe the step details of the multi-target attack on the KKW protocol. In contrast to our analysis of Picnic, we will not calculate the concrete (computational) complexity of the attack. In particular, we will reduce the second step of the attack to a known problem, but will not analyze the known algorithms for this problem in order to determine the best one for a given set of parameters.

[10] In general, κ bits are sufficient to uniquely determine the key on average. However, even if several candidate keys are recovered, they can be filtered against the public key.

Step 1: Deriving PRG Output of the Unopened Player. We focus on the additive secret sharing of $\lambda_{a,b} = \lambda_a \cdot \lambda_b$. We assume that the view of P_i is unopened for $i \neq n$, hence the attacker has all shares of $\lambda_{a,b}$, except for the i'th share that is computed using a PRG applied to the seed of unopened P_i. Observe that $\lambda_{a,b} = \lambda_a \cdot \lambda_b$ is not uniform, as it is equal to 0 with probability 3/4. Consequently, the attacker can compute a guess for P_i's share of $\lambda_{a,b}$, which is correct with probability 3/4 by XORing together all the known $n-1$ shares. Hence, in this case, the attacker does not obtain direct outputs of P_i's pseudo-random bits, but rather noisy bits with a noise of 1/4.

Step 2: Matching a Run and a PRG Seed. According to the previous step, finding a match between the 2^d runs and $2^{\kappa-d}$ PRG seeds reduces to finding a pair of highly correlated strings (with expected correlation of 3/4) among two groups of strings (which, other than the matching pair, are assumed to be independent and uniform). This is known as the nearest neighbor search problem. The trivial algorithm for this problem simply exhausts all string pairs and runs in time $2^{\kappa-d} \cdot 2^d = 2^\kappa$. However, there are more efficient algorithms for this well-studied problem (cf. [18,23]).

Step 3: Recovering the Secret Witness. Given a run and a seed for the unopened player, we can compute all random bits used by this player in the run.

In the KKW protocol, for each wire in the circuit, the players holds an n-out-of-n secret sharing of a random mask along with masked value of the wire, which is public (and given to the players in the online execution of the protocol). In particular, this applies to the input wires, whose value encodes the bits of the secret witness x. The randomness of the unopened player allows the attacker to compute the missing share for each wire α of x, and thus compute the random mask λ_α for this wire by summing together (XORing) all the n shares for the mask $[\lambda_\alpha]$. Finally, the attacker XORs the mask λ_α with the public masked value of the wire $\hat{z}_\alpha = z_\alpha \oplus \lambda_\alpha$, which gives the value of the corresponding bit of x, z_α.

5 The Multi-target Attack on Picnic

The attack on Picnic is a variant of the general attack of Sect. 4.1. In this section, we describe it in more detail and start with an overview below.

5.1 Overview of the Attack

Given $D = 2^d$ runs, our goal is to devise a concrete attack on Picnic by matching the PRG output of the unopened player in each run with output obtained by evaluating the PRG with arbitrary seeds (similarly to the generic attack described in Sect. 4). If each run would contain values about the same PRG output bits, we could sort these values and efficiently match each PRG evaluation with the runs. However, as we will see later each run contains data about different bits of the PRG output of the corresponding unopened player (and the number of known bits varies according to the run). Based on this fact, we describe below a more specific (yet still incomplete) outline of the steps, parameterized by κ, d, d', ℓ.

1. Out of 2^d runs, filter out ones that contain less data (about the PRG output of the unopened player) than some threshold.
2. For of each remaining run, $r \in [1, 2^{d'}]$: extract a prefix of ℓ bits that are output by the PRG of the unopened player (including possible unknown bits). Enumerate over all possible guesses for the unknown bits in the prefix, and store all the generated fully specified ℓ-bit *expanded strings* in a hash table (with a pointer to run r).
3. For each PRG seed $k \in [1, 2^{\kappa-d'}]$: derive an ℓ-bit PRG output string using the seed k, and search for it in the hash table. For each match: obtain the corresponding run r and compare the additional PRG output bits computed from this run with the PRG output. In case of equality, compute and output the corresponding secret key x.

Analysis Sketch. We briefly analyze the attack for the specific case where we wish to obtain the information theoretic complexity of $2^{\kappa-d}$ (assuming $d \leq \kappa/2$). In this case, we must have $d = d'$, i.e., we cannot use any filtering in Step 1.

We introduce another parameter $0 < \tau \leq 1$, which quantifies the fraction of bits that we can extract from each run about the ℓ-bit prefix of the PRG output of the unopened player. Namely, we assume that for an ℓ-bit prefix, we can determine $\tau\ell$ bits, while $(1-\tau)\ell$ are unknown.[11] Hence, for each run, we obtain $2^{(1-\tau)\ell}$ expanded strings in Step 2 and the hash table contains a total of $2^{d+(1-\tau)\ell}$ strings of ℓ bits. We refer to τ as the *information rate* that we can achieve.

Given a random ℓ-bit PRG output in Step 3, the expected number of matches with the hash table is $2^{-\ell} \cdot 2^{d+(1-\tau)\ell} = 2^{d-\tau\ell}$, hence the total number of matches tested in the attack (before the key is recovered) is $2^{\kappa-d} \cdot 2^{d-\tau\ell} = 2^{\kappa-\tau\ell}$.

Taking into account all the steps, the expected complexity of the attack is $max(2^{\kappa-d}, 2^{d+(1-\tau)\ell}, 2^{\kappa-\tau\ell})$. We balance the first and third terms by setting $\tau\ell = d$, or $\ell = d/\tau$. Then, the complexity becomes $max(2^{\kappa-d}, 2^{d/\tau})$, which implies that information theoretic complexity can be obtained as long as $d/\tau \leq \kappa - d$, or $d \leq \kappa \cdot (\tau/(1+\tau))$. The optimal complexity in this case is $2^{\kappa-d} = 2^{\kappa(1/(1+\tau))}$. When $d > \kappa \cdot (\tau/(1+\tau))$, the information theoretic complexity cannot be reached, and we will apply filtering to optimize the complexity.

Optimizations and Parameters for the Attack. Clearly, the complexity of the attack depends in a strong way on the information rate τ, namely, on the ability to extract as much information as possible from each run about the PRG output of the unopened player. The first part of the concrete analysis below (which is the most technical one) involves deriving methods that maximize the

[11] We assume here for that sake of simplicity that τ is constant and does not depend on the analyzed run (although as we will see later, this does not necessarily hold).

information rate. We first show that a naive method achieves $\tau = 1/4$, giving (optimal) complexity of $2^{\kappa(1/(1+\tau))} = 2^{4\kappa/5} \approx 2^{102}$ for $\kappa = 128$. We then utilize the design of Picnic (and the underlying LowMC circuit) in order to maximize the information rate. In particular, we obtain $\tau = 1/2$, which significantly improves the complexity to $2^{2\kappa/3} \approx 2^{85}$ for $\kappa = 128$. Finally, by applying filtering, we reduce the optimal complexity to about 2^{77}.

As a concrete example of the parameters, we note that our optimized attack has $\tau \geq 1/2$, hence we need to match $\ell = d/\tau < 2d$ PRG output bits. In this paper, we only consider data complexity of $d < 64+9 = 73$, hence $\ell < 2{\cdot}73 = 146$. These PRG output bits are used in the evaluation of $\lceil 146/3 \rceil = 49$ Sboxes, whereas all LowMC variants in Picnic have at least 200 Sboxes.

5.2 Deriving PRG Output of the Unopened Player

We start by describing a preliminary method to extract PRG output of the unopened player. We then present two optimized methods, exploiting the specific Sbox design of LowMC. In the extended version of this paper [9] we describe an additional method which is not directly used in our attack, but is interesting nevertheless.

Preliminary Extraction Method. We consider a binary multiplication gate $(w_c = w_a \cdot w_b)$. Recall that in ZKBoo (and ZKB++),

$$w_c^{(i)} = (w_a^{(i)} \cdot w_b^{(i)}) + (w_a^{(i+1)} \cdot w_b^{(i)}) + (w_a^{(i)} \cdot w_b^{(i+1)}) + R_i(c) - R_{i+1}(c). \quad (2)$$

Let us assume that the views and random seeds of players 2, 3 are revealed. Consider $i = 3$, for which the equation above reduces to:

$$w_c^{(3)} = (w_a^{(3)} \cdot w_b^{(3)}) + (w_a^{(1)} \cdot w_b^{(3)}) + (w_a^{(3)} \cdot w_b^{(1)}) + R_3(c) - R_1(c).$$

Moreover, we assume that

$$w_a^{(3)} = w_b^{(3)} = 0. \quad (3)$$

Note that since view 3 is revealed, the attacker knows when this event occurs. Conditioned on this event, the equation simplifies to $w_c^{(3)} = R_3(c) - R_1(c)$, or

$$R_1(c) = R_3(c) - w_c^{(3)}.$$

Since $R_3(c)$ and $w_c^{(3)}$ are known from random seed and view of player 3 (respectively), then the attacker can compute $R_1(c)$ with probability 1, conditioned on (3).

In Boolean circuits (as in Picnic), we expect (generalized) condition (3) to hold for 1/4 of the AND gates (the probability is over the randomness of the view of P_{e+1}). Consequently, the attacker knows about $\tau = 1/4$ of the output bits produced by the PRG (not including the ones used in the initial Share function). Note that the locations of these known output bits depend on $w_a^{(e+1)}$ and $w_b^{(e+1)}$, which are different for each run.

Below, we exploit the specific structure of the Picnic circuit in order to optimize the information rate.

Extraction Method 1. Recall that the AND operations performed by a LowMC Sbox are

$$S'(w_{a_1}, w_{a_2}, w_{a_3}) = (w_{a_2} \cdot w_{a_3}, w_{a_1} \cdot w_{a_3}, w_{a_1} \cdot w_{a_2}),$$

where S' denotes the function obtained from S by only considering AND operations. Denote the output wires of these 3 AND operations by $w_{c_1}, w_{c_2}, w_{c_3}$, respectively, and write the basic equation of (2) with $i = 3$ for the 3 AND gates:

$$\begin{aligned} w_{c_1}^{(3)} &= (w_{a_2}^{(3)} \cdot w_{a_3}^{(3)}) \oplus (w_{a_2}^{(1)} \cdot w_{a_3}^{(3)}) \oplus (w_{a_2}^{(3)} \cdot w_{a_3}^{(1)}) \oplus R_3(c_1) \oplus R_1(c_1), \\ w_{c_2}^{(3)} &= (w_{a_1}^{(3)} \cdot w_{a_3}^{(3)}) \oplus (w_{a_1}^{(1)} \cdot w_{a_3}^{(3)}) \oplus (w_{a_1}^{(3)} \cdot w_{a_3}^{(1)}) \oplus R_3(c_2) \oplus R_1(c_2), \\ w_{c_3}^{(3)} &= (w_{a_1}^{(3)} \cdot w_{a_2}^{(3)}) \oplus (w_{a_1}^{(1)} \cdot w_{a_2}^{(3)}) \oplus (w_{a_1}^{(3)} \cdot w_{a_2}^{(1)}) \oplus R_3(c_3) \oplus R_1(c_3). \end{aligned} \quad (4)$$

Assuming the views of players 2, 3 are revealed, the unknown values in these 3 equations are the view and randomness variables of player 1:

$$R_1(c_1), R_1(c_2), R_1(c_3), w_{a_1}^{(1)}, w_{a_2}^{(1)}, w_{a_3}^{(1)}.$$

Observe that for every value of the 3 known bits $w_{a_1}^{(3)}, w_{a_2}^{(3)}, w_{a_3}^{(3)}$, we obtain a linear equation system with 3 equations. Our goal is to perform Gaussian elimination on this system in order to eliminate the unknown variables $w_{a_1}^{(1)}, w_{a_2}^{(1)}, w_{a_3}^{(1)}$ and remain with linear relations in the 3 randomness variables of player 1, $R_1(c_1), R_1(c_2), R_1(c_3)$. We are thus interested in the rank of the equation system in $w_{a_1}^{(1)}, w_{a_2}^{(1)}, w_{a_3}^{(1)}$ as a function of the known variables $w_{a_1}^{(3)}, w_{a_2}^{(3)}, w_{a_3}^{(3)}$.

Since the equation system is symmetric, the rank depends only on the Hamming weight (HW) of $w_{a_1}^{(3)}, w_{a_2}^{(3)}, w_{a_3}^{(3)}$. It is easy to check that the following holds:

- If $HW = 0$ (i.e., $w_{a_1}^{(3)} = w_{a_2}^{(3)} = w_{a_3}^{(3)} = 0$), the rank is 0 and we obtain the 3 PRG output bits $R_1(c_1), R_1(c_2), R_1(c_3)$.
- If $HW > 0$, the rank is 2 and we obtain 1 PRG output bit (or a linear combination of output bits) according to the specific values of $w_{a_1}^{(3)}, w_{a_2}^{(3)}, w_{a_3}^{(3)}$.

The first case $w_{a_1}^{(3)} = w_{a_2}^{(3)} = w_{a_3}^{(3)} = 0$ occurs with probability 1/8 (over the randomness of the view of P_3). Note that the equation system is never of full rank and we can always obtain at least 1 bit of information about $R_1(c_1), R_1(c_2), R_1(c_3)$.

Example 1. If $w_{a_1}^{(3)} = w_{a_2}^{(3)} = w_{a_3}^{(3)} = 1$, we XOR together the 3 equations to eliminate $w_{a_1}^{(1)}, w_{a_2}^{(1)}, w_{a_3}^{(1)}$, and can compute the value of $R_1(c_1) \oplus R_1(c_2) \oplus R_1(c_3)$.

We performed the analysis assuming the views of P_2, P_3 were opened in the run, but similar analysis applies (with appropriate indexing modifications) regardless of which 2 views are opened. We summarize our findings below.

Proposition 1. *Given access to the open views and randomness of* P_e, P_{e+1} *in Picnic, for any triplet of wires that are input to a LowMC Sbox and its corresponding triplet of output wires* $w_{c_1}, w_{c_2}, w_{c_3}$*, with probability 1/8 (over the randomness of the view of* P_{e+1}*), we can easily compute the corresponding PRG output bits of* P_{e+2}*, namely,* $R_{e+2}(c_1), R_{e+2}(c_2), R_{e+2}(c_3)$*. Otherwise (with probability 7/8) we can compute one of seven possible linear equations on these bits, where each particular linear equation is obtained with probability 1/8 (over the randomness of the view of* P_{e+1}*).*

Hence, with high probability, for *most runs* we obtain 3 bits of information for at least 1/8 of the Sboxes and 1 bit of information for the remaining Sboxes. We therefore obtain at least $3/8 + 7 \cdot 1/8 = 10/8 = 5/4$ bits of information on average per 3-bit Sbox, or $\tau_1 = 5/12$ bits of information per PRG output bit (for most runs). This is significantly better than the ratio of 1/4 obtained in a generic manner above.

Extraction Method 2. Assume that P_2, P_3 are opened and reconsider the equation system of (4). If we guess the values of $w_{a_1}, w_{a_2}, w_{a_3}$, we can easily deduce the unknown PRG output bits $R_1(c_1), R_1(c_2), R_1(c_3)$ by first computing $w_{a_1}^{(1)}, w_{a_2}^{(1)}, w_{a_3}^{(1)}$. On its own, this is a useless observation since we could have directly guessed these 3 PRG output bits. However, let us assume that the analyzed Sbox is located in the first LowMC Sbox layer. This implies that each of $w_{a_1}, w_{a_2}, w_{a_3}$ is a linear function of the unknown LowMC secret key x corresponding to the run (as there is no non-linear function applied to compute these bits from the known plaintext). Therefore, the knowledge of the 3 bits $w_{a_1}, w_{a_2}, w_{a_3}$ input to the Sbox directly translates to knowledge of 3 linear equations on the LowMC secret key. More specifically, we have $w_{a_i} = l_{a_i}(x)$ for $i \in \{1, 2, 3\}$, where $l_{a_i}(x)$ is a linear equation on the secret key x. Recall that the Share function outputs 3 shares that sum to the LowMC secret key $x = x_1 \oplus x_2 \oplus x_3$. Therefore, for $i \in \{1, 2, 3\}$ we have

$$w_{a_i} = l_{a_i}(x) = l_{a_i}(x_1) \oplus l_{a_i}(x_2) \oplus l_{a_i}(x_3), \text{ or}$$

$$l_{a_i}(x_1) = w_{a_i} \oplus l_{a_i}(x_2) \oplus l_{a_i}(x_3). \tag{5}$$

We (assume to) know $w_{a_i}, l_{a_i}(x_2), l_{a_i}(x_3)$, and therefore, we can derive $l_{a_i}(x_1)$. The 3 bits $l_{a_i}(x_1)$ are linear combinations of PRG output bits of P_1 that are output by the Share function, and we have shown that they are directly deduced from the knowledge of w_{a_i}.

Altogether, we guess 3 bits and obtain 6 PRG output bits. Crucially for the attack, that indices of the computed 6 PRG (linear combinations of) output bits are fixed among all runs.

Proposition 2. *Given access to the open views and randomness of P_e, P_{e+1} in Picnic for $e \in \{2,3\}$, for any triplet of wires that are input to a LowMC Sbox in the first Sbox layer, a guess for the 3 bit values for these wires allows to easily compute a guess for values of 6 (linear combinations of) PRG output bits of P_{e+2}. The locations of these output bits depend only on the choice of Sbox. Moreover, the same holds for any Sbox in the i'th Sbox layer, given that we have a guess for the all the Sbox inputs in layers $1, 2, \ldots, i-1$.*

Note that the proposition only applies to $e \in \{2,3\}$, as for $e = 1$, the key share of $P_{e+2} = P_3$ (namely x_3) is not computed using a PRG. The last part of the proposition holds since each input wire to each Sbox in the i'th layer can be expressed as a linear combination of the key bits and the output bits of the Sboxes in previous layers. These output bits are (assumed to be) known.

Recall that in the previous extraction method we obtained an information rate of $\tau_1 = 5/12$ for most runs. Here, we guess 3 bits and obtain 6 PRG output bits, i.e. $\tau_2 = 1/2 > \tau_1$, and hence this method can be viewed as an improvement over the previous one. On the other hand, for specific runs which deviate from the average, the first method may yield a higher information rate, thus the methods are not always directly comparable. Indeed, our attack will combine these methods according to some parameter values.

5.3 Exploiting PRG Output of the Unopened Player

We focus on a single run that contains data about the PRG output of the unopened player (P_{e+2}). We only exploit data for runs with $e \in \{2,3\}$ in the attack.

We call the useful data extracted from a run a *target string* (TS). The target string is indexed according to triplets of bits (corresponding to the Sboxes in LowMC's circuit), where the relevant information about each triplet consists of the known view and randomness bits of P_{e+1}. For example, if $e = 2$ (i.e., players 2,3 are opened), then for each Sbox in LowMC's circuit, the TS contains all the known view and randomness bits of P_3 that appear in the equation system of (4). Recall from Proposition 1 that for each triplet, we may obtain all the 3 PRG output bits of the unopened player, and in this case, we call it a *full triplet.* Otherwise, we obtain 1 bit of a linear equation on the 3 PRG output bits (the linear equation itself depends on the view of P_{e+1}) and call the triplet a *partial triplet.*

The triplet data in each target string is sorted according to the indices of LowMC's Sboxes. Importantly, Sboxes in each layer i appear together before layer $i+1$ (the order within each layer is chosen arbitrarily, but is consistent among all target strings). Given a target string ts, we refer to the data of the i'th triplet by $ts[i]$, and to the data of the triplet sequence $i, i+1, \ldots, j$ as $ts[i, j]$.

For the purpose of exploiting Proposition 2, we also need auxiliary information from the Share function about the shares of P_e, P_{e+1}. We append this data to each TS.

Target String Expansion. We elaborate on Step 2 of the general attack of Sect. 5.1 by defining the expansion of a target string ts. Essentially, it is a set of strings that correspond to all possible PRG outputs of the unopened player (for some specific triplets) that match the partial information in ts.

Given parameters $t_1, t_2 \geq 0$, assume $ts[t_1+1, t_1+t_2]$ contains $t_3 \leq t_2$ full triplets (and $t_2 - t_3$ partial triplets). We can expand the t_2 triplets of $ts[t_1 + 1, t_1 + t_2]$ according to Proposition 1 into a set of $2^{2(t_2-t_3)}$ expanded strings, each of length $3t_2$ bits. Similarly, we can expand triplets $ts[1, t_1]$ according to Proposition 2 into a set of 2^{3t_1} expanded strings, each of length $6t_1$ bits.

Combining these two expansion methods into one expansion function (denoted $expand(ts)$), we obtain a set of $2^{3t_1+2(t_2-t_3)}$ strings. Each string is of length $\ell = 6t_1 + 3t_2$ bits and represents possible values for certain $6t_1 + 3t_2$ PRG output bits, which we call *matching bits* (mb). Given a PRG seed k, we denote by $PRG_k[mb]$ the $6t_1 + 3t_2$-bit PRG output value for the matching bits, when evaluated with k. Given a TS, ts, generated with seed k, only one of the $2^{3t_1+2(t_2-t_3)}$ strings in $expand(ts)$ is equal to $PRG_k[mb]$. We summarize below.

Proposition 3. *Given a target string ts, parameters $t_1, t_2 \geq 0$, and assuming $ts[t_1 + 1, t_1 + t_2]$ contains $t_3 \leq t_2$ full triplets, the expansion of ts with parameters t_1, t_2, t_3 is a set denoted $expand(ts)$ that contains $2^{3t_1+2(t_2-t_3)}$ expanded strings, each of length $6t_1+3t_2$ bits. Each string in $expand(ts)$ contains $6t_1+3t_2$ possible values for the matching bits, derived from ts by a guess for the missing information.*

5.4 The Multi-target Attack

It is obvious that a run with a large number of full triplets in $ts[t_1+1, t_1+t_2]$ is more useful for our purpose, as it contains more data about the PRG output of the unopened player (equivalently, its expanded set according to Proposition 3 is relatively small). Hence, we filter the target data strings, keeping those with a large value of t_3. Each remaining target string is then expanded and the resultant expanded strings are stored to be matched with data obtained from PRG evaluations. Based on Proposition 1, we derive the following proposition.

Proposition 4. *Given integer parameters $t_1 \geq 0$ and $0 \leq t_3 \leq t_2$, the probability (over the randomness of the view of P_{e+1}) that for an arbitrary target string ts, the t_2 triplets of $ts[t_1 + 1, t_1 + t_2]$ contain at least t_3 full triplets is*

$$\Gamma(t_2, t_3) \stackrel{\text{def}}{=} \sum_{i=t_3}^{t_2} \binom{t_2}{t_3} \left(\frac{1}{8}\right)^{i} \cdot \left(\frac{7}{8}\right)^{t_2-i}.$$

We describe the attack below, using positive integer parameters $\kappa, r, r', t_1, t_2, t_3$.

1. For each of the 2^d available runs: denote by ts the target string of the current run. If $e+2 = 3$ (P_{e+2} is the unopened player), or if $ts[t_1+1, t_1+t_2]$ contains less than t_3 full triplets, then discard the run.
2. For of each remaining $2^{d'}$ runs: compute $expand(ts)$, and store each of the $6t_1 + 3t_2$-bit expanded strings $s \in expand(ts)$ (along with a pointer to ts) in a hash table L, indexed by s.
3. For each PRG seed $k \in [1, 2^{\kappa-d'}]$:
   - Evaluate the PRG, derive $PRG_k[mb]$ and search for a match in L.
   - For each match with an expanded string s, recover the corresponding ts.
   - Continue to compute the PRG output on k and compare with ts.
   - If the PRG outputs match, we have guessed the correct seed k for the unopened player in ts with high probability. Derive and output the signing key x based on the Share function $x = x_1 \oplus x_2 \oplus x_3$ by computing the missing share using k.

Analysis. In order to analyze the attack, observe that on average, in 2/3 of the runs P_1 or P_2 are unopened. Out of the remaining runs, a fraction of $\Gamma(t_2, t_3)$ contains at least t_3 full triplets in $ts[t_1 + 1, t_1 + t_2]$ (according to Proposition 4). Consequently, we expect

$$2^{d'} = 2/3 \cdot \Gamma(t_2, t_3) \cdot 2^d. \tag{6}$$

Next, according to Proposition 3, L is expected to contain at most $2^{d'+3t_1+2(t_2-t_3)}$ expanded strings, which gives the memory complexity of the attack (and a lower bound on its time complexity).

Finally, the expected number of matches in Step 3 between a random $6t_1+3t_2$-bit string $PRG_k[mb]$ and one of the expanded $2^{d'+3t_1+2(t_2-t_3)}$ strings in L is at most $2^{d'+3t_1+2(t_2-t_3)} \cdot 2^{-6t_1-3t_2} = 2^{d'-3t_1-t_2-2t_3}$. Hence, the total expected number of matches that we need to test in Step 3 is upper bounded by $2^{\kappa-d'} \cdot 2^{d'-3t_1-t_2-2t_3} = 2^{\kappa-3t_1-t_2-2t_3}$.

Taking all steps into account, the total time complexity is upper bounded by $max(2^d, 2^{\kappa-d'}, 2^{d'+3t_1+2(t_2-t_3)}, 2^{\kappa-3t_1-t_2-2t_3})$. Plugging in the value of d' calculated in (6), we obtain

$$max(2^d, 3/2 \cdot 1/\Gamma(t_2, t_3) \cdot 2^{\kappa-d}, 2/3 \cdot \Gamma(t_2, t_3) \cdot 2^{d+3t_1+2(t_2-t_3)}, 2^{\kappa-3t_1-t_2-2t_3}). \tag{7}$$

We balance the second and fourth terms by setting

$$2^{3t_1+t_2+2t_3} = 2/3 \cdot \Gamma(t_2, t_3) \cdot 2^d, \tag{8}$$

i.e., $2/3 \cdot \Gamma(t_2, t_3) = 2^{3t_1+t_2+2t_3-d}$. Thus, the third term (which also represents the memory complexity) becomes $2^{6t_1+3t_2}$ and the time complexity upper bound is calculated as

$$max(2^d, 3/2 \cdot 1/\Gamma(t_2, t_3) \cdot 2^{\kappa-d}, 2^{6t_1+3t_2}), \tag{9}$$

under constraint (8). We now analyze the complexity of the attack for various choice of the free parameters t_1, t_2, t_3.

Achieving the Information Theoretic Complexity. If we want a time complexity close to the information theoretic complexity we apply a minimal amount of filtering. Based on the conclusion of Sect. 5.1, it is best to use the extraction method with has highest information rate. Thus, we use the second extraction method (summarized in Proposition 2), which has $\tau_2 = 1/2$, by setting $t_2 = t_3 = 0$ (and $\Gamma(t_2, t_3) = 1$). The analysis becomes similar to the one of Sect. 5.1, with the exception of the filtering constant 2/3 and rounding factors. We can come close to the information theoretic time complexity and obtain time complexity of $3/2 \cdot 2^{\kappa-d}$ as long as $3/2 \cdot 2^{\kappa-d} \geq 4/9 \cdot 2^{2d}$, or

$$d \leq \log 3/2 + \kappa/3$$

(the formula only holds for values of d that satisfy $2^d = 3/2 \cdot 2^{3t_1}$ for an integer t_1).

For example, if $\kappa = 128$, we can only exploit up to $2^d \leq 3/2 \cdot 2^{42}$ data (by setting $t_1 = 14$). The optimal time complexity is therefore about $2^{128-42} = 2^{86}$. For $\kappa = 192$, we can reach the information theoretic time complexity for almost the entire range of $D \leq 324 \cdot 2^{64}$, whereas for $\kappa = 256$, we obtain the information theoretic time complexity for the full range $D \leq 438 \cdot 2^{64}$.

General Time Complexity Optimization. When more than $3/2 \cdot 2^{\kappa/3}$ runs are available and our goal is to optimize time complexity, we could not accurately optimize the attack analytically as a function of the known parameters κ, d. Instead, we optimized the most precise original attack complexity Eq. (7) for several choices of κ, d by brute force. In particular, for $\kappa = 128$, if we restrict ourselves to 2^{64} signatures ($r = \log(219 \cdot 2^{64}) \approx 71$), then we (approximately) obtain $T = 2^{77}$ and $M = 2^{76}$ memory by setting $t_1 = 0, t_2 = 25, t_3 = 13$. This demonstrates the fact that when a large amount of data is available, we do not exploit the second extraction method in the optimal attack, namely, we set $t_1 = 0$. This is due to the fact that filtering a large amount of data results in a better information rate using the first extraction method.

We can also obtain some time-memory tradeoffs by applying more filtering. For example, we obtain $T = 2^{83}, M = 2^{57}$ by setting $t_1 = 0, t_2 = 19, t_3 = 13$.

6 Seed Collision Attack on Picnic

In this section we describe our seed collision attack on Picnic. Unlike our main multi-target attack, this attack (almost) reaches the information theoretic complexity of $2^\kappa = T \cdot D$ for $D = 2^{\kappa/2}$ (but only in the single-user setting for the particular point of $D = 2^{\kappa/2}$). The reason we can reach the information theoretic complexity here is that the matching step is much easier compared to that of our main multi-target attack.

Assume the attacker has access to two runs generated with the same private key. For run $i \in \{1, 2\}$, denote by $P^{(i)}_{e^{(i)}+2}$ the unopened player and assume that in both runs the unopened player uses the same seed $k^{(1)}_{e^{(1)}+2} = k^{(2)}_{e^{(2)}+2}$. Moreover, assume that $e^{(i)} \in \{2, 3\}$. Then, in both runs, the pseudo-random bits generated by the PRG of the unopened players are identical, and in particular, the outputs of their Share functions are identical, namely $x^{(1)}_{e^{(1)}+2} = x^{(2)}_{e^{(2)}+2}$. Hence, $x \oplus x^{(1)}_{e^{(1)}} \oplus x^{(1)}_{e^{(1)}+1} = x \oplus x^{(2)}_{e^{(2)}} \oplus x^{(2)}_{e^{(2)}+1}$ (since both runs are generated with the same private key x). Therefore, the attacker can easily detect this event by verifying the condition

$$x^{(1)}_{e^{(1)}} \oplus x^{(1)}_{e^{(1)}+1} = x^{(2)}_{e^{(2)}} \oplus x^{(2)}_{e^{(2)}+1}. \tag{10}$$

The key recovery process (once again) has to use the available data about the PRG outputs of the unopened players. Assuming that $k^{(1)}_{e^{(1)}+2} = k^{(2)}_{e^{(2)}+2}$, the attacker can recover the secret key x by exploiting the fact that the remaining PRG outputs of $P^{(1)}_{e^{(1)}+2}$ and $P^{(2)}_{e^{(2)}+2}$ are identical, assuming that the seeds of $P^{(1)}_{e^{(1)}+1}$ and $P^{(2)}_{e^{(2)}+1}$ are different (i.e., $k^{(1)}_{e^{(1)}+1} \neq k^{(2)}_{e^{(2)}+1}$), which occurs with high probability. This is done by independently analyzing each Sbox, observing the corresponding input triplets of bits in the two target strings for the runs. Assuming that the 3-bit randomness values of $P^{(1)}_{e^{(1)}+1}$ and $P^{(2)}_{e^{(2)}+1}$ are different for the Sbox (which occurs with probability 7/8), then the attacker can always obtain two linear equations on the secret inputs to the Sbox.

Example 2. Examine again the equation system (4), assuming for simplicity that player 1 is unopened in both runs, namely, $e^{(1)} + 2 = e^{(2)} + 2 = 1$. Moreover, assume that in the first run, the 3 relevant view bits of player 3 are equal to zero, i.e., $w^{(3)}_{a_1} = w^{(3)}_{a_2} = w^{(3)}_{a_3} = 0$. Then, we can compute the common random bits of unopened player 1, namely, $R_1(c_1), R_1(c_2), R_1(c_3)$. Furthermore, assume that for the second run, the 3 relevant view bits of player 3 satisfy $u^{(3)}_{a_1} = u^{(3)}_{a_2} = 0, u^{(3)}_{a_3} = 1$ (we index these bits with u, as they are different across the runs). Then, since the randomness of player 1 is known, we can deduce the share values $u^{(1)}_{a_1}, u^{(1)}_{a_2}$ in the second run, which reveal the wire values w_{a_1}, w_{a_2}.

Linearizing the Circuit. After analyzing all the Sboxes, the attacker knows 2 out of 3 (linear combinations of) input bits to a fraction of about 7/8 of the Sboxes. This makes the 3 output bits of each such Sbox linear functions of the inputs. We call these Sboxes *linearized Sboxes.* From the viewpoint of the attacker, the only non-linear operations that remain in the circuit involve the non-linearized Sboxes (whose expected fraction is 1/8). Based on this observation, we set up a linear equation system where the variables are the values of the κ key bits in addition to the 3 unknown output values of each non-linearized Sbox. Note that the value of every wire in the circuit can be expressed as a linear combination of these variables. Assuming that the circuit has K Sboxes, the expected number of variables is $\kappa + 3 \cdot K/8$.

In order to get the values of linear equations in the variables, we deduce values of specific wires (or linear combination of wires) in the circuit. First, note that the output of the circuit is known and gives rise to κ linear equations. We obtain additional equations based on the 2 known input bits of linearized

Sboxes. Hence, the expected number of equations is $\kappa + 2 \cdot 7K/8 = \kappa + 14K/8$. For Picnic, we expect to obtain many more equations than variables to the system, whose solution gives the secret signing key. For example, if $\kappa = 128$, then $K = 200$, implying that the expected number of variables is 203 and the number of equations is 478. A simple Chernoff bound shows that the numbers of variables and equations are close to their mean with high probability. For example, for $\kappa = 128, K = 200$, with probability (more than) 0.9999 the attacker knows 2 out of 3 input bits for at least $7/8 \cdot 200 - 25 = 150$ Sboxes. This implies that with probability 0.9999, the number of variables is at most $\kappa + 3 \cdot (200/8 + 25) = 278$ and the number of equations is at least $\kappa + 2 \cdot 150 = 428$.

In case the attacker is unlucky and still has to spend considerable effort in key recovery, it is possible to exploit several collisions for this purpose at the price of increased data complexity. As in typical collision attacks, the expected number of collisions grows quadratically as a function of the data. In particular, after obtaining 4 collisions (requiring about twice the amount of data), the expected fraction of Sboxes with 3 unknown input bits is $(1/8)^4 = 1/4096$. Since the LowMC instances only have a few hundreds of Sboxes, the system becomes completely linear in the key bits and the attacker directly solves for the key.

Details of the Attack. The seed collision attack is described below.

1. Store each of the $\approx 2/3 \cdot 2^d$ available runs (generated with the same private key) for which $e^{(i)} \in \{2, 3\}$ in a hash table L, with the value $x^{(i)}_{e^{(i)}} \oplus x^{(i)}_{e^{(i)}+1}$ as the index.
2. For each collision $x^{(i)}_{e^{(i)}} \oplus x^{(i)}_{e^{(i)}+1} = x^{(j)}_{e^{(j)}} \oplus x^{(j)}_{e^{(j)}+1}$ between runs i, j in L:
   - If $k^{(i)}_{e^{(i)}+1} = k^{(j)}_{e^{(j)}+1}$, discard the collision.
   - Otherwise, if the known PRG output bits of runs i, j do not match, discard the collision.
   - Otherwise, recover and output the secret key by solving a system of linear equations.

Analysis. Since the seeds are of a length κ bits, we need about $d = 1 + \kappa/2$ to have two runs i, j for which $k^{(i)}_{e^{(i)}+2} = k^{(j)}_{e^{(j)}+2}$ with probability larger than 1/2 (for these runs $k^{(i)}_{e^{(i)}+1} \neq k^{(j)}_{e^{(j)}+1}$ with probability $1 - 2^{-\kappa} \approx 1$). On the other hand, $x^{(i)}_{e^{(i)}} \oplus x^{(i)}_{e^{(i)}+1} = x^{(j)}_{e^{(j)}} \oplus x^{(j)}_{e^{(j)}+1}$ is a κ-bit condition that occurs for an arbitrary pair of runs with probability $2^{-\kappa}$. Hence, summing over all pairs of runs, the expected total number of collisions that we need to test in Step 2 is about $1/4 \cdot 2^{2d-\kappa} = 1$ and the complexity of the attack is $2^d = 2 \cdot 2^{\kappa/2}$. As the linear system of equations considered in the final step has several hundreds of variables, the complexity of solving it can be bounded by 2^{30} bit operations using Gaussian elimination, and this complexity can be neglected.

The Multiple User Setting. In the multiple user setting we independently run the attack on the data of each user. Assume that we have 2^u users with 2^d distinct runs available per user. Then, the success probability per user (assuming $d \leq 1 + \kappa/2$) is about $1/4 \cdot 2^{2d-\kappa}$, and the success probability across all users is roughly $1/4 \cdot 2^{u+2d-\kappa}$, implying that we need $d \approx 1 + (\kappa - u)/2$ in order to recover the key of one of the users with high (constant) probability. Therefore, we require a total of $2 \cdot 2^{u+d} = 2 \cdot 2^{(\kappa+u)/2}$ runs. More generally, if the number of available runs varies among the different users, i.e., we have 2^{d_i} available runs for user i, then the success probability is proportional to

$$1/4 \cdot 2^{-\kappa} \cdot \sum_{i=1}^{2^u} 2^{2d_i}.$$

The expression is minimized when all d_i's are equal, implying that a skewed distribution of data helps the attacker.

7 Multi-target Attacks on Additional Cryptosystems

In this section we give two examples of optimizations used in MPC protocols that weaken their resistance to multi-target attacks. We further give an example of a general public key scheme that is vulnerable to a multi-user single-target attack. For each example, we reference at least one vulnerable scheme that was proposed recently. We note that all of the described attacks can be prevented by appropriate use of salting. This results in some performance overhead, which depends on the underlying scheme.

We assume throughout this section that the desired security level is κ bits.

7.1 Hash-Based Commitments Optimization

We consider the widely used hash-based commitment scheme, which utilizes a hash function $H : \{0,1\}^* \rightarrow \{0,1\}^{2\kappa}$. In order to commit to a string W, one selects a sufficiently long random string *rand*, and the commitment is defined as $H(rand, W)$. The commitment is opened by revealing $rand, W$. Several MPC protocols such as [14,17] optimize this scheme by omitting *rand* and defining the commitment as $H(W)$, given that W has sufficient min-entropy of at least κ bits. This way, only W has to be sent when opening the commitment, thus saving communication. However, the optimization clearly exposes the protocol to multi-user multi-target attacks, as the attacker may try to derive a preimage to one out of many available commitments.

7.2 Seed Tree Optimization

We analyze an MPC protocol optimization that is used in the KKW protocol [14]. In the unoptimized protocol, the seeds of all n players are (essentially) independent and opening $n-1$ players out of n requires $\kappa \cdot (n-1)$ bits of communication.

As shown in Sect. 4.2, the unoptimized protocol is already vulnerable to multi-user multi-target attacks. We now consider an optimized version of the protocol described in [14], which further weakens its security against such attacks.

The optimization involves building a seed tree construction (cf. [19]) which generates seeds for the n players that participate in the MPC protocol in a way that reduces the communication required to reveal $n-1$ seeds. The seed tree is a binary tree, where each node has a label of κ bits. The label of the root is a randomly generated master seed of κ bits, and the two κ-bit labels of the 2 children of each node are defined recursively by running a PRG on the label of the parent and outputting 2κ bits. The tree is of depth $\log n$ and the seeds of the n players are defined as the labels of the n leaves. In order to reveal the seeds of all players but player i, it is sufficient to reveal the labels of the siblings of the path from the root to leaf i, which requires only $\kappa \cdot \log n$ communication.

Observe that in the original protocol, the attacker had only a single target per run, which was the seed of the unopened player. In contrast, in the optimized protocol, each node on the path from the root to the unopened leaf i is a target, as the attacker knows one of its κ-bit outputs from the $\log n$ revealed labels. Hitting one of these targets allows the attacker to easily compute the label of leaf i. The degradation in security is proportional to $\log n$, which is not large, but should still be noted.

7.3 Public Key Scheme Construction

Finally, we consider a public key scheme that uses a secret signing key $x \in \{0,1\}^\kappa$ and generates its public key as $pk = g(x)$, where g is a deterministic function. Typically, g involved invoking a PRF at least once on input x and additional (deterministically generated) inputs. An example of such a scheme is the recently proposed TACHYON signature algorithm [2] (which was presented at ACM CCS 2018). This described scheme is clearly vulnerable to a multi-user single-target attack, where the attacker obtains access to several public keys that belong to several users. The attacker attempts to recover the secret key of one (or several) of the users by iteratively guessing a value for x', computing $pk' = g(x')$, and comparing with the available public keys.

8 Conclusions

In this paper we described multi-target attacks on the Picnic signature scheme and on the related KKW protocol. Our attacks have two features that stem from Picnic's novel design and distinguish them from standard multi-target attacks:

1. The vulnerability of the cryptosystem (Picnic) to multi-target attacks is not evident, even when one carefully looks for it. As a result, it was missed by the designers.
2. Internally, the multi-target attacks cannot apply a typical sort-and-match algorithm, and efficient key recovery requires cryptanalytic effort.

The attacks expose a design weakness in the way Picnic uses a PRG during signing. Although our attacks are generally impractical, in some cases this design weakness could be leveraged in combination with an additional implementation weakness (such as generation of seeds with low entropy[12]) to mount a practical attack. Such an attack would have been harder to carry out had the PRG been appropriately salted.

Besides the short-term impact of our analysis on enhancing Picnic's security, we hope that it will contribute to the secure design of novel cryptosystems in the future.

Acknowledgements. The authors would like to thank the Picnic designers for helpful discussions about this work and anonymous reviewers for valuable suggestions. The first author was supported by the Israeli Science Foundation through grant No. 573/16 and by the European Research Council under the ERC starting grant agreement No. 757731 (LightCrypt).

References

1. Albrecht, M.R., Rechberger, C., Schneider, T., Tiessen, T., Zohner, M.: Ciphers for MPC and FHE. In: Oswald and Fischlin [21], pp. 430–454
2. Behnia, R., Ozmen, M.O., Yavuz, A.A., Rosulek, M.: TACHYON: fast signatures from compact knapsack. In: Lie et al. [16], pp. 1855–1867
3. Bernstein, D.J., et al.: SPHINCS: practical stateless hash-based signatures. In: Oswald and Fischlin [21], pp. 368–397
4. Boneh, D., Eskandarian, S., Fisch, B.: Post-quantum group signatures from symmetric primitives. IACR Cryptology ePrint Archive 2018:261 (2018)
5. Chase, M., et al.: Picnic: a family of post-quantum secure digital signature algorithms. https://microsoft.github.io/Picnic/
6. Chase, M., et al.: Post-quantum zero-knowledge and signatures from symmetric-key primitives. In: Thuraisingham, B.M., Evans, D., Malkin, T., Xu, D. (eds.) Proceedings of the 2017 ACM SIGSAC Conference on Computer and Communications Security, CCS 2017, Dallas, TX, USA, 30 October–03 November 2017, pp. 1825–1842. ACM (2017)
7. Checkoway, S., et al.: On the practical exploitability of dual EC in TLS implementations. In: Fu, K., Jung, J. (eds.) Proceedings of the 23rd USENIX Security Symposium, San Diego, CA, USA, 20–22 August 2014, pp. 319–335. USENIX Association (2014)
8. Derler, D., Ramacher, S., Slamanig, D.: Post-quantum zero-knowledge proofs for accumulators with applications to ring signatures from symmetric-key primitives. In: Lange, T., Steinwandt, R. (eds.) PQCrypto 2018. LNCS, vol. 10786, pp. 419–440. Springer, Cham (2018). https://doi.org/10.1007/978-3-319-79063-3_20
9. Dinur, I., Nadler, N.: Multi-target attacks on the picnic signature scheme and related protocols. IACR Cryptology ePrint Archive 2018:1212 (2018)
10. Fiat, A., Shamir, A.: How to prove yourself: practical solutions to identification and signature problems. In: Odlyzko, A.M. (ed.) CRYPTO 1986. LNCS, vol. 263, pp. 186–194. Springer, Heidelberg (1987). https://doi.org/10.1007/3-540-47721-7_12

[12] The Picnic specification recommends to generate the seeds based on the (high-entropy) private key, but does not (and cannot) enforce this.

11. Giacomelli, I., Madsen, J., Orlandi, C.: ZKBoo: faster zero-knowledge for Boolean circuits. In: Holz, T., Savage, S. (eds.) 25th USENIX Security Symposium, USENIX Security 2016, Austin, TX, USA, 10–12 August 2016, pp. 1069–1083. USENIX Association (2016)
12. Hülsing, A., Rijneveld, J., Song, F.: Mitigating multi-target attacks in hash-based signatures. In: Cheng, C.-M., Chung, K.-M., Persiano, G., Yang, B.-Y. (eds.) PKC 2016. LNCS, vol. 9614, pp. 387–416. Springer, Heidelberg (2016). https://doi.org/10.1007/978-3-662-49384-7_15
13. Ishai, Y., Kushilevitz, E., Ostrovsky, R., Sahai, A.: Zero-knowledge from secure multiparty computation. In: Johnson, D.S., Feige, U. (eds.) Proceedings of the 39th Annual ACM Symposium on Theory of Computing, San Diego, California, USA, 11–13 June 2007, pp. 21–30. ACM (2007)
14. Katz, J., Kolesnikov, V., Wang, X.: Improved non-interactive zero knowledge with applications to post-quantum signatures. In: Lie et al. [16], pp. 525–537
15. Lamport, L.: Constructing digital signatures from a one way function. Technical report SRI-CSL-98, SRI International Computer Science Laboratory (1979)
16. Lie, D., Mannan, M., Backes, M., Wang, X. (eds.): Proceedings of the 2018 ACM SIGSAC Conference on Computer and Communications Security, CCS 2018, Toronto, ON, Canada, 15–19 October 2018. ACM (2018)
17. Lindell, Y., Riva, B.: Blazing fast 2PC in the offline/online setting with security for malicious adversaries. In: Ray, I., Li, N., Kruegel, C. (eds.) Proceedings of the 22nd ACM SIGSAC Conference on Computer and Communications Security, Denver, CO, USA, 12–16 October 2015, pp. 579–590. ACM (2015)
18. May, A., Ozerov, I.: On computing nearest neighbors with applications to decoding of binary linear codes. In: Oswald and Fischlin [21], pp. 203–228
19. Naor, D., Naor, M., Lotspiech, J.: Revocation and tracing schemes for stateless receivers. In: Kilian, J. (ed.) CRYPTO 2001. LNCS, vol. 2139, pp. 41–62. Springer, Heidelberg (2001). https://doi.org/10.1007/3-540-44647-8_3
20. NIST's post-quantum cryptography project. https://csrc.nist.gov/Projects/Post-Quantum-Cryptography
21. Oswald, E., Fischlin, M. (eds.): EUROCRYPT 2015. LNCS, vol. 9056. Springer, Heidelberg (2015). https://doi.org/10.1007/978-3-662-46800-5
22. Unruh, D.: Quantum proofs of knowledge. In: Pointcheval, D., Johansson, T. (eds.) EUROCRYPT 2012. LNCS, vol. 7237, pp. 135–152. Springer, Heidelberg (2012). https://doi.org/10.1007/978-3-642-29011-4_10
23. Valiant, G.: Finding correlations in subquadratic time, with applications to learning parities and the closest pair problem. J. ACM **62**(2), 13:1–13:45 (2015)

Durandal: A Rank Metric Based Signature Scheme

Nicolas Aragon[1], Olivier Blazy[1], Philippe Gaborit[1], Adrien Hauteville[1(✉)], and Gilles Zémor[2]

[1] XLIM-DMI, University of Limoges,
123 Avenue Albert Thomas, 87060 Limoges Cedex, France
adrien.hauteville@inria.fr

[2] Université de Bordeaux, Institut de Mathématiques, UMR 5251,
351 cours de la Libération, 33400 Talence, France

Abstract. We describe a variation of the Schnorr-Lyubashevsky approach to devising signature schemes that is adapted to rank based cryptography. This new approach enables us to obtain a randomization of the signature, which previously seemed difficult to derive for code-based cryptography. We provide a detailed analysis of attacks and an EUF-CMA proof for our scheme. Our scheme relies on the security of the Ideal Rank Support Learning and the Ideal Rank Syndrome problems and a newly introduced problem: Product Spaces Subspaces Indistinguishability, for which we give a detailed analysis. Overall the parameters we propose are efficient and comparable in terms of signature size to the Dilithium lattice-based scheme, with a signature size of 4 kB for a public key of size less than 20 kB.

1 Introduction

During the last few years and especially since the 2017 call for proposals of the NIST for post-quantum cryptosystems, there has been a burst of activity in post-quantum cryptography and notably in code-based cryptography.

As far as encryption schemes are concerned, code-based cryptography has satisfactory solutions, in the form of cryptosystems whose security is reduced to well known problems: decoding random structured matrices like ideal or quasi-cyclic matrices [1,2,4]. However, the situation is very different for signature schemes.

Essentially there exist two types of signature schemes: hash-and-sign schemes and proof of knowledge based signatures.

For hash-and-sign schemes, signing consists in finding a small weight pre-image of a random syndrome, with a non-negligible probability. For instance: CFS in code based cryptography [7], GPV for lattices [19], Ranksign for rank metric [15], NTRUSign for lattices [23], pqsigRM [24]. The main drawback of this approach is that the system relies on hiding a trapdoor within the public key:

N. Aragon—This work was partially funded by French DGA.

Y. Ishai and V. Rijmen (Eds.): EUROCRYPT 2019, LNCS 11478, pp. 728–758, 2019.
https://doi.org/10.1007/978-3-030-17659-4_25

typically the secret is a decoding (or approximate decoding) algorithm which is hidden in the public matrix that describes the code. Whereas for lattices this type of masking can be efficiently randomized because of properties of the Euclidean distance [19], it has proved much more difficult for coding theory. In practice there exist two published code-based signature schemes: CFS [7] and RankSign [16] (see also SURF [8]), but for these schemes the public key can be distinguished from a random matrix [9,10]. Overall, for signature, this approach is similar to classical McEliece Encryption for which there is always a sword of Damocles lying over its head, namely the possible existence of a structural attack which recovers the hidden structure of and hence breaks the scheme. Relating the distinguishing problem to another well known problem seems a difficult feature to obtain. For the case of the RankSign scheme, a structural attack was recently found in [9]; it is always possible to repair and counter such attacks, like it was the case for all the sequels of NTRUSign [23], but this illustrates the difficulty of relying on this approach, when the secret trapdoor is not randomized.

The second approach for devising a signature scheme consists in proving that one knows a small weight vector associated to a given syndrome. It can done in two ways.

A first way consists in considering a zero-knowledge authentication algorithm and turning it into a signature scheme through the Fiat-Shamir transform. If the probability of cheating (associated to soundness) is very small, this approach can be efficient, but when the cheating probability is of order 1/2, it leads to very large signature sizes, since the number of necessary rounds is very large. It is typically the case for the Stern authentication protocol [31] for which the cheating probability is 2/3 (it was decreased to 1/2 in [3] and adapted to the rank metrix in [17]). Overall this approach is very interesting in terms of security reductions since one is reduced to generic problems *without* any masking, but rather inefficient in terms of signature size which can easily reach several hundred thousand bits, which is questionable in practice.

A second approach was proposed in a sequence of papers initiated by Lyubashevsky [26] in 2009. This approach is in the spirit of the Schnorr signature scheme [30] but adapted to the lattice context. The idea works as follows: for a public random matrix H, the secret is a matrix S of small weight vectors, to which one associates a matrix of syndromes HS^T. The signature consists in a proof of knowledge of the small weight matrix S from a sparse challenge c. The signature has the form $z = y + cS$, for y a random vector of moderate weight, typically of several orders of magnitude higher than the weights of cS. The idea of the proof of knowledge is that through z the verifier is convinced that the prover knows the secret matrix S because of the use of cS in the signature. At the same time, the vector y guarantees the randomization of the signature since its more noisy distribution enables it to absorb the less noisy distribution of cS. The main appeal of this approach is that it enables one to avoid the repetition related to zero-knowledge protocols with high probability of cheating, for instance the Dilithium [28] signature of NIST has a length of only 4 kB.

This previous approach can be straightforwardly adapted globally to code-based cryptography, but there is a problem is the randomization part: for the Hamming metric the randomization has to be considered on the whole length of the word, and not only on independent coordinates as when dealing with the Euclidean metric. In practice it means that it seems difficult to randomize the signature [11]: consequently, whenever a signature is produced, information leaks from the secret, so that after only a few signatures it becomes possible to recover the whole secret.

Overall, this second approach seems very promising but finding a good randomization strategy is a challenge.

Our contribution. We build upon the Schnorr-Lyubashevsky approach in a rank metric context and propose a way to efficiently randomize the signature. The main idea consists in extending the number of small weight secret vectors and adding another secret matrix S', so that the signature has the form $z = y + cS + pS'$ where p serves the purpose of providing extra randomization. In this way, the prover benefits from relaxed conditions that he uses to derive a randomization of the signature. We give a proof in the EUF-CMA security model, reducing the security of the scheme to the Rank Support Learning (RSL), the (ideal) Rank Syndrome Decoding (RSD) problem and a newly introduced problem, the Product Spaces Subspaces Indistinguishability (PSSI) problem for which we give a detailed analysis of a distinguisher. Our approach is developed for the rank metric and does not have an obvious Hamming metric counterpart. Overall our scheme is efficient in terms of signature sizes (a few kB) and of key sizes (of order 20 kB), with a security reduction to the ideal-RSD problem (a generalization of the quasi-cyclic RSD problem).

Roadmap. The paper is organized as follows: Sect. 2 recalls the required material from rank based cryptography, Sect. 3 gives a general overview and a precise description of the scheme, Sect. 4 is concerned with the security of the scheme. Finally, Sects. 5 and 6 describe the main practical attacks and examples of parameters for our scheme.

2 Presentation of Rank Metric Codes

Notation. In what follows, q denotes a power of a prime p. The finite field with q elements is denoted by $\mathbb{F}_q$ and for any positive integer m the finite field with q^m elements is denoted by $\mathbb{F}_{q^m}$. We will frequently view $\mathbb{F}_{q^m}$ as an m-dimensional vector space over $\mathbb{F}_q$. The Grassmannian $\mathbf{Gr}(k, \mathbb{F}_{q^m})$ represents the set of all subspaces of $\mathbb{F}_{q^m}$ of dimension k.

We use bold lowercase and capital letters to denote vectors and matrices respectively.

2.1 General Definitions

Definition 1 (Rank metric over $\mathbb{F}_{q^m}^n$). *Let $\boldsymbol{x} = (x_1, \ldots, x_n) \in \mathbb{F}_{q^m}^n$ and $(\beta_1, \ldots, \beta_m) \in \mathbb{F}_{q^m}^m$ be a basis of $\mathbb{F}_{q^m}$ viewed as an m-dimensional vector space* *over* $\mathbb{F}_q$. *Each coordinate* x_j *is associated to a vector of* $\mathbb{F}_q^m$ *in this basis:* $x_j = \sum_{i=1}^{m} m_{ij}\beta_i$. *The* $m \times n$ *matrix associated to* $\boldsymbol{x}$ *is given by* $\boldsymbol{M}(\boldsymbol{x}) = (m_{ij})_{\substack{1 \leqslant i \leqslant m \\ 1 \leqslant j \leqslant n}}$. *The rank weight* $\|\boldsymbol{x}\|$ *of* $\boldsymbol{x}$ *is defined as*

$$\|\boldsymbol{x}\| \stackrel{def}{=} \operatorname{Rank} \boldsymbol{M}(\boldsymbol{x}).$$

The associated distance $d(\boldsymbol{x}, \boldsymbol{y})$ *between elements* $\boldsymbol{x}$ *and* $\boldsymbol{y}$ *in* $\mathbb{F}_{q^m}^n$ *is defined by* $d(\boldsymbol{x}, \boldsymbol{y}) = \|\boldsymbol{x} - \boldsymbol{y}\|$.

Definition 2 ($\mathbb{F}_{q^m}$-linear code). *An* $\mathbb{F}_{q^m}$*-linear code* $\mathcal{C}$ *of dimension* k *and length* n *is a subspace of dimension* k *of* $\mathbb{F}_{q^m}^n$ *embedded with the rank metric. In this case we speak of an* $[n, k]_{q^m}$ *code. A code* $\mathcal{C}$ *can be represented in two equivalent ways:*

- *by a generator matrix* $\boldsymbol{G} \in \mathbb{F}_{q^m}^{k \times n}$. *Each row of* $\boldsymbol{G}$ *is an element of a basis of* $\mathcal{C}$,

$$\mathcal{C} = \{\boldsymbol{x}\boldsymbol{G}, \boldsymbol{x} \in \mathbb{F}_{q^m}^k\}$$

- *by a parity-check matrix* $\boldsymbol{H} \in \mathbb{F}_{q^m}^{(n-k) \times n}$. *Each row of* $\boldsymbol{H}$ *determines a parity-check equation satisfied by the elements of* $\mathcal{C}$:

$$\mathcal{C} = \{\boldsymbol{x} \in \mathbb{F}_{q^m}^n : \boldsymbol{H}\boldsymbol{x}^T = \boldsymbol{0}\}$$

We say that $\boldsymbol{G}$ *(respectively* $\boldsymbol{H}$*) is in systematic form if it is of the form* $(\boldsymbol{I}_k | \boldsymbol{A})$ *(respectively* $(\boldsymbol{I}_{n-k} | \boldsymbol{B})$*).*

As in the Hamming metric case, the notion of the support of a word is crucial to the rank metric. This notion appears very often in rank metric code-based cryptography, notably to compute the complexity of some algorithms.

Definition 3 (Support of a word). *Let* $\boldsymbol{x} = (x_1, \ldots, x_n) \in \mathbb{F}_{q^m}^n$. *The support* E *of* $\boldsymbol{x}$, *denoted* $\operatorname{Supp}(\boldsymbol{x})$, *is the* $\mathbb{F}_q$*-subspace of* $\mathbb{F}_{q^m}$ *generated by the coordinates of* $\boldsymbol{x}$:

$$E = \langle x_1, \ldots, x_n \rangle_{\mathbb{F}_q}$$

This definition is coherent with the definition of the rank weight since we have $\dim E = \|\boldsymbol{x}\|$.

The number of supports of dimension w of $\mathbb{F}_{q^m}$ is denoted by the Gaussian coefficient

$$\begin{bmatrix} m \\ w \end{bmatrix}_q = \prod_{i=0}^{w-1} \frac{q^m - q^i}{q^w - q^i} = \Theta(q^{w(m-w)})$$

We also need to define homogeneous matrices.

Definition 4 (Homogeneous matrices). *Let* $\boldsymbol{M} \in \mathbb{F}_{q^m}^{k \times n}$ *be a matrix over* $\mathbb{F}_{q^m}$. *The matrix* $\boldsymbol{M} = (m_{ij})$ *is said to be homogeneous of support* E *if the* $\mathbb{F}_q$*-subspace of* $\mathbb{F}_{q^m}$ *spanned by its coefficients* m_{ij} *is equal to* E. *If* $d = \dim E$, *then* $\boldsymbol{M}$ *is also said to be homogeneous of weight* d.

2.2 Double Circulant and Ideal Codes

To describe an $[n, k]_{q^m}$ linear code, we can give its systematic generator matrix or its systematic parity-check matrix. In both cases, the number of bits needed to represent such a matrix is $k(n-k)m\lceil \log_2 q \rceil$. To reduce the size of the representation of a code, we introduce double circulant codes.

First we need to define circulant matrices.

Definition 5 (Circulant matrix). *A square $n \times n$ matrix $\boldsymbol{M}$ is said to be circulant if it is of the form*

$$\boldsymbol{M} = \begin{pmatrix} m_0 & m_1 & \dots & m_{n-1} \\ m_{n-1} & m_0 & \ddots & m_{n-2} \\ \vdots & \ddots & \ddots & \vdots \\ m_1 & m_2 & \dots & m_0 \end{pmatrix}$$

We denote $\mathcal{M}_n(\mathbb{F}_{q^m})$ the set of circulant matrices of size $n \times n$ over $\mathbb{F}_{q^m}$.

The following proposition states an important property of circulant matrices.

Proposition 1. *$\mathcal{M}_n(\mathbb{F}_{q^m})$ is an $\mathbb{F}_{q^m}$-algebra isomorphic to $\mathbb{F}_{q^m}[X]/(X^n-1)$. The canonical isomorphism is given by*

$$\begin{aligned} \varphi : \mathbb{F}_{q^m}[X]/(X^n-1) &\longrightarrow \mathcal{M}_n(\mathbb{F}_{q^m}) \\ \sum_{i=0}^{n-1} m_i X^i &\longmapsto \begin{pmatrix} m_0 & m_1 & \dots & m_{n-1} \\ m_{n-1} & m_0 & \ddots & m_{n-2} \\ \vdots & \ddots & \ddots & \vdots \\ m_1 & m_2 & \dots & m_0 \end{pmatrix} \end{aligned}$$

In the following, in order to simplify notation, we will identify the polynomial $G(X) = \sum_{i=0}^{n-1} g_i X^i \in \mathbb{F}_{q^m}[X]$ with the vector $\boldsymbol{g} = (g_0, \dots, g_{n-1}) \in \mathbb{F}_{q^m}^n$. We will denote $\boldsymbol{ug} \mod P$ the vector of the coefficients of the polynomial

$$\left(\sum_{j=0}^{n-1} u_j X^j\right)\left(\sum_{i=0}^{n-1} g_i X^i\right) \mod P$$

or simply $\boldsymbol{ug}$ if there is no ambiguity in the choice of the polynomial P.

Definition 6 (Double circulant codes). *A $[2n, n]_{q^m}$ linear code $\mathcal{C}$ is said to be double circulant if it has a generator matrix $\boldsymbol{G}$ of the form $\boldsymbol{G} = (\boldsymbol{A}|\boldsymbol{B})$ where $\boldsymbol{A}$ and $\boldsymbol{B}$ are two circulant matrices of size n.*

With the previous notation, we have $\mathcal{C} = \{(\boldsymbol{xa}, \boldsymbol{xb}), \boldsymbol{x} \in \mathbb{F}_{q^m}^n\}$. If $\boldsymbol{a}$ is invertible in $\mathbb{F}_{q^m}[X]/(X^n-1)$, then $\mathcal{C} = \{(\boldsymbol{x}, \boldsymbol{xg}), \boldsymbol{x} \in \mathbb{F}_{q^m}^n\}$ where $\boldsymbol{g} = \boldsymbol{a}^{-1}\boldsymbol{b}$. In this case we say that $\mathcal{C}$ is generated by $\boldsymbol{g}$ (mod $X^n - 1$). Thus we only need $nm\lceil \log_2 q \rceil$ bits to describe a $[2n, n]_{q^m}$ double circulant code.

We can generalize double circulant codes by choosing another polynomial P, rather than $X^n - 1$, to define the quotient-ring $\mathbb{F}_{q^m}[X]/(P)$. These codes are called ideal codes.

Definition 7 (Ideal codes). *Let $P(X) \in \mathbb{F}_q[X]$ be a polynomial of degree n and $\boldsymbol{g}_1, \boldsymbol{g}_2 \in \mathbb{F}_{q^m}^n$. Let $G_1(X) = \sum_{i=0}^{n-1} g_{1i}X^i$ and $G_2(X) = \sum_{j=0}^{n-1} g_{1j}X^j$ be the polynomials associated respectively to $\boldsymbol{g}_1$ and $\boldsymbol{g}_2$.*

The $[2n, n]_{q^m}$ ideal code $\mathcal{C}$ with generator $(\boldsymbol{g}_1, \boldsymbol{g}_2)$ is the code with generator matrix

$$\boldsymbol{G} = \left(\begin{array}{c|c} G_1(X) \mod P & G_2(X) \mod P \\ XG_1(X) \mod P & XG_2(X) \mod P \\ \vdots & \vdots \\ X^{n-1}G_1(X) \mod P & X^{n-1}G_2(X) \mod P \end{array} \right)$$

More concisely, we have $\mathcal{C} = \{(\boldsymbol{x}\boldsymbol{g}_1 \mod P, \boldsymbol{x}\boldsymbol{g}_2 \mod P), \boldsymbol{x} \in \mathbb{F}_{q^m}^n\}$. We will often omit mentioning the polynomial P if there is no ambiguity. If $\boldsymbol{g}_1$ is invertible, we may express the code in systematic form, $\mathcal{C} = \{(\boldsymbol{x}, \boldsymbol{x}\boldsymbol{g}), \boldsymbol{x} \in \mathbb{F}_{q^m}^n\}$ with $\boldsymbol{g} = \boldsymbol{g}_1^{-1}\boldsymbol{g}_2 \mod P$.

The advantage of ideal codes over double circulant codes is that they are resistant to the folding attack of [22]. Such codes have been used for NIST propositions LAKE and LOCKER.

2.3 Difficult Problems in Rank Metric

In order to design rank metric code-based cryptosystems, we need to define difficult problems in rank metric. The first problem corresponds to the classical problem of syndrome decoding, adapted to the rank metric.

Problem 1. **Rank Support Decoding** (RSD). Let $\boldsymbol{H}$ be an $(n-k) \times n$ parity-check matrix of an $[n, k]$ $\mathbb{F}_{q^m}$-linear code, $\boldsymbol{s} \in \mathbb{F}_{q^m}^{n-k}$ and r an integer. The $\mathrm{RSD}_{q,m,n,k,r}$ problem is to find $\boldsymbol{e}$ such that $\|\boldsymbol{e}\| = r$ and $\boldsymbol{H}\boldsymbol{e}^T = \boldsymbol{s}^T$.

This problem is probabilistically reduced to the well-known NP-complete Syndrome Decoding problem in the Hamming metric [18].

The following problem was introduced in [12]. It is similar to the RSD problem, the difference is that instead of having one syndrome, we are given several syndromes of errors of same support and the goal is to find this support.

Problem 2. **Rank Support Learning (RSL)** [12]. Let $\boldsymbol{H}$ be a random full-rank $(n-k) \times n$ matrix over $\mathbb{F}_{q^m}$. Let $\mathcal{O}$ be an oracle which, given $\boldsymbol{H}$, gives samples of the form $\boldsymbol{H}\boldsymbol{s}_1^T, \boldsymbol{H}\boldsymbol{s}_2^T, \dots, \boldsymbol{H}\boldsymbol{s}_N^T$, with the vectors $\boldsymbol{s}_i$ randomly chosen from a space E^n, where E is a random subspace of $\mathbb{F}_{q^m}$ of dimension r. The $RSL_{q,m,n,k,r}$ problem is to recover E given only access to the oracle.

We denote $RSL_{q,m,n,k,r,N}$ the $RSL_{q,m,n,k,r}$ problem where we are allowed to make exactly N calls to the oracle, meaning we are given exactly N syndrome values $\boldsymbol{H}\boldsymbol{s}_i^T$. By an instance of the RSL problem, we shall mean a sequence

$$(\boldsymbol{H}, \boldsymbol{H}\boldsymbol{s}_1^T, \boldsymbol{H}\boldsymbol{s}_2^T, \dots, \boldsymbol{H}\boldsymbol{s}_N^T)$$

that we can also view as a pair of matrices $(\boldsymbol{H}, \boldsymbol{T})$, where $\boldsymbol{T}$ is the matrix whose columns are the $\boldsymbol{H}\boldsymbol{s}_i^T$.

The last problem we need before introducing our scheme is a variant of the RSD problem. Instead of searching for the error associated to a syndrome, this problem consists in finding an error associated to a syndrome which belongs to a given $\mathbb{F}_q$-affine subspace of $\mathbb{F}_{q^m}^{n-k}$. Formally:

Problem 3. **Affine Rank Syndrome Decoding** (ARSD). Let $\boldsymbol{H}$ be an $(n-k)\times n$ parity-check matrix of an $[n,k]$ $\mathbb{F}_{q^m}$-linear code, $\boldsymbol{H}'$ an $(n-k)\times n'$ random matrix over $\mathbb{F}_{q^m}$, F an $\mathbb{F}_q$-subspace of $\mathbb{F}_{q^m}$ of dimension r', $\boldsymbol{s}\in\mathbb{F}_{q^m}^{n-k}$ and r an integer. The $\mathrm{ARSD}_{q,m,n,k,r,n',F}$ problem is to find $\boldsymbol{e}\in\mathbb{F}_{q^m}^{n}$ and $\boldsymbol{e}'\in\mathbb{F}_{q^m}^{n'}$ such that

$$\begin{cases}\boldsymbol{H}\boldsymbol{e}^T+\boldsymbol{H}'\boldsymbol{e}'^T=\boldsymbol{s}\\ \|\boldsymbol{e}\|=r\\ \mathrm{Supp}(\boldsymbol{e}')\subseteq F\end{cases}$$

Remark: This problem can seen as that of finding a vector $\boldsymbol{x}$ of weight r such that $\boldsymbol{H}\boldsymbol{x}^T=\boldsymbol{s}'$ with $\boldsymbol{s}'\in\{\boldsymbol{s}-\boldsymbol{H}'\boldsymbol{x}'^T:\mathrm{Supp}(\boldsymbol{x}')\subseteq F\}$. This set is an $\mathbb{F}_q$-affine subspace of $\mathbb{F}_{q^m}^{n-k}$, which explains the name of the problem.

The following proposition shows that the ARSD problem in the worst case is as hard as the RSD problem for large values of m.

Proposition 2. *Let $\mathcal{A}$ be an algorithm which can solve the $\mathrm{ARSD}_{q,m,n,k,r,n',F}$ problem with $m\geqslant\frac{r(n-r)+n'\dim F}{n-k-r}$. Then $\mathcal{A}$ can be used to solve the $\mathrm{RSD}_{q,m,n,k,r}$ problem with non negligible probability.*

Proof. Let $\boldsymbol{H}\in\mathbb{F}_{q^m}^{(n-k)\times n}$, $\boldsymbol{s}\in\mathbb{F}_{q^m}^{n-k}$ such that $\boldsymbol{s}=\boldsymbol{H}\boldsymbol{e}^T$ with $\|\boldsymbol{e}\|=r$ be an instance of the $\mathrm{RSD}_{q,m,n,k,r}$ problem. First we need to transform this instance into an instance of the ARSD problem. Let $\boldsymbol{H}'\in\mathbb{F}_{q^m}^{(n-k)\times n'}$ and let F be a subspace of $\mathbb{F}_{q^m}$ of dimension r' such that $m\geqslant\frac{r(n-r)+n'\dim F}{n-k-r}$. Let $\boldsymbol{s}'=\boldsymbol{s}+\boldsymbol{H}'\boldsymbol{e}'^T$ with $\mathrm{Supp}(\boldsymbol{e}')=F$.

$(\boldsymbol{H},\boldsymbol{s}',r,\boldsymbol{H}',F)$ is an instance of the $\mathrm{ARSD}_{q,m,n,k,r,n',F}$ problem. Let $(\boldsymbol{x},\boldsymbol{x}')$ be a solution of this instance given by algorithm $\mathcal{A}$. Now we will prove that this solution is unique with a non negligible probability.

Let us consider the application f defined by

$$\begin{aligned} f:\mathcal{S}_{\mathbb{F}_{q^m}^n}(r)\times F^{n'} &\rightarrow \mathbb{F}_{q^m}^{n-k}\\ (\boldsymbol{x},\boldsymbol{x}') &\mapsto \boldsymbol{H}\boldsymbol{x}^T+\boldsymbol{H}'\boldsymbol{x}'^T\end{aligned}$$

where $\mathcal{S}_{\mathbb{F}_{q^m}^n}(r)$ is the set of words of $\mathbb{F}_{q^m}^n$ of rank r. By definition of the ARSD problem, we have $(\boldsymbol{x},\boldsymbol{x}')\in f^{-1}(\{\boldsymbol{s}'\})$.

Let $S(\mathbb{F}_{q^m}^n,r)$ denote the cardinality of $\mathcal{S}_{\mathbb{F}_{q^m}^n}(r)$. By definition of the rank metric, $S(\mathbb{F}_{q^m}^n,r)$ is equal to the number of matrices of $\mathbb{F}_q^{m\times n}$ of rank r and we have

$$S(\mathbb{F}_{q^m}^n,r)=\prod_{i=0}^{r-1}\frac{(q^m-q^i)(q^n-q^i)}{q^r-q^i}=\Theta\left(q^{r(m+n-r)}\right).$$

Thus the cardinality of the codomain of f is in $\Theta(q^{r(n+n-r)+n'r'})$ and the cardinality of its domain is equal to $q^{m(n-k)}$. We have $m \geqslant \frac{r(n-r)+n'r'}{n-k-r}$ which implies $m(n-k) \geqslant r(m+n-r)+n'r'$, hence $\boldsymbol{s}'$ has only one preimage with a non negligible probability. Thus $\boldsymbol{H}\boldsymbol{x}^T + \boldsymbol{H}'\boldsymbol{x}'^T = \boldsymbol{s}' = \boldsymbol{H}\boldsymbol{e}^T + \boldsymbol{H}\boldsymbol{e}'^T$ implies $(\boldsymbol{x}, \boldsymbol{x}') = (\boldsymbol{e}, \boldsymbol{e}')$ so $\boldsymbol{x}$ is a solution of the instance of the $\mathrm{RSD}_{q,m,n,k,r}$ problem. □

Remark: All these problems are defined for random codes but can straightforwardly be specialized to the families of double circulant codes or of ideal random codes. In this case, these problems are denoted I – RSD, I – ARSD and I – RSL respectively. The reductions are unchanged, the only difference being that the I – RSD problem is reduced to the Syndrome Decoding problem for ideal codes, which has not been proven NP-complete. However this problem is considered hard by the community since the best attacks stay exponential.

2.4 Bounds on Rank Metric Codes

One can define bounds on the size or the minimum distance of rank metric codes that are similar to well-known bounds for Hamming metric codes. The rank Gilbert-Varshamov bound (or rank Gilbert-Varshamov distance, denoted d_{RGV}) gives the maximum rank-weight for which the RSD problem has typically a unique solution.

Definition 8 (Rank Gilbert-Varshamov (RGV) bound). *Let $B(\mathbb{F}_{q^m}^n, t)$ be the size of the ball of radius t in rank rank metric. The quantity d_{RGV} is defined as the smallest t such that $B(\mathbb{F}_{q^m}^n, t) \geqslant q^{m(n-k)}$.*

Asymptotically we have

$$d_{RGV}(m,n,k) \sim \frac{m+n-\sqrt{(m-n)^2+4km}}{2}$$

$$d_{RGV}(n,n,k) \sim n\left(1-\sqrt{\frac{k}{n}}\right) \text{ when } m=n. \quad (1)$$

The quantity $q^{m(n-k)}$ corresponds to the number of syndromes $\boldsymbol{s} \in \mathbb{F}_{q^m}^{n-k}$ and by definition,

$$B(\mathbb{F}_{q^m}^n, t) = \sum_{i=0}^{t} S(\mathbb{F}_{q^m}^n, i)$$

where $S(\mathbb{F}_{q^m}^n, i)$ is the size of the sphere of radius i, which correspond to the number of matrices of size $m \times n$ and of rank i over $\mathbb{F}_q$. This quantity is equal to $\prod_{j=0}^{i-1} \frac{(q^m-q^j)(q^n-q^j)}{q^i-q^j} = \Theta\left(q^{i(m+n-i)}\right)$. The asymptotic expressions are obtained by solving for t the equation $t(m+n-t) = m(n-k)$ [25].

The rank Singleton bound gives the weight above which the RSD problem becomes polynomial.

Definition 9 (Rank Singleton bound). *The rank Singleton bound for an $\mathbb{F}_{q^m}$-linear $[n,k]$ code is defined as the quantity*

$$d_{RS}(m,n,k) = \frac{m(n-k)}{\max(m,n)}.$$

We can obtain this equality by counting the number of equations and unknowns over $\mathbb{F}_q$ of the RSD problem. Indeed, given a random support E of dimension r, we can express the error $\boldsymbol{e}$ in a basis of E with nr unknowns over $\mathbb{F}_q$ (r unknowns per coordinate). The parity-check equations gives us $(n-k)$ equations over $\mathbb{F}_{q^m}$, meaning $m(n-k)$ equations over $\mathbb{F}_q$. If $nr \geqslant m(n-k)$ then this instance of the RSD problem has a solution $\boldsymbol{e}$ of support E with a non-negligible probability. Such a solution can easily be found by solving a linear system. Therefore, the RSD problem becomes polynomial if $r \geqslant \left\lceil \frac{m(n-k)}{n} \right\rceil$.

3 A New Signature Scheme Based on the RSL Problem

3.1 General Overview

Our scheme consists of adapting to the rank metric the idea proposed in [27]. This idea can be viewed as a framework for an authentication scheme and can be loosely described as follows. Two matrices, over some fixed finite field, $\boldsymbol{H}$ and $\boldsymbol{T}$, are public, and a Prover wishes to prove that she is in possession of secret matrix $\boldsymbol{S}$ with "small" entries such that $\boldsymbol{T} = \boldsymbol{H}\boldsymbol{S}^T$. She chooses a random vector $\boldsymbol{y}$ of small norm (to be defined appropriately) according to some distribution $\mathcal{D}_y$. She computes the syndrome $\boldsymbol{x} = \boldsymbol{H}\boldsymbol{y}^T$ of $\boldsymbol{y}$ and sends it to the verifier. The verifier chooses a random vector $\boldsymbol{c}$ of the appropriate length and of small norm according to some distribution $\mathcal{D}_c$ and sends it as a challenge to the Prover. The Prover computes $\boldsymbol{z} = \boldsymbol{y} + \boldsymbol{c}\boldsymbol{S}$ and sends it to the Verifier. The verifier checks that $\boldsymbol{z}$ is of small norm, and that

$$\boldsymbol{H}\boldsymbol{z}^T - \boldsymbol{T}\boldsymbol{c}^T = \boldsymbol{x}.$$

This scheme is described on Fig. 1.

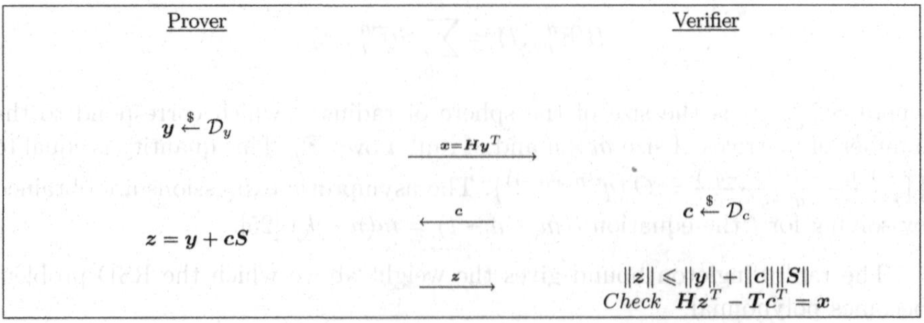

Prover		Verifier
$\boldsymbol{y} \xleftarrow{\$} \mathcal{D}_y$		
	$\xrightarrow{\boldsymbol{x}=\boldsymbol{H}\boldsymbol{y}^T}$	
	$\xleftarrow{\boldsymbol{c}}$	$\boldsymbol{c} \xleftarrow{\$} \mathcal{D}_c$
$\boldsymbol{z} = \boldsymbol{y} + \boldsymbol{c}\boldsymbol{S}$		
	$\xrightarrow{\boldsymbol{z}}$	$\lVert\boldsymbol{z}\rVert \leqslant \lVert\boldsymbol{y}\rVert + \lVert\boldsymbol{c}\rVert\lVert\boldsymbol{S}\rVert$ *Check* $\boldsymbol{H}\boldsymbol{z}^T - \boldsymbol{T}\boldsymbol{c}^T = \boldsymbol{x}$

Fig. 1. Overview of the authentication framework from [27]

The general idea is that cheating is difficult for the Prover because it requires finding a vector $\boldsymbol{z}$ of small norm such that $\boldsymbol{H}\boldsymbol{z}^T$ equals a prescribed value, and this is an instance of the decoding problem for a random code. Also, the vector $\boldsymbol{z}$ sent by the legitimate Prover should yield no useful information on the secret $\boldsymbol{S}$, because the noisy random vector $\boldsymbol{y}$ drowns out the sensitive quantity $\boldsymbol{cS}$.

If we instantiate this scheme in the rank metric, $\boldsymbol{H}$ would be a random matrix over $\mathbb{F}_{q^m}$, and for $\boldsymbol{S}$ to be a matrix of small norm would mean it to be homogeneous matrix of some small rank r. Requiring that the vectors $\boldsymbol{y}$ and $\boldsymbol{c}$ are also small will mean that they are taken in random subspaces of $\mathbb{F}_{q^m}$ of fixed dimensions respectively w and d.

The problem with this approach in the rank metric is that adding $\boldsymbol{y}$ to $\boldsymbol{cS}$ does not hide $\boldsymbol{cS}$ properly. Indeed, the verifier, or any witness to the protocol of Fig. 1, can recover the support of the secret matrix $\boldsymbol{S}$ even after a single instance of the protocol, using techniques from the decoding of LRPC codes [13]: since the verifier has $\boldsymbol{c}$ he can choose a basis $f_1, \dots, f_d$ of $\text{Supp}(\boldsymbol{c})$ and then with high probability it will occur that:

$$\bigcap_{i=1}^{d} f_i^{-1}\,\text{Supp}(\boldsymbol{z}) = \text{Supp}(\boldsymbol{S})$$

and with the support of $\boldsymbol{S}$ the verifier can compute $\boldsymbol{S}$ explicitely from the linear equations $\boldsymbol{H}\boldsymbol{S}^T = T$. A less efficient version of this attack, requiring multiple signatures, was described in [29].

To tackle this problem, we will modify the protocol of Fig. 1 by adding an other term to $\boldsymbol{z}$.

3.2 An Authentication Scheme

We will first describe our scheme as an authentication scheme. It calls upon the notion of product of $\mathbb{F}_q$-linear subspaces of $\mathbb{F}_{q^m}$.

Definition 10. *If E and F are two $\mathbb{F}_q$-linear subspaces of $\mathbb{F}_{q^m}$, their* product, *denoted EF, is defined as the $\mathbb{F}_q$-*subspace *consisting of the $\mathbb{F}_q$-linear span of the set of vectors*

$$\{ef, e \in E, f \in F\}$$

where ef stands for the product of e by f in the field $\mathbb{F}_{q^m}$. The product of E with itself will be denoted $E^{\langle 2 \rangle}$, so as not to confuse it with the cartesian product.

The public key consists of a random $(n-k) \times n$ matrix $\boldsymbol{H}$ over $\mathbb{F}_{q^m}$ and two matrices $\boldsymbol{T}$ and $\boldsymbol{T}'$, of size $(n-k) \times lk$ and $(n-k) \times l'k$ respectively, and such that $(\boldsymbol{H}, \boldsymbol{T}|\boldsymbol{T}')$ is an instance of the RSL problem, where $|$ denotes matrix concatenation. The private key consist of two homogeneous matrices $\boldsymbol{S}$ and $\boldsymbol{S}'$ of weight r such that $\boldsymbol{H}\boldsymbol{S}^T = \boldsymbol{T}$ and $\boldsymbol{H}\boldsymbol{S}'^T = \boldsymbol{T}'$. Accordingly, $\boldsymbol{S}$ and $\boldsymbol{S}'$ are $lk \times n$ and $l'k \times n$ matrices respectively. We denote by E the vector space spanned by the coordinates of $\boldsymbol{S}$ and $\boldsymbol{S}'$.

In the commitment step, we sample uniformly at random two vector spaces: $W \in \mathbf{Gr}(w, \mathbb{F}_{q^m})$ and $F \in \mathbf{Gr}(d, \mathbb{F}_{q^m})$. We then randomly choose $\boldsymbol{y} \in (W + EF)^n$. This vector will be used to mask the secret information in answer to the challenge. The commitment consists of $\boldsymbol{x} = \boldsymbol{H}\boldsymbol{y}^T$ together with the subspace F.

The verifier then chooses a challenge $\boldsymbol{c} \in F^{l'k}$.

To answer the challenge, the prover first computes $\boldsymbol{y} + \boldsymbol{c}\boldsymbol{S}'$. Since the entries of the vector $\boldsymbol{c}$ are in F and the entries of the matrix $\boldsymbol{S}'$ are in E, we have that $\boldsymbol{c}\boldsymbol{S}'$ has its entries in the product space EF, and the vector $\boldsymbol{y} + \boldsymbol{c}\boldsymbol{S}'$ has its entries in the space $W + EF$, like the vector $\boldsymbol{y}$. The linear span of the coordinates of $\boldsymbol{y} + \boldsymbol{c}\boldsymbol{S}'$ is typically equal, or very close to $W + EF$, and this yields too much information on the secret space E to be given to the verifier. To counter this, we add a vector $\boldsymbol{p}\boldsymbol{S}$. Coordinates of $\boldsymbol{p}$ are chosen in F, so that the coordinates of $\boldsymbol{p}\boldsymbol{S}$ fall in the product space EF, and through linear algebra the prover chooses $\boldsymbol{p}$ such that the linear span of the coordinates of the sum $\boldsymbol{z} = \boldsymbol{y} + \boldsymbol{c}\boldsymbol{S}' + \boldsymbol{p}\boldsymbol{S}$ is restricted to a smaller subspace: namely a subspace $W + U$ for U some subspace of EF of codimension λ inside EF. In other words, $\boldsymbol{z} = \boldsymbol{y} + \boldsymbol{c}\boldsymbol{S}' + \boldsymbol{p}\boldsymbol{S}$ is computed so as to be of rank at most $w + rd - \lambda$. The vector $\boldsymbol{z}$ is then sent to the verifier, together with the vector $\boldsymbol{p}$. This operation is at the heart of the present protocol and the derived signature scheme. More details are given about this in the following section and in Sect. 4.1.

The verifier accepts if $\|\boldsymbol{z}\| \leqslant rd + w - \lambda$ and $\boldsymbol{H}\boldsymbol{z}^T - \boldsymbol{T}'\boldsymbol{c}^T + \boldsymbol{T}\boldsymbol{p}^T = \boldsymbol{x}$. An overview of this protocol is given in Fig. 2.

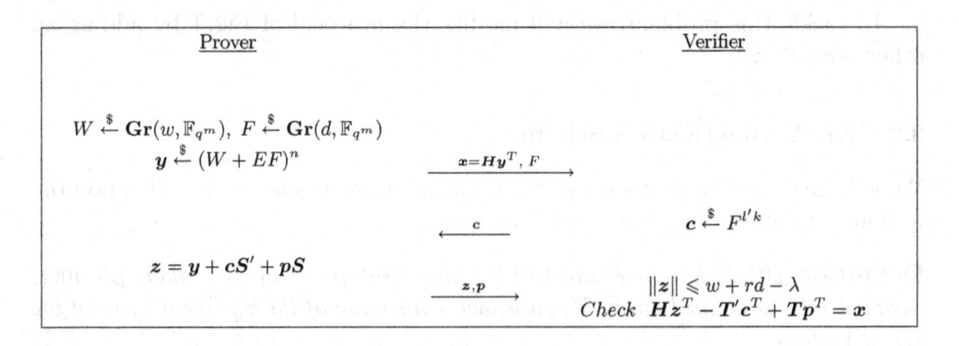

Fig. 2. Overview as an authentication scheme

Using the Fiat-Shamir heuristic, we turn this authentication scheme into a signature scheme.

3.3 Signature Scheme

Key generation

- Randomly choose an $(n - k) \times n$ ideal double circulant matrix $\boldsymbol{H}$ as in Definition 7 for an irreducible polynomial P, in practice we consider $k = \frac{n}{2}$

- Choose a random subspace E of dimension r of $\mathbb{F}_{q^m}$ and sample l vectors $\boldsymbol{s}_i$ and l' vectors $\boldsymbol{s}'_i$ of length n from the same support E of dimension r
- Set $\boldsymbol{t}_i = \boldsymbol{H}\boldsymbol{s}_i^T$ and $\boldsymbol{t'}_i = \boldsymbol{H}\boldsymbol{s'}_i^T$
- Output $(\boldsymbol{H}, \boldsymbol{t}_1, \ldots, \boldsymbol{t}_l, \boldsymbol{t'}_1, \ldots, \boldsymbol{t'}_{l'})$ as public key, and $(\boldsymbol{s}_1, \ldots, \boldsymbol{s}_l, \boldsymbol{s'}_1, \ldots, \boldsymbol{s'}_{l'})$ as secret key

Note that, since $\boldsymbol{H}$ has an ideal structure, each relation of the form $\boldsymbol{H}\boldsymbol{s}_i^T = \boldsymbol{t}_i$ can be shifted $\mod P$ to generate k syndrome relations. We denote $\boldsymbol{S}$ (respectively $\boldsymbol{S}'$) the matrix consisting of all $\boldsymbol{s}_i$ (respectively $\boldsymbol{s'}_i$) and their ideal shifts. Let $\boldsymbol{T} = \boldsymbol{H}\boldsymbol{S}^T$ and $\boldsymbol{T}' = \boldsymbol{H}\boldsymbol{S'}^T$: the public key consists of $(\boldsymbol{H}, \boldsymbol{T}, \boldsymbol{T}')$. $\boldsymbol{T}$ and $\boldsymbol{T}'$ are $\frac{n}{2} \times lk$ and $\frac{n}{2} \times l'k$ matrices respectively, but can be described using only the vectors $(\boldsymbol{t}_1, \ldots, \boldsymbol{t}_l)$ and $(\boldsymbol{t'}_1, \ldots, \boldsymbol{t'}_{l'})$. The secret key consists of the homogeneous matrices $\boldsymbol{S}$ and $\boldsymbol{S}'$ of rank r such that $\boldsymbol{H}\boldsymbol{S}^T = \boldsymbol{T}$ and $\boldsymbol{H}\boldsymbol{S}'^T = \boldsymbol{T}'$.

Figure 3 describes the key pair.

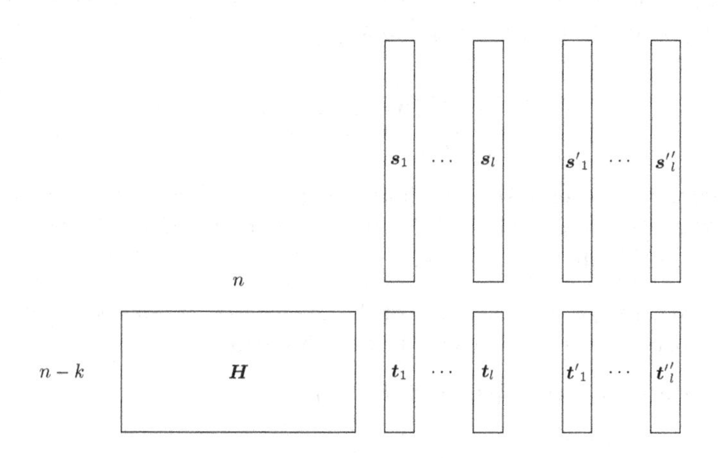

Fig. 3. Overview of public and secret key

Signature of a message μ

- Randomly choose W, a subspace of $\mathbb{F}_{q^m}$ of dimension w.
- Randomly choose F, a subspace of $\mathbb{F}_{q^m}$ of dimension d.
- Sample $\boldsymbol{y} \in (W + EF)^n$ and set $\boldsymbol{x} = \boldsymbol{H}\boldsymbol{y}^T$.
- For some hash function $\mathcal{H}$, set $\boldsymbol{c} = \mathcal{H}(\boldsymbol{x}, F, \mu)$, $\boldsymbol{c} \in F^{l'k}$. This is done by using the output of $\mathcal{H}$ as the coordinates of $\boldsymbol{c}$ in a basis of F.
- Choose U, a subspace of the product space EF, of dimension $rd - \lambda$, and such that U contains no non-zero elements of the form ef, for $e \in E$ and $f \in F$. More details on this process are given in Subsect. 3.4.
- Solve $\boldsymbol{z} = \boldsymbol{y} + \boldsymbol{c}\boldsymbol{S}' + \boldsymbol{p}\boldsymbol{S}$ with $\boldsymbol{p} \in F^{lk}$ as unknown, such that $\text{Supp}(\boldsymbol{z}) \subset W + U$: as mentioned in the previous section, $\boldsymbol{p}$ is computed through linear algebra. Specifically, we write $\boldsymbol{p} = (p_1, \ldots, p_{lk})$, and each coordinate $p_i \in F$ of $\boldsymbol{p}$ is decomposed as

$$p_i = \sum_{\ell=1}^{d} p_{i\ell} f_\ell$$

where $f_1, \ldots, f_d$ is a basis of F that will be used to describe the space F. The j-th coordinate of the vector $\boldsymbol{cS}$ is therefore equal to

$$(\boldsymbol{cS})_j = \sum_{i=1}^{lk} \sum_{\ell=1}^{d} p_{i\ell} f_\ell \boldsymbol{S}_{ij}. \tag{2}$$

Recall that $f_\ell \boldsymbol{S}_{ij}$ is in FE because $\boldsymbol{S}$ has support E. Choose a basis of EF of the form $u_1, \ldots, u_{rd-\lambda}, v_1, \ldots, v_\lambda$, where $u_1, \ldots, u_{rd-\lambda}$ is a basis of U (the typical dimension of EF is rd). Let $\pi_1, \ldots, \pi_\lambda$ be the projections of elements of EF on the last λ coordinates of the above basis. For $h = 1 \ldots \lambda$, applying π_h to all n coordinates of the vector $\boldsymbol{y} + \boldsymbol{cS}' + \boldsymbol{pS}$ and declaring the result to equal 0, we will obtain a linear system of λn equations in the variables p_{ij} by using linearity in (2) to express $\pi_h[(\boldsymbol{cS})_j]$ as

$$\sum_{i=1}^{lk} \sum_{\ell=1}^{d} p_{i\ell} \pi_h(f_\ell \boldsymbol{S}_{ij}). \tag{3}$$

Parameters are chosen so that this system has more variables than equations and typically has a solution. If it doesn't, another space U is sampled.
- Output $(\boldsymbol{z}, F, \boldsymbol{c}, \boldsymbol{p})$ as signature.

The signature consists therefore of the challenge $\boldsymbol{c}$, computed through a hash function, together with the answer to this challenge.

Verification of a signature $(\mu, \boldsymbol{z}, F, \boldsymbol{c}, \boldsymbol{p})$

- Check that $\|zv\| \leqslant rd + w - \lambda$,
- Verify that $\mathcal{H}(\boldsymbol{Hz}^T - \boldsymbol{T}'\boldsymbol{c}^T + \boldsymbol{Tp}^T, F, \mu) = \boldsymbol{c}$.

To verify the signature, we have to check the rank weight of $\boldsymbol{z}$ and the equality $\mathcal{H}(\boldsymbol{x}, F, \mu) = \boldsymbol{c}$. The vector $\boldsymbol{x}$ is recomputed using the answer to the challenge. The complete signature scheme is summarized on Fig. 4.

3.4 Filtering Vector Spaces

The goal of filtering U during the signature step is to ensure to there is no non-zero element of the form ef in the support of $\boldsymbol{z}$, for $e \in E$ and $f \in F$. This is to prevent an attack that would recover E through techniques for decoding LRPC codes [13]. Indeed, if there is an element of the form ef in $\mathrm{Supp}(\boldsymbol{z})$, then $e \in E \cap f^{-1}\,\mathrm{Supp}(\boldsymbol{z})$ which allows an attack against the secret key (moreover elements of this form can be used to distinguish between signatures and randomly generated vectors, as explained in Sect. 4.1). To achieve that, we need to find a pair (U, F) such that:

Key generation: $E \xleftarrow{\$} \mathbf{Gr}(r, \mathbb{F}_{q^m})$
Signing key: $\boldsymbol{S} \xleftarrow{\$} E^{n \times lk}$, $\boldsymbol{S}' \xleftarrow{\$} E^{n \times l'k}$
Verification key: $\boldsymbol{H} \xleftarrow{\$}$ ideal $\mathcal{M}^{\frac{n}{2} \times n}$, $\boldsymbol{T} = \boldsymbol{H}\boldsymbol{S}^T, \boldsymbol{T}' = \boldsymbol{H}\boldsymbol{S}'^T$

Sign$(\mu, \boldsymbol{S}, \boldsymbol{S}')$:

1. $W \xleftarrow{\$} \mathbf{Gr}(w, \mathbb{F}_{q^m})$,
   $F \xleftarrow{\$} \mathbf{Gr}(d, \mathbb{F}_{q^m})$
2. $\boldsymbol{y} \xleftarrow{\$} (W + EF)^n$, $\boldsymbol{x} = \boldsymbol{H}\boldsymbol{y}^T$
3. $\mathbf{c} = \mathcal{H}(\boldsymbol{x}, F, \mu)$, $\mathbf{c} \in F^{l'k}$
4. $U \xleftarrow{\$}$ filtered subspace of EF of dimension $rd - \lambda$
5. $\boldsymbol{z} = \boldsymbol{y} + \mathbf{c}\boldsymbol{S}' + \boldsymbol{p}\boldsymbol{S}$, $\boldsymbol{z} \in W + U$
6. Output $(\boldsymbol{z}, F, \mathbf{c}, \boldsymbol{p})$

Verify$(\mu, \boldsymbol{z}, F, \mathbf{c}, \boldsymbol{p}, \boldsymbol{H}, \boldsymbol{T}, \boldsymbol{T}')$:

1. Accept if and only if :
   $\|\boldsymbol{z}\| \leqslant w + rd - \lambda$ and
   $\mathcal{H}(\boldsymbol{H}\boldsymbol{z}^T - \boldsymbol{T}'\mathbf{c}^T + \boldsymbol{T}\boldsymbol{p}^T, F, \mu) = \mathbf{c}$

Fig. 4. The Durandal Signature scheme

- U is a subspace of EF of dimension $rd - \lambda$
- For every non-zero $x = ef$ with $e \in E$ and $f \in F$, we have that $x \notin U$.

We argue that, for a given F, finding the required space U is quite manageable. We use the following obvious proposition to check the second condition:

Proposition 3. *Let U be a subspace of EF of dimension $rd - \lambda$. Let $E/\mathbb{F}_q$ be a set of representatives of the equivalence relation $x \equiv y \iff \exists \alpha \in \mathbb{F}_q^*$ such that $x = \alpha y$. We have the following equivalence:*

$$\{ef : e \in E, f \in F\} \cap U = \{0\} \iff \forall e \in E/\mathbb{F}_q, eF \cap U = \{0\}.$$

Hence, the cost of this verification is $(q^r - 1)/(q - 1)$ intersections of subspaces of dimension d and $rd - \lambda$, that is to say $\frac{q^r-1}{q-1}(d + rd - \lambda)^2 m$ operations in $\mathbb{F}_q$.

We now briefly estimate the probability that a random U contains no element $x = ef$. For simplicity, we only deal with a typical practical case, namely $q = 2$ and $d = r$.

The subspace U is chosen randomly and uniformly of codimension λ inside the vector space EF: we study the probability that U contains no non-zero product ef. Let $x = ef$ be such a non-zero product. Let $\mathcal{U}_x$ be the event $\{x \in U\}$. We are interested in $1 - \mathbb{P}(\mathcal{U})$ where

$$\mathcal{U} = \bigcup_{x=ef,\, x \neq 0} \mathcal{U}_x.$$

Clearly, $\mathbb{P}(\mathcal{U}_x) = 2^{-\lambda}$. Our goal is to evaluate $\mathbb{P}(\mathcal{U})$ through inclusion-exclusion, i.e.

$$\mathbb{P}(\mathcal{U}) = \sum_x \mathbb{P}(\mathcal{U}_x) - \sum_{x,y} \mathbb{P}(\mathcal{U}_x \cap \mathcal{U}_y) + \cdots + (-1)^i \sum_{X \in \Pi, |X|=i} \mathbb{P}\left(\bigcap_{x \in X} \mathcal{U}_x\right) + \cdots \quad (4)$$

where Π denotes the set of non-zero elements of EF of the form $x = ef$. We have $|\Pi| = (2^r - 1)^2$. Note that whenever X is made up of linearly independent elements, then the events $\mathcal{U}_x, x \in X$ are independent in the sense of probability, so that

$$\mathbb{P}\left(\bigcap_{x \in X} \mathcal{U}_x\right) = 2^{-\lambda|X|}.$$

More generally, since any linear combination of vectors that are in U is also in U, we have

$$\mathbb{P}\left(\bigcap_{x \in X} \mathcal{U}_x\right) = 2^{-\lambda \text{rk}(X)}$$

where rk(X) denotes the rank of X.

For $\lambda = 2r - 1$, tedious computations show that the contribution of the non full-rank subsets X for a growing (with r) set of first terms of (4) is negligible, so that we have

$$\mathbb{P}(\mathcal{U}) \approx 2 - \frac{4}{2!} + \frac{8}{3!} + \cdots \approx 1 - e^{-2}$$

Giving $1 - \mathbb{P}(\mathcal{U}) \approx e^{-2}$.

3.5 Value of λ

In order to find U that contains no element $x = ef$, we need to take the highest value possible for λ. We denote $\boldsymbol{z}_1 = \boldsymbol{y} + \boldsymbol{c}\boldsymbol{S}'$. When $\boldsymbol{z}_1$ is written as a $rd \times n$ matrix over $\mathbb{F}_q$ by rewriting each of its coordinates in a basis of EF of the form $\{u_1, \ldots, u_{rd-\lambda}, v_1, \ldots v_\lambda\}$ such that $U = \{u_1, \ldots, u_{rd-\lambda}\}$, we want $\boldsymbol{pS}$ to be equal to z_1 on the last λ lines, corresponding to $\{v_1, \ldots v_\lambda\}$. This gives λn equations in the base field, and the system has dlk unknowns (the coordinates of $\boldsymbol{p}$ in a basis of F). This gives the following condition on λ:

$$\lambda n < dlk \Leftrightarrow \lambda < \frac{dlk}{n}.$$

Since we want to maximize the value of λ, we take $\lambda = \lfloor \frac{dlk}{n} \rfloor$.

3.6 Computational Cost

Key generation. The most costly operation of the key generation step is the multiplication of $\boldsymbol{H}$ and the syndromes $\boldsymbol{s}_i$. Each matrix-vector multiplication costs n^2 multiplications in $\mathbb{F}_{q^m}$, hence a total cost of $(l + l')n^2$ multiplications.

Signature of a message μ. The signature step splits naturally into two phases: an offline phase during which the signature support is computed (this is the most costly part) and an online phase to compute the actual signature. The two phases are as follows:

1. Offline phase
   - Choose the vector spaces W and F.
   - Sample $\boldsymbol{y} \in (W + EF)^n$ and set $\boldsymbol{x} = \boldsymbol{H}\boldsymbol{y}^T$.
   - Choose U, a random subspace of EF of dimension $rd - \lambda$. If U contains non-zero elements of the form ef, $e \in E$ and $f \in F$, choose another U.
   - Write the $\mathbb{F}_{q^m}$-coordinates of the vector $\boldsymbol{pS}$ in a basis of EF of the form $\{u_1, \ldots, u_{rd-\lambda}, v_1, \ldots, v_\lambda\}$ where $U = \langle u_1, \ldots, u_{rd-\lambda}\rangle$ to obtain linear expressions in the variables p_{ij} of the form (3). Compute a $\lambda n \times \lambda n$ matrix $\boldsymbol{D}$ that inverts this linear mapping of the p_{ij}. This will allow to compute $\boldsymbol{p}$ such that $\boldsymbol{z} \in U$ in the online phase with a matrix multiplication instead of an inversion. If the linear map cannot be inverted to produce the matrix $\boldsymbol{D}$, choose another random subspace U of EF.
2. Online phase
   - Set $\boldsymbol{c} = \mathcal{H}(\boldsymbol{x}, F, \mu)$, $\boldsymbol{c} \in F^{l'k}$
   - Solve $\boldsymbol{z} = \boldsymbol{y} + \boldsymbol{cS'} + \boldsymbol{pS}$ with $\boldsymbol{p} \in F^{lk}$, using the matrix $\boldsymbol{D}$ computed during the offline phase
   - Output $(\boldsymbol{z}, F, \boldsymbol{c}, \boldsymbol{p})$ as signature.

The most costly step in the offline phase is the computation of the matrix $\boldsymbol{D}$, which requires inverting a linear system over $\mathbb{F}_q$ with λn equations, hence the cost is $(\lambda n)^3$ multiplications in $\mathbb{F}_q$.

The online phase consists in the computation of $\boldsymbol{p}$ which costs $(\lambda n)^2$ multiplications in $\mathbb{F}_q$ as well as the computation of $\boldsymbol{z} = \boldsymbol{y} + \boldsymbol{cS'} + \boldsymbol{pS}$ which costs $(l'k)^2 + (lk)^2$ multiplications in $\mathbb{F}_{q^m}$ for computing the matrix/vector products.

Verification of a signature. The most costly step during the verification phase is the computation of $\boldsymbol{H}\boldsymbol{z}^T - \boldsymbol{T}'\boldsymbol{c}^T + \boldsymbol{T}\boldsymbol{p}^T$, which costs $n^2 + (l'k)^2 + (lk)^2$ multiplications in $\mathbb{F}_{q^m}$.

4 Security of the Scheme

4.1 Product Spaces Subspaces Indistinguishability (PSSI)

The PSSI problem is a new problem which appears naturally when we try to prove the indistinguishability of the signatures.

Problem 4. **Product Spaces Subspaces Indistinguishability.** Let E be a fixed $\mathbb{F}_q$-subspace of $\mathbb{F}_{q^m}$ of dimension r. Let F_i, U_i and W_i be subspaces defined as follows:

- $F_i \xleftarrow{\$} \mathbf{Gr}(d, \mathbb{F}_{q^m})$
- $U_i \xleftarrow{\$} \mathbf{Gr}(rd - \lambda, EF_i)$ such that $\{ef, e \in E, f \in F\} \cap U_i = \{0\}$

– $W_i \xleftarrow{\$} \mathbf{Gr}(w, \mathbb{F}_{q^m})$

The $\mathrm{PSSI}_{r,d,\lambda,w,\mathbb{F}_{q^m}}$ problem consists in distinguishing samples of the form $(\boldsymbol{z}_i, F_i)$ where $\boldsymbol{z}_i$ is a random vector of $\mathbb{F}_{q^m}^n$ of support $W_i + U_i$ from samples of the form $(\boldsymbol{z}'_i, F_i)$ where $\boldsymbol{z}'_i$ is a random vector of $\mathbb{F}_{q^m}^n$ of weight $w + rd - \lambda$.

In order to study the complexity of this problem, we first reduce it to the case where the samples are of the form (Z_i, F_i) with $Z_i = \mathrm{Supp}(\boldsymbol{z}_i)$. Let us suppose we have a distinguisher D for this last case. Then given N samples of the PSSI problem, it is easy to compute the supports Z_i of the vectors $\boldsymbol{z}_i$ and to use D to distinguish if Z_i is a random subspace of dimension $w + rd - \lambda$ or if it is of the form $W_i + U_i$ with U_i a subspace of the product space EF_i.

Conversely, let us suppose we have a distinguisher D' for the PSSI problem. We are given N samples of the form (Z_i, F_i). For each sample, we can compute a random vector $\boldsymbol{z}_i$ of support Z_i and use D' to distinguish whether $\boldsymbol{z}_i$ is a random vector of weight $w + rd - \lambda$ or whether its support is of the form $W_i + U_i$.

Thus we can consider the case when the samples are only couples of subspaces of $\mathbb{F}_{q^m}$.

This problem is related to the decoding of LRPC codes [13]. Indeed we can consider a subspace $Z = U + W$ as the noisy support of a syndrome for an LRPC code, the noise corresponding to W. Consequently, it is natural to try and apply techniques used for decoding LRPC codes in order to solve the PSSI problem. The first idea is to use the basic decoding algorithm (see [13]). It consists in computing intersections I of the form $f^{-1}Z \cap f'^{-1}Z$ with $(f, f') \in F^2$. If Z is of the form $U + W$ then the probability that $\dim I \neq 0$ is much higher than if Z were truly random. However, this technique cannot be used because we filter the subspace U.

The decoding algorithm for LRPC codes has been improved in [5]. The idea is to consider product spaces of the form ZF_i where F_i is a subspace of F of dimension 2. The probability that $\dim ZF_i = 2(w + rd - \lambda)$ depends on whether Z is random or not. We study in detail the advantage of this distinguisher in the following paragraphs.

Consider the product subspace EF inside $\mathbb{F}_{q^m}$ with $\dim F = \dim E = r$. Suppose E is unknown, the typical dimension of the product EF is then r^2, if we assume $r^2 \ll m$. We now suppose we are given a subspace Q of $\mathbb{F}_{q^m}$ of dimension r^2 that is either a product space EF or a randomly chosen one, and we wish to distinguish between the two events. One easy way to do so, if the dimension m of the ambient space $\mathbb{F}_{q^m}$ is large enough, is to multiply Q by F. If Q is random, we will get the typical product dimension $\dim FQ = r \dim Q = r^3$. Whereas if $Q = EF$, we will get $\dim FQ \leq \binom{r+1}{2} r < r^3$. In fact, to distinguish the two cases it is enough to multiply Q by any subspace A of F of dimension 2, since we will have $\dim AQ \leq 2r^2 - r$ when $Q = EF$ and $\dim AQ = 2r^2$ in the typical random Q case.

To make our two cases difficult to distinguish, our query space Q is actually chosen to be constructed in one of two ways, making up a *distinguishing problem*:

Distinguishing problem. Distinguish whether Q is of the form (i) or (ii) below:

(i) $Q = U + W$ where U is a subspace of EF of codimension λ. The space E is chosen randomly of dimension r as before. The subspace U is chosen in such a way so that, for any subspace A of F dimension 2, we have $\dim AU = 2\dim U$. The space W is chosen randomly of dimension w, so that $\dim Q = r^2 - \lambda + w$.

(ii) Q is a random subspace of dimension $r^2 - \lambda + w$. Equivalently we may think of Q of the form $Q = V + W$ where both V and W are random (and independent) of dimensions $r^2 - \lambda$ and w respectively.

The purpose of choosing such a subspace U of EF is to make the dimension of AU equal to that of AV for a random V. Adding the random space W to U should keep the probability distributions of $\dim(AQ)$ equal for both ways of constructing Q. The purpose of W is to make the dimension of Q sufficiently large with respect to the dimension m of the ambient space, so that multiplying Q by a space of dimension more than 2 will typically fill up the whole space $\mathbb{F}_{q^m}$ anyway. In this manner, the two ways of constructing Q will be indistinguishable by measuring dimensions of the product of Q by a subspace.

First, we give a criterion for a subspace U of EF to have the property that $\dim(AU) = 2\dim U$ for any subspace A of dimension 2 of F.

Let E, F be two subspaces of $\mathbb{F}_{q^m}$, both of dimension r over $\mathbb{F}_q$. Let us make the remark that the maximum possible dimension of $F^{\langle 2\rangle}$ is $\binom{r+1}{2}$, and the maximum possible dimension of $F^{\langle 2\rangle}E$ is therefore $r\binom{r+1}{2}$.

Let $f_1, f_2, \dots, f_r$ be a basis of F. Denote by F_2 the subspace of F generated by f_1, f_2, by F_3 the subspace of F generated by f_1, f_2, f_3, and so on.

Lemma 1. *Suppose* $\dim F^{\langle 2\rangle}E = r\binom{r+1}{2}$. *Then* $f_1FE \cap f_2FE = f_1f_2E$.

Proof. We have $F_2FE = f_1FE + f_2FE$, and $f_1FE \cap f_2FE \supset f_1f_2E$, therefore

$$\dim(F_2FE) \geq 2\dim(EF) - \dim F \tag{5}$$

by using the formula $\dim(A + B) = \dim A + \dim B - \dim(A \cap B)$. Similarly, $F_{i+1}FE = F_iFE + f_{i+1}FE$ and $F_iFE \cap f_{i+1}FE \supset f_jf_{i+1}E$ for all $j = 1, 2, \dots i$. From which we have

$$\dim(F_{i+1}FE) \geq \dim(F_iFE) + \dim(FE) - i\dim E.$$

Now $\dim F^{\langle 2\rangle}E = r\binom{r+1}{2}$ only occurs when we have equality in all the above inequalities, in particular we have equality in (5) which implies that the inclusion $f_1f_2E \subset f_1FE \cap f_2FE$ is also an equality. □

Lemma 2. *Let U be a subspace of EF. If we suppose* $\dim(F^{\langle 2\rangle}E) = r\binom{r+1}{2}$, *we have that there exists a subspace $A \subset F$ of dimension 2 such that,*

$$\dim(AU) < 2\dim U$$

if and only if U contains two non-zero elements of the form fe and $f'e$ $f, f' \in F$, $e \in E$ where f and f' are two linearly independent elements of F.

Proof. Let A be a subspace of F of dimension 2 generated by f_1, f_2. We have $AS = f_1U + f_2U$ so that $\dim(AU) < 2\dim U$ if and only if $f_1U \cap f_2U \neq \{0\}$. But we have

$$f_1U \cap f_2U \subset f_1FE \cap f_2FE$$

and under the hypothesis $\dim(F^{\langle 2\rangle}E) = r\binom{r+1}{2}$ we have that $\dim(AFE) = 2r^2 - r$ and $f_1FE \cap f_2FE = f_1f_2E$. Therefore $f_1U \cap f_2U$ contains a non-zero element if and only if U contains an element of the form f_2e and an element of the form f_1e, for $e \in E$, $e \neq 0$. □

Corollary 1. *Suppose* $\dim(F^{\langle 2\rangle}E) = r\binom{r+1}{2}$, *and that* U *is a subspace of* FE *such that for any two non-zero elements* $f \in F$ *and* $e \in E$, $ef \notin U$. *Then we have, for any subspace* $A \subset F$ *of dimension* 2,

$$\dim(AU) = 2\dim U.$$

Next, we study the probability distribution of the dimension of the product space $A(U+W)$, where W is random of dimension w, and U is either constructed as above or uniform random. We only focus on the binary extension field case $q = 2$, and from the previous discussion we only keep the property that $\dim(AU)$ is maximal. In other words, for the purpose of the following analysis, U is a fixed subspace of $\mathbb{F}_{2^m}$ with $\dim U = u$, A is a fixed subspace of $\mathbb{F}_{2^m}$ of dimension $\dim A = 2$ and we suppose that we have $\dim(AU) = 2u$. Let W be a *random* subspace of dimension $\dim W = w$ of $\mathbb{F}_{2^m}$. The space W is chosen by choosing $x_1, x_2, \ldots, x_w$ random independent (in the sense of probability) elements of $\mathbb{F}_{2^m}$ and W is taken to be the subspace generated by the x_i. Strictly speaking, $x_1, \ldots, x_w$ may turn out not to be linearly independent and not generate a space of dimension w. However, w will be taken to be much smaller than m, so that this event happens with negligible probability.

Our goal is to study the probability that $A(U+W)$ does not have dimension $2(u+w)$ and see how it may vary for two different spaces U_1 and U_2.

Consider the mapping:

$$A^w \xrightarrow{\Phi} \mathbb{F}_{2^m}$$

$$(a_1, a_2, \ldots, a_w) \mapsto a_1x_1 + a_2x_2 + \cdots + a_wx_w.$$

The product space $A(U+W)$ does not have maximal dimension, namely $2(u+w)$, if and only if there is a non-zero $\mathbf{a} = (a_1, a_2, \ldots, a_w)$ in A^w such that $\Phi(\mathbf{a}) \in AU$. This event $\mathcal{E}$, over all choices of $\mathbf{x} = (x_1, \ldots, x_w)$, can therefore be written as:

$$\mathcal{E} = \bigcup_{\substack{\mathbf{a}\in A^w \\ \mathbf{a}\neq 0}} \mathcal{E}_{\mathbf{a}}$$

where $\mathcal{E}_{\mathbf{a}}$ denotes the event $\Phi(\mathbf{a}) \in AU$. Since $\mathbb{P}(\mathcal{E}_{\mathbf{a}}) = \frac{4^u}{2^m}$, the union bound gives us

$$\mathbb{P}(\mathcal{E}) \leq (4^w - 1)\frac{4^u}{2^m}. \qquad (6)$$

We now study the lower bound

$$\mathbb{P}(\mathcal{E}) \geq \sum_{\substack{\mathbf{a}\in A^w \\ \mathbf{a}\neq 0}} \mathbb{P}(\mathcal{E}_{\mathbf{a}}) - \sum_{\mathbf{a},\mathbf{b}} \mathbb{P}(\mathcal{E}_{\mathbf{a}} \cap \mathcal{E}_{\mathbf{b}}) \tag{7}$$

where the second sum runs over all unordered pairs of distinct w-tuples $\mathbf{a}$ and $\mathbf{b}$. To evaluate this second sum we split the pairs $\mathbf{a}, \mathbf{b}$ into two disjoint sets:

1. linearly independent pairs $\mathbf{a}, \mathbf{b}$. In which case the two random variables $\mathbf{ax}$ and $\mathbf{bx}$ are independent, and we have

$$\mathbb{P}(\mathcal{E}_{\mathbf{a}} \cap \mathcal{E}_{\mathbf{b}}) = \mathbb{P}(\mathcal{E}_{\mathbf{a}})\,\mathbb{P}(\mathcal{E}_{\mathbf{b}}) = \left(\frac{4^u}{2^m}\right)^2.$$

2. linearly dependent pairs $\mathbf{a}, \lambda\mathbf{a}$, for some $\lambda \in \mathbb{F}_{2^m}$, $\lambda \neq 1$, such that $\lambda\mathbf{a} \in A^w$. In this case, we have

$$\mathbb{P}(\mathcal{E}_{\mathbf{a}} \cap \mathcal{E}_{\mathbf{b}}) = |AU \cap \lambda AU|\frac{1}{2^m} \leq \frac{4^u}{2^m}.$$

   We now estimate the number of such pairs $\mathbf{a}, \lambda\mathbf{a}$.
   Denote the non-zero elements of A by $a_1, a_2, a_3 = a_1 + a_2$ (recall that A is a vector space). Suppose we have $\lambda a_1 = a_2$ and $\lambda a_2 = a_3 = a_1 + a_2$ ($\lambda a_2 = a_2$ would imply $\lambda = 1$ and $\lambda a_2 = a_1$ would imply $\lambda^2 = 1$ hence $\lambda = 1$ in a field of characteristic 2). Then $a_2 a_1^{-1} = \lambda$ satisfies $\lambda^2 + \lambda + 1$ which is not possible for odd m and happens with negligible probability for even m. Assuming this does not happen we have therefore that any $\mathbf{a} \in A^w$ such that $\lambda\mathbf{a} \in A^w$ must have all non-zero coefficients equal. Hence the number of such pairs $\mathbf{a}, \lambda\mathbf{a}$ is at most 3.2^w. Inequality (7) gives us therefore:

$$\mathbb{P}(\mathcal{E}) \geq (4^w - 1)\frac{4^u}{2^m} - \binom{4^w - 1}{2}\frac{4^{2u}}{2^{2m}} - 3.2^w\frac{4^u}{2^m}. \tag{8}$$

From which we get:

Proposition 4. *If* U *and* V *are two spaces of dimension* u *such that* $\dim(AU) = \dim(AV) = 2u$ *then*

$$|\mathbb{P}(\dim[A(U + W)] < 2(u + w)) - \mathbb{P}(\dim[A(V + W)] < 2(u + w))| \leq \binom{4^w - 1}{2}\frac{4^{2u}}{2^{2m}} + 3.2^w\frac{4^u}{2^m}.$$

Product space distinguisher. We go back to our distinguishing problem defined above. As mentioned in the discussion leading up to the problem, it is natural to try and distinguish between (i) and (ii) by computing the dimension of some AQ for many instances of Q and basing the decision on the number of times an abnormal (less than $2\dim Q$) turns up. The consequence of Proposition 4 is that to distinguish confidently with this method requires a very large number of queries. Specifically, if the two probabilities of producing an abnormal dimension are p and $p(1 + \varepsilon)$, then the number of products AQ that one must produce is of the order $1/p\varepsilon^2$. Proposition 4 gives $p \approx 2^{2u+2w-m}$ and $\varepsilon = 2^{\log_2 3 - w}$.

Proposition 5. *By applying Proposition 4 to the* $\mathrm{PSSI}_{r,d,\lambda,w,\mathbb{F}_{q^m}}$ *problem, the advantage with which one may distinguish the two distributions is of the order of* $2^{m-2(rd-\lambda)}$.

Remark: One might also consider computing product spaces of the form ZE' where E' is a subspace of E of dimension larger than 2. However, we have chosen our parameters such that $3(w + rd - \lambda) > m$ so this idea cannot work.

4.2 New Problem: Advanced Product Subspaces Indistinguishability (PSSI$^+$)

The PSSI$^+$ problem is a generalization of the previous problem, with some extra side information. We need to consider this problem for our security proof.

Problem 5. **Advanced Product Spaces Subspaces Indistinguishability.** Let E be a fixed $\mathbb{F}_q$-subspace of $\mathbb{F}_{q^m}$ of dimension r. Let F_i, U_i and W_i be subspaces defined as before:

- $F_i \xleftarrow{\$} \mathbf{Gr}(d, \mathbb{F}_{q^m})$
- $U_i \xleftarrow{\$} \mathbf{Gr}(rd - \lambda, EF_i)$ such that $\{ef, e \in E, f \in F\} \cap U_i = \{0\}$
- $W_i \xleftarrow{\$} \mathbf{Gr}(w, \mathbb{F}_{q^m})$

Let $\boldsymbol{H}$ be a randomly chosen $(n-k) \times n$ ideal double circulant matrix as in Definition 7 for an irreducible polynomial P.

- Sample l vectors $\boldsymbol{s}_i$ and l' vectors $\boldsymbol{s}_i'$ of length n from the same support E of dimension r
- Set $\boldsymbol{S}$ (respectively $\boldsymbol{S}'$) the matrix consisting of all $\boldsymbol{s}_i$ (respectively $\boldsymbol{s'}_i$) and their ideal shifts. Let $\boldsymbol{T} = \boldsymbol{H}\boldsymbol{S}^T$ and $\boldsymbol{T}' = \boldsymbol{H}\boldsymbol{S}'^T$.

The $\mathrm{PSSI}^+(N)_{r,d,\lambda,w,\mathbb{F}_{q^m}}$ problem consists in distinguishing N samples of the form $(\boldsymbol{z}_i, F_i)$ where $\boldsymbol{z}_i$ is a random vector of $\mathbb{F}_{q^m}^n$ of support $W_i + U_i$ from samples of the form $(\boldsymbol{z}_i', F_i)$ where $\boldsymbol{z}_i'$ is a random vector of $\mathbb{F}_{q^m}^n$ of weight $w + rd - \lambda$ when additionally given $\boldsymbol{H}, \boldsymbol{T}, \boldsymbol{T}'$.

The PSSI$^+$ problem consists of an instance of the PSSI problem and an instance of the RSL problem that share the same secret support E. The question is to determine whether or not the instance of RSL can be used in order to reduce the difficulty of PSSI.

In general, two difficult problems taken together do not necessarily make up another hard problem. For example, two difficult instances of the factorization of large integers n, n' with $n = pq$ and $n' = pq'$ where p, q and q' are prime is a an easy problem.

In our case, the knowledge of an instance of RSL could be useful if it gives us some information on the support E. But, for our parameters, the best attacks on the RSL problem are based on the GRS$^+$ algorithm [6,9,12] and this algorithm recovers the whole support or nothing. Moreover, the main idea behind the

GRS^+ algorithm (which consists of looking for a subspace E' which contains E) cannot be applied to the PSSI^+ problem since E is "multiplied" by an F_i at each sample. Thus it appears that the knowledge of an instance of RSL that shares the same secret support E does not help to solve the PSSI problem and we will consider that the PSSI^+ problem is as hard to attack as the PSSI problem.

4.3 Security Model

One of the security models for signature schemes is existential unforgeability under an adaptive chosen message attack (EUF-CMA). Basically, it means that even if an adversary has access to a signature oracle, it cannot produce a valid signature for a new message with a non negligible probability.

Existential Unforgeability under Chosen Message Attacks [20] ($\mathsf{EUF-CMA}$). Even after querying N valid signatures on chosen messages (μ_i), an adversary should not be able to output a valid signature on a fresh message μ. To formalize this notion, we define a signing oracle OSign:

- $\mathsf{OSign}(\mathsf{vk}, \mu)$: This oracle outputs a signature on μ valid under the verification key vk. The requested message is added to the signed messages set $\mathcal{SM}$.

$\mathbf{Exp}^{\mathsf{euf}}_{\mathcal{S},\mathcal{A}}(\lambda)$
1. $\mathsf{param} \leftarrow \mathsf{Setup}(1^\lambda)$
2. $(\mathsf{vk}, \mathsf{sk}) \leftarrow \mathsf{KeyGen}(\mathsf{param})$
3. $(\mu^*, \sigma^*) \leftarrow \mathcal{A}(\mathsf{vk}, \mathsf{OSign}(\mathsf{vk}, \cdot))$
4. $b \leftarrow \mathsf{Verify}(\mathsf{vk}, \mu^*, \sigma^*)$
5. `IF` $\mu^* \in \mathcal{SM}$ `RETURN` 0
6. `ELSE RETURN` b

The probability of success against this game is denoted by

$$\mathsf{Succ}^{\mathsf{euf}}_{\mathcal{S},\mathcal{A}}(\lambda) = \mathbb{P}\left(\mathbf{Exp}^{\mathsf{euf}}_{\mathcal{S},\mathcal{A}}(\lambda) = 1\right), \quad \mathsf{Succ}^{\mathsf{euf}}_{\mathcal{S}}(\lambda, t) = \max_{\mathcal{A} \leq t} \mathsf{Succ}^{\mathsf{euf}}_{\mathcal{S},\mathcal{A}}(\lambda).$$

4.4 $\mathsf{EUF-CMA}$ Proof

To prove the $\mathsf{EUF-CMA}$ security of our scheme, we proceed in two steps. In the first step, we show that an adversary with access to N valid signatures has a negligible advantage over the same adversary with only access to the public keys. In other words, we prove that signatures do not leak information on the secret keys. In the second step, we show that if we only have access to the public keys, a valid signature allows us to solve an instance of the I − ARSD problem.

We will use the following technical Lemma.

Lemma 3. *Let $\mathcal{F}$ be a family of functions defined by*

$$\mathcal{F} = \left\{ \begin{array}{lll} f_{\boldsymbol{H}} : (W + EF)^n & \rightarrow & \mathbb{F}_{q^m}^{n-k} \\ \boldsymbol{y} & \mapsto & \boldsymbol{x} = \boldsymbol{y}\boldsymbol{H}^T \end{array} \right\}$$

Since $\boldsymbol{H}$ is chosen uniformly at random amongst the $(n-k) \times n$ ideal double circulant matrices, $\mathcal{F}$ is a pairwise independent family of functions.
The number of choices for $\boldsymbol{y}$ depends on W and F and on the choice of the coordinates of $\boldsymbol{y}$. Overall, the entropy of $\boldsymbol{y}$ is equal to

$$\Theta\left(\begin{bmatrix} m \\ w \end{bmatrix}_q \begin{bmatrix} m \\ d \end{bmatrix}_q q^{(w+rd)n}\right) = 2^{(w(m-w)+d(m-d)+(w+rd)n)\log q+O(1)}$$

Since $\|\boldsymbol{y}\| > d_{RGV}$, any vector of $\mathbb{F}_{q^m}^{n-k}$ can be reached, thus the entropy of $\boldsymbol{x}$ is equal to $2^{(n-k)m\log q}$. According to the Leftover Hash Lemma [21], we have

$$\Delta(\mathcal{D}_{\mathbf{G_0}}, \mathcal{U}) < \frac{\varepsilon}{2}$$

where $\Delta(X,Y)$ denotes the statistical distance between X and Y, $\mathcal{D}_{\mathbf{G_0}}$ denotes the distribution of $\boldsymbol{x}$ in game $\mathbf{G_0}$, $\mathcal{U}$ denotes the uniform distribution over $\mathbb{F}_{q^m}^{n-k}$ (the distribution of $\boldsymbol{x}'$ in game $\mathbf{G_1}$) and

$$\varepsilon = 2^{\frac{((n-k)m-w(m-w)+d(m-d)+(w+rd)n)\log q}{2}+O(1)}.$$

Proofs. For the first step, we proceed in a sequence of games. We denote $\mathbb{P}_{\mathbf{G_i}}$ the probability that the adversary returns 1 in the end of the game $\mathbf{G_i}$ and $\mathsf{Adv}(\mathbf{G_i}) = |\mathbb{P}_{\mathbf{G_i}} - \frac{1}{2}|$ the advantage of the adversary for the game $\mathbf{G_i}$.

- $\mathbf{G_0}$: this is the real $\mathsf{EUF-CMA}$ game for $\mathcal{S}$. The adversary has access to the signature oracle OSign to obtain valid signatures.

$$\mathbb{P}_{\mathbf{G_0}} = \mathsf{Succ}^{\mathsf{euf}}_{\mathcal{S},\mathcal{A}}(\lambda).$$

- $\mathbf{G_1}$: we replace $\boldsymbol{z}$ by a vector $\boldsymbol{z}'$ of the same weight chosen uniformly at random in the correct subspace U of $W + EF$, and sample $\boldsymbol{c}', \boldsymbol{p}'$ uniformly with support F.
  Now set $\boldsymbol{x}' = \boldsymbol{H}\boldsymbol{z}' - \boldsymbol{c}'\boldsymbol{T}' - \boldsymbol{p}'\boldsymbol{T}$ and use the Random Oracle to set $\boldsymbol{c} = \mathcal{H}(\boldsymbol{x}', F, \mu)$.
  In $\mathbf{G_0}$, $\boldsymbol{x}$ is the syndrome of the vector $\boldsymbol{y}$ of support of the form $EF + W$, while here $\boldsymbol{x}'$ is not necessarily. Under Lemma 3 we conclude

$$\mathsf{Adv}(\mathbf{G_1}) \leq \mathsf{Adv}(\mathbf{G_0}) + \varepsilon.$$

  The parameters of the signature are chosen such that ε is lower than the security parameter.
- $\mathbf{G_2}$: We now sample $\boldsymbol{z}$ at random in $\mathbb{F}_{q^m}^n$ with the same weight, and proceed as in $\mathbf{G_2}$.
  This corresponds to an instance of the PSSI$^+(N)$ Problem 5. Since the adversary can have access to at most N signatures, we have

$$|\mathsf{Adv}(\mathbf{G_2}) - \mathsf{Adv}(\mathbf{G_1})| \leqslant \mathsf{Adv}(\mathrm{PSSI}^+(N)).$$

- $\mathbf{G_3}$: We now pick $\boldsymbol{T}, \boldsymbol{T}'$ at random and proceed as before. The difference between $\mathbf{G_3}$ and $\mathbf{G_2}$ resides in the public key, specifically whether it was sampled using vectors in a given subspace or not.

$$|\mathsf{Adv}(\mathbf{G_3}) - \mathsf{Adv}(\mathbf{G_2})| \leqslant \mathsf{Adv}(\mathrm{DRSL}).$$

At this step, everything we send to the adversary is random, and independent from any secret keys. Hence the security of our scheme is reduced to the case where no signature is given to the attacker.

If he can compute a valid signature after game $\mathbf{G_3}$, then the challenger can compute a solution of the $\mathrm{I-ARSD}$ problem. Indeed, the couple $(\boldsymbol{z}, \boldsymbol{p})$ is a solution of the instance $(\boldsymbol{H}, -\boldsymbol{T}, F, \boldsymbol{x} + \boldsymbol{T}'\boldsymbol{c}^T, w + rd - \lambda)$ of the $\mathrm{I-ARSD}$ problem. According to Proposition 2, the $\mathrm{I-ARSD}$ problem is reduced to the $\mathrm{I-RSD}$ problem.

Finally, we can now give our main theorem:

Theorem 1 (EUF-CMA security). *Under the hypothesis of the hardness of the* PSSI^+ *Problem 4.1 and of the* $\mathrm{DRSL}, \mathrm{I-RSD}$ *Problem 1, our signature scheme is secure under the EUF-CMA model in the Random Oracle Model.*

5 Attacks

5.1 Attacks on the RSL Problem

In this section we will study the hardness of recovering the secret matrices $\boldsymbol{S}$ and $\boldsymbol{S}'$ from $\boldsymbol{H}, \boldsymbol{T}, \boldsymbol{T}'$. This is exactly an instance of the $\mathrm{RSL}_{q,m,n,k,w,N}$ problem.

We will use the setting proposed in [12]. First, we recall how the problem is reduced to searching for a codeword of weight w in a code containing q^N codewords of this form. We introduce the following $\mathbb{F}_q$-linear code:

$$C = \{\boldsymbol{x} \in \mathbb{F}_{q^m}^n : \boldsymbol{A}\boldsymbol{x} \in W_T\}$$

where W_T is the $\mathbb{F}_q$-linear subspace of $\mathbb{F}_{q^m}^{n-k}$ generated by the linear combinations of the elements of the public matrices $\boldsymbol{T}$ and $\boldsymbol{T}'$. As in Lemma 1 in [12], we define:

$$C' = \{\sum_i \alpha_i \boldsymbol{s}_i, \alpha_i \in \mathbb{F}_q\}.$$

We have:
- $\dim_{\mathbb{F}_q} C \leqslant km + N$
- $C' \subset C$
- the elements of C' are of weight $\leqslant w$.

Combinatorial attack. In [12], the authors search for a codeword of rank w in C by using information-set decoding techniques, using the fact that C' contains q^N words of weight w. As this codeword will very likely be a linear combinations of the vectors s_i, it will reveal the secret support E with high probability. Theorem 2 in [12] gives a complexity of $q^{min(e_-,e_+)}$, where:

$$e_- = \left(w - \left\lfloor \frac{N}{n} \right\rfloor\right)\left(\left\lfloor \frac{K}{n} \right\rfloor - \left\lfloor \frac{N}{n} \right\rfloor\right)$$
$$e_+ = \left(w - \left\lfloor \frac{N}{n} \right\rfloor - 1\right)\left(\left\lfloor \frac{K}{n} \right\rfloor - \left\lfloor \frac{N}{n} \right\rfloor - 1\right) + n\left(\left\lfloor \frac{K}{n} \right\rfloor - \left\lfloor \frac{N}{n} \right\rfloor - 1\right)$$

where $K = km + N$. See [12] for more details about this.

Algebraic attacks. We will now study how algebraic attacks can be used to find codewords of weight w in C.

We are looking for $X \in C$ such that $X \in E^n$. We can write X as:

$$X = \begin{matrix} \sum_{i=1}^{w} x_1^{(i)} y_1^{(i)} & \cdots & \sum_{i=1}^{w} x_1^{(i)} y_n^{(i)} \\ \vdots & \ddots & \vdots \\ \sum_{i=1}^{w} x_m^{(i)} y_1^{(i)} & \cdots & \sum_{i=1}^{w} x_m^{(i)} y_n^{(i)} \end{matrix}$$

where $(x^{(1)}, \ldots, x^{(w)})$ represent a basis of E, and the $(y_i^j), 1 \leqslant i \leqslant w, 1 \leqslant j \leqslant n$ represent the coordinates of X written in this basis.

C has length nm and dimension $N + km$ in $\mathbb{F}_q$, which gives $(n-k)m - N$ parity check equations, and $(n+m)w$ unknowns (the $x_i^{(j)}$ and the $(y_i^{(j)})$).

To decrease the number of unknowns, we will first write the basis of E in an echelon form, which removes w^2 unknowns:

$$\forall (i,j) \in [1,w]^2, i \neq j, x_i^{(j)} = 0$$
$$x_i^{(i)} = 1.$$

Then we will use the fact that for a fixed basis of E, the solution space has dimension N, which allows us to set N of the $(y_i^{(j)})$ to specialize one solution, as in [9]: for a random subset $I \subseteq [1,n] \times [1,w]$ of size $N-1$:

$$\forall (i,j) \in I, y_i^{(j)} = 0$$
$$y_{i_0}^{(j_0)} = 1, (i_0, j_0) \notin I$$

which removes N unknowns.

Proposition 6. *Using this setting we obtain:*

- *$(n-k)m - N$ equations*
- *$(n+m)w - w^2 - N$ unknowns.*

We implemented this approach in Magma to try it on small examples, and the combinatorial attacks become much more efficient than the algebraic approach when the number of samples is around kr, whereas this attack is faster when the number of samples is higher. Another drawback of this attack is the high memory cost, making parameters as small as $n = m = 30, k = 15, r = 3$ with kr samples too big for a computer using 16 GB of RAM.

For concrete parameters (Sect. 6), we chose N, the number of samples for the RSL problem, equal to either $k(r-1)$ or $k(r-2)$. Our experiments on smaller parameters showed that combinatorial attacks should be way faster for this number of samples. This also defeats the setting proposed in [9] since it needs at least kr samples.

The parameter set I gives 2117 unknowns for 23836 equations and the parameter set II gives 2809 unknowns for 29154 equations. Based on our experiments on smaller parameters this seems really hard to reach.

5.2 Attack on the ARSD Problem

As explained in the security proof in Sect. 4.4, a forgery attack consists in solving an instance of the ARSD Problem 3. In order to choose the parameters of our signature, we need to deal with the complexity of the attacks on this problem. These attacks are adapted from those against the RSD problem [6,14] to which ARSD is very similar.

The following proposition gives a bound beyond which the problem becomes polynomial.

Proposition 7. *Let* $(\boldsymbol{H}, \boldsymbol{H}', \boldsymbol{s}, F)$ *be an instance of the* $ARSD_{q,m,n,k,r,n',F}$ *problem. If* $\max(m,n)r + n'r' \geqslant m(n-k)$ *then the* ARSD *problem can be solved in polynomial time with a probabilistic algorithm.*

Proof. To prove this proposition, we will use the method used to compute the Singleton bound.

Let us begin with the case $n \geqslant m$. Let E be a subspace of $\mathbb{F}_{q^m}$ of dimension r and suppose that there exists a solution $(\boldsymbol{x}, \boldsymbol{x}')$ of the ARSD problem such that $\text{Supp}(\boldsymbol{x}) = E$. Then, we can express the coordinate x_j of $\boldsymbol{x}$ in a basis E_i of E:

$$\forall j \in \{1 \dots n\}, x_j = \sum_{i=1}^{r} \lambda_{ij} E_i$$

Likewise, we can express the coordinates x'_t of $\boldsymbol{x}'$ in a basis of F:

$$\forall t \in \{1 \dots n'\}, x'_t = \sum_{s=1}^{r'} \lambda'_{st} F_s$$

Let us write the linear system satisfied by the unknown $(\lambda_{ij}, \lambda'_{st})$:

$$\begin{aligned}
& & \boldsymbol{H}\boldsymbol{x}^T \quad &+ \quad \boldsymbol{H}'\boldsymbol{x}'^T = \boldsymbol{s} \\
\iff & \forall i \in \{1 \dots n-k\}, & \sum_{j=1}^{n} H_{ij} x_j \quad &+ \quad \sum_{t=1}^{n'} H'_{it} x'_t = s_i \\
\iff & \forall i \in \{1 \dots n-k\}, & \sum_{j=1}^{n} H_{ij} \sum_{i=1}^{r} \lambda_{ij} E_i &+ \sum_{t=1}^{n'} H'_{it} \sum_{s=1}^{r'} \lambda'_{st} F_s = s_i
\end{aligned} \tag{9}$$

The $(n-k)$ linear equations (9) over $\mathbb{F}_{q^m}$ can be projected on $\mathbb{F}_q$ to obtain $m(n-k)$ linear equations over $\mathbb{F}_q$. Since we have $nr + n'r \geqslant m(n-k)$, there are more unknowns than equations so the system admits at least a solution with a non negligible probability.

In the case $m > n$, we need to consider the matrix $\boldsymbol{M}(\boldsymbol{x})$ associated to $\boldsymbol{x}$ (cf Definition 1) and express its rows in a basis of a subspace E of dimension r of $\mathbb{F}_q^n$. Since the support of $\boldsymbol{x}'$ is fixed, its coordinates still give us $n'r'$ unknowns over $\mathbb{F}_q$. This gives us $mr + n'r'$ unknowns over $\mathbb{F}_q$ in total. Then we transform the equation $\boldsymbol{H}\boldsymbol{x}^T + \boldsymbol{H}'\boldsymbol{x}'^T = \boldsymbol{s}$ into a linear system over $\mathbb{F}_q$ as previously. This operation is not difficult but cumbersome and we do not give the explicit equations. The resulting linear system has $m(n-k)$ equations and $mr + n'r'$ unknowns over $\mathbb{F}_q$. It has a solution with a non negligible probability since $mr + n'r' \geqslant m(n-k)$.

In both cases, the solution of the system solves the ARSD problem. □

In the case where $\max(m,n)r + n'r' < m(n-k)$, we need to adapt the best attacks against the RSD problem [14] to the ARSD case. The general idea is to find a subspace E of dimension δ such that $\text{Supp}(\boldsymbol{x}) \subset E$ (in the case $n \geqslant m$). Then we can express the coordinates of $\boldsymbol{x}$ if $n \geqslant m$ or the rows of the matrix $\boldsymbol{M}(\boldsymbol{x})$ if $m > n$ in a basis of E exactly as in the previous proposition. We want δ as large as possible to increase the probability that $\text{Supp}(\boldsymbol{x}) \subset E$ but we have to take δ such that $\max(m,n)\delta + n'r' < m(n-k)$ in order to obtain an over-constrained linear system. Hence $\delta = \left\lfloor \frac{m(n-k)-n'r'}{\max(m,n)} \right\rfloor$. The probability that $E \supset \text{Supp}(\boldsymbol{x})$ depends on m and n:

$$\text{If } n \geqslant m, \text{ then } \quad \mathbb{P}(E \supset \text{Supp}(\boldsymbol{x})) = \begin{bmatrix} \delta \\ r \end{bmatrix}_q \begin{bmatrix} m \\ r \end{bmatrix}_q^{-1} = \Theta(q^{-r(m-\delta)}).$$

$$\text{If } n < m, \text{ then } \quad \mathbb{P}(E \supset \text{Supp}(\boldsymbol{x})) = \begin{bmatrix} \delta \\ r \end{bmatrix}_q \begin{bmatrix} n \\ r \end{bmatrix}_q^{-1} = \Theta(q^{-r(n-\delta)}).$$

In order to respect the constraints of our signature, we have to insure that the instance of the ARSD problem has several solutions. Thus, the average complexity of this attack is equal to the inverse of the probability $\mathbb{P}(E \supset \text{Supp}(\boldsymbol{x}))$ divided by the number of solutions times the cost of the linear algebra. The number of solutions is in $\Theta\left(q^{r(m+n-r)+n'r'-m(n-k)}\right)$ (see Proposition 2 for details).

Proposition 8. *In the case* $\max(m,n)r + n'r' < m(n-k)$*, the complexity of the best attack against the* $\text{ARSD}_{q,m,n,k,r,n',F}$ *problem is in*

$$\mathcal{O}\left(m^3(n-k)^3 q^{r\left\lceil \frac{km+n'r'}{\max(m,n)} \right\rceil - r(m+n-r) - n'r' + m(n-k)}\right).$$

Remark: We did not consider the improvement of the attack of the RSD problem in [6] because this attack does not fit the case where there are several solutions to the RSD problem.

6 Parameters

6.1 Constraints

In this section we recap the different constraints on our parameters.

Choice of l, l', r and d. First we need to choose l' such that the entropy of $\mathbf{c}$ is high enough. For our parameters, $l' = 1$ is always enough since $\mathbf{c} \in F^{l'dk}$ and $dk > 512$. In practice using less than dk coordinates for $\mathbf{c}$ is a possibility to make the parameters a little smaller.

We then need to choose r high enough so that the attacks on both the RSD and RSL problems are hard. d and l must be chosen such that $\lambda \geqslant r + d$: $d = r$ and $l = 4$ is a way to meet this condition. In the sets of parameters given below, this value of l leads to $N = k(r-1)$ and $N = k(r-2)$ respectively, which allows us to be pretty conservative with respect to the attacks on the RSL problem.

Choice of m. In order to avoid the distinguisher attack for a security parameter of 128, the relation $m - 2u \geqslant 128 + 64$ (we use Proposition 5, setting $u = rd - \lambda$), must be verified to fit the security proof: we consider that the adversary has access to 2^{64} signatures, so the probability of distinguishing signatures and random vectors must be lower than 2^{-192}. We choose a prime m (so there is no intermediate field between $\mathbb{F}_q$ and $\mathbb{F}_{q^m}$) such that $m \geqslant 192 + 2u$.

Choice of n, k and w. They must be chosen such that $3(u + w) > m$ to avoid the distinguisher attack using subspaces of dimension 3, and $(u+w) < (n-k)-\lambda$ in order to keep the weight of the signature below the Singleton bound $-\lambda$ (due to ARSD). k is taken prime for having access to really sparse polynomials to define the ideal codes.

6.2 Example of Parameters

The public key consists of:

- $\boldsymbol{H}$ which can be recovered from a seed (256 bits)
- $l(n-k)m\log(q)$ bits to describe the syndromes.

The signature consists of:

- $(rd + w - \lambda)(n + m - rd - w + \lambda)\log(q)$ bits to describe $\boldsymbol{z}$. We give $\text{Supp}(\boldsymbol{z})$ in echelon form as well as the coordinates in this basis
- A seed to describe F (256 bits)
- 512 bits to describe $\mathbf{c}$
- $dlk\log(q)$ bits to describe $\boldsymbol{p}$.

The complexity of the key recovery attack is computed using the complexity of the combinatorial attack given in Sect. 5.1.

For our parameters, the complexity of the forgery attack using the algorithm against ARSD described in Sect. 5.2 is disproportionately large compared to the key recovery attack. Parameter sets were chosen for a security of 128 bits.

	m	n	k	l	l'	d	r	w	λ	q	Public key size	Signature size	Key recovery attack	Distinguisher	Security
I	241	202	101	4	1	6	6	57	12	2	15.25 kB	4.06 kB	461	193	128
II	263	226	113	4	1	7	7	56	14	2	18.61 kB	5.02 kB	660	193	128

The implementation of our scheme on an Intel(R) Core(TM) i5-7440HQ CPU running at 2.80 GHz gives the following computation times:

Parameter	Keygen	Online signature phase	Verification
I	4 ms	4 ms	5 ms
II	5 ms	5 ms	6 ms

For the offline phase, the most costly step, the computation of the matrix $\boldsymbol{D}$, takes 350 ms for parameter I and 700 ms for parameter II.

Acknowledgements. This work has been supported in part by the French ANR projects CBCRYPT (ANR-17-CE39-0007) and ID-FIX (ANR-16-CE39-0004). The authors would like to thank Alain Couvreur for his insightful comments.

References

1. Aguilar Melchor, C., et al.: HQC 2017. NIST Round 1 submission for Post-Quantum Cryptography (2017)
2. Aguilar Melchor, C., et al.: RQC 2017. NIST Round 1 submission for Post-Quantum Cryptography (2017)
3. Aguilar Melchor, C., Gaborit, P., Schrek, J.: A new zero-knowledge code based identification scheme with reduced communication. In: Proceedings of the IEEE ITW (2011)
4. Aragon, N., et al.: BIKE 2017. NIST Round 1 submission for Post-Quantum Cryptography (2017)
5. Aragon, N., Gaborit, P., Hauteville, A., Ruatta, O., Zémor, G.: Low rank parity check codes: new decoding algorithms and application to cryptography. IEEE Trans. Inf. Theory (2019, submitted)
6. Aragon, N., Gaborit, P., Hauteville, A., Tillich, J.-P.: A new algorithm for solving the rank syndrome decoding problem. In: Proceedings of the IEEE ISIT (2018)
7. Courtois, N.T., Finiasz, M., Sendrier, N.: How to achieve a McEliece-based digital signature scheme. In: Boyd, C. (ed.) ASIACRYPT 2001. LNCS, vol. 2248, pp. 157–174. Springer, Heidelberg (2001). https://doi.org/10.1007/3-540-45682-1_10
8. Debris-Alazard, T., Sendrier, N., Tillich, J.-P.: The problem with the surf scheme. Preprint (2017). https://arxiv.org/abs/1706.08065
9. Debris-Alazard, T., Tillich, J.-P.: Two attacks on rank metric code-based schemes: Ranksign and an identity-based-encryption scheme. In: ASIACRYPT (2018)
10. Faugère, Je.-C., Gauthier, V., Otmani, A., Perret, L., Tillich, J.-P.: A distinguisher for high rate McEliece cryptosystems. IEEE Trans. Inf. Theory IT **59**(10), 6830–6844 (2013)

11. Fukushima, K., Sarathi Roy, P., Xu, R., Kiyomoto, S., Morozov, K., Takagi, T.: RaCoSS. NIST Round 1 submission for Post-Quantum Cryptography (2017)
12. Gaborit, P., Hauteville, A., Phan, D.H., Tillich, J.-P.: Identity-based encryption from codes with rank metric. In: Katz, J., Shacham, H. (eds.) CRYPTO 2017. LNCS, vol. 10403, pp. 194–224. Springer, Cham (2017). https://doi.org/10.1007/978-3-319-63697-9_7
13. Gaborit, P., Murat, G., Ruatta, O., Zémor, G.: Low rank parity check codes and their application to cryptography. In: Proceedings of the WCC (2013)
14. Gaborit, P., Ruatta, O., Schrek, J.: On the complexity of the rank syndrome decoding problem. IEEE Trans. Inf. Theory IT **62**(2), 1006–1019 (2016)
15. Gaborit, P., Ruatta, O., Schrek, J., Zémor, G.: New results for rank-based cryptography. In: Pointcheval, D., Vergnaud, D. (eds.) AFRICACRYPT 2014. LNCS, vol. 8469, pp. 1–12. Springer, Cham (2014). https://doi.org/10.1007/978-3-319-06734-6_1
16. Gaborit, P., Ruatta, O., Schrek, J., Zémor, G.: RankSign: an efficient signature algorithm based on the rank metric. In: Mosca, M. (ed.) PQCrypto 2014. LNCS, vol. 8772, pp. 88–107. Springer, Cham (2014). https://doi.org/10.1007/978-3-319-11659-4_6
17. Gaborit, P., Schrek, J., Zémor, G.: Full cryptanalysis of the chen identification protocol. In: Yang, B.-Y. (ed.) PQCrypto 2011. LNCS, vol. 7071, pp. 35–50. Springer, Heidelberg (2011). https://doi.org/10.1007/978-3-642-25405-5_3
18. Gaborit, P., Zémor, G.: On the hardness of the decoding and the minimum distance problems for rank codes. IEEE Trans. Inf. Theory IT **62**(12), 7245–7252 (2016)
19. Gentry, C., Peikert, C., Vaikuntanathan, V.: Trapdoors for hard lattices and new cryptographic constructions. In: STOC (2008)
20. Goldwasser, S., Micali, S., Rivest, R.: A digital signature scheme secure against adaptive chosen-message attacks. SIAM J. Comput. **17**(2), 281–308 (1988)
21. Håstad, J., Impagliazzo, R., Levin, L., Luby, M.: A pseudorandom generator from any one-way function. SIAM J. Comput. **28**(4), 1364–1396 (1999)
22. Hauteville, A., Tillich, J.-P.: New algorithms for decoding in the rank metric and an attack on the LRPC cryptosystem. In: Proceedings of the IEEE ISIT (2015)
23. Hoffstein, J., Howgrave-Graham, N., Pipher, J., Silverman, J., Whyte, W.: NTRUSign: digital signatures using the NTRU lattice. In: Joye, M. (ed.) CT-RSA 2003. LNCS, vol. 2612, pp. 122–140. Springer, Heidelberg (2003). https://doi.org/10.1007/3-540-36563-X_9
24. Lee, W., Kim, Y.-S., Lee, Y.-W., No, J.-S.: pqsigRM 2017. NIST Round 1 submission for Post-Quantum Cryptography (2017)
25. Loidreau, P.: On cellular code and their cryptographic applications. In: Proceedings of ACCT (2014)
26. Lyubashevsky, V.: Fiat-Shamir with aborts: applications to lattice and factoring-based signatures. In: Matsui, M. (ed.) ASIACRYPT 2009. LNCS, vol. 5912, pp. 598–616. Springer, Heidelberg (2009). https://doi.org/10.1007/978-3-642-10366-7_35
27. Lyubashevsky, V.: Lattice signatures without trapdoors. In: Pointcheval, D., Johansson, T. (eds.) EUROCRYPT 2012. LNCS, vol. 7237, pp. 738–755. Springer, Heidelberg (2012). https://doi.org/10.1007/978-3-642-29011-4_43
28. Lyubashevsky, V., et al.: CRYSTALS-DILITHIUM 2017. NIST Round 1 submission for Post-Quantum Cryptography (2017)
29. Persichetti, E.: Improving the efficiency of code-based cryptography. Ph.D. thesis, The University of Auckland (2012). https://persichetti.webs.com/Thesis%20Final.pdf

30. Schnorr, C.-P.: Efficient signature generation by smart cards. J. Cryptol. **4**, 161–174 (1991)
31. Stern, J.: A new identification scheme based on syndrome decoding. In: Stinson, D.R. (ed.) CRYPTO 1993. LNCS, vol. 773, pp. 13–21. Springer, Heidelberg (1994). https://doi.org/10.1007/3-540-48329-2_2

SeaSign: Compact Isogeny Signatures from Class Group Actions

Luca De Feo[1(✉)] and Steven D. Galbraith[2]

[1] Université Paris-Saclay – UVSQ, LMV, UMR CNRS 8100, Versailles, France
luca.de-feo@uvsq.fr
https://defeo.lu/

[2] Mathematics Department, University of Auckland, Auckland, New Zealand
s.galbraith@auckland.ac.nz

Abstract. We give a new signature scheme for isogenies that combines the class group actions of CSIDH with the notion of Fiat-Shamir with aborts. Our techniques allow to have signatures of size less than one kilobyte at the 128-bit security level, even with tight security reduction (to a non-standard problem) in the quantum random oracle model. Hence our signatures are potentially shorter than lattice signatures, but signing and verification are currently very expensive.

1 Introduction

Stolbunov [49] was the first to sketch a signature scheme based on isogeny problems. Stolbunov's scheme is in the framework of class group actions. However the scheme was not analysed in the post-quantum setting, and a naive implementation would leak the private key. Due to renewed interest in class group actions, especially CSIDH [13] (due to Castryck, Lange, Martindale, Panny and Renes) and the scheme by De Feo, Kieffer and Smith [22], it is of interest to develop a secure signature scheme in this setting. Our main contribution is to use Lyubashevsky's "Fiat-Shamir with aborts" strategy [40] to obtain a secure signature scheme. We also describe some methods to obtain much shorter signatures than in Stolbunov's original proposal.

Currently it is a major problem to get practical signatures from isogeny problems. Yoo *et al.* (see Table 1 of [53]) state signatures of over 100 KiB and signing/verification that take a few seconds on a PC. This can be reduced using some optimisations. For example [28] state approximately 12 KiB for this signature scheme (for classical 128-bit security level) and approximately 11 KiB for their main scheme. In contrast, in this paper we are able to get signatures smaller than a kilobyte, which is better even than lattice signatures. Unfortunately, signing and verification are very slow (the order of minutes), but we hope that future work (see for example [23]) will lead to more efficient schemes.

We now briefly summarise the main findings in the paper (for more details see Table 2). For the parameters $(n, B) = (74, 5)$ as used in CSIDH [13] we propose a signature scheme whose public key is 4 MiB, signature size is 978 bytes, and

Y. Ishai and V. Rijmen (Eds.): EUROCRYPT 2019, LNCS 11478, pp. 759–789, 2019.
https://doi.org/10.1007/978-3-030-17659-4_26

verification time is under 3 min (signing time is three times longer than this on average, since rejection sampling requires repeating the signing algorithm). For the same parameters we show that one can reduce the public key size to only 32 bytes, but this increases the signature size to around 3 KiB and does not add any significant additional cost to signing or verification time. One can obtain even shorter signatures by taking different choices of parameters, for example taking $(n, B) = (20, 3275)$ leads to signatures as small as 416 bytes, but we do not have an estimate of the verification time for these parameters.

The paper is organised as follows. Section 3 gives the basic signature scheme concept, that was proposed by Stolbunov, and our secure variant based on Fiat-Shamir with aborts. Section 4 explains how to get shorter signatures, at the expense of public key size, by using challenges that are more than just a single bit. This optimisation also leads to faster signing and verification. Section 5 shows how to retain the benefit of shorter signatures, while also having a short public key, by using modified Merkle trees. Section 7 shows how to use our scheme in the context of lossy keys, from which we obtain tight security in the quantum random oracle model via the results of Kiltz, Lyubashevsky and Schaffner [36] (and this security enhancement involves no increase in signature size, though the primes are larger so computations will be somewhat slower). This is the first time that lossy keys have been used in the isogeny setting. Section 8 explains that, if a quantum computer is available during parameter generation, then a much more practical signature scheme can be obtained by following the methods in Stolbunov's thesis.

The name "SeaSign" is a reference to the name CSIDH, which is pronounced "sea-side".

2 Background and Notation

We use the following notation: $\#X$ is the number of elements in a finite set X; log denotes the logarithm in base 2; KiB and MiB denote kilobytes and megabytes respectively; for $B \in \mathbb{N}$ we denote by $[-B, B]$ the set of integers u with $-B \leq u \leq B$.

2.1 Elliptic Curves, Isogenies, Ideal Class Groups

References for elliptic curves over finite fields and isogenies are Silverman [48], Washington [52], Galbraith [26], Sutherland [50] and De Feo [20]. A good reference for ideal class groups and class group actions is Cox [19].

Let E be an elliptic curve over a field K and let $P \in E(K)$ be a point of order m. Then there is a unique (up to isomorphism) elliptic curve E' and separable isogeny $\phi : E \to E'$ such that $\ker(\phi) = \langle P \rangle$. Vélu [51] gives an algorithm to compute an equation for E' and rational functions that enable to compute ϕ. The complexity of this algorithm is linear in m and requires field operations in K, so when K is a finite field it has cost $O(m \log(\#K)^2)$ bit operations using standard arithmetic. In the worst case (i.e., when m is large) this algorithm is exponential-time. In practice this computation is only feasible when m is relatively small (say $m < 1000$) and when the field K over which P is defined is not too large (say, at most a few thousand bits) For an elliptic curve E over a field K we define $\mathrm{End}(E)$ to be the ring of endomorphisms of E defined over the algebraic closure of K, and $\mathrm{End}_K(E)$ to be the ring of endomorphisms defined over K. Since we are mostly concerned with the CSIDH [13] approach, we will be interested in supersingular elliptic curves E such that $j(E) \in \mathbb{F}_p$, where p is a large prime. In this case $\mathrm{End}(E)$ is a maximal order in a quaternion algebra, while $\mathrm{End}_{\mathbb{F}_p}(E)$ is an order in the imaginary quadratic field $\mathbb{Q}(\sqrt{-p})$. Indeed, $\mathbb{Z}[\sqrt{-p}] \subseteq \mathrm{End}_{\mathbb{F}_p}(E)$.

We will be concerned with the ideal class group of the order $\mathcal{O} = \mathrm{End}_{\mathbb{F}_p}(E)$. This is the quotient of the group of fractional invertible ideals in $\mathcal{O}$ by the subgroup of principal fractional invertible ideals. The principal ideal $(1) = \mathcal{O}$ is the identity element of the ideal class group. Given two invertible $\mathcal{O}$-ideals $\mathfrak{a}, \mathfrak{b}$ we write $\mathfrak{a} \equiv \mathfrak{b}$ if $\mathfrak{a}$ and $\mathfrak{b}$ are equivalent (meaning that $\mathfrak{a}\mathfrak{b}^{-1}$ is a principal fractional $\mathcal{O}$-ideal).

2.2 Class Group Actions and Computational Problems

Let p be a prime. Let E be an ordinary elliptic curve over $\mathbb{F}_p$ with $\mathrm{End}(E) \cong \mathcal{O}$ or E a supersingular curve over $\mathbb{F}_p$ with $\mathrm{End}_{\mathbb{F}_p}(E) \cong \mathcal{O}$ where $\mathcal{O}$ is an order in an imaginary quadratic field. Let $\mathrm{Cl}(\mathcal{O})$ be the ideal class group of $\mathcal{O}$. One can define the action of an $\mathcal{O}$-ideal $\mathfrak{a}$ on the curve E as the image curve E' under the isogeny $\phi : E \to E'$ whose kernel is equal to the subgroup $E[\mathfrak{a}] = \{P \in E(\overline{\mathbb{F}}_p) : \alpha(P) = 0 \ \forall \alpha \in \mathfrak{a}\}$. We denote E' by $\mathfrak{a} * E$.

The set $\{j(E)\}$ of isomorphism classes of elliptic curves with $\mathrm{End}(E) \cong \mathcal{O}$ is a principal homogeneous space for $\mathrm{Cl}(\mathcal{O})$. Good references for the details are Couveignes [18] and Stolbunov [49]. The key exchange protocol proposed by Couveignes and Stolbunov is for Alice to send $\mathfrak{a} * E$ to Bob and Bob to send $\mathfrak{b} * E$ to Alice; the shared key is $(\mathfrak{a}\mathfrak{b}) * E$.

The difficulty is that if $\mathfrak{a} \subset \mathcal{O}$ is an arbitrary ideal then the subgroup $E[\mathfrak{a}]$ is typically defined over a very large field extension and the computation of $\mathfrak{a} * E$ has exponential complexity. For efficient computation it is necessary to work with ideals that are a product of powers of small prime ideals, so it is necessary to find a "smooth" ideal in the ideal class of $\mathfrak{a}$. Techniques for smoothing an ideal class in the context of isogeny computation were first proposed in [27] and developed further in [8,12,34]. The state of the art is [8] which computes $\mathfrak{a} * E$ for any ideal class in subexponential complexity in $\log(\#\mathrm{Cl}(\mathcal{O}))$.

Since subexponential complexity is not good enough for cryptographic applications it is necessary to choose ideals deliberately of the form $\mathfrak{a} = \prod_{i=1}^{n} \mathfrak{l}_i^{e_i}$ where $\mathfrak{l}_1, \ldots, \mathfrak{l}_n$ are split prime $\mathcal{O}$-ideals of small norm ℓ_i and where $(e_1, \ldots, e_n)$ is an appropriately chosen vector of exponents. Then, the action of $\mathfrak{a}$ can be computed as a composition of isogenies of degree ℓ_i. Throughout the paper we assume that $\{\mathfrak{l}_1, \ldots, \mathfrak{l}_n\}$ is a set of non-principal prime ideals in $\mathcal{O}$, generating $\mathrm{Cl}(\mathcal{O})$, of norm polynomial in the size of the class group. Theoretically we have the bounds $\#\mathrm{Cl}(\mathcal{O}) = O(\sqrt{p}\log(p))$ and, assuming a generalised Riemann hypothesis, $\ell_i = O(\log(p)^2)$. In practice one usually takes $\ell_i = O(\log(p))$ for efficiency reasons; heuristically, this is more than enough to generate the class group.

The basic computational assumption is to invert the action of an ideal. Couveignes called Problem 1 "vectorisation" and Stolbunov called it "Group Action Inverse Problem (GAIP)". The CSIDH paper speaks of hard homogeneous spaces and calls the problem "Key recovery".

Problem 1. Given two elliptic curves E and E_A tover the same field with $\mathrm{End}(E) = \mathrm{End}(E_A) = \mathcal{O}$. Find an ideal $\mathfrak{a}$ such that $j(E_A) = j(\mathfrak{a} * E)$.

The best classical algorithms for this problem in the general case have exponential time (at least $\sqrt{\#\mathrm{Cl}(\mathcal{O})}$ isogeny computations). Childs, Jao and Soukharev [15] were the first to point out that this problem can be formulated as a "hidden shift" problem, and so quantum algorithms for the hidden shift problem can be applied. Hence, there are subexponential-time quantum algorithms for Problem 1 based on the quantum algorithms of Kuperberg [38] and Regev [45]. It is still an active area of research to assess the exact quantum hardness of these problems; see the recent papers by Biasse-Iezzi-Jacobson [9], Bonnetain-Schrottenloher [11], Jao-LeGrow-Leonardi-Ruiz-Lopez [33] and Bernstein-Lange-Martindale-Panny [7]. But at the very least, Kuperberg's algorithm requires at least $\tilde{O}(2^{\sqrt{\log(p)/2}})$ quantum gates, thus taking

$$p > 2^{2\lambda^2}, \tag{1}$$

where λ is the security parameter, should be sufficient to make Problem 1 hard for a quantum computer.

If the ideals $\mathfrak{a}$ in Problem 1 are sampled uniformly at random then the problem admits a random self-reduction: given an instance (E, E_A) one can choose random ideal classes $\mathfrak{b}_1, \mathfrak{b}_2$ and construct the instance $(E_1, E_2) = (\mathfrak{b}_1 * E, \mathfrak{b}_2 * E_A)$, which is now uniformly distributed across the set of pairs of isomorphism classes of curves in the isogeny class. If $\mathfrak{a}'$ is a solution to the instance (E_1, E_2) then any ideal equivalent to the fractional ideal $\mathfrak{a}'\mathfrak{b}_1\mathfrak{b}_2^{-1}$ is a solution to the original instance. This is a nice feature for security proofs that is not shared by SIDH [32][1]; we use this idea in Sect. 4.2.

As already mentioned, when instantiating the group action in practice, one must choose parameters that make evaluating isogenies of degree ℓ_i as efficient as possible. This is done both by choosing the primes ℓ_i to be as small as possible, and also by arranging that the kernel subgroups $E[\ell_i]$ are defined over as small a field extension as possible (so that Vélu's formulas can be used). In the ordinary case, the best technique currently available to select parameters is due to De Feo, Kieffer and Smith [22]. Despite the optimisations described in [22], this technique requires years of CPU time to construct a good curve. Like [22], CSIDH [13] chooses a special prime of the form $p + 1 = 4\prod_{i=1}^{n} \ell_i$, but, instead of ordinary

[1] On the other hand, SIDH has the advantage that no subexponential-time algorithm is known to break it.

curves, it uses supersingular curves defined over $\mathbb{F}_p$. This makes the search for a suitable curve virtually instantaneous, and produces very efficient parameters; indeed note that the formula for $p+1$ implies that each prime ℓ_i splits in $\mathbb{Q}(\sqrt{-p})$ as a product $(\ell_i) = \mathfrak{l}_i\bar{\mathfrak{l}}_i$ of distinct prime ideals. For key exchange, CSIDH samples the exponent vectors $\mathbf{e} = (e_1, \ldots, e_n) \in [-B, B]^n \subseteq \mathbb{Z}^n$ for a suitable constant B.

This leads to a special case of Problem 1 where the ideals may not be uniformly distributed in the ideal class group. For further discussion see Definition 1 and the discussion that follows it. In this special case one can also consider a straightforward meet-in-the-middle attack: Let E and $\mathfrak{a} * E$ be given, where $\mathfrak{a} = \prod_{i=1}^{n} \mathfrak{l}_i^{e_i}$ over $e_i \in [-B, B]$. We compute lists (assume n is even)

$$L_1 = \left\{ (\prod_{i=1}^{n/2} \mathfrak{l}_i^{e_i}) * E : e_i \in [-B, B] \right\}, L_2 = \left\{ (\prod_{i=n/2+1}^{n} \mathfrak{l}_i^{e_i}) * E_A : e_i \in [-B, B] \right\}.$$

If $L_1 \cap L_2 \neq \varnothing$ then we have solved the isogeny problem. This attack is faster than general methods when the set of ideal classes generated is a small subset of $\mathrm{Cl}(\mathcal{O})$. Hence for security we may require

$$(2B + 1)^n > 2^{2\lambda}, \tag{2}$$

where λ is the security parameter. Further, there is a quantum algorithm due to Tani, which is straightforward to adapt to this problem (we refer to Sect. 5.2 of De Feo, Jao and Plût [21] for details). This means we might need to take $(2B + 1)^n > 2^{3\lambda}$ to have post-quantum security. However, recent analyses [2,35] question the pertinence of the complexity models of the meet-in-the-middle and Tani algorithms, and advocate for more relaxed bounds.

Choosing the best values of B, n, p for large choices of λ (e.g., satisfying the constraints of Eqs. (1) and (2)) is non-trivial, but will generally lead to sampling in a very small subset of the whole ideal class group.

We remark that Kuperberg's algorithm uses the entire class group, and there seems to be no way to improve the algorithm for the case where the "hidden shift" is sampled from a distribution far from the uniform distribution. We leave the study of this question to future work.

By taking into account the best known attacks, the CSIDH authors propose parameters for the three NIST categories [43], as summarised in Table 1. Note that in all CSIDH instances the set of sampled ideal classes is (heuristically) likely to cover the whole class group. Their implementation of the smallest parameter size CSIDH-512 computes one class group action in 40 ms on a 3.5 GHz processor.

For our signature schemes we may work with more general primes than considered in CSIDH [13]. For example, CSIDH takes $p+1 = 4\prod_{i=1}^{n} \ell_i$, whereas we may be able to use fewer primes and just multiply by a random co-factor to get a large enough p.

Table 1. Proposed parameters for CSIDH [13]. Effective parameters p, n and B for CSIDH-1024 and CSIDH-1792 were not given in the paper, and are produced here following their methodology. Message size is the number of bytes to re present a j-invariant, and private key size is the space required to store the exponent vector $\mathbf{e} \in \mathbb{Z}^n$.

	n	$\lfloor \log_2 p \rfloor$	B	NIST level	classical security	quantum security	message size	private key size
CSIDH-512	74	510	5	1	127 bits	62 qbits	64B	37B
CSIDH-1024	130	1019	8	3	257 bits	94 qbits	127B	82B
CSIDH-1792	208	1786	10	5	449 bits	129 qbits	223B	130B

2.3 Public Key Signature Schemes

One can describe Fiat-Shamir-type signatures in various ways, including the language of sigma protocols or identification schemes. In the main body of our paper we mostly work with the language of signatures, and give proofs directly in this formulation. In Sect. 7.1 we use the language of identification schemes, and introduce the terminology fully there.

A *canonical identification scheme* consists of algorithms $(\mathsf{KeyGen}, \mathsf{P}_1, \mathsf{P}_2, \mathsf{V})$ and a set ChSet. The randomised algorithm $\mathsf{KeyGen}(1^\lambda)$ outputs a key pair (pk, sk). The deterministic algorithm P_1 takes sk and randomness r_1 and computes $(W, \mathsf{st}) = P_1(sk, r_1)$. Here st denotes state information to be passed to P_2. A challenge c is sampled uniformly from ChSet. The deterministic algorithm P_2 then computes $Z = P_2(sk, W, c, \mathsf{st}, r_2)$ or $\perp$, where r_2 is the randomness. The output $\perp$ corresponds to an abort in the "Fiat-Shamir with aborts" paradigm. We require that $\mathsf{V}(pk, W, c, Z) = 1$ for a correctly formed transcript (W, c, Z).

A *public key signature scheme* consists of algorithms $\mathsf{KeyGen}, \mathsf{Sign}, \mathsf{Verify}$. The randomised algorithm $\mathsf{KeyGen}(1^\lambda)$ outputs a pair (pk, sk), where λ is a security parameter. The randomised algorithm Sign takes input the private key sk and a message msg, and outputs $\sigma = \mathsf{Sign}(sk, \mathsf{msg})$. The verification algorithm $\mathsf{Verify}(pk, \mathsf{msg}, \sigma)$ returns 0 or 1. We require $\mathsf{Verify}(pk, \mathsf{msg}, \mathsf{Sign}(sk, \mathsf{msg})) = 1$.

The *Fiat-Shamir transform* is a construction to turn a canonical identification scheme into a public key signature scheme. The main idea is to make the interactive identification scheme into a non-interactive scheme by replacing the challenge c by a hash $H(W, \mathsf{msg})$.

The standard notion of security is *unforgeability against chosen-message attack (UF-CMA)*. A UF-CMA adversary against the signature scheme is a randomised polynomial-time algorithm A that plays the following game against a challenger. The challenger runs KeyGen to get (pk, sk) and runs $A(pk)$. The adversary A sends messages msg to the challenger, and receives $\sigma = \mathsf{Sign}(sk, \mathsf{msg})$ in return. The adversary outputs $(\mathsf{msg}^*, \sigma^*)$ and wins if $\mathsf{Verify}(pk, \mathsf{msg}^*, \sigma^*) = 1$ and if msg^* was not one of the messages previously sent by the adversary to the challenger. A signature scheme is UF-CMA secure if there is no polynomial-time adversary that wins with non-negligible probability.

3 Basic Signature Scheme

This section contains our main ideas and presents a basic signature scheme. We focus in this section on classical adversaries and proofs in the random oracle model. Hence our signature is based on the traditional Fiat-Shamir transform. For schemes and analysis against a post-quantum adversary see Sect. 7.

For simplicity, we describe our schemes in the setting of a general class group action on a set of j-invariants of elliptic curves. In Sect. 3.3 we explain one small subtlety that arises when implementing the scheme in the setting of CSIDH.

3.1 Stolbunov's Scheme

Section 2.B of Stolbunov's PhD thesis [49] contains a sketch of a signature scheme based on isogeny problems (though his description is not complete and he does not give a proof of security). It is a Fiat-Shamir scheme based on an identification protocol. Section 4 of Couveignes [18] also sketches the identification protocol, but does not mention signature schemes.

The public key consists of E and $E_A = \mathfrak{a} * E$, where $\mathfrak{a} = \prod_{i=1}^{n} \mathfrak{l}_i^{e_i}$ is the private key. To construct the private key one uniformly chooses an exponent vector $\mathbf{e} = (e_1, \ldots, e_n) \in [-B, B]^n \subseteq \mathbb{Z}^n$ for some suitably chosen constant B. Stolbunov assumes the relation lattice for the ideal class group is known, and uses it in Sect. 2.6.1 to sample ideal classes uniformly at random. Section 2.6.2 of [49] suggests an approach to approximate the uniform distribution.

In the identification protocol the prover generates t random ideals $\mathfrak{b}_k = \prod_{i=1}^{n} \mathfrak{l}_i^{f_{k,i}}$ for $1 \leq k \leq t$ and computes $\mathcal{E}_k = \mathfrak{b}_k * E$. Here the exponent vectors $\mathbf{f}_k = (f_{k,1}, \ldots, f_{k,n})$ are uniformly and independently sampled in a region like $[-B, B]^n$ (Stolbunov assumes these ideal classes are uniformly sampled). The prover sends $(j(\mathcal{E}_k) : 1 \leq k \leq t)$ to the verifier. The verifier responds with t uniformly chosen challenge bits $b_1, \ldots, b_t \in \{0, 1\}$. If $b_k = 0$ the prover responds with $\mathbf{f}_k = (f_{k,1}, \ldots, f_{k,n})$ and the verifier checks that $j(\mathcal{E}_k) = j((\prod_{i=1}^{n} \mathfrak{l}_i^{f_{k,i}}) * E)$. If $b_k = 1$ the prover responds with a representation of $\mathfrak{b}_k \mathfrak{a}^{-1}$. When $b_k = 1$ the verifier checks that $j(\mathcal{E}_k) = j((\mathfrak{b}_k \mathfrak{a}^{-1}) * E_A)$. A cheating prover (who does not know the private key) can succeed with probability $1/2^t$.

The major problem with the above idea is how to represent the ideal class of $\mathfrak{b}_k \mathfrak{a}^{-1}$ in a way that does not leak $\mathfrak{a}$. Stolbunov notes that sending the vector $\mathbf{f}_k - \mathbf{e} = (f_{k,i} - e_i)_{1 \leq i \leq n}$ would not be secure as it would leak the private key. Instead, Stolbunov (and also Couveignes) work in the setting where the relation lattice in the ideal class group is known; we discuss this further in Sect. 8. A main contribution of our paper is to give solutions to this problem (using Fiat-Shamir with aborts) that do not require to know the relation lattice.

To obtain a signature scheme Stolbunov applies the Fiat-Shamir transform, and hence obtains the challenge bits b_k as the hash value $H(j(\mathcal{E}_1), \ldots, j(\mathcal{E}_t), \mathsf{msg})$ where H is a cryptographic hash function with t-bit output and msg is the message to be signed. The signature consists of the binary string $b_1 \cdots b_t$ and the representations of the ideal classes $\mathfrak{b}_k$ when $b_k = 0$ and $\mathfrak{b}_k \mathfrak{a}^{-1}$ when $b_k = 1$.

The verifier computes, for $1 \le k \le t$, $\mathcal{E}_k = \mathfrak{b}_k * E$ when $b_k = 0$ and $\mathcal{E}_k = \mathfrak{b}_k \mathfrak{a}^{-1} * E_A$ when $b_k = 1$. The verifier then computes $H(j(\mathcal{E}_1), \dots, j(\mathcal{E}_t), \mathsf{msg})$ and checks whether this is equal to the binary string $b_1 \cdots b_t$, and accepts the signature if and only if the strings agree.

We stress that neither Couveignes nor Stolbunov give a secure post-quantum signature scheme. Both authors assume that the relations in the ideal class group have been computed (Stolbunov needs this to prevent leakage). However the cost to compute the relations in the ideal class group on a classical computer is in essentially the same asymptotic complexity class as the cost to break the scheme on a quantum computer (using the Kuperberg or Regev algorithms). Hence it may not make sense to require the Key Generation algorithm of the scheme to compute the relations in the ideal class group. On the other hand, in the fully post-quantum setting where quantum computers are readily available then the relation lattice can be computed in polynomial time. We revisit this issue in Sect. 8.

3.2 Using Rejection Sampling

To prevent signatures from leaking the private key, we use rejection sampling in exactly the way proposed by Lyubashevsky [40] in the context of lattice signatures.

Let $B > 0$ be a constant. When generating the private key we sample uniformly $e_i \in [-B, B]$ for $1 \le i \le n$. Let $\mathbf{e} = (e_1, \dots, e_n)$. The value B may be chosen large enough that $\prod_{i=1}^n \mathfrak{l}_i^{e_i}$ covers most ideal classes and so that the output distribution is close to uniformly distributed in $\mathrm{Cl}(\mathcal{O})$, but we avoid any explicit requirement or assumption that this distribution is uniform. We refer to Definition 1 for more discussion of this issue, and in Sect. 7 we consider a variant where the ideals are definitely not distributed uniformly in $\mathrm{Cl}(\mathcal{O})$.

Exponents $f_{k,i}$ are sampled uniformly in $[-(nt+1)B, (nt+1)B]$, where t is the number of parallel rounds of the identification/signature protocol and n is the number of primes. Let $\mathbf{f}_k = (f_{k,1}, \dots, f_{k,n})$, $\mathfrak{b}_k = \prod_{i=1}^n \mathfrak{l}_i^{f_{k,i}}$ and define $\mathcal{E}_k = \mathfrak{b}_k * E$.

If the k-th challenge bit b_k is zero then the prover responds with $\mathbf{f}_k = (f_{k,1}, \dots, f_{k,n})$ and the verifier checks that $j(\mathcal{E}_k) = j((\prod_{i=1}^n \mathfrak{l}_i^{f_{k,i}}) * E)$ as in the basic scheme above.[2] If $b_k = 1$ then the prover is required to provide a representation of $\mathfrak{b}_k \mathfrak{a}^{-1}$, the idea is to compute the vector $\mathbf{z}_k = (z_{k,1}, \dots, z_{k,n})$ defined by $z_{k,i} = f_{k,i} - e_i$ for $1 \le i \le n$. As already noted, outputting $\mathbf{z}$ directly would potentially leak the secret. To prevent this leakage we only output $\mathbf{z}_k$ if all its entries satisfy $|z_{k,i}| \le ntB$. We give the signature scheme in Fig. 1. It remains to show that in the accepting case the vector leaks no information about the

[2] In the scheme and analysis we apply rejection sampling to the case $b_k = 0$ as well as the case $b_k = 1$. An alternative would be to only apply rejection sampling in the case $b_k = 1$. It doesn't really matter one way or the other, since in both settings we are able to simulate a signer in the random oracle model and so the security proof works.

Algorithm 1 KeyGen

Input: $B, \mathfrak{l}_1, \ldots, \mathfrak{l}_n, E$
Output: $sk = \mathbf{e}$ and $pk = E_A$

1: $\mathbf{e} \leftarrow [-B, B]^n$
2: $E_A = (\prod_{i=1}^n \mathfrak{l}_i^{e_i}) * E$
3: **return** $sk = \mathbf{e}, pk = E_A$

Algorithm 2 Sign

Input: msg, (E, E_A), $\mathbf{e}$
Output: $(\mathbf{z}_1, \ldots, \mathbf{z}_t, b_1, \ldots, b_t)$

1: **for** $k = 1, \ldots, t$ **do**
2: $\quad \mathbf{f}_k \leftarrow [-(nt+1)B, (nt+1)B]^n$
3: $\quad \mathcal{E}_k = (\prod_{i=1}^n \mathfrak{l}_i^{f_{k,i}}) * E$
4: **end for**
5: $b_1 \| \cdots \| b_t = H(j(\mathcal{E}_1), \ldots, j(\mathcal{E}_t), \mathsf{msg})$
6: **for** $k = 1, \ldots, t$ **do**
7: $\quad$ **if** $b_k = 0$ **then**
8: $\quad\quad \mathbf{z}_k = \mathbf{f}_k$
9: $\quad$ **else**
10: $\quad\quad \mathbf{z}_k = \mathbf{f}_k - \mathbf{e}$
11: $\quad$ **end if**
12: $\quad$ **if** $\mathbf{z}_k \notin [-ntB, ntB]^n$ **then**
13: $\quad\quad$ **return** $\perp$
14: $\quad$ **end if**
15: **end for**
16: **return** $\sigma = (\mathbf{z}_1, \ldots, \mathbf{z}_t, b_1, \ldots, b_t)$

Algorithm 3 Verify

Input: msg, (E, E_A), σ
Output: Valid/Invalid

1: Parse σ as $(\mathbf{z}_1, \ldots, \mathbf{z}_t, b_1, \ldots, b_t)$
2: **for** $k = 1, \ldots, t$ **do**
3: $\quad$ **if** $b_k = 0$ **then**
4: $\quad\quad \mathcal{E}_k = (\prod_{i=1}^n \mathfrak{l}_i^{z_{k,i}}) * E$
5: $\quad$ **else**
6: $\quad\quad \mathcal{E}_k = (\prod_{i=1}^n \mathfrak{l}_i^{z_{k,i}}) * E_A$
7: $\quad$ **end if**
8: **end for**
9: $b'_1 \| \cdots \| b'_t = H(j(\mathcal{E}_1), \ldots, j(\mathcal{E}_t), \mathsf{msg})$
10: **if** $(b'_1, \ldots, b'_t) = (b_1, \ldots, b_t)$ **then**
11: $\quad$ **return** Valid
12: **else**
13: $\quad$ **return** Invalid
14: **end if**

Fig. 1. The basic signature scheme using rejection sampling.

private key, and that the rejecting case occurs with low probability. We do this in the following two lemmas.

Lemma 1. *The distribution of vectors* $\mathbf{z}_k$ *output by the signing algorithm is the uniform distribution and therefore is independent of the private key* $\mathbf{e}$.

Proof. Let $U = [-(nt+1)B, (nt+1)B]$. Then $\#U = 2(nt+1)B+1$. If $e \in [-B, B]$ then

$$[-ntB, ntB] \subseteq U - e = \{f - e : f \in U\} \subseteq [-(nt+2)B, (nt+2)B].$$

Hence, when rejection sampling (only outputting values $f_{k,i} - e_i$ in the range $[-ntB, ntB]$) is applied then the output distribution of $\mathbf{z}_k$ is the uniform distribution on $[-ntB, ntB]^n$. This argument does not depend on the choice of $\mathbf{e}$, so the output distribution is independent of $\mathbf{e}$. □

Lemma 2. *The probability that the signing algorithm outputs a signature (i.e., does not output* $\perp$*) is at least* $1/e > 1/3$.

Proof. Let notation be as in the proof of Lemma 1. For fixed $e \in [-B, B]$ and uniformly sampled $f \in U = [-(nt+1)B, (nt+1)B]$, the probability that a value $f - e$ lies in $[-ntB, ntB]$ is

$$\frac{2ntB+1}{2(nt+1)B+1} = 1 - \frac{2B}{2(nt+1)B+1} \geq 1 - \frac{1}{nt+1}.$$

Hence, the probability that all of the values $z_{k,i}$ over $1 \leq k \leq t, 1 \leq i \leq n$ lie in $[-ntB, ntB]$ is at least $(1-1/(nt+1))^{nt}$. Using the inequality $1-1/(x+1) \geq e^{-1/x}$ for $x \geq 1$ it follows that the probability that all values are in the desired range is at least

$$\left(e^{-1/nt}\right)^{nt} = e^{-1}.$$

This completes the proof. □

We can therefore get a rough idea of parameters and efficiency for the scheme. Let λ be a security parameter (e.g., $\lambda = 128$ or $\lambda = 256$), for security we need at least $t = \lambda$ so that an attacker cannot guess the hash value or invert the hash function (see also the proof of Theorem 1). We also need a large enough set of private keys, so we need $(2B+1)^n$ large enough. The signature contains one hash value of t bits, plus t vectors $\mathbf{f}_k$ or $\mathbf{z}_k$ with entries of size bounded by $(nt+1)B$, for a total of $\lambda + t\lceil n\log(2(nt+1)B+1)\rceil$ bits (assuming each vector is represented optimally). If we take $t = \lambda = 128$, and $(n, B) = (74, 5)$ as in CSIDH-512, we obtain signatures of around 20 KiB (see also Table 2).

To sign/verify one needs to evaluate the action of either of $\mathfrak{b}_k$ and $\mathfrak{b}_k\mathfrak{a}^{-1}$ for every $1 \leq k \leq t$, which means that for each k and each prime $\mathfrak{l}_i$ one needs to compute up to ntB isogenies of degree ℓ_i. Hence, the total number of isogeny computations is upper bounded by $(nt)^2B$. The quadratic dependence on nt is a major inconvenience. For example, taking $(n, t, B) = (74, 128, 5)$ gives around 2^{28} isogeny computations in signature/verification. We can make t small using the techniques in later sections, but one needs n large unless B is going to get very large. So even going down to $t = 8$ still has signatures requiring around 2^{20} isogeny computations. The acceptance probability estimate from Lemma 2 is very close to the true value: for example, when $(n, t, B) = (74, 128, 5)$ then the true acceptance probability is approximately 0.36790, while $e^{-1} \approx 0.36788$.

3.3 CSIDH Implementation

The above description represents the isomorphism class of $\mathfrak{a} * E$ using a j-invariant. But, as explained in [13,24], in the case of supersingular curves over $\mathbb{F}_p$ there are two isomorphism classes for each j-invariant and so the j-invariant alone is not an adequate representation for $\mathfrak{a} * E$. Castryck *et al.* [13] observe that the Montgomery model for these curves provides an elegant solution to the dilemma. Instead of representing $\mathfrak{a} * E$ with a j-invariant one uses the "A coefficient" of the Montgomery equation. This works when choosing $p \equiv 3 \pmod 8$ and using curves whose endomorphism ring is on the "floor" of the 2-isogeny volcano; we refer to Proposition 8 of [13] for the details.

In short, when implementing our signature schemes using CSIDH one should choose $p \equiv 3 \pmod 8$ and replace the words "j-invariant" by "Montgomery coefficient". In terms of the security analysis, strictly speaking the security proofs use variants of the computational problems expressed in terms of Montgomery coefficients. It is a simple exercise to show that these problems are equivalent to problems expressed using j-invariants. Hence the theorem statements in our paper are all correct in the setting of CSIDH.

3.4 Security Proof

We now prove security of the basic scheme in the random oracle model against a classical adversary. The proof technique is the standard approach that uses the forking lemma. In this section we do not consider quantum adversaries, or give a proof in the quantum random oracle model (QROM). A proof in the QROM follows from the approach in Sect. 7.

First we need to discuss some subtleties about the distribution of ideal classes coming from the key generation and signing algorithms.

Definition 1. *Fix distinct ideals $\mathfrak{l}_1, \dots, \mathfrak{l}_n$. For $B \in \mathbb{N}$, consider the random variable $\mathfrak{a}$ which is the ideal class of $\prod_{i=1}^n \mathfrak{l}_i^{e_i}$ over a uniformly random $\mathbf{e} \in [-B, B]^n$. Define $\mathcal{D}_B$ to be the distribution on $\mathrm{Cl}(\mathcal{O})$ corresponding to this random variable. Define M_B to be an upper bound on the probability, over $\mathfrak{a}, \mathfrak{b}$ sampled from $\mathcal{D}_B$, that $\mathfrak{a} \equiv \mathfrak{b}$.*

In other words, $\mathcal{D}_B$ is the output distribution of the public key generation algorithm. Understanding the distribution $\mathcal{D}_B$ is non-trivial in general.[3] For small B and n (so that $(2B+1)^n \ll \#\mathrm{Cl}(\mathcal{O})$) we expect $\mathcal{D}_B$ to be the uniform distribution on a subset of $\mathrm{Cl}(\mathcal{O})$ of size $(2B+1)^n$. For fixed n and large enough B it should be the case that $\mathcal{D}_B$ is very close to the uniform distribution on $\mathrm{Cl}(\mathcal{O})$. A full study of the distribution $\mathcal{D}_B$ is beyond the scope of this paper, but is a good problem for future work.

For the isogeny problem to be hard for public keys we certainly need $M_B \le 1/2^\lambda$, where λ is the security parameter. In the proof we will need to use M_{ntB}, since the concern is about the auxiliary curves generated during the signing algorithm. We do not require these curves to be uniformly sampled, but in practice we can certainly assume that $M_{ntB} = O(1/\sqrt{p})$. In any case, it is negligible in the security parameter.

Problem 2. Let notation be as in the key generation protocol of the scheme. Given (E, E_A), where $E_A = \mathfrak{a} * E$ for some ideal $\mathfrak{a} = \prod_{i=1}^n \mathfrak{l}_i^{e_i}$ and where the exponent vector $\mathbf{e} = (e_1, \dots, e_n)$ is uniformly sampled in $[-B, B]^n \subseteq \mathbb{Z}^n$, to compute any ideal equivalent to $\mathfrak{a}$.

[3] Even the analogous problem of understanding the distribution of $\prod_i \ell_i^{e_i} \pmod q$, where ℓ_i are small primes and q is some integer, is an open problem in general.

Depending on how close to uniform is the distribution $\mathcal{D}_B$, this problem may or may not be equivalent to Problem 1 and may or may not have a random self-reduction. Nevertheless, we believe this is a plausible assumption.

We recall the forking lemma, in the formulation of Bellare and Neven [4].

Lemma 3 *(Bellare and Neven [4]). Fix an integer $Q \geq 1$. Let A be a randomised algorithm that takes as input $h_1, \dots, h_Q \in \{0,1\}^t$ and outputs (J, σ) where $1 \leq J \leq Q$ with probability $\wp$. Consider the following experiment: $h_1, \dots, h_Q$ are chosen uniformly at random in $\{0,1\}^t$; $A(h_1, \dots, h_Q)$ returns (I, σ) such that $I \geq 1$; $h'_I, \dots, h'_Q$ are chosen uniformly at random in $\{0,1\}^t$; $A(h_1, \dots, h_{I-1}, h'_I, \dots, h'_Q)$ returns (I', σ'). Then the probability that $I' = I$ and $h'_I \neq h_I$ is at least $\wp(\wp/Q - 1/2^t)$.*

Theorem 1. *In the random oracle model, the basic signature scheme of Fig. 1 is unforgeable under a chosen message attack under the assumption that Problem 2 is hard.*

Proof. Consider a polynomial-time adversary A against the signature scheme. So A takes a public key, makes queries to the hash function H and the signing oracle, and outputs a forgery of a signature with respect the public key.

Let $(E, E_A = \mathfrak{a} * E)$ be an instance of Problem 2. The simulator runs the adversary A with public key (E, E_A).

Suppose the adversary A makes at most Q (polynomial in the security parameter) queries in total to either the random oracle H or the signing oracle. We now explain how the simulator responds to these queries. The simulator maintains a list, initially empty, of pairs $(x, H(x))$ for each value of the random oracle that has been defined.

Sign queries: To answer a Sign query on message msg the simulator chooses t uniformly chosen bits $b_1, \dots, b_t \in \{0,1\}$. When $b_k = 0$ the simulator randomly samples $z_k \leftarrow [-ntB, ntB]^n$ and sets $\mathfrak{b}_k = \prod_{i=1}^n \mathfrak{l}_i^{z_{k,i}}$ and computes $\mathcal{E}_k = \mathfrak{b}_k * E$, just like in the real signing algorithm. When $b_k = 1$ the simulator chooses a random ideal $\mathfrak{c}_k = \prod_{i=1}^n \mathfrak{l}_i^{z_{k,i}}$ for $z_{k,i} \in [-ntB, ntB]$ and computes $\mathcal{E}_k = \mathfrak{c}_k * E_A$. By Lemma 1, the values $j(\mathcal{E}_k)$ and $\mathbf{z}_k$ are distributed exactly as in the real signing algorithm. We program the random oracle (update the hash list) so that $H(j(\mathcal{E}_1), \dots, j(\mathcal{E}_t), \mathsf{msg}) := b_1 \cdots b_t$, unless the random oracle has already been defined on this input in which case the simulation fails and outputs $\perp$. The probability of failure is at most Q/M_{ntB}^t, where M_{ntB} is defined in Definition 1 to be an upper bound on the probability of a collision in the sampling of ideal classes. Note that Q/M_{ntB}^t is negligible. Assuming the simulation does not fail, the output is a valid signature and is indistinguishable from signatures output by the real scheme in the random oracle model.

Hash queries: To answer a random oracle query on input x the simulator checks if (x, y) already appears in the list, and if so returns y. Otherwise the simulator chooses uniformly at random $y \in \{0,1\}^t$ and sets $H(x) := y$ and adds (x, y) to the list.

Eventually A outputs a forgery $(\mathsf{msg}, \sigma = (\mathbf{z}_1, \ldots, \mathbf{z}_t, b_1 \cdots b_t))$ that passes the verification equation. Define $\mathfrak{c}_k = \prod_i \mathfrak{l}_i^{z_{k,i}}$. The proof now invokes the Forking Lemma (see Bellare-Neven [4]). The adversary is replayed with the same random tape and the exact same simulation, except that one of the hash queries is answered with a different binary string. With non-negligible probability the adversary outputs a forgery $\sigma = (\mathbf{z}'_1, \ldots, \mathbf{z}'_t, b'_1 \cdots b'_t)$ for the same message msg and the same input $(j(\mathcal{E}_1), \ldots, j(\mathcal{E}_t), \mathsf{msg})$ to H, but a different output string $b'_1 \cdots b'_t$. Let k be an index such that $b_k \neq b'_k$ (without loss of generality $b_k = 0$ and $b'_k = 1$). Then the ideal classes $\mathfrak{c}_k$ and $\mathfrak{c}'_k$ in the two signatures are such that $j(\mathfrak{c}_k * E) = j(\mathfrak{c}'_k * E_A)$ and so $\mathfrak{c}'_k \mathfrak{c}_k^{-1} = \prod_i \mathfrak{l}_i^{z'_{k,i} - z_{k,i}}$ is a solution to the problem instance. □

We make two observations about the use of the forking lemma. First, as always, the proof is not tight since if the adversary succeeds with probability ϵ then the simulator solves the computational problem with probability proportional to ϵ^2. Second, the hash output length t in Lemma 3 only appears in the term $1/2^t$, so it suffices to take $t = \lambda$. There may be situations where a larger hash output is needed; for more discussion about hash output sizes we refer to Neven, Smart and Warinschi [44].

4 Smaller Signatures and Faster Signing/Verification

The signature size of the basic scheme is rather large (around 20 KiB), since the sigma protocol that underlies the identification scheme only has single bit challenges. In practice we need $t \geq 128$, which means signatures are very large (several megabytes). To get shorter signatures it is natural to try to increase the size of the challenges. In this section we sketch an approach to obtain s-bit challenge values for any small integer $s \in \mathbb{N}$, by trading the challenge size with the public key size. This optimisation also dramatically speeds up signing and verification. In the next section we explain how to shorten the public keys again.

The basic idea is to have public keys $(E_{A,0} = \mathfrak{a}_0 * E, \ldots, E_{A,2^s-1} = \mathfrak{a}_{2^s-1} * E)$. For each $0 \leq m < 2^s$ we choose $\mathbf{e}_m \leftarrow [-B, B]^n$ and set $E_{A,m} = (\prod_{i=1}^n \mathfrak{l}_i^{e_{m,i}}) * E$. The signing algorithm for user A chooses t random ideals $\mathfrak{b}_k = \prod_{i=1}^n \mathfrak{l}_i^{f_{k,i}}$ and computes $\mathcal{E}_k = \mathfrak{b}_k * E$, as before. Now we have s-bit challenges $b_1, \ldots, b_t \in \{0, 1, \ldots, 2^s - 1\}$. For each $1 \leq k \leq t$ the signer computes $\mathbf{z}_k = \mathbf{f}_k - \mathbf{e}_{b_k}$, which corresponds to the ideal class $\mathfrak{c}_k = \mathfrak{b}_k \mathfrak{a}_{b_k}^{-1}$ and the verifier can check that $j(\mathcal{E}_k) = j(\mathfrak{c}_k * E_{A,b_k})$.

A signature consists of one hash value, plus t vectors $\mathbf{z}_k$ with entries of size bounded by ntB, i.e., a total of $\lambda + t\lceil n \log(2ntB+1) \rceil$ bits, similar to the previous section. But now for security we only require $ts \geq \lambda$. Taking, say, $\lambda = 128$ and $s = 16$ can mean t as low as 8, and so only 8 vectors need to be transmitted as part of the signature, giving signatures of well under 1 KiB (see Table 3). Of course the public key now includes 2^{16} j-invariants (elements of $\mathbb{F}_p$) which would be around 4 MiB, and key generation is also 2^{16} times slower.

As far as we can tell, this idea cannot be applied to the schemes of Yoo *et al.* [53] or Galbraith *et al.* [28].

4.1 Security

A trivial modification to the proof of Theorem 1 can be applied in this setting. But note that the forking lemma produces two signatures such that $b_k \neq b'_k$ for some index k. Hence from a successful forger we derive two ideal classes $\mathfrak{c}_k$ and $\mathfrak{c}'_k$ such that $j(\mathfrak{c}_k * E_{A,b_k}) = j(\mathfrak{c}'_k * E_{A,b'_k})$. It follows that $(\mathfrak{c}'_k)^{-1}\mathfrak{c}_k$ is an ideal class corresponding to an isogeny $E_{A,b_k} \to E_{A,b'_k}$. Hence the computational assumption underlying the scheme is the following.

Problem 3. Let notation be as in the key generation protocol of the scheme. Consider a set of 2^s elliptic curves $\{E_{A,0}, \ldots, E_{A,2^s-1}\}$, all of the form $E_{A,m} = \mathfrak{a}_m * E$ for some ideal $\mathfrak{a}_m = \prod_{i=1}^{n} \mathfrak{l}_i^{e_{m,i}}$ where the exponent vectors $\mathbf{e}_m$ are uniformly sampled in $[-B, B]^n \subseteq \mathbb{Z}^n$. The problem is to compute an ideal corresponding to any isogeny $E_{A,m} \to E_{A,m'}$ for some $m \neq m'$.

We believe this problem is hard for classical and quantum computers. One can easily obtain a non-tight reduction of this problem to Problem 2. However, if the ideals $\mathfrak{a}_m$ are not sampled uniformly at random from $\mathrm{Cl}(\mathcal{O})$ then we do not know how to obtain a random-self-reduction for this problem, which prevents us from having a tight reduction to Problem 2.

Theorem 2. *In the random oracle model, the signature scheme of this section is unforgeable under a chosen message attack under the assumption that Problem 3 is hard.*

The proof of this theorem is almost identical to the proof of Theorem 1 and so is omitted.

4.2 Variant Based on a More Natural Problem

Problem 3 is a little un-natural. It would be more pleasing to prove security based on Problem 1 or Problem 2. We now explain that one can prove security based on Problem 1, under an assumption about uniform sampling of ideal classes.

Suppose in this section that the distribution $\mathcal{D}_B$ of Definition 1 has negligible statistical distance (Renyi divergence can also be used here) from the uniform distribution. This assumption is reasonable for bounded n and very large B; but we leave for future work to determine whether practical parameters for isogeny based cryptography can be obtained under this constraint.

Lemma 4. *Let parameters be such that the statistical distance between $\mathcal{D}_B$ and the uniform distribution on $\mathrm{Cl}(\mathcal{O})$ is negligible. Suppose that all the prime ideals $\mathfrak{l}_i$ have norm bounded as $O(\log(p))$ Then given an algorithm that runs in time T and solves Problem 3 with probability ϵ, there is an algorithm to solve Problem 1 with time $T + O(2^s \log(p)^5)$ and success probability $\epsilon/2$.*

Proof. Let A be an algorithm for Problem 3, and let $(E, E_A = \mathfrak{a} * E)$ be an instance of Problem 1.

Choose random ideal classes $\mathfrak{b}_0, \ldots, \mathfrak{b}_{2^s-1}$ (chosen as $\mathfrak{b}_m = \prod_{i=1}^n \mathfrak{l}_i^{u_{i,m}}$ for $0 \le m < 2^s$ and $u_{i,m} \in [-B, B]$) and compute $E'_{A,m} = \mathfrak{b}_m * E$ for $0 \le m < 2^{s-1}$ and $E'_{A,m} = \mathfrak{b}_m * E_A$ for $2^{s-1} \le m < 2^s$. Choose a random permutation π on $\{0, 1, \ldots, 2^s - 1\}$ and construct the sequence $E_{A,m} = E'_{A,\pi(m)}$. This computation takes $O(2^s \log(p)^5)$ bit operations, since n and B and the norm ℓ_i of $\mathfrak{l}_i$ are all $O(\log(p))$. Note that these curves are all uniformly sampled in the isogeny class, and so there is no way to distinguish whether any individual curve has been generated from E or E_A. This is where the subtlety about distributions appears: it is crucial that the curves derived from the pair (E, E_A) are indistinguishable from the curves in Problem 3.

Now run the algorithm A on this input. Since the input is indistinguishable from a real input, A runs in time T and succeeds with probability ϵ. In the case of success, we have an ideal $\mathfrak{c}$ corresponding to an isogeny $E_{A,m} \to E_{A,m'}$ for some $m \ne m'$. With probability $1/2$ we have that one of the curves, say $E_{A,m}$, is known to the simulator as $\mathfrak{b} * E$ and the other (i.e., $E_{A,m'}$) is known as $\mathfrak{b}' * E_A$. If this event occurs then we have $\mathfrak{c}\mathfrak{b} * E = \mathfrak{b}' * E_A$ (or vice versa) in which case $\mathfrak{c}\mathfrak{b}(\mathfrak{b}')^{-1}$ is a solution to the original instance. □

Note that this proof introduces an extra $1/2$ factor in the success probability, but this is not a serious issue since the security proof isn't tight anyway.

Using this result, the following theorem is an immediate consequence of Theorem 2.

Theorem 3. *Let parameters be such that the statistical distance between $\mathcal{D}_B$ and the uniform distribution on $\mathrm{Cl}(\mathcal{O})$ is negligible. In the random oracle model, the signature scheme of this section is unforgeable under a chosen message attack under the assumption that Problem 1 is hard.*

We have a tight proof in Sect. 7 based on a less standard assumption (see Problem 4). It is an open problem to have a tight proof and also the security based on Problem 1.

4.3 Reducing Storage for Private Keys

Rather than storing all the private keys $\mathfrak{a}_m$ for $0 \le m < 2^s$ one could have generated them using a pseudorandom function as $\mathsf{PRF}(\mathsf{seed}, i)$ where seed is a seed and i is used to generate the i-th private key (which is an integer exponent vector). The prover only needs to store seed and can then recompute the private keys as needed. Of course, during key generation one needs to compute all the public keys, but during signing one only needs to determine $t \approx 8$ private keys (although this adds a cost to the signing algorithm).

5 Smaller Public Keys

The approach of Sect. 4 gives signatures that are potentially quite small, but at the expense of very large public keys. In some settings (e.g., software signing or licence checks) large public keys can be easily accommodated, while in other settings (e.g., certificate chains) it makes no sense to shorten signatures at the expense of public key size. In this section we explain how to use techniques from hash-based signatures to compress the public key while also maintaining compact signatures. The key idea is to use a Merkle tree [41] with leaves the public curves $E_{A,0}, \dots, E_{A,2^s-1}$, and use the tree root (a single hash value) as public key. However, the security of plain Merkle trees depends on collision resistance of the underlying hash function, thus requiring hashes of size at least twice the security parameter. Instead, we use a modified Merkle tree, as introduced in the hash-based signatures XMSS-T [31] and SPHINCS+ [5], whose security relies on the second-preimage resistance of a keyed hash function.

Let λ be a security parameter, and let n, B, s, t, p be as in the previous sections; we assume that $\lceil \log p \rceil > 2\lambda$, as this is the case in any secure instantiation. Let the following (public) functions be given:

- $\mathsf{PRF}_{\mathrm{secret}} : \{0,1\}^\lambda \times \{0,1\}^s \to [-B, B]^n$,
- $\mathsf{PRF}_{\mathrm{key}} : \{0,1\}^\lambda \times \{0,1\}^{s+1} \to \{0,1\}^\lambda$,
- $\mathsf{PRF}_{\mathrm{mask}} : \{0,1\}^\lambda \times \{0,1\}^{s+1} \to \{0,1\}^{\lceil \log p \rceil}$ three pseudo-random functions, and
- $M : \{0,1\}^\lambda \times \{0,1\}^{\lceil \log p \rceil} \to \{0,1\}^\lambda$ a keyed hash function (where we think of the first λ bits as the key and the second $\lceil \log p \rceil$ bits as the input).

Finally, let PK.seed and SK.seed be two random seeds; as the names suggest, PK.seed is part of the public key, while SK.seed is part of the secret key. Like in Sect. 4.3, we define the secret ideals $\mathfrak{a}_m = \prod_{i=1}^n \mathfrak{l}_i^{e_{m,i}}$, where $\mathbf{e}_m = \mathsf{PRF}_{\mathrm{secret}}(\mathsf{SK.seed}, m)$, and the public curves $E_{A,m} = \mathfrak{a}_m * E$, for $0 \le m < 2^s$.

We set up a hash tree by defining $h_{l,u}$ for $0 \le l \le s$ and $0 \le u < 2^{s-l}$. First we set

$$h_{s,u} = M\big(\mathsf{PRF}_{\mathrm{key}}(\mathsf{PK.seed}, 2^s + u),\ j(E_{A,u}) \oplus \mathsf{PRF}_{\mathrm{mask}}(\mathsf{PK.seed}, 2^s + u)\big)$$

for $0 \le u < 2^s$, where $\oplus$ denotes bitwise XOR. Now, for any $0 \le l < s$, the rows of the hash tree are defined as

$$h_{l,u} = M\big(\mathsf{PRF}_{\mathrm{key}}(\mathsf{PK.seed}, 2^l + u),\ (h_{l+1,2u} \| h_{l+1,2u+1}) \oplus \mathsf{PRF}_{\mathrm{mask}}(\mathsf{PK.seed}, 2^l + u)\big).$$

Finally, the public key is set to the pair $(\mathsf{PK.seed}, h_{0,0})$.

To prove that a value $E_{A,u}$ is in the hash tree, we use its *authentication path.* That is the list of the hash values $h_{l,u'}$, for $1 \le l \le s$, occurring as siblings of the nodes on the path from $h_{s,u}$ to the root. The proof in [31, Appendix B] shows that having M output λ-bit hashes gives a (classical) security of approximately 2^λ. See [5,31] for more details.

Typically, in hash-based signatures the secret key would only contain SK.seed, since all secret and public values can be reconstructed from it at an acceptable cost. However, in our case recomputing the leaves of the hash tree (2^s class group actions) is much more expensive than recomputing the internal nodes ($2^s - 1$ hash function evaluations), thus we set the secret key to the tuple

$(\mathsf{SK.seed}, h_{s,0}, \dots, h_{s,2^s-1})$. This is a considerably large secret key, e.g., around 1 MiB when $\lambda = 128$ and $s = 16$, but it is offset by a more than tenfold gain in signing time. Also note that the values $h_{s,u}$ can (and will) be leaked without any loss in security, they are indeed part of the uncompressed public key, thus they are more formally treated as auxiliary signer data, rather than as part of the secret key.

To sign we proceed like in Sect. 4, but the signature now needs to contain additional information. The signer computes the random ideals $\mathfrak{b}_1, \dots, \mathfrak{b}_t$ and the associated curves $\mathcal{E}_1, \dots, \mathcal{E}_t$ to obtain the challenges $b_1, \dots, b_t$. Then, using $\mathsf{PRF}_{\mathrm{secret}}$, they obtain the secrets $\mathfrak{a}_{b_1}, \dots, \mathfrak{a}_{b_t}$, recompute the public curves $E_{A,b_1}, \dots, E_{A,b_t}$, and the ideals $\mathfrak{c}_i = \mathfrak{a}_{b_i}^{-1}\mathfrak{b}_i$. The signature is made of the ideals $\mathfrak{c}_1, \dots, \mathfrak{c}_t$, the curves $E_{A,b_1}, \dots, E_{A,b_t}$, and their authentication paths in the hash tree. The verifier computes $\mathcal{E}_i$ as $\mathfrak{c}_i * E_{A,b_i}$, obtains the challenges $b_1, \dots, b_t$, and uses them to verify the authentication paths. Hence, the signature contains t ideals represented as vectors in $[-ntB, ntB]^n$, t curves represented by their j-invariants, and t authentication paths of length s. The t authentication paths eventually merge before the root, thus some hash values will be repeated. We can save some space by only sending the hash values once, in some standardised order: the worst case happening when no path merges before level $\log(t)$, no more than $t(s - \log(t))$ hash values need to be sent as part of the signature. In total, a signature requires at most $t\lceil n\log(2ntB + 1)\rceil + t\log(p) + t\lambda(s - \log(t))$ bits. For our parameters $t = 8, s = 16$ and $\lambda = 128$, this adds about 2 KiB to the signature of Sect. 4. Note that this is still less than half the size of the best stateless hash-based signature schemes (the NIST candidate SPHINCS+ [5,6] has size-optimized signatures of 8080 bytes at the NIST security level 1), and is comparable in size to stateful hash-based signatures (e.g., the IETF draft XMSS [30, § 5.3.1]) and to the shortest known lattice-based signatures.

Concerning security, the proofs of the previous sections, and that of [31, Appendix B] can be combined to prove the following theorem.

Theorem 4. *The signature scheme of this section is unforgeable under a chosen message attack under the following assumptions:*

- *Problem 3 is hard;*
- *The* multi-function multi-target second-preimage resistance *of the keyed hash function M;*
- *The pseudo-randomness of* $\mathsf{PRF}_{\mathrm{secret}}$*;*

when the hash function H and the pseudo-random functions $\mathsf{PRF}_{\mathrm{key}}$ *and* $\mathsf{PRF}_{\mathrm{mask}}$ *are modelled as random oracles (ideal random functions).*

Like in the previous section, it is possible to replace Problem 3 with Problem 1, modulo some additional assumptions. Both proofs are straightforward adaptations, and we omit them for conciseness. As already noted, the proofs are not tight, however the part concerned with the second-preimage resistance of M is.

6 Performance

Table 2 gives some estimates of cost for the schemes presented in Sects. 3, 4, 5. The rows of the table are divided into three sections.

The first section of the table (under the heading "Exact") reports the parameter sizes, as a number of bits, already computed in each section, where λ is the security parameter, n, B and s are as described previously (in Sect. 3 we have $s = 1$). To simplify the expressions we assume that all hash functions have λ-bit outputs, and we set the parameter $t = \lambda/s$.

In all sections we give a rough lower bound for the performance of the keygen and sign/verify algorithms, in terms of $\mathbb{F}_p$-operations. The lower bound only takes into account the number of operations needed to compute and evaluate the isogeny path, and so the exact cost may be higher.

The operation count is based on the following estimates.

1. Based on [17,46], we estimate that computing/evaluating an isogeny of degree ℓ, when given a kernel point, costs $O(\ell)$ operations.
2. By the prime number theorem $\sum_{i=1}^{n} \ell_i \sim \frac{1}{2}n^2 \ln(n)$, and the estimate is very accurate already for $n > 3$.

Putting these estimates together, an ideal with exponent vector within $[-C, C]^n$ can be evaluated in $O(Cn^2 \log(n))$ operations on average and in the worst case. We note that the above estimate is not likely to be the dominant part in the computation, especially asymptotically, as scalar multiplications of elliptic points are likely to dominate. However, estimating this part of the algorithm is much more complex and dependent on specific optimisations, we thus leave a more precise analysis for future work.

The second section of rows in the table (under the heading "Asymptotic") gives asymptotic estimates in terms only of the security parameter λ, and the parameter of s of Sect. 4. We now give a brief justification for the parameter restrictions in terms of λ.

1. Kuperberg's algorithm is believed to require at least $2^{\sqrt{\log(N)}}$ operations in a group of size N. In our case $N > \sqrt{p}$. Taking $\log(p) > 2\lambda^2$ gives

$$\sqrt{\log(N)} > \sqrt{\tfrac{1}{2}\log(p)} > \sqrt{\tfrac{1}{2}2\lambda^2} = \lambda.$$

   So we choose $\log(p) \approx 2\lambda^2$.
2. To resist a classical meet-in-the-middle attack we need $(2B + 1)^n > 2^{2\lambda}$, although the work of Adj *et al.* [2] suggests this may be too cautious. For security against Tani's quantum algorithm we may require $(2B + 1)^n > 2^{3\lambda}$, and so $n \log(B) \sim 3\lambda$, though again this may not be necessary [35]. In any case, we have $n \log(B) = \Omega(\lambda)$.
3. Assuming that one wants to optimise for (asymptotic) performance, the best choice is then to take $B = O(1)$ and $n = \Omega(\lambda)$, which means that the prime ideals $\mathfrak{l}_i$ have norm $\ell_i = \Omega(n \log(n)) = \Omega(\lambda \log(\lambda))$. Note that this is compatible with the requirement $\log(p) > 2\lambda^2$, since $\sum_{i=1}^{n} \ln(\ell_i) \sim n \ln(n) \sim \lambda \log(\lambda)^2$.

Table 2. Parameter size and performance of the various signature protocols. Parameters taken in the asymptotic analysis are: $\log p \sim 2\lambda^2$, $n \log(B) \sim 3\lambda$, $B = O(1)$. The entry CSIDH is for parameters $(\lambda, n, B, \log(p)) = (128, 74, 5, 510)$ with $(s,t) = (1, 128)$ in the first column and $(s,t) = (16, 8)$ in the second two columns. All logarithms are in base 2. Signing time is on average 3 times the estimated verification time.

	Rejection sampling (Sect. 3.2)	Shorter signatures (Sect. 4)	Smaller public keys (Sect. 5)
Exact			
Sig size	$\lambda n \lceil \log(2n\lambda B + 1) \rceil + \lambda$	$\frac{\lambda}{s} n \lceil \log(2n\frac{\lambda}{s}B + 1) \rceil + \lambda$	$\frac{\lambda}{s}(n \lceil \log(2n\frac{\lambda}{s}B + 1) \rceil + \log p) + \lambda(\lambda - \frac{\lambda}{s} \log \frac{\lambda}{s})$
PK size	$\log p$	$2^s \log p$	2λ
SK size	$n \log(2B + 1)$	λ	$(2^s + 1)\lambda$
Performance ($\mathbb{F}_p$-ops)			
→ keygen	$\Omega(Bn^2 \log(n))$	$\Omega(2^s Bn^2 \log(n))$	$\Omega(2^s Bn^2 \log(n))$
→ sign/verify	$\Omega(\lambda^2 Bn^3 \log(n))$	$\Omega((\lambda/s)^2 Bn^3 \log(n))$	$\Omega((\lambda/s)^2 Bn^3 \log(n))$
Asymptotic			
Sig size	$O(\lambda^2 \log(\lambda))$	$O((\lambda^2/s) \log(\lambda))$	$O(\lambda^3/s)$
PK size	$2\lambda^2$	$2^{s+1}\lambda^2$	2λ
SK size	3λ	λ	$(2^s + 1)\lambda$
Performance (bits)			
→ keygen	$\Omega(\lambda^4 \log(\lambda)^2)$	$\Omega(2^s \lambda^4 \log(\lambda)^2)$	$\Omega(2^s \lambda^4 \log(\lambda)^2)$
→ sign/verify	$\Omega(\lambda^7 \log(\lambda)^2)$	$\Omega((\lambda^7/s^2) \log(\lambda)^2)$	$\Omega((\lambda^7/s^2) \log(\lambda)^2)$
CSIDH			
Sig size	20144 B	978 B	3136 B
PK size	64 B	4096 KiB	32 B
SK size	32 B	16 B	1024 KiB
Est. keygen time	0.03 s	1966 s	1966 s
Est. verify time	36372 s	142 s	142 s

4. Instead of measuring performance in terms of $\mathbb{F}_p$-operations, here we measure them in terms of bit-operations. After substituting B and n, this adds a factor $\lambda^2 \log(\lambda)$ in front of the lower bound if using fast (quasi-linear complexity) modular arithmetic.

Note that our asymptotic choices forbid the key space from covering the whole class group. If the conditions of Problem 1 are wanted, different choices must be made for n and B. In this case it is best to choose primes of the form $p + 1 = 4\prod_{i=1}^{n} \ell_i$, as in CSIDH [13]. Then, $n\log(n) \sim \log(p) \sim 2\lambda^2$ and so we have $n \sim \lambda^2/\log(\lambda)$. To have a distribution of ideal classes close to uniform we need $(2B+1)^n \gg \sqrt{p}$ and so $\log(B) > \log(\sqrt{p})/n \sim \log(\lambda)$. Hence $B > \sqrt{n}$, making all asymptotic bounds considerably worse.

The third block of rows (under the heading "CSIDH") gives concrete sizes obtained by fixing $\lambda = 128$ and $s = 16$ and using the CSIDH-512 primitive, i.e., $(n, B, \log(p)) = (74, 5, 510)$. We estimate these parameters to correspond to the NIST-1 security level. Note that we are able to get smaller signatures at similar cost, for example see the various options in Table 3 (and one can also potentially consider $s > 16$, such as $(s, t) = (21, 6)$). However, for Table 2 we choose the same parameters as [13] so that we are able to refer to their running-time computations. We estimate real-world performance, using as baseline the worst-case time for one isogeny action in CSIDH. In [13,42], for an exponent vector in $[-B, B]^n$, this time is reported to be around 30 ms. Accordingly, we multiply this time by the size of the exponent vector to obtain our estimates. Note that the estimates are very rough, as they purposely ignore other factors such as hash tree computations. However the results in [5,31] show that hash trees much larger than ours can be computed in a fraction of the time we need to compute isogenies.

Table 3. Parameter choices for small signatures, with $(s, t) = (16, 8)$, at around 128-bit classical security level. Signature size is $nt\lceil\log_2(2ntB + 1)\rceil + 128$ bits.

n	B	$\lceil\log_2(2ntB+1)\rceil$	Signature size (bytes)
20	3275	20	416
28	293	17	492
33	124	16	544
37	55	15	571
46	22	14	660

7 Tight Security Reduction Based on Lossy Keys

We now explain how to implement lossy keys in our setting. This allows us to use the methods of Kiltz, Lyubashevsky and Schaffner [36] (that build on work of Abdalla, Fouque, Lyubashevsky and Tibouchi [1]) to obtain signatures from lossy identification schemes. This approach gives a *tight reduction* in the *quantum random oracle model.*

Here's the basic idea to get a lossy scheme, using uniform distributions for simplicity (one can also use discrete Gaussians in this setting): Take a very large prime p so that the ideal class group is very large, but use relatively small values for n and B so that $\{\mathfrak{a} = \prod_{i=1}^{n} \mathfrak{l}_i^{e_i} : |e_i| \leq B\}$ is a very small subset of the class group.[4] The real key is $(E, E_A = \mathfrak{a} * E)$ for such an $\mathfrak{a}$. The lossy key is (E, E_A) where E_A is a uniformly random curve in the isogeny class. Further, choose parameters so that the $f_{k,i}$ are also such that $\{\mathfrak{b} = \prod_{i=1}^{n} \mathfrak{l}_i^{f_{k,i}} : |f_{k,i}| \leq (nt+1)B\}$ is a small subset of the ideal class group. In the case of a real key, the signatures define ideals that correspond to "short" paths from E or E_A to a curve $\mathcal{E}$. In the case of a lossy key, then such ideals do not exist, as for a curve $\mathcal{E}$ it is not the case that there is a short path from E to $\mathcal{E}$ AND a short path from E_A to $\mathcal{E}$.

In the remainder of this section we develop these ideas.

7.1 Background Definitions

We closely follow Kiltz, Lyubashevsky and Schaffner [36]. A *canonical identification scheme* consists of algorithms $(\mathsf{IGen}, \mathsf{P}_1, \mathsf{P}_2, \mathsf{V})$ and a set ChSet. The randomised algorithm $\mathsf{IGen}(1^\lambda)$ outputs a key pair (pk, sk). The deterministic algorithm P_1 takes sk and randomness r_1 and computes $(W, \mathsf{st}) = \mathsf{P}_1(sk, r_1)$. Here st denotes state information to be passed to P_2. A challenge c is sampled uniformly from ChSet. The deterministic algorithm P_2 then computes $Z = \mathsf{P}_2(sk, W, c, \mathsf{st}, r_2)$ or $\perp$, where r_2 is the randomness. The output $\perp$ corresponds to an abort in the "Fiat-Shamir with aborts" paradigm. We require that $\mathsf{V}(pk, W, c, Z) = 1$ for a correctly formed transcript (W, c, Z).

We assume, for each value of λ, there are well-defined sets $\mathcal{W}$ and $\mathcal{Z}$, such that $\mathcal{W}$ contains all W output by P_1 and $\mathcal{Z}$ contains all Z output by P_2. The scheme is *commitment recoverable* if, given c and $Z = \mathsf{P}_2(sk, W, c, \mathsf{st})$, there is a unique $W \in \mathcal{W}$ such that $\mathsf{V}(pk, W, c, Z) = 1$ and this W can be efficiently computed from (pk, c, Z)

A canonical identification scheme is ϵ_{zk}-naHVZK *non-abort honest verifier zero knowledge* if there is a simulator that given only pk outputs (W, c, Z) whose distribution has statistical distance at most ϵ_{zk} from the output distribution of the real protocol conditioned on $\mathsf{P}_2(sk, W, c, \mathsf{st}, r_2) \neq \perp$.

A *lossy identification scheme* is a canonical identification scheme as above together with a lossy key generation algorithm $\mathsf{LossIGen}$, which is a randomised algorithm that on input 1^λ outputs pk. An adversary against a lossy identification scheme is a randomised algorithm A that takes an input pk and returns 0 or 1. The advantage $\mathrm{Adv}^{\mathsf{LOSS}}(A)$ of an adversary against a lossy identification scheme is defined to be

$$\left|\Pr\left(A(pk) = 1 : pk \leftarrow \mathsf{LossIGen}(1^\lambda)\right) - \Pr\left(A(pk) = 1 : pk \leftarrow \mathsf{IGen}(1^\lambda)\right)\right|.$$

[4] It might even be possible to consider working with subgroups, in the quantum algorithm case where the class group structure is known. For example, private keys could be sampled from a large subgroup and lossy keys from a non-trivial coset.

The two security properties of a lossy identification scheme are:

1. There is no polynomial-time adversary that has non-negligible advantage $\mathsf{Adv}^{\mathsf{LOSS}}$ in distinguishing real and lossy keys.
2. The probability, over (pk, W, c) where pk is an output of the lossy key generation algorithm $\mathsf{LossIGen}$, $W \leftarrow \mathcal{W}$ and $c \leftarrow \mathsf{ChSet}$, that there is some $Z \in \mathcal{Z}$ with $\mathsf{V}(pk, W, c, Z) = 1$, is negligible.
   This will allow to show that no unbounded quantum adversary can pass the identification protocol (or, once we have applied Fiat-Shamir, forge a signature) with respect to a lossy public key, because with overwhelming probability no such signature exists.

7.2 Scheme

We can re-write our scheme in this setting, see Fig. 2. Here we are assuming that E is a supersingular elliptic curve with $j(E) \in \mathbb{F}_p$ where p satisfies the constraint

$$\sqrt{p} > (4(nt+1)B+1)^n 2^\lambda \tag{3}$$

This bound is sufficient for the keys to be lossy.

We use the generic deterministic signature construction from Kiltz, Lyubashevsky and Schaffner [36], and use the fact that signatures can be shortened because the identification protocol is commitment recoverable. We refer to the full version of the paper for details.

7.3 Proofs

We now explain that our identification scheme satisfies the required properties, from which the security of the signature scheme will follow from Theorem 3.1 of [36].

We make some heuristic assumptions.

Heuristic 1: There are at least $\sqrt{p}$ supersingular elliptic curves with j-invariant in $\mathbb{F}_p$.
This assumption, combined with the bound $\sqrt{p} \gg (4(nt+1)B)^n$ of Eq. (3), implies that the curves $\mathcal{E}_k$ constructed by algorithm P_1 are a negligibly small proportion of all such curves.

Heuristic 2: Each choice of $\mathbf{f}_k \in [-(nt+1)B, (nt+1)B]^n$ gives a unique value for $j(\mathcal{E}_k)$.
This is extremely plausible given Eq. (3). It implies that the min-entropy of the values W output by P_1 is extremely high (more than sufficient for the security proofs).

Under heuristic assumption 1, we now show that the keys are lossy. The lossy key generator outputs a pair (E, E_A) where E and E_A are randomly sampled supersingular elliptic curves with $j(E), j(E_A) \in \mathbb{F}_p$. To implement this one constructs a supersingular curve with j-invariant in $\mathbb{F}_p$ and then runs long pseudorandom walks in the isogeny graph until the uniform mixing bounds imply that E_A is uniformly distributed.

Algorithm 4 IGen

Input: $B, \mathfrak{l}_1, \ldots, \mathfrak{l}_n, E$
Output: $sk = \mathbf{e}$ and $pk = E_A$

1: $\mathbf{e} \leftarrow [-B, B]^n$
2: $E_A = (\prod_{i=1}^n \mathfrak{l}_i^{e_i}) * E$
3: **return** $sk = \mathbf{e}$, $pk = E_A$

Algorithm 5 P_1

Input: $(E, E_A), r_1$
Output: $W = (j(\mathcal{E}_1), \ldots, j(\mathcal{E}_t))$, $\mathsf{st} = (\mathbf{f}_1, \ldots, \mathbf{f}_t)$

1: **for** $k = 1, \ldots, t$ **do**
2: $\quad \mathbf{f}_k \leftarrow [-(nt+1)B, (nt+1)B]^n$ using $\mathsf{PRF}(r_1)$
3: $\quad \mathcal{E}_k = (\prod_{i=1}^n \mathfrak{l}_i^{f_{k,i}}) * E$
4: **end for**
5: **return** $(j(\mathcal{E}_1), \ldots, j(\mathcal{E}_t)), (\mathbf{f}_1, \ldots, \mathbf{f}_t)$

Algorithm 6 P_2

Input: $(E, E_A), \mathbf{e}, W, c, \mathsf{st}, r_2$
Output: $Z = (\mathbf{z}_1, \ldots, \mathbf{z}_t)$

1: Parse c as $b_1 \| \cdots \| b_t$
2: **for** $k = 1, \ldots, t$ **do**
3: $\quad$ **if** $b_k = 0$ **then**
4: $\quad\quad \mathbf{z}_k = \mathbf{f}_k$
5: $\quad$ **else**
6: $\quad\quad \mathbf{z}_k = \mathbf{f}_k - \mathbf{e}$
7: $\quad$ **end if**
8: $\quad$ **if** $\mathbf{z}_k \notin [-ntB, ntB]^n$ **then**
9: $\quad\quad$ **return** $\perp$
10: $\quad$ **end if**
11: **end for**
12: **return** $\sigma = (\mathbf{z}_1, \ldots, \mathbf{z}_t)$

Algorithm 7 V

Input: $(E, E_A), (W, c, Z)$
Output: Valid/Invalid

1: Parse W as $(j_1, \ldots, j_t)$
2: Parse c as $b_1 \| \cdots \| b_t$
3: Parse Z as $(\mathbf{z}_1, \ldots, \mathbf{z}_t)$
4: **for** $k = 1, \ldots, t$ **do**
5: $\quad$ **if** $b_k = 0$ **then**
6: $\quad\quad \mathcal{E}_k = (\prod_{i=1}^n \mathfrak{l}_i^{z_{k,i}}) * E$
7: $\quad$ **else**
8: $\quad\quad \mathcal{E}_k = (\prod_{i=1}^n \mathfrak{l}_i^{z_{k,i}}) * E_A$
9: $\quad$ **end if**
10: **end for**
11: **if** $(j_1, \ldots, j_t) = (j(\mathcal{E}_1), \ldots, j(\mathcal{E}_t))$ **then**
12: $\quad$ **return** Valid
13: **else**
14: $\quad$ **return** Invalid
15: **end if**

Fig. 2. The identification protocol. Note that P_1 does not need sk, while P_2 does not use r_2 (it really is deterministic) and does not use W. Also note that the scheme is commitment recoverable.

Lemma 5. *Let parameters satisfy the bound of Eq. (3) and suppose heuristic 1 holds. Let (E, E_A) be a key output by the lossy key generator. Then with overwhelming probability there is no ideal $\mathfrak{a} = \prod_{i=1}^n \mathfrak{l}_i^{f_i}$ such that $\mathbf{f} \in [-2(nt+1)B, 2(nt+1)B]^n$ and $j(E_A) = j(\mathfrak{a} * E)$.*

Proof. If $f_i \in [-2(nt+1)B, 2(nt+1)B]$ then there are $4(nt+1)B+1$ choices for each f_i and so at most $(4(nt+1)B+1)^n$ choices for $\mathfrak{a}$. Given E it means there are at most that many $j(\mathfrak{a} * E)$. Since E_A is uniformly and independently sampled from a set of size at least $\sqrt{p} > (4(nt+1)B+1)^n 2^\lambda$, the probability that

$j(E_A)$ lies in the set of all possible $j(\mathfrak{a} * E)$ is at most $1/2^\lambda$, which is negligible. □

We consider the following decisional problem. It is an open challenge to give a "search to decision" reduction in this context (showing that if one can solve Problem 4 then one can solve Problem 2). This seems to be non-trivial.

Problem 4. Consider two distributions on pairs (E, E_A) of supersingular elliptic curves over $\mathbb{F}_p$. Let $\mathcal{D}_1$ be the output distribution of the algorithm IGen. Let $\mathcal{D}_2$ be the uniform distribution (i.e., output distribution of the lossy key generation algorithm). The *decisional short isogeny problem* is to distinguish the two distributions when given one sample.

The next result shows the second part of the security property for lossy keys.

Lemma 6. *Assume heuristic 1. Let pk be an output of the lossy key generation algorithm* LossIGen. *Let $W \leftarrow \mathcal{W}$ be an output of* P_1. *Let $c \leftarrow$ ChSet be a uniformly chosen challenge. Then the probability that there is some $Z \in \mathcal{Z}$ with* $\mathsf{V}(pk, W, c, Z) = 1$, *is negligible.*

Proof. Let $pk = (E, E_A)$ be an output of $\mathsf{LossIGen}(1^\lambda)$. By Lemma 5 we have that with overwhelming probability $j(E_A) \neq j(\mathfrak{a} * E)$ for all ideals $\mathfrak{a}$ of the form in Lemma 5. Let $W = (j(\mathcal{E}_1), \ldots, j(\mathcal{E}_t))$ be an element of $\mathcal{W}$, so that each $\mathcal{E}_k$ is of the form $\mathfrak{a}_k * E$ where $\mathfrak{a}_k = \prod_i \mathfrak{l}_i^{f_{k,i}}$ for $f_{k,i} \in [-(nt+1)B, (nt+1)B]$.

Let $c \leftarrow$ ChSet be a uniformly chosen challenge, which means that $c \neq 0$ with overwhelming probability. Then there is some k with $c_k \neq 0$ and so if Z was to satisfy the verification algorithm $\mathsf{V}(pk, W, c, Z) = 1$ then it would follow that $\mathbf{z}_k$ gives an ideal $\mathfrak{c}_k$ such that $j(\mathcal{E}_k) = j(\mathfrak{c}_k * E_A)$. From $\mathfrak{a}_k * E \cong \mathcal{E}_k \cong \mathfrak{c}_k * E_A$ it follows that $E_A \cong (\mathfrak{c}_k^{-1}\mathfrak{a}_k) * E$. But $\mathfrak{c}_k^{-1}\mathfrak{a}_k = \prod_i \mathfrak{l}_i^{f_{k,i} - z_{k,i}}$, which violates the claim about E_A corresponding to Lemma 5. Hence with overwhelming probability Z does not exist, and the result is proved. □

Note that Heuristic 2 also shows that there are "unique responses" in the sense of Definition 2.7 of [36] (not just computationally unique, but actually unique). But we won't need this for the result we state.

We now discuss *no-abort honest verifier zero-knowledge* (naHVZK). This is simply the requirement that there is a simulator that produces transcripts (W, c, Z) that are statistically close to real transcripts output by the protocol.

Lemma 7. *The identification scheme (sigma protocol) of Fig. 2 has no-abort honest verifier zero-knowledge.*

Proof. This is simple to show in our setting (due to the rejection sampling): Instead of choosing $W = (j((\prod_i \mathfrak{l}_i^{f_{1,i}}) * E), \ldots, j((\prod_i \mathfrak{l}_i^{f_{k,i}}) * E))$, then c, and then $Z = (\mathbf{z}_1, \ldots, \mathbf{z}_k)$ the simulator chooses Z first, then c, and then sets, for $1 \leq k \leq t$, $j_k = j((\prod_i \mathfrak{l}_i^{z_{k,i}}) * E)$ when $c_k = 0$ and $j_k = j((\prod_i \mathfrak{l}_i^{z_{k,i}}) * E_A)$ when $c_k = 1$. Setting $W = (j_1, \ldots, j_k)$ it follows that (W, c, Z) is a transcript that satisfies the verification algorithm. Further, the distribution of triples (W, c, Z)

is identical to the distribution from the real protocol since, for any choice of the private key, this choice of W would have arisen for some choice of the original vectors $\mathbf{f}_k$. □

Theorem 5. *Assume Heuristic 1, and the hardness of Problem 2. Then the deterministic signature scheme of Kiltz, Lyubashevsky and Schaffner applied to Fig. 2 has UF-CMA security in the quantum random oracle model, with a tight security reduction.*

Proof. See Theorem 3.1 of [36]. In particular this theorem gives a precise statement of the advantage. □

One can then combine this proof with the optimisations of Sects. 4 and 5, to get a compact signature scheme with tight post-quantum security based on a merger of the assumptions corresponding to Problems 3 and 4.

8 Using the Relation Lattice

This section explains an alternative solution to the problem of representing an ideal class without leaking the private key of the signature scheme. This variant can be considered if a quantum computer is available during system setup. Essentially, this is the scheme from Stolbunov's thesis (see Sect. 3.1), which can be used securely once the relation lattice is known. Note that this section is about signatures that involve sampling ideal classes uniformly and so the techniques can't be used in the lossy keys setting.

Let $(\mathfrak{l}_1, \ldots, \mathfrak{l}_n)$ be a sequence of $\mathcal{O}$-ideals that generates $\mathrm{Cl}(\mathcal{O})$. Define

$$L = \left\{ (x_1, \ldots, x_n) \in \mathbb{Z}^n : \prod_{i=1}^{n} \mathfrak{l}_i^{x_i} \equiv (1) \right\}.$$

Then L is a rank n lattice with volume equal to $\#\mathrm{Cl}(\mathcal{O})$. Indeed, we have the exact sequence of Abelian groups

$$0 \to L \to \mathbb{Z}^n \to \mathrm{Cl}(\mathcal{O}) \to 1$$

where the map $f : \mathbb{Z}^n \to \mathrm{Cl}(\mathcal{O})$ is the group homomorphism $(x_1, \ldots, x_n) \mapsto \prod_i \mathfrak{l}_i^{x_i}$. We call L the *relation lattice.*

A basis for this lattice can be constructed in subexponential time using classical algorithms [8,29]. However, of interest to us is that a basis can be constructed in probabilistic polynomial time using quantum algorithms: the function $f : \mathbb{Z}^n \to \mathrm{Cl}(\mathcal{O})$ defined in the previous paragraph can be evaluated in polynomial time [16,47], and finding a basis for $L = \ker f$ is an instance of the Hidden Subgroup Problem for $\mathbb{Z}^n$, which can be solved in polynomial time using Kitaev's generalisation of Shor's algorithm [37]. The classical approach is not very interesting since the underlying computational assumption is only subexponentially hard for quantum computers, but it might make sense in a certain setting. The quantum case would make sense in a post-quantum world where a quantum computer can be used to set up the system parameters for the system and then is not required for further use. It might also be possible to construct (E, p) such that computing the relation lattice is efficient (e.g., constructing E so that $\mathrm{Cl}(\mathrm{End}(E))$ has smooth order), but we do not consider such approaches in this paper.

For the remainder of this section we assume that the relation lattice is known. Let $\{\mathbf{x}_1, \ldots, \mathbf{x}_n\}$ be a basis for L. Let $\mathcal{F} = \{\sum_{i=1}^{n} : u_i \mathbf{x}_i : -1/2 \leq u_i < 1/2\}$ be the centred fundamental domain of the basis of L. Then there is a one-to-one correspondence between $\mathcal{F} \cap \mathbb{Z}^n$ and $\mathrm{Cl}(\mathcal{O})$ by $(z_1, \ldots, z_n) \in \mathcal{F} \cap \mathbb{Z}^n \mapsto \prod_{i=1}^{n} \mathfrak{l}_i^{z_i}$.

Returning to Stolbunov's signature scheme, the solution to the problem is then straightforward: Given $\mathfrak{a} = \prod_{i=1}^{n} \mathfrak{l}_i^{e_i}$ and $\mathfrak{b}_k = \prod_{i=1}^{n} \mathfrak{l}_i^{f_{k,i}}$, a representation of $\mathfrak{b}_k \mathfrak{a}^{-1}$ is obtained by computing the vector $\mathbf{z}' = \mathbf{f}_k - \mathbf{e}$ and then using Babai rounding to get the unique vector $\mathbf{z}$ in $\mathcal{F} \cap (\mathbf{z}' + L)$. The vector $\mathbf{z}$ is sent as the response to the k-th challenge. Since $\mathfrak{b}_k$ is a uniformly chosen ideal class, the class $\mathfrak{b}_k \mathfrak{a}^{-1}$ is also uniformly distributed as an ideal class, and hence the vector $\mathbf{z} \in \mathcal{F} \cap \mathbb{Z}^n$ is uniformly distributed and carries no information about the private key.

Lemma 8. *If $\mathfrak{b}_k$ is a uniformly chosen ideal class then the vector $\mathbf{z} \in \mathcal{F} \cap \mathbb{Z}^n$ corresponding to $\mathbf{f}_k - \mathbf{e}$ is uniformly distributed.*

Proof. For fixed $\mathbf{e}$ the vector $\mathbf{z}$ depends only on the ideal class of $\mathfrak{b}_k$. But $\mathfrak{b}_k$ is uniform and independent of $\mathbf{e}$ and not known to verifier. □

The above discussion fixes a particular fundamental domain and uses Babai rounding to compute an element in it, but this may not lead to the most efficient signature scheme. One can consider different fundamental domains and different "reduction" algorithms to compute $\mathbf{z}$. Since the cost of signature verification depends on the size of the entries in $\mathbf{z}$, a natural computational problem is to efficiently compute a short vector $(z_1, \ldots, z_n)$ corresponding to a given ideal class; we discuss this problem in the next subsection.

8.1 Solving Close Vector Problems in the Relation Lattice

Let $\mathbf{w} = (w_1, \ldots, w_n) \in \mathbb{Z}^n$ be given and suppose we want to compute the isogeny $\mathfrak{a} * E$ where $\mathfrak{a} = \prod_{i=1}^{n} \mathfrak{l}_i^{w_i}$. Since the computation of the isogeny depends on the sizes of $|w_i|$ it is natural to first compute a short vector $(z_1, \ldots, z_n)$ that represents the same element of $\mathbb{Z}^n / L$. This can be done by solving a close vector problem in the lattice L. Namely, if $\mathbf{v} \in L$ is such that $\|\mathbf{w} - \mathbf{v}\|$ is short, then $\mathbf{z} = \mathbf{w} - \mathbf{v}$ is a short vector that can be used to compute $\mathfrak{a} * E$. Hence, the problem of interest is the close vector problem in the relation lattice.

Note that most literature and algorithms for solving close lattice vector problems are with respect to the Euclidean norm, whereas for isogeny problems the natural norms are the 1-norm $\|\mathbf{z}\|_1 = \sum_{i=1}^{n} |z_i|$ or the ∞-norm $\|\mathbf{z}\|_\infty = \max_i |z_i|$. The choice of norm depends on how the isogeny is computed. The algorithm for computing $\mathfrak{a} * E$ given in [13] depends mostly on the ∞-norm, since the Vélu formulae are used and a block of isogenies are handled together in each iteration. However, the intuitive cost of the isogeny (and this is appropriate when using modular polynomials to compute the isogenies) is given by the 1-norm. If the entries z_i are uniformly distributed in $[-\|\mathbf{z}\|_\infty, \|\mathbf{z}\|_\infty]$ then we have $\|\mathbf{z}\|_\infty \approx \sqrt{3/n}\|\mathbf{z}\|_2$ and $\|\mathbf{z}\|_1 \approx \frac{n}{2}\|\mathbf{z}\|_\infty \approx \sqrt{3n/4}\|\mathbf{z}\|_2$.

There are many approaches to solving the close vector problem. All methods start with pre-processing the lattice using some basis reduction, and in our case one can perform a major precomputation to produce a basis customised for solving close vector problems. Once the instance $\mathbf{w}$ is provided one can perform one of the following three approaches: the Babai nearest plane method (or an iterative version of it, as done by Lindner and Peikert [39]); enumeration; reducing to SVP (the Kannan embedding technique) and running a basis reduction algorithm. The choice of method depends on the quality of the original basis, the amount of time available to spend on solving CVP (note that a reduction in the sizes of the $|z_i|$ pays dividends in the time to compute $\mathfrak{a} * E$, and so it may be worth to devote more than a few cycles to this problem).

For this paper we focus on the Babai nearest plane algorithm. Let $\mathbf{b}_1, \ldots, \mathbf{b}_n$ be the (ordered) reduced lattice basis and $\mathbf{b}_1^*, \ldots, \mathbf{b}_n^*$ the Gram-Schmidt vectors. Equation (4.3) of Babai [3] shows that the nearest plane algorithm on input $\mathbf{w}$ outputs a vector $\mathbf{v} \in L$ with

$$\|\mathbf{w} - \mathbf{v}\|_2^2 \leq (\|\mathbf{b}_1^*\|_2^2 + \|\mathbf{b}_2^*\|_2^2 + \cdots + \|\mathbf{b}_n^*\|_2^2)/4. \tag{4}$$

Bounds on $\|\mathbf{b}_i^*\|$ are regularly discussed in the literature. For example, much work on the BKZ algorithm is devoted to understanding the sizes of these vectors; see Gama-Nguyen and Chen-Nguyen [14].

Fukase and Kashiwabara [25] have discussed lattice reduction algorithms that produce a basis that minimises the right hand size of Eq. (4) and hence are good for solving CVP using the nearest-plane algorithm. Blömer [10] has given a variant of the near-plane algorithm that efficiently solves CVP when given a dual-HKZ-reduced basis.

For our calculations we simply consider a BKZ-reduced lattice basis and, following Chen-Nguyen [14], assume that

$$\|\mathbf{b}_i^*\|_2 \approx \|\mathbf{b}_1\|_2^{1-0.0263(i-1)}.$$

Some similar calculations are given in [11].

8.2 Optimal Signature Size

We now use an idea that is implicit in the work of Couveignes [18] and Stolbunov [49] that gives signatures of optimal size when the relation lattice is known. Suppose the ideal class group is cyclic of order N and let $\mathfrak{g}$ be a generator (whose factorisation over $(\mathfrak{l}_1, \ldots, \mathfrak{l}_n)$ is known). Then one can choose the private key by uniformly sampling an integer $0 \leq x < N$ and letting $\mathfrak{a} = \mathfrak{g}^x$ in $\mathrm{Cl}(\mathcal{O})$. The public key is $E_A = \mathfrak{a} * E$ as before (this computation requires "smoothing" the ideal class using the relation lattice). When signing one chooses the t random ideals $\mathfrak{b}_k$ by choosing uniform integers y_k in $[0, N)$ and computing $\mathfrak{b}_k = \mathfrak{g}^{y_k}$. As before $\mathcal{E}_k = \mathfrak{b}_k * E$. Finally, in the scheme, when $b_k = 0$ we return y_k and when $b_k = 1$ we return $y_k - x \pmod{N}$. The verifier just sees a uniformly distributed integer modulo N, and uses this to recompute $\mathcal{E}_k$ from either E or E_A (again, this requires reducing a vector modulo the relation lattice and then computing the corresponding isogenies). This scheme is clearly optimal from the point of view of signature size, since one cannot represent a random element of a group of order N in fewer than $\log_2(N)$ bits.

The method used to compute the isogenies during verification is left for the verifier to decide. In practice all users will work with the same prime p (e.g., the 512-bit CSIDH prime) in which case the relation lattice can be precomputed and optimised. The verifier then solves the CVP instances using their preferred method and then computes the isogenies.

The full version of the paper contains a table of parameters for this scheme.

9 Conclusions

We have given a signature scheme suitable for the CSIDH isogeny setting. This solves an unresolved problem in Stolbunov's thesis. We have also shown how to get shorter signatures by increasing the public key size. We do not know how to obtain a similar trade-off between public key size and signature size for the schemes of Yoo *et al.* [53] or Galbraith *et al.* [28] based on the SIDH setting.

Acknowledgements. Thanks to Samuel Dobson for doing some experiments with discrete Gaussians. Thanks to Lorenz Panny for comments and suggestions. Thanks to Damien Stehlé for references about solving close vector problems. Thanks to the Eurocrypt referees for their careful reading of the paper.

Luca De Feo was supported by the French *Programme d'Investissements d'Avenir* under the national project RISQ n° P141580-3069086/DOS0044212.

References

1. Abdalla, M., Fouque, P.-A., Lyubashevsky, V., Tibouchi, M.: Tightly-secure signatures from lossy identification schemes. In: Pointcheval, D., Johansson, T. (eds.) EUROCRYPT 2012. LNCS, vol. 7237, pp. 572–590. Springer, Heidelberg (2012). https://doi.org/10.1007/978-3-642-29011-4_34
2. Adj, G., Cervantes-Vázquez, D., Chi-Domínguez, J.J., Menezes, A., Rodríguez-Henríquez, F.: On the cost of computing isogenies between supersingular elliptic curves. In: Cid, C., Jacobson Jr., M. (eds.) SAC 2018. LNCS, vol. 11349, pp. 322–343. Springer, Cham (2019). https://doi.org/10.1007/978-3-030-10970-7_15
3. Babai, L.: On Lovász lattice reduction and the nearest lattice point problem. Combinatorica **6**(1), 1–13 (1986)
4. Bellare, M., Neven, G.: Multi-signatures in the plain public-key model and a general forking lemma. In: Juels, A., Wright, R.N., di Vimercati, S.D.C. (eds.) ACM CCS 2006, pp. 390–399. ACM (2006)

5. Bernstein, D.J., et al.: SPHINCS+, November 2017. https://sphincs.org/data/sphincs+-submission-nist.zip
6. Bernstein, D.J., et al.: SPHINCS: practical stateless hash-based signatures. In: Oswald, E., Fischlin, M. (eds.) EUROCRYPT 2015. LNCS, vol. 9056, pp. 368–397. Springer, Heidelberg (2015). https://doi.org/10.1007/978-3-662-46800-5_15
7. Bernstein, D.J., Lange, T., Martindale, C., Panny, L.: Quantum circuits for the CSIDH: optimizing quantum evaluation of isogenies. In: Ishai, Y., Rijmen, V. (eds.) EUROCRYPT 2019. LNCS, vol. 11477, pp. 409–441. Springer, Cham (2019)
8. Biasse, J., Fieker, C., Jacobson, M.J.: Fast heuristic algorithms for computing relations in the class group of a quadratic order, with applications to isogeny evaluation. LMS J. Comput. Math. **19**(A), 371–390 (2016)
9. Biasse, J.-F., Iezzi, A., Jacobson Jr., M.J.: A note on the security of CSIDH. In: Chakraborty, D., Iwata, T. (eds.) INDOCRYPT 2018. LNCS, vol. 11356, pp. 153–168. Springer, Cham (2018). https://doi.org/10.1007/978-3-030-05378-9_9
10. Blömer, J.: Closest vectors, successive minima, and dual HKZ-bases of lattices. In: Montanari, U., Rolim, J.D.P., Welzl, E. (eds.) ICALP 2000. LNCS, vol. 1853, pp. 248–259. Springer, Heidelberg (2000). https://doi.org/10.1007/3-540-45022-X_22
11. Bonnetain, X., Schrottenloher, A.: Quantum security analysis of CSIDH and ordinary isogeny-based schemes. IACR Cryptology ePrint Archive 2018/537 (2018)
12. Bröker, R., Charles, D.X., Lauter, K.E.: Evaluating large degree isogenies and applications to pairing based cryptography. In: Galbraith, S.D., Paterson, K.G. (eds.) Pairing 2008. LNCS, vol. 5209, pp. 100–112. Springer, Heidelberg (2008). https://doi.org/10.1007/978-3-540-85538-5_7
13. Castryck, W., Lange, T., Martindale, C., Panny, L., Renes, J.: CSIDH: an efficient post-quantum commutative group action. In: Peyrin, T., Galbraith, S.D. (eds.) ASIACRYPT 2018. LNCS, vol. 11274, pp. 395–427. Springer, Cham (2018). https://doi.org/10.1007/978-3-030-03332-3_15
14. Chen, Y., Nguyen, P.Q.: BKZ 2.0: better lattice security estimates. In: Lee, D.H., Wang, X. (eds.) ASIACRYPT 2011. LNCS, vol. 7073, pp. 1–20. Springer, Heidelberg (2011). https://doi.org/10.1007/978-3-642-25385-0_1
15. Childs, A., Jao, D., Soukharev, V.: Constructing elliptic curve isogenies in quantum subexponential time. J. Math. Cryptol. **8**(1), 1–29 (2014)
16. Cohen, H.: A Course in Computational Algebraic Number Theory, vol. 138. Springer, New York (1993). https://doi.org/10.1007/978-3-662-02945-9
17. Costello, C., Hisil, H.: A simple and compact algorithm for SIDH with arbitrary degree isogenies. In: Takagi, T., Peyrin, T. (eds.) ASIACRYPT 2017. LNCS, vol. 10625, pp. 303–329. Springer, Cham (2017). https://doi.org/10.1007/978-3-319-70697-9_11
18. Couveignes, J.M.: Hard homogeneous spaces. eprint 2006/291 (2006)
19. Cox, D.A.: Primes of the Form x2 + ny2: Fermat, Class Field Theory, and Complex Multiplication. Wiley, Hoboken (1997)
20. De Feo, L.: Mathematics of isogeny based cryptography. Notes from a summer school on mathematics for post-quantum cryptography (2017). https://arxiv.org/abs/1711.04062
21. De Feo, L., Jao, D., Plût, J.: Towards quantum-resistant cryptosystems from supersingular elliptic curve isogenies. J. Math. Cryptol. **8**(3), 209–247 (2014)
22. De Feo, L., Kieffer, J., Smith, B.: Towards practical key exchange from ordinary isogeny graphs. In: Peyrin, T., Galbraith, S.D. (eds.) ASIACRYPT 2018. LNCS, vol. 11274, pp. 365–394. Springer, Cham (2018). https://doi.org/10.1007/978-3-030-03332-3_14

23. Decru, T., Panny, L., Vercauteren, F.: Faster SeaSign signatures through improved rejection sampling. To appear at PQCrypto 2019 (2019)
24. Delfs, C., Galbraith, S.D.: Computing isogenies between supersingular elliptic curves over F_p. Des. Codes Crypt. **78**(2), 425–440 (2016)
25. Fukase, M., Kashiwabara, K.: An accelerated algorithm for solving SVP based on statistical analysis. J. Inf. Process. **23**(1), 67–80 (2015)
26. Galbraith, S.D.: Mathematics of Public Key Cryptography. Cambridge University Press, Cambridge (2012)
27. Galbraith, S.D., Hess, F., Smart, N.P.: Extending the GHS weil descent attack. In: Knudsen, L.R. (ed.) EUROCRYPT 2002. LNCS, vol. 2332, pp. 29–44. Springer, Heidelberg (2002). https://doi.org/10.1007/3-540-46035-7_3
28. Galbraith, S.D., Petit, C., Silva, J.: Identification protocols and signature schemes based on supersingular isogeny problems. In: Takagi, T., Peyrin, T. (eds.) ASIACRYPT 2017. LNCS, vol. 10624, pp. 3–33. Springer, Cham (2017). https://doi.org/10.1007/978-3-319-70694-8_1
29. Hafner, J.L., McCurley, K.S.: A rigorous subexponential algorithm for computation of class groups. J. Am. Math. Soc. **2**(4), 837–850 (1989)
30. Huelsing, A., Butin, D., Gazdag, S.L., Rijneveld, J., Mohaisen, A.: XMSS: eXtended Merkle signature scheme. RFC 8391, May 2018
31. Hülsing, A., Rijneveld, J., Song, F.: Mitigating multi-target attacks in hash-based signatures. In: Cheng, C.-M., Chung, K.-M., Persiano, G., Yang, B.-Y. (eds.) PKC 2016. LNCS, vol. 9614, pp. 387–416. Springer, Heidelberg (2016). https://doi.org/10.1007/978-3-662-49384-7_15
32. Jao, D., De Feo, L.: Towards quantum-resistant cryptosystems from supersingular elliptic curve isogenies. In: Yang, B.-Y. (ed.) PQCrypto 2011. LNCS, vol. 7071, pp. 19–34. Springer, Heidelberg (2011). https://doi.org/10.1007/978-3-642-25405-5_2
33. Jao, D., LeGrow, J., Leonardi, C., Ruiz-Lopez, L.: A subexponential-time, polynomial quantum space algorithm for inverting the CM group action. To appear in proceedings of MathCrypt (2019)
34. Jao, D., Soukharev, V.: A subexponential algorithm for evaluating large degree isogenies. In: Hanrot, G., Morain, F., Thomé, E. (eds.) ANTS 2010. LNCS, vol. 6197, pp. 219–233. Springer, Heidelberg (2010). https://doi.org/10.1007/978-3-642-14518-6_19
35. Jaques, S., Schanck, J.M.: Quantum cryptanalysis in the RAM model: claw-finding attacks on SIKE. Cryptology ePrint Archive, Report 2019/103 (2019). https://eprint.iacr.org/2019/103
36. Kiltz, E., Lyubashevsky, V., Schaffner, C.: A concrete treatment of fiat-shamir signatures in the quantum random-oracle model. In: Nielsen, J.B., Rijmen, V. (eds.) EUROCRYPT 2018. LNCS, vol. 10822, pp. 552–586. Springer, Cham (2018). https://doi.org/10.1007/978-3-319-78372-7_18
37. Kitaev, A.Y.: Quantum measurements and the Abelian stabilizer problem. arXiv preprint quant-ph/9511026 (1995). https://arxiv.org/abs/quant-ph/9511026
38. Kuperberg, G.: A subexponential-time quantum algorithm for the dihedral hidden subgroup problem. SIAM J. Comput. **35**(1), 170–188 (2005)
39. Lindner, R., Peikert, C.: Better key sizes (and attacks) for LWE-based encryption. In: Kiayias, A. (ed.) CT-RSA 2011. LNCS, vol. 6558, pp. 319–339. Springer, Heidelberg (2011). https://doi.org/10.1007/978-3-642-19074-2_21
40. Lyubashevsky, V.: Fiat-Shamir with aborts: applications to lattice and factoring-based signatures. In: Matsui, M. (ed.) ASIACRYPT 2009. LNCS, vol. 5912, pp. 598–616. Springer, Heidelberg (2009). https://doi.org/10.1007/978-3-642-10366-7_35

41. Merkle, R.C.: A certified digital signature. In: Brassard, G. (ed.) CRYPTO 1989. LNCS, vol. 435, pp. 218–238. Springer, New York (1990). https://doi.org/10.1007/0-387-34805-0_21
42. Meyer, M., Reith, S.: A faster way to the CSIDH. In: Chakraborty, D., Iwata, T. (eds.) INDOCRYPT 2018. LNCS, vol. 11356, pp. 137–152. Springer, Cham (2018). https://doi.org/10.1007/978-3-030-05378-9_8
43. National Institute of Standards and Technology: Announcing request for nominations for public-key post-quantum cryptographic algorithms (2016). https://www.federalregister.gov/d/2016-30615
44. Neven, G., Smart, N.P., Warinschi, B.: Hash function requirements for Schnorr signatures. J. Math. Cryptol. **3**(1), 69–87 (2009)
45. Regev, O.: A subexponential time algorithm for the dihedral hidden subgroup problem with polynomial space. arXiv:quant-ph/0406151, June 2004
46. Renes, J.: Computing isogenies between Montgomery curves using the action of (0, 0). In: Lange, T., Steinwandt, R. (eds.) PQCrypto 2018. LNCS, vol. 10786, pp. 229–247. Springer, Cham (2018). https://doi.org/10.1007/978-3-319-79063-3_11
47. Shanks, D.: On Gauss and composition. In: Number Theory and Applications, pp. 163–204. NATO - Advanced Study Institute. Kluwer Academic Press (1989)
48. Silverman, J.H.: The Arithmetic of Elliptic Curves. GTM, vol. 106. Springer, New York (1986). https://doi.org/10.1007/978-1-4757-1920-8
49. Stolbunov, A.: Cryptographic schemes based on isogenies. Doctoral thesis, NTNU (2012)
50. Sutherland, A.: Elliptic curves. Lecture Notes from a Course (18.783). MIT (2017). http://math.mit.edu/classes/18.783/2017/lectures
51. Vélu, J.: Isogénies entre courbes elliptiques. Comptes Rendus de l'Académie des Sciences de Paris **273**, 238–241 (1971)
52. Washington, L.C.: Elliptic Curves: Number Theory and Cryptography, 2nd edn. CRC Press, Boca Raton (2008)
53. Yoo, Y., Azarderakhsh, R., Jalali, A., Jao, D., Soukharev, V.: A post-quantum digital signature scheme based on supersingular isogenies. In: Kiayias, A. (ed.) FC 2017. LNCS, vol. 10322, pp. 163–181. Springer, Cham (2017). https://doi.org/10.1007/978-3-319-70972-7_9

Author Index

Abusalah, Hamza II-277
Agarwal, Navneet II-381
Aggarwal, Divesh I-531, II-442
Agrawal, Shweta I-191
Aharonov, Dorit III-219
Alamati, Navid II-55
Albrecht, Martin R. II-717
Alwen, Joël I-129
Anand, Sanat II-381
Ananth, Prabhanjan II-532
Applebaum, Benny II-504, III-441
Aragon, Nicolas III-728
Asharov, Gilad II-214
Attrapadung, Nuttapong I-34
Aviram, Nimrod II-117

Backes, Michael III-281
Badrinarayanan, Saikrishna I-593
Băetu, Ciprian II-747
Ball, Marshall I-501
Barak, Boaz I-226
Bar-On, Achiya I-313
Bartusek, James III-636
Beimel, Amos III-441
Ben-Sasson, Eli I-103
Bernstein, Daniel J. II-409
Bitansky, Nir III-667
Blazy, Olivier III-728
Boyle, Elette II-3
Brakerski, Zvika II-504, III-219, III-619

Canteaut, Anne III-585
Chen, Hao II-34
Cheu, Albert I-375
Chiesa, Alessandro I-103
Chillotti, Ilaria II-34
Choudhuri, Arka Rai II-351, II-532
Chung, Kai-Min II-442, III-219
Coladangelo, Andrea III-247
Coretti, Sandro I-129
Couteau, Geoffroy II-473, II-562

Dachman-Soled, Dana I-501
De Feo, Luca III-759
Dinur, Itai I-343, III-699
Dodis, Yevgeniy I-129
Döttling, Nico I-531, II-292, III-281
Ducas, Léo II-717
Dunkelman, Orr I-313
Durak, F. Betül II-747
Dutta, Avijit I-437
Dziembowski, Stefan I-625

Eckey, Lisa I-625

Farràs, Oriol III-441
Faust, Sebastian I-625
Fehr, Serge III-472
Fisch, Ben II-324
Fuchsbauer, Georg I-657

Gaborit, Philippe III-728
Galbraith, Steven D. III-759
Ganesh, Chaya I-690
Garg, Sanjam III-33
Gay, Romain III-33
Gellert, Kai II-117
Genise, Nicholas II-655
Ghosh, Satrajit III-154
Goel, Aarushi II-532
Goyal, Vipul I-562, II-351
Green, Ayal III-219
Grilo, Alex B. III-247
Guan, Jiaxin III-500

Haitner, Iftach III-667
Hajiabadi, Mohammad III-33
Hamlin, Ariel II-244
Hanrot, Guillaume II-685
Hanzlik, Lucjan III-281
Hauck, Eduard III-345
Hauteville, Adrien III-728
Herold, Gottfried II-717
Hesse, Julia I-625
Hoang, Viet Tung II-85
Hofheinz, Dennis II-562
Hong, Cheng III-97
Hopkins, Samuel B. I-226

Hostáková, Kristina I-625
Hubert Chan, T.-H. I-720, II-214
Huguenin-Dumittan, Loïs II-747

Jaeger, Joseph I-467
Jager, Tibor II-117
Jain, Aayush I-226, I-251
Jain, Abhishek II-351, II-532
Jeffery, Stacey III-247
Jost, Daniel I-159

Kales, Daniel I-343
Kamara, Seny II-183
Kamath, Chethan II-277
Katsumata, Shuichi II-622, III-312
Katz, Jonathan III-97
Keller, Nathan I-313
Kiltz, Eike III-345
Kirshanova, Elena II-717
Klein, Karen II-277
Klooß, Michael I-68
Kluczniak, Kamil III-281
Kohl, Lisa II-3
Kolesnikov, Vladimir III-97
Kölsch, Lukas I-285
Komargodski, Ilan III-667
Kothari, Pravesh I-226
Kowalczyk, Lucas I-3
Kulkarni, Mukul I-501

Lai, Ching-Yi III-219
Lai, Russell W. F. II-292
Lallemand, Virginie III-585
Lange, Tanja II-409
Leander, Gregor III-585
Lehmann, Anja I-68
Lepoint, Tancrède III-636
Leurent, Gaëtan III-527
Li, Ting III-556
Lin, Han-Hsuan II-442
Lin, Huijia I-251, I-501
Liu, Qipeng III-189
Loss, Julian III-345
Lu, Wen-jie III-97
Lyubashevsky, Vadim III-619

Ma, Fermi III-636
Malavolta, Giulio II-292
Malkin, Tal I-501
Martindale, Chloe II-409
Matt, Christian I-251
Maurer, Ueli I-159
Micciancio, Daniele II-655, III-64
Miller, David II-85
Moataz, Tarik II-183
Montgomery, Hart II-55
Mularczyk, Marta I-159

Nadler, Niv III-699
Nandi, Mridul I-437
Nayak, Kartik II-214
Neumann, Patrick III-585
Nielsen, Jesper Buus I-531
Nilges, Tobias III-154
Nir, Oded III-441
Nishimaki, Ryo II-622

Obremski, Maciej I-531
Orlandi, Claudio I-690
Orrù, Michele I-657
Ostrovsky, Rafail II-244

Panny, Lorenz II-409
Pass, Rafael I-720, II-214
Patranabis, Sikhar II-55
Pellet-Mary, Alice II-685
Persiano, Giuseppe I-404
Peter, Naty III-441
Peyrin, Thomas III-527
Pietrzak, Krzysztof II-277
Pinkas, Benny III-122
Polyakov, Yuriy II-655
Postlethwaite, Eamonn W. II-717
Prabhakaran, Manoj II-381
Promitzer, Angela I-343
Purwanto, Erick I-531

Quach, Willy II-593

Ramacher, Sebastian I-343
Rechberger, Christian I-343
Ren, Ling II-214
Riabzev, Michael I-103
Rothblum, Ron D. II-593
Roy, Arnab II-55
Rupp, Andy I-68

Sahai, Amit I-226, I-251
Sattath, Or III-219

Schneider, Jonas III-281
Schneider, Thomas III-122
Scholl, Peter II-3
Seurin, Yannick I-657
Shi, Elaine I-720, II-214
Shumow, Dan II-151
Smith, Adam I-375
Song, Yifan I-562
Song, Yongsoo II-34
Spooner, Nicholas I-103
Srinivasan, Akshayaram I-593
Stehlé, Damien II-685
Stevens, Marc II-717
Sun, Yao III-556

Talayhan, Abdullah II-747
Talnikar, Suprita I-437
Tessaro, Stefano I-467
Tkachenko, Oleksandr III-122
Trieu, Ni II-85
Tsabary, Rotem II-504
Tschudi, Daniel I-690

Ullman, Jonathan I-375

Vaikuntanathan, Vinod III-619
Vaudenay, Serge II-747
Vidick, Thomas II-442, III-247
Virza, Madars I-103

Walter, Michael II-277
Wang, Xiao III-97
Ward, Nicholas P. I-103
Wee, Hoeteck I-3
Weiss, Mor II-244
Weizman, Ariel I-313
Wesolowski, Benjamin III-379
Wichs, Daniel II-244, II-593, III-619
Wiemer, Friedrich III-585
Woodage, Joanne II-151

Yamada, Shota II-622, III-312
Yamakawa, Takashi II-622
Yanai, Avishay III-122
Yeo, Kevin I-404
Yogev, Eylon III-667
Yuan, Chen III-472

Zeber, David I-375
Zémor, Gilles III-728
Zhandary, Mark III-500
Zhandry, Mark III-3, III-189, III-408, III-636
Zhilyaev, Maxim I-375

Printed in the United States
By Bookmasters